Advances in Experimental Medicine and Biology

Volume 962

More information about this series at http://www.springer.com/series/5584

Yoram Groner • Yoshiaki Ito
Paul Liu • James C. Neil
Nancy A. Speck • Andre van Wijnen
Editors

RUNX Proteins in Development and Cancer

Springer

Editors
Yoram Groner
Department of Molecular Genetics
Weizmann Institute of Science
Rehovot, Israel

Yoshiaki Ito
Cancer Science Institute of Singapore
National University of Singapore
Singapore, Singapore

Paul Liu
Genetics and Molecular Biology Branch
National Human Genome Research Institute
Bethesda, Maryland, USA

James C. Neil
Institute of Infection, Immunity and Inflammation
University of Glasgow
Glasgow, UK

Nancy A. Speck
Department of Cell and Development Biology
University of Pennsylvania
Philadelphia, Pennsylvania, USA

Andre van Wijnen
Mayo Clinic
Rochester, Minnesota, USA

Dr. Liu is working with US Government

ISSN 0065-2598 ISSN 2214-8019 (electronic)
Advances in Experimental Medicine and Biology
ISBN 978-981-10-3231-8 ISBN 978-981-10-3233-2 (eBook)
DOI 10.1007/978-981-10-3233-2

Library of Congress Control Number: 2017932394

Printed on acid-free paper

This Springer imprint is published by Springer Nature
The registered company is Springer Nature Singapore Pte Ltd.
The registered company address is: 152 Beach Road, #21-06/08 Gateway East, Singapore 189721, Singapore

Contents

About the Editors and Contributors

About the Editors

Yoram Groner is the Dr. Barnet Berris professor of cancer research at the Weizmann Institute of Science, Israel. Dr. Groner cloned the *RUNX3* gene in 1994 and since then has studied the biological functions of Runx3 and the regulation of its expression, primarily in immunology and neurogenesis.

Yoshiaki Ito M.D., Ph.D. is the Yong Loo Lin professor in medical oncology at the National University of Singapore. Dr. Ito, along with his longtime collaborator Dr. Katsuya Shigesada, discovered, purified, and cloned the RUNX and CBFβ proteins based on their binding to the polyomavirus enhancer in 1990 and 1993. Together, Ito and Shigesada contributed much of the early seminal work on the biochemistry of RUNX and CBFβ. Dr. Ito was the first to describe a role for Runx3 in cancer of the gastrointestinal tract, and he has continued to work extensively in this area.

P. Paul Liu M.D., Ph.D., is a senior investigator in the Genetics and Molecular Biology Branch at the National Human Genome Research Institute, USA. Dr. Liu identified the *CBFB* gene as a target of the inv Puig-Kroger and Corbi (2006) in acute myeloid leukemia in 1993. Dr. Liu studies the normal and oncogenic RUNX and CBFβ proteins in hematopoietic development and leukemia using both mouse and zebra fish models.

James C. Neil Ph.D., FRSE, FRSB, is a professor of virology and molecular oncology at the University of Glasgow, UK. He discovered that the *Runx2* gene was a frequent proviral insertion site for the Moloney murine leukemia virus in transgenic mouse models of thymic lymphoma in 1997. His current interests include the role of RUNX proteins in cancer survival and maintenance.

Nancy A. Speck Ph.D., is professor and chair of cell and developmental biology at the University of Pennsylvania, USA. Dr. Speck purified and cloned the RUNX and CBFβ proteins based on their binding to the Moloney murine leukemia virus enhancer in 1992–1993. Her work has focused on the role of Runx1 and CBFβ in hematopoiesis, particularly in the formation of hematopoietic stem cells in the embryo.

Andre van Wijnen Ph.D., is a professor of orthopedic surgery, biochemistry, and molecular biology at the Mayo Clinic, Rochester, Minnesota, USA. Dr. Wijnen works on Runx2 in bone biology and its role in cell cycle regulation.

About the Corresponding Authors

Suk-Chul Bae Ph.D., is a professor of biochemistry at Chungbuk National University and the director of the Institute for Tumor Research, Korea. Dr. Bae cloned murine *Runx3* and has shown that Runx3 is a suppressor of lung cancer.

Clara D. Bloomfield M.D., is a distinguished university professor and William G. Pace III professor of cancer research in the Department of Internal Medicine at the Ohio State University, USA. She was among the first investigators to show that older patients and even elderly patients with leukemia could be treated and cured with chemotherapy. Dr. Bloomfield has published extensively on the prognostic significance of chromosomal abnormalities in leukemia, including core-binding factor leukemia. She played a significant role in revising the World Health Organization classification of leukemia and lymphoma to include cytogenetics.

Karen Blyth Ph.D., is head of transgenic models at the Cancer Research UK Beatson Institute in Glasgow, UK. She studies the role of RUNX genes in promoting and suppressing tumor formation, with a particular focus on breast cancer.

Constanze Bonifer Ph.D., is a professor at the Institute of Cancer and Genomic Sciences and chair of experimental hematology at the University of Birmingham, UK. She has worked for many years in the area of epigenetics and chromatin biology, with a particular focus on the blood cell lineage. She has published extensively on chromatin regulation by RUNX1 and its oncogenic fusion protein, AML1-ETO.

John H. Bushweller Ph.D., is professor of molecular physiology and biological physics at the University of Virginia, USA. Dr. Bushweller published some of the first structures of the DNA-binding domain of RUNX1 and the CBFβ subunit. His laboratory uses structural and biophysical approaches to study normal and oncogenic forms of the RUNX1 and CBFβ transcription factors and is developing small molecule inhibitors of the oncogenic forms of the proteins.

Ewan R. Cameron Ph.D., MRCVS, is a professor of veterinary pathology, public health, and disease investigation and the head of the School of Veterinary Medicine, University of Glasgow, UK. He studies the dual role of the RUNX proteins in both promoting and suppressing tumor development.

Lucio Castilla Ph.D., is a professor of molecular, cell, and cancer biology at the University of Massachusetts Medical School, USA. Dr. Castilla was the first to develop a mouse model for CBFβ-SMMHC and showed that CBFβ-SMMHC induces the formation of preleukemic stem cells.

Marella F.T.R. de Bruijn Ph.D., is an associate professor of developmental hematopoiesis at the Weatherall Institute of Molecular Medicine, Oxford, UK. Dr. de Bruijn studies the role of Runx1 in hemogenic endothelium. She was the first to identify an enhancer in *Runx1* that marks hemogenic endothelium and hematopoietic stem cells in the embryo.

Paul J. Farrell Ph.D., FRCPath, FMedSci, is a professor of medicine at the Imperial College London, UK. He showed that Epstein-Barr virus regulates the transcription of the *RUNX1* and *RUNX3* genes and studies the role of RUNX proteins in B-cell proliferation.

Anthony M. Ford Ph.D., is a senior staff scientist in the Centre for Evolution and Cancer at the Institute of Cancer Research in London, UK. In collaboration with Dr. Mel Greaves, Dr. Ford showed that the initial steps in the development of pediatric cancer can occur in utero. He currently studies the clonal evolution of pediatric acute lymphoblastic leukemia associated with the *ETV6-RUNX*1 fusion gene.

Alan D. Friedman M.D., is the King Fahd professor of pediatric oncology and professor of pediatrics at the Johns Hopkins University School of Medicine, USA. He has published extensively on the role of the normal and oncogenic forms of RUNX1 and CBFβ in myeloid cell development and leukemia. Dr. Friedman was one of the first to show that the RUNX proteins regulate cell cycle progression.

Susumu Goyama M.D., Ph.D., is an associate professor of cellular therapy at the Institute of Medical Science, the University of Tokyo. He showed that the normal Runx1 protein promotes the growth of leukemia cells expressing certain oncogenic fusion proteins, including AML1-ETO. His current interests include developing antileukemia therapies targeting RUNX1.

Mel Greaves Ph.D., FMedSci, FRS, is a professor of cell biology at the Institute of Cancer Research in London, UK. Dr. Greaves is a pioneer in the study of pediatric leukemia. He was the first to show that the initial steps in the development of pediatric cancer can occur in utero. Specifically, he found that the t(12;21), the most common translocation in pediatric leukemia that creates the *ETV6-RUNX1* fusion gene, could be detected on the Guthrie cards of pediatric leukemia patients and hence was present at birth.

John D. Gross Ph.D., is principal investigator and director of the NRM Lab at the University of California, San Francisco, USA. He was one of the first to show that CBFβ is an essential cofactor of the HIV-1 protein Vif, which together regulate the APOBEC3 family of retroviral restriction factors.

Sunil R. Hingorani M.D., Ph.D., is a member in the Clinical Research Division of the Fred Hutchinson Cancer Research Center and a professor of medicine at the University of Washington, USA. Dr. Hingorani studies the molecular basis of pancreatic cancer. He showed that Runx3 promotes the metastasis of pancreatic cancer cells.

R. Katherine Hyde Ph.D., is an assistant professor of biochemistry and molecular biology at the University of Nebraska Medical Center, USA. As a postdoctoral fellow, Dr. Hyde studied the in vivo properties of CBFβ-SMMHC, with an emphasis on hematopoiesis in the mouse embryo.

Valerie Kouskoff Ph.D., is a group leader at the Cancer Research UK Manchester Institute, University of Manchester, UK. She studies mesoderm formation and the hemangioblast at the earliest stages of blood cell formation.

Toshihisa Komori M.D., Ph.D., is a professor of cell biology at the Nagasaki University School of Dentistry, Japan. Dr. Komori was one of the first investigators to demonstrate the essential role of RUNX2 in bone formation.

Georges Lacaud Ph.D., is the leader of the Stem Cell Group at the Cancer Research UK Manchester Institute, University of Manchester, UK. Dr. Lacaud studies blood cell development in the embryo and from embryonic stem cells. He was the first to show that Runx1 has a role in the hemangioblast. Dr. Lacaud currently studies Runx1 target genes and the activity of the *Runx1* promoters during hematopoiesis in the mouse embryo.

You Mie Lee Ph.D., is a professor of pathophysiology in the College of Pharmacy at Kyungpook National University, Korea. She showed that *RUNX3* expression is silenced through repressive histone modifications in response to hypoxia.

Klaus H. Metzeler M.D., is a principal investigator in the Department of Internal Medicine III at the Ludwig-Maximilians-Universität in Munich, Germany. Dr. Metzeler investigates the functional relevance of driver mutations in acute myeloid leukemia.

James C. Mulloy Ph.D., is a member of the Division of Experimental Hematology and Cancer Biology at Cincinnati Children's Hospital Medical Center and a professor of pediatrics at the University of Cincinnati, USA. He studies the activity of AML1-ETO in human hematopoietic stem cells and the biochemical mechanisms by which AML1-ETO promotes self-renewal and preleukemia.

Stephen D. Nimer M.D., is the director of the Sylvester Comprehensive Cancer Center and a professor of medicine at the University of Miami, USA. Dr. Nimer studies the AML1-ETO (RUNX1-RUNX1T1) fusion protein and its role in leukemogenesis. He was the first to show that lysine acetylation of AML1-ETO is essential for its leukemogenic activity.

Motomi Osato M.D., Ph.D., is an associate professor at the National University of Singapore and a professor at Kumamoto University, Japan. He was one of the first to identify somatic mutations in *RUNX1* in patients with myeloid malignancies. He has since studied the functions of RUNX proteins in hematopoiesis and leukemogenesis with an emphasis on stem cells. His current interests also include the identification of *cis*-regulatory elements in the RUNX family genes.

Antonino Passaniti Ph.D., is a professor of pathology at the University of Maryland in Baltimore, USA. He discovered that Runx2 collaborates with the YAP protein in promoting cell transformation. He currently studies the role of Runx2 in angiogenesis and of the TAZ gene in breast cancer.

Takaomi Sanda M.D., Ph.D., is a principal investigator at the Cancer Science Institute of Singapore, National University of Singapore. Dr. Sanda identified the transcriptional regulatory circuit involving the *RUNX1* gene in T-cell acute lymphoblastic leukemia when he was a postdoctoral fellow. He is currently studying transcriptional regulatory elements aberrantly activated by oncogenic transcription factors in T-cell leukemia.

Paul Shore, Ph.D is a senior lecturer in the Faculty of Life Sciences, University of Manchester, Manchester, UK. His research focuses on the role of Runx2 in the metastasis and invasion of breast and prostate cancer cells to secondary sites in the bone.

Stefano Stifani Ph.D., is a professor at McGill University, Montreal, Canada. Dr. Stifani identified Runx1 based on its interaction with the mammalian hairy and enhancer of split protein, Hes-1. Dr. Stefani studies the role of Runx1 in the development of the nervous system.

Tahir H. Tahirov Ph.D., is a professor at the Eppley Institute for Research in Cancer and Allied Diseases at the University of Nebraska Medical Center, USA. Dr. Tahirov published the first X-ray crystal structure of the ternary complex consisting of the DNA-binding domain of RUNX1 and CBFβ bound to DNA and more recently determined the structural basis for the cooperative DNA binding of RUNX1 and its frequent partner ETS1.

Ichiro Taniuchi M.D., Ph.D., is a group director at the Riken Center for Integrative Medical Sciences, Japan. Dr. Taniuchi first described the contribution of RUNX proteins to silencing of the *Cd4* gene. His research centers on the mechanism of lineage choice in T-cell development, with a continued focus on RUNX.

Jean Paul Thiery Ph.D., is the Toh Chin Chye visiting professor in the Yong Loo Lin School of Medicine at the National University of Singapore and a principal investigator in the Institute of Molecular and Cell Biology, A*STAR, in Singapore. Dr. Thiery was the first to propose the crucial contribution of the epithelial-mesenchymal transition (EMT) to the progression of carcinoma and has made seminal contributions to the study of minimal residual disease in breast cancer. He also discovered the first cell-cell adhesion molecule, N-CAM, and developed new physical approaches to measure the strength of intercellular adhesion.

Dominic Chih-Cheng Voon Ph.D., is an associate professor at the Institute for Frontier Science Initiative, Kanazawa University, Japan. Dr. Voon discovered that *Runx3* safeguards gastric epithelial cells against epithelial-mesenchymal transition-induced cellular plasticity and tumorigenicity. His current interests include the role of RUNX proteins in epithelial inflammation and plasticity.

Michelle J. West Ph.D., is a professor of tumor virology at the University of Sussex, UK. Her work focuses on the regulation of cellular and viral gene expression and B-cell transformation by Epstein-Barr virus.

Owen Williams Ph.D., is a principal investigator in the Developmental Biology and Cancer Programme at the University College London, UK. Dr. Williams studies the ETV6-RUNX1 fusion protein and the mechanism by which it contributes to the development of pediatric B-cell acute lymphoblastic leukemia.

A. Woollard Ph.D., is an associate professor in the Department of Biochemistry at the University of Oxford, UK. Dr. Woollard described a role for the *C. elegans RUNX1* and *CBFB* genes (*rnt-1* and *bro-1*) in stem cell proliferation.

Overview

The year 2016 is the 25th anniversary of the cloning of the first mammalian RUNX gene and its simultaneous association with human disease (Miyoshi et al. 1991). It is thus a fitting occasion to assemble a compendium of articles on the RUNX proteins. Although the RUNX genes gained their notoriety based on their involvement in human disease, they were first identified in *Drosophila* by Nüsslein-Volhard and Wieschaus in their historic screen for mutations that affected the basic body plan of the embryo (Nüsslein-Volhard and Wieschaus 1980). The mutant *runt*, named for the diminutive embryo, was categorized as one of six pair-rule mutants. Later cloning of the *Drosophila runt* gene by Gergen and colleagues revealed that it encoded a nuclear protein, with no known function (Kania 1990). The human *RUNX1* gene (originally named *AML1*) was the next to be discovered based on its location on chromosome 21 at the breakpoint of the 8;21 translocation in acute myeloid leukemia (Miyoshi et al. 1991). The homology of *RUNX1* to the *Drosophila runt* gene was noted, but the protein's function remained obscure. Only upon the purification and cloning of the mammalian RUNX proteins based on their biochemical properties, i.e., their ability to bind a specific sequence of DNA, did it become apparent that the RUNX proteins were transcription factors (Kamachi et al. 1990; Wang and Speck 1992; Wang et al. 1993; Meyers et al. 1993; Ogawa et al. 1993a, b). Also revealed by the biochemical studies was that the RUNX proteins constituted one subunit in a heterodimeric complex that also contained a non-DNA-binding partner, core-binding factor beta (CBFβ) (Kamachi et al. 1990; Wang et al. 1993; Ogawa et al. 1993a). At around the same time, the gene encoding CBFβ (*Cbfb*) was cloned; the human *CBFB* gene was identified as one of the genes disrupted by the inv(16) in acute myeloid leukemia (Liu et al. 1993). This convergence of information from multiple independent lines of inquiry and in particular the identification of translocations in *RUNX1* and *CBFB* in leukemia was an extremely exciting time in the field.

The field that began with a handful of laboratories has grown into a large and varied enterprise; as of today, a PubMed search for RUNX proteins yields >8000 references. It has become evident that the RUNX proteins are involved in many different areas of biology, ranging from basic cellular processes such as cell cycle and DNA repair, cell specification during development, stem cell biology, pathogenesis of solid and liquid tumors, and virology. This variety of functions is reflected by the diverse topics covered in this compendium.

Several investigators who were involved in the early days of discovery have contributed to this compendium, along with many other outstanding investigators who later joined the field and who have greatly broadened our knowledge about the RUNX proteins and their many biological and biochemical roles. It is impossible to cover all of the various developmental, biochemical, and disease processes in which these proteins play a part. Nevertheless, it is our hope that this volume will be a useful resource to those in the field or interested in the RUNX proteins and will stimulate further research on these fascinating molecules.

Genomic and Protein Structure of RUNX Family

For readers new to the field, here are the basics. There are three RUNX genes, *RUNX1*, *RUNX2*, and *RUNX3*, in mammals. Figure 1 shows genomic structure of three mammalian RUNX genes and the exon-intron structure of RUNX mRNAs including major splice variants. Note that there are two promoters in each gene, P1 (distal) and P2 (proximal). RUNX genes are evolutionarily very old, dating back to unicellular organisms (see Chap. 1 by Hughes and Woollard). Figure 2 shows an alignment of amino acid sequences of three human RUNX proteins. Amino acid sequences of RUNX proteins in many animal species are also well conserved. For example, Fig. 3 shows the amino acid sequence comparison between *Homo sapiens RUNX3* and *Caenorhabditis elegans Rnt-1*.

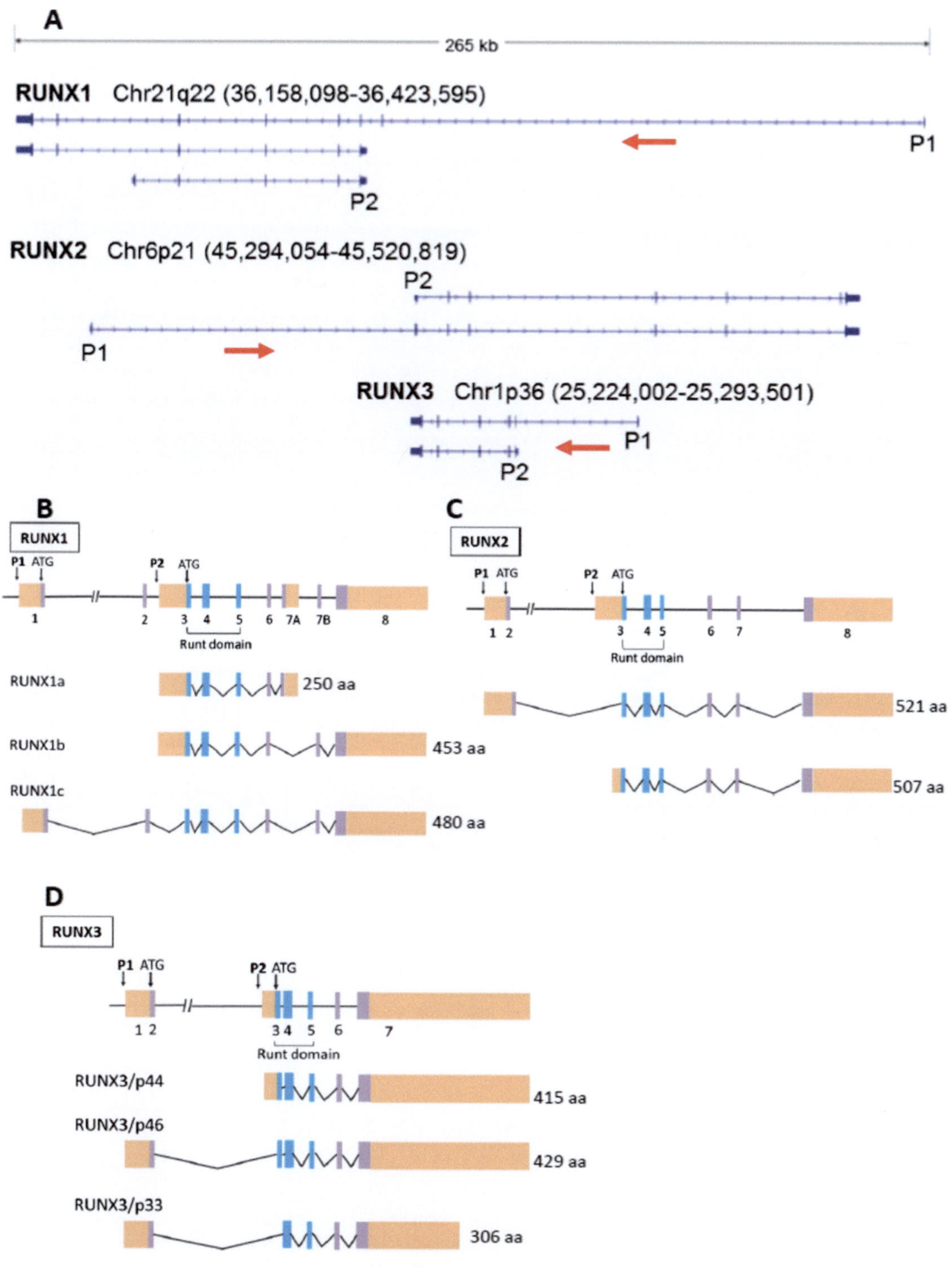

Fig. 1 Comparison of the genomic structure of and mRNA variants produced from the human *RUNX1*, *RUNX2*, and *RUNX3* genes. (**a**) Genomic structure of the three human RUNX genes. The positions of P1 and P2 promoters are indicated. *Red arrows* show the direction of transcription. Note the large size differences between the three genes. RUNX3, believed to be the ancestral gene, is the smallest (http://hgdownload.soe.ucsc.edu/downloads.html). (**b–d**) Exon-intron structure of *RUNX1* (**b**) (Levanon et al. 2001; Osato 2014), *RUNX2* (**c**) (Terry et al. 2004), and *RUNX3* (**d**) (Bangsow et al. 2001). The Runt domain is encoded in exons 3, 4, and 5 (*blue*). Sizes of exons and introns are not to scale. The splice variants of RUNX proteins suggest complex cellular mechanisms regulate and fine-tune RUNX activities

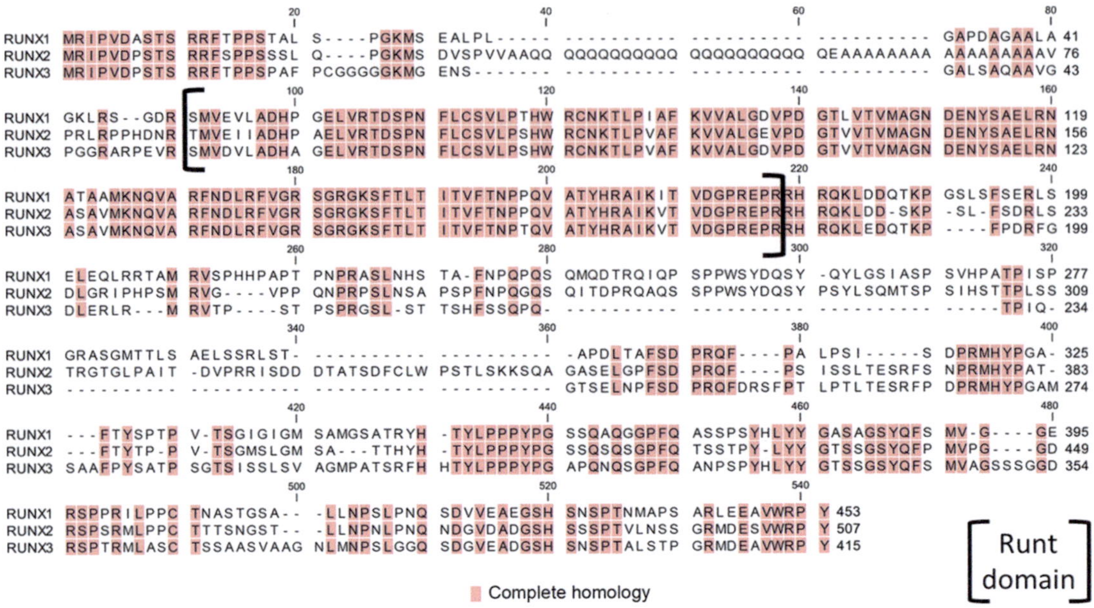

Fig. 2 Amino acid alignment of human RUNX1, RUNX2, and RUNX3. The highly conserved 128-amino acid region (*bracketed*) called the Runt domain is required for DNA binding and heterodimerization with CBFβ. The 5 amino acids at the C terminus, VWRPY, are also conserved. The VWRPY motif is required for binding to the corepressor Groucho (transducin-like enhancer of split 1)

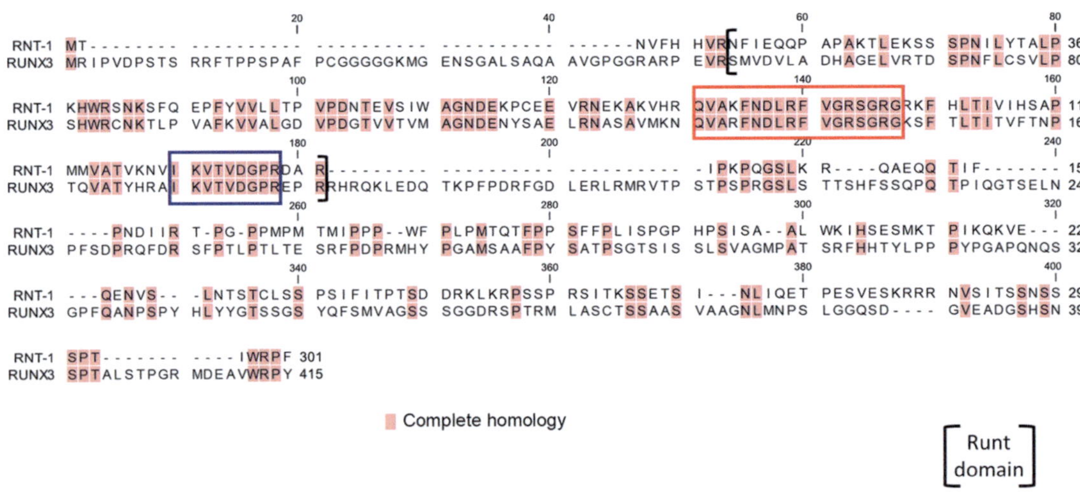

Fig. 3 Amino acid alignment of *Caenorhabditis elegans Rnt-1* (RNT-1: B0414.2) and *Homo sapiens RUNX3*. Two particularly highly conserved regions are indicated in boxes. The conserved sequences from aa131 to aa147 (*red box*) (the amino acid numbers correspond to *RUNX3*) contain a Walker motif A and AKT phosphorylation site, the significances of which are not known. The second conserved sequence is from aa170 to aa178 (*blue box*). These regions are likely to be important for evolutionarily conserved functions

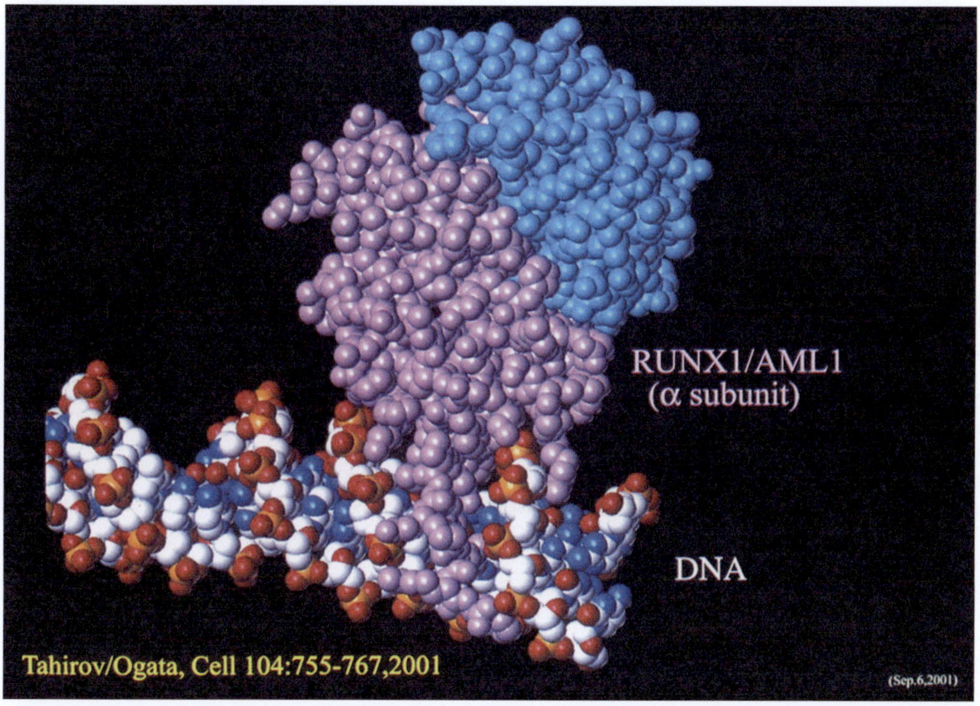

Fig. 4 Crystal structure of the Runt domain heterodimerized with a 134-amino acid region of CBFβ bound to DNA (Figure from Tahirov et al. 2001)

RUNX proteins have a conserved, obligate non-DNA-binding partner, CBFβ. CBFβ allosterically regulates DNA binding by the RUNX proteins (see Chap. 2 by Tahirov and Bushweller). Figure 4 shows crystal structure of the Runt domain together with CBFβ bound to DNA (courtesy of Dr. Tahir H. Tahirov).

Professor and Chair — Nancy A. Speck, Ph.D.
Department of Cell and Developmental Biology
University of Pennsylvania
Yong Loo Lin Professor of Medical Oncology — Yoshiaki Ito, M.D., Ph.D.
Cancer Science Institute of Singapore
National University of Singapore
September, 2016

References

Bangsow, C., Rubins, N., Glusman, G., Bernstein, Y., Negreanu, V., & Goldenberg, D., et al. (2001). The *RUNX3* gene – sequence, structure and regulated expression. *Gene, 279*, 221–232.

Kamachi, Y., Ogawa, E., Asano, M., Ishida, S., Murakami, Y., & Satake, M., et al. (1990). Purification of a mouse nuclear factor that binds to both the A and B cores of the polyomavirus enhancer. *Journal of virology, 64*, 4808–4819.

Kania, M.A., Bonner, A.S., Duffy, J.B., & Gergen, J.P (1990). The Drosophila segmentation gene *runt* encodes a novel nuclear regulatory protein that is also expressed in the developing nervous system. *Genes & development, 4*, 1701–1713.

Levanon, D., Glusman, G., Bangsow, T., Ben-Asher, E., Male, D. A., & Avidan, N., et al. (2001). Architecture and anatomy of the genomic locus encoding the human leukemia-associated transcription factor RUNX1/AML1. *Gene, 262*, 23–33.

Liu, P., Tarle, S. A., Hajra, A., Claxton, D. F., Marlton, P., & Freedman, M., et al. (1993). Fusion between transcription factor CBFb/PEBP2b and a myosin heavy chain in acute myeloid leukemia. *Science, 261*, 1041–1044.

Meyers, S., Downing, J. R., & Hiebert, S. W. (1993). Identification of AML-1 and the (8;21) translocation protein (AML-1/ETO) as sequence-specific DNA-binding proteins: The runt homology domain is required for DNA binding and protein-protein interactions. *Molecular and Cellular Biology, 13*, 6336–6345.

Miyoshi, H., Shimizu, K., Kozu, T., Maseki, N., Kaneko, Y., & Ohki, M. (1991). t(8;21) breakpoints on chromosome 21 in acute myeloid leukemia are clustered within a limited region of a single gene, *AML1*. *Proceedings of the National Academy of Sciences of the United States of America, 88*, 10431–10434.

Nüsslein-Volhard, C. & Wieschaus, E. (1980). Mutations affecting segment number and polarity in Drosophila. *Nature, 287*, 795–801.

Ogawa, E., Inuzuka, M., Maruyama, M., Satake, M., Naito-Fujimoto, M., & Ito, Y., et al. (1993a). Molecular cloning and characterization of PEBP2b, the heterodimeric partner of a novel Drosophila runt-related DNA binding protein PEBP2a. *Virology, 194*, 314–331.

Ogawa, E., Maruyama, M., Kagoshima, H., Inuzuka, M., Lu, J., & Satake, M., et al. (1993b). PEBP2/PEA2 represents a new family of transcription factor homologous to the products of the Drosophila *runt* and the human *AML1* gene. *Proceedings of the National Academy of Sciences of the United States of America, 90*, 6859–6863.

Osato, M. (2014). An unsung runt 6e isoform for HSC expansion. *Blood, 123*, 3684–3686.

Puig-Kroger, A. & Corbi, A. (2006). RUNX3: A new player in myeloid gene expression and immune response. *Journal of Cellular Biochemistry, 98*, 744–756.

Tahirov, T. H., Inoue-Bungo, T., Morii, H., Fujikawa, A., Sasaki, M., & Kimura, K., et al. (2001). Structural analyses of DNA recognition by the AML1/Runx-1 Runt domain and its allosteric control by CBFb. *Cell, 104*, 755–767.

Terry. A., Kilbey, A., Vaillant, F., Stewart, M., Jenkins, A., Cameron, E., & Neil, J. C. (2004). Conservation and expression of an alternative 3' exon of Runx2 encoding a novel proline-rich C-terminal domain. *Gene, 336*, 115–125.

Wang, S. & Speck, N. A. (1992). Purification of core-binding factor, a protein that binds the conserved core site in murine leukemia virus enhancers. *Molecular and Cellular Biology, 12*, 89–102.

Wang, S., Wang, Q., Crute, B. E., Melnikova, I. N., Keller, S. R., & Speck, N. A. (1993). Cloning and characterization of subunits of the T-cell receptor and murine leukemia virus enhancer core-binding factor. *Molecular and Cellular Biology, 13*, 3324–3339.

Part I

Evolution of RUNX Genes

RUNX in Invertebrates

1

S. Hughes and A. Woollard

Abstract

Runx genes have been identified in all metazoans and considerable conservation of function observed across a wide range of phyla. Thus, insight gained from studying simple model organisms is invaluable in understanding RUNX biology in higher animals. Consequently, this chapter will focus on the *Runx* genes in the diploblasts, which includes sea anemones and sponges, as well as the lower triploblasts, including the sea urchin, nematode, planaria and insect. Due to the high degree of functional redundancy amongst vertebrate *Runx* genes, simpler model organisms with a solo *Runx* gene, like *C. elegans*, are invaluable systems in which to probe the molecular basis of RUNX function within a whole organism. Additionally, comparative analyses of Runx sequence and function allows for the development of novel evolutionary insights. Strikingly, recent data has emerged that reveals the presence of a *Runx* gene in a protist, demonstrating even more widespread occurrence of *Runx* genes than was previously thought. This review will summarize recent progress in using invertebrate organisms to investigate RUNX function during development and regeneration, highlighting emerging unifying themes.

Keywords

Runx • Runt • *rnt-1* • *C. elegans* • Planarian • Sea urchin • *Drosophila*

1.1 Introduction

Although the triploblasts (which include mammals, insects, nematodes and sea urchins) and the diploblasts (corals and jellyfish) diverged very early in evolution, there are striking similarities between both groups, suggesting that a simple genetic "toolkit" directed the development of the

S. Hughes
Faculteit Techniek, Hogeschool van Arnhem en Nijmegen, Laan van Scheut 2, 6503 GL Nijmegen, The Netherlands

A. Woollard (✉)
Department of Biochemistry, University of Oxford, South Parks Road, Oxford OX1 3QU, UK
e-mail: alison.woollard@bioch.ox.ac.uk

Y. Groner et al. (eds.), *RUNX Proteins in Development and Cancer*, Advances in Experimental Medicine and Biology 962, DOI 10.1007/978-981-10-3233-2_1

common ancestor (Schierwater et al. 2009). Indeed, developmentally important transcription factors originated early in evolution and underwent a rapid expansion in number during early eumetazoan evolution (Coffman 2009; Degnan et al. 2009; Sebe-Pedros et al. 2011).

Transcription factors play crucial roles in development, as evidenced by the fact that a large proportion of developmentally impaired mutants in model organisms such as *Drosophila* and *C. elegans* have lesions in transcription factor genes. RUNX transcription factors are known for their involvement in several different embryonic and adult developmental processes, centered on controlling developmental decisions between cell proliferation and differentiation via interaction with various signal transduction pathways (Duffy et al. 1991; Coffman 2003, 2009; Nimmo and Woollard 2008). In almost all cases, RUNX function has been shown to be dependent on binding to CBFbeta, which acts to increase the affinity and specificity of DNA binding to target genes (Golling et al. 1996; Adya et al. 2000; Kaminker et al. 2001; Kagoshima et al. 2007). RUNX factors are also associated with context-dependent regulation via interaction with co-activators (e.g. Core Binding Factor, CBF and acetyltransferases e.g. p300) and co-repressors (e.g. Groucho) (Ito 1999; Speck 2001; Coffman 2003; Durst and Hiebert 2004; Chang et al. 2013).

Although *Runx* genes have been identified in all metazoans (Fig. 1.1), this review will focus on *Runx* in invertebrates. The RUNX family of transcription factors is defined by the presence of a highly conserved 128 amino acid Runt domain (Kagoshima et al. 1993; Crute et al. 1996). The Runt domain contains sites that are required for DNA binding, dimerization of Runx proteins with their binding partners and a C-terminal WRPY motif that is required for the interaction with the Groucho/TLE co-repressor (Kamachi et al. 1990; Kagoshima et al. 1993; Ogawa et al. 1993; Ito 1999). Although *Runx* genes have been identified in all metazoa, the core WRPY motif is absent in the *Runx* homologs of the dermosponge, *Amphimedon queenslandica*, and one of the two planarian *Schmidtea mediterranea Runx* (Robertson et al. 2009). Surprisingly, although *Runx* has until recently been considered to be specific to metazoa, two *Runx* homologs (*Co_Runx1* and *Co_Runx2*) have been identified in the unicellular amoeboid halozoan *Capsaspora owczarzaki*, (Sebe-Pedros et al. 2011). This suggests that *Runx* genes may actually have evolved prior to the divergence of protists from metazoans (Sebe-Pedros et al. 2011). Intriguingly, *Capsaspora* lacks any evidence of a CBFbeta homologue, suggesting RUNX may function independently in this organism. However, it is possible that sequence divergence makes the identification of a *Capsaspora* CBFbeta homologue particularly difficult, as CBFbeta homologues tend to be associated with a greater level of sequence divergence than *Runx* homologues. The functional significance of *Capsaspora Runx* genes remains to be elucidated. Likewise, very little functional information has been obtained from the solo sponge (*Amphimedon queenslandica* and *Oscarella carmela*) and sea squirt (*Ciona intestinalis*) *Runx* genes (Robertson et al. 2009), although these do provide valuable insights into the evolution of this important transcription factor family.

In contrast, several invertebrate phyla have *Runx* genes that have been subjected to extensive functional analysis, offering significant insights into molecular mechanism, functional conservation and possible links with human disease. The two premier model organisms for studying *Runx* are *Drosophila* and *C. elegans* although other useful insights have been gleaned from the sea urchin *Strongylocentrotus purpuratus* and more recently from the planarian flatworm *Schmidtea mediterranea*.

1.2 *Runx* Genes in the Fruit Fly, *Drosophila melanogaster*

Runx genes have been extensively studied in the fruit fly *Drosophila melanogaster*. In *Drosophila* as in other insects, four *Runx* genes have arisen as a consequence of gene duplication, independent of those that lead to the three vertebrate *Runx* genes (Rennert et al. 2003; Bao and Friedrich 2008). The first *Runx* family member to be

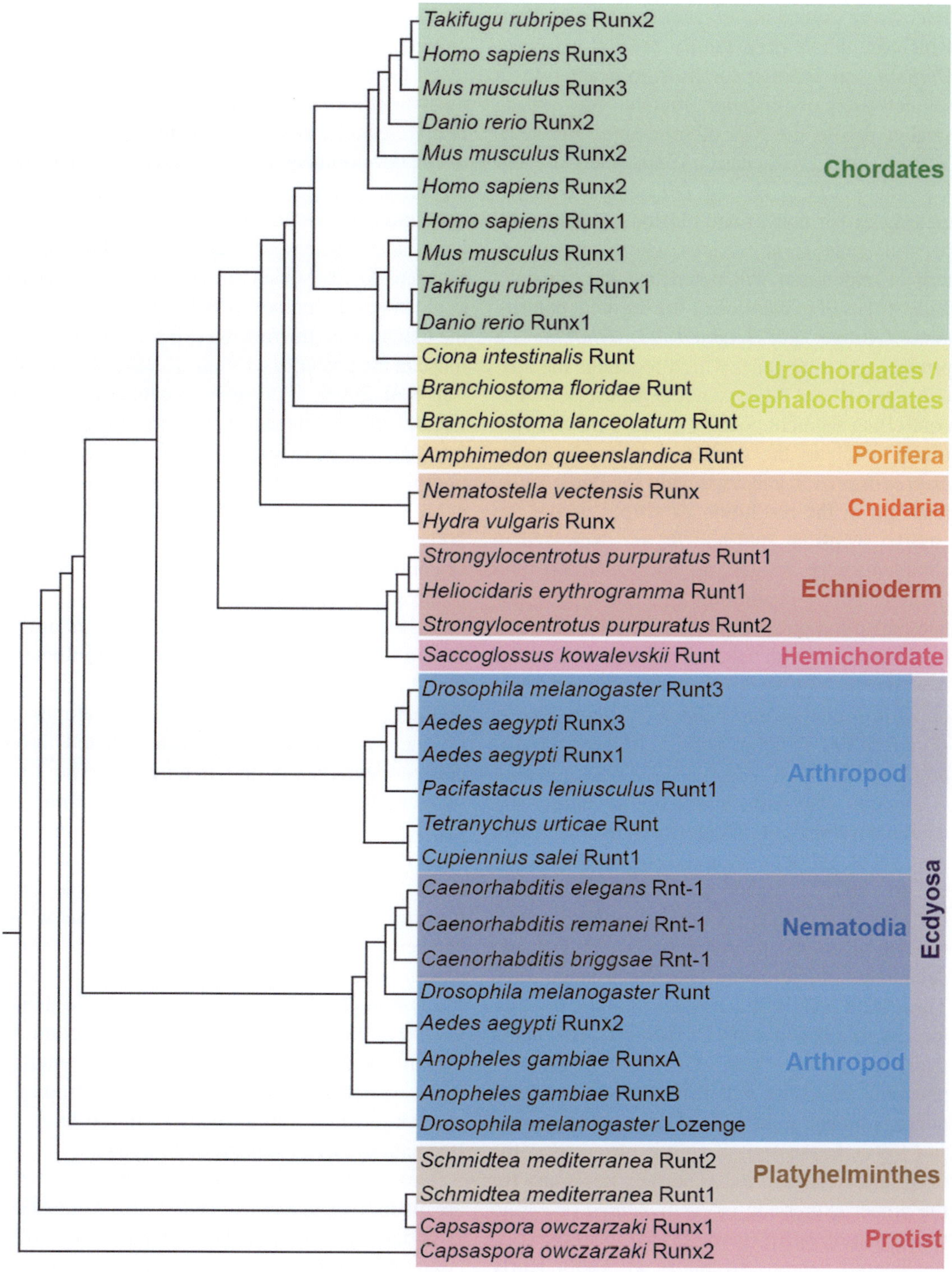

Fig. 1.1 *Runx* genes in the metazoa. *Runx* genes are represented in all major metazoan lineages, with a newly identified *Runx* gene in the unicellular protist *C. owczarzaki*. Alignments of whole Runx protein sequences were undertaken in MAFFT using Neighbor-joining, substitution model JTT and a bootstrap value of 1000 (Katoh et al. 2002)

extensively studied in *Drosophila* was *runt*, from which the whole gene family derived its name. *DmRunt* was isolated for its significant role in segmentation, with *runt* mutant flies being smaller due to the loss of segments (Nusslein-Volhard and Wieschaus 1980; Gergen and Wieschaus 1985). During *Drosophila* embryogenesis, at the mid-to-late blastoderm stage, the pair-rule genes form 7 stripes, whose precise pattern of expression will determine the one-cell-wide stripes of expression of the segment polarity genes (Klinger and Gergen 1993). *DmRunt* is a primary pair-rule gene, which regulates the spatial expression of other pair-rule genes, as well as controlling segment polarity genes. *DmRunt* positively regulates the secondary pair-rule genes, *fushi tarazu (ftz)*, and negatively regulates *hairy*, resulting in the resolution of stripes across the embryo such that *runt* and *ftz* are expressed in complementary stripes to *hairy* (Canon and Banerjee 2000). In addition, *runt* and *hairy* regulate each other independently of *ftz*. The result of this hierarchy, with *runt* at the top, is that the downstream segmentation genes convert positional information into patterns of gene expression, resulting in the generation of a regular and precise body plan.

DmRunt also plays a key role in embryonic neural development (Gergen and Butlet 1988; Kania et al. 1990; Duffy and Gergen 1991; Duffy et al. 1991; Canon and Banerjee 2000). *Drosophila* neurogenesis begins during embryogenesis when the neuroectoderm enlarges and delaminates to form the neuroblast stem cells. These stem cells will divide asymmetrically giving rise to a new neuroblast (self-renewal) and a ganglion mother cell, GMC (differentiated daughter cell), that will further divide to form neurons and/or glial cells (Campos-Ortega and Jan 1991). Expression of *runt* is observed in the GMC and neurons with its activity necessary for the proper expression of *even-skipped* (*eve*) and the formation of EL (*even skipped* (*eve*)-expressing lateral) neurons (Kania et al. 1990; Duffy et al. 1991). *runt* is necessary and sufficient to induce *eve* expression in the *Drosophila* nervous system, however the precise role for *runt* in the development of EL neurons is not fully understood.

Of the three other *Drosophila Runx* genes, the most significant is *lozenge*, *lz*, which was identified via genetic analysis through its contribution to eye development and its involvement in hematopoiesis. The eye develops from an epithelial structure (the eye imaginal disk) during the third larval stage, where an indentation in the epithelium marks the onset of differentiation (Daga et al. 1996). Precursor cells localized anterior to the indentation (the furrow) express *eyeless* while those in the posterior express *lz* (Daga et al. 1996; Yan et al. 2003). *lz* negatively regulates *seven-up* and *deadpan* while simultaneously up-regulating *bar* and *prospero* expression, resulting in the photoreceptors adopting their correct fate (Daga et al. 1996; Canon and Banerjee 2000; Yan et al. 2003). Thus, *lozenge* is crucial for the regulation of cell fate within the equivalence group of cells in the developing *Drosophila* eye.

lz is also a key regulator of cell fate and identity in *Drosophila* hematopoiesis. Multipotent blood cell progenitors are produced during two distinct time points in *Drosophila* development giving rise to three types of differentiated blood cell, collectively called hemocytes. The first wave of hematopoiesis occurs during embryogenesis, where prohemocytes arise from the head mesoderm and form two lateral clusters of cells, which will ultimately differentiate into plasmatocytes or crystal cells. The second wave of hematopoiesis comes during later larval stages, when blood cell progenitors arise from the lymph gland (Waltzer et al. 2010; Gold and Bruckner 2014). The final cell type that contributes to the blood cell population are lamellocytes, which are only produced upon immune challenge when foreign bodies are too large to be phagocytosed (Markus et al. 2009).

During the larval stage of hematopoiesis, there are distinct populations of cells with different differentiation potentials. The medullary zone (MZ) contains undifferentiated quiescent prohemocytes while the adjacent cortical zone (CZ) comprises of differentiated maturing hemocytes derived from the prohemocytes from the MZ

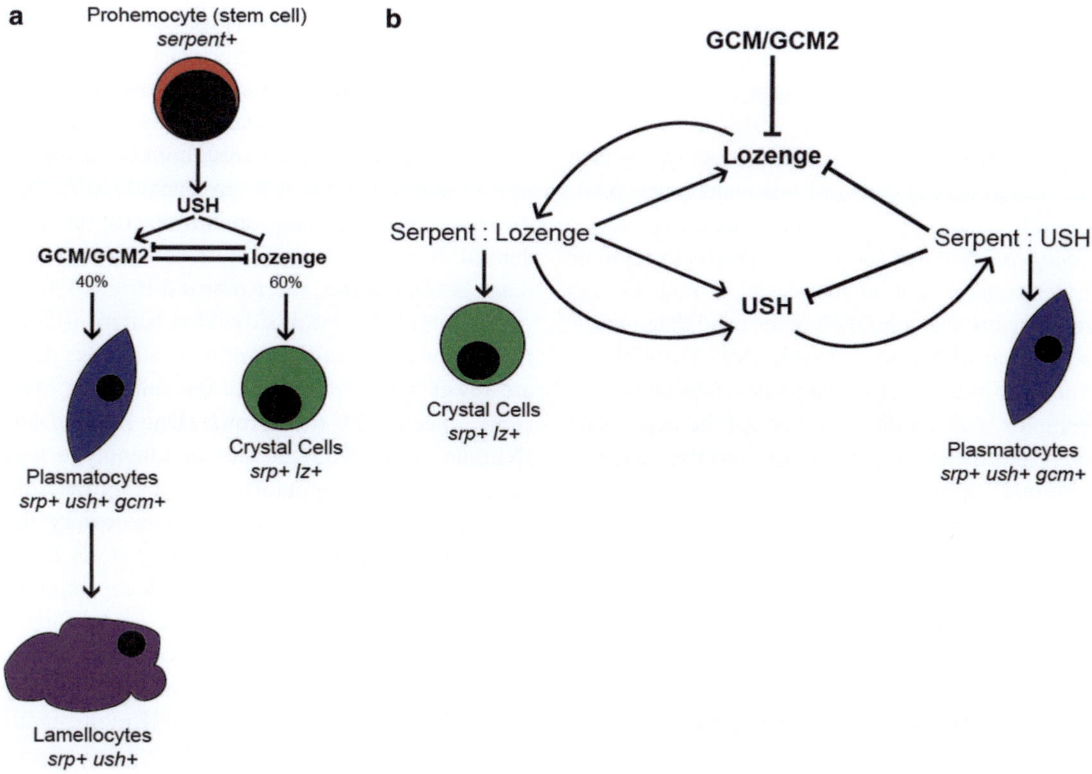

Fig. 1.2 Simplified diagram of the transcription factor network that controls cell fate in *Drosophila* hematopoiesis. (**a**) The prohemocytes are a stem cell population that express the GATA factor *serpent* (*srp*) that activates *ush* (*u-shaped*, friend of GATA (FOG) family) which will in turn function with *gcm/gcm2* (glial cells missing) to commit cells to the plasmatocyte lineage. In 60 % of the srp^+ prohemocytes, expression of *lozenge* (*lz*) will inhibit *gcm/gcm2*, and together with *srp*, will direct cells towards the crystal cell fate. (**b**) The regulation of cell differentiation by *lz/srp/ush* is dynamic, involving a bi-potential regulatory state that resolves two distinct cell populations; the crystal cells and the plasmatocytes. *srp* initiates and maintains *lz* expression. The SRP:LZ complex activates *ush* which will compete with LZ for binding to SRP. The SRP:USH complex negatively regulates both *lz* and *ush*, while GCM/GCM2 will independently suppress *lz* transcription (Adapted from Muratoglu et al. 2007; Braun and Woollard 2009; Wang et al. 2014)

(Jung et al. 2005). *lz* is only expressed in the CZ by prohemocytes adopting the crystal cell fate (Lebestky et al. 2000; Gajewski et al. 2007). Although *lz* expression is activated in all prohemocytes, only 60 % of these lz^+ cells will maintain *lz* expression via a feedback loop and differentiate into crystal cells while the remaining 40 % of cells are $lz^{\text{â}\check{}}$ and thus differentiate into plasmatocyes (Fig. 1.2a) (Bataille et al. 2005). The molecular mechanism by which *lz* expression translates to the lineage commitment of prohemocytes to either crystal cells or plasmatocytes involves a complex transcriptional circuit (Muratoglu et al. 2006, 2007). *lz* expression is regulated by a feedback loop involving the pan-hematopoietic GATA factor *serpent*, promoting crystal cell differentiation (Bataille et al. 2005), while expression of *ush* (friend-of-GATA family of transcription factors, *u-shaped*) in lz^+ prohaemocytes is required, together with *serpent*, to direct plasmatocyte cell fate (Fig. 1.2b) (Muratoglu et al. 2007). The complex regulation of *lz, srp* and *ush* is dynamic and results in two distinct cell populations, the plasmatocytes (srp^+ush^+) and crystal cells (srp^+lz^+). Several aspects of this circuitry remain to be elucidated, including the mechanism by which *ush* is turned off in crystal cells.

Additional antagonists of *lz* which direct crystal cell fate are the transcription factors *gcm* (glial cells missing) and its homologue *gcm2*, which act with reciprocal asymmetry with *lz* limiting the expression of *lz* and therefore reducing the production of crystal cells (Alfonso and Jones 2002; Bataille et al. 2005). The mechanism by which *gcm/gcm2* and *ush* act in combination to regulate *lz* expression and maintenance is unclear, but recent work has identified other candidates in the regulation of lineage commitment. Through the Salvador-Warts-Hippo pathway, *yorki* acts in a complex with *scalloped* to control the expression of *lz* and therefore regulate the proliferation and terminal differentiation of progenitor cells into crystal cells (Milton et al. 2014). Thus, *lz* is at the hub of an increasingly complex transcriptional network directing *Drosophila* hematopoiesis.

1.3 *Runx* Genes in the Nematode, *Caenorhabditis elegans*

The single *C. elegans Runx* homolog, *rnt-1*, is an important regulator of the balance between proliferation/self-renewal and differentiation in the lateral neuroectodermal seam cells (Kagoshima et al. 2005; Nimmo et al. 2005; Xia et al. 2007). The seam cells are a group of multipotent stem-cell like cells formed during embryogenesis that divide in a stereotypical pattern throughout larval development. Animals hatch with 10 seam cells per lateral side of the animal, most of which proceed through a re-iterative series of asymmetric divisions, interspersed by the odd symmetrical division in order to expand the number of progenitor cells. In this sense, the seam cells provide a useful paradigm for the stem cell mode of division. In general, at each larval molt there is an asymmetric division producing a posterior daughter cell that retains the ability to self-renew, and an anterior daughter cell that differentiates into either a hypodermal cell, a glial cell or a neuronal cell (Fig. 1.3a) (Sulston and Horvitz 1977). In addition, there is a single symmetrical (proliferative) division at the L2 stage whereby both daughter cells retain the proliferative ability and consequently expand the pool of seam cells so that adult worms have 16 seam cells per side (Fig. 1.3b). At the last larval stage (L4), after the final round of cell division, the seam cells terminally differentiate and fuse into a syncytium. However, although the terminal differentiation of the seam cells occurs at the start of adulthood, the cells are capable of further divisions under certain circumstances, as evidenced in heterochronic mutants (Nimmo and Slack 2009; Harandi and Ambros 2015).

The regulation of this division pattern is controlled by *rnt-1*. In *rnt-1* mutant animals, there are fewer seam cells due to the failure of divisions, specifically the symmetrical L2 division (Nimmo et al. 2005). A similar phenotype was observed in *bro-1* mutants, *bro-1* being the sole *C. elegans* homolog of CBFbeta necessary for correct RNT-1 function (Kagoshima et al. 2007; Xia et al. 2007). BRO-1 enhances the binding affinity and specificity of RNT-1, and is itself regulated by the GATA transcription factor, ELT-1 which acts as a direct activator of *bro-1* to promote seam cell proliferation (Brabin et al. 2011).

In contrast to the mutant phenotype of fewer seam cells at adulthood, overexpressing *rnt-1* and *bro-1* leads to seam cell hyperplasia at the expense of other differentiated cell types (Kagoshima et al. 2007). This is in large part due to the symmeterisation of normally asymmetric divisions, leading to the production of two proliferative daughters rather than a single one, and resulting in the tumourous appearance of the seam tissue (Nimmo et al. 2005; Kagoshima et al. 2007).

Expression of *rnt-1* is observed in the seam cells during embryogenesis and throughout larval development, where it is normally restricted to the proliferative (posterior, seam) daughter and not the hypodermal (anterior, differentiated) daughter cell (Kagoshima et al. 2005, 2007). Thus *rnt-1* expression is closely associated with, and crucial for, the promotion of the proliferative fate, at the expense of the differentiative fate. The molecular mechanism by which *rnt-1* promotes proliferation likely involves repression of the CIP/KIP CDK inhibitor *cki-1* in the posterior daughter destined to proliferate further (Nimmo et al. 2005).

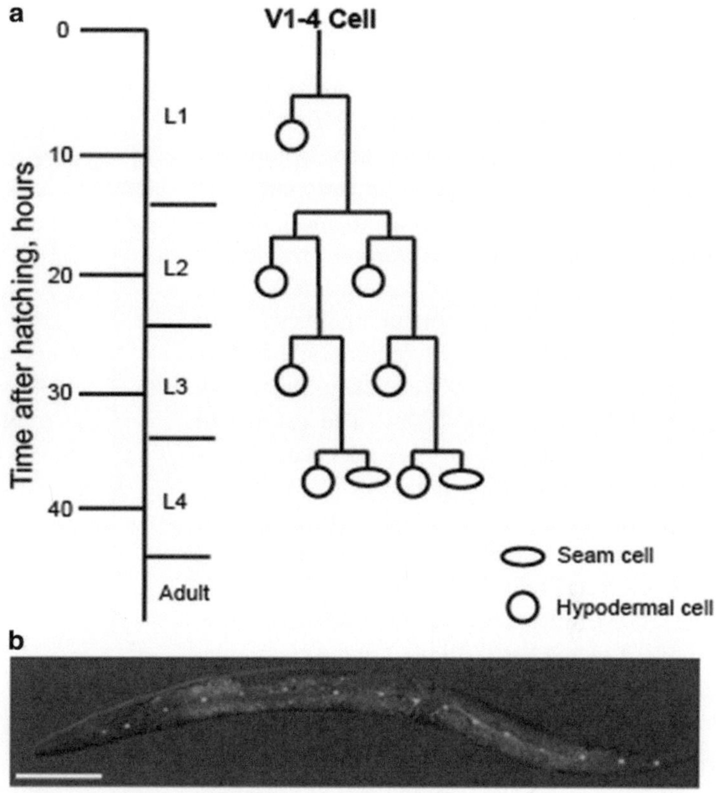

Fig. 1.3 Seam cells in *Caenorhabditis elegans*. (**a**) Lineage diagrams of the anterior V seam cells, which most obviously display the stem-like mode of division. The asymmetric divisions occur at each larval stage with an additional symmetric division at the L2 stage. In general, at adulthood, each V cell will have given rise to seven hypodermal nuclei and two seam cells that will terminally differentiate in adulthood. (**b**) An image of an early adult *C. elegans* which expresses a seam cell marker, *scm::gfp* (Strain, *JR667*). There are 16 seam cell nuclei running along each side of the animal at the end of development. Scale bar is 100 Î¼m

A major player in *rnt-1* regulation in *C. elegans* is the *ceh-20/unc-62* transcriptional partnership (homologous to the Pbx/Meis complex in mammals). Both *ceh-20* and *unc-62* mutants display seam cell hyperplasia, caused, like *rnt-1/bro-1* overexpression, by the symmetrisation of seam cell divisions such that both daughters adopt the proliferative fate (Hughes et al. 2013). *ceh-20/unc-62* seam hyperplasia is completely supressed in *rnt-1/bro-1* mutants, suggesting that *rnt-1* likely operates downstream of *ceh-20/unc-62* to promote proliferation. The fact that *rnt-1* expression appears to be de-repressed in anterior daughters (that would normally differentiate) when *ceh-20/unc-62* are silenced, suggests that *ceh-20/unc-62* function upstream to repress *rnt-1* expression in cells that normally quit the cell cycle in order to differentiate (Hughes et al. 2013).

The expression of *rnt-1* has also been observed in intestinal cells. Although RNT-1::GFP is undetectable in the intestine at adulthood, *rnt-1* mRNA is present in the adult intestine, suggestive of post-transcriptional regulation (Lee et al. 2012). Indeed, RNT-1 has been shown to be stabilized in the intestine following oxidative stress, with *rnt-1* mutants displaying increased sensitivity to these conditions (Lee et al. 2012). Given that the intestine is the first line of defence against the environment, it is possible that the post-transcriptional control of RNT-1 provides a mechanism for a rapid response to environmental changes. The p38 MAP kinase pathway plays an important function in stress response in *C. elegans* (Inoue et al. 2005) and acts to directly phosphorylate RNT-1, stabilising it via inhibition of degradation (Lee et al. 2012).

1.4 *Runx* Genes in the Sea Urchin, *Strongylocentrotus purpuratus*

Strongylocentrotus purpuratus has two *Runx* genes with the sole characterized *Runx*, *SpRunt-1*, expressed during embryogenesis and transiently expressed in adult coelomocytes as a consequence of immune challenge (Coffman et al. 1996; Pancer et al. 1999; Robertson et al. 2002). During embryogenesis, *SpRunt-1* promotes the expression of a number of zygotically induced *Wnt* genes, in particular *wnt6* and *wnt8* (Robertson et al. 2008). Indeed, morphillino-antisense silencing of *SpRunt-1* results in impaired cell proliferation during late blastula development and widespread apoptosis as a consequence of the down regulation of these *Wnts* (Coffman et al. 2004; Dickey-Sims et al. 2005; Robertson et al. 2008). The reverse of this, where *wnt6* and *wnt8* are silenced, phenocopies the proliferation defect of the *SpRunt-1* morphant. Evidence for the direct regulation of *Wnt* by Runt-1 comes from mutational analysis of a *wnt8 cis*-regulatory module (Minokawa et al. 2005). SpRunt-1 cooperates with the effectors Tcf/Lef and Krox/Blimp-1 at the *cis*-regulatory region ('module C') of *wnt8*, which is necessary for the beta-catenin dependent maintenance of *wnt8* activity in the endomesoderm (Minokawa et al. 2005; Robertson et al. 2008). Additionally, GSK-3beta (the sole sea urchin glycogen synthase kinase that targets mitogenic proteins for ubiquitination), which itself is negatively regulated by Wnt signaling, is able to stabilize SpRunt-1 when inhibited, highlighting the complex interplay between RUNX and Wnt (Fig. 1.4) (Robertson et al. 2008).

Recent evidence has implicated the serine/threonine kinase, AKT, as a key mediator of mitogenic RUNX function in sea urchin, via phosphorylation and inhibition of GSK-3 (Robertson et al. 2013), with *akt-2* morphant animals phenocopying *SpRunt-1* morphants (Dickey-Sims et al. 2005; Robertson et al. 2013). In a further complication it is thought that RUNX also activates PKC in a positive feedback loop to inhibit GSK-3beta (Dickey-Sims et al. 2005; Robertson et al. 2008, 2013). Overall, SpRunt-1 appears to have a number of distinct roles depending on developmental stage, but as in *C. elegans*, with an emphasis on promoting cell proliferation.

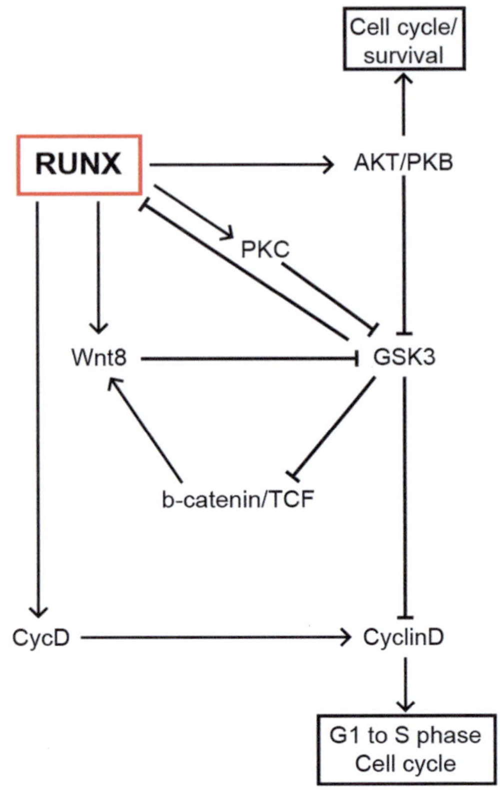

Fig. 1.4 Regulatory circuit through which *runx* regulates cell proliferation in the sea urchin embryo. The transcription factor Runx directly activates embryonic *wnt8* that is necessary for the beta-catenin dependence maintenance of *wnt8* activity. *SpRunt-1* is as an anti-apoptotic factor that, together with AKT functions through the direct regulation of PKC and GSK-3. RUNX and GSK-3 function in a mutually antagonistic regulatory pathway suggesting that, in sea urchin, RUNX promotes somatic cell proliferation by activating genes, including *pkc*, in a positive feedback loop to inhibit GSK-3 (Adapted from Robertson et al. 2002, 2008; Dickey-Sims et al. 2005, 2013)

1.5 *Runx* Genes in the Planarian Flatworm, *Schmidtea mediterranea*

Planarians are relatively simple free-living platyhelminthes that lie at an important juncture of the evolution of the basal metazoans (Newmark and Sanchez-Alvarado 2002). Planarians such as *Schmidtea mediterranea*, have amazing developmental plasticity due to the presence of a large population of pluripotent stem cells called neoblasts, with the striking ability to regenerate missing body parts following injury (Newmark and Sanchez-Alvarado 2002; Reddien and Sanchez-Alvarado 2004; Sanchez-Alvarado and Tsonis 2006; Forsthoefel and Newmark 2009; Salo et al. 2009; Wagner et al. 2011). After wounding, the neoblasts respond by undergoing proliferation, followed by migration to the wound site and finally local differentiation into the specific cell types required to generate new tissue (Eisenhoffer et al. 2008; Wenemoser and Reddien 2010; Lapan and Reddien 2011; Scimone et al. 2014).

Transcriptome analysis has revealed a number of genes that are significantly upregulated during the period of neoblast self-renewal as a response to damage (Sandmann et al. 2011; Wenemoser et al. 2012). *runt-1* is one such gene, being expressed within 30 min of wounding, likely as an immediate response to the injury. A second wave of *runt-1* expression is induced 3-12 hours post wounding (Wenemoser et al. 2012; Wurtzel et al. 2015). The role of *runt-1* in the planarian response to injury is to firstly direct the proliferation of cells, followed by the differentiation of these cells into lineage restricted precursors. Following wounding, knockdown of *Smed-Runt-1* by RNAi results in defects in cell positioning and photoreceptor phenotypes in the eye (Sandmann et al. 2011; Wenemoser et al. 2012), indicative of *Smed-Runt-1* promoting the formation of fate restricted neoblasts in the anterior of the animal following wounding to form eyes.

1.6 *Runx* Genes in the Cnidaria

A similar upregulation of *runt-1* has been observed following injury and during regeneration in the sea anemone (*Nematostella vecterisis*) (DuBuc et al. 2014) where *NvRunt-1* is localized to the pluripotent progenitors of the sensory neurons in ectodermal cells of the tentacle tips (Sullivan et al. 2008). Hydra, like sea anemones, are members of the phylum Cnidaria and are freshwater polyps with a symmetrical tubular body. As in *S. mediterranea*, a pool of heterogeneic stem cells have been identified in hydra (Govindasamy et al. 2014). These stem cells are quiescent until they become activated to enter the cell cycle following removal of the head (Govindasamy et al. 2014) with *runt-1* upregulated following decapitation (DuBuc et al. 2014; Petersen et al. 2015).

Thus, a role for *runt-1* in regeneration in planarians and cnidarians such as sea anemone and hydra appears to be associated with the stimulation of both cell proliferation and subsequent differentiation following injury. In this way, RUNX may play a key role in the transition of undifferentiated cells into committed lineage precursors, and therefore provide new insights into the control of regenerative processes.

1.7 Comparative Analysis Delineates Emerging Themes in RUNX Biology

Establishing functional relationships between genes in very diverse organisms is a daunting, yet appealing task, beset with problems of interpretation and translation between systems. Nevertheless, any systematic examination of RUNX biology throws up some immediate areas of commonality, both in terms of biological processes as well as molecular pathways, and it is these areas of commonality that may hold the key to unlocking a broader understanding of RUNX biology in increasingly complex organisms.

1.8 Conserved RUNX-Associated Biological Processes

1.8.1 Regulation of the Transition from Quiescence into Proliferation

Runx genes have an obvious role in promoting cell proliferation in many species. The function of *rnt-1* in *C. elegans* seam cells to promote proliferation bears a remarkable similarity to the role of mammalian *Runx1* in hair follicle stem cells (HFSC). Both stem cell systems are comprised of epidermal cells where divisions occur after long quiescent phases. In the worm, seam cells are quiescent until the molt preceding each larval transition when the cells divide in a *rnt-1*-dependent manner. Similarly, in mammals, Runx1 activates quiescent stem cells in the hair follicle, with *Runx1* mutant mice having an extended quiescent phase and defects in HFSC colony formation (Osoiro et al. 2008). Further invertebrate examples of *Runx* genes functioning in cell proliferation include the sea urchin, where inactivation of *SpRunt-1* is associated with proliferation defects and both hydra and planaria, where *runt-1* appears to be involved in promoting cell proliferation following injury (Sandmann et al. 2011; Wenemoser et al. 2012; DuBuc et al. 2014; Govindasamy et al. 2014; Petersen et al. 2015). These latter observations support the idea that *Runx* genes may have a general role to play in regeneration. An additional example of RUNX-dependent proliferation in mammals is in the nervous system, where Runx1 is required to sustain the proliferation of olfactory receptor neuron (ORN) precursors (Theriault et al. 2005). Indeed, in this example, over-expression of *Runx1* increased the number of proliferating cells, much like the over-expression of *rnt-1* in *C. elegans* seam cells, causing hyperplasia of this cell type (Nimmo et al. 2005; Kagoshima et al. 2007).

Moving from quiescence into proliferation involves transduction of growth factor signalling, and *Runx* genes appear generally to have an important role in this process. For example, in the HFSC system, *Runx1* mutants do not respond properly to a growth signal, thus proliferation fails. Intriguingly, in *C. elegans* which are starved, the *rnt-1* mutant phenotype is enhanced (Nimmo et al. 2005), and *rnt-1* was found to be one of the most highly up-regulated genes following re-feeding after starvation (Baugh et al. 2009), consistent with an important role for *Runx* genes in transducing environmental information to achieve properly coordinated growth and development. Furthermore, the role of *C. elegans rnt-1* in regulating stress response (Lee et al. 2012) is intriguing in the light of recent data suggesting that mammalian *Runx1* deficient hematopoietic stem cells (HSC) display increased stress resistance (Cai et al. 2015), together with lower rates of translation, attenuated p53 signalling and a decrease in ribosome biosynthesis. Understanding the molecular pathways that result in the altered metabolic profile of *Runx1*-deficient HSC will have significant implications in treating leukaemia.

Finally, the important role for *Runx* genes in controlling cell number in invertebrate models by promoting, or even repressing in some examples, (Kramer et al. 2006; Murthy et al. 2014) cell proliferation resonates strongly with the well-characterised role of *Runx* genes as oncogenes or tumour suppressors, depending on context (Strom et al. 2000; Cameron and Neil 2004; Ito 2004; Wotton et al. 2004; Keita et al. 2013; Wysokinski et al. 2015). This suggests that invertebrate model systems have useful contributions to make the field of Runx-associated carcinogenesis.

1.8.2 Lineage Commitment and Cell Fate Determination

Runx genes have been described as molecular switches coordinating the developmental balance between proliferation and differentiation (Nimmo and Woollard 2008). There are certainly many examples of *Runx* genes acting to promote proliferation, as we have seen, and there are several examples of *Runx* genes acting in lineage commitment and cell fate decisions; there are two examples of *Runx* genes being required for eye development (planaria and fly), two examples of a requirement during haematopoiesis (mammals and fly) and several examples of a role in neurogenesis (fly, worm, mammals, planaria). But whether these shared functions are orthologous, in the sense that they indicate an ancient origin, or whether they are examples of *Runx* genes being co-opted during evolution for different purposes, some common between different organisms and some not, is difficult to determine.

The most intriguing shared function is surely haematopoiesis. Runx1 has long been known to regulate the differentiation of hematopoietic stem cells (HSCs) from myeloid precursors in mammals (Tanaka et al. 1995; Ahn et al. 1998; Yokomizo et al. 2001). In fact, RUNX proteins are expressed throughout all hematopoietic lineages, being necessary for the emergence of the first HSCs through to their terminal differentiation. In *Drosophila*, the hemocytes formed during larval development (in a process resembling vertebrate definitive hematopoiesis) most closely resemble vertebrate myeloid linages (Waltzer et al. 2010), with the plasmatocytes having a similar function to vertebrate macrophages (Lanot et al. 2001; Wood and Jacinto 2007). The parallels between the complex network of transcription factors regulating lineage commitment in *Drosophila* crystal cells and human thymocytes are striking, with co-factors such as GATA factors figuring prominently in both cases. In recent years, there is increasing evidence for a role of *RUNX* in the immune system beyond haematopoiesis (Ito et al. 2008; Kitoh et al. 2009; Wong et al. 2011, 2012, 2014; Lotem et al. 2013). It has long been known that sea urchin *Runx* is expressed as a consequence of immune challenge, and more evidence is emerging for the function of *RUNX* in the mammalian immune system (reviewed in Voon et al. 2015) that may allow for future comparative analysis.

1.9 Conserved RUNX Molecular Pathways

Evaluating conservation of molecular mechanisms involving RUNX is perhaps even more difficult than assessing conserved processes. Firstly, experiments may be difficult to translate between organisms. Secondly, transcription factors can be co-opted into many different signalling pathways over the course of evolution, and adopt many different target genes depending on the context of their precise role. Finally, *Runx* genes have emerged, been lost, multiplied and diverged, so that evolutionary history presents many molecular fossils that are hard to interpret, and there is the additional confounding factor of convergent evolution. Nevertheless, certain similarities in the molecular architecture of RUNX function across highly divergent groups appear to stand out. One example is the interaction of RUNX with cell cycle genes, and other examples include interactions with Wnt signalling and GATA factors.

1.9.1 Interaction with Cell Cycle Genes

The role of *Runx* genes in the transition from quiescence to proliferation is associated in several cases with the direct regulation of the cell cycle. In *C. elegans rnt-1* mutants, expression of *cki-1* (cyclin dependent kinase inhibitor of the CIP/KIP family) is upregulated in seam cells, and depleting *cki-1* in these animals rescues the seam cell proliferation defect (Nimmo et al. 2005). RNT-1 is therefore acting (directly or indirectly) to repress the expression of *cki-1* in seam cells destined to divide. With striking similarity, RUNX1 and RUNX2 have been shown to repress the cyclin-dependent kinase inhibitor p21 in mammalian cell culture (Strom et al. 2000; Bernardin and Friedman 2002; Westendorf et al. 2002; Bernardin-Fried et al. 2004). Similarly, in sea urchin, RUNX induces cyclinD during embryogenesis leading to cell cycle progression (Coffman et al. 2004; Dickey-Sims et al. 2005; Robertson et al. 2008).

1.9.2 Interaction with Wnt Signalling

In the sea urchin, experiments show that SpRunt1 binds directly to *wnt6* and *wnt8* in the late blastula stage of embryogenesis (Robertson et al. 2008), and depletion of SpRunt1 is associated with a decrease in Wnt signalling. This is also the case in mammals where *wnt4* gene expression is reduced in *Runx1* knockout mice (Naillat et al. 2015), although the mechanism in this latter case likely involves TCF/LEF (T-cell factor/lymphoid enhancer factor) binding to RUNX1 in order to attenuate Wnt signalling. Indeed, there are several examples of TCF interactions with *Runx* genes, including the binding of TCF1 to RUNX2 during osteoblast development (Kahler and Westendorf 2003), the interaction of TCF7 and

RUNX1 in haematopoiesis (Wu et al. 2012) and the interaction of TCF4 with RUNX3 to regulate Wnt signalling, which has been linked to gastric cancer (Ito et al. 2008, 2011). Overall, RUNX, TCF/LEF and Wnt signalling have been shown to act together in a context dependent manner to activate or repress transcription of genes to control cell fate choice in a variety of tissues. However, although interactions between Wnt signaling and Runt have been demonstrated in sea urchin, there is little to support this in *Drosophila* or *C. elegans*. Indeed, in nematodes it is likely that, at least in the stem cell-like seam cells, *rnt-1* acts in a parallel pathway to Wnt (Gleason and Eisenmann 2010; Hughes et al. 2013; Gorrepati et al. 2015).

1.9.3 Interaction with GATA Factors

In the nematode RNT-1 and BRO-1 regulate the proliferation of seam cells, with the GATA transcription factor ELT-1 directly regulating *bro-1* (Brabin et al. 2011). The function of RNT-1, BRO-1 and ELT-1 in the worm directly reflect the roles of RUNX, CBFbeta and GATA in stem cells in other systems. The interaction of these transcription factors is reminiscent of the situation in *Drosophila* and mammalian haematopoiesis where GATA/Serpent, RUNX/Lozenge and CBF-beta/Brother tightly control cell fate choice (Li and Gergen 1999; Waltzer et al. 2003; 2010; Pencovich et al. 2011).

1.10 Conclusion

There are intriguing connections between RUNX functions in mammals and invertebrates, centering on the regulation of cell proliferation and lineage commitment. Invertebrate models such as *C. elegans*, *Drosophila* and the sea urchin are useful in the study of RUNX function because they offer unique options in relation to genetic manipulation and ease and speed of experimentation. Work in *C. elegans* offers the particular advantage of the lack of functional redundancy issues, as it contains a solo *Runx* homologue. However, it does not appear to be the case that research in invertebrate models will necessarily uncover a single ancestral function of *Runx* genes that explains the range of functions documented in mammals. On the contrary, different invertebrate models have proved invaluable to highlight and investigate specific functions of *Runx* genes reported in vertebrates. Taken together, studies of invertebrate RUNX biology provides a wealth of information that will be instrumental in our understanding of the importance of *Runx* genes in developmental control and in the fight against disease.

References

Adya, N., Castilla, L. H., & Liu, P. P. (2000). Function of CBFb/Bro proteins. *Seminars in Cell and Developmental Biology, 11*(5), 361–368.

Ahn, M. Y., Huang, G., Bae, S. C., Wee, H. J., Kim, W. Y., & Ito, Y. (1998). Negative regulation of granulocytic differentiation in the myeloid precursor cell line by 32DcI3 by *ear-2*, a mammalian homolog of *Drosophila seven-up*, and a chimeric leukemogenic gene, AML1/ETO. *PNAS, 95*(4), 1812–1817.

Alfonso, T. B., & Jones, B. W. (2002). *gcm2* promotes glial cell differentiation and is required with glial cells missing for macrophage development in *Drosophila*. *Developmental Biology, 248*(2), 369–383.

Bao, R., & Friedrich, M. (2008). Conserved cluster organisation of insect *Runx* genes. *Development Genes and Evolution, 218*(10), 567–574.

Bataille, L., Auge, B., Ferjoux, G., Haenlin, M., & Waltzer, L. (2005). Resolving embryonic blood cell fate choice in *Drosophila*: Interplay of GCM and RUNX factors. *Development, 132*, 4635–4644.

Baugh, L. R., Demodena, J., & Sternberg, P. W. (2009). RNA Pol II accumulates at promoters of growth genes during developmental arrest. *Science, 324*(5923), 92–94.

Bernardin, F., & Friedman, A. D. (2002). AML1 stimulates G1 to S progression via its transactivation domain. *Oncogene, 21*(20), 3247–3252.

Bernardin-Fried, F., Kummalue, T., Leijen, S., Collector, M. I., Ravid, K., & Friedman, A. D. (2004). AML1/RUNX1 increases during G1 to S cell cycle progression independent of cytokine-dependent phosphorylation and induces cyclin D3 gene expression. *Journal of Biological Chemistry, 279*(15), 15678–15687.

Brabin, C., Appleford, P. J., & Woollard, A. (2011). The *Caenorhabditis elegans* GATA factor ELT-1 works through the cell proliferation regulator BRO-1 and the fusogen EFF-1 to maintain the seam stem-like fate. *PLoS Genetics, 7*(8), e1002200.

Braun, T., & Woollard, A. (2009). RUNX factors in development: Lessons from invertebrate model systems. *Blood cells, Molecules and Diseases, 43*, 43–48.

Cai, X., Gao, L., Teng, L., Ge, J., Oo, Z. M., Kumar, A. R., et al. (2015). *Runx1* deficiency decrease ribosome biogenesis and confers stress resistance to hematopoietic stem and progenitor cells. *Cell Stem Cell, 17*(2), 165–177.

Cameron, E. R., & Neil, J. C. (2004). The *Runx* genes: Lineage-specific oncogenes and tumor suppressors. *Oncogene, 23*, 4308–4314.

Campos-Ortega, J. A., & Jan, Y. N. (1991). Genetic and molecular bases of neurogenesis in *Drosophila melanogaster. Annual Review of Neuroscience, 14*, 399–420.

Canon, J., & Banerjee, U. (2000). Runt and Lozenge function in *Drosophila* development. *Cell and Developmental Biology, 11*, 327–336.

Chang, L. S. H., Ito, K., & Ito, Y. (2013). RUNX family: Regulation and diversification of roles through interacting proteins. *International Journal of Cancer, 132*, 1260–1271.

Coffman, J. A. (2003). Runx transcription factors and the developmental balance between cell proliferation and differentiation. *Cell Biology International, 27*, 315–324.

Coffman, J. A. (2009). Is Runx a linchpin for developmental signalling in metazoans? *Journal of Cellular Biochemistry, 107*, 194–202.

Coffman, J. A., Kirchhamer, C. V., Harrington, M. G., & Davidson, E. H. (1996). *SpRunt-1*, a new member of the runt domain family of transcription factors, is a positive regulator of the aboral ectoderm-specific CyIIIA gene in sea urchin embryos. *Developmental Biology, 174*, 43–54.

Coffman, J. A., Dickey-Sims, C., Haug, J. S., McCarthy, J. J., & Robertson, A. J. (2004). Evaluation of developmental phenotypes produced by morpholino antisense targeting of a sea urchin *Runx* gene. *BMC Biology, 2*, 6–15.

Crute, B. E., Lewis, A. F., Wu, Z., Bushweller, J. H., & Speck, N. A. (1996). Biochemical and biophysical properties of the core-binding factor alpha2 (AML2) DNA binding domain. *Journal of Biological Chemistry, 271*, 26251–26260.

Daga, A., Karlovich, C. A., Dumstrei, K., & Banerjee, U. (1996). Patterning of cells in the *Drosophila* eye by *Lozenge*, which shares homologous domains with AML1. *Genes and Development, 10*, 1194–1205.

Degnan, B. M., Vervoor, M., Larroux, C., & Richards, G. S. (2009). Early evolution of metazoan transcription factors. *Current Opinion in Genetics and Development, 19*, 591–599.

Dickey-Sims, C., Robertson, A. J., Rupp, D. E., McCarthy, J. J., & Coffman, J. A. (2005). Runx-dependent expression of PKC is criticla for cell survival in the sea urchin embryo. *BMC Biology, 3*, 18–30.

DuBuc, T. Q., Traylor-Knowles, N., & Martindate, M. Q. (2014). Initiating a regenerative response; cellular and molecular features of wound healing in the cnidarian *Nematostella vectensis*. *BMC Biology, 12*, 24.

Duffy, J. B., & Gergen, J. P. (1991). The *Drosophila* segmentation gene *runt* acts as a position-specific numerator element necessary for the uniform expression of the sex-determining gene *Sex-lethal. Genes and Development, 5*(12A), 2176–2187.

Duffy, J. B., Kania, M. A., & Gergen, J. P. (1991). Expression and function of the *Drosophila* gene *runt* in early stages of neural development. *Development, 114*(4), 1223–1230.

Durst, K. L., & Hiebert, S. W. (2004). Role of RUNX family members in transcriptional repression and gene silencing. *Oncogene, 23*, 4220–4224.

Eisenhoffer, G. T., Kang, H., & Sanchez-Alvarado, A. (2008). A molecular analysis of stem cells and their descendants during cell turnover and regeneration in the planarian *Schmidtea mediterranea*. *Cell Stem Cell, 3*, 327–339.

Forsthoefel, D. J., & Newmark, P. J. (2009). Emerging patterns in planarian regeneration. *Current Opinion in Genetics and Development, 19*, 412–420.

Gajewski, K. M., Sorrentino, R. P., Lee, J. H., Zhang, Q., Russell, M., & Schulz, P. A. (2007). Identification of a crystal cell-specific enhancer of the black cells prophenoloxidase gene in *Drosophila*. *Genesis, 45*(4), 200–207.

Gergen, J. P., & Butlet, B. A. (1988). Isolation of the *Drosophila* segmentation gene *runt* and analysis of its expression during embryogenesis. *Genes and Development, 2*, 1179–1193.

Gergen, J. P., & Wieschaus, E. F. (1985). The localised requirements for a gene affecting segmentation in *Drosophila*: Analysis of larvae mosaic for *runt*. *Developmental Biology, 109*, 321–335.

Gleason, J. E., & Eisenmann, D. M. (2010). Wnt signalling controls the stem cell-like asymmetric division of the epithelial seam cells during *C. elegans* larval development. *Developmental Biology, 348*(1), 58–66.

Gold, K. S., & Bruckner, K. (2014). *Drosophila* as a model for the two myeloid blood cell systems in vertebrates. *Experimental Hematology, 42*, 717–727.

Golling, G., Li, L., Pepling, M., Stebbins, M., & Gergen, J. P. (1996). *Drosophila* homologs of the proto-oncogene product PEBP2/CBF beta regulate the DNA-binding properties of Runt. *Molecular and Cellular Biology, 16*(3), 932–942.

Gorrepati, L., Krause, M. W., Chen, W., Brodigan, T. M., Correa-Mendez, M., & Eisenmann, D. M. (2015). Identification of Wnt pathway target genes regulating the division and differentiation of larval seam cells and vulval precursor cells in *Caenorhabditis elegans*. *G3, 5*, 1551–1566.

Govindasamy, N., Murthy, S., & Ghanekar, Y. (2014). Slow-cycling stem cells in hydra contribute to head regeneration. *Biology Open, 3*, 1236–1244.

Harandi, O. F., & Ambros, V. R. (2015). Control of stem cell self-renewal and differentation by the heterochronic genes and the cellular asymmetry machinary in *Caenorhabditis elegans*. *PNAS, 112*, E287–E296.

Hughes, S., Brabin, C., Appleford, P. J., & Woollard, A. (2013). CEH-20/Pbx and UNC-62/Meis function upstream of *rnt-1*/Runx to regulate asymmetric divisions of the *C. elegans* stem-like seam cells. *Biology Open, 2*(7), 718–727.

Inoue, H., Hisamoto, N., An, J. H., Oliveira, R. P., Nishida, E., Blackwell, T. K., & Matsumoto, K. (2005). The *C. elegans* p38 MAPK pathway regulates nuclear localisation of the transcription factor SKN-1 in oxidative stress response. *Genes and Development, 19*(19), 2278–27783.

Ito, Y. (1999). Molecular basis of tissue-specific gene expression mediated by the Runt domain transcription factor PEBP2/CBF. *Genes and Cells, 4*, 685–696.

Ito, Y. (2004). Onocgenic potential of the RUNX gene family: 'Overview'. *Oncogene, 23*, 4198–4208.

Ito, K., Lim, A. C., Salto-Tellez, M., Motoda, L., Osato, M., Chuang, L. S., et al. (2008). RUNX3 attenuates beta-catenin/T cell factors in intestinal tumorigenesis. *Cancer Cell, 14*(3), 226–237.

Ito, K., Chuang, L. S., Ito, T., Chang, T. L., Fukamachi, H., Salto-Tellez, M., & Ito, Y. (2011). Loss of *Runx3* is a key event in inducing precancerous state of the stomach. *Gasterienterology, 140*(5), 1536–1546.

Jung, S. H., Evans, C. J., Uemura, C., & Banerjee, U. (2005). The *Drosophila* lymph gland as a developmental model of hematopoiesis. *Development, 132*, 2521–2533.

Kagoshima, H., Shigesada, K., Satake, M., Ito, Y., Miyoshi, H., Ohki, M., et al. (1993). The Runt domain identifies a new family of heteromeric transcriptional regulators. *Trends in Genetics, 9*(10), 338–341.

Kagoshima, H., Sawa, S., Mitnai, T. R., Burglin, K., Shigesada, K., & Kohara, Y. (2005). The *C. elegans* RUNX transcription factor RNT-1/MAB-2 is required for asymmetrical cell division of the T blast cell. *Developmental Biology, 287*, 262–273.

Kagoshima, H., Nimmo, R. A., Saad, N., Tanaka, J., Miwa, Y., Mitani, S., et al. (2007). The *C. elegans* CBFbeta homologue BRO-1 interacts with the Runx factor, RNT-1, to promote stem cell proliferation and self-renewal. *Development, 134*(21), 3905–3015.

Kahler, R. A., & Westendorf, J. J. (2003). Lymphoid enhancer factor-1 and beta-catenin inhibit Runx2-dependent transcriptional activation of the osteocalcin promoter. *Journal of Biological Chemistry, 278*(14), 11937–11944.

Kamachi, Y., Ogawa, E., Asano, M., Ishida, S., Murakami, Y., Satake, M., et al. (1990). Purification of a mouse nuclear factor that binds to both the A and B cores of the polyomavirus enhancer. *Journal of Virology, 64*, 4808–4819.

Kaminker, J. S., Singh, R., Lebestky, T., Yan, H., & Banerjee, U. (2001). Redendant function of Runt domain binding partners, Big Brother and Brother, during *Drosophila* development. *Development, 128*(14), 2639–2648.

Kania, M. A., Bonner, A. S., Duffy, J. B., & Gergen, J. P. (1990). The *Drosophila* segmentatio gene *runt* encodes a novel nuclear regulatory protein that is also expressed in the developing nervous system. *Genes and Development, 4*(10), 1701–1713.

Katoh, K., Misawa, K., Kuma, K., & Miyata, T. (2002). MAFFT: a novel method for rapid multiple seqeunce alignment based on fast Fourier transform. *Nucleic Acids Research, 30*, 3059–3066.

Keita, M., Bachvarova, M., Morin, C., Plante, M., Gregoire, J., Renaud, M.-C., et al. (2013). The RUNX1 transcription factor is expresed in serious epithelial ovarina carcinoma and contributes to cell proliferation, migration and invasion. *Cell Cycle, 12*(6), 972–986.

Kitoh, A., Ono, M., Naoe, Y., Ohkura, N., Yamaguchi, T., Yaguchi, H., et al. (2009). Indispensale role of the Runx1-Cbfbeta transcription complex for *in vivo*-suppressive function of FoxP3^{+} regulator T cells. *Immunity, 31*, 609–620.

Klinger, M., & Gergen, J. P. (1993). Regulation of *runt* transcription by *Drosophila* segmentation genes. *Mechanisms of Development, 43*, 3–19.

Kramer, I., Sigrist, M., de Nooij, J. C., Taniuchi, I., HJessell, T. M., & Arber, S. (2006). A role for Runx transcription factor signalling in dorsal root ganglion sensory neurone diversification. *Neurone, 49*(3), 379–393.

Lanot, R., Zachary, D., Holder, F., & Meister, M. (2001). Postembryonic hematopoiesis in *Drosophila*. *Developmental Biology, 230*(2), 243–257.

Lapan, S. W., & Reddien, P. W. (2011). *dlx* and *sp6-9* control optic cup regeneratin in prototypic eye. *PLoS Genetics, 7*(8), e1002226.

Lebestky, T., Chang, T., Hartenstein, V., & Banerjee, U. (2000). Specification of *Drosophila* hematopoietic lineage by conserved transcription factors. *Science, 288*, 146–149.

Lee, K., Shim, J., Bae, J., Kim, Y.-J., & Lee, J. (2012). Stabilization of RNT-1 protein, Runt-related transcription factor (RUNX) protein homolog of *Caenorhabditis elegans*, by oxaditive stress through mitogen-activated protein kinase pathway. *Journal of Biological Chemistry, 287*(13), 10444–10452.

Li, L.-H., & Gergen, J. P. (1999). Differential interactions between Brother proteins and Runt domain proteins in the *Drosophila* embryo and eye. *Development, 126*, 3313–3322.

Lotem, J., Levanon, D., Negreanu, V., Leshkowitz, D., Friedlander, G., & Groner, Y. (2013). Runx3-mediated transcriptional progeam in cytotoxic lymphocytes. *PLoS One, 8*(11), e80467.

Markus, R., Laurinyecz, B., Kurucz, E., Honti, V., Bajusz, I., Sipos, B., et al. (2009). Sessile hemocytes as a hematopoietic compartment in *Drosophila melanogaster*. *PNAS, 106*(12), 4805–4809.

Milton, C. C., Grusche, F. A., Degoutin, J. L., Yu, E., Dai, Q., Lai, E. C., & Harvey, K. F. (2014). The Hippo

pathway regulates hematopoiesis in *Drosophila melanogaster*. *Current Biology, 24*, 2673–2680.
Minokawa, T., Wikramanayake, A. H., & Davidson, E. H. (2005). *cis*-regulatory inputs of the *wnt8* gene in the sea urchin endomesoderm network. *Developmental Biology, 288*, 545–558.
Muratoglu, S., Garratt, G., Hyman, K., Gajewski, K., Schulz, R. A., & Fossett, N. (2006). Regulation of *Drosophila* friend of GATA gene, *u-shaped*, during hematopoiesis: a direct role for *serpent* and *lozenge*. *Developmental Biology, 296*, 561–579.
Muratoglu, S., Hough, B., Mon, S. T., & Fossett, N. (2007). The GATA factor Serpent cross-regulates *lozenge* and *u-shaped* expression during *Drosophila* blood cell development. *Developmental Biology, 311*, 636–649.
Murthy, M., Bocking, S., Verginelli, F., & Stifani, S. (2014). Transcription factor Runx1 inhibits proliferation and promotes developmental maturation in a selected population of inner olfactory nerve layer olfactory ensheathing cells. *Gene, 540*(2), 191–200.
Naillat, F., Yan, W., Karjalainen, R., Liakhovitskaia, A., Samoylenko, A., Xu, Q., et al. (2015). Indentification of the gene regulated by Wnt-4, a critical signal for commitment of the ovary. *Experimental Cell Research, 332*, 163–178.
Newmark, P. J., & Sanchez-Alvarado, A. (2002). Not your father's panarian: a classic model enters the era of functional genomics. *Nature Reviews Genetics, 3*, 210–219.
Nimmo, R. A., & Slack, F. J. (2009). An elegant miRror: microRNAs in stem cells, developmental timing and cancer. *Chromosoma, 118*, 405–418.
Nimmo, R. A., & Woollard, A. (2008). Worming out the biology of Runx. *Developmental Biology, 313*, 492–500.
Nimmo, R. A., Antebi, A., & Woollard, A. (2005). *mab-2* encodes RNT-1, a *C. elegans* Runx homologue essential for controlling cell proliferation in a stem cell-like developmental lineage. *Development, 132*, 5034–5054.
Nusslein-Volhard, C., & Wieschaus, E. F. (1980). Mutations affecting segment number and polarity in *Drosophila*. *Nature, 287*, 795–801.
Ogawa, E., Maruyama, M., Kagoshima, H., Inuzuka, M., Lu, J., Satake, M., et al. (1993). PEBP2/PEA2 represents a family of transcription factors homologous to the products of the *Drosophila runt* gene and the human AML1 gene. *PNAS, 90*, 6859–6863.
Osoiro, K. M., Lee, S. E., McDermitt, D. J., Waghmare, S. K., Zhang, Y. V., Woo, H. N., & Tumbar, T. (2008). Runx1 mediates developmental, but not injury-driven, hair follicle stem cell activation. *Development, 135*, 1059–1068.
Pancer, Z., Rast, J. P., & Davidson, E. H. (1999). Origins of immunity: transcription factors and homologues of effector genes of the vertebrate immune system expressed in sea urchin coelomocytes. *Immunogenetics, 49*, 773–786.
Pencovich, N., Jaschek, R., Tanay, A., & Groner, Y. (2011). Dynamic combinatorial interactions of RUNX1 and cooperating partners regulates megakaryocytic differentiation in cell line models. *Blood, 117*(1), e1–14.
Petersen, H. O., Hoger, S. K., Looso, M., Lengfeld, T., Khn, A., Warnken, U., et al. (2015). A comprehensive transcriptomic and proteomic analysis of hydra head regeneration. *Molecular Biology and Evolution, 32*(8), 1928–1947.
Reddien, P. W., & Sanchez-Alvarado, A. (2004). Fundamentals of planarian regeneration. *Annual Review of Cell and Developmental Biology, 20*, 725–757.
Rennert, J., Coffman, J. A., Mushegian, A. R., & Robertson, A. J. (2003). The evolution of *Runx* genes I. A comparative study of sequences from phylogenetically diverse model organisms. *BMC Evolutionary Biology, 3*, 4–15.
Robertson, A. J., Dickey, C. E., McCarthy, J. J., & Coffman, J. A. (2002). The expression of *SpRunt* during sea urchin embryogenesis. *Mechanisms of Development, 117*, 327–330.
Robertson, A. J., Coluccio, A., Knowlton, P., Dickey-Sims, C., & Coffman, J. A. (2008). *Runx* expression is mitogenic and mutually linked to Wnt activity in blastula-stage sea urchin embryos. *PLoS One, 3*(11), e3770.
Robertson, A. J., Larroux, C., Degnan, B. M., & Coffman, J. A. (2009). The evolution of *Runx* genes II. The C-terminal Groucho recruitment motif is present in both eumetazoans and homoscleromorphs but absent in a haplosclerid demosponge. *BMC Research Notes, 2*, 59–68.
Robertson, A. J., Coluccio, A., Jensen, S., Rydlizky, K., & Coffman, J. A. (2013). Sea urchin *akt* activity is Runx-dependent and required for post-cleavage stage cell division. *Biology Open, 2*, 472–478.
Salo, E., Abril, J. F., Cebria, F., Eckelt, K., Fernandez-Taboada, E., Handberg-Thorsager, M., et al. (2009). Planarian regeneration: achivements and future directions after 20 years of research. *International Journal of Developmental Biology, 58*(8-10), 1317–1327.
Sanchez-Alvarado, A., & Tsonis, P. A. (2006). Bridging the regeneration gap: genetic insights from diverse animal models. *Nature Reviews Genetics, 7*, 873–884.
Sandmann, T., Vogg, M. C., Owlarn, S., Boutros, M., & Bartscherer, K. (2011). The head-regeneration transcriptome of the planarian *Schmidtea mediterranea*. *Genome Biology, 12*, R76.
Schierwater, B., Eitel, M., Jakob, W., Osigus, H.-J., Hadrys, H., Dellaporta, S. L., et al. (2009). Concatenated analysis sheds light on early metazoan evolution and fuels a modern "Urmetazoon" hypothesis. *PLoS Biology, 7*(1), e1000020.
Scimone, M. L., Kravarik, K. M., Lapan, S. W., & Reddien, P. W. (2014). Neoblast specialisation in regeneration of the planarian *Schmidtea mediterranea*. *Stem Cell Reports, 3*, 339–352.

Sebe-Pedros, A., de Mendoza, A., Lang, B. F., Degnan, B. M., & Ruiz-Trillo, L. (2011). Unexpected repertoire of metazoan transcription factors in the unicellular holozoan *Capsaspora owczarzaki*. *Molecular Biology and Evolution, 28*(3), 1241–1254.

Speck, N. A. (2001). Core Binding Factor and its role in normal hematopoietic development. *Current Opinion in Hematology, 8*(4), 192–196.

Strom, D. K., Nip, J., Westendorf, J. J., Linggi, B., Lutterbach, B., Downing, J. R., et al. (2000). Expression of the AML-1 oncogene shortens the G(1) phase of the cell cycle. *Journal of Biological Chemistry, 275*, 3438–3445.

Sullivan, J. C., Sher, D., Eisenstein, M., Shigesada, K., Reitzek, A. M., Marlow, H., et al. (2008). The evolutionary origin of the Runx/CBFbeta transcription factors - Studies of the most basal metazoans. *BMC Evolutionary Biology, 8*, 228–248.

Sulston, J. E., & Horvitz, H. R. (1977). Post-embryonic cell lineages of the nematode, *Caenorhabditis elegans*. *Developmental Biology, 56*(1), 110–156.

Tanaka, T., Tanaka, K., Ogawa, S., Kurokawa, M., Mitani, K., Nishida, J., et al. (1995). An acute myeloid leukemia gene, AML1, regulates hemopoietci myeloid cell differentiation and transcriptional activation antagonistically by two alternative spliced forms. *EMBO Journal, 14*(2), 341–350.

Theriault, F. M., Nuthall, H. N., Dong, Z., Lo, R., Barnabe-Heider, F., Miller, F. D., & Stifani, S. (2005). Role for Runx1 in the proliferation and neuronal differentation of selected progenitor cells in the mammalian nervous system. *Journal of Neuroscience, 25*(8), 2050–2061.

Voon, D. C., Hor, Y. T., & Ito, Y. (2015). The RUNX complex: Reaching beyond haematopoiesis into immunity. *Immunology, 146*(4), 523–536.

Wagner, D. E., Wang, I. E., & Reddien, P. W. (2011). Clonogenic neoblasts are pluripotent adult stem cells that underlie planarian regeneration. *Science, 332*(6031), 811–816.

Waltzer, L., Ferjoux, G., Bataille, L., & Haenlin, M. (2003). Cooperation between the GATA and RUNX factors Serpent and Lozenge during *Drosophia* hematopoiesis. *EMBO Journal, 22*(24), 6516–6525.

Waltzer, L., Gobert, V., Osman, D., & Haenlin, M. (2010). Transcription factor interplay during *Drosophila* haemoatopoiesis. *International Journal of Developmental Biology, 54*, 1107–1115.

Wang, L., Kounatidis, I., & Ligoxygakis, P. (2014a). *Drosophila* as a model to study the role of blood cells in inflammation, innate immunity and cancer. *Frontiers in Cellular and Infection Microbiology, 3*(113), 1–17.

Wang, Y., Godec, J., Ben-Aissa, K., Cui, K., Zhao, K., Pucsek, A. C., et al. (2014b). The transcription factors T-bet and Runx are required for the ontogeny of pathogenic interferon-gamma-producing T helper 17 cells. *Immunity, 40*(3), 355–366.

Wenemoser, D., & Reddien, P. W. (2010). Planarian regeneration involves distinct stem cell responses to wounds and tissue absence. *Developmental Biology, 344*(2), 979–991.

Wenemoser, D., Lapan, S. W., Wilkinson, A. W., Bell, G. W., & Reddien, P. W. (2012). A molecular wound response program assocaited with regeneration initation in planarians. *Genes and Development, 26*, 988–1002.

Westendorf, J. J., Zaidi, S. K., Cascino, J. E., Kahler, R., van Wijen, A. J., Lian, J. B., et al. (2002). Runx2 (CbfaI, AML-3) interacts with histone deacetylase 6 and represses the p21 (CIP1/WAF1) promoter. *Molecular and Cellular Biology, 22*(22), 9782–7992.

Wong, F. W., Kurokawa, M., Satake, M., & Kohu, K. (2011). Down-regulation of *Runx1* expression by TCR signal involves an autoregulatory mechanism and contributes to IL-2 production. *The Journal of Biological Chemistry, 286*(13), 11110–11118.

Wong, F. W., Kohu, K., Nakamura, A., Ebina, M., Kikuchi, T., Tazawa, R., et al. (2012). *Runx1* deficiency in CD4^{+} T cells causes fatal autoimmune inflammatory lung disease due to spontaneous hyperactivation of cells. *The Journal of Immunology, 188*, 5408–5420.

Wood, W., & Jacinto, A. (2007). *Drosophila melanogaster* embryonic haemocytes: masters of multitasking. *Nature Reviews Molecular Cell Biology, 8*(7), 542–551.

Wotton, S., Blyth, K., Kilbey, A., Jenkins, A., Terry, A., Bernardin-Fried, F., et al. (2004). RUNX1 transformation of primary embryonic fibroblasts is revealed in the absence of p53. *Oncogene, 23*, 5476–5486.

Wu, J. Q., Seay, M., Schulz, V. P., Hariharan, M., Tuck, D., Lian, J., et al. (2012). Tcf7 is an important regulator of the switch of self-renewal and differentiation in a multipotential hematopoietic cell line. *PLoS Genetics, 8*(3), e1002565.

Wurtzel, O., Cote, L. E., Poririer, A., Satjia, R., Regev, A., & Reddien, P. W. (2015). A generic and cell-type specific wound response precedes regeneration in planarians. *Developmental Cell, 35*, 632–645.

Wysokinski, D., Pawlowska, E., & Blasiak, J. (2015). RUNX2: A master bone growth regulator that may be involved in the DNA damage response. *DNA and Cell Biology, 34*(5), 305–315.

Xia, D., Zhnag, Y., Huang, X., Sun, Y., & Zhnag, H. (2007). The *C. elegans* CBFbeta homolog, BRO-1, regulates the proliferation, differentation and specification of the stem cell-like seam cell lineages. *Developmental Biology, 309*(2), 259–272.

Yan, H., Canon, J., & Banerjee, U. (2003). A transcriptional chain linking eye specification to terminal determination of cone cells in the *Drosophila* eye. *Developmental Biology, 263*, 323–329.

Yokomizo, T., Ogawa, M., Osato, M., Kanno, T., Yoshida, H., Fujimoto, T., et al. (2001). Requirement of Runx1/AML1/PEBP2alphaB for the generation of haematopoietic cells from endothelial cells. *Genes and Cells, 6*(1), 13–23.

Part II

RUNX and CBFβ: Structure and Function

Structure and Biophysics of CBFβ/RUNX and Its Translocation Products

2

Tahir H. Tahirov and John Bushweller

Abstract

The core binding factor (CBF) transcription factor is somewhat unique in that it is composed of a DNA binding RUNX subunit (RUNX1, 2, or 3) and a non-DNA binding CBFβ subunit, which modulates RUNX protein activity by modulating the auto-inhibition of the RUNX subunits. Since the discovery of this fascinating transcription factor more than 20 years ago, there has been a robust effort to characterize the structure as well as the biochemical properties of CBF. More recently, these efforts have also extended to the fusion proteins that arise from the subunits of CBF in leukemia. This chapter highlights the work of numerous labs which has provided a detailed understanding of the structure and function of this transcription factor and its fusion proteins.

Keywords

CBFβ • RUNX • Structure • X-ray • NMR

2.1 Biochemical Properties of CBFβ and RUNX

The DNA binding domain of the RUNX proteins is termed the Runt domain. Much of the early biochemical characterization of CBF focused on studies of CBFβ, the Runt domain, and DNA. Sedimentation equilibrium measurements (Tang et al. 2000a) confirmed the 1:1:1 stiochiometry of the complex. Measurements of the affinity of the Runt domain for DNA and for CBFβ as well as of the CBFβ-Runt domain complex for DNA using both electromobility shift assays (EMSA) as well as isothermal titration calorimetry provided a detailed thermodynamic

T.H. Tahirov
Eppley Institute for Research in Cancer and Allied Diseases, Fred & Pamela Buffett Cancer Center, University of Nebraska Medical Center, Omaha, NE 68198, USA
e-mail: ttahirov@unmc.edu

J. Bushweller (✉)
Department of Molecular Physiology and Biological Physics, University of Virginia, Charlottesville, VA 22908, USA
e-mail: jhb4v@virginia.edu

Y. Groner et al. (eds.), *RUNX Proteins in Development and Cancer*, Advances in Experimental Medicine and Biology 962, DOI 10.1007/978-981-10-3233-2_2

box for this system (Crute et al. 1996; Huang et al. 1998; Tang et al. 2000a, b). Consistent with early predictions, these measurements indeed showed a significant enhancement (six to ten-fold) of Runt domain-DNA binding in the presence of the CBFβ subunit. In the context of full-length RUNX1, these effects are more substantive, with >40-fold enhancement of DNA binding (Gu et al. 2000), suggesting interactions between other regions of RUNX1 and the Runt domain to mediate this auto-inhibition.

2.2 Structures of the CBFβ Heterodimerization Domain and the Runt Domain

Early studies established that a 141 amino acid N-terminal fragment of CBFβ retains all of the binding determinants for interaction with the Runt domain (Ogawa et al. 1993; Huang et al. 1998). Deletion studies established the limits of the DNA binding domain, or Runt domain, of the RUNX proteins as the domain mediating both DNA and CBFβ binding. The first structures of the isolated domains were determined using NMR spectroscopy. The structures of CBFβ (Goger et al. 1999; Huang et al. 1999) showed it to be a unique α/β fold. Chemical shift changes upon binding of the Runt domain were used to map the contact surface on CBFβ and guide subsequent mutagenesis studies (see below). The structures of the Runt domain were solved using Runt domain-DNA complexes (Berardi et al. 1999; Nagata et al. 1999) as the isolated Runt domain is very poorly behaved in solution. The structures of the Runt domain established it as a member of the s-type Ig fold DNA binding domains which includes p53, NF-kB, NFAT, and STAT1. Again here, NMR approaches were used to map the contact surfaces on the Runt domain and guide subsequent mutagenesis studies (see below). Subsequently, a solution structure of the isolated Runt domain was also reported, using NMR spectroscopy under conditions of high salt (Perez-Alvarado et al. 2000). At a later time point, X-ray crystal structures of the isolated Runt domain were also reported (Backstrom et al. 2002).

2.3 Structures of CBFβ-Runt Domain-DNA Complexes and Mechanism of CBFβ Regulation of the Runt Domain

Structures of CBFβ-Runt domain-DNA complexes have been determined using X-ray crystallography (Bravo et al. 2001; Tahirov et al. 2001). In addition, structures of CBFβ-Runt domain and Runt domain-DNA complexes have also been determined (Warren et al. 2000; Tahirov et al. 2001). As shown in Fig. 2.1, these structures provide detailed insights into the interaction of the two proteins with one another and of the Runt domain with DNA. Like other Ig-fold DNA binding domains, the Runt domain makes contacts both in the major and minor grooves of the DNA using loops from one end of the barrel. Unlike other Ig-fold DNA binding domains, the Runt domain utilizes its C-terminus to make most of the observed sequence-specific DNA contacts. The DNA in the structures is significantly distorted from standard B-form, with a pronounced bend observed. It is clear from these structures that CBFβ does not make any contacts with the DNA and therefore must regulate DNA binding by the Runt domain via an allosteric mechanism. A large surface area is buried between the two proteins, explaining their high affinity for one another. With the availability of structures of the Runt domain alone, binary complexes, and the ternary CBFβ-Runt domain-DNA complex, comparisons among these have provided insights into the mechanism of CBFβ regulation of DNA binding. Significant structural changes were observed in the Runt domain for a number of the loop regions. The most significant change was observed in the βG-G' loop, termed the S-switch, mediated particularly by contacts from CBFβ to T161 as well as A165. The changes here are consistent with NMR studies that showed significant chemical shift changes for this region upon CBFβ binding (Tang et al. 2000a).

The structural studies show that either DNA or CBFβ can induce similar conformational changes in the Runt domain, suggesting that the effect of CBFβ is to shift an existing conformational equi-

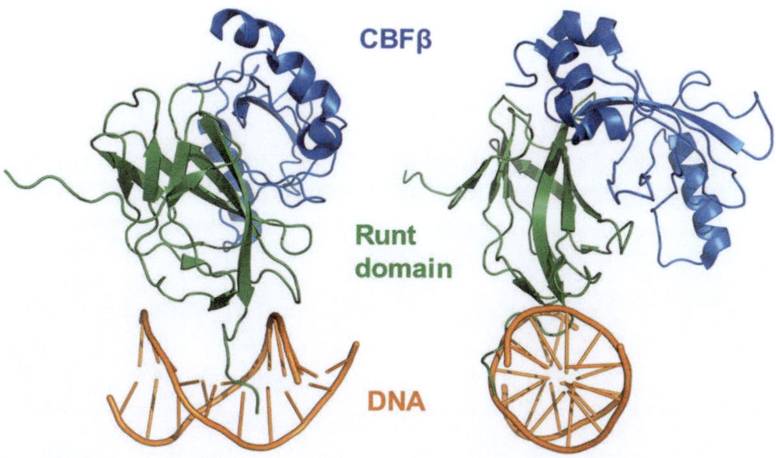

Fig. 2.1 Ribbon representation of the structure of the CBFβ:Runt domain:DNA ternary complex. CBFβ is shown in *blue*, the Runt domain in *green*, and the DNA in *orange*. For clarity the structure is shown in two different orientations, rotated by 90 ° relative to one another. Image was rendered from PDB code 1H9D

librium in the Runt domain. Indeed, NMR relaxation analysis of the Runt domain shows evidence of conformational exchange for the S-switch region which is quenched upon CBFβ binding (Yan et al. 2004), a phenomenon which is also observed for the DNA. This was subsequently confirmed by introduction of a mutation into the S-switch region which enhanced DNA binding and quenched the conformational exchange (Li et al. 2006).

2.4 Mutagenesis Studies of CBFβ and the Runt Domain, Including Disease Associated Mutations in the Runt Domain

Extensive mutagenesis studies of CBFβ as well as the Runt domain have been carried out, with the intention of identifying energetic hotspots for CBFβ-Runt domain as well as Runt domain-DNA binding. The structural studies of the ternary complexes included characterization of mutations as well (Bravo et al. 2001; Tahirov et al. 2001). The mutations of CBFβ identified N104, in particular, as a critical hotspot for binding to the Runt domain, consistent with the extensive contacts made by this residue with the Runt domain (Tang et al. 2000b). An extensive mutagenesis study of the Runt domain, including characterization of effects of the mutations on structural integrity, identified energetically critical residues for DNA binding (Li et al. 2003), consistent with the structures of the CBFβ-Runt domain-DNA complex. Interestingly, the most critical interactions are largely mediated by residues at the C-terminus of the Runt domain including T169, D171, R174, and R177.

Point mutations in the Runt domain of *RUNX1* are observed in familial platelet disorder with predisposition for acute myelogenous leukemia (FPD/AML), AML M0, radiation-associated and therapy-related myelodysplastic syndrome and AML, isolated cases of AML M2, M5a, M3 relapse, and in chronic myelogenous leukemia in blast phase (Fig. 2.2) (Blyth et al. 2005; Mangan and Speck 2011). Point mutations in the Runt domain of *RUNX2* are observed in the inherited skeletal disorder cleidocranial dysplasia (CCD) (Fig. 2.2) (Otto et al. 2002). A detailed characterization of these disease-associated mutations, including impact on CBFβ and DNA binding as well as impact on the structural integrity of the mutated Runt domains, provided important insights into the effects of these mutations (Matheny et al. 2007). Loss of DNA binding severely impaired RUNX1 function whereas loss

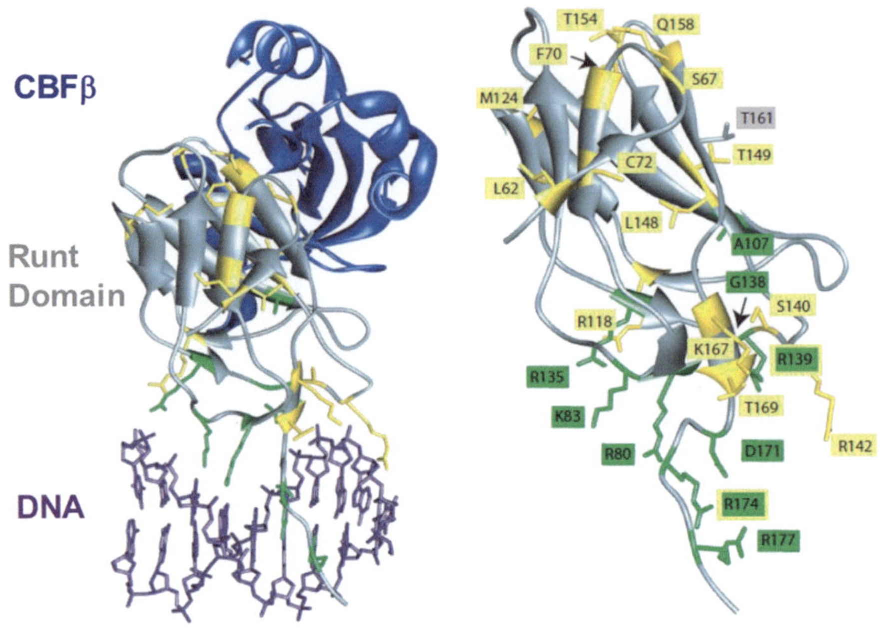

Fig. 2.2 Ribbon diagram of the Runt domain:CBFβ:DNA ternary complex. The Runt domain and CBFβ are shown in *grey* and *blue*, respectively, and DNA is *purple*. Amino acids mutated in RUNX2 in CCD are *yellow*, whereas *green* indicates amino acids mutated in RUNX1 in FPD/AML, AML M0 subtype, radiation-associated and therapy-related myelodysplastic syndrome and AML, AML M2, M5a, M3 relapse, and chronic myelogenous leukemia in blast phase. T161 (*grey*) is a hotspot for binding of the Runt domain to CBFβ

of CBFβ binding resulted in hypomorphic alleles. It was concluded that hypomorphic alleles in *RUNX2* can cause CCD but the more robust DNA binding mutations in *RUNX1* are necessary for leukemia development.

2.5 Structural Basis of Ets1 Activation by Runx1

Among the well-studied interaction partners of Runx1 is the Ets1 transcription factor. DNA binding of Ets1 is regulated by the inhibition regulatory module (IRM) flanking the DNA-binding Ets domain (ED) and the serine rich region (SRR) (Goetz et al. 2000) (Fig. 2.3a). The IRM is comprised of inhibitory helices HI1 and HI2 N-terminal to the ED (Skalicky et al. 1996; Lee et al. 2005). The IRM packs onto the Ets domain and reinforces the stability of the SRR, thereby reducing its DNA binding affinity (Lee et al. 2008). A notable feature of Ets1 is its ability to reinforce the autoinhibition by Ca^{2+} signaling-mediated phosphorylation of the serines in SRR which reduces the DNA-binding activity of Ets1 up to 50-fold (Cowley and Graves 2000; Pufall et al. 2005; Lee et al. 2008).

Runx1 cooperates with Ets1 on composite Ets1•Runx1 motifs found in a number of genes (Hollenhorst et al. 2007, 2009). Among the well-characterized motifs is an GGATGTGG motif of *T cell receptor alpha* (*TCRα*) and *beta* (*TCRβ*) gene enhancers (Wotton et al. 1994, Giese et al. 1995, Sun et al. 1995). Two features of Runx1-Ets1 cooperation are notable. First is the ability of Runx1 to activate the DNA binding of both phosphorylated and unphosphorylated Ets1, and, second is the highly concerted mechanism of cooperation in which DNA bound Runx1 stimulates Ets1 activity (Goetz et al. 2000; Shrivastava et al. 2014).

Runx1 overcomes Ets1 autoinhibition by direct physical interaction (Giese et al. 1995; Kim et al. 1999; Goetz et al. 2000; Gu et al. 2000), however, the details of their cooperation remained a mystery until the report of the crystal structures of Runx1 and Ets1 in complex with a

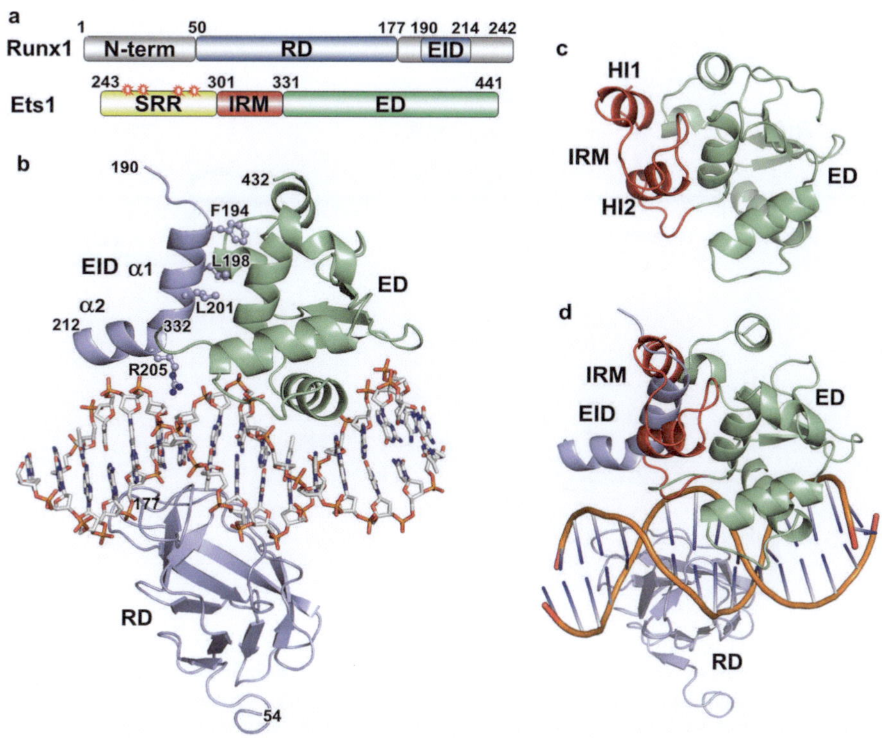

Fig. 2.3 Structure of $Runx1_{1-242}$•$Ets1_{296-441}$•TCRα enhancer (TCRαE) ternary complex and comparison with autoinhibited Ets1. (**a**) Runx1 and Ets1 DNA binding and inhibitory domains. The highlighted domains are: N-terminal, Runt domain (RD), Ets1 interacting domain (EID) for Runx1; and serine rich region (SRR), inhibition regulatory module (IRM) and ETS domain (ED) for Ets1. The red stars indicate the location of phosphorylation sites. (**b**) Crystal structure of $Runx1_{1-242}$•$Ets1_{296-441}$•TCRαE (PDB access code 4L0Z). EID residues interacting with ED and DNA are shown as balls and sticks. (**c**) NMR structure of autoinhibited Ets1 (PDB access code 1R36). (**d**) Comparison of EID position in $Runx1_{1-242}$•$Ets1_{296-441}$•TCRα with the position of IRM in autoinhibited Ets1 after the superimposition of EDs. ED of autoinhibited Ets1 is omitted for clarity

TCRα enhancer ($Runx1_{1-242}$•$Ets1_{296-441}$•TCRαE) (Fig. 2.3b) (Shrivastava et al. 2014). The structure revealed the Ets domain and the Runt domain (RD) are bound to their respective binding sites on the DNA. The helices α1 (194–203) and α2 (204–211) of the Ets1-interaction domain (EID) in Runx1 interact with the Ets domain and DNA, respectively. The major contributors to Ets1•Runx1 interaction are F194, L198 and L201 of the Runx1 EID helix α1, the side chains of which are packed into a wide hydrophobic depression at the Ets1 surface (Fig. 2.3b). Comparison of the autoinhibited Ets1 structure (Lee et al. 2005) (Fig. 2.3c) with the structure of Ets1 in $Runx1_{1-242}$•$Ets1_{296-441}$•TCRαE shows that the EID of Runx1 is bound in an area that is occupied by helices HI1 and HI2 in an autoinhibited Ets1 (Fig. 2.3d). This means that interaction of Runx1 EID with Ets1 repositions the inhibition regulatory module (IRM) of Ets1 together with the serine rich region (SRR) and results in elimination of the autoinhibitory effect of the SRR-IRM module on Ets1 (Fig. 2.4). Consistent with the crystal structures, deletion of Runx1 residues 191–214 or alanine substitutions of the three key Ets1-interacting residues in EID helix α1 individually or in combination resulted in a complete loss of Ets1 activation (Shrivastava et al. 2014). Similar loss of activation was achieved also by disrupting electrostatic interactions between the EID helix α2 of Runx1 and DNA through substitution of Arg205 with glutamate.

In addition to cooperation on composite Ets1•Runx1 motifs on *TCRα or TCRβ* enhancers

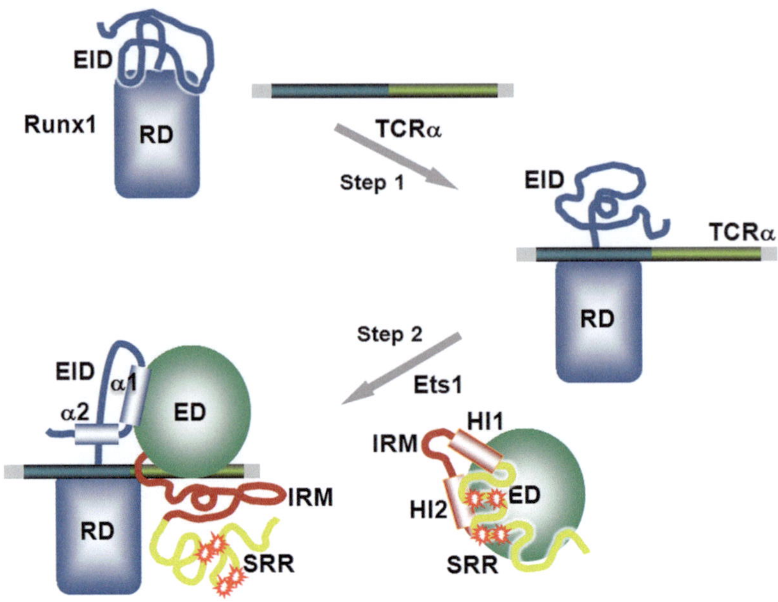

Fig. 2.4 Mechanism of Ets1 activation by Runx1. Step1 – Runx1 binding to TCRαE DNA exposes the disordered EID. Step2 – Upon approaching Ets1, EID of DNA-bound Runx1 binds to ED and DNA by forming helices α1 and α2, and displacing both IRM helices HI1 and HI2. This results in disorder of IRM and destabilization of the phosphorylated SRR inhibitory conformation, producing a fully active Ets1. The *red stars* indicate the presence of phosphorylated serines in SRR

(Kim et al. 1999; Shrivastava et al. 2014), Runx1-Ets1 cooperation was observed also on SC1/core DNA with a different spacing between the Runx1 and Ets1 binding sites (Goetz et al. 2000; Gu et al. 2000). The Runx1•Ets1 cooperation on composite sites with different spacing can be explained by flexibility of the linker between the RD and EID. Indeed, this linker is disordered in $Runx1_{1-242}$•$Ets1_{296-441}$•TCRα crystals (Shrivastava et al. 2014).

Ets1 activation by Runx1 was more prominent with phosphorylated forms of Ets1. Even with two phosphorylated serines (S282 and S285) out of several SSR serines in $Ets1_{276-441}$ ($Ets1_{276-441}$**) the DNA-binding affinity of $Ets1_{276-441}$** is 14-fold lower compared with a wild type $Ets_{276-441}$ (Shrivastava et al. 2014). Nevertheless, in presence of $Runx1_{1-242}$ the $Ets1_{276-441}$** binds DNA similar to a wild type $Ets1_{276-441}$ (Shrivastava et al. 2014). Normally, the activation of transcription factors that are inhibited by phosphorylation requires their dephosphorylation (Holmberg et al. 2002). Thus, phosphorylated Ets1 is unique since it is activated by Runx1 without dephosphorylation.

Trans-activation of the *TCRα* gene enhancer fragment by Runx1 and Ets1 has been studied in a series of transient transfection experiments using either wild-type or phosphorylated Ets1 (Shrivastava et al. 2014). The data show that trans-activation of the *TCRα* gene enhancer is synergistic with either form of Ets1. Consistent with the structure-based mechanism of Ets1 activation by Runx1, the mutations in Runx1 EID that disrupt the EID•Ets1 interaction also eliminate the synergistic trans-activation of *TCRα* gene.

2.6 Biochemical Properties of CBFβ-SMMHC

The *CBFB gene* is disrupted in ~10 % of AML by the inversion of chromosome 16 [inv(16)(p13q22)], and less frequently by the t(16;16)(p13q22), associated with 100 % of AML-M4Eo subtype (Liu et al. 1993). This inversion breaks and joins the *CBFB* and *MYH11* genes, encoding the fusion protein CBFβ-

SMMHC (Liu et al. 1993; Look 1997). Heterozygous knock-in mice for CBFβ-SMMHC lack definitive hematopoiesis, a similar phenotype to that seen for the complete knockout of Runx1 or Cbfβ (Castilla et al. 1996a). Neither *Runx1*$^{+/-}$ nor *Cbfb*$^{+/-}$ heterozygous mice exhibit the dramatic hematopoietic defects associated with the *CBFB-MYH11* knock-in allele (Castilla et al. 1996; Okuda et al. 1996; Sasaki et al. 1996; Wang et al. 1996a, b), indicating that CBFβ-SMMHC acts as a dominant negative on RUNX function. Using isothermal titration calorimetry, it was shown that CBFβ-SMMHC binds ~ 10-fold more tightly to the RUNX1 Runt domain than does wild type CBFβ (Lukasik et al. 2002). Similar results were obtained using EMSA (Huang et al. 2004). As patients present with one allele of the wildtype CBFβ and one allele of CBFβ-SMMHC, RUNX1 will be preferentially bound to the fusion protein in this situation. Other studies have shown decreased RUNX1 binding to the *MPO* promoter (Cao et al. 1997) and to the *INK4b* promoter (Markus et al. 2007) in the presence of CBFβ-SMMHC, clearly indicating that the fusion protein will cause a global change in RUNX1 driven gene expression by blocking its ability to bind to its target genes. A recent RNA-Seq and ChIP-Seq study has suggested that CBFβ-SMMHC can both increase and decrease expression at different loci (Mandoli et al. 2014). These studies clearly established that the binding of CBFβ-SMMHC to RUNX1 is essential for its oncogenic function and therefore established this as an appropriate target for therapeutic intervention.

2.7 Structure/Function of RUNX1-RUNX1T1 (AML1-ETO)

RUNX1-RUNX1T1 (or AML1-ETO) is the chimeric protein formed as a result of the t(8;21), which is among the most common chromosomal rearrangements in adult acute myeloid leukemia (AML) (Rubnitz and Look 1998). RUNX1-ETO contains the N-terminal 177 amino acids of Runx1 fused in frame to nearly all of ETO (Eight Twenty-One, encoded by *RUNX1T1*) (Erickson et al. 1992; Nucifora et al. 1993). AML1-ETO has five conserved domains: the Runt domain from RUNX1, and four from ETO (eTAFH, HHR, Nervy, MYND) (Davis et al. 2003; Hug and Lazar 2004) (Fig. 2.5). As described above, the Runt domain mediates both DNA binding and heterodimerization with CBFβ. The eTAFH (or

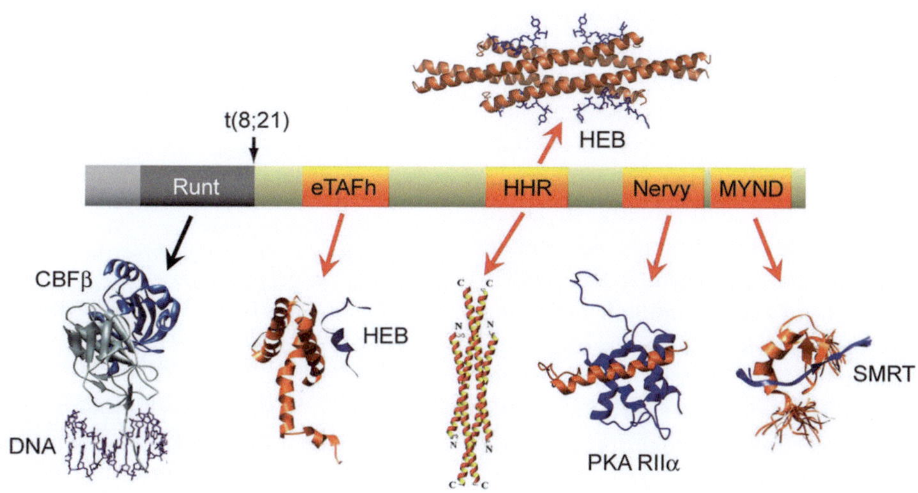

Fig. 2.5 AML1-ETO and its interacting proteins. Shown is a schematic diagram of AML1-ETO with RUNX1-derived sequences in *grey* and RUNX1T1 sequences in *gold* and *orange*. Structures of conserved domains (*dark grey* or *orange*) and their interacting proteins or peptides from those proteins (*blue*), and DNA (*purple*) are shown *above* and *below*

NHR1) domain interacts with the nuclear hormone receptor co-repressor (N-CoR) (Wei et al. 2007), and also with the activation domain of E proteins (E2A and HEB) (Zhang et al. 2004). The HHR (NHR2) domain forms an alpha-helical tetramer that mediates oligomerization of AML1-ETO with itself, with other ETO proteins (Kitabayashi et al. 1998), with Gfi1, and with the co-repressors Sin3 and histone deacetylases 1 and 3 (Lutterbach et al. 1998; Davis et al. 1999, 2003; Amann et al. 2001; McGhee et al. 2003; Liu et al. 2006). The HHR domain has also been shown to bind to E proteins (Sun et al. 2013). Nervy (NHR3) is an α-helical domain that binds the regulatory subunit of type II cyclic AMP-dependent protein kinase (Fukuyama et al. 2001; Yang et al. 2004). Myeloid-Nervy-DEAF-1 (MYND or NHR4) is a zinc-chelating domain structurally homologous to the PHD and RING finger domains, and mediates interactions with N-CoR, the silencing mediator of retinoid and thyroid hormone receptor (SMRT), and the DNA binding protein SON (Ahn et al. 1998; Gelmetti et al. 1998; Lutterbach et al. 1998; Wang et al. 1998; Liu et al. 2007).

Structures of all five conserved domains in RUNX1-ETO bound to interacting proteins or peptides from those proteins have been solved (Warren et al. 2000; Bravo et al. 2001; Tahirov et al. 2001; Liu et al. 2006, 2007; Plevin et al. 2006; Wei et al. 2007; Park et al. 2009; Corpora et al. 2010; Sun et al. 2013) (Fig. 2.5). The structural information was used to introduce amino acid substitutions into each of AML1-ETO's conserved domains that did not perturb their overall structure, but greatly impaired their interaction with other proteins. As a result, four interactions that are essential for AML1-ETO's activity have been identified: DNA and CBFβ binding by the Runt domain, oligomerization through the HHR domain, and E-protein binding by the HHR domain (Liu et al. 2006, 2007; Park et al. 2009; Roudaia et al. 2009; Corpora et al. 2010; Sun et al. 2013). The interaction of eTAFH with HEB, Nervy with PKA RIIα, and MYND with NCoR or SMRT, on the other hand, do not promote AML1-ETO's leukemogenic activity.

References

Ahn, M. Y., Huang, G., Bae, S. C., Wee, H. J., Kim, W. Y., & Ito, Y. (1998). Negative regulation of granulocytic differentiation in the myeloid precursor cell line 32Dcl3 by ear-2, a mammalian homolog of Drosophila seven-up, and a chimeric leukemogenic gene, AML1/ETO. *Proceedings of the National Academy of Sciences of the United States of America, 95*(4), 1812–1817.

Amann, J. M., Nip, J., Strom, D. K., Lutterbach, B., Harada, H., Lenny, N., et al. (2001). ETO, a target of t(8;21) in acute leukemia, makes distinct contacts with multiple histone deacetylases and binds mSin3A through its oligomerization domain. *Molecular and Cellular Biology, 21*(19), 6470–6483.

Backstrom, S., Wolf-Watz, M., Grundstrom, C., Hard, T., Grundstrom, T., & Sauer, U. H. (2002). The RUNX1 Runt domain at 1.25A resolution: A structural switch and specifically bound chloride ions modulate DNA binding. *Journal of Molecular Biology, 322*(2), 259–272.

Berardi, M. J., Sun, C., Zehr, M., Abildgaard, F., Peng, J., Speck, N. A., & Bushweller, J. H. (1999). The Ig fold of the core binding factor alpha Runt domain is a member of a family of structurally and functionally related Ig-fold DNA-binding domains. *Structure, 7*(10), 1247–1256.

Blyth, K., Cameron, E. R., & Neil, J. C. (2005). The RUNX genes: Gain or loss of function in cancer. *Nature Reviews. Cancer, 5*(5), 376–387.

Bravo, J., Li, Z., Speck, N. A., & Warren, A. J. (2001). The leukemia-associated AML1 (Runx1)—CBF beta complex functions as a DNA-induced molecular clamp. *Nature Structural Biology, 8*(4), 371–378.

Cao, W., Britos-Bray, M., Claxton, D. F., Kelley, C. A., Speck, N. A., Liu, P. P., & Friedman, A. D. (1997). CBF beta-SMMHC, expressed in M4Eo AML, reduced CBF DNA-binding and inhibited the G1 to S cell cycle transition at the restriction point in myeloid and lymphoid cells. *Oncogene, 15*(11), 1315–1327.

Castilla, L. H., Wijmenga, C., Wang, Q., Stacy, T., Speck, N. A., Eckhaus, M., et al. (1996a). Failure of embryonic hematopoiesis and lethal hemorrhages in mouse embryos heterozygous for a knocked-in leukemia gene CBFB-MYH11. *Cell, 87*(4), 687–696.

Corpora, T., Roudaia, L., Oo, Z. M., Chen, W., Manuylova, E., Cai, X., et al. (2010). Structure of the AML1-ETO NHR3-PKA(RIIalpha) complex and its contribution to AML1-ETO activity. *Journal of Molecular Biology, 402*(3), 560–577.

Cowley, D. O., & Graves, B. J. (2000). Phosphorylation represses Ets-1 DNA binding by reinforcing autoinhibition. *Genes & Development, 14*(3), 366–376.

Crute, B. E., Lewis, A. F., Wu, Z., Bushweller, J. H., & Speck, N. A. (1996). Biochemical and biophysical properties of the core-binding factor alpha2 (AML1) DNA-binding domain. *The Journal of Biological Chemistry, 271*(42), 26251–26260.

Davis, J. N., Williams, B. J., Herron, J. T., Galiano, F. J., & Meyers, S. (1999). ETO-2, a new member of the ETO-family of nuclear proteins. *Oncogene, 18*(6), 1375–1383.

Davis, J. N., McGhee, L., & Meyers, S. (2003). The ETO (MTG8) gene family. *Gene, 303*, 1–10.

Erickson, P., Gao, J., Chang, K. S., Look, T., Whisenant, E., Raimondi, S., et al. (1992). Identification of breakpoints in t(8;21) acute myelogenous leukemia and isolation of a fusion transcript, AML1/ETO, with similarity to Drosophila segmentation gene, runt. *Blood, 80*(7), 1825–1831.

Fukuyama, T., Sueoka, E., Sugio, Y., Otsuka, T., Niho, Y., Akagi, K., & Kozu, T. (2001). MTG8 proto-oncoprotein interacts with the regulatory subunit of type II cyclic AMP-dependent protein kinase in lymphocytes. *Oncogene, 20*(43), 6225–6232.

Gelmetti, V., Zhang, J., Fanelli, M., Minucci, S., Pelicci, P. G., & Lazar, M. A. (1998). Aberrant recruitment of the nuclear receptor corepressor-histone deacetylase complex by the acute myeloid leukemia fusion partner ETO. *Molecular and Cellular Biology, 18*(12), 7185–7191.

Giese, K., Kingsley, C., Kirshner, J. R., & Grosschedl, R. (1995). Assembly and function of a TCR alpha enhancer complex is dependent on LEF-1-induced DNA bending and multiple protein-protein interactions. *Genes & Development, 9*(8), 995–1008.

Goetz, T. L., Gu, T. L., Speck, N. A., & Graves, B. J. (2000). Auto-inhibition of Ets-1 is counteracted by DNA binding cooperativity with core-binding factor alpha2. *Molecular and Cellular Biology, 20*(1), 81–90.

Goger, M., Gupta, V., Kim, W. Y., Shigesada, K., Ito, Y., & Werner, M. H. (1999). Molecular insights into PEBP2/CBF beta-SMMHC associated acute leukemia revealed from the structure of PEBP2/CBF beta. *Nature Structural Biology, 6*(7), 620–623.

Gu, T. L., Goetz, T. L., Graves, B. J., & Speck, N. A. (2000). Auto-inhibition and partner proteins, core-binding factor beta (CBFbeta) and Ets-1, modulate DNA binding by CBFalpha2 (AML1). *Molecular and Cellular Biology, 20*(1), 91–103.

Hollenhorst, P. C., Shah, A. A., Hopkins, C., & Graves, B. J. (2007). Genome-wide analyses reveal properties of redundant and specific promoter occupancy within the ETS gene family. *Genes & Development, 21*(15), 1882–1894.

Hollenhorst, P. C., Chandler, K. J., Poulsen, R. L., Johnson, W. E., Speck, N. A., & Graves, B. J. (2009). DNA specificity determinants associate with distinct transcription factor functions. *PLoS Genetics, 5*(12), e1000778.

Holmberg, C. I., Tran, S. E., Eriksson, J. E., & Sistonen, L. (2002). Multisite phosphorylation provides sophisticated regulation of transcription factors. *Trends in Biochemical Sciences, 27*(12), 619–627.

Huang, X., Crute, B. E., Sun, C., Tang, Y. Y., Kelley 3rd, J. J., Lewis, A. F., et al. (1998). Overexpression, purification, and biophysical characterization of the heterodimerization domain of the core-binding factor beta subunit. *The Journal of Biological Chemistry, 273*(4), 2480–2487.

Huang, X., Peng, J. W., Speck, N. A., & Bushweller, J. H. (1999). Solution structure of core binding factor beta and map of the CBF alpha binding site. *Nature Structural Biology, 6*(7), 624–627.

Huang, G., Shigesada, K., Wee, H. J., Liu, P. P., Osato, M., & Ito, Y. (2004). Molecular basis for a dominant inactivation of RUNX1/AML1 by the leukemogenic inversion 16 chimera. *Blood, 103*(8), 3200–3207.

Hug, B. A., & Lazar, M. A. (2004). ETO interacting proteins. *Oncogene, 23*(24), 4270–4274.

Kim, W. Y., Sieweke, M., Ogawa, E., Wee, H. J., Englmeier, U., Graf, T., & Ito, Y. (1999). Mutual activation of Ets-1 and AML1 DNA binding by direct interaction of their autoinhibitory domains. *The EMBO Journal, 18*(6), 1609–1620.

Kitabayashi, I., Ida, K., Morohoshi, F., Yokoyama, A., Mitsuhashi, N., Shimizu, K., et al. (1998). The AML1-MTG8 leukemic fusion protein forms a complex with a novel member of the MTG8(ETO/CDR) family, MTGR1. *Molecular and Cellular Biology, 18*(2), 846–858.

Lee, G. M., Donaldson, L. W., Pufall, M. A., Kang, H. S., Pot, I., Graves, B. J., & McIntosh, L. P. (2005). The structural and dynamic basis of Ets-1 DNA binding autoinhibition. *The Journal of Biological Chemistry, 280*(8), 7088–7099.

Lee, G. M., Pufall, M. A., Meeker, C. A., Kang, H. S., Graves, B. J., & McIntosh, L. P. (2008). The affinity of Ets-1 for DNA is modulated by phosphorylation through transient interactions of an unstructured region. *Journal of Molecular Biology, 382*(4), 1014–1030.

Li, Z., Yan, J., Matheny, C. J., Corpora, T., Bravo, J., Warren, A. J., et al. (2003). Energetic contribution of residues in the Runx1 Runt domain to DNA binding. *The Journal of Biological Chemistry, 278*(35), 33088–33096.

Li, Z., Lukasik, S. M., Liu, Y., Grembecka, J., Bielnicka, I., Bushweller, J. H., & Speck, N. A. (2006). A mutation in the S-switch region of the Runt domain alters the dynamics of an allosteric network responsible for CBFbeta regulation. *Journal of Molecular Biology, 364*(5), 1073–1083.

Liu, P., Tarle, S. A., Hajra, A., Claxton, D. F., Marlton, P., Freedman, M., et al. (1993). Fusion between transcription factor CBF beta/PEBP2 beta and a myosin heavy chain in acute myeloid leukemia. *Science, 261*(5124), 1041–1044.

Liu, Y., Cheney, M. D., Gaudet, J. J., Chruszcz, M., Lukasik, S. M., Sugiyama, D., et al. (2006). The tetramer structure of the Nervy homology two domain, NHR2, is critical for AML1/ETO's activity. *Cancer Cell, 9*(4), 249–260.

Liu, Y., Chen, W., Gaudet, J., Cheney, M. D., Roudaia, L., Cierpicki, T., et al. (2007). Structural basis for recognition of SMRT/N-CoR by the MYND domain and its contribution to AML1/ETO's activity. *Cancer Cell, 11*(6), 483–497.

Look, A. T. (1997). Oncogenic transcription factors in the human acute leukemias. *Science, 278*(5340), 1059–1064.

Lukasik, S. M., Zhang, L., Corpora, T., Tomanicek, S., Li, Y., Kundu, M., et al. (2002). Altered affinity of CBFbeta-SMMHC for Runx1 explains its role in leukemogenesis. *Nature Structural Biology, 9*(9), 674–679.

Lutterbach, B., Sun, D., Schuetz, J., & Hiebert, S. W. (1998). The MYND motif is required for repression of basal transcription from the multidrug resistance 1 promoter by the t(8;21) fusion protein. *Molecular and Cellular Biology, 18*(6), 3604–3611.

Mandoli, A., Singh, A. A., Jansen, P. W. T. C., Wierenga, A. T. J., Riahi, H., Franci, G., et al. (2014). CBFB-MYH11/RUNX1 together with a compendium of hematopoietic regulators, chromatin modifiers and basal transcription factors occupies self-renewal genes in inv(16) acute myeloid leukemia. *Leukemia, 28*(4), 770–778.

Mangan, J. K., & Speck, N. A. (2011). RUNX1 mutations in clonal myeloid disorders: From conventional cytogenetics to next generation sequencing, a story 40 years in the making. *Critical Reviews in Oncogenesis, 16*(1–2), 77–91.

Markus, J., Garin, M. T., Bies, J., Galili, N., Raza, A., Thirman, M. J., et al. (2007). Methylation-independent silencing of the tumor suppressor INK4b (p15) by CBFbeta-SMMHC in acute myelogenous leukemia with inv(16). *Cancer Research, 67*(3), 992–1000.

Matheny, C. J., Speck, M. E., Cushing, P. R., Zhou, Y., Corpora, T., Regan, M., et al. (2007). Disease mutations in RUNX1 and RUNX2 create nonfunctional, dominant-negative, or hypomorphic alleles. *The EMBO Journal, 26*(4), 1163–1175.

McGhee, L., Bryan, J., Elliott, L., Grimes, H. L., Kazanjian, A., Davis, J. N., & Meyers, S. (2003). Gfi-1 attaches to the nuclear matrix, associates with ETO (MTG8) and histone deacetylase proteins, and represses transcription using a TSA-sensitive mechanism. *Journal of Cellular Biochemistry, 89*(5), 1005–1018.

Nagata, T., Gupta, V., Sorce, D., Kim, W. Y., Sali, A., Chait, B. T., et al. (1999). Immunoglobulin motif DNA recognition and heterodimerization of the PEBP2/CBF Runt domain. *Nature Structural Biology, 6*(7), 615–619.

Nucifora, G., Birn, D. J., Erickson, P., Gao, J., LeBeau, M. M., Drabkin, H. A., & Rowley, J. D. (1993). Detection of DNA rearrangements in the AML1 and ETO loci and of an AML1/ETO fusion mRNA in patients with t(8;21) acute myeloid leukemia. *Blood, 81*(4), 883–888.

Ogawa, E., Inuzuka, M., Maruyama, M., Satake, M., Naito-Fujimoto, M., Ito, Y., & Shigesada, K. (1993). Molecular cloning and characterization of PEBP2 beta, the heterodimeric partner of a novel Drosophila runt-related DNA binding protein PEBP2 alpha. *Virology, 194*(1), 314–331.

Okuda, T., van Deursen, J., Hiebert, S. W., Grosveld, G., & Downing, J. R. (1996). AML1, the target of multiple chromosomal translocations in human leukemia, is essential for normal fetal liver hematopoiesis. *Cell, 84*, 321–330.

Otto, F., Kanegane, H., & Mundlos, S. (2002). Mutations in the RUNX2 gene in patients with cleidocranial dysplasia. *Human Mutation, 19*(3), 209–216.

Park, S., Chen, W., Cierpicki, T., Tonelli, M., Cai, X., Speck, N. A., & Bushweller, J. H. (2009). Structure of the AML1-ETO eTAFH domain – HEB peptide complex and its contribution to AML1-ETO activity. *Blood, 113*, 3558–3567.

Perez-Alvarado, G. C., Munnerlyn, A., Dyson, H. J., Grosschedl, R., & Wright, P. E. (2000). Identification of the regions involved in DNA binding by the mouse PEBP2alpha protein. *FEBS Letters, 470*(2), 125–130.

Plevin, M. J., Zhang, J., Guo, C., Roeder, R. G., & Ikura, M. (2006). The acute myeloid leukemia fusion protein AML1-ETO targets E proteins via a paired amphipathic helix-like TBP-associated factor homology domain. *Proceedings of the National Academy of Sciences of the United States of America, 103*(27), 10242–10247.

Pufall, M. A., Lee, G. M., Nelson, M. L., Kang, H. S., Velyvis, A., Kay, L. E., et al. (2005). Variable control of Ets-1 DNA binding by multiple phosphates in an unstructured region. *Science, 309*(5731), 142–145.

Roudaia, L., Cheney, M. D., Manuylova, E., Chen, W., Morrow, M., Park, S., et al. (2009). CBF{beta} is critical for AML1-ETO and TEL-AML1 activity. *Blood, 113*, 3070–3079.

Rubnitz, J. E., & Look, A. T. (1998). Molecular basis of leukemogenesis. *Current Opinion in Hematology, 5*(4), 264–270.

Sasaki, K., Yagi, H., Bronson, R. T., Tominaga, K., Matsunashi, T., Deguchi, K., et al. (1996). Absence of fetal liver hematopoiesis in transcriptional co-activator, core binding factor b (*Cbfb*) deficient mice. *Proceedings of the National Academy of Sciences of the United States of America, 93*, 12359–12363.

Shrivastava, T., Mino, K., Babayeva, N. D., Baranovskaya, O. I., Rizzino, A., & Tahirov, T. H. (2014). Structural basis of Ets1 activation by Runx1. *Leukemia, 28*(10), 2040–2048.

Skalicky, J. J., Donaldson, L. W., Petersen, J. M., Graves, B. J., & McIntosh, L. P. (1996). Structural coupling of the inhibitory regions flanking the ETS domain of murine Ets-1. *Protein Science, 5*(2), 296–309.

Sun, W., Graves, B. J., & Speck, N. A. (1995). Transactivation of the Moloney murine leukemia virus and T-cell receptor beta-chain enhancers by cbf and ets requires intact binding sites for both proteins. *Journal of Virology, 69*(8), 4941–4949.

Sun, X. J., Wang, Z., Wang, L., Jiang, Y., Kost, N., Soong, T. D., et al. (2013). A stable transcription factor complex nucleated by oligomeric AML1-ETO controls leukaemogenesis. *Nature, 500*(7460), 93–97.

Tahirov, T. H., Inoue-Bungo, T., Morii, H., Fujikawa, A., Sasaki, M., Kimura, K., et al. (2001). Structural analyses of DNA recognition by the AML1/Runx-1 Runt domain and its allosteric control by CBFbeta. *Cell, 104*(5), 755–767.

Tang, Y. Y., Crute, B. E., Kelley, J. J., Huang, X., Yan, J., Shi, J., et al. (2000a). Biophysical characterization of interactions between the core binding factor alpha and beta subunits and DNA. *FEBS Letters, 470*(2), 167–172.

Tang, Y. Y., Shi, J., Zhang, L., Davis, A., Bravo, J., Warren, A. J., et al. (2000b). Energetic and functional contribution of residues in the core binding factor beta (CBFbeta) subunit to heterodimerization with CBFalpha. *The Journal of Biological Chemistry, 275*(50), 39579–39588.

Wang, Q., Stacy, T., Binder, M., Marín-Padilla, M., Sharpe, A. H., & Speck, N. A. (1996a). Disruption of the *Cbfa2* gene causes necrosis and hemorrhaging in the central nervous system and blocks definitive hematopoiesis. *Proceedings of the National Academy of Sciences of the United States of America, 93*, 3444–3449.

Wang, Q., Stacy, T., Miller, J. D., Lewis, A. F., Huang, X., Bories, J.-C., et al. (1996b). The CBFb subunit is essential for CBFa2 (AML1) function in vivo. *Cell, 87*, 697–708.

Wang, J., Hoshino, T., Redner, R. L., Kajigaya, S., & Liu, J. M. (1998). ETO, fusion partner in t(8;21) acute myeloid leukemia, represses transcription by interaction with the human N-CoR/mSin3/HDAC1 complex. *Proceedings of the National Academy of Sciences of the United States of America, 95*(18), 10860–10865.

Warren, A. J., Bravo, J., Williams, R. L., & Rabbitts, T. H. (2000). Structural basis for the heterodimeric interaction between the acute leukaemia-associated transcription factors AML1 and CBFbeta. *The EMBO Journal, 19*(12), 3004–3015.

Wei, Y., Liu, S., Lausen, J., Woodrell, C., Cho, S., Biris, N., et al. (2007). A TAF4-homology domain from the corepressor ETO is a docking platform for positive and negative regulators of transcription. *Nature Structural & Molecular Biology, 14*(7), 653–661.

Wotton, D., Ghysdael, J., Wang, S., Speck, N. A., & Owen, M. J. (1994). Cooperative binding of Ets-1 and core binding factor to DNA. *Molecular and Cellular Biology, 14*(1), 840–850.

Yan, J., Liu, Y., Lukasik, S. M., Speck, N. A., & Bushweller, J. H. (2004). CBFbeta allosterically regulates the Runx1 Runt domain via a dynamic conformational equilibrium. *Nature Structural & Molecular Biology, 11*(9), 901–906.

Yang, H. T., Wu, D. H., Xue, X. Y., Liang, W. X., Miao, X. Y., Pang, H., & Chen, S. J. (2004). Cloning, expression, purification and crystallization of NHR3 domain from acute myelogenous leukemia-related protein AML1-ETO. *Acta Biochimica et Biophysica Sinica (Shanghai), 36*(8), 566–570.

Zhang, J., Kalkum, M., Yamamura, S., Chait, B. T., & Roeder, R. G. (2004). E protein silencing by the leukemogenic AML1-ETO fusion protein. *Science, 305*(5688), 1286–1289.

3 Covalent Modifications of RUNX Proteins: Structure Affects Function

Ezra Blumenthal, Sarah Greenblatt, Guang Huang, Koji Ando, Ye Xu, and Stephen D. Nimer

Abstract

The RUNX family of transcription factors plays important roles in tissue-specific gene expression. Many of their functions depend on specific post-translational modifications (PTMs), and in this review, we describe how PTMs govern RUNX DNA binding, transcriptional activity, protein stability, cellular localization, and protein-protein interactions. We also report how these processes can be disrupted in disease settings. Finally, we describe how alterations of RUNX1, or the enzymes that catalyze its post-translational modifications, contribute to hematopoietic malignancies.

Keywords

RUNX1 • RUNX2 • RUNX3 • CBFβ • Post-Translational Modifications • Acetylation • Methylation • Phosphorylation • Ubiquitylation • Transcriptional Activation • Transcriptional Repression • Acute Myeloid Leukemia • FPD/AML • Cleidocranial Dysplasia

3.1 Contents of the Chapter (Including 1 Figure and 1 Table)

Post-translational modifications (PTMs) are key regulators of RUNX protein function. Together these PTMs govern the transcriptional activity of RUNX proteins by modulating their DNA binding, protein stability, cellular localization, and protein-protein interactions. The sensitivity of RUNX proteins to subtle changes in these properties allows extracellular signals to influence stem cell, progenitor cell, and differentiated cell biology. Furthermore, RUNX protein modifications help create scaffolds that facilitate

E. Blumenthal • S. Greenblatt • K. Ando
Sylvester Comprehensive Cancer Center, Miami, FL 33136, USA

G. Huang
Division of Experimental Hematology and Cancer Biology, Cincinnati Children's Hospital Medical Center, Cincinnati, OH, USA

Y. Xu • S.D. Nimer (✉)
Department of Medicine, University of Miami Miller School of Medicine, Miami, FL 33136, USA

Sylvester Comprehensive Cancer Center, Miami, FL 33136, USA
e-mail: snimer@med.miami.edu

Y. Groner et al. (eds.), *RUNX Proteins in Development and Cancer*, Advances in Experimental Medicine and Biology 962, DOI 10.1007/978-981-10-3233-2_3

the recruitment of proteins that either promote or inhibit transcription. Among RUNX proteins, the functional loss of RUNX1 activity in the myeloid malignancies, driven by genomic alterations, indicates its critical role as a suppressor of myeloid transformation. In this section, we describe how PTMs affect RUNX functions in normal and malignant cells.

3.2 RUNX1 Post-translational Modifications in Transcriptional Activation

Lineage specific gene expression requires the complex interplay between transcription factors with DNA binding ability and histone modifying enzymes (Fig. 3.1, Table 3.1). The transcription factor RUNX1 associates with chromatin modifiers, cofactors, and other transcription factors at the regulatory regions of target genes critical for myeloid and lymphoid differentiation, such as *SPI1* Zhao et al. (2008) and *EBF1* Seo et al. (2012). Genetic alterations involving *RUNX1* are common in acute myeloid leukemia (AML), occurring in approximately 15% of patients (The Cancer Genome Atlas Research Network 2013). They include point mutations, truncating mutations, amplifications, and chromosomal translocations that generate fusion proteins, most commonly RUNX1-ETO, also known as AML1-ETO. RUNX1 binds DNA with high affinity only when bound to its heterodimeric protein partner CBFβ, which is encoded by a gene that is also

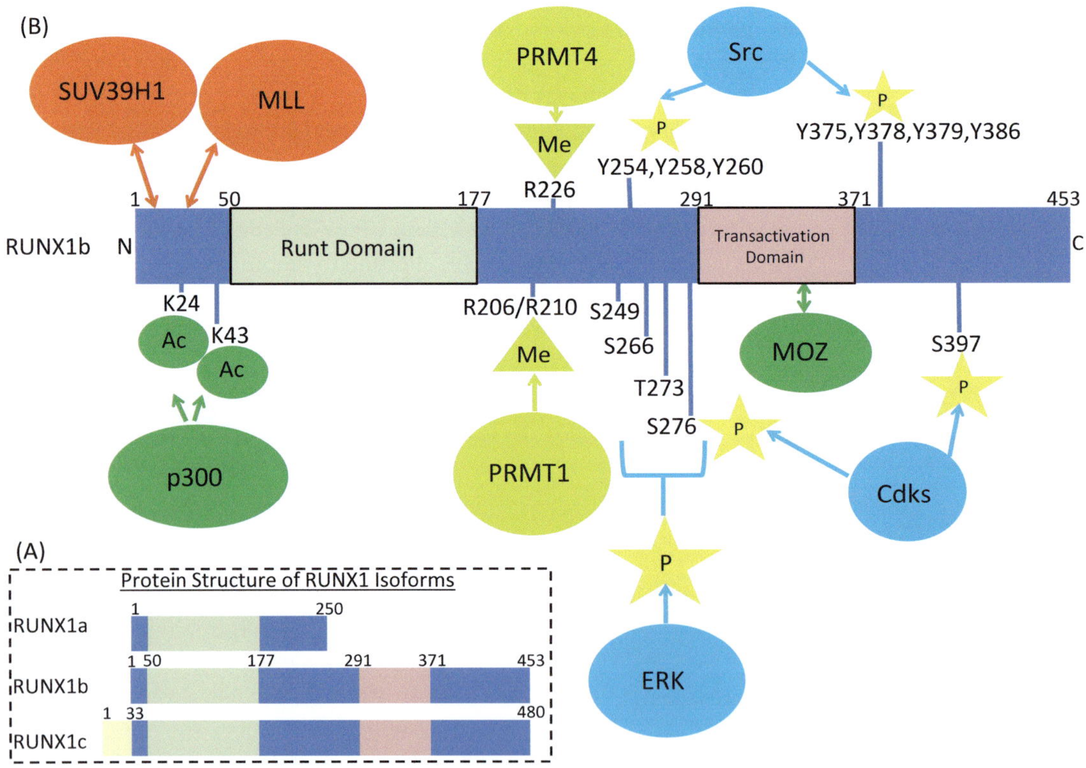

Fig. 3.1 (**a**) Protein structure of RUNX1 isoforms. The DNA-binding Runt domain (*light green*) is conserved across all RUNX1 isoforms. The transactivation domain (*red*) is found only in RUNX1b and RUNX1c. (**b**) RUNX1b: selected protein interactions and post-translational modifications. Kinases (*light blue*) interact with the RUNX1 C-terminus following cytokine stimulation and regulate cell cycle progression. The lysine acetyltransferases p300 and MOZ (*green*) generally promote transcriptional activation by RUNX1. Arginine methyltransferases and lysine methyltransferases (mustard and orange, respectively) both stimulate and inhibit transcription by RUNX1 and play important roles in hematopoietic differentiation

Table 3.1 List of RUNX interacting proteins

Protein	Interacting partner	Function	Reference
RUNX1	p300/CBP	Lysine Acetyltransferase	Yamaguchi et al. (2004)
	MOZ	Lysine Acetyltransferase	Kitabayashi et al. (2001)
	MLL	Lysine Methyltransferase	Huang et al. (2011), Koh et al. (2013)
	PRMT1	Arginine Methyltransferase	Zhao et al. (2008)
	PRMT4	Arginine Methyltransferase	Vu et al. (2013)
	PRMT6	Arginine Methyltransferase	Reed-Inderbitzin et al. (2006)
	SUV39H1	Lysine Methyltransferase	Herglotz et al. (2013) and Chakraborty et al. (2003)
	ERK	Kinase	Tanaka et al. (1996), Zhang et al. (2004) and Yoshimi et al. (2012)
	CDKs	Kinase	Biggs et al. (2006) and Zhang et al. (2008)
	PIM1	Kinase	Aho et al. (2006)
	HIPK2	Kinase	Wee et al. (2008) and Aikawa et al. (2006)
	IKK	Kinase	Nakagawa et al. (2011)
	HDACs	Deacetylase	Guo and Friedman (2011), Lutterbach et al. (2000) and Chakraborty et al. (2003)
	Sin3a	Transcriptional Regulator	Zhao et al. (2008), Imai et al. (2004) and Lutterbach et al. (2000)
	Groucho	Transcriptional Regulator	Lutterbach et al. (2000)
	BMI1	Polycomb Protein	Yu et al. (2012)
	SHP2	Tyrosine Phosphatase	Huang et al. (2012)
	Src	Kinase	Huang et al. (2012)
	APC/C	Ubiquitin Ligase	Biggs et al. (2006)
	SMADs	Transcription Factors	Coco Lo et al. (1997)
	CHIP/Stub1	Ubiquitin Ligase	Shang et al. (2009)
RUNX2	p300/CBP	Lysine Acetyltransferase	Sierra et al. (2003)
	MOZ	Lysine Acetyltransferase	Shang et al. (2009)
	MORF	Lysine Acetyltransferase	Pelletier et al. (2002)
	HDAC4, HDAC5	Deactetylase	Jeon et al. (2006)
	SMADs	Transcription Factor	Jeon et al. (2006) and Zhang et al. (2000)
	ERK	Kinase	Franceschi et al. (2003) and Qiao et al. (2004)
	PKA	Kinase	Selvamurugan et al. (2000)
	PKC	Kinase	Kim et al. (2006)
	GSK3β	Kinase	Kugimiya et al. (2007)
RUNX3	p300/CBP	Lysine Acetyltransferase	Jin et al. (2004)
	BRD2	Transcriptional Regulator	Li et al. (2002)
	SirT2	Deacetylase	Lee et al. (2013)
	Smurf	Ubiquitin Ligase	Jin et al. (2004)
	SMADs	Transcription Factors	Zaidi et al. (2002)
	HDAC1,-2,-4,-5	Deacetylases	Jin et al. (2004)

involved in a chromosomal translocation, one that generates the CBFβ-SMMHC fusion protein (Kamikubo et al. 2010). Taken together, AML1-ETO and CBFβ-SMMHC expressing leukemias are referred to as CBF-AML.

3.2.1 Acetylation

The recruitment of coactivator proteins to specific promoters is a critical step for transcriptional activation. Many of these transcriptional coactivators possess histone acetyltransferase activity. RUNX proteins are rich in lysine residues, and are modified by multiple members of the lysine acetyltransferase (KAT) family. In general, the acetylation of RUNX proteins stimulates their transcriptional activity. Members of the KAT family that bind and acetylate RUNX proteins include p300 and MOZ, while P/CAF and GCN5 can modify the RUNX1 fusion protein AML1/MDS1/EVI1 (Yamaguchi et al. 2004; Jin et al. 2004; Kitabayashi et al. 2001; Senyuk et al. 2003). The KAT proteins p300 and CBP are transcriptional coactivators with distinct roles in normal hematopoiesis (Rebel et al. 2002). The p300-mediated acetylation of lysines 24 and 43 on RUNX1 augments RUNX1 DNA binding and transcriptional activation. Mutation of these sites to arginine does not disrupt the interaction with p300 but rather impairs RUNX1 DNA binding (Yamaguchi et al. 2004). In t(8;21) AML, the AML1-ETO fusion protein is acetylated by p300 on lysine 24 and 43, which promotes its ability to activate key self-renewing genes; the absence of K43 acetylation significantly abrogates AML1-ETO mediated leukemogenesis in vivo (Wang et al. 2011).

The *Myst* acetyltransferase family member *Moz* is another coactivator of RUNX1-mediated transcription. RUNX1 binds MOZ through its C-terminal transactivation domain, which drives the expression of genes involved in monocyte/macrophage differentiation (Kitabayashi et al. 2001). The MYST domain of MOZ has KAT activity, however, it is neither required for interacting with RUNX1 nor essential for activating transcription, implying that other domains within MOZ regulate transcriptional activation by RUNX1. MOZ can also be found in a fusion protein that contains the p300 homolog CBP, and in fact MOZ, CBP and MOZ-CBP can all acetylate RUNX1 in vitro. Acetylation by MOZ or CBP has been shown to promote RUNX1 transactivation, while acetylation by MOZ-CBP attenuates RUNX1-driven gene expression (Kitabayashi et al. 2001). Although the mechanism remains unclear, it is possible that MOZ-CBP disrupts the interaction between RUNX1 and CBP or MOZ, or another cofactor necessary for transactivation. In sum, the KAT proteins p300, CBP and MOZ play important roles in the RUNX1-mediated transcriptional program, and their dysregulation could contribute to aberrant gene regulation in hematopoietic malignancies.

3.2.2 Methylation

Protein methyltransferases are another group of histone modifying enzymes that regulate RUNX1 transcriptional activity and recruitment to its target gene promoters. We have shown that the lysine methyltransferase MLL physically interacts with the N-terminal region of RUNX1, and promotes the deposition of H3K4me3 activating marks upstream of the critical RUNX1 target gene PU.1. This interaction also appears to stabilize RUNX1 by inhibiting its poly-ubiquitination (Huang et al. 2011). PRMT1, an arginine methyltransferase that targets histone H4R3, methylates RUNX1 at R206 and R210, which abrogates its association with the corepressor SIN3A, enhances its transcriptional activity, and facilitates the binding of RUNX1 to its target gene promoters including *ITGA2B* and *SPI1* (Zhao et al. 2008). Interestingly, knock-in mice with arginine-to-lysine mutations at R206/R210 primarily display impaired peripheral T-cell maintenance (Mizutani et al. 2015). PRMT1 also methylates AML1-ETO (and a truncated isoform AE9a) affecting its transcriptional activating properties (Shia et al. 2012). Thus, multiple methyltransferases are involved in the control of RUNX1-mediated transcriptional activation.

3.2.3 Phosphorylation

[Note: Amino acid positions in this section refer to RUNX1c].

RUNX1 is also subject to phosphorylation by kinase signaling cascades; these include kinases activated by hematopoietic cytokines and growth factors, and kinases that function as cell cycle regulatory proteins. In response to cytokine stimulation, extracellular signal-regulated kinase (ERK) phosphorylates the C-terminus of RUNX1 at S276, S293, T300, S303, and S462 (Tanaka et al. 1996; Zhang et al. 2004; Yoshimi et al. 2012), which increase RUNX1 mediated transactivation by preventing its interaction with Sin3A (Imai et al. 2004). Mutation of four RUNX1 ERK phosphorylation sites (S276/S293/T300/S303) impairs T-cell differentiation, although the mutants retain the ability to rescue early hematopoiesis (Yoshimi et al. 2012). The sensitivity of T-lymphocytes to specific changes in RUNX1 phosphorylation and methylation suggests a tissue-specific role of ERK signaling in RUNX1 function.

RUNX1 directly regulates the G1 to S transition, a process that is inhibited by the leukemia-associated CBFβ-SMMHC and AML1-ETO fusion proteins. While RUNX1 RNA levels remain constant throughout the cell cycle, RUNX1 protein levels increase at the G1 to S transition, and then decrease due to ubiquitin-mediated degradation during G2/M. Serine phosphorylation of RUNX1 occurs during G_2/M by cyclin B/Cdk1 and cyclin A/Cdk2 on S276, and S303, triggering RUNX1 ubiquitination by the APC-CDC20 complex (Biggs et al. 2006). Cyclin B/CDK1 and cyclin D3/CDK6 also phosphorylate RUNX1 on S48 and S424; while S48 is phosphorylated throughout the cell cycle, S303 and S424 are phosphorylated most prominently during G2/M and G1, respectively (Zhang et al. 2008). These modifications reduce the interaction of RUNX1 with HDAC1 and HDAC3, further promoting transcriptional activation (Guo and Friedman 2011). PIM1 kinase, another regulator of cell cycle progression, also interacts with RUNX1 to enhance its transactivation activity following cytokine stimulation (Aho et al. 2006).

Cross talk between serine/threonine phosphorylation and histone acetylation can synergistically enhance the transcriptional activity of RUNX1. Homeodomain-interacting protein kinase 2 (HIPK2) is a nuclear kinase that forms a complex with both RUNX1 and p300, and it initiates a phosphorylation cascade that stimulates transcriptional activation (Wee et al. 2008; Aikawa et al. 2006). *Hipk1/2* double-deficient mice show impaired phosphorylation of both p300 and RUNX1, and defects in definitive hematopoiesis (Aikawa et al. 2006).

Taken together, the phosphorylation of RUNX1 provides for its dynamic regulation in response to extracellular signals and cell cycle progression. Serine/threonine phosphorylation serves to increase RUNX1 transcriptional activity while also decreasing its protein stability. As we describe below, tyrosine phosphorylation of RUNX1 also regulates its protein-protein interactions and diminishes its DNA binding ability.

3.3 RUNX1 Post-translational Modifications in Transcriptional Repression

Covalent modifications of RUNX family members can also have negative effects on transcription. BMI1, a component of the polycomb repressive complex 1 (PRC1), has been shown to be recruited to chromatin at key RUNX1 binding sites in megakaryocytes and lymphocytes (Yu et al. 2012). Transcriptional repression by RUNX1 is critical for normal hematopoietic development. For example, RUNX1 and RUNX3 are required for CD4 silencing in vivo, a necessary process for cytotoxic T-cell development and maturation (Levanon et al. 2002). Analysis of *Runx1* knockout mice demonstrates that RUNX1 suppresses the nuclear translocation of NF-kB. Normally, NF-kB is bound to IkB in the cytoplasm. Phosphorylation of IkB by IkB kinase (IKK) degrades IkB, leading to nuclear import of NF-kB, which induces expression of NF-kB target genes. RUNX1 inhibits the enzymatic activity of IKK by binding to it in the cytoplasm, thereby suppressing the nuclear shuttling of NF-kB (Nakagawa et al. 2011).

3.3.1 Deacetylation

Histone deacetylase (HDAC) complexes classically participate in chromatin remodeling and gene repression, and several members of HDAC complexes have been found to interact with RUNX1 including HDAC1, SIN3A and Gro/TLE. While SIN3A also interacts with RUNX2, RUNX3, and RUNX1-ETO, mutation of the SIN3A binding domain in RUNX1 reduces the recruitment of histone deacetylases and impairs RUNX1-mediated repression of the CDKNIA promoter (Lutterbach et al. 2000). Trichostatin A, a broad-spectrum histone deacetylase inhibitor, impairs RUNX1-guided gene repression, further highlighting the contribution of histone deacetylases to the negative regulation of gene expression by RUNX1 (Lutterbach et al. 2000). Interestingly, chromosomal translocations involving *RUNX1* generate RUNX1 fusion proteins that show enhanced recruitment of co-repressors, compared to wildtype RUNX1. The ALL associated t(12;21) fusion protein TEL/AML1 binds SIN3A with higher affinity than RUNX1 alone. It appears that in pathologic settings, RUNX1 fusion proteins may more potently repress RUNX1 target gene expression, acting at least in part by suppressing wildtype RUNX1 function (Guidez et al. 2000).

3.3.2 Methylation

Methylation of RUNX1 can also promote its repression of transcription. Hematopoietic stem cells express high levels of PRMT4 (CARM1), a Type-I arginine methyltransferase that methylates RUNX1 on an evolutionarily conserved residue, arginine 223. RUNX1-R223me2 drives the assembly of a repressive complex that inhibits transcription of mir-223, a driver of myeloid differentiation, and other RUNX1 target genes. AML patient samples significantly overexpress PRMT4, suggesting that R223-methyl-RUNX1 contributes to the block in differentiation that is characteristic of human AML (Vu et al. 2013).

RUNX1 has also been shown to form a corepressor complex with PRMT6 (and SIN3A and HDAC1) to mediate repression of its target genes prior to megakaryocytic differentiation (Herglotz et al. 2013). The histone H3 lysine 9 specific methyltransferase SUV39H1 binds RUNX1 at its N-terminus. H3K9me is a histone mark that recruits heterochromatin protein-1 (HP1) to silence gene expression. The interaction between RUNX1 and SUV39H1 decreases the affinity of RUNX1 for DNA and also impairs transcription of the RUNX1 target gene CSF1R (Chakraborty et al. 2003; Reed-Inderbitzin et al. 2006).

3.3.3 Phosphorylation

Similar to methylation, phosphorylation of RUNX1 can alter its effects on gene expression. RUNX1 tyrosine phosphorylation was recently implicated in impairing megakaryocyte development, as the level of tyrosine phosphorylation inversely correlates with the extent of differentiation. RUNX1 interacts with the non-receptor tyrosine kinase c-Src and the tyrosine phosphatase SHP2; its tyrosyl phosphorylation decreases the interactions of RUNX1 with CBFβ and GATA1 and FLI1, but increases its affinity for the chromatin remodeling SWI/SNF subunits BRG1 and SNF5 (Huang et al. 2012; Neel and Speck 2012). Additionally, the phosphorylated tyrosine residues in RUNX1 impair its ability to bind and transactivate DNA. These findings suggest that while serine/threonine phosphorylation events promote transcriptional activation by RUNX1, tyrosine phosphorylation may promote gene repression.

3.4 Protein Stability and Localization of RUNX1

3.4.1 Ubiquitination

The levels of RUNX1 are tightly regulated at both the transcriptional and post-translational level. RUNX1 degradation is mediated by the ubiquitin-proteasome system, yet multiple protein-protein interactions can stabilize RUNX1 by shielding it from ubiquitin ligases. For example, the interaction of RUNX1 with its obligate binding partner CBFβ stabilizes RUNX1 through inhibition of ubiquitination (Huang et al. 2001). The mixed lineage leukemia (MLL) protein also binds RUNX1 to produce a similar stabilizing effect (Huang et al. 2011). The anaphase promoting complex (APC) is an E3 ubiquitin ligase known to regulate mitosis. The CDC20 subunit of APC targets RUNX1 for degradation following RUNX1 phosphorylation by CDK1 or CDK2 (Biggs et al. 2006). The heat shock protein-binding co-chaperone protein CHIP/STUB1 impairs protein stability by promoting the ubiquitination and degradation of misfolded or unfolded proteins. While reports have suggested that its E3 ligase activity depends on its interaction with HSP70/90 (Murata et al. 2003), CHIP/Stub1 is capable of promoting the degradation of RUNX1 independently of HSP70 (Shang et al. 2009).

It appears that several leukemia-associated RUNX1 mutant proteins are relatively resistant to ubiquitin-mediated degradation, allowing them to function as dominant negatives. For example, in-vitro assays demonstrate that the RUNX1 L117P and I150T mutants are ubiquitinated to a lesser extent than wildtype RUNX1. These runt domain mutants lack transcriptional activity; they can competitively bind to CBFβ and/or DNA and suppress transactivation by wildtype RUNX1 (Koh et al. 2013). The runt domain also harbors the nuclear localization signal, which is critical for the nuclear translocation of RUNX1. Wildtype RUNX1 is exclusively localized to the nucleus, but many leukemia-associated runt domain mutants exhibit both cytoplasmic and nuclear staining (I150T, P156A, R170Q), or distinct cytoplasmic staining (Y260X) (Koh et al. 2013). In summary, runt domain mutations impair the transcriptional activity of RUNX1 through stabilization of dominant-negative mutants or disruption of nuclear localization.

3.5 Post-translational Modifications of RUNX2 and RUNX3

Runx2 (*Aml3*) is a critical regulator of bone development and it also interacts with multiple acetyltransferases, including p300, MOZ, and MORF (Sierra et al. 2003; Pelletier et al. 2002). Acetylation of RUNX2 generally favors bone growth; for instance, following ERK activation (e.g. in response to BMP2), p300 acetylates RUNX2 which enhances its transcriptional activity and stability, triggering osteoblast differentiation and bone formation (Jeon et al. 2006). The interaction between p300 and RUNX2 is promoted by the SMAD proteins but blocked if the phosphorylation sites in the C-terminus of RUNX2 are mutated; these mutations inhibit responsiveness of cells to TGF-β/BMP signaling (Zhang et al. 2000). The H3K36 trimethyltransferase, WHSC1, also promotes the interaction between RUNX2 and p300, which drives expression of the *Spp1* and *Collagen type 1a* genes (Lee et al. 2014). Loss of one *Whsc1* allele results in Wolf-Hirschorn Syndrome (WHS), a disease characterized by skeletal abnormalities and hypo-ossification.

Histone deacetylases (HDACs) such as HDAC4 and HDAC5, can reverse RUNX2 acetylation and impair osteoblast differentiation, decreasing bone formation (Jeon et al. 2006). As predicted, the use of HDAC inhibitors can promote osteoblast maturation and bone growth, as can overexpression of miR-29a, which blocks HDAC4 expression, stabilizes acetyl-RUNX2 and rescues defects in osteoblast differentiation (Ko et al. 2015).

RUNX2 activity is modulated throughout osteoblast differentiation by multiple signaling pathways including MAPK/ERK, cAMP/PKA, and DAG/PKC (Franceschi et al. 2003). IGF-1 stimulates the PI3K and ERK pathways to enhance RUNX2 DNA binding and transactivation (Qiao et al. 2004). Similarly, FGF2 activates ERK signaling and RUNX2 phosphorylation to enhance expression of osteocalcin (Franceschi et al. 2003). FGF2 also activates PKC which phosphorylates S247 on RUNX2, promoting its transcriptional activity (Kim et al. 2006). PKA is stimulated by parathyroid hormone to phosphorylate RUNX2 on S347 which drives *Collagenase-3* transcription (Selvamurugan et al. 2000). Clearly, RUNX2 is a critical point of integration for a variety of signaling pathways that augment the expression of genes essential for osteoblast activity and bone development.

Similar to RUNX1, phosphorylation can also negatively regulate the transcriptional activity of RUNX2. Phosphorylation on serine 104 disrupts its binding to CBFβ, which subjects RUNX2 to proteolytic degradation. Notably, the S104R mutant, which mimics constitutive phosphorylation, is seen in patients with cleidocranial dysplasia, a hereditary congenital disorder characterized by underdeveloped bone and teeth (Wee et al. 2002). Phosphorylation of S451 occurs within the C-terminal inhibitory domain of RUNX2 to diminish transcriptional activity, although the mechanism remains unclear (Wee et al. 2002). GSK3β-mediated phosphorylation of RUNX2 inhibits its transactivation; therefore inhibiting GSK3β activity may be useful for those with catabolic bone disorders. Indeed, mice with heterozygous loss of GSK3β, or wildtype mice treated with lithium chloride, an inhibitor of GSK3β, exhibit increased bone mass (Kugimiya et al. 2007).

Evidently, PTMs and protein-protein interactions are crucial to RUNX2 activity and highlight its bone-specific functions. Indeed, the loss of key RUNX2 interactions result in profound dysostosis as seen in developmental disorders like WHS and cleidocranial dysplasia. Given its essential role in bone development, an expanded understanding of the PTMs that influence RUNX2-driven gene expression may have implications for treating other bone disorders in the future.

Runx3 appears to be essential for neurogenesis (Levanon et al. 2002; Inoue et al. 2002) and it appears to function as a tumor suppressor gene in bladder cancer, gastric cancer, and lung cancer development (Kim et al. 2005; Li et al. 2002; Lee et al. 2013). The stability of the RUNX3 protein results from a dynamic equilibrium of RUNX3 acetylation, deacetylation and ubiquitination. Upon TGF-β stimulation, p300 acetylates RUNX3, impairing its ubiquitination and promoting its stabilization (Jin et al. 2004). Acetyl-RUNX3 has a higher affinity for the bromodomain-containing protein, BRD2, than its unmodified counterpart and the RUNX3-BRD2 complex positively regulates transcription of *p21* and *ARF* genes, preventing K-Ras induced transformation (Lee et al. 2013). The HDAC SIRT2 deacetylates RUNX3, which allows RUNX3 to be ubiquitinated by E3 ligase SMURF and subsequently degraded (Jin et al. 2004; Kim et al. 2011). Since a number of cancers have abnormally low expression of p300 or inactivating mutations in p300 (Iyer et al. 2004), a reduction in acetyl-RUNX3 may contribute to a decline in transcription of tumor suppressor genes. As all of the RUNX proteins have been associated with cancer (Ito et al. 2015), further study of their protein-protein interactions and PTMs will clarify their contribution to normal and aberrant gene expression.

3.6 RUNX1 as a Tumor Suppressor

As mentioned above, *RUNX1* can be altered through chromosomal rearrangement, copy number variation, point mutation, and internal tandem duplication (Ferro et al. 2004). In general, loss of RUNX1 function impairs early hematopoiesis and differentiation. However, it is clear that RUNX1 can act as a tumor suppressor or tumor promoter depending on the cellular context and its expression level. Several oncogenic mouse model systems show a requirement for at least low levels of *Runx1* in the hematologic malignancies; indeed, complete knockout of *Runx1* inhibited leukemia development in an MLL-AF9 driven AML model (Goyama et al. 2013). Examining alterations in RUNX1 in the context of PTMs, and the enzymes that catalyze them, will provide insight into its context dependent roles.

Inactivating mutations of *RUNX1* are frequently found in patients with MDS and cytogenetically normal AML, and implicate its role as a canonical tumor suppressor. Heterozygous germ line mutations of *RUNX1* are associated with familial platelet disorder with predisposition to AML (FPD/AML), a disease where approximately 35% of carriers develop AML (Owen et al. 2008). In several RUNX1 fusion proteins, the gene rearrangement eliminates functional domains in RUNX1 that affect its transcriptional regulatory properties. Several of these fusion proteins have lost the C-terminal regulatory region of RUNX1 but retain their ability to bind to RUNX1 consensus sequences. As a result, these fusion proteins are able to compete with wildtype RUNX1 for target gene occupancy. The absent C-terminus of RUNX1 mediates critical interactions with the chromatin modifying enzymes MOZ and MLL, and includes the sites of extensive phosphorylation that follow extracellular signals. Therefore, point mutations that disrupt the C-terminal PTM code may phenocopy the physical loss of this region by attenuating essential protein interactions and the ability of RUNX1 to respond to mitogenic or antiproliferative signaling pathways.

Mutations in chromatin modifying enzymes may also contribute to inactivation of RUNX1 by modifying its protein level in the nucleus or affecting its overall protein stability. MLL fusion proteins induce abnormal RUNX1 ubiquitination and downregulate its expression (Zhao et al. 2014). Somatic RUNX1 mutations have also been identified in 15% of cytogenetically normal AML, with specific mutations correlated with mislocalization of the RUNX1 protein and defective ubiquitination (Licht 2001). Thus, it appears that there are multiple mechanisms for the disruption of RUNX1 activity in cancer cells, either by direct inactivating mutations, or by indirect post-translational regulation.

Taken in a larger context, many mechanisms exist to fine tune RUNX1 activity and there are many opportunities for disruption of its PTM code. While MOZ and MLL proteins normally promote transcription activation through interaction with RUNX1, the MOZ-CBP and MLL fusion proteins block RUNX1 mediated transcription. The CBFB-SMMHC translocation formed by inv(16) sequesters the transcription activating kinase HIPK2 in the cytoplasm, preventing the phosphorylation of RUNX1 and p300. Mutations have also been noted in HIPK2 itself in AML patients, leading to disrupted subcellular localization of RUNX1 (Wee et al. 2008; Calzado et al. 2007). Thus, the overall activity of RUNX1 in hematopoietic cells depends on multiple factors, including the enzymes that catalyze its PTMs.

Increased *RUNX1* gene copy numbers and protein levels have been cited as evidence for an oncogenic role of RUNX1 in human cancer. Patients with Trisomy 21 have an increased susceptibility to AML due to augmented gene dosage (De Vita et al. 2010), while amplification of RUNX1 occurs in a subset of patients with T-ALL and B-ALL (Grossmann et al. 2011). However, increased gene dosage may not be correlated with increased RUNX1 activity and may mask the contribution of other genes located in the same region. The mouse studies in which knockout of RUNX1 inhibited leukemia development can also be viewed in the context of overall RUNX1 regulation, namely that cancer cells retain low levels of RUNX1 even in the context of inactivating mutations and that this may be required for leukemogenesis. In summary, RUNX1 activity is highly regulated and often disrupted through genetic alteration and PTMs, generally suggesting a tumor suppressive role in hematopoietic malignancies.

3.7 Conclusion

Post-translational modifications regulate the function of RUNX proteins by affecting their DNA binding, cellular localization, stability, and protein-protein interactions. RUNX proteins often act as scaffolds to mediate the formation of activating or repressive complexes that regulate tissue-specific gene expression. As pharmacologic targeting of transcription factors has proven challenging, an understanding of the enzymes that catalyze RUNX modifications and their biological roles may have implications for developing new therapies for disorders such as cleidocranial dysplasia or t(8;21) RUNX1-ETO AML. For example, blocking RUNX2 phosphorylation through the inhibition of GSK3β may provide a paradigm by which RUNX2 activity can be partially restored in patients with catabolic bone disorders. Moreover, as the leukemogenicity of t(8;21) AML is dependent on the acetylation of AML1-ETO, the use of KAT inhibitors may overcome the aberrant gene expression and self-renewal that is seen in this context.

The roles of RUNX proteins in development and disease have been under intense study since the cloning of *AML1* in 1991 (Coco Lo et al. 1997). As the use of next generation sequencing continues to expand our knowledge of the RUNX transcriptome, we must continue to integrate these approaches with traditional biochemistry and cell biology to further our understanding of RUNX PTMs and RUNX-related diseases.

Acknowledgements We are grateful to Delphine Prou, Ph.D., for her helpful suggestions and comments.

References

Aho, T. L., Sandholm, J., Peltola, K. J., Ito, Y., & Koskinen, P. J. (2006). Pim-1 kinase phosphorylates RUNX family transcription factors and enhances their activity. *BMC Cell Biology, 7*, 21.

Aikawa, Y., et al. (2006). Roles of HIPK1 and HIPK2 in AML1- and p300-dependent transcription, hematopoiesis and blood vessel formation. *The EMBO Journal, 25*, 3955–3965.

Biggs, J. R., Peterson, L. F., Zhang, Y., Kraft, A. S., & Zhang, D.-E. (2006). AML1/RUNX1 phosphorylation by cyclin-dependent kinases regulates the degradation of AML1/RUNX1 by the anaphase-promoting complex. *Molecular and Cellular Biology, 26*, 7420–7429.

Calzado, M. A., Renner, F., Roscic, A., & Schmitz, M. L. (2007). HIPK2: A versatile switchboard regulating the transcription machinery and cell death. *Cell Cycle, 6*, 139–143.

Chakraborty, S., Sinha, K. K., Senyuk, V., & Nucifora, G. (2003). SUV39H1 interacts with AML1 and abrogates AML1 transactivity. AML1 is methylated in vivo. *Oncogene, 22*, 5229–5237.

Coco Lo, F., Pisegna, S., & Diverio, D. (1997). The AML1 gene: A transcription factor involved in the pathogenesis of myeloid and lymphoid leukemias. *Haematologica, 82*, 364–370.

De Vita, S., et al. (2010). Trisomic dose of several chromosome 21 genes perturbs haematopoietic stem and progenitor cell differentiation in Down's syndrome. *Oncogene, 29*, 6102–6114.

Ferro, M. T., et al. (2004). Chromosome 21 tandem repetition and AML1 (RUNX1) gene amplification. *Cancer Genetics and Cytogenetics, 149*, 11–16.

Franceschi, R. T., et al. (2003). Multiple signaling pathways converge on the Cbfa1/Runx2 transcription factor to regulate osteoblast differentiation. *Connective Tissue Research, 44*(Suppl 1), 109–116.

Goyama, S., et al. (2013). Transcription factor RUNX1 promotes survival of acute myeloid leukemia cells. *The Journal of Clinical Investigation, 123*, 3876–3888.

Grossmann, V., et al. (2011). Prognostic relevance of RUNX1 mutations in T-cell acute lymphoblastic leukemia. *Haematologica, 96*, 1874–1877.

Guidez, F., et al. (2000). Recruitment of the nuclear receptor corepressor N-CoR by the TEL moiety of the childhood leukemia-associated TEL-AML1 oncoprotein. *Blood, 96*, 2557–2561.

Guo, H., & Friedman, A. D. (2011). Phosphorylation of RUNX1 by cyclin-dependent kinase reduces direct interaction with HDAC1 and HDAC3. *The Journal of Biological Chemistry, 286*, 208–215.

Herglotz, J., et al. (2013). Histone arginine methylation keeps RUNX1 target genes in an intermediate state. *Oncogene, 32*, 2565–2575.

Huang, G., et al. (2001). Dimerization with PEBP2beta protects RUNX1/AML1 from ubiquitin-proteasome-mediated degradation. *The EMBO Journal, 20*, 723–733.

Huang, G., et al. (2011). The ability of MLL to bind RUNX1 and methylate H3K4 at PU.1 regulatory regions is impaired by MDS/AML-associated RUNX1/AML1 mutations. *Blood, 118*, 6544–6552.

Huang, H., et al. (2012). A Src family kinase-Shp2 axis controls RUNX1 activity in megakaryocyte and T-lymphocyte differentiation. *Genes & Development, 26*, 1587–1601.

Imai, Y., et al. (2004). The corepressor mSin3A regulates phosphorylation-induced activation, intranuclear location, and stability of AML1. *Molecular and Cellular Biology, 24*, 1033–1043.

Inoue, K.-I., et al. (2002). Runx3 controls the axonal projection of proprioceptive dorsal root ganglion neurons. *Nature Neuroscience, 5*, 946–954.

Ito, Y., Bae, S.-C., & Chuang, L. S. H. (2015). The RUNX family: Developmental regulators in cancer. *Nature Reviews. Cancer, 15*, 81–95.

Iyer, N. G., Özdag, H., & Caldas, C. (2004). p300/CBP and cancer. *Oncogene, 23*, 4225–4231.

Jeon, E.-J., et al. (2006). Bone morphogenetic protein-2 stimulates Runx2 acetylation. *The Journal of Biological Chemistry, 281*, 16502–16511.

Jin, Y.-H., et al. (2004). Transforming growth factor-beta stimulates p300-dependent RUNX3 acetylation, which inhibits ubiquitination-mediated degradation. *The Journal of Biological Chemistry, 279*, 29409–29417.

Kamikubo, Y., et al. (2010). Accelerated leukemogenesis by truncated CBFβ-SMMHC defective in high-affinity binding with RUNX1. *Cancer Cell, 17*, 455–468.

Kim, W.-J., et al. (2005). RUNX3 inactivation by point mutations and aberrant DNA methylation in bladder tumors. *Cancer Research, 65*, 9347–9354.

Kim, B. G., et al. (2006). Runx2 phosphorylation induced by fibroblast growth factor-2/protein kinase C pathways. *Proteomics, 6*, 1166–1174.

Kim, W.-J., et al. (2011). Nicotinamide inhibits growth of carcinogen induced mouse bladder tumor and human bladder tumor xenograft through up-regulation of RUNX3 and p300. *The Journal of Urology, 185*, 2366–2375.

Kitabayashi, I., Aikawa, Y., Nguyen, L. A., Yokoyama, A., & Ohki, M. (2001). Activation of AML1-mediated transcription by MOZ and inhibition by the MOZ-CBP fusion protein. *The EMBO Journal, 20*, 7184–7196.

Ko, J.-Y., et al. (2015). MicroRNA-29a mitigates glucocorticoid induction of bone loss and fatty marrow by rescuing Runx2 acetylation. *Bone, 81*, 80–88.

Koh, C. P., et al. (2013). RUNX1 meets MLL: Epigenetic regulation of hematopoiesis by two leukemia genes. *Leukemia, 27*, 1793–1802.

Kugimiya, F., et al. (2007). GSK-3β controls osteogenesis through regulating Runx2 activity. *PloS One, 2*, e837.

Lee, Y.-S., et al. (2013). Runx3 inactivation is a crucial early event in the development of lung adenocarcinoma. *Cancer Cell, 24*, 603–616.

Lee, Y. F., Nimura, K., Lo, W. N., Saga, K., & Kaneda, Y. (2014). Histone H3 lysine 36 methyltransferase Whsc1 promotes the association of Runx2 and p300 in the activation of bone-related genes. *PloS One, 9*, e106661.

Levanon, D., et al. (2002). The Runx3 transcription factor regulates development and survival of TrkC dorsal root ganglia neurons. *The EMBO Journal, 21*, 3454–3463.

Li, Q.-L., et al. (2002). Causal relationship between the loss of RUNX3 expression and gastric cancer. *Cell, 109*, 113–124.

Licht, J. D. (2001). AML1 and the AML1-ETO fusion protein in the pathogenesis of t(8;21) AML. *Oncogene, 20*, 1–20.

Lutterbach, B., et al. (2000). A mechanism of repression by acute myeloid leukemia-1, the target of multiple chromosomal translocations in acute leukemia. *The Journal of Biological Chemistry, 275*, 651–656.

Mizutani, S., et al. (2015). Loss of RUNX1/AML1 arginine-methylation impairs peripheral T cell homeostasis. *British Journal of Haematology, 170*, 859–873.

Murata, S., Chiba, T., & Tanaka, K. (2003). CHIP: A quality-control E3 ligase collaborating with molecular chaperones. *The International Journal of Biochemistry & Cell Biology, 35*, 572–578.

Nakagawa, M., et al. (2011). AML1/RUNX1 functions as a cytoplasmic attenuator of NF-κB signaling in the repression of myeloid tumors. *Blood, 118*, 6626–6637.

Neel, B. G., & Speck, N. A. (2012). Tyrosyl phosphorylation toggles a Runx1 switch. *Genes & Development, 26*, 1520–1526.

Owen, C. J., et al. (2008). Five new pedigrees with inherited RUNX1 mutations causing familial platelet disorder with propensity to myeloid malignancy. *Blood, 112*, 4639–4645.

Pelletier, N., Champagne, N., Stifani, S., & Yang, X.-J. (2002). MOZ and MORF histone acetyltransferases interact with the Runt-domain transcription factor Runx2. *Oncogene, 21*, 2729–2740.

Qiao, M., Shapiro, P., Kumar, R., & Passaniti, A. (2004). Insulin-like growth factor-1 regulates endogenous RUNX2 activity in endothelial cells through a phosphatidylinositol 3-kinase/ERK-dependent and Akt-independent signaling pathway. *The Journal of Biological Chemistry, 279*, 42709–42718.

Rebel, V. I., et al. (2002). Distinct roles for CREB-binding protein and p300 in hematopoietic stem cell self-renewal. *PNAS, 99*, 14789–14794.

Reed-Inderbitzin, E., et al. (2006). RUNX1 associates with histone deacetylases and SUV39H1 to repress transcription. *Oncogene, 25*, 5777–5786.

Selvamurugan, N., Pulumati, M. R., Tyson, D. R., & Partridge, N. C. (2000). Parathyroid hormone regulation of the rat collagenase-3 promoter by protein kinase A-dependent transactivation of core binding factor alpha1. *The Journal of Biological Chemistry, 275*, 5037–5042.

Senyuk, V., Sinha, K. K., Chakraborty, S., Buonamici, S., & Nucifora, G. (2003). P/CAF and GCN5 acetylate the AML1/MDS1/EVI1 fusion oncoprotein. *Biochemical and Biophysical Research Communications, 307*, 980–986.

Seo, W., Ikawa, T., Kawamoto, H., & Taniuchi, I. (2012). Runx1-Cbfβ facilitates early B lymphocyte development by regulating expression of Ebf1. *The Journal of Experimental Medicine, 209*, 1255–1262.

Shang, Y., et al. (2009). CHIP functions an E3 ubiquitin ligase of Runx1. *Biochemical and Biophysical Research Communications, 386*, 242–246.

Shia, W.-J., et al. (2012). PRMT1 interacts with AML1-ETO to promote its transcriptional activation and progenitor cell proliferative potential. *Blood, 119*, 4953–4962.

Sierra, J., et al. (2003). Regulation of the bone-specific osteocalcin gene by p300 requires Runx2/Cbfa1 and the vitamin D3 receptor but not p300 intrinsic histone acetyltransferase activity. *Molecular and Cellular Biology, 23*, 3339–3351.

Tanaka, T., et al. (1996). The extracellular signal-regulated kinase pathway phosphorylates AML1, an acute myeloid leukemia gene product, and potentially regulates its transactivation ability. *Molecular and Cellular Biology, 16*, 3967–3979.

The Cancer Genome Atlas Research Network. (2013). Genomic and epigenomic landscapes of adult de novo acute myeloid leukemia. *The New England Journal of Medicine, 368*, 2059–2074.

Vu, L. P., et al. (2013). PRMT4 blocks myeloid differentiation by assembling a methyl-RUNX1-dependent repressor complex. *Cell Reports, 5*, 1625–1638.

Wang, L., et al. (2011). The leukemogenicity of AML1-ETO is dependent on site-specific lysine acetylation. *Science, 333*, 765–769.

Wee, H. J., Huang, G., Shigesada, K., & Ito, Y. (2002). Serine phosphorylation of RUNX2 with novel potential functions as negative regulatory mechanisms. *EMBO Reports, 3*, 967–974.

Wee, H. J., Voon, D. C.-C., Bae, S.-C., & Ito, Y. (2008). PEBP2-beta/CBF-beta-dependent phosphorylation of RUNX1 and p300 by HIPK2: Implications for leukemogenesis. *Blood, 112*, 3777–3787.

Yamaguchi, Y., et al. (2004). AML1 is functionally regulated through p300-mediated acetylation on specific lysine residues. *The Journal of Biological Chemistry, 279*, 15630–15638.

Yoshimi, M., Goyama, S., & Kawazu, M. (2012). Multiple phosphorylation sites are important for RUNX1 activity in early hematopoiesis and T-cell differentiation. *European Journal of Immunology, 42*(4), 1044–1050.

Yu, M., et al. (2012). Direct recruitment of polycomb repressive complex 1 to chromatin by core binding transcription factors. *Molecular Cell, 45*, 330–343.

Zaidi, S. K., et al. (2002). Integration of Runx and Smad regulatory signals at transcriptionally active subnuclear sites. *PNAS, 99*, 8048–8053.

Zhang, Y. W., et al. (2000). A RUNX2/PEBP2alpha A/CBFA1 mutation displaying impaired transactivation and Smad interaction in cleidocranial dysplasia. *PNAS, 97*, 10549–10554.

Zhang, Y., Biggs, J. R., & Kraft, A. S. (2004). Phorbol ester treatment of K562 cells regulates the transcriptional activity of AML1c through phosphorylation. *The Journal of Biological Chemistry, 279*, 53116–53125.

Zhang, L., Fried, F. B., Guo, H., & Friedman, A. D. (2008). Cyclin-dependent kinase phosphorylation of RUNX1/AML1 on 3 sites increases transactivation potency and stimulates cell proliferation. *Blood, 111*, 1193–1200.

Zhao, X., et al. (2008). Methylation of RUNX1 by PRMT1 abrogates SIN3A binding and potentiates its transcriptional activity. *Genes & Development, 22*, 640–653.

Zhao, X., et al. (2014). Downregulation of RUNX1/CBFβ by MLL fusion proteins enhances hematopoietic stem cell self-renewal. *Blood, 123*, 1729–1738.

Part III

RUNX Genes and Mammalian Development

The Role of Runx1 in Embryonic Blood Cell Formation

4

Amanda D. Yzaguirre, Marella F.T.R. de Bruijn, and Nancy A. Speck

Abstract

The *de novo* generation of hematopoietic stem and progenitor cells (HSPC) occurs solely during embryogenesis from a population of epithelial cells called hemogenic endothelium (HE). During midgestation HE cells in multiple intra- and extraembryonic vascular beds leave the vessel wall as they transition into HSPCs in a process termed the endothelial to hematopoietic transition (EHT). Runx1 expression in HE cells orchestrates the transcriptional switch necessary for the transdifferentiation of endothelial cells into functional HSPCs. Runx1 is widely considered the master regulator of developmental hematopoiesis because it plays an essential function during specification of the hematopoietic lineage during embryogenesis. Here we review the role of Runx1 in embryonic HSPC formation, with a particular focus on its role in hemogenic endothelium.

Keywords

Runx1 • Hemogenic endothelium • Hematopoiesis • Hematopoietic stem cell

Almost all blood cells in the adult mammal differentiate from hematopoietic stem cells (HSCs) in the bone marrow. However HSCs do not originate in the bone marrow, and instead differentiate in the embryo before bone or bone marrow forms (Müller et al. 1994). The majority of HSCs, defined as cells that can engraft adult transplant recipients, differentiate from immature HSC precursors called pre-HSCs (Rybtsov et al. 2011; Taoudi et al. 2008). Pre-HSCs, in turn, differentiate from a small population of hemogenic endothelial cells (Zovein et al. 2008; Chen et al. 2009).

A.D. Yzaguirre • N.A. Speck (✉)
Department of Cell and Developmental Biology, Abramson Family Cancer Research Institute, Perelman School of Medicine, University of Pennsylvania, Philadelphia, PA 19103, USA
e-mail: nancyas@exchange.upenn.edu

M.F.T.R. de Bruijn
MRC Molecular Haematology Unit, Weatherall Institute of Molecular Medicine, Oxford OX3 9DS, UK

Y. Groner et al. (eds.), *RUNX Proteins in Development and Cancer*, Advances in Experimental Medicine and Biology 962, DOI 10.1007/978-981-10-3233-2_4

The maturation of pre-HSCs into HSCs predominantly takes place in the fetal liver, which is colonized by pre-HSCs via the circulation (Rybtsov et al. 2016; Kieusseian et al. 2012). Following birth, HSCs leave the fetal liver and settle in the bone marrow, where they remain for the rest of adult life.

HSCs develop at midgestation in the mouse embryo, and at 1 month of gestation in the human embryo (Ivanovs et al. 2011; Müller et al. 1994). However, before HSCs are present, several other primitive types of blood cells emerge that are essential for embryonic viability, and some of which contribute to tissue macrophages in the adult (Ginhaux and Jung 2014). Hematopoietic progenitors (cells that can produce differentiated blood cells, but do not possess long-term multilineage reconstitution potential) and HSCs form in three waves, as described below. Runx1 is important for the differentiation of all embryonic blood cell lineages, and is particularly essential for the differentiation of blood cells in the second two waves from hemogenic endothelium.

4.1 Primitive Hematopoiesis-The First Wave

Primitive hematopoietic cells are one of the earliest functional cell populations to appear during embryogenesis. They emerge in the extraembryonic yolk sac shortly after gastrulation, and prior to the onset of circulation, a functional vascular system, and the development of HSCs (Palis et al. 1999; Ferkowicz and Yoder 2005). Primitive hematopoietic cells in this first wave include unipotent primitive erythrocyte progenitors, bipotent erythrocyte/megakaryocyte progenitors, and primitive macrophages (Xu et al. 2001; Tober et al. 2007; Palis et al. 1999; Moore and Metcalf 1970; Haar and Ackerman 1971; Tracey et al. 1998). These primitive blood cells have distinct morphological and functional features compared to their "definitive" counterparts that form during the second and third waves of hematopoiesis. For example, primitive erythrocytes are larger than definitive erythrocytes, they express embryonic and adult globins, and they retain their nucleus when entering the circulation (Palis et al. 1999; Palis 2014; Kingsley et al. 2004). Primitive megakaryocytes have a lower ploidy class than definitive megakaryocytes, and more rapidly produce platelets that prevent hemorrhaging in the primitive vascular plexus as it develops into a functional cardiovascular system (Xu et al. 2001; Potts et al. 2014).

The mesodermal cells that give rise to primitive hematopoietic cells originate from a population of proximal epiblasts that migrate through the primitive streak and into the extraembryonic yolk sac early during gastrulation (Lawson et al. 1991). The mesoderm accumulates to form thickened regions called mesodermal masses that then differentiate into blood islands consisting of primitive erythroblasts, and angioblasts that will form the vascular plexus of the yolk sac (Ferkowicz and Yoder 2005; Haar and Ackerman 1971). Due to their parallel development and close physical association, it was initially hypothesized that blood and endothelial cells in the yolk sac shared a common progenitor called the hemangioblast (Murray 1932; Sabin 1920). This idea was supported by the demonstration that hemangioblast-like progenitors that gave rise to both blood and endothelial cells could be isolated from embryonic stem (ES) cell cultures (Choi et al. 1998; Zambidis et al. 2005). It later became apparent that the putative bi-potent hemangioblast was actually a tri-lineage progenitor that could also give rise to smooth muscle cells (Ema et al. 2003). In vivo clonal analyses provided further evidence against the existence of a bipotential hemangioblast in the yolk sac, and instead suggested that yolk sac endothelium and hematopoietic cells are derived from adjacent but independent regions of the epiblast, and are specified prior to entering the primitive streak (Padron-Barthe et al. 2014; Ueno and Weissman 2006).

Runx1 is expressed in the mesodermal masses in the yolk sac, and in the progenitors of primitive hematopoietic cells in the mouse embryo with the exception of primitive erythrocytes that initially express Runx1 but rapidly downregulate its expression shortly after emergence (North et al. 1999; Zeigler et al. 2006; Lacaud et al. 2002).

Two of the three primitive hematopoietic lineages, primitive erythrocytes and megakaryocytes can form in the absence of Runx1, however their normal development is affected by Runx1 loss. Runx1-deficient embryos produce numbers of primitive erythroid colonies comparable to littermate controls and do not appear anemic (Yokomizo et al. 2008; Lacaud et al. 2002). However, more detailed analysis revealed reduced expression of cell surface Ter119 and the hematopoietic transcription factors KLF1 and GATA1, and defective maturation of Runx1-deficient erythrocytes (Yokomizo et al. 2008; Castilla et al. 1996). Furthermore, about 30% of primitive erythrocytes derived from *Runx1*$^{-/-}$ embryos displayed a deformed shape characterized by a rough punctate surface (Yokomizo et al. 2008). Despite these abnormalities *Runx1*$^{-/-}$ primitive erythrocytes are functional, as indicated by normal levels of benzidine staining (hemoglobinization) and the fact that *Runx1*$^{-/-}$ embryos survive until E12.5, which is longer than GATA1-deficient embryos, which die by E10.5 with severe anemia due to the lack of functional primitive erythrocytes (Yokomizo et al. 2008; Fujiwara et al. 1996; Okuda et al. 1996; Wang et al. 1996a). Runx1 is not required for the formation of primitive diploid megakaryocytes, although their numbers were lower in Runx1 deficient yolk sacs (Potts et al. 2014). Primitive macrophages, on the other hand, absolutely require Runx1, as they are lacking in *Runx1*$^{-/-}$ embryonic stem cell differentiation cultures (Lacaud et al. 2002) and embryos (Li et al. 2006). In summary, in the absence of Runx1, primitive macrophages are absent, diploid megakaryocytes are reduced in number, and primitive erythropoiesis is abnormal. Although it is often stated that Runx1 is required for definitive, but not primitive hematopoiesis, this is inaccurate as Runx1 is strictly required for the development of one primitive blood cell lineage, and important for the normal development of two others.

Runx1 has also been shown to play a role during primitive hematopoiesis in zebrafish and Xenopus embryos. In Xenopus embryos, Runx1 is expressed in the ventral blood island (VBI), which is analogous to mouse yolk sac blood islands (Tracey et al. 1998). Inhibiting Runx1 function via the injection of a dominant negative form of Runx1 mRNA prior to the VBI stage drastically reduced the number of Benzidine^{+} primitive erythrocytes (Tracey et al. 1998). Similarly, in zebrafish embryos, morpholino knockdown of Runx1 expression at the one to eight cell stage resulted in fewer primitive erythrocytes (Kalev-Zylinska et al. 2002). The primitive macrophage and megakaryocyte populations were not examined in either species. The decrease in primitive erythrocytes in both zebrafish and Xenopus embryos is contrary to what is observed in the mouse and suggests that Runx1 plays a more essential role in primitive erythropoiesis during zebrafish and Xenopus development.

4.2 Definitive Hematopoiesis-The Second and Third Waves

The term "definitive" in the context of developmental hematopoiesis has several meanings, but was originally used to describe adult erythrocytes, which unlike primitive erythrocytes are small and concave, lose their nuclei before entering the circulation, and do not express embryonic globin (Palis et al. 1999; Palis 2014; Kingsley et al. 2004). Defined this way, definitive hematopoiesis encompasses two overlapping waves of blood development. Wave 2 is characterized by the generation of erythro-myeloid progenitors (EMPs) and lymphoid progenitors in the yolk sac and embryo proper (Yoder 2014). EMPs can be found as early as E8.25 in the murine yolk sac (Palis et al. 1999; Mcgrath et al. 2015) and heart (Nakano et al. 2013). The next wave 2 progenitor to appear are lymphoid progenitors, which are found at E9.5 in the yolk sac and the paired dorsal aorta, and by E10.5 in the umbilical artery (UA) and vitelline artery (VA) (Yoshimoto et al. 2011, 2012). Adult repopulating HSCs (wave 3) do not appear until E10.5; they are generated initially in the dorsal aorta (DA), UA, and VA, and can subsequently be found in the yolk sac, head and placental vasculature (Li et al. 2012, 2016; Rhodes et al. 2008; Gordon-Keylock et al. 2013; De Bruijn et al. 2000; Gekas et al. 2005). They are thought to arrive via circulation in these latter sites, instead of being generated *in situ* (Dieterlen-Lievre 1975; Cumano et al. 2001; Medvinsky and Dzierzak 1996; Iizuka et al. 2016).

Definitive hematopoietic cells are derived from a population of epithelial cells called hemogenic endothelium (HE) that are part of the interior lining of specific blood vessels in the embryo (Swiers et al. 2013b). HE is a transient population that gives rise to hematopoietic progenitors and stem cells in a process termed the endothelial to hematopoietic transition (EHT) (Kissa and Herbomel 2010). Live-imaging studies of HE cells *in vitro* and *in vivo* have captured this dynamic process (Kissa and Herbomel 2010; Boisset et al. 2010; Bertrand et al. 2010; Lancrin et al. 2009). However, scanning electron microscopy revealed that at the ultramicroscopic level HE cells are more oblong, with rounded cell bodies and filopodia-like protrusions of the membrane as compared to non-hemogenic endothelial cells (Bos et al. 2015). As the EHT progresses, the HE cell bends away from the vessel wall until it rounds up and detaches from the endothelial layer becoming a mobile hematopoietic cell (Kissa and Herbomel 2010; Boisset et al. 2010; Bertrand et al. 2010; Lancrin et al. 2009; Eilken et al. 2009).

In mouse embryos, HE is localized in the yolk sac, the large arteries of the embryo proper, the heart, and the chorionic plexus (Rhodes et al. 2008; Li et al. 2012; Nakano et al. 2013; Yzaguirre and Speck 2016). HE cells are distinguished from non-hemogenic endothelial cells based on Runx1 expression (North et al. 1999) (Fig. 5.1). Runx1 is a critical regulator of the EHT and as such, suppresses an endothelial transcriptional program and initiates a hematopoietic program in HE allowing the EHT to occur (Lancrin et al. 2012; Chen et al. 2009; North et al. 1999; Yokomizo et al. 2001). Transcriptional and

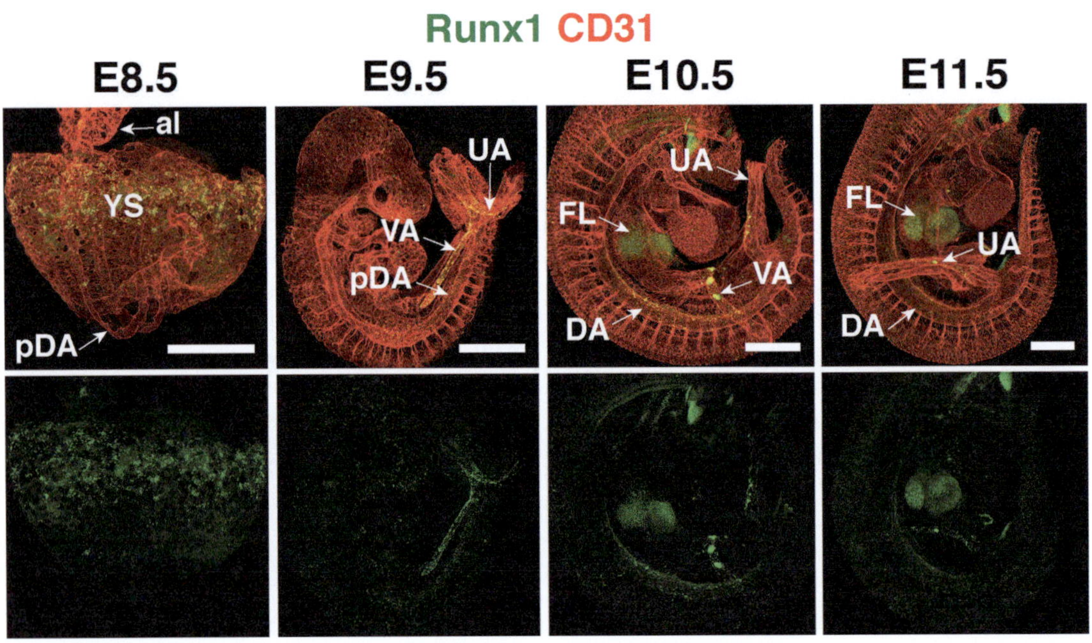

Fig. 5.1 Location of Runx1 expression and hemogenic endothelium in the mouse embryo. Confocal Z-projections of mouse embryos between embryonic day (E) 8.5 and E11.5 immunostained for the endothelial and hematopoietic marker CD31 (*red*) and Runx1 (*green*). Runx1 is expressed in HE and hematopoietic cells in the yolk sac (YS) at E8.5. At E9.5, Runx1 expression is prominent in the vitelline artery (VA) and umbilical artery (UA). An E10.5 embryo (head removed) shows Runx1 protein in the vitelline artery, umbilical artery, dorsal aorta (DA), and the site of colonization, the fetal liver (FL). At E11.5 Runx1 expression in the fetal liver intensifies as hematopoietic cells colonize it. Conversely expression of Runx1 in the large arteries at E11.5 diminishes as the hemogenic endothelium gives rise to hematopoietic cells that subsequently enter circulation. *al* allantois, *pDA* paired dorsal aortae. Scale bar = 500 μm

functional analyses demonstrated that HE cells derived from E8.5 mouse embryos preferentially form endothelial tubules in culture conditions that support both endothelial and hematopoietic cells (Swiers et al. 2013a). In contrast, E10.5 HE preferentially forms hematopoietic cells *in vitro*. The functional change that occurs between E8.5 and E10.5 was accompanied by a transcriptional shift characterized by the upregulation of hematopoietic factors such as Runx1, Meis1, Gata2, Gata3 and Myb suggesting that initially HE cells are functional endothelial cells, but as the hematopoietic program ramps up during midgestation HE loses endothelial function and gains hematopoietic potential (Swiers et al. 2013a).

In mammalian embryos, after the EHT occurs, newborn hematopoietic cells adhere to the vessel wall within the lumen forming clusters of hematopoietic cells. The peak of EHT in the mouse embryo (E10.5) is marked by the formation of hundreds of Kit^+ hematopoietic clusters within the lumens of the DA, VA and UA and dozens residing within the vascular plexus of the yolk sac (Yokomizo and Dzierzak 2010; Frame et al. 2015; Yzaguirre and Speck 2016). Analysis of the Kit^+ hematopoietic clusters within the embryo proper between E10.5 and E11.5 has revealed that they consist of lymphoid progenitors, a small number of myeloid progenitors, and pre-HSCs that can mature into HSCs capable of long-term multilineage reconstitution (Boisset et al. 2015; Li et al. 2014; Taoudi et al. 2008; Rybtsov et al. 2011). By E12.5 most hematopoietic cluster cells have entered the circulation and made their way to the fetal liver where they undergo maturation and proliferation, expanding the pool of HSCs and hematopoietic progenitors (Ema and Nakauchi 2000; Kieusseian et al. 2012; Rybtsov et al. 2016). Beginning at E17.5 HSCs migrate to the bone marrow where they will reside throughout the lifetime of the animal (Christensen et al. 2004).

In zebrafish embryos the EHT occurs away from the lumen of the dorsal aorta, and the newly formed hematopoietic cells must traverse through the subaortic space and enter circulation via the axial vein (Kissa et al. 2008). Once in circulation hematopoietic cells migrate to the caudal hematopoietic tissue that is akin to mammalian fetal liver where they differentiate and expand before traveling to definitive hematopoietic organs (Murayama et al. 2006).

4.3 Runx1 Is Required During Definitive Hematopoiesis

Runx1 is expressed in all sites of blood formation. During gastrulation Runx1 is expressed in the extraembryonic mesoderm that gives rise to primitive hematopoietic cells (Swiers et al. 2013a; Lacaud et al. 2002; Zeigler et al. 2006). During definitive hematopoiesis Runx1 is the most reliable marker of hemogenic endothelium and is expressed by all hematopoietic cells with the exception of erythrocytes (North et al. 2004, 1999, 2002; Lorsbach et al. 2004). In addition to hematopoietic tissues, Runx1 is expressed in the olfactory epithelium, spinal ganglia, maxillary processes and the mesenchyme that flanks the ventral length of the dorsal aorta (North et al. 1999; Levanon et al. 2001a). Germline deletion of Runx1 results in the elimination of all definitive hematopoietic cells and embryonic lethality by E12.5 (Wang et al. 1996a; Okuda et al. 1996). Embryonic lethality of Runx1 deficient embryos is due to hemorrhaging within the ventricle of the central nervous system, the pericardial space, and the peritoneal cavity (Okuda et al. 1996; Wang et al. 1996a). The hemorrhaging is likely secondary to the lack of definitive hematopoietic cells because hematopoietic cells are involved in vascular remodeling during embryogenesis. For example, hematopoietic cells express angiopoietin-1 (Ang-1), a chemoattractant that promotes blood vessel sprouting (Witzenbichler et al. 1998). Analysis of the vasculature of Runx1 deficient embryos revealed decreased branching in the head, pericardium and vitelline artery in the yolk sac (Takakura et al. 2000). When Runx1 deficient explants were supplemented with hematopoietic cells or Ang-1 the vascular defects were rescued, suggesting that the vascular defects that cause hemorrhaging in Runx1 deficient embryos are due to the loss of Ang-1 expressing definitive hematopoietic cells (Takakura et al. 2000).

Runx1 can bind DNA as a monomer *in vitro,* but when Runx1 heterodimerizes with its non-DNA binding subunit CBFβ, flexible DNA-recognition loops in Runx1 are stabilized and its binding affinity for DNA increases (see Bushweller and Tahirov, this volume). Embryos deficient for CBFβ died by E12.5 with hemorrhaging akin to Runx1 deficient embryos and had significantly fewer definitive hematopoietic progenitors in their fetal livers when compared to littermate controls (Sasaki et al. 1996; Bresciani et al. 2014; Wang et al. 1996b; Niki et al. 1997). Similar results were obtained in CBFβ deficient zebrafish, confirming that CBFβ is required for Runx1 to function during definitive hematopoiesis (Sasaki et al. 1996; Bresciani et al. 2014). Interestingly, definitive hematopoiesis is not blocked in CBFβ deficient embryos to the same extent as in Runx1 deficient embryos. For example, the hematopoietic-specific transcription factor c-Myb is not expressed at sites of definitive hematopoiesis in Runx1 deficient zebrafish embryos, but it is expressed in the dorsal aorta of CBFβ deficient zebrafish embryos (Bresciani et al. 2014). Furthermore, definitive erythroid and myeloid progenitors are never found within the fetal livers of Runx1 deficient embryos but small numbers (approximately 2.5% of wildtype controls) are present in CBFβ deficient fetal livers (Wang et al. 1996b). These studies suggest that the low-affinity binding of Runx1 to DNA in the absence of CBFβ is enough to initiate definitive hematopoiesis but is not sufficient to supply enough definitive hematopoietic cells to prevent embryonic lethality.

Unlike Runx1 and CBFβ deficient embryos, Runx1 heterozygous mutant embryos survive well into adulthood and have relatively minor defects in hematopoietic development. There are fewer erythroid/myeloid progenitors in the yolk sacs, fetal livers and aorta/gonad/mesonephros regions of *Runx1*$^{+/-}$ embryos compared to wild type littermate controls (Wang et al. 1996a, 1996b; Mukouyama et al. 2000). Unexpectedly, the development of HSCs in *Runx1*$^{+/-}$ embryos is accelerated and spatially shifted (Cai et al. 2000). Specifically, HSCs were readily detected in the E10.5 AGM, and could also be detected in the yolk sacs of *Runx1*$^{+/-}$ embryos at E10.5 (Cai et al. 2000). This is in contrast to wild type embryos in which very few HSCs are present in the E10.5 dorsal aorta and are found in the yolk sac approximately 24 h later (Müller et al. 1994). Therefore, reduced Runx1 dosage suppresses definitive hematopoiesis (wave 2) and changes the spatial and temporal development of HSCs (wave 3). The mechanism behind the temporal and spatial shift in HSC development associated with Runx1 haploinsufficiency is not known, but a subsequent study on the differentiation of Runx1$^{+/-}$ embryonic stem (ES) cells provided a clue. The commitment of *Runx1*$^{+/-}$ ES cells to hemangioblasts, and subsequently to hematopoietic lineages was found to be accelerated by approximately 12 h compared to that of wildtype ES cells (Lacaud et al. 2004). Therefore the acceleration in HSC formation may originate at a very early stage in hematopoietic development, in the formation of the tri-lineage hemangioblast, in which Runx1 is expressed (Lacaud et al. 2002).

4.4 Runx1 Is Required in Hemogenic Endothelium for the Development of Definitive Hematopoietic Cells

The studies of Runx1 and CBFβ knockout mice demonstrated that Runx1 and CBFβ are essential for definitive hematopoiesis but did not pinpoint when and in which cell population Runx1 is necessary. The observation that Runx1 is expressed in endothelial cells at all sites of hematopoietic cluster formation and was required for cluster formation led to the hypothesis that Runx1 is required for the transition from endothelial to definitive hematopoietic cells. To test this hypothesis Runx1 was ablated in endothelial cells via endothelial specific Cre-recombinase mediated excision, which led to the complete abrogation of definitive hematopoiesis and embryonic lethality by E13.5 (Li et al. 2006; Chen et al. 2009). Also, endothelium sorted from the yolk sac and embryo proper of E10.5 *Runx1*$^{-/-}$ mice could not generate hematopoietic cells when plated on an OP9 stromal cell layer in conditions that support EHT (Yokomizo et al. 2001). These findings suggest that Runx1 expression is required in endothelial cells for the *de novo* generation of definitive hematopoietic cells. This point was further supported by a study that took the reverse approach by restoring endogenous Runx1 expression in Tek-expressing endothelial cells in Runx1 reversible knockout mouse embryos (Liakhovitskaia et al. 2009). Restoration of Runx1 expression in endothelial cells was sufficient to rescue lymphoid lineages, myeloid lineages and HSCs, and prolonged the life of the embryos up until birth (Liakhovitskaia et al. 2009). Postnatal lethality of these mice likely resulted from the loss of Runx1 expression in non-hematopoietic tissues. In fact, Runx1 null mice in which Runx1 expression is restored in only endothelial/hematopoietic cells have defects in neuronal differentiation and mineralization of the skull and sternum, demonstrating additional roles of Runx1 during development (Kobayashi et al. 2012; Liakhovitskaia et al. 2010).

After HE cells transition into hematopoietic cells they continue to express Runx1, which led to the hypothesis that Runx1 remains essential even after the EHT. However, conditional deletion of Runx1 in hematopoietic cells via Vav1-Cre, did not result in the ablation of EMPs or HSCs, nor did it affect embryonic or adult viability, indicating that Runx1 is not essential in Vav1 expressing hematopoietic cells (Chen et al. 2009). However, Runx1 deletion in hematopoietic cells does cause defects that include thrombocytopenia and defective lymphopoiesis in adult mice (Chen et al. 2009; Ichikawa et al. 2004; Growney et al. 2005; Putz et al. 2006). Therefore, although Runx1 expression in hematopoietic cells is not essential for the generation and survival of definitive hematopoietic cells, it is required for lineage-specific differentiation and homeostasis. One caveat of this study is that Vav1-Cre is active in circulating and fetal liver hematopoietic cells but was not detected in hematopoietic cluster cells within the dorsal aorta, leaving open the possibility that Runx1 is required for a short period after the EHT and before fetal liver colonization.

To more precisely determine the temporal requirement of Runx1 expression in hemogenic endothelium, Tober et al. conditionally deleted Runx1 during 24-h intervals between E7.5 and E11.5 using a tamoxifen-inducible endothelial-specific Cre driven from vascular endothelial cadherin (*Cdh5*) regulatory sequences (*Cdh5*-Cre^{ERT}) then assessed hematopoiesis (Tober et al. 2013). They found that when Runx1 was deleted between E8.25 –E9.25 that EMP numbers were dramatically reduced, indicating that Runx1 is critical in that time frame for the formation of EMPs from hemogenic endothelium. On the other hand deletion between E9.0 – E10.0 had no effect on EMP numbers, indicating that by E10.0 the requirement for Runx1 in hemogenic endothelium for the majority of EMP formation has ended. In contrast, the *de novo* development of HSCs was dependent on Runx1 expression in the endothelium up until E11.5. Thus, the requirement for Runx1 expression in HE for the development of EMPs and HSCs is temporally uncoupled, which is consistent with the sequential development of EMPs and HSCs during embryogenesis. This study however, did not determine if Runx1 was required in hematopoietic cluster cells because although vascular endothelial cadherin protein (CD144) is expressed on the surface of hematopoietic cluster cells, *Cdh5* mRNA is approximately 80% lower in hematopoietic cluster cells (Tober et al. 2013; North et al. 2002; Jaffredo et al. 2005; Fraser et al. 2003). It was unclear whether Cre^{ERT} protein levels correlated with *Cdh5* mRNA or vascular endothelial cadherin protein levels, and was present and active in hematopoietic cluster cells.

The molecular basis for the transient requirement for Runx1 was explored in a mouse embryonic stem (ES) cell model (Hoogenkamp et al. 2009; Lichtinger et al. 2012). Using a *Runx1*$^{-/-}$ mouse ES cell line expressing inducible Runx1, Hoogenkamp et al. demonstrated that Runx1 bound to an upstream regulatory element (URE) of *Spi1*, which encodes a transcription factor required for hematopoiesis called Pu.1. Runx1 initiated chromatin unfolding in the *Spi1* URE at the onset of hematopoietic development (Hoogenkamp et al. 2009). Furthermore, using

ChIP and *in vivo* footprinting they found that weak and transient binding of Runx1 to the URE was sufficient to establish stable transcription factor complexes at *cis*-regulatory elements that could sustain *Spi1* expression even after removal of Runx1 (Hoogenkamp et al. 2009).

Genome-wide analysis by the same group using the same ES cell differentiation model compared HE cells before and after the induction of Runx1 (Lichtinger et al. 2012). They found that after Runx1 induction in HE, Runx1 bound to sites that contained little or no H3K9Ac and subsequently strongly increased H3K9Ac levels, illustrating that Runx1 does not require high levels of active chromatin marks to bind to its target sites, but once bound can induce chromatin activation. Furthermore, Runx1 was shown to recruit hematopoietic regulators, SCL/TAL1 and FLI1 to target sites in HE cells to activate a hematopoietic transcriptional program (Lichtinger et al. 2012). This study illustrates Runx1's ability to orchestrate a hematopoietic-specific program in HE by changing the binding profiles of hematopoietic regulators and insuring proper progression through the EHT.

As hemogenic endothelial cells begin to transition into hematopoietic cells, one of the earliest hematopoietic markers to be expressed is the αIIb integrin subunit CD41 (Mikkola et al. 2003). A subset of endothelial cells in the dorsal aorta of Runx1 deficient embryos express CD41, suggesting that in the absence of Runx1, hemogenic endothelium is at least partially specified and can switch on hematopoietic gene expression (Liakhovitskaia et al. 2014). To determine if Runx1 expression close to the onset of EHT is sufficient for generating definitive hematopoietic cells, Liakhovitskaia et al., restored Runx1 expression in $CD41^+$ cells in Runx1 deficient embryos via CD41 (*Itga2b*)-Cre (Liakhovitskaia et al. 2014). Restoring Runx1 expression in $CD41^+$ cells rescued the generation of HSCs, and the embryos survived until birth, suggesting that Runx1 is required and sufficient for the progression of $CD41^+$ cells into HSCs (Liakhovitskaia et al. 2014). $CD41^+$ cells isolated from wild type mouse embryos or embryonic stem cell cultures can give rise to hematopoietic cells but cannot generate endothelial progenitors, indicating that $CD41^+$ cells are committed to the hematopoietic lineage (Hashimoto et al. 2007; Li et al. 2005). The finding that restoring Runx1 expression in $CD41^+$ (*Itga2b*-Cre expressing) cells can rescue HSCs suggests that Runx1 is not required until the endothelial to hematopoietic transition is initiated and hematopoietic fate has been cemented. However, transcriptional analysis of hemogenic and non-hemogenic endothelial cells isolated from E8.5 embryos revealed that while CD41 protein at the surface of either cell population is low to non-existent at E8.5, both hemogenic and non-hemogenic endothelial cells express *Itga2b* mRNA (Swiers et al. 2013a). Therefore the *Itga2b*-Cre used by Liakhovitskaia et al. (Liakhovitskaia et al. 2014) may have restored Runx1 expression in all endothelium at E8.5 rather than specifically in HE cells initiating the EHT. Thus, it is formally possible that Runx1 expression in *Itga2b*-expressing endothelial cells earlier in development, prior to the onset of EHT, is necessary for the *de novo* generation of definitive hematopoietic cells.

4.5 Regulation of Runx1 Expression During the Specification of Hemogenic Endothelium

Although Runx1 is required for the successful transition of HE cells into hematopoietic cells it is not required for the specification of hemogenic endothelium. This was perhaps best illustrated in live-imaging studies of EHT in Runx1 deficient zebrafish embryos. In Runx1 morphant zebrafish embryos, HE cells bend away from the endothelial monolayer, initiating the EHT, but fragment before forming a hematopoietic cell (Kissa and Herbomel 2010; Zhen et al. 2013), a phenomenon that was also observed in *Runx1*$^{-/-}$ mouse ES-derived HE cells (Lancrin et al. 2009; Eilken et al. 2009). Furthermore, as mentioned above CD41 is expressed by HE cells in the DA of E10.5 *Runx1*$^{-/-}$ mouse embryos, indicating that the hematopoietic program is at least partially initiated in the absence of Runx1 expression (Liakhovitskaia et al. 2014).

Although Runx1 is not required for the specification of HE it was proposed to play a role in determining cell fate in mesoderm-derived progenitors. $Etv2^+$ $Flk1^+$ mesodermal cells give rise to both endothelial cells and blood (Kataoka et al. 2011; Wareing et al. 2012). Whether the $Etv2^+$ $Flk1^+$ mesodermal progenitor gives rise to a non-hemogenic endothelial cell or a HE cell was recently reported to be controlled, at least in part, by Runx1 (Eliades et al. 2016). At E7.5, $Runx1^+$ $Etv2^+$ $Flk1^+$ cells reside within the extraembryonic yolk sac and co-express mesodermal and endothelial specific markers. At E8.5, a subset of $Etv2^+$ cells migrate from the area at the boundary of the yolk sac and embryo proper into the embryo proper and downregulate mesoderm-specific genes (Eliades et al. 2016). A similar observation was made by Tanaka et al., who reported that between E7.5 and E8.5 $Runx1^+$ $Gata1^-$ cells located at the boundary between the extraembryonic yolk sac and the embryo proper, migrate to the embryo proper where they contribute to the intraembryonic vasculature and blood (Tanaka et al. 2014). Interestingly, the $Etv2^+$ population at E7.5 expresses Runx1 and has hemogenic potential, likely representing at least in part the yolk sac blood island cells. At E8.5, in contrast most $Etv2^+$ cells do not express Runx1, and lack hematopoietic potential, from which it was suggested that Runx1 is silenced in the majority of $Etv2^+$ cells between E7.5 and E8.5 (Eliades et al. 2016). The mechanism of silencing involves Bmi1, a member of the Polycomb Repressive Complex 1 (PRC1) (Eliades et al. 2016), which physically interacts with Runx1 (Yu et al. 2012). Ectopic expression of Runx1, or inhibition of PRC1 conferred hemogenic potential to the E8.5 $Etv2^+$ population, suggesting that the hemogenic potential of the E8.5 $Etv2^+$ population is restricted through Runx1 silencing (Eliades et al. 2016). These results demonstrate that the default program in $Etv2^+$ $Flk1^+$ progenitors may be the hematopoietic program, initiated by Runx1. Bmi1 then represses Runx1 expression at E8.5 to promote a vascular fate.

Silencing of Runx1 expression in endothelium is also mediated through the homeobox protein, HoxA3. During hematopoietic development the expression of Runx1 and HoxA3 in the endothelium is mutually exclusive, in part because HoxA3 directly interacts with and represses Runx1 expression (Iacovino et al. 2011). Ectopic expression of HoxA3 during ES cell differentiation and in cells derived from E10.5 mouse embryos resulted in the downregulation of hematopoietic markers and inhibited hematopoietic specification, and increased the expression of endothelial-specific genes, suggesting that HoxA3 reinforces an endothelial fate while suppressing the hematopoietic potential of endothelial progenitors (Iacovino et al. 2011). Interestingly, when Runx1 is ectopically expressed in HoxA3-induced ES-derived endothelial progenitor cells the expression of hematopoietic genes is rescued, indicating that high levels of Runx1 can override HoxA3 activity (Iacovino et al. 2011).

4.6 Does Runx1 Function as a Master Regulator of Hematopoiesis?

The term "master regulator" is often used to describe a gene that sits at the very top of a regulatory hierarchy. However a stringent test of a master regulator is whether it can reprogram one cell type into another (Chan and Kyba 2013). Logically, the most likely cells that would respond to direct reprogramming by Runx1 are endothelial cells. However direct reprogramming studies have shown that Runx1 alone is not sufficient to reprogram either human umbilical vein endothelial cells (HUVECs) or human adult dermal endothelial cells (hDMECs) into hematopoietic progenitor cells (Sandler et al. 2014). Only when Runx1 was combined with Spi1, Fosb and Gfi1 could relatively efficient reprogramming of endothelial cells be achieved (Sandler et al. 2014). Interestingly, both *Spi1* (Pu.1) and *Gfi1* are direct downstream targets of Runx1 (Lancrin et al. 2012; Huang et al. 2008; Hoogenkamp et al. 2009), but when they were individually removed from a transduction cocktail containing all four transcription factors the efficiency of reprogramming significantly decreased, suggesting that ectopic Runx1 alone was unable to efficiently drive their expression (Sandler et al. 2014). Therefore, by this strict definition Runx1 is not a master regulator, as it is not by itself sufficient to reprogram HUVECs or hDMECs into blood cells. The reason for this is unclear, but may be because Runx1 cannot access various downstream targets in specific endothelial subtypes. Endothelial cells of different tissues and developmental stages are diverse in function, phenotype, transcription and chromatin state (Nolan et al. 2013; Aird 2012; Chi et al. 2003; Casanello et al. 2014), therefore it would be interesting to determine if other endothelial subtypes are more permissive to respecification by Runx1. Runx1 can induce a hematopoietic program in E8.5 Etv2^{+} endothelial cells, therefore the ability of endothelial cells to respond to Runx1 activity may be lost as development proceeds (Eliades et al. 2016).

4.7 Downstream Targets of Runx1 that Regulate the EHT

In order to transition morphologically and functionally into hematopoietic cells, HE cells must extinguish their endothelial-specific transcriptional program and upregulate a hematopoietic program; a transcriptional switch that is largely orchestrated by Runx1. Two direct targets of Runx1, *Gfi1* and *Gfi1b*, encode nuclear zinc finger transcriptional repressors that inhibit expression of endothelial genes in HE during the EHT (Lancrin et al. 2012). In Gfi1/Gfi1b deficient mouse embryos, HE cells in the yolk sac fail to transition morphologically into hematopoietic cells and remain locked in the endothelial layer (Lancrin et al. 2012). However, dissociation of Gfi1/Gfi1b deficient yolk sac frees the hematopoietic cells, which can then form hematopoietic colonies in clonogenic assays, suggesting that Gfi1/Gfi1b deficient HE cells can form functional hematopoietic progenitors but are unable to physically transition into a morphological hematopoietic cell (Lancrin et al. 2012). Conversely, when Gfi1 and Gfi1b expression was induced in *Runx1*$^{-/-}$ embryonic stem cell derived-HE, the HE cells could undergo the morphological transition into rounded cells but the round cells could not form colonies in hematopoietic clonogenic assays, thus illustrating that during the EHT Gfi1 and Gfi1b repress an endothelial fate allowing for the morphological transition of flat HE cells into rounded hematopoietic cells (Lancrin et al. 2012). Interestingly, a subsequent study found that hematopoietic clusters did not form in the arteries of Gfi1/Gfi1b deficient embryos, and Gfi1 expressing cells remained within the endothelial layer. However, unlike in the yolk sac, dissociated cells from the arteries could not differentiate into hematopoietic colonies, indicating that Gfi1 and Gfi1b have functions in blood cell formation in the major arteries in addition to their requirement for the EHT (Thambyrajah et al. 2016).

Identifying the transcriptional program regulated by Runx1 in hemogenic endothelium is challenging because HE is a rare population that exists only transiently during midgestation. To overcome these challenges Lie-A-ling et al. (Lie-A-Ling et al. 2014) used an alternative technique to chromatin immunoprecipitation called DNA adenine methyltransferase identification (DamID). DamID relies on the fusion of a transcription factor (such as Runx1) to the *Escherichia coli* DNA adenine methyltransferase (Dam). When the transcription factor binds DNA the fused Dam protein adds stable methylation tags to adenines within nearby GATC sequences allowing for identification of transcription factor binding sites without the need for antibodies (Lie-A-Ling et al. 2014). To identify Runx1 targets in HE, Lie-A-Ling et al. established *Runx1*$^{-/-}$ ES cell lines containing doxycycline inducible Runx1-Dam, and then differentiated the ES cells into HE. Fortuitously, the inducible system was leaky, allowing for low levels of Runx1 expression in the absence of doxycycline that were not sufficient for EHT, but were sufficient for the detection of Runx1 occupancy by DamID (Lie-A-Ling et al. 2014). Comparison of the Runx1-DamID methylation and RNA-Seq datasets led to the identification of 235 genes that were both bound by Runx1 and differentially expressed in HE cells generated from wild type and *Runx1*$^{-/-}$ ES cells (Lie-A-Ling et al. 2014). The expression of 80 of the genes was negatively correlated with Runx1 occupancy and that of 155 genes was positively correlated (Lie-A-Ling et al. 2014), consistent with Runx1's ability to function as a transcriptional repressor or activator in the same cell type (Canon and Banerjee 2003). The target genes that positively correlated with Runx1 expression were associated with cell adhesion, integrin signaling, cellular movement and interaction with the extracellular matrix (Lie-A-Ling et al. 2014). Interestingly, very few hematopoietic genes were identified as Runx1 targets, suggesting that the HE was in an early stage of differentiation and had not yet initiated a hematopoietic-specific program. Thus an early function of Runx1 in HE is to regulate the expression of genes involved in the activation of migration and adhesion of HE cells prior to the EHT.

4.8 Transcriptional and Translational Regulation of Runx1 Expression During Embryonic Hematopoiesis

The spatio-temporal specific expression pattern of Runx1 during embryonic hematopoiesis is controlled, in part, through transcriptional regulation. In vertebrates, *Runx1* is transcribed from two alternative promoters, the distal (P1) promoter and the proximal (P2) promoter (Ghozi et al. 1996; Rennert et al. 2003; Levanon et al. 2001b; Bee et al. 2009b; Telfer and Rothenberg 2001). The P2 promoter differs from the P1 promoter in that it is associated with a large CpG island that may influence differential regulation of transcription from P1 versus P2 (Levanon et al. 2001b; Bee et al. 2009b). Furthermore, the conserved binding sites associated with each promoter are different; P1 contains a cAMP-responsive element, a CCAAT box, GATA, SMAD and RUNX motifs whereas P2 contains CCATT boxes, initiator sequences, a GC-box, OCT and ETS motifs (Ghozi et al. 1996; Bee et al. 2009a, 2010; Martinez et al. 2016). Differential promoter usage in addition to RNA splicing leads to a vast array of Runx1 isoforms. The full-length isoforms generated from the P1 and P2 promoters are referred to as Runx1c and Runx1b, respectively. Runx1c (465aa) is larger than Runx1b (451aa) due to a difference of 19aa at their N-termini, but there are no data that suggest these differences confer distinct properties to the Runx1c and Runx1b proteins (Fujita et al. 2001; Challen and Goodell 2010).

Differential promoter usage during hematopoiesis does, however, control the timing and level of Runx1 expression. Analysis of *Runx1* promoter activity in mouse embryos and ES cell differentiation models revealed that P2 is dominant early during primitive hematopoiesis and at the onset of definitive hematopoiesis, whereas P1 activity ramps up later in development during fetal liver and bone marrow hematopoiesis (Bee et al. 2009b, 2010; Fujita et al. 2001; Pozner et al. 2007; Sroczynska et al. 2009). In mice, abrogation of P2 via insertion of a neomycin resistance gene resulted in fewer hematopoietic clusters in the large arteries of the embryo proper, significantly fewer hematopoietic progenitors in the fetal liver and yolk sac, reduced thymopoiesis, and perinatal lethality (Pozner et al. 2007; Bee et al. 2010). The prolonged survival of P2-attenuated mice compared to Runx1 null mice (E12.5 lethality), is likely due to the overlap of P1 and P2 activity in hemogenic endothelium, and therefore P1 promoter activity alone promotes the *de novo* generation of sufficient numbers of definitive hematopoietic cells to prevent embryonic lethality (Sroczynska et al. 2009; Bee et al. 2009b). Loss of P1 activity, on the other hand, is less detrimental than P2 loss. P1-null mouse embryos have fewer hematopoietic clusters and produce fewer hematopoietic progenitors in the yolk sac and large arteries of the embryo proper compared to littermate controls, but the decrease in hematopoietic cells is not as severe as that caused by P2 attenuation, and loss of P1 is not lethal (Bee et al. 2010). However, the bone marrow and peripheral blood of P1-null adult mice does exhibit a significant decrease in white blood cells and platelets and an increase in the percentage of bone marrow HSCs and hematopoietic progenitors (Bee et al. 2010). Interestingly, one functional P2-deleted *Runx1* allele in the absence of P1-activity was sufficient to rescue embryonic lethality, but one functional P1 allele in the absence of P2 was not, suggesting that the dosage and timing of Runx1 expression is critical for the generation of definitive hematopoietic cells (Bee et al. 2010; Pozner et al. 2007).

P1 and P2 regulate the timing and dosage of Runx1 during hematopoiesis but they do not confer tissue specificity in mammalian embryos (Ghozi et al. 1996; Bee et al. 2009a). Hematopoietic specific expression is mediated by enhancers located within and upstream of the *Runx1* gene locus (Schuette et al. 2016). The best known of these is a 531 bp enhancer located between P1 and P2, 23.4 kb downstream of the ATG in exon 1 of *Runx1* (Nottingham et al. 2007; Ng et al. 2010). The +23 enhancer drives reporter expression at all sites of hematopoiesis in mouse embryos (Nottingham et al. 2007; Ng et al. 2010). Specifically, the +23 enhancer is active in HE, hematopoietic clusters and fetal liver hematopoietic cells. It is not, however, active in non-hematopoietic tissues that express Runx1, such as the mesenchyme beneath the dorsal aorta (Nottingham et al. 2007; Ng et al. 2010). ChIP analysis of the +23 enhancer demonstrated association with Gata2, Runx1, Ets transcription factors, and the SCL/Lmo2/Lbd-1 complex in a myeloid progenitor cell line (Nottingham et al. 2007). To determine if transcription factor binding was necessary for activity of the +23 enhancer in mouse embryos, Nottingham et al. assessed whether or not activity of the enhancer was disrupted after mutating the RUNX, ETS or GATA motifs. They found that the RUNX motif was not required for +23 enhancer activity but the ETS and GATA motifs were required, therefore the +23 enhancer confers hematopoietic specific expression of Runx1 and is regulated, in part, through interaction with Gata and Ets transcription factors (Nottingham et al. 2007).

Post-transcriptional control of Runx1 occurs through variations in translational efficiency and transcript attenuation via miRNAs. The translational efficiencies of P1 and P2-derived transcripts differ due to distinct 5′ untranslated regions (UTR). P1-derived transcripts have a relatively short 5′ UTR (452 bp) that directs efficient cap-dependent translation (Pozner et al. 2000). In contrast, P2-derived transcripts have a long 5′ UTR (1631 bp) containing an internal ribosomal entry site (IRES), which mediates cap-independent translation (Pozner et al. 2000). It has been proposed that P2-derived transcripts are poorly translated due to the length of the UTR and cis-acting elements within it, including the IRES as well as multiple upstream AUG codons and GC-rich islands (Pozner et al. 2000; Levanon et al. 1996). A possible explanation for the presence of both IRES and cap-dependent translation of Runx1 mRNA is that IRES-containing transcripts are translated during mitosis and under stress conditions when cap-dependent translation is impaired (Levanon and Groner 2004). Further post-transcriptional regulation of Runx1 occurs through miRNA transcript attenuation. In addition to distinct 5'UTRs, Runx1 mRNA isoforms have different 3′ UTRs that range in size from 150 to 4000 bp (Levanon et al. 2001b). Several putative miRNA binding sites were identified in the 3′ UTR of Runx1, and the length of the 3'UTR was shown to change the susceptibility to miRNA targeting and attenuation (Ben-Ami et al. 2009). Although the role that translational regulation of Runx1 plays during embryonic hematopoiesis has not been elucidated, it is plausible that it influences isoform, dose, timing and cell specific expression of Runx1 during development.

The past 10 years have seen the shaping of the roadmap of hematopoietic cell development from hemogenic endothelium: discrete cellular intermediates of the HSC lineage have been identified, along with the identification of distinct populations of HE giving rise to EMPs and HSCs. In addition, our understanding of the critical role Runx1 plays in this process has deepened with the identification of new target genes. The rapid developments in imaging and expression profiling technologies now enable taking the study of *de novo* hematopoietic cell generation to the single cell level where cell fate decisions are made.

This will no doubt lead to more exciting insights into the role of Runx1 in blood stem and progenitor cell generation.

References

Aird, W. C. (2012). Endothelial cell heterogeneity. *Cold Spring Harbor Perspectives in Medicine, 2*, a006429.

Bee, T., Ashley, E. L., Bickley, S. R., Jarratt, A., Li, P. S., Sloane-Stanley, J., et al. (2009a). The mouse Runx1 +23 hematopoietic stem cell enhancer confers hematopoietic specificity to both Runx1 promoters. *Blood, 113*, 5121–5124.

Bee, T., Liddiard, K., Swiers, G., Bickley, S. R., Vink, C. S., Jarratt, A., et al. (2009b). Alternative Runx1 promoter usage in mouse developmental hematopoiesis. *Blood Cells, Molecules & Diseases, 43*, 35–42.

Bee, T., Swiers, G., Muroi, S., Pozner, A., Nottingham, W., Santos, A. C., et al. (2010). Nonredundant roles for Runx1 alternative promoters reflect their activity at discrete stages of developmental hematopoiesis. *Blood, 115*, 3042–3050.

Ben-Ami, O., Pencovich, N., Lotem, J., Levanon, D., & Groner, Y. (2009). A regulatory interplay between miR-27a and Runx1 during megakaryopoiesis. *Proceedings of the National Academy of Sciences of the United States of America, 106*, 238–243.

Bertrand, J. Y., Chi, N. C., Santoso, B., Teng, S., Stainier, D. Y., & Traver, D. (2010). Haematopoietic stem cells derive directly from aortic endothelium during development. *Nature, 464*, 108–111.

Boisset, J. C., Van Cappellen, W., Andrieu-Soler, C., Galjart, N., Dzierzak, E., & Robin, C. (2010). In vivo imaging of haematopoietic cells emerging from the mouse aortic endothelium. *Nature, 464*, 116–120.

Boisset, J. C., Clapes, T., Klaus, A., Papazian, N., Onderwater, J., Mommaas-Kienhuis, M., et al. (2015). Progressive maturation toward hematopoietic stem cells in the mouse embryo aorta. *Blood, 125*, 465–469.

Bos, F. L., Hawkins, J. S., & Zovein, A. C. (2015). Single-cell resolution of morphological changes in hemogenic endothelium. *Development, 142*, 2719–2724.

Bresciani, E., Carrington, B., Wincovitch, S., Jones, M., Gore, A. V., Weinstein, B. M., et al. (2014). CBFβ and RUNX1 are required at 2 different steps during the development of hematopoietic stem cells in zebrafish. *Blood, 124*, 70–78.

Cai, Z., De Bruijn, M., Ma, X., Dortland, B., Luteijn, T., Downing, R. J., & Dzierzak, E. (2000). Haploinsufficiency of AML1 affects the temporal and spatial generation of hematopoietic stem cells in the mouse embryo. *Immunity, 13*, 423–431.

Canon, J., & Banerjee, U. (2003). In vivo analysis of a developmental circuit for direct transcriptional activation and repression in the same cell by a Runx protein. *Genes & Development, 17*, 838–843.

Casanello, P., Schneider, D., Herrera, E. A., Uauy, R., & Krause, B. J. (2014). Endothelial heterogeneity in the umbilico-placental unit: DNA methylation as an innuendo of epigenetic diversity. *Frontiers in Pharmacology, 5*, 49.

Castilla, L. H., Wijmenga, C., Wang, Q., Stacy, T., Speck, N. A., Eckhaus, M., et al. (1996). Failure of embryonic hematopoiesis and lethal hemorrhages in mouse embryos heterozygous for a knocked-in leukemia gene CBFB-MYH11. *Cell, 87*, 687–696.

Challen, G. A., & Goodell, M. A. (2010). Runx1 isoforms show differential expression patterns during hematopoietic development but have similar functional effects in adult hematopoietic stem cells. *Experimental Hematology, 38*, 403–416.

Chan, S. S., & Kyba, M. (2013). What is a master regulator? *Journal of Stem Cell Research and Therapy, 3*, 114.

Chen, M. J., Yokomizo, T., Zeigler, B. M., Dzierzak, E., & Speck, N. A. (2009). Runx1 is required for the endothelial to haematopoietic cell transition but not thereafter. *Nature, 457*, 887–891.

Chi, J. T., Chang, H. Y., Haraldsen, G., Jahnsen, F. L., Troyanskaya, O. G., Chang, D. S., et al. (2003). Endothelial cell diversity revealed by global expression profiling. *Proceedings of the National Academy of Sciences of the United States of America, 100*, 10623–10628.

Choi, K., Kennedy, M., Kazarov, A., Papadimitriou, J. C., & Keller, G. (1998). A common precursor for hematopoietic and endothelial cells. *Development, 125*, 725–732.

Christensen, J. L., Wright, D. E., Wagers, A. J., & Weissman, I. L. (2004). Circulation and chemotaxis of fetal hematopoietic stem cells. *PLoS Biology, 2*, E75.

Cumano, A., Ferraz, J. C., Klaine, M., Di Santo, J. P., & Godin, I. (2001). Intraembryonic, but not yolk sac hematopoietic precursors, isolated before circulation, provide long-term multilineage reconstitution. *Immunity, 15*, 477–485.

De Bruijn, M. F., Speck, N. A., Peeters, M. C., & Dzierzak, E. (2000). Definitive hematopoietic stem cells first develop within the major arterial regions of the mouse embryo. *The EMBO Journal, 19*, 2465–2474.

Dieterlen-Lievre, F. (1975). On the origin of haemopoietic stem cells in the avian embryo: An experimental approach. *Journal of Embryology and Experimental Morphology, 33*, 607–619.

Eilken, H. M., Nishikawa, S., & Schroeder, T. (2009). Continuous single-cell imaging of blood generation from haemogenic endothelium. *Nature, 457*, 896–900.

Eliades, A., Wareing, S., Marinopoulou, E., Fadlullah, M. Z., Patel, R., Grabarek, J. B., et al. (2016). The hemogenic competence of endothelial progenitors is restricted by Runx1 silencing during embryonic development. *Cell Reports, 15*, 2185–2199.

Ema, H., & Nakauchi, H. (2000). Expansion of hematopoietic stem cells in the developing liver of a mouse embryo. *Blood, 95*, 2284–2288.

Ema, M., Faloon, P., Zhang, W. J., Hirashima, M., Reid, T., Stanford, W. L., et al. (2003). Combinatorial effects of Flk1 and Tal1 on vascular and hematopoietic development in the mouse. *Genes & Development, 17*, 380–393.

Ferkowicz, M. J., & Yoder, M. C. (2005). Blood island formation: Longstanding observations and modern interpretations. *Experimental Hematology, 33*, 1041–1047.

Frame, J. M., Fegan, K. H., Conway, S. J., Mcgrath, K. E., & Palis, J. (2015). Definitive hematopoiesis in the yolk sac emerges from Wnt-responsive hemogenic endothelium independently of circulation and arterial identity. *Stem Cells, 34*, 431–444.

Fraser, S. T., Ogawa, M., Yokomizo, T., Ito, Y., & Nishikawa, S. (2003). Putative intermediate precursor between hematogenic endothelial cells and blood cells in the developing embryo. *Development, Growth & Differentiation, 45*, 63–75.

Fujita, Y., Nishimura, M., Taniwaki, M., Abe, T., & Okuda, T. (2001). Identification of an alternatively spliced form of the mouse AML1/RUNX1 gene transcript AML1c and its expression in early hematopoietic development. *Biochemical and Biophysical Research Communications, 281*, 1248–1255.

Fujiwara, Y., Browne, C. P., Cunniff, K., Goff, S. C., & Orkin, S. H. (1996). Arrested development of embryonic red cell precursors in mouse embryos lacking transcription factor GATA-1. *Proceedings of the National Academy of Sciences of the United States of America, 93*, 12355–12358.

Gekas, C., Dieterlen-Lievre, F., Orkin, S. H., & Mikkola, H. K. (2005). The placenta is a niche for hematopoietic stem cells. *Developmental Cell, 8*, 365–375.

Ghozi, M. C., Bernstein, Y., Negreanu, V., Levanon, D., & Groner, Y. (1996). Expression of the human acute myeloid leukemia gene AML1 is regulated by two promoter regions. *Proceedings of the National Academy of Sciences of the United States of America, 93*, 1935–1940.

Ginhaux, F., & Jung, S. (2014). Monocytes and macrophages: Developmental pathways and tissue homeostasis. *Nature Reviews Immunology, 14*, 392–404.

Gordon-Keylock, S., Sobiesiak, M., Rybtsov, S., Moore, K., & Medvinsky, A. (2013). Mouse extraembryonic arterial vessels harbor precursors capable of maturing into definitive HSCs. *Blood, 122*, 2338–2345.

Growney, J. D., Shigematsu, H., Li, Z., Lee, B. H., Adelsperger, J., Rowan, R., et al. (2005). Loss of Runx1 perturbs adult hematopoiesis and is associated with a myeloproliferative phenotype. *Blood, 106*, 494–504.

Haar, J. L., & Ackerman, G. A. (1971). A phase and electron microscopic study of vasculogenesis and erythropoiesis in the yolk sac of the mouse. *The Anatomical Record, 170*, 199–223.

Hashimoto, K., Fujimoto, T., Shimoda, Y., Huang, X., Sakamoto, H., & OGAWA, M. (2007). Distinct hemogenic potential of endothelial cells and CD41+ cells in mouse embryos. *Development, Growth & Differentiation, 49*, 287–300.

Hoogenkamp, M., Lichtinger, M., Krysinska, H., Lancrin, C., Clarke, D., Williamson, A., et al. (2009). Early chromatin unfolding by RUNX1: A molecular explanation for differential requirements during specification versus maintenance of the hematopoietic gene expression program. *Blood, 114*, 299–309.

Huang, G., Zhang, P., Hirai, H., Elf, S., Yan, X., Chen, Z., et al. (2008). PU.1 is a major downstream target of AML1 (RUNX1) in adult mouse hematopoiesis. *Nature Genetics, 40*, 51–60.

Iacovino, M., Chong, D., Szatmari, I., Hartweck, L., Rux, D., Caprioli, A., et al. (2011). HoxA3 is an apical regulator of haemogenic endothelium. *Nature Cell Biology, 13*, 72–78.

Ichikawa, M., Asai, T., Saito, T., Seo, S., Yamazaki, I., Yamagata, T., et al. (2004). AML-1 is required for megakaryocytic maturation and lymphocytic differentiation, but not for maintenance of hematopoietic stem cells in adult hematopoiesis. *Nature Medicine, 10*, 299–304.

Iizuka, K., Yokomizo, T., Watanabe, N., Tanaka, Y., Osata, M., Takaku, T., et al. (2016). Lack of phenotypical and morphological evidences of endothelial to hematopoietic transition in the murine embryonic head during hematopoietic stem cell emergence. *PLoS One, 11*, e0156427.

Ivanovs, A., Rybtsov, S., Welch, L., Anderson, R. A., Turner, M. L., & Medvinsky, A. (2011). Highly potent human hematopoietic stem cells first emerge in the intraembryonic aorta-gonad-mesonephros region. *The Journal of Experimental Medicine, 208*, 2417–2427.

Jaffredo, T., Bollerot, K., Sugiyama, D., Gautier, R., & Drevon, C. (2005). Tracing the hemangioblast during embryogenesis: Developmental relationships between endothelial and hematopoietic cells. *The International Journal of Developmental Biology, 49*, 269–277.

Kalev-Zylinska, M. L., Horsfield, J. A., Flores, M. V., Postlethwait, J. H., Vitas, M. R., Baas, A. M., et al. (2002). Runx1 is required for zebrafish blood and vessel development and expression of a human RUNX1-CBF2T1 transgene advances a model for studies of leukemogenesis. *Development, 129*, 2015–2030.

Kataoka, H., Hayashi, M., Nakagawa, R., Tanaka, Y., Izumi, N., Nishikawa, S., et al. (2011). Etv2/ER71 induces vascular mesoderm from Flk1+PDGFRα+ primitive mesoderm. *Blood, 118*, 6975–6986.

Kieusseian, A., Brunet De LA Grange, P., Burlen-Defranoux, O., Godin, I., & Cumano, A. (2012). Immature hematopoietic stem cells undergo maturation in the fetal liver. *Development, 139*, 3521–3530.

Kingsley, P. D., Malik, J., Fantauzzo, K. A., & Palis, J. (2004). Yolk sac-derived primitive erythroblasts enucleate during mammalian embryogenesis. *Blood, 104*, 19–25.

Kissa, K., & Herbomel, P. (2010). Blood stem cells emerge from aortic endothelium by a novel type of cell transition. *Nature, 464*, 112–115.

Kissa, K., Murayama, E., Zapata, A., Cortes, A., Perret, E., Machu, C., & Herbomel, P. (2008). Live imaging of emerging hematopoietic stem cells and early thymus colonization. *Blood, 111*, 1147–1156.

Kobayashi, A., Senzaki, K., Ozaki, S., Yoshikawa, M., & Shiga, T. (2012). Runx1 promotes neuronal differentiation in dorsal root ganglion. *Molecular and Cellular Neurosciences, 49*, 23–31.

Lacaud, G., Gore, L., Kennedy, M., Kouskoff, V., Kingsley, P., Hogan, C., et al. (2002). Runx1 is essential for hematopoietic commitment at the hemangioblast stage of development in vitro. *Blood, 100*, 458–466.

Lacaud, G., Kouskoff, V., Trumble, A., Schwantz, S., & Keller, G. (2004). Haploinsufficiency of Runx1 results in the acceleration of mesodermal development and hemangioblast specification upon in vitro differentiation of ES cells. *Blood, 103*, 886–889.

Lancrin, C., Sroczynska, P., Stephenson, C., Allen, T., Kouskoff, V., & Lacaud, G. (2009). The haemangioblast generates haematopoietic cells through a haemogenic endothelium stage. *Nature, 457*, 892–895.

Lancrin, C., Mazan, M., Stefanska, M., Patel, R., Lichtinger, M., Costa, G., et al. (2012). GFI1 and GFI1B control the loss of endothelial identity of hemogenic endothelium during hematopoietic commitment. *Blood, 120*, 314–322.

Lawson, K. A., Meneses, J. J., & Pedersen, R. A. (1991). Clonal analysis of epiblast fate during germ layer formation in the mouse embryo. *Development, 113*, 891–911.

Levanon, D., & Groner, Y. (2004). Structure and regulated expression of mammalian RUNX genes. *Oncogene, 23*, 4211–4219.

Levanon, D., Bernstein, Y., Negreanu, V., Ghozi, M. C., Bar-Am, I., Aloya, R., et al. (1996). A large variety of alternatively spliced and differentially expressed mRNAs are encoded by the human acute myeloid leukemia gene AML1. *DNA and Cell Biology, 15*, 175–185.

Levanon, D., Brenner, O., Negreanu, V., Bettoun, D., Woolf, E., Eilam, R., et al. (2001a). Spatial and temporal expression pattern of Runx3 (Aml2) and Runx1 (Aml1) indicates non-redundant functions during mouse embryogenesis. *Mechanisms of Development, 109*, 413–417.

Levanon, D., Glusman, G., Bangsow, T., Ben-Asher, E., Male, D. A., Avidan, N., et al. (2001b). Architecture and anatomy of the genomic locus encoding the human leukemia-associated transcription factor RUNX1/AML1. *Gene, 262*, 23–33.

Li, W., Ferkowicz, M. J., Johnson, S. A., Shelley, W. C., & Yoder, M. C. (2005). Endothelial cells in the early murine yolk sac give rise to CD41-expressing hematopoietic cells. *Stem Cells and Development, 14*, 44–54.

Li, Z., Chen, M. J., Stacy, T., & Speck, N. A. (2006). Runx1 function in hematopoiesis is required in cells that express Tek. *Blood, 107*, 106–110.

Li, Z., Lan, Y., He, W., Chen, D., Wang, J., Zhou, F., et al. (2012). Mouse embryonic head as a site for hematopoietic stem cell development. *Cell Stem Cell, 11*, 663–675.

Li, Y., Esain, V., Teng, L., Xu, J., Kwan, W., Frost, I. M., et al. (2014). Inflammatory signaling regulates embryonic hematopoietic stem and progenitor cell production. *Genes & Development, 28*, 2597–2612.

Li, Z., Vink, C. S., Mariani, S. A., & Dzierzak, E. (2016). Subregional localization and characterization of Ly6aGFP-expressing hematopoietic cells in the mouse embryonic head. *Developmental Biology, 416*, 34–41.

Liakhovitskaia, A., Gribi, R., Stamateris, E., Villain, G., Jaffredo, T., Wilkie, R., et al. (2009). Restoration of Runx1 expression in the Tie2 cell compartment rescues definitive hematopoietic stem cells and extends life of Runx1 knockout animals until birth. *Stem Cells, 27*, 1616–1624.

Liakhovitskaia, A., Lana-Elola, E., Stamateris, E., Rice, D. P., Van 'T Hof, R. J., & Medvinsky, A. (2010). The essential requirement for Runx1 in the development of the sternum. *Developmental Biology, 340*, 539–546.

Liakhovitskaia, A., Rybtsov, S., Smith, T., Batsivari, A., Rybtsova, N., Rode, C., et al. (2014). Runx1 is required for progression of CD41+ embryonic precursors into HSCs but not prior to this. *Development, 141*, 3319–3323.

Lichtinger, M., Ingram, R., Hannah, R., Müller, D., Clarke, D., Assi, S. A., et al. (2012). RUNX1 reshapes the epigenetic landscape at the onset of haematopoiesis. *The EMBO Journal, 31*, 4318–4333.

Lie-A-Ling, M., Marinopoulou, E., Li, Y., Patel, R., Stefanska, M., Bonifer, C., et al. (2014). RUNX1 positively regulates a cell adhesion and migration program in murine hemogenic endothelium prior to blood emergence. *Blood, 124*, e11–e20.

Lorsbach, R. B., Moore, J., Ang, S. O., Sun, W., Lenny, N., & Downing, J. R. (2004). Role of Runx1 in adult hematopoiesis: Analysis of Runx1-IRES-GFP knock-in mice reveals differential lineage expression. *Blood, 103*, 2522–2529.

Martinez, M., Hinojosa, M., Trombly, D., Morin, V., Stein, J., Stein, G., et al. (2016). Transcriptional Auto-Regulation of RUNX1 P1 Promoter. *PloS One, 11*, e0149119.

Mcgrath, K. E., Frame, J. M., Fegan, K. H., Bowen, J. R., Conway, S. J., Catherman, S. C., et al. (2015). Distinct sources of hematopoietic progenitors emerge before HSCs and provide functional blood cells in the mammalian embryo. *Cell Reports, 11*, 1892–1904.

Medvinsky, A., & Dzierzak, E. (1996). Definitive hematopoiesis is autonomously initiated by the AGM region. *Cell, 86*, 897–906.

Mikkola, H. K., Fujiwara, Y., Schlaeger, T. M., Traver, D., & Orkin, S. H. (2003). Expression of CD41 marks the initiation of definitive hematopoiesis in the mouse embryo. *Blood, 101*, 508–516.

Moore, M. A., & Metcalf, D. (1970). Ontogeny of the haemopoietic system: Yolk sac origin of in vivo and in vitro colony forming cells in the developing mouse embryo. *British Journal of Haematology, 18*, 279–296.

Mukouyama, Y., Chiba, N., Hara, T., Okada, H., Ito, Y., Kanamaru, R., et al. (2000). The AML1 transcription factor functions to develop and maintain hematogenic precursor cells in the embryonic aorta-gonad-mesonephros region. *Developmental Biology, 220*, 27–36.

Müller, A. M., Medvinsky, A., Strouboulis, J., Grosveld, F., & Dzierzak, E. (1994). Development of hematopoietic stem cell activity in the mouse embryo. *Immunity, 1*, 291–301.

Murayama, E., Kissa, K., Zapata, A., Mordelet, E., Briolat, V., Lin, H. F., et al. (2006). Tracing hematopoietic precursor migration to successive hematopoietic organs during zebrafish development. *Immunity, 25*, 963–975.

Murray, P. (1932). *The development in vitro of the blood of the early chick embryo*. London: Proceedings of the Royal Society.

Nakano, H., Liu, X., Arshi, A., Nakashima, Y., Van Handel, B., Sasidharan, R., et al. (2013). Haemogenic endocardium contributes to transient definitive haematopoiesis. *Nature Communications, 4*, 1564.

Ng, C. E., Yokomizo, T., Yamashita, N., Cirovic, B., Jin, H., Wen, Z., et al. (2010). A Runx1 intronic enhancer marks hemogenic endothelial cells and hematopoietic stem cells. *Stem Cells, 28*, 1869–1881.

Niki, M., Okada, H., Takano, H., Kuno, J., Tani, K., Hibino, H., et al. (1997). Hematopoiesis in the fetal liver is impaired by targeted mutagenesis of a gene encoding a non-DNA binding subunit of the transcription factor, polyomavirus enhancer binding protein 2/core binding factor. *Proceedings of the National Academy of Sciences of the United States of America, 94*, 5697–5702.

Nolan, D. J., Ginsberg, M., Israely, E., Palikuqi, B., Poulos, M. G., James, D., et al. (2013). Molecular signatures of tissue-specific microvascular endothelial cell heterogeneity in organ maintenance and regeneration. *Developmental Cell, 26*, 204–219.

North, T., Gu, T. L., Stacy, T., Wang, Q., Howard, L., Binder, M., et al. (1999). Cbfa2 is required for the formation of intra-aortic hematopoietic clusters. *Development, 126*, 2563–2575.

North, T. E., De Bruijn, M. F., Stacy, T., Talebian, L., Lind, E., Robin, C., et al. (2002). Runx1 expression marks long-term repopulating hematopoietic stem cells in the midgestation mouse embryo. *Immunity, 16*, 661–672.

North, T. E., Stacy, T., Matheny, C. J., Speck, N. A., & De Bruijn, M. F. (2004). Runx1 is expressed in adult mouse hematopoietic stem cells and differentiating myeloid and lymphoid cells, but not in maturing erythroid cells. *Stem Cells, 22*, 158–168.

Nottingham, W. T., Jarratt, A., Burgess, M., Speck, C. L., Cheng, J. F., Prabhakar, S., et al. (2007). Runx1-mediated hematopoietic stem-cell emergence is controlled by a Gata/Ets/SCL-regulated enhancer. *Blood, 110*, 4188–4197.

Okuda, T., Van Deursen, J., Hiebert, S. W., Grosveld, G., & Downing, J. R. (1996). AML1, the target of multiple chromosomal translocations in human leukemia, is essential for normal fetal liver hematopoiesis. *Cell, 84*, 321–330.

Padron-Barthe, L., Temino, S., Villa Del Campo, C., Carramolino, L., Isern, J., & Torres, M. (2014). Clonal analysis identifies hemogenic endothelium as the source of the blood-endothelial common lineage in the mouse embryo. *Blood, 124*, 2523–2532.

Palis, J. (2014). Primitive and definitive erythropoiesis in mammals. *Frontiers in Physiology, 5*, 3.

Palis, J., Robertson, S., Kennedy, M., Wall, C., & Keller, G. (1999). Development of erythroid and myeloid progenitors in the yolk sac and embryo proper of the mouse. *Development, 126*, 5073–5084.

Potts, K. S., Sargeant, T. J., Markham, J. F., Shi, W., Biben, C., Josefsson, E. C., et al. (2014). A lineage of diploid platelet-forming cells precedes polyploid megakaryocyte formation in the mouse embryo. *Blood, 124*, 2725–2729.

Pozner, A., Goldenberg, D., Negreanu, V., Le, S. Y., Elroy-Stein, O., Levanon, D., & Groner, Y. (2000). Transcription-coupled translation control of AML1/RUNX1 is mediated by cap- and internal ribosome entry site-dependent mechanisms. *Molecular and Cellular Biology, 20*, 2297–2307.

Pozner, A., Lotem, J., Xiao, C., Goldenberg, D., Brenner, O., Negreanu, V., et al. (2007). Developmentally regulated promoter-switch transcriptionally controls Runx1 function during embryonic hematopoiesis. *BMC Developmental Biology, 7*, 84.

Putz, G., Rosner, A., Nuesslein, I., Schmitz, N., & Buchholz, F. (2006). AML1 deletion in adult mice causes splenomegaly and lymphomas. *Oncogene, 25*, 929–939.

Rennert, J., Coffman, J. A., Mushegian, A. R., & Robertson, A. J. (2003). The evolution of Runx genes I. A comparative study of sequences from phylogenetically diverse model organisms. *BMC Evolutionary Biology, 3*, 4.

Rhodes, K. E., Gekas, C., Wang, Y., Lux, C. T., Francis, C. S., Chan, D. N., et al. (2008). The emergence of hematopoietic stem cells is initiated in the placental vasculature in the absence of circulation. *Cell Stem Cell, 2*, 252–263.

Rybtsov, S., Sobiesiak, M., Taoudi, S., Souilhol, C., Senserrich, J., Liakhovitskaia, A., et al. (2011). Hierarchical organization and early hematopoietic specification of the developing HSC lineage in the AGM region. *The Journal of Experimental Medicine, 208*, 1305–1315.

Rybtsov, S., Ivanovs, A., Zhao, S., & Medvinsky, A. (2016). Concealed expansion of immature precursors underpins acute burst of adult HSC activity in foetal liver. *Development, 143*, 1284–1289.

Sabin, F. R. (1920). Studies on the origin of blood-vessels and of red blood-corpuscles as seen in the living blastoderm of chicks during the second day of incubation. *Contributions to Embryology, 30 x 24 cm*, 213–262.

Sandler, V. M., Lis, R., Liu, Y., Kedem, A., James, D., Elemento, O., et al. (2014). Reprogramming human endothelial cells to haematopoietic cells requires vascular induction. *Nature, 511*, 312–318.

Sasaki, K., Yagi, H., Bronson, R. T., Tominaga, K., Matsunashi, T., Deguchi, K., et al. (1996). Absence of fetal liver hematopoiesis in mice deficient in transcriptional coactivator core binding factor beta. *Proceedings of the National Academy of Sciences of the United States of America, 93*, 12359–12363.

Schuette, J., Wang, H., Antoniou, S., Jarrat, A., Wilson, N. K., Riepsaame, J., et al. (2016). An experimentally validated network of nine haematopoietic transcription factors reveals mechanisms of cell state stability. *eLife, 5*, e11469.

Sroczynska, P., Lancrin, C., Kouskoff, V., & Lacaud, G. (2009). The differential activities of Runx1 promoters define milestones during embryonic hematopoiesis. *Blood, 114*, 5279–5289.

Swiers, G., Baumann, C., O'rourke, J., Giannoulatou, E., Taylor, S., Joshi, A., et al. (2013a). Early dynamic fate changes in haemogenic endothelium characterized at the single-cell level. *Nature Communications, 4*, 2924.

Swiers, G., Rode, C., Azzoni, E., & De Bruijn, M. F. (2013b). A short history of hemogenic endothelium. *Blood Cells, Molecules & Diseases, 51*, 206–212.

Takakura, N., Watanabe, T., Suenobu, S., Yamada, Y., Noda, T., Ito, Y., et al. (2000). A role for hematopoietic stem cells in promoting angiogenesis. *Cell, 102*, 199–209.

Tanaka, Y., Sanchez, V., Takata, N., Yokomizo, T., Yamanaka, Y., Kataoka, H., et al. (2014). Circulation-independent differentiation pathway from extraembryonic mesoderm toward hematopoietic stem cells via hemogenic angioblasts. *Cell Reports, 8*, 31–39.

Taoudi, S., Gonneau, C., Moore, K., Sheridan, J. M., Blackburn, C. C., Taylor, E., & Medvinsky, A. (2008). Extensive hematopoietic stem cell generation in the AGM region via maturation of VE-cadherin+CD45+ pre-definitive HSCs. *Cell Stem Cell, 3*, 99–108.

Telfer, J. C., & Rothenberg, E. V. (2001). Expression and function of a stem cell promoter for the murine CBFalpha2 gene: Distinct roles and regulation in natural killer and T cell development. *Developmental Biology, 229*, 363–382.

Thambyrajah, R., Mazan, M., Patel, R., Moignard, V., Stefanska, M., Marinopoulou, E., et al. (2016). GFI1 proteins orchestrate the emergence of haematopoietic stem cells through recruitment of LSD1. *Nature Cell Biology, 18*, 21–32.

Tober, J., Koniski, A., Mcgrath, K. E., Vemishetti, R., Emerson, R., De Mesy-Bentley, K. K., et al. (2007). The megakaryocyte lineage originates from hemangioblast precursors and is an integral component both of primitive and of definitive hematopoiesis. *Blood, 109*, 1433–1441.

Tober, J., Yzaguirre, A. D., Piwarzyk, E., & Speck, N. A. (2013). Distinct temporal requirements for Runx1 in hematopoietic progenitors and stem cells. *Development, 140*, 3765–3776.

Tracey, W. D., Pepling, M. E., Horb, M. E., Thomsen, G. H., & Gergen, J. P. (1998). A Xenopus homologue of aml-1 reveals unexpected patterning mechanisms leading to the formation of embryonic blood. *Development, 125*, 1371–1380.

Ueno, H., & Weissman, I. L. (2006). Clonal analysis of mouse development reveals a polyclonal origin for yolk sac blood islands. *Developmental Cell, 11*, 519–533.

Wang, Q., Stacy, T., Binder, M., Marin-Padilla, M., Sharpe, A. H., & Speck, N. A. (1996a). Disruption of the Cbfa2 gene causes necrosis and hemorrhaging in the central nervous system and blocks definitive hematopoiesis. *Proceedings of the National Academy of Sciences of the United States of America, 93*, 3444–3449.

Wang, Q., Stacy, T., Miller, J. D., Lewis, A. F., Gu, T. L., Huang, X., et al. (1996b). The CBFbeta subunit is essential for CBFalpha2 (AML1) function in vivo. *Cell, 87*, 697–708.

Wareing, S., Eliades, A., Lacaud, G., & Kouskoff, V. (2012). ETV2 expression marks blood and endothelium precursors, including hemogenic endothelium, at the onset of blood development. *Developmental Dynamics, 241*, 1454–1464.

Witzenbichler, B., Maisonpierre, P. C., Jones, P., Yancopoulos, G. D., & Isner, J. M. (1998). Chemotactic properties of angiopoietin-1 and -2, ligands for the endothelial-specific receptor tyrosine kinase Tie2. *The Journal of Biological Chemistry, 273*, 18514–18521.

Xu, M. J., Matsuoka, S., Yang, F. C., Ebihara, Y., Manabe, A., Tanaka, R., et al. (2001). Evidence for the presence of murine primitive megakaryocytopoiesis in the early yolk sac. *Blood, 97*, 2016–2022.

Yoder, M. C. (2014). Inducing definitive hematopoiesis in a dish. *Nature Biotechnology, 32*, 539–541.

Yokomizo, T., & Dzierzak, E. (2010). Three-dimensional cartography of hematopoietic clusters in the vasculature of whole mouse embryos. *Development, 137*, 3651–3661.

Yokomizo, T., Ogawa, M., Osato, M., Kanno, T., Yoshida, H., Fujimoto, T., et al. (2001). Requirement of Runx1/AML1/PEBP2alphaB for the generation of haematopoietic cells from endothelial cells. *Genes to Cells, 6*, 13–23.

Yokomizo, T., Hasegawa, K., Ishitobi, H., Osato, M., Ema, M., Ito, Y., et al. (2008). Runx1 is involved in primitive erythropoiesis in the mouse. *Blood, 111*, 4075–4080.

Yoshimoto, M., Montecino-Rodriguez, E., Ferkowicz, M. J., Porayette, P., Shelley, W. C., Conway, S. J., et al. (2011).

Embryonic day 9 yolk sac and intra-embryonic hemogenic endothelium independently generate a B-1 and marginal zone progenitor lacking B-2 potential. *Proceedings of the National Academy of Sciences of the United States of America, 108*, 1468–1473.

Yoshimoto, M., Porayette, P., Glosson, N. L., Conway, S. J., Carlesso, N., Cardoso, A. A., et al. (2012). Autonomous murine T-cell progenitor production in the extra-embryonic yolk sac before HSC emergence. *Blood, 119*, 5706–5714.

Yu, M., Mazor, T., Huang, H., Huang, H. T., Kathrein, K. L., Woo, A. J., et al. (2012). Direct recruitment of polycomb repressive complex 1 to chromatin by core binding transcription factors. *Molecular Cell, 45*, 330–343.

Yzaguirre, A. D., & Speck, N. A. (2016). Insights into blood cell formation from hemogenic endothelium in lesser-known anatomic sites. *Developmental Dynamics, 245*, 1011–1028.

Zambidis, E. T., Peault, B., Park, T. S., Bunz, F., & Civin, C. I. (2005). Hematopoietic differentiation of human embryonic stem cells progresses through sequential hematoendothelial, primitive, and definitive stages resembling human yolk sac development. *Blood, 106*, 860–870.

Zeigler, B. M., Sugiyama, D., Chen, M., Guo, Y., Downs, K. M., & Speck, N. A. (2006). The allantois and chorion, when isolated before circulation or chorio-allantoic fusion, have hematopoietic potential. *Development, 133*, 4183–4192.

Zhen, F., Lan, Y., Yan, B., Zhang, W., & Wen, Z. (2013). Hemogenic endothelium specification and hematopoietic stem cell maintenance employ distinct Scl isoforms. *Development, 140*, 3977–3985.

Zovein, A. C., Hofmann, J. J., Lynch, M., French, W. J., Turlo, K. A., Yang, Y., et al. (2008). Fate tracing reveals the endothelial origin of hematopoietic stem cells. *Cell Stem Cell, 3*, 625–636.

Runx1 Structure and Function in Blood Cell Development

5

Constanze Bonifer, Elena Levantini, Valerie Kouskoff, and Georges Lacaud

Abstract

RUNX transcription factors belong to a highly conserved class of transcriptional regulators which play various roles in the development of the majority of metazoans. In this review we focus on the founding member of the family, RUNX1, and its role in the transcriptional control of blood cell development in mammals. We summarize data showing that RUNX1 functions both as activator and repressor within a chromatin environment, a feature that requires its interaction with multiple other transcription factors and co-factors. Furthermore, we outline how RUNX1 works together with other factors to reshape the epigenetic landscape and the three-dimensional structure of gene loci within the nucleus. Finally, we review how aberrant forms of RUNX1 deregulate blood cell development and cause hematopoietic malignancies.

Keywords

RUNX1 • Blood cell development • Isoforms • Regulation of RUNX1 activity • Chromatin structure • Transcriptional networks • Leukaemia

C. Bonifer (✉)
Institute for Cancer and Genomic Sciences, University of Birmingham, Birmingham, UK
e-mail: C.Bonifer@bham.ac.uk

E. Levantini
Beth Israel Diaconess Medical Center, Harvard Medical School, Boston, MA, USA

Istituto di Tecnologie Biomediche, Consiglio Nazionale delle Richerche, Pisa, Italy

V. Kouskoff
Division of Developmental Biology & Medicine, The University of Manchester, Manchester, UK

G. Lacaud
Cancer Research UK Manchester Institute, University of Manchester, Manchester, UK

5.1 The RUNX Family

RUNX genes have been found in the majority of metazoan sequenced genomes and encode α-subunits of heterodimeric transcription factors, involved in the control of proliferation and differentiation during development. There are single copies of *RUNX* genes in bilaterians, while insects and vertebrates have at least three such genes, which result from gene duplications (Rennert et al. 2003). The major functions of all three mammalian RUNX transcription factors were initially revealed using knockout mice.

Y. Groner et al. (eds.), *RUNX Proteins in Development and Cancer*, Advances in Experimental Medicine and Biology 962, DOI 10.1007/978-981-10-3233-2_5

RUNX1 was shown to be necessary for definitive hematopoiesis, with its knockout leading to midgestation lethality (Okuda et al. 1996; Wang et al. 1996). RUNX2, in contrast, is implicated in skeletal development (Ducy et al. 1997; Komori et al. 1997; Otto et al. 1997). *Runx2*$^{-/-}$ mice lack osteoblast differentiation and bone ossification and die soon after birth (Otto et al. 1997). Finally RUNX3 deficient mice suffer from limb ataxia, caused by defects in dorsal root ganglia development (Inoue et al. 2002; Levanon et al. 2002). However the phenotypes of these knockout mice probably reflect just one aspect of the function of these genes, as they probably are likely to have multiple more subtle roles in different tissues and at different stages of differentiation and development. The details of the knock-out phenotypes will be described elsewhere in this book, here we will concentrate on the structure and function of the founding members of the RUNX family, RUNX1.

5.2 Structure and Versions of the RUNX1 Protein

The most common and specific feature of RUNX proteins is a highly conserved (90% between the 3 human RUNX proteins), 128 amino acid long, DNA-binding domain, called the runt homology domain (RHD), located near the N-terminus. Apart from recognition and binding to specific DNA sequences, it contributes to the nuclear localization of RUNX transcription factors. This domain is also necessary for binding to CBF-β the β subunit of core transcription factors, which does not interact with DNA, but increases the affinity of the α subunit for DNA binding and the stability of the complex (Huang et al. 2001). The C-terminal part is less conserved, and contains an activation domain, an inhibitory domain, a region riche in proline (PY) and a nuclear matrix-targeting signal. In addition all *Runx* genes encode proteins with a five amino acid C-terminal motif (VWRPY in most cases), known as the recruitment signal for Groucho/TLE co-repressors (Coffman 2003). The observation that expression of *Runx2* or *Runx3* expression in *Runx1* deficient P-Sp cells, or substitutions of the C-terminal part of RUNX1 by RUNX2 or RUNX3, can rescue definitive hematopoiesis (Fukushima-Nakase et al. 2005; Goyama et al. 2004) suggest an important functional conservation between the domains in the C-terminal part of the different RUNX proteins.

Apart from high similarity in the domain organization, the three mammalian *Runx* genes also share conservation of the transcription from two different promoters, distal P1 and proximal P2. Both promoters contain RUNX-binding sites, thus the expression of a given *Runx* gene can be potentially regulated by its own protein product and by the two other RUNX transcription factors (Levanon and Groner 2004). The P2 proximal 5′ UTR is exceptionally long and weakly translated *in vitro*, while the distal 5′ UTR mediates efficient, cap-dependent translation (Bae et al. 1995; Pozner et al. 2000). The major isoforms of *Runx1* gene transcripts were determined more than two decades ago (Miyoshi et al. 1995). The proximal P2 promoter drives the expression of *Runx1a* and *Runx1b* whereas the distal P1 promoter generates *Runx1c*. The only difference between the RUNX1c and RUNX1b protein, generated by the full length transcripts, is their N-terminal amino acid sequences (Fujita et al. 2001; Miyoshi et al. 1995), but it remains unclear to what extent this small disparity might confer them with different functions. The third isoform *Runx1a* encodes a truncated version of *Runx1b*. RUNX1a potentially acts as an inhibitor of RUNX1b and RUNX1c (Tanaka et al. 1995). Forced expression of *Runx1a* has been shown to enhance both in human and mouse expansion of HSC *in vitro* and *in vivo* (Liu et al. 2009a; Tsuzuki et al. 2007; Tsuzuki and Seto 2012). In addition, RUNX1a has also been shown to enhance hematopoietic lineage commitment from human ES and iPS cells (Ran et al. 2013). Interestingly some somatic mutations of *RUNX1* found in leukaemia create proteins that resemble RUNX1a (Osato 2004). Various other *RUNX1* mRNAs, differing in size and giving rise to a variety of protein isoforms were described (Aziz-Aloya et al. 1998;

Bae et al. 1993; Levanon et al. 1996; Miyoshi et al. 1995; Telfer 2001) but their functional relevance remains largely unknown.

The two major *Runx1* isoforms are sequentially expressed during mouse and human ES cell *in vitro* to hematopoiesis, whereby the proximal P2 isoforms are being generated prior to the distal P1 isoforms (Challen and Goodell 2010; Fujita et al. 2001; Zambidis 2005). Similarly, during mouse embryonic hematopoiesis the proximal P2 derived *Runx1* mRNAs are found dominantly until E10.5, after which stage the *Runx1* transcripts in the fetal liver are mainly distal P1 promoter-derived (Bee et al. 2009). Interestingly, in zebrafish the two *Runx1* promoters appear to be active at different sites of hematopoietic development, as P1-EGFP and P2-EGFP transgenic lines express EGFP in distinct populations of blood cells (Lam et al. 2009). To investigate the activities of distal and proximal *Runx1* promoters at the single-cell level and to be able to isolate and evaluate the biological potential defined by their activities, mouse ES-cell lines with fluorescent proteins or truncated human *CD4* reporter genes knocked into the distal and proximal *Runx1* promoters, were generated (Sroczynska et al. 2009). Studies of these ES cells revealed that early expression of the proximal *Runx1* isoform is associated with a hemogenic cell population, whereas the subsequent onset of distal *Runx1* transcription coincides with the loss of endothelial phenotype and the appearance of definitive hematopoietic progenitors. The distal-GFP–positive cell population is highly enriched in definitive progenitors during ES-cell differentiation in vitro, as well as in the mouse embryos. Intriguingly, the activities of the two RUNX1 promoters switch during fetal liver hematopoiesis with the distal P1 promoter becoming predominant (Sroczynska et al. 2009). In adult mice, the overwhelming majority of *Runx1* mRNAs detected by PCR in hematopoietic cells initiate from the distal P1 promoter (Bee et al. 2009; Telfer 2001). Using a dual reporter fluorescent mouse, P1 was confirmed as the dominant promoter in adult hematopoiesis, being active in all *Runx1*-expressing populations (Draper et al. 2016). In contrast, proximal P2 activity is far more heterogeneous, confined to progenitor subsets of the granulocyte/macrophage (GM) and lymphoid lineages and megakaryocytes. With the exception of megakaryocytes, it appears that downregulation of proximal P2 transcription is a prerequisite of terminal differentiation of these lineages. At least in myeloid lineages, P2-expression correlates with enhanced CFU-C activity and in the Pre-Megakaryocyte/Erythroid progenitor (PreMegE) with increased proliferation (Draper et al. 2016).

The functional differences between the different isoforms generated by the distal and proximal promoters remain unclear. *In vitro* translated RUNX1b binds more weakly to DNA than the RUNX1c protein and transduction with the proximal isoform results in delay in neutrophilic differentiation of 32Dcl.3 cell line (Telfer 2001). Several studies have evaluated the respective requirements for distal or proximal derived transcripts during the onset of blood cell development when RUNX1 was shown to be absolutely critical (Chen et al. 2009; Lancrin et al. 2009). Specific P2 proximal morpholinos in zebrafish (Lam et al. 2009) or a mouse model hypomorphic for the activity of the P2 proximal promoter have established the critical importance for proximal transcripts in blood cell emergence *in vivo*. However this could reflect that only the P2 promoter is active at this stage (Bee et al. 2009; Sroczynska et al. 2009) rather than a specific biochemical activity of RUNX1b. Indeed both RUNX1b and RUNX1c have been shown to be able to rescue the block in blood development observed in the absence of RUNX1. In vivo deletions of distal RUNX1c during hematopoietic development in mouse or zebrafish result in more subtle blood defects (Bee et al. 2009; Lam et al. 2009; Pozner et al. 2007; Sroczynska et al. 2009) a finding consistent with the observation that RUNX1 is dispensable to some extent after the emergence of blood cells (Chen et al. 2009). In adult mice, it was recently reported that inactivation of the expression of P1 distal RUNX1 transcripts results in numerous lineage specific defects reminiscent of total RUNX1 deficiency

(Draper et al. 2016), a finding consistent with the prominence of distal P1 transcripts at this stage.

5.3 Regulation of RUNX1 Activity

RUNX proteins are not in themselves strong transcriptional regulators. The transcriptional activity of RUNX1 is instead modulated by interactions with other regulators of transcription (transcription factors or transcriptional co-activators or co-repressors) or by post-translational modifications that affect stability, activity or cellular localization. RUNX1 has been shown to interact with multiple other transcription factors and we will give only a few examples here. The V-Ets Avian Erythroblastosis Virus E26 Oncogene Homolog 1 (ETS-1) transcription factor binds to DNA cooperatively with RUNX1, interacting with its N-terminal part, and significantly increasing its DNA binding affinity (Gu et al. 2000; Kim et al. 1999; Wotton et al. 1994). In myeloid cells, RUNX1 interacts with CCAAT/enhancer binding protein alpha (C/EBPα) and PU.1 to activate macrophage colony-stimulating factor receptor (M-CSFR) expression (Zhang et al. 1998). In erythroid cells RUNX1 is part of a large complex together with LIM domain binding 1 (LDB1), GATA binding protein 1 (GATA1), T-cell acute lymphocytic leukemia (TAL1/SCL) and core-binding factor, runt domain, alpha subunit 2; translocated to 3 (CBFA2T3/ETO2) (Meier et al. 2006) or interacts with lysine-specific demethylase 1A (KDM1A/LSD1) and myelin basic protein expression factor 2 (MYEF2) (van Riel et al. 2012). In blood stem/progenitor cells, RUNX1 has been shown to bind to DNA in combination with SCL, lymphoblastic leukemia 1 (LYL1), LIM domain only 2 (LMO2), GATA2, ERG, and Friend leukemia integration 1 (FLI1) and at least RUNX1, GATA2, SCL, and ERG have been shown to have direct protein-protein interactions (Wilson et al. 2010). Interestingly, expression of RUNX1 at the onset of hematopoiesis in hemogenic endothelium initiates a rapid global alteration of the binding of SCL/TAL1 and FLI1 to induce the acquisition of hematopoietic fate (Lichtinger et al. 2012). Finally the transcriptional activity of RUNX1 can also be inhibited by binding with transcription factors, for example forkhead box P3 (FOXP3) to suppress interleukin 2 (IL2) and interferon gamma (IFN-γ) expression in T regulatory cells (Ono et al. 2007).

RUNX1 has been shown to interact with a growing number of chromatin modifying and remodeling proteins. RUNX1 interacts with the lysine acetyl transferase MOZ (MYST3) and this interaction leads to increased RUNX1 transcriptional activity (Bristow and Shore 2003; Kindle et al. 2005; Kitabayashi et al. 2001). Similarly RUNX1 interacts with the transcriptional co-activators P300 and CBP to stimulate RUNX1 dependent transcription in myeloid cells (Kitabayashi 1998; Yamaguchi et al. 2004). Furthermore, RUNX1 has been shown to physically and functionally interact with the histone methyltransferase MLL to potentiate H3K4 methylation at the PU.1 locus and its transcription (Huang et al. 2011). In addition, MLL and RUNX1 association protects RUNX1 from proteasome degradation (Huang et al. 2011), similarly to RUNX1 association with CBFβ (Huang et al. 2001). Finally RUNX1 has been shown to collaborate with members of the SWI/SNF chromatin remodeling complex, such as BRG1 and INI1, to control the expression of hematopoietic genes (Bakshi et al. 2010). RUNX1 also interacts with epigenetic factors to repress transcription. RUNX1 recruits HDACS by interacting with several co-repressors such as Groucho/TLE as well as Sin3A (Levanon et al. 1998; Lutterbach et al. 2000; Wheeler et al. 2000). Treatment with the histone deacetylase inhibitor, Trichostatin A, impairs repression by RUNX1 of transcription of *p21* suggesting that histone deacetylases contribute to RUNX1-mediated repression (Lutterbach et al. 2000). RUNX1 has also been shown to associate with the histone H3 lysine 9 methyltransferase SUV39H1 (KMT1A) and HDAC1

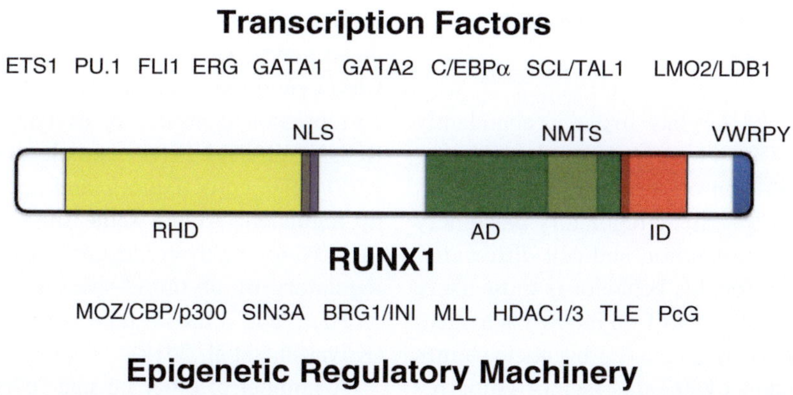

Fig. 5.1 **RUNX1 interacts with a large number of transcription factors and epigenetic modifiers.** The following functional domains of RUNX1 are indicated: runt homology domain (*RHD*), nuclear localisation signal (*NLS*), activation domain (*AD*), nuclear matrix targeting signal (*NMTS*), inhibitory signal (*ID*), and *VWRPY* sequence. Examples of hematopoietic transcription factors and chromatin modifier complexes interacting with RUNX1 are shown and include the histone acetylases (*MOZ, CBP, p300*), methylases (*MLL*) and deacetylases (*HDAC1 and 3*), the TLE co-repressors, chromatin remodelers such as SWI/SNF (*BRG INI*) and Polycomb group complexes (*PcG*)

and HDAC3 to repress transcription (Reed-Inderbitzin et al. 2006). Finally, the RUNX1/CBFβ complex was recently shown to directly recruit the Polycomb repressive complex 1 (PRC1) in megakaryocytic and lymphocytic cells (Ross et al. 2012; Yu et al. 2012) (Fig. 5.1).

As many other proteins, RUNX1 activity is also regulated by multiple post-translational modifications such as phosphorylation, methylation, acetylation and ubiquitination that reflect cell signaling and growth status. Phosphorylation of several serine and threonine residues was shown to regulate RUNX1 degradation, nuclear matrix association and transcriptional activity (Biggs et al. 2005). Activation of the extra cellular signal-regulated kinase (ERK) pathway lead to serine phosphorylation that results in increased RUNX1 transactivation potential (Tanaka et al. 1996). The ERK-mediated phosphorylation was also demonstrated to abolish RUNX1 interaction with Sin3A (Imai et al. 2004), potentiating its transcriptional activity but also targeting it for proteasome degradation through ubiquitination, creating a negative feedback loop (Biggs et al. 2006). Similarly, RUNX1 is also a substrate of cdk1/2/6 and phosphorylation of RUNX1 decreases interactions with HDAC, increases RUNX1 transcriptional activity but decreases its stability (Guo and Friedman 2011; Zhang et al. 2008). Finally, homeodomain-interacting protein kinase HIPK2 phosphorylates RUNX1 to promote activation of P300 and up-regulation of target gene transcription (Aikawa et al. 2006; Leong et al. 2016; Wee et al. 2008). As discussed above, RUNX1 can interact with lysine acetyl transferases to increase local histone acetylation and up regulate transcription. In addition, CBP, MOZ and P300 directly acetylate RUNX1 (Kitabayashi et al. 2001) and P300 acetylation of RUNX1 significantly increases its affinity for DNA (Yamaguchi et al. 2004). In contrast the HAT activity of MOZ seems to be mostly dispensable for its stimulation of RUNX1-dependent transcription (Kitabayashi et al. 2001). The transcriptional activity of RUNX1 is also enhanced by protein arginine methyltransferase 1 (PRMT1)–mediated methylation that abrogates its association with SIN3A (Zhao et al. 2008). Finally several E3-ubiquitin ligases were shown to participate in RUNX1 degradation. APC and the Skp1/cullin/F-box protein (SCF) complex degrade RUNX1 following its phosphorylation by cyclin dependent kinases (Biggs et al. 2006). Similarly the E3 ubiquitin ligase STUB1 (also called CHIP) has been shown to degrade RUNX1 (Shang et al. 2009).

5.4 Chromatin Programming by RUNX1

The previous chapter has made it abundantly clear that RUNX1 interacts with a bewildering number of other transcription factors as well as members of the epigenetic regulatory machinery to regulate gene expression and cell differentiation. The reason for this behavior is most likely due to the fact that RUNX1 on its own is incapable of binding to sites covered by nucleosomes and needs to interact with other factors (Gutierrez et al. 2000). This is exemplified by studies demonstrating that both at the T cell receptor alpha (*TCRa*) enhancer and the GM-CSF (*Csf2*) enhancer, RUNX1 was unable to bind to chromatin on its own, but required the cooperation with other factors to stably bind to DNA within chromatin (Bowers et al. 2010; Hernandez-Munain et al. 1998).

RUNX1 is capable of both activating and silencing gene expression depending on the genomic context, which, in turn dictates whether co-activators or co-repressors are recruited (Collins et al. 2009; Durst and Hiebert 2004). A good example of this dichotomy is the differential function of RUNX1 in the differentiation of T cells into $CD4^+$ and $CD8^+$ cells. CD8 development requires the repression of the CD4 locus which is mediated by a silencer element and silencing is initiated by the binding of RUNX1 with this element at early (double-negative) T cell developmental stages cells (Taniuchi et al. 2002) and which involves the DNA-methylation machinery (Sellars et al. 2015). However, RUNX1 is also required to activate the expression of $CD8^+$ gene locus by binding to crucial enhancer elements and recruiting co-activators, again indicating that it is the context in which RUNX1 acts that is decisive for how it influences transcriptional outcome. Another good example of how RUNX1 is able to differentially program chromatin is seen at the branch point of megakaryocyte versus erythroid differentiation. In human progenitor cells, RUNX1 recruits the arginine methyltransferase PRMT6 as well as HDACs to install inactive chromatin marks at megakaryocyte-specific RUNX1 target genes such as *CD41* and repress the expression of these genes (Herglotz et al. 2013). Once megakaryocyte differentiation is initiated, the RUNX1 co-repressor complex is exchanged against a RUNX1/GATA1/FOG complex that recruits p300/CBP co-activators and these genes are up-regulated. At the same time, RUNX1 uses PRMT6 to inactivate the chromatin of important regulators of erythroid-specific genes such as *KLF1*, and thus represses erythropoiesis (Kuvardina et al. 2015).

A number of other lineage-specific transcription factors, such as PAX5 or PU.1 (Nutt et al. 1999; Stopka et al. 2005) have been shown to be involved in both activating gene expression in one lineage, but at the same time repressing genes of alternate lineages. RUNX1 is widely expressed within the hematopoietic system and is thus unique in its ability to cooperate with multiple lineage-specific transcriptional regulators to help driving differential gene expression in multiple lineages. This feature is reflected in the phenotype of mice carrying a conditional *Runx1* allele and where the gene is ablated at the hematopoietic stem cell stage. Ablation does not dramatically impact on stem cell maintenance, but such mice display defects in multiple lineages whereby cell differentiation as such still takes place (Growney et al. 2005; Ichikawa et al. 2004; Putz et al. 2006). However, there is one stage where RUNX1 activity is crucially required: at the stage of stem cell emergence in the early embryo. RUNX1 is expressed from the hemangioblast stage onwards (Lacaud et al. 2002), first at a low level, followed by an up-regulation of expression, which is required to drive the endothelial hematopoietic transition (Chen et al. 2009; Lancrin et al. 2009). At this stage, the removal of RUNX1 leads to a complete absence of blood cell development. Moreover, in the germ-line knockout endothelial development was affected as well, and mice displayed extensive hemorrhaging (Okuda et al. 1996; Wang et al. 1996). The question now arises of the chromatin mechanisms that make the presence of RUNX1 essential for this particular developmental stage. This question was addressed by both single gene studies and global analyses. One study (Hoogenkamp et al.

2009) examined the establishment of an active chromatin structure at the *Spi1* locus by analyzing chromatin accessibility of different cell types representing different stages of hematopoietic specification. *Spi1* is the gene for the transcription factor PU.1 which is a target of RUNX1 (Huang et al. 2008) and which is essential for the development of myeloid cells (Scott et al. 1994). Using the differentiation of ES cells into hematopoietic cells as model, the study found that in the presence of RUNX1 chromatin of this gene was starting to become accessible already at the hemangioblast stage, which was accompanied by a selective demethylation of DNA. Employing an inducible version of RUNX1 in mouse embryonic stem cells with a *Runx1−/−* background (iRUNX1) (Lancrin et al. 2009) followed by *in vitro* differentiation they showed that the increase in chromatin accessibility was dependent on the presence of RUNX1. Thus suggested that even when expressed at very low levels and without being able to form stable complex, RUNX1 was capable of transiently interacting with its target and priming the gene for activation. That RUNX1 was truly capable of transiently binding to DNA was confirmed by experiments which employed a RUNX1-Dam-methylase fusion protein to measure the deposition of methyl groups in the vicinity of RUNX1 binding sites in the absence of the formation of a stable transcription factor complex (Hoogenkamp et al. 2009). Moreover, genome-wide studies using DamID in a wild-type genomic background showed that RUNX1 interaction with its targets at this early developmental stage was wide-spread, and more importantly, also involved endothelial genes (Lie-A-Ling et al. 2014), providing a possible molecular explanation for the defects in endothelial development seen in the knock-out mouse. Another genome-wide study investigated the mode of action of RUNX1 during the endothelial-hematopoietic transition (Lichtinger et al. 2012), again using the iRUNX1 system. At the same time, binding of other transcription factor (TAL1 and FLI1) was measured in the presence or absence of RUNX1. The experiments showed that (i) that the binding of TAL1/FLI1 did not determine where RUNX1 was binding, but that, once induced RUNX1 pulled in the other factors to new binding sites and (ii) this reorganization of the epigenetic landscape led to a wide-spread increase in histone acetylation at its binding sites. The alteration of the genomic neighborhood of TAL1 and FLI1 was also seen in genome-wide studies examining uninterrupted hematopoietic differentiation (Goode et al. 2016). Megakaryocyte differentiation is another example for the ability of RUNX1 to work with different partners to orchestrate transcription factor complex alterations in development. An interesting aspect of this work is the finding that RUNX1 cooperates with signaling responsive transcription factors such as AP-1 to open chromatin and establish stable transcription factor complexes (Pencovich et al. 2011). In summary, these experiments show that RUNX1 plays an important role of integrating differentiation and signaling dependent transcriptional responses by cooperating with multiple partners.

5.5 RUNX1 Function in the Context of Higher Order Chromatin Structure

Transcriptional activity occurs in discrete subnuclear foci and takes place within the 3-dimensional architecture of the nucleus. Recent studies have shown that genes are organized in topologically confined domains (TAD for topology associating domains) that determine the regulatory environment of genes (Dekker and Heard 2015). It is emerging that selective association of RUNX proteins to specific nuclear structures is another mechanism utilized to organize functional regulatory complexes within the nucleus. Throughout the interphase RUNX proteins are organized as transcriptionally active discrete subnuclear foci (Harrington et al. 2002; Zaidi et al. 2002, 2003). Transcriptionally active RUNX proteins associate with the nuclear matrix, whereas inactive C-terminally truncated RUNX proteins do not (Javed et al. 2000; Zaidi et al. 2001; Zeng et al. 1998). Further analysis of the molecular requirements for this interaction revealed that the association with the matrix is independent of DNA binding and requires a nuclear matrix targeting signal (NMTS), a 31-amino-acid segment near the C-terminus, that is distinct from nuclear localization signals (NLS) (Zaidi et al. 2001; Zeng et al. 1997). The biological importance of RUNX intranuclear distribution was demonstrated *in vivo* using mouse models expressing a RUNX protein with impaired subnuclear targeting (Dowdy et al. 2010). These mice show similar phenotypes to *Runx*-null mice, indicating the assembly and organization of RUNX-containing macromolecular complexes at subnuclear sites is intimately linked to RUNX biological activity. It is highly likely that such RUNX1 mutations disrupt crucial protein-protein interactions, however, the molecular nature of such interactions is currently unclear.

During mitosis the nuclear envelope is disassembled, chromatin condenses to form distinct chromosomes, and a large number of transcription factors are displaced from their target promoters or are degraded during mitotic gene silencing (Pande et al. 2009). In contrast, RUNX1 remains associated with the nucleolar organizing regions of acrocentric chromosomes (Bakshi et al. 2008). RUNX-foci persist throughout mitosis and undergo a spatio-temporal redistribution that results in equal partitioning of the protein into each of the progeny nuclei (Zaidi et al. 2003). A similar observation was made with the partner of RUNX1, CBFβ (Lopez-Camacho et al. 2014). Equal partitioning and a complete restoration of subnuclear organization of the foci during telophase could be a mechanism for the maintenance of epigenetic memory since it facilitates re-assembly of regulatory complexes after mitosis. Such "mitotic bookmarking" has now been described for several other transcription factors and chromatin associated proteins (Zaret 2014).

Understanding how transcription factors and *cis*-regulatory elements set up long-range interactions to orchestrate gene expression is a key issue in genomic biology. Since the initial observation reporting chromatin looping at the β-globin gene locus (Carter et al. 2002; Tolhuis et al. 2002), similar interactions have been shown to occur between promoters and 5′ and/or 3'UTRs of many genes, and even *trans*-interactions between loci on different chromosomes have been reported (Barnett et al. 2008; Boney-Montoya et al. 2010; Brown et al. 2002; Chavanas et al. 2008; Chen et al. 1998; Ling et al. 2006; Liu et al. 2009b; Marenduzzo et al. 2007; Theo Sijtse Palstra 2009; Vernimmen et al. 2007). A model in which intervening inactive chromatin lying between distant elements is looped out, and forms an active chromatin hub has been proposed (Patrinos et al. 2004; Theo Sijtse Palstra 2009). Structural proteins, transcription factors, or components of the pre-initiation complex have been implicated as candidate molecules mediating the chromatin looping (Deshane et al. 2010; Kim et al. 2007, 2009; Liu et al. 2009b; Marenduzzo et al. 2007; Theo Sijtse Palstra 2009; Williams et al. 2007).

A potential role of RUNX1 in mediating enhancer – promoter interactions in hematopoietic stem/progenitor cells was demonstrated in studies on the regulation of the human *CD34* and *Sp1(Pu.1)* gene loci. This work mapped the critical regulatory elements required for *hCD34* expression in LT-HSCs by generating different transgenic mouse lines carrying various combinations of *hCD34* genomic elements and identified a novel regulatory element located at +19 kb, the Downstream Regulatory Element (DRE), which was necessary for *hCD34* expression in LT-HSCs (Long term-HSCs) (Levantini et al. 2011). The DRE contains four binding sites for RUNX together with other stem-cell specific factor binding sites. Experiments with conditional RUNX1 knockout mice demonstrated that the presence of RUNX1 was essential for the activity of the DRE. The DRE physically interacted with the *hCD34* promoter through the RUNX1 binding sites as shown by chromosome conformation capture (3C) analysis. The frequency of promoter and DRE interaction was strongly decreased after mutating the RUNX binding sites, therefore providing *in vivo* evidence that these proteins are required to directly mediate chromatin looping between long-range regulatory elements in HSCs. At the *Spi1 (Pu.1)* locus it was shown that *PU.1* promoter interacts with a critical Upstream Regulatory Element (URE) (Ebralidze et al. 2008). Using a knock-in mouse model in which binding of all RUNX factors at the -14 kb upstream enhancer of PU.1 was abolished by mutating all four RUNX1 binding sites decreased interaction between the *PU.1* promoter and the URE in primary hematopoietic stem/progenitor cells (Staber et al. 2014). Outside of the stem cell compartment, RUNX proteins regulate the intranuclear positioning of the CD4 and CD8 loci (Collins et al. 2011), increasing the body of evidence showing that RUNX1 may be required for its ability to loop promoter-distal regulatory elements towards nuclear "transcription factories" (Jackson et al. 1993; Osborne et al. 2004), where genes are dynamically recruited during activation of transcription.

5.6 De-regulation of Transcriptional Networks by Aberrant Forms of RUNX1

Many of the transcription factors that we know today to be important for blood cell development and differentiation have also been described as leukaemic oncogenes. This is also true for RUNX1. Its previous name was AML1, which stands for Acute Myeloid Leukaemia, as it was one of the first transcription factor genes known to be mutated in human AML (Miyoshi et al. 1991). *RUNX1* is a frequent partner in chromosomal translocations, but a variety of disease related point mutations are also found which generally fall into two classes: one with point mutations in the DNA binding domain and one which only carry the DNA binding domain due to a nonsense mutation behind the RUNT domain (Michaud et al. 2002; Osato 2004). All of these versions of RUNX1 have in common that they either affect DNA binding of RUNX1 directly, or in one way or another disrupt or change the multitude of interactions with other factors that are required for undisturbed hematopoietic differentiation (Lam and Zhang 2012). This is reflected in the finding that the different types of mutations all cause diseases with different phenotypes where by translocations cause leukaemia, and the majority of point mutations tend to cause thrombocytopenia and platelet disorders that can later develop into AML (Song et al. 1999). These different phenotypes are also seen in mouse models (Matheny et al. 2007). However, another commonality of these different phenotypes is the fact that the majority of them are causes by heterozygous RUNX1 mutations. In fact, at least for the cells carrying two types of RUNX1 / Core binding factor translocations (t(8;21) and Inv(16), recent studies showed that they require a wild-type *RUNX1* copy to survive (Ben-Ami et al. 2013; Goyama et al. 2013). The study from the Groner lab also demonstrated that RUNX1 and RUNX1-ETO regulate different sets of genes within the same cells, which was underpinned by the finding that RUNX1 and RUNX1/ETO bind to overlapping, but distinct sites within the genome as shown by sequential ChIP and immunohistochemistry (Bakshi et al. 2008; Ptasinska et al. 2014). Currently it is unclear whether RUNX1 and RUNX1-ETO bind in a dynamic equilibrium to the same sites or whether binding of RUNX1 and RUNX1-ETO at different alleles is epigenetically stable.

In normal cells RUNX1 binds to its targets together with other transcription factors such as LMO2, TAL1, LYL1, FLI1 or ERG (Goode et al. 2016; Wilson et al. 2010). In the t(8;21) the RUNT domain of RUNX1 is fused to the transcriptional repressor ETO (MTG8) (Bae et al. 1993). Also RUNX1-ETO co-localizes with these factors in chromatin, and forms a stable complex with LMO2/LDB1 and the E-Box factor HEB, and altogether exists as a tetrameric complex that recruits the co-repressor SMRT/N-CoR both in vitro and in chromatin (Liu et al. 2007; Ptasinska et al. 2014; Reed-Inderbitzin et al. 2006; Sun et al. 2013; Trombly et al. 2015). The tetramer has a preference for binding to multiple RUNX binding motifs (Okumura et al. 2008). Importantly, these multiple interactions are required for the transforming activity of RUNX1-ETO (Liu et al. 2007; Martens et al. 2012; Sun et al. 2013). Depletion of RUNX1-ETO in t(8;21) cells using a knock-down approach leads to changes in binding of the transcription factors of the complex as well as in complex changes in gene expression (Ptasinska et al. 2012, 2014). Both up-regulation as well as down-regulation of gene expression was observed, whereby genes specific for stem cells (such as *CD34* or *ERG*), proliferation and cell cycle were down-regulated. Genes characteristic for myeloid differentiation were up-regulated, whereby direct target genes enriched for binding sites of the whole RUNX1-ETO complex of predominantly in the latter category. A common theme in hematopoiesis is the enforcement of commitment by lineage-determining transcription factors which have the dual function of activating genes of one lineage and repressing lineage inappropriate genes (Nutt et al. 1999; van Oevelen et al. 2015). Knock-down of RUNX1-ETO led to an increase in binding of RUNX1 at previous RUNX1-ETO binding sites (Ptasinska et al. 2014) together with a global increase in binding for the transcription factor C/EBPα which is essential for myeloid differentiation (Zhang et al. 1997). C/EBPα binding was also observed on RUNX1-ETO target genes that were down-regulated. This finding confirms previous results demonstrating that C/EBPα can act as a repressor (Zhang et al. 2013) and explains why in its absence lineage infidelity of gene expression is observed (Paul et al. 2015). An interesting finding was that the majority of changes in transcription factor binding occurred at pre-existing accessible chromatin sites, suggesting that RUNX1-ETO targets RUNX1 binding sites regions destined for differentiation-dependent factor exchange. The opposite was observed when an inducible version of RUNX1-ETO was expressed in murine progenitors (Regha et al. 2015). Induction led to a decrease in RUNX1 binding, an immediate down-regulation of myeloid regulators (*Cebpa, Cebpe, Spi (Pu.1, Irf8)* and an up-regulation of stem cell genes, including those encoding transcriptional regulators such as ERG. The precise molecular mechanism of how the expression of myeloid regulators feeds back on stem cell regulators is currently unclear. However it is clear that their direct binding is involved, as this has also been shown for PU.1 which associated with C/EBPα on stem cells genes in mature myeloid cells (Goode et al. 2016). In summary, these experiments demonstrate that RUNX1-ETO interferes with the function of RUNX1 as it hijacks the protein complexes it interacts with and represses *CEBPA* expression thus taking out one of the major drivers of myeloid differentiation. RUNX1-ETO on its own is unable to cause AML (Yergeau et al. 1997), however, cell differentiation is slowed down and cells are trapped in a cycle of self-renewal that awaits a second hit to become a full-blown leukemia. It is likely that a similar interference with normal RUNX1 function will also operate in other core binding factor leukemias such as Inv(16) or the t(12;21).

5.7 Future Directions

The studies described above picture RUNX1 as a transcription factor that is capable of integrating multiple inputs in form of interaction partners to reshape the epigenetic landscape within the three dimensional space of the nucleus. Importantly, many of these factors are either tissue-specifically expressed or (like AP-1) respond to signals and developmental cues. Currently the biochemical basis of this interactivity is unclear. We also do not understand, how RUNX1 directs cis-regulatory element interactions, i.e. whether it interacts with itself or with other factors at different cis-elements and which domains of the protein are involved. Such biochemical experiments will inform *in vivo* studies that should give us unprecedented insights into the function of one of the best-studied gene regulatory proteins and will serve as a model for many other lineage determining transcription factors. Last, but not least, the impact of the different disease-related mutations of *RUNX1* on the programming of chromatin and transcription at early stages of leukemogenesis prior to *in vivo* selection in mouse models of leukemia and in patients is poorly understood. The establishment of tractable cell line models carrying inducible versions of leukemic oncogenes offers a way by which such events can be studied at the systems level which will pave the way to identify signals and pathways that can be exploited to therapeutically target pre-leukemic cells (Goyama et al. 2016; Regha et al. 2015). Provided that scientists are still allowed to dig deep into the molecular mechanisms of gene regulation, interesting times lie ahead.

Acknowledgements Work in Constanze Bonifer's lab is supported by grants from Bloodwise, the BBSRC and the Kay Kendall Leukemia fund. Georges Lacaud and Valerie Kouskoff are funded by the BBSRC, MRC, Bloodwise and Cancer Research UK.

References

Aikawa, Y., Nguyen, L. A., Isono, K., Takakura, N., Tagata, Y., Schmitz, M. L., et al. (2006). Roles of HIPK1 and HIPK2 in AML1- and p300-dependent transcription, hematopoiesis and blood vessel formation. *The EMBO Journal, 25*, 3955–3965.

Aziz-Aloya, R. B., Levanon, D., Karn, H., Kidron, D., Goldenberg, D., Lotem, J., et al. (1998). Expression of AML1-d, a short human AML1 isoform, in embryonic stem cells suppresses in vivo tumor growth and differentiation. *Cell Death and Differentiation, 5*, 765–773.

Bae, S. C., Yamaguchi-Iwai, Y., Ogawa, E., Maruyama, M., Inuzuka, M., Kagoshima, H., et al. (1993). Isolation of PEBP2 alpha B cDNA representing the mouse homolog of human acute myeloid leukemia gene, AML1. *Oncogene, 8*, 809–814.

Bae, S. C., Takahashi, E., Zhang, Y. W., Ogawa, E., Shigesada, K., Namba, Y., et al. (1995). Cloning, mapping and expression of PEBP2 alpha C, a third gene encoding the mammalian Runt domain. *Gene, 159*, 245–248.

Bakshi, R., Zaidi, S. K., Pande, S., Hassan, M. Q., Young, D. W., Montecino, M., et al. (2008). The leukemogenic t(8;21) fusion protein AML1-ETO controls rRNA genes and associates with nucleolar-organizing regions at mitotic chromosomes. *Journal of Cell Science, 121*, 3981–3990.

Bakshi, R., Hassan, M. Q., Pratap, J., Lian, J. B., Montecino, M. A., van Wijnen, A. J., et al. (2010). The human SWI/SNF complex associates with RUNX1 to control transcription of hematopoietic target genes. *Journal of Cellular Physiology, 225*, 569–576.

Barnett, D. H., Sheng, S., Charn, T. H., Waheed, A., Sly, W. S., Lin, C. Y., et al. (2008). Estrogen receptor regulation of carbonic anhydrase XII through a distal enhancer in breast cancer. *Cancer Research, 68*, 3505–3515.

Bee, T., Liddiard, K., Swiers, G., Bickley, S. R. B., Vink, C. S., Jarratt, A., et al. (2009). Alternative Runx1 promoter usage in mouse developmental hematopoiesis. *Blood Cells, Molecules, and Diseases, 43*, 35–42.

Ben-Ami, O., Friedman, D., Leshkowitz, D., Goldenberg, D., Orlovsky, K., Pencovich, N., et al. (2013). Addiction of t(8;21) and inv(16) acute myeloid leukemia to native RUNX1. *Cell Reports, 4*, 1131–1143.

Biggs, J. R., Zhang, Y., Peterson, L. F., Garcia, M., Zhang, D. E., & Kraft, A. S. (2005). Phosphorylation of AML1/RUNX1 regulates its degradation and nuclear matrix association. *Molecular Cancer Research : MCR, 3*, 391–401.

Biggs, J. R., Peterson, L. F., Zhang, Y., Kraft, A. S., & Zhang, D. E. (2006). AML1/RUNX1 phosphorylation by cyclin-dependent kinases regulates the degradation of AML1/RUNX1 by the anaphase-promoting complex. *Molecular and Cellular Biology, 26*, 7420–7429.

Boney-Montoya, J., Ziegler, Y. S., Curtis, C. D., Montoya, J. A., & Nardulli, A. M. (2010). Long-range transcriptional control of progesterone receptor gene expression. *Molecular Endocrinology (Baltimore, MD), 24*, 346–358.

Bowers, S. R., Calero-Nieto, F. J., Valeaux, S., Fernandez-Fuentes, N., & Cockerill, P. N. (2010). Runx1 binds as a dimeric complex to overlapping Runx1 sites within a palindromic element in the human GM-CSF enhancer. *Nucleic Acids Research, 38*, 6124–6134.

Bristow, C. A. P., & Shore, P. (2003). Transcriptional regulation of the human MIP-1alpha promoter by RUNX1 and MOZ. *Nucleic Acids Research, 31*, 2735–2744.

Brown, R. C., Pattison, S., van Ree, J., Coghill, E., Perkins, A., Jane, S. M., & Cunningham, J. M. (2002). Distinct domains of erythroid Kruppel-like factor modulate chromatin remodeling and transactivation at the endogenous beta-globin gene promoter. *Molecular and Cellular Biology, 22*, 161–170.

Carter, D., Chakalova, L., Osborne, C. S., Dai, Y. F., & Fraser, P. (2002). Long-range chromatin regulatory interactions in vivo. *Nature Genetics, 32*, 623–626.

Challen, G. A., & Goodell, M. A. (2010). Runx1 isoforms show differential expression patterns during hematopoietic development but have similar functional effects in adult hematopoietic stem cells. *Experimental Hematology, 38*, 403–416.

Chavanas, S., Adoue, V., Mechin, M. C., Ying, S., Dong, S., Duplan, H., et al. (2008). Long-range enhancer associated with chromatin looping allows AP-1 regulation of the peptidylarginine deiminase 3 gene in differentiated keratinocyte. *PloS One, 3*, e3408.

Chen, X., Reitman, M., & Bieker, J. J. (1998). Chromatin structure and transcriptional control elements of the erythroid Kruppel-like factor (EKLF) gene. *The Journal of Biological Chemistry, 273*, 25031–25040.

Chen, M. J., Yokomizo, T., Zeigler, B. M., Dzierzak, E. A., & Speck, N. A. (2009). Runx1 is required for the endothelial to haematopoietic cell transition but not thereafter. *Nature, 457*, 887–891.

Coffman, J. A. (2003). Runx transcription factors and the developmental balance between cell proliferation and differentiation. *Cell Biology International, 27*, 315–324.

Collins, A., Littman, D. R., & Taniuchi, I. (2009). RUNX proteins in transcription factor networks that regulate T-cell lineage choice. *Nature Reviews Immunology, 9*, 106–115.

Collins, A., Hewitt, S. L., Chaumeil, J., Sellars, M., Micsinai, M., Allinne, J., et al. (2011). RUNX transcription factor-mediated association of Cd4 and Cd8 enables coordinate gene regulation. *Immunity, 34*, 303–314.

Dekker, J., & Heard, E. (2015). Structural and functional diversity of Topologically Associating Domains. *FEBS Letters, 589*, 2877–2884.

Deshane, J., Kim, J., Bolisetty, S., Hock, T. D., Hill-Kapturczak, N., & Agarwal, A. (2010). Sp1 regulates chromatin looping between an intronic enhancer and distal promoter of the human heme oxygenase-1 gene in renal cells. *The Journal of Biological Chemistry, 285*, 16476–16486.

Dowdy, C. R., Xie, R., Frederick, D., Hussain, S., Zaidi, S. K., Vradii, D., et al. (2010). Definitive hematopoiesis requires Runx1 C-terminal-mediated subnuclear targeting and transactivation. *Human Molecular Genetics, 19*, 1048–1057.

Draper, J. E., Sroczynska, P., Tsoulaki, O., Leong, H. S., Fadlullah, M. Z. H., Miller, C., et al. (2016). RUNX1B expression is highly heterogeneous and distinguishes megakaryocytic and erythroid lineage fate in adult mouse hematopoiesis. *PLoS Genetics, 12*, e1005814.

Ducy, P., Zhang, R., Geoffroy, V., Ridall, A. L., & Karsenty, G. (1997). Osf2/Cbfa1: A transcriptional activator of osteoblast differentiation. *Cell, 89*, 747–754.

Durst, K. L., & Hiebert, S. W. (2004). Role of RUNX family members in transcriptional repression and gene silencing. *Oncogene, 23*, 4220–4224.

Ebralidze, A. K., Guibal, F. C., Steidl, U., Zhang, P., Lee, S., Bartholdy, B., et al. (2008). PU.1 expression is modulated by the balance of functional sense and antisense RNAs regulated by a shared cis-regulatory element. *Genes & Development, 22*, 2085–2092.

Fujita, Y., Nishimura, M., Taniwaki, M., Abe, T., & Okuda, T. (2001). Identification of an alternatively spliced form of the mouse AML1/RUNX1 gene transcript AML1c and its expression in early hematopoietic development. *Biochemical and Biophysical Research Communications, 281*, 1248–1255.

Fukushima-Nakase, Y., Naoe, Y., Taniuchi, I., Hosoi, H., Sugimoto, T., & Okuda, T. (2005). Shared and distinct roles mediated through C-terminal subdomains of acute myeloid leukemia/Runt-related transcription factor molecules in murine development. *Blood, 105*, 4298–4307.

Goode, D. K., Obier, N., Vijayabaskar, M. S., Lie-a-Ling, M., Lilly, A. J., Hannah, R., et al. (2016). Dynamic gene regulatory networks drive hematopoietic specification and differentiation. *Developmental Cell, 36*, 572–587.

Goyama, S., Yamaguchi, Y., Imai, Y., Kawazu, M., Nakagawa, M., Asai, T., et al. (2004). The transcriptionally active form of AML1 is required for hematopoietic rescue of the AML1-deficient embryonic para-aortic splanchnopleural (P-Sp) region. *Blood, 104*, 3558–3564.

Goyama, S., Schibler, J., Cunningham, L., Zhang, Y., Rao, Y., Nishimoto, N., et al. (2013). Transcription factor RUNX1 promotes survival of acute myeloid leukemia cells. *The Journal of Clinical Investigation, 123*, 3876–3888.

Goyama, S., Schibler, J., Gasilina, A., Shrestha, M., Lin, S., Link, K. A., et al. (2016). UBASH3B/Sts-1-CBL axis regulates myeloid proliferation in human preleukemia induced by AML1-ETO. *Leukemia, 30*, 728–739.

Growney, J. D., Shigematsu, H., Li, Z., Lee, B. H., Adelsperger, J., Rowan, R., et al. (2005). Loss of Runx1 perturbs adult hematopoiesis and is associated with a myeloproliferative phenotype. *Blood, 106*, 494–504.

Gu, T. L., Goetz, T. L., Graves, B. J., & Speck, N. A. (2000). Auto-inhibition and partner proteins, core-binding factor beta (CBFbeta) and Ets-1, modulate DNA binding by CBFalpha2 (AML1). *Molecular and Cellular Biology, 20*, 91–103.

Guo, H., & Friedman, A. D. (2011). Phosphorylation of RUNX1 by cyclin-dependent kinase reduces direct interaction with HDAC1 and HDAC3. *Journal of Biological Chemistry, 286*, 208–215.

Gutierrez, J., Sierra, J., Medina, R., Puchi, M., Imschenetzky, M., van Wijnen, A., et al. (2000). Interaction of CBF alpha/AML/PEBP2 alpha transcription factors with nucleosomes containing promoter sequences requires flexibility in the translational positioning of the histone octamer and exposure of the CBF alpha site. *Biochemistry, 39*, 13565–13574.

Harrington, K. S., Javed, A., Drissi, H., McNeil, S., Lian, J. B., Stein, J. L., et al. (2002). Transcription factors RUNX1/AML1 and RUNX2/Cbfa1 dynamically associate with stationary subnuclear domains. *Journal of Cell Science, 115*, 4167–4176.

Herglotz, J., Kuvardina, O. N., Kolodziej, S., Kumar, A., Hussong, H., Grez, M., & Lausen, J. (2013). Histone arginine methylation keeps RUNX1 target genes in an intermediate state. *Oncogene, 32*, 2565–2575.

Hernandez-Munain, C., Roberts, J. L., & Krangel, M. S. (1998). Cooperation among multiple transcription factors is required for access to minimal T-cell receptor alpha-enhancer chromatin in vivo. *Molecular and Cellular Biology, 18*, 3223–3233.

Hoogenkamp, M., Lichtinger, M., Krysinska, H., Lancrin, C., Clarke, D., Williamson, A., et al. (2009). Early chromatin unfolding by RUNX1: A molecular explanation for differential requirements during specification versus maintenance of the hematopoietic gene expression program. *Blood, 114*, 299–309.

Huang, G., Shigesada, K., Ito, K., Wee, H.-J., Yokomizo, T., & Ito, Y. (2001). Dimerization with PEBP2beta protects RUNX1/AML1 from ubiquitin-proteasome-mediated degradation. *The EMBO Journal, 20*, 723–733.

Huang, G., Zhang, P., Hirai, H., Elf, S., Yan, X., Chen, Z., et al. (2008). PU.1 is a major downstream target of

AML1 (RUNX1) in adult mouse hematopoiesis. *Nature Genetics, 40*, 51–60.

Huang, G., Zhao, X., Wang, L., Elf, S., Xu, H., Zhao, X., et al. (2011). The ability of MLL to bind RUNX1 and methylate H3K4 at PU.1 regulatory regions is impaired by MDS/AML-associated RUNX1/AML1 mutations. *Blood, 118*, 6544–6552.

Ichikawa, M., Asai, T., Saito, T., Seo, S., Yamazaki, I., Yamagata, T., et al. (2004). AML-1 is required for megakaryocytic maturation and lymphocytic differentiation, but not for maintenance of hematopoietic stem cells in adult hematopoiesis. *Nature Medicine, 10*, 299–304.

Imai, Y., Kurokawa, M., Yamaguchi, Y., Izutsu, K., Nitta, E., Mitani, K., et al. (2004). The corepressor mSin3A regulates phosphorylation-induced activation, intranuclear location, and stability of AML1. *Molecular and Cellular Biology, 24*, 1033–1043.

Inoue, K.-I., Ozaki, S., Shiga, T., Ito, K., Masuda, T., Okado, N., et al. (2002). Runx3 controls the axonal projection of proprioceptive dorsal root ganglion neurons. *Nature Neuroscience, 5*, 946–954.

Jackson, D. A., Hassan, A. B., Errington, R. J., & Cook, P. R. (1993). Visualization of focal sites of transcription within human nuclei. *The EMBO Journal, 12*, 1059–1065.

Javed, A., Guo, B., Hiebert, S., Choi, J. Y., Green, J., Zhao, S. C., et al. (2000). Groucho/TLE/R-esp proteins associate with the nuclear matrix and repress RUNX (CBF(alpha)/AML/PEBP2(alpha)) dependent activation of tissue-specific gene transcription. *Journal of Cell Science, 113*(Pt 12), 2221–2231.

Kim, W. Y., Sieweke, M., Ogawa, E., Wee, H.-J., Englmeier, U., Graf, T., et al. (1999). Mutual activation of Ets-1 and AML1 DNA binding by direct interaction of their autoinhibitory domains. *The EMBO Journal, 18*, 1609–1620.

Kim, S. I., Bultman, S. J., Jing, H., Blobel, G. A., & Bresnick, E. H. (2007). Dissecting molecular steps in chromatin domain activation during hematopoietic differentiation. *Molecular and Cellular Biology, 27*, 4551–4565.

Kim, S. I., Bultman, S. J., Kiefer, C. M., Dean, A., & Bresnick, E. H. (2009). BRG1 requirement for long-range interaction of a locus control region with a downstream promoter. *Proceedings of the National Academy of Sciences of the United States of America, 106*, 2259–2264.

Kindle, K. B., Troke, P. J. F., Collins, H. M., Matsuda, S., Bossi, D., Bellodi, C., et al. (2005). MOZ-TIF2 inhibits transcription by nuclear receptors and p53 by impairment of CBP function. *Molecular and Cellular Biology, 25*, 988–1002.

Kitabayashi, I. (1998). Interaction and functional cooperation of the leukemia-associated factors AML1 and p300 in myeloid cell differentiation. *The EMBO Journal, 17*, 2994–3004.

Kitabayashi, I., Aikawa, Y., Nguyen, L. A., Yokoyama, A., & Ohki, M. (2001). Activation of AML1-mediated transcription by MOZ and inhibition by the MOZ-CBP fusion protein. *The EMBO Journal, 20*, 7184–7196.

Komori, T., Yagi, H., Nomura, S., Yamaguchi, A., Sasaki, K., Deguchi, K., et al. (1997). Targeted disruption of Cbfa1 results in a complete lack of bone formation owing to maturational arrest of osteoblasts. *Cell, 89*, 755–764.

Kuvardina, O. N., Herglotz, J., Kolodziej, S., Kohrs, N., Herkt, S., Wojcik, B., et al. (2015). RUNX1 represses the erythroid gene expression program during megakaryocytic differentiation. *Blood, 125*, 3570–3579.

Lacaud, G., Gore, L., Kennedy, M., Kouskoff, V., Kingsley, P., Hogan, C., et al. (2002). Runx1 is essential for hematopoietic commitment at the hemangioblast stage of development in vitro. *Blood, 100*, 458–466.

Lam, K., & Zhang, D. E. (2012). RUNX1 and RUNX1-ETO: Roles in hematopoiesis and leukemogenesis. *Frontiers in Bioscience, 17*, 1120–1139.

Lam, E. Y. N., Chau, J. Y. M., Kalev-Zylinska, M. L., Fountaine, T. M., Mead, R. S., Hall, C. J., et al. (2009). Zebrafish runx1 promoter-EGFP transgenics mark discrete sites of definitive blood progenitors. *Blood, 113*, 1241–1249.

Lancrin, C., Sroczynska, P., Stephenson, C., Allen, T., Kouskoff, V., & Lacaud, G. (2009). The haemangioblast generates haematopoietic cells through a haemogenic endothelium stage. *Nature, 457*, 892–895.

Leong, W. Y., Guo, H., Ma, O., Huang, H., Cantor, A. B., & Friedman, A. D. (2016). Runx1 phosphorylation by Src increases trans-activation via augmented stability, reduced histone deacetylase (HDAC) binding, and increased dna affinity, and activated Runx1 favors granulopoiesis. *The Journal of Biological Chemistry, 291*, 826–836.

Levanon, D., & Groner, Y. (2004). Structure and regulated expression of mammalian RUNX genes. *Oncogene, 23*, 4211–4219.

Levanon, D., Bernstein, Y., Negreanu, V., Ghozi, M. C., Bar-Am, I., Aloya, R., et al. (1996). A large variety of alternatively spliced and differentially expressed mRNAs are encoded by the human acute myeloid leukemia gene AML1. *DNA and Cell Biology, 15*, 175–185.

Levanon, D., Goldstein, R. E., Bernstein, Y., Tang, H., Goldenberg, D., Stifani, S., et al. (1998). Transcriptional repression by AML1 and LEF-1 is mediated by the TLE/Groucho corepressors. *Proceedings of the National Academy of Sciences of the United States of America, 95*, 11590–11595.

Levanon, D., Bettoun, D., Harris-Cerruti, C., Woolf, E., Negreanu, V., Eilam, R., et al. (2002). The Runx3 transcription factor regulates development and survival of TrkC dorsal root ganglia neurons. *The EMBO Journal, 21*, 3454–3463.

Levantini, E., Lee, S., Radomska, H. S., Hetherington, C. J., Alberich-Jorda, M., Amabile, G., et al. (2011). RUNX1 regulates the CD34 gene in haematopoietic stem cells by mediating interactions with a distal regulatory element. *The EMBO Journal, 30*, 4059–4070.

Lichtinger, M., Ingram, R., Hannah, R., Muller, D., Clarke, D., Assi, S.A., Lie-A-Ling, M., Noailles, L., Vijayabaskar, M.S., Wu, M., et al. (2012). RUNX1 reshapes the epigenetic landscape at the onset of haematopoiesis. *The EMBO Journal, 31*, 4318–4333.

Lie-A-Ling, M., Marinopoulou, E., Li, Y., Patel, R., Stefanska, M., Bonifer, C., et al. (2014). RUNX1 positively regulates a cell adhesion and migration program in murine hemogenic endothelium prior to blood emergence. *Blood, 124*, e11–e20.

Ling, J. Q., Li, T., Hu, J. F., Vu, T. H., Chen, H. L., Qiu, X. W., et al. (2006). CTCF mediates interchromosomal colocalization between Igf2/H19 and Wsb1/Nf1. *Science, 312*, 269–272.

Liu, Y., Chen, W., Gaudet, J., Cheney, M. D., Roudaia, L., Cierpicki, T., et al. (2007). Structural basis for recognition of SMRT/N-CoR by the MYND domain and its contribution to AML1/ETO's activity. *Cancer Cell, 11*, 483–497.

Liu, X., Zhang, Q., Zhang, D. E., Zhou, C., Xing, H., Tian, Z., et al. (2009a). Overexpression of an isoform of AML1 in acute leukemia and its potential role in leukemogenesis. *Leukemia, 23*, 739–745.

Liu, Z., Ma, Z., Terada, L. S., & Garrard, W. T. (2009b). Divergent roles of RelA and c-Rel in establishing chromosomal loops upon activation of the Igkappa gene. *Journal of Immunology, 183*, 3819–3830.

Lopez-Camacho, C., van Wijnen, A. J., Lian, J. B., Stein, J. L., & Stein, G. S. (2014). CBFbeta and the leukemogenic fusion protein CBFbeta-SMMHC associate with mitotic chromosomes to epigenetically regulate ribosomal genes. *Journal of Cellular Biochemistry, 115*, 2155–2164.

Lutterbach, B., Westendorf, J. J., Linggi, B., Isaac, S., Seto, E., & Hiebert, S. W. (2000). A mechanism of repression by acute myeloid leukemia-1, the target of multiple chromosomal translocations in acute leukemia. *The Journal of Biological Chemistry, 275*, 651–656.

Marenduzzo, D., Faro-Trindade, I., & Cook, P. R. (2007). What are the molecular ties that maintain genomic loops? *Trends in Genetics, 23*, 126–133.

Martens, J. H., Mandoli, A., Simmer, F., Wierenga, B. J., Saeed, S., Singh, A. A., et al. (2012). ERG and FLI1 binding sites demarcate targets for aberrant epigenetic regulation by AML1-ETO in acute myeloid leukemia. *Blood, 120*, 4038–4048.

Matheny, C. J., Speck, M. E., Cushing, P. R., Zhou, Y., Corpora, T., Regan, M., et al. (2007). Disease mutations in RUNX1 and RUNX2 create nonfunctional, dominant-negative, or hypomorphic alleles. *The EMBO Journal, 26*, 1163–1175.

Meier, N., Krpic, S., Rodriguez, P., Strouboulis, J., Monti, M., Krijgsveld, J., et al. (2006). Novel binding partners of Ldb1 are required for haematopoietic development. *Development (Cambridge, England), 133*, 4913–4923.

Michaud, J., Wu, F., Osato, M., Cottles, G. M., Yanagida, M., Asou, N., et al. (2002). In vitro analyses of known and novel RUNX1/AML1 mutations in dominant familial platelet disorder with predisposition to acute myelogenous leukemia: Implications for mechanisms of pathogenesis. *Blood, 99*, 1364–1372.

Miyoshi, H., Shimizu, K., Kozu, T., Maseki, N., Kaneko, Y., & Ohki, M. (1991). t(8;21) breakpoints on chromosome 21 in acute myeloid leukemia are clustered within a limited region of a single gene, AML1. *Proceedings of the National Academy of Sciences of the United States of America, 88*, 10431–10434.

Miyoshi, H., Ohira, M., Shimizu, K., Mitani, K., Hirai, H., Imai, T., et al. (1995). Alternative splicing and genomic structure of the AML1 gene involved in acute myeloid leukemia. *Nucleic Acids Research, 23*, 2762–2769.

Nutt, S. L., Heavey, B., Rolink, A. G., & Busslinger, M. (1999). Commitment to the B-lymphoid lineage depends on the transcription factor Pax5. *Nature, 401*, 556–562.

Okuda, T., van Deursen, J., Hiebert, S. W., Grosveld, G., & Downing, J. R. (1996). AML1, the target of multiple chromosomal translocations in human leukemia, is essential for normal fetal liver hematopoiesis. *Cell, 84*, 321–330.

Okumura, A. J., Peterson, L. F., Okumura, F., Boyapati, A., & Zhang, D. E. (2008). t(8;21)(q22;q22) Fusion proteins preferentially bind to duplicated AML1/RUNX1 DNA-binding sequences to differentially regulate gene expression. *Blood, 112*, 1392–1401.

Ono, M., Yaguchi, H., Ohkura, N., Kitabayashi, I., Nagamura, Y., Nomura, T., et al. (2007). Foxp3 controls regulatory T-cell function by interacting with AML1/Runx1. *Nature, 446*, 685–689.

Osato, M. (2004). Point mutations in the RUNX1/AML1 gene: Another actor in RUNX leukemia. *Oncogene, 23*, 4284–4296.

Osborne, C. S., Chakalova, L., Brown, K. E., Carter, D., Horton, A., Debrand, E., et al. (2004). Active genes dynamically colocalize to shared sites of ongoing transcription. *Nature Genetics, 36*, 1065–1071.

Otto, F., Thornell, A. P., Crompton, T., Denzel, A., Gilmour, K. C., Rosewell, I. R., et al. (1997). Cbfa1, a candidate gene for cleidocranial dysplasia syndrome, is essential for osteoblast differentiation and bone development. *Cell, 89*, 765–771.

Pande, S., Ali, S. A., Dowdy, C., Zaidi, S. K., Ito, K., Ito, Y., et al. (2009). Subnuclear targeting of the Runx3 tumor suppressor and its epigenetic association with mitotic chromosomes. *Journal of Cellular Physiology, 218*, 473–479.

Patrinos, G. P., de Krom, M., de Boer, E., Langeveld, A., Imam, A. M., Strouboulis, J., et al. (2004). Multiple interactions between regulatory regions are required to stabilize an active chromatin hub. *Genes & Development, 18*, 1495–1509.

Paul, F., Arkin, Y., Giladi, A., Jaitin, D. A., Kenigsberg, E., Keren-Shaul, H., et al. (2015). Transcriptional heterogeneity and lineage commitment in myeloid progenitors. *Cell, 163*, 1663–1677.

Pencovich, N., Jaschek, R., Tanay, A., & Groner, Y. (2011). Dynamic combinatorial interactions of RUNX1 and cooperating partners regulates megakaryocytic differentiation in cell line models. *Blood, 117*, e1–14.

Pozner, A., Goldenberg, D., Negreanu, V., Le, S. Y., Elroy-Stein, O., Levanon, D., & Groner, Y. (2000). Transcription-coupled translation control of AML1/RUNX1 is mediated by cap- and internal ribosome entry site-dependent mechanisms. *Molecular and Cellular Biology, 20*, 2297–2307.

Pozner, A., Lotem, J., Xiao, C., Goldenberg, D., Brenner, O., Negreanu, V., et al. (2007). Developmentally regulated promoter-switch transcriptionally controls Runx1 function during embryonic hematopoiesis. *BMC Developmental Biology, 7*, 84.

Ptasinska, A., Assi, S. A., Mannari, D., James, S. R., Williamson, D., Dunne, J., et al. (2012). Depletion of RUNX1/ETO in t(8;21) AML cells leads to genome-wide changes in chromatin structure and transcription factor binding. *Leukemia, 26*, 1829–1841.

Ptasinska, A., Assi, S. A., Martinez-Soria, N., Imperato, M. R., Piper, J., Cauchy, P., et al. (2014). Identification of a dynamic core transcriptional network in t(8;21) AML that regulates differentiation block and self-renewal. *Cell Reports, 8*, 1974–1988.

Putz, G., Rosner, A., Nuesslein, I., Schmitz, N., & Buchholz, F. (2006). AML1 deletion in adult mice causes splenomegaly and lymphomas. *Oncogene, 25*, 929–939.

Ran, D., Shia, W.-J., Lo, M.-C., Fan, J.-B., Knorr, D. A., Ferrell, P. I., et al. (2013). RUNX1a enhances hematopoietic lineage commitment from human embryonic stem cells and inducible pluripotent stem cells. *Blood, 121*, 2882–2890.

Reed-Inderbitzin, E., Moreno-Miralles, I., Vanden-Eynden, S. K., Xie, J., Lutterbach, B., Durst-Goodwin, K. L., et al. (2006). RUNX1 associates with histone deacetylases and SUV39H1 to repress transcription. *Oncogene, 25*, 5777–5786.

Regha, K., Assi, S. A., Tsoulaki, O., Gilmour, J., Lacaud, G., & Bonifer, C. (2015). Developmental-stage-dependent transcriptional response to leukaemic oncogene expression. *Nature Communications, 6*, 7203.

Rennert, J., Coffman, J. A., Mushegian, A. R., & Robertson, A. J. (2003). The evolution of Runx genes I. A comparative study of sequences from phylogenetically diverse model organisms. *BMC Evolutionary Biology, 3*, 4.

Ross, K., Sedello, A. K., Todd, G. P., Paszkowski-Rogacz, M., Bird, A. W., Ding, L., et al. (2012). Polycomb group ring finger 1 cooperates with Runx1 in regulating differentiation and self-renewal of hematopoietic cells. *Blood, 119*, 4152–4161.

Scott, E. W., Simon, M. C., Anastasi, J., & Singh, H. (1994). Requirement of transcription factor PU.1 in the development of multiple hematopoietic lineages. *Science, 265*, 1573–1577.

Sellars, M., Huh, J. R., Day, K., Issuree, P. D., Galan, C., Gobeil, S., et al. (2015). Regulation of DNA methylation dictates Cd4 expression during the development of helper and cytotoxic T cell lineages. *Nature Immunology, 16*, 746–754.

Shang, Y., Zhao, X., Xu, X., Xin, H., Li, X., Zhai, Y., et al. (2009). CHIP functions an E3 ubiquitin ligase of Runx1. *Biochemical and Biophysical Research Communications, 386*, 242–246.

Song, W. J., Sullivan, M. G., Legare, R. D., Hutchings, S., Tan, X., Kufrin, D., et al. (1999). Haploinsufficiency of CBFA2 causes familial thrombocytopenia with propensity to develop acute myelogenous leukaemia. *Nature Genetics, 23*, 166–175.

Sroczynska, P., Lancrin, C., Kouskoff, V., & Lacaud, G. (2009). The differential activities of Runx1 promoters define milestones during embryonic hematopoiesis. *Blood, 114*, 5279–5289.

Staber, P. B., Zhang, P., Ye, M., Welner, R. S., Levantini, E., Di Ruscio, A., et al. (2014). The Runx-PU.1 pathway preserves normal and AML/ETO9a leukemic stem cells. *Blood, 124*, 2391–2399.

Stopka, T., Amanatullah, D. F., Papetti, M., & Skoultchi, A. I. (2005). PU.1 inhibits the erythroid program by binding to GATA-1 on DNA and creating a repressive chromatin structure. *The EMBO Journal, 24*, 3712–3723.

Sun, X. J., Wang, Z., Wang, L., Jiang, Y., Kost, N., Soong, T. D., et al. (2013). A stable transcription factor complex nucleated by oligomeric AML1-ETO controls leukaemogenesis. *Nature, 500*, 93–97.

Tanaka, T., Tanaka, K., Ogawa, S., Kurokawa, M., Mitani, K., Nishida, J., et al. (1995). An acute myeloid leukemia gene, AML1, regulates hemopoietic myeloid cell differentiation and transcriptional activation antagonistically by two alternative spliced forms. *The EMBO Journal, 14*, 341–350.

Tanaka, T., Kurokawa, M., Ueki, K., Tanaka, K., Imai, Y., Mitani, K., et al. (1996). The extracellular signal-regulated kinase pathway phosphorylates AML1, an acute myeloid leukemia gene product, and potentially regulates its transactivation ability. *Molecular and Cellular Biology, 16*, 3967–3979.

Taniuchi, I., Osato, M., Egawa, T., Sunshine, M. J., Bae, S. C., Komori, T., et al. (2002). Differential requirements for Runx proteins in CD4 repression and epigenetic silencing during T lymphocyte development. *Cell, 111*, 621–633.

Telfer, J. (2001). Expression and function of a stem cell promoter for the murine CBFα2 gene: Distinct roles and regulation in natural killer and T cell development. *Developmental Biology, 229*, 363–382.

Theo Sijtse Palstra, R. J. (2009). Close encounters of the 3C kind: Long-range chromatin interactions and transcriptional regulation. *Briefings in Functional Genomics & Proteomics, 8*, 297–309.

Tolhuis, B., Palstra, R. J., Splinter, E., Grosveld, F., & de Laat, W. (2002). Looping and interaction between hypersensitive sites in the active beta-globin locus. *Molecular Cell, 10*, 1453–1465.

Trombly, D. J., Whitfield, T. W., Padmanabhan, S., Gordon, J. A., Lian, J. B., van Wijnen, A. J., et al. (2015). Genome-wide co-occupancy of AML1-ETO and N-CoR defines the t(8;21) AML signature in leukemic cells. *BMC Genomics, 16*, 309.

Tsuzuki, S., & Seto, M. (2012). Expansion of functionally defined mouse hematopoietic stem and progenitor cells by a short isoform of RUNX1/AML1. *Blood, 119*, 727–735.

Tsuzuki, S., Hong, D., Gupta, R., Matsuo, K., Seto, M., & Enver, T. (2007). Isoform-specific potentiation of stem and progenitor cell engraftment by AML1/RUNX1. *PLoS Medicine, 4*, e172.

van Oevelen, C., Collombet, S., Vicent, G., Hoogenkamp, M., Lepoivre, C., Badeaux, A., et al. (2015). C/EBPalpha activates pre-existing and de novo macrophage enhancers during induced Pre-B cell transdifferentiation and myelopoiesis. *Stem Cell Reports, 5*, 232–247.

van Riel, B., Pakozdi, T., Brouwer, R., Monteiro, R., Tuladhar, K., Franke, V., et al. (2012). A novel complex, RUNX1-MYEF2, represses hematopoietic genes in erythroid cells. *Molecular and Cellular Biology, 32*, 3814–3822.

Vernimmen, D., De Gobbi, M., Sloane-Stanley, J. A., Wood, W. G., & Higgs, D. R. (2007). Long-range chromosomal interactions regulate the timing of the transition between poised and active gene expression. *The EMBO Journal, 26*, 2041–2051.

Wang, Q., Stacy, T., Binder, M., Marín-Padilla, M., Sharpe, A. H., & Speck, N. A. (1996). Disruption of the Cbfa2 gene causes necrosis and hemorrhaging in the central nervous system and blocks definitive hematopoiesis. *Proceedings of the National Academy of Sciences of the United States of America, 93*, 3444–3449.

Wee, H.-J., Voon, D. C.-C., Bae, S. C., & Ito, Y. (2008). PEBP2-beta/CBF-beta-dependent phosphorylation of RUNX1 and p300 by HIPK2: Implications for leukemogenesis. *Blood, 112*, 3777–3787.

Wheeler, J. C., Shigesada, K., Gergen, J. P., & Ito, Y. (2000). Mechanisms of transcriptional regulation by Runt domain proteins. *Seminars in Cell and Developmental Biology, 11*, 369–375.

Williams, A. O., Isaacs, R. J., & Stowell, K. M. (2007). Down-regulation of human topoisomerase IIalpha expression correlates with relative amounts of specificity factors Sp1 and Sp3 bound at proximal and distal promoter regions. *BMC Molecular Biology, 8*, 36.

Wilson, N. K., Foster, S. D., Wang, X., Knezevic, K., Schutte, J., Kaimakis, P., et al. (2010). Combinatorial transcriptional control in blood stem/progenitor cells: Genome-wide analysis of ten major transcriptional regulators. *Cell Stem Cell, 7*, 532–544.

Wotton, D., Ghysdael, J., Wang, S., Speck, N. A., & Owen, M. J. (1994). Cooperative binding of Ets-1 and core binding factor to DNA. *Molecular and Cellular Biology, 14*, 840–850.

Yamaguchi, Y., Kurokawa, M., Imai, Y., Izutsu, K., Asai, T., Ichikawa, M., et al. (2004). AML1 is functionally regulated through p300-mediated acetylation on specific lysine residues. *The Journal of Biological Chemistry, 279*, 15630–15638.

Yergeau, D. A., Hetherington, C. J., Wang, Q., Zhang, P., Sharpe, A. H., Binder, M., et al. (1997). Embryonic lethality and impairment of haematopoiesis in mice heterozygous for an AML1-ETO fusion gene. *Nature Genetics, 15*, 303–306.

Yu, M., Mazor, T., Huang, H., Huang, H.-T., Kathrein, K. L., Woo, A. J., et al. (2012). Direct recruitment of polycomb repressive complex 1 to chromatin by core binding transcription factors. *Molecular Cell, 45*, 330–343.

Zaidi, S. K., Javed, A., Choi, J. Y., van Wijnen, A. J., Stein, J. L., Lian, J. B., & Stein, G. S. (2001). A specific targeting signal directs Runx2/Cbfa1 to subnuclear domains and contributes to transactivation of the osteocalcin gene. *Journal of Cell Science, 114*, 3093–3102.

Zaidi, S. K., Sullivan, A. J., van Wijnen, A. J., Stein, J. L., Stein, G. S., & Lian, J. B. (2002). Integration of Runx and Smad regulatory signals at transcriptionally active subnuclear sites. *Proceedings of the National Academy of Sciences of the United States of America, 99*, 8048–8053.

Zaidi, S. K., Young, D. W., Pockwinse, S. M., Javed, A., Lian, J. B., Stein, J. L., et al. (2003). Mitotic partitioning and selective reorganization of tissue-specific transcription factors in progeny cells. *Proceedings of the National Academy of Sciences of the United States of America, 100*, 14852–14857.

Zambidis, E. T. (2005). Hematopoietic differentiation of human embryonic stem cells progresses through sequential hematoendothelial, primitive, and definitive stages resembling human yolk sac development. *Blood, 106*, 860–870.

Zaret, K. S. (2014). Genome reactivation after the silence in mitosis: Recapitulating mechanisms of development? *Developmental Cell, 29*, 132–134.

Zeng, C., van Wijnen, A. J., Stein, J. L., Meyers, S., Sun, W., Shopland, L., et al. (1997). Identification of a nuclear matrix targeting signal in the leukemia and bone-related AML/CBF-alpha transcription factors. *Proceedings of the National Academy of Sciences of the United States of America, 94*, 6746–6751.

Zeng, C., McNeil, S., Pockwinse, S., Nickerson, J., Shopland, L., Lawrence, J. B., et al. (1998). Intranuclear targeting of AML/CBFalpha regulatory factors to nuclear matrix-associated transcriptional domains. *Proceedings of the National Academy of Sciences of the United States of America, 95*, 1585–1589.

Zhang, D. E., Zhang, P., Wang, N. D., Hetherington, C. J., Darlington, G. J., & Tenen, D. G. (1997). Absence of granulocyte colony-stimulating factor signaling and neutrophil development in CCAAT enhancer binding protein alpha-deficient mice. *Proceedings of the National Academy of Sciences of the United States of America, 94*, 569–574.

Zhang, D. E., Petrovick, M. S., Hiebert, S. W., Friedman, A. D., Hetherington, C. J., & Tenen, D. G. (1998). Multiple functional domains of AML1: PU.1 and C/EBPalpha synergize with different regions of AML1. *Molecular and Cellular Biology, 18*, 3915–3925.

Zhang, L., Fried, F. B., Guo, H., & Friedman, A. D. (2008). Cyclin-dependent kinase phosphorylation of RUNX1/AML1 on 3 sites increases transactivation potency and stimulates cell proliferation. *Blood, 111*, 1193–1200.

Zhang, H., Alberich-Jorda, M., Amabile, G., Yang, H., Staber, P. B., Di Ruscio, A., et al. (2013). Sox4 is a key oncogenic target in C/EBPalpha mutant acute myeloid leukemia. *Cancer Cell, 24*, 575–588.

Zhao, X., Jankovic, V., Gural, A., Huang, G., Pardanani, A., Menendez, S., et al. (2008). Methylation of RUNX1 by PRMT1 abrogates SIN3A binding and potentiates its transcriptional activity. *Genes & Development, 22*, 640–653.

Roles of Runx2 in Skeletal Development

6

Toshihisa Komori

Abstract

Runx2 is the most upstream transcription factor essential for osteoblast differentiation. It regulates the expression of *Sp7*, the protein of which is a crucial transcription factor for osteoblast differentiation, as well as that of bone matrix genes including *Spp1*, *Ibsp*, and *Bglap2*. Runx2 is also required for chondrocyte maturation, and Runx3 has a redundant function with Runx2 in chondrocyte maturation. Runx2 regulates the expression of *Col10a1*, *Spp1*, *Ibsp*, and *Mmp13* in chondrocytes. It also inhibits chondrocytes from acquiring the phenotypes of permanent cartilage chondrocytes. It regulates chondrocyte proliferation through the regulation of *Ihh* expression. Runx2 enhances osteoclastogenesis by regulating *Rankl*. Cbfb, which is a co-transcription factor for Runx family proteins, plays an important role in skeletal development by stabilizing Runx family proteins. In Cbfb isoforms, Cbfb1 is more potent than Cbfb2 in Runx2-dependent transcriptional regulation; however, the expression level of *Cbfb2* is three-fold higher than that of *Cbfb1*, demonstrating the requirement of Cbfb2 in skeletal development. The expression of *Runx2* in osteoblasts is regulated by a 343-bp enhancer located upstream of the P1 promoter. This enhancer is activated by an enhanceosome composed of Dlx5/6, Mef2, Tcf7, Ctnnb1, Sox5/6, Smad1, and Sp7. Thus, Runx2 is a multifunctional transcription factor that is essential for skeletal development, and Cbfb regulates skeletal development by modulating the stability and transcriptional activity of Runx family proteins.

Keywords

Runx2 • Osteoblast • Chondrocyte • Cbfb • Enhancer

T. Komori (✉)
Department of Cell Biology, Unit of Basic Medical Sciences, Nagasaki University Graduate School of Biomedical Sciences, Nagasaki 852-8588, Japan
e-mail: komorit@nagasaki-u.ac.jp

Y. Groner et al. (eds.), *RUNX Proteins in Development and Cancer*, Advances in Experimental Medicine and Biology 962, DOI 10.1007/978-981-10-3233-2_6

6.1 Introduction

Skeletal component cells including osteoblasts, chondrocytes, adipocytes, myoblasts, tendon cells, and fibroblasts, are derived from mesenchymal stem cells. Their lineages are determined by different transcription factors. Runx2 and Sp7 regulate osteoblast differentiation, the Sox family (Sox9, Sox5, and Sox6) regulates chondrocyte differentiation, the MyoD family (MyoD, Myf5, and myogenin) regulates myogenic differentiation, and the C/EBP family (C/EBPβ, C/EBPδ, and C/EBPα) and PPARγ2 regulate adipocyte differentiation (Komori 2006).

Bony skeletons are formed through intramembranous and endochondral ossification. In intramembranous bone development, mesenchymal cells differentiate into osteoblasts and bone is directly formed by osteoblasts. In endochondral ossification, cartilaginous skeletons are formed by chondrocytes, which acquire mature phenotypes at the diaphysis, in which vascular invasion occurs, and osteoclasts and mesenchymal cells invade the cartilage. Terminally differentiated chondrocytes die due to apoptosis, mesenchymal cells differentiate into osteoblasts, and bone is formed on the rudiments of cartilaginous structures. Cartilaginous structures are then completely replaced with bone (Inada et al. 1999; Marks Jr. and Odgren 2002).

Runx2, which belongs to the Runx family of proteins consisting of Runx1, Runx2, and Runx3, is a transcription factor that is essential for skeletal development. Runx family proteins have a runt domain, which directly binds to DNA. Runx2 is known to form a heterodimer with Cbfb and acquires an enhanced DNA-binding capacity (Komori 2005). A heterozygous mutation of RUNX2 has been shown to cause cleidocranial dysplasia, which is characterized by hypoplastic clavicles, open fontanelles, supernumerary teeth, and a short stature (Mundlos et al. 1997). In this review, a focus has been placed on the functions of Runx2 and Cbfb in the regulation of osteoblast and chondrocyte differentiation as well as transcriptional regulation of the *Runx2* gene.

6.2 Roles of Runx2 in Osteoblast Differentiation

Osteoblasts are completely absent in *Runx2*$^{-/-}$ mice, which indicates that Runx2 is an essential transcription factor for osteoblast differentiation (Komori et al. 1997; Otto et al. 1997) (Fig. 6.1). Canonical Wnt signaling and Sp7 are also crucial for osteoblast differentiation (Komori 2006). After committing to the osteoblastic lineage, osteoblasts express bone matrix protein genes at different expression levels depending on the maturation level of the cells. Immature mesenchymal cells and preosteoblasts weakly express *Col1a1*, the expression of which is up-regulated in immature osteoblasts (Inada et al. 1999). Immature osteoblasts have been shown to express *Spp1* and then *Ibsp*, while mature osteoblasts strongly express *Bglap2* (Maruyama et al. 2007; Aubin and Triffitt 2002). Mature osteoblasts are embedded into the bone matrix and ultimately become osteocytes, which express *Dmp1* (Toyosawa et al. 2001). Previous studies demonstrated that the expression of bone matrix protein genes including *Spp1*, *Ibsp*, and *Bglap2*, is virtually absent in *Runx2*$^{-/-}$ mice (Komori et al. 1997; Inada et al. 1999). Runx2 has the ability to up-regulate the expression of bone matrix protein genes including *Col1a1*, *Spp1*, *Ibsp*, *Bglap*, and *Fn1* (fibronectin 1) (Ducy et al. 1997; Sato et al. 1998; Harada et al. 1999; Lee et al. 2000), and Runx2 activates reporter activities including *Col1a1*, *Col1a2*, *Spp1*, and *Bglap2* promoters (Banerjee et al. 1997; Harada et al. 1999; Jimenez et al. 1999; Sato et al. 1998; Kern et al. 2001). However, the expression of *Ibsp* is reduced by Runx2 and HDAC3 *in vitro*, and Runx2 represses *Ibsp* promoter activity (Javed et al. 2001; Lamour et al. 2007).

The function of Runx2 in the early stage of osteoblast differentiation is very clear because osteoblast marker gene expression is absent in *Runx2*$^{-/-}$ mice, indicating that Runx2 is essential for the differentiation of mesenchymal stem cells into osteoblasts in an early stage (Komori et al. 1997; Otto et al. 1997) (Fig. 6.1). However, the

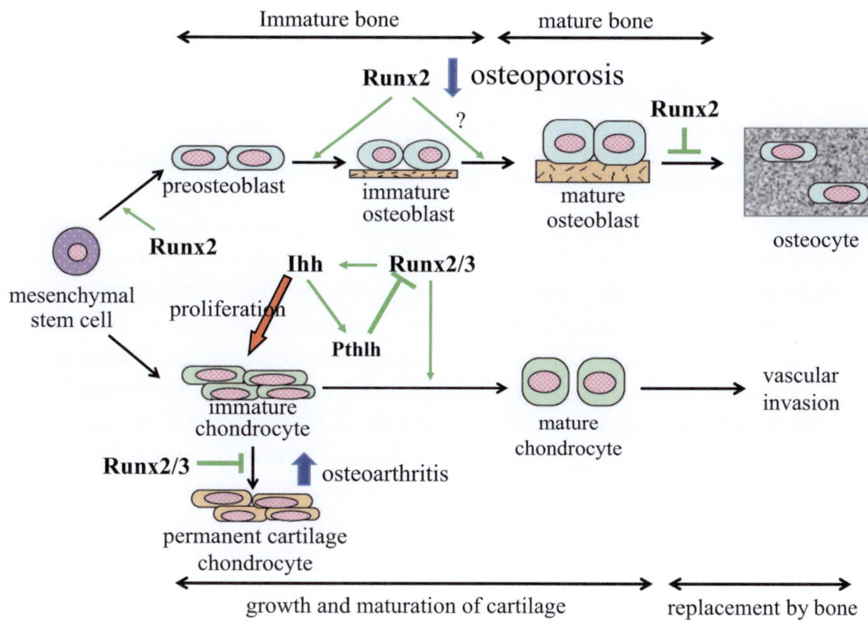

Fig. 6.1 Functions of Runx2 in osteoblast and chondrocyte differentiation. Runx2 directs the differentiation of mesenchymal stem cells to preosteoblasts and further differentiation to immature osteoblasts. The functions of Runx2 in committed osteoblasts are controversial. Runx2 inhibits the transition of osteoblasts to osteocytes. Although Runx2 is not required for the differentiation of mesenchymal stem cells to immature chondrocytes, it is necessary for the maturation of immature chondrocytes. Runx3 has a redundant function in chondrocyte maturation. Runx2 inhibits chondrocytes from acquiring the phenotypes of permanent cartilage chondrocytes. It also regulates chondrocyte proliferation by regulating *Ihh* expression. Ihh up-regulates the expression of *Pthlh*, the protein of which inhibits Runx2 and chondrocyte maturation. Pathologically, reductions in Runx2 expression and activity in the osteoblast lineage are associated with osteoporosis, while the up-regulation of Runx2 expression and activity in permanent cartilage chondrocytes is related to osteoarthritis

functions of Runx2 in committed osteoblasts are controversial. *Runx2* conditional knockout mice using Cre transgenic mice under the control of a 2.3-kb *Col1a1* promoter, which directs transgene expression to committed osteoblasts, were recently reported by two groups. The conditional deletion of exon 4, which contains a part of the runt domain, results in no overt phenotypes (Takarada et al. 2013), whereas mice with the conditional deletion of exon 8, which produces a truncated Runx2 protein, have been shown to develop osteopenia due to reduced bone formation (Adhami et al. 2015). Osteoclastogenesis is also reduced in the latter mice. Since these studies used the same Cre transgenic line, the expression level of the Cre transgene does not appear to be the cause of the difference in these phenotypes. However, the genetic backgrounds of their mice differed, which may explain the discrepancies observed. Further investigations are needed in order to clarify the functions of Runx2 in committed osteoblasts (Fig. 6.1).

We and others previously reported that the overexpression of *Runx2* using a 2.3-kb *Col1a1* promoter resulted in osteopenia due to reduced bone formation (Liu et al. 2001; Geoffroy et al. 2002; Kanatani et al. 2006). The expression of Runx2 is initially detected in preosteoblasts, increases in immature osteoblasts, and then decreases during osteoblast maturation. It is strongly expressed in embryos and young mice after birth, but gradually decreases and is low in adult mice (Maruyama et al. 2007). Therefore, the phenotypes of *Runx2* transgenic mice indicate that the maintenance of the strong expression of *Runx2* inhibits osteoblast maturation and keeps the osteoblasts in an immature stage. Furthermore, osteocytes are virtually absent in

Runx2 transgenic mice, indicating that Runx2 inhibits the transition of osteoblasts to osteocytes (Liu et al. 2001) (Fig. 6.1). Runx2 has also been shown to induce the expression of *Rankl*, which is essential for osteoclast differentiation, and enhances bone resorption (Enomoto et al. 2003; Geoffroy et al. 2002).

Cortical bone is reduced in dominant-negative (dn) *Runx2* transgenic mice under the control of a 2.3-kb *Col1a1* promoter. Bone formation in trabecular bone is marginally reduced in young adult dn-*Runx2* transgenic mice, but trabecular bone increases by 7 months of age. A previous study demonstrated that mineralization is increased in trabecular bone, urinary deoxypyridinoline, which is a marker for bone resorption, is reduced in dn-*Runx2* transgenic mice, and ovariectomy increases bone resorption in wild-type mice, but not in dn-*Runx2* transgenic mice (Maruyama et al. 2007). Therefore, osteoblast maturation appears to be accelerated in dn-*Runx2* transgenic mice and leads to the formation of mature bone, which is resistant to bone resorption, because cortical bone formed by mature osteoblasts is more resistant to bone resorption than trabecular bone formed by relatively immature osteoblasts. Furthermore, dn-*Runx2* inhibits osteoclastogenesis *in vitro*. Therefore, Runx2 regulates bone maturity and osteoclastogenesis and is involved in bone reductions after an estrogen deficiency (Maruyama et al. 2007).

6.3 Roles of Runx2 in Chondrocyte Differentiation

Although the entire skeleton of *Runx2*$^{-/-}$ mice is composed of cartilage, chondrocyte maturation is severely inhibited throughout most of the skeleton. *Col2a1*, which is expressed in immature chondrocytes, is expressed in whole *Runx2*$^{-/-}$ skeletons, whereas *Col10a1*, which is expressed in mature chondrocytes, is restrictedly expressed in the tibia, fibula, radius, and ulna. The expression of *Spp1*, *Ibsp*, and *Mmp13*, which are expressed in terminally differentiated chondrocytes, was found to be virtually absent in whole *Runx2*$^{-/-}$ skeletons (Inada et al. 1999). *Spp1* and *Mmp13* are directly regulated by Runx2 (Jimenez et al. 1999; Porte et al. 1999; Sato et al. 1998; Selvamurugan et al. 2000; Hess et al. 2001). These findings indicate that Runx2 is required for chondrocyte maturation (Fig. 6.1). Even in the restricted skeletons of *Runx2*$^{-/-}$ mice, in which chondrocyte maturation occurs, vascular invasion is absent, indicating that Runx2 is also required for vascular invasion into the cartilage (Zelzer et al. 2001; Lee et al. 2012; Himeno et al. 2002). In wild-type mice, osteoblast differentiation occurs in the perichondrium and the bone collar is formed. However, osteoblast differentiation is completely blocked in *Runx2*$^{-/-}$ mice and there are no osteoblasts in the perichondrium (Komori et al. 1997). Therefore, the absence of osteoblasts in the perichondrium may affect chondrocyte maturation. These findings indicate that Runx2 regulates chondrocyte maturation directly or indirectly through the regulation of osteoblast differentiation in the perichondrium.

In the prechondrogenic cell line, ATDC5, the expression of *Runx2* was found to be enhanced prior to differentiation to the hypertrophic phenotype, and a treatment with antisense oligonucleotides for *Runx2* inhibited chondrocyte maturation. The retrovirally forced expression of *Runx2* in chick immature chondrocytes also induced chondrocyte maturation (Enomoto et al. 2000). These findings indicate that Runx2 is an important regulatory factor in chondrocyte maturation (Komori 2000) (Fig. 6.1). The overexpression of *Runx2* in chondrocytes using a *Col2a1* promoter/enhancer has been shown to accelerate chondrocyte maturation and endochondral ossification (Takeda et al. 2001; Ueta et al. 2001). Chondrocyte maturation even occurs in permanent cartilage including articular cartilage, thyroid cartilage, cricoid cartilage, tracheal cartilage, and intervertebral discs, which are replaced with bone in *Runx2* transgenic mice. In contrast, the expression of dn-*Runx2* in chondrocytes decelerates chondrocyte maturation and endochondral ossification (Ueta et al. 2001). Since *Runx2* and dn-*Runx2* are only expressed in chondrocytes, these findings indicate that Runx2 directly regulates chondrocyte maturation (Fig. 6.1).

Tenascin is expressed in chondrocytes once cartilage tissue appears, but becomes limited to articular chondrocytes as cartilage development progresses. The expression of tenascin is absent in the presumptive joint regions of *Runx2* transgenic mice, while it is expressed in most chondrocytes in the skeleton of dn-*Runx2* transgenic mice. These findings indicate that Runx2 inhibits chondrocytes from acquiring the characteristics of permanent cartilage (Ueta et al. 2001) (Fig. 6.1). The mechanisms responsible for the specification of permanent cartilage have not yet been elucidated in detail. In *Runx2* transgenic mice, permanent cartilage undergoes endochondral ossification. Therefore, even permanent cartilage or cartilage fated to be permanent has the potential to be transient cartilage that enters the endochondral pathway. Furthermore, the lack of cell hypertrophy in permanent cartilage appears to be due to negative regulation by microenvironmental cues and mechanisms, which may down-regulate the expression of *Runx2*. The degeneration of permanent cartilage is a feature of the pathological changes occurring with osteoarthritis in articular joints. Osteoarthritis is frequently associated with the ectopic expression of a number of molecules such as *Col10a1* (von der Mark et al. 1992; Shlopov et al. 1997), *Spp1* (Pullig et al. 2000), and *Mmp13* (Shlopov et al. 1997), which are normally specific to hypertrophic chondrocytes and are encoded by the direct target genes of Runx2 (Li et al. 2011; Sato et al. 1998; Porte et al. 1999; Hess et al. 2001; Jimenez et al. 1999). Therefore, we previously proposed that the degradation of articular cartilage in osteoarthritis may be related to the uncontrolled behavior of permanent chondrocytes and abnormal expression of the hypertrophic phenotype, and also that Runx2 may be involved in osteoarthritis (Ueta et al. 2001) (Fig. 6.1). In accordance with these hypotheses, the degradation of articular cartilage in *Runx2*$^{+/-}$ mice was previously reported to be significantly reduced in an osteoarthritis mouse model (Kamekura et al. 2006).

Terminally differentiated chondrocytes have been detected in restricted parts of the skeleton in *Runx2*$^{-/-}$ mice, indicating that other transcription factors are also involved in chondrocyte maturation. Runx3 is expressed in prehypertrophic chondrocytes, and chondrocyte maturation is slightly disturbed at E15.5, but not in the newborn stage. Chondrocyte maturation was previously shown to be completely absent in the whole skeleton in *Runx2*$^{-/-}$ *Runx3*$^{-/-}$ mice, indicating that Runx2 and Runx3 have redundant functions in chondrocyte maturation and are essential for chondrocyte maturation (Yoshida et al. 2004).

In *Runx2*$^{-/-}$ mice, the lengths of the limbs are short, chondrocyte proliferation is reduced, and the expression of *Ihh*, which is expressed in prehypertrophic chondrocytes, is severely reduced. Several Runx2-binding motifs have been identified in the promoter region of *Ihh*, and Runx2 directly regulates *Ihh* expression (Yoshida et al. 2004). Therefore, Runx2 regulates not only chondrocyte maturation, but also chondrocyte proliferation through *Ihh* regulation (Komori 2005) (Fig. 6.1).

6.4 Roles of Cbfb in Skeletal Development

Runx1$^{-/-}$ mice and *Cbfb*$^{-/-}$ mice both die at midgestation due to the lack of fetal liver hematopoiesis, indicating that the heterodimerization of Runx1 and Cbfb is required for fetal liver hematopoiesis (Okuda et al. 1996; Sasaki et al. 1996; Wang et al. 1996a, b). Since *Cbfb*$^{-/-}$ mice die between E11.5–13.5, the involvement of Cbfb in skeletal development remains to be clarified. In an attempt to overcome lethality, we and others partially rescued the lack of fetal liver hematopoiesis in *Cbfb*$^{-/-}$ mice, showing the requirement of Cbfb in skeletal development (Kundu et al. 2002; Miller et al. 2002; Yoshida et al. 2002).

In order to precisely evaluate the functions of Cbfb in skeletal development, *Cbfb* was conditionally deleted using *Dermo1* Cre knock-in mice, in which Cre is expressed in mesenchymal cells, giving rise to chondrocyte and osteoblast lineages. The processes of endochondral and intramembranous ossification are both retarded in *Cbfb*$^{fl/fl/Cre}$ mice due to the deceleration of chondrocyte maturation and osteoblast differentiation. Chondrocyte proliferation was also shown to be reduced in *Cbfb*$^{fl/fl/Cre}$ mice (Qin et al. 2015). Similar findings have been reported in *Cbfb* conditional knockout mice using *Sp7*-Cre mice, *Col2a1* Cre mice, *Prrx1* Cre mice, and *Dermo1* Cre mice (Chen et al. 2014; Fei et al. 2014; Wu et al. 2014a, b; Lim et al. 2015). Although the development of endochondral bones is known to be severely affected in *Cbfb*$^{fl/fl/Cre}$ mice, but not in *Runx2*$^{+/-}$ mice, the development of calvariae and clavicles was affected less in *Cbfb*$^{fl/fl/Cre}$ mice than in *Runx2*$^{+/-}$ mice (Qin et al. 2015). Calvariae and the lateral parts of clavicles are formed through intramembranous ossification (Huang et al. 1997; Marks Jr. and Odgren 2002). Therefore, these findings indicate that Cbfb is vital for chondrocyte maturation and proliferation as well as osteoblast differentiation, and also that Cbfb is crucial for endochondral bone development, but is only partially required for intramembranous bone development.

Runx family protein levels are reduced in *Cbfb*$^{fl/fl/Cre}$ mice, indicating that Cbfb is required for the stability of Runx family proteins (Qin et al. 2015). Cbfb protects Runx2 from polyubiquitination (Lim et al. 2015). However, the levels of reduction differ among Runx family proteins and in cartilaginous limb skeletons and calvariae at E15.5 (Qin et al. 2015). As described above, Runx2 and Runx3 are essential for chondrocyte maturation, and Runx1 is involved in the development of the sternum, occipital bone, and palate by regulating chondrocyte differentiation (Kimura et al. 2010; Liakhovitskaia et al. 2010). Therefore, all Runx family proteins are involved in chondrocyte differentiation. In cartilaginous limb skeletons, the levels of all Runx family proteins are severely reduced in the order of Runx1>Runx3>Runx2, and exist at levels that are 3 %, 8 %, and 13 % those in wild-type mice, respectively. Although the function of Runx1 in the osteoblast lineage is unknown, Runx3 has been shown to play a role in the proliferation of osteoblast lineage cells (Bauer et al. 2015). In calvariae, Runx1 protein levels are the most severely reduced, at 7 % that in wild-type mice, whereas Runx2 and Runx3 protein levels in calvariae are 55 % and 25 %, respectively, those in wild-type mice. Therefore, protein stability differs among Runx family proteins in the absence of Cbfb, with the Runx2 protein being more stable in calvariae than in cartilaginous limb skeletons in the absence of Cbfb. Some unknown proteins may compensate for the lack of Cbfb in calvariae in order to protect against the degradation of the Runx2 protein. The relative stability of the Runx2 protein in the calvariae of *Cbfb*$^{fl/fl/Cre}$ mice explains why *Cbfb*$^{fl/fl/Cre}$ mice show a severe delay in endochondral ossification, but milder deformities in calvariae and the lateral parts of clavicles, which are formed through intramembranous ossification (Qin et al. 2015).

Two functional Cbfb isoforms have been identified: Cbfb1 and Cbfb2. They are formed by alternative splicing using donor splice sites located inside exon 5 and at the 3′ terminus of exon 5, respectively, and an acceptor splice site located at the 5′ terminus of exon 6 (Ogawa et al. 1993). *Cbfb1*$^{-/-}$ mice and *Cbfb2*$^{-/-}$ mice have been generated by mutating donor splice signals (Tachibana et al. 2011). *Cbfb1*$^{-/-}$ mice develop normally, whereas *Cbfb2*$^{-/-}$ mice show dwarfism and endochondral and intramembranous ossification is inhibited (Jiang et al. 2016). Although Cbfb1 and Cbfb2 exhibit similar activities for the stabilization of Runx family proteins, Cbfb1 is more potent at enhancing chondrocyte and osteoblast differentiation and the DNA binding of Runx2. However, the formation of the Cbfb1 isoform is strictly regulated in skeletal tissues, livers, and thymuses, in which Runx family transcription factors play important roles in osteoblast and chondrocyte differentiation, hematopoiesis, and T-cell development, respectively, and *Cbfb1* mRNA levels are one third those of *Cbfb2*. Therefore, Cbfb1 and Cbfb2 have redundant functions with different efficiencies, and modulations in the relative levels of the isoforms appear to adjust transcriptional activation by Runx2 to appropriate physiological levels (Jiang et al. 2016).

6.5 Transcriptional Regulation of the Runx2 Gene

Runx2 is expressed as two isoforms that possess different N-termini (type I Runx2 starting with the sequence MRIPV and type II Runx2 starting with the sequence MASNS), and are expressed under two promoters: the proximal (P2) and distal (P1) promoters, respectively (Fujiwara et al. 1999). Although both isoforms are expressed in osteoblasts and chondrocytes, the expression of *Runx2* in osteoblasts was found to be mainly transcribed from the P1 promoter (Park et al. 2001; Enomoto et al. 2000). The transcriptional regulation of *Runx2* was investigated in the P1 promoter (Fujiwara et al. 1999; Zambotti et al. 2002; Lee et al. 2005; Hassan et al. 2007; Zhang et al. 2009; Gaur et al. 2005), and the findings obtained showed that reporter mice under the control of the P1 promoter failed to express the reporter gene in osteoblasts (Lengner et al. 2002), suggesting the presence of an enhancer for osteoblast-specific expression.

GFP reporter mice using a 200-kb BAC clone of the *Runx2* gene locus, which includes the P1 and P2 promoters, recapitulate the endogenous expression of *Runx2* (Kawane et al. 2014). A 343-bp osteoblast-specific enhancer was identified following the serial deletion of the BAC clone. GFP is specifically expressed in the osteoblasts of GFP reporter mice driven by an enhancer and minimal promoter. The sequence of this enhancer is highly conserved among the mouse, human, dog, horse, opossum, and chicken. It is also highly enriched for histone H3 mono- and dimethylated at Lys4 and acetylated at Lys27 and Lys18, but depleted for histone H3 trimethylated at Lys4 in primary osteoblasts. Furthermore, the histone variant H2A.Z is enriched in the enhancer. These are typical chromatin modifications in enhancers. A 89 bp fragment in the 343-bp enhancer still retains the ability to direct the reporter gene to osteoblasts. Dlx5/6 and Mef2 have been shown to directly bind to the homeobox motif and Mef2-binding motif in the 89-bp core sequence, respectively. Dlx5/6 and Mef2 form an enhanceosome with Tcf7, Ctnnb1, Sox5/6, Smad1, and Sp7, which are integrated into the enhanceosome by a protein-protein interaction, and activate the enhancer. Since *Tcf7* and *Sp7* are known to be regulated by Runx2 (Mikasa et al. 2011; Yoshida et al. 2012), these transcription factors are reciprocally regulated in osteoblast differentiation. Although Msx2 and Dlx5 both bind to the homeobox motif in the enhancer, Msx2 inhibits enhancer activity. The binding of Msx2 to the homeobox motif is dominant in the uncommitted mesenchymal cell line, C3H10T1/2, whereas the binding of Dlx5 to it is dominant in osteoblastic MC3T3-E1 cells. Therefore, the switching of binding from Msx2 to Dlx5 in the homeobox motif is important for activation of the enhancer (Kawane et al. 2014). The 343-bp enhancer is useful for the screening of drugs for osteoporosis and bone regeneration by targeting *Runx2*, and is also important as a vector in gene therapy for bone diseases.

Acknowledgments This work was supported by the grant from the Japanese Ministry of Education, Culture, Sports, Science and Technology to TK (Grant number: 26221310).

References

Adhami, M. D., Rashid, H., Chen, H., Clarke, J. C., Yang, Y., & Javed, A. (2015). Loss of Runx2 in committed osteoblasts impairs postnatal skeletogenesis. *Journal of Bone and Mineral Research, 30*(1), 71–82. doi:10.1002/jbmr.2321.

Aubin, J., & Triffitt, J. (2002). Mesenchymal stem cells and osteoblast differentiation. In J. P. Bilezikian, L. G. Raisz, & G. A. Rodan (Eds.), *Principles of bone biology*. New York: Academic Press.

Banerjee, C., McCabe, L. R., Choi, J. Y., Hiebert, S. W., Stein, J. L., Stein, G. S., & Lian, J. B. (1997). Runt homology domain proteins in osteoblast differentiation: AML3/CBFA1 is a major component of a bone-specific complex. *Journal of Cellular Biochemistry, 66*(1), 1–8.

Bauer, O., Sharir, A., Kimura, A., Hantisteanu, S., Takeda, S., & Groner, Y. (2015). Loss of osteoblast Runx3 produces severe congenital osteopenia. *Molecular and Cellular Biology, 35*(7), 1097–1109. doi:10.1128/mcb.01106-14.

Chen, W., Ma, J., Zhu, G., Jules, J., Wu, M., McConnell, M., et al. (2014). Cbfβ deletion in mice recapitulates cleidocranial dysplasia and reveals multiple functions of Cbfβ required for skeletal development. *Proceedings of the National Academy of Sciences of the United States of America, 111*(23), 8482–8487. doi:10.1073/pnas.1310617111.

Ducy, P., Zhang, R., Geoffroy, V., Ridall, A. L., & Karsenty, G. (1997). Osf2/Cbfa1: A transcriptional activator of osteoblast differentiation. *Cell, 89*(5), 747–754.

Enomoto, H., Enomoto-Iwamoto, M., Iwamoto, M., Nomura, S., Himeno, M., Kitamura, Y., et al. (2000). Cbfa1 is a positive regulatory factor in chondrocyte maturation. *The Journal of Biological Chemistry, 275*(12), 8695–8702.

Enomoto, H., Shiojiri, S., Hoshi, K., Furuichi, T., Fukuyama, R., Yoshida, C. A., Kanatani, N., Nakamura, R., Mizuno, A., Zanma, A., Yano, K., Yasuda, H., Higashio, K., Takada, K., Komori, T. (2003). Induction of osteoclast differentiation by Runx2 through receptor activator of nuclear factor-κB ligand (RANKL) and osteoprotegerin regulation and partial rescue of osteoclastogenesis in Runx2$^{-/-}$ mice by RANKL transgene. *The Journal of Biological Chemistry, 278*(26), 23971–23977. doi:10.1074/jbc.M302457200 M302457200 [pii].

Fei, T., Mengrui, W., Lianfu, D., Guochun, Z., Junqing, M., Bo, G., et al. (2014). Core binding factor beta (Cbfβ) controls the balance of chondrocyte proliferation and differentiation by upregulating Indian hedgehog (Ihh) expression and inhibiting parathyroid hormone-related protein receptor (PPR) expression in postnatal cartilage and bone formation. *Journal of Bone and Mineral Research, 29*(7), 1564–1574. doi:10.1002/jbmr.2275.

Fujiwara, M., Tagashira, S., Harada, H., Ogawa, S., Katsumata, T., Nakatsuka, M., Komori, T., Takada, H. (1999). Isolation and characterization of the distal promoter region of mouse Cbfa1. *Biochimica et Biophysica Acta, 1446*(3), 265–272. doi:S0167–4781(99)00113-X [pii].

Gaur, T., Lengner, C. J., Hovhannisyan, H., Bhat, R. A., Bodine, P. V., Komm, B. S., et al. (2005). Canonical WNT signaling promotes osteogenesis by directly stimulating Runx2 gene expression. *The Journal of Biological Chemistry, 280*(39), 33132–33140. doi:10.1074/jbc.M500608200.

Geoffroy, V., Kneissel, M., Fournier, B., Boyde, A., & Matthias, P. (2002). High bone resorption in adult aging transgenic mice overexpressing Cbfa1/Runx2 in cells of the osteoblastic lineage. *Molecular and Cellular Biology, 22*(17), 6222–6233.

Harada, H., Tagashira, S., Fujiwara, M., Ogawa, S., Katsumata, T., Yamaguchi, A., et al. (1999). Cbfa1 isoforms exert functional differences in osteoblast differentiation. *The Journal of Biological Chemistry, 274*(11), 6972–6978.

Hassan, M. Q., Tare, R., Lee, S. H., Mandeville, M., Weiner, B., Montecino, M., et al. (2007). HOXA10 controls osteoblastogenesis by directly activating bone regulatory and phenotypic genes. *Molecular and Cellular Biology, 27*(9), 3337–3352. doi:10.1128/mcb.01544-06.

Hess, J., Porte, D., Munz, C., & Angel, P. (2001). AP-1 and Cbfa/runt physically interact and regulate parathyroid hormone-dependent MMP13 expression in osteoblasts through a new osteoblast-specific element 2/AP-1 composite element. *The Journal of Biological Chemistry, 276*(23), 20029–20038. doi:10.1074/jbc.M010601200.

Himeno, M., Enomoto, H., Liu, W., Ishizeki, K., Nomura, S., Kitamura, Y., & Komori, T. (2002). Impaired vascular invasion of Cbfa1-deficient cartilage engrafted in the spleen. *Journal of Bone and Mineral Research, 17*(7), 1297–1305. doi:10.1359/jbmr.2002.17.7.1297.

Huang, L. F., Fukai, N., Selby, P. B., Olsen, B. R., & Mundlos, S. (1997). Mouse clavicular development: Analysis of wild-type and cleidocranial dysplasia mutant mice. *Developmental Dynamics, 210*(1), 33–40. doi:10.1002/(sici)1097-0177(199709)210:1<33::aid-aja4>3.0.co;2-2.

Inada, M., Yasui, T., Nomura, S., Miyake, S., Deguchi, K., Himeno, M., et al. (1999). Maturational disturbance of chondrocytes in Cbfa1-deficient mice. *Developmental Dynamics, 214*(4), 279–290. doi:10.1002/(sici)1097-0177(199904)214:4<279::aid-aja1>3.0.co;2-w.

Javed, A., Barnes, G. L., Jasanya, B. O., Stein, J. L., Gerstenfeld, L., Lian, J. B., & Stein, G. S. (2001). runt homology domain transcription factors (Runx, Cbfa, and AML) mediate repression of the bone sialoprotein promoter: evidence for promoter context-dependent activity of Cbfa proteins. *Molecular and Cell Biology, 21*(8), 2891–2905 (0270-7306 (Print)).

Jiang, Q., Qin, X., Kawane, T., Komori, H., Matsuo, Y., Taniuchi, I., et al. (2016). Cbfb2 isoform dominates more potent Cbfb1 and is required for skeletal development. *Journal of Bone and Mineral Research.* doi:10.1002/jbmr.2814.

Jimenez, M. J., Balbin, M., Lopez, J. M., Alvarez, J., Komori, T., & Lopez-Otin, C. (1999). Collagenase 3 is a target of Cbfa1, a transcription factor of the runt gene family involved in bone formation. *Molecular and Cellular Biology, 19*(6), 4431–4442.

Kamekura, S., Kawasaki, Y., Hoshi, K., Shimoaka, T., Chikuda, H., Maruyama, Z., et al. (2006). Contribution of runt-related transcription factor 2 to the pathogenesis of osteoarthritis in mice after induction of knee joint instability. *Arthritis and Rheumatism, 54*(8), 2462–2470. doi:10.1002/art.22041.

Kanatani, N., Fujita, T., Fukuyama, R., Liu, W., Yoshida, C. A., Moriishi, T., Yamana, K., Miyazaki, T., Toyosawa, S., Komori, T. (2006). Cbfβ regulates Runx2 function isoform-dependently in postnatal bone development. *Developmental Biology, 296*(1), 48–61. doi:S0012–1606(06)00242–9 [pii]10.1016/j.ydbio.2006.03.039.

Kawane, T., Komori, H., Liu, W., Moriishi, T., Miyazaki, T., Mori, M., et al. (2014). Dlx5 and Mef2 regulate a novel Runx2 enhancer for osteoblast-specific expression. *Journal of Bone and Mineral Research, 29*(9), 1960–1969. doi:10.1002/jbmr.2240.

Kern, B., Shen, J., Starbuck, M., & Karsenty, G. (2001). Cbfa1 contributes to the osteoblast-specific expression of type I collagen genes. *The Journal of Biological*

Chemistry, 276(10), 7101–7107. doi:10.1074/jbc.M006215200.

Kimura, A., Inose, H., Yano, F., Fujita, K., Ikeda, T., Sato, S., et al. (2010). Runx1 and Runx2 cooperate during sternal morphogenesis. *Development, 137*(7), 1159–1167. doi:10.1242/dev.045005.

Komori, T. (2000). A fundamental transcription factor for bone and cartilage. *Biochemical and Biophysical Research Communications, 276*(3), 813–816. doi:10.1006/bbrc.2000.3460S0006-291X(00)93460-0 [pii].

Komori, T. (2005). Regulation of skeletal development by the Runx family of transcription factors. *Journal of Cellular Biochemistry, 95*(3), 445–453. doi:10.1002/jcb.20420.

Komori, T. (2006). Regulation of osteoblast differentiation by transcription factors. *Journal of Cellular Biochemistry, 99*(5), 1233–1239. doi:10.1002/jcb.20958.

Komori, T., Yagi, H., Nomura, S., Yamaguchi, A., Sasaki, K., Deguchi, K., Shimizu, Y., Bronson, R. T., Gao, Y. H., Inada, M., Sato, M., Okamoto, R., Kitamura, Y., Yoshiki, S., Kishimoto, T. (1997). Targeted disruption of Cbfa1 results in a complete lack of bone formation owing to maturational arrest of osteoblasts. *Cell, 89*(5), 755–764. doi:S0092-8674(00)80258-5 [pii].

Kundu, M., Javed, A., Jeon, J. P., Horner, A., Shum, L., Eckhaus, M., et al. (2002). Cbfβ interacts with Runx2 and has a critical role in bone development. *Nature Genetics, 32*(4), 639–644. doi:10.1038/ng1050.

Lamour, V., Detry, C., Sanchez, C., Henrotin, Y., Castronovo, V., & Bellahcene, A. (2007). Runx2- and histone deacetylase 3-mediated repression is relieved in differentiating human osteoblast cells to allow high bone sialoprotein expression. *The Journal of Biological Chemistry, 282*(50), 36240–36249. doi:10.1074/jbc.M705833200.

Lee, K. S., Kim, H. J., Li, Q. L., Chi, X. Z., Ueta, C., Komori, T., et al. (2000). Runx2 is a common target of transforming growth factor beta1 and bone morphogenetic protein 2, and cooperation between Runx2 and Smad5 induces osteoblast-specific gene expression in the pluripotent mesenchymal precursor cell line C2C12. *Molecular and Cellular Biology, 20*(23), 8783–8792.

Lee, M. H., Kim, Y. J., Yoon, W. J., Kim, J. I., Kim, B. G., Hwang, Y. S., et al. (2005). Dlx5 specifically regulates Runx2 type II expression by binding to homeodomain-response elements in the Runx2 distal promoter. *The Journal of Biological Chemistry, 280*(42), 35579–35587. doi:10.1074/jbc.M502267200.

Lee, S. H., Che, X., Jeong, J. H., Choi, J. Y., Lee, Y. J., Lee, Y. H., et al. (2012). Runx2 protein stabilizes hypoxia-inducible factor-1alpha through competition with von Hippel-Lindau protein (pVHL) and stimulates angiogenesis in growth plate hypertrophic chondrocytes. *The Journal of Biological Chemistry, 287*(18), 14760–14771. doi:10.1074/jbc.M112.340232.

Lengner, C. J., Drissi, H., Choi, J. Y., van Wijnen, A. J., Stein, J. L., Stein, G. S., & Lian, J. B. (2002). Activation of the bone-related Runx2/Cbfa1 promoter in mesenchymal condensations and developing chondrocytes of the axial skeleton. *Mechanisms of Development, 114*(1–2), 167–170.

Li, F., Lu, Y., Ding, M., Napierala, D., Abbassi, S., Chen, Y., et al. (2011). Runx2 contributes to murine Col10a1 gene regulation through direct interaction with its cis-enhancer. *Journal of Bone and Mineral Research, 26*(12), 2899–2910. doi:10.1002/jbmr.504.

Liakhovitskaia, A., Lana-Elola, E., Stamateris, E., Rice, D. P., van't Hof, R. J., & Medvinsky, A. (2010). The essential requirement for Runx1 in the development of the sternum. *Developmental Biology, 340*(2), 539–546. doi:10.1016/j.ydbio.2010.02.005.

Lim, K. E., Park, N. R., Che, X., Han, M. S., Jeong, J. H., Kim, S. Y., et al. (2015). Core binding factor β of osteoblasts maintains cortical bone mass via stabilization of Runx2 in mice. *Journal of Bone and Mineral Research, 30*(4), 715–722. doi:10.1002/jbmr.2397.

Liu, W., Toyosawa, S., Furuichi, T., Kanatani, N., Yoshida, C., Liu, Y., Himeno, M., Narai, S., Yamaguchi, A., Komori, T. (2001). Overexpression of Cbfa1 in osteoblasts inhibits osteoblast maturation and causes osteopenia with multiple fractures. *The Journal of Cell Biology, 155*(1), 157–166. doi:10.1083/jcb.200105052155/1/157 [pii].

Marks Jr., S., & Odgren, P. (2002). Structure and development of the skeleton. In J. P. Bilezikian, L. G. Raisz, & G. A. Rodan (Eds.), *Principles of bone biology*. New York: Academic Press.

Maruyama, Z., Yoshida, C. A., Furuichi, T., Amizuka, N., Ito, M., Fukuyama, R., et al. (2007). Runx2 determines bone maturity and turnover rate in postnatal bone development and is involved in bone loss in estrogen deficiency. *Developmental Dynamics, 236*(7), 1876–1890. doi:10.1002/dvdy.21187.

Mikasa, M., Rokutanda, S., Komori, H., Ito, K., Tsang, Y. S., Date, Y., et al. (2011). Regulation of Tcf7 by Runx2 in chondrocyte maturation and proliferation. *Journal of Bone and Mineral Metabolism, 29*(3), 291–299. doi:10.1007/s00774-010-0222-z.

Miller, J., Horner, A., Stacy, T., Lowrey, C., Lian, J. B., Stein, G., et al. (2002). The core-binding factor β subunit is required for bone formation and hematopoietic maturation. *Nature Genetics, 32*(4), 645–649. doi:10.1038/ng1049.

Mundlos, S., Otto, F., Mundlos, C., Mulliken, J. B., Aylsworth, A. S., Albright, S., et al. (1997). Mutations involving the transcription factor CBFA1 cause cleidocranial dysplasia. *Cell, 89*(5), 773–779.

Ogawa, E., Inuzuka, M., Maruyama, M., Satake, M., Naito-Fujimoto, M., Ito, Y., & Shigesada, K. (1993). Molecular cloning and characterization of PEBP2 β, the heterodimeric partner of a novel Drosophila runt-related DNA binding protein PEBP2α. *Virology, 194*(1), 314–331. doi:10.1006/viro.1993.1262.

Okuda, T., van Deursen, J., Hiebert, S. W., Grosveld, G., & Downing, J. R. (1996). AML1, the target of multiple chromosomal translocations in human leukemia, is essential for normal fetal liver hematopoiesis. *Cell, 84*(2), 321–330.

Otto, F., Thornell, A. P., Crompton, T., Denzel, A., Gilmour, K. C., Rosewell, I. R., et al. (1997). Cbfa1, a candidate gene for cleidocranial dysplasia syndrome, is essential for osteoblast differentiation and bone development. *Cell, 89*(5), 765–771.

Park, M. H., Shin, H. I., Choi, J. Y., Nam, S. H., Kim, Y. J., Kim, H. J., & Ryoo, H. M. (2001). Differential expression patterns of Runx2 isoforms in cranial suture morphogenesis. *Journal of Bone and Mineral Research, 16*(5), 885–892. doi:10.1359/jbmr.2001.16.5.885.

Porte, D., Tuckermann, J., Becker, M., Baumann, B., Teurich, S., Higgins, T., et al. (1999). Both AP-1 and Cbfa1-like factors are required for the induction of interstitial collagenase by parathyroid hormone. *Oncogene, 18*(3), 667–678. doi:10.1038/sj.onc.1202333.

Pullig, O., Weseloh, G., Gauer, S., & Swoboda, B. (2000). Osteopontin is expressed by adult human osteoarthritic chondrocytes: protein and mRNA analysis of normal and osteoarthritic cartilage. *Matrix Biology, 19*(3), 245–255.

Qin, X., Jiang, Q., Matsuo, Y., Kawane, T., Komori, H., Moriishi, T., et al. (2015). Cbfb regulates bone development by stabilizing Runx family proteins. *Journal of Bone and Mineral Research, 30*(4), 706–714. doi:10.1002/jbmr.2379.

Sasaki, K., Yagi, H., Bronson, R. T., Tominaga, K., Matsunashi, T., Deguchi, K., et al. (1996). Absence of fetal liver hematopoiesis in mice deficient in transcriptional coactivator core binding factor β. *Proceedings of the National Academy of Sciences of the United States of America, 93*(22), 12359–12363.

Sato, M., Morii, E., Komori, T., Kawahata, H., Sugimoto, M., Terai, K., et al. (1998). Transcriptional regulation of osteopontin gene in vivo by PEBP2αA/CBFA1 and ETS1 in the skeletal tissues. *Oncogene, 17*(12), 1517–1525. doi:10.1038/sj.onc.1202064.

Selvamurugan, N., Pulumati, M. R., Tyson, D. R., & Partridge, N. C. (2000). Parathyroid hormone regulation of the rat collagenase-3 promoter by protein kinase A-dependent transactivation of core binding factor α1. *The Journal of Biological Chemistry, 275*(7), 5037–5042.

Shlopov, B. V., Lie, W. R., Mainardi, C. L., Cole, A. A., Chubinskaya, S., & Hasty, K. A. (1997). Osteoarthritic lesions: Involvement of three different collagenases. *Arthritis and Rheumatism, 40*(11), 2065–2074. doi:10.1002/1529-0131(199711)40:11<2065::AID-ART20>3.0.CO;2-0.

Tachibana, M., Tenno, M., Tezuka, C., Sugiyama, M., Yoshida, H., & Taniuchi, I. (2011). Runx1/Cbfβ2 complexes are required for lymphoid tissue inducer cell differentiation at two developmental stages. *Journal of Immunology, 186*(3), 1450–1457. doi:10.4049/jimmunol.1000162.

Takarada, T., Hinoi, E., Nakazato, R., Ochi, H., Xu, C., Tsuchikane, A., et al. (2013). An analysis of skeletal development in osteoblast-specific and chondrocyte-specific runt-related transcription factor-2 (Runx2) knockout mice. *Journal of Bone and Mineral Research, 28*(10), 2064–2069. doi:10.1002/jbmr.1945.

Takeda, S., Bonnamy, J. P., Owen, M. J., Ducy, P., & Karsenty, G. (2001). Continuous expression of Cbfa1 in nonhypertrophic chondrocytes uncovers its ability to induce hypertrophic chondrocyte differentiation and partially rescues Cbfa1-deficient mice. *Genes & Development, 15*(4), 467–481. doi:10.1101/gad.845101.

Toyosawa, S., Shintani, S., Fujiwara, T., Ooshima, T., Sato, A., Ijuhin, N., & Komori, T. (2001). Dentin matrix protein 1 is predominantly expressed in chicken and rat osteocytes but not in osteoblasts. *Journal of Bone and Mineral Research, 16*(11), 2017–2026. doi:10.1359/jbmr.2001.16.11.2017.

Ueta, C., Iwamoto, M., Kanatani, N., Yoshida, C., Liu, Y., Enomoto-Iwamoto, M., et al. (2001). Skeletal malformations caused by overexpression of Cbfa1 or its dominant negative form in chondrocytes. *The Journal of Cell Biology, 153*(1), 87–100.

von der Mark, K., Kirsch, T., Nerlich, A., Kuss, A., Weseloh, G., Gluckert, K., & Stoss, H. (1992). Type X collagen synthesis in human osteoarthritic cartilage. Indication of chondrocyte hypertrophy. *Arthritis and Rheumatism, 35*(7), 806–811.

Wang, Q., Stacy, T., Binder, M., Marin-Padilla, M., Sharpe, A. H., & Speck, N. A. (1996a). Disruption of the Cbfa2 gene causes necrosis and hemorrhaging in the central nervous system and blocks definitive hematopoiesis. *Proceedings of the National Academy of Sciences of the United States of America, 93*(8), 3444–3449.

Wang, Q., Stacy, T., Miller, J. D., Lewis, A. F., Gu, T. L., Huang, X., et al. (1996b). The CBFβ subunit is essential for CBFα2 (AML1) function in vivo. *Cell, 87*(4), 697–708.

Wu, M., Li, C., Zhu, G., Wang, Y., Jules, J., Lu, Y., et al. (2014a). Deletion of core-binding factor β (Cbfβ) in mesenchymal progenitor cells provides new insights into Cbfβ/Runxs complex function in cartilage and bone development. *Bone, 65*, 49–59. doi:10.1016/j.bone.2014.04.031.

Wu, M., Li, Y. P., Zhu, G., Lu, Y., Wang, Y., Jules, J., et al. (2014b). Chondrocyte-specific knockout of Cbfβ reveals the indispensable function of Cbfβ in chondrocyte maturation, growth plate development and trabecular bone formation in mice. *International Journal of Biological Sciences, 10*(8), 861–872. doi:10.7150/ijbs.8521.

Yoshida, C. A., Furuichi, T., Fujita, T., Fukuyama, R., Kanatani, N., Kobayashi, S., Satake, M., Takada, K., Komori, T. (2002). Core-binding factor β interacts with Runx2 and is required for skeletal development.

Nature Genetics, 32(4), 633–638. doi:10.1038/ng1015ng1015 [pii].

Yoshida, C. A., Yamamoto, H., Fujita, T., Furuichi, T., Ito, K., Inoue, K., Yamana, K., Zanma, A., Takada, K., Ito, Y., Komori, T. (2004). Runx2 and Runx3 are essential for chondrocyte maturation, and Runx2 regulates limb growth through induction of Indian hedgehog. *Genes Developement, 18*(8), 952–963. doi:10.1101/gad.1174704 18/8/952 [pii].

Yoshida, C. A., Komori, H., Maruyama, Z., Miyazaki, T., Kawasaki, K., Furuichi, T., et al. (2012). SP7 inhibits osteoblast differentiation at a late stage in mice. *PloS One, 7*(3), e32364. doi:10.1371/journal.pone.0032364.

Zambotti, A., Makhluf, H., Shen, J., & Ducy, P. (2002). Characterization of an osteoblast-specific enhancer element in the CBFA1 gene. *The Journal of Biological Chemistry, 277*(44), 41497–41506. doi:10.1074/jbc.M204271200.

Zelzer, E., Glotzer, D. J., Hartmann, C., Thomas, D., Fukai, N., Soker, S., & Olsen, B. R. (2001). Tissue specific regulation of VEGF expression during bone development requires Cbfa1/Runx2. *Mechanisms of Development, 106*(1–2), 97–106.

Zhang, Y., Hassan, M. Q., Xie, R. L., Hawse, J. R., Spelsberg, T. C., Montecino, M., et al. (2009). Co-stimulation of the bone-related Runx2 P1 promoter in mesenchymal cells by SP1 and ETS transcription factors at polymorphic purine-rich DNA sequences (Y-repeats). *The Journal of Biological Chemistry, 284*(5), 3125–3135. doi:10.1074/jbc.M807466200.

Mitotic Gene Bookmarking: An Epigenetic Mechanism for Coordination of Lineage Commitment, Cell Identity and Cell Growth

7

Sayyed K. Zaidi, Jane B. Lian, Andre van Wijnen, Janet L. Stein, and Gary S. Stein

Abstract

Epigenetic control of gene expression contributes to dynamic responsiveness of cellular processes that include cell cycle, cell growth and differentiation. Mitotic gene bookmarking, retention of sequence-specific transcription factors at target gene loci, including the RUNX regulatory proteins, provide a novel dimension to epigenetic regulation that sustains cellular identity in progeny cells following cell division. Runx transcription factor retention during mitosis coordinates physiological control of cell growth and differentiation in a broad spectrum of biological conditions, and is associated with compromised gene expression in pathologies that include cancer.

Keywords

RUNX • Mitotic bookmarking • Epigenetic control • Gene expression

7.1 Introduction

Epigenetic regulation of gene expression is essential it enables cells to accommodate dynamic changes that take place during a variety of cellular processes, including cell cycle progression, cell growth, and cell differentiation(Chi et al. 2010; Sarkies and Sale 2012; Ptashne 2013; Cerase et al. 2014; Attar and Kurdistani 2014). DNA methylation and histone modifications are the most studied and well understood epigenetic mechanisms that contribute to spatial and temporal regulation of gene expression (Strahl and Allis 2000; He and Lehming 2003; Chi et al. 2010; Rivera and Bennett 2010; Schübeler 2015). We used RUNX proteins as a model for lineage commitment and maintenance to identify mitotic gene bookmarking – retention of sequence specific phenotypic transcription factors on target genes during mito-

S.K. Zaidi • J.B. Lian • J.L. Stein • G.S. Stein (✉)
Department of Biochemistry and University of Vermont Cancer Center, The Larner College of Medicine, University of Vermont, 89 Beaumont Avenue, Burlington, VT 05405, USA
e-mail: gary.stein@uvm.edu

A. van Wijnen
Departments of Orthopedic Surgery & Biochemistry and Molecular Biology, Mayo Clinic, 200 First Street SW, Rochester, MN 55905, USA

Y. Groner et al. (eds.), *RUNX Proteins in Development and Cancer*, Advances in Experimental Medicine and Biology 962, DOI 10.1007/978-981-10-3233-2_7

sis. This bookmarking represents a novel epigenetic mechanism that ensures maintenance of cellular identity across cell generations and coordinates cell growth and differentiation (Zaidi et al. 2003; Young et al. 2007a; Young et al. 2007b; Ali et al. 2008; Bakshi et al. 2008; Pande et al. 2009; Ali et al. 2010; Zaidi et al. 2010; Ali et al. 2012; Zaidi et al. 2014; Lopez-Camacho et al. 2014). Many independent studies (Xing 2005; Sarge and Park-Sarge 2005; Dey et al. 2009; Blobel et al. 2009; Zhao et al. 2011; Kadauke et al. 2012; Arampatzi et al. 2013; Caravaca et al. 2013; Kadauke and Blobel 2013; Lake et al. 2014; Zaret 2014; Lodhi et al. 2014; Wong et al. 2014; Lodhi et al. 2016; Lerner et al. 2016; Festuccia et al. 2016) have since confirmed that mitotic gene bookmarking is a prevalent epigenetic mechanism in a number of biological models and under physiological and pathological conditions. In this chapter, we present an overview of mitotic bookmarking as a key mechanistic, epigenetic dimension of RUNX control for multiple cellular processes.

7.2 Mitotic Bookmarking: A Historical Perspective

In the 1980s, Weintraub, Groudine, and Struhl found that a limited number of nuclease accessible sites on the condensed mitotic chromatin persist through the cell cycle (Struhl 1981; Groudine and Weintraub 1982; Weintraub 1985). In the 1990s, John and Workman proposed that these inheritable hypersensitive sites are putative accessible sites that provide a platform to place "bookmarks" for rapid activation of genes following mitosis (John and Workman 1998). This model explained observations by Levens and colleagues, as well as by Wu and colleagues that promoters of the *Myc*, *hsp70i, and βGlobin* genes each contain nuclease accessible sites that persist through mitosis (Martínez-Balbás et al. 1995; Michelotti et al. 1997). Wu and colleagues also examined several sequence-specific transcription factors during mitosis and found that these transcription factors are displaced from the condensed mitotic chromatin (Martínez-Balbás et al. 1995). In 2003, our group identified the osteogenic master regulator RUNX2 as the first sequence specific bookmark that remained associated with target genes through mitosis (Zaidi et al. 2003). Subsequent studies of RUNX family of phenotypic transcription factors from our group (Zaidi et al. 2003; Young et al. 2007a; Young et al. 2007b; Ali et al. 2008; Bakshi et al. 2008; Pande et al. 2009; Ali et al. 2010; Zaidi et al. 2010; Ali et al. 2012; Zaidi et al. 2014; Lopez-Camacho et al. 2014) and studies by other groups examining various transcription factors (Xing 2005; Sarge and Park-Sarge 2005; Dey et al. 2009; Blobel et al. 2009; Zhao et al. 2011; Kadauke et al. 2012; Arampatzi et al. 2013; Caravaca et al. 2013; Kadauke and Blobel 2013; Lake et al. 2014; Zaret 2014; Lodhi et al. 2014; Wong et al. 2014; Lodhi et al. 2016; Lerner et al. 2016; Festuccia et al. 2016) have identified mitotic bookmarking as a key epigenetic mechanism for regulation of genes that coordinate cell growth and lineage maintenance following mitosis.

7.3 Characteristics of Mitotic Bookmarks

Studies over the past decade have contributed to an emerging view of mitotic bookmarking and have revealed shared characteristics of mitotic bookmarks across biological models:

7.3.1 Properties of Genes That Are Bookmarked During Mitosis

Condensation of chromosomes during mitosis is a key event that leads to a brief pause in transcription (Hartl et al. 1993; Gottesfeld and Forbes 1997) and displacement of some sequence-specific transcription factors from their target genes (Martínez-Balbás et al. 1995). While reconfiguration of cellular architecture during mitosis requires coordination of several independent mechanisms (Pines 2006), studies have established that phosphorylation of histone H3 on serine residues at positions 10 and 28 plays a key role in condensation of mitotic chromosomes

(Kouzarides 2007; Margueron and Reinberg 2010). These phosphorylation events are mediated by the mitotic Aurora B kinase and result in displacement of several regulatory and structural proteins that include Heterochromatin Protein 1 (HP1), RNA Polymerase II, and chromatin remodeler B cell-specific Moloney murine leukemia virus integration site 1 (BMI1) from mitotic chromosomes (Sabbattini et al. 2007). It is noteworthy that earlier studies in the 1980s showed that mitotic chromosomes harbor nuclease-accessible regions that are inheritable (Struhl 1981; Groudine and Weintraub 1982; Weintraub 1985). Consistent with these observations, more recent studies, using nuclease accessibility assays combined with genome-wide approaches, have shown that the state of open chromatin remains preserved during mitosis, although it appears to be remodeled at the level of individual genes and regulatory elements (Hsiung et al. 2015). Studies focused on understanding posttranslational modifications of histone proteins show that mitotic chromosomes retain several histone modifications (Wang and Higgins 2013). Two of these modifications (Kouzarides 2007) – trimethylation of lysine 4 in histone 3 (H3K4me3), a modification associated with transcriptional activation, and a similar modification in lysine 27 (H3K27me3), a modification linked with transcriptional silencing – have been shown to be retained on promoters of genes that are rapidly reactivated following mitosis (Grandy et al. 2016). Bivalency of some genes, which represents the marking of gene regulatory regions with both activating and repressive histone marks, is emerging as a key mechanism that not only retains cellular memory, but also provides necessary plasticity for gene regulation in human embryonic stem cells, as well as in cancer cells.

In addition to histone modifications, variants of histone proteins and their nucleosomal distribution within the regulatory regions of certain genes during mitosis appear to contribute to transcriptional memory (Weber and Henikoff 2014). The histone variants H3.3 and H2A.Z are well-studied examples (Ng and Gurdon 2008; Kelly et al. 2010). For example, the H3.3 variant is predominantly distributed in actively transcribed genes during interphase, and this distribution is preserved during mitosis, indicating that incorporation of histone H3.3 in regulatory regions of genes that are reactivated immediately after mitosis may be a key hallmark of mitotically bookmarked genes. Consistent with a role of histone H3.3 in maintaining cellular memory, Gurdon and colleagues have found that incorporation of histone H3.3 into the Myogenic Differentiation 1 (*MyoD1*) gene promoter can maintain cellular transcription memory of the gene through 24 cell divisions (Ng and Gurdon 2007; Ng and Gurdon 2008). The histone H2A.Z variant appears to be retained during mitosis. Interestingly, genes that are active in the G2 phase of the cell cycle contain nucleosomes at the +1 position. These nucleosomes contain the H2A.Z variant, and slide onto the transcription start site (TSS) of specific genes during mitosis, thus resulting in a silenced state. It has been proposed that this sliding of H2A.Z-containing nucleosome at the +1 position may contribute to marking genes that require rapid reactivation following mitosis (Kelly et al. 2010). Together, these observations point to a central role for histone modifications and variants in making gene loci accessible in the condensed chromatin environment of mitotic chromosomes.

Studies using RUNX phenotypic transcription factors as a model to examine mitotic gene bookmarking have revealed two classes of genes that are bookmarked by RUNX proteins during mitosis (Zaidi et al. 2010). First, there are highly repetitive genes containing dozens of RUNX binding sites, thus providing a natural amplification of RUNX signal that is easily visualized by immunofluorescence microscopy. Ribosomal RNA genes are one example of such genes. Each of the rDNA repeats is tandemly organized as 200–300 copies on each of the 5 acrocentric chromosomes in humans and contains more than 40 binding sites for RUNX proteins. Collectively, these repeats provide a concentration of high affinity binding sites for RUNX proteins, resulting in allelic visualization of RUNX nuclear foci that co-localize with the RNA Polymerase I machinery and regulate the expression of rRNA genes post-mitotically (Young et al. 2007a; Ali et al. 2008).

Second, there are RUNX-responsive single copy genes with a limited number of RUNX binding sites (typically between 1 and 10) that must be coordinately regulated to maintain cellular identity, as well as proliferative and growth potential. For example, in osteoblasts, SMAD 2 and 4 genes – two key effectors of osteogenic BMP signaling – are localized on chromosome 18 and are bookmarked by RUNX2 during mitosis (Young et al. 2007b; Zaidi et al. 2011). This observation suggests that a subset of genes that are bookmarked by a transcription factor during mitosis may be organized in "mitotic enhancers" to coordinate post-mitotic gene expression (Fig. 7.1). It remains to be seen whether these "mitotic enhancers", in addition to containing multiple genes and transcriptional regulators, also share other properties of interphase enhancers, such as CTCF binding and presence of dimethylated H3K4 (H3K4me2) and acetylated H3K27 (H3K27ac) histone modifications.

7.3.2 Properties of Transcription Factors That Function as Mitotic Bookmarks

It is increasingly apparent that, in addition to interacting with target genes in a sequence specific manner, a transcription factor that occupies and bookmarks a subset of its target genes during mitosis is usually a phenotypic regulatory protein. Studies over the past decade have demonstrated that more than 20 transcription factors and chromatin regulatory proteins (including many lineage determining factors) are retained on mitotic chromosomes. Examples include proteins involved in genome organization (e.g., the chromatin remodeler BRD4 and the global enhancer binding protein CTCF), as well as lineage-restricted transcription factors that include the basic helix-loop-helix myogenic regulatory factors in muscle cell differentiation, CCAAT/enhancer-binding protein α in the adipocyte differentiation program, FoxA1 in liver cells, GATA1 and Runx1 in hematopoietic lineage differentiation, and Runx2 in osteoblast differentiation (Zaidi et al. 2003; Xing 2005; Sarge and Park-Sarge 2005; Young et al. 2007a; Young et al. 2007b; Ali et al. 2008; Bakshi et al. 2008; Pande et al. 2009; Dey et al. 2009; Blobel et al. 2009; Ali et al. 2010; Zaidi et al. 2010; Zhao et al. 2011; Ali et al. 2012; Kadauke et al. 2012; Arampatzi et al. 2013; Caravaca et al. 2013; Kadauke and Blobel 2013; Zaidi et al. 2014; Lake et al. 2014; Zaret 2014; Lodhi et al. 2014; Wong et al. 2014; Lopez-Camacho et al. 2014; Lodhi et al. 2016; Lerner et al. 2016; Festuccia et al. 2016). Each of these transcription factors occupy a subset of their target genes during mitosis in their respective lineages. Interference with mitotic bookmarking by these proteins results in deregulation of target genes following mitosis and compromised lineage identity.

Importantly, transcription factors that dictate lineage commitment of mesenchymal stem cell into muscle (MyoD), adipocytes (C/EBPa), or osteoblasts (RUNX2) not only bookmark RNA Pol II regulated genes during mitosis, but also the ribosomal RNA genes that are transcribed by RNA Pol I (Young et al. 2007a; Ali et al. 2008). Ribosomal RNA (rRNA) genes are intimately linked with cellular growth potential and provide a model system to study mechanistic ramifications of mitotic gene bookmarking. As discussed above, these tandemly repeated genes are naturally amplified regulatory units each with multiple binding sites for each member of a growing class of principal transcription factors. In addition to RUNX proteins, this class of mitotic transcriptional regulators that regulate both Pol I and Poll II transcribed genes include Myc, MyoD, and C/EBPa and RUNX proteins (Young et al. 2007a; Ali et al. 2008). In undifferentiated mesenchymal cells, rRNA genes are occupied by Myc, which is a transcriptional activator of these genes, and stimulates genes that are involved in cell proliferation. When mesenchymal stem cells are differentiated into myoblasts, adipocytes or osteoblasts, they exit the cell cycle and Myc is replaced by MyoD, Similarly, C/EBPa and RUNX2 each bookmark and downregulate rRNA genes, but also genes involved in cell cycle regulation. Concomitantly, these phenotypic proteins mitotically bookmark genes that are expressed immediately after mitosis and are critical for commitment to and maintenance of their respective lineages. (Ali et al. 2008). These findings

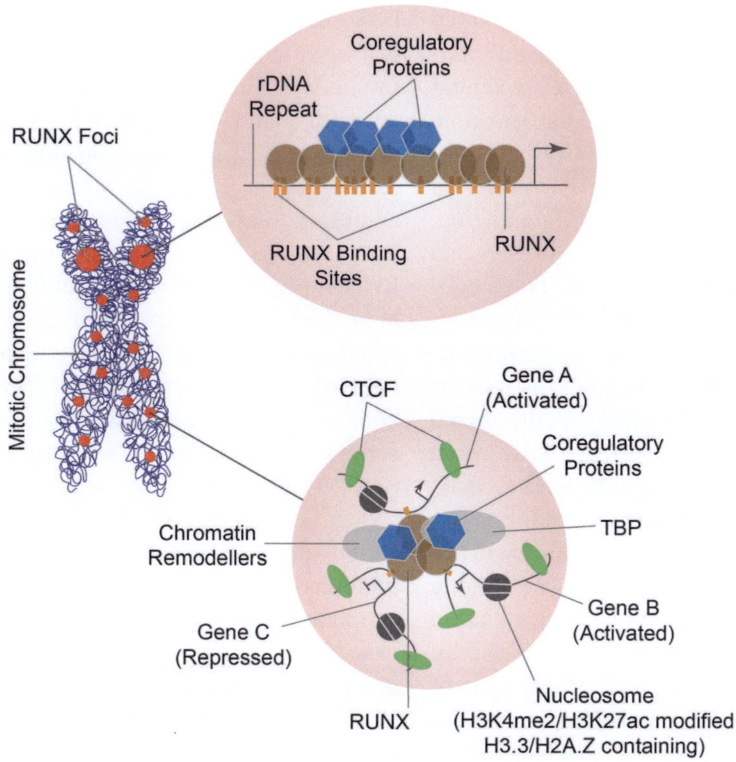

Fig. 7.1 Retention of RUNX proteins with target genes on mitotic chromosomes reveals two distinct mitotic microenvironments. RUNX proteins (shown in red circles) associate with target genes on mitotic chromosomes (shown in blue line). More than a decade of studies has revealed two distinct target genes that are occupied by RUNX proteins during mitosis: (1) Genes that normally have several hundred copies in human genome, with each copy carrying several dozen RUNX binding sites (depicted as orange vertical lines in the top left circle). These genes provide a physiological amplification of RUNX occupancy on mitotic chromosomes (organized as large, allelic foci that can be identified by immunofluorescence microscopy). A key example is ribosomal RNA genes that are intimately linked with cell growth, and – together with a subset of regulatory proteins (blue hexagons) – are regulated by RUNX proteins (brown circles). (2) Single copy genes that carry fewer RUNX binding sites, but are localized on the same chromosome (depicted in bottom left circle). These genes may also be occupied by CTCF proteins (green ovals), with specialized nucleosomes (dark gray circles) containing histone variants H3.3 and/or H2A.Z, and may be coordinately activated (e.g., Genes A and B, usually linked with phenotype maintenance) and repressed (e.g., Gene C, often linked with cell proliferation) by RUNX proteins. The collective outcome of mitotic gene bookmarking by RUNX proteins is a coordinate regulation of cell growth and proliferation as well as lineage maintenance in post-mitotic cells

highlight an important mechanistic aspect of mitotic bookmarking during lineage commitment: mechanistic coordination of cell proliferation, cell growth and cell identity. Genome-wide experimental approaches involving endogenous transcription factors in biologically relevant systems will provide further mechanistic insights into functional relevance of mitotic gene bookmarking in maintaining epigenetic cell memory in progeny cells.

7.4 Mitotic Bookmarking by Oncogenes as a Mechanism for Maintenance of Disease Phenotype

Recent studies also suggest that mitotic gene bookmarking has an important role in the onset, progression, and perpetuation of disease (Zaidi et al. 2014). A key example is provided by the leukemic fusion protein AML1-ETO (also known as RUNX1T1) that blocks myeloid cell differentiation and enhances proliferative potential (Bakshi et al. 2008). Interestingly, the leukemic AML1-ETO fusion protein mitotically bookmarks rRNA genes, as well as genes controlling cell proliferation and myeloid cell differentiation. Functionally, AML1-ETO upregulates rRNA and cell proliferation-related genes, but downregulates genes that mediate myeloid cell differentiation, promoting and/or supporting the transformed phenotype. Another recent example of cancer-related mitotic gene bookmarking is the mixed lineage leukemia protein (MLL). MLL is a chromatin-remodeling factor that is associated with leukemia and regulates transcription by recruiting chromatin modifying machinery to target genes (Gilliland et al. 2004). The mitotic retention of MLL with target genes favors rapid post-mitotic reactivation of target gene transcription required for the onset and progression of MLL post- mitotically (Blobel et al. 2009; Follmer et al. 2012). Another example of a link between human disease and mitotic gene bookmarking is provided by the examination of hepatocyte nuclear factor HNFβ1 (Lerner et al. 2016). HNFβ1 is frequently mutated in Congenital Abnormalities of Kidney and Urogenital Tract. Many of these mutations disrupt DNA binding activity of HNFβ1, and compromise its gene bookmarking capabilities. Whether disruption of mitotic gene bookmarking by HNFβ1 contributes to the observed congenital abnormalities remains to be seen. It will be informative to establish whether mitotic bookmarking of disease/cancer-related genes is a shared trait of all oncogenic proteins that interact with target genes in a sequence-specific manner.

7.5 Concluding Remarks

It is increasingly apparent that mitotic bookmarking is an essential epigenetic mechanism for maintenance of cellular memory through cell divisions. Emerging evidence indicate that phenotypic transcription factors mitotically bookmark a subset of target genes and that this bookmarking plays a key role in coordination of cell proliferation, growth and differentiation. Importantly, mitotic gene bookmarking by oncogenes in cancer cells appear to be necessary for maintenance of the tumor phenotype. There are several open ended questions that must be addressed to acquire comprehensive mechanistic insights into mitotic gene bookmarking by RUNX proteins and transcription factors in a broad biological context. First, do genes that are localized on the same chromosome and are regulated by the same transcription factors are also bookmarked during mitosis for coordinate transcriptional regulation post-mitotically? In addition, what extracellular signals regulate the switch between mitotic bookmarking of a gene by global transcriptional activators (e.g., Myc) in undifferentiated cells and by phenotypic transcription factors as cells commit to a specific lineage? Furthermore, what are the contributions of coregulatory proteins to reactivation of a bookmarked gene post-mitotically? Also, can the accumulation of transcription factors on mitotic chromosomes be therapeutically targeted in dividing cells? Finally, does mitotic bookmarking play any role in asymmetrically dividing cells? Additional studies will be necessary to functionally link mitotic gene bookmarking and maintenance of cellular memory within in relation to biological control and pathology.

References

Ali, S. A., Zaidi, S. K., Dacwag, C. S., et al. (2008). Phenotypic transcription factors epigenetically mediate cell growth control. *Proceedings of the National Academy of Sciences, 105*, 6632–6637. doi:10.1073/pnas.0800970105.

Ali, S. A., Zaidi, S. K., Dobson, J. R., et al. (2010). Transcriptional corepressor TLE1 functions with Runx2 in epigenetic repression of ribosomal RNA genes. *Proceedings of the National Academy of Sciences, 107*, 4165–4169. doi:10.1073/pnas.1000620107.

Ali, S. A., Dobson, J. R., Lian, J. B., et al. (2012). A Runx2-HDAC1 co-repressor complex regulates rRNA gene expression by modulating UBF acetylation. *Journal of Cell Science*. doi:10.1242/jcs.100909.

Arampatzi, P., Gialitakis, M., Makatounakis, T., & Papamatheakis, J. (2013). Gene-specific factors determine mitotic expression and bookmarking via alternate regulatory elements. *Nucleic Acids Research*. doi:10.1093/nar/gks1365.

Attar, N., & Kurdistani, S. K. (2014). Layered regulation of the epigenome and its links to cellular metabolism and physiology. *AACR Education book, 2014*(1), 31–36.

Bakshi, R., Zaidi, S. K., Pande, S., et al. (2008). The leukemogenic t(8;21) fusion protein AML1-ETO controls rRNA genes and associates with nucleolar-organizing regions at mitotic chromosomes. *Journal of Cell Science, 121*, 3981–3990. doi:10.1242/jcs.033431.

Blobel, G. A., Kadauke, S., Wang, E., et al. (2009). A reconfigured pattern of MLL occupancy within mitotic chromatin promotes rapid transcriptional reactivation following mitotic exit. *Molecular Cell, 36*, 970–983. doi:10.1016/j.molcel. 2009.12.001.

Caravaca, J. M., Donahue, G., Becker, J. S., et al. (2013). Bookmarking by specific and nonspecific binding of FoxA1 pioneer factor to mitotic chromosomes. *Genes & Development, 27*, 251–260. doi:10.1101/gad. 206458.112.

Cerase, A., Smeets, D., Tang, Y. A., et al. (2014). Spatial separation of Xist RNA and polycomb proteins revealed by superresolution microscopy. *Proceedings of the National Academy of Sciences of the United States of America, 111*, 2235–2240. doi:10.1073/pnas.1312951111.

Chi, P., Allis, C. D., & Wang, G. G. (2010). Covalent histone modifications--miswritten, misinterpreted and mis-erased in human cancers. *Nature Reviews. Cancer, 10*, 457–469. doi:10.1038/nrc2876.

Dey, A., Nishiyama, A., Karpova, T., et al. (2009). Brd4 marks select genes on mitotic chromatin and directs postmitotic transcription. *Molecular Biology of the Cell, 20*, 4899–4909. doi:10.1091/mbc.E09-05-0380.

Festuccia, N., Dubois, A., Vandormael-Pournin, S., et al. (2016). Mitotic binding of Esrrb marks key regulatory regions of the pluripotency network. *Nature Cell Biology*. doi:10.1038/ncb3418.

Follmer, N. E., Wani, A. H., & Francis, N. J. (2012). A polycomb group protein is retained at specific sites on chromatin in mitosis. *PLoS Genetics, 8*, e1003135. doi:10.1371/journal.pgen.1003135.s005.

Gilliland, D. G., Jordan, C. T., & Felix, C. A. (2004). The molecular basis of leukemia. *Hematology. American Society of Hematology Education Program*, 80–97. doi:10.1182/asheducation-2004.1.80.

Gottesfeld, J. M., & Forbes, D. J. (1997). Mitotic repression of the transcriptional machinery. *Trends in Biochemical Sciences, 22*, 197–202.

Grandy, R. A., Whitfield, T. W., Wu, H., et al. (2016). Genome-wide studies reveal that H3K4me3 modification in bivalent genes Is dynamically regulated during the pluripotent cell cycle and stabilized upon differentiation. *Molecular and Cellular Biology, 36*, 615–627. doi:10.1128/MCB.00877-15.

Groudine, M., & Weintraub, H. (1982). Propagation of globin DNAase I-hypersensitive sites in absence of factors required for induction: A possible mechanism for determination. *Cell, 30*, 131–139.

Hartl, P., Gottesfeld, J., & Forbes, D. J. (1993). Mitotic repression of transcription in vitro. *The Journal of Cell Biology, 120*, 613–624. doi:10.1083/jcb.120.3.613.

He, H., & Lehming, N. (2003). Global effects of histone modifications. *Briefings in Functional Genomics & Proteomics, 2*, 234–243.

Hsiung, C. C. S., Morrissey, C. S., Udugama, M., et al. (2015). Genome accessibility is widely preserved and locally modulated during mitosis. *Genome Research, 25*, 213–225. doi:10.1101/gr.180646.114.

John, S., & Workman, J. L. (1998). Bookmarking genes for activation in condensed mitotic chromosomes. *BioEssays, 20*, 275–279. doi:10.1002/(SICI)1521-1878(199804)20:4<275::AID-BIES1>3.0.CO;2-P.

Kadauke, S., & Blobel, G. A. (2013). Mitotic bookmarking by transcription factors. *Epigenetics & Chromatin, 6*, 1–1. doi:10.1186/1756-8935-6-6.

Kadauke, S., Udugama, M. I., Pawlicki, J. M., et al. (2012). Tissue-specific mitotic bookmarking by hematopoietic transcription factor GATA1. *Cell, 150*, 725–737. doi:10.1016/j.cell.2012.06.038.

Kelly, T. K., Miranda, T. B., Liang, G., et al. (2010). H2A.Z Maintenance during mitosis reveals nucleosome shifting on mitotically silenced genes. *Molecular Cell, 39*, 901–911. doi:10.1016/j.molcel.2010.08.026.

Kouzarides, T. (2007). Chromatin modifications and their function. *Cell, 128*, 693–705. doi:10.1016/j.cell.2007.02.005.

Lake, R. J., Tsai, P.-F., Choi, I., et al. (2014). RBPJ, the major transcriptional effector of notch signaling, remains associated with chromatin throughout mitosis, suggesting a role in mitotic bookmarking. *PLoS Genetics, 10*, e1004204–e1004215. doi:10.1371/journal.pgen.1004204.

Lerner, J., Bagattin, A., Verdeguer, F., et al. (2016). Human mutations affect the epigenetic/bookmarking function of HNF1B. *Nucleic Acids Research, 44*, 8097–8111. doi:10.1093/nar/gkw467.

Lodhi, N., Kossenkov, A. V., & Tulin, A. V. (2014). Bookmarking promoters in mitotic chromatin: Poly(ADP-ribose)polymerase-1 as an epigenetic mark. *Nucleic Acids Research, 42*, 7028–7038. doi:10.1093/nar/gku415.

Lodhi, N., Ji, Y., & Tulin, A. (2016). Mitotic bookmarking: Maintaining post-mitotic reprogramming of transcription reactivation. *Current Molecular Biology Reports, 2*, 10–15. doi:10.1007/s40610-016-0029-3.

Lopez-Camacho, C., van Wijnen, A. J., Lian, J. B., et al. (2014). CBFβ and the leukemogenic fusion protein CBFβ-SMMHC associate with mitotic chromosomes to epigenetically regulate ribosomal genes. *Journal of Cellular Biochemistry, 115*, 2155–2164. doi:10.1002/jcb.24892.

Margueron, R., & Reinberg, D. (2010). Chromatin structure and the inheritance of epigenetic information.

Nature Reviews. Genetics, 11, 285–296. doi:10.1038/nrg2752.

Martínez-Balbás, M. A., Dey, A., Rabindran, S. K., et al. (1995). Displacement of sequence-specific transcription factors from mitotic chromatin. *Cell, 83*, 29–38.

Michelotti, E. F., Sanford, S., & Levens, D. (1997). Marking of active genes on mitotic chromosomes. *Nature, 388*, 895–899. doi:10.1038/42282.

Ng, R. K., & Gurdon, J. B. (2007). Epigenetic memory of an active gene state depends on histone H3.3 incorporation into chromatin in the absence of transcription. *Nature Cell Biology, 10*, 102–109. doi:10.1038/ncb1674.

Ng, R. K., & Gurdon, J. B. (2008). Epigenetic memory of an active gene state depends on histone H3.3 incorporation into chromatin in the absence of transcription. *Nature Cell Biology, 10*, 102–109. doi:10.1038/ncb1674.

Pande, S., Ali, S. A., Dowdy, C. R., et al. (2009). Subnuclear targeting of the Runx3 tumor suppressor and its epigenetic association with mitotic chromosomes. *Journal of Cellular Physiology, 218*, 473–479. doi:10.1002/jcp.21630.

Pines, J. (2006). Mitosis: A matter of getting rid of the right protein at the right time. *Trends in Cell Biology, 16*, 55–63. doi:10.1016/j.tcb.2005.11.006.

Ptashne, M. (2013). Epigenetics: Core misconcept. *Proceedings of the National Academy of Sciences*. doi:10.1073/pnas.1305399110.

Rivera, R. M., & Bennett, L. B. (2010). Epigenetics in humans: An overview. *Current Opinion in Endocrinology, Diabetes, and Obesity, 17*, 493–499. doi:10.1097/MED.0b013e3283404f4b.

Sabbattini, P., Canzonetta, C., Sjoberg, M., et al. (2007). A novel role for the Aurora B kinase in epigenetic marking of silent chromatin in differentiated postmitotic cells. *The EMBO Journal, 26*, 4657–4669. doi:10.1038/sj.emboj.7601875.

Sarge, K. D., & Park-Sarge, O.-K. (2005). Gene bookmarking: Keeping the pages open. *Trends in Biochemical Sciences, 30*, 605–610. doi:10.1016/j.tibs.2005.09.004.

Sarkies, P., & Sale, J. E. (2012). Cellular epigenetic stability and cancer. *Trends in Genetics, 28*, 1–10. doi:10.1016/j.tig.2011.11.005.

Schübeler, D. (2015). Function and information content of DNA methylation. *Nature, 517*, 321–326. doi:10.1038/nature14192.

Strahl, B., & Allis, C. (2000). The language of covalent histone modifications. *Nature, 403*(6765), 41–45.

Struhl, G. (1981). A gene product required for correct initiation of segmental determination in Drosophila. *Nature, 293*, 36–41.

Wang, F., & Higgins, J. M. G. (2013). Histone modifications and mitosis: Countermarks, landmarks, and bookmarks. *Trends in Cell Biology, 23*, 175–184. doi:10.1016/j.tcb.2012.11.005.

Weber, C. M., & Henikoff, S. (2014). Histone variants: Dynamic punctuation in transcription. *Genes & Development, 28*, 672–682. doi:10.1101/gad.238873.114.

Weintraub, H. (1985). Assembly and propagation of repressed and depressed chromosomal states. *Cell, 42*, 705–711.

Wong, M. M., Byun, J. S., Sacta, M., et al. (2014). Promoter-bound p300 complexes facilitate post-mitotic transmission of transcriptional memory. *PloS One, 9*, e99989. doi:10.1371/journal.pone.0099989.

Xing, H. (2005). Mechanism of hsp70i gene bookmarking. *Science, 307*, 421–423. doi:10.1126/science.1106478.

Young, D. W., Hassan, M. Q., Pratap, J., et al. (2007a). Mitotic occupancy and lineage-specific transcriptional control of rRNA genes by Runx2. *Nature, 445*, 442–446. doi:10.1038/nature05473.

Young, D. W., Hassan, M. Q., Yang, X., et al. (2007b). Mitotic retention of gene expression patterns by the cell fate-determining transcription factor Runx2. *Proceedings of the National Academy of Sciences of the United States of America, 104*, 3189–3194. doi:10.1073/pnas.0611419104.

Zaidi, S. K., Young, D. W., Pockwinse, S. M., et al. (2003). Mitotic partitioning and selective reorganization of tissue-specific transcription factors in progeny cells. *Proceedings of the National Academy of Sciences of the United States of America, 100*, 14852–14857. doi:10.1073/pnas.2533076100.

Zaidi, S. K., Young, D. W., Montecino, M. A., et al. (2010). Mitotic bookmarking of genes: A novel dimension to epigenetic control. *Nature Reviews. Genetics, 11*, 583–589. doi:10.1038/nrg2827.

Zaidi, S. K., Young, D. W., Montecino, M., et al. (2011). Bookmarking the genome: Maintenance of epigenetic information. *Journal of Biological Chemistry, 286*, 18355–18361. doi:10.1074/jbc.R110.197061.

Zaidi, S. K., Grandy, R. A., Lopez-Camacho, C., et al. (2014). Bookmarking target genes in mitosis: A shared epigenetic trait of phenotypic transcription factors and oncogenes? *Cancer Research, 74*, 420–425. doi:10.1158/0008-5472.CAN-13-2837.

Zaret, K. S. (2014). Genome reactivation after the silence in mitosis: Recapitulating mechanisms of development? *Developmental Cell, 29*, 132–134. doi:10.1016/j.devcel.2014.04.019.

Zhao, R., Nakamura, T., Fu, Y., et al. (2011). Gene bookmarking accelerates the kinetics of post-mitotic transcriptional re-activation. *Nature Publishing Group, 13*, 1295–1304. doi:10.1038/ncb2341.

Roles of *Runx* Genes in Nervous System Development

8

Jae Woong Wang and Stefano Stifani

Abstract

Runt-related (Runx) transcription factors play essential roles during development and adult tissue homeostasis and are responsible for several human diseases. They regulate a variety of biological mechanisms in numerous cell lineages. Recent years have seen significant progress in our understanding of the functions performed by Runx proteins in the developing and postnatal mammalian nervous system. In both central and peripheral nervous systems, Runx1 and Runx3 display remarkably specific expression in mostly non-overlapping groups of postmitotic neurons. In the central nervous system, Runx1 is involved in the development of selected motor neurons controlling neural circuits mediating vital functions such as chewing, swallowing, breathing, and locomotion. In the peripheral nervous system, Runx1 and Runx3 play essential roles during the development of sensory neurons involved in circuits mediating pain, itch, thermal sensation and sense of relative position. Runx1 and Runx3 orchestrate complex gene expression programs controlling neuronal subtype specification and axonal connectivity. Runx1 is also important in the olfactory system, where it regulates the progenitor-to-neuron transition in undifferentiated neural progenitor cells in the olfactory epithelium as well as the proliferation and developmental maturation of specific glial cells termed olfactory ensheathing cells. Moreover, upregulated Runx expression is associated with brain injury and disease. Increasing knowledge of the functions of Runx proteins in the developing and postnatal nervous system is therefore expected to improve our understanding of nervous system development, homeostasis and disease.

J.W. Wang • S. Stifani (✉)
Montreal Neurological Institute, Department of Neurology and Neurosurgery, McGill University, Montreal, QC, H3A2B4, Canada
e-mail: stefano.stifani@mcgill.ca

Y. Groner et al. (eds.), *RUNX Proteins in Development and Cancer*, Advances in Experimental Medicine and Biology 962, DOI 10.1007/978-981-10-3233-2_8

Keywords

Astrocyte • Dorsal root ganglion • Hypoglossal motor neuron • Microglia • Motor neuron somatotopic map formation • Muscle innervation • Nociception • Olfactory ensheathing cells • Olfactory epithelium • Parabrachial nucleus • Proprioception • Runx1 • Runx2 • Runx3 • Sensory neuron development

8.1 Introduction

It was during the study of the expression and function of the *Drosophila runt* gene that it was first recognized that members of the *runt*-related gene family (hereafter collectively referred to as *Runx* unless otherwise indicated) are involved in the regulation of nervous system development. During *Drosophila* embryogenesis, *runt* is expressed in a specific subset of central nervous system (CNS) neurons termed *even-skipped*-expressing lateral (EL) neurons. *In vivo* studies showed that *runt* inactivation causes a selective loss of EL neurons (Duffy et al. 1991). Conversely, ectopic *runt* expression results in the formation of supernumerary EL neurons that can extend axons along the normal trajectory used by these cells (Dormand and Brand 1998). These observations provided the first *in vivo* evidence that *Runx* genes participate in context-restricted mechanisms regulating the specification of defined neuronal subtypes in the developing nervous system and they opened the way to the subsequent investigation of the roles of *Runx* genes in vertebrate neural development. This chapter will review some of the key functions performed by *Runx* genes during mammalian nervous system development, focusing mainly on their roles during the formation of neural circuits mediating somatosensory sensation, motor control, and olfaction.

8.2 Involvement of *Runx* Genes in Peripheral Nervous System Development

Dorsal root ganglion (DRG) sensory neurons in the peripheral nervous system (PNS) mediate somatosensory stimuli such as the sensations of pain (nociception), mechanical pressure (mechanoreception), or relative position (proprioception). Nociceptive neurons (nociceptors) are connected to the dorsal horn of the spinal cord and cutaneous structures and express neurotrophic factor receptors such as TrkA and Ret. Mechanoreceptive neurons (mechanoreceptors) also project their fibers to the spinal cord dorsal horn and the skin, but express other receptors for neurotrophic factors such as TrkB and TrkC. Proprioceptive neurons (proprioceptors) express TrkC and project to the ventral horn and intermediate zone of the spinal cord, as well as to muscle spindles and Golgi tendon organs (Inoue et al. 2002; Kramer et al. 2006; Yoshikawa et al. 2007; Honma et al. 2010; Abdo et al. 2011; Lallemend and Ernfors, 2012).

Both *Runx1* and *Runx3* are expressed in developing DRG sensory neurons, mostly in non-overlapping patterns. Mouse *Runx1* is preferentially expressed in several nociceptive neurons, including pain-related nociceptors and thermoceptors (Levanon et al. 2002; Chen et al. 2006a; Kramer et al. 2006; Marmigere et al. 2006; Lou et al. 2013, 2015). In contrast, *Runx3* expression marks for the most part proprioceptive neurons (Inoue et al. 2002; Levanon et al. 2002; Chen et al. 2006b; Kramer et al. 2006; Yoshikawa et al. 2007; Inoue et al. 2008; Lallemend and Ernfors, 2012). The next two sections will discuss evidence that Runx1 and Runx3 act during DRG sensory neuron development to regulate the acquisition of specific neuronal subtype identities, defined gene expression profiles, and formation of precise axonal innervations.

8.2.1 Runx1 Involvement in Establishment of Cutaneous Sensory Circuits

Most if not all DRG cutaneous sensory neurons mediating pain, itch and thermal sensation initially express TrkA during embryonic development. $TrkA^+$ DRG sensory neurons also express Runx1 in developing embryos (Levanon et al. 2002; Chen et al. 2006a; Kramer et al. 2006; Marmigere et al. 2006; Yoshikawa et al. 2007). At perinatal and postnatal stages, DRG expression of TrkA and Runx1 separates, resulting in the appearance of two main groups of sensory neurons characterized by the expression of TrkA or Runx1 (this latter group also expresses Ret). $TrkA^+$ sensory neurons acquire a 'peptidergic' phenotype characterized in part by the expression of the neuropeptide calcitonin-gene-related peptide (CGRP) and specific cell surface proteins. In contrast, $Runx1^+$ cells become 'non-peptidergic' sensory neurons defined by specific molecular traits and innervation of skin epidermis and hair follicle targets (Chen et al. 2006a; Kramer et al. 2006; Marmigere et al. 2006; Luo et al. 2007; Yoshikawa et al. 2007; Gascon et al. 2010; Yang et al. 2013; Lou et al. 2015). Runx1 is essential for both the initial separation of $TrkA^+$ and $TrkA^-$ sensory neuron lineages and the subsequent generation of cutaneous sensory neuron diversity, a process mediated in part by the transient or persistent nature of Runx1 expression after the separation of TrkA and Runx1 expression (Lou et al. 2015).

During non-peptidergic neuronal fate specification, Runx1 orchestrates transcriptional mechanisms that directly or indirectly regulate the expression of numerous genes defining the non-peptidergic phenotype. These include, to name only a few, genes encoding cold receptors TRPM8 and TRPA1, heat receptors TRPV1 and TRPV2, several Mrgpr class G protein-coupled receptors, and ATP-gated channels. Runx1 is also important to restrict the expression of peptidergic genes, including those encoding TrkA, CGRP, and mu-class opioid receptor (Chen et al. 2006a; Kramer et al. 2006; Yoshikawa et al. 2007; Liu et al. 2008; Ugarte et al. 2013). Runx1 is hypothesized to directly regulate *TrkA* expression because it can bind *in vitro* to a *TrkA* minimal enhancer containing putative Runx binding sites (Marmigere et al. 2006).

As mentioned, Runx1 also plays key roles in the further specification of several cutaneous sensory neuronal subtypes, including polymodal nociceptors, pruciceptors and other neurons associated with pain, as well as specific types of mechanoreceptors (Lou et al. 2013, 2015; Yang et al. 2013). As an example, Runx1 regulates the development of specific unmyelinated low-threshold mechanoreceptors that persistently express vesicular glutamate transporter 3 (VGLUT3) and mediate pleasant touch and/or pain (Lou et al. 2013). The Runx1-dependent transcription factor gene *Zfp521* is required to establish molecular features that define $VGLUT3^+$ mechanoreceptors. Runx1 and ZFP521 work in coordination to determine the molecular phenotype of $VGLUT3^+$ mechanoreceptors while suppressing traits typical of other types of sensory neurons such as polymodal nociceptors (Lou et al. 2015). Thus, Runx1 is a key player in the generation of DRG sensory neuron diversity by regulating various developmental processes leading to the separation of peptidergic *vs* non-peptidergic phenotypes as well as the specification of selected cutaneous sensory neuron subtypes.

Runx1 is also involved in regulating the connectivity of the DRG sensory neurons in which it is expressed. Non-peptidergic and peptidergic DRG neurons normally send their axons to separate targets in the dorsal spinal cord. In *Runx1*-deficient mice, the majority of nociceptive neurons in which *Runx1* would have been expressed had it not been inactivated send their axons to targets more typical of peptidergic neurons (Chen et al. 2006a; Yoshikawa et al. 2007). Conversely, ectopic Runx1 expression in DRG neurons is sufficient to cause axons of $TrkA^+$ neurons to project to layers of the spinal cord normally innervated by non-peptidergic neurons (Kramer et al. 2006). Consistent with these results, *in vitro* studies suggest that Runx1 might participate in mechanisms promoting axon growth and branching (Marmigere et al. 2006). These results are in agreement with the demonstration that *Drosophila runt* is involved in the control of photoreceptor neuron axonal targeting choices (Kaminker et al. 2002). Loss of mouse *Runx1* function also leads to the selective loss of sensory innervation to the epidermis. In contrast, sensory innervation of 'deep tissues', such as muscle and visceral organs, is not affected by *Runx1* inactivation, providing further evidence for a role of this gene in genetic programs controlling the differentiation of cutaneous pain pathways (Yang et al. 2013). Given these phenotypic manifestations, it is not surprising that mice lacking Runx1 in DRG sensory neurons display decreased thermal and mechanical pain perception, underscoring further the key role of Runx1 in the formation of functional nociceptive circuits (Chen et al. 2006a, b; Abdel Samad et al. 2010).

In summary, Runx1 acts at multiple levels during the formation of PNS neural circuits mediating pain, thermal, and itch sensations by controlling both sensory neuronal subtype specification and establishment of precise innervations.

8.2.2 Runx3 Involvement in Formation of Proprioceptive Circuits

During DRG development, $TrkC^+$ proprioceptive neurons are generated from transiently lived $TrkB^+/TrkC^+$ cells that also give rise to $TrkB^+$ mechanoreceptive neurons. Runx3 expression becomes detectable in $TrkC^+$ DRG cells at approximately the time when the latter arise from $TrkB^+/TrkC^+$ precursors (Inoue et al. 2002; Levanon et al. 2002; Kramer et al. 2006). More importantly, *Runx3* inactivation in *Runx3*-deficient mice results in decreased numbers of $TrkC^+$ cells and also decreased cells expressing Parvalbumin, a protein expressed preferentially in proprioceptive neurons. This phenotype is correlated with a concomitant increase in $TrkB^+$ neurons (Inoue et al. 2002; Levanon et al. 2002; Kramer et al. 2006; Inoue et al. 2007; Nakamura et al. 2008; Lallemend et al. 2012). The converse situation is observed after ectopic expression of Runx3 in all developing DRG neurons (Kramer et al. 2006). These findings show that Runx3 is important for the separation of $TrkC^+$ and $TrkB^+$ sensory neuronal lineages.

Runx3 may be directly involved in the extinction of *TrkB* expression in proprioceptive neurons because Runx3 binds to, and represses transcription from, a *TrkB* intronic gene regulatory element containing consensus Runx-binding sites (Inoue et al. 2007). Moreover, analysis of DRG and trigeminal ganglion development in mouse embryos lacking the gene *Brn3a* revealed that *Runx3* fails to be activated in *TrkC*$^+$ neurons in the absence of Brn3a (*Runx1* expression is also greatly attenuated in *TrkA*$^+$ nociceptors in *Brn3a*-deficient mice). These changes are accompanied by expanded expression of *TrkB*, followed by the loss of *TrkC* and *TrkA* expression (Dykes et al. 2010, 2011). Brn3a binds to a conserved upstream enhancer element within the *Runx3* locus, suggesting that Runx factors repress *TrkB* expression downstream of Brn3a (Dykes et al. 2010). It should be noted that separate studies suggest that the negative regulation of *TrkB* expression by Runx3 in sensory neurons mediating touch sensation is indirect and is mediated by another transcription factor, termed Shox2, which activates *TrkB* expression and is repressed by Runx3 (Abdo et al. 2011).

Perturbation of *Runx3* activity is also correlated with proprioceptive neuron axonal targeting defects. *Runx3*-deficient embryos display abnormal projections of proprioceptive DRG neurons to both peripheral and central targets. $TrkC^+$ proprioceptive neurons fail to establish proper connections in the ventral spinal cord and instead innervate more dorsal positions (Inoue et al. 2002; Chen et al. 2006b; Nakamura et al. 2008). Conversely, ectopic Runx3 expression in *TrkA*$^+$ DRG neurons results in innervation of more ventral sectors of the spinal cord, similar to the targeting of *TrkC*$^+$ neurons (Kramer et al. 2006). These observations provide evidence that, similar to Runx1 participation in nociceptive circuit formation, Runx3 is important for the connectivity of proprioceptive DRG neurons.

More recent studies have revealed an additional level of involvement of Runx3 during proprioceptive neuron axonal development. Proprioceptive neurons exhibit different rates of axon extension at different axial levels and these differences are determined by a segmental pattern of Runx3 levels at different axial positions. Runx3 is involved in proprioceptive neuron axonal extension at least in part by controlling the level of expression of genes encoding cytoskeletal proteins involved in axon growth (Lallemend et al. 2012). Taken together, these findings provide evidence that Runx3 plays important roles in proprioceptive neuron development by regulating both the specification of proprioceptive sensory neurons and the formation of proprioceptive neural circuits.

In summary, Runx1 and Runx3 mediate analogous developmental functions during the establishment of pathways mediating either cutaneous sensation or propioception, ranging from the regulation of specific *Trk* family gene expression and acquisition of cell-type specific transcriptional profiles to the control of neuronal connectivity. It should be mentioned that the presence of certain DRG neurons expressing both Runx1 and Runx3 has been observed during embryonic development and after birth. At postnatal stages, some of these Runx1⁺/Runx3⁺ cells also express TrkB, Ret, and TrkC, suggesting that these proteins are coexpressed in at least a particular group of mechanoreceptive DRG neurons (Nakamura et al. 2008; Yoshikawa et al. 2013). It remains to be determined whether Runx1 and Runx3 have overlapping or non-redundant roles in the specific sensory neuronal populations in which they are coexpressed at postnatal stages.

8.3 Roles of *Runx* Genes in Central Nervous System Development

Motor neurons are specialized cells controlling voluntary and involuntary functions, ranging from the contractile activity of multiple muscle groups to the activity of smooth and cardiac muscle fibers or glands. There are three main classes of motor neurons, termed somatic (innervating muscles that control functions such as speaking, swallowing, breathing, and locomotion), branchial (innervating muscles in the face and upper neck), and visceral (controlling components of the autonomic nervous system such as smooth muscle in the viscera). The next three sections will discuss evidence suggesting that Runx1 acts during the development of selected motor neurons to regulate the establishment of specific motor neuron identities and axonal innervations. Moreover, they will address the expression of *Runx1*, and other *Runx* genes, in other types of neural cells in the CNS.

8.3.1 *Runx1* Involvement in the Formation of Motor Circuits

Runx1 is expressed in selected types of postmitotic motor neurons, but not their undifferentiated mitotic progenitors. Runx1$^+$ neurons include visceral and somatic motor neurons in the murine brainstem and cervical spinal cord (Theriault et al. 2004; Stifani et al. 2008; Guizard et al. 2010; Chen et al. 2015; Yoshikawa et al. 2015). *Runx1* also displays a restricted expression in selected motor neuron subtypes in the chick cervical spinal cord (Dasen et al. 2005). The timing of Runx1 expression does not coincide with motor neuron generation and instead roughly correlates with the time when the Runx1$^+$ motor neurons are acquiring their specific molecular identities, cell body positions and/or axonal innervations (Stifani et al. 2008; Chen et al. 2015). Consistent with this finding, lack of *Runx1* activity in *Runx1*-deficient mice does not perturb the generation of the somatic motor neurons in which Runx1 would normally be expressed (Stifani et al. 2008).

The roles of Runx1 during motor neuron development are best characterized in brainstem somatic motor neurons located in the hypoglossal nucleus (12N). Motor neurons in 12N innervate muscles in the tongue controlling vital functions such as chewing, swallowing, and breathing. Specifically, 12N motor neurons innervate two main tongue muscle groups, anatomically defined as intrinsic or extrinsic based on their origin and location. Extrinsic and intrinsic tongue muscles can be further subdivided into two functional categories: muscles controlling either tongue protrusion ('protrusors') or tongue retraction ('retrusors') (Aldes 1995; Altschuler et al. 1994; McClung and Goldberg 1999, 2000). 12N motor neurons innervating these different tongue mus-

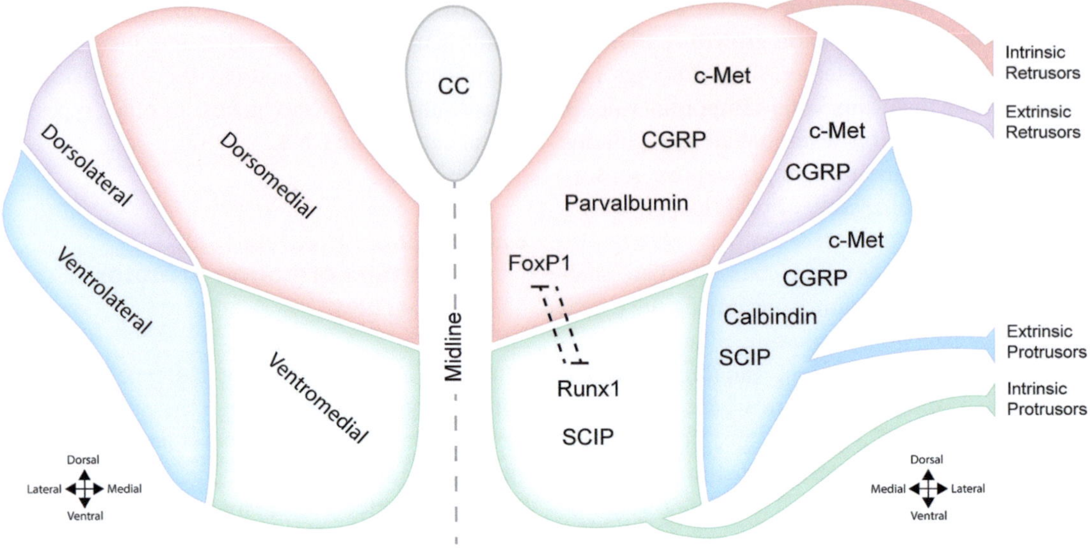

Fig. 8.1 Expression and proposed roles of Runx1 during hypoglossal motor neuron development. Schematic representation of the restricted expression of Runx1 in the ventromedial 12N anatomical quadrant at mediocaudal level. Runx1$^+$ 12N motor neurons coexpress the transcription factor SCIP but do not express the transcription factor FoxP1 nor the calcium-binding proteins Parvalbumin and Calbindin, the neurotransmitter CGRP, or the surface protein c-Met. The combinatorial expression of these proteins defines four separate motor neuron groups whose topology roughly corresponds to different 12N quadrants associated with innervation of different tongue muscle groups. Runx1 is thought to act in ventromedial 12N motor neurons to prevent the acquisition of molecular profiles defining dorsomedial and ventrolateral 12N motor neuron phenotypes. Runx1 and FoxP1 are proposed to cross-repress each other's expression, thereby defining the dorsoventral border of the medial 12N domain. Runx1 is also involved in mechanisms important for ventromedial 12N motor neuron axonal innervation of intrinsic protrusor muscles in the tongue. Abbreviations: *CC* central canal

cles are organized into a characteristic 'somatotopic map' in which motor neurons located in the dorsal half of the nucleus generally innervate retrusor muscles, whereas motor neurons located in the ventral part innervate protrusor muscles. Moreover, motor neurons innervating intrinsic muscles are generally located medially, with motor neurons innervating extrinsic muscles found more laterally (summarized in Fig. 8.1) (Aldes 1995; Chibuzo and Cummings 1982; Krammer et al. 1979).

During 12N development, Runx1 expression is mainly restricted to ventromedial 12N motor neurons characterized by a molecular profile distinct from other 12N motor neurons based on the expression of specific calcium-binding proteins, neurotransmitters, cell surface receptors and transcription factors (Fig. 8.1) (Chen et al. 2015; Yoshikawa et al. 2015). Importantly, the number of motor neurons exhibiting lateral or dorsal 12N motor neuron molecular traits is decreased when Runx1 is ectopically expressed in all developing 12N motor neurons. These results suggest that Runx1 acts to specifically promote a 'ventromedial' 12N motor neuron phenotype at the expense of 'ventrolateral' and 'dorsomedial' 12N motor neuron identities (Chen et al. 2015).

The Runx1$^+$ ventromedial 12N motor neurons send their axons to intrinsic tongue muscles mediating tongue protrusion (Fig. 8.1) (Chen et al. 2015; Yoshikawa et al. 2015). *In vivo Runx1* inactivation results in decreased 12N motor neuron axonal projections, and overall reduction of innervation density, to intrinsic protrusor muscles (Yoshikawa et al. 2015). *Runx1* inactivation also results in decreased 12N expression of the gene *Frizzled3*, which contributes to axonal pathfinding of 12N motor neurons (Hua et al. 2013; Yoshikawa et al. 2015). Together, these observations suggest that transcriptional programs involving Runx1 are involved in the establishment of motor circuits controlling tongue protrusion.

One of the consequences of ectopic Runx1 expression in 12N motor neurons is the detectable perturbation of the normal pattern of expression of other transcription factors involved in motor neuron development. A notable example is provided by FoxP1, which is a transcription factor that plays important roles in the regulation of spinal motor neuron subtype development (Dasen et al. 2008; Rousso et al. 2008; Palmesino et al. 2010). During 12N development, FoxP1 is expressed in dorsomedial 12N motor neurons neighboring the ventromedial Runx1^{+} cells (Fig. 8.1) (Chen et al. 2015). A number of observations suggest that Runx1 and FoxP1 may transcriptionally repress each other's expression, thereby defining their respective expression domains in the medial part of 12N. In potential agreement with this possibility, forced Runx1 expression in dorsal 12N motor neurons results in a significant decrease in the number of dorsomedial 12N neurons expressing FoxP1 (Chen et al. 2015). These observations raise the possibility that at least one of the functions of Runx1 during 12N development is to prevent FoxP1 expression in the ventral sector of the medial region, thereby contributing to the establishment of separate ventromedial (Runx1^{+}/FoxP1^{-}) and dorsomedial (FoxP1^{+}/Runx1^{-}) 12N motor neuron groups. As mentioned, ventromedial 12N motor neurons project to intrinsic protrusor muscles, whereas dorsomedial 12N motor neurons are believed to project to intrinsic retrusors. Thus, it is conceivable that the activities of Runx1 and FoxP1 are involved in the formation of neural circuits controlling either tongue protrusion or retraction, respectively (Fig. 8.1).

Runx1 is expressed in other selected groups of motor neurons, including spinal motor neurons involved in the control of posture and locomotion. For instance, certain Runx1^{+} motor neurons at cervical spinal cord level C1–C4 innervate the *anterior trapezius* muscle. Runx1 expression also marks defined motor neurons at level C4–C5 that innervate the *deltoideus* muscle (Stifani et al. 2008). It is possible that, similar to its roles in 12N motor neurons, Runx1 may also be important for the acquisition and/or maintenance of specific motor neuron gene expression patterns and axonal connectivity in other somatic motor neuron subtypes.

In summary, the involvement of Runx1 in motor neuron diversity generation and circuit formation shares common themes with the roles of Runx1 during PNS nociceptive neuron development. This situation in turn suggests that Runx1 may perform similar tasks during the formation of additional CNS circuits involving other types of neurons in which Runx1 is expressed.

8.3.2 *Runx1* Involvement in Other Central Neural Circuits

In addition to motor neurons, Runx1 is expressed in at least another neuronal population in the rostral brainstem, which was identified as part of the superior lateral subnucleus of the parabrachial nucleus (LPBS) on the basis of several anatomical and molecular properties (Zagami and Stifani 2010). Runx1 expression in these cells is first observed early during brain development and persists into the postnatal brain, similar to the situation observed in sensory and motor neurons. These observations suggest that Runx1 might be involved in mechanisms controlling the differentiation and/or target connectivity of LPBS neurons, as it does in other central and peripheral neurons.

Although the functional significance of the restricted expression of Runx1 in this selected neuronal population remains to be defined, it is worth mentioning that the LPBS has been implicated in the suppression of food intake in response to pain (Gibbs et al. 1973, 1976; Malick et al. 2001). Moreover, thermal and inflammatory noxious stimuli were shown to activate neurons in the LPBS in which Runx1 is expressed (Hermanson et al. 1998; Bester et al. 1995, 1997; Buritova et al. 1998). Taken together with the important roles of Runx1 in sensory nociceptor development and innervation discussed above, these observations raise the possibility that Runx1 might be involved in the formation of functional networks coordinating nociception and regulation of food intake.

8.3.3 *Runx2* Expression in the Central Nervous System

In contrast to Runx1, little is known about the involvement of Runx2 and Runx3 in the CNS. Previous studies have revealed the presence of *Runx2* transcripts in the adult mouse brain (Takarada and Yoneda 2009). Moreover, *Runx2*$^{LacZ/+}$ knock-in mice display restricted expression of βGalactosidase under the control of the *Runx2* promoter in the postnatal hippocampus and frontal lobe area (Jeong et al. 2008). More recent work has suggested that mouse *Runx2* is expressed under circadian control in specific brain regions including the paraventricular nucleus, olfactory bulb and suprachiasmatic nucleus (Reale et al. 2013). Runx2 expression is presumed to occur in neuronal cells in these brain areas, although it should be noted that separate studies suggest that at least some of the Runx2$^+$ cells in the brain correspond to glial cells (Takarada and Yeoneda 2009). *RUNX2* expression was also detected in the adult human hippocampus, and hippocampal *RUNX2* expression is decreased in bipolar disorder patients (Benes et al. 2007). It will be important to characterize further the brain cells in which Runx2 is expressed and the functions, and regulation, of Runx2 in these cells to determine whether of not at least some of the themes uncovered by the analysis of Runx1 and Runx3 in central and peripheral neurons also apply to Runx2.

8.4 Contribution of *Runx1* to Olfactory System Development

The previous sections have discussed the important roles performed by *Runx1* and *Runx3* in specific subtypes of central and peripheral neurons and have addressed how these functions occur mostly, if not entirely, during postmitotic neuronal development. This section will focus on the olfactory system, which offers a compelling example of the importance of Runx proteins in the biology of specific types of undifferentiated mitotic neural cells.

The olfactory system is an evolutionarily ancient sensory system mediating the sense of smell. It has a peripheral component, the olfactory epithelium (OE), located in the nasal cavity, and central elements, the olfactory bulb and olfactory cortex, in the brain. Both the OE and the olfactory bulb host populations of neurons that are constantly regenerated during adult life and thus the olfactory system is one of the few neural tissues with persistent renewal potential (Leinwand and Chalasani 2011; Takeuchi and Sakano 2014; Suzuki and Osumi 2015).

8.4.1 *Runx1* Involvement in Neural Progenitor Cell Proliferation in the Olfactory Epithelium

The developing OE was first identified as one of the sites of most robust *Runx1* expression in the murine nervous system almost two decades ago (Simeone et al. 1995; Levanon et al. 2001). Subsequent work demonstrated that, in contrast to motor and sensory neuronal lineages, *Runx1* is mainly expressed in mitotic olfactory sensory neuron (OSN) progenitor cells located on the basal side of the OE (Theriault et al. 2005). During embryonic development and postnatal life, these cells act as the neural stem/progenitor cell population in the OE, comprising the self-propagating and transit-amplifying pools that drive both development and persistent regeneration of OSNs (Kam et al. 2014; Takeuchi and Sakano 2014). In agreement with the ability of Runx proteins to regulate the balance between the undifferentiated and differentiated states in numerous cell lineages, *in vivo* loss-of-function studies showed that Runx1 acts in OSN progenitor cells to sustain cell proliferation and delay differentiation, thereby contributing to maintenance of the mitotic OSN progenitor pool and regulation of OSN generation. Importantly, forced exogenous Runx1 expression in primary cultures of self-propagating OE progenitor cells revealed that Runx1 is sufficient to enhance proliferation in this cellular context (Theriault et al. 2005). Thus, Runx1 acts to regulate the timing of neuronal differentiation in the OE at least in part by sustaining the proliferation of OSN progenitor cells.

8.4.2 *Runx1* Involvement in Proliferation and Developmental Maturation of Olfactory Ensheathing Cells

OSNs located in the OE send their axons to the olfactory bulb in the rostral part of the brain. The axons of OSNs are enveloped by a particular population of glial cells, termed olfactory ensheathing cells (OECs). Residing in both the OE and the olfactory bulb, OECs span the PNS and CNS and perform a remarkable array of functions during olfactory development and regeneration. They accompany and ensheath OSN axons, produce growth factors, cell adhesion molecules and extracellular matrix proteins that promote OSN axon growth and targeting, and act as phagocytic cells that engulf and remove apoptotic olfactory nerve debris (Su and He 2010; Chou et al. 2014; Roet and Verhaagen 2014).

Developmentally mature OECs derive from mitotic precursors that originate peripherally and give rise to different subtypes of differentiated OECs populating the olfactory bulb. Runx1 is expressed in at least some OEC precursor cells *en route* to the olfactory bulb and the proliferative ability of Runx1-expressing OECs is sensitive to Runx1 dosage. Specifically, decreased Runx1 levels are correlated with increased numbers of mitotic OECs, with a parallel decrease in the number of more differentiated OECs. In contrast, Runx1 overexpression results in reduced OEC proliferation (Murthy et al. 2014). Thus, Runx1 contributes to the development, and possibly regeneration, of the olfactory system by acting, at least in part, to either delay or promote the proliferation-to-differentiation transition in OSN or OEC precursors, respectively.

Runx1 expression is not limited to OEC precursors that are migrating to the olfactory bulb, but persists in a defined subgroup(s) of more developmentally mature OECs. These cells are specifically located in the inner portion of the olfactory bulb nerve layer (ONL) and exhibit characteristic molecular features, such as the expression of neuropeptide Y. In contrast, Runx1 expression is not detected in OECs that are located in the outer part of the ONL and express different sets of proteins, including the low-affinity NGF receptor p75NTR (Murthy et al. 2014). The expression of Runx1 in a particular subtype(s) of developmentally mature OECs with a specific topology and molecular profile is akin to the subtype-restricted expression of Runx1 in neuronal cells in the CNS and PNS, as discussed above. This observation raises the possibility that Runx1 may act to regulate specialized glial cell fate acquisition in the olfactory system, similar to its role in the acquisition of specific neuronal identities. Since inner and outer ONL OECs are believed to arise from common precursors, Runx1 may be involved in the specification of inner ONL OECs by promoting the expression of genes associated with this subtype and/or repressing genes associated with outer ONL OEC phenotype(s).

In summary, these findings suggest that Runx1 is involved in olfactory system development and regeneration by participating in multiple mechanisms controlling the transition from an immature proliferating state to more a developmentally mature phenotype in both neuronal and glial cell lineages.

8.5 *Runx* Expression in Glial Cells in the Central and Peripheral Nervous Systems

The expression of *Runx1* in OECs is not the only example of *Runx* gene expression in glial cells. Recent studies have provided evidence suggesting that Runx1 is expressed in Schwann cell progenitors in the PNS (Li et al. 2015). Schwann cells share a number of properties with OECs and are involved in many aspects of peripheral nerve biology, including myelination of axons (a property exhibited by many, but not all, Schwann cells), the secretion of growth factors supporting nerve development and regeneration, trophic support to neurons, and antigen presentation (Kidd et al. 2013).

A search for active enhancers in myelinating Schwann cells after peripheral nerve injury has recently identified a *Runx2* enhancer bound by c-Jun, a transcription factor required for Schwann cells to support nerve regeneration. These observations suggest that Runx2 might also be involved in Schwann cell biology, possibly by participating in the regulation of genes induced after peripheral nerve injury (Hung et al. 2015). The presence of *Runx2* transcripts was also detected in cultured rat forebrain astrocytes and astrocytic C6 glioma cells (Takarada and Yoneda 2009). Moreover, *RUNX2* is expressed in human gliomas, brain tumors of astrocytic origin (Vladimirova et al. 2008). Although the functional significance of these observations remains to be determined, they raise the possibility that Runx2 might be involved in mechanisms regulating proliferation and/or differentiation along the astrocyte lineage. In contrast to *RUNX2*, *RUNX3* expression is low or absent in glioma due to promoter hypermethylation (Mueller et al. 2007). *RUNX3* mRNA expression was observed in fetal and adult human brain, but whether this expression occurs in astrocytes and/or other glial cells remains to be determined (Mueller et al. 2007).

Together, these observations are suggesting that the involvement of Runx proteins in glial cell biology is not limited to OECs and extends to other glial cells in both the CNS and PNS. In these cells, Runx proteins may act to regulate the expression of genes important for the proliferation-to-differentiation transition and/or the acquisition of defined glial phenotypes. In this regard, it is important to note that Runx1 is upregulated in neurofibroma, a cancer of Schwann cell origin. Impairment of Runx1 activity in Schwann cell progenitor cells delays mouse neurofibroma formation *in vivo* as a result of decreased cell proliferation and increased cell apoptosis, suggesting that Runx1 may play a role in Schwann cell proliferation (Li et al. 2015). Future studies aimed at clarifying Runx expression and function in glial cells in the developing and postnatal CNS and PNS are expected to offer insight into gliogenesis and glial cell functions.

8.6 Conclusions and Perspectives

Much progress has been made in understanding the neural functions of *Runx* genes since the initial observation that *runt* is important for the differentiation of EL neurons in the *Drosophila* CNS. Our current knowledge points to essential roles for Runx1 and Runx3 during the development of specific neural circuits in the mammalian CNS and PNS. Moreover, Runx1 is involved in the development, and possibly regeneration, of the olfactory system, one of the few neural tissues with persistent renewal potential throughout life. In the future, it will be important to characterize further the molecular mechanisms underlying the roles of Runx proteins in the regulation of sensory and motor neuron subtype specific gene expression and axonal targeting choices, including the identity of additional Runx-regulated genes during these events. Moreover, little is known about the upstream mechanisms that establish the exquisite temporal and spatial specificity of *Runx* gene expression in both neuronal and glial cell lineages.

The possible involvement of Runx proteins in the adult CNS and/or PNS also remains to be determined. In the healthy brain, *Runx1* is not detectably expressed in neural stem/progenitor cells capable of supporting the genesis of new neural cells. However, endogenous *Runx1* expression becomes induced in a subpopulation of putative neural stem/progenitor cells after brain injury in adult mice (Logan et al. 2013, 2015). This effect is thought to be mediated at least in part by mechanisms involving FGF signaling, because culturing forebrain neural progenitor cells in the presence of basic FGF (bFGF) induces Runx1 expression, whereas removal of bFGF from the culture medium decreases Runx1 expression (Theriault et al. 2005; Logan et al. 2015). This possibility is also consistent with the observation that bFGF induces human *RUNX1* expression in olfactory neuroblastoma tumors (Nibu et al. 2000).

When expressed in CNS neural stem/progenitor cells as a result of injury or growth factor treatment, Runx1 can participate in mechanisms regulating proliferation and neuronal differentiation (Theriault et al. 2005; Logan et al. 2015), at least in part by enhancing proliferation through repression of cell cycle inhibitory genes such as *p21^Cip1^* (Theriault et al. 2005). Thus, although Runx1 does not appear to be physiologically expressed in neural stem/progenitor cells in the CNS, it has the potential to participate in neural stem/progenitor cell biology during processes associated with injury and repair of the adult nervous system.

Finally, it is important to mention that Runx1 can contribute to nervous system development and repair in an additional manner, namely by modulating the activity of microglia, the resident immune cells of the nervous system (Zusso et al. 2012). Microglia survey the nervous system for signs of infection, injury or disease and mediate immune responses during many neuropathological conditions (Ransohoff and El Khoury 2015; Shemer et al. 2015). They derive from primitive myeloid precursors that originate from the yolk sac and colonize the nervous system during embryonic development (Ginhoux et al. 2010; Prinz and Mildner 2011). In the developing brain, microglia are initially proliferative, amoeboid in shape and endowed with phagocytic activity. They mediate several important developmental functions including cell debris phagocytosis, guidance of axons in white matter tracts, and synaptic refinement. As development proceeds, microglia lose their amoeboid morphology and gradually progress to a surveillant, non-phagocytic state characterized by a highly ramified morphology (Schlegelmilch et al. 2010; Prinz and Mildner 2011; Wu et al. 2015). Runx1 is expressed in postnatal forebrain amoeboid microglia and its expression is downregulated as microglia progress to the ramified phenotype. Runx1 inhibits amoeboid microglia proliferation and promotes progression to the ramified state (Zusso et al. 2012). Thus, the involvement of Runx1 in microglia biology during brain development provides an additional example of this protein's ability to regulate proliferation and developmental maturation mechanisms in the developing and postnatal mammalian nervous system. The amoeboid-to-ramified morphological transition of microglia during development is almost recapitulated in reverse during the process of microglia activation in the adult brain, when surveillant microglia undergo a ramified-to-amoeboid transformation and become phagocytic in response to injury or disease (Ransohoff and El Khoury 2015; Shemer et al. 2015). Runx1 expression is upregulated in microglia following nerve injury in the adult nervous system (Zusso et al. 2012), implicating Runx1 in the regulation of at least certain functions of microglia in the injured/diseased adult nervous system. Runx2 expression has also been observed in microglia (Nakazato et al. 2014), but it is unknown whether Runx1 and Runx2 have overlapping or distinct functions in these cells.

In conclusion, increasing evidence implicates Runx proteins in the development, regeneration, and repair of the nervous system through the regulation of several important mechanisms in neurons, macroglia, and microglia. It is therefore anticipated that increased understanding of the roles of Runx proteins in neural cells and microglia will facilitate the study of nervous system development, homeostasis and disease.

References

Abdel Samad, O., Liu, Y., Yang, F. C., Kramer, I., Arber, S., & Ma, Q. (2010). Characterization of two Runx1-dependent nociceptor differentiation programs necessary for inflammatory versus neuropathic pain. *Molecular Pain, 6*, 45.

Abdo, H., Li, L., Lallemend, F., Bachy, I., Xu, X. J., Rice, F. L., & Ernfors, P. (2011). Dependence on the transcription factor Shox2 for specification of sensory neurons conveying discriminative touch. *The European Journal of Neuroscience, 34*, 1529–1541.

Aldes, L. D. (1995). Subcompartmental organization of the ventral (protrusor) compartment in the hypoglossal nucleus of the rat. *The Journal of Comparative Neurology, 353*, 89–108.

Altschuler, S. M., Bao, X., & Miselis, R. R. (1994). Dendritic architecture of hypoglossal motor neurons projecting to extrinsic tongue musculature in the rat. *The Journal of Comparative Neurology, 342*, 538–550.

Benes, F. M., Lim, B., Matzilevich, D., Walsh, J. P., Subbaraju, S., & Minns, M. (2007). Regulation of the GABA cell phenotype in hippocampus of schizophrenics and bipolars. *Proceedings of the National Academy of Sciences of the United States of America, 104*, 10164–10169.

Bester, H., Menendez, L., Besson, J. M., & Bernard, J. F. (1995). Spino (trigemino) parabrachiohypothalamic pathway: Electrophysiological evidence for an involvement in pain processes. *Journal of Neurophysiology, 73*, 568–585.

Bester, H., Matsumoto, N., Besson, J. M., & Bernard, J. F. (1997). Further evidence for the involvement of the spinoparabrachial pathway in nociceptive processes: A c-Fos study in the rat. *The Journal of Comparative Neurology, 383*, 439–458.

Buritova, J., Besson, J. M., & Bernard, J. F. (1998). Involvement of the spinoparabrachial pathway in inflammatory nociceptive processes: A c-Fos protein study in the awake rat. *The Journal of Comparative Neurology, 397*, 10–28.

Chen, C. L., Broom, D. C., Liu, Y., de Nooij, J. C., Li, Z., Cen, C., et al. (2006a). Runx1 determines nociceptive sensory neuron phenotype and is required for thermal and neuropathic pain. *Neuron, 49*, 365–377.

Chen, A. I., de Nooij, J. C., & Jessell, T. M. (2006b). Graded activity of transcription factor Runx3 specifies the laminar termination pattern of sensory axons in the developing spinal cord. *Neuron, 49*, 395–408.

Chen, X., Wang, J. W., Salin-Cantegrel, A., Dali, R., & Stifani, S. (2015, December 19). Transcriptional regulation of mouse hypoglossal motor neuron somatotopic map formation *Brain Structure and Function, 221*, 4187–4202. PMID: 26687503.

Chibuzo, G. A., & Cummings, J. F. (1982). An enzyme tracer study of the organization of the somatic motor center for the innervation of different muscles of the tongue: Evidence for two sources. *The Journal of Comparative Neurology, 205*, 273–281.

Chou, R. H., Lu, C. Y., Wei-Lee, F. J. R., Yu, Y. L., & Shyu, W. C. (2014). The potential therapeutic applications of olfactory ensheathing cells in regenerative medicine. *Cell Transplantation, 23*, 567–571.

Dasen, J. S., Tice, B. C., Brenner-Morton, S., & Jessell, T. M. (2005). A Hox regulatory network establishes motor neuron pool identity and target muscle connectivity. *Cell, 123*, 477–491.

Dasen, J. S., De Camilli, A., Wang, B., Tucker, P. W., & Jessell, T. M. (2008). Hox repertoires for motor neuron diversity and connectivity gated by a single accessory factor, FoxP1. *Cell, 134*, 304–316.

Dormand, E. L., & Brand, A. H. (1998). Runt determines cell fates in the Drosophila embryonic CNS. *Development, 125*, 1659–1667.

Duffy, J. B., Kania, M. A., & Gergen, J. P. (1991). Expression and function of the Drosophila gene runt in early stages of neural development. *Development, 113*, 1223–1230.

Dykes, I. M., Lanier, J., Eng, S. R., & Turner, E. E. (2010, January 22). Brn3a regulates neuronal subtype specification in the trigeminal ganglion by promoting Runx expression during sensory differentiation. *Neural Development, 5*, 3.

Dykes, I. M., Tempest, L., Lee, S. I., & Turner, E. E. (2011). Brn3a and Islet1 act epistatically to regulate the gene expression program of sensory differentiation. *The Journal of Neuroscience, 31*, 9789–9799.

Gascon, E., Gaillard, S., Malapert, P., Liu, Y., Rodat-Despoix, L., Samokhvalov, I. M., et al. (2010). Hepatocyte growth factor-Met signaling is required for Runx1 extinction and peptidergic differentiation in primary nociceptive neurons. *The Journal of Neuroscience, 30*, 12414–12423.

Gibbs, J., Young, R. C., & Smith, G. P. (1973). Cholecystokinin decreases food intake in rats. *Journal of Comparative and Physiological Psychology, 84*, 488–495.

Gibbs, J., Falasco, J. D., & McHugh, P. R. (1976). Cholecystokinin-decreased food intake in rhesus monkeys. *The American Journal of Physiology, 230*, 15–18.

Ginhoux, F., Greter, M., Leboeuf, M., Nandi, S., See, P., Gokhan, S., et al. (2010). Fate mapping analysis reveals that adult microglia derive from primitive macrophages. *Science, 330*, 841–845.

Guizard, N., Coupé, P., Stifani, N., Stifani, S., & Collins, D.L.. (2010). *Robust 3D reconstruction and mean-shift clustering of motoneurons from serial histological images*. In Proceedings of 5th international conference on Medical Imaging and Augmented Reality (MIAR) (pp. 191–199). 2010. Springer-Verlag Berlin/Heidelberg.

Hermanson, O., Larhammar, D., & Blomqvist, A. (1998). Preprocholecystokinin mRNA-expressing neurons in the rat parabrachial nucleus: Subnuclear localization, efferent projection, and expression of nociceptive-related intracellular signaling substances. *The Journal of Comparative Neurology, 400*, 255–270.

Honma, Y., Kawano, M., Kohsaka, S., & Ogawa, M. (2010). Axonal projections of mechanoreceptive dorsal root ganglion neurons depend on Ret. *Development, 137*, 2319–2328.

Hua, Z. L., Smallwood, P. M., & Nathans, J. (2013). Frizzled3 controls axonal development in distinct populations of cranial and spinal motor neurons. *eLife, 2*, e01482.

Hung, H. A., Sun, G., Keles, S., & Svaren, J. (2015). Dynamic regulation of Schwann cell enhancers after peripheral nerve injury. *The Journal of Biological Chemistry, 290*, 6937–6950.

Inoue, K., Ozaki, S., Shiga, T., Ito, K., Masuda, T., Okado, N., et al. (2002). Runx3 controls the axonal projection of proprioceptive dorsal root ganglion neurons. *Nature Neuroscience, 5*, 946–954.

Inoue, I. K., Osato, M., Lee, B., Bae, S. C., & Ito, Y. (2007). The transcription factor Runx3 represses the neurotrophin receptor TrkB during lineage commitment of dorsal root ganglion neurons. *The Journal of Biological Chemistry, 282*, 24175–24184.

Inoue, K., Shiga, T., & Ito, Y. (2008). Runx transcription factors in neuronal development. *Neural Development, 3*, 20.

Jeong, J. H., Jin, J. S., Kim, H. N., Kang, S. M., Liu, J. C., Lengner, C. J., et al. (2008). Expression of Runx2 transcription factor in non-skeletal tissues, sperm, and brain. *Journal of Cellular Physiology, 217*, 511–517.

Kam, J. W., Raja, R., & Cloutier, J. F. (2014). Cellular and molecular mechanisms regulating embryonic neurogenesis in the rodent olfactory epithelium. *International Journal of Developmental Neuroscience, 37*, 76–86.

Kaminker, J. S., Canon, J., Salecker, I., & Banerjee, U. (2002). Control of photoreceptor axon target choice by transcriptional repression of Runt. *Nature Neuroscience, 5*, 746–750.

Kidd, G. J., Ohno, N., & Trapp, B. D. (2013). Biology of Schwann cells. *Handbook of Clinical Neurology, 115*, 55–79.

Kramer, I., Sigrist, M., de Nooij, J. C., Taniuchi, I., Jessell, T. M., & Arber, S. (2006). A role for Runx transcription factor signaling in dorsal root ganglion sensory neuron diversification. *Neuron, 49*, 379–393.

Krammer, E. B., Rath, T., & Lischka, M. F. (1979). Somatotopic organization of the hypoglossal nucleus: A HRP study in the rat. *Brain Research, 170*, 533–537.

Lallemend, F., & Ernfors, P. (2012). Molecular interactions underlying the specification of sensory neurons. *Trends in Neurosciences, 35*, 373–381.

Lallemend, F., Sterzenbach, U., Hadjab-Lallemend, S., Aquino, J. B., Castelo-Branco, G., Sinha, I., et al. (2012). Positional differences of axon growth rates between sensory neurons encoded by Runx3. *The EMBO Journal, 31*, 3718–3729.

Leinwand, S. G., & Chalasani, S. H. (2011). Olfactory networks: From sensation to perception. *Current Opinion in Genetics & Development, 21*, 806–811.

Levanon, D., Brenner, O., Negreanu, V., Bettoun, D., Woolf, E., Eilam, R., et al. (2001). Spatial and temporal expression pattern of Runx3 (Aml2) and Runx1 (Aml1) indicates non-redundant functions during mouse embryogenesis. *Mechanisms of Development, 109*, 413–417.

Levanon, D., Bettoun, D., Harris-Cerruti, C., Woolf, E., Negreanu, V., Eilam, R., et al. (2002). The Runx3 transcription factor regulates development and survival of TrkC dorsal root ganglia neurons. *The EMBO Journal, 21*, 3454–3463.

Li, H., Zhao, X., Yan, X., Jessen, W. J., Kim, M. O., Dombi, E., Liu, P. P., Huang, G., Wu, J. (2015, June 15). Runx1 contributes to neurofibromatosis type 1 neurofibroma formation. *Oncogene*, *35*, 1468–1474. doi:10.1038/onc.2015.207.

Liu, Y., Yang, F., Okuda, T., Dong, X., Zylka, M. J., Chen, C., et al. (2008). Mechanisms of compartmentalized expression of Mrg class G protein-coupled sensory receptors. *The Journal of Neuroscience, 28*, 125–132.

Logan, T. T., Villapol, S., & Symes, A. J. (2013). TGF-β superfamily gene expression and induction of the Runx1 transcription factor in adult neurogenic regions after brain injury. *PloS One, 8*(3), e59250.

Logan, T. T., Rusnak, M., & Symes, A. J. (2015). Runx1 promotes proliferation and neuronal differentiation in adult mouse neurosphere cultures. *Stem Cell Research, 15*, 554–564.

Lou, S., Duan, B., Vong, L., Lowell, B. B., & Ma, Q. (2013). Runx1 controls terminal morphology and mechanosensitivity of VGLUT3-expressing C-mechanoreceptors. *The Journal of Neuroscience, 33*, 870–882.

Lou, S., Pan, X., Huang, T., Duan, B., Yang, F. C., Yang, J., et al. (2015). Incoherent feed-forward regulatory loops control segregation of C-mechanoreceptors, nociceptors, and pruriceptors. *The Journal of Neuroscience, 35*, 5317–5329.

Luo, W., Wickramasinghe, S. R., Savitt, J. M., Griffin, J. W., Dawson, T. M., & Ginty, D. D. (2007). A hierarchical NGF signaling cascade controls Ret-dependent and Ret-independent events during development of nonpetidergic DRG neurons. *Neuron, 54*, 739–754.

Malick, A., Jakubowski, M., Elmquist, J. K., Saper, C. B., & Burstein, R. (2001). A neurohistochemical blueprint for pain-induced loss of appetite. *Proceedings of the National Academy of Sciences of the United States of America, 98*, 9930–9935.

Marmigere, F., Montelius, A., Wegner, M., Groner, Y., Reichardt, L. F., & Ernfors, P. (2006). The Runx1/AML1 transcription factor selectively regulates development and survival of TrkA nociceptive sensory neurons. *Nature Neuroscience, 9*, 180–187.

McClung, J. R., & Goldberg, S. J. (1999). Organization of motoneurons in the dorsal hypoglossal nucleus that innervate the retrusor muscles of the tongue in the rat. *The Anatomical Record, 254*, 222–230.

McClung, J. R., & Goldberg, S. J. (2000). Functional anatomy of the hypoglossal innervated muscles of the rat tongue: A model for elongation and protrusion of the mammalian tongue. *The Anatomical Record, 260*, 378–386.

Mueller, W., Nutt, C. L., Ehrich, M., Riemenschneider, M. J., von Deimiling, A., van den Boom, D., & Louis, D. N. (2007). Downregulation of RUNX3 and TES by hypermethylation in glioblastoma. *Oncogene, 26*, 583–593.

Murthy, M., Bocking, S., Verginelli, F., & Stifani, S. (2014). Transcription factor Runx1 inhibits proliferation and promotes developmental maturation in a selected population of inner olfactory nerve layer olfactory ensheathing cells. *Gene, 540*, 191–200.

Nakamura, S., Senzaki, K., Yoshikawa, M., Nishimura, M., Inoue, K. I., Ito, Y., et al. (2008). Dynamic regulation of the expression of neurotrophin receptors by Runx3. *Development, 135*, 1703–17011.

Nakazato, R., Takarada, T., Watanabe, T., Nguyen, B. T., Ikeno, S., Hinoi, E., & Yoneda, Y. (2014). Constitutive and functional expression of runt-related transcription factor-2 by microglial cells. *Neurochemistry International, 74*, 24–35.

Nibu, K., Li, G., Kaga, K., & Rothstein, J. L. (2000). bFGF induces differentiation and death of olfactory neuroblastoma cells. *Biochemical and Biophysical Research Communications, 279*, 172–180.

Palmesino, E., Rousso, D. L., Kao, T.-J., Klar, A., Laufer, E., Uemura, O., et al. (2010). Foxp1 and Lhx1 coordinate motor neuron migration with axon trajectory choice by gating reelin signalling. *PLoS Biology, 8*(8), e1000446.

Prinz, M., & Mildner, A. (2011). Microglia in the CNS: Immigrants from another world. *Glia, 59*, 177–187.

Ransohoff RM, El Khoury J. 2015. Microglia in health and disease. *Cold Spring Harbor Perspectives in Biology*, *8*(1).

Reale, M. E., Webb, I. C., Wang, X., Baltazar, R. M., Coolen, L. M., & Lehman, M. N. (2013). The transcription factor Runx2 is under circadian control in the suprachiasmatic nucleus and functions in the control of rhythmic behavior. *PloS One, 8*(1), e54317.

Roet, K. C., & Verhaagen, J. (2014). Understanding the neural repair-promoting properties of olfactory ensheathing cells. *Experimental Neurology, 261*, 594–609.

Rousso, D. L., Gaber, Z. B., Wellik, D., Morrisey, E. E., & Novitch, B. G. (2008). Coordinated actions of the forkhead protein Foxp1 and Hox proteins in the columnar organization of spinal motor neurons. *Neuron, 59*, 226–240.

Schlegelmilch, T., Henke, K., & Peri, F. (2010). Microglia in the developing brain: From immunity to behavior. *Current Opinion in Neurobiology, 21*, 1–6.

Shemer, A., Erny, D., Jung, S., & Prinz, M. (2015). Microglia plasticity during health and disease: An immunological perspective. *Trends in Immunology, 36*, 614–624.

Simeone, A., Daga, A., & Calabi, F. (1995). Expression of runt in the mouse embryo. *Developmental Dynamics, 203*, 61–70.

Stifani, N., Freitas, A. R. O., Liakhovitskaia, A., Medvinsky, A., Kania, A., & Stifani, S. (2008). Suppression of interneuron programs and maintenance of selected spinal motor neuron fates by the transcription factor AML1/Runx1. *Proceedings of the National Academy of Sciences of the United States of America, 105*, 6451–6456.

Su, Z., & He, C. (2010). Olfactory ensheathing cells: Biology in neural development and regeneration. *Progress in Neurobiology, 92*, 517–532.

Suzuki, J., & Osumi, N. (2015). Neural crest and placode contributions to olfactory development. *Current Topics in Developmental Biology, 111*, 351–374.

Takarada, T., & Yoneda, Y. (2009). Transactivation by Runt-related factor-2 of matrix metalloproteinase-13 in astrocytes. *Neuroscience Letters, 451*, 99–104.

Takeuchi, H., & Sakano, H. (2014). Neural map formation in the mouse olfactory system. *Cellular and Molecular Life Sciences, 71*, 3049–3057.

Theriault, F. M., Roy, P., & Stifani, S. (2004). AML1/Runx1 is important for the development of hindbrain cholinergic branchiovisceral motor neurons and selected cranial sensory neurons. *Proceedings of the National Academy of Sciences of the United States of America, 101*, 10343–10348.

Theriault, F. M., Nuthall, H. N., Dong, Z., Lo, R., Barnabe-Heider, F., Miller, F. D., & Stifani, S. (2005). Role for Runx1 in the proliferation and neuronal differentiation of selected progenitor cells in the mammalian nervous system. *The Journal of Neuroscience, 25*, 2050–2061.

Ugarte, G. D., Diaz, E., Biscaia, M., Stehberg, J., Montecino, M., & van Zundert, B. (2013). Transcription of the pain-related TRPV1 gene requires Runx1 and C/EBPβ factors. *Journal of Cellular Physiology, 228*, 860–870.

Vladimirova, V., Waha, A., Luckerath, K., Pesheva, P., & Probmeister, R. (2008). Runx2 is expressed in human glioma cells and mediates the expression of galectin-3. *Journal of Neuroscience Research, 86*, 2450–2461.

Wu, Y., Dissing-Olesen, L., MacVicar, B. A., & Stevens, B. (2015). Microglia: Dynamic mediators of synapse development and plasticity. *Trends in Immunology, 36*(10), 605–613.

Yang, F. C., Tan, T., Huang, T., Christianson, J., Samad, O. A., Liu, Y., et al. (2013). Genetic control of the segregation of pain-related sensory neurons innervating the cutaneous versus deep tissues. *Cell Reports, 5*, 1353–1364.

Yoshikawa, M., Senzaki, K., Yokomizo, T., Takahashi, S., Ozaki, S., & Shiga, T. (2007). Runx1 selectively regulates cell fate specification and axonal projections of dorsal root ganglion neurons. *Developmental Biology, 303*, 663–674.

Yoshikawa, M., Murakami, Y., Senzaki, K., Masuda, T., Ozaki, S., Ito, Y., & Shiga, T. (2013). Co-expression of Runx1 and Runx3 in mechanoreceptive dorsal root ganglion neurons. *Developmental Neurobiology, 73*, 469–479.

Yoshikawa, M., Hirabayashi, M., Ito, R., Ozaki, S., Aizawa, S., Masuda, T., et al. (2015). Contribution of the Runx1 transcription factor to axonal pathfinding and muscle innervation by hypoglossal motoneurons. *Developmental Neurobiology, 75*, 1295–1314.

Zagami, C. J., & Stifani, S. (2010). Molecular characterization of the mouse superior lateral parabrachial nucleus through expression of the transcription factor Runx1. *PloS One, 5*(11), e13944.

Zusso, M., Methot, L., Lo, R., Grenhalgh, A. D., David, S., & Stifani, S. (2012). Regulation of postnatal forebrain amoeboid microglial cell proliferation and development by the transcription factor Runx1. *The Journal of Neuroscience, 32*, 11285–11298.

Runx Family Genes in Tissue Stem Cell Dynamics

9

Chelsia Qiuxia Wang, Michelle Meng Huang Mok, Tomomasa Yokomizo, Vinay Tergaonkar, and Motomi Osato

Abstract

The Runx family genes play important roles in development and cancer, largely via their regulation of tissue stem cell behavior. Their involvement in two organs, blood and skin, is well documented. This review summarizes currently known Runx functions in the stem cells of these tissues. The fundamental core mechanism(s) mediated by Runx proteins has been sought; however, it appears that there does not exist one single common machinery that governs both tissue stem cells. Instead, Runx family genes employ multiple spatiotemporal mechanisms in regulating individual tissue stem cell populations. Such specific Runx requirements have been unveiled by a series of cell type-, developmental stage- or age-specific gene targeting studies in mice. Observations from these experiments revealed that the regulation of stem cells by Runx family genes turned out to be far more complex than previously thought. For instance, although it has been reported that Runx1 is required for the endothelial-to-hematopoietic cell transition (EHT) but not thereafter, recent studies clearly demonstrated that Runx1 is also needed during the period subsequent to EHT, namely at perinatal stage. In addition, *Runx1* ablation in the

C.Q. Wang
Cancer Science Institute of Singapore, National University of Singapore, Singapore, Singapore

Institute of Molecular and Cell Biology, A*STAR, Singapore, Singapore

M.M.H. Mok
Cancer Science Institute of Singapore, National University of Singapore, Singapore, Singapore

T. Yokomizo
International Research Center for Medical Sciences, Kumamoto University, Chuo-ku, Kumamoto, Japan

V. Tergaonkar
Institute of Molecular and Cell Biology, A*STAR, Singapore, Singapore

M. Osato (✉)
Cancer Science Institute of Singapore, National University of Singapore, Singapore, Singapore

Institute of Bioengineering and Nanotechnology, A*STAR, Singapore, Singapore

Department of Paediatrics, National University of Singapore, Singapore, Singapore

International Research Center for Medical Sciences, Kumamoto University, Chuo-ku, Kumamoto, Japan
e-mail: csimo@nus.edu.sg

Y. Groner et al. (eds.), *RUNX Proteins in Development and Cancer*, Advances in Experimental Medicine and Biology 962, DOI 10.1007/978-981-10-3233-2_9

embryonic skin mesenchyme eventually leads to complete loss of hair follicle stem cells (HFSCs) in the adult epithelium, suggesting that Runx1 facilitates the specification of skin epithelial stem cells in a cell extrinsic manner. Further in-depth investigation into how Runx family genes are involved in stem cell regulation is warranted.

Keywords
Runx • Hmga2 • Hematopoietic stem cell • Hair follicle stem cell • Niche • Fetal liver • Leukemia • Skin cancer • Perinatal • Aging • DNA damage repair • Conditional knockout mice

9.1 Introduction

In multicellular organisms, terminally differentiated cells of most tissues are short-lived and therefore require constant replenishment from stem cells for homeostasis and tissue repair. Stem cells are functionally defined as cells that can self-renew and are multipotent. When required, stem cells undergo cell division to self-renew and/or provide for downstream progenitors and differentiated cells. Stem cells can also reside in a dormant state. Importantly, deregulation of the stem cell compartment often leads to organ failure or tumorigenesis.

It is well documented that the Runx family genes (*Runx1*, *Runx2*, *Runx3*, and *Cbfb*) play a fundamental role in controlling the stem cell populations of various tissues (Table 9.1). They are central players in the fine-tuning of the balance among cell proliferation, differentiation, and cell cycle exit (summarized in Wang et al. 2010). In this review, the role of Runx and Cbfβ in two well characterized murine tissue stem cells, namely hematopoietic stem cells (HSCs) and hair follicle stem cells (HFSCs), will be discussed.

9.2 Hematopoietic Stem Cells in the Blood

The hematopoietic system performs multiple functions, including oxygen transportation, blood clotting and providing immunity. Blood cells have a high turnover rate and are continuously replenished by HSCs and long term progenitors. HSCs, defined as cells that are multipotent and capable of long-term repopulating activity when transplanted into irradiated adult recipient mice (Cumano and Godin 2007), are generated during embryonic development in vertebrates (Fig. 9.1) (see chapter by Yzaguirre et al.). At E10.5-E11.5 of mouse development, the aorta-gonad-mesonephros (AGM) region is the primary intra-embryonic hemogenic territory of HSC

Table 9.1 Summary of the various tissue stem cells in which Runx family genes play a regulatory role

	Runx family gene involved				
Tissue stem cell	*Runx1*	*Runx2*	*Runx3*	*Cbfb*	References
Hematopoietic stem cell	✓		✓	✓	Yokomizo et al. (2001), Jacob et al. (2010). Wang et al. (2013), Wang et al. (2014), and Wang et al. (2015)
Hair follicle stem cell	✓				Osorio et al. (2011) and Hoi et al. (2010)
Skeletal stem cell		✓			Worthley et al. (2015)
Mammary stem cell	✓	✓			van Bragt et al. (2014) and Ferrari et al. (2015)
Gastric stem cell	✓				Matsuo et al. (2016)
Intestinal stem cell	✓?				Scheitz et al. (2012)
Neural crest stem cell	✓				Kanaykina et al. (2010)
Oral epithelial stem cell	✓?				Scheitz et al. (2012)

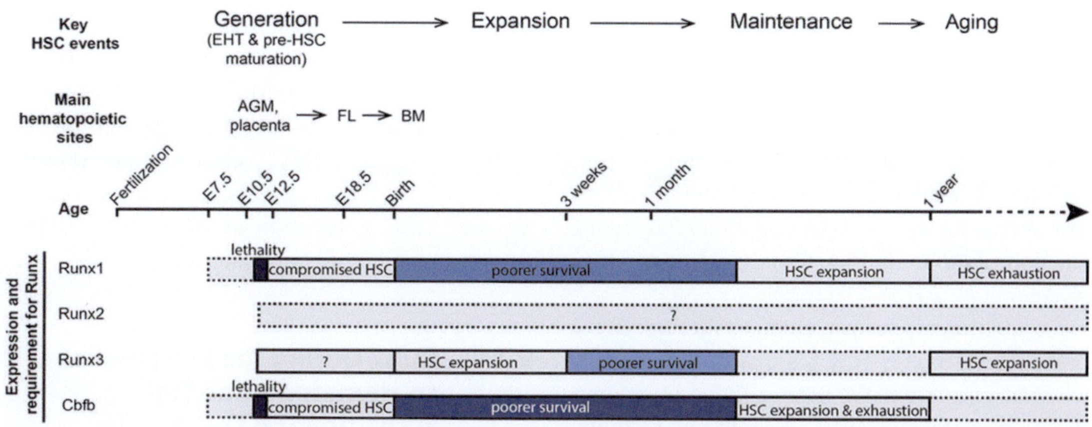

Fig. 9.1 Summary of the ontogeny of hematopoietic stem cells and requirement of Runx family genes. Timeline schematic diagram showing the key events which HSCs undergo and the main hematopoietic sites at which these events occur during ontogeny (*top* panel). Based on studies using knockout mouse models of *Runx1*, *Runx3* and *Cbfb*, the time windows during which the Runx proteins are required have been established (*bottom* panel). Intensity of the *blue boxes* represent the mortality ratio observed. The HSC-associated phenotypes are also described in boxes. Note that the indicated information for the time period E12.5 to approximately 2 months and the period thereafter is derived from observations in Runx-deficient mice using the Vav1-iCre and Mx1-Cre system, respectively. Dotted lines delineate the times at which the Runx family gene is expressed. Abbreviations: *HSC* hematopoietic stem cell, *EHT* endothelial-to-hematopoietic transition, *AGM* aorta-gonad-mesonephros, *FL* fetal liver, *BM* bone marrow

generation. Clusters of hematopoietic cells can be visualized emerging from the hemogenic endothelium into the lumen of the dorsal aorta—a phase known as endothelial-to-hematopoietic cell transition (EHT) (Jaffredo et al. 2005; Yokomizo and Dzierzak 2010; Kissa and Herbomel 2010). The HSC precursors, also known as pre-HSCs, that are generated from EHT must undergo maturation and acquire stem cell characteristics (Taoudi et al. 2008; Rybtsov et al. 2011) before becoming *bona fide* HSCs. The placenta is also capable of *de novo* HSC generation (Rhodes et al. 2008) and in fact harbors more HSCs than the AGM at its peak at E12.5–E13.5 (Gekas et al. 2005; Ottersbach and Dzierzak 2005).

Regardless of their sites of generation, the HSCs colonize the fetal liver (Dzierzak and Speck 2008; Cumano and Godin 2007), which not only provides a conducive environment for further maturation (Kieusseian et al. 2012), but which also serves as the predominant site for rapid HSC proliferation and differentiation to pools of various blood progenitors from E12 to E16 (Martinez-Agosto et al. 2007). Towards the end of the prenatal period, the HSCs proceed to colonize the bone marrow (Orkin and Zon 2008; Christensen et al. 2004), which will serve as the primary site of adult hematopoiesis throughout the postnatal life of the organism.

Interestingly, between 3 and 4 weeks after birth, most of the HSCs in the murine bone marrow abruptly transit from a predominantly proliferative state to a quiescent state, after which the HSC population size remains constant under steady state conditions (Bowie et al. 2006). Long-term progenitors, including classically defined short-term HSCs (ST-HSCs) and multipotent progenitors (MPPs), play a prime role in sustaining hematopoiesis under unperturbed conditions (Sun et al. 2014; Busch et al. 2015), while HSCs remain largely dormant in the lifetime of the organism and serves as a backup reservoir for the recovery from stressed hematopoiesis such as infection, cytotoxic drug treatment, transplantation or perhaps even aging.

As the organism ages, HSCs undergo a functional decline. Aged HSCs have a reduced self-renewal capability and a poorer ability to contribute to hematopoiesis compared to younger HSCs (Dykstra et al. 2011). As a compensatory mechanism to overcome their impaired function, their numbers are increased (Morrison et al. 1996). Aged HSCs also exhibit a biased differentiation potential towards the myeloid lineage (Dykstra et al. 2011). Previous studies have shown that accumulation of DNA damage, transcriptional changes, epigenetic modifications, and altered lineage contribution are factors which can contribute to the functional decline of HSCs (Geiger et al. 2013).

In addition to intrinsic changes through development and aging, HSCs are also subjected to external cues from the extracellular microenvironment. Although much still remains unknown about HSC niches during the developmental stage, two cell types that support HSC proliferation in the fetal liver are the Nestin$^+$NG2$^+$ pericytes located close to portal vessels (Khan et al. 2016) and lymphatic vessel endothelial receptor 1 (Lyve-1)$^+$ sinusoidal endothelial cells (Iwasaki et al. 2010). The vascular network in the placental labyrinth may provide external cues for HSC proliferation (Rhodes et al. 2008). In the extensively studied bone marrow niche, HSCs receives appropriate signals for their maintenance and differentiation from multiple niche-constituting cells (summarized in Birbrair and Frenette 2016). Such cells include osteolineage cells, Nestin$^+$ pericyte cells, CXCL12-abundant reticular (CAR) cells, leptin receptor (Lepr)$^+$ perivascular stromal cells, endothelial cells, glial fibrillary acidic protein (GFAP)$^+$ Schwann cells, sympathetic nerves, and even hematopoietic cells like CD169$^+$ macrophages and megakaryocytes. CXCR4-CXCL12, Tie2-angiopoietin1 (Ang1), CD44-osteopontin (OPN), integrin α_2 (or CD49b)-collagen, integrin $\alpha_4\beta_1$-Vcam1, and Robo4-Slit2 have been found to be important for HSC-niche interactions. It is increasingly apparent that changes in the HSC niche affects the age-related changes in HSCs (Arora et al. 2014).

Differential control of HSC behavior by genes at different stages of development has been described. For example, Sox17 (Kim et al. 2007) and Scl (Lecuyer and Hoang 2004) are required for HSC generation or maintenance during early embryonic development but dispensable for adult HSCs, while Bmi1 (Park et al. 2003), Gfi1 (Hock et al. 2004), Tie2 (Puri and Bernstein 2003) and C/ebpα (Ye et al. 2013) are critical for self-renewal in adult HSCs but not in fetal HSCs. Knowledge about the intrinsic molecular mechanisms and important signals supplied by the niche to regulate HSC behavior may provide clues into establishing successful protocols for *de novo* HSC generation *in vitro*, *ex vivo* HSC expansion and HSC rejuvenation.

9.2.1 Runx During the Endothelial-to-Hematopoietic Cell Transition (EHT)

During early development of mouse embryos, Runx1 and Cbfb are indispensable for the generation of HSCs, specifically at EHT. While *Runx1*-positive cells are found in hemogenic endothelial cells and hematopoietic cell clusters in the dorsal aorta of the AGM region at E9.5–E11.5 in wild-type embryos (Yokomizo and Dzierzak 2010; Yokomizo et al. 2001; Ng et al. 2010; North et al. 1999), *Runx1*$^{-/-}$ embryos showed lack of such hematopoietic cell clusters budding from the vessel wall (Yokomizo et al. 2001; North et al. 1999). As a consequence of the lack of EHT and thus the complete failure in generating HSCs and definitive hematopoiesis, both *Runx1*$^{-/-}$ and *Cbfb*$^{-/-}$ mice die between E11.5 and E12.5 (Okada et al. 1998; Niki et al. 1997; Ng et al. 2010; Okuda et al. 1996; Wang et al. 1996a; b; Sasaki et al. 1996) (Fig. 9.1).

Employing different developmental stage- or tissue-specific Cre system to ablate *Runx1* in mice, it was shown that the essential requirement for Runx1 function at this early developmental stage occurs within a specific time window. When *Runx1* is abrogated in endothelial cells by vascular endothelial cadherin (VEC) promoter-driven Cre, the knockout mice showed embryonic lethality due to the lack of intra-aortic hematopoietic cell clusters and HSCs (Chen et al. 2009). However, embryonic lethality was not observed when Cre driven by the promoter of a pan-hematopoietic *Vav1* gene, active starting from E10.5, was used to delete *Runx1*. Using temporal knockouts of *Runx1*$^{f/f}$*;Actb-Cre*^{ERT} and *Runx1*$^{f/f}$*;VEC-Cre*^{ERT} mice based on the tamoxifen-activated

Cre recombinase system, it was established that during the early developmental stage, Runx1 is required at least up to E11.5, after which removal of both *Runx1* alleles does not eliminate HSCs (Tober et al. 2013). These results appear to indicate that Runx1 is specifically required for HSC generation during EHT, but not thereafter.

In *Runx2*$^{-/-}$ and *Runx3*$^{-/-}$ mice, there are no inherent hematopoietic phenotypes at the early development stage including EHT. Yet, Runx2 was reported to be expressed in the AGM region, although Runx3 was not (Okada et al. 1998; Levanon et al. 2001). However, it is not clear whether the methods used for the examination of expression are sensitive and specific for individual Runx family genes. As such, the involvement of these genes during early development remains to be investigated. The roles of Runx1 in HSC generation during early development are described in detail elsewhere in this book (see chapter by Yzaguirre et al.).

9.2.2 Runx in Hematopoietic Stem Cells at the Peri-Natal Stage

Even after EHT is complete, it was reported that fetal livers from E14.5 *Runx1*$^{f/f}$;*Vav1-Cre* embryos contained fourfold fewer functional HSCs than control fetal livers in a limiting dilution transplantation assay (Cai et al. 2011) and *Cbfb*$^{f/f}$;*Vav1-Cre* fetal livers were completely unable to reconstitute recipient mice (Tober et al. 2013). Therefore, it seems likely that the Runx-deficient fetal HSCs are compromised. Probably due to such defects, *Cbfb*$^{f/f}$;*Vav1-iCre*$^{+}$ mice were not born at Mendelian ratios (Wang et al. 2015), although *Runx1*$^{f/f}$;*Vav1-iCre*$^{+}$ and *Runx3*$^{f/f}$;*Vav1-iCre*$^{+}$ mice were (data not shown). Interestingly, *Runx1*$^{f/f}$;*Runx3*$^{f/f}$;*Vav1-iCre*$^{+}$ mice could not be obtained from crosses of *Runx1*$^{f/+}$;*Runx3*$^{f/f}$;*Vav1-iCre*$^{+}$ with *Runx1*$^{f/f}$;*Runx3*$^{f/f}$;*Vav1-iCre*$^{-}$ mice (0 out of 73 mice genotyped) (Fig. 9.2). After birth, *Runx1*$^{f/f}$;*Vav1-iCre*$^{+}$ and *Cbfb*$^{f/f}$;*Vav1-iCre*$^{+}$ mice exhibit poorer survival than *Runx1*$^{f/f}$;*Mx1-Cre*$^{+}$ and *Cbfb*$^{f/f}$;*Mx1-Cre*$^{+}$ mice, respectively (Fig. 9.2a, d). Although the survival data do not directly translate to HSC defects, the most critical data supporting that the pronounced phenotypes in *Runx*-deficient *Vav1-iCre*$^{+}$ mice are due to functional HSC defects is derived from transplantation experiments. The ability of *Cbfb*$^{f/f}$;*Vav1-iCre*$^{+}$ adult HSCs to reconstitute recipient mice was severely impaired (Wang et al. 2015). Taken together, these results suggest that the perinatal absence of Runx family genes causes dysfunctional HSCs, even after EHT (Fig. 9.1).

Interestingly, a study involving reexpression of Runx1 in a reversible *Runx1* knockout mouse model using Tie2-Cre supports such a notion (Liakhovitskaia et al. 2009). Although restoration of Runx1 expression rescues the ability of the mice to generate HSCs as fetal liver cells isolated from such mice were able to reconstitute irradiated recipient mice, the mice still exhibit lethality at birth. Possible causes of lethality was attributed to the abnormal development of sternum (Liakhovitskaia et al. 2010), but defects in HSC migration from fetal liver to the bone marrow cannot be ruled out. Furthermore, a recent study investigating the involvement of Runx1 during early development revealed that Runx1 is required for the maturation of pre-HSCs (Liakhovitskaia et al. 2014).

Notably, the reconstitution ability of E14.5 *Cbfb*$^{f/f}$;*Vav1-Cre* fetal livers (Tober et al. 2013) was much worse than *Runx1*$^{f/f}$;*Vav1-Cre* fetal livers (Cai et al. 2011), indicating a possibility that Runx2 and Runx3 may be required after EHT. Supporting this possibility, it has been reported that there is a prominent increase in Runx2 levels when the intermediary pre-HSCs transit to become nascent HSCs (Zhou et al. 2016), suggesting that at least Runx2 may play a role at this transition stage. Furthermore, *Runx3*$^{f/f}$;*Vav1-iCre*$^{+}$ mice exhibited increased HSC compartment accompanied by leukocytosis, which was only observed in aged *Runx3*$^{f/f}$;*Mx1-Cre*$^{+}$ mice, but not the young ones (Fig. 9.2e, f). Surprisingly, *Runx3*$^{f/f}$;*Vav1-iCre*$^{+}$ mice showed poorer survival (Fig. 9.2b), which was not observed in *Runx3*$^{f/f}$;*Mx1-Cre*$^{+}$ mice (Wang et al. 2014).

The differences in phenotypes of *Runx*-deficient mice using the Vav1-iCre system versus Mx1-Cre system can be explained by the distinct requirement of a gene in developmental versus

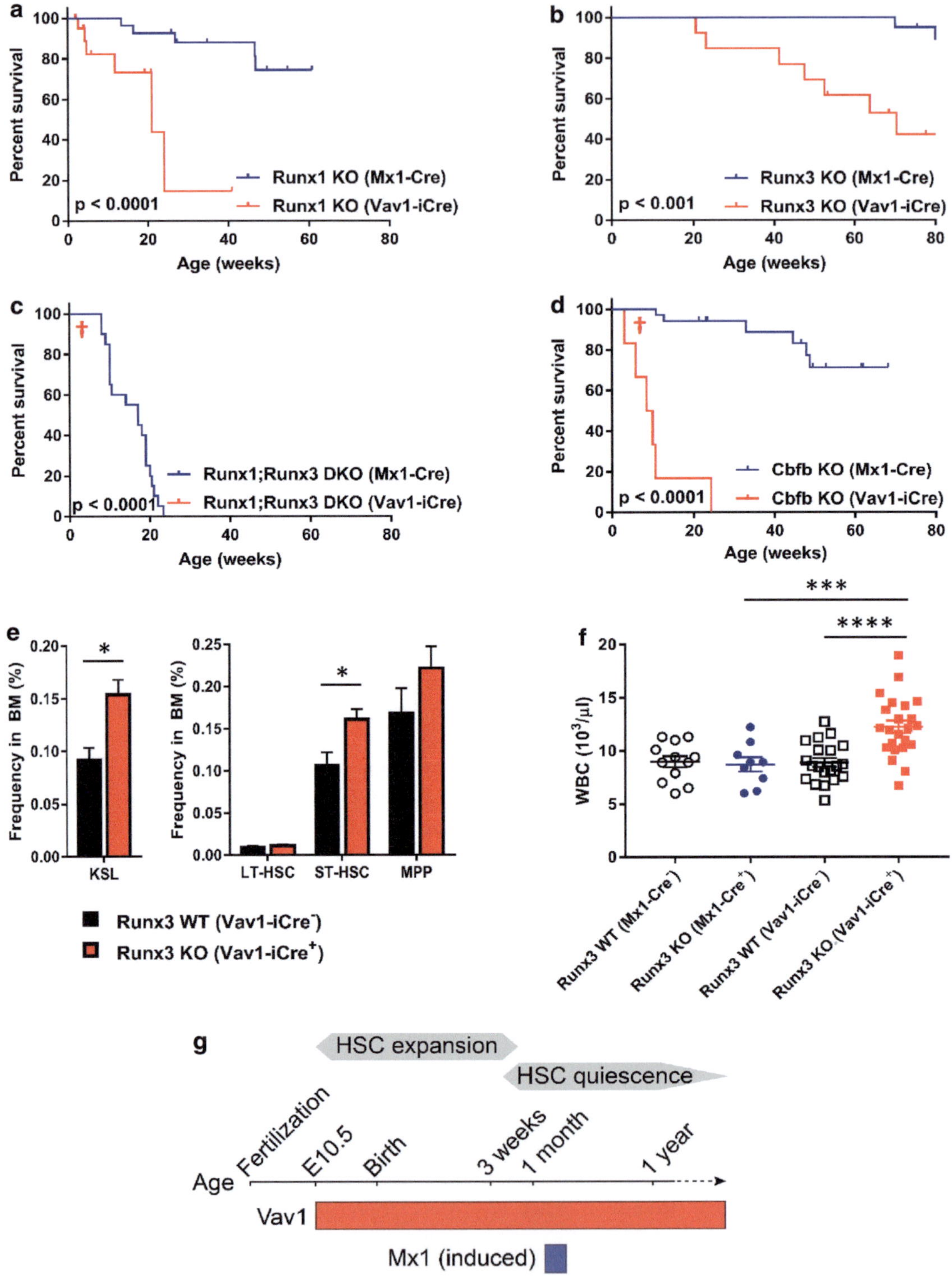

Fig. 9.2 Comparison of phenotypes of *Runx1*, *Runx3*, *Runx1*; *Runx3* and *Cbfb* conditional KO mice using different Cre-inducible systems. (**a–d**) Kaplan-Meier survival curves of (**a**) *Runx1*$^{f/f}$*;Mx1-Cre*$^{+}$ (*black line*, n = 27) versus *Runx1*$^{f/f}$*;Vav1-iCre*$^{+}$ (*red line*, n = 22), (**b**) *Runx3*$^{f/f}$*;Mx1-Cre*$^{+}$ (*black line*, n = 11) versus *Runx3*$^{f/f}$*;Vav1-iCre*$^{+}$ (*red line*, n = 12), (**c**) *Runx1*$^{f/f}$*;Runx3*$^{f/f}$*;Mx1-Cre*$^{+}$ (*black line*, n = 27) versus *Runx1*$^{f/f}$*;Runx3*$^{f/f}$*;Vav1-iCre*$^{+}$, and (**d**) *Cbfb*$^{f/f}$*;Mx1-Cre*$^{+}$ (*black line*, n = 35) versus *Cbfb*$^{f/f}$*;Vav1-iCre*$^{+}$ (*red line*, n = 6) mice. Hundred percent and

adult hematopoiesis (discussed in Koh et al. 2015): Vav1 promoter is activated at E10.5 onwards in all hematopoietic cells, including those during the maturation of pre-HSCs to nascent HSCs and the critical period of rapid HSC expansion, while Mx1 promoter is induced by polyinosinic–polycytidylic acid (pIpC) at the adult stage when HSCs are largely in quiescence (Fig. 9.2g). It is important to note that there are at least four distinct Vav1 promoter-driven Cre strains which could potentially delete in distinct but overlapping cell populations (Georgiades et al. 2002; Croker et al. 2004; Stadtfeld and Graf 2005; de Boer et al. 2003), and differences in phenotypes might be observed even amongst these four strains. Noteworthy is the enhanced expression of iCre (improved Cre) in the Vav1-iCre strain compared to the original Cre in the other Vav1-Cre strains (Shimshek et al. 2002; de Boer et al. 2003), which may cause some level of toxicity (Loonstra et al. 2001; Silver and Livingston 2001) and poorer survival.

Taken together, Runx may be required for one or more of the following processes during the perinatal stage: (1) maturation of pre-HSCs to nascent HSCs, (2) HSC expansion in the placenta, fetal liver or bone marrow, (3) migration of HSCs from placenta to fetal liver and/or from fetal liver to bone marrow, and/or (4) the tightly controlled transition when the abrupt change from proliferative status to quiescence in HSC occurs.

9.2.3 Runx in Hematopoietic Stem Cells of Young Adults

The Runx family genes are also important for the maintenance of HSCs in adults. In contrast to the complete lack of HSCs in *Runx1*$^{-/-}$ and *Cbfb*$^{-/-}$ mice, HSCs are still present in adult *Runx1* or *Cbfb* conditional knockout mice using the Mx1-Cre system, but showed disrupted regulation (Ichikawa et al. 2004; Growney et al. 2005; Wang et al. 2015). Phenotypic HSCs in *Runx1*$^{f/f}$*;Mx1-Cre*$^{+}$ mice exhibited a twofold expansion (Jacob et al. 2010). Recently, it has also been demonstrated that *Cbfb*$^{f/f}$*;Mx1-Cre*$^{+}$ show an even more pronounced expansion in the immunophenotypic HSC compartment (Wang et al. 2015), similar to the extent of expansion in *Runx1*$^{f/f}$*;Runx3*$^{f/f}$*;Mx1-Cre*$^{+}$ mice (Wang et al. 2014). These results suggest that both Runx1 and Runx3 serve to inhibit cell proliferation in HSCs, perhaps mediated by the following two molecular mechanisms.

First, Runx can directly inhibit cell cycle progression. Gene expression profiling of *Runx1*$^{f/f}$*;Runx3*$^{f/f}$*;Mx1-Cre*$^{+}$ HSC-enriched c-Kit^{+}Sca-1^{+} Lineage^{-} (KSL) cells showed significantly elevated levels of high-mobility group AT-hook 2 (Hmga2), a non-histone chromatin-interacting factor, which could have contributed to increased HSC self-renewal (Fig. 9.3). It has been shown that Runx1 plays a direct role in repressing the transcription of Hmga2 (Lam et al. 2014). The

Fig. 9.2 (continued) forty percent embryonic/perinatal lethality observed in *Runx1*$^{f/f}$*;Runx3*$^{f/f}$*;Vav1-iCre*$^{+}$ and *Cbfb*$^{f/f}$*;Vav1-iCre*$^{+}$, respectively, are marked in *red crosses* (†). The *p*-value from Mantel-Cox test is shown. *Vertical ticks* represent censored cases. Survival curves of wild-type controls are not shown for clarity purpose. (**e**) Frequency of hematopoietic stem/progenitor cell populations of *Runx3*$^{f/f}$*;Vav1-iCre*$^{-}$ (n = 6) and *Runx3*$^{f/f}$*;Vav1-iCre*$^{+}$ mice (n = 6) at 2 months old. Mean ± SEM of percentages in whole BM are shown. (**f**) White blood cell counts of *Runx3*$^{f/f}$*;Mx1-Cre*$^{-}$ (n = 12) and *Runx3*$^{f/f}$*;Mx1-Cre*$^{+}$ (n = 9) at 2 months post-induction, and *Runx3*$^{f/f}$*;Vav1-iCre*$^{-}$ (n = 20) and *Runx3*$^{f/f}$*;Vav1-iCre*$^{+}$ mice (n = 23) at 2 months old. Mean ± SEM of leukocyte counts are shown. (**g**) Diagram showing the time window of differential HSC properties (*top*) and the time when Vav1-iCre or Mx1-Cre are expressed in hematopoietic tissues (*bottom*). Abbreviations: *KSL* c-Kit^{+}Sca-1^{+}Lineage^{-}, *LT-HSC* long-term hematopoietic stem cell (c-Kit^{+}Sca-1^{+}Lineage^{-}CD34^{-}Flt3^{-}), *ST-HSC* short-term hematopoietic stem cell (c-Kit^{+}Sca-1^{+}Lineage^{-}CD34^{+}Flt3^{-}), *MPP* multipotent progenitor (c-Kit^{+}Sca-1^{+}Lineage^{-}CD34^{+}Flt3^{+}). *Asterisk*(s) represents significant difference (*$p < 0.05$, ***$p < 0.001$ and ****$p < 0.0001$, Student's t-test)

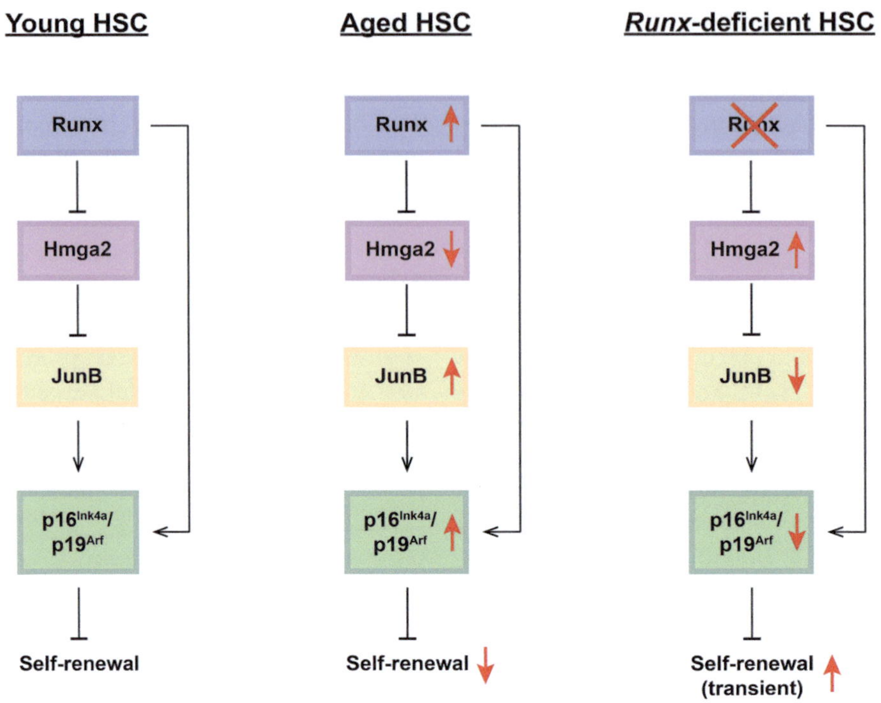

Fig. 9.3 Schematic diagram depicting the Runx-Hmga2 axis in controlling HSC behavior during aging. Runx has been shown to inhibit transcription of Hmga2, which in turn regulates transcription of age-related players, p16^{INK4a} and p19^{ARF} indirectly, possibly via JunB. p16^{INK4a} and p19^{ARF} inhibits HSC self-renewal during aging. Hence, increased Runx1 expression in aged HSCs culminates in reduced self-renewal and the absence of Runx causes increased HSC self-renewal, albeit transient, via this Runx-Hmga2 axis

transcriptional regulation of cell cycle regulatory proteins (such as D-type cyclins (Strom et al. 2000; Bernardin-Fried et al. 2004) and p21 (Strom et al. 2000; Galindo et al. 2005) by Runx family genes may also affect the cell cycle status of HSCs. For example, downregulation of Cdkn1a (p21) in *Runx1$^{f/f}$;Runx3$^{f/f}$;Mx1-Cre^{+}* KSL cells observed in gene expression profiling analysis may contribute to the expanded HSC population.

Additionally, Runx1 and Runx3 help to maintain stem cells in their quiescent state by regulation of the G_0/G_1 transition in a HSC niche-dependent manner, thereby controlling the HSC population size. Analysis of a panel of niche-related factors in *Runx1$^{f/f}$;Mx1-Cre^{+}* mice revealed that CXCR4 was downregulated. A luciferase reporter assay using the CXCR4 promoter, containing two Runx binding sites, showed that RUNX1 transactivates CXCR4 in a DNA-binding dependent manner, suggesting the direct transcriptional regulation of CXCR4 expression by RUNX1 (Jacob et al. 2010; Chin et al. 2016). Along with CXCR4, another niche-interacting factor, integrin α_2 (CD49b), was also downregulated in *Runx1$^{f/f}$;Mx1-Cre^{+}* HSCs. In the case of *Runx1$^{f/f}$;Runx3$^{f/f}$;Mx1-Cre^{+}* mice, the KSL population showed downregulation of several critical niche-related factors: Vcam-1, Cxcr4 and Robo4 (Wang et al. 2014). Such inadequate expression of niche-related factors in *Runx*-deficient HSCs led to a weakened niche interaction and, as a result, less quiescence and expansion of HSCs could be induced.

Although the numbers of HSCs are increased in *Runx1* and *Cbfb* conditional knockout mice, the HSCs are functionally impaired in the ability to repopulate hematopoiesis in recipient mice. While *Runx1$^{f/f}$;Mx1-Cre^{+}* HSCs can reconstitute recipient mice in a bone marrow transplantation experiment, the chimerism in *Runx1*$^{f/f}$;*Mx1-Cre*$^{+}$ recipient mice decreased with time, unlike in control recipient mice (Growney et al. 2005). Furthermore, young adult *Runx1*$^{f/f}$;*Runx3*$^{f/f}$;*Mx1-Cre*$^{+}$, *Cbfb*$^{f/f}$;*Mx1-Cre*$^{+}$ and surviving *Cbfb*$^{f/f}$;*Vav1-iCre*$^{+}$ mice exhibit a phenomenon called stem cell exhaustion (Wang et al. 2014). This is thought to be the result of the interplay of at least two mechanisms described below.

First, Runx proteins control DNA damage repair pathway(s) (Wang et al. 2014) (see chapter by Krishnan and Ito). In the absence of *Runx1* and *Runx3*, KSL cells showed impairment in DNA damage repair, accentuated by an inter-strand crosslinking agent. The increased DNA damage in Runx-deficient cells, coupled by the increased proliferation described earlier, renders high levels of replication stress to the cells, contributing to decreased HSC integrity and subsequent exhaustion of the stem cell pool. Second, Runx proteins regulate the transcription of several aforementioned niche-interacting factors. In *Runx*-deficient HSCs, there is compromised HSC-niche interactions important for maintaining HSC quiescence, which in turn causes the detachment of HSCs from the niche. Subsequently, the inability to maintain quiescence would result in the loss of long-term self-renewal capacity of *Runx*-deficient HSCs, eventually leading to stem cell exhaustion, despite the initial transient increase of HSCs.

The role of Runx2 in adult HSCs *per se* has yet to be studied, but Runx2 most probably affects HSC quiescence by generating the HSC niche components, namely the osteoblasts. Runx2 has been shown to be the principal transcriptional regulator of osteoblastic differentiation (Otto et al. 1997; Komori et al. 1997; Deguchi et al. 1999). This implies that the maintenance or replenishment of available HSC-niche sites during homeostatic bone turnover by Runx2 could potentially support stem cell quiescence, although more conclusive data is required. Additionally, Runx2 activates the transcription of OPN, an important factor for HSC quiescence, in osteoblasts (Sato et al. 1998). Contradictory, it has been recently demonstrated that Runx2 highly expressing immature osteoblasts were better than osteoblasts expressing lower levels of Runx2 for supporting proliferation and colony-forming capability of KSL population (Chitteti et al. 2010). Whether Runx2 is inhibitory or promoting for HSC quiescence remains to be further investigated.

9.2.4 Runx in Aging Hematopoietic Stem Cells

The phenotypes that distinguish aged HSCs from young HSCs include increased phenotypically defined HSC numbers, reduced self-renewal capacity, myeloid-biased differentiation, impaired homing to and enhanced mobilization from the bone marrow. Altered HSC-niche interactions can also play a role in HSC aging (Arora et al. 2014). Notably, the phenotypes commonly observed in various *Runx*-deficient mice seem to exhibit similarities to aging phenotypes (Fig. 9.1). When Runx family genes are ablated in the hematopoietic system, the HSC population as defined immunophenotypically is increased and there is a myeloid-biased differentiation. The motility of *Runx*-deficient HSCs is also increased. Hence, these features in *Runx*-deficient mice suggest a possibility that Runx family genes are essential to slow down the aging phenomenon.

Initial expansion and subsequent exhaustion of HSCs from *Runx1*$^{f/f}$;*Mx1-Cre*$^{+}$ mice (Jacob et al. 2010) and earlier death of these mice than the control mice may be attributable to progeria, or premature aging. Similarly, expansion of HSCs in aged *Runx3*$^{f/f}$;*Mx1-Cre*$^{+}$ mice may be caused by accelerated aging, although occurring to a lesser extent compared to *Runx1*$^{f/f}$;*Mx1-Cre*$^{+}$ mice (Wang et al. 2013). *Runx* deficiency may thus promote aging of HSCs and plausible mechanisms are discussed below.

Hmga2 was identified to be the only gene whose expression not only progressively declines with age in HSCs, but is preferentially expressed in HSCs but not in differentiated cells (Nishino et al. 2008). Hmga2 regulates the *Ink4a/Arf* locus via JunB, mediating its effects on aging via $p16^{Ink4a}$ and $p19^{Arf}$ (Fig. 9.3). It is well documented that $p16^{Ink4a}$ induction contributes to the decline of HSC function in aged mice, whereas its absence in HSCs rescues their compromised self-renewal capacity in transplantation assays (Janzen et al. 2006). Lin28b-let-7 has been reported to regulate Hmga2 (Copley et al. 2013). As discussed earlier, Runx1 transcriptionally represses Hmga2 in HSCs and is thus another regulator of Hmga2. Alternatively, Runx1 can induce the expression of $p19^{Arf}$ directly (Linggi et al. 2002).

Notably, *Ink4a/Arf* deletion in Hmga2 mutant mice does not completely rescue the stem cell defects, suggesting that there are other mechanisms mediated by Hmga2, in addition to its effects on HSC self-renewal via $p16^{Ink4a}$ and $p19^{Arf}$. Hmga2 is reported to be directly involved in DNA repair and aging-related DNA damage accumulation in HSCs might be due to the decrease of Hmga2 with age (Yu et al. 2014). Age-dependent accumulation of DNA damage in HSCs results in impaired self-renewal and thus decreased regenerative capacity of aged HSCs. As mentioned earlier, Runx proteins are also implicated in DNA repair and their roles in maintaining HSC genomic integrity could be via the regulation of Hmga2 (Wang et al. 2014). Taken together, it is imperative to determine if Runx family genes are indeed regulators of HSC ageing.

An alternative possibility for the ageing phenotype is that cytoplasmic sequestration of the IKK complex by Runx1 results in diminished NF-κB signaling (Nakagawa et al. 2011). In Runx-deficient cells, the derepression of NF-κB transcriptional targets can lead to increased senescence, and possibly the senescence-associated secretory phenotype (SASP) (Chien et al. 2011), resulting in a premature aging phenotype. Interestingly, the secretion of SASP factors can affect the cellular microenvironment, which may implicate Runx1 as having a role in influencing the HSCs and/or HSC niche via SASP.

FOXO3a is one of the few confirmed longevity genes (Flachsbart et al. 2009). SIRT1, a member of the sirtuin family of nicotinamide adenine dinucleotide (NAD^+)-dependent deacetylases, is another anti-aging gene that triggers FOXO3a transcription via its deacetylation activity (Brunet et al. 2004). SIRT1 is thought to contribute to increased longevity by acting as a guardian against cellular oxidative stress and DNA damage (Haigis and Guarente 2006). Intriguingly, Runx2 physically interacts with Sirt1 (Shakibaei et al. 2012) and RUNX3 physically interacts with FOXO3a (Yamamura et al. 2006). In addition, the transcription of Runx2 was found to be co-mediated by SIRT1 and FOXO3a (Tseng et al. 2011). Interestingly, Runx2 is reported to be downregulated in aged HSCs (Chambers et al. 2007), and this decrease may be linked to the downregulation of SIRT1 and FOXO3a during aging. The close relationship between Runx family genes with longevity factors, FOXO3a and SIRT1, warrants further study into the role of Runx in HSC aging.

Based on the discussion above, one would then expect Runx1 to be expressed at low levels in aged HSCs which are increased in number and have a myeloid-biased differentiation phenotype. Paradoxically, Runx1, a myeloid-related gene, is upregulated in aged HSCs (Rossi et al. 2005; Chambers et al. 2007). Such discrepancy could be due to differential expression of the Runx1 isoforms in young versus aged mice. Runx1bEx6e, functionally similar to RUNX1a in human, enhances expansion of HSCs (Komeno et al. 2014; Osato 2014) and may be upregulated in aged HSCs, while Runx1bEx6+ inhibits HSC proliferation and could be downregulated in aged HSCs.

As the heterodimerization partner of the Runx proteins, the significance of Cbfβ upregulation (Rossi et al. 2005; Chambers et al. 2007) in aged HSCs remains to be further investigated. In general, the role of Runx family genes in aging is not well studied and stands as the important theme for future research.

9.2.5 Runx and Leukemia

In general, myeloid related hematopoietic malignancies are more predominant at old age, whereas lymphoid related ones occur at younger age. However, human RUNX-related myeloid leukemia such as inv(16) and t(8;21) are prevalent in childhood, as well as adolescents and younger adults (AYAs) (Mrozek et al. 2012). As mentioned above, results obtained from the mouse model studies clearly demonstrated that Runx family genes play a development- and age-dependent requirement in HSCs. Hence, the importance of Runx at the perinatal stage may underlie the aforementioned young onset of Runx leukemias.

RUNX1 point mutations are more frequently found in elderly acute myeloid leukemia (AML) patients. Recurrent mutations in genes such as *DNMT3a*, *TET2* and *ASXL1* are detected singly in aged individuals of the general population who do not exhibit hematological disorders and result in clonal expansion of HSCs harbouring these mutations (Jaiswal et al. 2014; Genovese et al. 2014). Based on the genetic landscape of mutations found in elderly healthy individuals and AML patients, it is thought that Runx1 point mutations occur secondary to such preceding mutations found in pre-leukemic HSCs to cause progression to leukemia (Xie et al. 2014). Runx deficiency leading to stem cell exhaustion is counterintuitive to leukemogenesis. Yet, *RUNX1* mutations predispose cells to leukemogenesis by promoting a myeloid-biased cell status. This conundrum can be explained by the ability of the driver mutations to overcome the exhaustion conferred by *RUNX1* mutations. For this reason, the advantageous state of pre-leukemic HSCs in aged individuals could explain why *RUNX1* point mutations commonly occur in leukemias of older patients.

Notably, overexpression of Hmga2, repeatedly discussed in this chapter (Fig. 9.3), has been reported in leukemias (Fusco and Fedele 2007; Tan et al. 2016). As such, the pronounced increased Hmga2 levels in the absence of Runx family genes may be relevant to leukemogenesis in Runx-deficient status, perhaps by increasing the stem cell pool in which cooperative mutations may occur.

9.3 Hair Follicle Stem Cells in the Skin

The skin epidermis is important for protection against external environmental insults and prevention against dehydration. The skin epidermis is constantly renewing and consists of different compartments, including the hair follicle, its surrounding interfollicular epidermis and sebaceous glands. At least three stem cell populations have been documented to maintain the epidermis, one of which is the HFSC located in the outermost layer of the follicle known as the outer root sheath. While the HFSCs contributes permanently only to the hair follicle, the HFSCs can temporarily regenerate all types of cells in the epidermal layer upon injury.

The hair follicle is an accessory structure of the epidermis embedded deep in the dermis of the mammalian skin. Of the four different types of hair present in the mouse pelage, the guard hair is the well studied and the timelines provided in this chapter are representative of this particular hair type. The formation of hair follicles require constant molecular interactions between the mesenchymal and epithelial cells. The morphogenesis, or development, of the hair follicle occurs beneath the skin surface and begins at approximately E14.5 in the mouse (Fig. 9.4a). The hair follicles develop through a series of intermediate structures that starts with budding from the epidermis and is fully mature when it extends into the dermis with an inner root sheath and hair shaft. The emergence and maturation of HFSCs takes place concurrent with the development of the hair follicle (reviewed in Sennett and Rendl 2012; Forni et al. 2012). Induction of early HFSC precursors begins at about E12.5, and HFSCs with slow cycling properties emerge in hair placodes, an intermediate hair follicle structure, several days later (Nowak et al. 2008). By the end of the hair follicle morphogenesis, the adult HFSCs reside in a niche known as the bulge region, where they are kept in a quiescent state. The bulge is a hair follicle structure that is part of the upper outer root sheath and located below the sebaceous gland. Adult HFSCs in the bulge can be identified by CD34 expression concurrent with either keratin 14 (K14)-expressing cells (Malanchi et al. 2008) or integrin α_6 (Trempus et al. 2003).

At about 17 days after birth, development of the hair follicle is complete and morphogenesis ends with the first catagen stage when destruction of the hair bulb occurs. The first adult hair cycle is initiated around postnatal day 21 by activation

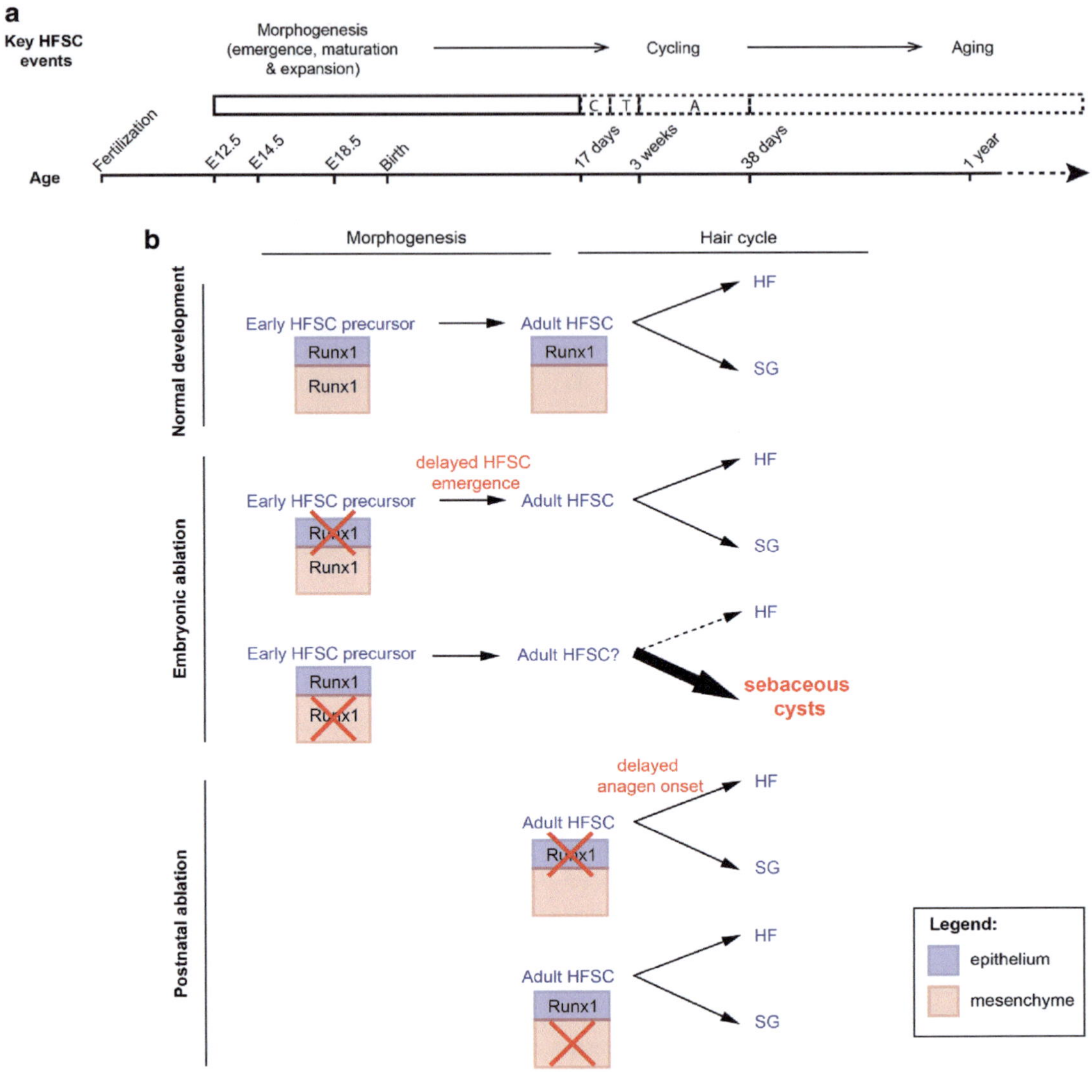

Fig. 9.4 Summary of the morphogenesis and cycling of hair follicle and requirement of Runx1 in hair follicle stem cells. (**a**) Timeline schematic diagram showing the key events of HFSCs/hair follicles during ontogeny. (**b**) Phenotypes exhibited by the *Runx1* conditional knockout mouse models used to decipher the Runx1 role in HFSC are summarized. *Colored boxes* indicate the tissue type in which Runx1 is ablated (*blue*, epithelium; *pink*, mesenchyme). Note that the absence of "Runx1" in mesenchymal tissue represents undetectable expression of Runx1 in this tissue at the adult stage. Abbreviations: *HFSC* hair follicle stem cell, *HF* hair follicle, *SG* sebaceous gland

and proliferation of HFSCs in the bulge. Structurally, each hair follicle is composed of an upper non-cycling region and a cycling region (Zhang et al. 2009). The structure of the non-cycling region includes the stem cell-containing bulge (Morris et al. 2004), oil-producing sebaceous gland and the infundibulum through which the hair shaft passes. To continuously produce new hair shafts during homeostasis, the cycling portion of each hair follicle, the bulb, repeatedly undergoes a 3-week long hair cycle that comprises anagen (growth phase), catagen (degeneration phase) and telogen (quiescent phase) (Alonso and Fuchs 2006). During anagen, HFSCs undergo self-renewal and then some of these HFSCs leave the bulge only at the next telogen-anagen transition. Once outside the niche, these cells then undergo further differentiation into transit amplifying progenitor cells, known as the matrix, throughout anagen (Zhang et al. 2009). The matrix encloses a group of mesenchymal cells, called the dermal papillae, which integrates signals from the bulge and elicits instructions to the epidermis to activate HFSC proliferation and induce differentiation for hair follicle regeneration.

The skin undergoes significant changes in terms of structure and function with age. Aged HFSCs maintain their numbers and gene signatures. However, the period of quiescence becomes progressively longer with age, suggesting that quiescent HFSCs become increasingly resistant to activation (Keyes et al. 2013).

9.3.1 Runx in Hair Follicle Stem Cells at the Developmental Stage

Runx1 is expressed in both epithelial and mesenchymal regions in the area around the hair follicle during morphogenesis (Raveh et al. 2006; Osorio et al. 2011). At placode stage, Runx1 is expressed in cells of the single layer of ectoderm fated to become adult HFSCs, and the Runx1-expressing cells ultimately contribute to all layers of hair follicle during morphogenesis (Osorio et al. 2011). Such early specification of stem cells is important for hair follicle morphogenesis (Nowak et al. 2008). Close examination of epithelial-specific *Runx1* knockout mice using *Runx1*$^{f/f}$*;K14-Cre*$^{+}$ mice revealed that there was a delay in the formation of mature hair follicles due to delayed HFSC emergence (Fig. 9.4b), which could be explained by reduced Lef1 levels and reduced Wnt signaling in the epidermis and the adjacent dermis cells (Osorio et al. 2011). However, this defect was overcome with time, suggesting that Runx1 is crucial for the timely emergence of HFSCs during embryogenesis, but is dispensable for this process (Osorio et al. 2011).

Runx1-expressing cells are also detected in the dermal layer just beneath the epidermis at E14.5. This dermal population increases by E17.5 (Raveh et al. 2006; Osorio et al. 2011). Deletion of *Runx1* by tamoxifen injections at E12.5, E13.5 and E14.5 in *Runx1*$^{CreER/f}$ mice resulted in very efficient deletion in the dermis, but not epidermis, generating a knockout mouse model with *Runx1* specifically ablated in mesenchymal tissue. In this mesenchymal knockout mouse model, the deletion of *Runx1* led to formation of hair follicles throughout development, but they are converted into enormous sebaceous cysts that lacked the bulb and bulge regions at the first hair cycle (Osorio et al. 2011) (Fig. 9.4b). This phenotype is due to defective maturation of the early HFSC precursors, leading to preferential differentiation towards sebaceous glands over hair bulb lineages. Hence, it is evident that mesenchymal loss of *Runx1* during embryogenesis affects hair follicle integrity much more than the epithelial loss of Runx1 does. Contrary to *Runx1* deletion in the epithelium, *Runx1* deficiency in the dermis resulted in upregulated Lef1 and Wnt signaling in both mesenchyme layer and epithelium, suggesting there are molecular interactions between the two compartments (Osorio et al. 2011).

Unlike Runx1 expression, the expression of Runx2 and Runx3 is much less pronounced in the epithelial layer of hair follicles. Runx2 is expressed in dermal papillae and bulb epithelium during hair follicle development, but is not expressed in other regions of the skin outside the hair follicles (Glotzer et al. 2008). Allografts of *Runx2* null skin at E18.5, compared to those of wild-type controls, showed that Runx2 is dispensable for normal hair follicle morphogenesis and cycling (Glotzer et al. 2008). However, there was a delay in hair follicle development as E18.5 *Runx2* null embryos had fewer follicles than the control embryos. This defect was shown to be due to decreased Sonic hedgehog expression.

Runx3 is expressed in the dermal layer just beneath the epidermal layer in placode and hair germ stages (Raveh et al. 2005). This is the region where the Runx3-expressing cells will form the dermal papillae. By birth, Runx3 is expressed in a cluster of cells that consisted of most, if not all, of the dermal papillae. The significance of this expression is worth further investigating as the involvement of Runx3 in HFSCs may act via its effects on the dermal papillae. *Runx3*$^{-/-}$ mice bred on heterogeneous ICR, MF1 background are able to survive until adulthood and these mice did not show major changes in the overall morphology of the skin and its appendages. This may be due to compensation by Runx2 as Runx2 is also expressed in the dermal papillae (Glotzer et al. 2008). A population of stem cells termed skin-derived precursors (SKPs) has been discovered in the dermal papillae (Toma et al. 2001). Whether Runx2 and Runx3 plays a role in specification of these SKPs remains to be studied.

9.3.2 Runx in Hair Follicle Stem Cells of Young Adults

Runx1 has a dynamic expression in the murine hair follicle throughout the hair cycle and when expressed, it is found only in specific hair follicle compartments. Runx1 marks adult HFSCs in the lower bulge during telogen (Scheitz et al. 2012) and is detected throughout the bulge during anagen when HFSCs self-renew. Dermal expression of Runx1 is undetected at the onset of the first hair cycle (Raveh et al. 2006). Consistent with the absence of Runx1 expression in the adult mesenchyme, dermal excision of *Runx1* by the induction of *Runx1*$^{CreER/f}$ mice with tamoxifen at PD21 resulted in normal skin (Fig. 9.4b). The overall expression patterns of Runx1 in human and mouse skin are largely similar (Hoi et al. 2010), suggesting that the functions of Runx in both organisms are also conserved.

Interestingly, Runx1 is dispensable during hair follicle development but is crucial for normal regulation of the hair cycle at the transition into adult skin homeostasis. Gene targeting of *Runx1* in the epithelial compartment in mice using K14-Cre system led to a delay in telogen-to-anagen transition of the first hair cycle (Osorio et al. 2008) (Fig. 9.4b). Immunofluorescence staining of epithelial *Runx1*-deficient hair follicles revealed the absence of specific hair lineage markers and progenitor matrix cell marker (Ephrin B1) characteristic of anagen phase at a time when wild-type hair follicles are in anagen. Using *Runx1*$^{f/f}$;β-actin-CreER mice injected with tamoxifen at various stages in the hair cycle, the requirement for Runx1 was found to be at the onset of anagen (Hoi et al. 2010). Clonogenic assays testing for skin stem cells resulted in cultured keratinocytes from *Runx1*$^{\Delta4/\Delta4}$;K14-Cre mice having fewer and smaller colonies that are unable to survive long term (Osorio et al. 2008). BrdU labeling experiment demonstrated that the loss of Runx1 at anagen onset impaired the proliferation of CD34^{+} bulge cells *in vivo* (Hoi et al. 2010). Taken together, loss of Runx1 impairs proliferation of HFSCs and leads to delayed HFSC activation into cell cycle. Notably, the telogen block is spontaneously overcome with age (Hoi et al. 2010) or injury (Osorio et al. 2008), suggesting that Runx1 is not completely indispensable for HFSCs to exit quiescence and that Runx1 loss did not affect the differentiation potential and fate decision of HFSCs. Runx1 may work, at least partly, by downregulating *Cdkn1a* (p21) (Hoi et al. 2010) and keeping HFSCs poised for receiving proliferation signals (Lee et al. 2014). However, whereas deletion of *Cdkn1a* rescued the proliferation defects in *Runx1* knockout keratinocytes *in vitro*, loss of *Cdkn1a* had the opposite effect in hair follicle cycling *in vivo*. *Runx1;p21* double knockout mice showed an even more prolonged telogen phase, thought to be due to the upregulation of compensatory CDK inhibitor expression, such as p15 (Lee et al. 2013).

As in the case of HSCs, Hmga2 was found to be an important factor in the maintenance of long-term self-renewal capability of HFSCs (Chen et al. 2012). An *in vitro* RNA interference screen coupled with serial passages of HFSCs led to the identification of Hmga2, and this hit was validated by shRNA transduction experiment. Intriguingly, Runx1 was also identified as the top hit—absence of Runx1 reduces the self-renewal ability of HFSCs. As such, the Runx1-Hmga2 axis that is apparent in HSCs may operate in HFSCs as well. In this case, however, Runx1 may act as a positive regulator of Hmga2 expression. The exact mechanism remains to be further studied.

Runx3 expression in the dermal papillae persists throughout all stages of the hair cycle (Raveh et al. 2005). Durations of hair cycle stages were found to be unaffected in *Runx3*$^{-/-}$ mice bred on heterogeneous ICR, MF1 background. As mesenchymal-epithelial interactions are important in the signals from dermal papillae to epidermis, it is possible that signals from Runx3-expressing cells in the dermal papillae may regulate HFSC function or numbers. Further investigation is warranted.

The expression of Runx2 during cycling is largely similar to that during development: Runx2 is expressed in dermal papillae and exhibits increasing asymmetric expression in bulb epithelium through anagen (Glotzer et al. 2008). Although Runx2 expression was detected at very low levels in freshly isolated CD34^{+}integrin α_6^{+} bulge cells (Hoi et al. 2010), its expression has not been detected in the bulge *in vivo* (Glotzer et al. 2008). Since there is some degree of overlap of expression of all three Runx genes, functional redundancy may operate in certain locations, such as the dermal papillae where both Runx2 and Runx3 are expressed. Hence, further in-depth analyses, such as analyses of *Runx2;Runx3* double knockout mice are required to clarify the role of Runx in HFSCs.

In general, the expression of Runx1 seems to be non-overlapping with those of Runx2 and Runx3. Runx1 seems to play a role in both epithelial and mesenchymal compartments, thus regulating HFSC in both cell intrinsic and extrinsic manner, whereas the roles of Runx2 and Runx3 in potentially regulating HFSCs seem confined to cell extrinsic mechanisms.

9.3.3 Runx in Aging Hair Follicle Stem Cells

Even after repeated stimulation of skin wounding over a period of 1 year, *Runx1*$^{\Delta4/\Delta4}$;K14-Cre hair follicles were able to continuously regenerate and result in new hair growth, suggesting that *Runx1*-deficient HFSCs maintained their long-term potential and did not result in the exhaustion of the stem cell pool (Osorio et al. 2008).

Although Runx proteins are not significantly differentially expressed with age in HFSCs, the transcription factor Nfatc1, implicated in hair follicle aging, was found to bind to the intronic regions of Runx1 (Keyes et al. 2013). In another study, the expression of type XVII collagen (COL17A1) decreased and/or its degradation or shedding by ELANE/elastase-2 (ELA2) increased with age in the HFSCs (Matsumura et al. 2016). As Runx1 has been demonstrated to transcriptionally activate ELA2 (Li et al. 2004; Lausen et al. 2006), Runx1 may function in aged HFSCs to promote aging. Since Runx1 poises HFSCs to respond rapidly to proliferation signals in young adult HFSCs, it will be interesting to find out if Runx is implicated in the prolonged telogen phase of aged HFSCs.

9.3.4 Runx and Skin Cancer

As HFSCs are a well appreciated source of skin appendage tumors such as basal cell carcinomas, Runx may be directly implicated in skin cancers (Lorz et al. 2009). Moreover, p21 encoded by *Cdkn1a* acts as a tumor suppressor in HFSCs (Topley et al. 1999) and Runx1 was found to negatively regulate *Cdkn1a* expression. As deletion of *Cdkn1a* only partially rescued the tumour impairment phenotype exhibited by *Runx1* KO mice, other Runx1 targets are required for full tumour growth (Lee et al. 2013). Runx1 is also shown to be a downstream factor of p63 (Ortt et al. 2008), which is expressed at a high level in skin squamous cell carcinomas (Wrone et al. 2004). The p63 protein is a member of the p53 family and an essential factor for proper development of stratified epithelium (Mills et al. 1999; Yang et al. 1999). Unlike *p53 (TP53)* which acts as a tumor suppressor and is frequently mutated in cancers, *p63 (TP63)* is typically associated with gene amplification in cancers (Romano and Sinha 2014). It was shown that p63 binds to a Runx1 intronic enhancer and positively regulates its expression (Ortt et al. 2008). For the above reasons, it is plausible that Runx1 can act as an oncogene in skin tumorigenesis.

Indeed, Runx1 is highly expressed in mouse skin papilloma and squamous cell carcinomas (Hoi et al. 2010). The conditional loss of *Runx1* in mouse epidermis impairs mouse skin tumorigenesis as mice in which *Runx1* is deleted using K14-Cre showed drastically delayed and reduced skin papilloma and squamous cell carcinoma formation in response to skin DMBA/TPA carcinogenic treatment (Hoi et al. 2010). The *Runx1* knockout bulge cells proliferated less than control cells even in response to stimulation by TPA, a strong proliferative-promoting agent (Hoi et al. 2010). Using timed deletion of *Runx1* at different stages of tumorigenesis (initiation, promotion and maintenance), *Scheitz* et al. nicely demonstrated that Runx1 is specifically required at the initiation stage and its expression in HFSCs is responsible for initiating squamous cell carcinomas in mice (Scheitz et al. 2012). In the absence of proliferative agents, however, Runx1 is also crucial for the maintenance of CD34-expressing papilloma cancer stem cells (Scheitz et al. 2012). Absence of Runx1 under such circumstances causes tumour regression, leading to shrinkage of tumor size. In addition, knockdown of *RUNX1* in the human skin cancer cell line, SCC13, renders the cells unable to grow (Scheitz et al. 2012). Further investigation revealed that RUNX1 acts to promote skin tumour growth by upregulating Stat3 activity via transcriptional repression of SOCS4 and potentially SOCS3 (Scheitz et al. 2012).

Overexpression of non-mutated RUNX3 has been reported in human basal cell carcinomas (Salto-Tellez et al. 2006). Whether the expression of RUNX3 is specifically required in the stem cell compartment for the development of this skin cancer subtype remains to be further investigated.

9.4 Future Perspectives

At least in HSCs and HFSCs, Runx seems to play converging roles, yet contrasting mechanisms operate. First, Runx1-expressing cells during early development mark stem cell precursors. Lack of Runx1 in the hematopoietic system led to complete failure in generating the blood system while lack of Runx1 in the skin led to impairment in generation of the mature hair follicle. Notably, the exact mechanism by which Runx1 affects blood and hair stem cells differs: *Runx1* deficiency impairs stem cell emergence in the hematopoietic system in a cell intrinsic manner, whereas *Runx1* ablation affects hair follicle development only via a cell extrinsic mechanism, indicating spatially different roles of Runx. Secondly, Runx1 works at the stem cell level to initiate adult stage of hematopoiesis and hair growth. Again, the mechanism differs: in young adults, Runx1 serves to promote proliferation in HFSCs, while it maintains quiescence in HSCs.

Such contrasting mechanisms in two different tissues, yet controlled by the same family of proteins, confound a proper understanding of fundamental core Runx functions. For example, while Hmga2 expression is upregulated in Runx-deficient HSCs, Runx seems to act in an opposite manner in HFSCs. In a similar scenario, p21 expression is positively regulated by Runx in HSCs but negatively in HFSCs. Possibly, intricate controls of the same pathway may be affected by other cell-type specific molecules, resulting in opposing effects.

In addition to HSCs and HFSCs, Runx function has also been detected in skeletal and mammary stem cells (Table 9.1). Our preliminary data suggest that Runx1 could be involved in stem cells of a wide spectrum of tissues, thus implying that Runx could potentially act as a global "master stem cell regulator". Hence, it will be imperative to examine how Runx functions globally in the various tissue stem cells. By extending our knowledge of Runx roles in other tissue stem cells, we may be able to stratify their roles based on common machinery in regulating the stem cell compartment.

Acknowledgements We thank members of MD2 Vivarium, NUS, for mouse husbandry. We thank Tumbar T., Nishimura E. and Speck NA for their careful and critical reading of our manuscript. This work was supported by National Medical Research Council, Biomedical Research Council, A*STAR (Agency of Science, Technology and Research), the National Research Foundation Singapore, the Singapore Ministry of Education under its Research Centres of Excellence initiative, and JSPS Kakenhi Grant Numbers JP15H04312 and JP16K14613, Japan.

References

Alonso, L., & Fuchs, E. (2006). The hair cycle. *Journal of Cell Science, 119*(Pt 3), 391–393. doi:10.1242/jcs02793.

Arora, N., Wenzel, P. L., McKinney-Freeman, S. L., Ross, S. J., Kim, P. G., Chou, S. S., et al. (2014). Effect of developmental stage of HSC and recipient on transplant outcomes. *Developmental Cell, 29*(5), 621–628. doi:10.1016/j.devcel.2014.04.013.

Bernardin-Fried, F., Kummalue, T., Leijen, S., Collector, M. I., Ravid, K., & Friedman, A. D. (2004). AML1/RUNX1 increases during G1 to S cell cycle progression independent of cytokine-dependent phosphorylation and induces cyclin D3 gene expression. *The Journal of Biological Chemistry, 279*(15), 15678–15687. doi:10.1074/jbc.M310023200.

Birbrair, A., & Frenette, P. S. (2016). Niche heterogeneity in the bone marrow. *Annals of the New York Academy of Sciences, 1370*(1), 82–96. doi:10.1111/nyas.13016.

Bowie, M. B., McKnight, K. D., Kent, D. G., McCaffrey, L., Hoodless, P. A., & Eaves, C. J. (2006). Hematopoietic stem cells proliferate until after birth and show a reversible phase-specific engraftment defect. *The Journal of Clinical Investigation, 116*(10), 2808–2816. doi:10.1172/JCI28310.

Brunet, A., Sweeney, L. B., Sturgill, J. F., Chua, K. F., Greer, P. L., Lin, Y., et al. (2004). Stress-dependent regulation of FOXO transcription factors by the SIRT1 deacetylase. *Science, 303*(5666), 2011–2015. doi:10.1126/science.1094637.

Busch, K., Klapproth, K., Barile, M., Flossdorf, M., Holland-Letz, T., Schlenner, S. M., et al. (2015). Fundamental properties of unperturbed haematopoiesis from stem cells in vivo. *Nature, 518*(7540), 542–546. doi:10.1038/nature14242.

Cai, X., Gaudet, J. J., Mangan, J. K., Chen, M. J., De Obaldia, M. E., Oo, Z., et al. (2011). Runx1 loss minimally impacts long-term hematopoietic stem cells. *PloS One, 6*(12), e28430. doi:10.1371/journal.pone.0028430.

Chambers, S. M., Shaw, C. A., Gatza, C., Fisk, C. J., Donehower, L. A., & Goodell, M. A. (2007). Aging hematopoietic stem cells decline in function and exhibit epigenetic dysregulation. *PLoS Biology, 5*(8), e201. doi:10.1371/journal.pbio.0050201.

Chen, M. J., Yokomizo, T., Zeigler, B. M., Dzierzak, E., & Speck, N. A. (2009). Runx1 is required for the endothelial to haematopoietic cell transition but not thereafter. *Nature, 457*(7231), 887–891. doi:10.1038/nature07619.

Chen, T., Heller, E., Beronja, S., Oshimori, N., Stokes, N., & Fuchs, E. (2012). An RNA interference screen uncovers a new molecule in stem cell self-renewal and long-term regeneration. *Nature, 485*(7396), 104–108. doi:10.1038/nature10940.

Chien, Y., Scuoppo, C., Wang, X., Fang, X., Balgley, B., Bolden, J. E., et al. (2011). Control of the senescence-associated secretory phenotype by NF-kappaB promotes senescence and enhances chemosensitivity. *Genes & Development, 25*(20), 2125–2136. doi:10.1101/gad.17276711.

Chin, D. W., Sakurai, M., Nah, G. S., Du, L., Jacob, B., Yokomizo, T., et al. (2016). RUNX1 haploinsufficiency results in granulocyte colony-stimulating factor hypersensitivity. *Blood Cancer Journal, 6*, e379. doi:10.1038/bcj.2015.105.

Chitteti, B. R., Cheng, Y. H., Streicher, D. A., Rodriguez-Rodriguez, S., Carlesso, N., Srour, E. F., & Kacena, M. A. (2010). Osteoblast lineage cells expressing high levels of Runx2 enhance hematopoietic progenitor cell proliferation and function. *Journal of Cellular Biochemistry, 111*(2), 284–294. doi:10.1002/jcb.22694.

Christensen, J. L., Wright, D. E., Wagers, A. J., & Weissman, I. L. (2004). Circulation and chemotaxis of fetal hematopoietic stem cells. *PLoS Biology, 2*(3), E75. doi:10.1371/journal.pbio.0020075.

Copley, M. R., Babovic, S., Benz, C., Knapp, D. J., Beer, P. A., Kent, D. G., et al. (2013). The Lin28b-let-7-Hmga2 axis determines the higher self-renewal potential of fetal haematopoietic stem cells. *Nature Cell Biology, 15*(8), 916–925. doi:10.1038/ncb2783.

Croker, B. A., Metcalf, D., Robb, L., Wei, W., Mifsud, S., DiRago, L., et al. (2004). SOCS3 is a critical physiological negative regulator of G-CSF signaling and emergency granulopoiesis. *Immunity, 20*(2), 153–165.

Cumano, A., & Godin, I. (2007). Ontogeny of the hematopoietic system. *Annual Review of Immunology, 25*, 745–785. doi:10.1146/annurev.immunol.25.022106.141538.

de Boer, J., Williams, A., Skavdis, G., Harker, N., Coles, M., Tolaini, M., et al. (2003). Transgenic mice with hematopoietic and lymphoid specific expression of Cre. *European Journal of Immunology, 33*(2), 314–325. doi:10.1002/immu.200310005.

Deguchi, K., Yagi, H., Inada, M., Yoshizaki, K., Kishimoto, T., & Komori, T. (1999). Excessive extramedullary hematopoiesis in Cbfa1-deficient mice with a congenital lack of bone marrow. *Biochemical and Biophysical Research Communications, 255*(2), 352–359. doi:10.1006/bbrc.1999.0163.

Dykstra, B., Olthof, S., Schreuder, J., Ritsema, M., & de Haan, G. (2011). Clonal analysis reveals multiple functional defects of aged murine hematopoietic stem cells. *The Journal of Experimental Medicine, 208*(13), 2691–2703. doi:10.1084/jem.20111490.

Dzierzak, E., & Speck, N. A. (2008). Of lineage and legacy: The development of mammalian hematopoietic stem cells. *Nature Immunology, 9*(2), 129–136. doi:10.1038/ni1560.

Ferrari, N., Riggio, A. I., Mason, S., McDonald, L., King, A., Higgins, T., et al. (2015). Runx2 contributes to the regenerative potential of the mammary epithelium. *Scientific Reports, 5*, 15658. doi:10.1038/srep15658.

Flachsbart, F., Caliebe, A., Kleindorp, R., Blanche, H., von Eller-Eberstein, H., Nikolaus, S., et al. (2009). Association of FOXO3A variation with human longevity confirmed in German centenarians. *Proceedings of the National Academy of Sciences of the United States of America, 106*(8), 2700–2705. doi:10.1073/pnas.0809594106.

Forni, M. F., Trombetta-Lima, M., & Sogayar, M. C. (2012). Stem cells in embryonic skin development. *Biological Research, 45*(3), 215–222. doi:10.4067/S0716-97602012000300003.

Fusco, A., & Fedele, M. (2007). Roles of HMGA proteins in cancer. *Nature Reviews. Cancer, 7*(12), 899–910. doi:10.1038/nrc2271.

Galindo, M., Pratap, J., Young, D. W., Hovhannisyan, H., Im, H. J., Choi, J. Y., et al. (2005). The bone-specific expression of Runx2 oscillates during the cell cycle to support a G1-related antiproliferative function in osteoblasts. *The Journal of Biological Chemistry, 280*(21), 20274–20285. doi:10.1074/jbc.M413665200.

Geiger, H., de Haan, G., & Florian, M. C. (2013). The ageing haematopoietic stem cell compartment. *Nature Reviews. Immunology, 13*(5), 376–389. doi:10.1038/nri3433.

Gekas, C., Dieterlen-Lievre, F., Orkin, S. H., & Mikkola, H. K. (2005). The placenta is a niche for hematopoietic stem cells. *Developmental Cell, 8*(3), 365–375. doi:10.1016/j.devcel.2004.12.016.

Genovese, G., Kahler, A. K., Handsaker, R. E., Lindberg, J., Rose, S. A., Bakhoum, S. F., et al. (2014). Clonal hematopoiesis and blood-cancer risk inferred from blood DNA sequence. *The New England Journal of Medicine, 371*(26), 2477–2487. doi:10.1056/NEJMoa1409405.

Georgiades, P., Ogilvy, S., Duval, H., Licence, D. R., Charnock-Jones, D. S., Smith, S. K., & Print, C. G. (2002). VavCre transgenic mice: A tool for mutagenesis in hematopoietic and endothelial lineages. *Genesis, 34*(4), 251–256. doi:10.1002/gene.10161.

Glotzer, D. J., Zelzer, E., & Olsen, B. R. (2008). Impaired skin and hair follicle development in Runx2 deficient mice. *Developmental Biology, 315*(2), 459–473. doi:10.1016/j.ydbio.2008.01.005.

Growney, J. D., Shigematsu, H., Li, Z., Lee, B. H., Adelsperger, J., Rowan, R., et al. (2005). Loss of Runx1 perturbs adult hematopoiesis and is associated with a myeloproliferative phenotype. *Blood, 106*(2), 494–504. doi:10.1182/blood-2004-08-3280.

Haigis, M. C., & Guarente, L. P. (2006). Mammalian sirtuins – emerging roles in physiology, aging, and calorie restriction. *Genes & Development, 20*(21), 2913–2921. doi:10.1101/gad.1467506.

Hock, H., Hamblen, M. J., Rooke, H. M., Schindler, J. W., Saleque, S., Fujiwara, Y., & Orkin, S. H. (2004). Gfi-1 restricts proliferation and preserves functional integrity of haematopoietic stem cells. *Nature, 431*(7011), 1002–1007. doi:10.1038/nature02994.

Hoi, C. S., Lee, S. E., Lu, S. Y., McDermitt, D. J., Osorio, K. M., Piskun, C. M., et al. (2010). Runx1 directly promotes proliferation of hair follicle stem cells and epithelial tumor formation in mouse skin. *Molecular and Cellular Biology, 30*(10), 2518–2536. doi:10.1128/MCB.01308-09.

Ichikawa, M., Asai, T., Saito, T., Seo, S., Yamazaki, I., Yamagata, T., et al. (2004). AML-1 is required for megakaryocytic maturation and lymphocytic differentiation, but not for maintenance of hematopoietic stem cells in adult hematopoiesis. *Nature Medicine, 10*(3), 299–304. doi:10.1038/nm997.

Iwasaki, H., Arai, F., Kubota, Y., Dahl, M., & Suda, T. (2010). Endothelial protein C receptor-expressing hematopoietic stem cells reside in the perisinusoidal niche in fetal liver. *Blood, 116*(4), 544–553. doi:10.1182/blood-2009-08-240903.

Jacob, B., Osato, M., Yamashita, N., Wang, C. Q., Taniuchi, I., Littman, D. R., et al. (2010). Stem cell exhaustion due to Runx1 deficiency is prevented by Evi5 activation in leukemogenesis. *Blood, 115*(8), 1610–1620. doi:10.1182/blood-2009-07-232249.

Jaffredo, T., Nottingham, W., Liddiard, K., Bollerot, K., Pouget, C., & de Bruijn, M. (2005). From hemangioblast to hematopoietic stem cell: an endothelial connection? *Experimental Hematology, 33*(9), 1029–1040. doi:10.1016/j.exphem.2005.06.005.

Jaiswal, S., Fontanillas, P., Flannick, J., Manning, A., Grauman, P. V., Mar, B. G., et al. (2014). Age-related clonal hematopoiesis associated with adverse outcomes. *The New England Journal of Medicine, 371*(26), 2488–2498. doi:10.1056/NEJMoa1408617.

Janzen, V., Forkert, R., Fleming, H. E., Saito, Y., Waring, M. T., Dombkowski, D. M., et al. (2006). Stem-cell ageing modified by the cyclin-dependent kinase inhibitor p16INK4a. *Nature, 443*(7110), 421–426. doi:10.1038/nature05159.

Kanaykina, N., Abelson, K., King, D., Liakhovitskaia, A., Schreiner, S., Wegner, M., & Kozlova, E. N. (2010). In vitro and in vivo effects on neural crest stem cell differentiation by conditional activation of Runx1 short isoform and its effect on neuropathic pain behavior. *Upsala Journal of Medical Sciences, 115*(1), 56–64. doi:10.3109/03009730903572065.

Keyes, B. E., Segal, J. P., Heller, E., Lien, W. H., Chang, C. Y., Guo, X., et al. (2013). Nfatc1 orchestrates aging in hair follicle stem cells. *Proceedings of the National Academy of Sciences of the United States of America, 110*(51), E4950–E4959. doi:10.1073/pnas.1320301110.

Khan, J. A., Mendelson, A., Kunisaki, Y., Birbrair, A., Kou, Y., Arnal-Estape, A., et al. (2016). Fetal liver hematopoietic stem cell niches associate with portal vessels. *Science, 351*(6269), 176–180. doi:10.1126/science.aad0084.

Kieusseian, A., Brunet de la Grange, P., Burlen-Defranoux, O., Godin, I., & Cumano, A. (2012). Immature hematopoietic stem cells undergo maturation in the fetal liver. *Development, 139*(19), 3521–3530. doi:10.1242/dev.079210.

Kim, I., Saunders, T. L., & Morrison, S. J. (2007). Sox17 dependence distinguishes the transcriptional regulation of fetal from adult hematopoietic stem cells. *Cell, 130*(3), 470–483. doi:10.1016/j.cell.2007.06.011.

Kissa, K., & Herbomel, P. (2010). Blood stem cells emerge from aortic endothelium by a novel type of cell transition. *Nature, 464*(7285), 112–115. doi:10.1038/nature08761.

Koh, C. P., Ng, C. E., Nah, G. S., Wang, C. Q., Tergaonkar, V., Matsumura, T., et al. (2015). Hematopoietic stem cell enhancer: a powerful tool in stem cell biology. *Histology Histopathology, 30*(6), 661–672. doi:10.14670/HH-30.661.

Komeno, Y., Yan, M., Matsuura, S., Lam, K., Lo, M. C., Huang, Y. J., et al. (2014). Runx1 exon 6-related alternative splicing isoforms differentially regulate hematopoiesis in mice. *Blood, 123*(24), 3760–3769. doi:10.1182/blood-2013-08-521252.

Komori, T., Yagi, H., Nomura, S., Yamaguchi, A., Sasaki, K., Deguchi, K., et al. (1997). Targeted disruption of Cbfa1 results in a complete lack of bone formation owing to maturational arrest of osteoblasts. *Cell, 89*(5), 755–764.

Lam, K., Muselman, A., Du, R., Harada, Y., Scholl, A. G., Yan, M., et al. (2014). Hmga2 is a direct target gene of RUNX1 and regulates expansion of myeloid progenitors in mice. *Blood, 124*(14), 2203–2212. doi:10.1182/blood-2014-02-554543.

Lausen, J., Liu, S., Fliegauf, M., Lubbert, M., & Werner, M. H. (2006). ELA2 is regulated by hematopoietic transcription factors, but not repressed by AML1-ETO. *Oncogene, 25*(9), 1349–1357. doi:10.1038/sj.onc.1209181.

Lecuyer, E., & Hoang, T. (2004). SCL: From the origin of hematopoiesis to stem cells and leukemia. *Experimental Hematology, 32*(1), 11–24.

Lee, J., Hoi, C. S., Lilja, K. C., White, B. S., Lee, S. E., Shalloway, D., & Tumbar, T. (2013). Runx1 and p21 synergistically limit the extent of hair follicle stem cell quiescence in vivo. *Proceedings of the National Academy of Sciences of the United States of America, 110*(12), 4634–4639. doi:10.1073/pnas.1213015110.

Lee, S. E., Sada, A., Zhang, M., McDermitt, D. J., Lu, S. Y., Kemphues, K. J., & Tumbar, T. (2014). High Runx1 levels promote a reversible, more-differentiated cell state in hair-follicle stem cells during quiescence. *Cell Reports, 6*(3), 499–513. doi:10.1016/j.celrep.2013.12.039.

Levanon, D., Brenner, O., Negreanu, V., Bettoun, D., Woolf, E., Eilam, R., et al. (2001). Spatial and temporal expression pattern of Runx3 (Aml2) and Runx1 (Aml1) indicates non-redundant functions during mouse embryogenesis. *Mechanisms of Development, 109*(2), 413–417.

Li, F. Q., Person, R. E., Takemaru, K., Williams, K., Meade-White, K., Ozsahin, A. H., et al. (2004). Lymphoid enhancer factor-1 links two hereditary leukemia syndromes through core-binding factor alpha regulation of ELA2. *The Journal of Biological Chemistry, 279*(4), 2873–2884. doi:10.1074/jbc.M310759200.

Liakhovitskaia, A., Gribi, R., Stamateris, E., Villain, G., Jaffredo, T., Wilkie, R., et al. (2009). Restoration of Runx1 expression in the Tie2 cell compartment rescues definitive hematopoietic stem cells and extends life of Runx1 knockout animals until birth. *Stem Cells, 27*(7), 1616–1624. doi:10.1002/stem.71.

Liakhovitskaia, A., Lana-Elola, E., Stamateris, E., Rice, D. P., van't Hof, R. J., & Medvinsky, A. (2010). The essential requirement for Runx1 in the development of the sternum. *Developmental Biology, 340*(2), 539–546. doi:10.1016/j.ydbio.2010.02.005.

Liakhovitskaia, A., Rybtsov, S., Smith, T., Batsivari, A., Rybtsova, N., Rode, C., et al. (2014). Runx1 is required for progression of CD41+ embryonic precursors into HSCs but not prior to this. *Development, 141*(17), 3319–3323. doi:10.1242/dev.110841.

Linggi, B., Muller-Tidow, C., van de Locht, L., Hu, M., Nip, J., Serve, H., et al. (2002). The t(8;21) fusion protein, AML1 ETO, specifically represses the transcription of the p14(ARF) tumor suppressor in acute myeloid leukemia. *Nature Medicine, 8*(7), 743–750. doi:10.1038/nm726.

Loonstra, A., Vooijs, M., Beverloo, H. B., Allak, B. A., van Drunen, E., Kanaar, R., et al. (2001). Growth inhibition and DNA damage induced by Cre recombinase in mammalian cells. *Proceedings of the National Academy of Sciences of the United States of America, 98*(16), 9209–9214. doi:10.1073/pnas.161269798.

Lorz, C., Segrelles, C., & Paramio, J. M. (2009). On the origin of epidermal cancers. *Current Molecular Medicine, 9*(3), 353–364.

Malanchi, I., Peinado, H., Kassen, D., Hussenet, T., Metzger, D., Chambon, P., et al. (2008). Cutaneous cancer stem cell maintenance is dependent on beta-catenin signalling. *Nature, 452*(7187), 650–653. doi:10.1038/nature06835.

Martinez-Agosto, J. A., Mikkola, H. K., Hartenstein, V., & Banerjee, U. (2007). The hematopoietic stem cell and its niche: A comparative view. *Genes & Development, 21*(23), 3044–3060. doi:10.1101/gad.1602607.

Matsumura, H., Mohri, Y., Binh, N. T., Morinaga, H., Fukuda, M., Ito, M., et al. (2016). Hair follicle aging is driven by transepidermal elimination of stem cells via COL17A1 proteolysis. *Science, 351*(6273), aad4395. doi:10.1126/science.aad4395.

Matsuo, J., Kimura, S., Yamamura, A., Koh, C., Hossain, M., Heng, D., Kohu, K., Voon, D., Hiai, H., Unno, M., So, J., Zhu, F., Srivastava, S., Meng, T., Yeoh, K., Osato, M., & Ito, Y. (2016) Identification of stem cells in the epithelium of the stomach corpus and antrum of mice. *Gastroenterology, 152*(1), 218–231. doi:10.1053/j.gastro.2016.09.018.

Mills, A. A., Zheng, B., Wang, X. J., Vogel, H., Roop, D. R., & Bradley, A. (1999). p63 is a p53 homologue required for limb and epidermal morphogenesis. *Nature, 398*(6729), 708–713. doi:10.1038/19531.

Morris, R. J., Liu, Y., Marles, L., Yang, Z., Trempus, C., Li, S., et al. (2004). Capturing and profiling adult hair follicle stem cells. *Nature Biotechnology, 22*(4), 411–417. doi:10.1038/nbt950.

Morrison, S. J., Wandycz, A. M., Akashi, K., Globerson, A., & Weissman, I. L. (1996). The aging of hematopoietic stem cells. *Nature Medicine, 2*(9), 1011–1016.

Mrozek, K., Marcucci, G., Nicolet, D., Maharry, K. S., Becker, H., Whitman, S. P., et al. (2012). Prognostic significance of the European LeukemiaNet standardized system for reporting cytogenetic and molecular alterations in adults with acute myeloid leukemia. *Journal of Clinical Oncology, 30*(36), 4515–4523. doi:10.1200/JCO.2012.43.4738.

Nakagawa, M., Shimabe, M., Watanabe-Okochi, N., Arai, S., Yoshimi, A., Shinohara, A., et al. (2011). AML1/RUNX1 functions as a cytoplasmic attenuator of NF-kappaB signaling in the repression of myeloid tumors. *Blood, 118*(25), 6626–6637. doi:10.1182/blood-2010-12-326710.

Ng, C. E., Yokomizo, T., Yamashita, N., Cirovic, B., Jin, H., Wen, Z., et al. (2010). A Runx1 intronic enhancer marks hemogenic endothelial cells and hematopoietic stem cells. *Stem Cells, 28*(10), 1869–1881. doi:10.1002/stem.507.

Niki, M., Okada, H., Takano, H., Kuno, J., Tani, K., Hibino, H., et al. (1997). Hematopoiesis in the fetal liver is impaired by targeted mutagenesis of a gene encoding a non-DNA binding subunit of the transcription factor, polyomavirus enhancer binding protein 2/core binding factor. *Proceedings of the National Academy of Sciences of the United States of America, 94*(11), 5697–5702.

Nishino, J., Kim, I., Chada, K., & Morrison, S. J. (2008). Hmga2 promotes neural stem cell self-renewal in young but not old mice by reducing p16Ink4a and p19Arf Expression. *Cell, 135*(2), 227–239. doi:10.1016/j.cell.2008.09.017.

North, T., Gu, T. L., Stacy, T., Wang, Q., Howard, L., Binder, M., et al. (1999). Cbfa2 is required for the formation of intra-aortic hematopoietic clusters. *Development, 126*(11), 2563–2575.

Nowak, J. A., Polak, L., Pasolli, H. A., & Fuchs, E. (2008). Hair follicle stem cells are specified and function in early skin morphogenesis. *Cell Stem Cell, 3*(1), 33–43. doi:10.1016/j.stem.2008.05.009.

Okada, H., Watanabe, T., Niki, M., Takano, H., Chiba, N., Yanai, N., et al. (1998). AML1(−/−) embryos do not express certain hematopoiesis-related gene transcripts including those of the PU.1 gene. *Oncogene, 17*(18), 2287–2293. doi:10.1038/sj.onc.1202151.

Okuda, T., van Deursen, J., Hiebert, S. W., Grosveld, G., & Downing, J. R. (1996). AML1, the target of multiple chromosomal translocations in human leukemia, is essential for normal fetal liver hematopoiesis. *Cell, 84*(2), 321–330.

Orkin, S. H., & Zon, L. I. (2008). Hematopoiesis: An evolving paradigm for stem cell biology. *Cell, 132*(4), 631–644. doi:10.1016/j.cell.2008.01.025.

Ortt, K., Raveh, E., Gat, U., & Sinha, S. (2008). A chromatin immunoprecipitation screen in mouse keratinocytes reveals Runx1 as a direct transcriptional target of DeltaNp63. *Journal of Cellular Biochemistry, 104*(4), 1204–1219. doi:10.1002/jcb.21700.

Osato, M. (2014). An unsung runt 6e isoform for HSC expansion. *Blood, 123*(24), 3684–3686. doi:10.1182/blood-2014-05-572891.

Osorio, K. M., Lee, S. E., McDermitt, D. J., Waghmare, S. K., Zhang, Y. V., Woo, H. N., & Tumbar, T. (2008). Runx1 modulates developmental, but not injury-driven, hair follicle stem cell activation. *Development, 135*(6), 1059–1068. doi:10.1242/dev.012799.

Osorio, K. M., Lilja, K. C., & Tumbar, T. (2011). Runx1 modulates adult hair follicle stem cell emergence and maintenance from distinct embryonic skin compartments. *The Journal of Cell Biology, 193*(1), 235–250. doi:10.1083/jcb.201006068.

Ottersbach, K., & Dzierzak, E. (2005). The murine placenta contains hematopoietic stem cells within the vascular labyrinth region. *Developmental Cell, 8*(3), 377–387. doi:10.1016/j.devcel.2005.02.001.

Otto, F., Thornell, A. P., Crompton, T., Denzel, A., Gilmour, K. C., Rosewell, I. R., et al. (1997). Cbfa1, a candidate gene for cleidocranial dysplasia syndrome, is essential for osteoblast differentiation and bone development. *Cell, 89*(5), 765–771.

Park, I. K., Qian, D., Kiel, M., Becker, M. W., Pihalja, M., Weissman, I. L., et al. (2003). Bmi-1 is required for maintenance of adult self-renewing haematopoietic stem cells. *Nature, 423*(6937), 302–305. doi:10.1038/nature01587.

Puri, M. C., & Bernstein, A. (2003). Requirement for the TIE family of receptor tyrosine kinases in adult but not fetal hematopoiesis. *Proceedings of the National Academy of Sciences of the United States of America, 100*(22), 12753–12758. doi:10.1073/pnas.2133552100.

Raveh, E., Cohen, S., Levanon, D., Groner, Y., & Gat, U. (2005). Runx3 is involved in hair shape determination. *Developmental Dynamics, 233*(4), 1478–1487. doi:10.1002/dvdy.20453.

Raveh, E., Cohen, S., Levanon, D., Negreanu, V., Groner, Y., & Gat, U. (2006). Dynamic expression of Runx1 in skin affects hair structure. *Mechanisms of Development, 123*(11), 842–850. doi:10.1016/j.mod.2006.08.002.

Rhodes, K. E., Gekas, C., Wang, Y., Lux, C. T., Francis, C. S., Chan, D. N., et al. (2008). The emergence of hematopoietic stem cells is initiated in the placental vasculature in the absence of circulation. *Cell Stem Cell, 2*(3), 252–263. doi:10.1016/j.stem.2008.01.001.

Romano, R. A., & Sinha, S. (2014). Family matters: Sibling rivalry and bonding between p53 and p63 in cancer. *Experimental Dermatology, 23*(4), 238–239. doi:10.1111/exd.12356.

Rossi, D. J., Bryder, D., Zahn, J. M., Ahlenius, H., Sonu, R., Wagers, A. J., & Weissman, I. L. (2005). Cell intrinsic alterations underlie hematopoietic stem cell aging. *Proceedings of the National Academy of Sciences of the United States of America, 102*(26), 9194–9199. doi:10.1073/pnas.0503280102.

Rybtsov, S., Sobiesiak, M., Taoudi, S., Souilhol, C., Senserrich, J., Liakhovitskaia, A., et al. (2011). Hierarchical organization and early hematopoietic specification of the developing HSC lineage in the AGM region. *The Journal of Experimental Medicine, 208*(6), 1305–1315. doi:10.1084/jem.20102419.

Salto-Tellez, M., Peh, B. K., Ito, K., Tan, S. H., Chong, P. Y., Han, H. C., et al. (2006). RUNX3 protein is overexpressed in human basal cell carcinomas. *Oncogene, 25*(58), 7646–7649. doi:10.1038/sj.onc.1209739.

Sasaki, K., Yagi, H., Bronson, R. T., Tominaga, K., Matsunashi, T., Deguchi, K., et al. (1996). Absence of fetal liver hematopoiesis in mice deficient in transcriptional coactivator core binding factor beta. *Proceedings of the National Academy of Sciences of the United States of America, 93*(22), 12359–12363.

Sato, M., Morii, E., Komori, T., Kawahata, H., Sugimoto, M., Terai, K., et al. (1998). Transcriptional regulation of osteopontin gene in vivo by PEBP2alphaA/CBFA1 and ETS1 in the skeletal tissues. *Oncogene, 17*(12), 1517–1525. doi:10.1038/sj.onc.1202064.

Scheitz, C. J., Lee, T. S., McDermitt, D. J., & Tumbar, T. (2012). Defining a tissue stem cell-driven Runx1/Stat3 signalling axis in epithelial cancer. *The EMBO Journal, 31*(21), 4124–4139. doi:10.1038/emboj.2012.270.

Sennett, R., & Rendl, M. (2012). Mesenchymal-epithelial interactions during hair follicle morphogenesis and cycling. *Seminars in Cell & Developmental Biology, 23*(8), 917–927. doi:10.1016/j.semcdb.2012.08.011.

Shakibaei, M., Shayan, P., Busch, F., Aldinger, C., Buhrmann, C., Lueders, C., & Mobasheri, A. (2012). Resveratrol mediated modulation of Sirt-1/Runx2 promotes osteogenic differentiation of mesenchymal stem cells: potential role of Runx2 deacetylation. *PloS One, 7*(4), e35712. doi:10.1371/journal.pone.0035712.

Shimshek, D. R., Kim, J., Hubner, M. R., Spergel, D. J., Buchholz, F., Casanova, E., et al. (2002). Codon-improved Cre recombinase (iCre) expression in the mouse. *Genesis, 32*(1), 19–26.

Silver, D. P., & Livingston, D. M. (2001). Self-excising retroviral vectors encoding the Cre recombinase overcome Cre-mediated cellular toxicity. *Molecular Cell, 8*(1), 233–243.

Stadtfeld, M., & Graf, T. (2005). Assessing the role of hematopoietic plasticity for endothelial and hepatocyte development by non-invasive lineage tracing. *Development, 132*(1), 203–213. doi:10.1242/dev.01558.

Strom, D. K., Nip, J., Westendorf, J. J., Linggi, B., Lutterbach, B., Downing, J. R., et al. (2000). Expression of the AML-1 oncogene shortens the G(1) phase of the cell cycle. *The Journal of Biological Chemistry, 275*(5), 3438–3445.

Sun, J., Ramos, A., Chapman, B., Johnnidis, J. B., Le, L., Ho, Y. J., et al. (2014). Clonal dynamics of native haematopoiesis. *Nature, 514*(7522), 322–327. doi:10.1038/nature13824.

Tan, L., Wei, X., Zheng, L., Zeng, J., Liu, H., Yang, S., & Tan, H. (2016). Amplified HMGA2 promotes cell growth by regulating Akt pathway in AML. *Journal of Cancer Research and Clinical Oncology, 142*(2), 389–399. doi:10.1007/s00432-015-2036-9.

Taoudi, S., Gonneau, C., Moore, K., Sheridan, J. M., Blackburn, C. C., Taylor, E., & Medvinsky, A. (2008). Extensive hematopoietic stem cell generation in the AGM region via maturation of VE-cadherin+CD45+ pre-definitive HSCs. *Cell Stem Cell, 3*(1), 99–108. doi:10.1016/j.stem.2008.06.004.

Tober, J., Yzaguirre, A. D., Piwarzyk, E., & Speck, N. A. (2013). Distinct temporal requirements for Runx1 in hematopoietic progenitors and stem cells. *Development, 140*(18), 3765–3776. doi:10.1242/dev.094961.

Toma, J. G., Akhavan, M., Fernandes, K. J., Barnabe-Heider, F., Sadikot, A., Kaplan, D. R., & Miller, F. D. (2001). Isolation of multipotent adult stem cells from the dermis of mammalian skin. *Nature Cell Biology, 3*(9), 778–784. doi:10.1038/ncb0901-778.

Topley, G. I., Okuyama, R., Gonzales, J. G., Conti, C., & Dotto, G. P. (1999). p21(WAF1/Cip1) functions as a suppressor of malignant skin tumor formation and a determinant of keratinocyte stem-cell potential. *Proceedings of the National Academy of Sciences of the United States of America, 96*(16), 9089–9094.

Trempus, C. S., Morris, R. J., Bortner, C. D., Cotsarelis, G., Faircloth, R. S., Reece, J. M., & Tennant, R. W. (2003). Enrichment for living murine keratinocytes from the hair follicle bulge with the cell surface marker CD34. *The Journal of Investigative Dermatology, 120*(4), 501–511. doi:10.1046/j.1523-1747.2003.12088.x.

Tseng, P. C., Hou, S. M., Chen, R. J., Peng, H. W., Hsieh, C. F., Kuo, M. L., & Yen, M. L. (2011). Resveratrol promotes osteogenesis of human mesenchymal stem cells by upregulating RUNX2 gene expression via the SIRT1/FOXO3A axis. *Journal of Bone and Mineral Research, 26*(10), 2552–2563. doi:10.1002/jbmr.460.

van Bragt, M. P., Hu, X., Xie, Y., & Li, Z. (2014). RUNX1, a transcription factor mutated in breast cancer, controls the fate of ER-positive mammary luminal cells. *eLife, 3*, e03881. doi:10.7554/eLife.03881.

Wang, Q., Stacy, T., Binder, M., Marin-Padilla, M., Sharpe, A. H., & Speck, N. A. (1996a). Disruption of the Cbfa2 gene causes necrosis and hemorrhaging in the central nervous system and blocks definitive hematopoiesis. *Proceedings of the National Academy of Sciences of the United States of America, 93*(8), 3444–3449.

Wang, Q., Stacy, T., Miller, J. D., Lewis, A. F., Gu, T. L., Huang, X., et al. (1996b). The CBFbeta subunit is essential for CBFalpha2 (AML1) function in vivo. *Cell, 87*(4), 697–708.

Wang, C. Q., Jacob, B., Nah, G. S., & Osato, M. (2010). Runx family genes, niche, and stem cell quiescence. *Blood Cells, Molecules & Diseases, 44*(4), 275–286. doi:10.1016/j.bcmd.2010.01.006.

Wang, C. Q., Motoda, L., Satake, M., Ito, Y., Taniuchi, I., Tergaonkar, V., & Osato, M. (2013). Runx3 deficiency results in myeloproliferative disorder in aged mice. *Blood, 122*(4), 562–566. doi:10.1182/blood-2012-10-460618.

Wang, C. Q., Krishnan, V., Tay, L. S., Chin, D. W., Koh, C. P., Chooi, J. Y., et al. (2014). Disruption of Runx1 and Runx3 leads to bone marrow failure and leukemia predisposition due to transcriptional and DNA repair defects. *Cell Reports, 8*(3), 767–782. doi:10.1016/j.celrep.2014.06.046.

Wang, C. Q., Chin, D. W., Chooi, J. Y., Chng, W. J., Taniuchi, I., Tergaonkar, V., & Osato, M. (2015). Cbfb deficiency results in differentiation blocks and stem/progenitor cell expansion in hematopoiesis. *Leukemia, 29*(3), 753–757. doi:10.1038/leu.2014.316.

Worthley, D. L., Churchill, M., Compton, J. T., Tailor, Y., Rao, M., Si, Y., et al. (2015). Gremlin 1 identifies a skeletal stem cell with bone, cartilage, and reticular stromal potential. *Cell, 160*(1–2), 269–284. doi:10.1016/j.cell.2014.11.042.

Wrone, D. A., Yoo, S., Chipps, L. K., & Moy, R. L. (2004). The expression of p63 in actinic keratoses, seborrheic keratosis, and cutaneous squamous cell carcinomas. *Dermatologic Surgery, 30*(10), 1299–1302. doi:10.1111/j.1524-4725.2004.30403.x.

Xie, M., Lu, C., Wang, J., McLellan, M. D., Johnson, K. J., Wendl, M. C., et al. (2014). Age-related mutations associated with clonal hematopoietic expansion and malignancies. *Nature Medicine, 20*(12), 1472–1478. doi:10.1038/nm.3733.

Yamamura, Y., Lee, W. L., Inoue, K., Ida, H., & Ito, Y. (2006). RUNX3 cooperates with FoxO3a to induce apoptosis in gastric cancer cells. *The Journal of Biological Chemistry, 281*(8), 5267–5276. doi:10.1074/jbc.M512151200.

Yang, A., Schweitzer, R., Sun, D., Kaghad, M., Walker, N., Bronson, R. T., et al. (1999). p63 is essential for regenerative proliferation in limb, craniofacial and epithelial development. *Nature, 398*(6729), 714–718. doi:10.1038/19539.

Ye, M., Zhang, H., Amabile, G., Yang, H., Staber, P. B., Zhang, P., et al. (2013). C/EBPa controls acquisition and maintenance of adult haematopoietic stem cell quiescence. *Nature Cell Biology, 15*(4), 385–394. doi:10.1038/ncb2698.

Yokomizo, T., & Dzierzak, E. (2010). Three-dimensional cartography of hematopoietic clusters in the vasculature of whole mouse embryos. *Development, 137*(21), 3651–3661. doi:10.1242/dev.051094.

Yokomizo, T., Ogawa, M., Osato, M., Kanno, T., Yoshida, H., Fujimoto, T., et al. (2001). Requirement of Runx1/AML1/PEBP2alphaB for the generation of haematopoietic cells from endothelial cells. *Genes to Cells, 6*(1), 13–23.

Yu, H., Lim, H. H., Tjokro, N. O., Sathiyanathan, P., Natarajan, S., Chew, T. W., et al. (2014). Chaperoning HMGA2 protein protects stalled replication forks in stem and cancer cells. *Cell Reports, 6*(4), 684–697. doi:10.1016/j.celrep.2014.01.014.

Zhang, Y. V., Cheong, J., Ciapurin, N., McDermitt, D. J., & Tumbar, T. (2009). Distinct self-renewal and differentiation phases in the niche of infrequently dividing hair follicle stem cells. *Cell Stem Cell, 5*(3), 267–278. doi:10.1016/j.stem.2009.06.004.

Zhou, F., Li, X., Wang, W., Zhu, P., Zhou, J., He, W., et al. (2016). Tracing haematopoietic stem cell formation at single-cell resolution. *Nature, 533*(7604), 487–492. doi:10.1038/nature17997.

Roles of the *RUNX1* Enhancer in Normal Hematopoiesis and Leukemogenesis

10

Wei-Siang Liau, Phuong Cao Thi Ngoc, and Takaomi Sanda

Abstract

Enhancers are regulatory elements in genomic DNA that contain specific sequence motifs that are bound by DNA-binding transcription factors. The activity of enhancers is tightly regulated in an integrated and combinatorial manner, thus yielding complex patterns of transcription in different tissues. Identifying enhancers is crucial to understanding the physiological and pathogenic roles of their target genes. The *RUNX1* intronic enhancer, eR1, acts in *cis* to regulate *RUNX1* gene expression in hematopoietic stem cells (HSCs) and hemogenic endothelial cells. RUNX1 and other hematopoietic transcription factors TAL1/SCL, GATA2, PU.1, LMO2 and LDB1 bind at this region. Interestingly, recent studies have revealed that this region is involved in a large cluster of enhancers termed a super-enhancer. The *RUNX1* super-enhancer is observed in normal HSCs and T-cell acute lymphoblastic leukemia cells. In this review, we describe the discovery of eR1 and its roles in normal development and leukemogenesis, as well as its potential applications in stem cell research.

Keywords

RUNX1 • eR1 • Super-enhancer • Hematopoietic stem cells • T-ALL

W.-S. Liau • P.C. Ngoc
Cancer Science Institute of Singapore, National University of Singapore, Singapore 117599, Singapore

T. Sanda, M.D., Ph.D. (✉)
Cancer Science Institute of Singapore, National University of Singapore, Singapore 117599, Singapore

Department of Medicine, Yong Loo Lin School of Medicine, National University of Singapore, Singapore 117599, Singapore
e-mail: takaomi_sanda@nus.edu.sg

10.1 Genetic Alterations of the *RUNX1* Gene in Human Acute Leukemia

Throughout the 1970s–1990s, a number of oncogenes and tumor suppressors were identified in hematological malignancies arising from breakpoints of chromosomal translocations, in which regulatory elements or coding regions are rearranged, resulting in the abnormal expression of

Y. Groner et al. (eds.), *RUNX Proteins in Development and Cancer*, Advances in Experimental Medicine and Biology 962, DOI 10.1007/978-981-10-3233-2_10

translocation partner genes or the creation of fusion genes (Look 1997; Rowley 2001; Speck and Gilliland 2002). The former mechanism predominantly occurs in lymphoid malignancies in which translocation partner genes are under the control of a potent regulatory element of a gene that is highly expressed in lymphoid cells, such as immunoglobulin (*IG*) or T-cell receptor (*TCR*). In these cases, the translocation partner genes (e.g., *MYC*, *TAL1* and *LMO2*) are structurally intact but are aberrantly overexpressed due to translocation, and they contribute to leukemogenesis. In contrast, chromosomal translocations that generate fusion genes are often present in acute myeloid leukemia (AML) and B-cell acute lymphoblastic leukemia (B-ALL). One of the most frequently involved genes is *RUNX1*, which was first identified by Miyoshi et al. in 1991 as a gene located at the breakpoint of chromosomal translocation t(8;21) (Miyoshi et al. 1991). This translocation creates the *RUNX1-RUNX1T1* fusion gene (also called *RUNX1-ETO* or *AML1-ETO*). Many other chromosomal translocations involving the *RUNX1* gene, such as the *RUNX1-MECOM* fusion (also called *RUNX1-EVI1* or *AML1-EVI1*), have been reported in AML (Mitani et al. 1994; Nucifora et al. 1994; Speck and Gilliland 2002). In those translocations, the breakpoints are located between exons 5 and 6 of the *RUNX1* gene. Consequently, the fusion proteins retain the N-terminal portion of the RUNX1 protein, which contains a RUNT domain that binds the RUNX DNA-binding motif, but are defective in transcriptional regulation and therefore act as dominant negative proteins against wild-type RUNX1 protein. Point mutations are also frequently found in sporadic and familial AML cases (Michaud et al. 2002; Osato et al. 1999; Preudhomme et al. 2000; Song et al. 1999). Most of these mutations are clustered in the RUNT domain, resulting in defective DNA binding. Notably, chromosomal translocations that rearrange the *IG* or *TCR* regulatory elements to the *RUNX1* gene locus have not been reported. Hence, these early studies have indicated that alteration of the transcription factor activity of RUNX1 but not overexpression of this protein is pathogenic, thus implicating *RUNX1* as a tumor suppressor.

In contrast to the t(8;21) translocation, the t(12;21) translocation that creates the *ETV6-RUNX1* fusion gene (also called *TEL-AML1* or *TEL-RUNX1*), is specific to B-ALL and occurs in approximately 25 % of cases (Golub et al. 1995; Greaves and Wiemels 2003; Wiemels et al. 2000). This translocation arises prenatally and also occurs in normal fetal hematopoiesis at a relatively high frequency (~1 %) (Greaves and Wiemels 2003; Mori et al. 2002). The breakpoints are predominantly located in intron 1 of the *RUNX1* gene, with some in intron 2. Thus, only the first few exons and the upstream sequence are defective (Wiemels et al. 2000). In these cases, the native enhancer that regulates the *RUNX1* gene may be lost or disrupted. Why does the translocation occur in intron 1? Unlike in the case of *IG* and *TCR* translocations or the *KMT2A (MLL)* gene fusion, the *ETV6-RUNX1* translocation is not associated with V(D)J recombination or topoisomerase treatment. Intron 1 could be an open chromatin region that is structurally "fragile". This question has only recently been addressed, owing the emergence of technologies that can provide a comprehensive view of enhancers (Nottingham et al. 2007; Ng et al. 2010; Kwiatkowski et al. 2014; Hnisz et al. 2016).

10.2 Discovery of the *Runx1* Intronic Enhancer (eR1)

RUNX1 is one of the best-characterized hematopoietic transcription factors and is crucial for the maintenance of hematopoiesis (Speck and Gilliland 2002) (please see other sections for details). RUNX1 expression is persistently high in embryonic and postnatal adult HSCs. RUNX1 is essential for HSC generation and maintenance (Jacob et al. 2010; Motoda et al. 2007; Okuda et al. 1996). The *RUNX1* gene is extremely large (>250 kb) and through two independent promoters ("P1" and "P2") in combination with exon skipping and alternative 3′ exon usage, it generates diverse mRNAs and proteins (Ghozi et al. 1996; Telfer and Rothenberg 2001) (please see other sections for details). Hence, the regulatory elements of the *RUNX1* gene have not been well characterized until recently, owing to technical challenges.

In the late 2000s, two groups identified the *RUNX1* enhancer, which is located in intron 1 between the P1 and P2 promoters of the mouse *Runx1* gene (Nottingham et al. 2007; Bee et al. 2009; Ng et al. 2010). Next generation sequencing-based methodologies, such as chromatin immunoprecipitation sequencing (ChIP-seq), were not readily available at that time. Thus, the researchers utilized targeted approaches to identify regulatory elements, including DNaseI hypersensitive site analysis and retrovirus insertional mutagenesis assays, which can predict open chromatin configurations that potentially contain *cis*-regulatory elements. By combining these profiles with a comparison of the genomic sequences among different species, the two groups pinpointed the element that is activated in HSCs and controls *RUNX1* gene expression.

The first discovery was made by a group at Oxford University in 2007. Using a combination of comparative genomics and chromatin analysis, Nottingham et al. analyzed conserved non-coding sequence elements (CNEs) across various species (Nottingham et al. 2007). They identified a highly-conserved CNE that was mapped to the DNaseI hypersensitive site within *RUNX1* intron 1. Reporter assays revealed that this element possesses enhancer activity in hematopoietic tissues of mouse embryos as well as in cell lines. The authors therefore termed this element the "+23" *RUNX1* enhancer because it is located 23.5 kb downstream of the ATG in exon 1. The same group subsequently demonstrated that this enhancer confers hematopoietic specificity to both the P1 and P2 *RUNX1* promoters (Bee et al. 2009). Both the P1 and P2 promoters are able to drive reporter gene expression when inserted upstream of the reporter with the +23 *RUNX1* enhancer in the hematopoietic cells of mouse embryos.

A group at the National University of Singapore also identified the same element through a different approach and has demonstrated that the *RUNX1* enhancer marks hemogenic endothelial cells (ECs) and HSCs (Ng et al. 2010). Similarly to Nottingham et al., Ng et al. used a combinatorial approach utilizing comparative genomic alignment and retroviral integration site (RIS) mapping. A region located in *RUNX1* intron 1, which has been frequently identified as an RIS is conserved among species. Using the EGFP reporter system, they demonstrated that this element could drive EGFP expression in mouse ECs and HSCs as well as in zebrafish HSCs. This study further confirmed the findings of Nottingham et al. and established the *RUNX1* intronic enhancer as a conserved *cis*-regulatory element that regulates *RUNX1* expression. They termed this element "+24" *RUNX1* enhancer, because it is located 24 kb downstream from the distal P1 promoter. The discrepancy in the naming ("+23" or "+24") was due to different definitions of the +1 site. The term "eR1" has recently been proposed by the same group (Koh et al. 2015). Of note, two other intronic enhancers, the mouse +25 and the human intron 5.2 enhancers, have also been identified (Markova et al. 2012; Ng et al. 2010) but do not seem to be active in HSCs.

10.3 Activity of the eR1 in Normal Hematopoiesis

In general, enhancers evolve quite rapidly and are rarely conserved among species due to positive evolutionary selection (Villar et al. 2015). In contrast, eR1 is highly conserved among eukaryotes, implying essential roles of this element and the RUNX1 protein in normal development. Indeed, eR1 is active in HSCs in mice and zebrafish (Nottingham et al. 2007; Bee et al. 2009; Ferrell et al. 2015; Ng et al. 2010). The central element of the eR1, which is highly conserved among species, is enriched in DNA-binding motifs, including E-box, Gata, Ets, Runx, Myb and Cebp sites (Nottingham et al. 2007; Ng et al. 2010). The hematopoietic transcription factors TAL1/SCL, GATA2, RUNX1, PU.1, LMO2 and LDB1 bind at this region (Nottingham et al. 2007; Bee et al. 2009). Recent ChIP-seq analyses have also confirmed the binding of these factors at the eR1 in mouse and human hematopoietic cells (Tijssen et al. 2011; Sanda et al. 2012).

Importantly, Ng et al. have demonstrated that when the eR1 is placed upstream of an *EGFP* reporter gene, the EGFP signal is specifically observed in the aorta-gonad-mesonephros (AGM) region (Ng et al. 2010), where HSCs emerge during definitive hematopoiesis (Muller et al. 1994). EGFP expression was observed in hemogenic ECs and CD45+ HSCs in mice. Notably, EGFP highly-positive cells are detected in greatest abundance in mouse long-term HSCs (LT-HSCs); however, EGFP expression is significantly reduced in short-term HSCs (ST-HSCs) and progenitor cells (Ng et al. 2010). In contrast, the EGFP signal was not observed in differentiated myeloid or lymphoid cells, which normally express the *Runx1* gene, suggesting that an additional regulatory element is required for activation of the *RUNX1* gene in these adult hematopoietic compartments. The eR1 is not active in erythroid cells, consistent with the pattern of endogenous *RUNX1* expression (Lorsbach et al. 2004; North et al. 1999). Using an independent system, Swiers et al. also demonstrated that eR1 marks hemogenic ECs at the single cell level (Swiers et al. 2013). These findings highlight the specificity of eR1 in the regulation of *RUNX1* expression during hematopoiesis. Interestingly, eR1 can also be activated in the AGM in zebrafish, although the enhancer sequence is not conserved between mice and zebrafish (Ng et al. 2010). This finding indicates that the element can still "capture" transcription factors endogenously expressed in zebrafish HSCs.

10.4 Super-Enhancer in Normal Development and Pathogenesis

The term "super-enhancer" has recently been defined (Hnisz et al. 2013; Loven et al. 2013; Pott and Lieb 2015; Whyte et al. 2013). Super-enhancers are clusters of enhancers marked by a high level of mediator binding and acetylation of histone H3 lysine 27 (H3K27Ac), as typically determined by ChIP-seq analysis. A super-enhancer spans a large genomic region containing a high degree of enrichment of transcription factors. Super-enhancers are also associated with high levels of p300 co-activator binding and monomethylation of histone H3 lysine 4 (H3K4me1). Super-enhancers typically exhibit an open chromatin configuration, as confirmed by DNaseI hypersensitive site sequencing (DNase-seq) (Hnisz et al. 2013; Pott and Lieb 2015).

Rapidly accumulating evidence demonstrates the importance of super-enhancers in the regulation of genes that are crucial in normal development or pathogenesis. In embryonic stem cells, for example, super-enhancers are regulated by stem cell factors, namely Oct4, Sox2 and Nanog (Whyte et al. 2013). Comprehensive analysis revealed that super-enhancers are associated with genes that control and define cell identity during the developmental process (Hnisz et al. 2013). Importantly, super-enhancers correlate with more robust transcriptional changes than regular enhancers (Huang et al. 2016; Kwiatkowski et al. 2014). Huang et al. reported that the expression of genes associated with super-enhancers more dramatically changes during hematopoietic cell differentiation than the expression of genes with typical enhancers (Huang et al. 2016). Super-enhancer-associated genes are more sensitive to transcriptional inhibition by small molecules such as the CDK7 inhibitor THZ1 in leukemia cells (Kwiatkowski et al. 2014). Notably, several groups reported that landscapes of super-enhancers are dynamic. For example, macrophages isolated from different tissues exhibit different landscapes of super-enhancers (Gosselin et al. 2014; Lavin et al. 2014). Adam et al. revealed that super-enhancers and their dense clusters of transcription factor binding sites ("epicenters") undergo remodeling upon lineage progression or in response to changes in the microenvironment (Adam et al. 2015). Transcription factor genes that are specifically expressed in hair follicle stem cells lost super-enhancers in follicle stem cell progeny (transit-amplifying cells), while transcription factor genes that are specifically expressed in transit-amplifying cells acquired new super-enhancers (Adam et al. 2015). Epicenters in some key super-enhancers shift upon environmental changes. Although the functional relationship between super-enhancers and epicenters is not yet clear, these studies indicate that dynamic remodeling of super-enhancers may be associated with stem cell plasticity and lineage choice.

Super-enhancers are also involved in pathogenesis, including cancer. A number of groups have demonstrated that super-enhancers are often enriched at critical cancer genes. For example, in T-cell acute lymphoblastic leukemia (T-ALL), super-enhancers are located in the oncogene *TAL1* and its regulatory partners (*GATA3, RUNX1* and *MYB*) (Kwiatkowski et al. 2014). In neuroblastoma, large super-enhancers have been detected near the oncogenes *ALK* and *MYCN* (Chipumuro et al. 2014). In small cell lung cancer, super-enhancers are associated with the *MYC* or *MYCN* oncogenes (Christensen et al. 2014). In diffuse large B-cell lymphoma (DLBCL), the *BCL6*, *IRF4* and *PAX5* gene loci are associated with super-enhancers (Chapuy et al. 2013). Interestingly, transcriptional inhibition by small molecules such as the CDK7 inhibitor THZ1 or the BRD inhibitor JQ1 result in a marked decrease in the expression of super-enhancer-associated genes and led to growth inhibition in these cancer cells (Kwiatkowski et al. 2014; Chipumuro et al. 2014; Chapuy et al. 2013). These results indicate that super-enhancers may be required to maintain high levels of expression of critical cancer genes. Most recently, Hnisz et al. reported that in T-ALL cells, tumor cell genomes harbor recurrent microdeletions that eliminate the boundary sites of insulated neighborhoods constrained within the CTCF cohesin-mediated loops (Hnisz et al. 2016). Somatic mutations of neighborhood boundaries are also found in many types of cancers. This study indicates that disruption of chromosome neighborhood boundaries leads to activation of proto-oncogenes that are possibly associated with super-enhancers.

10.5 The eR1 is Involved in a Large Super-Enhancer

The eR1 was originally discovered through observations of chromatin structure (Nottingham et al. 2007; Ng et al. 2010). Early studies demonstrated that intron 1 is a hotspot of the breakpoint of *ETV6-RUNX1* translocations (Golub et al. 1995; Greaves and Wiemels 2003; Wiemels et al. 2000). These findings indicated that the chromatin conformation around this region might be widely open. In normal HSCs, RUNX1 regulates gene expression in coordination with other hematopoietic transcription factors such as TAL1/SCL and GATA2. These factors also regulate their own genes by forming auto-regulatory and feed-forward loops (Novershtern et al. 2011; Wilkinson and Gottgens 2013). TAL1, GATA2 and other transcription factors bind at the *RUNX1* enhancer in HSCs (Nottingham et al. 2007; Ng et al. 2010).

Interestingly, we have reported a similar finding in *TAL1*-positive T-ALL cells (Sanda et al. 2012). *TAL1* is normally expressed in HSCs but is silenced during thymocyte development. In contrast, this gene is ectopically expressed in 40–60 % of T-ALL cases due to chromosomal translocations, intrachromosomal rearrangements (called "*SIL-TAL1* deletion" or "*SIL-TAL1* fusion") or mutations in non-coding elements (Look 1997; Mansour et al. 2014). In T-ALL cells, TAL1 frequently co-occupies genomic elements with GATA3 and RUNX1, which coordinately regulate gene expression (Sanda et al. 2012). In addition, TAL1, GATA3 and RUNX1 regulate each other by forming an interconnected auto-regulatory loop, in a manner similar to the structure identified earlier in normal HSCs. These three factors bind at the eR1 (Fig. 10.1) and positively regulate *RUNX1* expression in T-ALL cells. This work established that T-ALL cells possess a stem cell-like transcriptional circuitry. Interestingly, knockdown of *RUNX1* induces growth inhibition and cell death in *TAL1*-positive T-ALL cells. Thus, overexpression of RUNX1 appears to contribute to thymocyte transformation as a major component of the transcriptional circuitry in *TAL1*-positive T-ALL, in marked contrast to its role as a tumor suppressor in AML.

Notably, Kwiatkowski et al. have discovered that in T-ALL cells, the *RUNX1* gene is associated with an extremely large super-enhancer (>170 kb) covering all of intron 1 including the eR1 element (Kwiatkowski et al. 2014) (Fig. 10.1a). This region corresponds to the breakpoint of t(12;21) translocation. Although the

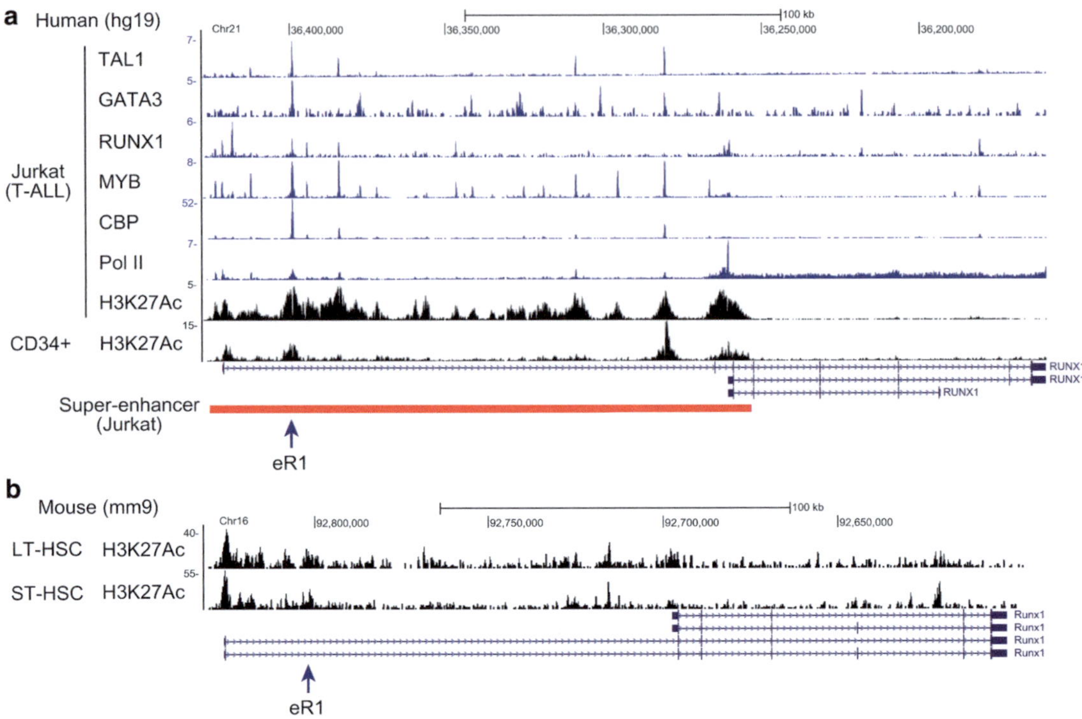

Fig. 10.1 Super-enhancer at the *RUNX1* gene locus in normal and malignant hematopoietic cells. (**a**) ChIP-seq gene track represents binding of transcription factors [TAL1, GATA3, RUNX1, MYB, CBP and RNA polymerase 2 (Pol II)] and histone H3 acetylation at lysine 27 (H3K27ac) in a human T-ALL cell line (Jurkat). H3K27ac in human CD34+ cells is also shown. The *x-axis* indicates the linear sequence of genomic DNA, and the *y-axis* indicates the total number of mapped reads. The *black* horizontal bar indicates the genomic scale in kilobases (kb). *Black* boxes in the gene map represent exons, and arrows indicate the location and direction of the transcriptional start site. The super-enhancer identified in Jurkat cells is indicated as a *red* box. The arrowhead shows the *RUNX1* intronic enhancer (eR1). The original ChIP-seq dataset can be found at http://www.ncbi.nlm.nih.gov/geo/ under superseries accession numbers GSE29181, GSE59657, GSE50625 and GSE17312 (Bernstein et al. 2010; Kwiatkowski et al. 2014; Mansour et al. 2014; Sanda et al. 2012). (**b**) ChIP-seq gene track represents H3K27ac in mouse long-term (LT) and short-term (ST)-HSCs (The original dataset can be found at http://www.ncbi.nlm.nih.gov/geo/under superseries accession number GSE60103 (Lara-Astiaso et al. 2014))

landscape is different, extensive H3K27Ac marks are also observed in primary human CD34+ hematopoietic cells (Fig. 10.1a) and mouse HSCs (Fig. 10.1b) (Bernstein et al. 2010; Khan and Zhang 2016; Lara-Astiaso et al. 2014), consistent with the activity of eR1 that is high in mouse LT-HSCs (Ng et al. 2010). This result suggests that the *RUNX1* super-enhancer is associated with the activity of eR1. A possible mechanism is that eR1 may function as an "epicenter" (Adam et al. 2015) that is crucial for controlling the dynamics of this super-enhancer. Activation of the eR1, for example, by an overexpression of TAL1, may lead to remodeling of the super-enhancer, thus contributing to a high-level of *RUNX1* expression in normal HSCs and malignant T-cells. In a recent study by Hnisz et al., the *RUNX1* super-enhancer locus displayed densely-connected enhancer-promoter interactions that had end points within the CTCF-CTCF loops in T-ALL cells (Hnisz et al. 2016). Thus, another possible mechanism by which the *RUNX1* super-enhancer is activated is through alteration of the chromatin neighborhood boundary which would modulate enhancer-promoter interactions. Further investigation is necessary to elucidate the functional relationship between eR1 and the super-enhancer.

10.6 Future Prospective: eR1 as a Powerful tool in Stem Cell Biology

Studies of the *RUNX1* intronic enhancer, eR1, have "rediscovered" the importance of the *RUNX1* gene in hematopoiesis. Analysis of the super-enhancer has also provided a broader picture of this enhancer than initially expected. This super-enhancer may be required to maintain the high expression level of the *RUNX1* gene in normal development and leukemogenesis. Importantly, the activity of eR1 is very potent and specific to HSCs. These characteristics are important for applications in stem cell biology because the developmental processes governing HSC generation, maintenance and cell differentiation are of great interest. Recent studies have clearly indicated that eR1 can be used as a marker for tracking hemogenic ECs and LT-HSCs with a reporter system *in vivo* (Ng et al. 2010; Swiers et al. 2013). In particular, unraveling HSC behavior in AGM and bone marrow by single live-cell imaging is feasible. Recent research has demonstrated that HSCs remain in a dormant phase, thus preventing stem cell exhaustion (Trumpp et al. 2010). It would therefore be interesting to determine whether eR1 actually marks dormant or active HSCs.

Because HSCs and leukemia stem cells (LSCs) share common features to maintain their stemness, the HSC marking approach may also be applied to the study of LSCs. In addition, eR1 can be used to express a transgene specifically in the HSC population. One can overexpress a potential oncogene or knock out the endogenous gene by expressing Cre recombinase in the HSC compartment under the control of eR1. This system is advantageous because many HSC enhancers used for transgenesis, such as *Tal1/Scl* and the *Gata2* enhancer, are also activated in differentiated cells (Koh et al. 2015). Hence, eR1 is considered the most specific HSC enhancer and can be used as a powerful tool in stem biology.

Acknowledgments We thank Dr. Motomi Osato for discussions and critical review. We thank the Nature Publishing Group Language Editing for editing the manuscript. This research is supported by the U.S. National Cancer Institute (1K99CA157951) and the National Research Foundation (NRF), Prime Minister's Office, Singapore under its Competitive Research Programme (Award No. NRF-NRFF2013-02).

References

Adam, R. C., et al. (2015). Pioneer factors govern super-enhancer dynamics in stem cell plasticity and lineage choice. *Nature, 521*, 366–370. doi:10.1038/nature14289.

Bee, T., et al. (2009). The mouse Runx1 +23 hematopoietic stem cell enhancer confers hematopoietic specificity to both Runx1 promoters. *Blood, 113*, 5121–5124. doi:10.1182/blood-2008-12-193003.

Bernstein, B. E., et al. (2010). The NIH roadmap epigenomics mapping consortium. *Nature Biotechnology, 28*, 1045–1048. doi:10.1038/nbt1010-1045.

Chapuy, B., et al. (2013). Discovery and characterization of super-enhancer-associated dependencies in diffuse large B cell lymphoma. *Cancer Cell, 24*, 777–790. doi:10.1016/j.ccr.2013.11.003.

Chipumuro, E., et al. (2014). CDK7 inhibition suppresses super-enhancer-linked oncogenic transcription in MYCN-driven cancer. *Cell, 159*, 1126–1139. doi:10.1016/j.cell.2014.10.024.

Christensen, C. L., et al. (2014). Targeting transcriptional addictions in small cell lung cancer with a covalent CDK7 inhibitor. *Cancer Cell, 26*, 909–922. doi:10.1016/j.ccell.2014.10.019.

Ferrell, P. I., Xi, J., Ma, C., Adlakha, M., & Kaufman, D. S. (2015). The RUNX1 +24 enhancer and P1 promoter identify a unique subpopulation of hematopoietic progenitor cells derived from human pluripotent stem cells. *Stem Cells, 33*, 1130–1141. doi:10.1002/stem.1940.

Ghozi, M. C., Bernstein, Y., Negreanu, V., Levanon, D., & Groner, Y. (1996). Expression of the human acute myeloid leukemia gene AML1 is regulated by two promoter regions. *Proceedings of the National Academy of Sciences of the United States of America, 93*(5), 1935–1940.

Golub, T. R., et al. (1995). Fusion of the TEL gene on 12p13 to the AML1 gene on 21q22 in acute lymphoblastic leukemia. *Proceedings of the National Academy of Sciences of the United States of America, 92*, 4917–4921.

Gosselin, D., et al. (2014). Environment drives selection and function of enhancers controlling tissue-specific macrophage identities. *Cell, 159*, 1327–1340. doi:10.1016/j.cell.2014.11.023.

Greaves, M. F., & Wiemels, J. (2003). Origins of chromosome translocations in childhood leukaemia. *Nature Reviews Cancer, 3*, 639–649. doi:10.1038/nrc1164.

Hnisz, D., et al. (2013). Super-enhancers in the control of cell identity and disease. *Cell, 155*, 934–947. doi:10.1016/j.cell.2013.09.053.

Hnisz, D., et al. (2016). Activation of proto-oncogenes by disruption of chromosome neighborhoods. *Science, 351*, 1454–1458. doi:10.1126/science.aad9024.

Huang, J., et al. (2016). Dynamic control of enhancer repertoires drives lineage and stage-specific transcription during hematopoiesis. *Developmental Cell, 36*, 9–23. doi:10.1016/j.devcel.2015.12.014.

Jacob, B., et al. (2010). Stem cell exhaustion due to Runx1 deficiency is prevented by Evi5 activation in leukemogenesis. *Blood, 115*, 1610–1620. doi:10.1182/blood-2009-07-232249.

Khan, A., & Zhang, X. (2016). dbSUPER: A database of super-enhancers in mouse and human genome. *Nucleic Acids Research, 44*, D164–D171. doi:10.1093/nar/gkv1002.

Koh, C. P., et al. (2015). Hematopoietic stem cell enhancer: A powerful tool in stem cell biology. *Histology and Histopathology, 30*, 661–672. doi:10.14670/HH-30.661.

Kwiatkowski, N., et al. (2014). Targeting transcription regulation in cancer with a covalent CDK7 inhibitor. *Nature, 511*, 616–620. doi:10.1038/nature13393.

Lara-Astiaso, D., et al. (2014). Immunogenetics. Chromatin state dynamics during blood formation. *Science, 345*, 943–949. doi:10.1126/science.1256271.

Lavin, Y., et al. (2014). Tissue-resident macrophage enhancer landscapes are shaped by the local microenvironment. *Cell, 159*, 1312–1326. doi:10.1016/j.cell.2014.11.018.

Look, A. T. (1997). Oncogenic transcription factors in the human acute leukemias. *Science, 278*, 1059–1064.

Lorsbach, R. B., Moore, J., Ang, S. O., Sun, W., Lenny, N., & Downing, J. R. (2004). Role of RUNX1 in adult hematopoiesis: analysis of RUNX1-IRES-GFP knock-in mice reveals differential lineage expression. *Blood, 103*(7), 2522–2529. doi:10.1182/blood-2003-07-2439.

Loven, J., et al. (2013). Selective inhibition of tumor oncogenes by disruption of super-enhancers. *Cell, 153*, 320–334. doi:10.1016/j.cell.2013.03.036.

Mansour, M. R., et al. (2014). Oncogene regulation. An oncogenic super-enhancer formed through somatic mutation of a noncoding intergenic element. *Science, 346*, 1373–1377. doi:10.1126/science.1259037.

Markova, E. N., Razin, S. V., & Kantidze, O. L. (2012). Fragment of intron 5.2 of the human RUNX1 gene important for transcription activation is neither enhancer nor MAR-element. *Doklady Biochemistry and Biophysics, 442*, 26–29. doi:10.1134/S1607672912010085.

Michaud, J., et al. (2002). In vitro analyses of known and novel RUNX1/AML1 mutations in dominant familial platelet disorder with predisposition to acute myelogenous leukemia: Implications for mechanisms of pathogenesis. *Blood, 99*, 1364–1372.

Mitani, K., et al. (1994). Generation of the AML1-EVI-1 fusion gene in the t(3;21)(q26;q22) causes blastic crisis in chronic myelocytic leukemia. *The EMBO Journal, 13*, 504–510.

Miyoshi, H., Shimizu, K., Kozu, T., Maseki, N., Kaneko, Y., & Ohki, M. (1991). t(8;21) breakpoints on chromosome 21 in acute myeloid leukemia are clustered within a limited region of a single gene, AML1. *Proceedings of the National Academy of Sciences of the United States of America, 88*, 10431–10434.

Mori, H., et al. (2002). Chromosome translocations and covert leukemic clones are generated during normal fetal development. *Proceedings of the National Academy of Sciences of the United States of America, 99*, 8242–8247. doi:10.1073/pnas.112218799.

Motoda, L., et al. (2007). Runx1 protects hematopoietic stem/progenitor cells from oncogenic insult. *Stem Cells, 25*, 2976–2986. doi:10.1634/stemcells.2007-0061.

Muller, A. M., Medvinsky, A., Strouboulis, J., Grosveld, F., & Dzierzak, E. (1994). Development of hematopoietic stem cell activity in the mouse embryo. *Immunity, 1*(4), 291–301.

Ng, C. E., et al. (2010). A Runx1 intronic enhancer marks hemogenic endothelial cells and hematopoietic stem cells. *Stem Cells, 28*, 1869–1881. doi:10.1002/stem.507.

North, T., Gu, T. L., Stacy, T., Wang, Q., Howard, L., Binder, M., Marin-Padilla, M., & Speck, N. A. (1999) Cbfa2 is required for the formation of intra-aortic hematopoietic clusters. *Development, 126*(11), 2563–2575.

Nottingham, W. T., et al. (2007). Runx1-mediated hematopoietic stem-cell emergence is controlled by a Gata/Ets/SCL-regulated enhancer. *Blood, 110*, 4188–4197. doi:10.1182/blood-2007-07-100883.

Novershtern, N., et al. (2011). Densely interconnected transcriptional circuits control cell states in human hematopoiesis. *Cell, 144*, 296–309. doi:10.1016/j.cell.2011.01.004.

Nucifora, G., et al. (1994). Consistent intergenic splicing and production of multiple transcripts between AML1 at 21q22 and unrelated genes at 3q26 in (3;21)(q26;q22) translocations. *Proceedings of the National Academy of Sciences of the United States of America, 91*, 4004–4008.

Okuda, T., van Deursen, J., Hiebert, S. W., Grosveld, G., & Downing, J. R. (1996). AML1, the target of multiple chromosomal translocations in human leukemia, is essential for normal fetal liver hematopoiesis. *Cell, 84*, 321–330.

Osato, M., et al. (1999). Biallelic and heterozygous point mutations in the runt domain of the AML1/PEBP2alphaB gene associated with myeloblastic leukemias. *Blood, 93*, 1817–1824.

Pott, S., & Lieb, J. D. (2015). What are super-enhancers? *Nature Genetics, 47*, 8–12. doi:10.1038/ng.3167.

Preudhomme, C., et al. (2000). High incidence of biallelic point mutations in the Runt domain of the AML1/PEBP2 alpha B gene in Mo acute myeloid leukemia and in myeloid malignancies with acquired trisomy 21. *Blood, 96*, 2862–2869.

Rowley, J. D. (2001). Chromosome translocations: Dangerous liaisons revisited. *Nature Reviews Cancer, 1*, 245–250. doi:10.1038/35106108.

Sanda, T., et al. (2012). Core transcriptional regulatory circuit controlled by the TAL1 complex in human T cell acute lymphoblastic leukemia. *Cancer Cell, 22*, 209–221. doi:10.1016/j.ccr.2012.06.007.

Song, W. J., et al. (1999). Haploinsufficiency of CBFA2 causes familial thrombocytopenia with propensity to develop acute myelogenous leukaemia. *Nature Genetics, 23*, 166–175. doi:10.1038/13793.

Speck, N. A., & Gilliland, D. G. (2002). Core-binding factors in haematopoiesis and leukaemia. *Nature Reviews Cancer, 2*, 502–513. doi:10.1038/nrc840.

Swiers, G., Baumann, C., O'Rourke, J., Giannoulatou, E., Taylor, S., Joshi, A., Moignard, V., Pina, C., Bee, T., Kokkaliaris, K. D., Yoshimoto, M., Yoder, M. C., Frampton, J., Schroeder, T., Enver, T., Gottgens, B., & de Bruijn, M. F. (2013) Early dynamic fate changes in haemogenic endothelium characterized at the single-cell level. *Nature Communications, 4*, 2924. doi:10.1038/ncomms3924.

Telfer, J. C., & Rothenberg, E. V. (2001). Expression and function of a stem cell promoter for the murine CBFalpha2 gene: Distinct roles and regulation in natural killer and T cell development. *Developmental Biology, 229*(2), 363–382. doi:10.1006/dbio.2000.9991.

Tijssen, M. R., et al. (2011). Genome-wide analysis of simultaneous GATA1/2, RUNX1, FLI1, and SCL binding in megakaryocytes identifies hematopoietic regulators. *Developmental Cell, 20*, 597–609. doi:10.1016/j.devcel.2011.04.008.

Trumpp, A., Essers, M., & Wilson, A. (2010). Awakening dormant haematopoietic stem cells. *Nature Reviews Immunology, 10*, 201–209. doi:10.1038/nri2726.

Villar, D., et al. (2015). Enhancer evolution across 20 mammalian species. *Cell, 160*, 554–566. doi:10.1016/j.cell.2015.01.006.

Whyte, W. A., et al. (2013). Master transcription factors and mediator establish super-enhancers at key cell identity genes. *Cell, 153*, 307–319. doi:10.1016/j.cell.2013.03.035.

Wiemels, J. L., Alexander, F. E., Cazzaniga, G., Biondi, A., Mayer, S. P., & Greaves, M. (2000). Microclustering of TEL-AML1 translocation breakpoints in childhood acute lymphoblastic leukemia. *Genes, Chromosomes & Cancer, 29*, 219–228.

Wilkinson, A. C., & Gottgens, B. (2013). Transcriptional regulation of haematopoietic stem cells. *Advances in Experimental Medicine and Biology, 786*, 187–212. doi:10.1007/978-94-007-6621-1_11.

Part IV

RUNX1 and CBFβ Fusion Proteins in Leukemia

RUNX1-ETO Leukemia

11

Shan Lin, James C. Mulloy, and Susumu Goyama

Abstract

AML1-ETO leukemia is the most common cytogenetic subtype of acute myeloid leukemia, defined by the presence of t(8;21). Remarkable progress has been achieved in understanding the molecular pathogenesis of AML1-ETO leukemia. Proteomic surveies have shown that AML-ETO forms a stable complex with several transcription factors, including E proteins. Genome-wide transcriptome and ChIP-seq analyses have revealed the genes directly regulated by AML1-ETO, such as CEBPA. Several lines of evidence suggest that AML1-ETO suppresses endogenous DNA repair in cells to promote mutagenesis, which facilitates acquisition of cooperating secondary events. Furthermore, it has become increasingly apparent that a delicate balance of AML1-ETO and native AML1 is important to sustain the malignant cell phenotype. Translation of these findings into the clinical setting is just beginning.

Keywords

AML1-ETO (RUNX1-ETO) • AML1(RUNX1) • Acute myeloid leukemia • Transcription factor complex • Epigenetics • Signal transduction • DNA repair

11.1 Introduction

The t(8;21)(q22;q22) translocation is one of the most frequent genetic alterations in acute myeloid leukemia (AML), and is typically present in French-American-British (FAB)-M2 subtype of AML (Rowley 1984). Leukemic blasts with the t(8;21) translocation show granulocytic maturation and a distinct immunophenotype characterized by a prevalent positivity for CD19, CD56, CD13, and

S. Lin • J.C. Mulloy
Division of Experimental Hematology and Cancer Biology, Cincinnati Children's Hospital Medical Center, University of Cincinnati College of Medicine, Cincinnati, OH, USA

S. Goyama (✉)
Division of Cellular Therapy, The Institute of Medical Science, The University of Tokyo, Tokyo, Japan
e-mail: goyama@ims.u-tokyo.ac.jp

Y. Groner et al. (eds.), *RUNX Proteins in Development and Cancer*, Advances in Experimental Medicine and Biology 962, DOI 10.1007/978-981-10-3233-2_11

CD34 (Bitter et al. 1987). Molecular cloning of the translocation breakpoints revealed rearrangement of the *AML1* gene (also referred to as *RUNX1*) on chromosome 21q22 and the *ETO* gene (also referred to as *RUNX1T1* or *MTG8*) on chromosome 8q22 (Miyoshi et al. 1993; Miyoshi et al. 1991; Erickson et al. 1992; Nisson et al. 1992). The rearrangement results in the generation of the AML1-ETO fusion protein. Since the discovery of t(8;21) and the AML1-ETO fusion, numerous studies have elucidated the clinical features and molecular basis of this type of leukemia. Patients with t(8;21) AML have a relatively favorable prognosis compared with other types of AML patients. However, substantial numbers of t(8;21) AML patients eventually relapse, indicating the heterogeneity within AML1-ETO leukemia. Furthermore, the outcome of older patients with t(8;21) AML who are not eligible for intensive chemotherapy remains dismal. A significant unmet medical need is therefore apparent for AML1-ETO leukemia. In this chapter, we summarize the current knowledge of the pathophysiologic mechanisms of AML1-ETO leukemia, on which will be the foundation to develop better therapies.

11.2 The Properties of AML1-ETO

11.2.1 Structure

The chromosomal breakpoints of the t(8;21) cluster within *AML1* intron 5 and *ETO* intron 1 to generate similar *AML1-ETO* chimeric genes in most cases (Erickson et al. 1992; Maseki et al. 1993; Nucifora et al. 1993). Mechanisms that drive the generation of the t(8;21) translocation have been unclear, but a recent report showed that Wnt/β-catenin signaling enhances transcription and genomic proximity of *AML1* and *ETO* genes, which appears to promote the generation of the *AML1-ETO* fusion gene (Ugarte et al. 2015) (Fig. 11.1a). AML1-ETO contains the N-terminal 177 amino acids of AML1 fused in frame with almost the entire ETO protein. The RUNT domain in AML1 mediates DNA binding as well as heterodimerization with a cofactor CBFβ. ETO contains four evolutionarily conserved domains termed nervy homology regions (NHR) 1–4 (Davis et al. 2003). Among the NHR domains, NHR2 contributes many of the biochemical properties of ETO. In particular, the NHR2-mediated oligomerization of AML1-ETO has been shown to be critical for leukemogenesis (Kwok et al. 2009; Liu et al. 2006; Yan et al. 2009). ETO also contains a nuclear localization signal (NLS) between NHR1 and NHR2, which is sufficient and necessary for nuclear localization of AML1-ETO (Odaka et al. 2000; Barseguian et al. 2002) (Fig. 11.1b).

11.2.2 Interacting Proteins

AML1-ETO regulates gene expression through interactions with diverse proteins. CBFβ is a cofactor of AML1 and also binds to AML1-ETO through the RUNT domain. The AML1/CBFβ complex is essential for emergence of hematopoietic cells, and homozygous loss of either AML1 or CBFβ alleles resulted in embryonic death with lack of definitive hematopoiesis. Similar to *AML1* gene, *CBFβ* gene is also frequently altered in AML patients with an inversion in chromosome 16, resulting in the fusion gene CBFβ-MYH11. CBFβ potentiates AML1-ETO binding to DNA as it does for AML1, and the RUNT/CBFβ interaction has been considered important for the development of AML1-ETO leukemia (Roudaia et al. 2009). However, a study reported that the transforming activity of AML1-ETO is independent of CBFβ interaction and instead requires the NHR2-mediated oligomerization (Kwok et al. 2009). Therefore, the precise role of CBFβ in AML1-ETO-induced leukemogenesis remains unclear. NCOR and SMRT were originally shown to interact with the NHR4 region (Lutterbach et al. 1998a; Gelmetti et al. 1998; Wang et al. 1998), but a later study found that other portions of ETO also contributed to the interactions with them (Zhang et al. 2001). Histone deacetylases (HDACs) 1–3 and mSin3A bind to AML1-ETO mainly through NHR2 (Lutterbach et al. 1998b; Wang et al. 1998; Gelmetti et al. 1998; Amann et al. 2001; Hildebrand et al. 2001) (reviewed in (Davis et al. 2003)). These findings suggest that

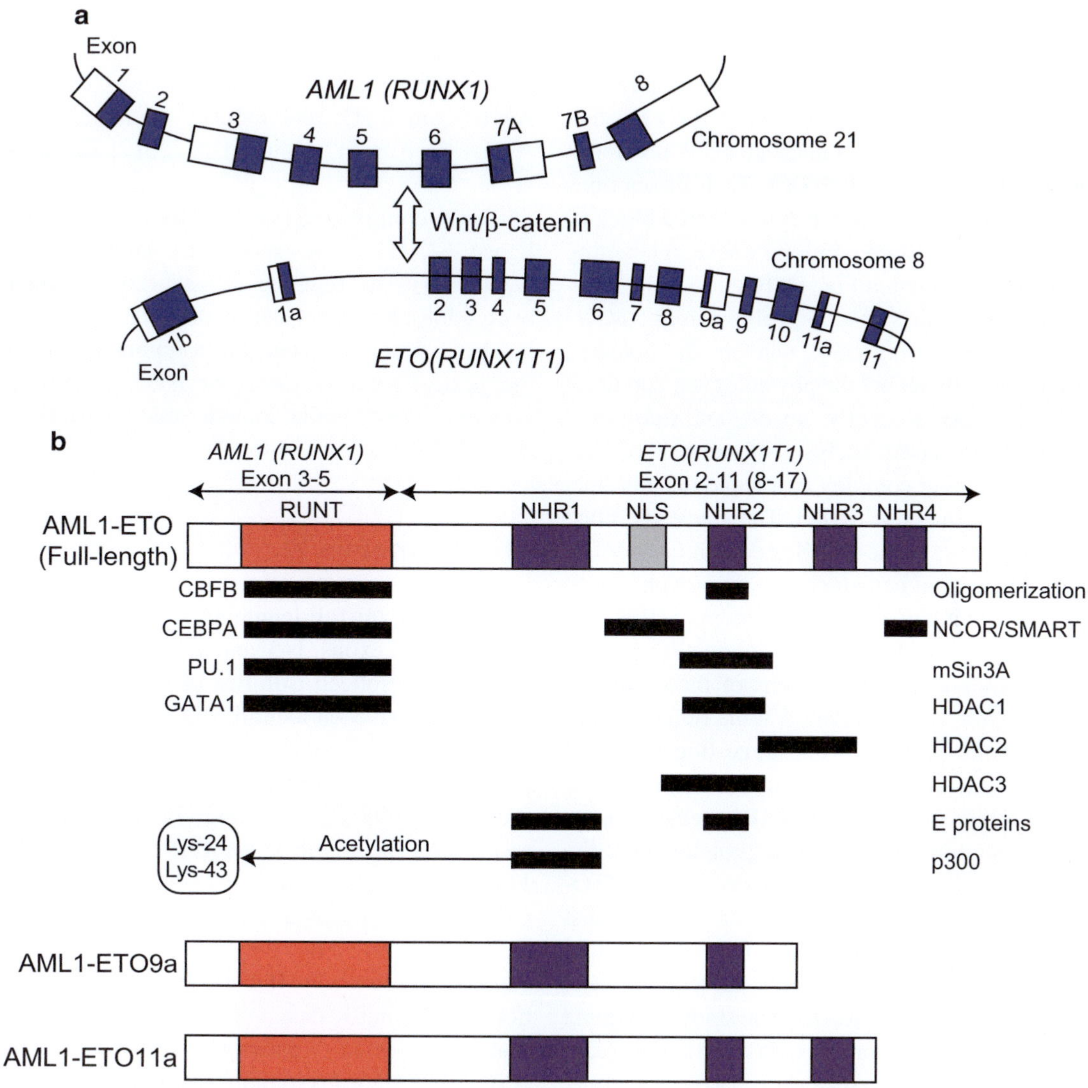

Fig. 11.1 Structure, isoforms, and interacting proteins of AML1-ETO. (**a**) Genomic structure of *AML1 (RUNX1)* on chromosome 21 and *ETO (RUNX1T1)* on chromosome 8. Wnt/β-catenin was shown to induce spatial proximity and translocation of AML1 and ETO, which led to the generation of the *AML1-ETO* fusion gene. Exons are depicted as *boxes*, and exons of ETO are numbered according to previous reports (Yan et al. 2006; Kozu et al. 2005). (**b**) Diagrams of full-length AML1-ETO and two shorter isoforms, illustrating the regions involved in interactions with other proteins. AML1-ETO has the RUNT domain and four Nervy homology regions (NHR1 – 4). The location of the nuclear localization signal (NLS) is also indicated. AML1-ETO9a and AML1-ETO11a lack NHR3/NHR4 and NHR4, respectively. Note that the information on interacting proteins was obtained mainly using AML1-ETO-overexpressing cells, and may not reflect physiological condition. Although E proteins were shown to interact with NHR1 (Zhang et al. 2004), a recent study that analyzed the natural AML1-ETO-containing complex found that oligomerized NHR2 created a novel protein-binding interface for E proteins (Sun et al. 2013)

AML1-ETO forms a corepressor complex with the NCOR/mSin3A/HDACs to repress expressions of target genes. AML1-ETO also interacts with transcription coactivators, such as p300 and PRMT1. In addition, E proteins, HEB and E2A, were shown to interact with AML1-ETO mainly through NHR1 (Zhang et al. 2004). The stable interaction between AML1-ETO and E proteins leads to inhibition of E protein-induced transcriptional activation. Furthermore, several hematopoietic transcription factors, including CEBPA, PU.1, and GATA1, bind to AML1-ETO through the RUNT domain (Vangala et al. 2003; Pabst et al. 2001; Choi et al. 2006). Such interactions are important for AML1-ETO-mediated block of hematopoietic differentiation. These interacting proteins are shown in Fig. 11.1b.

As most studies described above used AML1-ETO-overexpressing cells, whether the obtained results actually reflect the physiological condition was not clear. Recently, an elegant study provided an unbiased analysis of natural AML1-ETO-containing complex in a leukemic cell line, Kasumi-1 (Sun et al. 2013). The study identified an endogenous stable complex containing AML1-ETO and several hematopoietic transcription (co) factors, including CBFβ, E proteins (HEB and E2A), LYL1, LMO2 and its interacting partner LDB1. Individual knockdowns of these components significantly decreased some other components at the protein level, suggesting a mutual stabilization mechanism within the complex. Interestingly, oligomerized NHR2 created a novel protein-binding interface to E proteins, and disruption of this interaction abrogated AML1-ETO-induced leukemogenesis. These data suggest that AML1-ETO resides in the transcription factor complex, including E proteins, to regulate gene expression in leukemia cells. The study also found that AML1-ETO interacts with coactivators (e.g., p300) and corepressors (e.g., HDACs) only weakly in physiologic conditions (Sun et al. 2013), indicating the context-dependent exchange of these coregulators to achieve dynamic regulation of gene expression.

11.2.3 Posttranslational Modifications

As described above, AML1-ETO interacts with many epigenetic modifiers that also modulate posttranslational modifications of AML1-ETO itself. p300 interacts with AML1-ETO through NHR1 and induces acetylation of two lysine residues of AML1 (Lys-24 and Lys-43) (Wang et al. 2011a; Goyama et al. 2015a). PRMT1 binds to AML1-ETO through multiple regions of both AML1 and ETO, and induces methylation of AML1 at Arg-142 (Shia et al. 2012; Goyama et al. 2015a). As for ubiquitination, an E2 ubiquitin-conjugating enzyme UbcH8 and an E3 ligase SIAH1 were shown to interact with AML1-ETO to regulate its stability (Kramer et al. 2008). Phosphorylation of AML1-ETO has not been described well, but a report identified the phosphorylated forms of ETO protein in human $CD34^{+}$ hematopoietic cells (Erickson et al. 1996).

11.2.4 Isoforms

In addition to the full-length *AML1-ETO* that contains AML1 exons 1–5 and ETO exons 2–11, alternatively spliced isoforms of the *AML1-ETO* transcript have been identified in t(8;21) patients. The *ETO* gene is comprised of 14 exons, including 2 alternative exons 9a and 11a (Wolford and Prochazka 1998). Exon 9a and 11a provide a stop codon after the amino acid encoded by exon 8 or 11, respectively. *AML1-ETO9a* encodes a C-terminally truncated AML1-ETO protein that lack NHR3 and NHR4 domains, and has reduced capacity to repress AML1-mediated gene activation. Interestingly, despite the reduced transcription repression activity, expression of AML1-ETO9a leads to rapid development of leukemia in a mouse retroviral transduction-transplantation model (Yan et al. 2006). Because of the stronger leukemogenic potential, AML1-ETO9a has been widely used experimentally in mouse models for AML1-ETO leukemia. However, physiologic functions of AML1-ETO9a in human leukemia cells remain to be demonstrated. *AML1-ETO11a* produces a truncated protein lacking NHR4 domain with reduced transcriptional repressor activity (Kozu et al. 2005). The leukemogenic activity of AML1-ETO11a has not been examined. The scheme of these isoforms is also shown in Fig. 11.1b.

11.3 The Role of Native AML1 in AML1-ETO Leukemia

Dominant inhibition of native AML1 function has been considered as a central mechanism for AML-ETO-induced leukemogenesis. It has been shown that AML1-ETO recruits corepressors through the ETO moiety to repress AML1-induced gene activation (Gelmetti et al. 1998, Lutterbach et al. 1998b; Wang et al. 1998; Amann et al. 2001; Zhang et al. 2001; Hiebert et al. 2001). Furthermore, mice heterozygous for an AML1-ETO allele display early embryonic lethality and hematopoietic defects similar to those observed in AML1 null mice (Yergeau et al. 1997; Okuda et al. 1998). These observations gave rise to an important but perhaps oversimplified conclusion: AML1-ETO blocks AML1 function, thereby interfering with normal hematopoietic differentiation and inducing a preleukemic condition. However, increasing evidence suggests that the relationship between AML1-ETO and native AML1 may be more complex. Clinical data have shown that no inactivating mutations of AML1 have been found in t(8;21) AML while it is frequently mutated in other types of AML, indicating that AML1-ETO needs some activity of native AML1 to promote leukemogenesis (Goyama and Mulloy 2011). This implication has been reinforced by experimental data showing that knockdown of AML1 inhibits the growth and survival of AML1-ETO leukemia cells (Ben-Ami et al. 2013; Goyama et al. 2013). Mechanistically, a recent genome-wide ChIP-Seq and RNA-Seq analyses revealed that AML1 is actually a member of the transcription factor complex containing AML1-ETO, and the relative binding signals of AML1 and AML1-ETO on chromatin determine which genes are repressed or activated by AML1-ETO (Li et al. 2016). Thus, these new findings indicate that a delicate balance of AML1-ETO and native AML1 is important to sustain the malignant cell phenotype of AML1-ETO leukemia.

11.4 Mechanism of AML1-ETO-Mediated Transcriptional Regulation

Transcriptome analysis revealed that t(8;21) AML is associated with distinct gene expression profiles (Ross et al. 2004; Valk et al. 2004). To understand AML1-ETO-mediated transcriptional regulation, two essential questions need to be answered: (1) How AML1-ETO selects its targets or how AML1-ETO decides where to bind in the genome; (2) How AML1-ETO alters expression of its target genes.

11.4.1 Target Selection by AML1-ETO

AML1-ETO retains the intact RUNT domain and shares many target genes with native AML1. Genome-wide ChIP-seq analyses have shown that about 60–80 % of AML-ETO binding sites overlap with those of native AML1 in AML1-ETO-harboring leukemia cell lines (Kasumi-1 and SKNO-1) and t(8;21) patient samples (Gardini et al. 2008; Ptasinska et al. 2012; Li et al. 2016; Ben-Ami et al. 2013). Overall gene expression profile in response to knockdown of AML1-ETO or AML1 was inversely correlated, indicating an opposing role of them in the regulation of gene expression (Ben-Ami et al. 2013). However, genome-wide distributions of AML1-ETO and native AML1 are not identical. AML1-ETO preferentially binds to intergenic and intron regions, while AML1-binding sites are relatively enriched in promoter regions (Ben-Ami et al. 2013; Okumura et al. 2008). In addition, AML1-ETO has a selective preference for certain target genes that contain multimerized AML1 consensus sites in their regulatory elements (Okumura et al. 2008). A recent study also showed that AML1-ETO binds to the canonical short motifs of AML1 [5′-TG(T/C)GGT-3′] more efficiently than AML1, whereas AML1 prefers a longer motif (5′-TGTGGTTT-3′; with 2 additional thymidines to the short motif at the 3′ position) than AML1-ETO (Li et al. 2016). Intriguingly, AML1-ETO and AML1 often bind to adjacent sites in the majority of the overlap regions, and the ratio of AML1-ETO- or AML1-binding in regulatory regions affects up- or down-regulation of each gene. AML1-ETO-repressed genes tended to show higher binding signals for AML1-ETO, while AML-ETO-activated genes tended to exhibit higher binding signals for AML1 (Li et al. 2016). Thus, AML1-ETO binds to DNA mainly through RUNT domain, but has slightly altered DNA binding properties as compared with native AML1.

A portion of AML1-ETO binding sites do not contain the AML1 motifs (Ptasinska et al. 2012; Maiques-Diaz et al. 2012), indicating the presence of other cofactors linking the AML1-ETO protein to DNA. As described above, AML1-ETO forms a complex with E proteins (HEB and E2A), and ChIP-Seq analyses revealed the over-representation of E-boxes in AML1-ETO-binding regions (Gardini et al. 2008; Zhang et al. 2004; Sun et al. 2013). Another genome-wide analysis revealed that ETS family of transcription factors (ERG and FLI1) occupy similar genomic regions as AML1-ETO in t(8;21) AML cells and identified ERG/FLI1 as proteins that facilitate binding of AML1-ETO to genomic regions (Martens et al. 2012). SP1, which was shown to interact with AML1-ETO (Wei et al. 2008), also appears to have a similar function. Some promoter regions bound by AML1-ETO do not have an AML1-binding motif and only contain SP1 sites, and SP1 inhibition can reverse AML1-ETO-mediated transcriptional repression of the target genes (Maiques-Diaz et al. 2012). These data suggest that AML1-ETO also binds to DNA indirectly through the interacting cofactors, including E proteins and ETS family of transcription factors.

Chromatin accessibility is also associated with the binding of AML1-ETO to specific regions. It was shown that AML1-ETO binding sites are enriched in accessible chromatin regions that are marked with p300 and low/intermediate levels of acetylation (Saeed et al. 2012; Maiques-Diaz et al. 2012). In line with this, a recent study showed that AML1-ETO binding sites shift along the developmental stages from embryonic stem cells to myeloid progenitors (Regha et al. 2015), indicating that the cellular environment will affect AML1-ETO's choice of target genes. It remains unclear whether pre-existing accessible chromatin regions shape the binding profiles of AML1-ETO, or AML1-ETO acts as a pioneer factor to regulate the chromatin status as native AML1 does (Hoogenkamp et al. 2009).

11.4.2 Regulation of Gene Expression by AML1-ETO

In general, AML1-ETO is considered to act as a transcription repressor by recruiting co-repressors HDACs, NCOR and mSin3A via the ETO portion with a dominant negative effect over native AML1 (Gelmetti et al. 1998, Wang et al. 1998; Hiebert et al. 2001; Lutterbach et al. 1998b; Amann et al. 2001; Hildebrand et al. 2001). Cross analyses of ChIP-seq and gene expression data have supported this concept, showing that direct binding of AML1-ETO is significantly associated with target gene repression but not activation (Ptasinska et al. 2014; Maiques-Diaz et al. 2012). Furthermore, a genome wide co-occupancy of AML1-ETO and NCOR is validated in a recent study, and NCOR-bound genes are enriched in downregulated genes by AML1-ETO (Trombly et al. 2015). DNA methyltrasferase DNMT1 also interacts with AML1-ETO and is involved in AML1-ETO-mediated transcriptional repression (Liu et al. 2005). DNMT3A, which is upregulated by AML1-ETO in cooperation with HIF1α (Gao et al. 2015), is likely to have similar function, but whether DNMT3A and AML1-ETO forms a complex needs to be tested experimentally.

AML1-ETO is not always a transcriptional repressor, but can induce upregulation of several target genes through interactions with epigenetic factors, including p300, PRMT1, and JMJD1C. Histone acetyltransferases p300 promotes acetylation of Lysines (Lysine-24 and 43) in AML1-ETO and enhances its transcription activation ability, likely by recruiting transcriptional pre-initiation complex (Wang et al. 2011a).

p300 may also contribute to the recruitment of AML1-ETO to specific chromatin regions (Saeed et al. 2012), or may facilitate AML1-ETO-mediated gene activation by promoting acetylation of surrounding histones, but these hypotheses need to be demonstrated in future studies. PRMT1 is a protein arginine methyltransferase that promotes Histone 4 arginine 3 (H4R3) methylation, which is generally associated with gene activation. AML1-ETO9a recruits PRMT1 to promoters of target genes, resulting in enrichment of H4R3 methylation and transcription activation. PRMT1 also weakly methylates Arg-142 of AML1-ETO, which may increase the transcriptional potency (Shia et al. 2012). Similarly, a histone demethylase JMJD1C is directly recruited by AML1-ETO to its target genes and promotes AML1-ETO-induced transcriptional upregulation by maintaining low H3K9 dimethyl levels (Chen et al. 2015). In addition to these epigenetic factors, the AML1-ETO/AML complex recruits AP-1 transcription factor c-Jun for activating target genes but not for repression (Li et al. 2016). In line with this, several genes were shown to be upregulated by AML1-ETO in a JNK-signaling-dependent manner (Gao et al. 2007; Elsasser et al. 2003).

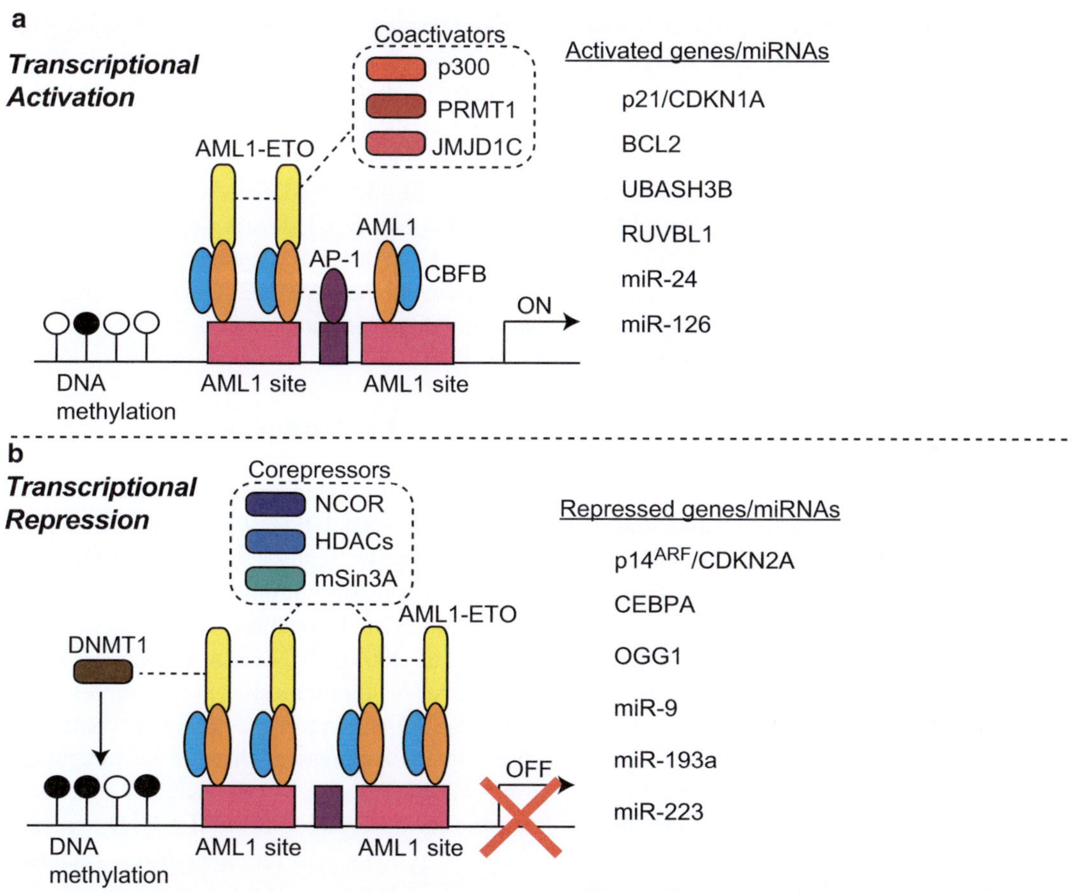

Fig. 11.2 Mechanisms of transcriptional activation and repression mediated by AML1-ETO. For transcriptional activation, AML1-ETO and native AML1 (together with its cofactor CBFB) often bind to adjacent AML1-motifs and the AML1-ETO/AML1 complex recruits AP-1. AML1-ETO also interacts with coactivators, including p300, PRMT1, and JMJD1C to activate target gene expression. For transcriptional repression, AML1-ETO often displaces native AML1 and recruits corepressors, including NCOR, HDACs, and mSin3A. AML1-ETO also interacts with DNMT1 to promote DNA methylation and to repress target gene expression. Representative target genes/miRNAs regulated by AML1-ETO are also shown

Thus, AML1-ETO alters gene expression through dynamic exchange of interacting proteins (Fig. 11.2). Future studies should address the following questions: (1) Do the AML-ETO-interacting coregulators work cooperatively or do they have their unique targets? (2) What determines the recruitment of repressor versus activator to the AML-ETO containing complex?

In addition to the direct transcriptional regulation, AML1-ETO also affects gene expression by interfering with the function of other transcription factors. CEBPA and PU.1 are transcription factors that play crucial roles in myeloid differentiation. AML1-ETO interacts with them and reduces their DNA binding activity (Pabst et al. 2001; Vangala et al. 2003). GATA1 is a major erythroid transcription factor, and AML1-ETO hampers transcriptional activity of GATA1 by preventing its acetylation (Choi et al. 2006). Functional suppression of these hematopoietic transcription factors represents a mechanism involved in the differentiation block in AML1-ETO leukemia. Furthermore, AML1-ETO may cause a genome-wide change in chromatin structure. Studies have shown that expression of AML1-ETO in U937 cells or removal of AML1-ETO in Kasumi-1 cells leads to a global redistribution of AML1 and HEB binding even in the regions without AML1-ETO (Gardini et al. 2008, Ptasinska et al. 2012). Thus, AML1-ETO alters the gene expression both locally and remotely through AML1-ETO multiple mechanisms, leading to a systematic reprogramming of transcription networks.

11.5 Dysregulated Genes in AML1-ETO Leukemia

Several AML1-ETO target genes that play important roles in leukemogenesis have been identified. Those include (1) myeloid transcription factors, (2) tumor suppressors, and (3) anti-apoptosis genes. AML1-ETO also regulates miRNAs, DNA repair genes, and genes in signal transduction pathways, which are summarized below.

11.5.1 Myeloid Transcription Factors

As described above, AML1-ETO binds to and inhibits functions of several myeloid transcription factors including native AML1, CEBPA, and PU.1, which leads to global suppression of myeloid gene expression. In particular, several independent studies have shown that CEBPA is a direct downstream target of AML1-ETO (Pabst et al. 2001; Ptasinska et al. 2014). Blasts in t(8;21) AML showed relatively low CEBPA expression compared with other subgroups of AML (Pabst et al. 2001). Experimentally, expression of AML1-ETO in U937 cells induced downregulation of CEBPA mRNA, protein and DNA binding activity through inhibition of positive autoregulation in the CEBPA promoter (Pabst et al. 2001). Conversely, depletion of AML1-ETO resulted in CEBPA upregulation and establishes a differentiation-associated transcriptional network dominated by de novo binding of CEBPA (Ptasinska et al. 2014). Thus, CEBPA is a critical target gene repressed by AML1-ETO, whose downregulation contributes to the AML1-ETO-mediated block of myeloid differentiation.

11.5.2 Tumor Suppressors

Dysregulation of tumor suppressors has been observed in AML1-ETO leukemia. AML1-ETO transcriptionally repressed $p14^{ARF}$ (CDKN2A) and NF1 expression through dominant inhibitory effects on AML1 function (Linggi et al. 2002; Yang et al. 2005). RUNX3, a well-known tumor suppressor in solid tumors, was also repressed by AML1-ETO (Cheng et al. 2008). In contrast to these tumor suppressors that are downregulated by AML1-ETO, $p21^{WAF1}$ (CDKN1A) expression is robustly upregulated in AML1-ETO leukemia at RNA, protein, and promoter levels (Peterson et al. 2007, Berg et al. 2008). The role of $p21^{waf1}$ in AML1-ETO-induced leukemogenesis is under debate; a study reported that the $p21^{waf1}$ pathway is involved in blocking leukemogenesis by AML1-ETO (Peterson et al. 2007), while another argued that the activated $p21^{waf1}$ is critical in preventing exhaustion of leukemic stem cells in AML (Viale et al. 2009). Different cell types and approaches were used in the two studies, which may have contributed to the discrepant findings.

11.5.3 Anti-apoptosis Genes

It was shown that BCL2 transcription was upregulated by AML1-ETO (Klampfer et al. 1996). However, t(8;21) patient samples do not uniformly show increased BCL2 expression, and this upregulation may depend on the status of p53(Banker et al. 1998). Another anti-apoptotic protein Bcl-xL was also upregulated in AML1-ETO-expressing leukemia cells and plays an essential role in their survival and self-renewal (Chou et al. 2012). However, Bcl-xL is not likely a direct transcriptional target of AML1-ETO. Instead, Bcl-xL expression was induced by enhanced thrombopoietin (TPO)/MPL signaling in AML1-ETO cells (Chou et al. 2012).

11.5.4 Other Target Genes

CD48 was recently shown as a transcriptional repression target of AML1-ETO. The AML1-ETO-mediated downregulation of CD48 is HDAC-dependent, and treatment with HDAC inhibitors restores the expression of CD48. CD48 downregulation appears to contribute to immune evasion of AML1-ETO cells by decreasing NK cell-mediated killing (Elias et al. 2014). AAA+ family member *RUVBL1* was also identified as a transcription activation target of AML1-ETO in a *Drosophila* model (Breig et al. 2014).

11.6 Dysregulated MicroRNAs in AML1-ETO Leukemia

Recent studies have demonstrated that t(8;21) AML patients showed a distinct miRNA expression profile (Li et al. 2008; Jongen-Lavrencic et al. 2008). Similar to target genes of AML1-ETO, some miRNAs are upregulated, while others are downregulated in AML1-ETO leukemia. Those miRNAs have been shown to have functional relevance in leukemogenesis, with some miRNAs acting as oncogenes and others as tumor suppressors.

11.6.1 Downregulated miRNAs

AML1-ETO has been shown to repress several miRNAs mainly by recruiting corepressors. miR-9 is downregulated in AML1-ETO leukemia compared with other subtypes of AML, and was shown to act as a tumor suppressor in AML1-ETO leukemia. Forced expression of miR-9 reduced leukemic growth and induced monocytic differentiation of t(8;21) AML cell lines (Emmrich et al. 2014). Studies have identified LIN28B/let-7/HMGA2 axis and a protein phosphatase UBSHASH3B as targets of miR-9 in AML1-ETO leukemia (Emmrich et al. 2014, Goyama et al. 2015b). miR-223, which is involved in myelopoiesis, was also shown to be a transcriptional target of AML1-ETO. The repressive effect of AML1-ETO on miR-223 gene included both histone deacetylation and DNA methylation, and demethylating treatment with 5-azacytidine enhanced miR-223 expression (Fazi et al. 2007). In addition, AML1-ETO repressed miR-193a expression by binding at AML1-binding sites and recruiting chromatin-remodeling enzymes, such as DNMTs and HDACs. Conversely, miR-193a repressed expression of multiple targets in the AML1-ETO-containing complex, such as AML1-ETO itself, DNMT3a and HDAC3. The negative feedback circuitry involving AML1-ETO and miR-193a appears to be important for leukemogenesis (Li et al. 2013). Thus, these downregulated miRNAs act as tumor suppressors in AML1-ETO leukemia.

11.6.2 Upregulated miRNAs

Several miRNAs are upregulated in AML1-ETO leukemia. AML1-ETO induced upregulation of miR-24, and miR-24 downregulates mitogen-activated protein kinase (MAPK) phosphatase-7 (a negative regulator of MAPK signaling) to stimulate myeloid proliferation of AML1-ETO leukemia (Zaidi et al. 2009). miR-126 is a critical regulator of both normal hematopoietic stem cells (HSCs)

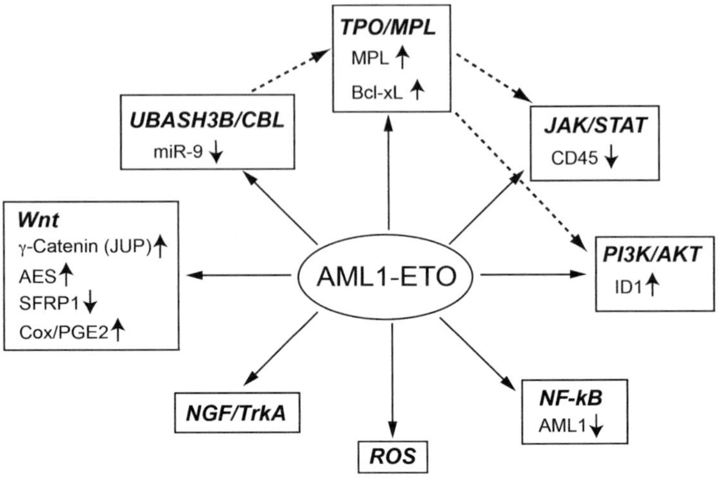

Fig. 11.3 Aberrantly activated signaling pathways in AML1-ETO leukemia. AML1-ETO directly modulates multiple signaling pathways. Dysregulated genes in AML1-ETO leukemia involved in the corresponding pathways are also shown

and leukemia stem cells (LSCs) (Lechman et al. 2012; de Leeuw et al. 2014, Lechman et al. 2016; Dorrance et al. 2015), and is highly expressed in t(8;21) AML (Li et al. 2008). Interestingly, a recent study showed that both knockout and overexpression of miR-126 promote leukemogenesis induced by AML1-ETO (Li et al. 2015). In agreement with other studies showing the critical role of miR-126 in LSCs, miR-126 overexpression enhanced LSC activity of AML1-ETO9a-expressing cells, partly by repressing ERRFI1 and SPRED1 (Li et al. 2015). Findings from other studies indicate that miR-126 negatively regulates PI3K/AKT pathway in HSCs and LSCs (Lechman et al. 2012, Lechman et al. 2016). Knockout of miR-126 also accelerates leukemogenesis, but miR-126-deficient leukemia cells became more sensitive to standard chemotherapy (Li et al. 2015). These data suggest that miR-126 plays a pivotal role in the regulation of LSCs and therapeutic resistance in AML1-ETO leukemia.

11.7 Dysregulated Signaling Pathways in AML1-ETO Leukemia

AML1-ETO has been shown to modulate the status of various signaling pathways mainly through transcriptional regulation of genes involved in specific signaling pathways. In this section, we summarize the pathways that are aberrantly activated in AML1-ETO leukemia (Fig. 11.3).

11.7.1 TPO/MPL Pathway

MPL is highly expressed in t(8;21) AML, and two independent studies identified TPO/MPL signaling as a key pathway to increase survival and leukemogenesis induced by AML1-ETO (Pulikkan et al. 2012; Chou et al. 2012). Enhanced TPO/MPL signaling led to upregulation of anti-apoptotic protein Bcl-xL (Chou et al. 2012) and activated PI3K/AKT and JAK/STAT pathways (Pulikkan et al. 2012). In addition, loss of CBL function was shown to enhance the TPO-mediated proliferation of AML1-ETO cells (Goyama et al. 2015b), which further indicated the important role of TPO/MPL pathway in AML1-ETO leukemia.

11.7.2 JAK/STAT Pathway

JAK/STAT signaling is a known downstream pathway of TPO/MPL signaling, and is therefore likely to be activated in AML1-ETO leukemia. In addition, a study showed that downregulation of CD45 in AML1-ETO leukemia also contributes to the enhanced JAK/STAT signaling (Lo et al. 2012). CD45 is a protein tyrosine phosphatase and negatively regulates the JAK/STAT pathway. Importantly, several reports have shown the substantial effect of JAK inhibitors to suppress AML1-ETO-induced leukemogenesis (Lo et al. 2013; Goyama et al. 2015b) (see the section of "Therapeutic strategies for AML1-ETO leukemia").

11.7.3 Wnt Pathway

Two studies have shown that AML1-ETO induced upregulation of γ-Catenin (JUP) to activate Wnt signaling (Muller-Tidow et al. 2004; Zheng et al. 2004). In addition, Groucho-related amino-terminal enhancer of split (AES), which acts as an enhancer of Wnt signaling, was upregulated by AML1-ETO (Steffen et al. 2011). Another study identified *SFRP1*, an antagonist of the Wnt signaling, as a transcriptional repression target of AML1-ETO (Cheng et al. 2011). Furthermore, AML1-ETO was shown to induce upregulation of the *Cox-2* gene, which in turn activated Wnt/β-catenin signaling (Zhang et al. 2013; Yeh et al. 2009). Cox inhibitors, deletion of β-catenin or γ-catenin, and overexpression of a dominant negative TCF (coactivator of β-catenin) resulted in suppression of the clonogenicity and leukemogenecity of AML1-ETO cells. Together, these findings indicate a critical role of Wnt pathway to increase/maintain LSCs in AML1-ETO leukemia.

11.7.4 PI3K/AKT Pathway

As described above, the enhanced TPO/MPL signaling in AML1-ETO cells was shown to activate PI3K/AKT pathway, and the activated PI3K/AKT mediates the MPL-directed antiapoptotic function in AML1-ETO cells (Pulikkan et al. 2012). In addition, a recent report showed that acetylated AML1-ETO upregulated ID1, which interacts with AKT1 to promote its phosphorylation. Inhibition of ID1 function induced apoptosis and prevented AML1-ETO-induced leukemogenesis (Wang et al. 2015). These data suggest that the PI3K/AKT pathway promotes the survival and proliferation of AML1-ETO cells. However, miR-126, which is highly expressed in AML1-ETO leukemia, was shown to target multiple components of the PI3K/AKT pathway to preserve quiescence, increase self-renewal, and promote chemotherapy resistance in normal and malignant stem cells (Lechman et al. 2016; Lechman et al. 2012). It is therefore possible that PI3K/AKT signaling is attenuated in LSCs of AML1-ETO leukemia.

11.7.5 Other Pathways

AML1-ETO was shown to induce TrkA (NTRK1) expression, a receptor for nerve growth factor (NGF). The upregulated TrkA allowed NGF-induced expansion of AML1-ETO-expressing human hematopoietic cells (Mulloy et al. 2005). A study using Drosophila as a model showed that AML1-ETO-expressing precursor cells expressed high levels of reactive oxygen species (ROS), and that ROS was a signaling factor promoting maintenance of the leukemic precursors (Sinenko et al. 2010). NF-κB signaling was inhibited by native AML1 through interaction with IκB kinase complex, and AML1-ETO lost this ability. Consequently, NF-κB signaling was activated in AML1-ETO cells (Nakagawa et al. 2011). A protein phosphatase UBASH3B, which is a known negative regulator of CBL, was upregulated in AML1-ETO cells through transcriptional and miR-9-mediated regulation (Goyama et al. 2015b). UBASH3B depletion impairs proliferation of AML1-ETO cells, and the growth inhibition caused by UBASH3B depletion can be rescued by ectopic expression of CBL mutants. Thus, UBASH3B/CBL pathway supports the growth of AML1-ETO cells, partly by activating TPO/MPL signaling (Goyama et al. 2015b).

11.8 AML1-ETO and DNA Damage

It has been considered that early/initiating oncogenic mutations, such as AML1-ETO, may promote mutagenesis in tumor cells. Increasing evidence suggests that AML1-ETO indeed has such function. Expression of AML1-ETO in U937 hematopoietic precursor cells as well as in primary human cord blood cells led to downregulation of genes involved in multiple DNA repair pathways, including *OGG1*, *POLE, BRCA2, and ATM* (Krejci et al. 2008; Alcalay et al. 2003). Consequently, expression of AML1-ETO in cord blood cells increased the frequency of γH2A.X foci (a marker of DNA double-strand breaks), activated the p53 pathway, and increased the mutation frequency *in vivo* (Krejci et al. 2008). These observations were confirmed by a recent study using a Pig-a assay that evaluates mutation frequency of *PIGA* gene (Forster et al. 2016). Somatic mutations of *PIGA* are growth neutral and can be determined using flow cytometry analysis measuring CD55 and CD59 expression. This assay demonstrated that the mutation frequency of *PIGA* in AML1-ETO-expressing cells was significantly higher than that in control cells. The study also showed that *OGG1* is a repressive target of AML-ETO, indicating an important role of OGG1 that is involved in base excision repair in AML1-ETO-mediated mutagenesis (Forster et al. 2016). Another study showed that AML1-ETO-transduced mouse bone marrow cells and Kasumi-1 cells displayed higher levels of DNA damage in part owing to their suppressed expression of key homologous recombination (HR)-associated genes including *RAD51*, *ATM*, *BRCA1* and *BRCA2* (Esposito et al. 2015). Thus, AML1-ETO suppresses endogenous DNA repair in cells, which probably facilitates acquisition of cooperating secondary events. This property of AML1-ETO leukemia may provide therapeutic opportunities, as described below (see the section of "Therapeutic strategies for AML1-ETO leukemia").

11.9 Collaborative Genetic Alterations in AML1-ETO Leukemia

It has been shown that AML1-ETO alone is not sufficient for leukemogenic transformation. Transgenic or conditional expression of the fusion protein was not able to induce AML in mice (Yuan et al. 2001; Higuchi et al. 2002). Retroviral expression of AML1-ETO in human CD34+ cells did not cause leukemia in immunodeficient mice (Mulloy et al. 2002, 2003). Thus, it is now widely accepted that AML1-ETO needs additional genetic alterations to develop full-blown leukemia. The collaborative genetic alterations can be classified into three categories: (1) chromosomal aberrations, (2) signaling pathway genes, (3) epigenetic genes (Table 11.1). It is an

Table 11.1 Cooperating genetic events in AML1-ETO leukemia

Gene name	Frequency (%)	Impact on prognosis	References of experimental confirmation for the cooperativity
Chromosomal aberrations			
LOS (X or Y)	50	Good	111
Del(9q)	15		112
Trisomy 8	5	Poor	
Mutations in signaling pathways			
KIT	20	Poor (D816)	103,104,107
NRAS	10		106
KRAS	5		105
FLT3-ITD/TKD	10		102
CBL	5		71
JAK2	3		
Mutations in epigenetic genes			
ASXL1	10	Poor	
ASXL2	20	Poor	
IDH1/IDH2	5		

LOS Loss of sex chromosome

open question how many mutations are required for the development of AML1-ETO leukemia. The classic '2-hit' model of AML-ETO leukemia was proposed based on experimental data in mouse transplantation assays (Gilliland et al. 2004). Expression of KIT, FLT3-ITD, NRAS or KRAS mutant together with AML1-ETO in mouse hematopoietic progenitors was sufficient to produce *in vivo* leukemia (Schessl et al. 2005; Wang et al. 2011b; Nick et al. 2012; Zhao et al. 2014). However, studies using primary human cord blood cells have suggested that human cells are more resistant to oncogene-induced transformation, and more than '2-hits' may be required to generate human AML-ETO leukemia. Expression of KIT, NRAS or CBL mutant with AML1-ETO in human cord blood cells was shown to increase the growth of AML1-ETO cells *in vitro*, but none of these combinations induce overt leukemia in xenograft models (Chou et al. 2011; Goyama et al. 2015b; Wichmann et al. 2015). Developing human AML1-ETO leukemia models *in vivo* with defined sets of mutations is an important future challenge, which may require the improvement of host environment in recipient mice with immunodeficiency (Goyama et al. 2015c).

11.9.1 Chromosomal Aberrations

Chromosomal aberrations are frequently detected (70 %) in AML1-ETO leukemia (Mrozek et al. 2008; Krauth et al. 2014). Most frequent aberrations are loss of sex chromosomes (50 %), followed by 9q deletion [del(9q); 15 %] and trisomy 8 (+8; 5 %) (Krauth et al. 2014). Loss of sex chromosome (− X or − Y) is usually present as the dominant clone and persists at relapse, suggesting an important role of this aberration in AML-ETO leukemia. Because loss of sex chromosome in t(8;21) patients involves both X and Y chromosomes at similar frequencies, candidate genes affecting AML1-ETO leukemia are likely to be located on the common regions of sex chromosomes known as the pseudoautosomal regions (PARs, PAR1, 2, and 3). Indeed, a recent report identified a GM-CSF receptor α subunit (*CSF2RA*), which is located in PAR1, as a tumor suppressor gene in a murine transplantation model of AML1-ETO-induced leukemia (Matsuura et al. 2012). For del(9q), TLE1 and TLE4 were suggested as candidate tumor suppressors residing in the genomic region lost in t(8;21) AML patients (Dayyani et al. 2008).

11.9.2 Mutations in Signaling Pathway Genes

Activating mutations in signaling pathways are the best studied collaborating mutations in AML-ETO leukemia. *KIT* mutations are quite prevalent (around 20 %) in t(8;21) patients (Krauth et al. 2014; Boissel et al. 2006; Shih et al. 2008), and cooperativity between AML1-ETO and activating KIT mutants in the induction of AML has been demonstrated using mouse transplantation models and a human cord blood cell culture assay (Wichmann et al. 2015; Nick et al. 2012; Wang et al. 2011b). In addition to KIT, mutations in signaling pathway genes (NRAS, KRAS, FLT3-ITD, FLT3-TKD, CBL and JAK2) were found in 30 % of t(8;21) patients. Experimental models have demonstrated the significance of mutations in *FLT3-ITD*, *NRAS*, *KRAS*, and *CBL* in AML1-ETO-induced leukemia (Goyama et al. 2015b; Nick et al. 2012; Wang et al. 2011b; Wichmann et al. 2015; Zhao et al. 2014; Chou et al. 2011; Schessl et al. 2005). Furthermore, a recent study identified *PTPN11* (a protein tyrosine phosphatase) as a collaborative mutation in AML1-ETO leukemia (Hatlen et al. 2016).

11.9.3 Mutations in Epigenetic Genes

Recent sequencing studies have revealed the frequent mutation of epigenetic genes, *ASXL1* (10 %) and *ASXL2* (20 %) in AML1-ETO leukemia (Micol et al. 2014; Krauth et al. 2014). *ASXL1* and *ASXL2* mutations are mutually exclusive with one another, suggesting the shared mechanisms of leukemic transformation of AML1-ETO cells induced by these mutations. The functional basis for *ASXL1* and *ASXL2* mutations in AML1-

ETO-induced leukemogenesis needs to be investigated in future studies. Mutations in *IDH1* and *IDH2* genes have been found in approximately 5 % of patients with t(8;21) AML (Krauth et al. 2014). Loss of *Tet2*, another epigenetic gene that is frequently mutated in myeloid neoplasms, was shown to promote leukemogenesis in concert with AML1-ETO in a mouse transplantation model (Hatlen et al. 2016).

11.9.4 Impact of the Collaborating Mutations on Prognosis

Survival analyses have revealed that KIT-D816 mutations (present in exon 17) had adverse prognostic impact in AML1-ETO leukemia, while the impact of other KIT mutations (mutations in exon 8 or 11) was less significant (Qin et al. 2014; Tokumasu et al. 2015; Krauth et al. 2014). D816 is in the activation loop domain of KIT, and gain-of-function mutations in this region are known to cause SCF-independent activation of KIT (Lennartsson and Ronnstrand 2012). Consistent with this clinical observation, it was experimentally shown that co-expression of AML1-ETO with a mutated KIT activation loop domain induces a more aggressive AML phenotype than co-expression with a mutated KIT extracellular domain (Nick et al. 2012). A report suggested that *FLT3* mutations were also associated with a shorter event free survival (EFS) (Boissel et al. 2006), but this observation should be confirmed in a large cohort of t(8;21) patients. In addition, mutations in ASXL1 and ASXL2 showed a significant negative impact on EFS, although these mutations did not affect overall survival (Krauth et al. 2014; Micol et al. 2014). For the prognostic impact of chromosome aberrations, t(8;21) AML patients with loss of sex chromosomes showed a better prognosis, whereas those with trisomy 8 had shorter EFS (Krauth et al. 2014). Recently, a study analyzed mutational landscape of matched diagnosis and relapse DNA samples from patients with AML1-ETO leukemia. Although the relapse-specific mutations are mostly in the genes that have not been previously linked to leukemia, the data indicated that *GATA2* haploinsufficiency may play a role for AML1-ETO leukemia relapse (Sood et al. 2016).

11.10 Therapeutic Strategies for AML1-ETO Leukemia

Almost all patients with AML1-ETO leukemia achieve a complete remission (CR) after the anthracycline- and cytarabine-based induction chemotherapy. Furthermore, the incorporation of high-dose cytarabine for postremission therapy has substantially improved the outcome, resulting in favorable long-term outcome in many t(8;21) patients (Schlenk et al. 2004; Kayser et al. 2015; Paschka and Dohner 2013). However, a substantial proportion of patients, especially older patients who are unable to receive intensive chemotherapy, cannot be cured by the current treatment. Consequently, approximately half of the patients relapse, with a median time to relapse of 2.5 years after achieving CR (Marcucci et al. 2005; Hospital et al. 2014). Risk stratification based on genetic information may help to improve the outcome. As mentioned above, KIT-D816 and ASXL1 mutations have adverse prognostic impact in AML1-ETO leukemia patients (Krauth et al. 2014). Therefore, the patients with these mutations may benefit from more intensive therapy or allogeneic hematopoietic stem cell transplantation. In addition, monitoring AML1-ETO transcript levels during the course of therapy appears to be useful to predict outcome (Kayser et al. 2015). Additional therapies directed at minimum residual disease (MRD) could reduce relapse rate in patients with AML-ETO leukemia. Perhaps the most promising approach to improve the patient's outcome will be incorporating new molecular drugs into the standard of care. Below, we summarize novel therapeutic approaches for AML1-ETO leukemia that show promise in clinical or experimental investigations.

11.10.1 Gemtuzumab Ozogamicin

Gemtuzumab Ozogamicin (GO, marketed as Mylotarg) links an antibody directed against the CD33 antigen with calicheamicin, a DNA-damaging toxin. This drug-linked monoclonal antibody provides an efficient drug delivery into the cell expressing high level of CD33, including AML cells. GO was used as a drug for AML, but was withdrawn from the market in 2010 because of the increased patient death and limited benefit over standard therapies. However, subset analyses revealed a significant survival benefit in favorable-risk AML, and subsequent clinical studies have confirmed the efficacy of GO in AML1-ETO leukemia (Laszlo et al. 2014) (Fig. 11.4).

11.10.2 Signal Transduction Inhibitors

Signaling pathways that are activated in AML1-ETO leukemia can be therapeutic targets. The high expression level of KIT and the high incidence of activating KIT mutations in AML1-ETO leukemia make KIT signaling an excellent target, and the efficacy of a KIT inhibitor Dasatinib is currently being tested in clinical trials. One concern is that suppressing the pathway activated by a cooperating mutation may induce clonal evolution of subclones with other types of mutation. In theory, pathways that are directly activated by AML1-ETO, not by the collaborating mutations, should be targeted. From this viewpoint, the JAK/STAT pathway may be an interesting target to treat AML1-ETO leukemia. As described above, the JAK/STAT pathway is activated in AML1-ETO leukemia, either through enhanced TPO/MPL signaling or decreased expression of the tyrosine phosphatase CD45. Furthermore, several studies have already shown the substantial effect of JAK inhibitors (INCB018424 and TG101209) on AML1-ETO cells. Both drugs inhibited proliferation and promoted apoptosis of AML1-ETO-expressing mouse bone marrow progenitors and human cord blood cells (Goyama et al. 2015b; Lo et al. 2013). Furthermore, TG101209 treatment in AML1-ETO9a-induced leukemia mice significantly prolonged survival (Lo et al. 2013). These data demonstrate the potential therapeutic efficacy of JAK inhibitors in treating t(8;21) AML.

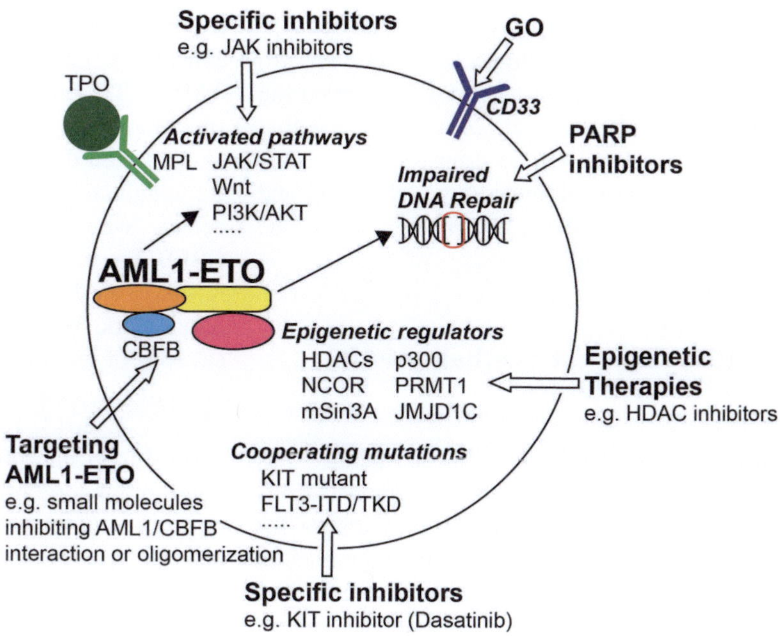

Fig. 11.4 Ongoing and future therapeutic approach for AML1-ETO leukemia. Molecular pathology of AML1-ETO leukemia provides potential therapeutic targets. Those include epigenetic regulators, aberrantly activated signaling pathways, impaired DNA repair machinery, and cooperating mutations. In addition, AML1-ETO fusion itself will be a great target to develop curative therapies. *GO* Gemtuzumab Ozogamicin

11.10.3 Epigenetic Therapies

Epigenetic aberrations, unlike genetic mutations, are potentially reversible and can be restored by epigenetic therapies. Given that AML1-ETO interacts with several epigenetic modifying enzymes, such as HDACs, p300, and PRMT1, targeting these epigenetic enzymes has great promise to treat AML1-ETO leukemia. Indeed, previous studies have shown the therapeutic potential of a HDAC inhibitor valproic acid (VPA) for AML1-ETO leukemia by inducing differentiation and/or apoptosis (Gottlicher et al. 2001; Insinga et al. 2005; Yang et al. 2007; Liu et al. 2007). A recent study used Panobinostat, a potent HDAC inhibitor, and showed that Panobinostat caused a significant antileukemic response in AML1-ETO9a-driven leukemia *in vivo* by triggering terminal myeloid differentiation (Bots et al. 2014). Interestingly, HDAC inhibitors promoted proteasomal degradation of AML1-ETO fusion protein (Bots et al. 2014; Yang et al. 2007), which can partly account for the robust antileukemic effect of these inhibitors. In addition to HDAC inhibitors, p300 inhibitors (Lys-CoA-Tat or C646) and Prmt1 knockdown delayed the development of AML1-ETO9a-driven leukemia in mouse transplantation models (Wang et al. 2011a; Shia et al. 2012). A recent study also showed that inhibitors of lysine specific demethylase 1 (KDM1A) effectively suppressed in vitro and in vivo growth of AML1-ETO-harbouring cell lines (McGrath et al. 2016). The potential effect of other epigenetic drugs, such as DNA methyltransferase inhibitors (e.g., azacitidine and decitabine) on AML1-ETO leukemia should be tested in future studies.

11.10.4 Synthetic Lethality: PARP Inhibitors

Synthetic lethality-based cancer therapy has attracted much attention since the demonstrations of substantial efficacy of PARP inhibitors for breast and ovarian cancers carrying mutations in *BRCA1* and *BRCA2* (Farmer et al. 2005). The BRCA genes are frequently mutated in cancers and are important for repairing double-strand breaks in DNA. Inhibition of another DNA repair enzyme called PARP is selectively lethal to the BRCA-mutated cancer cells, since the cells cannot repair DNA damage with simultaneous inhibition of BRCA and PARP pathways. This strategy can be applied for treating AML1-ETO leukemia cells in which the endogenous DNA repair pathway is substantially impaired (Esposito et al. 2015; Forster et al. 2016; Krejci et al. 2008; Alcalay et al. 2003). Indeed, a recent report clearly showed that AML1-ETO-transformed mouse bone marrow cells are extremely sensitive to PARP inhibition (Esposito et al. 2015), indicating the potential utility of PARP inhibitor-induced synthetic lethality for AML1-ETO leukemia.

11.10.5 Targeting AML1-ETO Itself

Recent studies have shown the dynamic changes of mutation patterns in leukemia cells between diagnosis and relapse. In AML1-ETO leukemia, AML1-ETO fusion protein shows 100 % stability during the course of disease, indicating the essential role of AML1-ETO to sustain the leukemogenecity (Krauth et al. 2014; Sood et al. 2016). Therefore, targeting AML1-ETO fusion protein itself has great promise to develop curative therapies. RUNT/CBFB interaction has been considered important for the development of AML1-ETO leukemia (Roudaia et al. 2009), and small molecules that inhibit this interaction have already been developed (Gorczynski et al. 2007; Cunningham et al. 2012). However, there is a debate whether AML1-ETO actually needs CBFB for leukemogenesis (Park et al. 2009; Kwok et al. 2010), which should be clarified in future work. Several studies have shown the critical role of the oligomerization of AML1-ETO through the NHR2 domain for leukemogenesis (Sun et al. 2013; Kwok et al. 2009; Yan et al. 2009; Liu et al. 2006). Therefore, blocking the homo-oligomeric properties of AML1-ETO will be an attractive therapeutic approach. Alternatively, agents that induce selective degradation of AML1-ETO protein can be developed as a specific drug for t(8;21) AML.

11.11 Closing Remarks

It was a common belief that AML1-ETO induces leukemogenesis by suppressing the functions of native AML1. However, recent evidence suggests that AML1-ETO regulates gene expression in concert with AML1, instead of simply suppressing its function. AML1-ETO forms a complex with several transcription factors including E proteins, and this AML1-ETO-containing transcription factor complex dynamically interacts with epigenetic factors to repress/activate target genes. Because AML1-ETO needs additional genetic alterations to induce leukemogenesis, targeting the collaborating mutations will be an efficient approach if they are druggable proteins. However, such an approach may promote the clonal evolution of AML1-ETO cells that lack the targeted mutation. Therefore, future studies should investigate ways to target AML1-ETO-containing transcription factor complex or pathways modulated directly by AML1-ETO. As summarized in this chapter, remarkable progress has been achieved in understanding the molecular pathogenesis of AML1-ETO leukemia. Translation of these findings into the clinical setting is just beginning.

References

Alcalay, M., Meani, N., Gelmetti, V., Fantozzi, A., Fagioli, M., Orleth, A., et al. (2003). Acute myeloid leukemia fusion proteins deregulate genes involved in stem cell maintenance and DNA repair. *The Journal of Clinical Investigation, 112*, 1751–1761.

Amann, J. M., Nip, J., Strom, D. K., Lutterbach, B., Harada, H., Lenny, N., et al. (2001). ETO, a target of t(8;21) in acute leukemia, makes distinct contacts with multiple histone deacetylases and binds mSin3A through its oligomerization domain. *Molecular and Cellular Biology, 21*, 6470–6483.

Banker, D. E., Radich, J., Becker, A., Kerkof, K., Norwood, T., Willman, C., & Appelbaum, F. R. (1998). The t(8;21) translocation is not consistently associated with high Bcl-2 expression in de novo acute myeloid leukemias of adults. *Clinical Cancer Research, 4*, 3051–3062.

Barseguian, K., Lutterbach, B., Hiebert, S. W., Nickerson, J., Lian, J. B., Stein, J. L., et al. (2002). Multiple subnuclear targeting signals of the leukemia-related AML1/ETO and ETO repressor proteins. *Proceedings of the National Academy of Sciences of the United States of America, 99*, 15434–15439.

Ben-Ami, O., Friedman, D., Leshkowitz, D., Goldenberg, D., Orlovsky, K., Pencovich, N., et al. (2013). Addiction of t(8;21) and inv(16) acute myeloid leukemia to native RUNX1. *Cell Reports, 4*, 1131–1143.

Berg, T., Fliegauf, M., Burger, J., Staege, M. S., Liu, S., Martinez, N., et al. (2008). Transcriptional upregulation of p21/WAF/Cip1 in myeloid leukemic blasts expressing AML1-ETO. *Haematologica, 93*, 1728–1733.

Bitter, M. A., le Beau, M. M., Rowley, J. D., Larson, R. A., Golomb, H. M., & Vardiman, J. W. (1987). Associations between morphology, karyotype, and clinical features in myeloid leukemias. *Human Pathology, 18*, 211–225.

Boissel, N., Leroy, H., Brethon, B., Philippe, N., de Botton, S., Auvrignon, A., et al. (2006). Incidence and prognostic impact of c-Kit, FLT3, and Ras gene mutations in core binding factor acute myeloid leukemia (CBF-AML). *Leukemia, 20*, 965–970.

Bots, M., Verbrugge, I., Martin, B. P., Salmon, J. M., Ghisi, M., Baker, A., et al. (2014). Differentiation therapy for the treatment of t(8;21) acute myeloid leukemia using histone deacetylase inhibitors. *Blood, 123*, 1341–1352.

Breig, O., Bras, S., Martinez Soria, N., Osman, D., Heidenreich, O., Haenlin, M., & Waltzer, L. (2014). Pontin is a critical regulator for AML1-ETO-induced leukemia. *Leukemia, 28*, 1271–1279.

Chen, M., Zhu, N., Liu, X., Laurent, B., Tang, Z., Eng, R., et al. (2015). JMJD1C is required for the survival of acute myeloid leukemia by functioning as a coactivator for key transcription factors. *Genes & Development, 29*, 2123–2139.

Cheng, C. K., Li, L., Cheng, S. H., Lau, K. M., Chan, N. P., Wong, R. S., et al. (2008). Transcriptional repression of the RUNX3/AML2 gene by the t(8;21) and inv(16) fusion proteins in acute myeloid leukemia. *Blood, 112*, 3391–3402.

Cheng, C. K., Li, L., Cheng, S. H., Ng, K., Chan, N. P., Ip, R. K., et al. (2011). Secreted-frizzled related protein 1 is a transcriptional repression target of the t(8;21) fusion protein in acute myeloid leukemia. *Blood, 118*, 6638–6648.

Choi, Y., Elagib, K. E., Delehanty, L. L., & Goldfarb, A. N. (2006). Erythroid inhibition by the leukemic fusion AML1-ETO is associated with impaired acetylation of the major erythroid transcription factor GATA-1. *Cancer Research, 66*, 2990–2996.

Chou, F. S., Wunderlich, M., Griesinger, A., & Mulloy, J. C. (2011). NRASG12D induces stepwise transformation in preleukemic human umbilical cord blood cultures expressing the AML1-ETO fusion gene. *Blood, 117*, 2237–2240.

Chou, F. S., Griesinger, A., Wunderlich, M., Lin, S., Link, K. A., Shrestha, M., et al. (2012). The thrombopoietin/MPL/Bcl-xL pathway is essential for survival and self-renewal in human preleukemia induced by AML1-ETO. *Blood, 120*, 709–719.

Cunningham, L., Finckbeiner, S., Hyde, R. K., Southall, N., Marugan, J., Yedavalli, V. R., et al. (2012). Identification of benzodiazepine Ro5-3335 as an inhibitor of CBF leukemia through quantitative high throughput screen against RUNX1-CBFbeta interaction. *Proceedings of the National Academy of Sciences of the United States of America, 109*, 14592–14597.

Davis, J. N., Mcghee, L., & Meyers, S. (2003). The ETO (MTG8) gene family. *Gene, 303*, 1–10.

Dayyani, F., Wang, J., Yeh, J. R., Ahn, E. Y., Tobey, E., Zhang, D. E., et al. (2008). Loss of TLE1 and TLE4 from the del(9q) commonly deleted region in AML cooperates with AML1-ETO to affect myeloid cell proliferation and survival. *Blood, 111*, 4338–4347.

de Leeuw, D. C., Denkers, F., Olthof, M. C., Rutten, A. P., Pouwels, W., Schuurhuis, G. J., et al. (2014). Attenuation of microRNA-126 expression that drives CD34+38- stem/progenitor cells in acute myeloid leukemia leads to tumor eradication. *Cancer Research, 74*, 2094–2105.

Dorrance, A. M., Neviani, P., Ferenchak, G. J., Huang, X., Nicolet, D., Maharry, K. S., et al. (2015). Targeting leukemia stem cells in vivo with antagomiR-126 nanoparticles in acute myeloid leukemia. *Leukemia, 29*, 2143–2153.

Elias, S., Yamin, R., Golomb, L., Tsukerman, P., Stanietsky-Kaynan, N., Ben-Yehuda, D., & Mandelboim, O. (2014). Immune evasion by oncogenic proteins of acute myeloid leukemia. *Blood, 123*, 1535–1543.

Elsasser, A., Franzen, M., Kohlmann, A., Weisser, M., Schnittger, S., Schoch, C., et al. (2003). The fusion protein AML1-ETO in acute myeloid leukemia with translocation t(8;21) induces c-jun protein expression via the proximal AP-1 site of the c-jun promoter in an indirect, JNK-dependent manner. *Oncogene, 22*, 5646–5657.

Emmrich, S., Katsman-Kuipers, J. E., Henke, K., Khatib, M. E., Jammal, R., Engeland, F., et al. (2014). miR-9 is a tumor suppressor in pediatric AML with t(8;21). *Leukemia, 28*, 1022–1032.

Erickson, P., Gao, J., Chang, K. S., Look, T., Whisenant, E., Raimondi, S., et al. (1992). Identification of breakpoints in t(8;21) acute myelogenous leukemia and isolation of a fusion transcript, AML1/ETO, with similarity to Drosophila segmentation gene, runt. *Blood, 80*, 1825–1831.

Erickson, P. F., Dessev, G., Lasher, R. S., Philips, G., Robinson, M., & Drabkin, H. A. (1996). ETO and AML1 phosphoproteins are expressed in CD34+ hematopoietic progenitors: Implications for t(8;21) leukemogenesis and monitoring residual disease. *Blood, 88*, 1813–1823.

Esposito, M. T., Zhao, L., Fung, T. K., Rane, J. K., Wilson, A., Martin, N., et al. (2015). Synthetic lethal targeting of oncogenic transcription factors in acute leukemia by PARP inhibitors. *Nature Medicine, 21*, 1481–1490.

Farmer, H., Mccabe, N., Lord, C. J., Tutt, A. N., Johnson, D. A., Richardson, T. B., et al. (2005). Targeting the DNA repair defect in BRCA mutant cells as a therapeutic strategy. *Nature, 434*, 917–921.

Fazi, F., Racanicchi, S., Zardo, G., Starnes, L. M., Mancini, M., Travaglini, L., et al. (2007). Epigenetic silencing of the myelopoiesis regulator microRNA-223 by the AML1/ETO oncoprotein. *Cancer Cell, 12*, 457–466.

Forster, V. J., Nahari, M. H., Martinez-Soria, N., Bradburn, A. K., Ptasinska, A., Assi, S. A., et al. (2016). The leukemia-associated RUNX1/ETO oncoprotein confers a mutator phenotype. *Leukemia, 30*, 251–254.

Gao, F. H., Wang, Q., Wu, Y. L., Li, X., Zhao, K. W., & Chen, G. Q. (2007). c-Jun N-terminal kinase mediates AML1-ETO protein-induced connexin-43 expression. *Biochemical and Biophysical Research Communications, 356*, 505–511.

Gao, X. N., Yan, F., Lin, J., Gao, L., Lu, X. L., Wei, S. C., et al. (2015). AML1/ETO cooperates with HIF1alpha to promote leukemogenesis through DNMT3a transactivation. *Leukemia, 29*, 1730–1740.

Gardini, A., Cesaroni, M., Luzi, L., Okumura, A. J., Biggs, J. R., Minardi, S. P., et al. (2008). AML1/ETO oncoprotein is directed to AML1 binding regions and co-localizes with AML1 and HEB on its targets. *PLoS Genetics, 4*, e1000275.

Gelmetti, V., Zhang, J., Fanelli, M., Minucci, S., Pelicci, P. G., & Lazar, M. A. (1998). Aberrant recruitment of the nuclear receptor corepressor-histone deacetylase complex by the acute myeloid leukemia fusion partner ETO. *Molecular and Cellular Biology, 18*, 7185–7191.

Gilliland, D. G., Jordan, C. T. & Felix, C. A. (2004). The molecular basis of leukemia. *Hematology. American Society Hematology Education Program*, 80–97.

Gorczynski, M. J., Grembecka, J., Zhou, Y., Kong, Y., Roudaia, L., Douvas, M. G., et al. (2007). Allosteric inhibition of the protein-protein interaction between the leukemia-associated proteins Runx1 and CBFbeta. *Chemistry & Biology, 14*, 1186–1197.

Gottlicher, M., Minucci, S., Zhu, P., Kramer, O. H., Schimpf, A., Giavara, S., et al. (2001). Valproic acid defines a novel class of HDAC inhibitors inducing differentiation of transformed cells. *The EMBO Journal, 20*, 6969–6978.

Goyama, S., & Mulloy, J. C. (2011). Molecular pathogenesis of core binding factor leukemia: Current knowledge and future prospects. *International Journal of Hematology, 94*, 126–133.

Goyama, S., Schibler, J., Cunningham, L., Zhang, Y., Rao, Y., Nishimoto, N., et al. (2013). Transcription factor RUNX1 promotes survival of acute myeloid leukemia cells. *The Journal of Clinical Investigation, 123*, 3876–3888.

Goyama, S., Huang, G., Kurokawa, M., & Mulloy, J. C. (2015a). Posttranslational modifications of RUNX1 as potential anticancer targets. *Oncogene, 34*, 3483–3492.

Goyama, S., Schibler, J., Gasilina, A., Shrestha, M., Lin, S., Link, K. A., et al. (2015b). UBASH3B/Sts-1-CBL axis regulates myeloid proliferation in human preleukemia induced by AML1-ETO. *Leukemia, 30*(3), 728–739.

Goyama, S., Wunderlich, M., & Mulloy, J. C. (2015c). Xenograft models for normal and malignant stem cells. *Blood, 125*, 2630–2640.

Hatlen, M. A., Arora, K., Vacic, V., Grabowska, E. A., Liao, W., Riley-Gillis, B., et al. (2016). Integrative genetic analysis of mouse and human AML identifies cooperating disease alleles. *The Journal of Experimental Medicine, 213*, 25–34.

Hiebert, S. W., Lutterbach, B., & Amann, J. (2001). Role of co-repressors in transcriptional repression mediated by the t(8;21), t(16;21), t(12;21), and inv(16) fusion proteins. *Current Opinion in Hematology, 8*, 197–200.

Higuchi, M., O'brien, D., Kumaravelu, P., Lenny, N., Yeoh, E. J., & Downing, J. R. (2002). Expression of a conditional AML1-ETO oncogene bypasses embryonic lethality and establishes a murine model of human t(8;21) acute myeloid leukemia. *Cancer Cell, 1*, 63–74.

Hildebrand, D., Tiefenbach, J., Heinzel, T., Grez, M., & Maurer, A. B. (2001). Multiple regions of ETO cooperate in transcriptional repression. *The Journal of Biological Chemistry, 276*, 9889–9895.

Hoogenkamp, M., Lichtinger, M., Krysinska, H., Lancrin, C., Clarke, D., Williamson, A., et al. (2009). Early chromatin unfolding by RUNX1: A molecular explanation for differential requirements during specification versus maintenance of the hematopoietic gene expression program. *Blood, 114*, 299–309.

Hospital, M. A., Prebet, T., Bertoli, S., Thomas, X., Tavernier, E., Braun, T., et al. (2014). Core-binding factor acute myeloid leukemia in first relapse: A retrospective study from the French AML Intergroup. *Blood, 124*, 1312–1319.

Insinga, A., Monestiroli, S., Ronzoni, S., Gelmetti, V., Marchesi, F., Viale, A., et al. (2005). Inhibitors of histone deacetylases induce tumor-selective apoptosis through activation of the death receptor pathway. *Nature Medicine, 11*, 71–76.

Jongen-Lavrencic, M., Sun, S. M., Dijkstra, M. K., Valk, P. J., & Lowenberg, B. (2008). MicroRNA expression profiling in relation to the genetic heterogeneity of acute myeloid leukemia. *Blood, 111*, 5078–5085.

Kayser, S., Schlenk, R. F., Grimwade, D., Yosuico, V. E., & Walter, R. B. (2015). Minimal residual disease-directed therapy in acute myeloid leukemia. *Blood, 125*, 2331–2335.

Klampfer, L., Zhang, J., Zelenetz, A. O., Uchida, H., & Nimer, S. D. (1996). The AML1/ETO fusion protein activates transcription of BCL-2. *Proceedings of the National Academy of Sciences of the United States of America, 93*, 14059–14064.

Kozu, T., Fukuyama, T., Yamami, T., Akagi, K., & Kaneko, Y. (2005). MYND-less splice variants of AML1-MTG8 (RUNX1-CBFA2T1) are expressed in leukemia with t(8;21). *Genes, Chromosomes & Cancer, 43*, 45–53.

Kramer, O. H., Muller, S., Buchwald, M., Reichardt, S., & Heinzel, T. (2008). Mechanism for ubiquitylation of the leukemia fusion proteins AML1-ETO and PML-RARalpha. *The FASEB Journal, 22*, 1369–1379.

Krauth, M. T., Eder, C., Alpermann, T., Bacher, U., Nadarajah, N., Kern, W., et al. (2014). High number of additional genetic lesions in acute myeloid leukemia with t(8;21)/RUNX1-RUNX1T1: Frequency and impact on clinical outcome. *Leukemia, 28*, 1449–1458.

Krejci, O., Wunderlich, M., Geiger, H., Chou, F. S., Schleimer, D., Jansen, M., et al. (2008). p53 signaling in response to increased DNA damage sensitizes AML1-ETO cells to stress-induced death. *Blood, 111*, 2190–2199.

Kwok, C., Zeisig, B. B., Qiu, J., Dong, S., & So, C. W. (2009). Transforming activity of AML1-ETO is independent of CBFbeta and ETO interaction but requires formation of homo-oligomeric complexes. *Proceedings of the National Academy of Sciences of the United States of America, 106*, 2853–2858.

Kwok, C., Zeisig, B. B., Dong, S., & So, C. W. (2010). The role of CBFbeta in AML1-ETO's activity. *Blood, 115*, 3176–3177.

Laszlo, G. S., Estey, E. H., & Walter, R. B. (2014). The past and future of CD33 as therapeutic target in acute myeloid leukemia. *Blood Reviews, 28*, 143–153.

Lechman, E. R., Gentner, B., Van Galen, P., Giustacchini, A., Saini, M., Boccalatte, F. E., et al. (2012). Attenuation of miR-126 activity expands HSC in vivo without exhaustion. *Cell Stem Cell, 11*, 799–811.

Lechman, E. R., Gentner, B., Ng, S. W., Schoof, E. M., Van Galen, P., Kennedy, J. A., et al. (2016). miR-126 regulates distinct self-renewal outcomes in normal and malignant hematopoietic stem cells. *Cancer Cell, 29*, 214–228.

Lennartsson, J., & Ronnstrand, L. (2012). Stem cell factor receptor/c-Kit: From basic science to clinical implications. *Physiological Reviews, 92*, 1619–1649.

Li, Z., Lu, J., Sun, M., Mi, S., Zhang, H., Luo, R. T., et al. (2008). Distinct microRNA expression profiles in acute myeloid leukemia with common translocations. *Proceedings of the National Academy of Sciences of the United States of America, 105*, 15535–15540.

Li, Y., Gao, L., Luo, X., Wang, L., Gao, X., Wang, W., et al. (2013). Epigenetic silencing of microRNA-193a contributes to leukemogenesis in t(8;21) acute myeloid leukemia by activating the PTEN/PI3K signal pathway. *Blood, 121*, 499–509.

Li, Z., Chen, P., Su, R., Li, Y., Hu, C., Wang, Y., et al. (2015). Overexpression and knockout of miR-126 both promote leukemogenesis. *Blood, 126*, 2005–2015.

Li, Y., Wang, H., Wang, X., Jin, W., Tan, Y., Fang, H., et al. (2016). Genome-wide studies identify a novel interplay between AML1 and AML1/ETO in t(8;21) acute myeloid leukemia. *Blood, 127*, 233–242.

Linggi, B., Muller-Tidow, C., van de Locht, L., Hu, M., Nip, J., Serve, H., et al. (2002). The t(8;21) fusion protein, AML1 ETO, specifically represses the transcription of the p14(ARF) tumor suppressor in acute myeloid leukemia. *Nature Medicine, 8*, 743–750.

Liu, S., Shen, T., Huynh, L., Klisovic, M. I., Rush, L. J., Ford, J. L., et al. (2005). Interplay of RUNX1/MTG8 and DNA methyltransferase 1 in acute myeloid leukemia. *Cancer Research, 65*, 1277–1284.

Liu, Y., Cheney, M. D., Gaudet, J. J., Chruszcz, M., Lukasik, S. M., Sugiyama, D., et al. (2006). The tetramer structure of the Nervy homology two domain, NHR2, is critical for AML1/ETO's activity. *Cancer Cell, 9*, 249–260.

Liu, S., Klisoviec, R. B., Vukosavljevic, T., Yu, J., Paschka, P., Huynh, L., et al. (2007). Targeting AML1/ETO-histone deacetylase repressor complex: A novel mechanism for valproic acid-mediated gene expression and cellular differentiation in AML1/ETO-positive acute myeloid leukemia cells. *The Journal of Pharmacology and Experimental Therapeutics, 321*, 953–960.

Lo, M. C., Peterson, L. F., Yan, M., Cong, X., Jin, F., Shia, W. J., et al. (2012). Combined gene expression and DNA occupancy profiling identifies potential therapeutic targets of t(8;21) AML. *Blood, 120*, 1473–1484.

Lo, M. C., Peterson, L. F., Yan, M., Cong, X., Hickman, J. H., Dekelver, R. C., et al. (2013). JAK inhibitors suppress t(8;21) fusion protein-induced leukemia. *Leukemia, 27*, 2272–2279.

Lutterbach, B., Sun, D., Schuetz, J., & Hiebert, S. W. (1998a). The MYND motif is required for repression of basal transcription from the multidrug resistance 1 promoter by the t(8;21) fusion protein. *Molecular and Cellular Biology, 18*, 3604–3611.

Lutterbach, B., Westendorf, J. J., Linggi, B., Patten, A., Moniwa, M., Davie, J. R., et al. (1998b). ETO, a target of t(8;21) in acute leukemia, interacts with the N-CoR and mSin3 corepressors. *Molecular and Cellular Biology, 18*, 7176–7184.

Maiques-Diaz, A., Chou, F. S., Wunderlich, M., Gomez-Lopez, G., Jacinto, F. V., Rodriguez-Perales, S., et al. (2012). Chromatin modifications induced by the AML1-ETO fusion protein reversibly silence its genomic targets through AML1 and Sp1 binding motifs. *Leukemia, 26*, 1329–1337.

Marcucci, G., Mrozek, K., Ruppert, A. S., Maharry, K., Kolitz, J. E., Moore, J. O., et al. (2005). Prognostic factors and outcome of core binding factor acute myeloid leukemia patients with t(8;21) differ from those of patients with inv(16): A cancer and leukemia group B study. *Journal of Clinical Oncology, 23*, 5705–5717.

Martens, J. H., Mandoli, A., Simmer, F., Wierenga, B. J., Saeed, S., Singh, A. A., et al. (2012). ERG and FLI1 binding sites demarcate targets for aberrant epigenetic regulation by AML1-ETO in acute myeloid leukemia. *Blood, 120*, 4038–4048.

Maseki, N., Miyoshi, H., Shimizu, K., Homma, C., Ohki, M., Sakurai, M., & Kaneko, Y. (1993). The 8;21 chromosome translocation in acute myeloid leukemia is always detectable by molecular analysis using AML1. *Blood, 81*, 1573–1579.

Matsuura, S., Yan, M., Lo, M. C., Ahn, E. Y., Weng, S., Dangoor, D., et al. (2012). Negative effects of GM-CSF signaling in a murine model of t(8;21)-induced leukemia. *Blood, 119*, 3155–3163.

Mcgrath, J. P., Williamson, K. E., Balasubramanian, S., Odate, S., ARORA, S., Hatton, C., et al. (2016). Pharmacological inhibition of the histone lysine demethylase KDM1A suppresses the growth of multiple acute myeloid leukemia subtypes. *Cancer Research, 76*, 1975–1988.

Micol, J. B., Duployez, N., Boissel, N., Petit, A., Geffroy, S., Nibourel, O., et al. (2014). Frequent ASXL2 mutations in acute myeloid leukemia patients with t(8;21)/RUNX1-RUNX1T1 chromosomal translocations. *Blood, 124*, 1445–1449.

Miyoshi, H., Shimizu, K., Kozu, T., Maseki, N., Kaneko, Y., & Ohki, M. (1991). t(8;21) breakpoints on chromosome 21 in acute myeloid leukemia are clustered within a limited region of a single gene, AML1. *Proceedings of the National Academy of Sciences of the United States of America, 88*, 10431–10434.

Miyoshi, H., Kozu, T., Shimizu, K., Enomoto, K., Maseki, N., Kaneko, Y., et al. (1993). The t(8;21) translocation in acute myeloid leukemia results in production of an AML1-MTG8 fusion transcript. *The EMBO Journal, 12*, 2715–2721.

Mrozek, K., Marcucci, G., Paschka, P., & Bloomfield, C. D. (2008). Advances in molecular genetics and treatment of core-binding factor acute myeloid leukemia. *Current Opinion in Oncology, 20*, 711–718.

Muller-Tidow, C., Steffen, B., Cauvet, T., Tickenbrock, L., Ji, P., Diederichs, S., et al. (2004). Translocation products in acute myeloid leukemia activate the Wnt signaling pathway in hematopoietic cells. *Molecular and Cellular Biology, 24*, 2890–2904.

Mulloy, J. C., Cammenga, J., Mackenzie, K. L., Berguido, F. J., Moore, M. A., & Nimer, S. D. (2002). The AML1-ETO fusion protein promotes the expansion of human hematopoietic stem cells. *Blood, 99*, 15–23.

Mulloy, J. C., Cammenga, J., Berguido, F. J., Wu, K., Zhou, P., Comenzo, R. L., et al. (2003). Maintaining the self-renewal and differentiation potential of human CD34+ hematopoietic cells using a single genetic element. *Blood, 102*, 4369–4376.

Mulloy, J. C., Jankovic, V., Wunderlich, M., Delwel, R., Cammenga, J., Krejci, O., et al. (2005). AML1-ETO fusion protein up-regulates TRKA mRNA expression in human CD34+ cells, allowing nerve growth factor-induced expansion. *Proceedings of the National Academy of Sciences of the United States of America, 102*, 4016–4021.

Nakagawa, M., Shimabe, M., Watanabe-Okochi, N., Arai, S., Yoshimi, A., Shinohara, A., et al. (2011). AML1/RUNX1 functions as a cytoplasmic attenuator of

NF-kappaB signaling in the repression of myeloid tumors. *Blood, 118*, 6626–6637.

Nick, H. J., Kim, H. G., Chang, C. W., Harris, K. W., Reddy, V., & Klug, C. A. (2012). Distinct classes of c-Kit-activating mutations differ in their ability to promote RUNX1-ETO-associated acute myeloid leukemia. *Blood, 119*, 1522–1531.

Nisson, P. E., Watkins, P. C., & Sacchi, N. (1992). Transcriptionally active chimeric gene derived from the fusion of the AML1 gene and a novel gene on chromosome 8 in t(8;21) leukemic cells. *Cancer Genetics and Cytogenetics, 63*, 81–88.

Nucifora, G., Birn, D. J., Erickson, P., Gao, J., Lebeau, M. M., Drabkin, H. A., & Rowley, J. D. (1993). Detection of DNA rearrangements in the AML1 and ETO loci and of an AML1/ETO fusion mRNA in patients with t(8;21) acute myeloid leukemia. *Blood, 81*, 883–888.

Odaka, Y., Mally, A., Elliott, L. T., & Meyers, S. (2000). Nuclear import and subnuclear localization of the proto-oncoprotein ETO (MTG8). *Oncogene, 19*, 3584–3597.

Okuda, T., Cai, Z., Yang, S., Lenny, N., Lyu, C. J., Van Deursen, J. M., et al. (1998). Expression of a knocked-in AML1-ETO leukemia gene inhibits the establishment of normal definitive hematopoiesis and directly generates dysplastic hematopoietic progenitors. *Blood, 91*, 3134–3143.

Okumura, A. J., Peterson, L. F., Okumura, F., Boyapati, A., & Zhang, D. E. (2008). t(8;21)(q22;q22) Fusion proteins preferentially bind to duplicated AML1/RUNX1 DNA-binding sequences to differentially regulate gene expression. *Blood, 112*, 1392–1401.

Pabst, T., Mueller, B. U., Harakawa, N., Schoch, C., Haferlach, T., Behre, G., et al. (2001). AML1-ETO downregulates the granulocytic differentiation factor C/EBPalpha in t(8;21) myeloid leukemia. *Nature Medicine, 7*, 444–451.

Park, S., Speck, N. A., & Bushweller, J. H. (2009). The role of CBFbeta in AML1-ETO's activity. *Blood, 114*, 2849–2850.

Paschka, P., & Dohner, K. (2013). Core-binding factor acute myeloid leukemia: Can we improve on HiDAC consolidation? *Hematology. American Society of Hematology. Education Program, 2013*, 209–219.

Peterson, L. F., Yan, M., & Zhang, D. E. (2007). The p21Waf1 pathway is involved in blocking leukemogenesis by the t(8;21) fusion protein AML1-ETO. *Blood, 109*, 4392–4398.

Ptasinska, A., Assi, S. A., Mannari, D., James, S. R., Williamson, D., Dunne, J., et al. (2012). Depletion of RUNX1/ETO in t(8;21) AML cells leads to genome-wide changes in chromatin structure and transcription factor binding. *Leukemia, 26*, 1829–1841.

Ptasinska, A., Assi, S. A., Martinez-Soria, N., Imperato, M. R., Piper, J., Cauchy, P., et al. (2014). Identification of a dynamic core transcriptional network in t(8;21) AML that regulates differentiation block and self-renewal. *Cell Reports, 8*, 1974–1988.

Pulikkan, J. A., Madera, D., Xue, L., Bradley, P., Landrette, S. F., Kuo, Y. H., et al. (2012). Thrombopoietin/MPL participates in initiating and maintaining RUNX1-ETO acute myeloid leukemia via PI3K/AKT signaling. *Blood, 120*, 868–879.

Qin, Y. Z., Zhu, H. H., Jiang, Q., Jiang, H., Zhang, L. P., Xu, L. P., et al. (2014). Prevalence and prognostic significance of c-KIT mutations in core binding factor acute myeloid leukemia: A comprehensive large-scale study from a single Chinese center. *Leukemia Research, 38*, 1435–1440.

Regha, K., Assi, S. A., Tsoulaki, O., Gilmour, J., Lacaud, G., & Bonifer, C. (2015). Developmental-stage-dependent transcriptional response to leukaemic oncogene expression. *Nature Communications, 6*, 7203.

Ross, M. E., Mahfouz, R., Onciu, M., Liu, H. C., Zhou, X., Song, G., et al. (2004). Gene expression profiling of pediatric acute myelogenous leukemia. *Blood, 104*, 3679–3687.

Roudaia, L., Cheney, M. D., Manuylova, E., Chen, W., Morrow, M., Park, S., et al. (2009). CBFbeta is critical for AML1-ETO and TEL-AML1 activity. *Blood, 113*, 3070–3079.

Rowley, J. D. (1984). Biological implications of consistent chromosome rearrangements in leukemia and lymphoma. *Cancer Research, 44*, 3159–3168.

Saeed, S., Logie, C., Francoijs, K. J., Frige, G., Romanenghi, M., Nielsen, F. G., et al. (2012). Chromatin accessibility, p300, and histone acetylation define PML-RARalpha and AML1-ETO binding sites in acute myeloid leukemia. *Blood, 120*, 3058–3068.

Schessl, C., Rawat, V. P., Cusan, M., Deshpande, A., Kohl, T. M., Rosten, P. M., et al. (2005). The AML1-ETO fusion gene and the FLT3 length mutation collaborate in inducing acute leukemia in mice. *The Journal of Clinical Investigation, 115*, 2159–2168.

Schlenk, R. F., Benner, A., Krauter, J., Buchner, T., Sauerland, C., Ehninger, G., et al. (2004). Individual patient data-based meta-analysis of patients aged 16 to 60 years with core binding factor acute myeloid leukemia: A survey of the German Acute Myeloid Leukemia Intergroup. *Journal of Clinical Oncology, 22*, 3741–3750.

Shia, W. J., Okumura, A. J., Yan, M., Sarkeshik, A., Lo, M. C., Matsuura, S., et al. (2012). PRMT1 interacts with AML1-ETO to promote its transcriptional activation and progenitor cell proliferative potential. *Blood, 119*, 4953–4962.

Shih, L. Y., Liang, D. C., Huang, C. F., Chang, Y. T., Lai, C. L., Lin, T. H., et al. (2008). Cooperating mutations of receptor tyrosine kinases and Ras genes in childhood core-binding factor acute myeloid leukemia and a comparative analysis on paired diagnosis and relapse samples. *Leukemia, 22*, 303–307.

Sinenko, S. A., Hung, T., Moroz, T., Tran, Q. M., Sidhu, S., Cheney, M. D., et al. (2010). Genetic manipulation of AML1-ETO-induced expansion of hematopoietic precursors in a Drosophila model. *Blood, 116*, 4612–4620.

Sood, R., Hansen, N. F., Donovan, F. X., Carrington, B., Bucci, D., Maskeri, B., et al. (2016). Somatic mutational landscape of AML with inv(16) or t(8;21) identifies patterns of clonal evolution in relapse leukemia. *Leukemia, 30*, 501–504.

Steffen, B., Knop, M., Bergholz, U., Vakhrusheva, O., Rode, M., Kohler, G., et al. (2011). AML1/ETO induces self-renewal in hematopoietic progenitor cells via the Groucho-related amino-terminal AES protein. *Blood, 117*, 4328–4337.

Sun, X. J., Wang, Z., Wang, L., Jiang, Y., Kost, N., Soong, T. D., et al. (2013). A stable transcription factor complex nucleated by oligomeric AML1-ETO controls leukaemogenesis. *Nature, 500*, 93–97.

Tokumasu, M., Murata, C., Shimada, A., Ohki, K., Hayashi, Y., Saito, A. M., et al. (2015). Adverse prognostic impact of KIT mutations in childhood CBF-AML: The results of the Japanese Pediatric Leukemia/Lymphoma Study Group AML-05 trial. *Leukemia, 29*, 2438–2441.

Trombly, D. J., Whitfield, T. W., Padmanabhan, S., Gordon, J. A., Lian, J. B., van Wijnen, A. J., et al. (2015). Genome-wide co-occupancy of AML1-ETO and N-CoR defines the t(8;21) AML signature in leukemic cells. *BMC Genomics, 16*, 309.

Ugarte, G. D., Vargas, M. F., Medina, M. A., Leon, P., Necunir, D., Elorza, A. A., et al. (2015). Wnt signaling induces transcription, spatial proximity, and translocation of fusion gene partners in human hematopoietic cells. *Blood, 126*, 1785–1789.

Valk, P. J., Verhaak, R. G., Beijen, M. A., Erpelinck, C. A., Barjesteh Van Waalwijk Van Doorn-Khosrovani, S., Boer, J. M., et al. (2004). Prognostically useful gene-expression profiles in acute myeloid leukemia. *The New England Journal of Medicine, 350*, 1617–1628.

Vangala, R. K., Heiss-Neumann, M. S., Rangatia, J. S., Singh, S. M., Schoch, C., Tenen, D. G., et al. (2003). The myeloid master regulator transcription factor PU.1 is inactivated by AML1-ETO in t(8;21) myeloid leukemia. *Blood, 101*, 270–277.

Viale, A., De Franco, F., Orleth, A., Cambiaghi, V., Giuliani, V., Bossi, D., et al. (2009). Cell-cycle restriction limits DNA damage and maintains self-renewal of leukaemia stem cells. *Nature, 457*, 51–56.

Wang, J., Hoshino, T., Redner, R. L., Kajigaya, S., & Liu, J. M. (1998). ETO, fusion partner in t(8;21) acute myeloid leukemia, represses transcription by interaction with the human N-CoR/mSin3/HDAC1 complex. *Proceedings of the National Academy of Sciences of the United States of America, 95*, 10860–10865.

Wang, L., Gural, A., Sun, X. J., Zhao, X., Perna, F., Huang, G., et al. (2011a). The leukemogenicity of AML1-ETO is dependent on site-specific lysine acetylation. *Science, 333*, 765–769.

Wang, Y. Y., Zhao, L. J., Wu, C. F., Liu, P., Shi, L., Liang, Y., et al. (2011b). C-KIT mutation cooperates with full-length AML1-ETO to induce acute myeloid leukemia in mice. *Proceedings of the National Academy of Sciences of the United States of America, 108*, 2450–2455.

Wang, L., Man, N., Sun, X. J., Tan, Y., Cao, M. G., Liu, F., et al. (2015). Regulation of AKT signaling by Id1 controls t(8;21) leukemia initiation and progression. *Blood, 126*, 640–650.

Wei, H., Liu, X., Xiong, X., Wang, Y., Rao, Q., Wang, M., & Wang, J. (2008). AML1-ETO interacts with Sp1 and antagonizes Sp1 transactivity through RUNT domain. *FEBS Letters, 582*, 2167–2172.

Wichmann, C., Quagliano-Lo Coco, I., Yildiz, O., Chen-Wichmann, L., Weber, H., Syzonenko, T., et al. (2015). Activating c-KIT mutations confer oncogenic cooperativity and rescue RUNX1/ETO-induced DNA damage and apoptosis in human primary CD34+ hematopoietic progenitors. *Leukemia, 29*, 279–289.

Wolford, J. K., & Prochazka, M. (1998). Structure and expression of the human MTG8/ETO gene. *Gene, 212*, 103–109.

Yan, M., Kanbe, E., Peterson, L. F., Boyapati, A., Miao, Y., Wang, Y., et al. (2006). A previously unidentified alternatively spliced isoform of t(8;21) transcript promotes leukemogenesis. *Nature Medicine, 12*, 945–949.

Yan, M., Ahn, E. Y., Hiebert, S. W., & Zhang, D. E. (2009). RUNX1/AML1 DNA-binding domain and ETO/MTG8 NHR2-dimerization domain are critical to AML1-ETO9a leukemogenesis. *Blood, 113*, 883–886.

Yang, G., Khalaf, W., Van De Locht, L., Jansen, J. H., Gao, M., Thompson, M. A., et al. (2005). Transcriptional repression of the Neurofibromatosis-1 tumor suppressor by the t(8;21) fusion protein. *Molecular and Cellular Biology, 25*, 5869–5879.

Yang, G., Thompson, M. A., Brandt, S. J., & Hiebert, S. W. (2007). Histone deacetylase inhibitors induce the degradation of the t(8;21) fusion oncoprotein. *Oncogene, 26*, 91–101.

Yeh, J. R., Munson, K. M., Elagib, K. E., Goldfarb, A. N., Sweetser, D. A., & Peterson, R. T. (2009). Discovering chemical modifiers of oncogene-regulated hematopoietic differentiation. *Nature Chemical Biology, 5*, 236–243.

Yergeau, D. A., Hetherington, C. J., Wang, Q., Zhang, P., Sharpe, A. H., Binder, M., et al. (1997). Embryonic lethality and impairment of haematopoiesis in mice heterozygous for an AML1-ETO fusion gene. *Nature Genetics, 15*, 303–306.

Yuan, Y., Zhou, L., Miyamoto, T., Iwasaki, H., Harakawa, N., Hetherington, C. J., et al. (2001). AML1-ETO expression is directly involved in the development of acute myeloid leukemia in the presence of additional mutations. *Proceedings of the National Academy of Sciences of the United States of America, 98*, 10398–10403.

Zaidi, S. K., Dowdy, C. R., Van Wijnen, A. J., Lian, J. B., Raza, A., Stein, J. L., et al. (2009). Altered Runx1 subnuclear targeting enhances myeloid cell proliferation and blocks differentiation by activating a miR-24/

MKP-7/MAPK network. *Cancer Research, 69*, 8249–8255.

Zhang, J., Hug, B. A., Huang, E. Y., Chen, C. W., Gelmetti, V., Maccarana, M., et al. (2001). Oligomerization of ETO is obligatory for corepressor interaction. *Molecular and Cellular Biology, 21*, 156–163.

Zhang, J., Kalkum, M., Yamamura, S., Chait, B. T., & Roeder, R. G. (2004). E protein silencing by the leukemogenic AML1-ETO fusion protein. *Science, 305*, 1286–1289.

Zhang, Y., Wang, J., Wheat, J., chen, X., Jin, S., Sadrzadeh, H., et al. (2013). AML1-ETO mediates hematopoietic self-renewal and leukemogenesis through a COX/beta-catenin signaling pathway. *Blood, 121*, 4906–4916.

Zhao, S., Zhang, Y., Sha, K., Tang, Q., Yang, X., Yu, C., et al. (2014). KRAS (G12D) cooperates with AML1/ETO to initiate a mouse model mimicking human acute myeloid leukemia. *Cellular Physiology and Biochemistry, 33*, 78–87.

Zheng, X., Beissert, T., Kukoc-Zivojnov, N., Puccetti, E., Altschmied, J., Strolz, C., et al. (2004). Gamma-catenin contributes to leukemogenesis induced by AML-associated translocation products by increasing the self-renewal of very primitive progenitor cells. *Blood, 103*, 3535–3543.

Clinical Relevance of *RUNX1* and *CBFB* Alterations in Acute Myeloid Leukemia and Other Hematological Disorders

12

Klaus H. Metzeler and Clara D. Bloomfield

Abstract

The translocation t(8;21), leading to a fusion between the *RUNX1* gene and the *RUNX1T1* locus, was the first chromosomal translocation identified in cancer. Since the first description of this balanced rearrangement in a patient with acute myeloid leukemia (AML) in 1973, *RUNX1* translocations and point mutations have been found in various myeloid and lymphoid neoplasms. In this chapter, we summarize the currently available data on the clinical relevance of core binding factor gene alterations in hematological disorders. In the first section, we discuss the prognostic implications of the core binding factor translocations *RUNX1-RUNX1T1* and *CBFB-MYH11* in AML patients. We provide an overview of the cooperating genetic events in patients with CBF-rearranged AML and their clinical implications, and review current treatment approaches for CBF AML and the utility of minimal residual disease monitoring. In the next sections, we summarize the available data on rare *RUNX1* rearrangements in various hematologic neoplasms and the role of *RUNX1* translocations in therapy-related myeloid neoplasia. The final three sections of the chapter cover the spectrum and clinical significance of *RUNX1* point mutations in AML and myelodysplastic syndromes, in familial platelet disorder with associated myeloid malignancy, and in acute lymphoblastic leukemia.

K.H. Metzeler
Laboratory for Leukemia Diagnostics, Department of Internal Medicine III, University of Munich, Munich, Germany

C.D. Bloomfield (✉)
The Ohio State University Comprehensive Cancer Center, Columbus, OH, USA
e-mail: Clara.Bloomfield@osumc.edu

Y. Groner et al. (eds.), *RUNX Proteins in Development and Cancer*, Advances in Experimental Medicine and Biology 962, DOI 10.1007/978-981-10-3233-2_12

Keywords
RUNX1 • Acute myeloid leukemia • Core binding factor leukemia • Myelodysplastic syndromes • Chromosomal translocations • Point mutations • Prognosis

12.1 Introduction

The balanced translocation t(8;21)(q22;q22), initially described by Janet D. Rowley in 1973 in a patient with acute myeloid leukemia (AML), was the first reciprocal chromosomal translocation identified in cancer cells (Rowley 1973). In 1991, Miyoshi and co-workers cloned and sequenced a novel gene, located on the breakpoint on chromosome 21, in AML patients with this translocation (Miyoshi et al. 1991). The gene was initially called *AML1*, and is now named Runt-related transcription factor 1 (*RUNX1*) due to its homology with the Drosophila gene Runt. Soon afterwards, it was established that on the molecular level, t(8;21)(q22;q22) leads to the formation of a chimeric fusion transcript that today is named *RUNX1-RUNX1T1* (Erickson et al. 1992; Miyoshi et al. 1993). The RUNX1 protein is part of a heterodimeric transcription factor called the "core binding factor", or CBF. Today, three different DNA-binding CBFα subunits (*RUNX1, RUNX2* and *RUNX3)* and one common non-DNA-binding CBFβ subunit (encoded by the *CBFB* gene) have been identified in humans (Speck and Gilliland 2002). In 1993, Liu and co-workers identified *CBFB* and the myosin heavy chain gene *MYH11* as the fusion partners in another recurrent balanced chromosomal rearrangement in AML, namely inv(16)(p13q22) and its variant, t(16;16)(p13;q22) (Liu et al. 1993). These seminal discoveries pointed towards an important role of CBF genes not only in normal hematopoiesis, but also in leukemia, and opened the road for further studies revealing that the *RUNX1* gene is frequently altered in myeloid malignancies and other hematological disorders through various mechanisms including chromosomal translocations, point mutations and deletions. In this chapter, we will review the clinical and prognostic significance of *RUNX1* alterations and the *CBFB-MYH11* fusion in AML and other hematological disorders.

12.2 Balanced Translocations Involving the Core Binding Factor Subunits in AML: t(8;21)(q22;q22); *RUNX1-RUNX1T1* and inv(16)(p13q22)/t(16;16)(p13;q22); *CBFB-MYH11*

12.2.1 Background

Between 1978 and 1984, the International Workshops on Chromosomes in Leukemia established that t(8;21)(q22;q22) is a recurrent event in AML, and closely linked to M2 morphology according to the French-American-British (FAB) classification (i.e., acute myelogenous leukemia with maturation) (Rowley and de la Chapelle 1978; Rowley 1980; Bloomfield et al. 1984). In subsequent studies, the incidence of the *RUNX1-RUNX1T1* rearrangement in adult AML was 4–8% (Slovak et al. 2000; Byrd et al. 2002; Mrózek 2004), and it was 7% in a very large cohort of 5876 patients aged 16–59 years (Grimwade et al. 2010). *RUNX1-RUNX1T1* is extremely rare in infants but occurs in 11–14% of children and adolescents, and represents the single most common balanced translocation in pediatric AML (Leverger et al. 1988; Raimondi et al. 1999; Mrózek et al. 2004; von Neuhoff et al. 2010; Harrison et al. 2010). The incidence of *RUNX1-RUNX1T1* decreases with age, and the translocation is less frequent in patients aged ≥60 years (Bloomfield et al. 1984; Byrd et al. 2002; Grimwade et al. 2010).

The pericentric inversion inv(16)(p13q22) was first described as a recurrent abnormality in AML in 1983. The original publications described an association with myelomonocytic leukemia with abnormal eosinophils (FAB M4eo), and reported that affected patients had favorable response to treatment (Le Beau et al. 1983; Bloomfield et al. 1984). The incidence of *CBFB-*

MYH11 in adults was 5% in the large British Medical Research Council (MRC) cohort, and 2–9% in other series, and it is found in 6–9% of pediatric AML (Raimondi et al. 1999; Mrózek et al. 2004; Grimwade et al. 2010; von Neuhoff et al. 2010; Harrison et al. 2010). Similar to *RUNX1-RUNX1T1*, the *CBFB-MYH11* fusion is less frequent in older adults (i.e., ≥60 years) (Byrd et al. 2002).

12.2.2 Prognosis of AML Patients with t(8;21)(q22;q22); *RUNX1-RUNX1T1* and inv(16)(p13q22) /t(16;16)(p13;q22); *CBFB-MYH11*

The reports from the International Workshops on Chromosomes in Leukemia established that karyotype is an important prognostic factor in AML, and revealed that both CBF rearrangements, t(8;21)(q22;q22) and inv(16)(p13q22), associate with relatively favorable outcomes. This finding was confirmed by long-term follow-up of the initial cohorts, although treatment was not uniform in these early series (Rowley 1980; Larson et al. 1983; Bloomfield et al. 1984; Swansbury et al. 1994). Later studies in patients who received more standardized, cytarabine-based induction and consolidation chemotherapy on cooperative group trials consistently showed that patients with CBF rearrangements had higher complete remission (CR) rates compared to patients with cytogenetically normal AML, and longer disease-free survival (DFS) and overall survival (OS) (Keating et al. 1987; Fenaux et al. 1989; Slovak et al. 2000). In a study of 1213 patients enrolled on 5 consecutive Cancer and Leukemia Group B (CALGB) treatment protocols, Byrd and colleagues demonstrated that the *RUNX1-RUNX1T1* and *CBFB-MYH11* fusions associated with high CR rates (91% and 85%, respectively), a low rate of primary refractory disease, and favorable DFS and OS (Byrd et al. 2002). In a more recent analysis of 5876 younger adults (<60 years) treated on trials of the MRC, *RUNX1-RUNX1T1* positive patients had a CR rate of 97% and a 10-year OS of 61%, while those with *CBFB-MYH11* had a CR rate of 92% and a 10-year OS of 55% (Fig. 12.1) (Grimwade et al. 2010). When patients with acute promyelocytic leukemia (APL) are excluded, t(8;21)(q22;q22) and inv(16)(p13q22) / t(16;16)(p13;q22) represent the most favorable cytogenetic subset in this very large cohort. In older adults (≥60 years) with CBF leukemias who received at least one cycle of induction chemotherapy, the CR rate was 88%, but 5-year OS was only 31% (Prébet et al. 2009). Pediatric AML patients with CBF rearrangements have excellent outcomes with reported OS rates of ~90% at 5 years and ~80% at 10 years (von Neuhoff et al. 2010; Harrison et al. 2010). Based on these data, it is generally accepted today that the CBF rearrangements, t(8;21)(q22;q22), inv(16)(p13q22) and t(16;16)(p13;q22), define a favorable cytogenetic subset of adult and pediatric AML patients, and this is reflected by current risk stratification systems and international guidelines (Grimwade et al. 2010; Döhner et al. 2010; Creutzig et al. 2012).

12.2.3 Role of Cooperating Cytogenetic Changes and Gene Mutations in AML with CBF Gene Rearrangement

Additional chromosomal alterations are found in most AML patients with the translocation t(8;21). Loss of a sex chromosome was identified as the most common secondary alteration in adults and children, followed by deletions in the long arm of chromosome 9 (band 9q22) (Larson et al. 1983; Bloomfield et al. 1984; Raimondi et al. 1999; Kuchenbauer et al. 2006). In a series of 111 adult patients, additional chromosomal abnormalities were found in 70%, including loss of a sex chromosome in 47%, 9q deletion in 15% and trisomy 8 in 6% (Krauth et al. 2014). Loss of a sex chromosome associated with favorable, and trisomy 8 with unfavorable event-free survival (EFS) in this series, although treatment was heterogeneous and the number of patients with +8 was small. Other reports also suggested that specific additional cytogenetic abnormalities, including

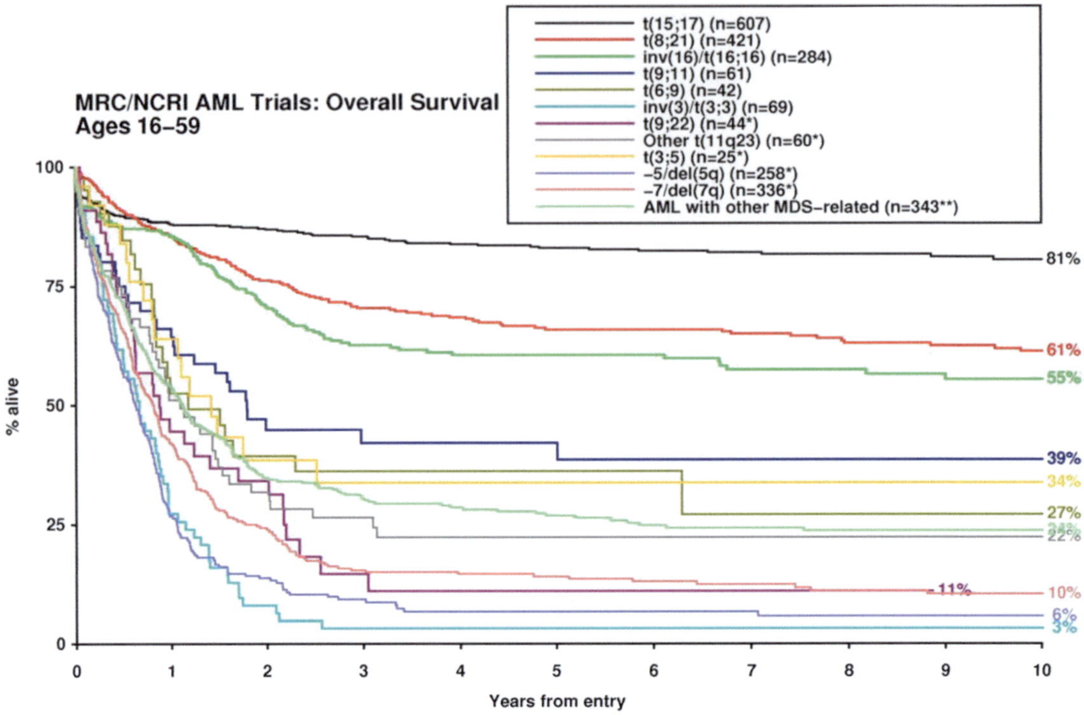

Fig. 12.1 Prognosis of AML patients with t(8;21)(q22;q22); *RUNX1-RUNX1T1* and inv(16)(p13q22)/t(16;16)(p13;q22); *CBFB-MYH11*. Overall survival of younger patients (age, 16–59) treated on British Medical Research Council trials, stratified according to cytogenetic findings. Survival for patients with t(8;21) is shown in red and survival for patients with inv(16) is shown in green (Figure reproduced from Grimwade et al. (2010) with permission)

del(9q) and loss of the Y chromosome in male patients, or *RUNX1-RUNX1T1* rearrangement in the context of a complex karyotype, adversely affect the outcomes of *RUNX1-RUNX1T1*-positive patients (Schoch et al. 1996; Schlenk et al. 2004; Appelbaum et al. 2006). In contrast, several large studies in children and adults demonstrated that the presence of secondary cytogenetic alterations including loss of a sex chromosome, del(9q), or trisomy 8 have no adverse impact on outcomes (Fenaux et al. 1989; Byrd et al. 2002; Grimwade et al. 2010; Harrison et al. 2010). Importantly, patients with *RUNX1-RUNX1T1* and *CBF-MYH11* have favorable outcomes even when these abnormalities occur within a complex karyotype (Byrd et al. 2002; Grimwade et al. 2010).

In AML patients with inv(16) or t(16;16), the most frequently identified additional cytogenetic alterations are trisomy 22, trisomy 11, and deletions on the short arm of chromosome 7, which are found in 14–19%, 10–16%, and 5–6% of patients, respectively (Schlenk et al. 2004; Marcucci et al. 2005; Grimwade et al. 2010; Paschka et al. 2013). At least three independent studies reported that patients with *CBFB-MYH11* and an additional chromosome 22 have a particularly low risk of relapse and favorable survival (Grimwade et al. 2010; Schlenk et al. 2004; Marcucci et al. 2005).

More recent analyses also include information on molecular gene mutations. Data from the German AML Study Group (AMLSG) show that 56% of *RUNX1-RUNX1T1* rearranged AML and 84% of *CBF-MYH11* rearranged AML harbor mutations in *KIT*, *FLT3*, *NRAS* or *KRAS* (Paschka and Döhner 2013). The frequency of *KIT* and *FLT3* gene mutations was similar in both subsets of CBF leukemias. *KIT* mutations were detected in 30% of *RUNX1-RUNX1T1* positive cases and 37% *CBF-MYH11* positive cases, and *FLT3* mutations were present in 13% and 17%, respectively. *RAS* mutations, however, were more common in AML with *CBF-MYH11* (53% vs. 21% in *RUNX1-RUNX1T1* rearranged AML). In another analysis of 11 different genes in 139 *RUNX1-RUNX1T1*-positive AML patients, at least one gene mutation was found in 50%, with *KIT*, *NRAS* and *ASXL1* being most commonly affected (Krauth et al. 2014).

In 2014, Micol and colleagues discovered novel mutations in the additional sex combs-like 2 (*ASXL2*) gene in almost a quarter of patients with *RUNX1-RUNX1T1* fusion. Notably, mutations in this gene were absent in patients with *CBFB-MYH11* rearrangement or mutated *RUNX1* (Micol et al. 2014). More comprehensive genetic analyses have revealed that mutations affecting epigenetic modifiers, including *ASXL1*, *ASXL2*, *EZH2* and *KDM6A*, the cohesin complex, and the zink finger transcription factor *ZBTB7A* are common in *RUNX1-RUNX1T1* rearranged AML, but rare or absent in patients with *CBFB-MYH11* (Hartmann et al. 2016; Duployez et al. 2016; Lavallée et al. 2016; Sood et al. 2016). The prognostic relevance of these mutations in CBF AML remains to be determined. SNP-array studies revealed that submicroscopic copy number alterations are rare in CBF rearranged AML (Kühn et al. 2012). Overall, these data indicate that, while activation of receptor tyrosine kinase signaling pathways is a common mechanism in CBF AML, other cooperating pathways may be specific to patients with either *RUNX1-RUNX1T1* or *CBFB-MYH1*.

12.2.3.1 Clinical Relevance of *KIT* Gene Mutations in CBF AML

Mutations in the *KIT* receptor tyrosine kinase in AML were first identified by Beghini and colleagues (Beghini et al. 1998), and were subsequently confirmed to be recurrent events in CBF leukemias, while they are rare in other cytogenetic subsets (Gari et al. 1999; Beghini et al. 2000; Schnittger et al. 2006). In some patients, *KIT* mutations become undetectable in CR while the *RUNX1-RUNX1T1* fusion remains detectable using similarly sensitive methods, suggesting that *KIT* mutations constitute a secondary hit that provides a growth and/or survival advantage to the leukemic cells (Wang et al. 2005). In several cohorts of CBF-rearranged adolescents and adults, *KIT* mutations were found in 20–47% of *RUNX1-RUNX1T1* and 30–45% of *CBFB-MYH11* positive patients and tended to associate with higher white blood counts (Care et al. 2003; Beghini et al. 2004; Wang et al. 2005; Cairoli et al. 2006; Paschka et al. 2006; Allen et al. 2013). In *RUNX1-RUNX1T1* rearranged AML, most of the mutations are activating missense mutations in the tyrosine kinase domain (exon 17), while mutations in the extracellular domain (exon 8) and the transmembrane and juxtamembrane domains (exons 10 and 11) occur more rarely and have not been analyzed in all studies (Allen et al. 2013). In contrast, exon 8 mutations are more common in patients with *CBFB-MYH11* (Paschka et al. 2013; Allen et al. 2013).

In several moderately-sized retrospective series of *RUNX1-RUNX1T1*-positive AML, *KIT*-mutated patients had a higher incidence of relapse (70–100%) compared to *KIT* wild-type patients (~35%), while the results in patients with *CBFB-MYH11* were discordant (Schnittger et al. 2006; Cairoli et al. 2006; Paschka et al. 2006; Nanri et al. 2005b; Boissel et al. 2006). In the largest cohort reported so far by the MRC study group, *KIT* mutations were found in 23% of 199 patients with *RUNX1-RUNX1T1* rearrangement. Only "high-level" *KIT* mutations with a mutant-to-wild type ratio of ≥25% associated with higher relapse risk (41% compared to 25% for *KIT* wild-type patients), while *KIT* mutations present at lower levels had no impact on relapse. Of note, *FLT3*-internal tandem duplications, but not *KIT* mutations, associated with shorter OS in this cohort (Allen et al. 2013). In contrast, 35% of 155 *CBFB-MYH11* rearranged patients in the same study had mutated *KIT*, and mutation status did not affect RFS or OS. In a large series of *CBFB-MYH11* patients from the German AMLSG, *KIT* mutations negatively affected RFS, but not OS (Paschka et al. 2013).

Collectively, these results establish *KIT* mutations as a predictor of higher relapse risk in adult *RUNX1-RUNX1T1* rearranged AML. *KIT* mutations were linked to shorter OS in some studies (Schnittger et al. 2006; Cairoli et al. 2006;

Boissel et al. 2006) but not in others (Paschka et al. 2006; Allen et al. 2013; Nanri et al. 2005b), suggesting that *KIT*-mutated patients may respond favorably to salvage therapy. The prognostic relevance of KIT mutations in AML with *CBFB-MYH11* is less well established, and there are conflicting reports on the prognostic relevance of *KIT* mutations in pediatric CBF AML patients (Paschka and Döhner 2013; Pollard et al. 2010).

12.2.4 Treatment of AML with CBF Leukemias

12.2.4.1 Chemotherapy and the Role of High-Dose Cytarabine

As outlined above, approximately 90% of adult *RUNX1-RUNX1T1*-positive AML patients achieve CR with cytarabine- and anthracycline-based ('7+3'-like) induction chemotherapy. A study from CALGB demonstrated that consolidation therapy with 4 cycles of high-dose cytarabine (HDAC, 3 g/m^2 twice daily on days 1,3 and 5), compared to cytarabine doses of 100 or 400 mg/m^2/day for 5 days, resulted in prolonged DFS particularly in patients with CBF leukemias (Bloomfield et al. 1998). The same group later showed that patients with *RUNX1-RUNX1T1* rearrangement who received three or four cycles of HDAC had superior 5-year DFS (71%) and OS (76%) compared to patients who received only one HDAC course (5 year DFS, 37%; 5-year OS, 44%). In an extended series of 96 *RUNX1-RUNX1T1*-positive patients, those receiving multiple HDAC courses had a 10 year survival of 56%, compared to 43% for those receiving only a single course (Marcucci et al. 2005). In patients with *CBFB-MYH11*, exposure to 3–4 HDAC cycles resulted in improved 5-year DFS compared to those receiving only one cycle (57% vs. 30%), with no improvement in OS (Byrd et al. 2004). A favorable impact of 3 cycles of HDAC consolidation, compared to 4 cycles of multiagent chemotherapy with lower-dose cytarabine, on DFS in CBF leukemias was also confirmed by a Japanese trial (Miyawaki et al. 2011). Daunorubicin dose escalation from 45 to 90 mg/m^2 during induction resulted in a trend towards improved EFS and OS in older patients (≥60 years) with CBF leukemia (Löwenberg et al. 2009). In a large British randomized trial of mostly younger patients and in a retrospective analysis of two French trials, daunorubicin dose escalation from 60 to 90 mg/m^2 during induction was not associated with improved survival in CBF leukemias (Prébet et al. 2014; Burnett et al. 2015). Therefore, '7+3'-like induction chemotherapy (preferentially with a daunorubicin dose of 60 mg/m^2) followed by 3–4 cycles of HDAC consolidation currently can be considered the standard treatment for adult AML patients with CBF rearrangement, although one study suggested that lower cumulative doses of cytarabine may be sufficient (Löwenberg et al. 2011).

12.2.4.2 Gemtuzumab Ozogamicin

Gemtuzumab ozogamicin (GO) is an anti-CD33 antibody coupled to the cytotoxic drug calicheamicin. In a subgroup analysis of the British MRC AML15 trial for patients <60 years, addition of a single dose of GO to induction chemotherapy led to significantly improved OS in patients with CBF leukemias (Burnett et al. 2011). A beneficial effect of GO in CBF leukemia patients was confirmed in a large meta-analysis of five randomized trials, which found a 5-year OS of 78% for patients receiving GO compared to 55% for those not receiving GO (Hills et al. 2014). The optimal dose and schedule of GO administration in CBF AML are unknown. GO was withdrawn from the US market due to concerns about early mortality in one trial (Petersdorf et al. 2013), and the drug is currently approved in Japan, but not in the US or Europe.

12.2.4.3 Allogeneic Stem Cell Transplantation for CBF AML

In a retrospective comparison of 118 AML patients with *RUNX1-RUNX1T1* translocation in first CR who underwent allogeneic stem cell transplantation (alloSCT) from a HLA-identical sibling to 132 patients receiving consolidation chemotherapy on German multicenter trials, patients receiving alloSCT had a reduced relapse risk, however, this was offset by increased treatment-related mortality. In summary, relapse-free survival (RFS) tended to be better for those patients receiving chemotherapy (Schlenk et al. 2008). For patients with *CBFB-MYH11* rearrangement, a German meta-analysis of 170 patients in first CR also found no RFS benefit of allogeneic SCT over consolidation chemotherapy (Schlenk et al. 2004). Results of a donor-versus-no donor analysis and a large-meta analysis of 24 trials confirmed that AML patients with CBF leukemias do not profit from alloSCT in first CR (Cornelissen et al. 2007; Koreth et al. 2009). While *KIT* mutations predict a higher relapse risk in *RUNX1-RUNX1T1*-positive AML, it is unknown whether alloSCT ameliorates this increased risk. Consequently, there is currently no consensus whether *KIT*-mutated patients should undergo alloSCT in first CR (Allen et al. 2013).

12.2.4.4 Tyrosine Kinase Inhibitors

Functional analyses of *KIT* mutations showed that they lead to constitutive activation of the receptor (Cammenga et al. 2005). Furthermore, *KIT* is overexpressed in *RUNX1-RUNX1T1*-rearranged AML patients irrespective of its mutation status (Bullinger et al. 2004; Valk et al. 2004). Wild-type and mutant *KIT* isoforms can be inhibited by various tyrosine kinase inhibitors (TKIs), providing a rationale for therapeutic use of TKIs in t(8;21) AML (Growney et al. 2005; Nanri et al. 2005a; Schittenhelm et al. 2006; Chevalier et al. 2010; Paschka and Döhner 2013). Mutated *KIT* isoforms exhibit variable sensitivity to different inhibitors. Clinical responses were observed in single patients or small series of patients with advanced disease receiving TKI, including imatinib and dasatinib (Nanri et al. 2005a; Chevalier et al. 2010). However, in a study of 26 high-risk patients with minimal residual disease (MRD) persistence or recurrence, 12 months of dasatinib maintenance did not avert hematological relapse in patients with molecular recurrence, or improve DFS in those with suboptimal MRD response (Boissel et al. 2015). Several prospective studies of dasatinib in CBF leukemias are ongoing.

12.2.4.5 Treatment of Relapsed Disease

Data from the 6th International Workshop on Chromosomes in Leukemia suggested that relapsed *RUNX1-RUNX1T1*-positive AML is relatively sensitive to repeated chemotherapy, and second CRs can be achieved in a considerable fraction of patients (Garson et al. 1989). In a retrospective analysis of 59 patients in first relapse, the rate of second CR after salvage chemotherapy was 88%, and 5-year survival after relapse was 51%. Addition of GO to salvage chemotherapy appeared to be beneficial, with a 5 year OS of 65% compared to 44% for those receiving chemotherapy without GO (Hospital et al. 2014). An analysis by the MRC group showed that relapsed *RUNX1-RUNX1T1* or *CBFB-MYH11*-positive AML patients who received salvage chemotherapy without alloSCT had a 5-year OS of 41% and 47%, respectively, compared to 29 % and 39 % for those receiving an alloSCT, yet this was not a randomized comparison and survival estimates may be biased. Nevertheless, these data indicate that CBF AML frequently remains chemoresponsive at the time of relapse, in contrast to relapsed non-CBF AML which generally is considered incurable without alloSCT (Burnett et al. 2013).

12.2.5 Minimal Residual Disease Monitoring in CBF AML

The *RUNX1-RUNX1T1* and *CBFB-MYH11* fusion transcripts can be detected with high sensitivity by RT-PCR, and this technique may be used to detect persisting leukemic cells in patients in clinical remission. However, it has been shown that some patients who have been in morphological CR for up to 8 years still have detectable *RUNX1-RUNX1T1* transcripts in the bone marrow (BM) and/or blood, although some studies reported that transcript levels tended to decrease and become undetectable over time (Nucifora et al. 1993; Nucifora and Rowley 1994; Kusec et al. 1994; Satake et al. 1995). Clonogenic progenitor assays revealed that the *RUNX1-RUNX1T1* fusion persisted in multipotent hematopoetic progenitor cells that were able to differentiate into mature trilineage myeloid cells and mature B cells *in vitro* and *in vivo* (Miyamoto et al. 1996, 2000). These studies indicate that *RUNX1-RUNX1T1*-positive pre-leukemic stem cells capable of self-renewal and differentiation can persist in the BM during CR, although their frequency gradually decreases over time. Miyamoto studied *RUNX1-RUNX1T1*-positive patients who had been in CR for 1–12.5 years using a nested RT-PCR assay with a sensitivity of $1{:}10^{-7}$. They found *RUNX1-RUNX1T1* transcripts in the BM of all 18 patients treated with chemotherapy only, but in none of the 4 patients who had undergone alloSCT (Miyamoto et al. 1996). Another study, however, reported that *RUNX1-RUNX1T1* transcripts were also detectable in 9 of 10 patients in CR after alloSCT (Jurlander et al. 1996). Taken together, these studies establish that *RUNX1-RUNX1T1*-positive cells can persist at low levels in t(8;21) AML patients who achieve long-term remissions. They also demonstrate that the *RUNX1-RUNX1T1* fusion alone is not sufficient to initiate AML, and secondary genetic lesions are needed.

Low-level persistence of *RUNX1-RUNX1T1* transcripts in patients who may be cured limits the utility of qualitative (end-point) RT-PCR assays for the detection of clinically meaningful residual disease. Nevertheless, a French multicenter study of 51 patients suggested that many patients in long-term remission ultimately become PCR-negative. Using a less sensitive one-step qualitative PCR technique, this study showed that patients who achieved PCR negativity during follow-up had a relapse rate of 15%, while all patients with persistently positive PCR results relapsed. This study also suggested the possibility of early MRD-based response assessment, since patients who became PCR-negative after induction and before consolidation chemotherapy had a relapse rate of 11%, compared to 72% for the remaining patients (Morschhauser et al. 2000).

The development of quantitative PCR (qPCR) techniques allowed serial monitoring of *RUNX1-RUNX1T1* and *CBFB-MYH11* transcript levels over time, and establishing critical threshold levels that are predictive of imminent hematological relapse (Tobal and Yin 1996; Marcucci et al. 1998; Krauter et al. 1999; Tobal et al. 2000; Krauter et al. 2003; Buonamici et al. 2002; Leroy et al. 2005). If quantitative MRD monitoring is to be used for clinical decision-making, careful standardization of methods and cut-offs is necessary to ensure comparable results from different laboratories. RNA-based assays for quantitative detection of *RUNX1-RUNX1T1, CBFB-MYH11* and other fusion transcripts have been established and validated by multinational consortia including the "Europe against Cancer" (EAC) initiative (van Dongen et al. 1999; Gabert et al. 2003). Of note, since the genomic breakpoints in the *RUNX1* locus are distributed over a region of ~25 kilobases, MRD monitoring on genomic DNA requires development of patient-specific assays. While this approach is feasible and offers the conceptual advantage of quantifying the proportion of leukemic cells more directly, it suffers from variable sensitivity and greatly increased complexity, and is thus not widely used (Duployez et al. 2014).

The clinical relevance of MRD measurements by qPCR was demonstrated in several large, uniformly treated patient cohorts analyzed according to the EAC recommendations. The British MRC group studied 278 CBF-AML patients aged 15–70 years, and found that a >3 log reduction of *RUNX1-RUNX1T1* transcript levels in BM after the first induction cycle was associated with a cumulative incidence of relapse of only 4% at 5 years, while patients with a lesser reduction had relapse rates exceeding 30%, although this did not translate into significant survival differences. Similarly, detection of <10 *CBFB-MYH11* copies per 10^5 copies of *ABL* in peripheral blood after induction 1 associated with a relatively low 5-year incidence of relapse (21%) and favorable survival after CR. After completion of therapy, BM MRD levels of over 500 *RUNX1-RUNX1T1* copies per 10^5 *ABL* copies were also highly predictive of relapse (relapse rate, 100% versus 7% for those with persistently lower levels) and inferior OS (5-year survival, 57% vs. 94%). For patients with *CBFB-MYH11*, detection of >10 copies in the peripheral blood associated with a

97% risk of relapse and 57% 5-year-survival, compared with a 7% relapse risk and 91% survival in those with MRD levels <10 copies. The median time from qPCR positivity to hematologic relapse was about 5 months, leading the authors to recommend MRD monitoring from BM every 3 months during the first 18 months of follow-up (Yin et al. 2012).

The French AML Intergroup reported data on 198 CBF-AML patients aged 18–60 years. A ≥3 log reduction of BM fusion transcripts after the first consolidation course associated with a lower relapse risk (hazard ratio, 0.31), while the risk of death was not significantly lower (hazard ratio, 0.51). Importantly, the prognostic significance of early MRD reduction with regard to relapse outweighed the impact of *KIT* and *FLT3* gene mutations in a multivariate analysis (Jourdan et al. 2013). Among *RUNX1-RUNX1T1* rearranged patients in this cohort, persistent MRD positivity or molecular relapse in blood after the end of therapy predicted hematological relapse in 21 of 28 patients, while persistent *RUNX1-RUNX1T1* MRD positivity in the BM at 2 years was found in 9% of patients who maintained long-term remissions (Willekens et al. 2016).

The German AMLSG group studied a cohort of 53 *CBFB-MYH11* rearranged patients and identified criteria for risk stratification. Patients who achieved qPCR negativity in at least one BM sample during consolidation therapy had favorable DFS, while qPCR negativity in at least two BM or PB samples during consolidation therapy and early follow-up predicted for superior DFS and OS. Conversion from PCR negativity to PCR positivity after consolidation therapy occurred in 10 patients, and 6 of them relapsed (Corbacioglu et al. 2010). In summary, these studies establish that MRD measurements by qPCR, and particularly early response kinetics during therapy, are strong prognostic markers in CBF AML. However, it remains unclear whether treatment modification in response to unfavorable MRD results is beneficial.

A Chinese study addressed this question and studied the role of MRD-directed treatment in 116 *RUNX1-RUNX1T1*-rearranged AML patients. Patients who did not sustain a 3-log reduction of *RUNX1-RUNX1T1* transcript levels after 2 cycles of intermediate-dose cytarabine-based consolidation chemotherapy were considered high risk and were recommended to undergo alloSCT, while those with better responses were scheduled for 6 cycles of consolidation chemotherapy. The trial was not randomized, and about 40 % of patients crossed over between the two arms for various reasons. In this cohort, alloSCT improved DFS and OS of high-risk, but not of low-risk patients, but this result requires confirmation from controlled trials (Zhu et al. 2013). A follow-up study from the same group investigated the prognostic relevance of MRD detection in the posttransplantation setting. Patients who achieved a >3 log reduction of BM *RUNX1-RUNX1*T1 transcript levels during the first 3 months after alloSCT had significantly lower relapse rates and longer DFS, compared to those with higher transcript levels. A multivariate analysis suggested that MRD levels outweigh *KIT* mutation status as a stronger predictor of post-transplant relapse risk, although this analysis is limited by the relatively small patient cohort (Wang et al. 2014).

12.3 Other Balanced Translocations Involving *RUNX1* in Myeloid Malignancies

Besides the t(8;21)(q22;q22), several other recurrent chromosomal translocations involving the *RUNX1* locus have been described in myeloid neoplasms and are discussed in the following sections.

12.3.1 AML with t(3;21)(q26;q22); *RUNX1-MECOM*

The balanced translocation t(3;21)(q26;q22) was initially identified in patients with chronic myeloid leukemia (CML) in blast crisis, and subsequently found in 3.6% of patients with therapy-related AML or myelodysplastic syndromes (t-AML/t-MDS) (Rubin et al. 1987; Rubin et al. 1990). In a cohort of 6515 adult AML patients, the translocation occurred in only 0.14%. More recently, this translocation was also observed in CML evolving into myeloid blast crisis after TKI treatment (Paquette et al. 2011). According to the 2016 WHO classification of haematopoietic neoplasms, detection of t(3;21)(q26;q22) is sufficient to establish a diagnosis of "AML with myelodysplasia-related changes" in patients with ≥20% blasts.

The t(3;21)(q26;q22) was shown to lead to the formation of *RUNX1-EVI1* and *RUNX1-MDS1* fusion transcripts (Nucifora et al. 1994; Mitani et al. 1994). The *MDS1* and *EVI1* genes are located closely to each other in chromosome band 3q26, and splicing of the second exon of *MDS1* to the second exon of *EVI1* can lead to the formation of a chimeric *MDS1/EVI1* transcript. Due to this close relation, *MDS1* and *EVI1* now are designated the '*MDS1* and *EVI1* complex locus' (*MECOM*), and the fusion gene in t(3;21)(q26;q22) has thus been named *RUNX1-MECOM*. High *EVI1* expression is found in most patients with t(3;21)(q26;q22). *MDS1/EVI1* levels were also high in some patients with t(3;21), but absent in others, indicating that *RUNX1-MDS1/EVI1* as well as *RUNX1-EVI1* fusions may occur depending on the location of the breakpoint in band 3q26 relative to the *MECOM* locus (Lugthart et al. 2010). With regard to outcomes, t-AML with t(3;21) associated with shorter OS compared to t-AML with t(8;21) in one series (Slovak et al. 2002).

12.3.2 AML with t(16;21)(q24;q22); *RUNX1-CBFA2T3*

The t(16;21)(q24;q22) is a rare, but recurrent chromosomal alteration found in therapy-related myeloid neoplasms. Gamou and colleagues reported that in this translocation, *RUNX1* is fused to *CBFA2T3* (previously called *MTG16*), a member of the conserved ETO family of transcriptional corepressors that shares a high degree of homology with *RUNX1T1*, the *RUNX1* translocation partner in t(8;21) (Gamou et al. 1998; Davis et al. 2003). Only 24 patients with t(16;21)(q24;q22) are currently reported in the Mitelman Database of Chromosome Aberrations and Gene Fusions in Cancer, including 12 who also had trisomy 8, suggesting a possible association between the two alterations (Mitelman et al. 2016). The clinical significance of this translocation in AML is unknown.

12.3.3 Rare Recurrent Translocations in AML Involving *RUNX1*

A number of additional, very rare but recurrent translocations involving *RUNX1* have been described in AML. In t(1;21)(p36;q22), *RUNX1* is fused to the *PRDM16* gene, a member of the positive regulatory (PR) domain gene family with similarity to *MECOM* (Sakai et al. 2005). In t(1;21)(p22;q22), *RUNX1* is fused to the *CLCA2* calcium channel gene (Giguère and Hébert 2010). In the t(11;21)(p14;q22), the fusion partner is *KIAA1549L*, a poorly characterized gene with unknown function (Abe et al. 2012). Finally, the t(20;21)(q13.2;q22.12) results in a *ZFP64-RUNX1* fusion involving the zinc finger protein *ZFP6* (Richkind et al. 2000). The clinical significance of these alterations is unknown due to their rarity.

12.4 Association of *RUNX1* Translocations with Therapy-Related Neoplasia

Petersen-Biergard and colleagues first reported an association between chromosomal rearrangements involving chromosome band 21q22 and t-MDS or t-AML (Pedersen-Bjergaard and Philip 1991). In 2002, an international workshop identified balanced 21q22 translocations in 15.5% of patients with t-MDS or therapy-related acute leukemias (Slovak et al. 2002). The most common primary diseases were breast cancer, Hodgkin disease and non-Hodgkin lymphoma, and most patients had received topoisomerase II inhibitors and alkylating agents with or without radiotherapy. The median latency of the secondary hematologic disorder was 39 months, significantly longer than for therapy-induced neoplasms with rearrangements involving *KMT2A* (chromosome band 11q23) or *CBFB* (16q22). A t(8;21) was present in 56% of these patients, and 22 additional translocations with documented involvement of the *RUNX1* locus were found, including t(3;21) in 20% and t(16;21) in 5% of patients. In a small series of 13 patients with t-AML and t(8;21), a CR rate of >90% was observed; however 10 of the 13 patients died after a median of 19 months (Gustafson et al. 2009). Likewise, Krauth and colleagues reported that among patients with t(8;21), those with t-AML had shorter OS compared to *de novo* patients (Krauth et al. 2014). The limited data available from retrospective case series suggests that the prognosis of t-AML patients with *RUNX1* rearrangements other than t(8;21) is relatively poor, with a median survival of less than 1 year (Slovak et al. 2002).

12.5 *RUNX1* Point Mutations in Myeloid Malignancies

12.5.1 *RUNX1* Mutations in AML

When the *RUNX1* gene was initially identified in 1991, Miyoshi et al. described a transcript encoding a 250-amino acid (AA) protein that was later named isoform AML1a, and today is known as transcript variant 3 (Miyoshi et al. 1991). Subsequently, the same group identified two additional transcript variants encoding proteins of 453 and 480 AA, which were designated AML1b (transcript variant 2) and AML1c (transcript variant 1), respectively (Miyoshi et al. 1995). The N-terminus of AML1c differs from that of AML1a and AML1b due to the use of an alternative promoter. All 3 proteins share a highly conserved, 128-AA Runt domain, a protein motif responsible for both DNA binding and heterodimerization. AML1b and AML1c contain a large C-terminal transactivation domain. Currently, the NCBI Gene database lists 13 exons, and 10 alternatively spliced RefSeq transcript isoforms, while the Ensembl database lists 9 protein-coding isoforms.

In 1999, Osato and colleagues were the first to identify somatically acquired *RUNX1* point mutations in 8 of 160 patients with myeloid leukemia (7 AML and 1 CML in blast crisis) (Osato et al. 1999). These mutations, located in the Runt domain, either disturb DNA binding and/or lead to weakened nuclear expression of *RUNX1*. The Runt domain is located in exons 3–5 and ranges from position 50–178 in the 453 AA transcript (position 77–205 in the 480 AA transcript). Early *RUNX1* mutation screening studies therefore often focused on exons 3–5, and did not include exons 1 and 2 or the C-terminal exons encoding the transactivation domain. Aggregate data from multiple cohorts available through the Catalogue of Somatic Mutations in Cancer (Forbes et al. 2015; Schnittger et al. 2011; Tang et al. 2009), and data from our own patients (Metzeler et al. 2016) indicate that *RUNX1* missense mutations cluster in the Runt homology domain (spanning exons 3–5) and are predicted to interfere with DNA binding, while truncating (nonsense and frame shift) mutations are distributed along the entire coding sequence (Fig. 12.2).

The reported incidence of *RUNX1* mutations in AML varies widely between studies (Tang et al. 2009), ranging from 3% in a series of pediatric AML patients (Taketani et al. 2003) to 33% in a cohort of adults with non-complex karyotypes (Schnittger et al. 2011). This large variability may be due to different baseline characteristics of the patient populations under study (e.g., age range, ethnicity, selection of cytogenetic subgroups, and *de novo* vs. secondary AML), and differences in the methods and target regions for mutation analyses. In recent, relatively large adult AML cohorts, the incidence of *RUNX1* mutations generally was in the range of 5–15% (Osato et al. 2001; Tang et al. 2009; Gaidzik et al. 2011; Patel et al. 2012; The Cancer Genome Atlas Research Network 2013; Kihara et al. 2014). The Cancer Genome Atlas Research Network (TCGA) consortium identified *RUNX1* mutations in 10% of 200 AML adult patients studied by whole-genome or whole-exome sequencing (The Cancer Genome Atlas Research Network 2013). Notably, analyses of clonal hierarchies in this cohort suggested that *RUNX1* mutations always were part of the "founding clone" that initiated the disease (Miller et al. 2013).

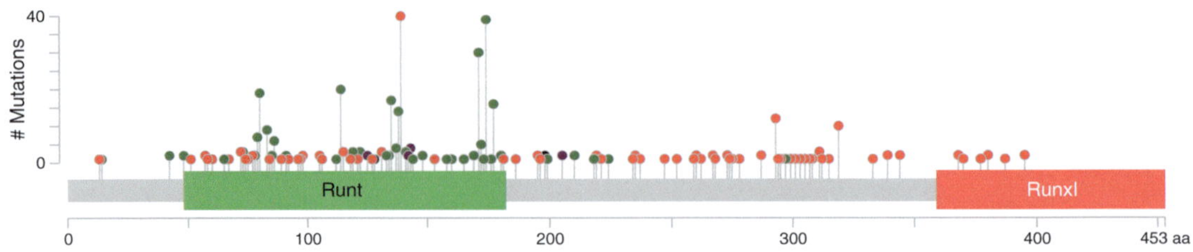

Fig. 12.2 Spectrum of somatic *RUNX1* point mutations in AML patients. Distribution of somatic *RUNX1* point mutations along the coding sequence of transcript variant 2 (NCBI accession number, NM_001001890.2). Truncating mutations (i.e., nonsense and frame shift changes) are shown in *red*, and missense variants are shown in *green* (Data are from the Catalogue of Somatic Mutations in Cancer (COSMIC) (Forbes et al. 2015) and from Metzeler et al. (2016))

12.5.2 Clinical Characteristics of AML Patients with Mutated *RUNX1*

Early studies indicated an association of somatic *RUNX1* mutations with FAB M0 morphology, and with secondary or treatment-related myeloid neoplasia (Asou 2003; Osato 2004). For example, Preudhomme and co-workers identified *RUNX1* mutations in ~10% of AML patients, and in 22% of patients with minimally differentiated (i.e., M0) AML. Twenty-one of 34 AML M0 patients in this series had biallelic *RUNX1* mutations, where a point mutation on one allele was accompanied by another point mutation or deletion of the second allele, while patients with non-M0 AML had monoallelic mutations (Preudhomme et al. 2000; Roumier et al. 2003). The association of *RUNX1* mutations with minimally differentiated AML was confirmed in larger series showing that 24–65% of patients with AML M0 carry *RUNX1* mutations (Dicker et al. 2007; Tang et al. 2009; Schnittger et al. 2011; Kao et al. 2014). Of note, in one large cohort that only included patients younger than 60 years, only 15% of M0 patients had mutated *RUNX1*, but the mutation was still enriched in this subgroup (Gaidzik et al. 2011). Tang and co-workers initially reported an association of *RUNX1* mutation with older age, an association that has been confirmed by multiple subsequent studies (Tang et al. 2009; Schnittger et al. 2011; Mendler et al. 2012; Greif et al. 2012). An association with male sex was also noted in some, but not all of these studies.

12.5.3 Cooperating Genetic Lesions in *RUNX1*-Mutated AML

RUNX1 mutations are found in patients with intermediate-risk (including cytogenetically normal AML) or unfavorable karyotypes, but are absent in those with favorable karyotypes (i.e., CBF leukemias including *RUNX1-RUNX1T1* rearranged AML, and APL) and in patients with balanced translocations involving 11q23 (*KMT2A; MLL*) (Tang et al. 2009; Gaidzik et al. 2011; The Cancer Genome Atlas Research Network 2013). *RUNX1* mutations are particularly common in patients with isolated trisomy 13, a rare cytogenetic subgroup with a *RUNX1* mutation frequency of 75–90% (Dicker et al. 2007; Schnittger et al. 2011; Herold et al. 2014). In several series, an association of *RUNX1* mutations with trisomy 8 was noted (Tang et al. 2009; Gaidzik et al. 2011; Alpermann et al. 2015). Accordingly, in two studies of adult AML patients with sole trisomy 8, *RUNX1* mutations were the most or second most common molecular alterations, occurring in 28% and 32% of patients, respectively (Becker et al. 2014; Alpermann et al. 2015). Finally, two studies reported an association between somatically acquired trisomy 21 and *RUNX1* mutations in AML (Preudhomme et al. 2000; Taketani et al. 2003). In one of these studies, the mutated *RUNX1* allele was present on two of the three copies of chromosome 21 in all 4 patients analyzed (Preudhomme et al. 2000). This indicates that trisomy 21 occurred as a secondary change after the *RUNX1* mutation, possibly acting by increasing the mutant-to-wild type allelic ratio.

With regard to coexisting molecular genetic alterations, *RUNX1* mutations were shown to be almost mutually exclusive with *NPM1* and *CEBPA* mutations in multiple AML cohorts (Tang et al. 2009; Schnittger et al. 2011; Gaidzik et al. 2011; Mendler et al. 2012; Greif et al. 2012; The Cancer Genome Atlas Research Network 2013). On the other hand, a positive association was observed between mutated *RUNX1* and presence of *KMT2A* (*MLL*) partial tandem duplications (*KMT2A*-PTD) (Tang et al. 2009; Schnittger et al. 2011; Gaidzik et al. 2011; Greif et al. 2012). *RUNX1* has subsequently been shown to interact directly with *KMT2A*. This interaction mediates histone H3K4 tri-methylation in the promoter region of the *SPI1* (PU.1) transcription factor that is involved in hematopoietic stem cell maintenance (Koh et al. 2013). Several groups also found a close association of mutations in *ASXL1* and *RUNX1*, with 22–44% of *RUNX1*-mutated patients carrying *ASXL1* mutations (Mendler et al. 2012; Schnittger et al. 2013; Paschka et al. 2015). Moreover, an association of *RUNX1* mutations with mutated *IDH2* was identified in two large studies (Gaidzik et al. 2011; The Cancer Genome Atlas Research Network 2013). While these associations suggest functional synergism between the *RUNX1* and frequently co-mutated genes, this has not yet been proven experimentally.

12.5.4 *RUNX1* Mutations and Prognosis in AML

Several relatively large patient cohorts provide information on the prognostic relevance of *RUNX1* gene mutations. Tang and colleagues identified *RUNX1* mutations in 13% of 470 patients with non-M3 AML (Tang et al. 2009). In the 330 patients who received standard induction and consolidation chemotherapy, multivariate analyses showed that *RUNX1* mutations associated with lower CR rates and shorter OS, but not with differences in DFS. In another cohort of 449 patients with normal or non-complex abnormal karyotypes who received non-uniform treatment, mutated *RUNX1* associated with inferior EFS and OS (Schnittger et al. 2011).

These two reports both showed an unfavorable impact of mutated *RUNX1* in subgroup analyses of patients with cytogenetically normal AML (CN-AML), and several studies focused on this cytogenetic subset. Mendler and colleagues identified *RUNX1* mutations in 12.5% of *de novo* CN-AML patients, and found an association with lower CR rates and inferior RFS, EFS and OS (Mendler et al. 2012). Similar results were obtained in multivariate analyses, and in subgroup analyses of younger (<60 years) and older (≥60 years) patients. Updated survival data are shown in Fig. 12.3. Of note, patients did not undergo alloSCT in first CR in this series. In another, smaller study focusing on CN-AML, an association between mutated *RUNX1* and inferior OS was observed in the entire cohort and in the subgroups of patients aged ≥60 years, and those within the European LeukemiaNet (ELN) Intermediate-I genetic group (Greif et al. 2012).

In the largest cohort published so far, Gaidzik and colleagues studied 945 patients aged 18–60 years treated on trials of the AMLSG. Only 5.6% of the patients carried *RUNX1* mutations, potentially due to the exclusion of older patients. *RUNX1* mutations associated with lower CR rate, and shorter RFS and OS. In multivariate analyses, *RUNX1* mutations were a significant predictor of shorter EFS and shorter RFS censored at the time of alloSCT, but not for CR rate, RFS without censoring, or OS (Gaidzik et al. 2011). In another relatively large cohort of 664 patients aged 18–86 years, *RUNX1* mutations associated with unfavorable OS only in patients younger than 60 years, and particularly in those with intermediate-risk cytogenetics (Metzeler et al. 2016).

At least two groups have addressed the impact of postremission therapy on outcomes of *RUNX1*-mutated patients. In the study by Tang and colleagues, *RUNX1* mutations were not associated with OS in the subset of patients who underwent alloSCT, suggesting that allografting might ameliorate the unfavorable prognostic impact of the mutation (Tang et al. 2009). In agreement with

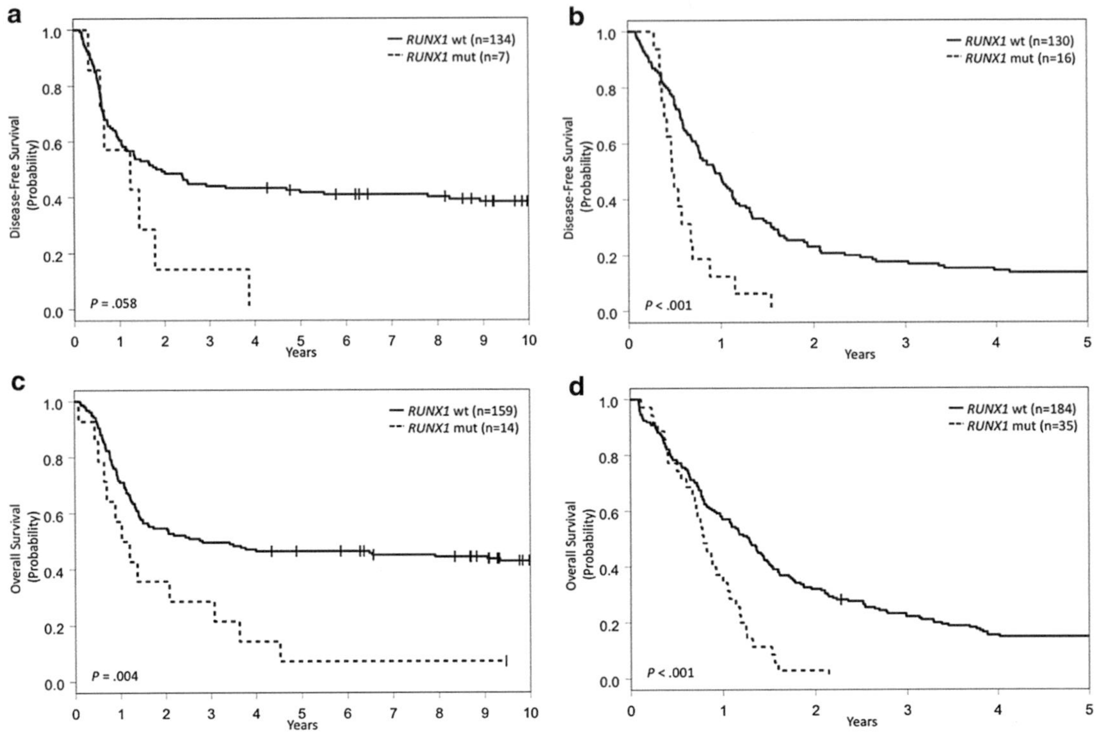

Fig. 12.3 *RUNX1* mutations and prognosis of patients with *de novo* cytogenetically normal AML. *Top*: Disease-free survival of patients with *de novo* cytogenetically normal AML (**a**) aged <60 years and (**b**) aged ≥60 years, according to *RUNX1* mutation status. *Bottom*: Overall survival of patients with *de novo* cytogenetically normal AML (**c**) aged <60 years and (**d**) aged ≥60 years, according to *RUNX1* mutation status (Bloomfield et al. unpublished data)

these results, Gaidzik et al. found that the RFS of *RUNX1* mutated patients who underwent alloSCT was comparable to *RUNX1*-wild type patients, while all patients who did not receive a transplant uniformly relapsed, most within 1 year (Gaidzik et al. 2011).

In summary, while univariate analyses consistently showed an unfavorable prognosis of *RUNX1*-mutated adult AML patients, the results of multivariate analyses adjusting for potential confounders are less clear. These discrepancies may be due to different baseline characteristics, for example regarding the age range of included patients, differences in treatment regimens including the use of alloSCT, and the effects of other genetic alterations that are considered in the multivariate models. With regard to the last point, several groups have recently tried to combine the prognostic information conveyed by various genetic alterations into integrative risk stratification algorithms. In two of these models, *RUNX1* mutations emerged as a factor associated with relatively unfavorable OS (Kihara et al. 2014; Grossmann et al. 2012), while in a third study, *RUNX1* mutations were found in only 5% of patients and were not included in the proposed risk stratification system (Patel et al. 2012).

Besides their prognostic relevance at baseline, *RUNX1* mutations could also serve as novel markers for MRD detection, yet the heterogeneity of the mutations make monitoring via conventional PCR assays difficult. This issue may be solved through the use of next-generation sequencing (NGS) techniques, as demonstrated in a cohort of 103 intensively treated, *RUNX1*-mutated patients with available follow-up samples (Kohlmann et al. 2014). Although the sensitivity of the NGS assay was relatively limited, residual disease was detected in 46 of the 103 patients at time points ranging from 60 to

198 days after initial diagnosis. Detectable *RUNX1* mutation associated with shorter EFS and OS. Notably, *RUNX1* mutations detected at the time of initial diagnosis were stable in relapsed disease in 51 of 57 evaluable patients (89%), while in 6 patients (11%), mutations were lost.

12.5.5 Reasons for the Different Outcomes of AML with Mutations and Balanced Translocations Involving *RUNX1*

Two major modes of *RUNX1* gene alteration, point mutations and balanced translocations leading to chimeric fusion genes, are found in AML. As outlined above, the clinical consequences of these two types of alterations are remarkably different. *RUNX1* point mutations generally associate with inferior outcomes and FAB M0 morphology, while the *RUNX1-RUNX1T1* gene fusion associates with favorable outcomes and a more differentiated (FAB M2) phenotype. The causes for these discrepant effects are not well understood. Among the potential factors that have been implicated are differences in the spectrum of co-mutated partner genes, differences in the residual *RUNX1* activity of the mutant allele, variable dominant-negative effects on the intact second allele, and effects of the translocation fusion partner in the case of balanced translocations (Osato et al. 2001).

12.5.6 *RUNX1* Point Mutations in Myelodysplastic Syndromes and Other Myeloid Malignancies

RUNX1 mutations in patients with MDS were first described in 2000 (Imai et al. 2000). Harada and colleagues subsequently found *RUNX1* mutations in 24% of MDS patients with elevated BM blasts (refractory anemia with excess blasts, RAEB) or post-MDS AML, while mutations were rarely observed in low-risk MDS without increased blast count (Harada et al. 2004). Of note, the frequency of *RUNX1* mutations was particularly high (50%) in a cohort of patients who developed MDS or AML after chemotherapy, radiotherapy or radiation exposure due to the atomic bombs used against Japan (Harada et al. 2003; Harada et al. 2004). A link between exposure to ionizing radiation and *RUNX1*-mutated myelodysplasia was confirmed by a study of radiation-exposed residents near a former Soviet nuclear test site (Zharlyganova et al. 2008). In three more recently published, larger series of MDS patients, *RUNX1* mutations were found in 8–11% of individuals (Bejar et al. 2011; Papaemmanuil et al. 2013; Haferlach et al. 2014). In two of these studies, analyses of recurrently mutated genes by targeted NGS revealed that mutated *RUNX1* frequently co-occur with mutations in *SRSF2*, *ASXL1*, *EZH2*, and *STAG2* (Papaemmanuil et al. 2013; Haferlach et al. 2014).

In MDS, *RUNX1* mutations are relatively rare in patients with low-risk disease and are found more frequently in patients with increased BM blasts (RAEB) and those transforming to post-MDS secondary AML (s-AML) (Papaemmanuil et al. 2013; Haferlach et al. 2014). In the light of this association, it is not surprising that mutated *RUNX1* associated with inferior leukemia-free survival in one cohort (Dicker et al. 2010). Moreover, *RUNX1* mutations associated with shorter OS even after adjustment for the International Prognostic Scoring System (IPSS) risk group and age in another series (Bejar et al. 2011). Finally, *RUNX1* mutations were included in a recently proposed 14-gene score that identifies MDS patients with shorter OS (Haferlach et al. 2014). In a study of 38 paired samples from MDS patients who later progressed to s-AML, 9 (24%) had *RUNX1* mutations already at the MDS stage, and only one *RUNX1*-wild type patient gained a mutation at the time of progression. Although the median time between MDS diagnosis and progression to s-AML was only 9 months, these data suggest that *RUNX1* mutations are a predisposing factor for s-AML transformation that is already present during the MDS phase, and not a marker that is acquired at the time of progression (Flach et al. 2011).

Mutations in *RUNX1* were also detected in 9–15% of chronic myelomonocytic leukemia (CMML), but do not seem to be prognostically relevant in this entity (Itzykson et al. 2013; Kohlmann et al. 2010). Finally, in a study of 70 patients with advanced, *KIT*-mutated systemic mastocytosis, *RUNX1* mutations were found in 23 % and associated with shorter OS (Jawhar et al. 2016).

12.6 Familial Platelet Disorder with Associated Myeloid Malignancy

Familial platelet disorder with associated myeloid malignancy (FPDMM; also known as familial platelet disorder with propensity to acute myelogenous leukemia, FPD/AML; Online Mendelian Inheritance in Man [OMIM] identifier, #601399) is an extremely rare, heritable condition caused by heterozygous germline *RUNX1* mutations. This syndrome was initially described in 1985 by Dowton and colleagues as an autosomal dominant disorder of platelet production and function in a large family with bleeding diathesis, and 6 members of the same family developed hematologic neoplasms (Dowton et al. 1985). Through linkage analysis, a critical region on chromosome 21 was identified in several affected families, and mutation analysis of regional candidate genes revealed mutations in *RUNX1* in six of the seven families (Ho et al. 1996; Song et al. 1999).

In the meantime, at least 20 affected kindreds have been described in the literature, and 19 of them were found to carry diverse types of *RUNX1* mutations including missense, frameshift and nonsense mutations as well as large deletions affecting the *RUNX1* locus (Preudhomme et al. 2009). Truncating changes lead to loss of the C-terminal transactivation domain resulting in haploinsufficiency of *RUNX1*. Missense mutations frequently affect conserved residues in the Runt domain that are involved in DNA binding, and may exert a dominant-negative effect on the remaining, intact allele through heterodimerization (Michaud et al. 2002). Patients with FPDMM typically present with mild thrombocytopenia, an "aspirin-like" platelet aggregation defect with abnormal response to epinephrine and arachidonic acid, a dense granule storage pool deficiency, and prolonged bleeding time. The lifetime incidence of leukemia among affected individuals is reported to be 20–50% (Osato 2004). In a series of asymptomatic individuals with germline *RUNX1* mutations aged <50 years, clonal hematopoiesis was detected in 67%, a proportion that is much higher than expected during normal aging (Churpek et al. 2015). Progression to AML is often accompanied by somatically acquired "second hits" (mutations or deletions) involving the second *RUNX1* allele, as well as gains of additional mutations in genes recurrently mutated in sporadic AML (Antony-Debré et al. 2016). Recently, somatic mutations in the *CDC25C* gene, which is not known to be mutated in sporadic AML, were reported in 7 of 13 FPDMM patients from Japan, including 4 of 7 patients who had developed AML. This finding was not reproduced in a European cohort (Yoshimi et al. 2014; Antony-Debré et al. 2016).

12.7 *RUNX1* Gene Alterations in Acute Lymphoblastic Leukemia

Although *RUNX1* translocations were first detected in AML, they are also found in acute leukemias of lymphoid lineage, particularly in childhood B-cell precursor acute lymphoblastic leukemia (BCP-ALL). The t(12;21)(p13;q22), leads to rearrangement of *RUNX1* with the *ETV6* gene. The resulting *ETV6-RUNX1* fusion transcript (previously designated *TEL-AML1*) is controlled by the *ETV6* promoter and contains the N-terminal "helix-loop-helix" (HLH) domains of *ETV6* fused to a large C-terminal part of the *RUNX1* coding sequence, including the Runt and transactivation domains. This is in contrast to *RUNX1* fusion genes found in AML, which are under the control of the *RUNX1* promoter and lack the *RUNX1* C-terminus including the transactivation domain (Golub et al. 1995; Romana et al. 1995a).

The t(12;21)(p13;q22) is commonly cryptic and missed by metaphase cytogenetics, and thus was initially considered to be a rare event.

However, using fluorescence-*in situ* hybridization, the *ETV6-RUNX1* rearrangement can be detected in approximately 15–35% of pediatric BCP-ALL, particularly in patients aged 1–9 years, while it is rare (<3%) in adult BCP-ALL (Romana et al. 1995b; Fears et al. 1996; Zelent et al. 2004). Thus, t(12;21)(p13;q22) is the most common structural chromosomal alteration in pediatric cancer. The translocation is accompanied by a deletion of the second *ETV6* allele on the other chromosome in >50% of patients, suggesting that loss of *ETV6* function plays a role in this disease (Raynaud et al. 1996; Schwab et al. 2013). Overall, *ETV6-RUNX1*-rearranged childhood BCP-ALL patients seem to harbor a relatively high number of copy number alterations including deletions of *CDKN2A*/B, *PAX5* and *BTG1*, each occurring in 15–20% of patients (Kim et al. 1996; Mullighan et al. 2007; Schwab et al. 2013). Furthermore, mutations in the histone H3K36 methyltransferase *NSD2* are found in 20% of *ETV6-RUNX1* rearranged childhood ALL (Jaffe et al. 2013). Several studies conclusively demonstrated that children with the *ETV6-RUNX1* rearrangement have excellent treatment outcomes (Shurtleff et al. 1995; Borkhardt et al. 1997; Moorman et al. 2010; Bhojwani et al. 2012), although in some series a high frequency of late relapses (≥5 years after diagnosis) was noted (Forestier et al. 2008). In one large study of 1725 children and adolescents with BCP-ALL, those with *ETV6-RUNX1* had ~50% reduced risk of relapse or death compared to other genetic subsets, with no late relapses. These associations persisted in multivariate analyses adjusting for other known risk factors (Moorman et al. 2010).

Studies of monozygotic twins and neonatal blood spots (Guthrie cards) revealed that the *RUNX1-ETV6* rearrangement is frequently acquired before birth, and BCP-ALL can develop in affected children with a reported latency of up to 14 years (Ford et al. 1998; Wiemels et al. 1999a, b). Notably, using highly sensitive assays, *RUNX1-ETV6* fusion transcripts can be found in up to 1% of cord blood samples from healthy newborns (Mori et al. 2002). In these children, the rearrangement is present in 1 of 10^3–10^4 mononuclear cells, indicating that the offspring of the single cell that initially acquired the translocation gained a proliferative advantage and underwent clonal expansion. On the other hand, the proportion of newborns with detectable *RUNX1-ETV6* transcripts in cord blood samples exceeds the incidence of *RUNX1-ETV6*-positive childhood BCP-ALL by a factor of 100, indicating that affected newborns have a low absolute risk of developing ALL, and that the acquisition of secondary genetic lesions is necessary for the development of overt leukemia. Although more rare, *RUNX1-RUNX1T1* fusion transcripts have also been detected in healthy newborns (Mori et al. 2002). Quantitative PCR assays have been developed for the detection of MRD in children with t(12;21) (Pallisgaard et al. 1999; Seeger et al. 2001; Drunat et al. 2001).

Besides the t(12;21), *RUNX1* point mutations have been found in sporadic childhood ALL (Song et al. 1999) and affect about 15% of children with early T-cell precursor (ETP)-ALL (Zhang et al. 2012). *RUNX1* mutations also occur in T-ALL developing in patients with FPDMM (Owen et al. 2008; Preudhomme et al. 2009; Prébet et al. 2013).

References

Abe, A., Katsumi, A., Kobayashi, M., et al. (2012). A novel RUNX1-C11orf41 fusion gene in a case of acute myeloid leukemia with a t(11;21)(p14;q22). *Cancer Genetics, 205*(11), 608–611.

Allen, C., Hills, R. K., Lamb, K., et al. (2013). The importance of relative mutant level for evaluating impact on outcome of KIT, FLT3 and CBL mutations in core-binding factor acute myeloid leukemia. *Leukemia, 27*(9), 1891–1901.

Alpermann, T., Haferlach, C., Eder, C., et al. (2015). AML with gain of chromosome 8 as the sole chromosomal abnormality (+8sole) is associated with a specific molecular mutation pattern including ASXL1 mutations in 46.8 % of the patients. *Leukemia Research, 39*(3), 265–272.

Antony-Debré, I., Duployez, N., Bucci, M., et al. (2016). Somatic mutations associated with leukemic progression of familial platelet disorder with predisposition to acute myeloid leukemia. *Leukemia, 30*(4), 999–1002.

Appelbaum, F. R., Kopecky, K. J., Tallman, M. S., et al. (2006). The clinical spectrum of adult acute myeloid leukaemia associated with core binding factor translocations. *British Journal of Haematology, 135*(2), 165–173.

Asou, N. (2003). The role of a runt domain transcription factor AML1/RUNX1 in leukemogenesis and its clinical implications. *Critical Reviews in Oncology/Hematology, 45*(2), 129–150.

Becker, H., Maharry, K., Mrózek, K., et al. (2014). Prognostic gene mutations and distinct gene- and microRNA-expression signatures in acute myeloid leukemia with a sole trisomy 8. *Leukemia, 28*(8), 1754–1758.

Beghini, A., Larizza, L., Cairoli, R., & Morra, E. (1998). c-kit activating mutations and mast cell proliferation in human leukemia. *Blood, 92*(2), 701–702.

Beghini, A., Peterlongo, P., Ripamonti, C. B., et al. (2000). C-kit mutations in core binding factor leukemias. *Blood, 95*(2), 726–728.

Beghini, A., Ripamonti, C. B., Cairoli, R., et al. (2004). KIT activating mutations: Incidence in adult and pediatric acute myeloid leukemia, and identification of an internal tandem duplication. *Haematologica, 89*(8), 920–925.

Bejar, R., Stevenson, K., Abdel-Wahab, O., et al. (2011). Clinical effect of point mutations in myelodysplastic syndromes. *The New England Journal of Medicine, 364*(26), 2496–2506.

Bhojwani, D., Pei, D., Sandlund, J. T., et al. (2012). ETV6-RUNX1-positive childhood acute lymphoblastic leukemia: Improved outcome with contemporary therapy. *Leukemia, 26*(2), 265–270.

Bloomfield, C. D., Goldman, A., Hassfeld, D., & de la Chapelle, A. (1984). Fourth international workshop on chromosomes in leukemia 1982: Clinical significance of chromosomal abnormalities in acute nonlymphoblastic leukemia. *Cancer Genetics and Cytogenetics, 11*(3), 332–350.

Bloomfield, C. D., Lawrence, D., Byrd, J. C., et al. (1998). Frequency of prolonged remission duration after high-dose cytarabine intensification in acute myeloid leukemia varies by cytogenetic subtype. *Cancer Research, 58*(18), 4173–4179.

Boissel, N., Leroy, H., Brethon, B., et al. (2006). Incidence and prognostic impact of c-Kit, FLT3, and Ras gene mutations in core binding factor acute myeloid leukemia (CBF-AML). *Leukemia, 20*(6), 965–970.

Boissel, N., Renneville, A., Leguay, T., et al. (2015). Dasatinib in high-risk core binding factor acute myeloid leukemia in first complete remission: A French acute myeloid leukemia intergroup trial. *Haematologica, 100*(6), 780–785.

Borkhardt, A., Cazzaniga, G., Viehmann, S., et al. (1997). Incidence and clinical relevance of TEL/AML1 fusion genes in children with acute lymphoblastic leukemia enrolled in the German and Italian multicenter therapy trials. Associazione Italiana Ematologia Oncologia Pediatrica and the Berlin-Frankfurt-Münster Study Group. *Blood, 90*(2), 571–577.

Bullinger, L., Döhner, K., Bair, E., et al. (2004). Use of gene-expression profiling to identify prognostic subclasses in adult acute myeloid leukemia. *The New England Journal of Medicine, 350*(16), 1605–1616.

Buonamici, S., Ottaviani, E., Testoni, N., et al. (2002). Real-time quantitation of minimal residual disease in inv(16)-positive acute myeloid leukemia may indicate risk for clinical relapse and may identify patients in a curable state. *Blood, 99*(2), 443–449.

Burnett, A. K., Hills, R. K., Milligan, D., et al. (2011). Identification of patients with acute myeloblastic leukemia who benefit from the addition of gemtuzumab ozogamicin: Results of the MRC AML15 trial. *Journal of Clinical Oncology, 29*(4), 369–377.

Burnett, A. K., Goldstone, A., Hills, R. K., et al. (2013). Curability of patients with acute myeloid leukemia who did not undergo transplantation in first remission. *Journal of Clinical Oncology, 31*(10), 1293–1301.

Burnett, A. K., Russell, N. H., Hills, R. K., et al. (2015). A randomized comparison of daunorubicin 90 mg/m2 vs 60 mg/m2 in AML induction: Results from the UK NCRI AML17 trial in 1206 patients. *Blood, 125*(25), 3878–3885.

Byrd, J. C., Mrózek, K., Dodge, R. K., et al. (2002). Pretreatment cytogenetic abnormalities are predictive of induction success, cumulative incidence of relapse, and overall survival in adult patients with de novo acute myeloid leukemia: Results from Cancer and Leukemia Group B (CALGB 8461). *Blood, 100*(13), 4325–4336.

Byrd, J. C., Ruppert, A. S., Mrózek, K., Carroll, A. J., et al. (2004). Repetitive cycles of high-dose cytarabine benefit patients with acute myeloid leukemia and inv(16)(p13q22) or t(16;16)(p13;q22): Results from CALGB 8461. *Journal of Clinical Oncology, 22*(6), 1087–1094.

Cairoli, R., Beghini, A., Grillo, G., et al. (2006). Prognostic impact of c-KIT mutations in core binding factor leukemias: An Italian retrospective study. *Blood, 107*(9), 3463–3468.

Cammenga, J., Horn, S., Bergholz, U., et al. (2005). Extracellular KIT receptor mutants, commonly found in core binding factor AML, are constitutively active and respond to imatinib mesylate. *Blood, 106*(12), 3958–3961.

Cancer Genome Atlas Research Network. (2013). Genomic and epigenomic landscapes of adult de novo acute myeloid leukemia. *The New England Journal of Medicine, 368*(22), 2059–2074.

Care, R. S., Valk, P. J., Goodeve, A. C., et al. (2003). Incidence and prognosis of c-KIT and FLT3 mutations in core binding factor (CBF) acute myeloid leukaemias. *British Journal of Haematology, 121*(5), 775–777.

Chevalier, N., Solari, M. L., Becker, H., et al. (2010). Robust in vivo differentiation of t(8;21)-positive acute myeloid leukemia blasts to neutrophilic granulocytes induced by treatment with dasatinib. *Leukemia, 24*(10), 1779–1781.

Churpek, J. E., Pyrtel, K., Kanchi, K. L., et al. (2015). Genomic analysis of germ line and somatic variants in familial myelodysplasia/acute myeloid leukemia. *Blood, 126*(22), 2484–2490.

Corbacioglu, A., Scholl, C., Schlenk, R. F., et al. (2010). Prognostic impact of minimal residual disease in CBFB-MYH11-positive acute myeloid leukemia. *Journal of Clinical Oncology, 28*(23), 3724–3729.

Cornelissen, J. J., van Putten, W. L., Verdonck, L. F., et al. (2007). Results of a HOVON/SAKK donor versus no-donor analysis of myeloablative HLA-identical sibling stem cell transplantation in first remission acute myeloid leukemia in young and middle-aged adults: Benefits for whom? *Blood, 109*(9), 3658–3666.

Creutzig, U., van den Heuvel-Eibrink, M. M., Gibson, B., et al. (2012). Diagnosis and management of acute myeloid leukemia in children and adolescents: Recommendations from an international expert panel. *Blood, 120*(16), 3187–3205.

Davis, J. N., McGhee, L., & Meyers, S. (2003). The ETO (MTG8) gene family. *Gene, 303*, 1–10.

Dicker, F., Haferlach, C., Kern, W., et al. (2007). Trisomy 13 is strongly associated with AML1/RUNX1 mutations and increased FLT3 expression in acute myeloid leukemia. *Blood, 110*(4), 1308–1316.

Dicker, F., Haferlach, C., Sundermann, J., et al. (2010). Mutation analysis for RUNX1, MLL-PTD, FLT3-ITD, NPM1 and NRAS in 269 patients with MDS or secondary AML. *Leukemia, 24*(8), 1528–1532.

Döhner, H., Estey, E. H., Amadori, S., et al. (2010). Diagnosis and management of acute myeloid leukemia in adults: Recommendations from an international expert panel, on behalf of the European LeukemiaNet. *Blood, 115*(3), 453–474.

Dowton, S. B., Beardsley, D., Jamison, D., et al. (1985). Studies of a familial platelet disorder. *Blood, 65*(3), 557–563.

Drunat, S., Olivi, M., Brunie, G., et al. (2001). Quantification of TEL-AML1 transcript for minimal residual disease assessment in childhood acute lymphoblastic leukaemia. *British Journal of Haematology, 114*(2), 281–289.

Duployez, N., Nibourel, O., Marceau-Renaut, A., et al. (2014). Minimal residual disease monitoring in t(8;21) acute myeloid leukemia based on RUNX1-RUNX1T1 fusion quantification on genomic DNA. *American Journal of Hematology, 89*(6), 610–615.

Duployez, N., Marceau-Renaut, A., Boissel, N., et al. (2016). Comprehensive mutational profiling of core binding factor acute myeloid leukemia. *Blood, 127*(20), 2451–2459.

Erickson, P., Gao, J., Chang, K. S., et al. (1992). Identification of breakpoints in t(8;21) acute myelogenous leukemia and isolation of a fusion transcript, AML1/ETO, with similarity to Drosophila segmentation gene, runt. *Blood, 80*(7), 1825–1831.

Fears, S., Vignon, C., Bohlander, S. K., et al. (1996). Correlation between the ETV6/CBFA2 (TEL/AML1) fusion gene and karyotypic abnormalities in children with B-cell precursor acute lymphoblastic leukemia. *Genes, Chromosomes and Cancer, 17*(2), 127–135.

Fenaux, P., Preudhomme, C., Lai, J.-L., et al. (1989). Cytogenetics and their prognostic value in de novo acute myeloid leukaemia: A report on 283 cases. *British Journal of Haematology, 73*(1), 61–67.

Flach, J., Dicker, F., Schnittger, S., et al. (2011). An accumulation of cytogenetic and molecular genetic events characterizes the progression from MDS to secondary AML: An analysis of 38 paired samples analyzed by cytogenetics, molecular mutation analysis and SNP microarray profiling. *Leukemia, 25*(4), 713–718.

Forbes, S. A., Beare, D., Gunasekaran, P., et al. (2015). COSMIC: Exploring the world's knowledge of somatic mutations in human cancer. *Nucleic Acids Research, 43*(Database issue), D805–D811.

Ford, A. M., Bennett, C. A., Price, C. M., et al. (1998). Fetal origins of the TEL-AML1 fusion gene in identical twins with leukemia. *Proceedings of the National Academy of Sciences USA, 95*(8), 4584–4588.

Forestier, E., Heyman, M., Andersen, M. K., et al. (2008). Outcome of ETV6/RUNX1-positive childhood acute lymphoblastic leukaemia in the NOPHO-ALL-1992 protocol: Frequent late relapses but good overall survival. *British Journal of Haematology, 140*(6), 665–672.

Gabert, J., Beillard, E., van der Velden, V. H. J., et al. (2003). Standardization and quality control studies of "real-time" quantitative reverse transcriptase polymerase chain reaction of fusion gene transcripts for residual disease detection in leukemia – A Europe Against Cancer program. *Leukemia, 17*(12), 2318–2357.

Gaidzik, V. I., Bullinger, L., Schlenk, R. F., et al. (2011). RUNX1 mutations in acute myeloid leukemia: Results from a comprehensive genetic and clinical analysis from the AML study group. *Journal of Clinical Oncology, 29*(10), 1364–1372.

Gamou, T., Kitamura, E., Hosoda, F., et al. (1998). The partner gene of AML1 in t(16;21) myeloid malignancies is a novel member of the MTG8(ETO) family. *Blood, 91*(11), 4028–4037.

Gari, M., Goodeve, A., Wilson, G., et al. (1999). c-kit proto-oncogene exon 8 in-frame deletion plus insertion mutations in acute myeloid leukaemia. *British Journal of Haematology, 105*(4), 894–900.

Garson, O. M., Hagemeijer, A., Sakurai, M., et al. (1989). Cytogenetic studies of 103 patients with acute myelogenous leukemia in relapse. *Cancer Genetics and Cytogenetics, 40*(2), 187–202.

Giguère, A., & Hébert, J. (2010). CLCA2, a novel RUNX1 partner gene in a therapy-related leukemia with t(1;21)(p22;q22). *Cancer Genetics and Cytogenetics, 202*(2), 94–100.

Golub, T. R., Barker, G. F., Bohlander, S. K., et al. (1995). Fusion of the TEL gene on 12p13 to the AML1 gene on 21q22 in acute lymphoblastic leukemia. *Proceedings of the National Academy of Sciences USA, 92*(11), 4917–4921.

Greif, P. A., Konstandin, N. P., Metzeler, K. H., et al. (2012). RUNX1 mutations in cytogenetically normal acute myeloid leukemia are associated with a poor prognosis and up-regulation of lymphoid genes. *Haematologica, 97*(12), 1909–1915.

Grimwade, D., Hills, R. K., Moorman, A. V., et al. (2010). Refinement of cytogenetic classification in acute myeloid leukemia: Determination of prognostic significance of rare recurring chromosomal abnormalities among 5876 younger adult patients treated in the United Kingdom Medical Research Council trials. *Blood, 116*(3), 354–365. doi:10.1182/blood-2009-11-254441.

Grossmann, V., Schnittger, S., Kohlmann, A., et al. (2012). A novel hierarchical prognostic model of AML solely based on molecular mutations. *Blood, 120*(15), 2963–2972.

Growney, J. D., Clark, J. J., Adelsperger, J., et al. (2005). Activation mutations of human c-KIT resistant to imatinib mesylate are sensitive to the tyrosine kinase inhibitor PKC412. *Blood, 106*(2), 721–724.

Gustafson, S. A., Lin, P., Chen, S. S., et al. (2009). Therapy-related acute myeloid leukemia with t(8;21)(q22;q22) shares many features with de novo acute myeloid leukemia with t(8;21)(q22;q22) but does not have a favorable outcome. *American Journal of Clinical Pathology, 131*(5), 647–655.

Haferlach, T., Nagata, Y., Grossmann, V., et al. (2014). Landscape of genetic lesions in 944 patients with myelodysplastic syndromes. *Leukemia, 28*(2), 241–247.

Harada, H., Harada, Y., Tanaka, H., et al. (2003). Implications of somatic mutations in the AML1 gene in radiation-associated and therapy-related myelodysplastic syndrome/acute myeloid leukemia. *Blood, 101*(2), 673–680.

Harada, H., Harada, Y., Niimi, H., et al. (2004). High incidence of somatic mutations in the AML1/RUNX1 gene in myelodysplastic syndrome and low blast percentage myeloid leukemia with myelodysplasia. *Blood, 103*(6), 2316–2324.

Harrison, C. J., Hills, R. K., Moorman, A. V., et al. (2010). Cytogenetics of childhood acute myeloid leukemia: United Kingdom Medical Research Council Treatment trials AML 10 and 12. *Journal of Clinical Oncology, 28*(16), 2673–2681.

Hartmann, L., Dutta, S., Opatz, S., et al. (2016). ZBTB7A mutations in acute myeloid leukemia with t(8;21) translocation. *Nature Communications, 7*, 11733.

Herold, T., Metzeler, K. H., Vosberg, S., et al. (2014). Isolated trisomy 13 defines a homogeneous AML subgroup with high frequency of mutations in spliceosome genes and poor prognosis. *Blood, 124*(8), 1304–1311.

Hills, R. K., Castaigne, S., Appelbaum, F. R., et al. (2014). Addition of gemtuzumab ozogamicin to induction chemotherapy in adult patients with acute myeloid leukaemia: A meta-analysis of individual patient data from randomised controlled trials. *Lancet Oncology, 15*(9), 986–996.

Ho, C. Y., Otterud, B., Legare, R. D., et al. (1996). Linkage of a familial platelet disorder with a propensity to develop myeloid malignancies to human chromosome 21q22.1-22.2. *Blood, 87*(12), 5218–5224.

Hospital, M. A., Prebet, T., Bertoli, S., et al. (2014). Core-binding factor acute myeloid leukemia in first relapse: A retrospective study from the French AML intergroup. *Blood, 124*(8), 1312–1319.

Imai, Y., Kurokawa, M., Izutsu, K., et al. (2000). Mutations of the AML1 gene in myelodysplastic syndrome and their functional implications in leukemogenesis. *Blood, 96*(9), 3154–3160.

Itzykson, R., Kosmider, O., Renneville, A., et al. (2013). Prognostic score including gene mutations in chronic myelomonocytic leukemia. *Journal of Clinical Oncology, 31*(19), 2428–2436.

Jaffe, J. D., Wang, Y., Chan, H. M., et al. (2013). Global chromatin profiling reveals NSD2 mutations in pediatric acute lymphoblastic leukemia. *Nature Genetics, 45*(11), 1386–1391.

Jawhar, M., Schwaab, J., Schnittger, S., et al. (2016). Additional mutations in SRSF2, ASXL1 and/or RUNX1 identify a high-risk group of patients with KIT D816V(+) advanced systemic mastocytosis. *Leukemia, 30*(1), 136–143.

Jourdan, E., Boissel, N., Chevret, S., et al. (2013). Prospective evaluation of gene mutations and minimal residual disease in patients with core binding factor acute myeloid leukemia. *Blood, 121*(12), 2213–2223.

Jurlander, J., Caligiuri, M. A., Ruutu, T., et al. (1996). Persistence of the AML1/ETO fusion transcript in patients treated with allogeneic bone marrow transplantation for t(8;21) leukemia. *Blood, 88*(6), 2183–2191.

Kao, H. W., Liang, D. C., Wu, J. H., et al. (2014). Gene mutation patterns in patients with minimally differentiated acute myeloid leukemia. *Neoplasia, 16*(6), 481–488.

Keating, M. J., Cork, A., Broach, Y., et al. (1987). Toward a clinically relevant cytogenetic classification of acute myelogenous leukemia. *Leukemia Research, 11*(2), 119–133.

Kihara, R., Nagata, Y., Kiyoi, H., et al. (2014). Comprehensive analysis of genetic alterations and their prognostic impacts in adult acute myeloid leukemia patients. *Leukemia, 28*(8), 1586–1595.

Kim, D. H., Moldwin, R. L., Vignon, C., et al. (1996). TEL-AML1 translocations with TEL and CDKN2 inactivation in acute lymphoblastic leukemia cell lines. *Blood, 88*(3), 785–794.

Koh, C. P., Wang, C. Q., Ng, C. E. L., et al. (2013). RUNX1 meets MLL: Epigenetic regulation of hematopoiesis by two leukemia genes. *Leukemia, 27*(9), 1793–1802.

Kohlmann, A., Grossmann, V., Klein, H. U., et al. (2010). Next-generation sequencing technology reveals a characteristic pattern of molecular mutations in 72.8 % of chronic myelomonocytic leukemia by detecting frequent alterations in TET2, CBL, RAS, and RUNX1. *Journal of Clinical Oncology, 28*(24), 3858–3865.

Kohlmann, A., Nadarajah, N., Alpermann, T., et al. (2014). Monitoring of residual disease by next-generation deep-sequencing of RUNX1 mutations can identify acute myeloid leukemia patients with resistant disease. *Leukemia, 28*(1), 129–137.

Koreth, J., Schlenk, R., Kopecky, K. J., et al. (2009). Allogeneic stem cell transplantation for acute myeloid leukemia in first complete remission: Systematic review and meta-analysis of prospective clinical trials. *JAMA, 301*(22), 2349–2361.

Krauter, J., Wattjes, M. P., Nagel, S., et al. (1999). Real-time RT-PCR for the detection and quantification of AML1/MTG8 fusion transcripts in t(8;21)-positive AML patients. *British Journal of Haematology, 107*(1), 80–85.

Krauter, J., Gorlich, K., Ottmann, O., et al. (2003). Prognostic value of minimal residual disease quantification by real-time reverse transcriptase polymerase chain reaction in patients with core binding factor leukemias. *Journal of Clinical Oncology, 21*(23), 4413–4422.

Krauth, M.-T., Eder, C., Alpermann, T., et al. (2014). High number of additional genetic lesions in acute myeloid leukemia with t(8;21)/RUNX1-RUNX1T1: Frequency and impact on clinical outcome. *Leukemia, 28*(7), 1449–1458.

Kuchenbauer, F., Schnittger, S., Look, T., et al. (2006). Identification of additional cytogenetic and molecular genetic abnormalities in acute myeloid leukaemia with t(8;21)/AML1-ETO. *British Journal of Haematology, 134*(6), 616–619.

Kühn, M. W., Radtke, I., Bullinger, L., et al. (2012). High-resolution genomic profiling of adult and pediatric core-binding factor acute myeloid leukemia reveals new recurrent genomic alterations. *Blood, 119*(10), e67–e75.

Kusec, R., Laczika, K., Knöbl, P., et al. (1994). AML1/ETO fusion mRNA can be detected in remission blood samples of all patients with t(8;21) acute myeloid leukemia after chemotherapy or autologous bone marrow transplantation. *Leukemia, 8*(5), 735–739.

Larson, R. A., Le Beau, M. M., Vardiman, J. W., et al. (1983). The predictive value of initial cytogenetic studies in 148 adults with acute nonlymphocytic leukemia: A 12-year study (1970–1982). *Cancer Genetics and Cytogenetics, 10*(3), 219–236.

Lavallée, V.-P., Lemieux, S., Boucher, G., et al. (2016). RNA-sequencing analysis of core binding factor AML identifies recurrent ZBTB7A mutations and defines RUNX1-CBFA2T3 fusion signature. *Blood, 127*(20), 2498–2501.

Le Beau, M. M., Larson, R. A., Bitter, M. A., et al. (1983). Association of an inversion of chromosome 16 with abnormal marrow eosinophils in acute myelomonocytic leukemia. A unique cytogenetic-clinicopathological association. *The New England Journal of Medicine, 309*(11), 630–636.

Leroy, H., de Botton, S., Grardel-Duflos, N., et al. (2005). Prognostic value of real-time quantitative PCR (RQ-PCR) in AML with t(8;21). *Leukemia, 19*(3), 367–372.

Leverger, G., Bernheim, A., Daniel, M. T., Flandrin, G., et al. (1988). Cytogenetic study of 130 childhood acute nonlymphocytic leukemias. *Medical and Pediatric Oncology, 16*(4), 227–232.

Liu, P., Tarlé, S. A., Hajra, A., et al. (1993). Fusion between transcription factor CBF beta/PEBP2 beta and a myosin heavy chain in acute myeloid leukemia. *Science, 261*(5124), 1041–1044.

Löwenberg, B., Ossenkoppele, G. J., van Putten, W., et al. (2009). High-dose daunorubicin in older patients with acute myeloid leukemia. *The New England Journal of Medicine, 361*(13), 1235–1248.

Löwenberg, B., Pabst, T., Vellenga, E., et al. (2011). Cytarabine dose for acute myeloid leukemia. *The New England Journal of Medicine, 364*(11), 1027–1036.

Lugthart, S., Gröschel, S., Beverloo, H. B., et al. (2010). Clinical, molecular, and prognostic significance of WHO type inv(3)(q21q26.2)/t(3;3)(q21;q26.2) and various other 3q abnormalities in acute myeloid leukemia. *Journal of Clinical Oncology, 28*(24), 3890–3898.

Marcucci, G., Livak, K. J., Bi, W., et al. (1998). Detection of minimal residual disease in patients with AML1/ETO-associated acute myeloid leukemia using a novel quantitative reverse transcription polymerase chain reaction assay. *Leukemia, 12*(9), 1482–1489.

Marcucci, G., Mrózek, K., Ruppert, A. S., et al. (2005). Prognostic factors and outcome of core binding factor acute myeloid leukemia patients with t(8;21) differ from those of patients with inv(16): A cancer and leukemia group B study. *Journal of Clinical Oncology, 23*(24), 5705–5717.

Mendler, J. H., Maharry, K., Radmacher, M. D., et al. (2012). RUNX1 mutations are associated with poor outcome in younger and older patients with cytogenetically normal acute myeloid leukemia and with distinct gene and MicroRNA expression signatures. *Journal of Clinical Oncology, 30*(25), 3109–3118.

Metzeler, K. H., Herold, T., Rothenberg-Thurley, M., et al. (2016). Spectrum and prognostic relevance of driver gene mutations in acute myeloid leukemia. *Blood, 128*(5), 686–698.

Michaud, J., Wu, F., Osato, M., et al. (2002). In vitro analyses of known and novel RUNX1/AML1 mutations in dominant familial platelet disorder with predisposition to acute myelogenous leukemia: Implications for mechanisms of pathogenesis. *Blood, 99*(4), 1364–1372.

Micol, J. B., Duployez, N., Boissel, N., et al. (2014). Frequent ASXL2 mutations in acute myeloid leukemia patients with t(8;21)/RUNX1-RUNX1T1 chromosomal translocations. *Blood, 124*(9), 1445–1449.

Miller, C. A., Wilson, R. K., & Ley, T. J. (2013). Genomic landscapes and clonality of de novo AML. *The New England Journal of Medicine, 369*(15), 1473.

Mitani, K., Ogawa, S., Tanaka, T., et al. (1994). Generation of the AML1-EVI-1 fusion gene in the t(3;21)(q26;q22) causes blastic crisis in chronic myelocytic leukemia. *EMBO Journal, 13*(3), 504–510.

Mitelman, F., Johansson, B., & Mertens, F. (Eds.). (2016). *Mitelman database of chromosome aberrations and gene fusions in cancer [Online]*. http://cgap.nci.nih.gov/Chromosomes/Mitelman.

Miyamoto, T., Nagafuji, K., Akashi, K., et al. (1996). Persistence of multipotent progenitors expressing

AML1/ETO transcripts in long-term remission patients with t(8;21) acute myelogenous leukemia. *Blood, 87*(11), 4789–4796.

Miyamoto, T., Weissman, I. L., & Akashi, K. (2000). AML1/ETO-expressing nonleukemic stem cells in acute myelogenous leukemia with 8;21 chromosomal translocation. *Proceedings of the National Academy of Sciences USA, 97*(13), 7521–7526.

Miyawaki, S., Ohtake, S., Fujisawa, S., et al. (2011). A randomized comparison of 4 courses of standard-dose multiagent chemotherapy versus 3 courses of high-dose cytarabine alone in postremission therapy for acute myeloid leukemia in adults: The JALSG AML201 study. *Blood, 117*(8), 2366–2372.

Miyoshi, H., Shimizu, K., Kozu, T., et al. (1991). t(8;21) breakpoints on chromosome 21 in acute myeloid leukemia are clustered within a limited region of a single gene, AML1. *Proceedings of the National Academy of Sciences USA, 88*(23), 10431–10434.

Miyoshi, H., Kozu, T., Shimizu, K., et al. (1993). The t(8;21) translocation in acute myeloid leukemia results in production of an AML1-MTG8 fusion transcript. *The EMBO Journal, 12*(7), 2715–2721.

Miyoshi, H., Ohira, M., SHimizu, K., et al. (1995). Alternative splicing and genomic structure of the AML1 gene involved in acute myeloid leukemia. *Nucleic Acids Research, 23*(14), 2762–2769.

Moorman, A. V., Ensor, H. M., Richards, S. M., et al. (2010). Prognostic effect of chromosomal abnormalities in childhood B-cell precursor acute lymphoblastic leukaemia: results from the UK Medical Research Council ALL97/99 randomised trial. *Lancet Oncology, 11*(5), 429–438.

Mori, H., Colman, S. M., Xiao, Z., et al. (2002). Chromosome translocations and covert leukemic clones are generated during normal fetal development. *Proceedings of the National Academy of Sciences USA, 99*(12), 8242–8247.

Morschhauser, F., Cayuela, J. M., Martini, S., et al. (2000). Evaluation of minimal residual disease using reverse-transcription polymerase chain reaction in t(8;21) acute myeloid leukemia: A multicenter study of 51 patients. *Journal of Clinical Oncology, 18*(4), 788–794.

Mrózek, K., Heerema, N. A., & Bloomfield, C. D. (2004). Cytogenetics in acute leukemia. *Blood Reviews, 18*(2), 115–136.

Mullighan, C. G., Goorha, S., Radtke, I., et al. (2007). Genome-wide analysis of genetic alterations in acute lymphoblastic leukaemia. *Nature, 446*(7137), 758–764.

Nanri, T., Matsuno, N., Kawakita, T., et al. (2005a). Imatinib mesylate for refractory acute myeloblastic leukemia harboring inv(16) and a C-KIT exon 8 mutation. *Leukemia, 19*(9), 1673–1675.

Nanri, T., Matsuno, N., Kawakita, T., et al. (2005b). Mutations in the receptor tyrosine kinase pathway are associated with clinical outcome in patients with acute myeloblastic leukemia harboring t(8;21)(q22;q22). *Leukemia, 19*(8), 1361–1366.

Nucifora, G., & Rowley, J. D. (1994). The AML1 and ETO genes in acute myeloid leukemia with a t(8;21). *Leukemia & Lymphoma, 14*(5–6), 353–362.

Nucifora, G., Larson, R. A., & Rowley, J. D. (1993). Persistence of the 8;21 translocation in patients with acute myeloid leukemia type M2 in long-term remission. *Blood, 82*(3), 712–715.

Nucifora, G., Begy, C. R., Kobayashi, H., et al. (1994). Consistent intergenic splicing and production of multiple transcripts between AML1 at 21q22 and unrelated genes at 3q26 in (3;21)(q26;q22) translocations. *Proceedings of the National Academy of Sciences USA, 91*(9), 4004–4008.

Osato, M. (2004). Point mutations in the RUNX1/AML1 gene: Another actor in RUNX leukemia. *Oncogene, 23*(24), 4284–4296.

Osato, M., Asou, N., Abdalla, E., et al. (1999). Biallelic and Heterozygous Point Mutations in the runt domain of the AML1/PEBP2alphaB gene associated eith myeloblastic leukemias. *Blood, 93*(6), 1817–1824.

Osato, M., Yanagida, M., Shigesada, K., & Ito, Y. (2001). Point mutations of the *RUNX1/AML1* gene in sporadic and familial myeloid leukemias. *International Journal of Hematology, 74*(3), 245–251.

Owen, C. J., Toze, C. L., Koochin, A., et al. (2008). Five new pedigrees with inherited RUNX1 mutations causing familial platelet disorder with propensity to myeloid malignancy. *Blood, 112*(12), 4639–4645.

Pallisgaard, N., Clausen, N., Schroder, H., & Hokland, P. (1999). Rapid and sensitive minimal residual disease detection in acute leukemia by quantitative real-time RT-PCR exemplified by t(12;21) TEL-AML1 fusion transcript. *Genes, Chromosomes and Cancer, 26*(4), 355–365.

Papaemmanuil, E., Gerstung, M., Malcovati, L., et al. (2013). Clinical and biological implications of driver mutations in myelodysplastic syndromes. *Blood, 122*(22), 3616–3627 -quiz 3699.

Paquette, R. L., Nicoll, J., Chalukya, M., et al. (2011). Frequent EVI1 translocations in myeloid blast crisis CML that evolves through tyrosine kinase inhibitors. *Cancer Genetics, 204*(7), 392–397.

Paschka, P., & Döhner, K. (2013). Core-binding factor acute myeloid leukemia: Can we improve on HiDAC consolidation? *Hematology. American Society of Hematology. Education Program, 2013*(1), 209–219.

Paschka, P., Marcucci, G., Ruppert, A. S., et al. (2006). Adverse prognostic significance of KIT mutations in adult acute myeloid leukemia with inv(16) and t(8;21): A cancer and leukemia group B study. *Journal of Clinical Oncology, 24*(24), 3904–3911.

Paschka, P., Du, J., Schlenk, R. F., et al. (2013). Secondary genetic lesions in acute myeloid leukemia with inv(16) or t(16;16): A study of the German-Austrian AML study group (AMLSG). *Blood, 121*(1), 170–177.

Paschka, P., Schlenk, R. F., Gaidzik, V. I., et al. (2015). ASXL1 mutations in younger adult patients with acute myeloid leukemia: A study by the German-Austrian acute myeloid leukemia study group. *Haematologica, 100*(3), 324–330.

Patel, J. P., Gönen, M., Figueroa, M. E., et al. (2012). Prognostic relevance of integrated genetic profiling in acute myeloid leukemia. *The New England Journal of Medicine, 366*(12), 1079–1089.

Pedersen-Bjergaard, J., & Philip, P. (1991). Balanced translocations involving chromosome bands 11q23 and 21q22 are highly characteristic of myelodysplasia and leukemia following therapy with cytostatic agents targeting at DNA-topoisomerase II. *Blood, 78*(4), 1147–1148.

Petersdorf, S. H., Kopecky, K. J., Slovak, M., et al. (2013). A phase 3 study of gemtuzumab ozogamicin during induction and postconsolidation therapy in younger patients with acute myeloid leukemia. *Blood, 121*(24), 4854–4860.

Pollard, J. A., Alonzo, T. A., Gerbing, R. B., et al. (2010). Prevalence and prognostic significance of KIT mutations in pediatric patients with core binding factor AML enrolled on serial pediatric cooperative trials for de novo AML. *Blood, 115*(12), 2372–2379.

Prébet, T., Boissel, N., Reutenauer, S., et al. (2009). Acute myeloid leukemia with translocation (8;21) or inversion (16) in elderly patients treated with conventional chemotherapy: A collaborative study of the French CBF-AML intergroup. *Journal of Clinical Oncology, 27*(28), 4747–4753.

Prébet, T., Carbuccia, N., Raslova, H., et al. (2013). Concomitant germ-line RUNX1 and acquired ASXL1 mutations in a T-cell acute lymphoblastic leukemia. *European Journal of Haematology, 91*(3), 277–279.

Prébet, T., Bertoli, S., Delaunay, J., et al. (2014). Anthracycline dose intensification improves molecular response and outcome of patients treated for core binding factor acute myeloid leukemia. *Haematologica, 99*(10), e185–e187.

Preudhomme, C., Warot-Loze, D., Roumier, C., et al. (2000). High incidence of biallelic point mutations in the Runt domain of the AML1/PEBP2 alpha B gene in Mo acute myeloid leukemia and in myeloid malignancies with acquired trisomy 21. *Blood, 96*(8), 2862–2869.

Preudhomme, C., Renneville, A., Bourdon, V., et al. (2009). High frequency of RUNX1 biallelic alteration in acute myeloid leukemia secondary to familial platelet disorder. *Blood, 113*(22), 5583–5587.

Raimondi, S. C., Chang, M. N., Ravindranath, Y., et al. (1999). Chromosomal abnormalities in 478 children with acute myeloid leukemia: Clinical characteristics and treatment outcome in a cooperative pediatric oncology group study-POG 8821. *Blood, 94*(11), 3707–3716.

Raynaud, S., Cave, H., Baens, M., et al. (1996). The 12;21 translocation involving TEL and deletion of the other TEL allele: Two frequently associated alterations found in childhood acute lymphoblastic leukemia. *Blood, 87*(7), 2891–2899.

Richkind, K., Hromas, R., Lytle, C., et al. (2000). Identification of two new translocations that disrupt the AML1 gene. *Cancer Genetics, 122*(2), 141–143.

Romana, S. P., Mauchauffé, M., Le Coniat, M., et al. (1995a). The t(12;21) of acute lymphoblastic leukemia results in a tel-AML1 gene fusion. *Blood, 85*(12), 3662–3670.

Romana, S. P., Poirel, H., Leconiat, M., et al. (1995b). High frequency of t(12;21) in childhood B-lineage acute lymphoblastic leukemia. *Blood, 86*(11), 4263–4269.

Roumier, C., Fenaux, P., Lafage, M., et al. (2003). New mechanisms of AML1 gene alteration in hematological malignancies. *Leukemia, 17*(1), 9–16.

Rowley, J. D. (1973). Identificaton of a translocation with quinacrine fluorescence in a patient with acute leukemia. *Annales de Génétique, 16*(2), 109–112.

Rowley, J. D. (1980). General report on the second international workshop on chromosomes in leukemia. *International Journal of Cancer, 26*(5), 531–533.

Rowley, J. D., & de la Chapelle, A. (1978). General report on the first international workshop on chromosomes in leukemia. *International Journal of Cancer, 21*(3), 307–308.

Rubin, C. M., Larson, R. A., Bitter, M. A., et al. (1987). Association of a chromosomal 3;21 translocation with the blast phase of chronic myelogenous leukemia. *Blood, 70*(5), 1338–1342.

Rubin, C. M., Larson, R. A., Anastasi, J., et al. (1990). t(3;21)(q26;q22): A recurring chromosomal abnormality in therapy-related myelodysplastic syndrome and acute myeloid leukemia. *Blood, 76*(12), 2594–2598.

Sakai, I., Tamura, T., Narumi, H., et al. (2005). Novel RUNX1-PRDM16 fusion transcripts in a patient with acute myeloid leukemia showing t(1;21)(p36;q22). *Genes, Chromosomes and Cancer, 44*(3), 265–270.

Satake, N., Maseki, N., Kozu, T., et al. (1995). Disappearance of AML1-MTG8(ETO) fusion transcript in acute myeloid leukaemia patients with t(8;21) in long-term remission. *British Journal of Haematology, 91*(4), 892–898.

Schittenhelm, M. M., Shiraga, S., Schroeder, A., et al. (2006). Dasatinib (BMS-354825), a dual SRC/ABL kinase inhibitor, inhibits the kinase activity of wild-type, juxtamembrane, and activation loop mutant KIT isoforms associated with human malignancies. *Cancer Research, 66*(1), 473–481.

Schlenk, R. F., Benner, A., Krauter, J., et al. (2004). Individual patient data-based meta-analysis of patients aged 16 to 60 years with core binding factor acute myeloid leukemia: A survey of the German acute myeloid leukemia intergroup. *Journal of Clinical Oncology, 22*(18), 3741–3750.

Schlenk, R. F., Pasquini, M. C., Pérez, W. S., et al. (2008). HLA-identical sibling allogeneic transplants versus chemotherapy in acute myelogenous leukemia with t(8;21) in first complete remission: Collaborative study between the German AML intergroup and CIBMTR. *Biology of Blood and Marrow Transplantation, 14*(2), 187–196.

Schnittger, S., Kohl, T. M., Haferlach, T., et al. (2006). KIT-D816 mutations in AML1-ETO-positive AML are associated with impaired event-free and overall survival. *Blood, 107*(5), 1791–1799.

Schnittger, S., Dicker, F., Kern, W., et al. (2011). RUNX1 mutations are frequent in de novo AML with noncomplex karyotype and confer an unfavorable prognosis. *Blood, 117*(8), 2348–2357.

Schnittger, S., Eder, C., Jeromin, S., et al. (2013). ASXL1 exon 12 mutations are frequent in AML with intermediate risk karyotype and are independently associated with an adverse outcome. *Leukemia, 27*(1), 82–91.

Schoch, C., Haase, D., Haferlach, T., et al. (1996). Fifty-one patients with acute myeloid leukemia and translocation t(8;21)(q22;q22): An additional deletion in 9q is an adverse prognostic factor. *Leukemia, 10*(8), 1288–1295.

Schwab, C. J., Chilton, L., Morrison, H., et al. (2013). Genes commonly deleted in childhood B-cell precursor acute lymphoblastic leukemia: Association with cytogenetics and clinical features. *Haematologica, 98*(7), 1081–1088.

Seeger, K., Kreuzer, K. A., Lass, U., et al. (2001). Molecular quantification of response to therapy and remission status in TEL-AML1-positive childhood ALL by real-time reverse transcription polymerase chain reaction. *Cancer Research, 61*(6), 2517–2522.

Shurtleff, S. A., Buijs, A., Behm, F. G., et al. (1995). TEL/AML1 fusion resulting from a cryptic t(12;21) is the most common genetic lesion in pediatric ALL and defines a subgroup of patients with an excellent prognosis. *Leukemia, 9*(12), 1985–1989.

Slovak, M. L., Kopecky, K. J., Cassileth, P. A., et al. (2000). Karyotypic analysis predicts outcome of preremission and postremission therapy in adult acute myeloid leukemia: A Southwest Oncology Group/Eastern Cooperative Oncology Group Study. *Blood, 96*(13), 4075–4083.

Slovak, M. L., Bedell, V., Popplewell, L., et al. (2002). 21q22 balanced chromosome aberrations in therapy-related hematopoietic disorders: Report from an international workshop. *Genes, Chromosomes and Cancer, 33*(4), 379–394.

Song, W. J., Sullivan, M. G., Legare, R. D., et al. (1999). Haploinsufficiency of CBFA2 causes familial thrombocytopenia with propensity to develop acute myelogenous leukaemia. *Nature Genetics, 23*(2), 166–175.

Sood, R., Hansen, N. F., Donovan, F. X., et al. (2016). Somatic mutational landscape of AML with inv(16) or t(8;21) identifies patterns of clonal evolution in relapse leukemia. *Leukemia, 30*(2), 501–504.

Speck, N. A., & Gilliland, D. G. (2002). Core-binding factors in haematopoiesis and leukaemia. *Nature Reviews Cancer, 2*(7), 502–513.

Swansbury, G. J., Lawler, S. D., Alimena, G., et al. (1994). Long-term survival in acute myelogenous leukemia: A second follow-up of the fourth international workshop on chromosomes in leukemia. *Cancer Genetics and Cytogenetics, 73*(1), 1–7.

Taketani, T., Taki, T., Takita, J., et al. (2003). AML1/RUNX1 mutations are infrequent, but related to AML-M0, acquired trisomy 21, and leukemic transformation in pediatric hematologic malignancies. *Genes, Chromosomes and Cancer, 38*(1), 1–7.

Tang, J. L., Hou, H. A., Chen, C. Y., et al. (2009). AML1/RUNX1 mutations in 470 adult patients with de novo acute myeloid leukemia: Prognostic implication and interaction with other gene alterations. *Blood, 114*(26), 5352–5361.

Tobal, K., & Yin, J. A. (1996). Monitoring of minimal residual disease by quantitative reverse transcriptase-polymerase chain reaction for AML1-MTG8 transcripts in AML-M2 with t(8; 21). *Blood, 88*(10), 3704–3709.

Tobal, K., Newton, J., Macheta, M., et al. (2000). Molecular quantitation of minimal residual disease in acute myeloid leukemia with t(8;21) can identify patients in durable remission and predict clinical relapse. *Blood, 95*(3), 815–819.

Valk, P. J., Verhaak, R. G., Beijen, M. A., et al. (2004). Prognostically useful gene-expression profiles in acute myeloid leukemia. *The New England Journal of Medicine, 350*(16), 1617–1628.

van Dongen, J. J. M., Macintyre, E. A., Gabert, J. A., et al. (1999). Standardized RT-PCR analysis of fusion gene transcripts from chromosome aberrations in acute leukemia for detection of minimal residual disease. Report of the BIOMED-1 Concerted Action: Investigation of minimal residual disease in acute leukemia. *Leukemia, 13*(12), 1901–1928.

von Neuhoff, C., Reinhardt, D., Sander, A., et al. (2010). Prognostic impact of specific chromosomal aberrations in a large group of pediatric patients with acute myeloid leukemia treated uniformly according to trial AML-BFM 98. *Journal of Clinical Oncology, 28*(16), 2682–2689.

Wang, Y. Y., Zhou, G. B., Yin, T., et al. (2005). AML1-ETO and C-KIT mutation/overexpression in t(8;21) leukemia: Implication in stepwise leukemogenesis and response to Gleeve. *Proceedings of the National Academy of Sciences USA, 102*(4), 1104–1109.

Wang, Y., Wu, D.-P., Liu, Q.-F., et al. (2014). In adults with t(8;21)AML, posttransplant RUNX1/RUNX1T1-based MRD monitoring, rather than c-KIT mutations, allows further risk stratification. *Blood, 124*(12), 1880–1886.

Wiemels, J. L., Cazzaniga, G., Daniotti, M., et al. (1999a). Prenatal origin of acute lymphoblastic leukaemia in children. *Lancet, 354*(9189), 1499–1503.

Wiemels, J. L., Ford, A. M., Van Wering, E. R., et al. (1999b). Protracted and variable latency of acute lymphoblastic leukemia after TEL-AML1 gene fusion in utero. *Blood, 94*(3), 1057–1062.

Willekens, C., Blanchet, O., Renneville, A., et al. (2016). Prospective long-term minimal residual disease monitoring using RQ-PCR in RUNX1-RUNX1T1-positive acute myeloid leukemia: Results of the French CBF-2006 trial. *Haematologica, 101*(3), 328–335.

Yin, J. A., O'Brien, M. A., Hills, R. K., et al. (2012). Minimal residual disease monitoring by quantitative RT-PCR in core binding factor AML allows risk stratification and predicts relapse: Results of the United Kingdom MRC AML-15 trial. *Blood, 120*(14), 2826–2835.

Yoshimi, A., Toya, T., Kawazu, M., et al. (2014). Recurrent CDC25C mutations drive malignant transformation in FPD/AML. *Nature Communications, 5*, 4770.

Zelent, A., Greaves, M., & Enver, T. (2004). Role of the TEL-AML1 fusion gene in the molecular pathogenesis of childhood acute lymphoblastic leukaemia. *Oncogene, 23*(24), 4275–4283.

Zhang, J., Ding, L., Holmfeldt, L., et al. (2012). The genetic basis of early T-cell precursor acute lymphoblastic leukaemia. *Nature, 481*(7380), 157–163.

Zharlyganova, D., Harada, H., Harada, Y., et al. (2008). High frequency of AML1/RUNX1 point mutations in radiation-associated myelodysplastic syndrome around semipalatinsk nuclear test site. *Journal of Radiation Research, 49*(5), 549–155.

Zhu, H. H., Zhang, X. H., Qin, Y. Z., et al. (2013). MRD-directed risk stratification treatment may improve outcomes of t(8;21) AML in the first complete remission: Results from the AML05 multicenter trial. *Blood, 121*(20), 4056–4062.

Mechanism of ETV6-RUNX1 Leukemia

13

Aishwarya Sundaresh and Owen Williams

Abstract

The t(12;21)(p13;q22) translocation is the most frequently occurring single genetic abnormality in pediatric leukemia. This translocation results in the fusion of the *ETV6* and *RUNX1* genes. Since its discovery in the 1990s, the function of the *ETV6-RUNX1* fusion gene has attracted intense interest. In this chapter, we will summarize current knowledge on the clinical significance of *ETV6-RUNX1*, the experimental models used to unravel its function in leukemogenesis, the identification of co-operating mutations and the mechanisms responsible for their acquisition, the function of the encoded transcription factor and finally, the future therapeutic approaches available to mitigate the associated disease.

Keywords

TEL-AML1 • Oncogene • Leukemia • Transcription factor • Signalling

13.1 Introduction

Chromosomal translocations are frequently the initiating genetic event in leukemogenesis, arising prenatally to generate a fusion gene encoding chimeric proteins. The resultant fusion genes often involve genes that encode transcription factors, which play key roles in the regulation of normal hematopoiesis.

A. Sundaresh • O. Williams (✉)
Cancer section, Developmental Biology and Cancer Programme, UCL Institute of Child Health, London, UK
e-mail: owen.williams@ucl.ac.uk

The t(12;21)(p13;q22) translocation results in the fusion of the *ETV6* (*TEL*) and *RUNX1* (*AML1*) genes, giving rise to the *ETV6-RUNX1* (*TEL-AML1*) fusion gene. These genes encode transcription factors that play important roles in hematopoiesis and deficiency in either results in failure of embryonic hematopoiesis (Okuda et al. 1996; Wang et al. 1996, 1997). In the adult, however, whereas *Etv6* is required for HSC survival (Hock et al. 2004) and establishment of bone marrow hematopoiesis (Wang et al. 1998), *Runx1* is predominantly necessary for maturation of more committed progenitors (Ichikawa et al. 2004; Growney et al. 2005; Putz et al. 2006). Germline variants and/or mutations in both genes

Y. Groner et al. (eds.), *RUNX Proteins in Development and Cancer*, Advances in Experimental Medicine and Biology 962, DOI 10.1007/978-981-10-3233-2_13

have been linked with predisposition to leukemia (Osato et al. 1999; Song et al. 1999; Moriyama et al. 2015; Noetzli et al. 2015; Topka et al. 2015; Zhang et al. 2015), and both genes are frequently disrupted by numerous different translocations in lymphoblastic and myeloid leukemias (Bohlander 2005; De Braekeleer et al. 2011).

ETV6-RUNX1 is the most common genetic alteration in acute lymphoblastic leukemia, exclusively associated with pre-B cell leukemia and present in up to 25% of cases (Golub et al. 1995). In patients, the t(12;21) translocation occurs during B cell differentiation prior to the onset of immunoglobulin gene rearrangement, giving rise to leukemic blasts that appear to be blocked at the pre-B cell stage (Romana et al. 1995; Hotfilder et al. 2002; Pine et al. 2003; Panzer-Grumayer et al. 2005). The most immature ETV6-RUNX1 population is identified by the aberrant $CD34^+CD38^-CD19^+$ phenotype, associated with an early stage of B cell lineage commitment (Castor et al. 2005; Hong et al. 2008).

Despite the prevalence of the t(12;21) translocation in childhood pre-B ALL, it is unlikely that expression of the *ETV6-RUNX1* fusion gene alone is responsible for the resulting leukemia. This is consistent with the substantial variation in onset of leukemia in identical twins with concordant t(12;21) ALL (Wiemels et al. 1999) and the high frequency of *ETV6-RUNX1* fusions found in unselected normal cord blood samples (Mori et al. 2002). Interestingly, estimation of the frequency of $ETV6\text{-}RUNX1^+$ cells in these cord blood samples suggested that acquisition of the fusion was associated with clonal expansion of progenitor cells (Mori et al. 2002). These studies have led to the hypothesis that the *ETV6-RUNX1* fusion results in the generation and expansion of a covert pre-leukemic clone that can persist for many years before acquiring secondary mutations and producing overt leukemia (Greaves 1999; Greaves and Wiemels 2003). More recently, whereas one study arrived at a similar estimate (Zuna et al. 2011), another found a much lower incidence of the fusion in normal cord blood samples (Lausten-Thomsen et al. 2011). Although there are many potential explanations for these differences (Brown 2011), whatever the frequency in the normal population, experimental models also indicate that ETV6-RUNX1 is insufficient to induce overt leukemia.

13.2 Experimental Models

Substantial insight into the role of ETV6-RUNX1 in leukemia induction has been provided by *in vitro* and *in vivo* experimental models. In 2001, Andreasson et al. cloned the *ETV6-RUNX1* fusion gene into a retroviral vector and used it to transduce growth factor dependent murine hematopoietic cell lines. Although expression of the fusion transcript was observed, ETV6-RUNX1 was unable to induce growth factor independence. Furthermore, transgenic mice expressing the fusion gene in lymphoid cells under the control of a B cell-specific immunoglobulin heavy chain enhancer failed to develop any malignant disorder or hematopoietic abnormalities (Andreasson et al. 2001). In line with this finding, transduction of adult bone marrow cells with a retroviral vector expressing the fusion gene and transplantation of these into syngeneic mice, only resulted in a low incidence of leukemia after a long latency (Bernardin et al. 2002). Furthermore, deficiency in $p16^{INK4a}p19^{ARF}$ was shown to cooperate with ETV6-RUNX1 in leukemia induction, providing the first evidence that secondary genetic aberrations may play a critical role in t(12;21) associated leukemia. Similarly, expression of a *ETV6-RUNX1* transgene in mice heterozygous for *Cdkn2a* was found to predispose to radiation-induced B malignancies (Li et al. 2013).

Since studies in twins originally suggested that t(12;21) pre-B ALL had a fetal origin (Ford et al. 1998), we examined the effect of ETV6-RUNX1 expression on the function of fetal liver hematopoietic progenitor cells (HPC) (Morrow et al. 2004). In this model, ETV6-RUNX1 enhanced the self-renewal capacity of B cell precursors *in vitro*, and hematopoietic reconstitution by transduced fetal liver HPC *in vivo*, with no apparent block in B cell development. Two further studies also used retroviral transduction to examine the effect of ETV6-RUNX1 expression on hematopoiesis *in vivo*, this time using adult bone marrow derived HPC (Tsuzuki et al. 2004; Fischer et al. 2005). Both studies found that whereas ETV6-RUNX1 did not affect B progenitor cell colony formation *in vitro*, it did cause a block in B cell maturation *in vivo*, resulting in a reduced number of ETV6-RUNX1$^+$ mature B cells. Taken together, these studies provided the first evidence that ETV6-RUNX1 perturbs normal hematopoiesis. The differences between the models regarding *in vitro* B progenitor self-renewal and the presence of a block in B cell development *in vivo* most likely result from differences in levels of ETV6-RUNX1 protein expressed by the different retroviral constructs used in each study and/or the nature of HPC targeted. In this respect, the FMEV vector used in the study by Fischer et al. (2005) has been optimized for high level expression in HSC (Schwieger et al. 2002). This possibility was recently tested using retroviral vectors expressing high and low levels of ETV6-RUNX1 (Tsuzuki and Seto 2013). Interestingly, high levels of ETV6-RUNX1 expression in mouse fetal liver HPC or pro-B cells were found to compromise *in vitro* B cell development, whereas low level expression promoted significant B progenitor self-renewal. The latter was not evident in transduced pro-B cells derived from adult bone marrow. Upon transplantation, low level ETV6-RUNX1 expression in fetal liver pro-B cells resulted in a B cell differentiation block, although this block was incomplete and mature ETV6-RUNX1$^+$ B cells did eventually emerge (Tsuzuki and Seto 2013).

In order to recapitulate expression of the ETV6-RUNX1 fusion gene more accurately, two studies introduced human *RUNX1* into the murine *Etv6* locus, such that the resulting *Etv6-RUNX1* fusion gene was expressed under the control of the endogenous *Etv6* locus (Schindler et al. 2009; van der Weyden et al. 2011). Although ETV6-RUNX1 expression alone did not induce leukemia in either model, both studies demonstrated that it predisposed to leukemia and lymphoma, either upon chemical (Schindler et al. 2009) or transposon-mediated (van der Weyden et al. 2011) mutagenesis. In the former model, conditional expression of the fusion gene in HSC increased the incidence of progenitor and T cell leukemias, and shortened disease latencies, induced by N-Nitroso-N-ethylurea exposure (Schindler et al. 2009). Interestingly, in van der Weyden's model, transposon-mediated mutagenesis resulted in increased incidence of pre-B ALL, as well as AML and T cell leukemias (van der Weyden et al. 2011). The differences in overall leukemia incidence and the subtypes of leukemia induced between these two models is most likely explained by the nature of the mutagen in each case, although differences in genetic backgrounds and fusion gene expression pattern (conditional versus constitutive) may also play a role.

In exploring how ETV6-RUNX1 alters normal hematopoiesis, Schindler et al. noted that expression of *Etv6-RUNX1* in adult bone marrow resulted in increased numbers of quiescent hematopoietic stem cells (HSC) and a loss of lymphoid committed progenitors, and transplantation of ETV6-RUNX1$^+$ fetal liver cells into adult mice failed to reconstitute even the most immature B lineage compartment. In contrast the fusion gene did not affect B cell maturation in the embryo but, consistent with data described above (Morrow et al. 2004), induced a transient increase in the self-renewal capacity of fetal liver-derived ETV6-RUNX1$^+$ B cell progenitors *in vitro* (Schindler et al. 2009). Indeed, such a direct effect of ETV6-RUNX1 on B cell progenitors may explain the observed increase in frequency of ETV6-RUNX1$^+$ cells detected in the absence of disease (Mori et al. 2002).

Interestingly, human cord blood HPC transduced with a lentivirus vector expressing ETV6-RUNX1 were found to generate the aberrant

$CD34^+CD38^-CD19^+$ B lineage population previously found to be associated with t(12;21) ALL (Castor et al. 2005), in transplanted mice (Hong et al. 2008). Furthermore, this aberrant B progenitor population demonstrated enhanced *in vitro* self-renewal as well as the capacity to regenerate itself as well as more mature B cell populations *in vivo*. However, consistent with the murine models, ETV6-RUNX1 expressing human HPC did not induce leukemia in transplanted mice, leading to the suggestion that $CD34^+CD38^-CD19^+$ cells represent a pre-leukemic B lineage population (Hong et al. 2008). This study was extended to show that ETV6-RUNX1 was only able to promote engraftment of this pre-leukemic population when expressed in HSC and not more committed progenitors (Fan et al. 2015), suggesting that the fusion maintains rather than induces self-renewal programs in the cell of origin.

In summary, all the experimental models described above favor the hypothesis that ETV6-RUNX1 enhances self-renewal of B cell progenitors but requires co-operating secondary genetic abnormalities for these aberrant cells to transform into overt leukemia.

13.3 Co-operating Genetic Lesions

Early studies demonstrated that deletions affecting the normal untranslocated *ETV6* allele were prevalent in t(12;21) ALL (Golub et al. 1995; Kim et al. 1996; Raynaud et al. 1996; Romana et al. 1996). Indeed, a recent study reported that in 143 ETV6-RUNX1 patient samples analyzed for additional genetic lesions, over half showed complete deletion of the normal *ETV6* allele (Stams et al. 2006). The frequent occurrence of *ETV6* deletions in t(12;21) patients suggested that loss of *ETV6* expression could be the genetic event that co-operates with ETV6-RUNX1 to induce leukemia. This could be attributed to loss of the tumor suppressor function of the normal ETV6 protein (Fenrick et al. 2000; Rompaey et al. 2000; Irvin et al. 2003). However, these deletions were often found to be sub-clonal (Romana et al. 1996; Maia et al. 2001), distinct in identical twins with concordant leukemia (Maia et al. 2001) and discordant in late-relapse versus diagnostic samples (Ford et al. 2001), suggesting they are most likely secondary genetic lesions associated with leukemia progression, rather than being essential for leukemic transformation.

To identify co-operating secondary events in ALL, Mullighan et al. performed a genome-wide screen of leukemic cells from 242 pediatric ALL cases (Mullighan et al. 2007). Interestingly, as well as detecting frequent abnormalities affecting genes involved in G1/S cell cycle progression, this study demonstrated that 40% of the ALL cases studied had lesions affecting genes involved in B cell maturation. These data were confirmed by an independent study of 40 pediatric ALL cases (Kuiper et al. 2007). The *PAX5* gene, which encodes the B cell lineage-specific activator protein (BSAP) essential for B cell commitment, was found to be one of the most frequently affected genes, altered by both deletions and mutations (Mullighan et al. 2007). In addition, deletions were found in genes such as *EBF1*, *IKZF1*, *E2A*, *LEF1* and *IKZF3*, which encode transcription factors directly involved in B lineage maturation pathways. In ETV6-RUNX1$^+$ ALL, deletions were more frequent than amplifications and were present at an average of 6 deletions per case. Interestingly, 28 % contained focal mono-allelic *PAX5* deletions without any apparent mutations in the remaining allele (Mullighan et al. 2007). In an independent study, focal deletions in *PAX5* were detected in 20% of ETV6-RUNX1$^+$ cases (Lilljebjorn et al. 2010). The relevance of these lesions is underlined by a recent study that demonstrated an increased incidence of pre-B ALL induced by viral and chemical mutagenesis in mice heterozygous for a loss-of-function *Pax5* allele (Dang et al. 2015), suggesting that the loss of one *PAX5* allele may indeed synergize with ETV6-RUNX1 to induce leukemia. More evidence for this possibility came from the demonstration that loss of one *Pax5* allele increased the proportion of B lineage ALL induced by transposon mediated mutagenesis in *Etv6-RUNX1* mice (van der Weyden et al. 2015).

Previous studies have hypothesized that many of the translocations and deletions commonly found in lymphoid malignancies result from the recognition of cryptic recombination signal sequences (RSS), close to the breakpoints of affected loci, by the RAG complex and subsequent illegitimate recombination events (Marculescu et al. 2002; Raschke et al. 2005; Mullighan et al. 2008a; Zhang and Swanson 2008; Iacobucci et al. 2009; Numata et al. 2010; Waanders et al. 2012). The RAG recombinase, encoded by the recombination activating genes 1 and 2 (*RAG1* and *RAG2*), is required for the sequential rearrangement of immunoglobulin and T cell receptor *V(D)J* loci, during B and T cell development, respectively. Although expression of these genes is tightly controlled during both B and T cell differentiation, most pre-B ALL express both genes constitutively (Bories et al. 1991), and elevated *RAG-1* gene expression is specifically associated with the t(12;21) ALL subtype (Ross et al. 2003; Haferlach et al. 2010).

Recently, whole genome and exome sequencing was used to systematically examine the association between RSS and the genomic breakpoints of structural variations found in t(12;21) ALL (Papaemmanuil et al. 2014). Interestingly, canonical RSS were found at 12%, truncated RSS motifs at 40 % and non-templated nucleotide sequences, characteristic of terminal deoxynucleotidyl transferase (TdT) activity normally associated with *V(D)J* recombination, at 70% of these breakpoints. Most of the genomic structural variants were found at the promoter and enhancer regions of active genes, including genes frequently affected by genomic alterations in pre-B ALL such as *ETV6*, *BTG1*, *TBL1XR1*, *RAG2* and *CDKN2A-CDKN2B*. Indeed, ETV6-RUNX1 associated *BTG1* deletions were previously shown to contain RSS at both ends of almost every deletion examined (Waanders et al. 2012). This suggests that the secondary deletion events driving leukemic transformation in the presence of *ETV6-RUNX1* result mainly from genomic rearrangements mediated by continuous aberrant RAG activity (Papaemmanuil et al. 2014).

Another enzyme linked with leukemia-associated acquired lesions is the activation-induced cytidine deaminase (AID), normally involved in immunoglobulin class switching and somatic hypermutation in germinal center B cells (Di Noia and Neuberger 2007). Low level AID expression in early B cells has been suggested to contribute to chromosomal translocations in combination with the RAG complex (Tsai et al. 2008). By virtue of its additional function as a structure specific endonuclease, the RAG complex was proposed to nick mismatches created by AID-mediated deamination of methylcytosine residues within CpG hotspots, resulting in lesion-specific double strand breaks. A more recent study found that many of the genes recurrently mutated in pre-B ALL are potential targets of AID (Swaminathan et al. 2015). In normal pre-B cells, AID expression was found to increase at the developmental stage associated with pre-BCR signaling-mediated IL7R downregulation, concomitant with increased RAG expression. Interestingly, stimulation of pre-B cells with inflammatory lipopolysaccharide (LPS) in the absence of IL7 induced further increases in AID expression. Since ETV6-RUNX1 was shown to induce expression of RAG complex genes in transduced mouse pre-B cells, this suggested an additional mechanism for mutagenesis in t(12;21) ALL. Indeed, further experiments demonstrated that cycles of LPS treatment, in the context of IL7 withdrawal, co-operated with ETV6-RUNX1 to induce AID- and RAG-dependent pre-B ALL (Swaminathan et al. 2015). This data is consistent with the hypothesis suggesting that abnormal or delayed infections of children with common pathogens may play a causative role in leukemogenesis (Greaves 1988). A previous study demonstrated that ETV6-RUNX1 confers resistance to TGF-β signaling in B cell precursors, providing a potential mechanism for clonal expansion of ETV6-RUNX1^{+} pre-leukemic cells in the presence of systemic TGF-β production (Ford et al. 2009). The study by Swaminathan et al. suggests that independent of whether expansion of the pre-leukemic ETV6-RUNX1^{+} cells is caused directly by the fusion gene itself (Morrow et al. 2004; Schindler et al. 2009) or by suppressing responses to immunomodulatory mediators (Ford et al. 2009), pathogen exposure may induce accumulation of secondary mutations and progression to overt leukemia (Swaminathan et al. 2015).

Genome-wide analysis of acquired copy number alterations in paired diagnostic and relapse pediatric ALL found that in some cases the relapse clone did not contain lesions present in the diagnostic clone, suggesting that both were derived from an ancestral clone present at diagnosis (Mullighan et al. 2008b). Anderson et al. recently used multiplexed fluorescence *in situ* hybridization to examine acquired copy number alterations in *ETV6*, *PAX5*, *CDKN2A*, as well as presence of the *ETV6-RUNX1* fusion gene, in single cells from 30 pediatric t(12;21) ALL cases (Anderson et al. 2011). This study found significant heterogeneity in the sub-clonal composition of these leukaemias at both diagnosis and relapse, suggesting that the clonal architecture of the disease is inherently dynamic and continually evolving. Interestingly, acquisition of the secondary lesions in sub-clones was not found to occur in any preferential order (Anderson et al. 2011). Single-cell genomic analysis was more recently used to dissect the heterogeneity of genetic abnormalities in t(12;21) ALL (Gawad et al. 2014). This study demonstrated significant heterogeneity at the level of point mutations in these leukemias, as well as the previously described structural lesions. Interestingly the point mutations were proposed to occur at later stages of clonal evolution and have hallmarks of apolipoprotein B mRNA editing enzyme, catalytic polypeptide-like (APOBEC) mutagenesis (Gawad et al. 2014).

These studies are consistent with the hypothesis that continuous illegitimate RAG- and AID-mediated recombination is a major driver of sub-clonal diversity (Papaemmanuil et al. 2014; Swaminathan et al. 2015). Thus, continued expression of ETV6-RUNX1 and associated induction of *RAG-1/2* gene expression (Ross et al. 2003; Haferlach et al. 2010), may be the underlying cause of the independent and recurrent acquisition of these genetic lesions. In this case, whether disease relapses resulted from expansion of minor sub-clones, already present at diagnosis, or from covert pre-leukemic clones following acquisition of new driver mutations, in all cases the *ETV6-RUNX1* fusion gene would be present in all of the sub-clones. This underlines the importance of studying the function of ETV6-RUNX1 in ALL cells and whether they require its continued expression for survival and progression.

13.4 ETV6-RUNX1 Function in t(12;21) Leukemia

The *ETV6-RUNX1* transcript encodes the N-terminal non-DNA binding moiety of ETV6 fused to almost the entire RUNX1 protein (Romana et al. 1995; Shurtleff et al. 1995). RUNX1 was originally thought to function as a transcriptional activator, positively regulating genes essential for normal hematopoiesis (Hernandez-Munain and Krangel 1994; Uchida et al. 1997; Kitabayashi et al. 1998; Okada et al. 1998). Furthermore, it was shown to co-operate with other transcription factors to drive target gene expression (Wotton et al. 1994; Petrovick et al. 1998). However, RUNX1 has also been shown to associate with a number of different co-repressor complexes and in certain contexts to mediate transcriptional repression (Takahashi et al. 1995; Aronson et al. 1997; Imai et al. 1998; Levanon et al. 1998; Fenrick et al. 1999; Javed et al. 2000; Lutterbach et al. 2000; Durst and Hiebert 2004). Therefore, the effect of RUNX1 on target gene transcription may depend on the balance between associated positive and negative cofactor complexes (Brettingham-Moore et al. 2015). The ETV6-RUNX1 fusion protein retains the ability to bind to RUNX1 target genes, by virtue of the retained RUNT domain, and inhibit RUNX1-dependent transcription, most likely by ETV6-mediated recruitment of nuclear receptor co-repressor (N-CoR)/histone deacetylase (HDAC) and mSin3A/HDAC complexes (Hiebert et al. 1996; Fears et al. 1997; Chakrabarti and Nucifora 1999; Fenrick et al. 2000; Guidez et al.

2000). Interestingly, as well as recruiting co-repressor complexes, the oligomerization properties of the ETV6 pointed domain have also been found to be essential for transcriptional repression (Lopez et al. 1999), although recent experiments suggest that the induced oligomerization is not sufficient to explain ETV6-RUNX1 activity in primary hematopoietic cells (Fischer et al. 2005). The ETV6-RUNX1 fusion has therefore been suggested to change the context dependent regulation of transcription by RUNX1 into constitutive repression of target gene expression (Zelent et al. 2004).

However, recent data from *Etv6-RUNX1* knock-in models point to a more complex function. Previous experiments demonstrated that heterozygous *Runx1-Eto* knock-in mice die during embryogenesis (Yergeau et al. 1997; Okuda et al. 1998). This is reminiscent of the constitutive *Runx1* deficient mice (Okuda et al. 1996; Wang et al. 1996) and suggests that the AML1-ETO fusion does act as a dominant negative inhibitor of wild-type RUNX1 function *in vivo*. In contrast, heterozygous *Etv6-RUNX1* mice displayed limited embryonal lethality, dependent on the background mouse strain used (Schindler et al. 2009; van der Weyden et al. 2011). Conditional expression of *Etv6-RUNX1* in adult mice (Schindler et al. 2009) revealed some similarities with the effects of conditional *Runx1* deletion on adult hematopoiesis (Ichikawa et al. 2004; Growney et al. 2005; Putz et al. 2006). However, whereas the *Etv6-RUNX1* allele was found to increase the number of adult HSC, conditional *Runx1* deletion resulted in their loss (Schindler et al. 2009). Thus, rather than recapitulating Runx1 loss in HSC, ETV6-RUNX1 would appear to have opposite effects. This suggests that the function of ETV6-RUNX1 cannot be restricted to its inhibition of RUNX1 activity.

An additional mechanism of action was suggested by the demonstration that ETV6-RUNX1 is able to dimerize with wild-type ETV6 (McLean et al. 1996), through interactions between the pointed domains of both proteins, and disrupt its activity (Gunji et al. 2004). This suggested that although the wild-type *ETV6* allele is lost in many t(12;21) ALL cases, interaction of ETV6 with the fusion protein could potentially play a role in disease initiation. However, the differences between the *Etv6-RUNX1* knock-in (Schindler et al. 2009) and *Etv6* deficient mice regarding embryonic lethality, yolk sac angiogenesis and HSC survival (Wang et al. 1997, 1998; Hock et al. 2004) suggest that ETV6-RUNX1 function is also not limited to inhibition of wild-type ETV6 function. Interestingly, mutation of the RUNX1 RUNT domain demonstrated that ETV6-RUNX1 requires both the ability to bind DNA (Morrow et al. 2007) and the CBFβ cofactor (Roudaia et al. 2009), which increases the affinity of RUNX1 for DNA, to promote pre-B cell self-renewal *in vitro*. This is consistent with ETV6-RUNX1 acting directly as a transcription factor, whatever its effects on RUNX1 target gene expression. A recent study examined changes in gene and protein expression in Ba/F3 cells conditionally expressing the ETV6-RUNX1 fusion, combining global gene expression analysis and stable isotope labelling by amino acids in cell culture (SILAC) (Linka et al. 2013). Using a novel anti-ETV6-RUNX1 antibody, they then used chromatin immunoprecipitation (ChIP) to determine which of the regulated genes was bound by the fusion protein. Although the bound promoters were found to be enriched in consensus RUNX1 binding motifs, most of these were not bound by either RUNX1 or ETV6 in these cells. This study indicates that although ETV6-RUNX1 binds DNA at potential RUNX1 binding sequences and regulates transcription, many of the genes it regulates are distinct to those regulated by RUNX1 and ETV6.

One way used to investigate ETV6-RUNX1 function is to examine gene expression patterns associated with the t(12;21) ALL patient samples. For example, high erythropoietin receptor (*EPOR*) gene expression is specifically associated with t(12;21) ALL (Fine et al. 2004). Interestingly, expression of ETV6-RUNX1 in Ba/F3 cells was found to induce *EPOR* (Inthal et al. 2008). Furthermore, EPO was found to increase proliferation and viability of ETV6-RUNX1⁺ leukemia cells. A more recent study confirmed that the *EPOR* gene is a direct transcriptional target of ETV6-RUNX1 and suggested that the effect of EPO on ETV6-RUNX1⁺ cell survival was via STAT5 mediated upregulation of BCL-XL expression (Torrano et al. 2011). These studies suggest that EPO induced signalling may be involved in the maintenance of ETV6-RUNX1⁺ pre-leukaemia and leukaemia cells *in vivo*.

Although ETV6-RUNX1 expression alone is not sufficient to cause leukemia, data from a number of recent studies have established that maintenance fusion protein expression is required for t(12;21) leukemia cell survival. In the first of these studies, silencing fusion gene expression, using *ETV6-RUNX1*-specific small interfering RNA (siRNA), induced apoptosis in the REH t(12;21) ALL cell line (Diakos et al. 2007). This was accompanied by decreased expression of HSP90 and survivin. Further experiments demonstrated that the effects on survivin expression were mediated by direct repression of micro RNA (miRNA) 494 and miRNA-320a expression by ETV6-RUNX1 (Diakos et al. 2010). Both miRNA were found to be under-expressed in t(12;21) ALL patients.

A modified approach, using short hairpin RNA (shRNA) directed against the *ETV6-RUNX1* fusion, was then used to silence fusion gene expression in REH cells. Global gene expression analysis revealed that the fusion protein was actively involved in regulating a variety of essential cellular pathways, including cell survival and stem cell self-renewal pathways (Fuka et al. 2011). Fusion gene silencing by shRNA was also shown to induce apoptosis in t(12;21) ALL cell lines and to impair leukemia engraftment and disease progression in recipient immunodeficient mice (Fuka et al. 2012). Cell death was accompanied by decreased phosphoinositide 3-kinase (PI3K)/AKT/mammalian target of rapamycin (mTOR) signalling, indicating that ETV6-RUNX1 maintains PI3K/AKT/mTOR activation. Pharmacological inhibition of the latter preferentially induced apoptosis in t(12;21) ALL cell lines and patient samples. This model was also used to demonstrate that *MDM2*, encoding the negative regulator of p53, is a direct transcriptional target of ETV6-RUNX1 and that p53 target genes were de-repressed following fusion gene silencing in t(12;21) cells (Kaindl et al. 2014). Elevated expression of *MDM2* was also found to be associated with the t(12;21) subtype of ALL. Finally, reactivation of p53 signalling by treatment of ETV6-RUNX1⁺ cells with Nutlin-3, a small molecule inhibitor of the interaction between MDM2 and p53, induced growth arrest and apoptosis.

We used *ETV6*-specific shRNA to target *ETV6-RUNX1* expression in REH cells, which lack the wild type *ETV6* allele (Mangolini et al. 2013). *ETV6-RUNX1* silencing resulted in growth inhibition and reduced colony formation. This study demonstrated that ETV6-RUNX1⁺ leukemia cells are dependent on aberrant STAT3 activity for survival and *in vivo* leukemia progression, and that STAT3 activation is mediated through ETV6-RUNX1 induced RAC1 activation. Furthermore, t(12;21) primary ALL cells were also susceptible to pharmacological STAT3 inhibition.

13.5 Therapeutic Implications

In general, ETV6-RUNX1⁺ ALL patients have been found to have a favorable prognosis (Loh et al. 2006; Rubnitz et al. 2008; Gandemer et al. 2012; Moorman et al. 2014; Pui et al. 2014), t(12;21) ALL cells being particularly susceptible to steroid, vincristine and L-asparaginase treatment *in vitro* (Ramakers-van Woerden et al. 2000; Krishna Narla et al. 2001). In contrast, other studies failed to demonstrate any independent significant effect of the translocation on treatment outcome (Hann et al. 2001; Seeger et al. 2001). Indeed, this subtype has been associated with a high incidence of late disease relapse (Harbott et al. 1997; Seeger et al. 1998, 1999; Forestier et al. 2008; Krentz et al. 2013) and ETV6-RUNX1⁺ patients have been found to make up a substantial proportion of relapsed ALL cases, questioning the prognostic significance of this abnormality at diagnosis (Bokemeyer et al. 2014). It has been suggested that differences in risk stratification, and consequent intensity of chemotherapy, as well as the type of treatment received may explain some of these differences in treatment outcome and relapse incidence (Loh et al. 2006). This has been used as a cautionary argument against reducing treatment intensity solely based on ETV6-RUNX1 status, the danger being that this may lead to an increased incidence of disease relapse (Loh et al. 2006). In addition to the prognosis of the group as a whole, within the t(12;21) ALL subtype, presence of the reciprocal *AML1-TEL* fusion (Stams et al. 2005) and additional copy number alterations (Stams et al. 2006; Bokemeyer et al. 2014) were found to affect treatment outcome. Relapsed ETV6-RUNX1$^+$ ALL cases frequently had deletions affecting genes associated with glucocorticoid signaling, and deletions involving the glucocorticoid receptor gene *NR3C1* were in particular associated with poor response to induction therapy (Kuster et al. 2011; Bokemeyer et al. 2014; Grausenburger et al. 2015).

Some light on the origin of disease relapse in t(12;21) ALL came from comparison of genomic TE-AML1 breakpoints, normal *ETV6* allele deletion and immunoglobulin/T cell receptor *V(D)J* rearrangements in diagnosis versus relapse. These studies suggested that in some cases, although both clones had identical *ETV6-RUNX1* breakpoints, the relapse clone did not emerge from the major clone at diagnosis (Ford et al. 2001; Konrad et al. 2003). This led to the hypothesis that relapse resulted from the incomplete elimination of pre-leukemic cells during initial therapy and their subsequent progression into leukemia upon acquisition of secondary transforming genetic aberrations (Ford et al. 2001; Konrad et al. 2003). This would fit with the length of time taken for many relapses to become manifest. It is also consistent with the slow response to therapy of minor clones detected at diagnosis, presumably representing pre-leukemic cells, and their much faster elimination by therapy at relapse, at which stage the pre-leukemic cells would have evolved into overt ALL (Konrad et al. 2003).

Taken together, these studies suggest that ETV6-RUNX1$^+$ ALL would benefit from novel, less toxic and more effective therapies. Although acquisition of particular secondary abnormalities has been associated with a poorer prognosis in ETV6-RUNX1$^+$ ALL, the significant degree of sub-clonal heterogeneity in this disease suggests that minor clones may be spared by therapies targeting pathways deregulated by these additional lesions. Furthermore, therapies that spare preleukemic cells may give rise to higher relapse rates. One solution would be to identify therapies that target the one indispensable abnormality common to preleukemic cells and all leukemia sub-clones, the ETV6-RUNX1 fusion protein itself. This may be particularly important in the case of ETV6-RUNX1, which is thought to actively contribute to acquisition of further genetic abnormalities by induction of aberrant *RAG* gene expression (Swaminathan et al. 2015). Significant progress has been made in elucidating the critical transcriptional and signaling pathways regulated by ETV6-RUNX1. Thus, studies indicating that ETV6-RUNX1$^+$ ALL cells appear to be particularly susceptible to PI3K/AKT/mTOR (Fuka et al. 2012), MDM2 (Kaindl et al. 2014) and STAT3 (Mangolini et al. 2013) inhibition, suggest that promising therapeutic approaches for this disease may be derived from targeting these pathways (Fig. 13.1).

13.6 Conclusions

From its first description over 20 years ago as one of the most common genetic aberrations in pediatric leukemia, the *ETV6-RUNX1* fusion has been the subject of intensive research. As outlined above, this work has given important insights into our understanding of this disease, but also it has served as a paradigm for leukemia and, in some aspects, cancer in general (Greaves 2009). This highlights the value of examining the detailed molecular mechanisms operating in this leukemia, both to our understanding of disease etiology and to future clinical advances in patient care.

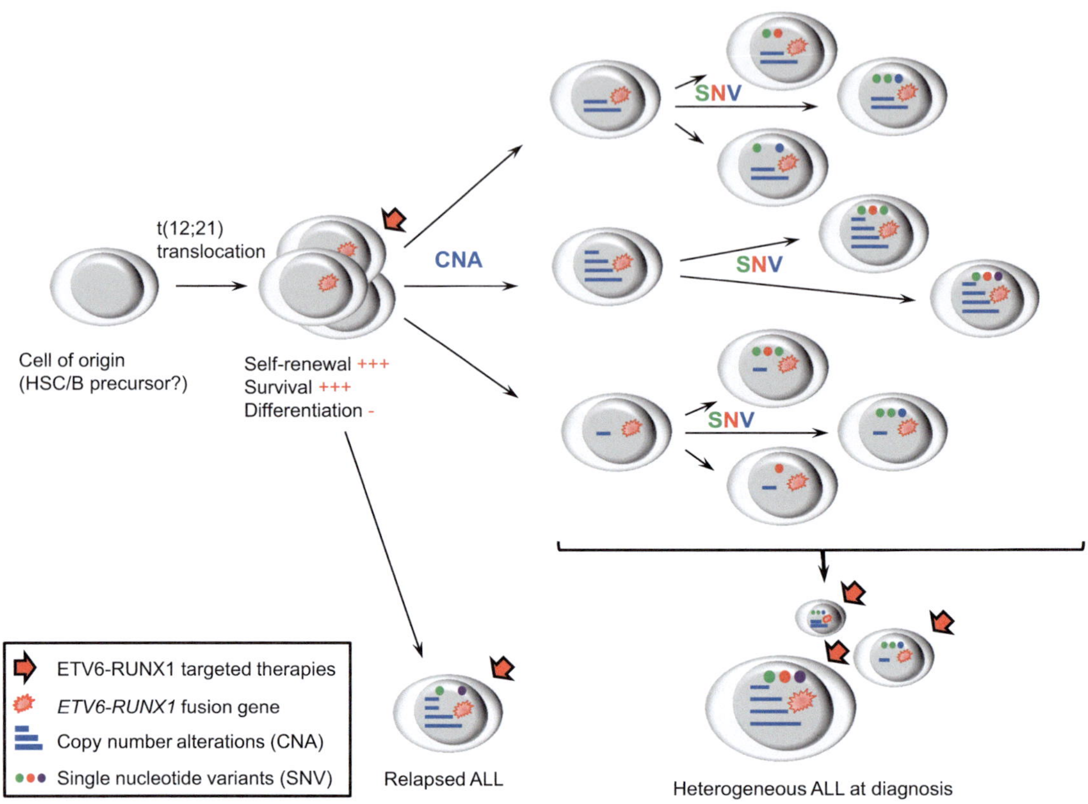

Fig. 13.1 Etiology, progression and therapy of t(12;21) ALL. The TEL-AML1 fusion enhances self-renewal and survival, while impairing differentiation, resulting in an expanded pool of pre-leukemic B progenitor cells. These progress into overt leukemia upon acquisition of additional copy number alterations (CNA), mediated by RAG and RAG/AID complexes and resulting from TEL-AML1 associated *RAG-1/2* expression and pathogen induced *AID* expression. Sub-clonal complexity is further increased by acquired single nucleotide variants (SNV). The resulting ALL at diagnosis may consist of a number of major and minor sub-clones. Leukemia elimination by ALL chemotherapy can be followed by late disease relapse, potentially arising from surviving pre-leukemic cells which acquire new secondary transforming genetic abnormalities. Novel therapies targeting TEL-AML1 function have the potential to eliminate all TEL-AML1 expressing cells, both leukemic and pre-leukemic, improving frontline therapy and reducing relapse incidence

References

Anderson, K., Lutz, C., van Delft, F. W., Bateman, C. M., Guo, Y., Colman, S. M., et al. (2011). Genetic variegation of clonal architecture and propagating cells in leukaemia. *Nature, 469*, 356–361.

Andreasson, P., Schwaller, J., Anastasiadou, E., Aster, J., & Gilliland, D. G. (2001). The expression of ETV6/CBFA2 (TEL/AML1) is not sufficient for the transformation of hematopoietic cell lines in vitro or the induction of hematologic disease in vivo. *Cancer Genetics and Cytogenetics, 130*, 93–104.

Aronson, B. D., Fisher, A. L., Blechman, K., Caudy, M., & Gergen, J. P. (1997). Groucho-dependent and -independent repression activities of Runt domain proteins. *Molecular and Cellular Biology, 17*, 5581–5587.

Bernardin, F., Yang, Y., Cleaves, R., Zahurak, M., Cheng, L., Civin, C. I., et al. (2002). TEL-AML1, expressed from t(12;21) in human acute lymphocytic leukemia, induces acute leukemia in mice. *Cancer Research, 62*, 3904–3908.

Bohlander, S. K. (2005). ETV6: A versatile player in leukemogenesis. *Seminars in Cancer Biology, 15*, 162–174.

Bokemeyer, A., Eckert, C., Meyr, F., Koerner, G., von Stackelberg, A., Ullmann, R., et al. (2014). Copy number genome alterations are associated with treatment response and outcome in relapsed childhood ETV6/RUNX1-positive acute lymphoblastic leukemia. *Haematologica, 99*, 706–714.

Bories, J. C., Cayuela, J. M., Loiseau, P., & Sigaux, F. (1991). Expression of human recombination activating genes (RAG1 and RAG2) in neoplastic lymphoid cells: Correlation with cell differentiation and antigen receptor expression. *Blood, 78*, 2053–2061.

Brettingham-Moore, K. H., Taberlay, P. C., & Holloway, A. F. (2015). Interplay between transcription factors and the epigenome: Insight from the role of RUNX1 in leukemia. *Frontiers in Immunology, 6*, 499.

Brown, P. (2011). TEL-AML1 in cord blood: 1 % or 0.01 %? *Blood, 117*, 2–4.

Castor, A., Nilsson, L., Astrand-Grundstrom, I., Buitenhuis, M., Ramirez, C., Anderson, K., et al. (2005). Distinct patterns of hematopoietic stem cell involvement in acute lymphoblastic leukemia. *Nature Medicine, 11*, 630–637.

Chakrabarti, S. R., & Nucifora, G. (1999). The leukemia-associated gene TEL encodes a transcription repressor which associates with SMRT and mSin3A. *Biochemical and Biophysical Research Communications, 264*, 871–877.

Dang, J., Wei, L., de Ridder, J., Su, X., Rust, A. G., Roberts, K. G., et al. (2015). PAX5 is a tumor suppressor in mouse mutagenesis models of acute lymphoblastic leukemia. *Blood, 125*, 3609–3617.

De Braekeleer, E., Douet-Guilbert, N., Morel, F., Le Bris, M. J., Ferec, C., & De Braekeleer, M. (2011). RUNX1 translocations and fusion genes in malignant hemopathies. *Future Oncology, 7*, 77–91.

Di Noia, J. M., & Neuberger, M. S. (2007). Molecular mechanisms of antibody somatic hypermutation. *Annual Review of Biochemistry, 76*, 1–22.

Diakos, C., Krapf, G., Gerner, C., Inthal, A., Lemberger, C., Ban, J., et al. (2007). RNAi-mediated silencing of TEL/AML1 reveals a heat-shock protein- and survivin-dependent mechanism for survival. *Blood, 109*, 2607–2610.

Diakos, C., Zhong, S., Xiao, Y., Zhou, M., Vasconcelos, G. M., Krapf, G., et al. (2010). TEL-AML1 regulation of survivin and apoptosis via miRNA-494 and miRNA-320a. *Blood, 116*, 4885–4893.

Durst, K. L., & Hiebert, S. W. (2004). Role of RUNX family members in transcriptional repression and gene silencing. *Oncogene, 23*, 4220–4224.

Fan, D., Zhou, X., Li, Z., Li, Z. Q., Duan, C., Liu, T., et al. (2015). Stem cell programs are retained in human leukemic lymphoblasts. *Oncogene, 34*, 2083–2093.

Fears, S., Gavin, M., Zhang, D. E., Hetherington, C., Ben-David, Y., Rowley, J. D., et al. (1997). Functional characterization of ETV6 and ETV6/CBFA2 in the regulation of the MCSFR proximal promoter. *Proceedings of the National Academy of Sciences of the United States of America, 94*, 1949–1954.

Fenrick, R., Amann, J. M., Lutterbach, B., Wang, L., Westendorf, J. J., Downing, J. R., et al. (1999). Both TEL and AML-1 contribute repression domains to the t(12;21) fusion protein. *Molecular and Cellular Biology, 19*, 6566–6574.

Fenrick, R., Wang, L., Nip, J., Amann, J. M., Rooney, R. J., Walker-Daniels, J., et al. (2000). TEL, a putative tumor suppressor, modulates cell growth and cell morphology of ras-transformed cells while repressing the transcription of stromelysin-1. *Molecular and Cellular Biology, 20*, 5828–5839.

Fine, B. M., Stanulla, M., Schrappe, M., Ho, M., Viehmann, S., Harbott, J., et al. (2004). Gene expression patterns associated with recurrent chromosomal translocations in acute lymphoblastic leukemia. *Blood, 103*, 1043–1049.

Fischer, M., Schwieger, M., Horn, S., Niebuhr, B., Ford, A., Roscher, S., et al. (2005). Defining the oncogenic function of the TEL/AML1 (ETV6/RUNX1) fusion protein in a mouse model. *Oncogene, 24*, 7579–7591.

Ford, A. M., Bennett, C. A., Price, C. M., Bruin, M. C., Van Wering, E. R., & Greaves, M. (1998). Fetal origins of the TEL-AML1 fusion gene in identical twins with leukemia. *Proceedings of the National Academy of Sciences of the United States of America, 95*, 4584–4588.

Ford, A. M., Fasching, K., Panzer-Grumayer, E. R., Koenig, M., Haas, O. A., & Greaves, M. F. (2001). Origins of "late" relapse in childhood acute lymphoblastic leukemia with TEL-AML1 fusion genes. *Blood, 98*, 558–564.

Ford, A. M., Palmi, C., Bueno, C., Hong, D., Cardus, P., Knight, D., et al. (2009). The TEL-AML1 leukemia fusion gene dysregulates the TGF-beta pathway in early B lineage progenitor cells. *The Journal of Clinical Investigation, 119*, 826–836.

Forestier, E., Heyman, M., Andersen, M. K., Autio, K., Blennow, E., Borgstrom, G., et al. (2008). Outcome of ETV6/RUNX1-positive childhood acute lymphoblastic leukaemia in the NOPHO-ALL-1992 protocol: frequent late relapses but good overall survival. *British Journal of Haematology, 140*, 665–672.

Fuka, G., Kauer, M., Kofler, R., Haas, O. A., & Panzer-Grumayer, R. (2011). The leukemia-specific fusion gene ETV6/RUNX1 perturbs distinct key biological functions primarily by gene repression. *PloS One, 6*, e26348.

Fuka, G., Kantner, H. P., Grausenburger, R., Inthal, A., Bauer, E., Krapf, G., et al. (2012). Silencing of ETV6/RUNX1 abrogates PI3K/AKT/mTOR signaling and impairs reconstitution of leukemia in xenografts. *Leukemia, 26*, 927–933.

Gandemer, V., Chevret, S., Petit, A., Vermylen, C., Leblanc, T., Michel, G., et al. (2012). Excellent prognosis of late relapses of ETV6/RUNX1-positive childhood acute lymphoblastic leukemia: lessons from the FRALLE 93 protocol. *Haematologica, 97*, 1743–1750.

Gawad, C., Koh, W., & Quake, S. R. (2014). Dissecting the clonal origins of childhood acute lymphoblastic leukemia by single-cell genomics. *Proceedings of the National Academy of Sciences of the United States of America, 111*, 17947–17952.

Golub, T. R., Barker, G. F., Bohlander, S. K., Hiebert, S. W., Ward, D. C., Bray-Ward, P., et al. (1995). Fusion of the TEL gene on 12p13 to the AML1 gene on 21q22 in acute lymphoblastic leukemia. *Proceedings of the National Academy of Sciences of the United States of America, 92*, 4917–4921.

Grausenburger, R., Bastelberger, S., Eckert, C., Kauer, M., Stanulla, M., Frech, C., et al. (2015). Genetic alterations in glucocorticoid signaling pathway components are associated with adverse prognosis in children with relapsed ETV6/RUNX1-positive acute lymphoblastic leukemia. *Leukemia and Lymphoma*, 1–11.

Greaves, M. F. (1988). Speculations on the cause of childhood acute lymphoblastic leukemia. *Leukemia, 2*, 120–125.

Greaves, M. (1999). Molecular genetics, natural history and the demise of childhood leukaemia. *European Journal of Cancer, 35*, 173–185.

Greaves, M. (2009). Darwin and evolutionary tales in leukemia. The Ham-Wasserman Lecture. *Hematology American Society Hematology Education Program*, 3–12.

Greaves, M. F., & Wiemels, J. (2003). Origins of chromosome translocations in childhood leukaemia. *Nature Reviews. Cancer, 3*, 639–649.

Growney, J. D., Shigematsu, H., Li, Z., Lee, B. H., Adelsperger, J., Rowan, R., et al. (2005). Loss of Runx1 perturbs adult hematopoiesis and is associated with a myeloproliferative phenotype. *Blood, 106*, 494–504.

Guidez, F., Petrie, K., Ford, A. M., Lu, H., Bennett, C. A., MacGregor, A., et al. (2000). Recruitment of the nuclear receptor corepressor N-CoR by the TEL moiety of the childhood leukemia-associated TEL-AML1 oncoprotein. *Blood, 96*, 2557–2561.

Gunji, H., Waga, K., Nakamura, F., Maki, K., Sasaki, K., Nakamura, Y., et al. (2004). TEL/AML1 shows dominant-negative effects over TEL as well as AML1. *Biochemical and Biophysical Research Communications, 322*, 623–630.

Haferlach, T., Kohlmann, A., Wieczorek, L., Basso, G., Kronnie, G. T., Bene, M. C., et al. (2010). Clinical utility of microarray-based gene expression profiling in the diagnosis and subclassification of leukemia: Report from the international microarray innovations in leukemia study group. *Journal of Clinical Oncology, 28*, 2529–2537.

Hann, I., Vora, A., Harrison, G., Harrison, C., Eden, O., Hill, F., et al. (2001). Determinants of outcome after intensified therapy of childhood lymphoblastic leukaemia: Results from medical research council United Kingdom acute lymphoblastic leukaemia XI protocol. *British Journal of Haematology, 113*, 103–114.

Harbott, J., Viehmann, S., Borkhardt, A., Henze, G., & Lampert, F. (1997). Incidence of TEL/AML1 fusion gene analyzed consecutively in children with acute lymphoblastic leukemia in relapse. *Blood, 90*, 4933–4937.

Hernandez-Munain, C., & Krangel, M. S. (1994). Regulation of the T-cell receptor delta enhancer by functional cooperation between c-Myb and core-binding factors. *Molecular and Cellular Biology, 14*, 473–483.

Hiebert, S. W., Sun, W., Davis, J. N., Golub, T., Shurtleff, S., Buijs, A., et al. (1996). The t(12;21) translocation converts AML-1B from an activator to a repressor of transcription. *Molecular and Cellular Biology, 16*, 1349–1355.

Hock, H., Meade, E., Medeiros, S., Schindler, J. W., Valk, P. J., Fujiwara, Y., et al. (2004). Tel/Etv6 is an essential and selective regulator of adult hematopoietic stem cell survival. *Genes & Development, 18*, 2336–2341.

Hong, D., Gupta, R., Ancliff, P., Atzberger, A., Brown, J., Soneji, S., et al. (2008). Initiating and cancer-propagating cells in TEL-AML1-associated childhood leukemia. *Science, 319*, 336–339.

Hotfilder, M., Rottgers, S., Rosemann, A., Jurgens, H., Harbott, J., & Vormoor, J. (2002). Immature CD34+CD19- progenitor/stem cells in TEL/AML1-positive acute lymphoblastic leukemia are genetically and functionally normal. *Blood, 100*, 640–646.

Iacobucci, I., Storlazzi, C. T., Cilloni, D., Lonetti, A., Ottaviani, E., Soverini, S., et al. (2009). Identification and molecular characterization of recurrent genomic deletions on 7p12 in the IKZF1 gene in a large cohort of BCR-ABL1-positive acute lymphoblastic leukemia patients: On behalf of Gruppo Italiano Malattie Ematologiche dell'Adulto Acute Leukemia Working Party (GIMEMA AL WP). *Blood, 114*, 2159–2167.

Ichikawa, M., Asai, T., Saito, T., Seo, S., Yamazaki, I., Yamagata, T., et al. (2004). AML-1 is required for megakaryocytic maturation and lymphocytic differentiation, but not for maintenance of hematopoietic stem cells in adult hematopoiesis. *Nature Medicine, 10*, 299–304.

Imai, Y., Kurokawa, M., Tanaka, K., Friedman, A. D., Ogawa, S., Mitani, K., et al. (1998). TLE, the human homolog of groucho, interacts with AML1 and acts as a repressor of AML1-induced transactivation. *Biochemical and Biophysical Research Communications, 252*, 582–589.

Inthal, A., Krapf, G., Beck, D., Joas, R., Kauer, M. O., Orel, L., et al. (2008). Role of the erythropoietin receptor in ETV6/RUNX1-positive acute lymphoblastic leukemia. *Clinical Cancer Research, 14*, 7196–7204.

Irvin, B. J., Wood, L. D., Wang, L., Fenrick, R., Sansam, C. G., Packham, G., et al. (2003). TEL, a putative tumor suppressor, induces apoptosis and represses transcription of Bcl-XL. *The Journal of Biological Chemistry, 278*, 46378–46386.

Javed, A., Guo, B., Hiebert, S., Choi, J. Y., Green, J., Zhao, S. C., et al. (2000). Groucho/TLE/R-esp proteins associate with the nuclear matrix and repress RUNX (CBF(alpha)/AML/PEBP2(alpha)) dependent activation of tissue-specific gene transcription. *Journal of Cell Science, 113*, 2221–2231.

Kaindl, U., Morak, M., Portsmouth, C., Mecklenbrauker, A., Kauer, M., Zeginigg, M., et al. (2014). Blocking ETV6/RUNX1-induced MDM2 overexpression by Nutlin-3 reactivates p53 signaling in childhood leukemia. *Leukemia, 28*, 600–608.

Kim, D. H., Moldwin, R. L., Vignon, C., Bohlander, S. K., Suto, Y., Giordano, L., et al. (1996). TEL-AML1 translocations with TEL and CDKN2 inactivation in acute lymphoblastic leukemia cell lines. *Blood, 88*, 785–794.

Kitabayashi, I., Yokoyama, A., Shimizu, K., & Ohki, M. (1998). Interaction and functional cooperation of the leukemia-associated factors AML1 and p300 in myeloid cell differentiation. *The EMBO Journal, 17*, 2994–3004.

Konrad, M., Metzler, M., Panzer, S., Ostreicher, I., Peham, M., Repp, R., et al. (2003). Late relapses evolve from slow-responding subclones in t(12;21)-positive acute lymphoblastic leukemia: evidence for the persistence of a preleukemic clone. *Blood, 101*, 3635–3640.

Krentz, S., Hof, J., Mendioroz, A., Vaggopoulou, R., Dorge, P., Lottaz, C., et al. (2013). Prognostic value of genetic alterations in children with first bone marrow relapse of childhood B-cell precursor acute lymphoblastic leukemia. *Leukemia, 27*, 295–304.

Krishna Narla, R., Navara, C., Sarquis, M., & Uckun, F. M. (2001). Chemosensitivity of TEL-AML1 fusion transcript positive acute lymphoblastic leukemia cells. *Leukemia & Lymphoma, 41*, 615–623.

Kuiper, R. P., Schoenmakers, E. F., van Reijmersdal, S. V., Hehir-Kwa, J. Y., van Kessel, A. G., van Leeuwen, F. N., et al. (2007). High-resolution genomic profiling of childhood ALL reveals novel recurrent genetic lesions affecting pathways involved in lymphocyte differentiation and cell cycle progression. *Leukemia, 21*, 1258–1266.

Kuster, L., Grausenburger, R., Fuka, G., Kaindl, U., Krapf, G., Inthal, A., et al. (2011). ETV6/RUNX1-positive relapses evolve from an ancestral clone and frequently acquire deletions of genes implicated in glucocorticoid signaling. *Blood, 117*, 2658–2667.

Lausten-Thomsen, U., Madsen, H. O., Vestergaard, T. R., Hjalgrim, H., Nersting, J., & Schmiegelow, K. (2011). Prevalence of t(12;21)[ETV6-RUNX1]-positive cells in healthy neonates. *Blood, 117*, 186–189.

Levanon, D., Goldstein, R. E., Bernstein, Y., Tang, H., Goldenberg, D., Stifani, S., et al. (1998). Transcriptional repression by AML1 and LEF-1 is mediated by the TLE/Groucho corepressors. *Proceedings of the National Academy of Sciences of the United States of America, 95*, 11590–11595.

Li, M., Jones, L., Gaillard, C., Binnewies, M., Ochoa, R., Garcia, E., et al. (2013). Initially disadvantaged, TEL-AML1 cells expand and initiate leukemia in response to irradiation and cooperating mutations. *Leukemia, 27*, 1570–1573.

Lilljebjorn, H., Soneson, C., Andersson, A., Heldrup, J., Behrendtz, M., Kawamata, N., et al. (2010). The correlation pattern of acquired copy number changes in 164 ETV6/RUNX1-positive childhood acute lymphoblastic leukemias. *Human Molecular Genetics, 19*, 3150–3158.

Linka, Y., Ginzel, S., Kruger, M., Novosel, A., Gombert, M., Kremmer, E., et al. (2013). The impact of TEL-AML1 (ETV6-RUNX1) expression in precursor B cells and implications for leukaemia using three different genome-wide screening methods. *Blood cancer Journal, 3*, e151.

Loh, M. L., Goldwasser, M. A., Silverman, L. B., Poon, W. M., Vattikuti, S., Cardoso, A., et al. (2006). Prospective analysis of TEL/AML1-positive patients treated on Dana-Farber Cancer Institute Consortium protocol 95-01. *Blood, 107*, 4508–4513.

Lopez, R. G., Carron, C., Oury, C., Gardellin, P., Bernard, O., & Ghysdael, J. (1999). TEL is a sequence-specific transcriptional repressor. *The Journal of Biological Chemistry, 274*, 30132–30138.

Lutterbach, B., Westendorf, J. J., Linggi, B., Isaac, S., Seto, E., & Hiebert, S. W. (2000). A mechanism of repression by acute myeloid leukemia-1, the target of multiple chromosomal translocations in acute leukemia. *The Journal of Biological Chemistry, 275*, 651–656.

Maia, A. T., Ford, A. M., Jalali, G. R., Harrison, C. J., Taylor, G. M., Eden, O. B., et al. (2001). Molecular tracking of leukemogenesis in a triplet pregnancy. *Blood, 98*, 478–482.

Mangolini, M., de Boer, J., Walf-Vorderwulbecke, V., Pieters, R., den Boer, M. L., & Williams, O. (2013). STAT3 mediates oncogenic addiction to TEL-AML1 in t(12;21) acute lymphoblastic leukemia. *Blood, 122*, 542–549.

Marculescu, R., Le, T., Simon, P., Jaeger, U., & Nadel, B. (2002). V(D)J-mediated translocations in lymphoid neoplasms: A functional assessment of genomic instability by cryptic sites. *The Journal of Experimental Medicine, 195*, 85–98.

McLean, T. W., Ringold, S., Neuberg, D., Stegmaier, K., Tantravahi, R., Ritz, J., et al. (1996). TEL/AML-1 dimerizes and is associated with a favorable outcome in childhood acute lymphoblastic leukemia. *Blood, 88*, 4252–4258.

Moorman, A. V., Enshaei, A., Schwab, C., Wade, R., Chilton, L., Elliott, A., et al. (2014). A novel integrated cytogenetic and genomic classification refines risk stratification in pediatric acute lymphoblastic leukemia. *Blood, 124*, 1434–1444.

Mori, H., Colman, S. M., Xiao, Z., Ford, A. M., Healy, L. E., Donaldson, C., et al. (2002). Chromosome translocations and covert leukemic clones are generated during normal fetal development. *Proceedings of the National Academy of Sciences of the United States of America, 99*, 8242–8247.

Moriyama, T., Metzger, M. L., Wu, G., Nishii, R., Qian, M., Devidas, M., et al. (2015). Germline genetic variation in ETV6 and risk of childhood acute lymphoblastic leukaemia: A systematic genetic study. *The Lancet Oncology, 16*, 1659–1666.

Morrow, M., Horton, S., Kioussis, D., Brady, H. J., & Williams, O. (2004). TEL-AML1 promotes development of specific hematopoietic lineages consistent with preleukemic activity. *Blood, 103*, 3890–3896.

Morrow, M., Samanta, A., Kioussis, D., Brady, H. J., & Williams, O. (2007). TEL-AML1 preleukemic activity requires the DNA binding domain of AML1 and the dimerization and corepressor binding domains of TEL. *Oncogene, 26*, 4404–4414.

Mullighan, C. G., Goorha, S., Radtke, I., Miller, C. B., Coustan-Smith, E., Dalton, J. D., et al. (2007). Genome-wide analysis of genetic alterations in acute lymphoblastic leukaemia. *Nature, 446*, 758–764.

Mullighan, C. G., Miller, C. B., Radtke, I., Phillips, L. A., Dalton, J., Ma, J., et al. (2008a). BCR-ABL1 lymphoblastic leukaemia is characterized by the deletion of Ikaros. *Nature, 453*, 110–114.

Mullighan, C. G., Phillips, L. A., Su, X., Ma, J., Miller, C. B., Shurtleff, S. A., et al. (2008b). Genomic analysis of the clonal origins of relapsed acute lymphoblastic leukemia. *Science, 322*, 1377–1380.

Noetzli, L., Lo, R. W., Lee-Sherick, A. B., Callaghan, M., Noris, P., Savoia, A., et al. (2015). Germline mutations in ETV6 are associated with thrombocytopenia, red cell macrocytosis and predisposition to lymphoblastic leukemia. *Nature Genetics, 47*, 535–538.

Numata, M., Saito, S., & Nagata, K. (2010). RAG-dependent recombination at cryptic RSSs within TEL-AML1 t(12;21)(p13;q22) chromosomal translocation region. *Biochemical and Biophysical Research Communications, 402*, 718–724.

Okada, H., Watanabe, T., Niki, M., Takano, H., Chiba, N., Yanai, N., et al. (1998). AML1(−/−) embryos do not express certain hematopoiesis-related gene transcripts including those of the PU.1 gene. *Oncogene, 17*, 2287–2293.

Okuda, T., van Deursen, J., Hiebert, S. W., Grosveld, G., & Downing, J. R. (1996). AML1, the target of multiple chromosomal translocations in human leukemia, is essential for normal fetal liver hematopoiesis. *Cell, 84*, 321–330.

Okuda, T., Cai, Z., Yang, S., Lenny, N., Lyu, C. J., van Deursen, J. M., et al. (1998). Expression of a knocked-in AML1-ETO leukemia gene inhibits the establishment of normal definitive hematopoiesis and directly generates dysplastic hematopoietic progenitors. *Blood, 91*, 3134–3143.

Osato, M., Asou, N., Abdalla, E., Hoshino, K., Yamasaki, H., Okubo, T., et al. (1999). Biallelic and heterozygous point mutations in the runt domain of the AML1/PEBP2alphaB gene associated with myeloblastic leukemias. *Blood, 93*, 1817–1824.

Panzer-Grumayer, E. R., Cazzaniga, G., van der Velden, V. H., del Giudice, L., Peham, M., Mann, G., et al. (2005). Immunogenotype changes prevail in relapses of young children with TEL-AML1-positive acute lymphoblastic leukemia and derive mainly from clonal selection. *Clinical Cancer Research, 11*, 7720–7727.

Papaemmanuil, E., Rapado, I., Li, Y., Potter, N. E., Wedge, D. C., Tubio, J., et al. (2014). RAG-mediated recombination is the predominant driver of oncogenic rearrangement in ETV6-RUNX1 acute lymphoblastic leukemia. *Nature Genetics, 46*, 116–125.

Petrovick, M. S., Hiebert, S. W., Friedman, A. D., Hetherington, C. J., Tenen, D. G., & Zhang, D. E. (1998). Multiple functional domains of AML1: PU.1 and C/EBPalpha synergize with different regions of AML1. *Molecular and Cellular Biology, 18*, 3915–3925.

Pine, S. R., Wiemels, J. L., Jayabose, S., & Sandoval, C. (2003). TEL-AML1 fusion precedes differentiation to pre-B cells in childhood acute lymphoblastic leukemia. *Leukemia Research, 27*, 155–164.

Pui, C. H., Pei, D., Campana, D., Cheng, C., Sandlund, J. T., Bowman, W. P., et al. (2014). A revised definition for cure of childhood acute lymphoblastic leukemia. *Leukemia, 28*, 2336–2343.

Putz, G., Rosner, A., Nuesslein, I., Schmitz, N., & Buchholz, F. (2006). AML1 deletion in adult mice causes splenomegaly and lymphomas. *Oncogene, 25*, 929–939.

Ramakers-van Woerden, N. L., Pieters, R., Loonen, A. H., Hubeek, I., van Drunen, E., Beverloo, H. B., et al. (2000). TEL/AML1 gene fusion is related to in vitro drug sensitivity for L-asparaginase in childhood acute lymphoblastic leukemia. *Blood, 96*, 1094–1099.

Raschke, S., Balz, V., Efferth, T., Schulz, W. A., & Florl, A. R. (2005). Homozygous deletions of CDKN2A caused by alternative mechanisms in various human cancer cell lines. *Genes, Chromosomes & Cancer, 42*, 58–67.

Raynaud, S., Cave, H., Baens, M., Bastard, C., Cacheux, V., Grosgeorge, J., et al. (1996). The 12;21 translocation involving TEL and deletion of the other TEL allele: two frequently associated alterations found in childhood acute lymphoblastic leukemia. *Blood, 87*, 2891–2899.

Romana, S. P., Mauchauffe, M., Le Coniat, M., Chumakov, I., Le Paslier, D., Berger, R., et al. (1995). The t(12;21) of acute lymphoblastic leukemia results in a tel-AML1 gene fusion. *Blood, 85*, 3662–3670.

Romana, S. P., Le Coniat, M., Poirel, H., Marynen, P., Bernard, O., & Berger, R. (1996). Deletion of the short arm of chromosome 12 is a secondary event in acute lymphoblastic leukemia with t(12;21). *Leukemia, 10*, 167–170.

Rompaey, L. V., Potter, M., Adams, C., & Grosveld, G. (2000). Tel induces a G1 arrest and suppresses Ras-induced transformation. *Oncogene, 19*, 5244–5250.

Ross, M. E., Zhou, X., Song, G., Shurtleff, S. A., Girtman, K., Williams, W. K., et al. (2003). Classification of pediatric acute lymphoblastic leukemia by gene expression profiling. *Blood, 102*, 2951–2959.

Roudaia, L., Cheney, M. D., Manuylova, E., Chen, W., Morrow, M., Park, S., et al. (2009). CBFbeta is critical for AML1-ETO and TEL-AML1 activity. *Blood, 113*, 3070–3079.

Rubnitz, J. E., Wichlan, D., Devidas, M., Shuster, J., Linda, S. B., Kurtzberg, J., et al. (2008). Prospective analysis of TEL gene rearrangements in childhood acute lymphoblastic leukemia: A Children's Oncology Group study. *Journal of Clinical Oncology, 26*, 2186–2191.

Schindler, J. W., Van Buren, D., Foudi, A., Krejci, O., Qin, J., Orkin, S. H., et al. (2009). TEL-AML1 corrupts hematopoietic stem cells to persist in the bone marrow and initiate leukemia. *Cell Stem Cell, 5*, 43–53.

Schwieger, M., Lohler, J., Friel, J., Scheller, M., Horak, I., & Stocking, C. (2002). AML1-ETO inhibits maturation of multiple lymphohematopoietic lineages and induces myeloblast transformation in synergy with

ICSBP deficiency. *The Journal of Experimental Medicine, 196*, 1227–1240.

Seeger, K., Adams, H. P., Buchwald, D., Beyermann, B., Kremens, B., Niemeyer, C., et al. (1998). TEL-AML1 fusion transcript in relapsed childhood acute lymphoblastic leukemia. The Berlin-Frankfurt-Munster Study Group. *Blood, 91*, 1716–1722.

Seeger, K., Buchwald, D., Peter, A., Taube, T., von Stackelberg, A., Schmitt, G., et al. (1999). TEL-AML1 fusion in relapsed childhood acute lymphoblastic leukemia. *Blood, 94*, 374–376.

Seeger, K., von Stackelberg, A., Taube, T., Buchwald, D., Korner, G., Suttorp, M., et al. (2001). Relapse of TEL-AML1–positive acute lymphoblastic leukemia in childhood: A matched-pair analysis. *Journal of Clinical Oncology, 19*, 3188–3193.

Shurtleff, S. A., Buijs, A., Behm, F. G., Rubnitz, J. E., Raimondi, S. C., Hancock, M. L., et al. (1995). TEL/AML1 fusion resulting from a cryptic t(12;21) is the most common genetic lesion in pediatric ALL and defines a subgroup of patients with an excellent prognosis. *Leukemia, 9*, 1985–1989.

Song, W. J., Sullivan, M. G., Legare, R. D., Hutchings, S., Tan, X., Kufrin, D., et al. (1999). Haploinsufficiency of CBFA2 causes familial thrombocytopenia with propensity to develop acute myelogenous leukaemia. *Nature Genetics, 23*, 166–175.

Stams, W. A., den Boer, M. L., Beverloo, H. B., Meijerink, J. P., van Wering, E. R., Janka-Schaub, G. E., et al. (2005). Expression levels of TEL, AML1, and the fusion products TEL-AML1 and AML1-TEL versus drug sensitivity and clinical outcome in t(12;21)-positive pediatric acute lymphoblastic leukemia. *Clinical Cancer Research, 11*, 2974–2980.

Stams, W. A., Beverloo, H. B., den Boer, M. L., de Menezes, R. X., Stigter, R. L., van Drunen, E., et al. (2006). Incidence of additional genetic changes in the TEL and AML1 genes in DCOG and COALL-treated t(12;21)-positive pediatric ALL, and their relation with drug sensitivity and clinical outcome. *Leukemia, 20*, 410–416.

Swaminathan, S., Klemm, L., Park, E., Papaemmanuil, E., Ford, A., Kweon, S. M., et al. (2015). Mechanisms of clonal evolution in childhood acute lymphoblastic leukemia. *Nature Immunology, 16*, 766–774.

Takahashi, A., Satake, M., Yamaguchi-Iwai, Y., Bae, S. C., Lu, J., Maruyama, M., et al. (1995). Positive and negative regulation of granulocyte-macrophage colony-stimulating factor promoter activity by AML1-related transcription factor, PEBP2. *Blood, 86*, 607–616.

Topka, S., Vijai, J., Walsh, M. F., Jacobs, L., Maria, A., Villano, D., et al. (2015). Germline ETV6 mutations confer susceptibility to acute lymphoblastic leukemia and thrombocytopenia. *PLoS Genetics, 11*, e1005262.

Torrano, V., Procter, J., Cardus, P., Greaves, M., & Ford, A. M. (2011). ETV6-RUNX1 promotes survival of early B lineage progenitor cells via a dysregulated erythropoietin receptor. *Blood, 118*, 4910–4918.

Tsai, A. G., Lu, H., Raghavan, S. C., Muschen, M., Hsieh, C. L., & Lieber, M. R. (2008). Human chromosomal translocations at CpG sites and a theoretical basis for their lineage and stage specificity. *Cell, 135*, 1130–1142.

Tsuzuki, S., & Seto, M. (2013). TEL (ETV6)-AML1 (RUNX1) initiates self-renewing fetal pro-B cells in association with a transcriptional program shared with embryonic stem cells in mice. *Stem Cells, 31*, 236–247.

Tsuzuki, S., Seto, M., Greaves, M., & Enver, T. (2004). Modeling first-hit functions of the t(12;21) TEL-AML1 translocation in mice. *Proceedings of the National Academy of Sciences of the United States of America, 101*, 8443–8448.

Uchida, H., Zhang, J., & Nimer, S. D. (1997). AML1A and AML1B can transactivate the human IL-3 promoter. *Journal of Immunology, 158*, 2251–2258.

van der Weyden, L., Giotopoulos, G., Rust, A. G., Matheson, L. S., van Delft, F. W., Kong, J., et al. (2011). Modeling the evolution of ETV6-RUNX1-induced B-cell precursor acute lymphoblastic leukemia in mice. *Blood, 118*, 1041–1051.

van der Weyden, L., Giotopoulos, G., Wong, K., Rust, A. G., Robles-Espinoza, C. D., Osaki, H., et al. (2015). Somatic drivers of B-ALL in a model of ETV6-RUNX1; Pax5(+/−) leukemia. *BMC Cancer, 15*, 585.

Waanders, E., Scheijen, B., van der Meer, L. T., van Reijmersdal, S. V., van Emst, L., Kroeze, Y., et al. (2012). The origin and nature of tightly clustered BTG1 deletions in precursor B-cell acute lymphoblastic leukemia support a model of multiclonal evolution. *PLoS Genetics, 8*, e1002533.

Wang, Q., Stacy, T., Binder, M., Marin-Padilla, M., Sharpe, A. H., & Speck, N. A. (1996). Disruption of the Cbfa2 gene causes necrosis and hemorrhaging in the central nervous system and blocks definitive hematopoiesis. *Proceedings of the National Academy of Sciences of the United States of America, 93*, 3444–3449.

Wang, L. C., Kuo, F., Fujiwara, Y., Gilliland, D. G., Golub, T. R., & Orkin, S. H. (1997). Yolk sac angiogenic defect and intra-embryonic apoptosis in mice lacking the Ets-related factor TEL. *The EMBO Journal, 16*, 4374–4383.

Wang, L. C., Swat, W., Fujiwara, Y., Davidson, L., Visvader, J., Kuo, F., et al. (1998). The TEL/ETV6 gene is required specifically for hematopoiesis in the bone marrow. *Genes & Development, 12*, 2392–2402.

Wiemels, J. L., Ford, A. M., Van Wering, E. R., Postma, A., & Greaves, M. (1999). Protracted and variable latency of acute lymphoblastic leukemia after TEL-AML1 gene fusion in utero. *Blood, 94*, 1057–1062.

Wotton, D., Ghysdael, J., Wang, S., Speck, N. A., & Owen, M. J. (1994). Cooperative binding of Ets-1 and core binding factor to DNA. *Molecular and Cellular Biology, 14*, 840–850.

Yergeau, D. A., Hetherington, C. J., Wang, Q., Zhang, P., Sharpe, A. H., Binder, M., et al. (1997). Embryonic lethality and impairment of haematopoiesis in mice heterozygous for an AML1-ETO fusion gene. *Nature Genetics, 15*, 303–306.

Zelent, A., Greaves, M., & Enver, T. (2004). Role of the TEL-AML1 fusion gene in the molecular pathogenesis of childhood acute lymphoblastic leukaemia. *Oncogene, 23*, 4275–4283.

Zhang, M., & Swanson, P. C. (2008). V(D)J recombinase binding and cleavage of cryptic recombination signal sequences identified from lymphoid malignancies. *The Journal of Biological Chemistry, 283*, 6717–6727.

Zhang, M. Y., Churpek, J. E., Keel, S. B., Walsh, T., Lee, M. K., Loeb, K. R., et al. (2015). Germline ETV6 mutations in familial thrombocytopenia and hematologic malignancy. *Nature Genetics, 47*, 180–185.

Zuna, J., Madzo, J., Krejci, O., Zemanova, Z., Kalinova, M., Muzikova, K., et al. (2011). ETV6/RUNX1 (TEL/AML1) is a frequent prenatal first hit in childhood leukemia. *Blood, 117*, 368–369.

ETV6-RUNX1^{+} Acute Lymphoblastic Leukaemia in Identical Twins

14

Anthony M. Ford and Mel Greaves

Abstract

Acute leukaemia is the major subtype of paediatric cancer with a cumulative risk of 1 in 2000 for children up to the age of 15 years. Childhood acute lymphoblastic leukaemia (ALL) is a biologically and clinically diverse disease with distinctive subtypes; multiple chromosomal translocations exist within the subtypes and each carries its own prognostic relevance. The most common chromosome translocation observed is the t(12;21) that results in an in-frame fusion between the first five exons of *ETV6 (TEL)* and almost the entire coding region of *RUNX1 (AML1)*.

The natural history of childhood ALL is almost entirely clinically silent and is well advanced at the point of diagnosis. It has, however, been possible to backtrack this process through molecular analysis of appropriate clinical samples: (i) leukaemic clones in monozygotic twins that are either concordant or discordant for ALL; (ii) archived neonatal blood spots or Guthrie cards from individuals who later developed leukaemia; and (iii) stored, viable cord blood cells.

Here, we outline our studies on the aetiology and pathology of childhood ALL that provide molecular evidence for a monoclonal, prenatal origin of *ETV6-RUNX1*+ leukaemia in monozygotic identical twins. We provide mechanistic support for the concept that altered patterns of infection during early childhood can deliver the necessary promotional drive for the progression of *ETV6-RUNX1*+ pre-leukaemic cells into a postnatal overt leukaemia.

Keywords

Leukemia • TEL-AML1 • ETV6-RUNX1 • RUNX • Twins • In utero • Infection

A.M. Ford, Ph.D. (✉) • M. Greaves, FRS
Centre for Evolution and Cancer, The Institute of Cancer Research, London SM2 5NG, UK
e-mail: tony.ford@icr.ac.uk

Y. Groner et al. (eds.), *RUNX Proteins in Development and Cancer*, Advances in Experimental Medicine and Biology 962, DOI 10.1007/978-981-10-3233-2_14

14.1 Introduction

Acute Lymphoblastic Leukaemia (ALL), a disease of the bone marrow, accounts for about 30 % of cancer diagnosed in children under the age of 15 years (Dickinson 2005). The disease is biologically and clinically diverse with distinctive subtypes, each characterized by an association between age at presentation of overt leukaemia and various recurrent genetic alterations. Multiple chromosomal translocations exist within the subtypes and each carries its own prognostic relevance (reviewed in (Rowley et al. 2015)).

The most common chromosome translocation observed in ALL is the t(12;21) (Golub et al. 1995; Romana et al. 1995). The translocation results in an in-frame fusion between the first five exons of *ETV6* and almost the entire coding region of *RUNX1*; bringing together the PTD and repression domains of ETV6 and the DNA binding (RHD), repression and transactivation domains of RUNX1 (Golub et al. 1995; Romana et al. 1995), Fig. 14.1. Both RUNX1 and ETV6 are important transcription factors required for normal haematopoiesis (Okuda et al. 1996; Wang et al. 1996).

Although cryptic at the level of karyotype, both FISH and RT-PCR studies have shown the *ETV6-RUNX1* fusion to be present in around 25 % of cases of B-cell precursor ALL (BCP-ALL), with an age related distribution peak of 2–5 years that matches the peak of incidence of the leukaemia (Shurtleff et al. 1995).

To further our understanding of the aetiology and natural history of childhood ALL, two key questions have been addressed; (1) precisely when and how is the *ETV6-RUNX1* fusion gene generated in the development and clonal evolution of overt leukaemia and (2) whether occurrence of the fusion gene is a leukaemia initiating event sufficient for overt leukaemia.

14.2 Identical Twins with Concordant Leukaemia

The notion that genetic changes necessary for overt leukaemia might occur before birth was raised over 50 years ago and based on studies of concordant leukaemia in identical (monozygotic) twins (Clarkson and Boyse 1971). Clarkson and Boyse suggested that a demonstration of shared, non-constitutive, cytogenetic abnormalities in leukaemic cells isolated from such twin pairs might provide a prenatal, monoclonal explanation for the concordant leukaemia.

Monozygotic identical twins occur when a single egg is fertilized by a single sperm to form one zygote. Subsequently, the zygote will divide into two separate embryos, the timing of which is critical to the formation of the placenta(s) and amniotic sac(s), Fig. 14.2. If the zygote splits within the first 3 days, two separate placentas and amniotic sacs are formed (dichorionic and diamniotic). If the split occurs between days four and nine after fertilization then the twins will share

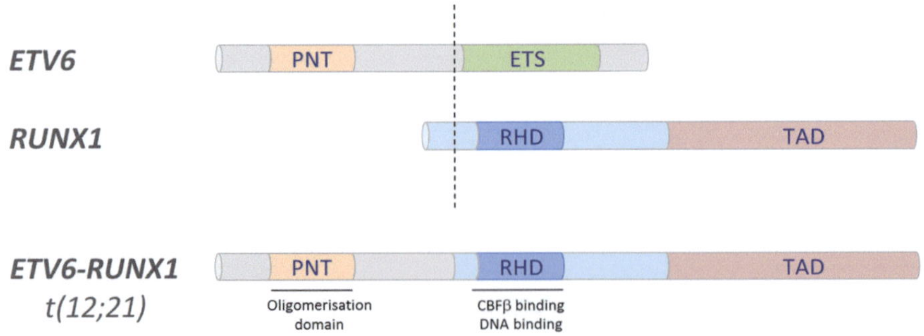

Fig. 14.1 Functional domains in the ETV6-RUNX1 fusion. A schematic representation of the full length ETV6, RUNX1 and ETV6-RUNX1 proteins. The fusion protein retains the oligomerization domain of ETV6 (PNT) and the DNA binding (RHD), repressor and activation (TAD) domains of RUNX1

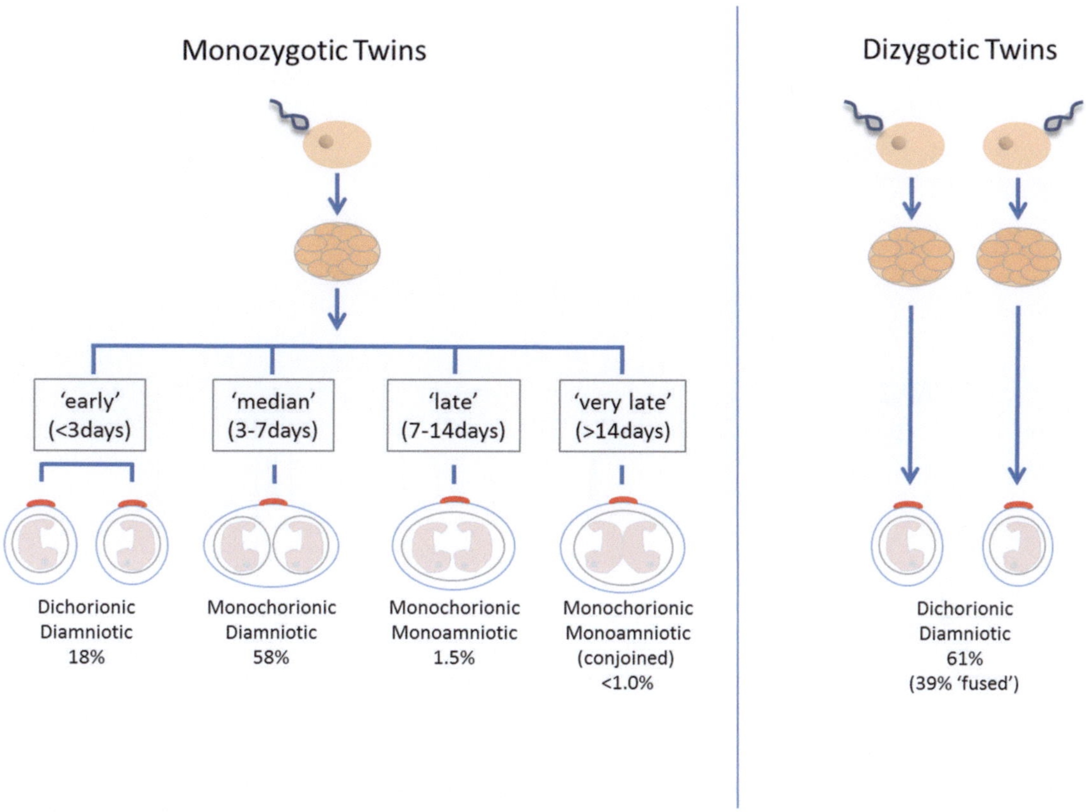

Fig. 14.2 Placental status in twin embryos. The schematic shows the placenta as a *red* oval, the amnion a *grey* oval and the chorion in *blue* (Frequency data is taken from Strong and Corney 1967)

one placenta with separate sacs (monochorionic and diamniotic) and notably will share their supply of blood. 60 % of monozygotic twins are in this category. If the split occurs after 9 days then the twins will share a single placenta and sac (monochorionic and monoamniotic). Non-identical or fraternal twins result from the fertilization of two separate eggs by two separate sperm (dizygotic) and consequently do not share their blood supplies. Monozygotic twins are genetically identical unless there has been a mutation in development.

Over seventy pairs of monozygotic twins with concordant acute leukaemia have been recorded in the literature (Greaves et al. 2003) and such cases usually share the same morphological and immunological subtype of leukaemia and development of their clinical symptoms usually occurs within a short time of each other (Greaves et al. 2003). The concordance rate for leukaemia in infant twins (<1 year) is almost 100 %, while that for older identical twins, including those with *ETV6-RUNX1*+ ALL, is less at 10–15 % (Greaves et al. 2003) suggesting the occurrence of additional, postnatal, genetic events.

14.3 Molecular Evidence for a Monoclonal, Prenatal Origin of *ETV6-RUNX1+* Leukaemia in Identical Twins

Chimeric fusion genes are formed by normal, error-prone repair of DNA double-strand breaks (DSBs) (Wiemels and Greaves 1999). Gene fusions between *ETV6* and *RUNX1* involve the noncoding introns of each gene and the breaks are both scattered and diverse within the respective breakpoint cluster regions (Golub et al. 1995). The breakpoints on chromosome 12 cluster

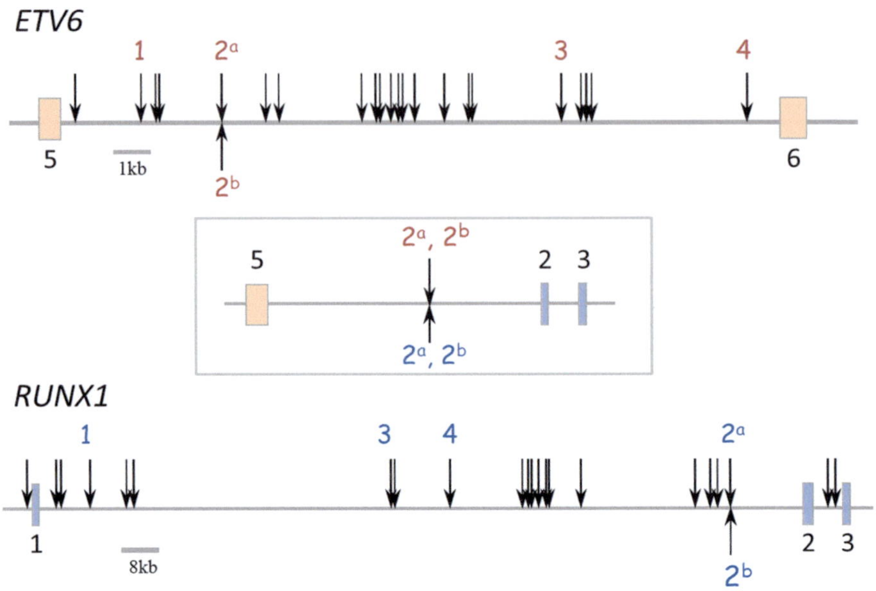

Fig. 14.3 Clonotypic genomic breakpoints of *ETV6* and *RUNX1* in monozygotic twins and singletons with ALL. Individual breakpoints are shown for *ETV6* and *RUNX1* in singletons (1,3 and 4 respectively) and identical breakpoints are shown for monozygotic twins with concordant ALL (2a, 2b)

within a single 12 kb intron of *ETV6* whereas those on chromosome 21 occur mainly within the large (~150 kb) first intron of *RUNX1* (Wiemels and Greaves 1999) and Fig. 14.3). As a consequence, each break and subsequent fusion junction is both clonotypic and patient specific at the DNA level and therefore the genomic fusion sequence provides a unique marker of clonal identity and a stable imprint of single cell origin. We reasoned that cloning and sequencing of the *ETV6-RUNX1* fusion region in twin pairs with concordant t(12;21) childhood ALL should provide unambiguous evidence for any clonal relationship as well as provide clues to the mechanism of recombination. We first cloned the *ETV6-RUNX1* fusion gene from a pair of monozygotic twins who were diagnosed at ages 3 years 6 months and 4 years 10 months respectively (Ford et al. 1998). Sequence analysis of twin 1 identified nucleotides within intron 5 of the *ETV6* gene and intron 1 of *RUNX1*. An identical fusion sequence in twin 2 confirmed that the twin leukaemias were derivatives of the same single cell or clone in which the unique and non-constitutive *ETV6-RUNX1* fusion had first arisen. Clonal identity was further supported by the finding that the leukaemic cells in the two twins shared an identical rearranged immunoglobulin heavy chain gene (*IGH*) allele (Ford et al. 1998). The most reasonable explanation for this finding was a single cell origin of the *ETV6-RUNX1* fusion in one foetus *in utero*, followed by an intraplacental metastasis of clonal progeny to the other twin via the shared vascular anastomoses.

Further unequivocal evidence to support the pre-natal origin of childhood *ETV6-RUNX1*+ leukaemia was provided by the scrutiny of neonatal blood spots, or Guthrie cards, taken at birth from a second pair of identical twins with concordant leukaemia. Guthrie cards are prepared by heel prick in the first days of life and are usually used for detection of inherited mutations and in screening for inborn errors of metabolism such as phenylketonuria. Given the natural history of childhood leukaemia, the assumption was that concordant identical twins with *ETV6-RUNX1*+ ALL might have cells with fusion gene sequences already present in their blood at birth. A simple way of testing this idea was through a backtracking analysis of the Guthrie cards of such patients. We studied a pair of identical twins diagnosed with concordant *ETV6-RUNX1*+ ALL at age 4 and for whom Guthrie cards were still available (Wiemels et al. 1999a). Diagnostic DNA was first used to establish that the sequence of the *ETV6-RUNX1* fusion was identical between the twin pairs and then individual segments of Guthrie card were used to confirm the presence of the fusion gene in the blood at birth and consequently the in *utero* clonal origin of the leukaemia.

A third twin pair provided new and unexpected insight into the time frame necessary for critical sequential events to occur. Unusually, these twins were diagnosed with *ETV6-RUNX1+* ALL over 8 years apart; at ages five and fourteen (Wiemels et al. 1999b). Cloning and sequencing of the *ETV6-RUNX1* fusion present in each twin showed perfect identity, again indicative of a single cell origin. However at the time when the first twin was diagnosed, the bone marrow of the second twin was haematologically normal and remained so for 8 years. Retrospective analysis by PCR of an archived bone marrow smear from the then 'unaffected' twin showed the presumptive *ETV6-RUNX1+* pre-leukaemic clone to be present 8 years before clinical diagnosis of ALL. These data suggest that subsequent to initiation of a prenatal, pre-leukaemic clone, almost certainly as a result of *ETV6-RUNX1* fusion alone, the period required for appearance of overt leukaemia can be both extremely variable and protracted, with latency of up to 14 years (Wiemels et al. 1999b).

Since *ETV6-RUNX1+* leukaemia in twins is no different, biologically or clinically from that seen in single children, at least some singletons are also likely to have prenatal initiation of leukaemia. We used nine sets of diagnostic samples with paired blood spots to backtrack the fusion gene to birth in non-twinned children with *ETV6-RUNX1+* ALL and provided more direct evidence that this disease can at least initiate *in utero* (Wiemels et al. 1999a).

14.4 Is an *ETV6-RUNX1* Fusion Gene Sufficient for Overt Leukaemia?

Taken together, these studies provide strong evidence, in most cases, for a prenatal origin of *ETV6-RUNX1+* leukaemia. However, it is now clear that not all individuals with a *ETV6-RUNX1* fusion gene go on to develop overt disease. In a retrospective study of over 600 normal new born cord bloods, we showed the frequency of fusion gene positive cord bloods to be 1 %; approximately 100 times the collective frequency of overt, clinically diagnosed leukaemia with *ETV6-RUNX1* fusion (Mori et al. 2002). The data, along with the modest rate of twin concordance (5–10 %), supports the view that detectable *ETV6-RUNX1+* cells in healthy children represent expanded clones of pre-leukaemic cells that can remain pathologically and clinically silent or covert in the absence of additional, postnatal genetic hits, perhaps for up to 14 years. However, a postnatal fusion of ETV6 and RUNX1 in some cases, cannot be ruled out.

14.5 *ETV6-RUNX1* as an Initiating or 'Founder' Event in ALL

A number of studies on singletons and pairs of monozygotic twins with *ETV6-RUNX1+* leukaemia have now been described that shed light on the important genetic events 'secondary' to gene fusion (Ford et al. 1998; Wiemels et al. 1999b; Broadfield et al. 2004; Teuffel et al. 2004; Maia et al. 2004; Bateman et al. 2010; Bungaro et al. 2008; Alpar et al. 2015). FISH analyses at diagnosis of *ETV6-RUNX1+* ALL show the fusion gene to be present in every leukaemic cell (Anderson et al. 2011) and the majority of cases also show some sub clonal deletion of the non-translocated 'normal' *ETV6* allele (Raynaud et al. 1996; Kempski and Sturt 2000). The deletions vary in size between patients and both FISH and loss of heterozygosity (LOH) studies show that 73 % of *ETV6-RUNX1+* cases have a partially or fully deleted second *ETV6* allele (Patel et al. 2003). Although the second *ETV6* allele was identified in the remaining patients, no ETV6 expression was detected. Taken together, these findings support the hypothesis that loss of ETV6 expression may be a critical secondary event for leukaemogenesis in *ETV6-RUNX1+* ALL and the assumption that ETV6 can act as a tumor suppressor gene.

Recurrent copy number alterations (CNAs) are the likely "driver" events that contribute critically to clonal diversification and selection. In *ETV6-RUNX1*+ ALL they typically include deletions of genes involved in B-cell development and differentiation such as *PAX5*, *BTG1*, the *RAG* family and the wild-type copy of *ETV6* (Mullighan 2012). If deletions of the normal copy of *ETV6* and indeed all other recurrent 'driver' CNAs are consistently secondary to *ETV6-RUNX1* fusion and therefore postnatal, then a testable prediction would be that these deletions should be distinct or different within monozygotic twin pairs. To address this idea we first used paired interphase FISH and SNP array information to identify recurrent CNAs in 5 pairs of twins with *ETV6-RUNX1*+ ALL. Sporadic CNAs classified as non-functional "passengers" were either identical (4/19) in the twin pairs and thought to precede the *ETV6-RUNX1* fusion event, or were distinct (15/19) (Bateman et al. 2010). Significantly, all 32 CNAs identified between the twin pairs that were regarded as being 'drivers' of leukaemia were discordant (Fig. 14.4). As expected, this discordance was further reflected by singletons and twin pairs that shared the same *ETV6* deletion but harboured different deletion boundaries (Ford et al. 2001; Maia et al. 2001; Bateman et al. 2010).

In a second study on a single set of monozygotic twins with *ETV6-RUNX1*+ ALL we used a whole genome sequencing approach to better determine the developmental timing of these events. We identified the *ETV6-RUNX1* translocation to be the only recognised fusion product shared by the twins and, despite the presence of LOH in 27 and 41 cytoband regions respectively, we found the only other mutation in common to be an inactivating germline mutation of neurofibromatosis type 1 (Ma et al. 2013). Despite a paucity of single base and indel 'driver' mutations within the leukaemia clones, none of the mutations identified were found to be shared, further supporting the concept that these genetic changes are both secondary to *ETV6-RUNX1* fusion and post-natal.

14.6 A Candidate Pre-leukaemic Stem Cell Population with an Early B Lineage Phenotype

Childhood ALL is associated with a rare population of CD34$^+$, CD38$^{-/\ low}$, CD19$^+$ cells not usually detectable in normal bone marrow

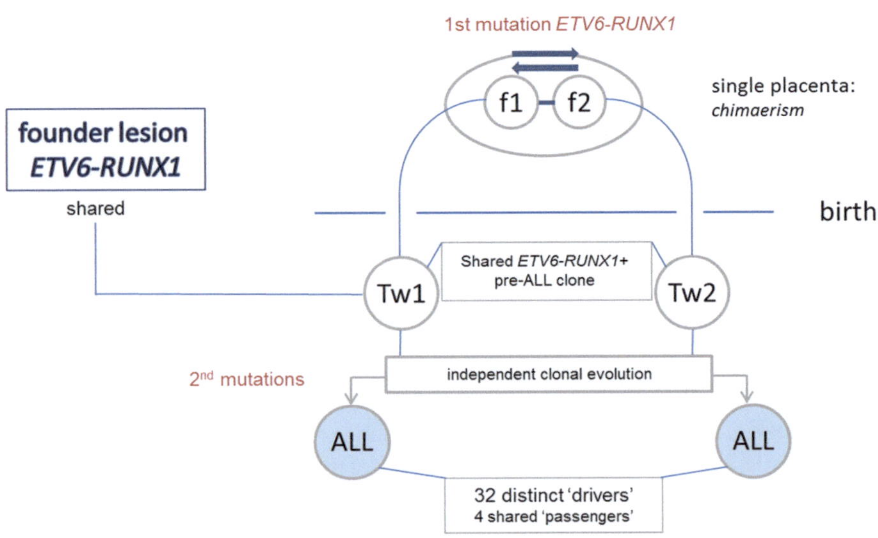

Fig. 14.4 Genomics of *ETV6-RUNX1* ALL in monozygotic twins. Combined data from 5 sets of twins with concordant *ETV6-RUNX1*+ ALL (Data taken from Bateman et al. 2015)

(Hotfilder et al. 2002; Castor et al. 2005; Hong et al. 2008; le Viseur et al. 2008) and is accompanied by clonal rearrangement of the *IGH* genes, indicative of a pre-B cell phenotype. In addition, ALLs characterised by *ETV6-RUNX1* fusion maintain a phenotype of $CD10^+$, $CD19^+$, along with recombinase gene activity (RAG) and expression of TdT. The pre-B cell however, is not necessarily the cell in which the functional impact of the *ETV6-RUNX1* fusion gene is first observed. In two early studies, distinct and specific B-cell receptor gene rearrangements were identified in one set of twins (Teuffel et al. 2004) suggesting that separate pre-leukaemic clones were already present at birth. However, Bungaro and colleagues (Bungaro et al. 2008) identified both shared and distinct rearrangements at diagnosis in a set of monozygotic twins with dichorionic placentas. Not only was this suggestive of a common clonal origin *in utero* but, in this case, is also indicative of the passage of cells from one foetus to the other via the blood system of the Mother. In a more detailed screen of IG/TCR rearrangements in 5 pairs of twins with concordant *ETV6-RUNX1*+ ALL, we revealed the pre-leukaemic initiating function of the *ETV6-RUNX1* fusion to be associated with clonal expansion of an early foetal B-cell (Alpar et al. 2015). In all pairs of twins studied, the cells carried identical incomplete or complete *IGH* variable-diversity-joining (VDJ) regions together with substantial, sub-clonal and divergent rearrangements. In addition, most descendent cells with stem cell (self-renewal) activity were shared and maintained in both twins after birth and provided an opportunity for necessary postnatal, secondary genetic hits to occur (Fig. 14.5).

The notion that *ETV6-RUNX1* fusion is an initiating event, insufficient itself for overt leukaemia, was also confirmed in a seminal study of identical twins with ***dis***cordant *ETV6-RUNX1*+ leukaemia (Hong et al. 2008). One twin was diagnosed with *ETV6-RUNX1*+ pre-B cell ALL at age 2, while the other twin has remained healthy for over 10 years. The bone marrow of the leukaemic child contained the $CD34^+$, $CD38^{-/low}$,

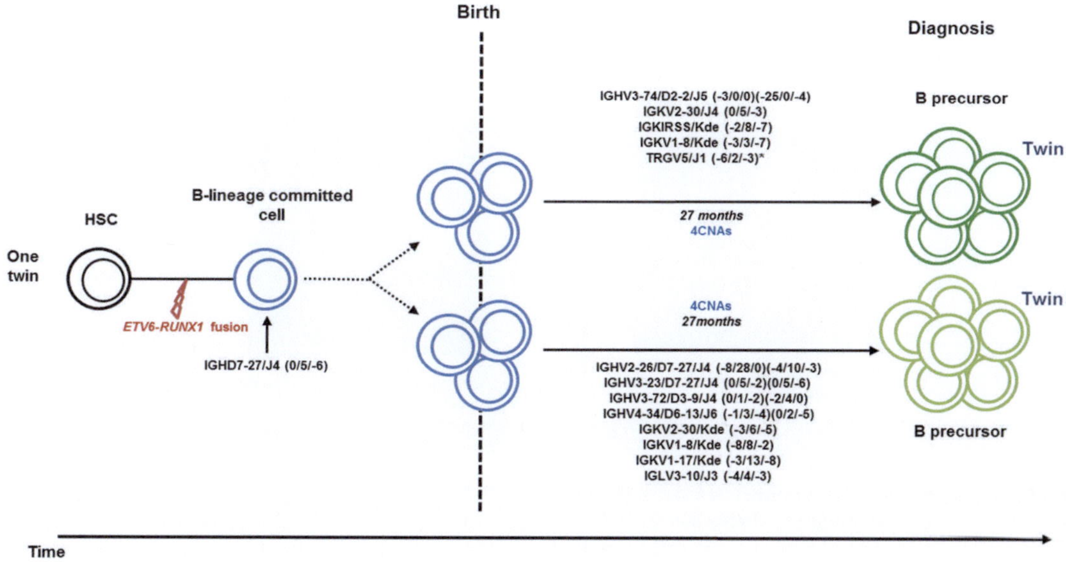

Fig. 14.5 Evolutionary sequence of clonal immunogenotypic markers identified in a pair of monozygotic twins with concordant *ETV6-RUNX1*+ ALL. Rearrangements are depicted both in the context of an *in utero* origin of pre-leukaemia, perhaps occurring in a progenitor cell already committed to the B cell lineage and the diverging postnatal evolution of the overt leukaemia (Alpar et al. 2015) (CNA data was adopted from Bateman et al. (2015). *HSC* hematopoietic stem cell)

CD19⁺ cancer-propagating population while the blasts, as well as presenting the *ETV6-RUNX1* fusion, showed additional loss of the nontranslocated *ETV6* allele, loss of one copy of *PAX5* and gain of 10p (Bateman et al. 2010), Fig. 14.6). Immuno-FISH for CD19 protein and the *ETV6-RUNX1* fusion in the peripheral blood pre-B cells of the healthy child detected the fusion gene at a frequency of ~0.1 %, but these cells did not show additional chromosome aberrations. Cloning and sequencing of the respective *ETV6-RUNX1* fusion junctions revealed their complete identity and added further support to the *in utero* origin of the pre-leukaemic clone (AF unpublished).

14.7 An Infectious Aetiology for Childhood ALL?

Given the age related peak incidence of 2–5 years for BCP-ALL, it has long been proposed that infection(s) in childhood might accelerate the transformation of *ETV6-RUNX1+* pre-leukaemic cells to overt leukaemia (Kinlen 1988; Greaves 1988). Although we and others (MacKenzie et al. 1999, 2001; McNally and Eden 2004) have not identified any specific exogenous viral sequences, epidemiological studies support the notion that a relationship exists between improved social conditions and childhood ALL (Dockerty et al.

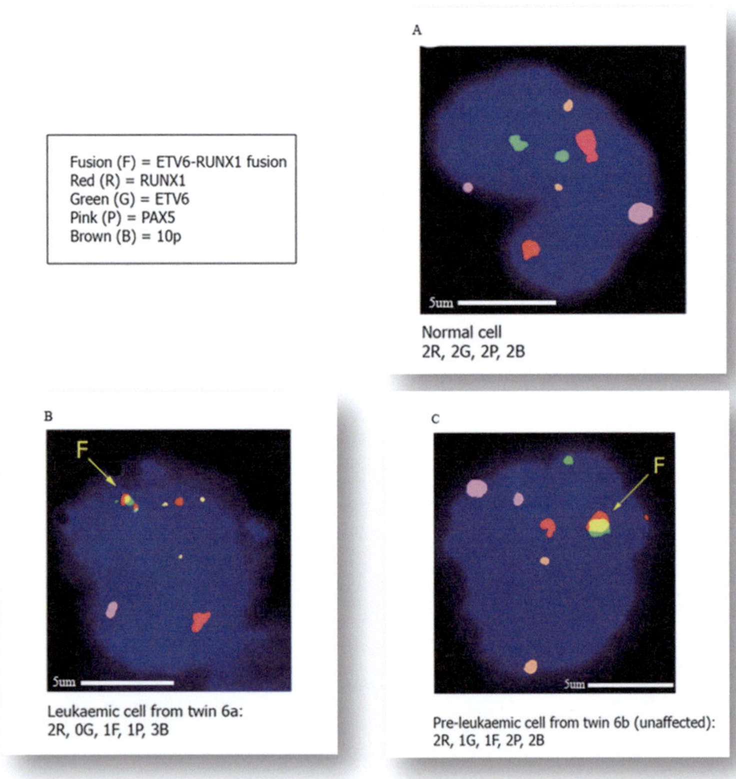

Fig. 14.6 Interphase FISH to confirm CNA status in cells from monozygotic twins clinically discordant for *ETV6-RUNX1+* ALL. (**a**) A normal bone marrow cell from the healthy twin (6b) shows 2 *red* (*RUNX1* gene), 2 *green* (*ETV6* gene), 2 *pink* (*PAX5* gene), and 2 *brown* (chromosome 10p) signals, respectively. (**b**) A leukaemic cell from twin 6a shows the *ETV6-RUNX1* gene fusion, 1 normal *RUNX1* and the remnant from the rearranged *RUNX1* gene, 1 copy of *PAX5* and 3 signals for 10p. (**c**) A *ETV6-RUNX1* fusion gene positive cell in the unaffected twin (6b) shows 1 *ETV6-RUNX1* gene fusion, the normal *RUNX1* and the remnant from rearranged *RUNX1*, 1 copy of normal *ETV6*, 2 copies of *PAX5* and 2 signals for chromose 10p to show that loss of *ETV6* and *PAX5* and gain of 10p are not observed in the preleukaemic, *ETV6-RUNX1+* cells of the unaffected twin. Five cells with the *ETV6-RUNX1* fusion were identified in the unaffected twin (twin 6b), of a total of 4251 scored (Bateman et al. 2015)

2001). In a highly protective unchallenged or 'hygienic' environment a delayed exposure of infants to an otherwise common infection may trigger a rare abnormal immune response by selection and expansion of pre-leukaemic (*ETV6-RUNX1*+) clones (Greaves 2006, 1988). In this context, we have shown that altered cytokine environments in the context of inflammation, such as TGFβ, can eliminate normal pre-B cell clones from the repertoire and support the selective outgrowth of pre-B cell clones that already harbour *ETV6-RUNX1* (Ford et al. 2009).

Somatic recombination and mutation of *IGH* genes to create antibody diversity in B cells requires the proteins encoded by the *RAG1* and *RAG2* genes that introduce DNA double-strand breaks and the subsequent recombination of V(D)J gene segments (Oettinger et al. 1990). The enzyme activation-induced cytidine deaminase (AICDA) subsequently then enables somatic hypermutation of V region genes followed by class switching (Li et al. 2004). During normal B cell ontogeny the activities of these enzymes are kept strictly segregated (Hardy and Hayakawa 2001).

14.8 Genetic Changes that Complement *ETV6-RUNX1* Fusion

The presence of V(D)J recombination signal sequences (RSS) close to CNAs commonly deleted in *ETV6-RUNX1*+ ALL has suggested a role for aberrant RAG endonuclease targeting at these loci (Zhang and Swanson 2008; Mullighan et al. 2008). To obtain a more detailed picture of these secondary genetic events we carried out genome analysis of diagnostic samples from 57 cases of *ETV6-RUNX1*+ ALL paired with matched remission (constitutive) DNA. We performed low-depth whole-genome sequencing and structural variation analysis on the leukaemic samples of 51 cases and used exome sequencing of 56 cases to search for recurrent somatic variants (Papaemmanuil et al. 2014).

We observed a paucity of recurrent coding-region mutations but resolved 354 of 523 structural variations at breakpoint sites to base-pair resolution. We searched for conserved RSS, along with proposed recognition motifs for the APOBEC family of enzymes that deaminate cytosine to uracil and for the presence of CpGs (Tsai et al. 2008). Although we did not find conserved RSS motifs near the breakpoints of the founding *ETV6-RUNX1* rearrangement, consistent with this rearrangement arising in a very early B-lineage progenitor cell, enrichment for RSS was particularly prominent at gene deletions targeting known B-cell ALL genes such as *ETV6* and *BTG1*. We did not observe specific enrichment of CpGs or any of the proposed AICDA recognition motifs at breakpoint junctions relative to other cancers; however another comprehensive analysis of translocation breakpoints did reveal a breakage mechanism that involved the RAG complex acting at AICDA-deaminated methyl-CpGs (Tsai et al. 2008). Nonetheless, 43 % of the 354 deletion breakpoints in our *ETV6-RUNX1* study showed conserved RAG recognition motifs: CACACTG–spacer–ACAAAAACC; compared to a complete absence of RSS at over 12,000 breakpoints examined in breast, pancreatic and prostate cancer. These observations signify the existence of a lymphoid-specific endogenous mutagenesis programme (Papaemmanuil et al. 2014; Swaminathan et al. 2015).

To test the hypothesis that, in the context of inflammation, RAGs and AICDA can cooperate to induce secondary genetic lesions and accelerate transformation of a *ETV6-RUNX1* pre-leukaemic clone to overt leukaemia, we subjected *ETV6-RUNX1* expressing pre-B cells to consecutive rounds of stimulation with the inflammatory mimic lipopolysaccharide (LPS) in the presence or absence of IL7. Signalling through the IL7-R has been shown to safeguard human pre-B cells from premature activation of AICDA (Swaminathan et al. 2015). We noted both upregulation of RAG1/RAG2 mRNA levels in the presence of *ETV6-RUNX1* alone and the subsequent upregulation of AICDA by over 20 fold in the presence of LPS and absence of IL7 (Swaminathan et al. 2015). Intravenous injection of these vulnerable cells into NOD/SCID recipients triggered development of leukaemia within 3 weeks. In contrast, pre-B cells isolated from

Aicda −/− or *Rag1−/−* animals respectively delayed or abrogated leukaemia development (Swaminathan et al. 2015). These data provide sound genetic evidence that, in the context of inflammatory/repetitive infectious stimulation, clonal evolution of pre-leukaemic *ETV6-RUNX1+* pre-B cells requires both RAG and AICDA activities.

14.9 Summary

Our studies on the aetiology and pathology of childhood ALL have provided the molecular evidence for a monoclonal, prenatal origin of *ETV6-RUNX1+* leukaemia in monozygotic identical twins and provide mechanistic support for the concept that altered patterns of infection during early childhood can deliver the necessary promotional drive for the progression of *ETV6-RUNX1+* pre-leukaemic cells into a postnatal overt leukaemia.

References

Alpar, D., Wren, D., Ermini, L., Mansur, M. B., van Delft, F. W., Bateman, C. M., et al. (2015). Clonal origins of ETV6-RUNX1(+) acute lymphoblastic leukemia: Studies in monozygotic twins. *Leukemia, 29*, 839–846.

Anderson, K., Lutz, C., van Delft, F. W., Bateman, C. M., Guo, Y., Colman, S. M., et al. (2011). Genetic variegation of clonal architecture and propagating cells in leukaemia. *Nature, 469*, 356–361.

Bateman, C. M., Colman, S. M., Chaplin, T., Young, B. D., Eden, T. O., Bhakta, M., et al. (2010). Acquisition of genome-wide copy number alterations in monozygotic twins with acute lymphoblastic leukemia. *Blood, 115*, 3553–3558.

Bateman, C. M., Alpar, D., Ford, A. M., Colman, S. M., Wren, D., Morgan, M., et al. (2015). Evolutionary trajectories of hyperdiploid ALL in monozygotic twins. *Leukemia, 29*, 58–65.

Broadfield, Z. J., Hain, R. D., Harrison, C. J., Reza Jalali, G., McKinley, M., Michalova, K., et al. (2004). Complex chromosomal abnormalities in utero, 5 years before leukaemia. *British Journal of Haematology, 126*, 307–312.

Bungaro, S., Irving, J., Tussiwand, R., Mura, R., Minto, L., Molteni, C., et al. (2008). Genomic analysis of different clonal evolution in a twin pair with t(12;21) positive acute lymphoblastic leukemia sharing the same prenatal clone. *Leukemia, 22*, 208–211.

Castor, A., Nilsson, L., Astrand-Grundstrom, I., Buitenhuis, M., Ramirez, C., Anderson, K., et al. (2005). Distinct patterns of hematopoietic stem cell involvement in acute lymphoblastic leukemia. *Nature Medicine, 11*, 630–637.

Clarkson, B. D., & Boyse, E. A. (1971). Possible explanation of the high concoddance for acute leukaemia in monozygotic twins. *Lancet, 1*, 699–701.

Dickinson, H. O. (2005). The causes of childhood leukaemia. *BMJ, 330*, 1279–1280.

Dockerty, J. D., Draper, G., Vincent, T., Rowan, S. D., & Bunch, K. J. (2001). Case-control study of parental age, parity and socioeconomic level in relation to childhood cancers. *International Journal of Epidemiology, 30*, 1428–1437.

Ford, A. M., Bennett, C. A., Price, C. M., Bruin, M. C., Van Wering, E. R., & Greaves, M. (1998). Fetal origins of the TEL-AML1 fusion gene in identical twins with leukemia. *Proceedings of the National Academy of Sciences of the United States of America, 95*, 4584–4588.

Ford, A. M., Fasching, K., Panzer-Grumayer, E. R., Koenig, M., Haas, O. A., & Greaves, M. F. (2001). Origins of "late" relapse in childhood acute lymphoblastic leukemia with TEL-AML1 fusion genes. *Blood, 98*, 558–564.

Ford, A. M., Palmi, C., Bueno, C., Hong, D., Cardus, P., Knight, D., et al. (2009). The TEL-AML1 leukemia fusion gene dysregulates the TGF-beta pathway in early B lineage progenitor cells. *The Journal of Clinical Investigation, 119*, 826–836.

Golub, T. R., Barker, G. F., Bohlander, S. K., Hiebert, S. W., Ward, D. C., Bray-Ward, P., et al. (1995). Fusion of the TEL gene on 12p13 to the AML1 gene on 21q22 in acute lymphoblastic leukemia. *Proceedings of the National Academy of Sciences of the United States of America, 92*, 4917–4921.

Greaves, M. F. (1988). Speculations on the cause of childhood acute lymphoblastic leukemia. *Leukemia, 2*, 120–125.

Greaves, M. (2006). Infection, immune responses and the aetiology of childhood leukaemia. *Nature Reviews. Cancer, 6*, 193–203.

Greaves, M. F., Maia, A. T., Wiemels, J. L., & Ford, A. M. (2003). Leukemia in twins: Lessons in natural history. *Blood, 102*, 2321–2333.

Hardy, R. R., & Hayakawa, K. (2001). B cell development pathways. *Annual Review of Immunology, 19*, 595–621.

Hong, D., Gupta, R., Ancliff, P., Atzberger, A., Brown, J., Soneji, S., et al. (2008). Initiating and cancer-propagating cells in TEL-AML1-associated childhood leukemia. *Science, 319*, 336–339.

Hotfilder, M., Rottgers, S., Rosemann, A., Jurgens, H., Harbott, J., & Vormoor, J. (2002). Immature CD34+CD19- progenitor/stem cells in TEL/AML1-positive acute lymphoblastic leukemia are genetically and functionally normal. *Blood, 100*, 640–646.

Kempski, H. M., & Sturt, N. T. (2000). The TEL-AML1 fusion accompanied by loss of the untranslocated TEL allele in B-precursor acute lymphoblastic leukaemia of childhood. *Leukemia & Lymphoma, 40*, 39–47.

Kinlen, L. (1988). Evidence for an infective cause of childhood leukaemia: Comparison of a Scottish new town with nuclear reprocessing sites in Britain. *Lancet, 2*, 1323–1327.

le Viseur, C., Hotfilder, M., Bomken, S., Wilson, K., Rottgers, S., Schrauder, A., et al. (2008). In childhood acute lymphoblastic leukemia, blasts at different stages of immunophenotypic maturation have stem cell properties. *Cancer Cell, 14*, 47–58.

Li, Z., Woo, C. J., Iglesias-Ussel, M. D., Ronai, D., & Scharff, M. D. (2004). The generation of antibody diversity through somatic hypermutation and class switch recombination. *Genes & Development, 18*, 1–11.

Ma, Y., Dobbins, S. E., Sherborne, A. L., Chubb, D., Galbiati, M., Cazzaniga, G., et al. (2013). Developmental timing of mutations revealed by whole-genome sequencing of twins with acute lymphoblastic leukemia. *Proceedings of the National Academy of Sciences of the United States of America, 110*, 7429–7433.

MacKenzie, J., Perry, J., Ford, A. M., Jarrett, R. F., & Greaves, M. (1999). JC and BK virus sequences are not detectable in leukaemic samples from children with common acute lymphoblastic leukaemia. *British Journal of Cancer, 81*, 898–899.

MacKenzie, J., Gallagher, A., Clayton, R. A., Perry, J., Eden, O. B., Ford, A. M., et al. (2001). Screening for herpesvirus genomes in common acute lymphoblastic leukemia. *Leukemia, 15*, 415–421.

Maia, A. T., Ford, A. M., Jalali, G. R., Harrison, C. J., Taylor, G. M., Eden, O. B., & Greaves, M. F. (2001). Molecular tracking of leukemogenesis in a triplet pregnancy. *Blood, 98*, 478–482.

Maia, A. T., Koechling, J., Corbett, R., Metzler, M., Wiemels, J. L., & Greaves, M. (2004). Protracted postnatal natural histories in childhood leukemia. *Genes, Chromosomes & Cancer, 39*, 335–340.

McNally, R. J., & Eden, T. O. (2004). An infectious aetiology for childhood acute leukaemia: A review of the evidence. *British Journal of Haematology, 127*, 243–263.

Mori, H., Colman, S. M., Xiao, Z., Ford, A. M., Healy, L. E., Donaldson, C., et al. (2002). Chromosome translocations and covert leukemic clones are generated during normal fetal development. *Proceedings of the National Academy of Sciences of the United States of America, 99*, 8242–8247.

Mullighan, C. G. (2012). Molecular genetics of B-precursor acute lymphoblastic leukemia. *The Journal of Clinical Investigation, 122*, 3407–3415.

Mullighan, C. G., Phillips, L. A., Su, X., Ma, J., Miller, C. B., Shurtleff, S. A., & Downing, J. R. (2008). Genomic analysis of the clonal origins of relapsed acute lymphoblastic leukemia. *Science, 322*, 1377–1380.

Oettinger, M. A., Schatz, D. G., Gorka, C., & Baltimore, D. (1990). RAG-1 and RAG-2, adjacent genes that synergistically activate V(D)J recombination. *Science, 248*, 1517–1523.

Okuda, T., van Deursen, J., Hiebert, S. W., Grosveld, G., & Downing, J. R. (1996). AML1, the target of multiple chromosomal translocations in human leukemia, is essential for normal fetal liver hematopoiesis. *Cell, 84*, 321–330.

Papaemmanuil, E., Rapado, I., Li, Y., Potter, N. E., Wedge, D. C., Tubio, J., et al. (2014). RAG-mediated recombination is the predominant driver of oncogenic rearrangement in ETV6-RUNX1 acute lymphoblastic leukemia. *Nature Genetics, 46*, 116–125.

Patel, N., Goff, L. K., Clark, T., Ford, A. M., Foot, N., Lillington, D., et al. (2003). Expression profile of wild-type ETV6 in childhood acute leukaemia. *British Journal of Haematology, 122*, 94–98.

Raynaud, S., Cave, H., Baens, M., Bastard, C., Cacheux, V., Grosgeorge, J., et al. (1996). The 12;21 translocation involving TEL and deletion of the other TEL allele: Two frequently associated alterations found in childhood acute lymphoblastic leukemia. *Blood, 87*, 2891–2899.

Romana, S. P., Mauchauffe, M., Le Coniat, M., Chumakov, I., Le Paslier, D., Berger, R., & Bernard, O. A. (1995). The t(12;21) of acute lymphoblastic leukemia results in a tel-AML1 gene fusion. *Blood, 85*, 3662–3670.

Rowley, J. D., Le Beau, M. M., & Rabbitts, T. H. (2015). *Chromosomal translocations and genome rearrangements in cancer*. Cham: Springer.

Shurtleff, S. A., Buijs, A., Behm, F. G., Rubnitz, J. E., Raimondi, S. C., Hancock, M. L., et al. (1995). TEL/AML1 fusion resulting from a cryptic t(12;21) is the most common genetic lesion in pediatric ALL and defines a subgroup of patients with an excellent prognosis. *Leukemia, 9*, 1985–1989.

Strong, S. J., & Corney, G. (1967). *The placenta in twin pregnancy*. Oxford: Pergamon Press.

Swaminathan, S., Klemm, L., Park, E., Papaemmanuil, E., Ford, A., Kweon, S. M., et al. (2015). Mechanisms of clonal evolution in childhood acute lymphoblastic leukemia. *Nature Immunology, 16*, 766–774.

Teuffel, O., Betts, D. R., Dettling, M., Schaub, R., Schafer, B. W., & Niggli, F. K. (2004). Prenatal origin of separate evolution of leukemia in identical twins. *Leukemia, 18*, 1624–1629.

Tsai, A. G., Lu, H., Raghavan, S. C., Muschen, M., Hsieh, C. L., & Lieber, M. R. (2008). Human chromosomal translocations at CpG sites and a theoretical basis for their lineage and stage specificity. *Cell, 135*, 1130–1142.

Wang, Q., Stacy, T., Binder, M., Marin-Padilla, M., Sharpe, A. H., & Speck, N. A. (1996). Disruption of the Cbfa2 gene causes necrosis and hemorrhaging in the central nervous system and blocks definitive hematopoiesis. *Proceedings of the National Academy of Sciences of the United States of America, 93*, 3444–3449.

Wiemels, J. L., Cazzaniga, G., Daniotti, M., Eden, O. B., Addison, G. M., Masera, G., et al. (1999a). Prenatal origin of acute lymphoblastic leukaemia in children. *Lancet, 354*, 1499–1503.

Wiemels, J. L., Ford, A. M., Van Wering, E. R., Postma, A., & Greaves, M. (1999b). Protracted and variable latency of acute lymphoblastic leukemia after TEL-AML1 gene fusion in utero. *Blood, 94*, 1057–1062.

Wiemels, J. L., & Greaves, M. (1999). Structure and possible mechanisms of TEL-AML1 gene fusions in childhood acute lymphoblastic leukemia. *Cancer Research, 59*, 4075–4082.

Zhang, M., & Swanson, P. C. (2008). V(D)J recombinase binding and cleavage of cryptic recombination signal sequences identified from lymphoid malignancies. *The Journal of Biological Chemistry, 283*, 6717–6727.

Molecular Basis and Targeted Inhibition of CBFβ-SMMHC Acute Myeloid Leukemia

15

Lucio H. Castilla and John H. Bushweller

Abstract

Acute myeloid leukemia (AML) is characterized by recurrent chromosomal rearrangements that encode for fusion proteins which drive leukemia initiation and maintenance. The inv(16)(p13q22) rearrangement is a founding mutation and the associated CBFβ-SMMHC fusion protein is essential for the survival of inv(16) AML cells. This Chapter will discuss our understanding of the function of this fusion protein in disrupting hematopoietic homeostasis and creating pre-leukemic blasts, in its cooperation with other co-occurring mutations during leukemia initiation, and in leukemia maintenance. In addition, this chapter will discuss the current approaches used for the treatment of inv(16) AML and the recent development of AI-10-49, a selective targeted inhibitor of CBFβ-SMMHC/RUNX1 binding, the first candidate targeted therapy for inv(16) AML.

Keywords

CBF • CBFb-MYH11 • CBFbeta-SMMHC • RUNX1 • AML • inv(16) • Leukemia • AI-10-49 • Targeted therapies • Protein-protein interaction inhibitor • PPI

15.1 CBFβ-SMMHC Leukemia Oncoprotein in Inv(16) Acute Myeloid Leukemia (AML)

The core-binding factor is a heterodimeric transcription factor composed of a DNA binding subunit (RUNX) and a stabilizing subunit (CBFβ), which regulates the expression of genes associated with differentiation of hematopoietic cells. The genes encoding for RUNX1 and CBFβ are frequent targets of mutations in human leukemia.

L.H. Castilla (✉)
Department of Molecular, Cell and Cancer Biology, University of Massachusetts Medical School, Worcester, MA 01605, USA
e-mail: Lucio.Castilla@umassmed.edu

J.H. Bushweller
Molecular Physiology and Biophysics, University of Virginia Medical School, Charlottesville, VA 22908, USA
e-mail: jhb4v@virginia.edu

Y. Groner et al. (eds.), *RUNX Proteins in Development and Cancer*, Advances in Experimental Medicine and Biology 962, DOI 10.1007/978-981-10-3233-2_15

The *RUNX1* gene is disrupted by a variety of chromosomal translocations that encode oncogenic fusion proteins and by point mutations (Blyth et al. 2005). The *CBFB* gene is disrupted in approximately 10 % of AML by the inversion of chromosome 16 [inv(16)(p13q22)], and less frequently by the translocation t(16;16)(p13q22), henceforth called inv(16) (Le Beau et al. 1983). Acute myeloid leukemia with inv(16) is characterized by precursor cells with myelomonocytic morphology, eosinophils and abnormal basophilic granules. The World Health Organization defines inv(16) AML in the category of "*AML with recurrent genetic abnormalities*" (Swerdlow et al. 2008). The French-American-British AML classification system, based on morphology features, places practically all inv(16) AML cases in the subtype M4-with eosinophilia (M4-eo) (Bennett et al. 1976). Sporadically, inv(16) AML cases can be found in other AML subtypes. In addition, inv(16) has been reported in a small number of BCR-ABL–positive chronic myelogenous leukemia cases transitioning to blast crisis (Enright et al. 1992; Secker-Walker et al. 1992; Wu et al. 2006; Han et al. 2014).

The inv(16) q-arm breakpoint is in intron 5 of *CBFB* and the p-arm breakpoint is in an intron of the *MYH11* gene, and generates the *CBFB-MYH11* fusion gene (Liu et al. 1993a). The *MYH11* gene encodes for the smooth muscle myosin heavy chain (SMMHC) protein, and *CBFB-MYH11* encodes the fusion protein CBFβ-SMMHC. The expected reciprocal fusion, *MYH11-CBFB*, is thought to be irrelevant in leukemia because the *MYH11* promoter is specific for smooth muscle cells and inactive in hematopoietic cells (Liu et al. 1993b; Miano et al. 1994). Ten *CBFB-MYH11* isoforms (type-A to J) have been reported in inv(16) AML samples, with variations of the breakpoint primarily at the *MYH11* locus, which always preserve the open reading frame (Liu et al. 1996; Monma et al. 2007; Park et al. 2009). Over 80 % of inv(16) AML cases show the isoform type-A, which includes the first 5 exons of *CBFB*, joining at nucleotide position 1921 of the *MYH11* transcript sequence. (Claxton et al. 1994, Liu et al. 1995). The CBFβ-SMMHC fusion integrates protein domains from the native proteins with essential functions for its leukemic activity (Fig. 15.1). RUNX1 binds to the fusion protein through two interacting sequences, the RUNX binding domain (RBD) in the CBFβ region, and the high-affinity binding domain (HABD) near the N-terminal region of SMMHC (Lukasik et al. 2002).

A region overlapping with the RBD and HABD was also called the hyper-hetero-dimerization domain (Shigesada et al. 2004). The assembly-competence domain (ACD) near the C-terminus of SMMHC is responsible for dimerization and multimerization of the fusion protein into filament structures (Ikebe et al. 2001; Sohn et al. 1997). The ACD consists of nine hepta-amino acid α-helical rods in exons 40 and 41, preceding the non-helical tail (D'Costa et al. 2005; Zhang et al. 2006). Functionally, the ACD-directed multimerization is critical for the localization of CBFβ-SMMHC to the nucleus, and for its activity in proliferation and transformation (Kummalue et al. 2002). In addition, the ACD is

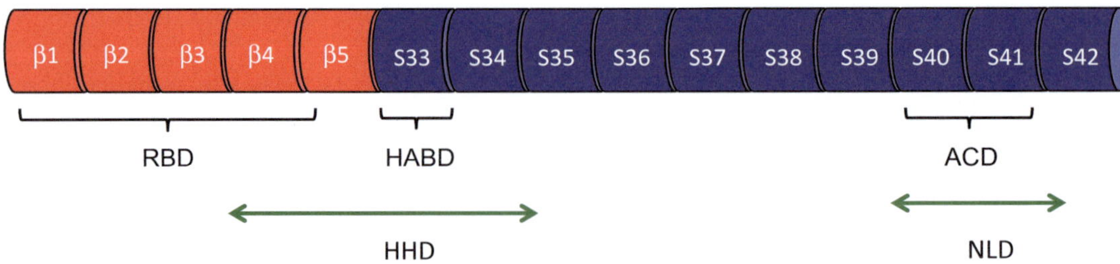

Fig. 15.1 Functional domains in CBFβ-SMMHC fusion protein. Schematic illustrating the exons (*boxes*) of CBFβ (*red*) and SMMHC (*blue*) amino acid sequences. The RUNX binding domain (RBD), high affinity binding domain (HABD), hetero-hyperdimerization domain (HHD), assembly domain (ACD) and nuclear localization domain (NLD) are shown

associated with binding to HIPK2 and HDAC8, and repression of RUNX activity, although it is not clear if these factors bind to the ACD or if dimerization is needed for their association with the fusion protein.

15.2 Mechanism of CBFβ-SMMHC Mediated Leukemogenesis

The inv(16) is thought to be a founding event in AML, and only considered a secondary event in a chronic myelogenous leukemia. Initially, the inv(16) rearrangement occurs in the hematopoietic stem cells (HSCs) and can remain undiagnosed for several years before AML presentation (McHale et al. 2003). The *CBFB-MYH11* transcript can be detected in 3 % of bone marrow samples from healthy individuals (Song et al. 2011), and the CD34+CD38- cells from inv(16) AML samples (Haase et al. 1995; Mehrotra et al. 1995).

The CBFβ-SMMHC fusion protein is a dominant repressor of RUNX function in hematopoiesis. Heterozygous CBFβ-SMMHC expression, using the $Cbfb^{MYH11/+}$ heterozygous knock-ins, blocks embryo definitive hematopoiesis and causes lethality at midgestation (Castilla et al. 1996), a similar phenotype as in homozygous $Runx1^{-/-}$ and $Cbfb^{-/-}$ knock-out embryos (Okuda et al. 1996; Sasaki et al. 1996; Wang et al. 1996a, b).

Several lines of evidence indicate that expression of RUNX proteins is essential for CBFβ-SMMHC associated leukemia. First, isothermal calorimetry revealed that RUNX1 binds ~10 fold more tightly to CBFβ-SMMHC than to wild type CBFβ, indicating that the fusion protein may directly reduce RUNX function (Lukasik et al. 2002). In line with these findings, RUNX1 binding to the promoters of RUNX targets *MPO* and *INK4b* is decreased in the presence of CBFβ-SMMHC (Cao et al. 1997; Markus et al. 2007). Second, CBFβ-SMMHC may bind to and sequester the RUNX1 kinase HIPK2, rendering RUNX1 dephosphorylated/inactive and unable to associate with p300 (Wee et al. 2008). Third, the viability of human inv(16) AML cell lines depend on RUNX1 expression (Ben-Ami et al. 2013). Fourth, the CBFβ-SMMHC function in leukemia development and maintenance requires basal levels of Runx1 and Runx2 in mice. A retroviral mutagenesis screen in $Cbfb^{+/MYH11}$ knock in mice identified retroviral insertions at the *Runx2* locus (Castilla et al. 2004). The leukemia median latency in $Cbfb^{+/MYH11}$ mice was accelerated with overexpression of *Runx2* and delayed in *Runx2*-heterozygous background (Kuo et al. 2009). Similarly, the hematopoietic block in $Cbfb^{+/MYH11}$ mice was partially rescued and leukemia development was delayed in a *Runx1*-null background (Hyde et al. 2015). The RUNX1/HABD binding may provide an explanation for the dominant negative phenotype associated with CBFβ-SMMHC and a rationale for the leukemia-associated dysregulation of hematopoietic development. Importantly, expression of a CBFβ-SMMHC lacking the HABD, using $Cbfb^{+/MYH11d179-221}$ knock-in mice, confirmed the expected partial rescue of definitive hematopoiesis and decrease in RUNX1 affinity to CBFβ-SMMHC (Kamikubo et al. 2010). Surprisingly $Cbfb^{+/MYH11d179-221}$ knock-in mice developed leukemia with accelerated median latency, arguing that high RUNX1-affinity may be required for differentiation block but not for leukemia development. This finding has important clinical relevance because the *CBFB-MYH11* isoform type-I lacks the HABD domain. The underlining mechanism driving leukemia activity of HABD-null fusion protein is not well understood. It is possible that the HABD may form part of a leukemia inhibitory domain, or that it may repress the leukemia promoting activity of a cooperating oncoprotein, such as MN1. The data may also suggest that differentiation block is unlinked to leukemogenesis, and leukemia may have developed faster in this model due to a larger pool of leukemia-initiating cells (L-ICs, also called leukemia stem cells).

15.3 CBFβ-SMMHC Induces Myeloid Pre-leukemia

The inv(16) rearrangement occurs in human HSCs and induces myeloid but not lymphoid leukemia in human and in mouse models. An explanation for this lineage bias emerged from

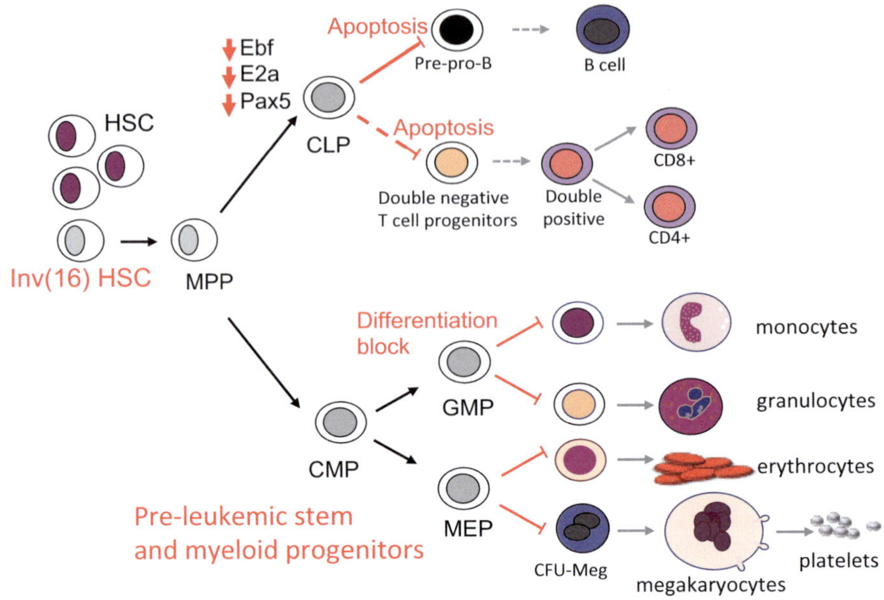

Fig. 15.2 Model of inv(16) associated pre-leukemia, depicting major events altered in hematopoietic differentiation from HSCs that have acquired an inv(16) rearrangement

studies on the effects of CBFβ-SMMHC expression in early hematopoiesis, utilizing conditional *Cbfb*$^{+/MYH11}$ knock-in mice. The differentiation of CBFβ-SMMHC-expressing HSCs induces the emergence of pre-leukemic myeloid progenitors and apoptosis in early stages of lymphoid differentiation (Fig. 15.2).

These pre-leukemic myeloid progenitors are restricted to the bone marrow, have a megakaryocyte/erythroid progenitor (MEP) immunophenotype [Lin(−),kit(+)CD34(low), FcγrII/III(low)], have a predominant morphology of blast/myeloblast and promyelocyte and are blocked in differentiation capacity (Kuo et al. 2006; Xue et al. 2014). These pre-leukemic cells transform to full blown leukemia with a median latency of 5 months (Kuo et al. 2006). Conversely, CBFβ-SMMHC-expressing HSCs produce normal numbers of common lymphoid progenitors but with reduced expression of key factors responsible for directing B and T cell differentiation, including *Ebf*, *E2a* and *Pax5*. Their commitment to pre-pro B and pre-B cell progenitors, the early stages of B cell differentiation is associated with increased apoptosis and reduced number of progenitors (Kuo et al. 2008). Similarly, commitment to double-negative T cell progenitors is associated with increased apoptosis and reduced numbers (Zhao et al. 2007). These defects in B and T cell differentiation were associated with increased levels in pro-apoptotic protein Bim, and are consistent with its dominant negative regulation of RUNX1/CBFβ function at these compartments, since competitive repopulation assays with donor HSCs carrying conditional *Runx1-null* or CBFβ-SMMHC activation result in loss of T and B cells (Growney et al. 2005; Kuo et al. 2006). In addition, *Runx1-null* and *Cbfb-null* in early lymphoid compartments have demonstrated that Runx1/Cbfβ function regulates early lymphoid differentiation (Egawa et al. 2007; Seo et al. 2012). Recent studies analyzing the clonal complexity of patient AML samples demonstrated that at diagnosis, the samples consist of multiple clones, all sharing the founding mutation and differing on the composition of "cooperating" mutations (Ding et al. 2012). In addition, the AML samples also include pre-leukemic cells, which are not sensitive to chemotherapeutic drugs, and that may serve as precursors for the expansion of resistant clones at relapse (Corces-Zimmerman et al. 2014; Shlush et al. 2014). A recent study determined the somatic mutation landscape in diagnosis/relapse-matched inv(16) AML samples confirmed the clonal evolution model (Sood et al. 2016). This study revealed that (a) the inv(16) rearrangement is found in all leukemia clones at both stages of disease, (b) a fraction of "cooperating" mutations were present at both stages, and (c) a group of mutations detected at diagnosis were lost at relapse, and a group of "new" mutations were detected only at relapse.

The inv(16) rearrangement is a driver mutation in AML and CBFβ-SMMHC expression is necessary but not sufficient to cause leukemia in mice. Previous studies using retroviral transduction, transgenic and knock-in approaches for inv(16) were recently reviewed elsewhere (Chin et al. 2015). Chimeric mice carrying *Cbfb*$^{+/MYH11}$ knock-in mouse embryonic stem cells were highly predisposed to AML upon treatment with chemical (ENU) or retroviral mutagenesis (Castilla et al. 1999; Castilla et al. 2004), validating the role of CBFβ-SMMHC and the need for "cooperating" mutations for AML transformation.

In most cases, the inv(16) is the only chromosomal rearrangement, and approximately 10 to 40 % of inv(16) AML samples show secondary chromosomal rearrangements, including trisomy of chromosome 8 and 22, and 7q deletion (Duployez et al. 2016; Grimwade et al. 2010). The inv(16) AML samples frequently show a small number of other mutations, which are predominantly associated with activation of tyrosine kinase pathways. Recent efforts to identify the spectrum of mutations, by targeted deep-sequencing of inv(16) AML samples revealed overall a lower cumulative mutation rate in inv(16) AML than AML samples with normal karyotype (Duployez et al. 2016; Kihara et al. 2014). The mutations were focused on genes encoding components of the tyrosine-kinase pathway, with approximately 90 % of activating mutations in the receptors *KIT* and *FLT3* and in the small GTPases of the RAS family *NRAS* and *KRAS* as previously reported (Boissel et al. 2006; Goemans et al. 2005; Haferlach et al. 2010; Paschka et al. 2006, 2013). The predominant *KIT* mutations are exon-8 mutations or D816 mutations. In *FLT3*, mutations include internal tandem duplications (ITD) of 3 to 400 amino acids in the juxtamembrane (JM) to tyrosine-kinase domain (TKD) 1 region, or point mutations at amino acid V596 in JM, N676 in the TKD1 domain or D835 and I836 in TKD 2 (Opatz et al. 2013). In addition, truncating mutations in the adaptor protein CBL and null mutations in the GTPase activating protein NF1 are found at lower frequency. Contrary to AML cases with normal karyotype or translocation t(8;21), the inv(16) AML are not associated with mutations in genes encoding components of cohesion (RAD21, SMC1A, SMC3 and STAG2), spliceosome (SRSF2) or DNA methylation complexes (TET2, IDH1 and IDH2).

The functional significance of several inv(16) AML cooperating mutations identified in patient samples have been validated in mice. The expression of oncogenic *FLT3-ITD* and *KIT* mutations were shown to cooperate with CBFβ-SMMHC in mouse leukemia development (Kim et al. 2008; Zhao et al. 2012). Importantly, the *KIT* mutations are present in 27 % of inv(16) AML with type-A *CBFB-MYH11* isoform but not in non-type A inv(16) AML, and confer poor clinical prognosis (Care et al. 2003; Paschka et al. 2006; Schwind et al. 2013). Expression of the Nras-G12D mutation increases survival of CBFβ-SMMHC-expressing pre-leukemic cells and cooperated in AML in *Cbfb*$^{Cbfb\text{-}MYH11/+}$ and *Nras*$^{LSL\text{-}G12D/+}$ knock-in mice (Xue et al. 2014). In this model, Nras-G12D expression accelerated the median latency of leukemia to 5 weeks, and increased the L-IC activity 10–200 fold.

Mutations in the tumor suppressor *TP53* are associated with complex karyotype AML or therapy related AML but are rare (< 3 %) with inv(16) AML (Haferlach et al. 2008; Seifert et al. 2009). The induction of p53 in response to irradiation or Etoposide is reduced by CBFβ-SMMHC activity, suggesting an alternative mechanism for interfering with p53-dependent apoptosis (Britos-Bray et al. 1998). This regulation could result from the binding of the kinase HIPK2 to CBFβ-SMMHC and its sequestration from its target, p53, thereby leaving p53 in its dephosphorylated/inactive form (Wee et al. 2008). Alternatively, the histone deacetylase HDAC8, which binds to the SMMHC region of CBFβ-SMMHC, has recently been shown to inhibit p53-acetylation (K379), thereby inhibiting its activity (Qi et al. 2015). Under either scenario, inhibition of p53 function is critical for L-IC maintenance in mice and p53 reactivation could have anti-leukemic activity.

15.4 Current Model for Inv(16) Acute Myeloid Leukemia

Based previous studies in the leukemia field, our current model is that inv(16) AML develops from HSCs that undergo somatic mutations and selective pressure, resulting in the clonal evolution to leukemia and, in some cases, relapse (see Fig. 15.3). Initially, the HSCs of AML development and progression is that the inv(16) is a founding mutation in HSCs, which produces a preleukemic cell population in the bone marrow with impaired multi-lineage differentiation (Fig. 15.3, grey nucleus). The preleukemic stem and myeloid progenitors are leukemia precursors with reduced cell death and probably impaired DNA-damage checkpoints, which can remain asymptomatic for several years.

These precursors induce AML by gaining "cooperating" mutations, with predominance of activating mutations in components of the receptor tyrosine kinase pathway. The inv(16) AML sample at diagnosis is composed of several leukemic clones that share inv(16) rearrangement, and each with a different combination of "cooperating" mutations (Fig. 15.3, colored nuclei), each represented at different frequency. In addition, diagnosis samples present pre-leukemic clones and few normal stem/progenitor cells. Most AML clones are sensitive to standard chemotherapy (cytarabine/daunorubicin) treatment and patients achieve remission at high frequency. The evolution to relapse and resistance found in a fraction of inv(16) cases could emerge through two non-exclusive possibilities: The preleukemic cells have low proliferation rate, are less sensitive to treatment, thereby persisting during remission and gaining treatment-associated mutations to produce resistant clones at relapse. Similarly, persistent inv(16) AML clones that are not completely eliminated by treatment can gain mutations and expand during relapse. Finally, we cannot rule out that the detection of "new" mutations in relapse inv(16) AML are leukemic clones present in the diagnosis AML sample at a frequency below detection by deep-sequencing methods (below 0.1–1 %), and which expand after the elimination of dominant clones during treatment.

15.5 Transcription Profile in Inv(16) AML

The expression of CBFβ-SMMHC defines a specific transcriptional cluster in AML, with most samples presenting the expected inv(16) (Valk et al. 2004). High expression of *MYH11* is the most discriminatory feature in the cluster and predictor of inv(16), followed by overexpression of *CLIPR59* and *ST18* transcripts. Interestingly, the cluster is characterized by downregulation of *CBFB* expression, probably as a direct result of

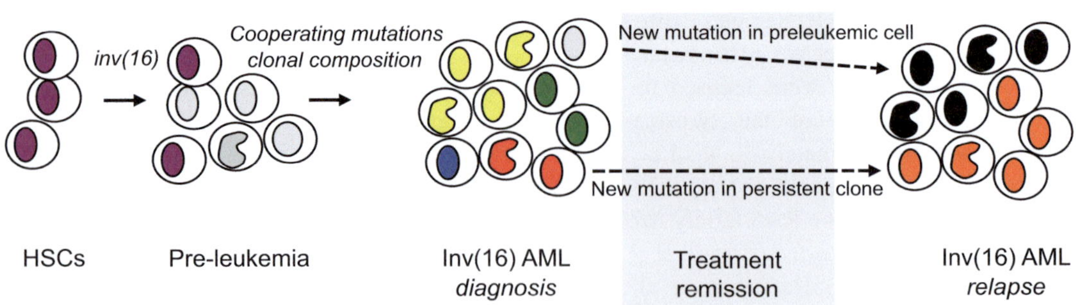

Fig. 15.3 Model of inv(16) AML development and clonal evolution to relapse

the rearrangement. Within the inv(16) AML cluster, the presence of non-type A *CBFB-MYH11* isoforms marks distinct transcription profiles, suggesting functional differences (Schwind et al. 2013). Non-type A inv(16) samples correlate with overexpression of myeloid differentiation gene *GFI1*, the DNA-methyltransferase *DNMT3B* and the apoptosis related gene *CYCS*, and downregulation of cytokine signaling transcripts *CD9* and *CD52*. The inv(16) AML cluster does not define expression of specific microRNAs (Li et al. 2008). However, the microRNA 126 is upregulated in core binding factor leukemia by promoter demethylation. The most discriminating miRNAs are upregulation of miRNA-424 and downregulation of miR-10a and miR-10b (Jongen-Lavrencic et al. 2008).

15.6 Current Approach to Treatment of Inv(16) AML

Patients with inv(16) AML usually undergo aggressive chemotherapy regimens involving a remission "induction" phase with high-dose cytarabine (Ara-C) and daunorubicin, followed by a "consolidation" phase with cytarabine. This treatment regime is generally well tolerated by young patients showing a 5 year overall survival of 45–65 % (Pulsoni et al. 2008, Ravandi et al. 2007). However, most patients are older and the 5 year overall survival for patients older than 60 years old is approximately 20 % (Farag et al. 2006). Emerging literature suggests that inability to cure AML with current therapies, including cytotoxic chemotherapy, kinase inhibitors, or monoclonal antibodies, may be attributed to a population of L-ICs that are resistant to treatment, are quiescent, have long term self-renewal potential, and can fully recapitulate tumor phenotype at time of relapse (Guzman and Allan 2014). An example of such failure is that despite aggressive treatments inv(16) AML patients display a 3 year relapse incidence of 29 % in younger patients and 55 % in older patients (Delaunay et al. 2003). At time of relapse, these patients present with the inv(16) rearrangement, although other mutations may have been lost at relapse (Kottaridis et al. 2002; Nakano et al. 1999; Shih et al. 2008; Sood et al. 2016), strongly suggesting that the inv(16) driving clones (L-ICs) have not been effectively eliminated by treatment. Surprisingly, there has not been a new agent approved by the FDA for relapsed AML in decades. These data clearly indicate that the identification of targeted therapies that can improve the therapeutic response and decrease the risk of relapse for inv(16) AML patients is essential. Importantly, there is no targeted therapy currently in use in the clinic for treatment of inv(16) AML.

15.7 Development of Potent Selective Small Molecule Inhibitor of the CBFβ-SMMHC/RUNX Function

The increasing understanding of CBFβ-SMMHC function in the past years has prompted the emergence of proof-of-principle inhibitors for inv(16) AML. A fluorescence resonance energy transfer (FRET) screen identified the benzodiazepine Ro5–3335 as an efficient small molecule inhibitor of RUNX1/CBFβ binding (Cunningham et al. 2012). This inhibitor preferentially eliminated leukemic cells expressing CBFβ-SMMHC and RUNX1-RUNX1T1, and reduced leukemia burden in mice and fish. A recent study demonstrated that Ro5–3335 binds to and inhibits the SWI/SNF members SMARCA2 and WDR9 (Illendula et al. 2016). The SWI/SNF complex associates with RUNX1 to control expression of its targets (Bakshi et al. 2010; Huang et al. 2012), suggesting that Ro5–3335 may be repressing RUNX1 activity indirectly via inhibition of SMARCA2. As mentioned above, Kuo and co-workers reported a functional role for the SMMHC portion of the CBFβ-SMMHC fusion protein by way of recruitment of HDAC8 and p53 to deacetylate p53 and decrease its activity (Qi et al. 2015). They showed that inhibition of HDAC8 with a small molecule inhibitor had significant effects on inv(16) cells including reducing the L-IC frequency as well as reactivating the gene expression program regulated by p53. The reactivation of p53 should increase the sensitivity of the inv(16)

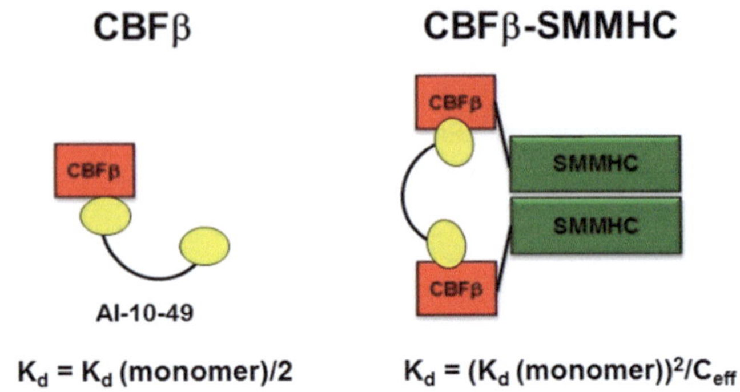

Fig. 15.4 Schematic illustrating the use of poly-valency to enhance potency and selectivity of inhibitor for CBFβ-SMMHC versus wild type CBFβ. Equations below indicate calculated values for K_d for a bi-valent inhibitor for CBFβ and CBFβ-SMMHC

cells to traditional cytotoxic chemotherapy, suggesting the use of HDAC8 inhibitors in combination with the daunorubicin and cytarabine currently used for treatment.

We recently described the development of a small molecule inhibitor of the CBFβ-SMMHC/RUNX interaction (Illendula et al. 2015). By coupling the GFP derivatives CFP (cyan fluorescent protein) and YFP (yellow fluorescent protein) to the Runt domain and CBFβ-SMMHC, respectively, we developed a FRET based assay for this protein-protein interaction. Screening using this assay followed by verification of binding to the CBFβ portion of CBFβ-SMMHC using NMR spectroscopy resulted in the identification of a modest potency inhibitor of this interaction (IC_{50} = 22 μM). Because wild type CBFβ is monomeric in solution whereas CBFβ-SMMHC oligomerizes (Lukasik et al. 2002), we utilized the principles of poly-valency to enhance both the potency and the selectivity of this inhibitor class (Fig. 15.4). Specifically, bi-valent derivatives of the original lead molecule were synthesized with polyethylene glycol based linkers of differing lengths. Such bi-valent molecules can make a bi-valent interaction with CBFβ-SMMHC whereas they can only make a mono-valent interaction with wild type CBFβ. This resulted in inhibitors with IC_{50}'s in the 300–400 nM range depending on the length of the linker. This class of inhibitors was further modified to improve activity and address an *in vivo* metabolic liability to generate a final optimized inhibitor labeled AI-10-49.

AI-10-49 has an IC_{50} of 260 nM in the FRET assay, a K_d for CBFβ-SMMHC of 170 nM as measured by isothermal titration calorimetry (ITC), and a half-life in mice > 6 h. Co-immunoprecipitation studies in ME-1 cells (an inv(16) AML cell line) showed that AI-10-49 inhibits binding of CBFβ-SMMHC to RUNX1 while having limited to no effect on the binding of wild type CBFβ to RUNX1 (Fig. 15.4), confirming our design principles and clearly establishing the outstanding selectivity of this inhibitor.

15.8 Mechanism of Action of CBFβ-SMMHC Inhibitor

The specificity of AI-10-49 in disrupting endogenous RUNX1 binding to CBFβ-SMMHC versus CBFβ was assessed in ME-1 cells. AI-10-49 effectively dissociated RUNX1 from CBFβ-SMMHC, with 90 % dissociation after a 6-h treatment, while having only a modest effect on CBFβ-RUNX1 association (Fig. 15.5a). Expression of the RUNX1-regulated genes *RUNX3*, *CSF1R*, and *CEBPA* is repressed by CBFβ-SMMHC in inv(16) AML (Cheng et al. 2008; Guo et al. 2012; Zhang et al. 1996). Previous studies have shown decreased RUNX1 binding to target genes in the presence of CBFβ-SMMHC (Cao et al. 1997; Markus et al. 2007), suggesting that CBFβ-SMMHC represses RUNX1 target genes by blocking binding of RUNX1 to target DNA sites. Consistent with this

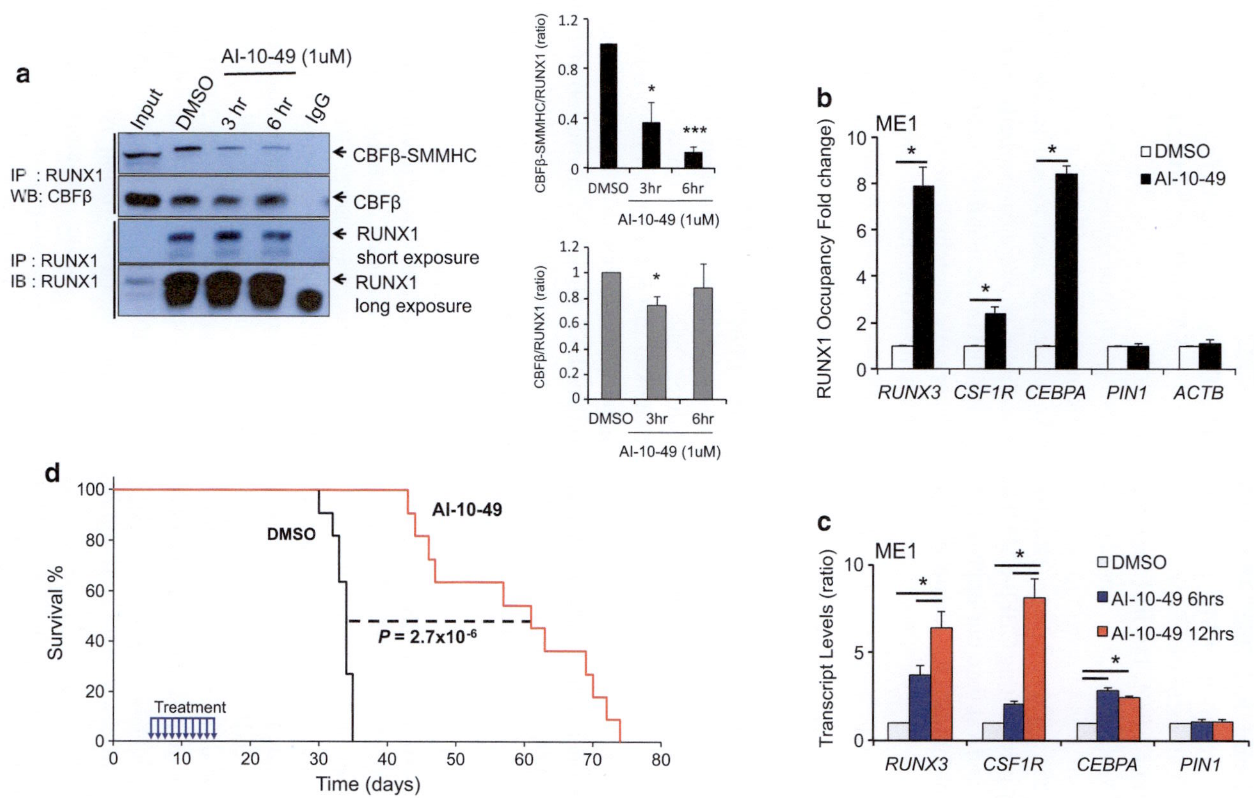

Fig. 15.5 Efficacy of AI-10-49 in cells and in mice. (**a**) Results of co-IP in ME-1 cells after AI-10-49 treatment. (**b**) Effects of AI-10-49 on RUNX1 occupancy on target and non-target genes. (**c**) Effect of AI-10-49 on expression of target and non-target genes. (**d**) Effect of 10 days treatment with AI-10-49 on leukemia latency in a mouse model of inv(16) leukemia

model, chromatin-immunoprecipitation (ChIP) assays showed that treatment of ME-1 cells for 6 h with AI-10-49 increased RUNX1 occupancy 8, 2.2, and 8 fold at the *RUNX3*, *CSF1R* and *CEBPA* promoters, respectively, whereas no enrichment was observed at control loci (Fig. 15.5b). In accordance with this, treatment of ME-1 cells for 6 or 10 h with AI-10-49 increased expression of *RUNX3*, *CSF1R*, and *CEBPA*, but had no effect on control gene *PIN1* (Fig. 15.5c). Importantly, neither of these effects was observed in inv(16)-negative U937 cells. These data establish AI-10-49 selectivity in inhibiting CBFβ-SMMHC binding to RUNX1 and validate our approach of using bivalent inhibitors to achieve this specificity.

15.9 Efficacy of CBFβ-SMMHC Inhibitor in a Mouse Model of inv.(16) Leukemia and Against Human Patient Derived Inv(16) Leukemia Cells

As described earlier, oncogenic NRAS mutations are a frequent "cooperating" event in inv(16) AML, and we have developed a knock-in mouse model of inv(16) AML by combining the conditional $Nras^{LSL\text{-}G12D}$ and $Cbfb^{MYH11}$ alleles (Xue et al. 2014), which is suitable for pharmacologic studies. To test the effects of AI-10-49 administration *in vivo*, we transplanted mice with $Cbfb^{+/MYH11}$:$Ras^{+/G12D}$ leukemic cells, waited 5 days for leukemia engraftment and then treated mice with vehicle (DMSO) or AI-10-49 for 10 days and assessed the effect on disease latency. As shown in Fig. 15.5d, vehicle treated mice succumbed to leukemia with a median latency of 33.5 days, whereas AI-10-49 treated mice survived significantly longer (median latency = 61 days; $P = 2.7 \times 10^{-6}$). These results demonstrate that transient treatment with AI-10-49 induces significant reduction in leukemia expansion in vivo. Importantly, examination of the hematopoietic compartment as well as all major organs shows no evidence of toxicity for AI-10-49. In addition, in work carried out in the laboratory of Dr. Monica Guzman, we have also demonstrated efficacy against inv(16)+ patient cells and no activity against normal karyotype AML patient samples.

15.10 Future Plans for CBFβ-SMMHC Inhibitor

Dysregulation of gene expression is a hallmark of all cancers. It is critical for conferring stem cell like properties, such as self-renewal and chemoresistance, on these cancer cells. Such properties contribute to the inability to completely eradicate cancer cells, thereby leading to relapse. The specific gene expression program that confers these properties derives from aberrant activity of specific transcription factors, which are drivers of disease. There are numerous examples of such transcription factor drivers in cancer such as fusion proteins of RUNX1 and CBFβ in leukemia (Look 1997), fusion proteins of ERG in prostate cancer and Ewing's sarcoma (Hessels and Schalken 2013; Ladanyi 1995), ETV-1 in melanoma (Jane-Valbuena et al. 2010), and members of the ETS family of transcription factors in a variety of cancers (Oh et al. 2012). Transcription factors have traditionally been viewed as "undruggable" due to the need to target more challenging protein-protein or protein nucleic acid interactions through which these proteins act. There are still relatively few examples of such agents for cancer treatment, with the MDM2-p53 inhibitors being one example of such an agent that has progressed to the clinic (Fotouhi and Graves 2005; Miyazaki et al. 2013; Zhang et al. 2014; Zhao et al. 2013). As there are very few such agents, our development of an effective inhibitor targeting the CBFβ-SMMHC transcription factor fusion is encouraging and should spur further development along these lines.

Strikingly, no new agent for AML patients who relapse with current therapy has been approved by the US FDA in decades. The development of this targeted agent could provide a new treatment option for inv(16) AML patients, particularly for those who have relapsed.

AI-10-49 and its analogs have been patented in the US (PCT/US2010/034748), patent is in process in the EU, and efforts to commercialize this agent or related analogs are ongoing. Approximately 50 % of inv(16) leukemia patients will relapse within 5 years (Delaunay et al. 2003). When they do, they invariably retain the inv(16) whereas the cooperating mutations are altered. This strongly implies the CBFβ-SMMHC fusion protein is a key driver of the disease as well as an early event. The minimal residual disease that persists after current treatment approaches clearly provides a pool of cells from which relapse can occur (Guzman and Allan 2014). The targeting of inv(16) containing cells in the context of minimal residual disease provides a unique potential therapeutic approach for deploying these inhibitors.

15.11 Small Molecule Inhibitors of Wildtype CBFβ-RUNX Function

Starting from the chemical scaffold identified for inhibition of the CBFβ-SMMHC/RUNX protein-protein interaction, we have also recently developed optimized inhibitors of wildtype CBFβ-RUNX binding (Illendula et al. 2016).

The RUNX proteins are auto-inhibited and this auto-inhibition is relieved by CBFβ, so inhibitors of CBFβ binding to RUNX proteins should leave RUNX proteins in an auto-inhibited state where they cannot bind DNA effectively and therefore their occupancy in the genome decreases (Fig. 15.6). The small molecule inhibitors we have developed have low micromolar IC_{50} values for inhibition of CBFβ-RUNX binding, reduce the occupancy of RUNX1 on target genes, and alter the expression of RUNX1 regulated genes (Illendula et al. 2016).

Interestingly, we have also shown that the inhibitors act allosterically by altering the dynamic behavior of key hotspot residues for the binding of CBFβ to RUNX. These compounds show activity against a number of different leukemia cell lines, but are most effective against the inv(16) cell line ME-1. As mentioned earlier, inv(16) AML blasts are completely dependent in wildtype RUNX activity for survival. This addiction has also been reported for t(8;21) AML expressing RUNX1-RUNX1T1 and t(9;11) AML expressing MLL-fusion proteins (Ben-Ami et al.

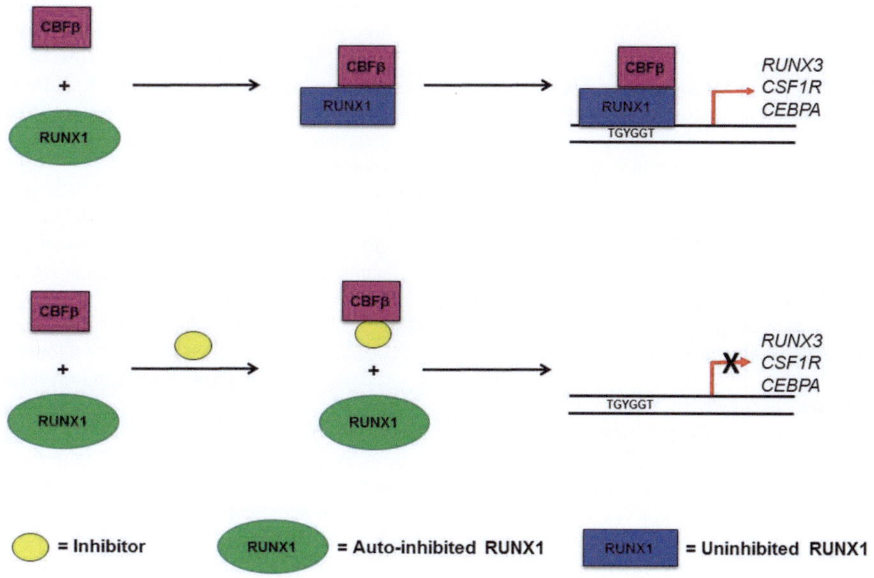

Fig. 15.6 Schematic illustrating mechanism of action of the CBFβ-RUNX inhibitors

2013; Goyama et al. 2013). As the affinity of wild type CBFβ for RUNX1 is ~10-fold weaker than CBFβ-SMMHC binding to RUNX1, it is likely that these inhibitors are reducing the binding of wild type CBFβ to RUNX1 thereby abrogating the remaining normal RUNX1 function and leading to cell death in several AML subtypes.

References

Bakshi, R., Hassan, M. Q., Pratap, J., Lian, J. B., Montecino, M. A., Van Wijnen, A. J., et al. (2010). The human SWI/SNF complex associates with RUNX1 to control transcription of hematopoietic target genes. *Journal of Cellular Physiology, 225*, 569–576.

Ben-Ami, O., Friedman, D., Leshkowitz, D., Goldenberg, D., Orlovsky, K., Pencovich, N., et al. (2013). Addiction of t(8;21) and inv(16) acute myeloid leukemia to native RUNX1. *Cell Reports, 4*, 1131–1143.

Bennett, J. M., Catovsky, D., Daniel, M. T., Flandrin, G., Galton, D. A., Gralnick, H. R., & Sultan, C. (1976). Proposals for the classification of the acute leukaemias. French-American-British (FAB) co-operative group. *British Journal of Haematology, 33*, 451–458.

Blyth, K., Cameron, E. R., & Neil, J. C. (2005). The RUNX genes: Gain or loss of function in cancer. *Nature Reviews Cancer, 5*, 376–387.

Boissel, N., Leroy, H., Brethon, B., Philippe, N., De Botton, S., Auvrignon, A., et al. (2006). Incidence and prognostic impact of c-Kit, FLT3, and Ras gene mutations in core binding factor acute myeloid leukemia (CBF-AML). *Leukemia, 20*, 965–970.

Britos-Bray, M., Ramirez, M., Cao, W., Wang, X., Liu, P. P., Civin, C. I., & Friedman, A. D. (1998). CBFbeta-SMMHC, expressed in M4eo acute myeloid leukemia, reduces p53 induction and slows apoptosis in hematopoietic cells exposed to DNA-damaging agents. *Blood, 92*, 4344–4352.

Cao, W., Britos-Bray, M., Claxton, D. F., Kelley, C. A., Speck, N. A., Liu, P. P., & Friedman, A. D. (1997). CBF beta-SMMHC, expressed in M4Eo AML, reduced CBF DNA-binding and inhibited the G1 to S cell cycle transition at the restriction point in myeloid and lymphoid cells. *Oncogene, 15*, 1315–1327.

Care, R. S., Valk, P. J., Goodeve, A. C., Abu-Duhier, F. M., Geertsma-Kleinekoort, W. M., wilson, G. A., et al. (2003). Incidence and prognosis of c-KIT and FLT3 mutations in core binding factor (CBF) acute myeloid leukaemias. *British Journal of Haematology, 121*, 775–777.

Castilla, L. H., Wijmenga, C., Wang, Q., Stacy, T., Speck, N. A., Eckhaus, M., et al. (1996). Failure of embryonic hematopoiesis and lethal hemorrhages in mouse embryos heterozygous for a knocked-in leukemia gene CBFB-MYH11. *Cell, 87*, 687–696.

Castilla, L. H., Garrett, L., Adya, N., Orlic, D., Dutra, A., Anderson, S., et al. (1999). The fusion gene Cbfb-MYH11 blocks myeloid differentiation and predisposes mice to acute myelomonocytic leukaemia. *Nature Genetics, 23*, 144–146.

Castilla, L. H., Perrat, P., Martinez, N. J., Landrette, S. F., Keys, R., Oikemus, S., et al. (2004). Identification of genes that synergize with Cbfb-MYH11 in the pathogenesis of acute myeloid leukemia. *Proceedings of the National Academy of Sciences of the United States of America, 101*, 4924–4929.

Cheng, C. K., Li, L., Cheng, S. H., Lau, K. M., Chan, N. P., Wong, R. S., et al. (2008). Transcriptional repression of the RUNX3/AML2 gene by the t(8;21) and inv(16) fusion proteins in acute myeloid leukemia. *Blood, 112*, 3391–3402.

Chin, D. W., Watanabe-Okochi, N., Wang, C. Q., Tergaonkar, V., & Osato, M. (2015). Mouse models for core binding factor leukemia. *Leukemia, 29*, 1970–1980.

Claxton, D. F., Liu, P., Hsu, H. B., Marlton, P., Hester, J., Collins, F., et al. (1994). Detection of fusion transcripts generated by the inversion 16 chromosome in acute myelogenous leukemia. *Blood, 83*, 1750–1756.

Corces-Zimmerman, M. R., Hong, W. J., Weissman, I. L., Medeiros, B. C., & Majeti, R. (2014). Preleukemic mutations in human acute myeloid leukemia affect epigenetic regulators and persist in remission. *Proceedings of the National Academy of Sciences of the United States of America, 111*, 2548–2553.

Cunningham, L., Finckbeiner, S., Hyde, R. K., Southall, N., Marugan, J., Yedavalli, V. R., et al. (2012). Identification of benzodiazepine Ro5-3335 as an inhibitor of CBF leukemia through quantitative high throughput screen against RUNX1-CBFbeta interaction. *Proceedings of the National Academy of Sciences of the United States of America, 109*, 14592–14597.

D'costa, J., Chaudhuri, S., Civin, C. I., & Friedman, A. D. (2005). CBFbeta-SMMHC slows proliferation of primary murine and human myeloid progenitors. *Leukemia, 19*, 921–929.

Delaunay, J., Vey, N., Leblanc, T., Fenaux, P., Rigal-Huguet, F., Witz, F., et al. (2003). Prognosis of inv(16)/t(16;16) acute myeloid leukemia (AML): A survey of 110 cases from the French AML Intergroup. *Blood, 102*, 462–469.

Ding, L., Ley, T. J., Larson, D. E., Miller, C. A., Koboldt, D. C., Welch, J. S., et al. (2012). Clonal evolution in relapsed acute myeloid leukaemia revealed by whole-genome sequencing. *Nature, 481*, 506–510.

Duployez, N., Marceau-Renaut, A., Boissel, N., Petit, A., Bucci, M., Geffroy, S., et al. (2016). Comprehensive mutational profiling of core binding factor acute myeloid leukemia. *Blood, 127*, 2451–2459.

Egawa, T., Tillman, R. E., Naoe, Y., Taniuchi, I., & Littman, D. R. (2007). The role of the Runx transcription factors in thymocyte differentiation and in homeostasis of naive T cells. *The Journal of Experimental Medicine, 204*, 1945–1957.

Enright, H., Weisdorf, D., Peterson, L., Rydell, R. E., Kaplan, M. E., & Arthur, D. C. (1992). Inversion of chromosome 16 and dysplastic eosinophils in accelerated phase of chronic myeloid leukemia. *Leukemia, 6*, 381–384.

Farag, S. S., Archer, K. J., Mrozek, K., Ruppert, A. S., Carroll, A. J., Vardiman, J. W., et al. (2006). Pretreatment cytogenetics add to other prognostic factors predicting complete remission and long-term outcome in patients 60 years of age or older with acute myeloid leukemia: Results from Cancer and Leukemia Group B 8461. *Blood, 108*, 63–73.

Fotouhi, N., & Graves, B. (2005). Small molecule inhibitors of p53/MDM2 interaction. *Current Topics in Medicinal Chemistry, 5*, 159–165.

Goemans, B. F., Zwaan, C. M., Miller, M., Zimmermann, M., Harlow, A., Meshinchi, S., et al. (2005). Mutations in KIT and RAS are frequent events in pediatric core-binding factor acute myeloid leukemia. *Leukemia, 19*, 1536–1542.

Goyama, S., Schibler, J., Cunningham, L., Zhang, Y., Rao, Y., Nishimoto, N., et al. (2013). Transcription factor RUNX1 promotes survival of acute myeloid leukemia cells. *The Journal of Clinical Investigation, 123*, 3876–3888.

Grimwade, D., Hills, R. K., Moorman, A. V., Walker, H., Chatters, S., Goldstone, A. H., et al. (2010). Refinement of cytogenetic classification in acute myeloid leukemia: Determination of prognostic significance of rare recurring chromosomal abnormalities among 5876 younger adult patients treated in the United Kingdom Medical Research Council trials. *Blood, 116*, 354–365.

Growney, J. D., Shigematsu, H., Li, Z., Lee, B. H., Adelsperger, J., Rowan, R., et al. (2005). Loss of Runx1 perturbs adult hematopoiesis and is associated with a myeloproliferative phenotype. *Blood, 106*, 494–504.

Guo, H., Ma, O., Speck, N. A., & Friedman, A. D. (2012). Runx1 deletion or dominant inhibition reduces Cebpa transcription via conserved promoter and distal enhancer sites to favor monopoiesis over granulopoiesis. *Blood, 119*, 4408–4418.

Guzman, M. L., & Allan, J. N. (2014). Concise review: Leukemia stem cells in personalized medicine. *Stem Cells, 32*, 844–851.

Haase, D., Feuring-Buske, M., Konemann, S., Fonatsch, C., Troff, C., Verbeek, W., et al. (1995). Evidence for malignant transformation in acute myeloid leukemia at the level of early hematopoietic stem cells by cytogenetic analysis of CD34+ subpopulations. *Blood, 86*, 2906–2912.

Haferlach, C., Dicker, F., Herholz, H., Schnittger, S., Kern, W., & Haferlach, T. (2008). Mutations of the TP53 gene in acute myeloid leukemia are strongly associated with a complex aberrant karyotype. *Leukemia, 22*, 1539–1541.

Haferlach, C., Dicker, F., Kohlmann, A., Schindela, S., Weiss, T., Kern, W., et al. (2010). AML with CBFB-MYH11 rearrangement demonstrate RAS pathway alterations in 92 % of all cases including a high frequency of NF1 deletions. *Leukemia, 24*, 1065–1069.

Han, E., Lee, H., Kim, M., Kim, Y., Han, K., Lee, S. E., et al. (2014). Characteristics of hematologic malignancies with coexisting t(9;22) and inv(16) chromosomal abnormalities. *Blood Research, 49*, 22–28.

Hessels, D., & Schalken, J. A. (2013). Recurrent gene fusions in prostate cancer: Their clinical implications and uses. *Current Urology Reports, 14*, 214–222.

Huang, H., Woo, A. J., Waldon, Z., Schindler, Y., Moran, T. B., Zhu, H. H., et al. (2012). A Src family kinase-Shp2 axis controls RUNX1 activity in megakaryocyte and T-lymphocyte differentiation. *Genes & Development, 26*, 1587–1601.

Hyde, R. K., Zhao, L., Alemu, L., & Liu, P. P. (2015). Runx1 is required for hematopoietic defects and leukemogenesis in Cbfb-MYH11 knock-in mice. *Leukemia, 29*, 1771–1778.

Ikebe, M., Komatsu, S., Woodhead, J. L., Mabuchi, K., Ikebe, R., Saito, J., et al. (2001). The tip of the coiled-coil rod determines the filament formation of smooth muscle and nonmuscle myosin. *The Journal of Biological Chemistry, 276*, 30293–30300 Epub 2001 Jun 6.

Illendula, A., Pulikkan, J. A., Zong, H., Grembecka, J., Xue, L., Sen, S., et al. (2015). Chemical biology. A small-molecule inhibitor of the aberrant transcription factor CBFbeta-SMMHC delays leukemia in mice. *Science, 347*, 779–784.

Illendula, A., Gilmour, J., Grembecka, J., Srimath Tirumala, V., Boulton, A., Kuntimaddi, A., et al. (2016). Small molecule inhibitor of CBFβ-RUNX binding for RUNX transcription factor driven cancers. *eBioMedicine, 8*, 117–131.

Jane-Valbuena, J., Widlund, H. R., Perner, S., Johnson, L. A., Dibner, A. C., lin, W. M., et al. (2010). An oncogenic role for ETV1 in melanoma. *Cancer Research, 70*, 2075–2084.

Jongen-Lavrencic, M., Sun, S. M., Dijkstra, M. K., Valk, P. J., & Lowenberg, B. (2008). MicroRNA expression profiling in relation to the genetic heterogeneity of acute myeloid leukemia. *Blood, 111*, 5078–5085.

Kamikubo, Y., Zhao, L., Wunderlich, M., Corpora, T., Hyde, R. K., Paul, T. A., et al. (2010). Accelerated leukemogenesis by truncated CBF beta-SMMHC defective in high-affinity binding with RUNX1. *Cancer Cell, 17*, 455–468.

Kihara, R., Nagata, Y., Kiyoi, H., Kato, T., Yamamoto, E., Suzuki, K., et al. (2014). Comprehensive analysis of genetic alterations and their prognostic impacts in adult acute myeloid leukemia patients. *Leukemia, 28*, 1586–1595.

Kim, H. G., Kojima, K., Swindle, C. S., Cotta, C. V., Huo, Y., Reddy, V., & Klug, C. A. (2008). FLT3-ITD cooperates with inv(16) to promote progression to acute myeloid leukemia. *Blood, 111*, 1567–1574.

Kottaridis, P. D., Gale, R. E., Langabeer, S. E., Frew, M. E., Bowen, D. T., & Linch, D. C. (2002). Studies of FLT3 mutations in paired presentation and relapse samples from patients with acute myeloid leukemia: Implications for the role of FLT3 mutations in leukemogenesis, minimal residual disease detection, and possible therapy with FLT3 inhibitors. *Blood, 100*, 2393–2398.

Kummalue, T., Lou, J., & Friedman, A. D. (2002). Multimerization via its myosin domain facilitates nuclear localization and inhibition of core binding factor (CBF) activities by the CBFbeta-smooth muscle myosin heavy chain myeloid leukemia oncoprotein. *Molecular and Cellular Biology, 22*, 8278–8291.

Kuo, Y. H., Landrette, S. F., Heilman, S. A., Perrat, P. N., Garrett, L., Liu, P. P., et al. (2006). Cbf beta-SMMHC induces distinct abnormal myeloid progenitors able to develop acute myeloid leukemia. *Cancer Cell, 9*, 57–68.

Kuo, Y. H., Gerstein, R. M., & Castilla, L. H. (2008). Cbfbeta-SMMHC impairs differentiation of common lymphoid progenitors and reveals an essential role for RUNX in early B-cell development. *Blood, 111*, 1543–1551.

Kuo, Y. H., Zaidi, S. K., Gornostaeva, S., Komori, T., Stein, G. S., & Castilla, L. H. (2009). Runx2 induces acute myeloid leukemia in cooperation with Cbfbeta-SMMHC in mice. *Blood, 113*, 3323–3332.

Ladanyi, M. (1995). The emerging molecular genetics of sarcoma translocations. *Diagnostic Molecular Pathology, 4*, 162–173.

Le Beau, M. M., Larson, R. A., Bitter, M. A., Vardiman, J. W., Golomb, H. M., & Rowley, J. D. (1983). Association of an inversion of chromosome 16 with abnormal marrow eosinophils in acute myelomonocytic leukemia. A unique cytogenetic-clinicopathological association. *New England Journal of Medicine, 309*, 630–636.

Li, Z., Lu, J., Sun, M., Mi, S., Zhang, H., Luo, R. T., et al. (2008). Distinct microRNA expression profiles in acute myeloid leukemia with common translocations. *Proceedings of the National Academy of Sciences of the United States of America, 105*, 15535–15540.

Liu, P., Claxton, D. F., Marlton, P., Hajra, A., Siciliano, J., Freedman, M., et al. (1993a). Identification of yeast artificial chromosomes containing the inversion 16 p-arm breakpoint associated with acute myelomonocytic leukemia. *Blood, 82*, 716–721.

Liu, P., Tarle, S. A., Hajra, A., Claxton, D. F., Marlton, P., Freedman, M., et al. (1993b). Fusion between transcription factor CBF beta/PEBP2 beta and a myosin heavy chain in acute myeloid leukemia. *Science, 261*, 1041–1044.

Liu, P. P., Hajra, A., Wijmenga, C., & Collins, F. S. (1995). Molecular pathogenesis of the chromosome 16 inversion in the M4Eo subtype of acute myeloid leukemia. [see comments]. [erratum appears in Blood 1997 Mar 1;89(5):1842]. *Blood, 85*, 2289–2302.

Liu, P. P., Wijmenga, C., Hajra, A., Blake, T. B., Kelley, C. A., Adelstein, R. S., Bagg, A., Rector, J., Cotelingam, J., Willman, C. L., & Collins, F. S. (1996). Identification of the chimeric protein product of the CBFB-MYH11 fusion gene in inv(16) leukemia cells. [erratum appears in Genes Chromosomes Cancer 1997 Jan;18(1):71]. *Genes, Chromosomes & Cancer, 16*, 77–87.

Look, A. T. (1997). Oncogenic transcription factors in the human acute leukemias. *Science, 278*, 1059–1064.

Lukasik, S. M., Zhang, L., Corpora, T., Tomanicek, S., Li, Y., Kundu, M., et al. (2002). Altered affinity of CBF beta-SMMHC for Runx1 explains its role in leukemogenesis. *Nature Structural Biology, 9*, 674–679.

Markus, J., Garin, M. T., Bies, J., Galili, N., Raza, A., Thirman, M. J., et al. (2007). Methylation-independent silencing of the tumor suppressor INK4b (p15) by CBFbeta-SMMHC in acute myelogenous leukemia with inv(16). *Cancer Research, 67*, 992–1000.

Mchale, C. M., Wiemels, J. L., Zhang, L., Ma, X., Buffler, P. A., Feusner, J., et al. (2003). Prenatal origin of childhood acute myeloid leukemias harboring chromosomal rearrangements t(15;17) and inv(16). *Blood, 101*, 4640–4641.

Mehrotra, B., George, T. I., Kavanau, K., Avet-Loiseau, H., Moore 2nd, D., Willman, C. L., et al. (1995). Cytogenetically aberrant cells in the stem cell compartment (CD34+lin-) in acute myeloid leukemia. *Blood, 86*, 1139–1147.

Miano, J. M., Cserjesi, P., Ligon, K. L., Periasamy, M., & Olson, E. N. (1994). Smooth muscle myosin heavy chain exclusively marks the smooth muscle lineage during mouse embryogenesis. *Circulation Research, 75*, 803–812.

Miyazaki, M., Naito, H., Sugimoto, Y., Yoshida, K., Kawato, H., Okayama, T., et al. (2013). Synthesis and evaluation of novel orally active p53-MDM2 interaction inhibitors. *Bioorganic & Medicinal Chemistry, 21*, 4319–4331.

Monma, F., Nishii, K., Shiga, J., Sugahara, H., Lorenzo, F. T., Watanabe, Y., et al. (2007). Detection of the CBFB/MYH11 fusion gene in de novo acute myeloid leukemia (AML): A single-institution study of 224 Japanese AML patients. *Leukemia Research, 31*, 471–476.

Nakano, Y., Kiyoi, H., Miyawaki, S., Asou, N., Ohno, R., Saito, H., & Naoe, T. (1999). Molecular evolution of acute myeloid leukaemia in relapse: Unstable N-ras and FLT3 genes compared with p53 gene. *British Journal of Haematology, 104*, 659–664.

Oh, S., Shin, S., & Janknecht, R. (2012). ETV1, 4 and 5: An oncogenic subfamily of ETS transcription factors. *Biochimica et Biophysica Acta, 1826*, 1–12.

Okuda, T., Van Deursen, J., Hiebert, S. W., Grosveld, G., & Downing, J. R. (1996). AML1, the target of multiple chromosomal translocations in human leukemia, is essential for normal fetal liver hematopoiesis. *Cell, 84*, 321–330.

Opatz, S., Polzer, H., Herold, T., Konstandin, N. P., Ksienzyk, B., Zellmeier, E., et al. (2013). Exome sequencing identifies recurring FLT3 N676 K mutations in core-binding factor leukemia. *Blood, 122*, 1761–1769.

Park, T. S., Lee, S. T., Song, J., Lee, K. A., Lee, J. H., Kim, J., et al. (2009). Detection of a novel CBFB/MYH11 variant fusion transcript (K-type) showing partial insertion of exon 6 of CBFB gene using two commercially available multiplex RT-PCR kits. *Cancer Genetics and Cytogenetics, 189*, 87–92.

Paschka, P., Marcucci, G., Ruppert, A. S., Mrozek, K., Chen, H., Kittles, R. A., et al. (2006). Adverse prognostic significance of KIT mutations in adult acute myeloid leukemia with inv(16) and t(8;21): A Cancer and Leukemia Group B Study. *Journal of Clinical Oncology, 24*, 3904–3911.

Paschka, P., Du, J., Schlenk, R. F., Gaidzik, V. I., Bullinger, L., Corbacioglu, A., et al. (2013). Secondary genetic lesions in acute myeloid leukemia with inv(16) or t(16;16): A study of the German-Austrian AML Study Group (AMLSG). *Blood, 121*, 170–177.

Pulsoni, A., Iacobelli, S., Bernardi, M., Borgia, M., Camera, A., Cantore, N., et al. (2008). M4 acute myeloid leukemia: The role of eosinophilia and cytogenetics in treatment response and survival. The GIMEMA experience. *Haematologica, 93*, 1025–1032.

Qi, J., Singh, S., Hua, W. K., Cai, Q., Chao, S. W., Li, L., et al. (2015). HDAC8 inhibition specifically targets inv(16) acute myeloid leukemic stem cells by restoring p53 acetylation. *Cell Stem Cell, 17*, 597–610.

Ravandi, F., Burnett, A. K., Agura, E. D., & Kantarjian, H. M. (2007). Progress in the treatment of acute myeloid leukemia. *Cancer, 110*, 1900–1910.

Sasaki, K., Yagi, H., Bronson, R. T., Tominaga, K., Matsunashi, T., Deguchi, K., et al. (1996). Absence of fetal liver hematopoiesis in mice deficient in transcriptional coactivator core binding factor beta. *Proceedings of the National Academy of Sciences of the United States of America, 93*, 12359–12363.

Schwind, S., Edwards, C. G., Nicolet, D., Mrozek, K., Maharry, K., wu, Y. Z., et al. (2013). inv(16)/t(16;16) acute myeloid leukemia with non-type A CBFB-MYH11 fusions associate with distinct clinical and genetic features and lack KIT mutations. *Blood, 121*, 385–391.

Secker-Walker, L. M., Morgan, G. J., Min, T., Swansbury, G. J., Craig, J., Yamada, T., et al. (1992). Inversion of chromosome 16 with the Philadelphia chromosome in acute myelomonocytic leukemia with eosinophilia. Report of two cases. *Cancer Genetics and Cytogenetics, 58*, 29–34.

Seifert, H., Mohr, B., Thiede, C., Oelschlagel, U., Schakel, U., Illmer, T., et al. (2009). The prognostic impact of 17p (p53) deletion in 2272 adults with acute myeloid leukemia. *Leukemia, 23*, 656–663.

Seo, W., Ikawa, T., Kawamoto, H., & Taniuchi, I. (2012). Runx1-Cbfbeta facilitates early B lymphocyte development by regulating expression of Ebf1. *The Journal of Experimental Medicine, 209*, 1255–1262.

Shigesada, K., Van De Sluis, B., & Liu, P. P. (2004). Mechanism of leukemogenesis by the inv(16) chimeric gene CBFB/PEBP2B-MHY11. *Oncogene, 23*, 4297–4307.

Shih, L. Y., Liang, D. C., Huang, C. F., Chang, Y. T., Lai, C. L., Lin, T. H., et al. (2008). Cooperating mutations of receptor tyrosine kinases and Ras genes in childhood core-binding factor acute myeloid leukemia and a comparative analysis on paired diagnosis and relapse samples. *Leukemia, 22*, 303–307.

Shlush, L. I., Zandi, S., Mitchell, A., Chen, W. C., Brandwein, J. M., Gupta, V., et al. (2014). Identification of pre-leukaemic haematopoietic stem cells in acute leukaemia. *Nature, 506*, 328–333.

Sohn, R. L., Vikstrom, K. L., Strauss, M., Cohen, C., Szent-Gyorgyi, A. G., & leinwand, L. A. (1997). A 29 residue region of the sarcomeric myosin rod is necessary for filament formation. *Journal of Molecular Biology, 266*, 317–330.

Song, J., Mercer, D., Hu, X., Liu, H., & Li, M. M. (2011). Common leukemia- and lymphoma-associated genetic aberrations in healthy individuals. *The Journal of Molecular Diagnostics, 13*, 213–219.

Sood, R., Hansen, N. F., Donovan, F. X., Carrington, B., Bucci, D., Maskeri, B., et al. (2016). Somatic mutational landscape of AML with inv(16) or t(8;21) identifies patterns of clonal evolution in relapse leukemia. *Leukemia, 30*, 501–504.

Swerdlow, S. H., Campo, E., Harris, N. L., Jaffe, E. S., Pileri, S. A., Stein, H., et al. (2008). WHO classification of tumours of haematopoietic and lymphoid tissues. In OMS (Ed.), *WHO classification of tumours of haematopoietic and lymphoid tissues* (4th ed.). Lyon: World Health Organization.

Valk, P. J., Verhaak, R. G., Beijen, M. A., Erpelinck, C. A., Barjesteh Van Waalwijk Van Doorn-Khosrovani, S., Boer, J. M., et al. (2004). Prognostically useful gene-expression profiles in acute myeloid leukemia. *The New England Journal of Medicine, 350*, 1617–1628.

Wang, Q., Stacy, T., Binder, M., Marin-Padilla, M., Sharpe, A. H., & Speck, N. A. (1996a). Disruption of the Cbfa2 gene causes necrosis and hemorrhaging in the central nervous system and blocks definitive hematopoiesis. *Proceedings of the National Academy of Sciences of the United States of America, 93*, 3444–3449.

Wang, Q., Stacy, T., Miller, J. D., Lewis, A. F., Gu, T. L., Huang, X., et al. (1996b). The CBFbeta subunit is essential for CBFalpha2 (AML1) function in vivo. *Cell, 87*, 697–708.

Wee, H. J., Voon, D. C., Bae, S. C., & Ito, Y. (2008). PEBP2-beta/CBF-beta-dependent phosphorylation of RUNX1 and p300 by HIPK2: Implications for leukemogenesis. *Blood, 112*, 3777–3787.

Wu, Y., Slovak, M. L., Snyder, D. S., & Arber, D. A. (2006). Coexistence of inversion 16 and the Philadelphia chromosome in acute and chronic myeloid leukemias: Report of six cases and review of literature. *American Journal of Clinical Pathology, 125*, 260–266.

Xue, L., Pulikkan, J. A., Valk, P. J., & Castilla, L. H. (2014). NrasG12D oncoprotein inhibits apoptosis of preleukemic cells expressing Cbfbeta-SMMHC via activation of MEK/ERK axis. *Blood, 124*, 426–436.

Zhang, D. E., Hetherington, C. J., Meyers, S., Rhoades, K. L., Larson, C. J., Chen, H. M., et al. (1996). CCAAT enhancer-binding protein (C/EBP) and AML1 (CBF alpha2) synergistically activate the macrophage colony-stimulating factor receptor promoter. *Molecular and Cellular Biochemistry, 16*, 1231–1240.

Zhang, L., D'costa, J., Kummalue, T., Civin, C. I., & Friedman, A. D. (2006). Identification of a region on the outer surface of the CBFbeta-SMMHC myeloid oncoprotein assembly competence domain critical for multimerization. *Oncogene, 25*, 7289–7296.

Zhang, Z., Ding, Q., Liu, J. J., Zhang, J., Jiang, N., Chu, X. J., et al. (2014). Discovery of potent and selective spiroindolinone MDM2 inhibitor, RO8994, for cancer therapy. *Bioorganic & Medicinal Chemistry, 22*, 4001–4009.

Zhao, L., Cannons, J. L., Anderson, S., Kirby, M., Xu, L., Castilla, L. H., et al. (2007). CBFB-MYH11 hinders early T-cell development and induces massive cell death in the thymus. *Blood, 109*, 3432–3440.

Zhao, L., Melenhorst, J. J., Alemu, L., Kirby, M., Anderson, S., Kench, M., et al. (2012). KIT with D816 mutations cooperates with CBFB-MYH11 for leukemogenesis in mice. *Blood, 119*, 1511–1521.

Zhao, Y., Yu, S., Sun, W., Liu, L., Lu, J., Mceachern, D., et al. (2013). A potent small-molecule inhibitor of the MDM2-p53 interaction (MI-888) achieved complete and durable tumor regression in mice. *Journal of Medicinal Chemistry, 56*, 5553–5561.

Part V

RUNX and CBFβ in Cancer and Immune Function

The *RUNX* Genes as Conditional Oncogenes: Insights from Retroviral Targeting and Mouse Models

16

James C. Neil, Kathryn Gilroy, Gillian Borland, Jodie Hay, Anne Terry, and Anna Kilbey

Abstract

The observation that the *Runx* genes act as targets for transcriptional activation by retroviral insertion identified a new family of dominant oncogenes. However, it is now clear that *Runx* genes are 'conditional' oncogenes whose over-expression is growth inhibitory unless accompanied by another event such as concomitant over-expression of MYC or loss of p53 function. Remarkably, while the oncogenic activities of either MYC or RUNX over-expression are suppressed while p53 is intact, the combination of both neutralises p53 tumour suppression *in vivo* by as yet unknown mechanisms. Moreover, there is emerging evidence that endogenous, basal RUNX activity is important to maintain the viability and proliferation of MYC-driven lymphoma cells. There is also growing evidence that the human *RUNX* genes play a similar conditional oncogenic role and are selected for over-expression in end-stage cancers of multiple types. Paradoxically, reduced RUNX activity can also predispose to cell immortalisation and transformation, particularly by mutant Ras. These apparently conflicting observations may be reconciled in a stage-specific model of RUNX involvement in cancer. A question that has yet to be fully addressed is the extent to which the three *Runx* genes are functionally redundant in cancer promotion and suppression.

Keywords

Cancer • Lymphoma • Retroviral mutagenesis • Senescence • Oncogene • Tumour suppressor

J.C. Neil (✉) • K. Gilroy • G. Borland • J. Hay
A. Terry • A. Kilbey
Molecular Oncology Laboratory, Centre for Virus Research, University of Glasgow,
Bearsden, Glasgow G61 1QH, UK
e-mail: James.Neil@glasgow.ac.uk

Y. Groner et al. (eds.), *RUNX Proteins in Development and Cancer*, Advances in Experimental Medicine and Biology 962, DOI 10.1007/978-981-10-3233-2_16

16.1 The Murine *Runx* Genes as Targets for Insertional Mutagenesis

The discovery that the murine *Runx* genes can act as targets for transcriptional activation by murine leukaemia viruses (MLVs) provided evidence that they belong to the class of proto-oncogenes: genes normally involved in growth control that can be activated to play a dominant oncogenic role in cancer. MLVs are members of the gamma-retrovirus genus and induce tumours in mice primarily by insertional mutagenesis. Strong promoter/enhancer sequences in the long terminal repeats of these viruses affect the expression of genes near, or even at a considerable distance from, the insertion site. MLV induced tumours may display multiple insertions at complementing oncogenes; gene activation events predominate although tumour suppressor inactivation can also occur (reviewed in (Uren et al. 2005)). Completion of the mouse genome sequence has greatly increased the ease of mapping retroviral insertions in tumours, greatly extending the reach of mutagenesis screens for cancer –relevant genes (Hwang et al. 2002; Mikkers et al. 2002; Suzuki et al. 2002). While the process of retroviral integration into the host genome was previously thought to be random with regard to base sequence and location, it is now clear that some retroviruses display very significant and distinct biases. For MLV and related gamma-retroviruses, intrinsic integration preference arises at least in part from interaction of the viral integrase protein with the BET (bromodomain and extraterminal) host chromatin tethering factors Brd2,3 and 4 (De Rijck et al. 2013; Gupta et al. 2013; Sharma et al. 2013). This specificity is of interest with regard to oncogenic potential as BET binding is also a feature of 'super-enhancers' – highly cell-type specific tandem clusters of enhancer elements that appear to define cell identity and the cancer phenotype (Whyte et al. 2013). This intrinsic bias has to be allowed for, particularly when analysing and interpreting high throughput datasets (Cattoglio et al. 2010; LaFave et al. 2014). These findings led us to propose a two-stage model of gamma-retrovirus oncogenesis: selective integration at 'dangerous' sites, followed by clonal selection of collaborating mutations (Huser et al. 2014).

The first recorded example of *Runx* gene targeting by MLV arose from an early screen where a single case of Akv MLV insertion was recorded close to the P1 promoter of *Runx1* in a case of myeloid leukaemia in a BXH2 mouse (Suzuki et al. 2002). This appears to have been a fortuitous observation, as subsequent studies have shown a relatively low frequency of targeting of the *Runx* genes in end-stage tumours of wild-type mice. However, frequent activation of *Runx* genes has been observed in mice where predisposition to tumour development is conferred, for example, by a germ-line *MYC* oncogene over-expressed under tissue-specific transcriptional controls. While all three *Runx* genes have been shown to be capable of acting as targets in *MYC* transgenic mice, the frequency varies according to model. *Runx1* and *Runx3* are targeted in B-cell lymphomas accelerated by neonatal infection with Moloney MLV in the Eμ-Myc model (Mikkers et al. 2002; Uren et al. 2008). All three *Runx* genes have been observed as targets in virus-accelerated T-cell lymphomas of CD2-*MYC* but with frequency of targeting *Runx2*>*Runx3*>*Runx1* (Mikkers et al. 2002; Stewart et al. 2002; Stewart et al. 1997; Wotton et al. 2002). Moreover, a recent high throughput/NGS study confirmed the relative rarity of *Runx* gene insertions in clonally expanded cell populations in wild-type mice, and indicated that, in contrast, activation of a *Runx* family member is virtually obligatory in virus-accelerated CD2-*MYC* lymphomas (Huser et al. 2014).

A diagram summarising the location and orientation of recorded proviral insertions at the murine *Runx* genes is presented in Fig. 16.1. We have included only those examples where significant clonal expansion has provided corroborating evidence that these insertions played a causal role in tumour outgrowth. A track showing H3K27Ac 'enhancer' marks in mouse thymus is included for comparison. While there is substantial correspondence between H3K27Ac marks and the peaks of insertion, this overlap is not seen at the *Runx2* P1 promoter. However, it should be noted

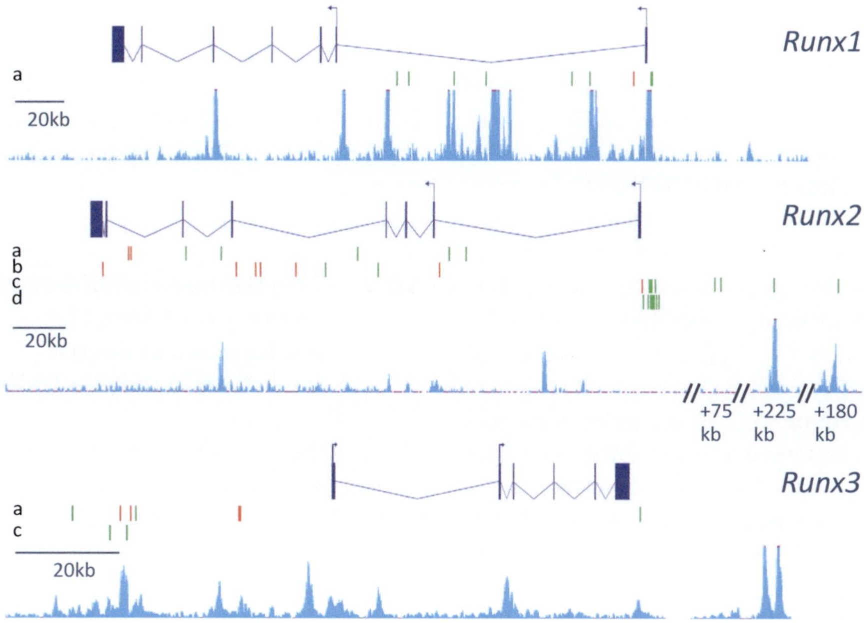

Fig. 16.1 *In vivo* clonally expanded insertions of MLV and transposons at the *Runx* gene loci in murine tumours. The results of multiple studies have been collated. The *top* of each track shows gene structure, with *solid vertical bars* representing exons and *lines* representing introns, with *arrows* showing transcriptional start sites. Proviral insertions in the forward direction are coloured *green* while those in the reverse direction are coloured *red*. Sources of data for tracks (**a–d**) are as follows: (**a**) CD2-*MYC* transgenic T-cell lymphomas analysed by high throughput/NGS analysis (read number >100) (Huser et al. 2014). (**b**) CD2-*MYC* transgenic T-cell lymphomas from restriction mapping and direct sequence analyses (Cameron et al. 2003; Stewart et al. 2007; Wotton et al. 2002) (**c**) Common insertion sites from the retroviral tagged cancer gene database (ref) (**d**) Sleeping beauty transposon insertions in leukaemia/lymphomas from *Rassf1a* deficient mice (van der Weyden et al. 2012). Insertions with more than 100 copies are shown. The *bottom* of each track shows H3K27ac intensity in adult mouse thymus, obtained from the mouse ENCODE repository

that available ChIP-Seq datasets may not reflect the target cell at the stage of development when retroviral integration occurred.

In the CD2-*MYC* system it is clear that MLV induces transcriptional activation of *Runx* family genes, as cell lines established from these lymphomas show very high level expression of the targeted gene with no mutational changes in coding sequence (Stewart et al. 2002; Stewart et al. 1997; Wotton et al. 2002). Moreover, the strong statistical bias of proviral orientation where insertions have been observed at the *Runx2* P1 promoter fits with the 'enhancer mode' of retroviral insertion, where the viral long terminal repeat is upstream and backwards with regard to the targeted promoter (Huser et al. 2014). Expanded clonal insertions far upstream of *Runx2* in CD2-*MYC* lymphomas suggest that regulation of this gene in early T-cell development also involves distant cis-acting enhancer elements that serve as targets for long-range activation by retroviral insertions (Huser et al. 2014).

The pattern is somewhat less clear for *Runx1*. In the 6i lymphoma cell line derived from CD2-*MYC* lymphoma, MLV insertion is in the 'promoter insertion' mode and the result has been established as over-expression of the P1 isoform of RUNX1 (Wotton et al. 2002). Other examples from the RTCGD (retroviral tagged cancer gene database; http://variation.osu.edu/rtcgd/index.html) show insertions between the P1 and P2 promoters mainly in the opposite orientation to the gene, and the consequences for activation of either promoter have not been investigated.

In the case of *Runx3*, insertions appear to cluster at an upstream 'super-enhancer.' In the 1i

CD2-*MYC* lymphoma cell line MLV insertion at this site is in the upstream and backwards mode and is associated with over-expression of the P1 isoform of RUNX3 (Stewart et al. 2002) and a similar location and orientation is evident in other lymphomas from the CD2-*MYC* series (Fig. 16.1). While other RTCGD database examples are less clearly biased in orientation, multiple insertions were reported at the homologue of this site in MLV-induced lymphomas in the rat. This site was originally designated *Dsi1* before the discovery of the *Runx* genes, and it was noted that all insertions were in the same orientation, which we can now read as upstream and backwards with respect to *Run*x3. Notably, in the index case from which Dsi1 was cloned, there was also a proviral insertion at rat c-*Myc* (Vijaya et al. 1987). The *Runx3* gene was also identified as a target for MLV insertion in two independent transplanted B-ALLs in a mouse model of *BCR-ABL* under selection for imatinib resistance. In this system over-expression of either RUNX3 or RUNX1 was shown to diminish imatinib-induced apoptosis (Miething et al. 2007).

The *Runx2* gene has also been reported as a target for retroviral insertional mutagenesis in myeloid leukaemias of *Cbfb-MYH11* (Inv16) mice where it was initially considered a candidate tumour suppressor on the basis that the insertions were intragenic and potentially disruptive (Castilla et al. 2004). However, further study revealed that reduced dosage of *Runx2* suppressed disease in this model, while ectopic expression of full-length *Runx2* cooperated with *Cbfb-MYH11* in transplantation assays (Kuo et al. 2009), suggesting that the intragenic insertions may have been activating events. This conclusion appears to conflict with a recent report of intragenic *Runx2* insertions of the transposon *Sleeping Beauty* in leukaemias/lymphomas of *Rassfs1a*$^{-/-}$ mice which were also interpreted as inactivating events (van der Weyden et al. 2012). However, in view of the location and orientation of these insertions in *Runx2* (Fig. 16.1) and the lack of corroborating evidence that the gene is up-regulated by these insertions, it is difficult to judge which interpretation is correct. As discussed later (Fig. 16.6), other lines of evidence suggest that both may have credence, if reduced Runx expression is favoured in the early tumour development while high expression drives later stages.

16.2 The Human *RUNX* Genes Act as Frequent Targets for Retroviral Vector Insertion in CD34+ Cells

The development of vector-induced leukaemias in gene therapy trials (Hacein-Bey-Abina et al. 2008; Howe et al. 2008) demonstrated the mutagenic potential of murine gamma-retroviruses in human subjects and focused attention on the factors affecting this adverse event (Scobie et al. 2009). High throughput studies of MLV vector integration in human CD34+ cells revealed pronounced hot-spots or hyper-clusters, including the *LMO2* gene that was the most frequently activated target in gene therapy-related leukaemias (Cattoglio et al. 2010). The selective targeting of MLV integration to regions of active chromatin marked by acetylated histones due to direct interaction of the viral integrase protein with Brd/BET chromatin tethers sheds light on these observations (Gupta et al. 2013; Sharma et al. 2013). As shown in Fig. 16.2, the human *RUNX* genes also serve as preferential targets for integration in CD34+ and K562 cells *in vitro*, with a distribution that mirrors H3K27Ac chromatin marks. There is an evident difference between the targeted sites for each *RUNX* gene. For *RUNX3*, the upstream 'super-enhancer' is most frequently targeted, while most insertions in *RUNX1* and *RUNX2* are intragenic. The lack of orientation bias of these insertions is consistent with the interpretation that the clustered pattern results from preferential integration and that no significant post-integration clonal selection has occurred during the limited culture period prior to harvesting for analysis (Huser et al. 2014).

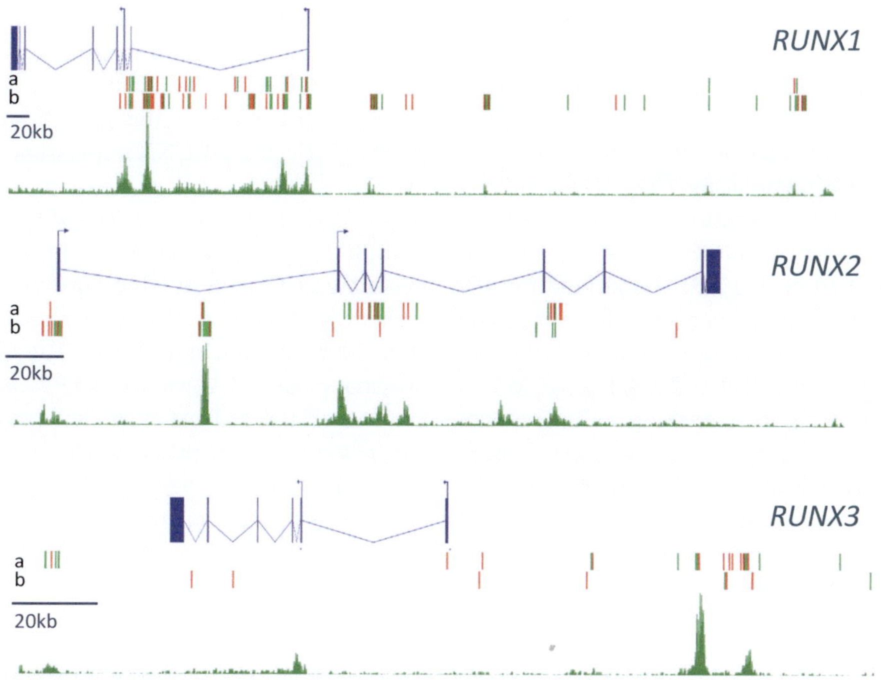

Fig. 16.2 Insertions at the *RUNX* gene loci in human haematopoietic cells transduced with MLV vector or infected with MLV/VSV pseudotypes. The *top* of each track shows gene structure, with *solid vertical bars* representing exons and lines representing introns, with *arrows* showing transcriptional start sites. Insertions in the forward direction are coloured *green* while those in the reverse direction are coloured *red*. Tracks (**a–b**) represent datasets: (**a**) MLV vector insertions in CD34+ cells (Cattoglio et al. 2010). (**b**) MLV Insertions in K562 cells (LaFave et al. 2014). The *bottom* of each track shows H3K27ac intensity in CD34+ primary cells, obtained from the human Epigenetics Roadmap

The frequent targeting of all three *RUNX* genes by retroviral vectors in CD34+ cells is also consistent with the fact that they are all transcriptionally active in early haematopoiesis, including *RUNX2*, providing a parallel with the targeting of *Runx2* by retroviral insertion in the Inv16 leukaemia model (Castilla et al. 2004). Moreover, the sites of preferred integration overlap to some extent with those selected in clonal end-stage murine lymphomas induced by MLV (Fig. 16.1). Despite these common features, end-stage patient leukaemias have not yet revealed the *RUNX* genes as targets, nor have the *RUNX* genes emerged from monitoring of insertion sites in the blood of healthy trial subjects (Hacein-Bey-Abina et al. 2008; Schwarzwaelder et al. 2007). This is perhaps not surprising in light of the requirement for activation of MYC or loss of p53 to facilitate the oncogenic effects of RUNX overexpression in murine models.

16.3 Switching the *Runx* Genes from Growth Suppressors to Oncogenes In Vivo: The Roles of MYC and p53

The potent synergy between RUNX2 and MYC in transgenic mice over-expressing both genes in the T-cell compartment (CD2-*MYC*/CD2-*Runx2* mice) confirmed the dominant oncogenic potential of RUNX2 (Vaillant et al. 1999). While the generality of the MYC/RUNX synergy was later confirmed by synergistic induction of B-cell lymphomas in *vav-Runx1*/Eμ-*Myc* mice (Blyth et al. 2009), most studies to date have been conducted with the CD2-*Runx2* model. Under the control of the CD2 LCR, RUNX2 was shown to accelerate lymphoma development in Eμ-Pim1 and CD2-v-Myb transgenic mice as well as in $p53^{null}$ mice (Blyth et al. 2001; Cameron et al. 2003), indicating a non-redundant, unique role for RUNX in tumour development. However, the selective targeting of *c-Myc* and *N-Myc* in virus-accelerated tumours of CD2-*Runx2* mice was a further indication of a special relationship between RUNX and MYC oncogenic functions (Blyth et al. 2001).

An unusual feature of the CD2-*MYC* model is the undetectable expression of the CD2-*MYC* transgene in the majority of mice that remain healthy (Stewart et al. 1993). To account for this phenotype it has been suggested that variegated expression of the hCD2 locus in early lymphoid development (Williams et al. 2008) leads to MYC-induced apoptosis in expressing cells and that compensatory expansion of transgene non-expressing cells leads to replenishment and generation of an apparently normal lymphoid compartment (Blyth et al. 2006). It seems likely that the potent oncogenicity of the CD2-*MYC*/CD2-*Runx2* combination is at least partially explained by the simultaneous activation of both oncogenes at the same stage of early lymphoid development. Similarly, the efficiency of virus acceleration in CD2-*MYC* mice may be explained by the activation of collaborating genes by viral infection and integration in lymphoid progenitor cells that have yet to activate the CD2-*MYC* transgene (Stewart et al. 1993). The remarkably low apoptotic index of spontaneous CD2-*Runx2* lymphomas and reduced apoptosis in compound transgenics compared to CD2-*MYC* alone in this setting indicates that the major selective advantage of RUNX2 in this context is survival and ablation of MYC-induced apoptosis (Blyth et al. 2006).

In the CD2-*Runx2* model, where RUNX2 expression is initiated in early lymphoid development, there is a gene dose-dependent predisposition to thymic lymphoma development (Vaillant et al. 1999). However, analysis of thymic development in the transgenic mice revealed a marked deficit in fetal thymocyte numbers rather than a preneoplastic expansion, indicating that the initial response to ectopic RUNX2 expression is growth suppressive. Moreover, analysis of the immature CD8+ subset (CD8ISP) which was expanded in these mice showed that these were small cells with a markedly lower proliferation rate than CD8ISP from healthy mice (Vaillant et al. 2002). A similar though less marked phenotype was observed in *vav-Runx1* mice which displayed reduced proliferation in haematopoietic stem/progenitor cells and B-cells, although with enhanced survival (Blyth et al. 2009). The ability of MYC to counteract the growth suppressive potential of ectopic RUNX expression is evident from the rapid onset of tumours in vav-Runx1/Eμ*Myc* and in CD2-*Runx2*/CD2-*MYC*, where the CD8ISP population observed in CD2-*Runx2* mice is transformed to large blastic cells in the premalignant thymus (Blyth et al. 2009; Blyth et al. 2006; Vaillant et al. 1999). The rapid onset of lymphomas in CD2-*Runx2*/p53^{null} mice shows that loss of the p53 pathway can also synergise with ectopic Runx expression (Blyth et al. 2001). A diagram summarising the collaboration of *Myc* and *Runx* genes in the context of T-cell lymphomas under the influence of CD2 LCR-driven gene expression is summarised in Fig. 16.3.

16.4 Over-Expression of RUNX and MYC Combine to Disarm the p53 Response In Vivo: A Dual Signal Hypothesis

While *MYC* and *Runx* transgenes are potently synergistic in lymphomagenesis, they also collaborate independently with germline inactivation of p53 (Blyth et al. 1995; Hsu et al. 1995). In accord with this observation, both MYC and RUNX induce p53 responses when ectopically expressed (Wolyniec et al. 2009; Zindy et al. 1998). Remarkably, the combination of transgenic MYC and RUNX2 appears to neutralise p53 activation, abolishing the selection to lose the wild-type *Trp53* allele in primary and transplanted lymphomas on a CD2-*MYC*/ *Runx2*/Trp53+/− background. The fact that the wild-type allele is rapidly lost on *in vitro* establishment of cell lines argues that it was intact but not active in the primary lymphomas (Blyth et al. 2006). These observations suggest that, *in vivo*, RUNX expression modifies the effects of MYC on the p53 pathway and vice versa. While the underlying mechanism has yet to be uncovered (Fig. 16.3b), it is notable that other oncogenes identified as potent MYC collaborators in retroviral mutagenesis screens impinge on the p53 pathway, including Bmi1, a repressor of Ink4a/Arf (Jacobs

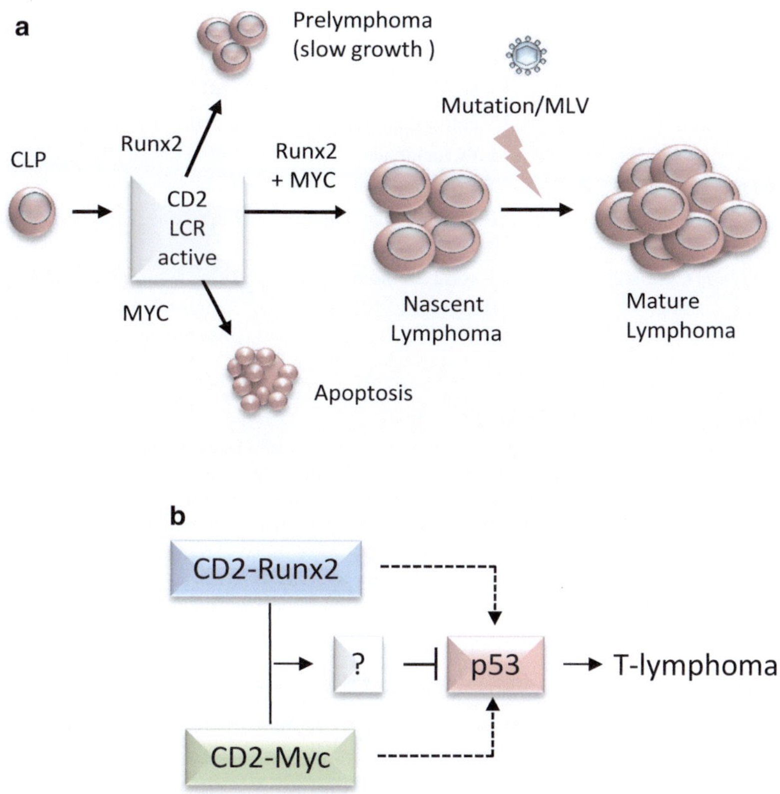

Fig. 16.3 Model of RUNX/MYC collaboration in lymphomagenesis. See text for explanation and supporting references. Panel *A* shows the key features of premalignant phenotype and disease in CD2-*MYC*, CD2-*Runx2* and compound transgenic mice. Panel *B* shows the interaction between RUNX2, MYC and p53 in the early T-cell compartment. Both CD2-*Runx2* and CD2-*MYC* transgenes are independently synergistic with germ-line inactivation of p53, suggesting that the oncogenic activity of both genes is antagonised by the tumour suppressor function of p53, while the combination of both oncogenes appears to overcome p53. We hypothesise that this phenomenon entails co-activation one or more genes that neutralises p53 function in T-lymphoma cells. CLP: common lymphoid progenitor

et al. 1999), and Gfi1, an indirect regulator of p53 activity via LSD1 mediated demethylation (Khandanpour and Moroy 2013).

Induction of MYC expression in primary fibroblasts leads to apoptosis unless pre-empted by a ‘dual signal’ provided by survival factors (Harrington et al. 1994) or relieved by inactivation of p53 (Hermeking and Eick 1994). It is interesting to compare this phenomenon to dual signalling hypotheses that evolved to account for shaping of the T-cell repertoire by positive and negative selection, and the proliferation of T-cells in response to foreign antigens (Zinkernagel et al. 1978). Notably, MYC expression enhances positive selection of thymocytes with functional T-cell receptors (Rudolph et al. 2000), while early studies showed that ectopic expression of RUNX1 could block TCR-signalling induced apoptosis of T-cell hybridomas (Fujii et al. 1998). It is tempting to suggest that the extremely potent synergy between *Runx* and *Myc* genes in driving lymphoma development is not merely a disordered response in the context of cancer, but an inherent feature of a signalling network that normally licenses lymphoid cells to proliferate in response to exogenous signals. The occurrence of autoimmune disease and hypersensitivity in *Runx*-deficient mice also provides indirect support for this hypothesis (Brenner et al. 2004; de Bruijn and Speck 2004; Wong et al. 2012).

16.5 Lymphoma Progression: Identification of Third Hit Genes in *MYC/Runx2* Lymphomas

Analysis of rapid onset tumours arising in CD2-*MYC*/CD2-*Runx2* transgenic mice showed that these are clonal outgrowths as indicated by their unique patterns of T-cell receptor rearrangement (Vaillant et al. 1999). This observation suggested that a further selective step is required to drive the end-stage lymphomas and led us to conduct further retroviral mutagenesis screens to identify the key target genes. Neonatal infection of CD2-

Table 16.1 Analysis of MYC/RUNX collaboration by retroviral insertional mutagenesis

Transgene	Disease −MLV	Disease +MLV	Preferred RIM targets
CD2-MYC	T-cell lymphoma Low incidence	T-cell lymphoma Rapid onset 100 %	*Runx2, Runx3, Runx1*
CD2-*Runx2*	T-cell lymphoma Low incidence, later onset	T-cell lymphoma Rapid onset 100 %	*C-Myc, N-Myc, Ikzf1*
CD2-*MYC/Runx2*	T-cell lymphoma Rapid onset 100 %	T-cell lymphoma 100 % Accelerated onset and dissemination	Multiple: TCR, PI3K, JAK-STAT signalling pathways, chemokine receptors, G1 checkpoint controls

Huser et al. (2014) and Stewart et al. (2007)
T-cell lymphoma onset is accelerated markedly in CD2-*MYC* and CD2-*Runx2* mice by neonatal infection with Moloney MLV, but with complementary patterns of insertional mutagenic targets that show a strong reciprocal relationship between *Myc* and *Runx* oncogenes. *Ikzf1* is a target for intragenic insertions that can generate dominant negative isoforms that may relieve MYC repression as an alternative mechanism of activation. Retroviral acceleration of tumour onset in highly tumour-prone CD2-*Runx2*/CD2-*MYC* mice reveals multiple target genes. A common selective advantage predicted for these insertions is the ability to grow in the absence of exogenous growth signals

MYC/CD2-*Runx2* mice with Moloney MLV leads to even more rapid tumour onset, increased clonal complexity and increased dissemination of primary thymic lymphomas to peripheral lymphoid tissues (Stewart et al. 2007). As outlined in Table 16.1, analysis of the preferred RIM targets in the progressing lymphomas showed a strong bias towards G1 checkpoint genes (D cyclins) and other genes that overcome the requirement for exogenous growth factor signals (e.g. Pim kinases). A further deep profiling analysis of many thousands of insertion sites in these tumours (Huser et al. 2014) confirmed these core progression genes as part of a broader set enriched for T-cell receptor, PI3K and JAK-STAT pathways along with selected chemokine receptors involved in T-cell homing to thymus (*Ccr7*, *Ccr9*). Most of these genes were frequently targeted in tumours of wild type mice suggesting that this gene set is frequently recruited in the normal course of viral infection where unscheduled proliferation of the target cell provides a selective advantage. While the consequence in otherwise normal cells might be self-limiting proliferation, activation of these genes in a lymphoma stem cell transformed by MYC and RUNX over-expression is sustained proliferation in the absence of exogenous signals (Huser et al. 2014).

Potential parallels with this 'three-hit' model are emerging from recent studies on human cancer where a recent review highlighted evidence for RUNX2 synergy with PI3K/AKT signalling in multiple cancer types (Cohen-Solal et al. 2015). Moreover, it may be interesting to re-evaluate evidence for similar oncogene combinations that may have been overlooked e.g. in osteosarcomas where *RUNX2* and *CCND3* on chromosome 6p21 are frequently co-amplified and over-expressed (Lu et al. 2008) and *MYC* is also frequently over-expressed (Gamberi et al. 1998).

16.6 Evidence of RUNX Addiction in Lymphoma Development

Evidence that endogenous RUNX activity is important for lymphoma development was provided by the delayed onset of T-cell lymphomas in *Runx1*$^{+/-}$ mice, whether these are induced by Moloney MLV infection or by the potent CD2-*MYC*/CD2-*Runx2* combination. Moreover, the lack of evidence of loss of heterozygosity in these lymphomas argued strongly that the *Runx1* gene was acting to promote tumour development rather than as a tumour suppressor (Wotton et al. 2002). Even more strikingly, the frequently occurring T-cell lymphomas in *Trp53*^{null} mice are virtually abolished on a *Runx1*$^{+/-}$ background (Shimizu et al. 2013). While these observations suggested an important pro-oncogenic role for basal RUNX1 expression, it could not be excluded that the gene plays an essential role in T-cell development and influences lymphomagenesis only indirectly by controlling the size of the target cell population.

A recent study in Eμ-*Myc*/*Runx1cKO* mice has provided direct evidence of addiction to RUNX1 in primary lymphoma cells which, in contrast to normal splenic lymphocytes, resist even mono-allelic deletion *in vivo*. While established lymphoma cell lines that have lost p53 function become permissive for complete loss of *Runx1*, the *Runx1*^{null} cells display a proliferative disadvantage and become markedly more sensitive to chemotherapeutics including doxorubicin and dexamethasone (Borland et al. 2016). These findings validate the *Runx* genes and their downstream effectors as targets for lymphoma therapies. Another notable feature of this study is that the transcriptional signature conferred by deletion of *Runx1* in these cells is enriched for genes involved in B-cell survival, proliferation and differentiation but does not include the 'ribosomal biogenesis' signature seen in *Runx1*^{null} haematopoietic progenitors (Cai et al. 2015) or the mitotic checkpoint signature observed in human AML cells after RUNX1 knockdown (Ben-Ami et al. 2013). While these findings again emphasise the cell context-dependent roles of RUNX, it should be noted that there are other players that may influence the outcome of loss of RUNX expression. In Eμ-*Myc* lymphomas, the cells overexpress MYC, a major driver of ribosome biogenesis that may be able to rescue loss of RUNX1 expression (Borland et al. 2016), while Kasumi AML cells express RUNX1-ETO, a potential antagonist of RUNX-dependent gene expression (Ben-Ami et al. 2013).

16.7 RUNX, p53 and Senescence: Insights from Primary Cells

Primary fibroblasts have provided many useful insights into the activities of cancer-relevant genes in the absence of the many genetic and epigenetic changes that affect the responses of established cancer cell lines (Etzold et al. 2016; Serrano et al. 1997). In primary murine embryonic fibroblasts (Fig. 16.4) or human foreskin

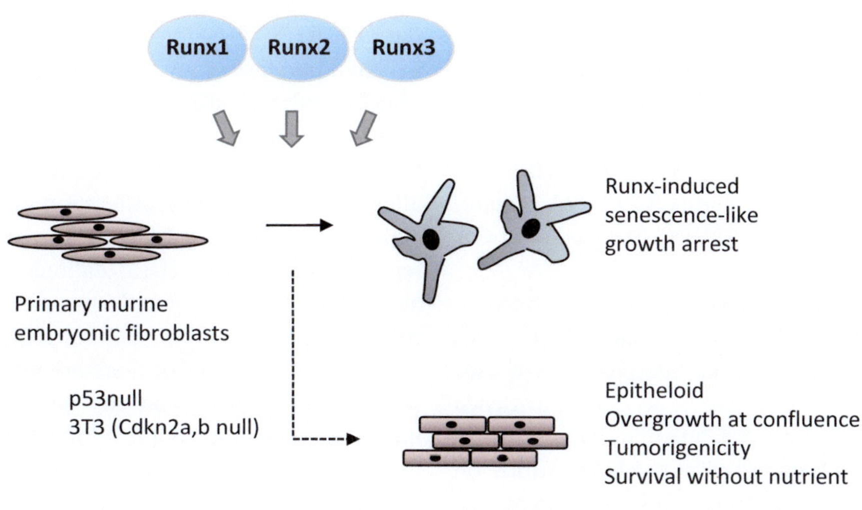

Fig. 16.4 Contrasting effects of ectopic RUNX expression in wild-type and established mouse fibroblasts. In primary mouse embryonic fibroblasts, RUNX overexpression leads to a profound growth arrest with flattened morphology and accumulation of senescence-associated β-galactosidase. In p53 deficient MEFs or NIH3T3 cells, ectopic RUNX expression leads instead to a morphological change resembling mesenchymal to epithelial transition along with enhanced growth and/or survival (Kilbey et al. 2007; Wotton et al. 2008, 2004)

fibroblasts (Hs68), ectopic expression of any of the three *Runx* genes (P1 or P2 isoforms) leads to senescence-like growth arrest (SLGA) (Kilbey et al. 2007; Wotton et al. 2004). Unlike cells undergoing Ras-induced SLGA which develops as a response to unscheduled proliferation and DNA damage (Di Micco et al. 2006), cells with enforced RUNX expression do not display nuclear DNA damage foci, and have a phenotype that instead resembles the effects of ectopic p53 expression. The requirements for induction of this failsafe response are not fully understood, but appear to engage both p53 and p16 (Cdkn2a) pathways. In primary MEFs lacking p53, the effect of ectopic RUNX1 is not growth arrest, but over-proliferation at confluence and increased tumorigenicity in nude mice (Wotton et al. 2004). In NIH3T3 cells, which have an abnormal p53 pathway response and lack p16 expression due to a large genomic deletion encompassing *Cdkn2a/Cdkn2b* (Wotton et al. 2004), the effect of ectopic RUNX expression is to promote an epitheloid morphology resembling mesenchymal to epithelial transition, and markedly enhanced survival under stress conditions (Wotton et al. 2008). This survival phenotype is associated with direct RUNX modulation of multiple enzymes involved in sphingolipid metabolism (Sgpp1, Ucgc, St3gal5), shifting the 'rheostat' from pro-apoptotic ceramides to pro-survival sphingosine-1-phospate (Kilbey et al. 2010). In human Hs68 cells, RUNX1 expression induces p53 despite the absence of detectable p14ARF expression, while the induction of SLGA is blocked by HPV E6. Moreover, Leiden fibroblasts (CDKN2A mutant) are resistant to RUNX1-induced SLGA (Wolyniec et al. 2009).

The strongly growth suppressive effect of RUNX over-expression in normal cells provides an important caveat to the interpretation of cancer cell inhibition by ectopic RUNX as evidence of a tumour suppressor function. This problem potentially compromises many publications on the *RUNX* genes and their putative roles in human cancer where in general only established cell lines are available for study. Moreover, the fact that the RUNX1-ETO fusion oncoprotein is also a potent inducer of senescence-like growth arrest in fibroblasts (Wolyniec et al. 2009) argues against a yin-yang interpretation of its functional relationship to RUNX1 mediated by regulation of p14/p19ARF (Linggi et al. 2002).

16.8 Translational Relevance: *RUNX* Genes as Dominant Oncogenes in Human Cancer

The relative rarity of copy number gains affecting the *RUNX* genes in human cancer and the widespread assumption that they are predominantly tumour suppressors has until recently diverted attention from their capacity to act as dominant oncogenes. Amplification of a large domain of chromosome 21 that encompasses *RUNX1* has been observed in a poor prognostic subset of B-ALL, presenting an exception to this rule (Harrison et al. 2014). Although it has been argued that these leukaemias do not significantly over-express *RUNX1* mRNA compared to ALLs lacking *RUNX1* amplification (Strefford et al. 2006) it is conceivable that other leukaemias over-express RUNX1 by different mechanisms, as *RUNX1* mRNA is highly elevated in many ALLs (Niini et al. 2002). Notably, gains of chromosome 21 and *RUNX1* copies are also evident in progressing t(12;21) leukaemias that express the TEL-RUNX1 fusion oncoprotein, in contrast to the frequent loss of the normal, non-translocated *TEL* allele (Lilljebjorn et al. 2010). The requirement for activity of the RUNX1 protein expressed from the untranslocated *RUNX1* allele for survival and proliferation of leukaemia cell lines harbouring RUNX1 fusion oncoproteins (Ben-Ami et al. 2013; Zaliova et al. 2011) also argues against a simple tumour suppressor role /dominant negative inhibitor relationship. It should also be noted that there are many ways in which RUNX expression can be dysregulated, including post-translational modification and translational controls via miRNA. Examples of RUNX oncogenic activity apparently mediated by such mechanisms have emerged from recent studies on human cancer cells of multiple types (Bledsoe et al. 2014; Browne et al. 2016; Shin et al. 2016).

16.9 The *Runx* Genes as Tumour Suppressors in Haematopoietic Cancers: Evidence from Mouse Models

The severe development defects resulting from germ-line deletion of the *Runx* genes has delayed assessment of their tumour suppressor activity *in vivo*, requiring the development of conditional knockout strains. As reviewed recently (Chin et al. 2015), conditional knockouts of *Runx1*, *Runx3* or *Cbfb* have revealed mainly myeloproliferative or myelodysplastic disease and/or haematopoietic stem cell expansions of varying degree. Deletion of *Runx1* in HSPC mediated by vav-Cre was shown to lead to reduced cell size as a result of diminished ribosome biogenesis, along with reduced apoptosis and resistance to genotoxic and ER stress, and it was suggested that this phenotype provides a selective advantage for null cells (Cai et al. 2015). Dual deletion of *Runx1* and *Runx3* using the Mx1-Cre system resulted mainly in bone marrow failure although this was preceded by expansion of haematopoietic stem/progenitor cells (HSPC) and almost 20 % of the mice developed fatal myeloproliferative disorder (Wang et al. 2014). While there are few reports of the development of spontaneous malignant disease in knockouts (Chin et al. 2015), *Runx1* inactivation collaborates strongly with other oncogenic insults such as FLT3-ITD or N-Ras to induce AML-like disease (Mead et al. 2013; Motoda et al. 2007). The predisposition towards MDS/AML rather than other malignancies in familial platelet disorder due to *RUNX1* mutation (Owen et al. 2008) suggests that the unique sensitivity of the myeloid lineage to *RUNX1* mutation and loss of function is conserved from mouse to human.

However, *RUNX1* mutations have also been found in some human lymphoid malignancies, notably in early T-ALLs (18 %) and a small subset of B-ALL also carrying BCR-ABL (Grossmann et al. 2011). Moreover, on the basis of a systems biology approach dubbed 'reverse engineering' of transcription networks, RUNX1 was predicted to act as a tumour suppressor in this lineage (Della Gatta et al. 2012). Evidence from mouse models in support of this designation is rather limited although one study of Mx1-Cre mediated deletion of *Runx1* reported thymic lymphoma in a proportion of *Runx1KO* mice. However, the major phenotype observed in these mice was myelodysplasia and a block in T-cell development and it is unclear from the report whether the lymphomas actually arose from *Runx1*^{null} cells (Putz et al. 2006). Another early study in chimeric mice showed that *Runx1KO* cells were preferentially targeted in T-cell lymphomas induced by chemical mutagenesis (Kundu et al. 2005). As highlighted earlier, reduction to a single functional allele in *Runx1*$^{+/-}$ slows onset of MLV-induced T-cell lymphomas (Wotton et al. 2002) and virtually ablates spontaneous T-cell lymphomas in p53-deficient mice (Shimizu et al. 2013), arguing for a pro-oncogenic role. Moreover, a recent study has provided evidence that primary B-cell lymphomas in the Eμ-*Myc* model are addicted to RUNX1, while established cell lines lacking p53 become permissive to Cre-mediated deletion and display *Rag* gene de-repression, providing a potential explanation for the apparent oncogene/tumour suppressor paradox in the lymphoid compartment (Borland et al. 2016).

16.10 RUNX2 and Oncogene-Induced Senescence: A Temporal Model for RUNX Function in Cancer

Murine primary embryonic fibroblasts and osteoblasts lacking RUNX2 are prone to spontaneous immortalisation and tumorigenic conversion (Kilbey et al. 2007; Zaidi et al. 2007). Both cell types display reduced basal expression of a number of negative regulators of cell cycle progression that have been implicated as effectors of oncogene-induced failsafe responses (p16Ink4a, p19Ink4a, p53 and p21Waf1). These observations provide a rationale for the failure of primary fibroblasts and osteoblasts to undergo early growth arrest in the oxidative conditions of cell culture (Parrinello et al. 2003) and suggest a non-redundant role for RUNX2, which is perhaps not surprising as RUNX2 is the predominant expressed family member in both cell types. *Runx2* null MEFs also resist H-Ras oncogene-induced senescence and become tumorigenic (Kilbey et al. 2007). However, despite their failure to arrest in response to mutant H-Ras, failsafe effectors are induced in *Runx2* null fibroblasts at levels comparable to wild-type cells. The basis of their continued proliferation in the presence of failsafe effector expression is not fully understood, but is associated with altered expression of chromatin remodelling factors that regulate cyclin gene expression (Kilbey et al. 2008). It has also been reported that loss of *Runx1* impairs N-Ras-induced failsafe responses in haematopoietic progenitors (Motoda et al. 2007), while *Runx3*cKO mice show accelerated lung tumour development in a K-Ras knock-in model (Lee et al. 2013) suggesting that this may be a wider feature of Runx oncogenesis (Fig. 16.5).

However, at this point knockout mouse models have supported a pro-oncogenic rather than a suppressor role for RUNX2 *in vivo* (Ferrari et al. 2015) as have many recent studies of human cancer. A rare co-occurrence of cleido-cranial dysplasia and AML suggested a possible loss-of-function scenario but instead the authors of that study found that RUNX2 was actually over-expressed, a phenomenon they suggested might be explained by compensatory up-regulation of the wild-type allele (Schnerch et al. 2014). As the *RUNX* genes can cross-regulate (Brady et al. 2009) it should also be kept in mind that functional loss of one family member may have consequences for other members, and that the point mutations of RUNX1 in AML and a handful of other cancers may affect more than merely RUNX1. These observations also invite us to propose a temporal model to explain the dualistic behaviour of the *Runx* genes in cancer. We hypothesise that reduced levels of RUNX expression may act early to promote cancer development in a number of ways; by facilitating the growth of cells carrying initiator mutations (including mutant Ras alleles), preventing exit

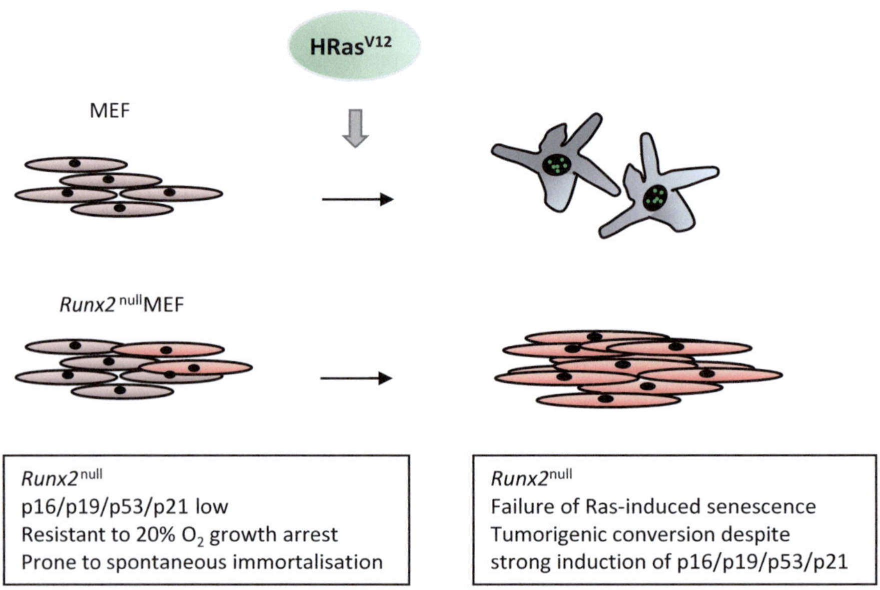

Fig. 16.5 *Runx2*^{null} mouse fibroblasts are prone to spontaneous immortalisation and resist Ras-induced senescence. *RUNX2* deficient primary embryonic fibroblasts express low levels of markers of aging and failsafe senescence (p16/p19^{Cdkn2a}, p53, p21^{Cdkn1a}), resist early growth arrest in normoxic culture and have an increased propensity for spontaneous immortalisation after 3T3 passage. Introduction of HRas^{V12} results in senescence-like growth arrest with nuclear DNA damage foci (*green*) in wild-type MEFs while *Runx2*^{null} cells proliferate and become tumorigenic, despite apparent induction of failsafe mediators (Kilbey et al. 2007)

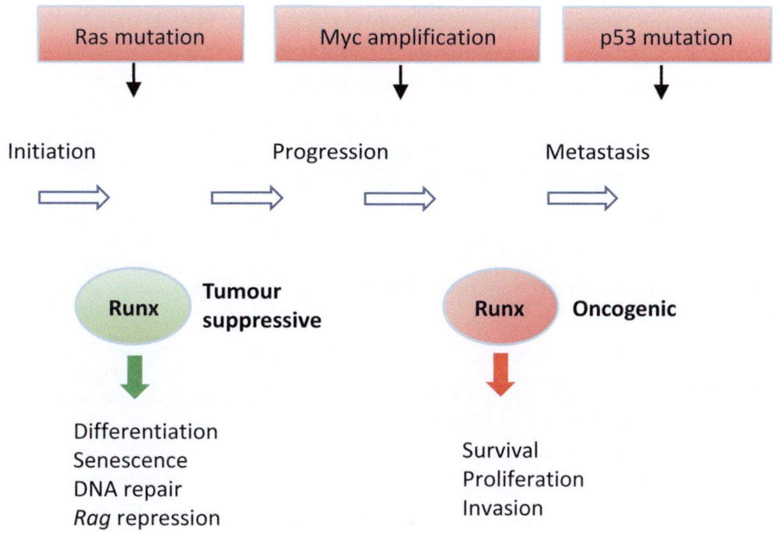

Fig. 16.6 A temporal model for RUNX function in cancer. In this model, RUNX expression is required for tumour suppressive cell fate decisions and protection against mutational damage in early tumorigenesis where compromised expression increases the probability of transformation. At later stages, over-expression of MYC and/or loss of p53 function unmask the latent oncogenic potential of RUNX and increased expression is selected in end-stage tumours

from stem cell/progenitor compartments, promoting genomic instability, impairing DNA repair and finally by de-repressing potentially mutagenic *Rag* genes. In contrast, in the later stages of cancer where Myc drives proliferation and the p53 pathway is compromised, *Runx* gene activity drives tumour cell growth and metastatic potential. This temporal model is also compatible with the apparent addiction of AML cells with RUNX1-ETO to expression of the wild-type allele of *RUNX1* (Ben-Ami et al. 2013) and the copy number gains of the unaffected *RUNX1* allele in progressing TEL-RUNX1 leukaemias (Lilljebjorn et al. 2012) (Fig. 16.6).

16.11 The *RUNX* Genes: Isoforms or Functionally Divergent Genes?

For historical reasons, the field has focused heavily on *RUNX1* in haemato-oncology, *RUNX2* in bone development, and *RUNX3* in immune cell function and tumour suppression. However, given the evidence of functional overlap as well as cross-regulation between family members (Spender et al. 2005) maintaining a singular focus on one family member while ignoring its relatives appears myopic. The fact that only *RUNX1* has emerged as a common target for chromosomal translocation events in human leukaemia appears suggestive of a unique function for this family member. However, it should be noted that this bias could arise instead due to the relatively high expression of *RUNX1* in haematopoietic progenitors where initiating events occur and/or to specific features of the *RUNX1* locus on chromosome 21 that confer susceptibility to rearrangement (Levanon et al. 2001).

Despite their unique roles in specific biological niches, indications of functional overlap between family members can be seen in the T-cell lineage where RUNX1 and RUNX3 act sequentially to silence CD4 (Taniuchi et al. 2002) and in bone where both RUNX1 and RUNX3 are required in addition to the bone 'master regulator' RUNX2 for full osteoblast function and skeletal development (Bauer et al. 2015; Liakhovitskaia et al. 2010). These observations prompt the question of redundancy and whether the products of all *Runx* genes should be regarded as isoforms that play unique roles in development only because of their tissue-specific expression patterns. An analogy may be drawn from the Myc gene family where the entire *N-Myc* coding sequence can replace *c-Myc* in murine development despite their significant sequence divergence (Malynn et al. 2000). While this type of experiment has not been fully recapitulated for the *Runx* family, substitution of the C-terminus of RUNX1 with equivalent domains of RUNX2 and RUNX3 to create chimeric proteins led to at least partial rescue of haematopoietic development *in vivo* (Fukushima-Nakase et al. 2005). Moreover, all three genes appeared equally efficient in rescuing haematopoietic development of *Runx1*^{null} cells in an *in vitro* co-culture system (Goyama et al. 2004).

Direct comparison of all three genes by ectopic expression in murine fibroblasts and gene expression microarray analysis showed a very high degree of redundancy, with a strong overlap in the signature gene expression changes and no examples of opposing regulation. However, there were clearly differences with regard to the relative potency of regulation for individual target genes that could conceivably translate into functional differences in specific niches *in vivo* (Wotton et al. 2008). In our view the degree of RUNX redundancy and the biological contexts in which it may operate largely remains to be addressed.

16.12 Conclusions and Prospects

There is growing evidence that the oncogenic potential of the *Runx* gene family revealed by their powerful co-operation with MYC overexpression or p53 loss in mouse models is highly relevant to human cancer, where a growing body of literature attests to the important roles that RUNX family members play in supporting the oncogenic phenotypes of end-stage cancers and cell lines. The tumour suppressor features of the *Runx* genes have been less amenable to dissection in *in vivo* models, but are now being elucidated using conditional knockout models. Evidence that the *Runx* genes operate in a complex integrated regulatory network suggests that future studies should address effects on all three genes where any single gene is affected.

Acknowledgements Our work was supported by a joint programme grant from Bloodwise (grant number 13046) and Cancer Research UK (grant number A11951).

References

Bauer, O., Sharir, A., Kimura, A., Hantisteanu, S., Takeda, S., & Groner, Y. (2015). Loss of osteoblast Runx3 produces severe congenital osteopenia. *Molecular and Cellular Biology, 35*, 1097–1109.

Ben-Ami, O., Friedman, D., Leshkowitz, D., Goldenberg, D., Orlovsky, K., Pencovich, N., et al. (2013). Addiction of t(8;21) and inv(16) acute myeloid leukemia to native RUNX1. *Cell Reports, 4*, 1131–1143.

Bledsoe, K. L., McGee-Lawrence, M. E., Camilleri, E. T., Wang, X., Riester, S. M., van Wijnen, A. J., et al. (2014). RUNX3 facilitates growth of Ewing sarcoma cells. *Journal of Cellular Physiology, 229*, 2049–2056.

Blyth, K., Terry, A., O'Hara, M., Baxter, E. W., Campbell, M., Stewart, M., et al. (1995). Synergy between a human c-myc transgene and p53 null genotype in murine thymic lymphomas: Contrasting effects of homozygous and heterozygous p53 loss. *Oncogene, 10*, 1717–1723.

Blyth, K., Terry, A., Mackay, N., Vaillant, F., Bell, M., Cameron, E. R., et al. (2001). Runx2: A novel oncogenic effector revealed by in vivo complementation and retroviral tagging. *Oncogene, 20*, 295–302.

Blyth, K., Vaillant, F., Hanlon, L., Mackay, N., Bell, M., Jenkins, A., et al. (2006). Runx2 and MYC collaborate in lymphoma development by suppressing apoptotic and growth arrest pathways in vivo. *Cancer Research, 66*, 2195–2201.

Blyth, K., Slater, N., Hanlon, L., Bell, M., Mackay, N., Stewart, M., et al. (2009). Runx1 promotes B-cell survival and lymphoma development. *Blood Cells, Molecules & Diseases, 43*, 12–19.

Borland, G., Kilbey, A., Hay, J., Gilroy, K., Terry, A., Mackay, N., et al. (2016). Addiction to Runx1 is partially attenuated by loss of p53 in the Emu-Myc lymphoma model. *Oncotarget, 7*, 22973–22987.

Brady, G., Whiteman, H. J., Spender, L. C., & Farrell, P. J. (2009). Downregulation of RUNX1 by RUNX3 requires the RUNX3 VWRPY sequence and is essential for Epstein-Barr virus-driven B-cell proliferation. *Journal of Virology, 83*, 6909–6916.

Brenner, O., Levanon, D., Negreanu, V., Golubkov, O., Fainaru, O., Woolf, E., & Groner, Y. (2004). Loss of Runx3 function in leukocytes is associated with spontaneously developed colitis and gastric mucosal hyperplasia. *Proceedings of the National Academy of Sciences of the United States of America, 101*, 16016–16021.

Browne, G., Dragon, J. A., Hong, D., Messier, T. L., Gordon, J. A., Farina, N. H., et al. (2016). MicroRNA-378-mediated suppression of Runx1 alleviates the aggressive phenotype of triple-negative MDA-MB-231 human breast cancer cells. *Tumour Biology, 37*, 8825–8839.

Cai, X., Gao, L., Teng, L., Ge, J., Oo, Z. M., Kumar, A. R., et al. (2015). Runx1 deficiency decreases ribosome biogenesis and confers stress resistance to hematopoietic stem and progenitor cells. *Cell Stem Cell, 17*, 165–177.

Cameron, E. R., Blyth, K., Hanlon, L., Kilbey, A., Mackay, N., Stewart, M., et al. (2003). The Runx genes as dominant oncogenes. *Blood Cells, Molecules & Diseases, 30*, 194–200.

Castilla, L. H., Perrat, P., Martinez, N. J., Landrette, S. F., Keys, R., Oikemus, S., et al. (2004). Identification of genes that synergize with Cbfb-MYH11 in the pathogenesis of acute myeloid leukemia. *Proceedings of the National Academy of Sciences of the United States of America, 101*, 4924–4929.

Cattoglio, C., Pellin, D., Rizzi, E., Maruggi, G., Corti, G., Miselli, F., et al. (2010). High-definition mapping of retroviral integration sites identifies active regulatory elements in human multipotent hematopoietic progenitors. *Blood, 116*, 5507–5517.

Chin, D. W., Watanabe-Okochi, N., Wang, C. Q., Tergaonkar, V., & Osato, M. (2015). Mouse models for core binding factor leukemia. *Leukemia, 29*, 1970–1980.

Cohen-Solal, K. A., Boregowda, R. K., & Lasfar, A. (2015). RUNX2 and the PI3K/AKT axis reciprocal activation as a driving force for tumor progression. *Molecular Cancer, 14*, 137.

de Bruijn, M. F., & Speck, N. A. (2004). Core-binding factors in hematopoiesis and immune function. *Oncogene, 23*, 4238–4248.

De Rijck, J., de Kogel, C., Demeulemeester, J., Vets, S., El Ashkar, S., Malani, N., et al. (2013). The BET family of proteins targets moloney murine leukemia virus integration near transcription start sites. *Cell Reports, 5*, 886–894.

Della Gatta, G., Palomero, T., Perez-Garcia, A., Ambesi-Impiombato, A., Bansal, M., Carpenter, Z. W., et al. (2012). Reverse engineering of TLX oncogenic transcriptional networks identifies RUNX1 as tumor suppressor in T-ALL. *Nature Medicine, 18*, 436–440.

Di Micco, R., Fumagalli, M., Cicalese, A., Piccinin, S., Gasparini, P., Luise, C., et al. (2006). Oncogene-induced senescence is a DNA damage response triggered by DNA hyper-replication. *Nature, 444*, 638–642.

Etzold, A., Galetzka, D., Weis, E., Bartsch, O., Haaf, T., Spix, C., et al. (2016). CAF-like state in primary skin fibroblasts with constitutional BRCA1 epimutation sheds new light on tumor suppressor deficiency-related changes in healthy tissue. *Epigenetics, 11*, 1–12.

Ferrari, N., Riggio, A. I., Mason, S., McDonald, L., King, A., Higgins, T., et al. (2015). Runx2 contributes to the regenerative potential of the mammary epithelium. *Scientific Reports, 5*, 15658.

Fujii, M., Hayashi, K., Niki, M., Chiba, N., Meguro, K., Endo, K., et al. (1998). Overexpression of AML1 renders a T hybridoma resistant to T cell receptor-mediated apoptosis. *Oncogene, 17*, 1813–1820.

Fukushima-Nakase, Y., Naoe, Y., Taniuchi, I., Hosoi, H., Sugimoto, T., & Okuda, T. (2005). Shared and distinct roles mediated through C-terminal subdomains of acute myeloid leukemia/Runt-related transcription factor molecules in murine development. *Blood, 105*, 4298–4307.

Gamberi, G., Benassi, M. S., Bohling, T., Ragazzini, P., Molendini, L., Sollazzo, M. R., et al. (1998). C-myc and c-fos in human osteosarcoma: Prognostic value of mRNA and protein expression. *Oncology, 55*, 556–563.

Goyama, S., Yamaguchi, Y., Imai, Y., Kawazu, M., Nakagawa, M., Asai, T., et al. (2004). The transcriptionally active form of AML1 is required for hematopoietic rescue of the AML1-deficient embryonic para-aortic splanchnopleural (P-Sp) region. *Blood, 104*, 3558–3564.

Grossmann, V., Kern, W., Harbich, S., Alpermann, T., Jeromin, S., Schnittger, S., et al. (2011). Prognostic relevance of RUNX1 mutations in T-cell acute lymphoblastic leukemia. *Haematologica, 96*, 1874–1877.

Gupta, S. S., Maetzig, T., Maertens, G. N., Sharif, A., Rothe, M., Weidner-Glunde, M., et al. (2013). Bromo- and extraterminal domain chromatin regulators serve as cofactors for murine leukemia virus integration. *Journal of Virology, 87*, 12721–12736.

Hacein-Bey-Abina, S., Garrigue, A., Wang, G. P., Soulier, J., Lim, A., Morillon, E., et al. (2008). Insertional oncogenesis in 4 patients after retrovirus-mediated gene therapy of SCID-X1. *The Journal of Clinical Investigation, 118*, 3132–3142.

Harrington, E. A., Bennett, M. R., Fanidi, A., & Evan, G. I. (1994). c-Myc-induced apoptosis in fibroblasts is inhibited by specific cytokines. *The EMBO Journal, 13*, 3286–3295.

Harrison, C. J., Moorman, A. V., Schwab, C., Carroll, A. J., Raetz, E. A., Devidas, M., et al. (2014). An international study of intrachromosomal amplification of chromosome 21 (iAMP21): Cytogenetic characterization and outcome. *Leukemia, 28*, 1015–1021.

Hermeking, H., & Eick, D. (1994). Mediation of c-Myc-induced apoptosis by p53. *Science, 265*, 2091–2093.

Howe, S. J., Mansour, M. R., Schwarzwaelder, K., Bartholomae, C., Hubank, M., Kempski, H., et al. (2008). Insertional mutagenesis combined with acquired somatic mutations causes leukemogenesis following gene therapy of SCID-X1 patients. *The Journal of Clinical Investigation, 118*, 3143–3150.

Hsu, B., Marin, M. C., el-Naggar, A. K., Stephens, L. C., Brisbay, S., & McDonnell, T. J. (1995). Evidence that c-myc mediated apoptosis does not require wild-type p53 during lymphomagenesis. *Oncogene, 11*, 175–179.

Huser, C. A., Gilroy, K. L., de Ridder, J., Kilbey, A., Borland, G., Mackay, N., et al. (2014). Insertional mutagenesis and deep profiling reveals gene hierarchies and a Myc/p53-dependent bottleneck in lymphomagenesis. *PLoS Genetics, 10*, e1004167.

Hwang, H. C., Martins, C. P., Bronkhorst, Y., Randel, E., Berns, A., Fero, M., & Clurman, B. E. (2002). Identification of oncogenes collaborating with p27Kip1 loss by insertional mutagenesis and high-throughput insertion site analysis. *Proceedings of the National Academy of Sciences of the United States of America, 99*, 11293–11298.

Jacobs, J. J., Scheijen, B., Voncken, J. W., Kieboom, K., Berns, A., & van Lohuizen, M. (1999). Bmi-1 collaborates with c-Myc in tumorigenesis by inhibiting c-Myc-induced apoptosis via INK4a/ARF. *Genes & Development, 13*, 2678–2690.

Khandanpour, C., & Moroy, T. (2013). Growth factor independence 1 (Gfi1) as a regulator of p53 activity and a new therapeutical target for ALL. *Oncotarget, 4*, 374–375.

Kilbey, A., Blyth, K., Wotton, S., Terry, A., Jenkins, A., Bell, M., et al. (2007). Runx2 disruption promotes immortalization and confers resistance to oncogene-induced senescence in primary murine fibroblasts. *Cancer Research, 67*, 11263–11271.

Kilbey, A., Terry, A., Cameron, E. R., & Neil, J. C. (2008). Oncogene-induced senescence: An essential role for Runx. *Cell Cycle, 7*, 2333–2340.

Kilbey, A., Terry, A., Jenkins, A., Borland, G., Zhang, Q., Wakelam, M. J., et al. (2010). Runx regulation of sphingolipid metabolism and survival signaling. *Cancer Research, 70*, 5860–5869.

Kundu, M., Compton, S., Garrett-Beal, L., Stacy, T., Starost, M. F., Eckhaus, M., et al. (2005). Runx1 deficiency predisposes mice to T-lymphoblastic lymphoma. *Blood, 106*, 3621–3624.

Kuo, Y. H., Zaidi, S. K., Gornostaeva, S., Komori, T., Stein, G. S., & Castilla, L. H. (2009). Runx2 induces acute myeloid leukemia in cooperation with Cbfbeta-SMMHC in mice. *Blood, 113*, 3323–3332.

LaFave, M. C., Varshney, G. K., Gildea, D. E., Wolfsberg, T. G., Baxevanis, A. D., & Burgess, S. M. (2014). MLV integration site selection is driven by strong enhancers and active promoters. *Nucleic Acids Research, 42*, 4257–4269.

Lee, Y. S., Lee, J. W., Jang, J. W., Chi, X. Z., Kim, J. H., Li, Y. H., et al. (2013). Runx3 inactivation is a crucial early event in the development of lung adenocarcinoma. *Cancer Cell, 24*, 603–616.

Levanon, D., Glusman, G., Bangsow, T., Ben-Asher, E., Male, D. A., Avidan, N., et al. (2001). Architecture and anatomy of the genomic locus encoding the human leukemia-associated transcription factor RUNX1/AML1. *Gene, 262*, 23–33.

Liakhovitskaia, A., Lana-Elola, E., Stamateris, E., Rice, D. P., van't Hof, R. J., & Medvinsky, A. (2010). The essential requirement for Runx1 in the development of the sternum. *Developmental Biology, 340*, 539–546.

Lilljebjorn, H., Soneson, C., Andersson, A., Heldrup, J., Behrendtz, M., Kawamata, N., et al. (2010). The correlation pattern of acquired copy number changes in 164 ETV6/RUNX1-positive childhood acute lymphoblastic leukemias. *Human Molecular Genetics, 19*, 3150–3158.

Lilljebjorn, H., Rissler, M., Lassen, C., Heldrup, J., Behrendtz, M., Mitelman, F., et al. (2012). Whole-exome sequencing of pediatric acute lymphoblastic leukemia. *Leukemia, 26*, 1602–1607.

Linggi, B., Muller-Tidow, C., van de Locht, L., Hu, M., Nip, J., Serve, H., et al. (2002). The t(8;21) fusion protein, AML1 ETO, specifically represses the transcription of the p14(ARF) tumor suppressor in acute myeloid leukemia. *Nature Medicine, 8*, 743–750.

Lu, X. Y., Lu, Y., Zhao, Y. J., Jaeweon, K., Kang, J., Xiao-Nan, L., et al. (2008). Cell cycle regulator gene CDC5L, a potential target for 6p12-p21 amplicon in osteosarcoma. *Molecular Cancer Research, 6*, 937–946.

Malynn, B. A., de Alboran, I. M., O'Hagan, R. C., Bronson, R., Davidson, L., DePinho, R. A., & Alt, F. W. (2000). N-myc can functionally replace c-myc in murine development, cellular growth, and differentiation. *Genes & Development, 14*, 1390–1399.

Mead, A. J., Kharazi, S., Atkinson, D., Macaulay, I., Pecquet, C., Loughran, S., et al. (2013). FLT3-ITDs instruct a myeloid differentiation and transformation bias in lymphomyeloid multipotent progenitors. *Cell Reports, 3*, 1766–1776.

Miething, C., Grundler, R., Mugler, C., Brero, S., Hoepfl, J., Geigl, J., et al. (2007). Retroviral insertional mutagenesis identifies RUNX genes involved in chronic myeloid leukemia disease persistence under imatinib treatment. *Proceedings of the National Academy of Sciences of the United States of America, 104*, 4594–4599.

Mikkers, H., Allen, J., Knipscheer, P., Romeijn, L., Hart, A., Vink, E., & Berns, A. (2002). High-throughput retroviral tagging to identify components of specific signaling pathways in cancer. *Nature Genetics, 32*, 153–159.

Motoda, L., Osato, M., Yamashita, N., Jacob, B., Chen, L. Q., Yanagida, M., et al. (2007). Runx1 protects hematopoietic stem/progenitor cells from oncogenic insult. *Stem Cells, 25*, 2976–2986.

Niini, T., Vettenranta, K., Hollmen, J., Larramendy, M. L., Aalto, Y., Wikman, H., et al. (2002). Expression of myeloid-specific genes in childhood acute lymphoblastic leukemia – A cDNA array study. *Leukemia, 16*, 2213–2221.

Owen, C. J., Toze, C. L., Koochin, A., Forrest, D. L., Smith, C. A., Stevens, J. M., et al. (2008). Five new pedigrees with inherited RUNX1 mutations causing familial platelet disorder with propensity to myeloid malignancy. *Blood, 112*, 4639–4645.

Parrinello, S., Samper, E., Krtolica, A., Goldstein, J., Melov, S., & Campisi, J. (2003). Oxygen sensitivity severely limits the replicative lifespan of murine fibroblasts. *Nature Cell Biology, 5*, 741–747.

Putz, G., Rosner, A., Nuesslein, I., Schmitz, N., & Buchholz, F. (2006). AML1 deletion in adult mice causes splenomegaly and lymphomas. *Oncogene, 25*, 929–939.

Rudolph, B., Hueber, A. O., & Evan, G. I. (2000). Reversible activation of c-Myc in thymocytes enhances positive selection and induces proliferation and apoptosis in vitro. *Oncogene, 19*, 1891–1900.

Schnerch, D., Lausch, E., Becker, H., Felthaus, J., Pfeifer, D., Mundlos, S., et al. (2014). Up-regulation of RUNX2 in acute myeloid leukemia in a patient with an inherent RUNX2 haploinsufficiency and cleidocranial dysplasia. *Leukemia & Lymphoma, 55*, 1930–1932.

Schwarzwaelder, K., Howe, S. J., Schmidt, M., Brugman, M. H., Deichmann, A., Glimm, H., et al. (2007).

Gammaretrovirus-mediated correction of SCID-X1 is associated with skewed vector integration site distribution in vivo. *The Journal of Clinical Investigation, 117*, 2241–2249.

Scobie, L., Hector, R. D., Grant, L., Bell, M., Nielsen, A. A., Meikle, S., et al. (2009). A novel model of SCID-X1 reconstitution reveals predisposition to retrovirus-induced lymphoma but no evidence of gammaC gene oncogenicity. *Molecular Therapy, 17*, 1031–1038.

Serrano, M., Lin, A. W., McCurrach, M. E., Beach, D., & Lowe, S. W. (1997). Oncogenic ras provokes premature cell senescence associated with accumulation of p53 and p16INK4a. *Cell, 88*, 593–602.

Sharma, A., Larue, R. C., Plumb, M. R., Malani, N., Male, F., Slaughter, A., et al. (2013). BET proteins promote efficient murine leukemia virus integration at transcription start sites. *Proceedings of the National Academy of Sciences of the United States of America, 110*, 12036–12041.

Shimizu, K., Yamagata, K., Kurokawa, M., Mizutani, S., Tsunematsu, Y., & Kitabayashi, I. (2013). Roles of AML1/RUNX1 in T-cell malignancy induced by loss of p53. *Cancer Science, 104*, 1033–1038.

Shin, M. H., He, Y., Marrogi, E., Piperdi, S., Ren, L., Khanna, C., et al. (2016). A RUNX2-mediated epigenetic regulation of the survival of p53 defective cancer cells. *PLoS Genetics, 12*, e1005884.

Spender, L. C., Whiteman, H. J., Karstegl, C. E., & Farrell, P. J. (2005). Transcriptional cross-regulation of RUNX1 by RUNX3 in human B cells. *Oncogene, 24*, 1873–1881.

Stewart, M., Cameron, E., Campbell, M., McFarlane, R., Toth, S., Lang, K., et al. (1993). Conditional expression and oncogenicity of c-myc linked to a CD2 gene dominant control region. *International Journal of Cancer, 53*, 1023–1030.

Stewart, M., Terry, A., Hu, M., O'Hara, M., Blyth, K., Baxter, E., et al. (1997). Proviral insertions induce the expression of bone-specific isoforms of PEBP2alphaA (CBFA1): Evidence for a new myc collaborating oncogene. *Proceedings of the National Academy of Sciences of the United States of America, 94*, 8646–8651.

Stewart, M., MacKay, N., Cameron, E. R., & Neil, J. C. (2002). The common retroviral insertion locus Dsi1 maps 30 kilobases upstream of the P1 promoter of the murine Runx3/Cbfa3/Aml2 gene. *Journal of Virology, 76*, 4364–4369.

Stewart, M., Mackay, N., Hanlon, L., Blyth, K., Scobie, L., Cameron, E., & Neil, J. C. (2007). Insertional mutagenesis reveals progression genes and checkpoints in MYC/Runx2 lymphomas. *Cancer Research, 67*, 5126–5133.

Strefford, J. C., van Delft, F. W., Robinson, H. M., Worley, H., Yiannikouris, O., Selzer, R., et al. (2006). Complex genomic alterations and gene expression in acute lymphoblastic leukemia with intrachromosomal amplification of chromosome 21. *Proceedings of the National Academy of Sciences of the United States of America, 103*, 8167–8172.

Suzuki, T., Shen, H., Akagi, K., Morse, H. C., Malley, J. D., Naiman, D. Q., et al. (2002). New genes involved in cancer identified by retroviral tagging. *Nature Genetics, 32*, 166–174.

Taniuchi, I., Osato, M., Egawa, T., Sunshine, M. J., Bae, S. C., Komori, T., et al. (2002). Differential requirements for Runx proteins in CD4 repression and epigenetic silencing during T lymphocyte development. *Cell, 111*, 621–633.

Uren, A. G., Kool, J., Berns, A., & van Lohuizen, M. (2005). Retroviral insertional mutagenesis: Past, present and future. *Oncogene, 24*, 7656–7672.

Uren, A. G., Kool, J., Matentzoglu, K., de Ridder, J., Mattison, J., van Uitert, M., et al. (2008). Large-scale mutagenesis in p19(ARF)- and p53-deficient mice identifies cancer genes and their collaborative networks. *Cell, 133*, 727–741.

Vaillant, F., Blyth, K., Terry, A., Bell, M., Cameron, E. R., Neil, J., & Stewart, M. (1999). A full-length Cbfa1 gene product perturbs T-cell development and promotes lymphomagenesis in synergy with myc. *Oncogene, 18*, 7124–7134.

Vaillant, F., Blyth, K., Andrew, L., Neil, J. C., & Cameron, E. R. (2002). Enforced expression of Runx2 perturbs T cell development at a stage coincident with beta-selection. *Journal of Immunology, 169*, 2866–2874.

van der Weyden, L., Papaspyropoulos, A., Poulogiannis, G., Rust, A. G., Rashid, M., Adams, D. J., et al. (2012). Loss of RASSF1A synergizes with deregulated RUNX2 signaling in tumorigenesis. *Cancer Research, 72*, 3817–3827.

Vijaya, S., Steffen, D. L., Kozak, C., & Robinson, H. L. (1987). Dsi-1, a region with frequent proviral insertions in Moloney murine leukemia virus-induced rat thymomas. *Journal of Virology, 61*, 1164–1170.

Wang, C. Q., Krishnan, V., Tay, L. S., Chin, D. W., Koh, C. P., Chooi, J. Y., et al. (2014). Disruption of Runx1 and Runx3 leads to bone marrow failure and leukemia predisposition due to transcriptional and DNA repair defects. *Cell Reports, 8*, 767–782.

Whyte, W. A., Orlando, D. A., Hnisz, D., Abraham, B. J., Lin, C. Y., Kagey, M. H., et al. (2013). Master transcription factors and mediator establish super-enhancers at key cell identity genes. *Cell, 153*, 307–319.

Williams, A., Harker, N., Ktistaki, E., Veiga-Fernandes, H., Roderick, K., Tolaini, M., et al. (2008). Position effect variegation and imprinting of transgenes in lymphocytes. *Nucleic Acids Research, 36*, 2320–2329.

Wolyniec, K., Wotton, S., Kilbey, A., Jenkins, A., Terry, A., Peters, G., et al. (2009). RUNX1 and its fusion oncoprotein derivative, RUNX1-ETO, induce senescence-like growth arrest independently of replicative stress. *Oncogene, 28*, 2502–2512.

Wong, W. F., Kohu, K., Nakamura, A., Ebina, M., Kikuchi, T., Tazawa, R., et al. (2012). Runx1 deficiency in CD4+ T cells causes fatal autoimmune inflammatory lung disease due to spontaneous hyperactivation of cells. *Journal of Immunology, 188*, 5408–5420.

Wotton, S., Stewart, M., Blyth, K., Vaillant, F., Kilbey, A., Neil, J. C., & Cameron, E. R. (2002). Proviral insertion indicates a dominant oncogenic role for Runx1/AML-1 in T-cell lymphoma. *Cancer Research, 62*, 7181–7185.

Wotton, S. F., Blyth, K., Kilbey, A., Jenkins, A., Terry, A., Bernardin-Fried, F., et al. (2004). RUNX1 transformation of primary embryonic fibroblasts is revealed in the absence of p53. *Oncogene, 23*, 5476–5486.

Wotton, S., Terry, A., Kilbey, A., Jenkins, A., Herzyk, P., Cameron, E., & Neil, J. C. (2008). Gene array analysis reveals a common Runx transcriptional programme controlling cell adhesion and survival. *Oncogene, 27*, 5856–5866.

Zaidi, S. K., Pande, S., Pratap, J., Gaur, T., Grigoriu, S., Ali, S. A., et al. (2007). Runx2 deficiency and defective subnuclear targeting bypass senescence to promote immortalization and tumorigenic potential. *Proceedings of the National Academy of Sciences of the United States of America, 104*, 19861–19866.

Zaliova, M., Madzo, J., Cario, G., & Trka, J. (2011). Revealing the role of TEL/AML1 for leukemic cell survival by RNAi-mediated silencing. *Leukemia, 25*, 313–320.

Zindy, F., Eischen, C. M., Randle, D. H., Kamijo, T., Cleveland, J. L., Sherr, C. J., & Roussel, M. F. (1998). Myc signaling via the ARF tumor suppressor regulates p53-dependent apoptosis and immortalization. *Genes & Development, 12*, 2424–2433.

Zinkernagel, R. M., Callahan, G. N., Althage, A., Cooper, S., Klein, P. A., & Klein, J. (1978). On the thymus in the differentiation of "H-2 self-recognition" by T cells: Evidence for dual recognition? *The Journal of Experimental Medicine, 147*, 882–896.

RUNX1 and CBFβ Mutations and Activities of Their Wild-Type Alleles in AML

17

R. Katherine Hyde, Paul Liu, and Alan D. Friedman

Abstract

Mutations in *RUNX1* and *CBFB* have long been recognized as important in hematological malignancies. Point mutations and deletions of *RUNX1* are frequently found in myelodysplastic syndrome, myeloproliferative disease, and acute myeloid leukemia. Germline mutations in *RUNX1* are associated with familial platelet disorder with predisposition to AML. In addition, as will be discussed in other chapters, both *RUNX1* and *CBFB* are involved in recurrent chromosomal rearrangements in leukemia. More recently, roles for the non-mutated *RUNX1* and *CBFB* genes have been identified in multiple leukemia subtypes. This chapter will discuss the roles of *RUNX1* and *CBFB*, both in diseases caused by their mutations or deletions, as well as in the context of chromosomal rearrangements.

Keywords

AML • MDS • MPD • *RUNX1* • *CBFB* • Inv(16) • t(8;21) • MLL

17.1 Introduction

RUNX1 and *CBFB* are common targets of chromosomal translocations and inversions in human leukemia, which lead to the formation of fusion genes between *RUNX1* or *CBFB* and other partner genes (De Braekeleer et al. 2011; Liu et al. 1993). These fusion genes are the initiating events and they play critical roles during leukemogenesis (Lam and Zhang 2012; Shigesada et al. 2004). These fusion genes are covered in other chapters in this book. In this chapter we will discuss other roles that *RUNX1* and *CBFB* play in leukemia. In addition to fusion genes

R.K. Hyde (✉)
Department of Biochemistry and Molecular Biology, and Fred and Pamela Buffet Cancer Center, University of Nebraska Medical Center, Omaha, NE, USA
e-mail: kate.hyde@unmc.edu

P. Liu
Division of Intramural Research, National Human Genome Research Institute, NIH, Bethesda, MD, USA

A.D. Friedman
Division of Pediatric Oncology, Johns Hopkins University School of Medicine, Baltimore, MD, USA

Y. Groner et al. (eds.), *RUNX Proteins in Development and Cancer*, Advances in Experimental Medicine and Biology 962, DOI 10.1007/978-981-10-3233-2_17

from chromosome translocations, *RUNX1* is frequently mutated (point mutations, small indels, and large deletions) in myelodysplastic syndromes and acute myeloid leukemia (Osato 2004). *RUNX1* mutations are also associated with certain inherited blood diseases that predispose patients to leukemia, such as the familial platelet disorder with predisposition to leukemia (West et al. 2014). Moreover, accumulating data suggest that normal *RUNX1* and *CBFB* are required for full leukemic transformation by the *RUNX1* and *CBFB* fusion genes, and may be required for survival by other types of leukemia as well (Goyama et al. 2013; Roudaia et al. 2009). These findings have led to more comprehensive understanding of how *RUNX1* and *CBFB* contribute to various types of leukemia and other related hematologic malignancies, and may open up new directions for the development of targeted therapy of leukemia.

17.2 RUNX1 Point Mutations During Myeloid Transformation

A useful model of acute myeloid leukemia (AML) postulates requirement for Type I mutations that drive proliferation, e.g. Bcr-Abl, Ras activation, or FLT3-ITD, and Type II mutations that block differentiation, e.g. PML-RARα or *CEBPA* mutation (Gilliland and Griffin 2002). Further, Type I mutations in isolation lead to myeloproliferative disease (MPD), e.g. Bcr-Abl in CML or JAK2(V617F) in polycythemia vera, with advent of a Type II mutation then leading to blast crisis with progression to AML, and Type II mutations in isolation contribute to myelodysplastic syndromes (MDS), the latter often associated with cytopenias and lineage dysplasia.

RUNX1 gene mutations are associated with *de novo* AML, with MPD progression to AML, with MDS, and with Familial Platelet Disorder with Predisposition to AML (FPD/AML), which might itself be considered an MDS subset. In this section we review *RUNX1* alterations, aside from those involving chromosomal translocation, associated with these myeloid disorders and data implicating RUNX1 as a key mediator of myeloid differentiation.

17.2.1 RUNX1 Mutation in AML

The first study to identify *RUNX1* gene intrinsic mutations in myeloid malignancy included evaluation of 109 AML and 8 CML blast phase patients using an RNA:cDNA cleavage assay as the initial screen (Osato et al. 1999). Six patients harbored *RUNX1* mutations, including 3 of the 9 FAB M0 cases evaluated. Two of these M0 cases harbored biallelic *RUNX1* alterations each leading to early translational termination, suggesting that complete absence of RUNX1 may play a common role in this subset of AML that is characterized by a very early block in myeloid differentiation. In addition, the eight variant *RUNX1* alleles in these six cases involved the DNA-binding, N-terminal Runt homology domain (RHD) or residues just upstream. With one exception each had reduced capacity for DNA-binding and no ability to trans-activate the MCSFR promoter and in fact dominantly inhibited trans-activation by wild-type (WT) RUNX1. A second early study evaluated the RHD alone in 131 AML patients (Preudhomme et al. 2000). Twelve manifested *RUNX1* mutations, including 9 of 41 within the FAB M0 subset, and as in the prior study the large majority (8 of 9) of these FAB M0 cases containing bialleleic *RUNX1* alterations, with a subset containing the same alteration of both alleles suggestive of gene conversion with LOH. In addition this study identified several AML cases with *RUNX1* mutations in the setting of trisomy 21, with the two of three *RUNX1* alleles harboring the identical point mutation as a consequence of large-scale recombination. Neither of these studies identified *RUNX1* mutation in cases harboring t(8;21) or inv(16), chromosomal abnormalities that involve the genes encoding RUNX1 or its DNA-binding partner, CBFβ, respectively. In addition, no mutations in the *CBFB* gene were identified in 100 AML or 30 MDS cases (Leroy et al. 2002).

A third study evaluated 470 *de novo*, non-FAB M3 AML cases (Tang et al. 2009). Sixty three

RUNX1 mutations were identified, 28 missense or in-frame and 35 nonsense or frameshift. Thirty two were in the RHD (exons 3–5) and 31 in or surrounding the downstream trans-activation domain (TAD; exons 6–8). Most missense mutations were in the RHD, with TAD-region mutations mainly frameshift. Eleven of the frameshift mutations generate premature stop codons, whereas 17 are predicted to generate elongated proteins. *RUNX1* mutations were present in 40% of FAB M0, 18% of M1, 6% of M2, and 15% of M4 cases. There was a 3:1 male:female ratio and older age (median 62 vs 48 years) amongst *RUNX1*-mutant cases. Thirteen percent of intermediate risk patients with normal karyotype (NK) or NK with simple chromosome additions or deletions and 9% of high-risk patients with complex karyotypes harbored *RUNX1* mutations, and no low-risk patients with t(8;21) or inv(16) manifested *RUNX1* gene mutations. Presence of a *RUNX1* mutation led to worse prognosis, both within the total cohort and the NK subset.

A more recent study focused on 449 *de novo* AML patients with NK or NK with simple chromosomal additions/deletions (Schnittger et al. 2011). One hundred and forty seven of these patients (33%) harbored *RUNX1* mutations; 101 were heterozygous, with the remainder either having lost the WT allele or having two different, biallelic mutations. Sixty one percent of the mutations were located in the RHD, with the remainder in or surrounding the C-terminal trans-activation domain. The large majority of mutations downstream of the Runt domain were again frameshift, whereas frameshift and missense mutations were again similarly represented within the Runt domain. *RUNX1*-mutated cases showed no male or female predilection but were again associated with a higher mean age (70.5 vs 67.1 years). Within this NK cohort, *RUNX1* mutation was found in 65% of FAB M0, 30% of M1, 32% of M2, and 20% of M4 cases. Of note, 90% of patients with trisomy 13 also harbored *RUNX1* mutations, consistent with an earlier study associating these genetic features (Dicker et al. 2007). Similar to earlier findings (Tang et al. 2009), only two *CEBPA* and one *NPM1* mutation occurred in *RUNX1*-mutated cases; lack of *RUNX1* overlap with *CEBPA* gene mutation is consistent with a role for RUNX1 in regulating normal myeloid differentiation, including via activation of the *CEBPA* locus as discussed below. FLT3 tyrosine kinase receptor activating mutations were found in 42% and activating N-Ras mutations in an additional 18% of *RUNX1*-mutant cases, representing co-expressed Type I proliferative alterations. In addition, MLL partial tandem duplication (MLL-PTD) showed 37% overlap. *RUNX1* mutation negatively impacted prognosis, reducing event-free survival (EFS) from approximately 40 to 25% and overall survival (OS) from 58 to 37% at 3 years. This difference was true for both the total cohort, for those with age < 60 years, for those with NK without chromosomal lose or gain, and for those lacking the high-risk FLT3-ITD or MLL-PTD alterations. Prognosis was worse in the presence of co-expressed FLT3-ITD but was not affected by location of the *RUNX1* mutation in the RHD versus downstream or by presence of a remaining wild-type *RUNX1* allele.

A fifth study evaluated 945 AML patients, ages 18–60 (Gaidzik et al. 2011). *RUNX1* mutations were found in 53 (5.6%) of these cases, and were again correlated with presence of MLL-PTD and also *IDH1* or *IDH2* mutation, and had little overlap with *CEBPA* or *NPM1* mutation. *RUNX1* mutation again negatively impacted prognosis, with 26% versus 44% 4-year EFS. Microarray RNA expression analysis identified a *RUNX1*-mutated signature amongst 14 cases also present in other high-risk cases, e.g. those with monosomy 7, inv(3);t(3;3), or complex karyotype, lacking *RUNX1* mutation. This signature included high expression of pro-survival genes, including Bcl-xL. An earlier study had compared global mRNA expression amongst 35 FAB M0 cases, including 14 with *RUNX1* mutations, with 253 additional AML cases (Silva et al. 2009). Amongst the 35 M0 cases, those with *RUNX1*-mutation manifested a unique gene expression pattern. Both the *RUNX1* wild-type and mutant M0 patients manifested decreased *MPO*, *CEBPA*, *CEBPD*, *SPI1/PU.1*, *ETV6/TEL*, and *JAK2* and up-regulation of *RARA*, *GATA3*, and *AKT3*. In addition, the *RUNX1*-mutant cohort contained increased expression of B cell genes, including *EBF1, RUNX3, FLT3, TDT, BCR, IGHM, IGL, BLNK*, and *SYK*.

17.2.2 RUNX1 Mutation in MPD and Congenital Neutropenia

Patients with MPD include those with polycythemia vera (PV), essential thrombocytopenia (ET), and primary myelofibrosis (PMF), each often associated with the activating JAK2(V617F) mutation. Amongst a cohort of 417 MPD patients, 18 progressed to acute leukemia (Ding et al. 2009). Five patients harbored *RUNX1* mutations (two missense, three frameshift) in the RHD after leukemic transformation, and four of these possessed coincident JAK2(V617F). The two PV and one ET patients had received prior busulfan and/or hydroxyurea, whereas the two PMF patients had not received prior chemotherapy.

Chronic myeloid leukemia (CML) cases, commonly associated with Bcr-Abl, invariably progresses from a chronic myelo-proliferative phase to blast crisis if not treated. Amongst a cohort of 85 accelerated phase or blast crisis CML patients, 11 (13%) harbored heterozygous *RUNX1* gene mutations (Zhao et al. 2012). Eight were in the RHD (six missense, two frameshift truncating), and three in the TAD (one frameshift truncating, two frameshift elongating). Each of these 11 cases manifested myeloid rather than lymphoid blast morphology. Two RHD variants studied, H78Q and V91fs-ter94, inhibited granulocytic differentiation of the 32Dcl3 cell line in response to G-CSF. And upon co-transduction with Bcr-Abl into murine marrow cells, these variants induced progression from chronic phase to blast crisis or accelerated phase, respectively. Deep sequence analysis of an additional cohort of 39 CML patients identified 10/29 with myeloid or unspecified and 3/10 with lymphoid blast crisis with heterozygous *RUNX1* alterations, making RUNX1 the most frequently mutated transcription factor in this group of leukemias (Kanno et al. 1998).

Four RHD missense variants expressed in CML blast crisis patients lacked DNA-binding and trans-activation potential, but could still bind CBFβ and inhibit trans-activation by WT RUNX1, potentially via CBFβ sequestration (Zhao et al. 2012). RUNX1(R293X), truncated just upstream of the TAD, strongly bound both CBFβ and DNA and also inhibited activation by WT RUNX1, potentially reflecting both competition for RUNX1 *cis* elements and increased affinity for CBFβ, similar to C-terminally truncated RUNX1 variants (Grossmann et al. 2011). RUNX1(G381fs-ter570) retains trans-activation activity; however, this elongated protein interfered with the ability of WT RUNX1 to form high molecular weight complexes, and did not form such complexes itself, perhaps thereby compromising RUNX1 activities *in vivo*.

At least 20% of patients with congenital neutropenia (CN) progress to AML, often associated with mutations of the *CSFR3* gene leading to expression of truncated GCSFR proteins capable of transducing proliferative and survival but not maturation signals into myeloid cells. A recent study identified *RUNX1* mutations in 14 of 18 AML and 5 of 12 MDS cases arising in CN patients (Skokowa et al. 2014). Among the 14 AML patients with *RUNX1* mutations, twelve had a coincident *CSF3R* mutation. *RUNX1* mutations arose subsequent to *CSF3R* mutations, and co-expression of either of two RUNX1 RHD mutants with truncated GCSFR(d715) in human marrow CD34$^+$ cells led to increased proliferation and impaired myeloid differentiation compared to GCSFR(d715) alone. These same investigators identified *RUNX1* mutations in 9 of 307 pediatric AML patients (3%), with a preponderance of FAB M1 cases, RHD alteration, and association with complex karyotype.

17.2.3 RUNX1 Mutation in MDS

Amongst 37 patients evaluated for *RUNX1* exons 3–6 alterations, mutations were identified in a chronic myelomonocytic leukemia (CMML) patient and in a patient with AML post-MDS (MDS/AML), in particular V105ter and R139G (Imai et al. 2000). Neither of these RUNX1 variants bound DNA, and the latter manifested increased CBFβ affinity and inhibited WT DNA-binding. In a second study, 0/94 MDS patients harbored *RUNX1* mutations (Preudhomme et al. 2000), and in a third *RUNX1* exon 3–5 mutations were detected in 2/74 sporadic, 6/13 atomic bomb-associated, and 2/6 therapy-related MDS patients, suggesting a key role for DNA damage for inducing *RUNX1* mutations in MDS cases (Harada et al. 2003). This group then extended their analysis to the entire *RUNX1* coding sequence and to additional patients, finding *RUNX1* mutations in 15/88 sporadic and 11/22 of secondary MDS/AML patients (Harada et al. 2004). All mutations were heterozygous with retention of a wild-type *RUNX1* allele. A subsequent study of 140 therapy-related MDS/AML patients identified 22 (16%) with heterozygous *RUNX1* mutations, 13 in the RHD and 9 C-terminal. Nineteen had received alkylating agents and two radiation, and presence of *RUNX1* mutations in MDS correlated highly with subsequent transformation to AML (Christiansen et al. 2004). *RUNX1* mutations were also present in 21% of 57 patients with Fanconi anemia (FA)-associated MDS or MDS/AML (Quentin et al. 2011). FA patients harbor genetic defects that interfere with their ability to repair DNA breaks, suggesting shared pathogenesis with patients who develop MDS/AML after exposure to DNA-damaging agents. Notably, *Runx1*$^{-/-}$/*Runx3*$^{-/-}$ mice manifest a related DNA repair defect, with impaired recruitment of monoubiquitinated-FANCD2 to DNA damage foci (Wang et al. 2014).

A study of 81 patients with CMML identified 30 with 32 *RUNX1* mutations, 28 heterozygous; 23 mutations were in the RHD and 9 C-terminal (Kuo et al. 2009). Presence of *RUNX1* mutation in these patients did not alter survival, but those with C-terminal mutations more rapidly progressed to AML. A recent study that evaluated the entire *RUNX1* coding sequence identified mutations in 20/143 or 14% of sporadic MDS and 27/84 (32%) of CMML patients (Tsai et al. 2015). These authors found that MDS or CMML patients with RUNX1 variants possessing markedly reduce trans-activation activity progressed more rapidly to AML; these included frameshift mutations in the RHD or TAD and missense RHD mutations that prevent DNA-binding.

The majority of RUNX1 RHD missense mutations associated with MDS or AML occur in residues within three Runt domain loops that contact DNA; among ten such variants, six prevented DNA-binding without reducing CBFβ affinity while the others prevented both interactions, suggesting global Runt domain disruption in those instances (Matheny et al. 2007). Transduction of either of two RHD variants, K83N or R135G, into murine marrow followed by transplantation did not affect lymphoid, megakaryocytic, granulocytic, or monocytic differentiation; however, marrow Ter119^{+} erythroid cells were reduced twofold, and a similar specific effect on the erythroid lineage was evident after transduction into human cord blood progenitors (Cammenga et al. 2007). In addition, these variants increased murine colony-forming unit (CFU) replating capability over at least six generations, indicating impairment of myeloid maturation in this assay. In a related study, transduction of RUNX1(D171N), a RHD mutation commonly associated with MDS that prevents DNA-binding, or RUNX1(S291fs), a frameshift mutation downstream of the RHD, into murine marrow followed by transplantation led to MDS-RAEB and MDS/AML. The S291fs variant showed a greater propensity for erythroid dysplasia and pancytopenia (Watanabe-Okochi et al. 2008). The D171N variant increased human CFU replating and markedly increased human long-term colony-initiating cells (LTC-IC), an *in vitro* assay reflective of immature marrow stem cells; in addition, D171N reduced the number and size of erythroid BFU-E but increased the number and size of myeloid CFU-GM compared with WT RUNX1 when introduced into human marrow CD34^{+} cells (Harada et al. 2013). In addition, while D171N transduction alone did not stimulate CD34^{+} cell proliferation, proliferation induced by the BMI1 Polycomb protein, itself commonly over-expressed in high-risk MDS cases, was further increased by co-expressed RUNX1(D171N).

17.2.4 RUNX1 Mutations in FPD/AML

FPD/AML is an autosomal dominant disorder associated with reduced megakaryocyte (Meg) ploidy, thrombocytopenia, platelet dysfunction, and progression to AML in approximately 35% of patients. Analysis of six pedigrees identified heterozygous nonsense *RUNX1* RHD mutations in three, intragenic *RUNX1* deletion in one, and RHD missense mutations in two, with reduced marrow CFU-Meg (Song et al. 1999). A subsequent study of three additional pedigrees identified a RHD missense and two RHD nonsense/frameshift mutations (Michaud et al. 2002). Studies of additional pedigrees identified additional examples of monoallelic *RUNX1* deletions, as well as C-terminal *RUNX1* mutations, with the majority of cases manifesting heterozygous RHD mutations (Ganly et al. 2004; Preudhomme et al. 2009).

Runx1$^{+/-}$ mice have only a mild ~15% decrease in platelet count with a normal number of marrow Megs (Sun and Downing 2004), suggesting that lack of a single *RUNX1* allele does not have consequences as severe as expression of a dominant-interfering RUNX1 missense or nonsense/frameshift variant, the latter associated with the large majority of FPD/AML cases and with those *de novo* AML, MPD/AML, or MDS/AML cases that harbor *RUNX1* point mutations. Consistent with this idea, among the five RUNX1 RHD variants identified in the initial two FPD/AML family studies, each had markedly reduced DNA affinity, and five retained the ability to bind CBFβ and inhibited reporter trans-activation by WT RUNX1, the latter effect rescued by exogenous CBFβ (Michaud et al. 2002). This latter study also noted that families with RUNX1 variants that function in a dominant-negative manner appear to have a higher incidence of AML than those with haplo-insufficiency and that at least in two FPD/AML cases evaluated the second *RUNX1* allele was preserved.

Defects in megakaryopoiesis in FPD/AML may in part reflect reduced RUNX1 activation of the *MYL9* and *MYH9* promoters and repression of the *MYH10* gene, leading to deregulation of myosin II expression (Bluteau et al. 2012).

An induced pluripotent stem cells (iPSC) line developed from an FPD/AML patient with RUNX1(Y260X) was subjected to gene editing using a WT *RUNX1* exon 5–8 donor vector and a zinc finger nuclease pair targeted to exon 5 to stimulate homologous recombination. Uncorrected and corrected lines generated similar numbers of CD34$^+$CD45$^+$ hematopoietic progenitors, with the corrected lines manifesting a twofold increase in Meg potential, indicating a key role for RUNX1 alteration in the FPD/AML megakaryopoiesis defect (Connelly et al. 2014). iPSC lines developed from FPD/AML patients harboring G172E, G143W, or N233fsX283 RUNX1 variants manifested two to threefold reduced CD45$^+$ or CD34$^+$ cells and reduced erythroid, Meg, and granulocyte-monocyte (GM) CFUs compared to three iPSC control lines, with each of these defects rescued by exogenous RUNX1 (Sakurai et al. 2014). iPSCs generated from an FPD/AML patient with RUNX1(R174Q) manifested reduced Meg and erythroid colony formation with increased CFU-GM, whereas iPSC derived from an FPD/AML patient with *RUNX1* haploinsufficiency only manifested the Meg/erythroid defect, and both the reduced Meg/erythroid potential and elevated GM potential returned towards baseline after introduction of exogenous RUNX1 (Antony-Debre et al. 2015).

17.2.5 Summary of RUNX1 Variants and Implications for the Pathogenesis of Myeloid Transformation

Investigation of *RUNX1* mutations associated with the various myeloid disorders considered in the preceding sections finds that in each the majority of mutations are found in the RHD and markedly reduce DNA binding affinity without affecting CBFβ interaction, implicating CBFβ sequestration as a potential mechanism for dominant inhibition of the remaining WT *RUNX1* allele. Downstream of the RHD, the majority of alterations generate truncated proteins via nonsense or frameshift alterations, predicted to inhibit normal RUNX1 (and perhaps RUNX2

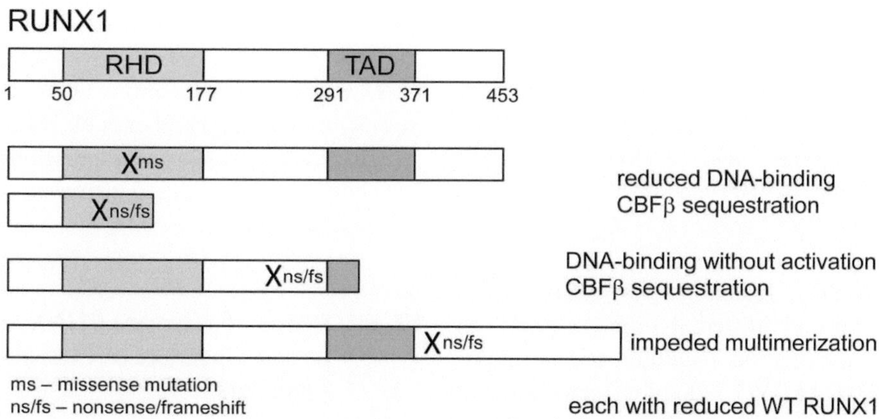

Fig. 17.1 Diagram depicting general categories of mutant RUNX1 variants and associated mechanisms accounting for reduced RUNX1 activity consequent to *RUNX1* point mutation in myeloid malignancies. RUNX1 RHD mutation impairs DNA-binding and facilitates CBFβ sequestration. RUNX1 C-terminal mutation allows competitive DNA-binding while impeding trans-activation and generates elongated variants that interfere with RUNX1 multimerization. These mutations each also lead to overall reduction in RUNX1 expression

and RUNX3) activity both via competition for RUNX1 *cis* DNA elements and for CBFβ. Additional, elongated variants might interfere with RUNX1 multimerization. In addition, *RUNX1* allele deletion or mutation at a minimum reduces RUNX1 activity secondary to reduced WT RUNX1 expression (Fig. 17.1). Amplification of mutant alleles by trisomy 21 or the presence of biallelic mutations, as seen often in FAB M0 AML cases, would further increase the dominant-inhibitory effect of RUNX1 variants.

Several clinical and laboratory observations implicate *RUNX1* mutation as a Type II alteration during myeloid transformation, contributing to impaired differentiation. First, clustering of biallelic *RUNX1* mutation in M0 cases, an AML subset associated with a complete block in granulocyte/monocyte maturation, suggests that residual RUNX1 activity in the other AML subsets enables their partial maturation. Second, the very low coincidence of *RUNX1* and *CEBPA* mutations in AML cases combined with the key role played by C/EBPα in myeloid differentiation (Friedman 2015) further supports a role for RUNX1 mutations as Type II alterations. Consistent with this premise, RUNX1 apparently directly regulates C/EBPα expression – conditional *Runx1* gene deletion in adult mice reduces *Cebpa* mRNA expression in hematopoietic progenitors, and Runx1 binds and regulates the *Cebpa* gene via four cis elements present in an evolutionarily conserved +37 kb *Cebpa* enhancer located at +42 kb in the *CEBPA* locus (Guo et al. 2012, 2014; Cooper et al. 2015). Among a cohort of 150 human NK AML cases, *CEBPA* expression was 1.4-fold lower in *RUNX1*-mutated compared with *RUNX1* WT cases, and almost two fold lower in cases with biallelic *RUNX1* alterations, and the degree of reduction was similar amongst RHD or C-terminal variants and amongst missense compared with nonsense/frameshift *RUNX1* variants (Grossmann et al. 2012). RUNX1 also directly regulates expression of the gene encoding PU.1, another key transcriptional regulator of myeloid differentiation (Huang et al. 2008), as well as those for myeloperoxidase, neutrophil elastase, MCSFR and additional markers of early myelopoiesis (Friedman 2007).

In contrast to *CEBPA*, mutated in ~10% of human AML cases, *SPI1/PU.1* gene mutations are rare (Dohner et al. 2003). Reduced levels of murine PU.1 consequent to deletion of its +14 kb enhancer impacts monopoiesis while sparing granulopoiesis (Rosenbauer et al. 2004), whereas the opposite is true upon *Cebpa* knockdown or +37 kb enhancer deletion (Ma et al. 2014; Guo et al. 2016). *Runx1* gene deletion in adult mice reduces *Cebpa* more than *Pu.1* in marrow stem/

progenitor cells, and in this setting granulopoiesis is reduced whereas monopoiesis is increased (Guo et al. 2014), further implicating *CEBPA* as the more critical RUNX1 target relevant to both normal myelopoiesis and myeloid transformation. In addition, C/EBPα and PU.1 activate their respective myeloid enhancers (Cooper et al. 2015; Yeamans et al. 2007).

While *RUNX1* mutations may well impact pathways beyond differentiation relevant to myeloid transformation, including cell proliferation, DNA repair, p53 induction, cell survival, and ribosome biogenesis (Wang et al. 2014; Friedman 2009; Britos-Bray et al. 1998; Goyama et al. 2013; Cai et al. 2015), a framework wherein *RUNX1* mutations synergize with Type I pro-proliferation/pro-survival alterations is likely central to the majority of AML cases associated with *RUNX1* point mutations (Fig. 17.2). Of note,

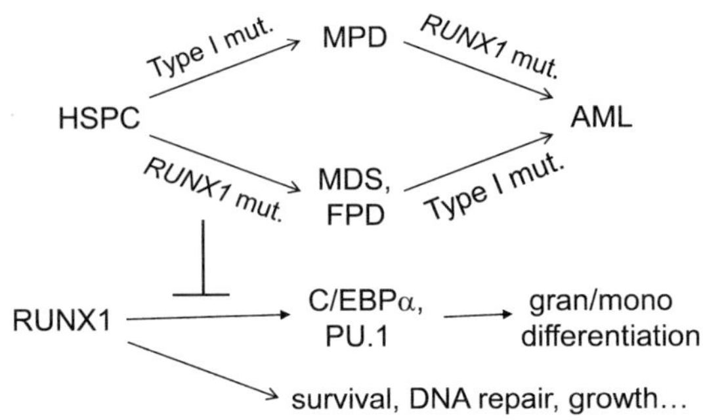

Fig. 17.2 Model of myeloid transformation wherein RUNX1 point mutants interfere with myeloid differentiation. Type I mutations such as FLT3ITD, Bcr-abl, activated N-Ras, GCSFR(d715), or JAK2(V617F) stimulate hematopoietic stem/progenitor cell (HSPC) proliferation and survival. Type I mutations in isolation can lead to myeloproliferative diseases (MPD), *RUNX1* mutation alone can contribute to a subset of myelodysplastic syndrome (MDS) cases and when inherited through the germline lead to Familial Platelet Disorder (FPD), and the combination of a Type I alteration and monoallelic or biallelic *RUNX1* mutation can led to acute myeloid leukemia (AML). Other Type II mutations, e.g. PML-RARα, RUNX1-ETO, CBFβ-SMMHC, or *CEBPA* mutation, can substitute for RUNX1 alterations in this model. RUNX1 mutant proteins inhibit RUNX1 induction of C/EBPα and PU.1 to impede granulocytic and monocytic myeloid differentiation as well as additional pathways regulated by RUNX1, including those that favor cell survival, DNA repair, or cell growth. In addition to Type I and Type II alterations, additional mutations, e.g. affecting splicing, metabolism, or cell senescence, may be required for full transformation

approximately 2% of pediatric B-lineage acute lymphoblastic leukemia (ALL) is associated with *RUNX1* gene amplification without apparent RUNX1 mutations (Penther et al. 2002). Perhaps in these cases RUNX1 over-expression acts like a Type I oncogene, leading to increased cell growth and proliferation.

17.3 Roles of Normal *RUNX1/CBFB* in the Context of Fusion Proteins

As discussed above, mutations that cause loss of RUNX1 activity are common in several types of hematological malignancies, indicating that RUNX1 has tumor suppressor qualities. Surprisingly, RUNX1 and CBFβ mutations are rarely found in patients with recurrent chromosomal abnormalities implying that, in the context of leukemic fusion proteins, RUNX1 may have important growth promoting or pro-survival activities (Tang et al. 2009; Schnittger et al. 2011; Patel et al. 2012). In this section, we will discuss the emerging roles for RUNX1 and CBFβ in leukemia with inversion of chromosome 16, the t(8;21) translocation, or rearrangements involving the Mixed-Lineage Leukemia protein, MLL.

17.3.1 Inversion 16 AML

Inversion 16 (inv(16)(p13q22) or the less common t(16;16)(p13q22) translocation is associated with AML subtype M4 with Eosinophilia (M4Eo) according to the French-American-British (FAB) classification (Le Beau et al. 1983). Both chromosomal abnormalities generate a fusion between *CBFB* and the *MYH11* gene, which encodes Smooth Muscle Myosin Heavy Chain (SMMHC) (Liu et al. 1993, 1996). The fusion gene, *CBFB-MYH11*, encodes the fusion protein CBFβ-SMMHC, and is causative for the development of AML (Castilla et al. 1999).

As will be discussed in more detail in subsequent chapters, the CBFβ-SMMHC fusion protein retains the ability to bind RUNX1 with the CBFβ half of the protein, but also contains a second, high affinity RUNX binding domain in the SMMHC tail which allows CBFβ-SMMHC to bind RUNX1 with a higher affinity than wild-type CBFβ (Lukasik et al. 2002). This has led to a model of CBFβ-SMMHC activity in which the fusion protein dominantly represses RUNX1 (Lukasik et al. 2002). Results in tissue culture and mouse models have provided further support that the fusion protein has dominant repressor activity (Wijmenga et al. 1996; Adya et al. 1998; Lutterbach et al. 1999). Knockin mice in which *Cbfb-MYH11* is expressed from the endogenous *Cbfb* locus (*Cbfb*^{MYH11}) also show evidence of the fusion protein's dominant repressor activity. *Cbfb*$^{+/MYH11}$ embryos have a complete block in definitive hematopoiesis, develop severe central nervous system hemorrhaging, and die between embryonic day 12.5 and 13.5 (Table 17.1) (Castilla et al. 1996). This phenotype is strikingly similar to what is observed with mice homozygous for null alleles of either *Runx1* (*Runx1*$^{-/-}$) or *Cbfb* (*Cbfb*$^{-/-}$) (Okuda et al. 1996; Sasaki et al. 1996; Wang et al. 1996a, b; Niki et al. 1997). This indicates that, in vivo, CBFβ-SMMHC can act as a dominant repressor of RUNX1.

Comparison of the different mouse models during the earlier phase of blood development, primitive hematopoiesis, indicates that CBFβ-SMMHC also has RUNX1-repression independent activities. *Cbfb*$^{+/MYH11}$ embryos have defects in the differentiation of primitive blood cells that are significantly more severe than seen in either *Runx1*$^{-/-}$ or *Cbfb*$^{-/-}$ embryos (Castilla et al. 1996; Hyde and Liu 2010). In addition, gene expression analysis of the peripheral blood from *Cbfb*$^{+/MYH11}$, *Cbfb*$^{-/-}$, and their respective wildtype littermate embryos shows that expression of the fusion protein and loss of *Cbfb* induce distinct changes in gene expression, with relatively few genes found to be deregulated in both genotypes. The majority of the gene expression changes in *Cbfb*$^{+/MYH11}$ embryos are increases in expression, which is the opposite of what would be expected if the fusion protein's only activity was the repression of the transcriptional activator, RUNX1 (Hyde and Liu 2010). Together, these data indicate that *Cbfb-MYH11* induces both differentiation defects and gene expression changes that are independent of RUNX repression.

Analysis of inv(16) patient data also supports the conclusion that CBFβ-SMMHC has important RUNX1-repression independent activities. Many of the genes deregulated in *Cbfb*$^{+/MYH11}$ embryos, but not in *Cbfb*$^{-/-}$ embryos are also expressed in samples from inv(16) patients (Hyde and Liu 2010). In addition, if CBFβ-SMMHC were acting solely as a dominant repressor of RUNX1, one would expect inv(16) AML to have similar blast cell morphology and prognosis to AML with homozygous deletion of *RUNX1*. Rather, these two subtypes have very different clinical presentations. Inv(16) is associated with myelomonocytic, M4 AML with good prognosis (Le Beau et al. 1983), while deletion of *RUNX1* is found almost exclusively in stem cell-like, M0 AML with poor prognosis (Tang et al. 2009;

Table 17.1 Summary of *Cbfb-MYH11* knockin models

Genotype	Description	Phenotype
Cbfb$^{+/MYH1}$ Castilla et al. (1999, 1996)	Traditional knockin, expresses full length CBFβ-SMMHC	Embryonic lethal, chimeras develop leukemia
Mx1-Cre^{+}, *Cbfb*$^{+/56M}$ Kuo et al. (2006)	Inducible expression of full length CBFβ- SMMHC	Develop leukemia similar to traditional knockin chimeras
Mx1-Cre^{+}, *Cbfb*$^{+/56M}$, *Runx1*$^{+/lz}$ Hyde et al. (2015)	Inducible expression of full length CBFβ-SMMHC in a *Runx1* deficient background	Delayed leukemia development
Cbfb$^{+/MYH11d179-231}$ Kamikubo et al. (2010)	Traditional knockin, expresses CBFβ-SMMHC lacking the HABD	Accelerated leukemia development
Mx1-Cre^{+}, *Cbfb*$^{-/56M}$ Heilman et al. (2006)	Inducible expression of full length CBFβ-SMMHC and lacking wildtype *Cbfb* expression	Accelerated leukemia development

Schnittger et al. 2011; Patel et al. 2012). These clinical observations provide further support that CBFβ-SMMHC has important RUNX1-repression independent activities.

Although the fusion protein has activities other than repression of RUNX1, it should not be assumed that RUNX1 is dispensable in CBFβ-SMMHC driven leukemia. In mouse embryos expressing *Cbfb-MYH11*, but lacking RUNX1 activity due to either homozygous null alleles (*Runx1*$^{-/-}$) or a hetozygous dominant negative allele (*Runx1*$^{+/lz}$), there is a rescue of the fusion gene associated differentiation defects (Table 17.1) (Hyde et al. 2015). In addition, adult mice expressing the fusion protein from an inducible allele, but lacking RUNX1 activity (*Mx1-Cre*$^{+}$, *Cbfb*$^{+/56M}$, *Runx1*$^{+/lz}$) have delayed leukemia development as compared to mice expressing CBFβ-SMMHC in a *Runx1* sufficient background (*Mx1-Cre*$^{+}$, *Cbfb*$^{+/56M}$, *Runx1*$^{+/+}$). Strikingly, approximately 20% of *Mx1-Cre*$^{+}$, *Cbfb*$^{+/56M}$, *Runx1*$^{+/lz}$ mice failed to develop leukemia within 1 year, in contrast to *Mx1-Cre*$^{+}$, *Cbfb*$^{+/56M}$, *Runx1*$^{+/+}$ mice, all of whom developed leukemia by 6 months (Hyde et al. 2015). These results demonstrate that RUNX1 activity is required for CBFβ-SMMHC activity during leukemia development.

RUNX1 activity is also required for the continued survival and propagation of CBFβ-SMMHC expressing leukemia cells. Knockdown of RUNX1 expression by short hairpin RNA (shRNA) in the inv(16) patient derived cell line, ME-1 induces significant apoptosis (Ben-Ami et al. 2013). Similarly, an inhibitor of RUNX1, Ro5–3335, causes decreased viability of ME-1 cells in culture; and decreased leukemic burden and increased survival when given to leukemic, CBFβ-SMMHC expressing mice (Cunningham et al. 2012). These results further demonstrate the important role of RUNX1 in CBFβ-SMMHC driven leukemia.

In the experiments described above, it is not possible to discern whether RUNX1 is required as part of a complex with the fusion protein, or with wildtype CBFβ. Chromatin immunoprecipitation (ChIP) experiments in ME-1 cells demonstrate that RUNX1 co-localizes with both CBFβ-SMMHC and wildtype CBFβ, binding distinct DNA sequences with each dimerization partner (Mandoli et al. 2013). These results imply that RUNX1 is involved in transcriptionally active complexes with both forms of CBFβ, each of which may be important. Interestingly, RUNX1 was also found associated with genes that are not bound by either the fusion protein or CBFβ, implying that RUNX1 may also have CBFβ- independent activities, as has been suggested previously (Pabst et al. 2001; Wheeler et al. 2002; Cammenga et al. 2007). Currently, the relative importance of each complex is not understood. It is possible that RUNX1 activity in each of the three configurations is important in CBFβ-SMMHC driven leukemia, and that balance among the different RUNX containing complexes influences leukemia development. This idea is supported by the finding that expression of a mutant *Cbfb-MYH11* allele with decreased RUNX1 binding affinity (*Cbfb*$^{+/MYH11d179-231}$), which is expected to allow more RUNX1 to function with wildtype CBFβ or in CBFβ-independent complexes, causes accelerated leukemogenesis (Kamikubo et al. 2010). In addition, mice expressing *Cbfb-MYH11*, but lacking wildtype *Cbfb*, which is expected to favor both CBFβ-SMMHC:RUNX1 and CBFβ-independent complexes also show accelerated leukemogenesis (Table 17.1) (Heilman et al. 2006). While more experimentation is required, these studies imply that RUNX1 contributes to CBFβ-SMMHC induced leukemogenesis through multiple mechanisms.

17.3.2 t(8;21) AML

The chromosomal translocation 8q22;21q22 (t(8;21)) is associated with M2 AML, and is classified as a good prognosis AML (Rowley 1973). The translocation generates a fusion protein between RUNX1 and the transcriptional repressor, RUNX1T1 (formerly ETO) (Erickson et al. 1992). The resulting fusion protein, RUNX1-RUNX1T1 (AML1-ETO), contains the DNA binding domain of RUNX1 and the protein interaction domains of RUNX1T1 which interact with a number of transcriptional repressors (Peterson and Zhang 2004). Consequently, RUNX1-RUNX1T1 is thought to act by repressing expression of its target genes.

In patients with t(8;21) AML, the second *RUNX1* allele is very rarely deleted or mutated, leading to speculation that wildtype RUNX1 is required in leukemia cells expressing the RUNX1-RUNX1T1 fusion protein (Tang et al. 2009; Schnittger et al. 2011; Patel et al. 2012). This conclusion is supported by work in two different models of human AML: the t(8;21) AML patient derived cell line, Kasumi-1, and cord blood derived $CD34^+$ cells transduced to express the RUNX1-RUNX1T1. In both models, loss of wildtype RUNX1 activity either by shRNA knockdown or overexpression of a dominant negative RUNX1 mutant results in cell cycle progression defects and apoptosis (Ben-Ami et al. 2013; Goyama et al. 2013). Consistent with these observations, Kasumi-1 cells treated with the RUNX1 inhibitor, Ro5–3335, show decreased survival, indicating that wildtype RUNX1 is required in t(8;21) AML cells (Cunningham et al. 2012).

Presumably wildtype RUNX1 activity is needed for the transcription of a subset of target genes that are important for cell survival. Indeed, multiple studies demonstrate that, although wildtype RUNX1 and RUNX1-RUNX1T1 both bind the majority of target genes, each protein also uniquely regulates the expression of a subset of genes (Okumura et al. 2008; Ben-Ami et al. 2013; Shimada et al. 2000; Gardini et al. 2008; Ptasinska et al. 2012, 2014; Li et al. 2016; Trombly et al. 2015). In addition, knockdown of RUNX1 in Kasumi-1 cells causes decreased expression of genes known to regulate the mitotic checkpoint implying that requirement for wildtype RUNX1 is rooted in its ability to transactivate critical cell cycle progression genes (Ben-Ami et al. 2013).

The role of RUNX1 at genes bound by both the wildtype protein and the fusion protein is less clear. Based on chromatin immunoprecipitation experiments, it has been proposed that RUNX1 and RUNX1-RUNX1T1 compete for mutually exclusive binding to these target genes, establishing a dynamic equilibrium that allows for fine-tuned control of target gene expression (Ptasinska et al. 2014). However, similar experiments using different antibodies indicate that RUNX1 and RUNX1-RUNX1T1 physically interact and simultaneously bind adjacent DNA motifs in target gene promoters, regulating target gene expression as a single complex (Li et al. 2016). Further experimentation will be needed to resolving these conflicting observations.

The role of CBFβ in the RUNX1-RUNX1T1 transcriptional complex is similarly complicated. RUNX1-RUNX1T1 retains the ability to interact with CBFβ, and forms a complex with the β subunit in vivo (Sun et al. 2013). It is known that the fusion protein's DNA binding ability is required for its leukemogenic activity (Kwok et al. 2009; Yan et al. 2009; Roudaia et al. 2009). Presumably, interaction with CBFβ stabilizes RUNX1-RUNX1T1's interaction with DNA, just as it does for wildtype RUNX1. Studies testing this possibility have produced contradictory conclusions. Kwok et al. used mutant RUNX1-RUNXT1 with point mutations that severely inhibited CBFβ binding, as well as shRNA knockdown of *CBFB*, to test the role of CBFβ in RUNX1-RUNX1T1 activity in mouse bone marrow cells. They found that loss of CBFβ binding by the fusion protein or knockdown of CBFβ expression impaired colony forming and serial replating ability, but did not prevent the growth of RUNX1-RUNX1T1 expressing cells in vitro (Kwok et al. 2009, 2010). These findings led the authors to conclude that CBFβ contributes, but is not strictly required for RUNX1-RUNX1T1 activity. In contrast, Roudaia et al. found that the same double point mutant expressed in mouse bone marrow cells eliminated RUNX1-RUNX1T1's ability to induce leukemia in cooperation with TEL-PDGFβR, an activated tyrosine kinase receptor (Park et al. 2009; Roudaia et al. 2009). This observation led to the conclusion that CBFβ is required for RUNX1-RUNX1T1's activity. The most likely explanation for these contradictory results is that the ability to induce leukemia in vivo is a more stringent test of RUNX1-RUNX1T1 activity than colony forming ability in vitro, so may have less tolerance for the loss of CBFβ. Although it will require further experimentation to resolve the requirement for

CBFβ by RUNX1-RUNX1T1, it is clear from both groups' work that CBFβ contributes to the fusion protein's activity and that disrupting their interaction may have therapeutic benefit for t(8;21) AML patients.

17.3.3 MLL Rearranged AML

The *Mixed-Lineage Leukemia* (*MLL*, also known as *KMT2A*) gene encodes a histone methyl transferase that is a critical regulator of transcription during hematopoietic development (Hanson et al. 1999; Yagi et al. 1998; Hess et al. 1997; Muntean and Hess 2012). MLL is also a common target of translocations found in both AML and ALL. These translocations involve the N-terminus of MLL fused to components of the transcriptional elongation complex. To date, over 60 different fusion partners for MLL have been identified (de Boer et al. 2013; Muntean and Hess 2012). Regardless of the fusion partner involved, MLL rearranged leukemia is classified as poor prognosis, and all fusion proteins are assumed to function in a similar manner (Neff and Armstrong 2013). Recent work, as described below, has identified potentially important differences in the molecular mechanisms of individual fusion proteins, especially as they relate to RUNX1.

Like *Runx1*, *Mll* is required for normal fetal liver hematopoiesis. As both factors are transcriptional regulators, it has been proposed that they may act together to regulate normal blood development (Hanson et al. 1999; Yagi et al. 1998; Hess et al. 1997; Muntean and Hess 2012). In fact, RUNX1 and MLL have recently been shown to physically interact (Huang et al. 2011). The C-terminal region of MLL directly binds the N-terminus of RUNX1, and protects RUNX1 from proteosomal degradation. The MLL:RUNX1 complex directly regulates another important hematopoietic transcription factor, PU.1. Interestingly, some RUNX1 mutations found in patients with MDS and AML affect MLL binding. N-terminal frameshift mutations cause increased binding of RUNX1 to MLL, leading to increased PU.1 expression. In contrast, point mutations in RUNX1's MLL binding domain significantly decrease interaction with MLL, resulting in decreased PU.1 expression (Huang et al. 2011). These results imply that abrogation of the RUNX1:MLL interaction, and subsequent deregulated target gene expression, contributes to MDS and leukemia development in some patients with RUNX1 mutations.

In MLL fusions, the C-terminal, RUNX1 binding domain of MLL is lost. Consequently, most fusion proteins are not expected to directly interact with RUNX1. However, genetic experiments have revealed that RUNX1 still plays an important role in MLL rearranged leukemias. Surprisingly, that role appears to be different in the context of different fusion proteins.

Translocation (11;19)(q23;p13.3) generates a fusion between MLL and the eleven nineteen leukemia protein, ENL (also called MLLT1). In overexpression studies, all MLL-fusion proteins tested induce decreased expression of RUNX1, with MLL-ENL causing the most severe reduction in RUNX1 protein (Zhao et al. 2014). In a mouse model of MLL-ENL leukemia, reduced RUNX1 expression cooperates with the fusion protein during leukemia development. When MLL-ENL is expressed from a retrovirus, mice lacking *Runx1* due to genetic deletion show increased proliferation and accelerated leukemia development as compared to mice expressing MLL-ENL in a *Runx1* sufficient background (Nishimoto et al. 2011). In this model, loss of *Runx1* is associated with decreased expression of cell cycle regulatory genes (p19ARF, p21) and apoptotic genes (p53, Bax), which is reversed by overexpression of *Runx1* (Nishimoto et al. 2011). These results indicate that decreased RUNX1 activity facilitates MLL-ENL induced leukemia initiation, likely due to decreased expression of RUNX1 target genes.

In contrast to its role during leukemia initiation, the role of RUNX1 after leukemic transformation may be different. In MLL-ENL expressing patient leukemia cells transduced with shRNAs, knockdown of RUNX1 causes decreased growth in culture, implying that RUNX1 is required for the maintenance of MLL-ENL leukemia, in contrast to its role during initiation (Goyama et al. 2013). An alternative explanation for the different outcomes observed between Nishimoto et al. and Goyama et al. is that RUNX1 has similar roles in both pre-leuekmic and leukemic cells, but that decreased RUNX1 expression has different consequences than complete loss of the gene (Nishimoto et al. 2011; Goyama et al. 2013).

A similar role for RUNX1 has been reported in leukemia driven by the MLL-AF9 fusion protein, which is the product of the t9;11(p22;q23) translocation. Expression of the MLL-AF9 protein causes a repression of RUNX1 (Goyama et al. 2013; Zhao et al. 2014). Further reduction of RUNX activity in pre-leuekmic MLL-AF9 cells by heterozygous loss of *Runx1* and *Cbfb* alleles (*Runx1*$^{+/-}$, *Cbfb*$^{+/-}$) accelerates leukemia development, while over expression of *Runx1* in pre-leukemic or leukemic cells expressing a MLL-AF9 knockin allele induced terminal differentiation and reduced colony forming ability (Zhao et al. 2014). In contrast, knockdown of RUNX1 or pharmacological inhibition of its activity in cordblood cells transformed by MLL-AF9 expression reduced cell growth (Goyama et al. 2013). Similarly, deletion of both *Runx1* and *Cbfb* in MLL-AF9 leukemia cells also reduced colony forming ability (Goyama et al. 2013). These differences could be due to a dosage effect of RUNX activity: decreased RUNX activity accelerates leukemia development, but below a certain threshold level, MLL-AF9 cells no longer survive, as proposed by Zhao et al. (2014). Alternatively, RUNX1 could have different roles during initiation and maintenance. During initiation, it is possible that decreased RUNX1 induces expansion of the hematopoietic stem and progenitor cell population, resulting in a larger pool of pre-leukemic cells, and increasing the odds of leukemic transformation. However, after transformation RUNX1 activity may be more important for continued survival of the leukemic cells, so loss of RUNX1 results in decreased growth (Zhao et al. 2014; Goyama et al. 2013). Further experimentation will be required to resolve this issue.

Translocation t(4;11)(q21;q23) fuses MLL with the AF4 gene, producing the MLL-AF4 fusion protein, as well as the reciprocal product, AF4-MLL. Experiments in mouse and human cells indicate the expression of both fusion genes is required for leukemia development, indicating that both translocation products have functional roles (Bursen et al. 2004, 2010; Montes et al. 2011).

RUNX1 is more highly expressed in t(4;11) cell lines and patient samples than in those with other MLL rearrangements, implying that RUNX1 expression is uniquely regulated in MLL-AF4 leukemias (Zhao et al. 2014; Wilkinson et al. 2013). Consistent with this observation, chromatin immunoprecipitation (ChIP) and knockdown experiments in cell lines with the t(4;11) rearrangement indicate that RUNX1 is a direct target of the MLL-AF4 protein's transactivation activity (Wilkinson et al. 2013). Interestingly, by ChIP, RUNX1 co-localizes with the AF4-MLL fusion protein. Knockdown experiments show that RUNX1 is required for the expression of their shared target genes, as well as the colony forming ability of t(4;11) cell lines. Collectively, this work supports a model in which MLL-AF4 induces expression of RUNX1, which then forms a complex with AF4-MLL to activate target gene expression, and promote leukemic cell survival (Wilkinson et al. 2013). Based on these observations, RUNX1 appears to be similarly required in t(4;11) leukemias as in other MLL rearranged leukemias, although the mechanism of RUNX1 activity may be unique.

17.4 Perspective

Mutations that reduce RUNX1 activity are common in MDS and AML, including biallelic *RUNX1* mutations in FAB M0 AML cases. On the other hand, AMLs expressing RUNX1-ETO or CBFβ-SMMHC, as well as leukemias expressing MLL-ENL or MLL-AF9, require presence of wild-type RUNX1. Perhaps in RUNX1 mutant cases, RUNX2 and/or RUNX3 provide requisite survival signals, with RUNX1-ETO and CBFβ-SMMHC inhibiting all three RUNX isoforms necessitating residual RUNX1 activity. Maneuvers that increase RUNX1 activity may prove useful therapeutically in RUNX1 mutant

MDS/AML cases, whereas inactivation of the wild-type RUNX1 allele may prove useful in those expressing RUNX1-ETO, CBFβ-SMMHC, or MLL fusion oncoproteins.

References

Adya, N., Stacy, T., Speck, N. A., & Liu, P. P. (1998). The leukemic protein core binding factor beta (CBFbeta)-smooth-muscle myosin heavy chain sequesters CBFalpha2 into cytoskeletal filaments and aggregates. *Molecular and Cellular Biology, 18*, 7432–7443.

Antony-Debre, I., Manchev, V. T., Balayn, N., Bluteau, D., Tomowiak, C., Legrand, C., et al. (2015). Level of RUNX1 activity is critical for leukemic predisposition but not for thrombocytopenia. *Blood, 125*, 930–940.

Ben-Ami, O., Friedman, D., Leshkowitz, D., Goldenberg, D., Orlovsky, K., Pencovich, N., et al. (2013). Addiction of t(8;21) and inv(16) acute myeloid leukemia to native RUNX1. *Cell Reports, 4*, 1131–1143.

Bluteau, D., Glembotsky, A. C., Raimbault, A., Balayn, N., Gilles, L., Rameau, P., et al. (2012). Dysmegakaryopoiesis of FPD/AML pedigrees with constitutional RUNX1 mutations is linked to myosin II deregulated expression. *Blood, 120*, 2708–2718.

Britos-Bray, M., Ramirez, M., Cao, W., Wang, X., Liu, P. P., Civin, C. I., & Friedman, A. D. (1998). CBFbeta-SMMHC, expressed in M4eo acute myeloid leukemia, reduces p53 induction and slows apoptosis in hematopoietic cells exposed to DNA- damaging agents. *Blood, 92*, 4344–4352.

Bursen, A., Moritz, S., Gaussmann, A., Moritz, S., Dingermann, T., & Marschalek, R. (2004). Interaction of AF4 wild-type and AF4.MLL fusion protein with SIAH proteins: Indication for t(4;11) pathobiology? *Oncogene, 23*, 6237–6249.

Bursen, A., Schwabe, K., Ruster, B., Henschler, R., Ruthardt, M., Dingermann, T., & Marschalek, R. (2010). The AF4.MLL fusion protein is capable of inducing ALL in mice without requirement of MLL. AF4. *Blood, 115*, 3570–3579.

Cai, X., Gao, L., Teng, L., Ge, J., Oo, Z. M., Kumar, A. R., et al. (2015). Runx1 deficiency decreases ribosome biogenesis and confers stress resistance to hematopoietic stem and progenitor cells. *Cell Stem Cell, 17*, 165–177.

Cammenga, J., Niebuhr, B., Horn, S., Bergholz, U., Putz, G., Buchholz, F., et al. (2007). RUNX1 DNA-binding mutants, associated with minimally differentiated acute myelogenous leukemia, disrupt myeloid differentiation. *Cancer Research, 67*, 537–545.

Castilla, L. H., Wijmenga, C., Wang, Q., Stacy, T., Speck, N. A., Eckhaus, M., et al. (1996). Failure of embryonic hematopoiesis and lethal hemorrhages in mouse embryos heterozygous for a knocked-in leukemia gene CBFB-MYH11. *Cell, 87*, 687–696.

Castilla, L. H., Garrett, L., Adya, N., Orlic, D., Dutra, A., Anderson, S., et al. (1999). The fusion gene Cbfb-MYH11 blocks myeloid differentiation and predisposes mice to acute myelomonocytic leukaemia [letter]. *Nature Genetics, 23*, 144–146.

Christiansen, D. H., Andersen, M. K., & Pedersen-Bjergaard, J. (2004). Mutations of AML1 are common in therapy-related myelodysplasia following therapy with alkylating agents and are significantly associated with deletion or loss of chromosome arm 7q and with subsequent leukemic transformation. *Blood, 104*, 1474–1481.

Connelly, J. P., Kwon, E. M., Gao, Y., Trivedi, N. S., Elkahloun, A. G., Horwitz, M. S., et al. (2014). Targeted correction of RUNX1 mutation in FPD patient-specific induced pluripotent stem cells rescues megakaryopoietic defects. *Blood, 124*, 1926–1930.

Cooper, S., Guo, H., & Friedman, A. D. (2015). The +37 kb Cebpa enhancer is critical for Cebpa myeloid gene expression and contains functional sites that bind SCL, GATA2, C/EBPalpha, PU.1, and additional Ets factors. *PLoS One, 10*, e0126385.

Cunningham, L., Finckbeiner, S., Hyde, R. K., Southall, N., Marugan, J., Yedavalli, V. R., et al. (2012). Identification of benzodiazepine Ro5-3335 as an inhibitor of CBF leukemia through quantitative high throughput screen against RUNX1-CBFbeta interaction. *Proceedings of the National Academy of Sciences of the United States of America, 109*, 14592–14597.

de Boer, J., Walf-Vorderwulbecke, V., & Williams, O. (2013). In focus: MLL-rearranged leukemia. *Leukemia, 27*, 1224–1228.

De Braekeleer, E., Douet-Guilbert, N., Morel, F., Le Bris, M. J., Ferec, C., & De Braekeleer, M. (2011). RUNX1 translocations and fusion genes in malignant hemopathies. *Future Oncology, 7*, 77–91.

Dicker, F., Haferlach, C., Kern, W., Haferlach, T., & Schnittger, S. (2007). Trisomy 13 is strongly associated with AML1/RUNX1 mutations and increased FLT3 expression in acute myeloid leukemia. *Blood, 110*, 1308–1316.

Ding, Y., Harada, Y., Imagawa, J., Kimura, A., & Harada, H. (2009). AML1/RUNX1 point mutation possibly promotes leukemic transformation in myeloproliferative neoplasms. *Blood, 114*, 5201–5205.

Dohner, K., Tobis, K., Bischof, T., Hein, S., Schlenk, R. F., Frohling, S., & Dohner, H. (2003). Mutation analysis of the transcription factor PU.1 in younger adults (16 to 60 years) with acute myeloid leukemia: a study of the AML Study Group Ulm (AMLSG ULM). *Blood, 102*, 3850; author reply 3850–1.

Erickson, P., Gao, J., Chang, K. S., Look, T., Whisenant, E., Raimondi, S., et al. (1992). Identification of breakpoints in t(8;21) acute myelogenous leukemia and isolation of a fusion transcript, AML1/ETO, with similarity to Drosophila segmentation gene, runt. *Blood, 80*, 1825–1831.

Friedman, A. D. (2007). Transcriptional control of granulocyte and monocyte development. *Oncogene, 26*, 6816–6828.

Friedman, A. D. (2009). Cell cycle and developmental control of hematopoiesis by Runx1. *Journal of Cellular Physiology, 219*, 520–524.

Friedman, A. D. (2015). C/EBPalpha in normal and malignant myelopoiesis. *International Journal of Hematology, 101*, 330–341.

Gaidzik, V. I., Bullinger, L., Schlenk, R. F., Zimmermann, A. S., Rock, J., Paschka, P., et al. (2011). RUNX1 mutations in acute myeloid leukemia: Results from a comprehensive genetic and clinical analysis from the AML study group. *Journal of Clinical Oncology, 29*, 1364–1372.

Ganly, P., Walker, L. C., & Morris, C. M. (2004). Familial mutations of the transcription factor RUNX1 (AML1, CBFA2) predispose to acute myeloid leukemia. *Leukemia & Lymphoma, 45*, 1–10.

Gardini, A., Cesaroni, M., Luzi, L., Okumura, A. J., Biggs, J. R., Minardi, S. P., et al. (2008). AML1/ETO oncoprotein is directed to AML1 binding regions and co-localizes with AML1 and HEB on its targets. *PLoS Genetics, 4*, e1000275.

Gilliland, D. G., & Griffin, J. D. (2002). The roles of FLT3 in hematopoiesis and leukemia. *Blood, 100*, 1532–1542.

Goyama, S., Schibler, J., Cunningham, L., Zhang, Y., Rao, Y., Nishimoto, N., et al. (2013). Transcription factor RUNX1 promotes survival of acute myeloid leukemia cells. *The Journal of Clinical Investigation, 123*, 3876–3388.

Grossmann, V., Kohlmann, A., Zenger, M., Schindela, S., Eder, C., Weissmann, S., et al. (2011). A deep-sequencing study of chronic myeloid leukemia patients in blast crisis (BC-CML) detects mutations in 76.9% of cases. *Leukemia, 25*, 557–560.

Grossmann, V., Bacher, U., Kohlmann, A., Butschalowski, K., Roller, A., Jeromin, S., et al. (2012). Expression of CEBPA is reduced in RUNX1-mutated acute myeloid leukemia. *Blood Cancer Journal, 2*, e86.

Guo, H., Ma, O., Speck, N. A., & Friedman, A. D. (2012). Runx1 deletion or dominant inhibition reduces Cebpa transcription via conserved promoter and distal enhancer sites to favor monopoiesis over granulopoiesis. *Blood, 119*, 4408–4418.

Guo, H., Ma, O., & Friedman, A. D. (2014). The Cebpa +37-kb enhancer directs transgene expression to myeloid progenitors and to long-term hematopoietic stem cells. *Journal of Leukocyte Biology, 96*, 419–426.

Guo, H., Cooper, S., & Friedman, A. D. (2016). In vivo deletion of the Cebpa +37 kb enhancer markedly reduces Cebpa mRNA in myeloid progenitors but not in non-hematopoietic tissues to impair granulopoiesis. *PloS One, 11*, e0150809.

Hanson, R. D., Hess, J. L., Yu, B. D., Ernst, P., van Lohuizen, M., Berns, A., et al. (1999). Mammalian trithorax and polycomb-group homologues are antagonistic regulators of homeotic development. *Proceedings of the National Academy of Sciences of the United States of America, 96*, 14372–14377.

Harada, H., Harada, Y., Tanaka, H., Kimura, A., & Inaba, T. (2003). Implications of somatic mutations in the AML1 gene in radiation-associated and therapy-related myelodysplastic syndrome/acute myeloid leukemia. *Blood, 101*, 673–680.

Harada, H., Harada, Y., Niimi, H., Kyo, T., Kimura, A., & Inaba, T. (2004). High incidence of somatic mutations in the AML1/RUNX1 gene in myelodysplastic syndrome and low blast percentage myeloid leukemia with myelodysplasia. *Blood, 103*, 2316–2324.

Harada, Y., Inoue, D., Ding, Y., Imagawa, J., Doki, N., Matsui, H., et al. (2013). RUNX1/AML1 mutant collaborates with BMI1 overexpression in the development of human and murine myelodysplastic syndromes. *Blood, 121*, 3434–3446.

Heilman, S. A., Kuo, Y. H., Goudswaard, C. S., Valk, P. J., & Castilla, L. H. (2006). Cbfbeta reduces Cbfbeta-SMMHC-associated acute myeloid leukemia in mice. *Cancer Research, 66*, 11214–11218.

Hess, J. L., Yu, B. D., Li, B., Hanson, R., & Korsmeyer, S. J. (1997). Defects in yolk sac hematopoiesis in Mll-null embryos. *Blood, 90*, 1799–1806.

Huang, G., Zhang, P., Hirai, H., Elf, S., Yan, X., Chen, Z., et al. (2008). PU.1 is a major downstream target of AML1 (RUNX1) in adult mouse hematopoiesis. *Nature Genetics, 40*, 51–60.

Huang, G., Zhao, X., Wang, L., Elf, S., Xu, H., Zhao, X., et al. (2011). The ability of MLL to bind RUNX1 and methylate H3K4 at PU.1 regulatory regions is impaired by MDS/AML-associated RUNX1/AML1 mutations. *Blood, 118*, 6544–6552.

Hyde, R. K., & Liu, P. P. (2010). RUNX1 repression-independent mechanisms of leukemogenesis by fusion genes CBFB-MYH11 and AML1-ETO (RUNX1-RUNX1T1). *Journal of Cellular Biochemistry, 110*, 1039–1045.

Hyde, R. K., Zhao, L., Alemu, L., & Liu, P. P. (2015). Runx1 is required for hematopoietic defects and leukemogenesis in Cbfb-MYH11 knock-in mice. *Leukemia, 29*, 1771–1778.

Imai, Y., Kurokawa, M., Izutsu, K., Hangaishi, A., Takeuchi, K., Maki, K., et al. (2000). Mutations of the AML1 gene in myelodysplastic syndrome and their functional implications in leukemogenesis. *Blood, 96*, 3154–3160.

Kamikubo, Y., Zhao, L., Wunderlich, M., Corpora, T., Hyde, R. K., Paul, T. A., et al. (2010). Accelerated leukemogenesis by truncated CBF beta-SMMHC defective in high-affinity binding with RUNX1. *Cancer Cell, 17*, 455–468.

Kanno, T., Kanno, Y., Chen, L. F., Ogawa, E., Kim, W. Y., & Ito, Y. (1998). Intrinsic transcriptional activation-inhibition domains of the polyomavirus enhancer binding protein 2/core binding factor alpha subunit revealed in the presence of the beta subunit. *Molecular and Cellular Biology, 18*, 2444–2454.

Kuo Y. H., Landrette, S. F., Heilman, S. A., Perrat, P. N., Garrett, L, Liu, P. P., Le Beau, M. M., Kogan, S.C., & Castilla, L. H. (2006). Cbf beta-SMMHC induces distinct abnormal myeloid progenitors able to develop acute myeloid leukemia. *Cancer Cell, 1*, 57–68.

Kuo, M. C., Liang, D. C., Huang, C. F., Shih, Y. S., Wu, J. H., Lin, T. L., & Shih, L. Y. (2009). RUNX1 mutations are frequent in chronic myelomonocytic leukemia and mutations at the C-terminal region might predict acute myeloid leukemia transformation. *Leukemia, 23*, 1426–1431.

Kwok, C., Zeisig, B. B., Qiu, J., Dong, S., & So, C. W. (2009). Transforming activity of AML1-ETO is independent of CBFbeta and ETO interaction but requires formation of homo-oligomeric complexes. *Proceedings of the National Academy of Sciences of the United States of America, 106*, 2853–2858.

Kwok, C., Zeisig, B. B., Dong, S., & So, C. W. (2010). The role of CBFbeta in AML1-ETO's activity. *Blood, 115*, 3176–3177.

Lam, K., & Zhang, D. E. (2012). RUNX1 and RUNX1-ETO: Roles in hematopoiesis and leukemogenesis. *Frontiers in Bioscience (Landmark Ed), 17*, 1120–1139.

Le Beau, M. M., Larson, R. A., Bitter, M. A., Vardiman, J. W., Golomb, H. M., & Rowley, J. D. (1983). Association of an inversion of chromosome 16 with abnormal marrow eosinophils in acute myelomonocytic leukemia. A unique cytogenetic-clinicopathological association. *The New England Journal of Medicine, 309*, 630–636.

Leroy, H., Roumier, C., Grardel-Duflos, N., Macintyre, E., Lepelley, P., Fenaux, P., & Preudhomme, C. (2002). Unlike AML1, CBFbeta gene is not deregulated by point mutations in acute myeloid leukemia and in myelodysplastic syndromes. *Blood, 99*, 3848–3550.

Li, Y., Wang, H., Wang, X., Jin, W., Tan, Y., Fang, H., et al. (2016). Genome-wide studies identify a novel interplay between AML1 and AML1/ETO in t(8;21) acute myeloid leukemia. *Blood, 127*, 233–242.

Liu, P., Tarl, E. S. A., Hajra, A., Claxton, D. F., Marlton, P., Freedman, M., et al. (1993). Fusion between transcription factor CBF beta/PEBP2 beta and a myosin heavy chain in acute myeloid leukemia. *Science, 261*, 1041–1044.

Liu, P. P., Wijmenga, C., Hajra, A., Blake, T. B., Kelley, C. A., Adelstein, R. S., et al. (1996). Identification of the chimeric protein product of the CBFB-MYH11 fusion gene in inv(16) leukemia cells. *Genes, Chromosomes & Cancer, 16*, 77–87.

Lukasik, S. M., Zhang, L., Corpora, T., Tomanicek, S., Li, Y., Kundu, M., et al. (2002). Altered affinity of CBF beta-SMMHC for Runx1 explains its role in leukemogenesis. *Nature Structural Biology, 9*, 674–679.

Lutterbach, B., Hou, Y., Durst, K. L., & Hiebert, S. W. (1999). The inv(16) encodes an acute myeloid leukemia 1 transcriptional corepressor. *Proceedings of the National Academy of Sciences of the United States of America, 96*, 12822–12827.

Ma, O., Hong, S., Guo, H., Ghiaur, G., & Friedman, A. D. (2014). Granulopoiesis requires increased C/EBPalpha compared to monopoiesis, correlated with elevated Cebpa in immature G-CSF receptor versus M-CSF receptor expressing cells. *PloS One, 9*, e95784.

Mandoli, A., Singh, A. A., Jansen, P. W., Wierenga, A. T., Riahi, H., Franci, G., et al. (2013). CBFB-MYH11/RUNX1 together with a compendium of hematopoietic regulators, chromatin modifiers and basal transcription factors occupies self-renewal genes in inv(16) acute myeloid leukemia. *Leukemia, 228*, 770–778.

Matheny, C. J., Speck, M. E., Cushing, P. R., Zhou, Y., Corpora, T., Regan, M., et al. (2007). Disease mutations in RUNX1 and RUNX2 create nonfunctional, dominant-negative, or hypomorphic alleles. *The EMBO Journal, 26*, 1163–1175.

Michaud, J., Wu, F., Osato, M., Cottles, G. M., Yanagida, M., Asou, N., et al. (2002). In vitro analyses of known and novel RUNX1/AML1 mutations in dominant familial platelet disorder with predisposition to acute myelogenous leukemia: Implications for mechanisms of pathogenesis. *Blood, 99*, 1364–1372.

Montes, R., Ayllon, V., Gutierrez-Aranda, I., Prat, I., Hernandez-Lamas, M. C., Ponce, L., et al. (2011). Enforced expression of MLL-AF4 fusion in cord blood CD34+ cells enhances the hematopoietic repopulating cell function and clonogenic potential but is not sufficient to initiate leukemia. *Blood, 117*, 4746–4758.

Muntean, A. G., & Hess, J. L. (2012). The pathogenesis of mixed-lineage leukemia. *Annual Review of Pathology, 7*, 283–301.

Neff, T., & Armstrong, S. A. (2013). Recent progress toward epigenetic therapies: The example of mixed lineage leukemia. *Blood, 121*, 4847–4853.

Niki, M., Okada, H., Takano, H., Kuno, J., Tani, K., Hibino, H., et al. (1997). Hematopoiesis in the fetal liver is impaired by targeted mutagenesis of a gene encoding a non-DNA binding subunit of the transcription factor, polyomavirus enhancer binding protein 2/core binding factor. *Proceedings of the National Academy of Sciences of the United States of America, 94*, 5697–5702.

Nishimoto, N., Arai, S., Ichikawa, M., Nakagawa, M., Goyama, S., Kumano, K., et al. (2011). Loss of AML1/Runx1 accelerates the development of MLL-ENL leukemia through down-regulation of p19ARF. *Blood, 118*, 2541–2550.

Okuda, T., van Deursen, J., Hiebert, S. W., Grosveld, G., & Downing, J. R. (1996). AML1, the target of multiple chromosomal translocations in human leukemia, is essential for normal fetal liver hematopoiesis. *Cell, 84*, 321–330.

Okumura, A. J., Peterson, L. F., Okumura, F., Boyapati, A., & Zhang, D. E. (2008). t(8;21)(q22;q22) fusion proteins preferentially bind to duplicated AML1/RUNX1 DNA-binding sequences to differentially regulate gene expression. *Blood, 112*, 1392–1401.

Osato, M. (2004). Point mutations in the RUNX1/AML1 gene: Another actor in RUNX leukemia. *Oncogene, 23*, 4284–4296.

Osato, M., Asou, N., Abdalla, E., Hoshino, K., Yamasaki, H., Okubo, T., et al. (1999). Biallelic and heterozygous point mutations in the runt domain of the AML1/PEBP2alphaB gene associated with myeloblastic leukemias. *Blood, 93*, 1817–1824.

Pabst, T., Mueller, B. U., Harakawa, N., Schoch, C., Haferlach, T., Behre, G., et al. (2001). AML1-ETO downregulates the granulocytic differentiation factor C/EBPalpha in t(8;21) myeloid leukemia. *Nature Medicine, 7*, 444–451.

Park, S., Speck, N. A., & Bushweller, J. H. (2009). The role of CBFbeta in AML1-ETO's activity. *Blood, 114*, 2849–2850.

Patel, J. P., Gonen, M., Figueroa, M. E., Fernandez, H., Sun, Z., Racevskis, J., et al. (2012). Prognostic relevance of integrated genetic profiling in acute myeloid leukemia. *The New England Journal of Medicine, 366*, 1079–1089.

Penther, D., Preudhomme, C., Talmant, P., Roumier, C., Godon, A., Mechinaud, F., et al. (2002). Amplification of AML1 gene is present in childhood acute lymphoblastic leukemia but not in adult, and is not associated with AML1 gene mutation. *Leukemia, 16*, 1131–1134.

Peterson, L. F., & Zhang, D. E. (2004). The 8;21 translocation in leukemogenesis. *Oncogene, 23*, 4255–4262.

Preudhomme, C., Warot-Loze, D., Roumier, C., Grardel-Duflos, N., Garand, R., Lai, J. L., et al. (2000). High incidence of biallelic point mutations in the Runt domain of the AML1/PEBP2 alpha B gene in Mo acute myeloid leukemia and in myeloid malignancies with acquired trisomy 21. *Blood, 96*, 2862–2869.

Preudhomme, C., Renneville, A., Bourdon, V., Philippe, N., Roche-Lestienne, C., Boissel, N., et al. (2009). High frequency of RUNX1 biallelic alteration in acute myeloid leukemia secondary to familial platelet disorder. *Blood, 113*, 5583–5587.

Ptasinska, A., Assi, S. A., Mannari, D., James, S. R., Williamson, D., Dunne, J., et al. (2012). Depletion of RUNX1/ETO in t(8;21) AML cells leads to genome-wide changes in chromatin structure and transcription factor binding. *Leukemia, 26*, 1829–1841.

Ptasinska, A., Assi, S. A., Martinez-Soria, N., Imperato, M. R., Piper, J., Cauchy, P., et al. (2014). Identification of a dynamic core transcriptional network in t(8;21) AML that regulates differentiation block and self-renewal. *Cell Reports, 8*, 1974–1988.

Quentin, S., Cuccuini, W., Ceccaldi, R., Nibourel, O., Pondarre, C., Pages, M. P., et al. (2011). Myelodysplasia and leukemia of Fanconi anemia are associated with a specific pattern of genomic abnormalities that includes cryptic RUNX1/AML1 lesions. *Blood, 117*, e161–e170.

Rosenbauer, F., Wagner, K., Kutok, J. L., Iwasaki, H., Le Beau, M. M., Okuno, Y., et al. (2004). Acute myeloid leukemia induced by graded reduction of a lineage-specific transcription factor, PU.1. *Nature Genetics, 36*, 624–630.

Roudaia, L., Cheney, M. D., Manuylova, E., Chen, W., Morrow, M., Park, S., et al. (2009). CBFbeta is critical for AML1-ETO and TEL-AML1 activity. *Blood, 113*, 3070–3079.

Rowley, J. D. (1973). Identificaton of a translocation with quinacrine fluorescence in a patient with acute leukemia. *Annales de Génétique, 16*, 109–112.

Sakurai, M., Kunimoto, H., Watanabe, N., Fukuchi, Y., Yuasa, S., Yamazaki, S., et al. (2014). Impaired hematopoietic differentiation of RUNX1-mutated induced pluripotent stem cells derived from FPD/AML patients. *Leukemia, 28*, 2344–2354.

Sasaki, K., Yagi, H., Bronson, R. T., Tominaga, K., Matsunashi, T., Deguchi, K., et al. (1996). Absence of fetal liver hematopoiesis in mice deficient in transcriptional coactivator core binding factor beta. *Proceedings of the National Academy of Sciences of the United States of America, 93*, 12359–12363.

Schnittger, S., Dicker, F., Kern, W., Wendland, N., Sundermann, J., Alpermann, T., et al. (2011). RUNX1 mutations are frequent in de novo AML with noncomplex karyotype and confer an unfavorable prognosis. *Blood, 117*, 2348–2357.

Shigesada, K., van de Sluis, B., & Liu, P. P. (2004). Mechanism of leukemogenesis by the inv(16) chimeric gene CBFB/PEBP2B-MHY11. *Oncogene, 23*, 4297–4307.

Shimada, H., Ichikawa, H., Nakamura, S., Katsu, R., Iwasa, M., Kitabayashi, I., & Ohki, M. (2000). Analysis of genes under the downstream control of the t(8;21) fusion protein AML1-MTG8: Overexpression of the TIS11b (ERF-1, cMG1) gene induces myeloid cell proliferation in response to G-CSF. *Blood, 96*, 655–663.

Silva, F. P., Swagemakers, S. M., Erpelinck-Verschueren, C., Wouters, B. J., Delwel, R., Vrieling, H., et al. (2009). Gene expression profiling of minimally differentiated acute myeloid leukemia: M0 is a distinct entity subdivided by RUNX1 mutation status. *Blood, 114*, 3001–3007.

Skokowa, J., Steinemann, D., Katsman-Kuipers, J. E., Zeidler, C., Klimenkova, O., Klimiankou, M., et al. (2014). Cooperativity of RUNX1 and CSF3R mutations in severe congenital neutropenia: A unique pathway in myeloid leukemogenesis. *Blood, 123*, 2229–2237.

Song, W. J., Sullivan, M. G., Legare, R. D., Hutchings, S., Tan, X., Kufrin, D., et al. (1999). Haploinsufficiency of CBFA2 causes familial thrombocytopenia with propensity to develop acute myelogenous leukaemia. *Nature Genetics, 23*, 166–175.

Sun, W., & Downing, J. R. (2004). Haploinsufficiency of AML1 results in a decrease in the number of LTR-HSCs while simultaneously inducing an increase in more mature progenitors. *Blood, 104*, 3565–3572.

Sun, X. J., Wang, Z., Wang, L., Jiang, Y., Kost, N., Soong, T. D., et al. (2013). A stable transcription factor complex nucleated by oligomeric AML1-ETO controls leukaemogenesis. *Nature, 500*, 93–97.

Tang, J. L., Hou, H. A., Chen, C. Y., Liu, C. Y., Chou, W. C., Tseng, M. H., et al. (2009). AML1/RUNX1 mutations in 470 adult patients with de novo acute myeloid leukemia: prognostic implication and interaction with other gene alterations. *Blood, 114*, 5352–5361.

Trombly, D. J., Whitfield, T. W., Padmanabhan, S., Gordon, J. A., Lian, J. B., van Wijnen, A. J., et al. (2015). Genome-wide co-occupancy of AML1-ETO and N-CoR defines the t(8;21) AML signature in leukemic cells. *BMC Genomics, 16*, 309.

Tsai, S. C., Shih, L. Y., Liang, S. T., Huang, Y. J., Kuo, M. C., Huang, C. F., et al. (2015). Biological activities of RUNX1 mutants predict secondary acute leukemia transformation from chronic myelomonocytic leukemia and myelodysplastic syndromes. *Clinical Cancer Research, 21*, 3541–3551.

Wang, Q., Stacy, T., Binder, M., Marin-Padilla, M., Sharpe, A. H., & Speck, N. A. (1996a). Disruption of the Cbfa2 gene causes necrosis and hemorrhaging in the central nervous system and blocks definitive hematopoiesis. *Proceedings of the National Academy of Sciences of the United States of America, 93*, 3444–3449.

Wang, Q., Stacy, T., Miller, J. D., Lewis, A. F., Gu, T. L., Huang, X., et al. (1996b). The CBFbeta subunit is essential for CBFalpha2 (AML1) function in vivo. *Cell, 87*, 697–708.

Wang, C. Q., Krishnan, V., Tay, L. S., Chin, D. W., Koh, C. P., Chooi, J. Y., et al. (2014). Disruption of Runx1 and Runx3 leads to bone marrow failure and leukemia predisposition due to transcriptional and DNA repair defects. *Cell Reports, 8*, 767–782.

Watanabe-Okochi, N., Kitaura, J., Ono, R., Harada, H., Harada, Y., Komeno, Y., et al. (2008). AML1 mutations induced MDS and MDS/AML in a mouse BMT model. *Blood, 111*, 4297–4308.

West, A. H., Godley, L. A., & Churpek, J. E. (2014). Familial myelodysplastic syndrome/acute leukemia syndromes: A review and utility for translational investigations. *Annals of the New York Academy of Sciences, 1310*, 111–118.

Wheeler, J. C., Vanderzwan, C., Xu, X., Swantek, D., Tracey, W. D., & Gergen, J. P. (2002). Distinct in vivo requirements for establishment versus maintenance of transcriptional repression. *Nature Genetics, 32*, 206–210.

Wijmenga, C., Gregory, P. E., Hajra, A., Schrock, E., Ried, T., Eils, R., et al. (1996). Core binding factor beta-smooth muscle myosin heavy chain chimeric protein involved in acute myeloid leukemia forms unusual nuclear rod-like structures in transformed NIH 3 T3 cells. *Proceedings of the National Academy of Sciences of the United States of America, 93*, 1630–1635.

Wilkinson, A. C., Ballabio, E., Geng, H., North, P., Tapia, M., Kerry, J., et al. (2013). RUNX1 is a key target in t(4;11) leukemias that contributes to gene activation through an AF4-MLL complex interaction. *Cell Reports, 3*, 116–127.

Yagi, H., Deguchi, K., Aono, A., Tani, Y., Kishimoto, T., & Komori, T. (1998). Growth disturbance in fetal liver hematopoiesis of Mll-mutant mice. *Blood, 92*, 108–117.

Yan, M., Ahn, E. Y., Hiebert, S. W., & Zhang, D. E. (2009). RUNX1/AML1 DNA-binding domain and ETO/MTG8 NHR2-dimerization domain are critical to AML1-ETO9a leukemogenesis. *Blood, 113*, 883–886.

Yeamans, C., Wang, D., Paz-Priel, I., Torbett, B. E., Tenen, D. G., & Friedman, A. D. (2007). C/EBPalpha binds and activates the PU.1 distal enhancer to induce monocyte lineage commitment. *Blood, 110*, 3136–3142.

Zhao, L. J., Wang, Y. Y., Li, G., Ma, L. Y., Xiong, S. M., Weng, X. Q., et al. (2012). Functional features of RUNX1 mutants in acute transformation of chronic myeloid leukemia and their contribution to inducing murine full-blown leukemia. *Blood, 119*, 2873–2882.

Zhao, X., Chen, A., Yan, X., Zhang, Y., He, F., Hayashi, Y., et al. (2014). Downregulation of RUNX1/CBFbeta by MLL fusion proteins enhances hematopoietic stem cell self-renewal. *Blood, 123*, 1729–1738.

Roles of RUNX in B Cell Immortalisation

18

Michelle J. West and Paul J. Farrell

Abstract

RUNX1 and *RUNX3* are the main *RUNX* genes expressed in B lymphocytes. Both are expressed throughout B-cell development and play key roles at certain key developmental transitions. The tumour-associated Epstein-Barr virus (EBV) has potent B-cell transforming ability and manipulates *RUNX3* and *RUNX1* transcription through novel mechanisms to control B cell growth. In contrast to resting mature B cells where *RUNX1* expression is high, in EBV-infected cells *RUNX1* levels are low and *RUNX3* levels are high. Downregulation of *RUNX1* in these cells results from cross-regulation by RUNX3 and serves to relieve RUNX1-mediated growth repression. *RUNX3* is upregulated by the EBV transcription factor (TF) EBNA2 and represses *RUNX1* transcription through RUNX sites in the *RUNX1* P1 promoter. Recent analysis revealed that EBNA2 activates *RUNX3* transcription through an 18 kb upstream super-enhancer in a manner dependent on the EBNA2 and Notch DNA-binding partner RBP-J. This super-enhancer also directs *RUNX3* activation by two further RBP-J-associated EBV TFs, EBNA3B and 3C. Counter-intuitively, EBNA2 also hijacks RBP-J to target a super-enhancer region upstream of *RUNX1* to maintain some *RUNX1* expression in certain cell backgrounds, although the dual functioning EBNA3B and 3C proteins limit this activation. Interestingly, the B-cell genome binding sites of EBV TFs overlap extensively with RUNX3 binding sites and show enrichment for RUNX

M.J. West (✉)
School of Life Sciences, University of Sussex, Falmer, Brighton BN1 9QG, UK
e-mail: m.j.west@sussex.ac.uk

P.J. Farrell
Section of Virology, Faculty of Medicine, Imperial College London, Norfolk Place, London W2 1PG, UK

Y. Groner et al. (eds.), *RUNX Proteins in Development and Cancer*, Advances in Experimental Medicine and Biology 962, DOI 10.1007/978-981-10-3233-2_18

motifs. Therefore in addition to B-cell growth manipulation through the long-range control of *RUNX* transcription, EBV may also use RUNX proteins as co-factors to deregulate the transcription of many B cell genes during immortalisation.

Keywords
RUNX1 • RUNX3 • Enhancer • Transcription • Super-enhancer • Epstein-Barr virus • EBNA2 • EBNA3B • EBNA3C • Notch • RBP-J

18.1 Role of *RUNX* in B Cell Development

Both *RUNX3* and *RUNX1* are expressed throughout the haemopoietic system including B lymphocytes (Levanon et al. 1994; Bangsow et al. 2001; North et al. 2004) and their tightly regulated expression plays an important role in defining key developmental transitions (Whiteman and Farrell 2006; Imperato et al. 2015). For example, *RUNX1* upregulation is crucial for the endothelial-to-hematopoietic transition (Chen et al. 2009) and *RUNX3* upregulation by TGF-beta1 drives class switching to IgA in splenic B cells through the activation of Ig alpha gene transcription (Shi and Stavnezer 1998).

Studies in mice have confirmed a role for *Runx3* in IgA class-switching (Fainaru et al. 2004) and delineated the role of *Runx1* in embryonic and adult haematopoiesis (Imperato et al. 2015). Homozygous disruption of *Runx1* leads to embryonic lethally and a block to definitive haematopoiesis (Wang et al. 1996) and embryos from *Runx1* knock-out mice lack both myeloid and erythroid cells in their yolk sacs and liver (Okuda et al. 1996). Numerous studies of the role of *Runx1* in adult blood cell development using knock-out mice have led to the conclusion that loss of *Runx1* does not disrupt haematopoiesis completely (as haematopoietic stem cells are maintained), but impairs the development of particular cell lineages. The bone marrow of adult mice lacking functional *Runx1* showed impaired megakaryocytic maturation and defective T- and B-lymphocyte development (Ichikawa et al. 2004). A further similar study documented a reduction in the percentage of mature B cells ($CD19^+B220^+$) in the bone-marrow, with an almost complete lack of pre-B and pro-B cell precursors (Growney et al. 2005). A reduction in the number of mature B cells expressing IgM and the B-cell marker B220 in the spleens of *Runx1* knock-out adult mice, with no effect on the number of immature B cells was also reported, consistent with an early stage block in B-cell development (Putz et al. 2006). Taken together these studies indicate that *Runx1* plays a key role in B-cell maturation.

Consistent with a role for *Runx1* in B cell development, expression of a deregulated *Runx1* transgene in mice results in increased cell survival and an increase in the number of IgM+ and CD19+ cells in the bone marrow (Blyth et al. 2009). Although *Runx1* expression in foetal liver cells *in vitro* led to reduced total cell and B cell survival, an increased proportion of the cells that survived were B220+. These data suggest that, whilst the survival effects of *Runx1* may be context dependent, its expression induces a drive towards the B-cell lineage.

18.2 Cross Regulation of *RUNX1* by *RUNX3* in B Cells

The structure of the mammalian *RUNX* genes is highly conserved and their transcription is controlled from two distinct promoters, P1 (distal) and P2 (proximal), that give rise to different protein isoforms (Figs. 18.1 and 18.2). These promoters are differentially active in different cell-types and at different stages of differentiation (Bangsow et al. 2001; Levanon et al. 1996). In resting mature B cells *RUNX1* transcripts are

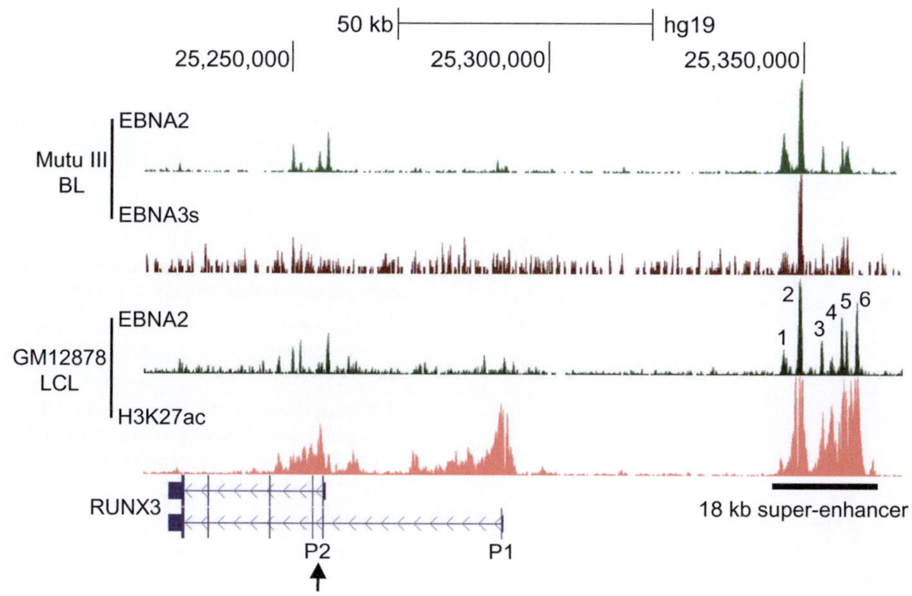

Fig. 18.1 Binding of EBV transcription factors at human *RUNX3* on chromosome 1. ChIP-sequencing reads for EBNA2 and EBNA3 proteins (3A, 3B and 3C) in Mutu III BL cells and the GM12878 LCL and H3K27ac signals in GM12878 from ENCODE. *RUNX3* runs right to left in the human genome. The *arrow* indicates the promoter active in B cells. Numbering indicates the major EBNA2 binding sites in the 18 kb upstream super-enhancer. The super-enhancer is bound by EBNA2 and is active in both cell lines

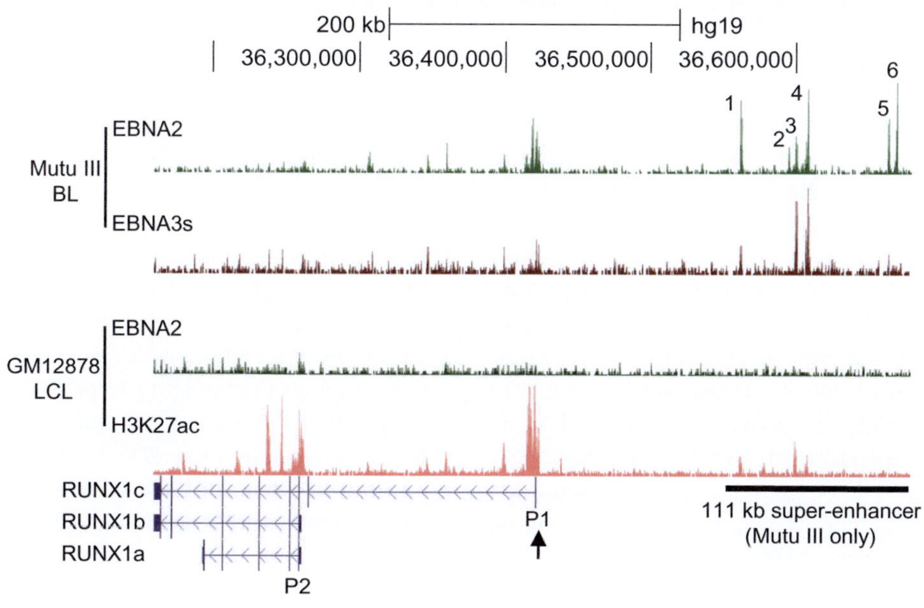

Fig. 18.2 Binding of EBV transcription factors at human *RUNX1* on chromosome 21. The P1 and P2 promoters and the RUNX1a, RUNX1b and RUNX1c isoforms are shown. Labelling as in Fig. 18.1. Numbering indicates the major EBNA2 binding sites in the 111 kb upstream super-enhancer, which is bound by EBNA2 and is active in Mutu III BL cells but not GM12878 cells

Table 18.1 EBV latent proteins and their roles and expression in EBV immortalised B cells

EBV latent protein	Essential for immortalisation	latency I expression	latency III expression	Function
EBNA1	+	+	+	Viral genome maintenance and segregation, latent replication, transcription activation.
EBNA2	+	–	+	Transcriptional activation of all viral latent genes and many cell genes. Binds RBP-J and PU.1.
EBNA3A	+	–	+	Transcriptional activation and repression. Competes with EBNA2 to bind RBP-J. Polycomb repressor complex-mediated cell gene silencing.
EBNA3B	–	–	+	Transcriptional activation and repression. Competes with EBNA2 to bind RBP-J. Tumour suppressor.
EBNA3C	+	–	+	Transcriptional activation and repression. Competes with EBNA2 to bind RBP-J. Polycomb repressor complex-mediated cell gene silencing. Deregulates the cell-cycle.
EBNA-LP	+/–	–	+	Coactivates genes with EBNA2.
LMP1	+	–	+	Transforming oncogene. Constitutively active CD40-like receptor. Activates NF-κB, JNK and p38 signalling pathways. Enhances cell survival and growth.
LMP2A	–	–	+	BCR mimic and decoy. Activates PI3-K/AKT. Regulates autophagy and blocks viral reactivation.
LMP2B	–	–	+	Blocks viral reactivation. Impairs activation of LMP2A.

Epstein-Barr nuclear antigens (EBNAs) are found in the nucleus and Latent membrane proteins (LMP) are found in the cell membrane. Genes essential for transformation of peripheral resting B cells are indicated (+). EBNA-LP (leader protein) plays an accessory role (+/–). Burkitt's lymphoma tumour cells display a latency I gene expression pattern but drift in culture to the latency III pattern seen in initially transformed cells and LCLs. For more information on latent gene function see (Kempkes and Robertson 2015; Cen and Longnecker 2015)

derived from the P1 promoter, leading to high-level expression of the RUNX1c isoform. RUNX1c expression drops dramatically when B cells are immortalised though latent infection with the tumour-associated herpesvirus EBV (Spender et al. 2002, 2005). In EBV-immortalised B cells, RUNX1c downregulation occurs concomitantly with the upregulation of *RUNX3* transcription from the P2 promoter (Spender et al. 2002). *RUNX3* is activated by the EBV TF EBNA2 (Spender et al. 2002), the master regulator of latent EBV gene transcription and a key inducer of hundreds of host genes including the growth regulators *MYC*, *CD23* and *FGR* (Kaiser et al. 1999; Wang et al. 1987; Knutson 1990; Spender et al. 2002; Maier et al. 2006; Zhao et al. 2006) (Table 18.1). Infection of resting B cells with EBV drives cell proliferation and this depends on the initial and continued expression of EBNA2, indicating a critical requirement for cell gene activation by EBNA2.

The differential expression of *RUNX1* and *RUNX3* distinguished between B-cell lines displaying different patterns of EBV latent gene expression (known as latency 1 and latency III). Only those latency III cell-lines expressing EBNA2 (in association with the other 8 EBV latent proteins) (Table 18.1) were characterised by high *RUNX3* expression and low *RUNX1* expression (Spender et al. 2002). Latency I cell-lines derived from Burkitt's lymphoma tumour biopsies, that express only the EBV genome maintenance and replication factor EBNA-1, maintained high RUNX1c expression and lacked detectable *RUNX3* expression. A similar induction of *RUNX3* and consequent repression of *RUNX1* occurs when B cells are activated by phorbol ester treatment and an inverse correlation between *Runx3* and *Runx1* expression is observed in the I.29μ mouse B-cell line upon *Runx3* induction by TGF-β (Shi and Stavnezer 1998).

Given this inverse correlation between *RUNX3* and *RUNX1* expression in B cells and the role of RUNX proteins as activators and repressors, a plausible explanation for the downregulation of *RUNX1* was found in direct repression by *RUNX3*. Following the identification of two adjacent RUNX motifs located between positions −2 and +12 in the *RUNX1* P1 promoter, mutagenesis revealed that *RUNX1* transcription was indeed repressed by RUNX3 through these sites (Spender et al. 2005). The ability of the RUNX proteins to activate or repress gene expression is determined by the higher complexes they form with activators or repressors of transcription in a gene- and context-specific manner (Durst and Hiebert 2004). Mammalian RUNX proteins contain both a transactivation region and a highly-conserved repression domain near the C-terminus, the VWRPY sequence. This VWRPY sequence recruits TLE family proteins, which mediate repression of RUNX target genes through their association with histone deacetylases (Imai et al. 1998; Levanon et al. 1998; McLarren et al. 2001; Wang et al. 2004). The repression of *RUNX1* by RUNX3 in B cells was shown to be dependent on the VWRPY sequence, since its removal prevented RUNX3-mediated repression without affecting RUNX3 binding to the *RUNX1* promoter (Brady et al. 2009). TLE3 was implicated as the co-repressor involved, since it was the only TLE family member expressed in B cell-lines where cross-repression was observed.

18.3 The Regulation of B Cell Growth by *RUNX1*

Given the role of EBNA2 in driving B cell growth, the upregulation of *RUNX3* by EBNA2 could represent an important step in B-cell immortalisation by EBV. Because of the interdependence of *RUNX1* and *RUNX3* levels, a key question to address was whether altered levels of RUNX1, RUNX3 or both proteins contributed to cell proliferation. The dependence of *RUNX1* repression on RUNX3 was confirmed under normal physiological conditions through siRNA depletion of endogenous RUNX3 in EBV transformed lymphoblastoid cell-lines (LCLs) (Spender et al. 2005). In these experiments, knock-down of *RUNX3* expression led to increased *RUNX1* expression as expected. Importantly, this reversal of cross regulation impaired the growth of the LCLs. These growth inhibitory effects were attributed to *RUNX1* when it was shown that ectopic expression of RUNX1c in an LCL blocked cell growth (Brady et al. 2009, 2013). Interestingly, the RUNX1c isoform (expressed in resting B cells) was more potent in inhibiting growth than RUNX1b, suggesting that the extra 30 N-terminal amino acids in RUNX1c play a role in the growth inhibitory function of this isoform (Brady et al. 2013). These data suggest that a switch from *RUNX1* to *RUNX3* may be a critical step in the activation of resting B cells; *RUNX1* must be removed for B cell proliferation to occur. The effects of *RUNX1* on B cell growth do however seem to be cell-type specific (see later). The experiments conducted to date do not rule out an additional role for *RUNX3* in promoting proliferation. For example, the RUNX3 target gene encoding the survival factor B-cell maturation antigen was induced in RUNX3 expressing B cells (Brady et al. 2009).

Functional differences between RUNX1b and RUNX1c in the regulation of cell proliferation were noted previously when RUNX1b but not RUNX1c was found to delay mitotic arrest in differentiating myeloid cells (Telfer et al. 2004). There are also other examples of RUNX1 and RUNX3 regulating growth differently in the same cell type. RUNX1 and RUNX3 have opposing roles in regulating the survival of cells expressing the BCR-ABL oncoprotein through transcriptional control of the 24p3R cell surface receptor (Sheng et al. 2009). BCR-ABL induces expression of the secreted receptor ligand 24p3, which induces apoptosis in normal cells expressing 24p3R, but in BCR-ABL+ cells 24p3R expression is repressed providing apoptosis resistance. In murine 32D myeloid BCR-ABL expressing cells, *24p3R* gene repression involves the displacement of RUNX3 from the *24p3R* gene promoter by RUNX1 and the recruitment of the Sin3A-HDAC co-repressor complex. This switch in RUNX binding is induced by BCL-ABL signalling through the Ras/MAP kinase pathway resulting in an increase in RUNX1 expression. Inhibition of BCR-ABL with Imatinib reversed this switch, promoting RUNX3 occupancy of the *24p3R* promoter, increased 24p3R expression and increased apoptosis sensitivity in these cells.

18.4 The Mechanism of *RUNX3* Activation by EBV

The original discovery of *RUNX3* as a target gene for the EBV transactivator EBNA2 came from experiments using an EBV transformed LCL expressing a conditionally active EBNA2-oestrogen receptor fusion protein. EBNA2 activation through β-estradiol addition to cells cultured in its absence for 5 days led to an induction of *RUNX3* mRNA (Spender et al. 2002). The upregulation of *RUNX3* by EBV was attributed to the direct actions of EBNA2 rather than the induced synthesis of another TF since the induction was observed in the presence of protein synthesis inhibitors. Subsequent investigations into the mechanism of *RUNX3* activation by EBNA2 using *RUNX3* promoter constructs containing promoter sequences close to the transcription start site failed to detect any reproducible effects of EBNA2 on the *RUNX3* P2 promoter, the promoter active in these cells (Spender et al. 2005; Gunnell et al. 2016).

The development of ChIP-sequencing assays for EBNA2 has shed considerable light on the potential mechanism of EBNA2 activation of a plethora of cellular genes, including many whose promoters had proved to be EBNA2 unresponsive (McClellan et al. 2013; Zhao et al. 2011). These genome-wide binding analyses revealed that a significant proportion (75 % or 86 %) of EBNA2 binding sites were located distal to gene transcription start sites and coincided with regions of the genome that display the chromatin signature of active or poised enhancers (H3K4 monomethylation with or without H3K27 acetylation) (McClellan et al. 2013). EBNA2 binding data generated in the latency III Burkitt's lymphoma (BL) cell-line Mutu III and new data from the GM12878 EBV transformed LCL revealed the presence of a large 18 kb enhancer region centred −97 kb upstream of the *RUNX3* P2 promoter (chr1:25,344,250-25,363,078, GRCh37/hg19) (Gunnell et al. 2016). This region displays high-level H3K27 acetylation in GM12878 and chromatin landscape analysis (dbSUPER, http://bioinfo.au.tsinghua.edu.cn/dbsuper/ (Khan and Zhang 2015)) classifies this region as a super-enhancer, a highly-active and characteristically large TF binding site cluster that drives transcription of lineage-specific genes (Whyte et al. 2013). An independent study also reported the super-enhancer characteristics of this EBNA2 binding region in EBV-infected cells (Zhou et al. 2015). Interestingly, based on H3K27 acetylation spread, this region is classified as a super-enhancer in GM12878 cells, early haematopoietic and T cell lineages but not CD19+ primary and CD20+ B cells. This suggests that the binding of EBNA2 drives its activation in EBV-infected B cells.

EBNA2 binding was also detected at the *RUNX3* P2 promoter, but the lack of EBNA2 responsiveness of this region in reporter assays indicates that these peaks may represent 'shadow' peaks of EBNA2 binding as a result of looping interactions between the EBNA2-bound super-enhancer region and the P2 promoter. This would result in the capturing of promoter DNA in cross-linked EBNA2 ChIP assays. In fact an adaptation of the genome-wide chromosome conformation capture technique that enriches for chromosome interactions involving promoters (capture Hi-C) (Mifsud et al. 2015) demonstrated that the *RUNX3* EBNA2 super-enhancer interacts with *RUNX* promoters in GM12878 cells, supporting its regulatory function (Gunnell et al. 2016).

The main six EBNA2 binding peaks from the *RUNX3* super-enhancer were cloned upstream of the *RUNX3* P2 promoter and used in reporter assays to determine which regions of the super-enhancer were responsible for directing EBNA2 activation. These experiments identified the 1.3 kb super-enhancer region 2 as the main EBNA2-responsive component of the super-enhancer.

Since EBNA2 does not bind DNA directly and relies on the hijacking of cellular TFs to control both viral and cell genes, cellular TFs bound to this region must be responsible for mediating the effects of EBNA2 on *RUNX3* transcription. As expected, this region contained binding sites for multiple cell TFs, but the use of a knock-out B cell line identified the Notch DNA binding partner RBP-J as the main TF responsible (Gunnell et al. 2016). This is consistent with the known association of EBNA2 with RBP-J via a conserved WϕP motif found in Notch family members and the essential nature of the EBNA2-RBPJ interaction for B-cell immortalisation (Ling and Hayward 1995; Yalamanchili et al. 1994).

A separate study identified a role for NF-κB binding to the *RUNX3* super-enhancer in maintaining *RUNX3* mRNA levels in EBV-infected cells, since *RUNX3* gene expression was reduced upon NF-κB inactivation (Zhou et al. 2015). Investigation of the role of an NF-κB binding site in enhancer region 2 supported these observations, demonstrating that this site was required for the maintenance of basal activity of the enhancer, but its mutation did not impair EBNA2 activation. These data are consistent with a role for NF-κB in EBNA2-independent activation of the *RUNX3* super-enhancer. Since the EBV oncogene and CD40 mimic, Latent membrane protein 1, activates NF-κB in EBV-infected cells (Rowe et al. 1994), this may represent an additional route exploited by EBV to maintain high *RUNX3* levels in infected cells.

ChIP-sequencing analysis of other EBV TFs expressed in immortalised B cells (EBNA3A, 3B and 3C) also detected binding of EBNA3B and 3C to enhancer region 2 of the *RUNX3* super-enhancer (Gunnell et al. 2016; McClellan et al. 2012, 2013). Microarray analysis of B cell lines infected with recombinant EBVs lacking all three EBNA3 proteins provided evidence for their role in activating *RUNX3* transcription (White et al. 2010) and this was confirmed at the protein level (Gunnell et al. 2016), a role that would not have been previously identified on the background of EBNA2 activation. Since these two factors also bind RBP-J on chromatin in a competitive manner with EBNA2 (Harth-Hertle et al. 2013; Wang et al. 2015a), the activation of *RUNX3* by EBNA3 proteins may also be mediated through this TF, with EBNA2, 3B and 3C binding independently to the super-enhancer.

It therefore appears that at least two B cell TFs play key roles in *RUNX3* super-enhancer control in immortalised B-cells; NF-kB and RBP-J. Further experiments will be required to determine whether the *RUNX3* super-enhancer does indeed play a role in the regulation of *RUNX3* transcription in other cell-types as predicted and whether additional TFs are required to direct cell-type specificity in these contexts.

18.5 Cell-Type Specific Feed-Forward Activation of *RUNX1* by EBV

Because *RUNX1* transcription is repressed in EBV-infected cells as a result of the induction of *RUNX3* by EBNA2, it was surprising that genome-wide binding analyses also identified binding sites for EBNA2 in a >100 kb region 139–250 kb upstream of *RUNX1* P1 (chr21:36,561,018-36,672,471, GRCh37/hg19) (Gunnell et al. 2016) (Fig. 18.2). These binding sites coincide with regions assigned as super-enhancers in specific cell-types (e.g. Diffuse large B cell lymphoma cell-lines, a breast cancer line, mammary epithelial cells and skeletal muscle (dbSUPER, http://bioinfo.au.tsinghua.edu.cn/dbsuper/ (Khan and Zhang 2015)). EBNA2 binds at six main sites within this region in a cell-type specific manner, with binding detected in Mutu III BL cells and not in LCLs. The lack of activity of the *RUNX1* superenhancer in LCLs is supported by the lack of H3K27 acetylation in this region in the GM12878 LCL (Fig. 18.2). Reporter assays examining EBNA2 responsiveness of the 6 main enhancer binding peaks identified enhancer region 4 as the main mediator of EBNA2 activation (Gunnell et al. 2016). As observed for *RUNX3*, EBNA2 activation of the *RUNX1* super-enhancer was dependent on the expression of RBP-J. The ability to access gene regulatory elements through association with RBP-J therefore appears be central in the control

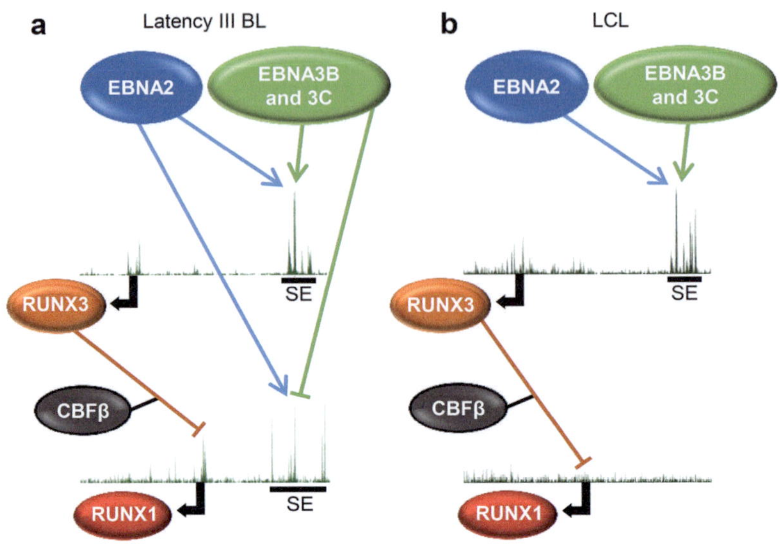

Fig. 18.3 Model of EBV control of the *RUNX* transcriptional network in different B cell backgrounds. EBNA2, 3B and 3C bind to the *RUNX3* super-enhancer (SE) in EBV-infected B cells to activate *RUNX3* transcription. RUNX3 then binds to the *RUNX1* P1 promoter to repress *RUNX1* transcription with its binding partner CBFβ. (**a**) In EBV-infected BL cell-lines expressing the latency III pattern of gene expression (Table 18.1), RUNX1 is also regulated positively through EBNA2 binding to a *RUNX1* SE. EBNA3B and 3C attenuate this EBNA2 activation, but higher RUNX1 expression is maintained. (**b**) In EBV-infected LCLs the *RUNX1* SE is inactive and not bound by EBNA2, 3B or 3C and *RUNX1* is repressed by RUNX3/CBFβ

of *RUNX1* and *RUNX3* expression by EBV (Fig. 18.3).

Although EBNA2 has the ability to target *RUNX1* and activate its transcription, it is unclear why EBNA2 would target *RUNX1* for activation, when its upregulation of *RUNX3* serves to repress *RUNX1* transcription (Spender et al. 2005). It is possible that the cell-type specificity of *RUNX1* targeting is important here. Experiments in LCLs highlight the importance of *RUNX1* repression for LCL growth, but EBNA2 activation of the *RUNX1* super-enhancer was observed in BL cells and not in LCLs. In BL cells, the characteristic *IG-MYC* translocation leads to growth deregulation through *MYC* overexpression. In this background, the repressive effects of *RUNX1* on growth may be overridden by the effects of *MYC*. It is also possible that the expression of some *RUNX1* might provide some advantage for the survival or growth of BL cells.

EBNA3B and 3C also bind *RUNX1* super-enhancer regions in the same cell-type specific manner as EBNA2, indicating multiple points of control by EBV (Gunnell et al. 2016) (Fig. 18.3). However, rather than contributing to *RUNX* upregulation as observed at *RUNX3*, these proteins appear to attenuate *RUNX1* expression. These dual roles of the EBNA3s are entirely consistent with their known roles as activators and repressors of transcription and their association with both co-activators and co-repressors (Knight et al. 2003; Cotter and Robertson 2000; Subramanian et al. 2002; Paschos et al. 2012) (Table 18.1).

18.6 *RUNX1* Regulation by Intronic Super-Enhancer Elements in Haematopoietic Cells

RUNX1 expression is controlled by enhancer elements in mouse and human cells, but unlike the intergenic upstream enhancers targeted by EBV, these enhancers are located in the region between the *RUNX1* P1 and P2 promoters. Comparative genomics combined with DNase hypersensitivity site (DHS) analysis identified a 194 bp enhancer in the mouse genome located 23.5 kb downstream of the *RUNX1* P1 transcript ATG (Nottingham et al. 2007). This +23 enhancer drives early haemopoietic-specific gene expression in mouse embryos in a pattern consistent with that of endogenous *RUNX1*. *In vitro* analysis of conserved TF binding sites in the enhancer identified conserved RUNX, Gata2 and Ets sites as crucial for its activity and RUNX1, Gata2 and the Ets TFs Fli-1, Elf1 and PU.1 bound the enhancer *in vivo* (Nottingham et al. 2007). Binding of the Gata-associated SCL, Lmo2 and Ldb1 complex was also detected by ChIP. Transgenic enhancer-reporter embryos confirmed that mutation of the Gata and Ets motifs severely disrupted enhancer activity in the embryo, although mutation of the RUNX motif had little effect (Nottingham et al. 2007). Unlike in zebrafish, where *RUNX1* P1 contains sites that control tissue-specific expression, the mouse P1 and P2 promoters alone lack tissue-specific activity, but tissue-specific expression of either promoter can be controlled by +23 kb enhancer (Bee et al. 2009).

In human cells, P1 and P2 display some cell-type specific activity and an equivalent 500 bp +23 kb enhancer (RE1) in intron 1 was found to regulate *RUNX1* expression in a haemopoietic cell-specific manner (Markova et al. 2011). The RE1 element was able to activate transcription from *RUNX1* P1 and P2 in reporter assays in Jurkat T leukaemic cells, from P2 in K562 erythroleukaemic cells, but failed to activate either promoter in HEK293 epithelial cells. A further 400 bp element identified in *RUNX1* intron 5.2 based on the presence of lymphoid-specific DHSs (RE2) did not activate *RUNX1* expression in reporter assays in any cell background. Chromosome conformation analysis detected haemopoietic cell-specific interactions between both P1 and P2 and RE1 (Markova et al. 2011) and showed that the promoters also interact with one another, forming a chromatin hub resembling that observed at *RUNX3* (Gunnell et al. 2016). RE2 also interacted with both P1 and P2, but separately, and this was not cell-type specific (Markova et al. 2011). Given the lack of detectable enhancer activity for RE2, this region may play an architectural role in the organisation of the locus, although a role as a matrix attachment region was ruled out (Markova et al. 2012).

Recent analysis confirmed the importance of the region between P1 and P2 in the regulation of *RUNX1* expression in Jurkat cells (Kwiatkowski et al. 2014). T-cell acute lymphoblastic leukaemia (T-ALL) cell-lines, including Jurkat, were found to be highly sensitive to treatment with a selective covalent inhibitor of the TFIIH kinase subunit CDK7 (Kwiatkowski et al. 2014). Genome-wide analysis in Jurkat cells identified *RUNX1* as one of a subset of genes that were selectively downregulated on inhibitor treatment. A *RUNX1* super-enhancer covering the entire 173 kb region between P1 and P2, containing the previously described RE1 element, was identified through examination of H3K27 acetylation profiles. The association of CDK7 with this entire super-enhancer region and the susceptibility of super-enhancers to inhibition likely explains the enhanced suppression of *RUNX1* transcription in these cells. T-ALL cell-line sensitivity to low-dose drug treatment is also probably enhanced as a result of the loss of expression of *TAL1* and *GATA3*, two other key components of the auto and cross-regulatory *RUNX1* transcriptional network (Kwiatkowski et al. 2014).

It is therefore clear that cell-type and developmental stage specific expression of *RUNX1* can be controlled by at least two extremely large super-enhancer regions, one located upstream of the gene locus that is exploited by EBV TFs in B cells and one located in an intronic region between the P1 and P2 promoters. Given the prediction that the upstream region has H3K27 acetylation levels consistent with superenhancer function in other cell-types, it is likely that this region has further cell-type specific and potentially disease specific roles in controlling *RUNX1* expression.

18.7 Notch Control of *RUNX* Gene Expression

The mechanism of RUNX super-enhancer control by EBV in immortalised B cells requires association of viral TFs with RBP-J. This reinforces the known interplay between the Notch and *RUNX* pathways in many developmental systems. For example, haematopoiesis in *Drosophila* is associated with Notch signalling-dependent induction of *RUNX1* (Lozenge) in specific precursor cells (Lebestky et al. 2003). *RUNX1* was also identified as a direct target gene induced by activated Notch1 signalling in murine mesodermal cells, but not embryonic stem cells (Meier-Stiegen et al. 2010). In mouse NIH3T3 cells Notch1 upregulates *Runx 1* transcription (Nakagawa et al. 2006) and Notch1-RBP-J null mice have impaired haematopoietic potential and reduced *Runx1* expression in specific compartments (Nakagawa et al. 2006). Notch signalling also controls *RUNX1* expression and haematopoietic stem cell development in Zebrafish (Burns et al. 2005). Although most studies have concentrated on *RUNX1*, *RUNX3* has also been shown to be a direct target of Notch in murine endothelial cells (Fu et al. 2011).

18.8 The Role of RUNX Proteins as Transcriptional Regulators in EBV-Immortalised B Cells

In addition to targeting *RUNX* genes via their super-enhancers to control *RUNX* transcription in EBV-infected cells, EBV TFs may also regulate cellular genes in association with RUNX proteins. Evidence for a role for RUNX proteins in EBV-directed gene control in B cells is based on the presence of RUNX motifs and/or RUNX3 binding at human genome binding sites for EBNA2 and EBNA3 proteins. Motif analysis of EBNA2 binding sites in the IB4 LCL found that 43 % had significant enrichment for a RUNX motif (Zhao et al. 2011). K-means cluster analysis of cell TF motifs at EBNA2 sites was used to examine whether specific sets of TFs co-bound at EBNA2 binding sites. These analyses identified a cluster of EBNA2 binding sites with RUNX motifs that also contained motifs for early B cell factor (EBF) and RBP-J (Zhao et al. 2011). A separate cluster of EBNA2 sites was also identified that contained RUNX motifs in association with RBP-J motifs, but lacked EBF motifs. Interestingly, the RUNX-EBF-RBP-J motif-containing sites had high H3K4 monomethylation levels and 37 % were bound at significant levels by the histone acetyl transferase p300, properties indicative of active enhancer regions. In contrast, the EBNA2 sites with RUNX-RBP-J motifs showed the least overlap with p300 binding (5%) and had low H3K4 monomethylation levels (Zhao et al. 2011), indicating that they may not be active regulatory regions. The most highly expressed RUNX protein in these LCLs is RUNX3 so these sites would be expected to be bound by RUNX3, but motif analyses cannot distinguish between RUNX1 and RUNX3 binding. These observations are interesting, however, in the context of a previous report where RUNX3 was found to suppress Notch activation of RBP-J dependent transcription through its association with the Notch intracellular domain (NIC) (Gao et al. 2010). Given that EBNA2 mimics NIC in its association with RBP-J, it is possible that RUNX3 might associate with EBNA2 to repress transcription at RBP-J bound sites and that the presence of EBF counteracts this repression. An EBF site in the viral LMP1 promoter has been shown to play a in role in EBNA2 activation of the promoter, although the involvement of RUNX3 in LMP1 transcription control is unknown (Zhao et al. 2011).

The potential involvement of RUNX proteins in EBV-mediated gene control also extends to the EBNA3 proteins. ChIP-sequencing analysis of epitope tagged EBNA3C found that similar to EBNA2, 44 % of EBNA3C sites were enriched for RUNX motifs (Jiang et al. 2014). Availability of RUNX3 ChIP-seq data for this later study also allowed analysis of binding overlap, which revealed that an even higher proportion (64 %) of EBNA3C sites in the human genome were also bound by RUNX3 (Jiang et al. 2014). EBNA3C sites that were co-bound by RUNX3 were found in 5 different clusters with different sets of cell TFs, some of which did not contain RBP-J, indicating that other TFs play a role in EBNA3C binding to DNA. Interestingly one

EBNA3C-RUNX3 co-bound site cluster was associated with chromatin marks found at strong promoters, but lacked RBP-J binding or any enrichment for RUNX motifs, consistent with some RUNX binding in the absence of a consensus site. Despite the 'active' chromatin profile, these sites were also co-bound by the Sin3A co-repressor, highlighting a potential role for RUNX3 in gene repression, in this case in the 'shutting down' of active promoter sites. Three other EBNA3C-RUNX3 clusters had mostly strong enhancer chromatin marks and RUNX motifs and included either PU.1, RBP-J and EBNA2, or BATF and IRF4. EBNA3C binding was enhanced in the presence of RUNX3 and RUNX3 has higher binding signals at sites bound by EBNA3C indicating some mutual enhancement or stabilisation of binding.

A high degree of overlap with RUNX3 binding was also detected in LCLs at binding sites for epitope tagged EBNA3A (63 %) or EBNA3B (83 %) (Schmidt et al. 2015; Wang et al. 2015a). For EBNA3A, 7 clusters of cell TF co-association were identified with 5 of them showing enrichment for RUNX3 binding and the presence of RUNX motifs (Schmidt et al. 2015). All of these 5 site clusters displayed chromatin marks associated with either weak or strong enhancers. Interestingly, in contrast to EBNA3C and its cell TF clusters, there were no detected associations of RUNX3 with EBNA3A in TF clusters containing co-repressors, indicating that RUNX3 association with different EBV TFs may be distinct and have different roles in different contexts. However, analysis of these datasets is hindered by the extensive overlap between the binding sites of EBNA2 and the EBNA3s (McClellan et al. 2013; Wang et al. 2015a), making it difficult to assign specificity to any interactions or co-operation. Interestingly, analysis of unique EBNA2, EBNA3A, 3B and 3C binding sites revealed that RUNX motifs are only enriched at sites bound by EBNA2 alone or EBNA3C alone and not for EBNA3A and EBNA3B, indicating that there may be specificity in the potential association between EBV TFs and RUNX3 (Wang et al. 2015a).

Although there is no evidence to date[1] of any direct or indirect interactions between EBV TFs and RUNX proteins, taken together these genome-wide binding studies strongly implicate RUNX3 in EBV-directed cellular gene regulation, through direct or indirect interactions at gene regulatory elements. This is consistent with the immortalisation of B cells by EBV TFs through the hijacking of cell TFs that drive the B-cell transcription program.

It is noteworthy that there is significant RUNX3 binding at the *RUNX3* super-enhancer and multiple RUNX motifs are present, including in enhancer 2, the most EBNA2 responsive region (Fig. 18.4) (Gunnell et al. 2016), so it will be interesting to determine the effects of mutating the RUNX binding sites in this region on EBNA2 responsiveness of the superenhancer. RUNX3 also bound to the *RUNX3* P1 promoter (Fig. 18.4). This is interesting given the known role of RUNX3 binding to two conserved RUNX sites in the *RUNX1* P1 promoter in the repression of *RUNX1* transcription in B cells (Spender et al. 2005) and the auto-repression of *RUNX2* via the same sites in rat (Drissi et al. 2000) . The tandem *RUNX1* P1 RUNX sites are completely conserved in the human *RUNX3* P1 promoter so it is possible that binding of RUNX3 to these sites may auto-repress its own P1 promoter. Consistent with this, *RUNX3* transcripts are derived from P2, not P1 in EBV immortalised B cells (Spender et al. 2005). Further analysis is required to determine the role played by RUNX3 in the regulation of its own enhancer and promoters, and indeed how RUNX3 binding to RUNX1 P1 may influence *RUNX1* enhancer activity.

[1]While this review was in press, an interaction between EBNA3C and EBNA3B and the RUNX binding partner CBFβ was demonstrated (Paschos et al. 2016). Depletion of RUNX3 or CBFβ reduced EBNA3B and EBNA3C binding to genomic sites and the activation or repression of cell genes by EBNA3C was reduced by CBFβ depletion. These data therefore provide good evidence for the role of RUNX complexes in the targeting of certain EBV EBNA3 proteins to DNA and in EBV-directed gene regulation in B cells.

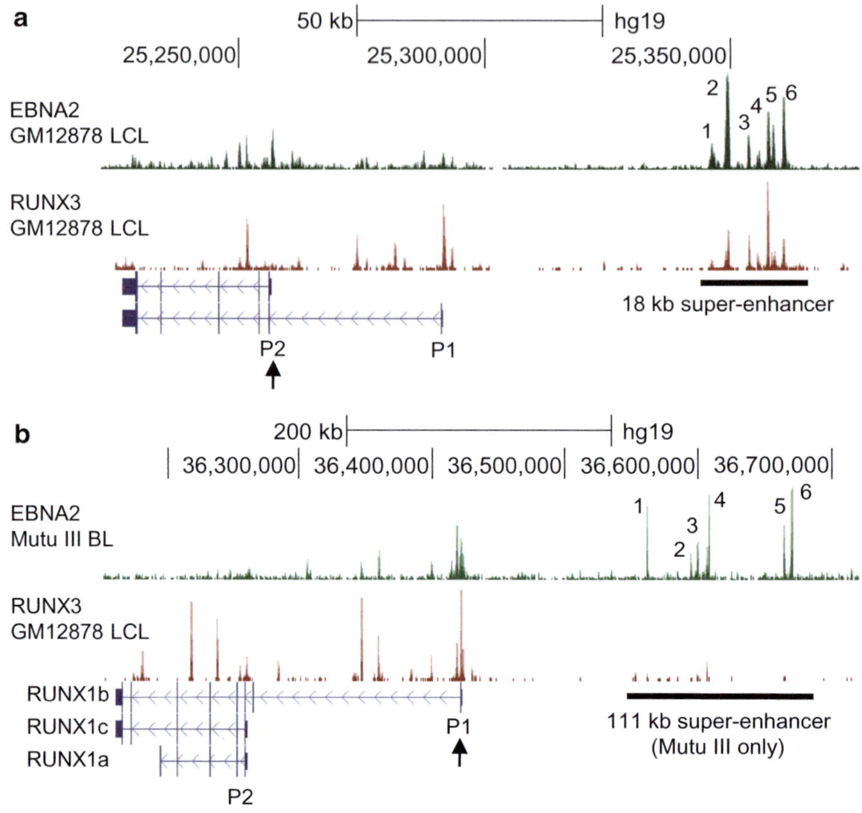

Fig. 18.4 RUNX3 binding at *RUNX1* and *RUNX3*. (**a**) Binding of RUNX3 at human *RUNX3* on chromosome 1. ChIP-sequencing reads for RUNX3 in GM12878 cells from ENCODE are shown alongside ChIP-sequencing reads for EBNA2 in GM12878. Labelling as in Fig. 18.1. (**b**) Binding of RUNX3 at human *RUNX1* on chromosome 21. ChIP-sequencing reads for RUNX3 in GM12878 cells from ENCODE are shown alongside ChIP-sequencing reads for EBNA2 in Mutu III BL cells. Labelling as in Fig. 18.2

18.9 Future Directions and Outstanding Questions

Studies of *RUNX* gene control in B cells immortalised by the tumour-associated Epstein-Barr virus have led to the identification of super-enhancers that play a key role in controlling the cell-type specific expression of both *RUNX3* and *RUNX1*. Further investigation is now needed to address the potential role of these super-enhancers in the spatial and temporal control of *RUNX* gene expression in the haemopoietic and other developmental systems and their contribution to *RUNX* gene deregulation in other disease states. The known susceptibility of super-enhancers to perturbation and their key role in lineage-specific growth regulation has already made them attractive targets for new transcription inhibitory molecules that have been shown to block T-ALL, small cell lung carcinoma, neuroblastoma and triple negative breast cancer cell growth (Chipumuro et al. 2014; Christensen et al. 2014; Kwiatkowski et al. 2014; Wang et al. 2015b). The perturbation of these EBV-driven super-enhancers may offer new therapeutic possibilities for the treatment of EBV-associated malignancies, given the key role played by RUNX proteins in the regulation of EBV-infected cell growth.

In spite of all this progress, there are still many questions regarding the role of *RUNX3* and *RUNX1* in B-cell gene regulation and growth control. One important priority is to determine in detail which genes are targeted by *RUNX3* and *RUNX1* in B cells and how combination with other cell TFs determines the dual roles of these proteins in gene activation and repression. It is also notable that the *RUNX1* super-enhancer is only active and bound by EBV TFs in Burkitt's lymphoma cells and not in the LCLs examined to date. Repression by the higher levels of RUNX3 in LCLs may be one possible explanation (Gunnell et al. 2016), but there may well be other effects of the high levels of MYC in Burkitt's lymphoma cells on the super-enhancer, and MYC binding sites that are bound by MYC/MAX are present in this region. Bringing together the rapidly increasing ChIP-sequencing information with functional studies of promoter/enhancer elements and further functional TF interactions is likely to give an even more comprehensive understanding of the roles of *RUNX* genes in B cell immortalisation in the near future.

References

Bangsow, C., Rubins, N., Glusman, G., Bernstein, Y., Negreanu, V., Goldenberg, D., et al. (2001). The RUNX3 gene – sequence, structure and regulated expression. *Gene, 279*(2), 221–232.

Bee, T., Ashley, E. L., Bickley, S. R., Jarratt, A., Li, P. S., Sloane-Stanley, J., et al. (2009). The mouse Runx1 +23 hematopoietic stem cell enhancer confers hematopoietic specificity to both Runx1 promoters. *Blood, 113*(21), 5121–5124. doi:10.1182/blood-2008-12-193003.

Blyth, K., Slater, N., Hanlon, L., Bell, M., Mackay, N., Stewart, M., et al. (2009). Runx1 promotes B-cell survival and lymphoma development. *Blood Cells, Molecules & Diseases, 43*(1), 12–19. doi:10.1016/j.bcmd.2009.01.013.

Brady, G., Whiteman, H. J., Spender, L. C., & Farrell, P. J. (2009). Downregulation of RUNX1 by RUNX3 requires the RUNX3 VWRPY sequence and is essential for Epstein-Barr virus-driven B-cell proliferation. *Journal of Virology, 83*(13), 6909–6916.

Brady, G., Elgueta Karstegl, C., & Farrell, P. J. (2013). Novel function of the unique N-terminal region of RUNX1c in B cell growth regulation. *Nucleic Acids Research, 41*(3), 1555–1568. doi:10.1093/nar/gks1273.

Burns, C. E., Traver, D., Mayhall, E., Shepard, J. L., & Zon, L. I. (2005). Hematopoietic stem cell fate is established by the Notch-Runx pathway. *Genes & Development, 19*(19), 2331–2342. doi:10.1101/gad.1337005.

Cen, O., & Longnecker, R. (2015). Latent Membrane Protein 2 (LMP2). *Current Topics in Microbiology and Immunology, 391*, 151–180. doi:10.1007/978-3-319-22834-1_5.

Chen, M. J., Yokomizo, T., Zeigler, B. M., Dzierzak, E., & Speck, N. A. (2009). Runx1 is required for the endothelial to haematopoietic cell transition but not thereafter. *Nature, 457*(7231), 887–891. doi:10.1038/nature07619.

Chipumuro, E., Marco, E., Christensen, C. L., Kwiatkowski, N., Zhang, T., Hatheway, C. M., et al. (2014). CDK7 inhibition suppresses super-enhancer-linked oncogenic transcription in MYCN-driven cancer. *Cell, 159*(5), 1126–1139. doi:10.1016/j.cell.2014.10.024.

Christensen, C. L., Kwiatkowski, N., Abraham, B. J., Carretero, J., Al-Shahrour, F., Zhang, T., et al. (2014). Targeting transcriptional addictions in small cell lung cancer with a covalent CDK7 inhibitor. *Cancer Cell, 26*(6), 909–922. doi:10.1016/j.ccell.2014.10.019.

Cotter 2nd, M. A., & Robertson, E. S. (2000). Modulation of histone acetyltransferase activity through interaction of epstein-barr nuclear antigen 3C with prothymosin alpha. *Molecular and Cellular Biology, 20*(15), 5722–5735.

Drissi, H., Luc, Q., Shakoori, R., Chuva De Sousa Lopes, S., Choi, J. Y., Terry, A., et al. (2000). Transcriptional autoregulation of the bone related CBFA1/RUNX2 gene. *Journal of Cellular Physiology, 184*(3), 341–350. doi:10.1002/1097-4652(200009)184:3<341::AID-JCP8>3.0.CO;2-Z.

Durst, K. L., & Hiebert, S. W. (2004). Role of RUNX family members in transcriptional repression and gene silencing. *Oncogene, 23*(24), 4220–4224. doi:10.1038/sj.onc.1207122.

Fainaru, O., Woolf, E., Lotem, J., Yarmus, M., Brenner, O., Goldenberg, D., et al. (2004). Runx3 regulates mouse TGF-beta-mediated dendritic cell function and its absence results in airway inflammation. *The EMBO Journal, 23*(4), 969–979. doi:10.1038/sj.emboj.7600085.

Fu, Y., Chang, A. C., Fournier, M., Chang, L., Niessen, K., & Karsan, A. (2011). RUNX3 maintains the mesenchymal phenotype after termination of the Notch signal. *The Journal of Biological Chemistry, 286*(13), 11803–11813. doi:10.1074/jbc.M111.222331.

Gao, J., Chen, Y., Wu, K. C., Liu, J., Zhao, Y. Q., Pan, Y. L., et al. (2010). RUNX3 directly interacts with intracellular domain of Notch1 and suppresses Notch signaling in hepatocellular carcinoma cells. *Experimental Cell Research, 316*(2), 149–157. doi:10.1016/j.yexcr.2009.09.025.

Growney, J. D., Shigematsu, H., Li, Z., Lee, B. H., Adelsperger, J., Rowan, R., et al. (2005). Loss of Runx1 perturbs adult hematopoiesis and is associated with a myeloproliferative phenotype. *Blood, 106*(2), 494–504. doi:10.1182/blood-2004-08-3280.

Gunnell, A., Webb, H. M., Wood, C. D., McClellan, M. J., Wichaidit, B., Kempkes, B., et al. (2016). RUNX super-enhancer control through the notch pathway by

Epstein-Barr virus transcription factors regulates B cell growth. *Nucleic Acids Research, 44*(10), 4636–4650. doi:10.1093/nar/gkw085.

Harth-Hertle, M. L., Scholz, B. A., Erhard, F., Glaser, L. V., Dolken, L., Zimmer, R., & Kempkes, B. (2013) Inactivation of intergenic enhancers by EBNA3A initiates and maintains polycomb signatures across a chromatin domain encoding CXCL10 and CXCL9. *PLoS Pathogens, 9*(9), e1003638. doi:10.1371/journal.ppat.1003638 PPATHOGENS-D-13-00854 [pii]

Ichikawa, M., Asai, T., Saito, T., Seo, S., Yamazaki, I., Yamagata, T., et al. (2004). AML-1 is required for megakaryocytic maturation and lymphocytic differentiation, but not for maintenance of hematopoietic stem cells in adult hematopoiesis. *Nature Medicine, 10*(3), 299–304. doi:10.1038/nm997.

Imai, Y., Kurokawa, M., Tanaka, K., Friedman, A. D., Ogawa, S., Mitani, K., et al. (1998). TLE, the human homolog of groucho, interacts with AML1 and acts as a repressor of AML1-induced transactivation. *Biochemical and Biophysical Research Communications, 252*(3), 582–589. doi:10.1006/bbrc.1998.9705.

Imperato, M. R., Cauchy, P., Obier, N., & Bonifer, C. (2015). The RUNX1-PU.1 axis in the control of hematopoiesis. *International Journal of Hematology, 101*(4), 319–329. doi:10.1007/s12185-015-1762-8.

Jiang, S., Willox, B., Zhou, H., Holthaus, A. M., Wang, A., Shi, T. T., Maruo, S., Kharchenko, P. V., Johannsen, E. C., Kieff, E., & Zhao, B. (2014). Epstein-Barr Virus nuclear antigen 3C binds to BATF/IRF4 or SPI1/IRF4 composite sites and recruits Sin3A to repress CDKN2A. *Proceedings of the National Academy of Sciences of the United States of America, 111*(1), 421–426. doi:1321704111 [pii] 10.1073/pnas.1321704111

Kaiser, C., Laux, G., Eick, D., Jochner, N., Bornkamm, G. W., & Kempkes, B. (1999). The proto-oncogene c-myc is a direct target gene of Epstein-Barr virus nuclear antigen 2. *Journal of Virology, 73*(5), 4481–4484.

Kempkes, B., & Robertson, E. S. (2015). Epstein-Barr virus latency: Current and future perspectives. *Current Opinion in Virology, 14*, 138–144. doi:10.1016/j.coviro.2015.09.007.

Khan, A., & Zhang, X. (2015). dbSUPER: A database of super-enhancers in mouse and human genome. *Nucleic Acids Research, 44*, D164–D171. doi:10.1093/nar/gkv1002.

Knight, J. S., Lan, K., Subramanian, C., & Robertson, E. S. (2003). Epstein-Barr virus nuclear antigen 3C recruits histone deacetylase activity and associates with the corepressors mSin3A and NCoR in human B-cell lines. *Journal of Virology, 77*(7), 4261–4272.

Knutson, J. C. (1990). The level of c-fgr RNA is increased by EBNA-2, an Epstein-Barr virus gene required for B-cell immortalization. *Journal of Virology, 64*(6), 2530–2536.

Kwiatkowski, N., Zhang, T., Rahl, P. B., Abraham, B. J., Reddy, J., Ficarro, S. B., et al. (2014). Targeting transcription regulation in cancer with a covalent CDK7 inhibitor. *Nature, 511*(7511), 616–620. doi:10.1038/nature13393.

Lebestky, T., Jung, S. H., & Banerjee, U. (2003). A serrate-expressing signaling center controls Drosophila hematopoiesis. *Genes & Development, 17*(3), 348–353. doi:10.1101/gad.1052803.

Levanon, D., Negreanu, V., Bernstein, Y., Bar-Am, I., Avivi, L., & Groner, Y. (1994). AML1, AML2, and AML3, the human members of the runt domain gene-family: cDNA structure, expression, and chromosomal localization. *Genomics, 23*(2), 425–432. doi:10.1006/geno.1994.1519.

Levanon, D., Bernstein, Y., Negreanu, V., Ghozi, M. C., Bar-Am, I., Aloya, R., et al. (1996). A large variety of alternatively spliced and differentially expressed mRNAs are encoded by the human acute myeloid leukemia gene AML1. *DNA and Cell Biology, 15*(3), 175–185.

Levanon, D., Goldstein, R. E., Bernstein, Y., Tang, H., Goldenberg, D., Stifani, S., et al. (1998). Transcriptional repression by AML1 and LEF-1 is mediated by the TLE/Groucho corepressors. *Proceedings of the National Academy of Sciences of the United States of America, 95*(20), 11590–11595.

Ling, P. D., & Hayward, S. D. (1995). Contribution of conserved amino acids in mediating the interaction between EBNA2 and CBF1/RBPJk. *Journal of Virology, 69*(3), 1944–1950.

Maier, S., Staffler, G., Hartmann, A., Hock, J., Henning, K., Grabusic, K., Mailhammer, R., Hoffmann, R., Wilmanns, M, Lang, R., Mages, J., Kempkes, B. (2006). Cellular target genes of Epstein-Barr virus nuclear antigen 2. *Journal of Virology, 80*(19), 9761–9771. doi:80/19/9761 [pii] 10.1128/JVI.00665–06

Markova, E. N., Kantidze, O. L., & Razin, S. V. (2011). Transcriptional regulation and spatial organisation of the human AML1/RUNX1 gene. *Journal of Cellular Biochemistry, 112*(8), 1997–2005. doi:10.1002/jcb.23117.

Markova, E. N., Razin, S. V., & Kantidze, O. L. (2012). Fragment of intron 5.2 of the human RUNX1 gene important for transcription activation is neither enhancer nor MAR-element. *Doklady Biochemistry and Biophysics, 442*, 26–29. doi:10.1134/S1607672912010085.

McClellan, M. J., Khasnis, S., Wood, C. D., Palermo, R. D. , Schlick, S. N., Kanhere, A. S., Jenner, R. G., & West, M. J. (2012). Downregulation of integrin receptor-signaling genes by Epstein-Barr virus EBNA 3C via promoter-proximal and -distal binding elements. *Journal of Virology, 86*(9), 5165–5178. doi:JVI.07161–11 [pii] 10.1128/JVI.07161-11

McClellan, M. J., Wood, C. D., Ojeniyi, O., Cooper, T. J., Kanhere, A., Arvey, A., Webb, H. M., Palermo, R. D., Harth-Hertle, M. L., Kempkes, B., Jenner, R. G., & West, M. J. (2013) Modulation of enhancer looping and differential gene targeting by Epstein-Barr virus transcription factors directs cellular reprogramming.

PLoS Pathogens, 9(9), e1003636. doi:10.1371/journal.ppat.1003636 PPATHOGENS-D-13-00786 [pii]

McLarren, K. W., Theriault, F. M., & Stifani, S. (2001). Association with the nuclear matrix and interaction with Groucho and RUNX proteins regulate the transcription repression activity of the basic helix loop helix factor Hes1. *The Journal of Biological Chemistry, 276*(2), 1578–1584. doi:10.1074/jbc.M007629200.

Meier-Stiegen, F., Schwanbeck, R., Bernoth, K., Martini, S., Hieronymus, T., Ruau, D., et al. (2010). Activated Notch1 target genes during embryonic cell differentiation depend on the cellular context and include lineage determinants and inhibitors. *PloS One, 5*(7), e11481. doi:10.1371/journal.pone.0011481.

Mifsud, B., Tavares-Cadete, F., Young, A. N., Sugar, R., Schoenfelder, S., Ferreira, L., et al. (2015). Mapping long-range promoter contacts in human cells with high-resolution capture Hi-C. *Nature Genetics, 47*(6), 598–606. doi:10.1038/ng.3286.

Nakagawa, M., Ichikawa, M., Kumano, K., Goyama, S., Kawazu, M., Asai, T., et al. (2006). AML1/Runx1 rescues Notch1-null mutation-induced deficiency of para-aortic splanchnopleural hematopoiesis. *Blood, 108*(10), 3329–3334. doi:10.1182/blood-2006-04-019570.

North, T. E., Stacy, T., Matheny, C. J., Speck, N. A., & de Bruijn, M. F. (2004). Runx1 is expressed in adult mouse hematopoietic stem cells and differentiating myeloid and lymphoid cells, but not in maturing erythroid cells. *Stem Cells, 22*(2), 158–168. doi:10.1634/stemcells.22-2-158.

Nottingham, W. T., Jarratt, A., Burgess, M., Speck, C. L., Cheng, J. F., Prabhakar, S., et al. (2007). Runx1-mediated hematopoietic stem-cell emergence is controlled by a Gata/Ets/SCL-regulated enhancer. *Blood, 110*(13), 4188–4197. doi:10.1182/blood-2007-07-100883.

Okuda, T., van Deursen, J., Hiebert, S. W., Grosveld, G., & Downing, J. R. (1996). AML1, the target of multiple chromosomal translocations in human leukemia, is essential for normal fetal liver hematopoiesis. *Cell, 84*(2), 321–330.

Paschos, K., Parker, G. A., Watanatanasup, E., White, R. E., & Allday, M. J. (2012) BIM promoter directly targeted by EBNA3C in polycomb-mediated repression by EBV. *Nucleic Acids Research, 40*(15), 7233–7246. doi:gks391 [pii] 10.1093/nar/gks391.

Paschos, K., Bazot, Q., Ho, G., Parker, G. A., Lees, J., Barton, G., & Allday, M. J. (2016). Core binding factor (CBF) is required for Epstein-Barr virus EBNA3 proteins to regulate target gene expression. *Nucleic Acids Research*. doi:10.1093/nar/gkw1167.

Putz, G., Rosner, A., Nuesslein, I., Schmitz, N., & Buchholz, F. (2006). AML1 deletion in adult mice causes splenomegaly and lymphomas. *Oncogene, 25*(6), 929–939. doi:10.1038/sj.onc.1209136.

Rowe, M., Peng-Pilon, M., Huen, D. S., Hardy, R., Croom-Carter, D., Lundgren, E., & Rickinson, A. B. (1994). Upregulation of bcl-2 by the Epstein-Barr virus latent membrane protein LMP1: A B-cell-specific response that is delayed relative to NF-kappa B activation and to induction of cell surface markers. *Journal of Virology, 68*(9), 5602–5612.

Schmidt, S. C., Jiang, S., Zhou, H., Willox, B., Holthaus, A. M., Kharchenko, P. V., et al. (2015). Epstein-Barr virus nuclear antigen 3A partially coincides with EBNA3C genome-wide and is tethered to DNA through BATF complexes. *Proceedings of the National Academy of Sciences of the United States of America, 112*(2), 554–559. doi:10.1073/pnas.1422580112.

Sheng, Z., Wang, S. Z., & Green, M. R. (2009). Transcription and signalling pathways involved in BCR-ABL-mediated misregulation of 24p3 and 24p3R. *The EMBO Journal, 28*(7), 866–876. doi:10.1038/emboj.2009.35.

Shi, M. J., & Stavnezer, J. (1998). CBF alpha3 (AML2) is induced by TGF-beta1 to bind and activate the mouse germline Ig alpha promoter. *Journal of Immunology, 161*(12), 6751–6760.

Spender, L. C., Cornish, G. H., Sullivan, A., & Farrell, P. J. (2002). Expression of transcription factor AML-2 (RUNX3, CBF(alpha)-3) is induced by Epstein-Barr virus EBNA-2 and correlates with the B-cell activation phenotype. *Journal of Virology, 76*(10), 4919–4927.

Spender, L. C., Whiteman, H. J., Karstegl, C. E., & Farrell, P. J. (2005). Transcriptional cross-regulation of RUNX1 by RUNX3 in human B cells. *Oncogene, 24*(11), 1873–1881.

Subramanian, C., Hasan, S., Rowe, M., Hottiger, M., Orre, R., & Robertson, E. S. (2002). Epstein-Barr virus nuclear antigen 3C and prothymosin alpha interact with the p300 transcriptional coactivator at the CH1 and CH3/HAT domains and cooperate in regulation of transcription and histone acetylation. *Journal of Virology, 76*(10), 4699–4708.

Telfer, J. C., Hedblom, E. E., Anderson, M. K., Laurent, M. N., & Rothenberg, E. V. (2004). Localization of the domains in Runx transcription factors required for the repression of CD4+ in thymocytes. *Journal of Immunology, 172*(7), 4359–4370.

Wang, F., Gregory, C. D., Rowe, M., Rickinson, A. B., Wang, D., Birkenbach, M., et al. (1987). Epstein-Barr virus nuclear antigen 2 specifically induces expression of the B-cell activation antigen CD23. *Proceedings of the National Academy of Sciences of the United States of America, 84*(10), 3452–3456.

Wang, Q., Stacy, T., Binder, M., Marin-Padilla, M., Sharpe, A. H., & Speck, N. A. (1996). Disruption of the Cbfa2 gene causes necrosis and hemorrhaging in the central nervous system and blocks definitive hematopoiesis. *Proceedings of the National Academy of Sciences of the United States of America, 93*(8), 3444–3449.

Wang, W., Wang, Y. G., Reginato, A. M., Glotzer, D. J., Fukai, N., Plotkina, S., et al. (2004). Groucho homologue Grg5 interacts with the transcription factor Runx2-Cbfa1 and modulates its activity during postnatal growth in mice. *Developmental Biology, 270*(2), 364–381. doi:10.1016/j.ydbio.2004.03.003.

Wang, A., Welch, R., Zhao, B., Ta, T., Keles, S., & Johannsen, E. (2015a) EBNA3 proteins regulate EBNA2 binding to distinct RBPJ genomic sites. *Journal of Virology, 90*, 2906–2919. doi:10.1128/JVI.02737–02715

Wang, Y., Zhang, T., Kwiatkowski, N., Abraham, B. J., Lee, T. I., Xie, S., et al. (2015b). CDK7-dependent transcriptional addiction in triple-negative breast cancer. *Cell, 163*(1), 174–186. doi:10.1016/j.cell.2015.08.063.

White, R. E., Groves, I. J., Turro, E., Yee, J., Kremmer, E., & Allday, M. J. (2010). Extensive co-operation between the Epstein-Barr virus EBNA3 proteins in the manipulation of host gene expression and epigenetic chromatin modification. *PloS One, 5*(11), e13979. doi:10.1371/journal.pone.0013979.

Whiteman, H. J., & Farrell, P. J. (2006). RUNX expression and function in human B cells. *Critical Reviews in Eukaryotic Gene Expression, 16*(1), 31–44.

Whyte, W. A., Orlando, D. A., Hnisz, D., Abraham, B. J., Lin, C. Y., Kagey, M. H., et al. (2013). Master transcription factors and mediator establish super-enhancers at key cell identity genes. *Cell, 153*(2), 307–319. doi:10.1016/j.cell.2013.03.035.

Yalamanchili, R., Tong, X., Grossman, S., Johannsen, E., Mosialos, G., & Kieff, E. (1994). Genetic and biochemical evidence that EBNA 2 interaction with a 63-kDa cellular GTG-binding protein is essential for B lymphocyte growth transformation by EBV. *Virology, 204*(2), 634–641. doi:S0042-6822(84)71578–9 [pii] 10.1006/viro.1994.1578.

Zhao, B., Maruo, S., Cooper, A., Chase, M. R., Johannsen, E., Kieff, E., & Cahir-McFarland, E. (2006) RNAs induced by Epstein-Barr virus nuclear antigen 2 in lymphoblastoid cell lines. *Proceedings of the National Academy of Sciences of the United States of America, 103*(6), 1900–1905. doi:0510612103 [pii] 10.1073/pnas.0510612103.

Zhao, B., Zou, J., Wang, H., Johannsen, E., Peng, C. W., Quackenbush, J., Mar, J. C., Morton, C. C., Freedman, M. L., Blacklow, S. C., Aster, J. C., Bernstein, B. E., & Kieff, E. (2011) Epstein-Barr virus exploits intrinsic B-lymphocyte transcription programs to achieve immortal cell growth. *Proceedings of the National Academy of Sciences of the United States of America, 108*(36), 14902–14907. doi:1108892108 [pii] 10.1073/pnas.1108892108

Zhou, H., Schmidt, S. C., Jiang, S., Willox, B., Bernhardt, K., Liang, J., et al. (2015). Epstein-Barr Virus oncoprotein super-enhancers control B cell growth. *Cell Host & Microbe, 17*(2), 205–216. doi:10.1016/j.chom.2014.12.013.

Roles of RUNX in Solid Tumors

19

Linda Shyue Huey Chuang, Kosei Ito,
and Yoshiaki Ito

Abstract

All *RUNX* genes have been implicated in the development of solid tumors, but the role each RUNX gene plays in the different tumor types is complicated by multiple interactions with major signaling pathways and tumor heterogeneity. Moreover, for a given tissue type, the specific role of each RUNX protein is distinct at different stages of differentiation. A regulatory function for RUNX in tissue stem cells points sharply to a causal effect in tumorigenesis. Understanding how RUNX dysregulation in cancer impinges on normal biological processes is important for identifying the molecular mechanisms that lead to malignancy. It will also indicate whether restoration of proper RUNX function to redirect cell fate is a feasible treatment for cancer. With the recent advances in RUNX research, it is time to revisit the many mechanisms/pathways that RUNX engage to regulate cell fate and decide whether cells proliferate, differentiate or die.

Keywords

RUNX • Solid tumors • Tumor suppressor • Oncogene • Wnt • TGFβ • RAS • Senescence • Protein-protein interaction • Stress-inducible gene • Precancerous state • Intestinal metaplasia

19.1 Introduction

The past decade has seen considerable progress in our understanding of the RUNX family of transcription factors in solid tumors. There are three mammalian *RUNX* genes and all have been directly implicated at various stages of tumor development, including initiation, progression and invasion. Animal knockout models of individual *Runx* genes revealed distinct developmental

L.S.H. Chuang • Y. Ito (✉)
Cancer Science Institute of Singapore,
Center for Translational Medicine, National
University of Singapore, 14 Medical Drive #12-01,
Singapore 117599, Singapore
e-mail: csiitoy@nus.edu.sg

K. Ito
Graduate School of Biomedical Sciences, Nagasaki
University, 1-7-1 Sakamoto, Nagasaki 852-8588, Japan

Y. Groner et al. (eds.), *RUNX Proteins in Development and Cancer*, Advances in Experimental Medicine and Biology 962, DOI 10.1007/978-981-10-3233-2_19

defects in hematopoiesis, bone development, the gastronintestinal tract and neurogenesis. While these findings reflect the tissue-specific expression and function of RUNX1 in the hematopoietic system, RUNX2 in bone, and RUNX3 in the gastrointestinal tract, lymphocytes as well as the dorsal root ganglion (DRG) neuron, it is noteworthy that all RUNX proteins are expressed in a broad range of tissues (Ito et al. 2015). Moreover, RUNX expression patterns are extremely dynamic and depend on the stage of differentiation as well as developmental and environmental cues. Because of the conserved Runt and the divergent C-terminal domains, RUNX proteins act redundantly in some cellular contexts and exert unique effects in others. This review summarizes the role of RUNX in solid tumors. This is by no means an exhaustive review for many types of cancer. Because of the focus of our laboratory, we will describe in detail our analyses of *RUNX3* function in tumors of the gastrointestinal tract. Through key examples, we will discuss how RUNX proteins engage different signaling pathways and biological processes to regulate proliferation, determine cell identity and influence tumor progression in solid tumors.

19.2 Ancient Origin of RUNX

RUNX genes have been uncovered in most, if not all, metazoans. Although *RUNX* was previously believed to be metazoan-specific, the discovery of two *Runx* genes in the holozoan *Capsaspora owczarzaki* indicates that *RUNX* originated in a unicellular organism, well before the emergence of multicellular metazoans (Sebe-Pedros et al. 2011). While the primordial role of *RUNX* remains unclear, it is reasonable to hypothesize that *RUNX* controls cell growth by orchestrating transcriptional programs in response to environmental cues. All other attributes of *RUNX* (eg. lineage specification) that were acquired later in evolution would likely reflect this original role. The extent to which the roles of multiple *RUNX* family members overlap, are inter-dependent or antagonistic is however unclear.

19.3 Dysregulation of RUNX in Solid Tumors

Although *RUNX1* aberrations are heavily implicated in leukemogenesis, recent studies have indicated causative roles for *RUNX1* in solid tumors.

RUNX1 is one of the significantly mutated genes in luminal breast cancer (Banerji et al. 2012; Ellis et al. 2012). Missense mutations at the Runt domain of RUNX1 and its binding partner, *CBFB*, are clear indications that the DNA binding ability/transcriptional activity of RUNX1, and perhaps other RUNX proteins, influence breast cancer growth (Banerji et al. 2012; Ellis et al. 2012). Yet, RUNX1 is highly expressed in a broad range of epithelial tumors, such as those of the skin, oesophageal, lung, colon, and interestingly, the breast (Scheitz et al. 2012). It has been proposed that leukemia, breast and skin cancers are stem cell disorders. Tight regulation of *RUNX1* expression appears to be necessary for proper stem cell function and differentiation.

RUNX2 has not been shown to be significantly mutated in cancer. Rather, RUNX2 mutation is identified as the main cause of the heritable dominant skeletal disorder cleidocranial dysplasia (CCD). Overexpression of RUNX2 is frequently observed in bone, breast and prostate cancers, suggesting that enhanced RUNX2 activity contributes to oncogenic growth in such tissues (Pratap et al. 2008; Akech et al. 2010) (Fig. 19.1). For example, human tissue microarray revealed that RUNX2 expression is elevated in triple negative (i.e. oestrogen receptor (ER)/progesterone receptor (PR)/HER2 negative) breast cancers and associated with a poor survival rate (McDonald et al. 2014). The chromosomal region *6p12-p21* where RUNX2 is located has been shown to be amplified in osteosarcoma (Sadikovic et al. 2010).

Unlike *RUNX1* and *RUNX2*, no familial disorder has been linked to *RUNX3* inactivation. *RUNX3* is located at *1p36*, a chromosomal region that is frequently deleted in a diverse range of cancers, including breast, lung, colorectal, neuroblastomas, hepatocellular carcinoma and pancreatic cancer (Nomoto et al. 2000; Mori et al. 2005; Nomoto et al. 2008; Henrich et al.

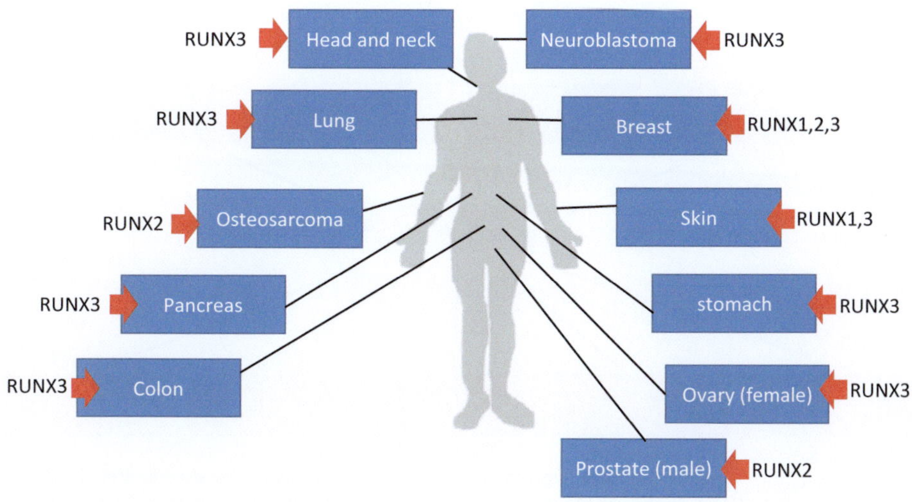

Fig. 19.1 The links of RUNX to solid tumors. Solid tumors with alterations of *RUNX* expression and activity are shown

2012). Moreover, *RUNX3* is silenced by hypermethylation of its promoter in cancers such as colon, lung, bladder, lung and bone (Chuang and Ito 2010) (Fig. 19.1). In fact, aberrant hypermethylation/inactivation of *RUNX3* is a very frequent event in the CpG island methylator phenotype (CIMP) subtype of colon cancer (Weisenberger et al. 2006). Mislocalisation of the RUNX3 protein in the cytoplasm has also been reported in gastric and breast cancer (Ito et al. 2005; Lau et al. 2006). The cytoplasmic localization of RUNX3 has been attributed to Src tyrosine kinase activation in the cancer cell lines (Goh et al. 2010) as well as defective TGFβ signaling (Ito et al. 2005). The crosstalk between epithelial cells and the microenvironment is a strong determinant of epithelial cancer initiation and progression. RUNX3 is one of the genes represented in a stroma-derived prognostic predictor – its expression in breast tumor stroma is associated a good clinical outcome (Finak et al. 2008).

19.4 RUNX in Developmental and Oncogenic Signaling Pathways

A central question is why the dysregulation of different *RUNX* genes led to distinct cancer types. For example, *RUNX1* haploinsufficiency caused predisposition to leukemia but its overexpression is necessary for skin tumors; RUNX2 overexpression is associated with bone cancer, as well as breast and prostate metastasis; RUNX3 inactivation is associated with multiple solid tumors, yet it is overexpressed in ovarian cancer (Nevadunsky et al. 2009; Lee et al. 2011). RUNX dosage therefore plays a critical role in determining proper cell growth. RUNX genes control cell fate through their ability to modulate the signaling outputs of major developmental pathways such as TGFβ, Wnt, Hippo, Hedgehog, Notch and Receptor tyrosine kinases (Ito et al. 2015) (Fig. 19.2).

If we could harness this potential by regulating RUNX expression, it would be a major advancement in cancer prevention or treatment.

19.5 Gastric Cancer: Proliferation, Survival and Invasion

A comprehensive molecular evaluation of gastric adenocarcinoma revealed key dysregulated pathways and putative drivers of various subtypes of gastric cancer (Cancer Genome Atlas Research 2014). The list of significantly mutated genes included those in the KRAS, β-catenin, TGFβ signaling, p53, Fanconi anemia and mitotic pathways (Cancer Genome Atlas Research 2014). As

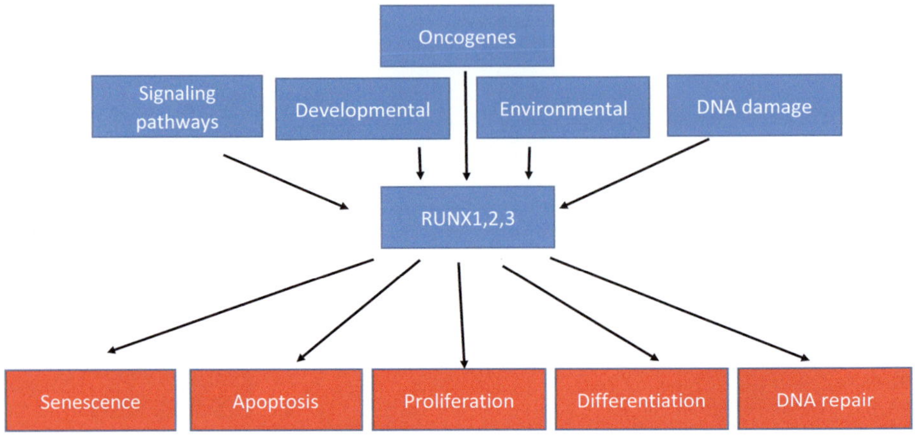

Fig. 19.2 RUNX integrates stimuli from the environment to influence cell fate. The *RUNX* family is regulated by signaling pathways, developmental stimuli, environmental cues, DNA damage, and oncogenic/hyperproliferative signals. In response to the diverse signals, RUNX proteins initiate transcription programs leading to senescence, DNA repair, apoptosis, differentiation or proliferation

discussed below, RUNX3 is a direct participant in most of these pathways.

In normal mouse and human stomach corpus, RUNX3 is expressed in the lower part of the epithelium, primarily in chief cells but not parietal cells (Ito et al. 2005; Ogasawara et al. 2009) (Figs. 19.3 and 19.4).

It is notable that 45–60% of human gastric cancer do not express RUNX3 – this is mainly due to RUNX3 promoter methylation/epigenetic silencing (Li et al. 2002). In addition, hemizygous deletion of the RUNX3 gene has also been identified. Mislocalisation of the RUNX3 protein in the cytoplasm was also reported to be a major form of RUNX3 inactivation (Fig. 19.3c) (Ito et al. 2005).

Helicobacter pylori infection is considered to be the primary cause of human gastric cancer. Chronic inflammation of stomach caused by *H. pylori* infection is associated with the loss of acid-producing parietal cells and pepsinogen-producing chief cells in the main body of the stomach (as known as corpus). Loss of parietal cells after *H pylori* infection is the main cause of oxyntic atrophy, a precancerous stomach condition (Weis and Goldenring 2009). The host inflammatory response, coupled with *H. pylori* virulence factors,resulted in promoter methylation and silencing of many tumor suppressor genes. *RUNX3* is one of them. The *CpG* island at the 5′ end of the P2 (proximal) promoter of *RUNX3* is frequently methylated in *H. pylori*-infected stomach and gastric cancer tissues (Katayama et al. 2009). The silencing of the *RUNX3* gene is therefore key epigenetic event during the development of gastric cancer.

Fig. 19.3 Comparison of RUNX3 expression in normal human gastric epithelial cells with cancer cells (Figures adapted from Ito et al. 2005, with permission from the American Association for Cancer Research). (**a**) Immunodetection of RUNX3 in the corpus and pyloric antrum of normal gastric epithelial cells. *Upper* and *lower boxed* regions are enlarged on the *right*. Counterstaining was done with hematoxylin. *Far right*, immunostaining without counterstaining in a section similar to the *lower* enlarged region is shown. *Arrows* indicate parietal cells with weaker immunoreactivity than the adjacent chief cells. (**b**) Detection of *RUNX3* mRNA by *in situ* hybridization with a *RUNX3* anti-sense or sense probe in normal gastric epithelial cells. (**c**) Immunodetection of RUNX3 in gastric cancer cells. Sections were prepared from differentiated (intestinal) gastric cancers. Three types of staining patterns for RUNX3 were observed: negative (44%; n = 43/97), positive (18%; n = 17/97), and cytoplasmic-positive (38%; n = 37/97). The *boxed* regions in the *upper panels* are enlarged *below*. Counterstaining was done with hematoxylin. Scale bars, 200 μm

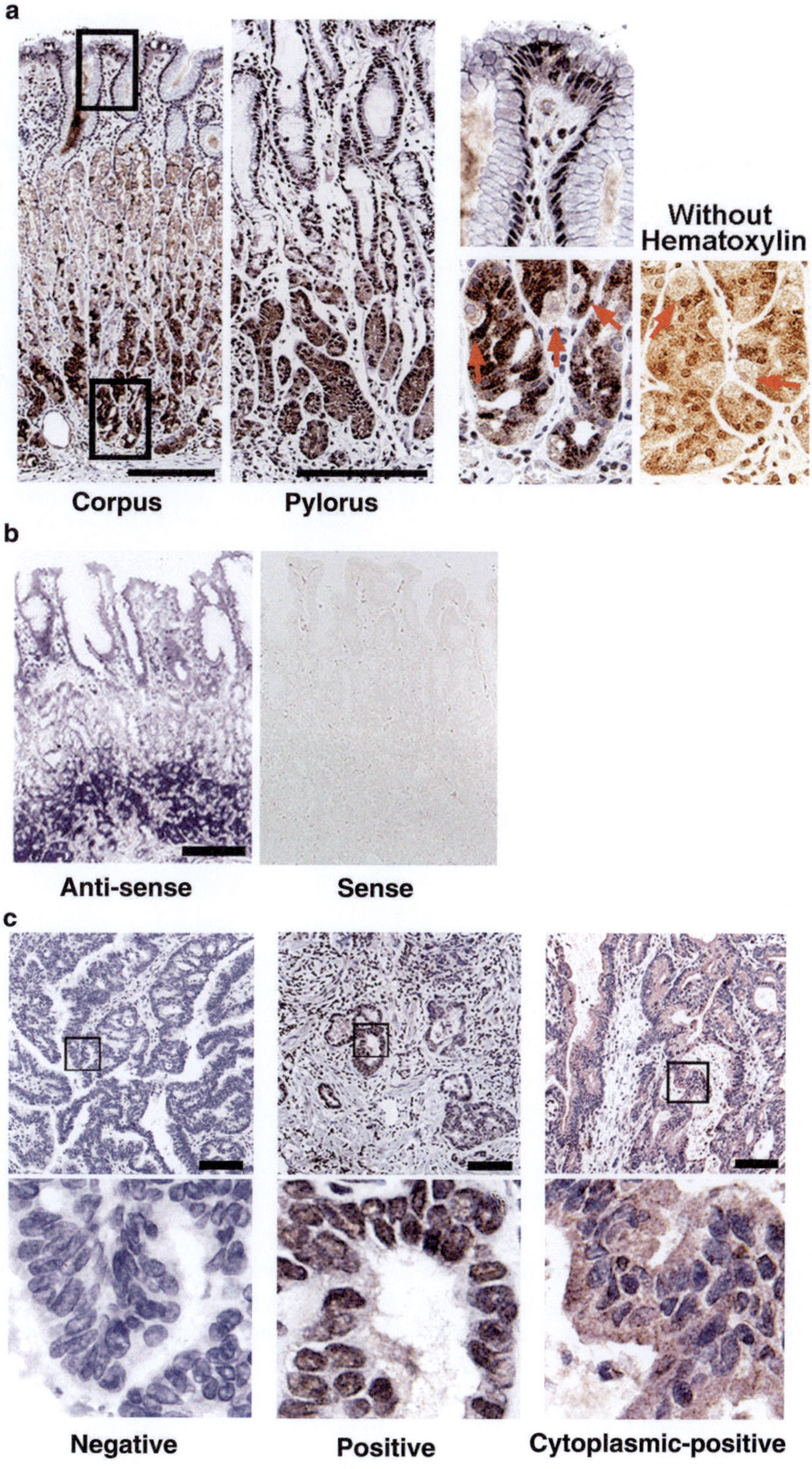
a
Without Hematoxylin
Corpus
Pylorus
b
Anti-sense
Sense
c
Negative
Positive
Cytoplasmic-positive

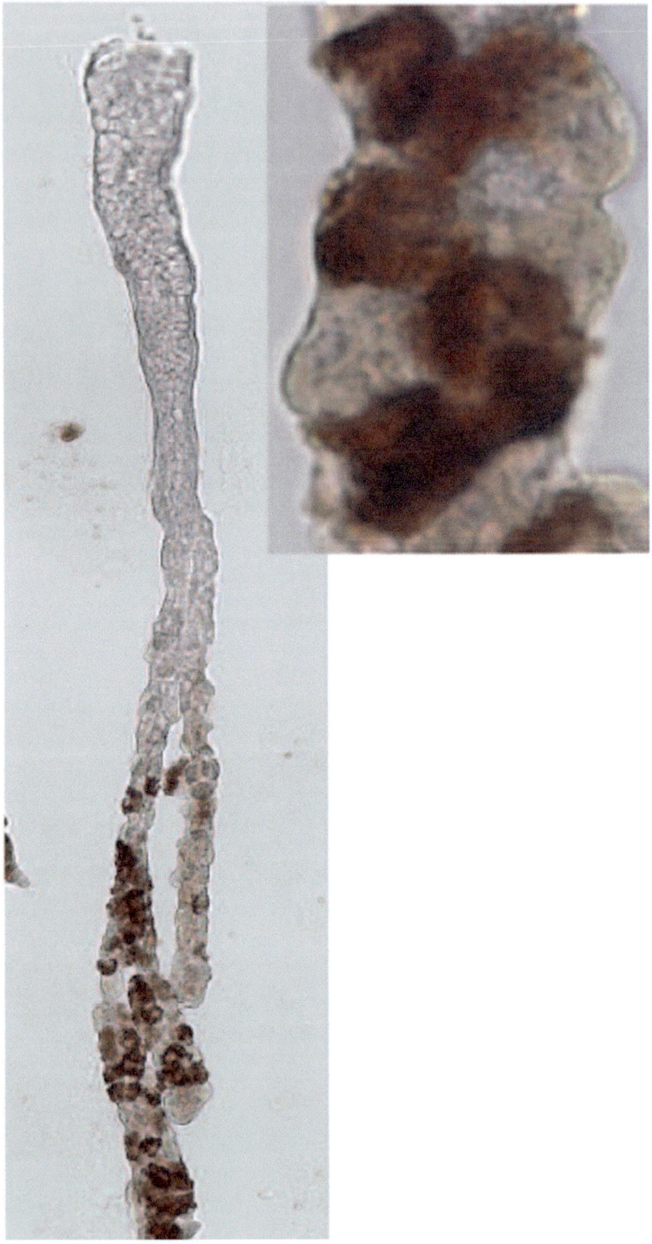

Fig. 19.4 RUNX3 is localized in the basal region of isolated fundic glands. Glands were isolated from the fundic mucosa of normal portions of resected human epithelium and evaluated for RUNX3 expression by immunohistochemistry. Original magnification: 100x; *inset*, 640x. (Figure adapted from Ogasawara et al. 2009, with permission from Histology and Histopathology)

From the gastric cancer-derived tissue, we found a *RUNX3* mutation, R122C, within the highly conserved Runt domain (Li et al. 2002). The oncogenic mutation R122C was also found in head and neck cancer (based upon data generated by the TCGA Research Network: http://cancergenome.nih.gov/) and thus represented a major breakthrough in understanding RUNX3's anti-tumor properties (Li et al. 2002). When we exogenously expressed RUNX3 in cultured cell lines, strong growth inhibition is often observed. Similar phenomenon has been reported for well-known tumor suppressors such as p53. We therefore suspected that RUNX3 may have tumor suppressor activity during the early stages of our studies on RUNX3. Likewise, when wild type RUNX3 is stably expressed in gastric cancer cell line, MKN28, a RUNX3-deficient gastric cancer cell line, tumorigenicity in immune-compromised mice is strongly inhibited compared with parental cell line. This is a strong indication that RUNX3 functions as a tumor suppressor in the stomach. When similar experiments were performed with the RUNX3^{R122C} mutant, tumor growth was not inhibited. Rather tumorigenic activity was enhanced beyond the level of the parental gastric cancer cell line: apparently, a single amino acid substitution R122C was sufficient to convert a strong tumor suppressor to an oncogene (Li et al. 2002). As we begin to understand the molecular mechanisms underlying growth inhibitory functions of wild-type RUNX3, how R122C provoked oncogenic activities becomes clear (see below).

In addition to clinical samples or molecular characterization, mouse models have also linked RUNX3 inactivation to gastric cancer. As with most animal models, we found interesting phenotypic variations among different mouse strains used to generate *Runx3* knockout. Conventional knockout of *Runx3* in C57BL/6 mice led to multiple developmental defects and early death – mostly within 24 h after birth. This is primarily due to a defective glosso-pharyngeal nervous system which resulted in an inability to control tongue and pharynx action to suck milk (unpublished data). In other strains such as BALB/c and ICR, a significant number of mice survived until adulthood. Groner's group reported that although *Runx3*$^{-/-}$ mice bred on ICR or MF1 background exhibited defects such as congenital sensory ataxia, reduced growth, and high mortality rates during the first 2 weeks after birth, some knockout mice survived and lived till old age (24 months) (Levanon et al. 2002; Brenner et al. 2004).

The strong influence of genetic variation on tumor development in epithelial tissues was also observed for different strains of Runx3 knockout mice. The gastric mucosa of the *Runx3*-null mice exhibited hyperplasia, due to increased proliferation and decreased sensitivity to transforming growth factor 1 (TGF-1) mediated growth suppression and apoptosis (Li et al. 2002). We had earlier reported that direct interaction with the effectors of TGFβ pathway, the SMAD proteins, is a common feature of all mammalian RUNX proteins (Hanai et al. 1999) (Fig. 19.5).

In response to TGFβ, RUNX3 directly activated the transcription of growth inhibitor *CDKN1A* (also known as *p21*^{CIP}) (Chi et al. 2005) and pro-apoptotic *BIM* in the stomach (Yano et al. 2006). Besides *BIM*, re-introduction of RUNX3 in a gastric cancer cell line also induced other genes involved in apoptosis, including those for Fas-associated death domain (FADD) and death-receptor mitochondria mediated apoptosis (Nagahama et al. 2008).

We investigated how *Runx3*-deficiency contributed to the distinct morphological changes in precancerous gastric epithelium. Conventional knockout of *Runx3* was generated with the BALB/c strain (Ito et al. 2011). These mice survived for 10–12 months and could be studied at the adult stage. The gastric epithelia in *Runx3*$^{-/-}$ mice show gradual loss of chief cells that express pepsinogen and the stomachs of 6-month-old *Runx3*$^{-/-}$ mice show nearly complete loss of pepsinogen expression, suggesting the loss of chief cells. Importantly, the population of Muc6- and TFF2-positive mucous neck cells was significantly elevated (Ito et al. 2011). Since chief cells are known to be derived from mucous neck cells, it is likely that chief cells trans-differentiated or retro-differentiated back to the precursor cells. This phenomenon is very similar to Spasmolytic Polypeptide Expressing Metaplasia (SPEM), a precancerous condition associated with 90% of resected gastric cancers (Weis and Goldenring 2009).

The *Runx3*$^{-/-}$ mouse stomach also showed intestinal metaplasia, characterized by the presence of intestinal goblet cells (indicated by the expression of intestine specific mucin Muc2) and Alcian Blue staining, which normally marks intestinal epithelium.

Stem cells in the stomach corpus epithelium are normally located in the isthmus. In the case of *Runx3*$^{-/-}$ stomach epithelium, bromodeoxyuridine (BrdU) incorporation – indicating the presence of rapidly growing cells – is observed throughout the epithelium suggesting that stem cell activity is enhanced. Organization or differentiation of epithelial cells might also be dysregulated

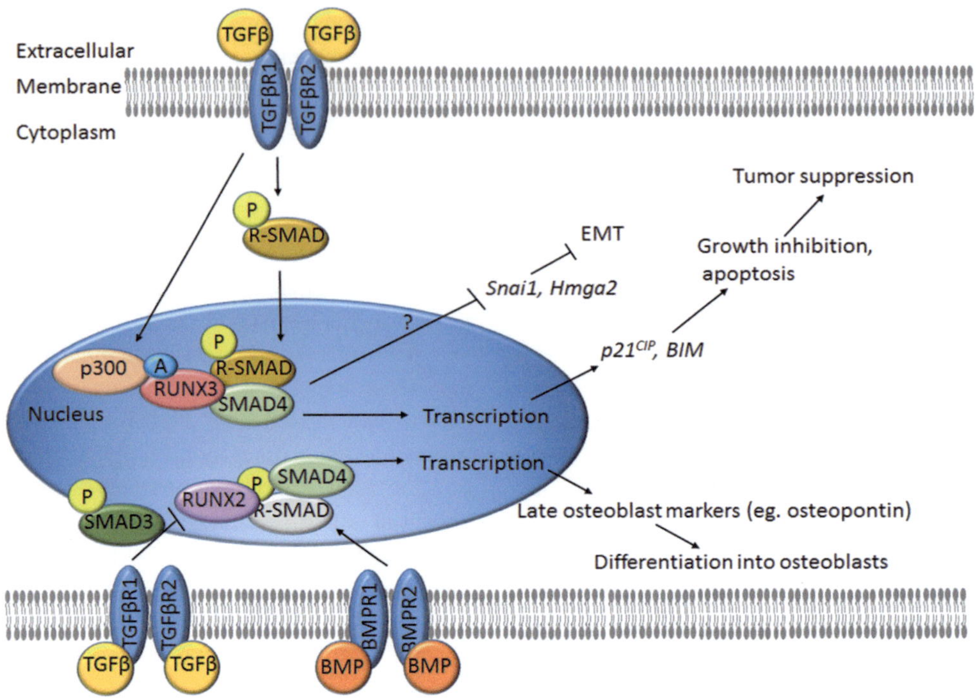

Fig. 19.5 Reciprocal interactions between RUNX and key elements of the TGFβ pathway. Activation of TGFβ receptors 1 and 2 (TGFβR1 and R2) by the TGFβ cytokine lead to phosphorylation of R-SMAD proteins. Phosphorylated R-SMAD translocates into the nucleus and forms a multiprotein complex with SMAD4 and RUNX3. This synergistic cooperation with RUNX3 and the SMAD proteins – key effectors of TGFb pathway – strongly induces transcription of $p21^{CIP}$ and *BIM*, which are respectively associated with growth inhibition and apoptosis as well as tumor suppression. RUNX3-SMAD complex also inhibits *Snail*, *Hmga2* and EMT by unknown mechanisms. Activation of TGFβR1 and R2 also enhance p300-mediated acetylation of RUNX3, resulting in increased stability of RUNX3, thereby promoting complex formation with SMAD effectors. Conversely, activation of TGFβR1 and R2 phosphorylates SMAD3 which inhibits RUNX2. Instead, the BMP pathway promotes interaction of RUNX2 with SMAD proteins, leading to transcription of late osteoblast markers (eg. Osteopontin) and differentiation to osteoblasts

(Ito et al. 2011). This observation unequivocally indicates strong tumor suppressor activity of Runx3 in the stomach epithelium.

Importantly, we detected a marked elevation of Wnt activity in the *Runx3*$^{-/-}$ gastric epithelium, suggesting that oncogenic Wnt drives the development of intestinal metaplasia in *Runx3*$^{-/-}$ stomach. This observation is consistent with the finding that RUNX3 inhibits the DNA binding activity of the TCF4/β-catenin complex (Ito et al. 2008) (see intestinal cancer section). The Runt domain of RUNX3 protein interacted with the DNA binding domain of TCF4, thereby inhibiting the DNA binding of both proteins. Indeed, Wnt activity is up-regulated in *Runx3*$^{-/-}$ mouse stomach (Ito et al. 2011).

Although reported mouse models of gastric cancer rarely develops invasive carcinoma, a low dose of the chemical carcinogen N-methyl-N-nitrosourea (MNU), which did not affect the stomach of wild-type mice, induced invasive gastric cancer in *Runx3*$^{-/-}$ mice at 52 weeks of age (Fig. 19.6c–e) (Ito et al. 2011). This observation suggests that RUNX3 protects against DNA damage-induced stress and tumorigenesis in the stomach. Indeed, a non-transciptional role for RUNX3 in the Fanconi anemia DNA repair pathway has been described (Wang et al. 2014) and is discussed in detail elsewhere in this book.

The susceptibility to carcinogen-induced gastric cancer may also indicate mitotic defects in *Runx3*$^{-/-}$ mice. Key regulators of mitosis Aurora

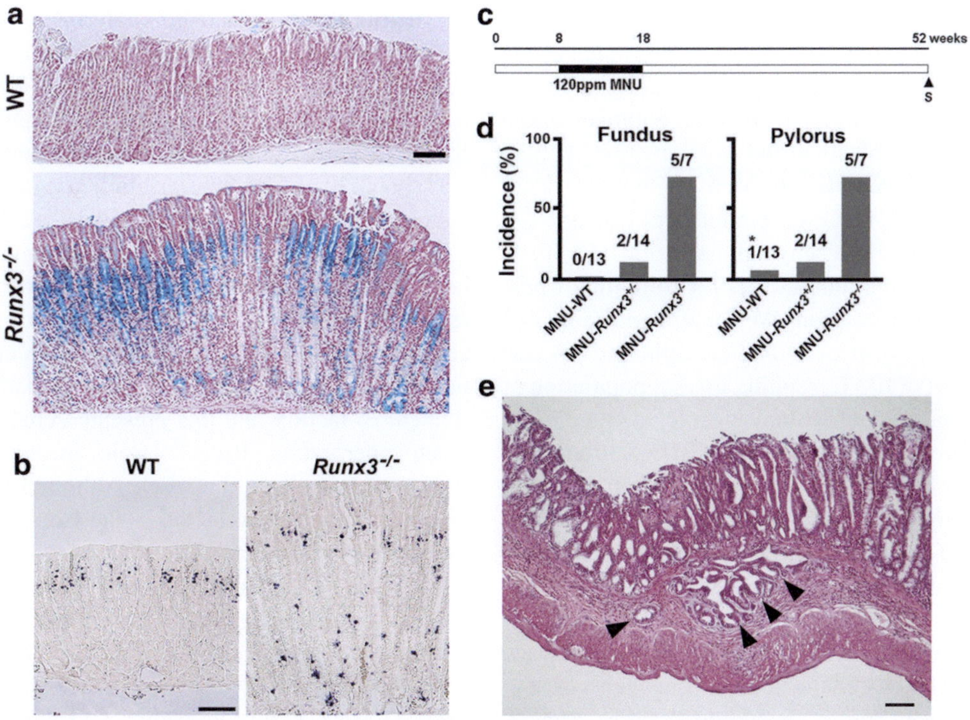

Fig. 19.6 Low dose of MNU induced cancer development in precancerous *Runx3*$^{-/-}$ stomach (Figures adapted from Ito et al. 2011, with permission from the Gastroenterology, Elsevier). (**a**) Alcian blue staining of wild-type (WT) and *Runx3*$^{-/-}$ fundic glands. (**b**) Detection of replicating cells by BrdU incorporation in WT and *Runx3*$^{-/-}$ fundic glands. (**c**) Mice were treated with 120 p.p.m. *N*-methyl-*N*-nitrosourea (MNU) through their drinking water for 10 weeks (from 8 to 18 weeks of age) and were sacrificed at 52 weeks (1 year) of age. (**d**) Frequency of adenocarcinoma development in the fundus and pylorus of MNU-treated WT, *Runx3*$^{+/-}$, and *Runx3*$^{-/-}$ mice at 52 weeks of age. All adenocarcinomas except one in WT (*) migrated into the submucosa. (**e**) Morphology of fundic glands in MNU-treated *Runx3*$^{-/-}$ mice at 52 weeks of age, stained by hematoxylin and eosin. Cancerous glands that invaded the submucosa are indicated by arrow heads. Scale bars, 100 μm

kinases induced phosphorylation of RUNX3 at G2/M transition (Chuang et al. 2016). It is possible that phosphorylation of RUNX triggers mitotic entry. Phosphorylation of threonine 173 in the Runt domain of RUNX3 (and its equivalent in RUNX1 and RUNX2) detaches RUNX from the DNA and promotes its localization at the mitotic apparatus such as the centrosome and midbody (Chuang et al. 2016). Currently, the biological consequence of T173 phosphorylation is unclear but not likely to involve DNA binding or transcription regulation. The identification of T173I mutation in colon cancer suggests the importance of this residue in cancer development (cancer.sanger.ac.uk/cosmic). As indicated above, RUNX protein can exert its tumor suppressor activity through protein-protein interaction without binding to DNA. The involvement of other RUNX proteins in mitosis has been reported. RUNX2 ensures transmission of parental epigenetic memory to daughter cells (Young et al. 2007a, b); RUNX1 contributes to the spindle assembly checkpoint (Ben-Ami et al. 2013). How much functional overlap of RUNX activities in mitosis remains to be seen.

Inflammation is well established as an oncogenic factor in the stomach. RUNX3 cooperates with TNF-α/NF-kB pathway and *Helicobacter pylori* infection to directly upregulate *IL23A* in gastric epithelial cells (Hor et al. 2014), However, *IL23B* is not produced and the Interleukin-23 (IL-23) heterodimeric cytokine cannot be formed. Why IL23A alone is secreted from gastric epithelial cells is unknown. However, the ability of RUNX3 to induce *IL23A* is strongly suggestive of a RUNX3 role in the innate immunity of gastric epithelial cells, where it enhances pathogen clearance during infection and inflammation. It may also be protecting stomach epithelium from inflammation, hence development of cancer.

Epithelial-mesenchymal transition (EMT) is one of the important factors for solid tumor progression and invasion. We found that the loss of Runx3 in gastric epithelial cells led to the induction of EMT, resulting in a subpopulation of cells which acquired tumorigenic, stem cell-like properties (Voon et al. 2012). RUNX3 therefore protects gastric epithelial cells from aberrant TGFβ signaling and subsequent reprogramming by EMT (Fig. 19.5). Involvement of RUNX in EMT is described in more detail elsewhere in this book.

The nuclear effector of the Hippo pathway, YAP-TEAD4, functions as an oncogene in several cancer types including gastric cancer. Increased TEAD-YAP expression is significantly correlated with poor survival of gastric cancer patients. We found that RUNX3 negatively regulates the oncogenic TEAD–YAP complex in gastric cancer (Qiao et al. 2015). The Runt domain of RUNX3 interacts with the DNA binding domain of TEAD4, resulting in attenuation of TEAD4 DNA binding activity and downregulation of TEAD-YAP mediated transcription. Various YAP-TEAD target genes (eg. *collagen type XII* and *calpain 6*) that were involved in metastasis and apoptosis were suppressed by RUNX3 (Qiao et al. 2015).

As discussed earlier, the cancer-derived mutation R122C has oncogenic effects. The $RUNX3^{R122C}$ mutant showed severe defects in DNA binding and could not induce $p21^{CIP1}$ transcription (Chi et al. 2005). $RUNX3^{R122C}$ is also defective in tumor suppressor activity mediated by protein-protein interactions: $RUNX3^{R122C}$ did not interact with Wnt effector TCF4 and failed to suppress oncogenic Wnt (Ito et al. 2011). $RUNX3^{R122C}$ also failed to interact with TEAD4 and, therefore, cannot suppress the oncogenic activity of TEAD4-YAP complex (Qiao et al. 2015).

The tumor suppressor function of RUNX3 in gastrointestinal tumors is much debated for many years, because of the low expression level of Runx3 in normal gastric epithelium. Expression of RUNX3 in normal human stomach epithelium was shown by in situ hybridization and immunohistochemistry (Fig. 19.3a, b) (Ito et al. 2005). Co–expression of RUNX3 and pepsinogen in isolated human gastric unit of corpus was also reported (Ogasawara et al. 2009). At issue is the expression level of Runx3 in the stomach epithelial cells of normal mice kept in specific pathogen free (SPF) facility and this presented challenges when interpreting Runx3's role in epithelial tumors (Levanon et al. 2011). Although *Runx3* expression is easily detected in the stomach epithelium of wild mice (ie. mice caught in the field), *Runx3* expression in mice kept in modern mouse facilities are generally quite low, requiring highly sensitive detection techniques. However, it has become increasingly apparent that Runx3 expression is highly dynamic (Whittle et al. 2015) and changes according to environmental cues. In response to oxidative and osmotic stress, the *Caenorhabditis elegans* RUNX homolog RNT-1 protein is rapidly stabilized in the intestine (Lee et al. 2012). In humans, chemotherapeutic and DNA damaging agent doxorubicin induced RUNX3 expression in different cultured cell lines (Yamada et al. 2010). Oncogenic stress, such as expression of mutant *K-Ras*, induced Runx3 expression in human embryonic kidney HEK293 cells (Lee et al. 2013). This observation reinforced the notion that stress response is a fundamental, as well as evolutionarily conserved, function of RUNX3. In this study, RUNX3 was found to be transcriptionally activated by oncogenic K-Ras to mediate the expression of p53. Conceptually, RUNX3 may serve as a monitor of the level of K-Ras activity and other oncogenic signals (Lee et al. 2013). The fact that RUNX2 was strongly induced in Ha-ras transformed NIH3T3 indicates a close relationship between RUNX induction and proliferation (Ogawa et al. 1993). It might be that all *RUNX* genes are activated by stress or other oncogenic signals to serve a fundamental function – protect normal cells from tumorigenesis.

The molecular identification of tissue stem cells, which can give rise to stomach cancer, will offer mechanistic insights on how gastric tumor is initiated, sustained or metastatic. We have successfully generated mice with *Runx1* or *Runx3* knockout in hematopoietic stem cells (Wang et al. 2014) and will be able to target tissue stem cells in the near future. Stem cell specific knockout of *Runx* genes singly or in combination will reveal more precise roles of individual *Runx* genes in the stomach epithelium.

As discussed above, completely different approaches have converged on the tumor suppressor function of RUNX3 in the stomach. Its abilities to engage multiple signaling pathways to suppress growth, attenuate oncogenic signaling, induce apoptosis and antagonize EMT have important implications in cancer treatment. A cell permeable RUNX protein was developed with promising results – locally administered RUNX3 suppressed the growth of subcutaneous human gastric tumor xenografts with increased $p21^{CIP1}$ and decreased VEGF expression – consistent with the interaction of RUNX3 with TGFβ signaling (Lim et al. 2013). Moreover, *RUNX3* is frequently epigenetically silenced in cancer, its expression, and perhaps anti-tumor activity, can be pharmacologically restored by inhibitors of DNA methyltransferases and histone deacetylases.

Given the prominent role of RUNX3 in the stomach, a pertinent question would be whether RUNX1 and RUNX2 functionally compensate for the anti-tumor activity of RUNX3. However, mutational analysis of RUNX1 in laser-captured gastric cancer cells of 44 patients did not reveal any significant mutation. Moreover, RUNX1 mRNA was detected in many gastric cancer cell lines and cancer tissues, suggesting that RUNX1 might not play a major role in suppression of most gastric cancers (Usui et al. 2006). It is important to note that while all RUNX proteins have the ability to regulate $p21^{CIP1}$ transcription through the multiple RUNX consensus binding sequence in the $p21^{CIP1}$ promoter, the downstream effects are different. RUNX1 regulates the $p21^{CIP1}$ promoter in a cell type dependent manner, transactivating the $p21^{CIP1}$ promoter in myeloid leukemia cells and repressing the $p21^{CIP1}$ promoter in NIH3T3 fibroblasts (Lutterbach et al. 2000). RUNX2 repressed the *CDKN1A* promoter and attenuated TGFβ1-mediated growth inhibition and apoptosis in vascular endothelial cells (Sun et al. 2004).

19.6 Intestinal Cancers: Interaction with Wnt Signaling

Studies on genetic alterations have overwhelmingly implicated the Wnt signaling pathway as the main player in colon cancer pathogenesis. Adult $Runx3^{-/-}$ BALB/c mice exhibited increased proliferation and hyperplasia in the epithelia of the jejunum and colon (Ito et al. 2008) (Fig. 19.7a, b).

Furthermore, Wnt target genes such as *CD44*, *cyclin D1*, *c-Myc*, *conductin* and *EphB2* were upregulated in the intestine of $Runx3^{-/-}$ mice. The adenomatous polyposis coli (APC) gene is well established as a major regulator of Wnt signaling in colorectal cancers. We found that at 65 weeks of age, $Runx3^{+/-}$ mice developed small intestinal adenomas at a frequency similar to that of $Apc^{Min/+}$ mice with the same BALB/c background (Fig. 19.7c) (Ito et al. 2008). Strikingly, adenomatous polyps were obtained from $Runx3^{+/-}$ $Apc^{Min/+}$*mice*, suggesting that the combined effects of defective *Runx3* and *Apc* genes drive progression from adenoma to adenocarcinoma (Fig. 19.7c). Since the analysis of the very small adenomas from $Runx3^{+/-}Apc^{Min/+}$ mice revealed either downregulation *Runx3* expression or nuclear accumulation of β-catenin, but not both phenotypes (Fig. 19.7d), it is possible that the adenomas developed because of biallelic inactivation of either *Apc* or *Runx3*. Interestingly, the large adenomas or adenocarcinomas of the $Runx3^{+/-}Apc^{Min/+}$ mice showed β-catenin accumulation suggesting that defects in both genes contribute to heightened activation of the oncogenic Wnt pathway. Mechanistically, RUNX3 formed a ternary complex with the Wnt effector complex TCF4-β-catenin, which resulted in attenuation of

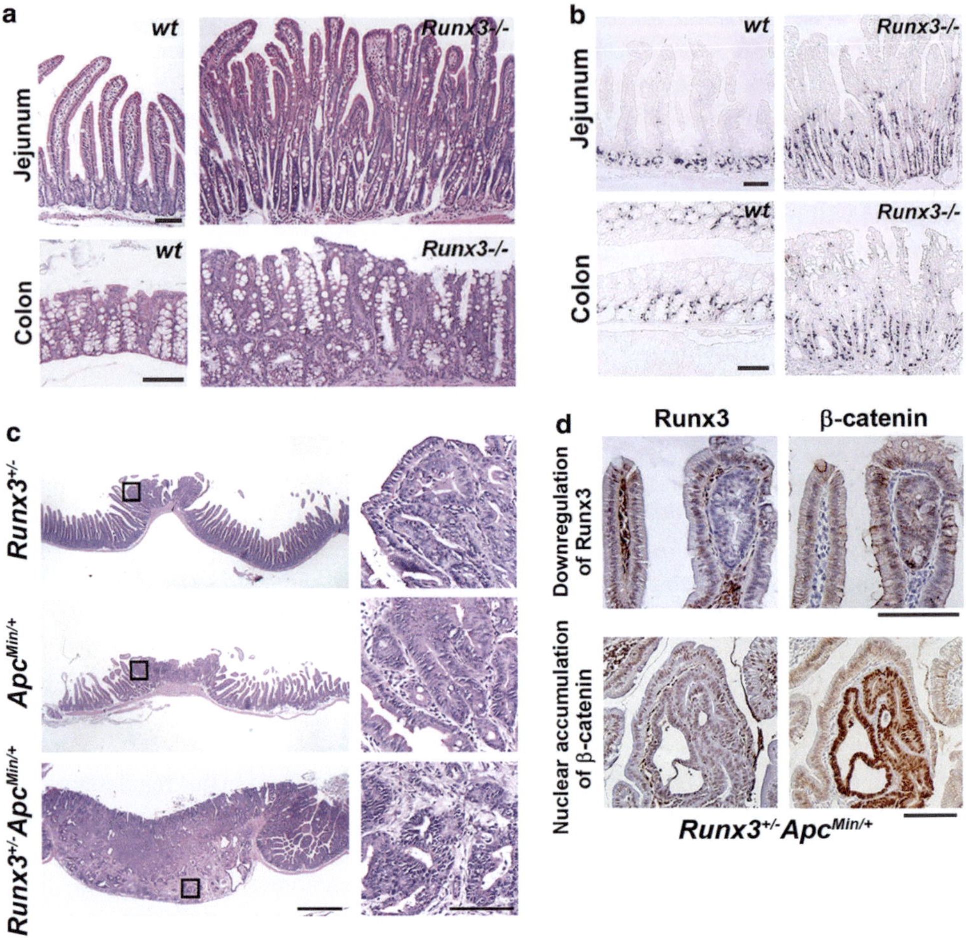

Fig. 19.7 Development of adenomatous polyps in *Runx3*$^{+/-}$ and adenocarcinoma in *Runx3*$^{+/-}$*Apc*$^{Min/+}$ BALB/c mice (Figures adapted from Ito et al. 2008, with permission from the Cell Press, Elsevier). (**a**) Morphology of jejunum and proximal colon of wild-type (WT) and *Runx3*$^{-/-}$ mice at 40 weeks of age. Tissues were stained by hematoxylin and eosin. (**b**) Detection of proliferating cells in WT and *Runx3*$^{-/-}$ jejunum and proximal colon by BrdU incorporation (adult mice at 40 weeks of age). (**c**) *Left*, Hematoxylin and eosin staining of small intestines in *Runx3*$^{+/-}$, *Apc*$^{Min/+}$ and *Runx3*$^{+/-}$*Apc*$^{Min/+}$ mice at 65 weeks of age. Adenomatous polyps in *Runx3*$^{+/-}$ and *Apc*$^{Min/+}$ mice were compared with adenocarcinomas in *Runx3*$^{+/-}$*Apc*$^{Min/+}$. *Right*, Boxed regions in *left* panel are enlarged. (**d**) Analysis of very small adenomas formed in the jejunum of *Runx3*$^{+/-}$*Apc*$^{Min/+}$ mice. Downregulation of Runx3 was not associated with activation of β-catenin (*upper panels*). In contrast, nuclear accumulation and activation of β-catenin were not associated with downregulation of Runx3 (*lower panels*). Counterstaining was done with hematoxylin. Scale bars correspond to 100 μm (**a–d**) and 1 mm (*left* in **b**)

TCF4-β-catenin DNA binding ability and suppression of oncogenic Wnt mediated transcription (Ito et al. 2008) (Fig. 19.8).

Moreover, the adenomatous polyps of *Runx3*$^{+/-}$ mice acquired *CpG* island methylation of the *Runx3* promoter, suggesting that DNA methylation is responsible for the downregulation the remaining *Runx3* gene. This reflects the human scenario and points to DNA methylation as major mechanism of *Runx3* inactivation in cancer. Given that Wnt signaling drives proliferation of intestinal epithelial cells, and that overactive Wnt has a causal role in intestinal tumor formation, it is not surprising that RUNX3, through its modulation of the Wnt pathway, is critical for the homeostatic regulation of growth and differentiation in the intestinal epithelia.

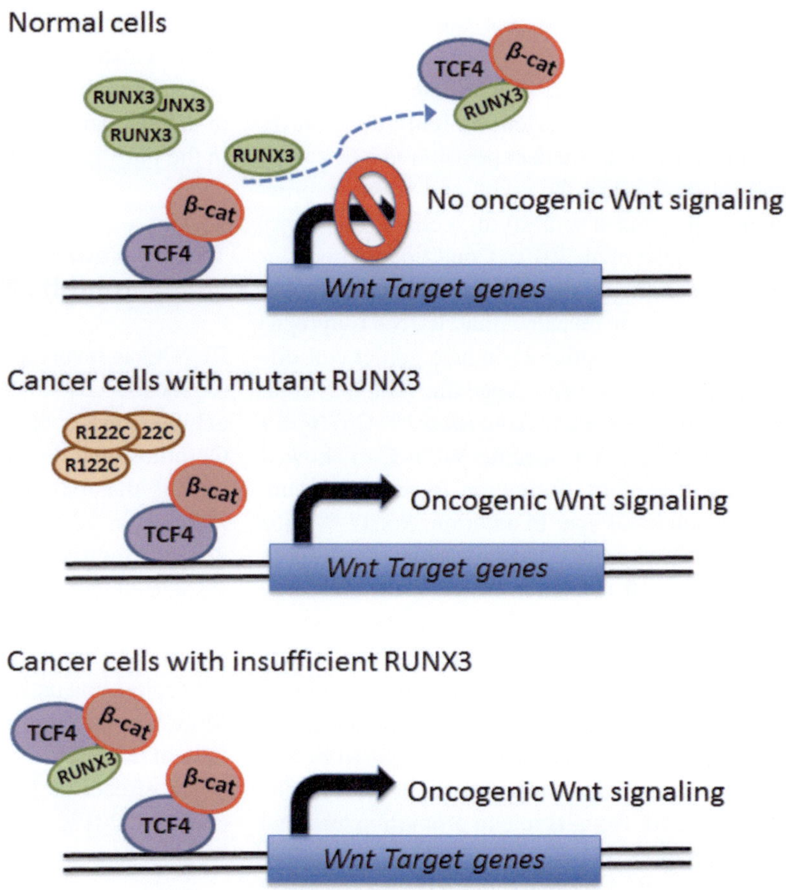

Fig. 19.8 Dysregulation of RUNX3 affects oncogenic Wnt signaling. *Top*, in normal cells with sufficient amounts of RUNX3, RUNX3 interacts with the TCF4-β-catenin complex and prevents the Wnt effector from binding to DNA; transcription of downstream genes of Wnt signaling is attenuated. *Middle*, in cancer cells with *RUNX3*^{R122C} mutation, the mutant RUNX3 fails to bind to TCF4-β-catenin. TCF4-β-catenin induces transcription of oncogenic Wnt downstream genes to promote tumorigenesis. *Bottom*, in cancer cells with insufficient RUNX3 levels (e.g. due to epigenetic silencing of *RUNX3*), excess TCF4-β-catenin proteins induce oncogenic Wnt transcriptional program

Recently, a different aspect of RUNX-TCF4-β-catenin interaction in the *Drosophila* embryonic midgut was reported (Fiedler et al. 2015). Instead of displacing TCF4 from the DNA, Runt functioned as a key component of the multi-protein Wnt enhanceosome that was tethered to TCF enhancers. Simultaneous interactions of RUNX with Groucho/TLE (through its C-terminal WRPY motif), TCF and a protein complex called ChiLS [composed of Chip/LDB ((Lin-11 Isl-1 Mec-3-) LIM-domain-binding protein) and single-stranded DNA binding protein (SSDP)] fine-tune the Wnt output (Fiedler et al. 2015). Human RUNX1/2/3 directly interacted with ChiLS, thereby suggesting the conservation of an ancient RUNX-Wnt link for a pivotal role in gut development. Whether RUNX3 suppress or activate the Wnt signaling pathway is thus context-dependent. Because the ChiLS complex also regulates Notch responsive enhancers, defining the factors that determine the function of the RUNX in this complex might help explain the interplay of RUNX, Notch and Wnt in intestinal cancer.

Bacteria *Citrobacter rodentium* infection is associated with epithelial cell hyperproliferation in mice. Moreover, it induced mucosal hyperplasia and adenoma formation in the colon of *Apc*$^{Min/+}$ mice; it has thus been proposed that infectious colitis is a risk factor for colon cancer (Newman 2001). RUNX3 is involved in the development of innate lymphoid cells (ILC), which reside on the intestinal mucosa to direct immune defenses against pathogenic infection (Ebihara et al. 2015). *Runx3* knockout in ILC cells resulted in impaired immune response against *Citrobacter rodentium* infection – the mice exhibited prolonged epithelial injury, crypt hyperplasia and increased inflammatory cell infiltration (Ebihara et al. 2015). It is possible that RUNX3 deficiency and ensuing defective immune response are risk factors to tumorigenesis.

RUNX1 also contributes to intestinal homeostasis. RUNX1 expression is highly dynamic in the intestine, with weak expression in a few stem cells at the base of the crypt, and strong expression in the transit amplifying cells at the upper crypt (Scheitz et al. 2012). Conditional knockout of *Runx1* in the mouse colon using the inducible Mx-Cre system revealed that RUNX1 upregulates *Klf4* transcription to induce goblet cell differentiation in the mouse intestine (Buchert et al. 2009). Conditional knockout *Runx1*$^{-/-}$ C57BL/6 J mice (with intestine-specific Villin-Cre) showed increased adenoma formation in the duodenum by 12 months of age. In addition, *Runx1*$^{-/-}$ mice in *Apc*^{min} background led to further increases in tumor size and formation in the small and large intestines (Fijneman et al. 2012). This suggests that RUNX1 is a tumor suppressor in the intestine. Knockdown of *RUNX1* in colon cancer cell lines, however, did not reveal any changes in cell growth (Scheitz et al. 2012). It is possible that RUNX1 tumor suppressor activity in the intestine stems, in part, from its role in promoting terminal differentiation.

In contrast to RUNX1 and RUNX3, RUNX2 promoted oncogenic activity in the human colon carcinoma cells; depletion of RUNX2 resulted in decreased proliferation, migration and invasion abilities of colorectal cell lines SW480 and DLD1 (Sase et al. 2012). Mechanistically, RUNX2 directly upregulated the metastatic gene osteopontin by binding to the RUNX binding sites in its promoter (Wai et al. 2006). In human colon carcinoma patients, elevated RUNX2 expression was significantly correlated with tumor stage and liver metastasis, further indicating its oncogenicity and as well as potential as prognostic factor (Sase et al. 2012).

A role for RUNX in gut development is conserved from *Caenorhabditis elegans* to *Drosophila* to human (Nam et al. 2002). One RUNX gene in the *Caenorhabditis elegans* is sufficient to dictate gut development and acquisition of multiple RUNX genes, while advantageous for complex organisms, might be inherently dangerous. Clearly, the imbalance of the RUNX1, RUNX2 and RUNX3 expression contributes to gastrointestinal cancers in humans – how RUNX2 activity impinges on that of RUNX1 and RUNX3 to activate an oncogenic transcriptional program in the intestine remains unknown.

19.7 Neuroblastoma: RUNX3 Inhibits MYCN

RUNX3 is involved in dorsal root ganglion neurogenesis. 20–40% of neuroblastoma cases exhibit loss of heterozygosity at *1p36*. Chromosomal deletion at *1p36* is reported to be one of the reasons why *RUNX3* expression is reduced in advanced neuroblastoma (Yu et al. 2013). Amplification of *MYCN* was significantly associated with reduced *RUNX3* expression in neuroblastoma patients. High *RUNX3* expression is associated with a more favorable prognosis in neuroblastoma patients. RUNX3 binds directly to MYCN to promote ubiquitin-mediated degradation of MYCN, thereby suppressing the oncogenic effects of MYCN transcriptional activity (Yu et al. 2013).

19.8 Dualistic Roles of RUNX3 in Pancreatic Cancer

The ability of RUNX3 to suppress proliferation and the discovery of a cancer mutation RUNX3^{R122C} that promotes oncogenic growth, would reasonably have led to the view of RUNX3 as a classical tumor suppressor. This is however a simplistic assumption. Increasingly, we note that precise expression levels of RUNX proteins are critical for proper growth. A recent study on pancreatic ductal adenocarcinoma (PDA) revealed that the regulation of RUNX3 expression is highly complex and dependent on multiple inputs from the gene expression landscape of the cancer cells (see relevant chapter in book). Hingorani's group explored RUNX3's activity in the context of key genetic changes such as activating *KRAS* mutation, *p53* point mutation and loss of *SMAD4* expression. They showed that SMAD4 regulates *RUNX3* expression in a biphasic, dose-dependent manner. Total loss of SMAD4 is correlated with elevated RUNX3 expression, which in turn promotes cell migration and metastasis of PDA cells (Whittle et al. 2015). Mechanistically, overexpression of RUNX3 upregulated osteopontin transcription to facilitate distant colonization. And yet, RUNX3 still possesses the ability to stimulate p21^{CIP1} expression and suppress proliferation in PDA cells, as previously observed in gastric cell lines. The role of RUNX3 in PDA is therefore dualistic. Depending on inputs from TGFβ and possibly p53, RUNX3 can function as both tumor suppressor and metastasis promoter in PDA (Whittle et al. 2015).

19.9 Skin Cancer: Stem Cells and Differentiation

It is not surprising, therefore, that RUNX genes can also function as potent oncogenes. An interesting example is skin cancer. All three *Runx* genes are expressed in the hair follicle (Levanon et al. 2001; Raveh et al. 2006). While some functional compensation of the three genes is likely, mouse knockout studies showed obvious phenotypes for single *Runx* knockout mice. Depletion of *Runx1* affects hair structure. *Runx1* expression is highly dynamic as cells progress through the different stages of hair follicle development, indicating that RUNX1 dosage is necessary for normal differentiation. *Runx2* knockout mice are impaired in follicle maturation. It was suggested that the dynamic expression of *Runx2* in the hair follicle cooperates with the hedgehog pathway to regulate skin thickness and hair follicle development (Glotzer et al. 2008). *Runx3* knockout mice showed changes in hair type composition and intrinsic shape of the hair (Raveh et al. 2005). In the human, immunohistochemistry revealed that RUNX3 is expressed all epidermal layers of normal skin – in particular, the number of RUNX3 expressing cells is prominent in the basal cell layer and hair shaft (Salto-Tellez et al. 2006).

Several lines of study have implicated RUNX proteins in skin tumor. Immunohistochemistry revealed that RUNX3 is overexpressed in basal cell carcinoma (BCC), when compared to normal epidermis. No mutation was found in the *RUNX3* gene, suggesting that it is fully functional and contributes as an oncogene to BCC pathogenesis (Salto-Tellez et al. 2006). Perhaps the most compelling evidence for a RUNX role in skin tumor come from the adult hair follicle stem cells (HFSC) in mice with conditional knockout of *Runx1* (using keratin 14 driven Cre on BL6 and/or CD1 backgrounds). These mice showed defective de novo production of hair shafts and differentiated hair lineages, which was attributed to a prolonged quiescent phase during the first hair cycle (Osorio et al. 2008). RUNX1 promoted proliferation of the HFSC, in part through transcriptional repression of cell cycle inhibitor *CDKN1A* and STAT inhibitors *SOCS3/4* (Hoi et al. 2010; Scheitz et al. 2012). RUNX1-mediated stimulation of Stat3 signaling is likely to be a major factor for cancer growth and survival in the skin. In fact, strong RUNX1 expression was observed in the skin epithelium, papillomas and squamous cell carcinomas derived from chemically induced skin tumors (Hoi et al. 2010). A two-step carcinogenic treatment on the *Runx1* conditional knockout mice revealed that RUNX1 expression was required for tumor initiation (Scheitz et al. 2012). The finding that RUNX1 is frequently overexpressed in human epithelial cancers, and necessary for the growth and survival of skin squamous cell carcinoma (SCC) and oral SCC (Scheitz et al. 2012), suggests that overexpression of Runx1 plays a key role in solid tumor initiation.

19.10 Bone Cancer and Metastasis

The frequent overexpression of RUNX2 in osteosarcoma suggests that it might promote bone cancer (Martin et al. 2011). RUNX2 expression is dynamically regulated during bone development. Its levels are low in mesenchymal progenitor cells, where it suppresses growth. During osteoblast differentiation, RUNX2 levels are upregulated and its synergistic cooperation with Smad proteins promotes bone specific gene transcription. The interaction of RUNX2 with the TGFβ/BMP pathway is critical for bone formation

(Fig. 19.5). Bone formation also requires the crosstalk of TGFβ/BMP with other pathways such as Wnt (Zhou et al. 2008), MAPK, Indian Hedgehog (Yoshida et al. 2004), Notch and Akt/mTOR – these signaling inputs have been linked to RUNX2 (Martin et al. 2011). While it is conceivable that dysregulated RUNX2 levels adversely affect the output of the above signaling pathways, the mechanistic link between RUNX2 and osteosarcoma remains unclear.

RUNX2 is also heavily implicated in metastasis to the bone from breast and prostate cancer. Runx2 is a direct upstream activator of genes involved in angiogenesis, survival, invasion and metastasis (Akech et al. 2010). These genes which include *VEGF*, *survivin* and *osteopontin* are associated with EMT. While normal prostate epithelial cells show negligible RUNX2 expression, advanced prostate tumors and metastatic prostate cancer cell lines are associated with high RUNX2 levels (Akech et al. 2010). The Runx2-Smad complex was shown to promote metastasis to distal sites; Runx2 expressing prostate cancer cells generated mixed osteolytic and osteoblastic lesions, which further metastasized to the lung (Zhang et al. 2015).

19.11 Breast Cancer: An Imbalance of RUNX Dosage

Since the involvement of RUNX in breast cancer is reviewed in detail elsewhere in this book, below is the brief comparison of the phenotypes of mice with altered Runx expression in the breast. The identification of RUNX1 as a significantly mutated gene in human luminal breast cancer suggests that RUNX1 plays a causal role in the pathogenesis of breast cancer (Ellis et al. 2012). In the mammary gland, the two major epithelial cell types are the luminal and the myoepithelial lineages. In adult virgin mice, the luminal and basal cells both show predominant *Runx1* expression, as compared to *Runx2* and *Runx3* (van Bragt et al. 2014). During pregnancy, *Runx1* expression is found mainly in the myoepithelial cells and absent from the alveolar luminal cells. Conditional knockout (KO) of *Runx1* using MMTV-Cre transgenic mice, which mainly targeted luminal epithelial cells, did not exhibit any gross morphological abnormalities. However, lactating Runx1 KO mice were distinguished by milk stasis and reduction in the luminal population. *Runx1* KO luminal cells exhibited a gene expression signature resembling progenitor cells in the luminal lineage (van Bragt et al. 2014). Elf5, a transcription factor which marks luminal progenitor cells and is critical for the alveolar cell lineage, was upregulated in *Runx1* KO luminal cells; conversely, expression of ductal luminal transcription factors, such as Foxa1, ERα and Cited1, were reduced (van Bragt et al. 2014). Importantly, the ER+ mature luminal cell population was decreased in the Runx1 KO mice and this phenotype that could be rescued if either *p53* or *RB1* was mutated. Moreover, *RUNX1* mutations frequently co-occur with *p53* or *RB1* mutations in breast cancer (van Bragt et al. 2014). Therefore, *RUNX1* mutation, in conjunction with the acquisition of oncogenic mutations in *p53* or *RB1*, are likely to play important roles in the pathogenesis of ER+ luminal breast cancer.

RUNX2 was also shown to play an important role in epithelial cancers originating from the breast. Mammary-specific Runx2 transgenic mouse models studies directly implicated RUNX2 in breast differentiation and cancer progression (McDonald et al. 2014; Owens et al. 2014; Ferrari et al. 2015). Runx2 is required for adult mammary stem/progenitor cell function (Owens et al. 2014; Ferrari et al. 2015) and is expressed in the basal as well as luminal cell lineages. Runx2 expression in the breast epithelium of pregnant mice is regulated in a temporal and hormonal manner (Owens et al. 2014). Ectopic expression of Runx2 disrupted lobular alveolar differentiation during pregnancy. Moreover, overexpression of Runx2 led to EMT-like changes, suggesting that Runx2 can promote metastasis (Owens et al. 2014). Conversely, *Runx2* deficiency was associated with reduced proliferation, delayed onset of breast cancer and better survival rates. The fact that RUNX2 expression is dynamically regulated and that its overexpression leads to impaired differentiation and cancer formation indicate that breast tumorigenesis stems, in part, from defective control of RUNX2 expression and deregulated differentiation in breast progenitor cells.

20% of *Runx3*$^{+/-}$ BALB/c female mice developed mammary ductal carcinoma (Huang et al. 2012). Expression of RUNX3 in ERα-positive MCF-7 cells resulted in inhibition of estrogen-dependent proliferation and transformation potential. This is due to the ability of RUNX3 to induce proteasome-specific degradation of ERα (Huang et al. 2012). Importantly, RUNX3 expression inversely correlates with the levels of ERα in human breast cancer tissues (Huang et al. 2012). RUNX3 is therefore associated with suppression of tumorigenesis of ERα-positive breast cancer cells.

It is clear that all RUNX proteins are involved in normal growth of breast tissues and it is tempting to speculate that RUNX1 and/or RUNX3 suppress the oncogenic tendencies of RUNX2. In other words, an imbalance of RUNX1/2/3 activities contribute to breast cancer progression.

19.12 Lung Cancer: RUNX3-p53-p14ARF Axis

The varied oncogenic outputs of mutated KRAS includes enhanced cell proliferation, suppression of apoptosis and modulation of the tumor microenvironment (eg. promotion of the angiogenesis and alteration of host immune response) (Pylayeva-Gupta et al. 2011). RAS mutations are frequently observed in lung cancer. We next investigated the relationship between oncogenic RAS activation and RUNX in the lung. Loss of *Runx3* resulted in hyperproliferation of bronchiolar epithelial cells and development of lung adenomas (Lee et al. 2013), suggesting that Runx3 deficiency might predispose lung epithelial cells to tumorigenesis. Using an oncogenic *KRAS*^{G12D} mouse cancer model, we demonstrated that targeted inactivation of *Runx3* in the lung resulted in accelerated lung adenocarcinoma formation. Furthermore, RUNX3 protects against oncogenic *KRAS* by collaborating with co-activator BRD2 to activate the p53-p14^{ARF} pathway (Lee et al. 2013). Intriguingly, the interaction of BRD2 with RUNX3 is enhanced by acetylation of the lysine 171 residue, which at the −2 position of the phosphorylation site T173, is part of the Aurora kinase b consensus motif (Chuang et al. 2016). How acetylation affects the phosphorylation of T173 remains unclear. Nevertheless, this finding draws attention to the exciting notion that post-translational modification of ancient, highly conserved motifs in the Runt domain is necessary for modulating the contact of the Runt domain with DNA (Bravo et al. 2001; Tahirov et al. 2001).

In normal cells, the persistent mitogenic stimulation induced by mutant RAS results in an irreversible cell cycle arrest also known as oncogene induced senescence – a crucial fail-safe mechanism that is activated by p53 and retinoblastoma protein. The p53-p14^{ARF} pathway protects against the consequences of replicative stress – namely genomic instability and malignant transformation – induced by oncogenic *RAS* mutations (Pylayeva-Gupta et al. 2011). Similarly, all RUNX family members induced senescence in mouse embryonic fibroblasts (Kilbey et al. 2007; Wolyniec et al. 2009). Ectopic expression of RUNX1 and RUNX1-ETO are associated with increases in reactive oxygen species, suggesting that the induction of senescence might be due in part to oxidative stress. In addition, RUNX proteins possess the capability to act upstream of p53. *P14*^{ARF}, which functions to stabilize the p53 protein, possesses RUNX binding sites in its promoter and is directly induced by RUNX1 as well as RUNX3 (Linggi et al. 2002; Lee et al. 2013).

19.13 Direct Transcription Regulation Versus Protein-Protein Interactions

RUNX genes function as tumor suppressors in some cancer types but act as oncogenes in other cases (Ito et al. 2015). As discussed above, the multiple mechanisms underlying the tumor suppressor activity of RUNX3 involve direct transcriptional regulation of growth inhibitory genes and/or disruption of DNA binding ability of oncogenic effectors through protein-protein interaction. Transcriptional regulation represents the classical RUNX3 tumor suppression scenario:

RUNX3 interacts with SMAD proteins and functions as an integral component of TGFβ tumor suppressor pathway at early stages of carcinogenesis (Ito et al. 2015). The other mechanism – tumor suppression through protein-protein interaction – has not been appreciated until recently. In addition to interacting with the DNA binding domains of TCF4 and TEAD4 and disrupting their activities, the Runt domain also binds the DNA binding domain of STAT5 (Ogawa et al. 2008). Complex formation between STAT5 and RUNX nullifies DNA binding of both proteins. Therefore, depending on the concentration of each protein, the level of oncogenic or tumor suppressive output would be significantly affected. For example, at elevated RUNX3 proteins levels (relative to the other three proteins), DNA binding activities of TCF4, TEAD4 and STAT5 would be significantly reduced. On the other hand, if TCF4 protein level is elevated due to enhanced Wnt activity, the amount of RUNX3 that interacts with DNA would be significantly reduced. This means that RUNX3 may simultaneously inhibit multiple oncogenic pathways and depending on cell context, the inactivation of RUNX3 would be pleiotropic. Another interesting aspect of this mechanism is that the DNA binding domains of RUNX3, TCF4, TEAD4, and STAT5 are highly conserved in the respective family. Therefore, RUNX1, RUNX2 and RUNX3 may interact with most, if not all, members of TCF, TEAD and STAT family. This possibility is extremely interesting and further studies will yield insights on the crosstalk of RUNX proteins with oncogenic effectors in cancer. In this light, the behavior of the oncogenic mutant RUNX3^{R122C} becomes clear. Not only is RUNX3^{R122C} unable to activate the transcription of growth inhibitory genes, it fails to interact with members of TCF and TEAD family and "resists" any existing oncogenic stimuli.

19.14 Conclusion

RUNX proteins are involved in many diverse mechanisms that cells employ to thwart malignant transformation. The interaction between RUNX and the TGFβ superfamily indicates how RUNX interact with a developmental pathway to direct differentiation and adult homeostasis (Fig. 19.5). The varied responses of RUNX to oncogenic signals (eg. Wnt, c-myc and mutant RAS) indicate how RUNX might influence or react to different oncogenic activities. Since many of these pathways crosstalk, a central question is how RUNX coordinate their crosstalk and integrate the signals to reach a cell fate decision. Aside from its classical role as transcription regulator, emerging evidence has indicated that the RUNX proteins directly participate in fundamental biological processes such as mitosis, centrosome function and DNA repair (Ito et al. 2015).

It is likely that the ability of RUNX to respond to diverse stimuli and regulate cell fate led to its evolutionary recruitment as developmental regulator as well as potent cancer gene. This is a double-edged sword. Depending on interacting proteins and post-translational modification, RUNX functions as either a transcriptional activator or repressor; if RUNX can function as a tumor suppressor by limiting proliferation and regulating differentiation, it should also be capable of promoting tumorigenesis in other cellular contexts. Understanding how RUNX dysregulation in cancer impinges on normal biological processes is important for identifying the molecular mechanisms that lead stepwise to malignancy.

Mounting evidence showed that tight regulation of RUNX expression is important for normal differentiation whereas dysregulated RUNX expression can lead to deregulated differentiation, tumor initiation and progression. Expression of RUNX or its downstream targets might therefore serve as biomarkers for early cancer detection and prognosis. Moreover, a pertinent question is whether we can augment RUNX tumor suppressor activity while decreasing its oncogenic potential. This will indicate whether restoration of proper RUNX expression to redirect cell fate or differentiation pathway is a feasible treatment for cancer. For example, the pro-tumorigenic activity of RUNX1 in stem cells might be suppressed by the restoration of RUNX3. In other words, the antidote to RUNX-induced tumors might well be its own family members.

This is an exciting period for RUNX research. The RUNX field is rapidly expanding into previously unanticipated directions – in the coming years we will gain a clearer understanding of the major theme underlying RUNX's pleiotropic properties for implementation in cancer therapy.

Acknowledgement This work was supported by the National Research Foundation Singapore, and the Singapore Ministry of Education under its Research Centres of Excellence initiative and also by the Singapore National Research Foundation under its Translational and Clinical Research Flagship Programme.

References

Akech, J., Wixted, J. J., Bedard, K., van der Deen, M., Hussain, S., Guise, T. A., et al. (2010). Runx2 association with progression of prostate cancer in patients: Mechanisms mediating bone osteolysis and osteoblastic metastatic lesions. *Oncogene, 29*, 811–821.

Banerji, S., Cibulskis, K., Rangel-Escareno, C., Brown, K. K., Carter, S. L., Frederick, A. M., et al. (2012). Sequence analysis of mutations and translocations across breast cancer subtypes. *Nature, 486*, 405–409.

Ben-Ami, O., Friedman, D., Leshkowitz, D., Goldenberg, D., Orlovsky, K., Pencovich, N., et al. (2013). Addiction of t(8;21) and inv(16) acute myeloid leukemia to native RUNX1. *Cell Reports, 4*, 1131–1143.

Bravo, J., Li, Z., Speck, N. A., & Warren, A. J. (2001). The leukemia-associated AML1 (Runx1) – CBF beta complex functions as a DNA-induced molecular clamp. *Nature Structural Biology, 8*, 371–378.

Brenner, O., Levanon, D., Negreanu, V., Golubkov, O., Fainaru, O., Woolf, E., & Groner, Y. (2004). Loss of Runx3 function in leukocytes is associated with spontaneously developed colitis and gastric mucosal hyperplasia. *Proceedings of the National Academy of Sciences of the United States of America, 101*, 16016–16021.

Buchert, M., Darido, C., Lagerqvist, E., Sedello, A., Cazevieille, C., Buchholz, F., et al. (2009). The symplekin/ZONAB complex inhibits intestinal cell differentiation by the repression of AML1/Runx1. *Gastroenterology, 137*, 156–164 164 e151-153.

Cancer Genome Atlas Research N. (2014). Comprehensive molecular characterization of gastric adenocarcinoma. *Nature, 513*, 202–209.

Chi, X. Z., Yang, J. O., Lee, K. Y., Ito, K., Sakakura, C., Li, Q. L., et al. (2005). RUNX3 suppresses gastric epithelial cell growth by inducing p21(WAF1/Cip1) expression in cooperation with transforming growth factor {beta}-activated SMAD. *Molecular and Cellular Biology, 25*, 8097–8107.

Chuang, L. S., & Ito, Y. (2010). RUNX3 is multifunctional in carcinogenesis of multiple solid tumors. *Oncogene, 29*, 2605–2615.

Chuang, L. S., Khor, J. M., Lai, S. K., Garg, S., Krishnan, V., Koh, C. G., et al. (2016). Aurora kinase-induced phosphorylation excludes transcription factor RUNX from the chromatin to facilitate proper mitotic progression. *Proceedings of the National Academy of Sciences of the United States of America, 113*, 6490–6495.

Ebihara, T., Song, C., Ryu, S. H., Plougastel-Douglas, B., Yang, L., Levanon, D., et al. (2015). Runx3 specifies lineage commitment of innate lymphoid cells. *Nature Immunology, 16*, 1124–1133.

Ellis, M. J., Ding, L., Shen, D., Luo, J., Suman, V. J., Wallis, J. W., et al. (2012). Whole-genome analysis informs breast cancer response to aromatase inhibition. *Nature, 486*, 353–360.

Ferrari, N., Riggio, A. I., Mason, S., McDonald, L., King, A., Higgins, T., et al. (2015). Runx2 contributes to the regenerative potential of the mammary epithelium. *Scientific Reports, 5*, 15658.

Fiedler, M., Graeb, M., Mieszczanek, J., Rutherford, T. J., Johnson, C. M., & Bienz, M. (2015). An ancient Pygo-dependent Wnt enhanceosome integrated by Chip/LDB-SSDP. *eLife, 4*, e09073.

Fijneman, R. J., Anderson, R. A., Richards, E., Liu, J., Tijssen, M., Meijer, G. A., et al. (2012). Runx1 is a tumor suppressor gene in the mouse gastrointestinal tract. *Cancer Science, 103*, 593–599.

Finak, G., Bertos, N., Pepin, F., Sadekova, S., Souleimanova, M., Zhao, H., et al. (2008). Stromal gene expression predicts clinical outcome in breast cancer. *Nature Medicine, 14*, 518–527.

Glotzer, D. J., Zelzer, E., & Olsen, B. R. (2008). Impaired skin and hair follicle development in Runx2 deficient mice. *Developmental Biology, 315*, 459–473.

Goh, Y. M., Cinghu, S., Hong, E. T., Lee, Y. S., Kim, J. H., Jang, J. W., et al. (2010). Src kinase phosphorylates RUNX3 at tyrosine residues and localizes the protein in the cytoplasm. *The Journal of Biological Chemistry, 285*, 10122–10129.

Hanai, J., Chen, L. F., Kanno, T., Ohtani-Fujita, N., Kim, W. Y., Guo, W. H., et al. (1999). Interaction and functional cooperation of PEBP2/CBF with Smads. Synergistic induction of the immunoglobulin germline Calpha promoter. *The Journal of Biological Chemistry, 274*, 31577–31582.

Henrich, K. O., Schwab, M., & Westermann, F. (2012). 1p36 tumor suppression – a matter of dosage? *Cancer Research, 72*, 6079–6088.

Hoi, C. S., Lee, S. E., Lu, S. Y., McDermitt, D. J., Osorio, K. M., Piskun, C. M., et al. (2010). Runx1 directly promotes proliferation of hair follicle stem cells and epithelial tumor formation in mouse skin. *Molecular and Cellular Biology, 30*, 2518–2536.

Hor, Y. T., Voon, D. C., Koo, J. K., Wang, H., Lau, W. M., Ashktorab, H., et al. (2014). A role for RUNX3 in inflammation-induced expression of IL23A in gastric epithelial cells. *Cell Reports, 8*, 50–58.

Huang, B., Qu, Z., Ong, C. W., Tsang, Y. H., Xiao, G., Shapiro, D., et al. (2012). RUNX3 acts as a tumor suppressor in breast cancer by targeting estrogen receptor alpha. *Oncogene, 31*, 527–534.

Ito, K., Liu, Q., Salto-Tellez, M., Yano, T., Tada, K., Ida, H., et al. (2005). RUNX3, a novel tumor suppressor, is frequently inactivated in gastric cancer by protein mislocalization. *Cancer Research, 65*, 7743–7750.

Ito, K., Lim, A. C., Salto-Tellez, M., Motoda, L., Osato, M., Chuang, L. S., et al. (2008). RUNX3 attenuates beta-catenin/T cell factors in intestinal tumorigenesis. *Cancer Cell, 14*, 226–237.

Ito, K., Chuang, L. S., Ito, T., Chang, T. L., Fukamachi, H., Salto-Tellez, M., & Ito, Y. (2011). Loss of Runx3 is a key event in inducing precancerous state of the stomach. *Gastroenterology, 140*, 1536–1546 e1538.

Ito, Y., Bae, S. C., & Chuang, L. S. (2015). The RUNX family: Developmental regulators in cancer. *Nature Reviews Cancer, 15*, 81–95.

Katayama, Y., Takahashi, M., & Kuwayama, H. (2009). Helicobacter pylori causes runx3 gene methylation and its loss of expression in gastric epithelial cells, which is mediated by nitric oxide produced by macrophages. *Biochemical and Biophysical Research Communications, 388*, 496–500.

Kilbey, A., Blyth, K., Wotton, S., Terry, A., Jenkins, A., Bell, M., et al. (2007). Runx2 disruption promotes immortalization and confers resistance to oncogene-induced senescence in primary murine fibroblasts. *Cancer Research, 67*, 11263–11271.

Lau, Q. C., Raja, E., Salto-Tellez, M., Liu, Q., Ito, K., Inoue, M., et al. (2006). RUNX3 is frequently inactivated by dual mechanisms of protein mislocalization and promoter hypermethylation in breast cancer. *Cancer Research, 66*, 6512–6520.

Lee, C. W., Chuang, L. S., Kimura, S., Lai, S. K., Ong, C. W., Yan, B., et al. (2011). RUNX3 functions as an oncogene in ovarian cancer. *Gynecologic Oncology, 122*, 410–417.

Lee, K., Shim, J., Bae, J., Kim, Y. J., & Lee, J. (2012). Stabilization of RNT-1 protein, runt-related transcription factor (RUNX) protein homolog of *Caenorhabditis elegans*, by oxidative stress through mitogen-activated protein kinase pathway. *The Journal of Biological Chemistry, 287*, 10444–10452.

Lee, Y. S., Lee, J. W., Jang, J. W., Chi, X. Z., Kim, J. H., Li, Y. H., et al. (2013). Runx3 inactivation is a crucial early event in the development of lung adenocarcinoma. *Cancer Cell, 24*, 603–616.

Levanon, D., Brenner, O., Negreanu, V., Bettoun, D., Woolf, E., Eilam, R., et al. (2001). Spatial and temporal expression pattern of Runx3 (Aml2) and Runx1 (Aml1) indicates non-redundant functions during mouse embryogenesis. *Mechanisms of Development, 109*, 413–417.

Levanon, D., Bettoun, D., Harris-Cerruti, C., Woolf, E., Negreanu, V., Eilam, R., et al. (2002). The Runx3 transcription factor regulates development and survival of TrkC dorsal root ganglia neurons. *The EMBO Journal, 21*, 3454–3463.

Levanon, D., Bernstein, Y., Negreanu, V., Bone, K. R., Pozner, A., Eilam, R., et al. (2011). Absence of Runx3 expression in normal gastrointestinal epithelium calls into question its tumour suppressor function. *EMBO Molecular Medicine, 3*, 593–604.

Li, Q. L., Ito, K., Sakakura, C., Fukamachi, H., Inoue, K., Chi, X. Z., et al. (2002). Causal relationship between the loss of RUNX3 expression and gastric cancer. *Cell, 109*, 113–124.

Lim, J., Duong, T., Do, N., Do, P., Kim, J., Kim, H., et al. (2013). Antitumor activity of cell-permeable RUNX3 protein in gastric cancer cells. *Clinical Cancer Research, 19*, 680–690.

Linggi, B., Muller-Tidow, C., van de Locht, L., Hu, M., Nip, J., Serve, H., et al. (2002). The t(8;21) fusion protein, AML1 ETO, specifically represses the transcription of the p14(ARF) tumor suppressor in acute myeloid leukemia. *Nature Medicine, 8*, 743–750.

Lutterbach, B., Westendorf, J. J., Linggi, B., Isaac, S., Seto, E., & Hiebert, S. W. (2000). A mechanism of repression by acute myeloid leukemia-1, the target of multiple chromosomal translocations in acute leukemia. *The Journal of Biological Chemistry, 275*, 651–656.

Martin, J. W., Zielenska, M., Stein, G. S., van Wijnen, A. J., & Squire, J. A. (2011). The Role of RUNX2 in Osteosarcoma Oncogenesis. *Sarcoma, 2011*, 282745.

McDonald, L., Ferrari, N., Terry, A., Bell, M., Mohammed, Z. M., Orange, C., et al. (2014). RUNX2 correlates with subtype-specific breast cancer in a human tissue microarray, and ectopic expression of Runx2 perturbs differentiation in the mouse mammary gland. *Disease Models & Mechanisms, 7*, 525–534.

Mori, T., Nomoto, S., Koshikawa, K., Fujii, T., Sakai, M., Nishikawa, Y., et al. (2005). Decreased expression and frequent allelic inactivation of the RUNX3 gene at 1p36 in human hepatocellular carcinoma. *Liver International, 25*, 380–388.

Nagahama, Y., Ishimaru, M., Osaki, M., Inoue, T., Maeda, A., Nakada, C., et al. (2008). Apoptotic pathway induced by transduction of RUNX3 in the human gastric carcinoma cell line MKN-1. *Cancer Science, 99*, 23–30.

Nam, S., Jin, Y. H., Li, Q. L., Lee, K. Y., Jeong, G. B., Ito, Y., et al. (2002). Expression pattern, regulation, and biological role of runt domain transcription factor, run, in *Caenorhabditis elegans*. *Molecular and Cellular Biology, 22*, 547–554.

Nevadunsky, N. S., Barbieri, J. S., Kwong, J., Merritt, M. A., Welch, W. R., Berkowitz, R. S., & Mok, S. C. (2009). RUNX3 protein is overexpressed in human epithelial ovarian cancer. *Gynecologic Oncology, 112*, 325–330.

Newman, D. K. (2001). Microbiology. How bacteria respire minerals. *Science, 292*, 1312–1313.

Nomoto, S., Haruki, N., Tatematsu, Y., Konishi, H., Mitsudomi, T., & Takahashi, T. (2000). Frequent allelic imbalance suggests involvement of a tumor suppressor gene at 1p36 in the pathogenesis of human lung cancers. *Genes, Chromosomes & Cancer, 28*, 342–346.

Nomoto, S., Kinoshita, T., Mori, T., Kato, K., Sugimoto, H., Kanazumi, N., et al. (2008). Adverse prognosis of epigenetic inactivation in RUNX3 gene at 1p36 in human pancreatic cancer. *British Journal of Cancer, 98*, 1690–1695.

Ogasawara, N., Tsukamoto, T., Mizoshita, T., Inada, K. I., Ban, H., Kondo, S., et al. (2009). RUNX3 expression correlates with chief cell differentiation in human gastric cancers. *Histology and Histopathology, 24*, 31–40.

Ogawa, E., Maruyama, M., Kagoshima, H., Inuzuka, M., Lu, J., Satake, M., et al. (1993). PEBP2/PEA2 represents a family of transcription factors homologous to the products of the Drosophila runt gene and the human AML1 gene. *Proceedings of the National Academy of Sciences of the United States of America, 90*, 6859–6863.

Ogawa, S., Satake, M., & Ikuta, K. (2008). Physical and functional interactions between STAT5 and Runx transcription factors. *Journal of Biochemistry, 143*, 695–709.

Osorio, K. M., Lee, S. E., McDermitt, D. J., Waghmare, S. K., Zhang, Y. V., Woo, H. N., & Tumbar, T. (2008). Runx1 modulates developmental, but not injury-driven, hair follicle stem cell activation. *Development, 135*, 1059–1068.

Owens, T. W., Rogers, R. L., Best, S. A., Ledger, A., Mooney, A. M., Ferguson, A., et al. (2014). Runx2 is a novel regulator of mammary epithelial cell fate in development and breast cancer. *Cancer Research, 74*, 5277–5286.

Pratap, J., Wixted, J. J., Gaur, T., Zaidi, S. K., Dobson, J., Gokul, K. D., et al. (2008). Runx2 transcriptional activation of Indian Hedgehog and a downstream bone metastatic pathway in breast cancer cells. *Cancer Research, 68*, 7795–7802.

Pylayeva-Gupta, Y., Grabocka, E., & Bar-Sagi, D. (2011). RAS oncogenes: Weaving a tumorigenic web. *Nature Reviews. Cancer, 11*, 761–774.

Qiao, Y., Lin, S. J., Chen, Y., Voon, D. C., Zhu, F., Chuang, L. S., et al. (2015). RUNX3 is a novel negative regulator of oncogenic TEAD-YAP complex in gastric cancer. *Oncogene, 35*, 2664.

Raveh, E., Cohen, S., Levanon, D., Groner, Y., & Gat, U. (2005). Runx3 is involved in hair shape determination. *Developmental Dynamics, 233*, 1478–1487.

Raveh, E., Cohen, S., Levanon, D., Negreanu, V., Groner, Y., & Gat, U. (2006). Dynamic expression of Runx1 in skin affects hair structure. *Mechanisms of Development, 123*, 842–850.

Sadikovic, B., Thorner, P., Chilton-Macneill, S., Martin, J. W., Cervigne, N. K., Squire, J., & Zielenska, M. (2010). Expression analysis of genes associated with human osteosarcoma tumors shows correlation of RUNX2 overexpression with poor response to chemotherapy. *BMC Cancer, 10*, 202.

Salto-Tellez, M., Peh, B. K., Ito, K., Tan, S. H., Chong, P. Y., Han, H. C., et al. (2006). RUNX3 protein is overexpressed in human basal cell carcinomas. *Oncogene, 25*, 7646–7649.

Sase, T., Suzuki, T., Miura, K., Shiiba, K., Sato, I., Nakamura, Y., et al. (2012). Runt-related transcription factor 2 in human colon carcinoma: A potent prognostic factor associated with estrogen receptor. *International Journal of Cancer, 131*, 2284–2293.

Scheitz, C. J., Lee, T. S., McDermitt, D. J., & Tumbar, T. (2012). Defining a tissue stem cell-driven Runx1/Stat3 signalling axis in epithelial cancer. *The EMBO Journal, 31*, 4124–4139.

Sebe-Pedros, A., de Mendoza, A., Lang, B. F., Degnan, B. M., & Ruiz-Trillo, I. (2011). Unexpected repertoire of metazoan transcription factors in the unicellular holozoan Capsaspora owczarzaki. *Molecular Biology and Evolution, 28*, 1241–1254.

Sun, L., Vitolo, M. I., Qiao, M., Anglin, I. E., & Passaniti, A. (2004). Regulation of TGFbeta1-mediated growth inhibition and apoptosis by RUNX2 isoforms in endothelial cells. *Oncogene, 23*, 4722–4734.

Tahirov, T. H., Inoue-Bungo, T., Morii, H., Fujikawa, A., Sasaki, M., Kimura, K., et al. (2001). Structural analyses of DNA recognition by the AML1/Runx-1 Runt domain and its allosteric control by CBFbeta. *Cell, 104*, 755–767.

Usui, T., Aoyagi, K., Saeki, N., Nakanishi, Y., Kanai, Y., Ohki, M., et al. (2006). Expression status of RUNX1/AML1 in normal gastric epithelium and its mutational analysis in microdissected gastric cancer cells. *International Journal of Oncology, 29*, 779–784.

van Bragt, M. P., Hu, X., Xie, Y., & Li, Z. (2014). RUNX1, a transcription factor mutated in breast cancer, controls the fate of ER-positive mammary luminal cells. *eLife, 3*, e03881.

Voon, D. C., Wang, H., Koo, J. K., Nguyen, T. A., Hor, Y. T., Chu, Y. S., et al. (2012). Runx3 protects gastric epithelial cells against epithelial-mesenchymal transition-induced cellular plasticity and tumorigenicity. *Stem Cells, 30*, 2088–2099.

Wai, P. Y., Mi, Z., Gao, C., Guo, H., Marroquin, C., & Kuo, P. C. (2006). Ets-1 and runx2 regulate transcription of a metastatic gene, osteopontin, in murine colorectal cancer cells. *The Journal of Biological Chemistry, 281*, 18973–18982.

Wang, C. Q., Krishnan, V., Tay, L. S., Chin, D. W., Koh, C. P., Chooi, J. Y., et al. (2014). Disruption of Runx1 and Runx3 leads to bone marrow failure and leukemia predisposition due to transcriptional and DNA repair defects. *Cell Reports, 8*, 767–782.

Weis, V. G., & Goldenring, J. R. (2009). Current understanding of SPEM and its standing in the preneoplastic process. *Gastric Cancer, 12*, 189–197.

Weisenberger, D. J., Siegmund, K. D., Campan, M., Young, J., Long, T. I., Faasse, M. A., et al. (2006). CpG island methylator phenotype underlies sporadic microsatellite instability and is tightly associated with BRAF mutation in colorectal cancer. *Nature Genetics, 38*, 787–793.

Whittle, M. C., Izeradjene, K., Rani, P. G., Feng, L., Carlson, M. A., DelGiorno, K. E., et al. (2015). RUNX3 controls a metastatic switch in pancreatic ductal adenocarcinoma. *Cell, 161*, 1345–1360.

Wolyniec, K., Wotton, S., Kilbey, A., Jenkins, A., Terry, A., Peters, G., et al. (2009). RUNX1 and its fusion oncoprotein derivative, RUNX1-ETO, induce senescence-like growth arrest independently of replicative stress. *Oncogene, 28*, 2502–2512.

Yamada, C., Ozaki, T., Ando, K., Suenaga, Y., Inoue, K., Ito, Y., et al. (2010). RUNX3 modulates DNA damage-mediated phosphorylation of tumor suppressor p53 at Ser-15 and acts as a co-activator for p53. *The Journal of Biological Chemistry, 285*, 16693–16703.

Yano, T., Ito, K., Fukamachi, H., Chi, X. Z., Wee, H. J., Inoue, K., et al. (2006). The RUNX3 tumor suppressor upregulates Bim in gastric epithelial cells undergoing transforming growth factor beta-induced apoptosis. *Molecular and Cellular Biology, 26*, 4474–4488.

Yoshida, C. A., Yamamoto, H., Fujita, T., Furuichi, T., Ito, K., Inoue, K., et al. (2004). Runx2 and Runx3 are essential for chondrocyte maturation, and Runx2 regulates limb growth through induction of Indian hedgehog. *Genes & Development, 18*, 952–963.

Young, D. W., Hassan, M. Q., Pratap, J., Galindo, M., Zaidi, S. K., Lee, S. H., et al. (2007a). Mitotic occupancy and lineage-specific transcriptional control of rRNA genes by Runx2. *Nature, 445*, 442–446.

Young, D. W., Hassan, M. Q., Yang, X. Q., Galindo, M., Javed, A., Zaidi, S. K., et al. (2007b). Mitotic retention of gene expression patterns by the cell fate-determining transcription factor Runx2. *Proceedings of the National Academy of Sciences of the United States of America, 104*, 3189–3194.

Yu, F., Gao, W., Yokochi, T., Suenaga, Y., Ando, K., Ohira, M., et al. (2013). RUNX3 interacts with MYCN and facilitates protein degradation in neuroblastoma. *Oncogene, 33*, 2601.

Zhang, X., Akech, J., Browne, G., Russell, S., Wixted, J. J., Stein, J. L., et al. (2015). Runx2-Smad signaling impacts the progression of tumor-induced bone disease. *International Journal of Cancer, 136*, 1321–1332.

Zhou, H., Mak, W., Zheng, Y., Dunstan, C. R., & Seibel, M. J. (2008). Osteoblasts directly control lineage commitment of mesenchymal progenitor cells through Wnt signaling. *The Journal of Biological Chemistry, 283*, 1936–1945.

RUNX3 and p53: How Two Tumor Suppressors Cooperate Against Oncogenic Ras?

20

Jung-Won Lee, Andre van Wijnen, and Suk-Chul Bae

Abstract

RUNX family members play pivotal roles in both normal development and neoplasia. In particular, *RUNX1* and *RUNX2* are essential for determination of the hematopoietic and osteogenic lineages, respectively. *RUNX3* is involved in lineage determination of various types of epithelial cells. Analysis of mouse models and human cancer specimens revealed that *RUNX3* acts as a tumor suppressor via multiple mechanisms. p53-related pathways play central roles in tumor suppression through the DNA damage response and oncogene surveillance, and RUNX3 is involved in both processes. In response to DNA damage, RUNX3 facilitates p53 phosphorylation by the ATM/ATR pathway and p53 acetylation by p300. When oncogenes are activated, RUNX3 induces ARF, thereby stabilizing p53. Here, we summarize the molecular mechanisms underlying the p53-mediated tumor-suppressor activity of RUNX3.

Keywords

RUNX • p53 • ARF • DNA damage • Oncogene surveillance • Tumor suppressor

20.1 Tumor-Suppressive Role of RUNX3

RUNX3 was first described two decades ago by several independent groups (Levanon et al. 1994; Bae et al. 1995; Wijmenga et al. 1995). The phenotypes of the *Runx3*-knockout mouse revealed that this gene is involved in various pathways, including TGF-β, Wnt, and the p53-mediated oncogene surveillance pathway. In the C57BL/6 strain background, *Runx3*$^{-/-}$ mice die within 24 h

J.-W. Lee • S.-C. Bae (✉)
Department of Biochemistry, School of Medicine, and Institute for Tumor Research, Chungbuk National University, Cheongju 28644, South Korea
e-mail: scbae@chungbuk.ac.kr

A. van Wijnen (✉)
Departments of Orthopedic Surgery & Biochemistry and Molecular Biology, Mayo Clinic, 200 First Street SW, Rochester, MN 55905, USA
e-mail: vanwijnen.andre@mayo.edu

Y. Groner et al. (eds.), *RUNX Proteins in Development and Cancer*, Advances in Experimental Medicine and Biology 962, DOI 10.1007/978-981-10-3233-2_20

of birth. The gastric mucosa of *Runx3*$^{-/-}$ mice exhibit hyperplasia as a result of elevated proliferation and diminished apoptosis in epithelial cells (Li et al. 2002). These cells are resistant to the growth-inhibitory and apoptosis-inducing effect of TGF-β1, indicating that *Runx3* is a major growth regulator of gastric epithelial cells and that its activity is controlled by TGF-β signaling. *Runx3*$^{-/-}$;*p53*$^{-/-}$ gastric epithelial cells are tumorigenic in nude mice, whereas *Runx3*$^{+/+}$;*p53*$^{-/-}$ gastric epithelial cells are not. Furthermore, some of the resultant tumors exhibit goblet cell character, a hallmark of intestinal metaplasia associated with gastric cancer (Fukamachi et al. 2004), strongly suggesting that the hyperplasia and tumorigenicity of *Runx3*$^{-/-}$;*p53*$^{-/-}$ gastric epithelia are cell-autonomous. About 60 % of human gastric cancers express lower levels of *RUNX3* due to hemizygous deletion or hypermethylation of the *RUNX3* promoter region (Li et al. 2002). Together, these results demonstrate that RUNX3 is a tumor suppressor.

Runx3 forms a ternary complex with β-catenin/Tcf4 and attenuates Wnt signaling activity. In addition, *Runx3* inactivation results in formation of spontaneous intestinal adenoma formation, as in the adenomatous polyposis coli (*Apc*) mutation in mice (Ito et al. 2008). Consistent with this, a significant fraction of human sporadic colorectal adenomas exhibit inactivation of *RUNX3*, implying that loss of *RUNX3* function is responsible for intestinal adenomas (Ito et al. 2008).

Runx3$^{-/-}$ mice, which die soon after birth, exhibit lung epithelial hyperplasia, whereas *Runx3*$^{+/-}$ mice spontaneously develop lung adenomas around the age of 18 months (Lee et al. 2010). Targeted deletion of *Runx3* in adult mouse lung results in deregulation of lung epithelial cell differentiation and development of lung adenoma (Lee et al. 2013). Together, these findings indicate that *Runx3* is essential for lung epithelial cell differentiation, and that downregulation of the gene is causally linked to lung cancer.

Analyses of human tumor samples revealed that downregulation of *RUNX3* is frequently associated with various types of cancers of the lung, colon, pancreas, liver, prostate, bile duct, breast, larynx, esophagus, uterine cervix, and testicular yolk sac. *RUNX3* expression is downregulated more than 50 % in these tumors (Goel et al. 2004; Ito et al. 2015; Kang et al. 2004, 2005; Kato et al. 2003; Kim et al. 2004; Ku et al. 2004; Li et al. 2004; Mori et al. 2005; Nakase et al. 2005; Oshimo et al. 2004; Sakakura et al. 2005; Schulmann et al. 2005; Tamura 2004; Wada et al. 2004; Xiao and Liu 2004; Yanagawa et al. 2003).

20.2 *p53*, a Master Guardian Against Cellular Transformation

Mutation of *p53* is common in a wide variety of human tumor cells. Indeed, *p53* is the gene most frequently mutated in human tumor cell genomes, and mutations in *p53* are present in almost half of human tumors (Weinberg 2007). These observations imply that *p53* functions as a major tumor suppressor. Functional analysis by gene targeting analysis confirmed that *p53* plays a distinctive tumor-suppressive role. In typical tumor-suppressor genes, homozygous inactivation in the mouse germ line disrupts embryonic development by dysregulating the proliferation of a variety of cell types. By contrast, deletion of *p53* from the mouse germ line causes no developmental defects in the majority of *p53*$^{-/-}$ embryos (Donehower et al. 1992). However, *p53*$^{-/-}$ mice have a short lifespan (about 5 months) because they develop early-onset cancer (Donehower et al. 1992). This observation provides strong evidence that p53 protein does not transduce anti-proliferative signals in normal mice. Instead, p53 is specialized to prevent the appearance of abnormal cells, specifically, those that are capable of inducing tumors.

This concept raised questions regarding how p53 recognizes abnormalities. Subsequent studies identified a variety of mechanisms that can activate p53. These observations made it clear that a diverse array of sensors is involved in monitoring the integrity and functioning of various cellular systems. When these sensors detect abnormalities that are capable of initiating tumors, they send signals to p53 and its regulators, resulting in p53 activation. The two most extensively studied sensors are activated by DNA damage and oncogene activity.

20.3 p53 Activation by the DNA Damage – Activated Sensor

DNA damage was the first type of stress shown to activate p53. Based on this observation, p53 is widely regarded as the guardian of the genome. The signaling routes that connect DNA damage with p53 are triggered by DNA damage – sensing kinases, including ataxia telangiectasia mutated kinase (ATM) and ATM- and RAD3-related kinase (ATR), which activate Chk1 and Chk2 (Bartek and Lukas 2003; Kruse and Gu 2009; Reinhardt et al. 2007); the Chk proteins in turn phosphorylate p53 (Kruse and Gu 2009; Shieh et al. 2000). In normal cells, the p53 level is usually very low, and the protein has a half-life of only 20 min due to rapid degradation. Mdm2 (in mouse) and Hdm2 (in human) are primarily responsible for the negative regulation of p53 in normal cells (Haupt et al. 1997; Lakin and Jackson 1999; Levine 1997). Mdm2 negatively regulates p53 through two main mechanisms. First, the direct binding of Mdm2 to the p53 N-terminus inhibits the transcriptional activation function of p53 (Momand et al. 1992; Oliner et al. 1993). Second, Mdm2 ubiquitinates p53 to facilitate its degradation by the 26S proteasome (Haupt et al. 1997; Kubbutat et al. 1997). When cells are subjected to certain types of damage, p53 is protected from Mdm2-mediated degradation, resulting in p53 accumulation. p53 phosphorylation by the DNA damage – induced protein kinase cascade prevents Mdm2 from binding p53 by altering its recognition domain (Lakin and Jackson 1999; Shieh et al. 1997; Smith and Jackson 1999). Consequently, Mdm2 fails to ubiquitinate p53, which is accordingly stabilized. Mdm2 itself is regulated by p53 via p53-response elements located in the *Mdm2* promoter (Barak et al. 1993; Perry et al. 1993). Thus, a negative-feedback loop exists between p53 and Mdm2. The intimate relationship between these two proteins was clearly revealed in studies in which *Mdm2* was knocked out in the mouse germline. Homozygous deletion of *Mdm2* results in lethality at the blastocyst stage due to inappropriately high levels of apoptosis. Remarkably, however, deletion of p53 completely rescues this phenotype (Montes de Oca Luna et al. 1995).

20.4 Involvement of RUNX Family Members in p53 Activation by DNA Damage

Recent studies identified a critical role for *RUNX3* in p53 activation by DNA damage. Yamada et al. showed that *RUNX3* is induced and forms a complex with p53 in response to the DNA-damaging agent adriamycin (Yamada et al. 2010). The RUNX3–p53 complex induces phosphorylation of p53 at Ser-15, thereby inhibiting Mdm2-mediated p53 degradation and promoting p53-dependent apoptosis. Conversely, *RUNX3* knockdown diminishes transcription of p53 target genes (Yamada et al. 2010). Therefore, RUNX3 is involved in DNA damage – dependent phosphorylation of p53 and acts as a co-activator of p53 (Ozaki et al. 2013; Yamada et al. 2010).

RUNX1 also forms a complex with p53 in response to DNA damage (Wu et al. 2013). Treatment with adriamycin induces formation of the RUNX1-p53 complex, which activates the p53 target genes *CDKN1A* and *BAX*. The transcriptional activity of the RUNX1–p53 complex is increased by p300-mediated acetylation of p53 (Wu et al. 2013). Knockdown of *RUNX1* significantly decreases the level of p53–p300 complex following adriamycin exposure. Consistent with these observations, RUNX1 and p53 synergistically up-regulate transcription of the stress sensor *GADD45A* (Satoh et al. 2012). Considering the physical interaction between RUNX family members and p300 (Jeon et al. 2006; Jin et al. 2004; Kitabayashi et al. 1998), the p53–p300 interaction might be bridged by RUNXs. The involvement of RUNXs in phosphorylation and acetylation of p53 is summarized in Fig. 20.1.

Aside from the RUNX–p53 association during DNA damage, RUNX proteins also have p53-inde-

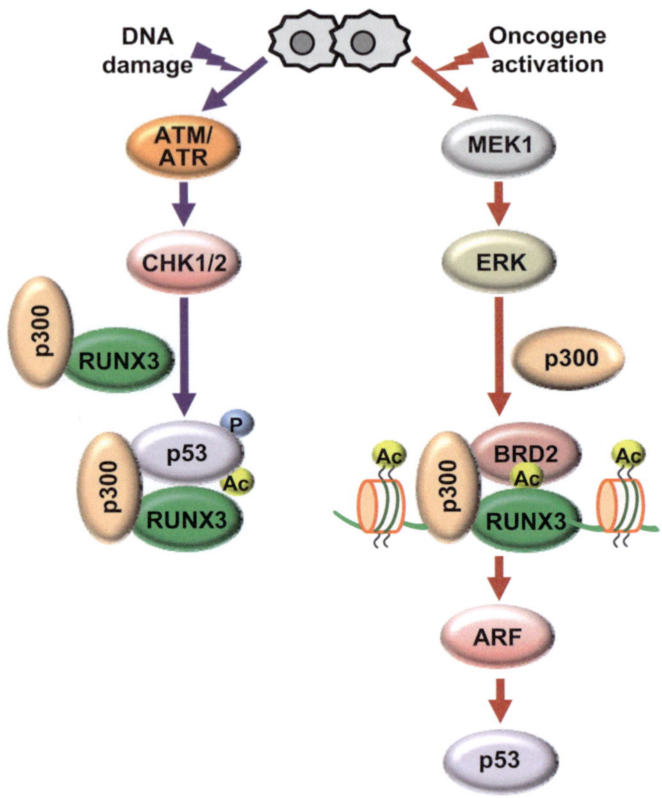

Fig. 20.1 Involvement of RUNX family members in p53 pathways triggered by DNA damage and oncogene activation. In response to DNA-damaging agents, p53 forms a complex with RUNX1 or RUNX3. The resultant RUNX–p53 complex induces phosphorylation of p53 at Ser-15, and acetylation at K373/382, thereby inhibiting Mdm2-mediated p53 degradation (Yamada et al. 2010; Wu et al. 2013). When *RAS* is activated, RUNX3 is activated by the MAPK pathway and acetylated by p300 at lysine residues; the modified and active protein then forms a complex with BRD2. The RUNX3–BRD2 complex induces *ARF*, which in turn stabilizes p53 (Lee et al. 2013)

pendent functions. *Runx1;Runx3* double-knockout mice exhibit bone marrow failure and myeloproliferative disorders that resemble the human disease Fanconi anemia (Wang et al. 2014). In addition, molecular biological analysis showed that RUNX1 and RUNX3 play pivotal roles in recruiting monoubiquitinated FANCD2 to sites of DNA damage. This direct link between RUNXs and the Fanconi anemia pathway indicates that RUNX family members are involved in directing DNA repair (Wang et al. 2014).

20.5 p53 Activation by the Oncogene-Activated Sensor

Paradoxically, many oncogenes that contribute to cell-cycle progression also play essential roles in apoptosis. This phenomenon can be explained by the observation that p53 is stabilized in response to abnormal oncogene activity. For example, mammalian cells are highly sensitive to constitutive activated oncogenic Ras. When oncogenic *K-Ras* (*K-Ras*^{G12V}) is ectopically expressed in wild-type mouse embryonic fibroblast cells (MEFs), the cells are driven toward senescence rather than cell-cycle progression (Serrano et al. 1997). By contrast, in *p53*$^{-/-}$ MEFs, expression of oncogenic *K-Ras* stimulates proliferation (Serrano et al. 1997). Similarly, the *E2F1* oncoprotein exerts both potent mitogenic and pro-apoptotic functions. The pro-apoptotic effects of the gene are stronger, so it is likely that most cells that express a high level of *E2F1* are rapidly eliminated via the *p53*-dependent apoptotic program (Wu and Levine 1994). However, if the oncogene-bearing cell happens to harbor inactivated *p53*, this oncogene-induced apoptosis is attenuated (Wu and Levine 1994).

Subsequent work revealed that the oncogene-induced *p53* activation pathway is mediated by an Mdm2 antagonist called p19^{Arf} in mouse cells and p14^{ARF} in human cells (hereafter, ARF)

(Kruse and Gu 2009; Lowe and Sherr 2003; Vogelstein et al. 2000; Zhang et al. 1998). Expression of oncogenic *Ras* induces proliferation in *Arf*$^{-/-}$ MEFs, whereas it induces senescence in normal MEFs. Similarly, ectopic expression of *Arf* in wild-type rodent cells strongly inhibits their proliferation. However, this inhibition is greatly reduced in cells lacking *p53* (Kruse and Gu 2009; Sherr 2006). Expression of *Arf* causes a rapid increase in the p53 level by sequestering Mdm2 in the nucleolus, the nuclear structure devoted to synthesis of ribosomal RNAs (Weber et al. 1999). Therefore, the *Arf* level is critical for cell fate decisions, i.e., whether to proliferate or die, in response to oncogene activation. Indeed, during tumorigenesis in humans ARF is frequently inactivated, conferring the same benefits as *p53* inactivation, via a variety of mechanisms (Ozenne et al. 2010; Sherr 1998).

The role of *ARF* in increasing the p53 level raised an important question: how do cells sense oncogene activation to induce *ARF* expression. Because oncogene activation normally occurs during cell-cycle progression, whereas DNA damage does not, the sensing mechanism must be distinct from that of DNA damage. Cells may monitor the levels of oncogene activity to determine whether they are growing according to the normal program. The sensing mechanism can be explained by the finding that the *ARF* promoter contains an E2F recognition sequence. Consequently, unusually high levels of E2F1 or E2F2 activity induce *ARF* transcription; as a result, Mdm2 is inhibited and p53 accumulates, ultimately inducing apoptosis (Komori et al. 2005; Lowe and Sherr 2003; Ozenne et al. 2010). This sensing mechanism suggests that the *ARF* promoter may be activated when E2F1 or E2F2 activity is higher than a threshold level that is not reached in normal cells.

20.6 Can Cells Sense Persistent Oncogene Activity?

Although the threshold-sensing mechanism for *ARF* expression can explain various kinds of oncogene-induced apoptosis, it is not sufficient to sense oncogenic *Ras* mutations, which are among the most frequent types of oncogene activation in multiple tumor types. Approximately 30 % of human tumors harbor *RAS* mutations (Bos 1989; Karnoub and Weinberg 2008; Pylayeva-Gupta et al. 2011). Mitogenic signaling activates the GTPase activity of RAS family proteins, which decreases to the basal level soon after the signal is transduced to downstream kinase pathways. Oncogenic *RAS* alleles harbor single point mutations, mostly in codon 12, 13, or 61 of the gene, resulting in constitutive activation of the encoded protein (Bos 1989; Bos et al. 1984; Pylayeva-Gupta et al. 2011; Verlaan-de Vries et al. 1986). The roles of oncogenic *RAS* in tumorigenesis have been most extensively studied in lung cancer, in which *K-RAS* is very frequently mutated (Bos 1989; Pylayeva-Gupta et al. 2011).

Johnson et al. showed that mice of the *K-Ras*^{LA} strain, in which expression of *K-Ras*^{G12D} can be spontaneously activated by random recombination, develop lung adenomas (Johnson et al. 2001). Similarly, in a *LoxP-Stop-LoxP-K-Ras* conditional mouse strain (*K-Ras*$^{LSL\text{-}G12D}$), expression of oncogenic *K-Ras* is controlled by a removable transcriptional termination *Stop* element; when the expression of endogenous *K-Ras*^{G12D} is triggered in lung by *Adeno-Cre* – mediated deletion of the *Stop* sequence, the mice develop adenomas after few months of stimulation (Jackson et al. 2001; Kim et al. 2005). Because expression of oncogenic *K-Ras* in the knock-in mice is driven by the endogenous promoter, these results demonstrate that persistent activation of *K-Ras* by heterozygous mutation is sufficient to induce tumorigenesis. Hence, it is important to determine the role of the ARF–p53 pathway in oncogenic *K-Ras* – induced lung cancer. Indeed, simultaneous activation of oncogenic *K-Ras* and inactivation of the *p53* tumor suppressor in mouse lung significantly accelerates the malignancy of the resultant cancer, suggesting that the p53 pathway can suppress *K-Ras* – induced lung adenocarcinoma (DuPage et al. 2009).

Although these experiments demonstrate the critical roles of oncogenic *K-Ras* and p53 in lung tumorigenesis, it remains unclear whether all, or merely some, *K-Ras* – activated cells are tumorigenic. To address this question, Guerra et al. generated *K-Ras*$^{LSL\text{-}G12D}$*-IRES-geo* mice harboring a *K-Ras*$^{LSL\text{-}G12D}$ allele containing *IRES-geo*, which enables tracing of targeted cells. They targeted *K-Ras*$^{LSL\text{-}G12D}$*-IRES-geo* throughout the whole body with tamoxifen-inducible *Cre-ERT*, and then analyzed the effect on tumor development (Guerra et al. 2003). Notably, systematic tamoxifen treatment targeted *K-Ras*$^{LSL\text{-}G12D}$*-IRES-geo* in 5–15 % of cells in most tissues, including stem cells. However, expression of *K-Ras*^{G12D} throughout the body failed to induce unscheduled proliferation or other growth abnormalities for up to 8 months. Only a subset of *K-Ras*^{G12D} – expressing lung epithelial cells underwent malignant transformation several months after inducer treatment. These results suggested that only a very small proportion of cells in a given cellular context are transformed by oncogenic *K-Ras* (Guerra et al. 2003).

Palmero et al. demonstrated that overexpression of oncogenic *K-Ras* activates the Arf–p53 pathway in primary cells and effectively induces senescence or apoptosis (Palmero et al. 1998). By contrast, Johnson et al. showed that oncogenic *K-Ras* mutation is sufficient to induce tumors even in the absence of p53 mutation (Johnson et al. 2001). Why, then, does the Arf–p53 pathway fail to protect mice against oncogenic *K-Ras* – induced transformation? Is it due to an inherent limit of the Arf–p53 pathway in specific cells, or instead to inactivation of the Arf–p53 pathway by an unknown mechanism? To answer this question, Junttila et al. and Feldser et al. induced lung adenocarcinoma by simultaneous inactivation of *p53* and activation of *K-Ras*, and then restored *p53* activity. Importantly, restoration of *p53* resulted in the regression of adenocarcinoma, but had no effect on adenoma (Feldser et al. 2010; Junttila et al. 2010). The Arf–p53 pathway was retained in MEFs expressing *K-Ras*^{G12D} (Palmero et al. 1998), indicating that this pathway was not activated by oncogenic *K-Ras* in mouse models. Palmero et al. showed that p53 was activated by overexpression of oncogenic *Ras* in cell lines, whereas Junttila et al. and Feldser et al. did not detect activation of p53 by endogenous levels of oncogenic *Ras* in mice. This discrepancy could be explained in two ways: (1) The ARF–p53 pathway may be activated only by Ras above some threshold that is higher than the endogenous level. (2) Alternatively, the p53 pathway may be able to sense persistent activation of endogenous Ras, but this mechanism can be abrogated via some unknown mechanism. Although several lines of evidence support the first possibility (Feldser et al. 2010; Junttila et al. 2010), other studies report that activation of *Ras* alone in normal cells is not sufficient to induce tumors, supporting the second possibility (Rauen 2007; Sugio et al. 1994). For example, Costello syndrome patients, who carry the *H-RAS*^{G12A} mutant alleles in their germ line, do not develop tumors at young age. Only 24 % of Costello syndrome patients develop malignancy, and even then only after several decades of life (Rauen 2007). Therefore, it is likely that normal cells are resistant to transformation by *Ras* activation alone, and that other genetic or epigenetic alterations are also required for tumorigenesis

20.7 Sensing Mechanism for the Persistence of Ras Activity

To determine whether cells can sense persistent activation of endogenous oncogenic Ras (the second possibility outlined above), it is important to identify the gene(s) responsible for abrogating the ARF–p53 pathway and the underlying sensing mechanism. It is worth mentioning that the activity level of oncogenic RAS is the same as that of mitogen-stimulated normal RAS. Therefore, heterozygous *RAS* mutation results in maintenance of 50 % of mitogen-stimulated RAS activity (Fig. 20.2). If cells possess a mechanism that can respond to this level of persistent RAS activity, it might also respond to 100 % of the mitogen-stimulated Ras activity level.

An important hint regarding the sensing mechanism was reported by Michieli et al. They showed in MEFs that transcription of *p21*, a cell-cycle inhibitor, is induced 1 h after serum stimulation in a p53-independent manner, and then

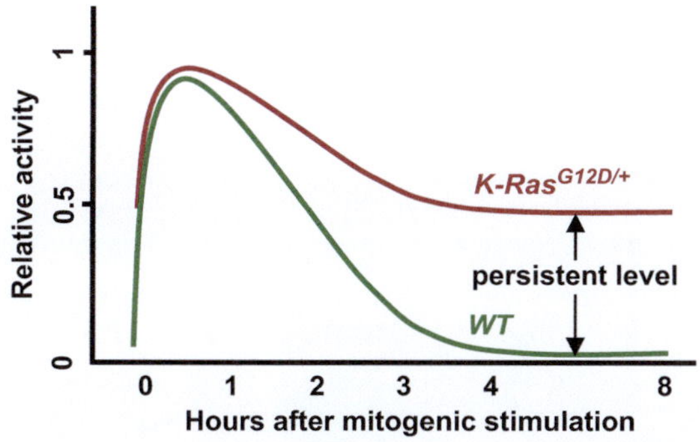

Fig. 20.2 Comparison between mitogen-stimulated normal RAS activity and oncogenic RAS activity. The activity level of oncogenic RAS is the same as that of mitogen-stimulated normal RAS; however, oncogenic RAS activity is constitutive (*red line*) whereas normal RAS activity is transient (*green line*). Therefore, heterozygous *RAS* mutation results in maintenance of 50 % of the activity level of mitogen-stimulated RAS

decreases to basal level 3 h after stimulation (Michieli et al. 1994). For a long time, the biological meaning of this apparent paradox (induction of a cell-cycle inhibitor by mitogenic stimulation) was not understood. Recently, Lee et al., found that, not only *p21*, but also *ARF*, is induced by serum stimulation at the same time points in HEK293 cells (Lee et al. 2013). The level of transient induction of *ARF* was sufficient to stabilize p53 when it was induced. The time points for the inductions of *p21* and *ARF* and the accumulation of p53 overlapped with those for RAS activation; accordingly, when oncogenic *RAS* was expressed, the induced levels of these proteins were maintained in cells. Mitogen-stimulated transient activation of the ARF–p53 pathway does not affect the cell cycle because it only occurs 1–3 h after mitogenic simulation, and is subsequently silenced at the G_1/S checkpoint. However, prolonged induction of the ARF–p53 pathway by oncogenic RAS induced cell-cycle arrest and apoptosis (Lee et al. 2013). Hence, cells evolve a mechanism for monitoring persistent RAS activity to protect cells from inappropriate proliferation (Lee et al. 2013). Thus, the immediate-early induction of *p21* was not a paradox, but instead the tip of an iceberg, i.e., a cellular monitoring system for sensing the persistent activation of oncogenic *RAS*.

How, then, does the *RAS* pathway induce *ARF* expression? Lee et al. also demonstrated that serum-stimulated transient induction of *ARF* was abolished by knockdown of *RUNX3* (Lee et al. 2013). When *RAS* is activated by normal mitogenic stimulation, RUNX3 is activated by the MAPK pathway and acetylated by p300 at lysine residues. It then forms a complex with BRD2, which contains bromodomains that bind to acetylated lysine. The RUNX3–BRD2 complex transiently induces *ARF*, which in turn transiently stabilizes p53. Soon after the mitogenic surge, MAPK activity is reduced. In this situation, the RUNX3–BRD2 complex dissociates, and *ARF* expression is repressed. On the other hand, when *K-RAS* is constitutively activated, the RUNX3–BRD2 complex is maintained, and expression of *ARF* and *p53* continues until the G_1/S checkpoint, leading to cell death. The regulation of the RAS-activated RUNX3–ARF–p53 pathway is depicted in Fig. 20.3. These results explain how cells can sense persistent RAS activity and demonstrate that the RUNX3–BRD2 complex functions as a sensor for persistently abnormal RAS activation. Therefore, cells that acquire a *K-RAS* mutation could be selected only when this sensing mechanism is abrogated.

Indeed, although the ARF–p53 pathway plays a major role in the cellular protection mechanisms that prevent pathologic consequences of abnormal oncogenic stimulation, it often fails to respond to oncogene activation during cancer development. For example, *K-Ras* codon 12 mutation is frequently found in lung adenocarcinoma, but *p53*-null mutations are rarely found at the preneoplastic stages that precede such cancers (atypical adenomatous hyperplasia [AAH] or bronchio-alveolar carcinoma [BAC]). Instead, *RUNX3* expression is downregulated in the majority of the preneoplastic stages by DNA hypermethylation (Lee et al. 2010, 2013; Lee and Bae 2016). This result suggests that loss of *RUNX3* expression prevents adequate oncogene surveillance during the preneoplastic stages of lung epithelial cells. Because *RUNX3* expression is not ubiquitous, other mechanisms must exist for monitoring aberrant RAS activation. However, frequent inactivation of *RUNX3* in various tumors

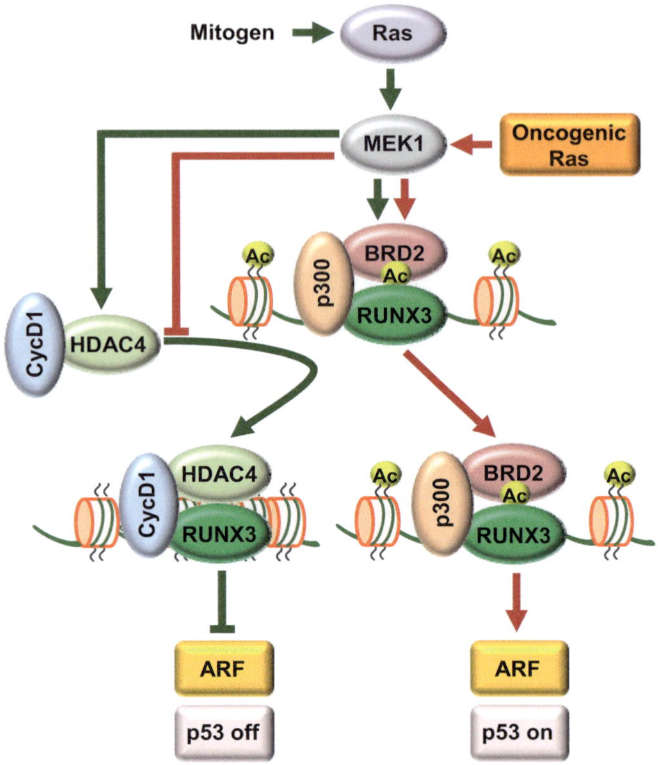

Fig. 20.3 Sensing mechanism for persistent Ras activity. When *RAS* is activated by normal mitogenic stimulation, RUNX3 is activated by the MAPK pathway and acetylated by p300 at lysine residues; the modified and activated protein then forms a complex with BRD2. The RUNX3–BRD2 complex induces *ARF*, which in turn stabilizes p53. Soon after the mitogenic surge, the Runx3–BRD2 complex is dissociated by the Cyclin D1–HDAC4 complex, and *ARF* expression is repressed (*green lines*). However, when *K-RAS* is constitutively activated, the RUNX3–BRD2 complex is maintained, and expression of *ARF* and *p53* continues until the G_1/S checkpoint, ultimately leading to cell cycle arrest or cell death (*red lines*)

suggests that the sensing mechanism mediated by the RUNX3–BRD2 complex plays a major role in surveillance of *RAS* activation.

20.8 Regulation of the RUNX3 Level by the p53–Mdm2 Pathway

Because accumulation of p53 leads to cell death, this protein must be quickly degraded in normal cells. When p53 is activated, it stimulates the synthesis of Mdm2, the agent of its own degradation. This relationship constitutes a negative-feedback loop that ensures that p53 is maintained at a very low steady-state level in normal cells. The connection between RAS and RUNX family members was first demonstrated by an early observation that the RUNX2 protein level is elevated in *H-Ras-V12* – transformed NIH3T3 cells (Ito 2004; Kamachi et al. 1990). Similarly, the RUNX3 protein level is also increased by ectopic expression of *K-Ras-V12* in HEK293 cells (Chi et al. 2009; Lee et al. 2013). Further study revealed that RUNX3, as well as p53, is a target of Mdm2. Mdm2 directly binds RUNX3 and ubiquitinates it on key lysine residues, leading to proteasomal degradation (Chi et al. 2009). The Mdm2-mediated degradation of RUNX3 is alleviated by expression of *ARF*. Therefore, both RUNX3 and p53 are stabilized by RAS activation through the ARF–Mdm2 signaling pathway. The connection between Mmd2, p53, and RUNX3 suggests that p53 and RUNX3 mutually control each other's expression in normal cells, ensuring that the levels of both proteins remain low. A schematic diagram of this feedback regulation is shown in Fig. 20.4.

20.9 Prospects

It is somewhat surprising that a regulator of differentiation, *RUNX3*, plays a key role in cellular defenses against DNA damage and oncogene activation. However, considering the machinery

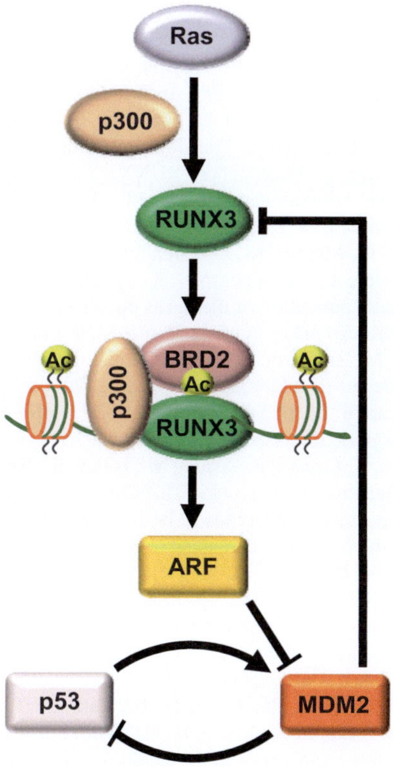

Fig. 20.4 Regulation of RUNX3 level by p53–Mdm2 pathway. RAS signaling triggers p300-mediated RUNX3 acetylation, and acetylated RUNX3 forms a complex with BRD2. The RUNX3–p300–BRD2 complex activates ARF, which inhibits Mdm2, thereby stabilizing p53. p53, in turn, stimulates the synthesis of Mdm2. RUNX3 and p53 are degraded by Mdm2-mediated ubiquitination (Chi et al. 2009). This relationship constitutes a negative-feedback loop that ensures the very low steady-state level of p53 and RUNX3 in normal cells

involved in cell-cycle decision making, contact points between the differentiation and cell-cycle programs would be necessary to ensure genome stability and proper balance of mitogenic signaling. When stimulated by mitogens, cells decide whether they will remain in G_1, retreat from the active cycle into G_0, or progress through the cell cycle. Similarly, cells that acquire DNA damage must halt the cell cycle to allow time for the repair machineries to restore genome integrity. The critical decision for cell-cycle progression or arrest is made 2–3 h after mitogenic stimulation, at the so-called restriction point (R-point) (Blagosklonny and Pardee 2002; Pardee 1974). A growing body of evidence indicates that deregulation of the R-point decision-making machinery accompanies the formation of most types of cancer (Weinberg 2007). To make an appropriate decision, cells must be aware of their own differentiation status; from this standpoint, *RUNX3* might serve to report cellular differentiation status to the R-point 'committee'. Accordingly, it would be interesting to investigate whether the tumor-suppressor activity of *RUNX3* is associated with R-point commitment.

As noted above, a recent effort to restore *p53* in *K-Ras*–activated mouse tumors failed to eliminate adenoma: *p53* restoration killed only malignant adenocarcinomas, leaving adenomas untouched (Feldser et al. 2010; Junttila et al. 2010). Most cancers are ultimately fatal due to the very high rate of recurrence (Kobayashi et al. 2005). Such recurrence occurs mainly due to persistent early lesions that are resistant to anti-cancer drugs. Therefore, to eradicate cancers, it is necessary to eliminate both early and malignant lesions. Recent findings demonstrated that *RUNX3* inactivation leads to inactivation of the *p53* pathway and induces development of adenomas in mouse and human (Ito et al. 2008; Lee et al. 2013). Therefore, *RUNX3* could represent a potential therapeutic target for the eradication of early-stage cancers.

Acknowledgements S-C. Bae is supported by a Creative Research Grant (2014R1A3A2030690) through the National Research Foundation (NRF) of Korea and a Research Grant of Chungbuk National University (2014). J-W. Lee is supported by the Basic Science Research Program (2014R1A1A2007204) through the NRF funded by the Ministry of Education of Korea.

References

Bae, S. C., Takahashi, E., Zhang, Y. W., Ogawa, E., Shigesada, K., Namba, Y., et al. (1995). Cloning, mapping and expression of PEBP2 alpha C, a third gene encoding the mammalian Runt domain. *Gene, 159*, 245–248.

Barak, Y., Juven, T., Haffner, R., & Oren, M. (1993). mdm2 expression is induced by wild type p53 activity. *The EMBO Journal, 12*, 461–468.

Bartek, J., & Lukas, J. (2003). Chk1 and Chk2 kinases in checkpoint control and cancer. *Cancer Cell, 3*, 421–429.

Blagosklonny, M. V., & Pardee, A. B. (2002). The restriction point of the cell cycle. *Cell Cycle, 1*, 103–110.

Bos, J. L. (1989). Ras oncogenes in human cancer: A review. *Cancer Research, 49*, 4682–4689.

Bos, J. L., Verlaan-de Vries, M., Jansen, A. M., Veeneman, G. H., van Boom, J. H., & van der Eb, A. J. (1984). Three different mutations in codon 61 of the human N-ras gene detected by synthetic oligonucleotide hybridization. *Nucleic Acids Research, 12*, 9155–9163.

Chi, X. Z., Kim, J., Lee, Y. H., Lee, J. W., Lee, K. S., Wee, H., et al. (2009). Runt-related transcription factor RUNX3 is a target of MDM2-mediated ubiquitination. *Cancer Research, 69*, 8111–8119.

Donehower, L. A., Harvey, M., Slagle, B. L., McArthur, M. J., Montgomery Jr., C. A., Butel, J. S., & Bradley, A. (1992). Mice deficient for p53 are developmentally normal but susceptible to spontaneous tumours. *Nature, 356*, 215–221.

DuPage, M., Dooley, A. L., & Jacks, T. (2009). Conditional mouse lung cancer models using adenoviral or lentiviral delivery of Cre recombinase. *Nature Protocols, 4*, 1064–1072.

Feldser, D. M., Kostova, K. K., Winslow, M. M., Taylor, S. E., Cashman, C., Whittaker, C. A., et al. (2010). Stage-specific sensitivity to p53 restoration during lung cancer progression. *Nature, 468*, 572–575.

Fukamachi, H., Ito, K., & Ito, Y. (2004). Runx3−/− gastric epithelial cells differentiate into intestinal type cells. *Biochemical and Biophysical Research Communications, 321*, 58–64.

Goel, A., Arnold, C. N., Tassone, P., Chang, D. K., Niedzwiecki, D., Dowell, J. M., et al. (2004). Epigenetic inactivation of RUNX3 in microsatellite unstable sporadic colon cancers. *International Journal of Cancer, 112*, 754–759.

Guerra, C., Mijimolle, N., Dhawahir, A., Dubus, P., Barradas, M., Serrano, M., et al. (2003). Tumor induction by an endogenous K-ras oncogene is highly dependent on cellular context. *Cancer Cell, 4*, 111–120.

Haupt, Y., Maya, R., Kazaz, A., & Oren, M. (1997). Mdm2 promotes the rapid degradation of p53. *Nature, 387*, 296–299.

Ito, Y. (2004). Oncogenic potential of the RUNX gene family: 'Overview'. *Oncogene, 23*, 4198–4208.

Ito, K., Lim, A. C., Salto-Tellez, M., Motoda, L., Osato, M., Chuang, L. S., et al. (2008). RUNX3 attenuates beta-catenin/T cell factors in intestinal tumorigenesis. *Cancer Cell, 14*, 226–237.

Ito, Y., Bae, S. C., & Chuang, L. S. (2015). The RUNX family: Developmental regulators in cancer. *Nature Reviews. Cancer, 15*, 81–95.

Jackson, E. L., Willis, N., Mercer, K., Bronson, R. T., Crowley, D., Montoya, R., et al. (2001). Analysis of lung tumor initiation and progression using conditional expression of oncogenic K-ras. *Genes & Development, 15*, 3243–3248.

Jeon, E. J., Lee, K. Y., Choi, N. S., Lee, M. H., Kim, H. N., Jin, Y. H., et al. (2006). Bone morphogenetic protein-2 stimulates Runx2 acetylation. *The Journal of Biological Chemistry, 281*, 16502–16511.

Jin, Y. H., Jeon, E. J., Li, Q. L., Lee, Y. H., Choi, J. K., Kim, W. J., et al. (2004). Transforming growth factor-beta stimulates p300-dependent RUNX3 acetylation, which inhibits ubiquitination-mediated degradation. *The Journal of Biological Chemistry, 279*, 29409–29417.

Johnson, L., Mercer, K., Greenbaum, D., Bronson, R. T., Crowley, D., Tuveson, D. A., & Jacks, T. (2001). Somatic activation of the K-ras oncogene causes early onset lung cancer in mice. *Nature, 410*, 1111–1116.

Junttila, M. R., Karnezis, A. N., Garcia, D., Madriles, F., Kortlever, R. M., Rostker, F., et al. (2010). Selective activation of p53-mediated tumour suppression in high-grade tumours. *Nature, 468*, 567–571.

Kamachi, Y., Ogawa, E., Asano, M., Ishida, S., Murakami, Y., Satake, M., et al. (1990). Purification of a mouse nuclear factor that binds to both the A and B cores of the polyomavirus enhancer. *Journal of Virology, 64*, 4808–4819.

Kang, G. H., Lee, S., Lee, H. J., & Hwang, K. S. (2004). Aberrant CpG island hypermethylation of multiple genes in prostate cancer and prostatic intraepithelial neoplasia. *The Journal of Pathology, 202*, 233–240.

Kang, S., Kim, J. W., Kang, G. H., Park, N. H., Song, Y. S., Kang, S. B., & Lee, H. P. (2005). Polymorphism in folate- and methionine-metabolizing enzyme and aberrant CpG island hypermethylation in uterine cervical cancer. *Gynecologic Oncology, 96*, 173–180.

Karnoub, A. E., & Weinberg, R. A. (2008). Ras oncogenes: Split personalities. *Nature Reviews. Molecular Cell Biology, 9*, 517–531.

Kato, N., Tamura, G., Fukase, M., Shibuya, H., & Motoyama, T. (2003). Hypermethylation of the RUNX3 gene promoter in testicular yolk sac tumor of infants. *The American Journal of Pathology, 163*, 387–391.

Kim, T. Y., Lee, H. J., Hwang, K. S., Lee, M., Kim, J. W., Bang, Y. J., & Kang, G. H. (2004). Methylation of RUNX3 in various types of human cancers and premalignant stages of gastric carcinoma. *Laboratory Investigation, 84*, 479–484.

Kim, C. F., Jackson, E. L., Woolfenden, A. E., Lawrence, S., Babar, I., Vogel, S., et al. (2005). Identification of bronchioalveolar stem cells in normal lung and lung cancer. *Cell, 121*, 823–835.

Kitabayashi, I., Yokoyama, A., Shimizu, K., & Ohki, M. (1998). Interaction and functional cooperation of the leukemia-associated factors AML1 and p300 in myeloid cell differentiation. *The EMBO Journal, 17*, 2994–3004.

Kobayashi, S., Boggon, T. J., Dayaram, T., Janne, P. A., Kocher, O., Meyerson, M., et al. (2005). EGFR mutation and resistance of non-small-cell lung cancer to gefitinib. *The New England Journal of Medicine, 352*, 786–792.

Komori, H., Enomoto, M., Nakamura, M., Iwanaga, R., & Ohtani, K. (2005). Distinct E2F-mediated transcriptional program regulates p14ARF gene expression. *The EMBO Journal, 24*, 3724–3736.

Kruse, J. P., & Gu, W. (2009). Modes of p53 regulation. *Cell, 137*, 609–622.

Ku, J. L., Kang, S. B., Shin, Y. K., Kang, H. C., Hong, S. H., Kim, I. J., et al. (2004). Promoter hypermethylation downregulates RUNX3 gene expression in colorectal cancer cell lines. *Oncogene, 23*, 6736–6742.

Kubbutat, M. H., Jones, S. N., & Vousden, K. H. (1997). Regulation of p53 stability by Mdm2. *Nature, 387*, 299–303.

Lakin, N. D., & Jackson, S. P. (1999). Regulation of p53 in response to DNA damage. *Oncogene, 18*, 7644–7655.

Lee, Y. S., & Bae, S. C. (2016). How do K-RAS-activated cells evade cellular defense mechanisms? *Oncogene, 35*, 827–832.

Lee, K. S., Lee, Y. S., Lee, J. M., Ito, K., Cinghu, S., Kim, J. H., et al. (2010). Runx3 is required for the differentiation of lung epithelial cells and suppression of lung cancer. *Oncogene, 29*, 3349–3361.

Lee, Y. S., Lee, J. W., Jang, J. W., Chi, X. Z., Kim, J. H., Li, Y. H., et al. (2013). Runx3 inactivation is a crucial early event in the development of lung adenocarcinoma. *Cancer Cell, 24*, 603–616.

Levanon, D., Negreanu, V., Bernstein, Y., Bar-Am, I., Avivi, L., & Groner, Y. (1994). AML1, AML2, and AML3, the human members of the runt domain gene-family: cDNA structure, expression, and chromosomal localization. *Genomics, 23*, 425–432.

Levine, A. J. (1997). p53, the cellular gatekeeper for growth and division. *Cell, 88*, 323–331.

Li, Q. L., Ito, K., Sakakura, C., Fukamachi, H., Inoue, K., Chi, X. Z., et al. (2002). Causal relationship between the loss of RUNX3 expression and gastric cancer. *Cell, 109*, 113–124.

Li, Q. L., Kim, H. R., Kim, W. J., Choi, J. K., Lee, Y. H., Kim, H. M., et al. (2004). Transcriptional silencing of the RUNX3 gene by CpG hypermethylation is associated with lung cancer. *Biochemical and Biophysical Research Communications, 314*, 223–228.

Lowe, S. W., & Sherr, C. J. (2003). Tumor suppression by Ink4a-Arf: Progress and puzzles. *Current Opinion in Genetics & Development, 13*, 77–83.

Michieli, P., Chedid, M., Lin, D., Pierce, J. H., Mercer, W. E., & Givol, D. (1994). Induction of WAF1/CIP1 by a p53-independent pathway. *Cancer Research, 54*, 3391–3395.

Momand, J., Zambetti, G. P., Olson, D. C., George, D., & Levine, A. J. (1992). The mdm-2 oncogene product forms a complex with the p53 protein and inhibits p53-mediated transactivation. *Cell, 69*, 1237–1245.

Montes de Oca Luna, R., Wagner, D. S., & Lozano, G. (1995). Rescue of early embryonic lethality in mdm2-deficient mice by deletion of p53. *Nature, 378*, 203–206.

Mori, T., Nomoto, S., Koshikawa, K., Fujii, T., Sakai, M., Nishikawa, Y., et al. (2005). Decreased expression and frequent allelic inactivation of the RUNX3 gene at 1p36 in human hepatocellular carcinoma. *Liver International, 25*, 380–388.

Nakase, Y., Sakakura, C., Miyagawa, K., Kin, S., Fukuda, K., Yanagisawa, A., et al. (2005). Frequent loss of RUNX3 gene expression in remnant stomach cancer and adjacent mucosa with special reference to topography. *British Journal of Cancer, 92*, 562–569.

Oliner, J. D., Pietenpol, J. A., Thiagalingam, S., Gyuris, J., Kinzler, K. W., & Vogelstein, B. (1993). Oncoprotein MDM2 conceals the activation domain of tumour suppressor p53. *Nature, 362*, 857–860.

Oshimo, Y., Oue, N., Mitani, Y., Nakayama, H., Kitadai, Y., Yoshida, K., et al. (2004). Frequent loss of RUNX3 expression by promoter hypermethylation in gastric carcinoma. *Pathobiology, 71*, 137–143.

Ozaki, T., Nakagawara, A., & Nagase, H. (2013). RUNX family participates in the regulation of p53-dependent DNA damage response. *International Journal of Genomics, 2013*, 271347.

Ozenne, P., Eymin, B., Brambilla, E., & Gazzeri, S. (2010). The ARF tumor suppressor: Structure, functions and status in cancer. *International Journal of Cancer, 127*, 2239–2247.

Palmero, I., Pantoja, C., & Serrano, M. (1998). p19ARF links the tumour suppressor p53 to Ras. *Nature, 395*, 125–126.

Pardee, A. B. (1974). A restriction point for control of normal animal cell proliferation. *Proceedings of the National Academy of Sciences of the United States of America, 71*, 1286–1290.

Perry, M. E., Piette, J., Zawadzki, J. A., Harvey, D., & Levine, A. J. (1993). The mdm-2 gene is induced in response to UV light in a p53-dependent manner. *Proceedings of the National Academy of Sciences of the United States of America, 90*, 11623–11627.

Pylayeva-Gupta, Y., Grabocka, E., & Bar-Sagi, D. (2011). RAS oncogenes: Weaving a tumorigenic web. *Nature Reviews. Cancer, 11*, 761–774.

Rauen, K. A. (2007). HRAS and the Costello syndrome. *Clinical Genetics, 71*, 101–108.

Reinhardt, H. C., Aslanian, A. S., Lees, J. A., & Yaffe, M. B. (2007). p53-deficient cells rely on ATM- and ATR-mediated checkpoint signaling through the p38MAPK/MK2 pathway for survival after DNA damage. *Cancer Cell, 11*, 175–189.

Sakakura, C., Hasegawa, K., Miyagawa, K., Nakashima, S., Yoshikawa, T., Kin, S., et al. (2005). Possible involvement of RUNX3 silencing in the peritoneal metastases of gastric cancers. *Clinical Cancer Research, 11*, 6479–6488.

Satoh, Y., Matsumura, I., Tanaka, H., Harada, H., Harada, Y., Matsui, K., et al. (2012). C-terminal mutation of RUNX1 attenuates the DNA-damage repair response in hematopoietic stem cells. *Leukemia, 26*, 303–311.

Schulmann, K., Sterian, A., Berki, A., Yin, J., Sato, F., Xu, Y., et al. (2005). Inactivation of p16, RUNX3, and

HPP1 occurs early in Barrett's-associated neoplastic progression and predicts progression risk. *Oncogene, 24*, 4138–4148.

Serrano, M., Lin, A. W., McCurrach, M. E., Beach, D., & Lowe, S. W. (1997). Oncogenic ras provokes premature cell senescence associated with accumulation of p53 and p16INK4a. *Cell, 88*, 593–602.

Sherr, C. J. (1998). Tumor surveillance via the ARF-p53 pathway. *Genes & Development, 12*, 2984–2991.

Sherr, C. J. (2006). Divorcing ARF and p53: An unsettled case. *Nature Reviews. Cancer, 6*, 663–673.

Shieh, S. Y., Ikeda, M., Taya, Y., & Prives, C. (1997). DNA damage-induced phosphorylation of p53 alleviates inhibition by MDM2. *Cell, 91*, 325–334.

Shieh, S. Y., Ahn, J., Tamai, K., Taya, Y., & Prives, C. (2000). The human homologs of checkpoint kinases Chk1 and Cds1 (Chk2) phosphorylate p53 at multiple DNA damage-inducible sites. *Genes & Development, 14*, 289–300.

Smith, G. C., & Jackson, S. P. (1999). The DNA-dependent protein kinase. *Genes & Development, 13*, 916–934.

Sugio, K., Kishimoto, Y., Virmani, A. K., Hung, J. Y., & Gazdar, A. F. (1994). K-ras mutations are a relatively late event in the pathogenesis of lung carcinomas. *Cancer Research, 54*, 5811–5815.

Tamura, G. (2004). Promoter methylation status of tumor suppressor and tumor-related genes in neoplastic and non-neoplastic gastric epithelia. *Histology and Histopathology, 19*, 221–228.

Verlaan-de Vries, M., Bogaard, M. E., van den Elst, H., van Boom, J. H., van der Eb, A. J., & Bos, J. L. (1986). A dot-blot screening procedure for mutated ras oncogenes using synthetic oligodeoxynucleotides. *Gene, 50*, 313–320.

Vogelstein, B., Lane, D., & Levine, A. J. (2000). Surfing the p53 network. *Nature, 408*, 307–310.

Wada, M., Yazumi, S., Takaishi, S., Hasegawa, K., Sawada, M., Tanaka, H., et al. (2004). Frequent loss of RUNX3 gene expression in human bile duct and pancreatic cancer cell lines. *Oncogene, 23*, 2401–2407.

Wang, C. Q., Krishnan, V., Tay, L. S., Chin, D. W., Koh, C. P., Chooi, J. Y., et al. (2014). Disruption of Runx1 and Runx3 leads to bone marrow failure and leukemia predisposition due to transcriptional and DNA repair defects. *Cell Reports, 8*, 767–782.

Weber, J. D., Taylor, L. J., Roussel, M. F., Sherr, C. J., & Bar-Sagi, D. (1999). Nucleolar Arf sequesters Mdm2 and activates p53. *Nature Cell Biology, 1*, 20–26.

Weinberg, R. A. (2007). *The biology of cancer*. New York: Garland Science.

Wijmenga, C., Speck, N. A., Dracopoli, N. C., Hofker, M. H., Liu, P., & Collins, F. S. (1995). Identification of a new murine runt domain-containing gene, Cbfa3, and localization of the human homolog, CBFA3, to chromosome 1p35-pter. *Genomics, 26*, 611–614.

Wu, X., & Levine, A. J. (1994). p53 and E2F-1 cooperate to mediate apoptosis. *Proceedings of the National Academy of Sciences of the United States of America, 91*, 3602–3606.

Wu, D., Ozaki, T., Yoshihara, Y., Kubo, N., & Nakagawara, A. (2013). Runt-related transcription factor 1 (RUNX1) stimulates tumor suppressor p53 protein in response to DNA damage through complex formation and acetylation. *The Journal of Biological Chemistry, 288*, 1353–1364.

Xiao, W. H., & Liu, W. W. (2004). Analysis of methylation and loss of heterozygosity of RUNX3 gene in hepatocellular carcinoma and its clinical significance. *Zhonghua Gan Zang Bing Za Zhi, 12*, 227–230.

Yamada, C., Ozaki, T., Ando, K., Suenaga, Y., Inoue, K., Ito, Y., et al. (2010). RUNX3 modulates DNA damage-mediated phosphorylation of tumor suppressor p53 at Ser-15 and acts as a co-activator for p53. *The Journal of Biological Chemistry, 285*, 16693–16703.

Yanagawa, N., Tamura, G., Oizumi, H., Takahashi, N., Shimazaki, Y., & Motoyama, T. (2003). Promoter hypermethylation of tumor suppressor and tumor-related genes in non-small cell lung cancers. *Cancer Science, 94*, 589–592.

Zhang, Y., Xiong, Y., & Yarbrough, W. G. (1998). ARF promotes MDM2 degradation and stabilizes p53: ARF-INK4a locus deletion impairs both the Rb and p53 tumor suppression pathways. *Cell, 92*, 725–734.

Runx3 and Cell Fate Decisions in Pancreas Cancer

21

Martin C. Whittle and Sunil R. Hingorani

Abstract

The RUNX family transcription factors are critical regulators of development and frequently dysregulated in cancer. RUNX3, the least well characterized of the three family members, has been variously described as a tumor promoter or suppressor, sometimes with conflicting results and opinions in the same cancer and likely reflecting a complex role in oncogenesis. We recently identified *RUNX3* expression as a crucial determinant of the predilection for pancreatic ductal adenocarcinoma (PDA) cells to proliferate locally or promulgate throughout the body. High RUNX3 expression induces the production and secretion of soluble factors that support metastatic niche construction and stimulates PDA cells to migrate and invade, while simultaneously suppressing proliferation through increased expression of cell cycle regulators such *as CDKN1A/p21*$^{WAF1/CIP1}$. RUNX3 expression and function are coordinated by numerous transcriptional and post-translational inputs, and interactions with diverse cofactors influence whether the resulting RUNX3 complexes enact tumor suppressive or tumor promoting programs. Understanding these exquisitely context-dependent tumor cell behaviors has the potential to inform clinical decision-making including the most appropriate timing and sequencing of local vs. systemic therapies.

Keywords

Tumor promoter • Tumor suppressor • Metastasis • Pancreatic ductal adenocarcinoma • RUNX3

M.C. Whittle
Clinical Research Division, Fred Hutchinson Cancer Research Center, 1100 Fairview Ave N, M5-C800, Seattle, WA 98109-1024, USA

S.R. Hingorani, M.D., Ph.D. (✉)
Clinical Research Division, Fred Hutchinson Cancer Research Center, 1100 Fairview Ave N, M5-C800, Seattle, WA 98109-1024, USA

Public Health Sciences Division, Fred Hutchinson Cancer Research Center, 1100 Fairview Ave N, M5-C800, Seattle, WA 98109-1024, USA

Division of Medical Oncology, University of Washington School of Medicine, Seattle, WA 98195, USA
e-mail: srh@fhcrc.org

Y. Groner et al. (eds.), *RUNX Proteins in Development and Cancer*, Advances in Experimental Medicine and Biology 962, DOI 10.1007/978-981-10-3233-2_21

21.1 *RUNX3*: Tumor Promoter or Suppressor?

The Runt-related (*RUNX*) family of transcription factors comprise three evolutionarily conserved proteins bearing a highly homologous Runt domain that, along with the obligatory cofactor core binding factor β (CBFβ), facilitates DNA binding to a consensus *5′* TGTGGT *3′* motif (Levanon and Groner 2004; Wu et al. 2014). *RUNX3* is the least studied and perhaps most enigmatic of this gene family. During normal development, *RUNX3* is expressed primarily in the hematopoietic system, in TrkC-positive neurons of the dorsal root ganglia, and in chondrocytes and odontoblasts/osteoblasts. In adult tissues, stable *RUNX3* expression persists in the hematopoietic system (Wu et al. 2014; Levanon et al. 2001, 2002; Yamashiro et al. 2002; Bauer et al. 2015) and, more controversially, in distinct epithelial compartments of the aerodigestive tract (Levanon et al. 2011; Ito 2012). *RUNX3* regulates signaling in critical developmental and homeostatic pathways, including TGFβ, WNT and MST, but almost certainly also acts outside of these defined programs (Hanai et al. 1999; Min et al. 2012; Ju et al. 2014).

The precise roles of *RUNX3* in tumor initiation and progression remain a matter of debate, with sometimes conflicting lines of evidence implicating *RUNX3* on either side of the cancer divide. The debate is confounded by a number of factors including the highly context-dependent nature of the outputs alluded to above, the specific manifestation of the malignant phenotype in question, and differences in reagents used and potentially their reliability in reporting expression levels in normal and neoplastic tissues. For example, a number of reports have identified *RUNX3* expression in the normal gastric epithelium and its subsequent loss during progression to invasive cancer, suggesting a tumor suppressive function (Li et al. 2002; Guo et al. 2002). Studies building upon this idea have proposed similar tumor suppressive roles in other solid tumors (Goel et al. 2004; Kim et al. 2005; Mori et al. 2005; Lee et al. 2010; Huang et al. 2012; Chen et al. 2013; Jin et al. 2013). Many studies have suggested that hypermethylation of the P2 promoter region of *RUNX3* serves as a primary mechanism for *RUNX3* silencing during cancer progression, but the relevance of this methylation event is not universally acknowledged (Kurklu et al. 2015). Chromosomal deletion of *RUNX3* has also been observed, although the nearby tumor suppressor, *CHD5*, and a number of other potential candidate tumor suppressor genes (TSG) have also been postulated as causes for loss of this locus (Wada et al. 2004; Carvalho et al. 2005; Bagchi et al. 2007; Okawa et al. 2008; Lotem et al. 2015). Perhaps the most provocative data supporting *RUNX3* as a TSG are found in studies of targeted deletion of *Runx3* in the murine lung epithelium (Lee et al. 2013). Biallelic excision of floxed *Runx3* alleles by inhalation of an adenoviral Cre recombinase-expressing vector elicited lung adenomas. These preinvasive lesions did not progress to adenocarcinomas, however, without the additional introduction of conditional mutant *Kras* expression. These data would seem to significantly bolster the designation of *Runx3* as an important TSG, however, some caveats remain: Runx3 expression was apparently undetectable in normal mouse lung epithelia ((Lee et al. 2013) and our own unpublished observations); the precise mechanism of adenoma induction in this model remains ambiguously defined; and important features of the adenomas caused by *Runx3* deletion were not described (e.g. rate of adenoma incidence in $Runx3^{fl/fl}$ and normal mice, proliferation and apoptotic rates within adenomas, advancement of cellular and molecular atypia with age, and so on).

A distinct body of work has shown that *RUNX3* expression is acquired during the course of cancer progression and that it therefore plays a tumor promoting or oncogenic role. For instance, acquired *RUNX3* expression correlates with poor histologic differentiation, invasion and metastasis in head and neck squamous cell carcinoma (HNSCC) (Tsunematsu et al. 2009). High *RUNX3* expression has also been observed in ovarian cancer (Nevadunsky et al. 2009; Barghout et al. 2015; Lee et al. 2011a), basal cell carcinoma (Salto-Tellez et al. 2006) and other skin cancers (Lee et al. 2011b), gastric cancer (Carvalho et al.

2005), inflammatory breast cancer (Fredika et al. 2012), Ewing sarcoma (Bledsoe et al. 2014), childhood AML (Cheng et al. 2008) and pancreatic ductal adenocarcinoma (PDA) (Whittle et al. 2015). RUNX3 has additionally been shown to mediate resistance to conventional and targeted therapies in ovarian cancer and chronic myeloid leukemia, respectively (Barghout et al. 2015; Miething et al. 2007). The debate over *RUNX3* expression in normal tissues remains central to its purported role and underlies the uncertainty of whether *RUNX3* is purposefully lost or gained in cancer (Lotem et al. 2015; Levanon et al. 2003). It should be acknowledged that undetectable levels in normal tissue that then rise during neoplastic transformation is not *de facto* evidence of a tumor-promoting role; indeed, the most classic example of such an expression pattern might be *TP53*, now known to be the quintessential gatekeeper and guardian of the genome but which, ironically, was first described as an oncogene until the subsequent discovery that the initial forms being studied contained point mutations that stabilized the resulting protein and permitted detection (DeLeo et al. 1979; Linzer and Levine 1979; Finlay et al. 1988). The cell cycle inhibitor and tumor suppressor CDKN2A/p16 is similarly undetectable unless elevated in response to genotoxic or oxidative stress and/or advanced cellular age (Kim and Sharpless 2006). Nevertheless, with considerable evidence on either side and undeniable complexity in its myriad functions, the most parsimonious explanation may be that RUNX3 can exhibit both tumor-promoting and suppressive behaviors depending on the context and the cellular phenotype in question (Table 21.1). We explore this idea in more detail below.

Table 21.1 Defining the functional outcome of *RUNX3* across cancer types

Cancer type	*RUNX3* designation	References
Acute myelogenous leukemia	Tumor promoting	Cheng et al. (2008), Damdinsuren et al. (2015), and Lacayo et al. (2004)
	Tumor suppressing	Estecio et al. (2015)
Basal cell carcinoma	Tumor promoting	Salto-Tellez et al. (2006) and Lee et al. (2011b)
Breast cancer	Tumor suppressing	Huang et al. (2012), Bai et al. (2013), Lau et al. (2006), and Subramaniam et al. (2009)
Ewing sarcoma	Tumor promoting	Bledsoe et al. (2014)
Gastric cancer	Tumor suppressing	Li et al. (2002), Guo et al. (2002), Chi et al. (2005), and Wei et al. (2005)
Head and neck cancer	Tumor promoting	Tsunematsu et al. (2009)
Hepatocellular carcinoma	Tumor suppressing	Mori et al. (2005), Nakanishi et al. (2011), Yang et al. (2014), and Zhang et al. (2015)
Lung adenocarcinoma	Tumor suppressing	Lee et al. (2010), Lee et al. (2013), Omar et al. (2012), and Sato et al. (2006)
Melanoma	Tumor promoting	Lee et al. (2011b)
	Tumor suppressing	Kitago et al. (2009) and Zhang et al. (2011)
Ovarian cancer	Tumor promoting	Nevadunsky et al. (2009), Barghout et al. (2015), and Lee et al. (2011a)
Pancreatic ductal adenocarcinoma	Tumor promoting	Li et al. (2004)
	Tumor suppressing	Wada et al. (2004), Xue et al. (2014), and Horiguchi et al. (2013)
	Tumor promoting/suppressing	Whittle et al. (2015)
Prostate cancer	Tumor suppressor	Chen et al. (2014)
Renal cell carcinoma	Tumor suppressor	Chen et al. (2013)

It may be instructive to revisit the classical definitions of a TSG and an oncogene. TSG have been classically described as inhibitors of tumorigenesis whose mutation or disruption acts recessively to unleash tumor growth. Stringent definitions additionally demand that inherited inactivating mutations in the putative TSG increase cancer susceptibility, by priming cells with one of two "hits" in that gene; that sporadic inactivation is observed frequently in cancer; and that restoration of gene function has the potential to reverse tumorigenicity (Sherr 2004; Knudson 1971). The National Cancer Institute (NCI) more generally defines a tumor suppressor protein as one that "helps control cell growth," which infers that TSG act broadly to inhibit diverse aspects of both normal and neoplastic physiology, such as regulation of the cell cycle, DNA damage repair and response to mitogenic signaling and stress. Indeed, TSG are often categorized as "gatekeepers", "caretakers" and/or "landscapers" of the genome, depending upon their normal function(s) and the results of their respective inactivation (Kinzler and Vogelstein 1997; Michor et al. 2004). In contrast, oncogenes are genes activated by mutation or overexpression that act dominantly to induce tumorigenesis. Oncogenes transform cells in classical experiments that test their proliferative and colony-forming capacity, by sustaining growth and maintaining survival signals in defiance of normal cues and restraints on cell behavior (Tucker et al. 1977; Pulciani et al. 1982; Tabin et al. 1982; Der et al. 1982). There is a pervasive temptation, often indulged, to rush to label proteins contributing to cancer formation as "oncogenes" and those opposing cancer progression as "tumor suppressors" before a full reckoning of the gene and its impact have occurred. In truth, a terminological gray area also exists between "tumor suppressor" and "oncogene" wherein lie proteins with diverse functions affecting a spectrum of cellular outcomes that can ultimately enhance or suppress tumor pathogenesis, and sometimes both. Using conventional definitions, these proteins are not accurately described by either designation while undoubtedly affecting tumor pathology and patient outcomes. Among these ambiguously defined proteins is RUNX3.

RUNX3 almost certainly does help to suppress tumor cell proliferation, but does not fit the mold of the "classical" tumor suppressor; disruptive RUNX3 mutations rarely occur in cancer and acquired transcriptional inactivation is equivocal in most epithelial cancers. RUNX3 cannot yet be described as a classical oncogene either, although it can promote metastasis and drug resistance, hallmarks of cancer on par with sustained proliferation and perhaps even more relevant to patient prognosis (Hanahan and Weinberg 2000, 2011). Setting semantics aside, these observations beget the question: can a protein be simultaneously a tumor suppressor and oncoprotein? The strict definitions for tumor suppressor and oncogene are mutually exclusive. Nevertheless, RUNX3 exemplifies a class of proteins that exhibit tumor-suppressing and tumor-promoting characteristics, which, to avoid unnecessary conflict and confusion, may require an extension of our cancer lexicon to include critical modulators of cancer behavior that defy strict categorization. We propose to describe RUNX3 and similarly acting proteins as "tumor modifiers" to reflect a strong influence on tumor behavior but with an ambipotent and context-dependent effect on pathobiology.

21.2 *RUNX3* Functions as a Metastatic Switch in PDA

We recently identified *RUNX3* as an important modifier of metastatic potential in PDA (Whittle et al. 2015). The extreme proclivity of PDA for metastatic spread is the primary cause of mortality from this disease. Even after successful surgical resection of the primary tumor, five-year survival is less than 20 %, and most of these patients ultimately succumb to disseminated disease rather than local relapse despite no clinically overt metastases at presentation (Allison et al. 1998).

The study of this extremely lethal carcinoma has been greatly aided in the past decade or so with the advent of highly faithful GEMM of the disease (Mazur and Siveke 2012; Perez-Mancera et al. 2012; Guerra and Barbacid 2013). Indeed, a number of models now exist that reliably phenocopy the distinct presentations seen in pancreas cancer patients beginning with the clinical syndrome of disease which includes jaundice and cachexia; same sites and frequencies of metastases to distant organs including malignant ascites; the same histologic progression scheme arising from the stochastic acquisition of preinvasive ductal lesions that progress spontaneously to invade and metastasize; and the aberrant expression and activation of signaling pathways that contribute to the malignant phenotype (Hruban et al. 2006).

The most frequent genetic mutations encountered in PDA activate the *KRAS* proto-oncogene and are thought to represent critical initiating events (Hruban et al. 1993). Additional mutations in key TSG including notably *P16, TP53* and *SMAD4/DPC4* shape the behavior of the resulting disease (Aguirre et al. 2003; Hingorani et al. 2003, 2005; Izeradjene et al. 2007; Caldas et al. 1994; van Es et al. 1995; Hahn et al. 1996). Targeted endogenous pancreatic expression of $Kras^{G12D}$ and $Trp53^{R172H}$, in particular, recapitulates the widely metastatic presentation encountered in the majority of PDA patients with great fidelity, and this murine $Kras^{LSL\text{-}G12D/+};Trp53^{LSL\text{-}R172H/+};p48^{Cre/+}$ (*KPC*) model is used extensively in translational studies of the disease (Olive et al. 2009; Provenzano et al. 2012; Courtin et al. 2013; Stromnes et al. 2014, 2015; Neesse et al. 2013; Sherman et al. 2014). During the course of our investigations on the influences of these cardinal genetic events in PDA pathogenesis, we observed that heterozygous mutation of *Smad4/Dpc4* could alter the differentiation state of the prototypical microscopic precursor lesions initiated by oncogenic *Kras*, termed pancreatic intraepithelial neoplasms (PanIN), to a less aggressive cystic precursor, mucinous cystic neoplasms (MCN). Each of these preinvasive neoplasms can ultimately progress to invasive cancers through mutation of additional TSG; however, to our surprise, the chronological sequence in which the mutations were acquired determined the ensuing pathology and prognosis of the resultant disease (Izeradjene et al. 2007; Crippa et al. 2008). We subsequently engineered targeted heterozygous deletion of *Smad4* into the highly metastatic *KPC* model to generate $Kras^{LSL\text{-}G12D/+};Trp53^{LSL\text{-}R172H/+};Dpc4^{flox/+};p48^{Cre/+}$ (*KPDC*) mice (Whittle et al. 2015). In this context of point mutation of *Trp53*, heterozygous inactivation of *Smad4* was no longer able to alter the differentiation state of the precursor lesions, as predicted (Izeradjene et al. 2007). However, despite the expected initiation and progression along the more conventional PanIN-to-PDA trajectory, the cancers that arose in *KPDC* mice were far less likely to metastasize than their *KPC* counterparts. This was reflected in the attenuated migration and invasion of *KPDC* cells *in vitro* together with an impaired ability to form pulmonary metastases even after direct intravenous inoculation of the purified primary tumor cells. An integrated series of genomic, transcriptomic and proteomic studies identified *Runx3*, a critical mediator of receptor-Smad dependent TGFβ signaling, as significantly overexpressed in primary invasive *KPC* tumor cells compared to the essentially undetectable levels in *KPDC* cells; normal pancreatic ductal cells and preinvasive *KPC* ductal cells similarly lacked *Runx3* expression. Acquired expression occurred stochastically within Runx3-positive tumors, consistent with previous reports (Li et al. 2004). Where metastases in *KPDC* mice were present, subpopulations of cells with elevated *Runx3* levels were observed in the primary tumor; this, in turn, was also typically accompanied by complete loss of *Smad4* expression (i.e. loss of heterozygosity (LOH)) in the same cells (discussed further below). Respective knockdown and overexpression of *Runx3* in *KPC* and *KPDC* cells reversed these phenotypes, compromising the metastatic potential of *KPC* cells and enhancing that of *KPDC* cells.

Runx3 drove metastasis primarily through the induced expression of numerous secreted factors, including versican, Sparc, osteopontin (*Spp1*) and collagen type VI alpha 1 (*Col6a1*). Of these, *Spp1* stands out for its well-known roles in tumor progression and metastasis (Urquidi et al. 2002; Wai and Kuo 2008) and has been shown to be upregulated by *Runx* family members, including *Runx3*, in a variety of contexts (Whittle et al. 2015; Ducy et al. 1997; Colla et al. 2005; Pratap et al. 2006; Bauer et al. 2014). The human and mouse *SPP1* promoters each contain two consensus *RUNX* binding sites within 1500 base pairs of the transcription start site, with particularly high homology in the site most proximal to transcription initiation. This conserved proximal site has been demonstrated to bind RUNX proteins (Inman and Shore 2003). *SPP1* encodes a secreted phosphoprotein in bone, kidney and endothelial cells, as well as sites of wound healing, and is also readily found in plasma and other bodily fluid compartments. *SPP1* is also highly expressed in many cancers, including lung, gastric, colorectal, hepatocellular and pancreas cancers (Chambers et al. 1996; Pan et al. 2003; Kolb et al. 2005; Higashiyama et al. 2007; Ng et al. 2015).

In pancreas cancer, Runx3 coordinates a secretory program, including Spp1 as a principal factor, that stimulates migration and invasion of cancer cells and preconditions a metastatic niche for successful colonization of disseminated tumor cells (Whittle et al. 2015). Spp1 enhances invasion and migration by promoting cell-matrix adhesion through interactions with integrin and CD44 receptors that are frequently overexpressed in malignancies (Katagiri et al. 1999; Furger et al. 2003). The increased tumoral expression and secretion of Spp1 can be detected in the plasma and presumably enables accumulation in other tissues. Elevated circulating SPP1 has been independently proposed as a marker for PDA and other cancers (Koopmann et al. 2004; Fredriksson et al. 2008). SPP1 plasma levels may also prognosticate the likelihood and site(s) of relapse after surgical resection of the primary tumor (Whittle et al. 2015; Rud et al. 2013; Poruk et al. 2013). The detailed mechanisms by which SPP1 supports survival of and distant colonization by circulating tumor cells remains to be fully elucidated, but may include stimulation of angiogenesis and disruption of immune cell action – two major barriers impeding successful metastatic outgrowth (Colla et al. 2005; Hirama et al. 2003; Rowe et al. 2014; Kim et al. 2014; Sangaletti et al. 2014). Thus, by increasing the migration of primary tumor cells and laying the foundation for distant colonization, *RUNX3* serves to both propel the "seed" and provide the "soil" for metastatic dissemination (Paget 1889; Fidler 2003).

Interestingly, *KPDC* mice succumbed at approximately the same age as *KPC* animals, despite the significantly lower metastatic disease burden. It became clear that tumors lacking *Runx3* expression tended to grow faster than those with high levels, an effect attributable to increased proliferation rather than decreased apoptosis; primary *KPDC* cells also proliferated faster in culture than *KPC* cells (Whittle et al. 2015). Induced overexpression and knockdown of *Runx3* in these primary cells confirmed its ability to suppress proliferation, providing a potential mechanism to explain the earlier designation of *Runx3* as a tumor suppressor. Increased expression of the cell cycle inhibitor *Cdkn1a* (*p21*), a known Runx3 target, drove this effect (Whittle et al. 2015; Chi et al. 2005). Nevertheless, a dramatic increase in metastasis outweighed the attenuated proliferation: primary tumor growth was slowed but not stopped and, in the bargain, the cells gained the ability to disseminate. Spontaneous loss of the remaining *Smad4/Dpc4* allele (i.e. LOH) in *KPDC* tumors restored *Runx3* levels and the metastatic potential of these cells; proliferation was also further unfettered, creating a uniquely lethal combination. These results were confirmed by engineering biallelic loss of *Dpc4* into the *KPC* background to generate $Kras^{LSL\text{-}G12D/+};Trp53^{LSL\text{-}R172H/+};Dpc4^{flox/flox};p48^{Cre/+}$ (*KPDDC*) tumors, which demonstrated metastases and accelerated primary tumor growth (Whittle et al. 2015). Taken together, these data revealed that acquired epithelial cell expression of *Runx3/RUNX3* manifests a "metastatic switch" that shifts the focus of the cell from division to dissemination (Fig. 21.1). Thus, in PDA at least, RUNX3 behaves as a "tumor modifier" by concomitantly suppressing cancer cell proliferation and enhancing metastatic potential.

21.3 Mechanisms of *RUNX3* Regulation in Cancer

The current state of the literature provides a fragmented understanding of *RUNX3* gene regulation. However, scattered clues have revealed potential prerequisites and mediators of *RUNX3* induction.

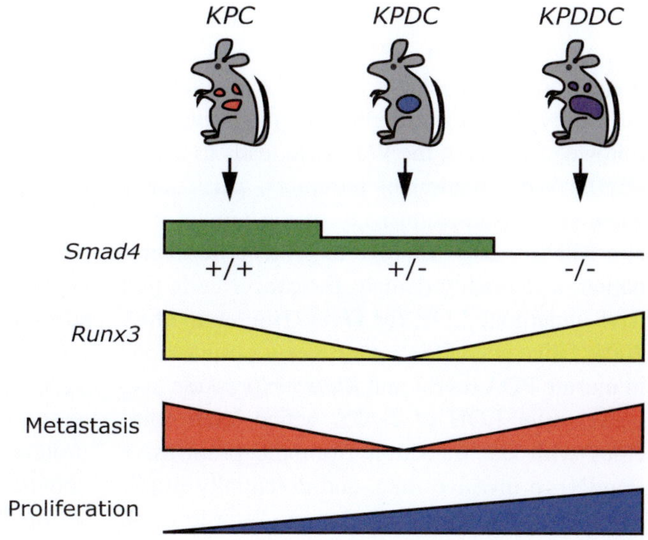

Fig. 21.1 ***Runx3* and PDA phenotypes.** Distinct patterns of primary tumor growth and metastatic disease are observed in *KPC*, *KPDC* and *KPDDC* GEMM of PDA. *Runx3* expression is observed in *KPC* and *KPDDC*, but not *KPDC*, tumors and correlates with metastatic potential

Acquired *RUNX3* expression generally occurs as a stochastic event and is typically not present in all cancer cells of RUNX3-positive tumors (Whittle et al. 2015; Li et al. 2004). This suggests that multiple regulatory mechanisms govern *RUNX3* induction. Increased transcription, increased translation, decreased degradation of message or protein, or some combination of these possibilities can all contribute to elevating RUNX3 levels. *RUNX3* promoter regions possess consensus binding sites for ETS1, SP1 and even RUNX proteins themselves (Bangsow et al. 2001). The initial upregulation of *RUNX3* may occur downstream of activated TGFβ receptors, though the mechanism for this is poorly understood (Haley et al. 2014). We have postulated a feed-forward mechanism involving sustained *RUNX3* transcription by RUNX3 itself – once a critical threshold of expression is achieved – that may be sufficient to maintain high expression levels (Whittle et al. 2015).

Enhanced protein stability may occur through dynamic RUNX3 protein complexes with differential susceptibilities to proteasomal degradation. Known interactions of RUNX3 with CBFβ, MDM2, SMAD proteins and/or TP53 may influence RUNX3 turnover in cancer, as well as modulate its specificity and mode of transcriptional regulation (Chi et al. 2005, 2009; Bae et al. 1995; Yamada et al. 2010). Disruptions to *SMAD4* and *TP53*, genetic events that form the foundation of oncogenic transformation in the epithelium of the pancreas and other organs, also appear to perturb regulation of RUNX3 levels (see below). Thus, a confluence of mechanisms including epigenetic promoter demethylation, genetic mutations, dysregulated signaling pathway activation and altered protein stability all conspire to precipitate *RUNX3* overexpression in cancer. The specific factors modulating *RUNX3* gene expression and protein levels appear to differ by cancer type and perhaps even by stage, a level of complexity perhaps befitting a transcription factor with the ability to enact powerful programs governing cell behavior.

21.4 RUNX3 and *TP53*

Inactivation of the *TP53* tumor suppressor gene is one of the most common events in cancers and occurs principally through genomic loss and/or point mutation (Takahashi et al. 1989; Baker et al. 1990; Prosser et al. 1990; Barton et al. 1991). In PDA, point mutation of one allele and subsequent LOH represents the most common pattern of disrupted *TP53* function, findings that have been substantiated in the *KPC* model

(Hingorani et al. 2005; Rozenblum et al. 1997; Yamano et al. 2000). TP53 assembles as a functional homotetramer that is rapidly degraded in the wildtype (WT) configuration; point mutation followed by loss of the WT allele leads to a substantial accumulation of tetramers, particularly when comprised exclusively of the mutant protein which is less sensitive to proteolytic degradation, and readily demarcates cancer cells that have undergone LOH for *TP53* (Hingorani et al. 2005; Olive et al. 2004). Our recent observations in murine PDA reveal that *Runx3* expression also occurs after LOH of *Trp53*, which itself coincides with the transition from the preinvasive (PanIN) to invasive state, and potentially implicates *Trp53* in *Runx3* regulation (Whittle et al. 2015). The timing of these events also suggests that Runx3 may profoundly contribute to the enhanced invasiveness and metastatic potential that accompanies the gain-of-function properties of stabilized mutant Trp53.

There are several canonical types of *TP53* mutation, but we focus here primarily on "conformational" point mutations exemplified by the R175H amino acid substitution (R172H in mice). The arginine 175 substitution produces a conformational change that alters the DNA-binding specificity of *TP53* such that the structure of DNA, rather than the typical consensus sequence, now drives the recruitment of TP53 to chromosomal targets (Gohler et al. 2005). This causes not only a loss of TP53 regulation of its canonical tumor-suppressive target genes, but a gain of function through regulation of a distinct cohort of TP53-mutant target genes, possibly to include *RUNX3* (Bossi et al. 2008; Freed-Pastor and Prives 2012). Gain-of-function effects can also become manifest when accumulated mutant $TP53^{R175H}$ protein interacts with *de novo* protein complexes that compound the loss of the canonical tumor-suppressive capacity of WT TP53. For example, interaction of mutant TP53 with TP63 and/or TP73 inhibits the tumor suppressive functions of these related family members (Bergamaschi et al. 2003; Irwin et al. 2003; Lang et al. 2004). We and others have found that RUNX3 is another such TP53 interactor that is potentially impacted by the functional gains of accumulated $TP53^{R175H}$ (Yamada et al. 2010; Whittle et al. 2015).

RUNX3 has been shown to directly interact with and contribute to the functional activity of WT TP53 (Yamada et al. 2010). In lung cancer, it can also contribute indirectly to TP53 protein stability by stimulating expression of $p14^{ARF}$/ $p19^{Arf}$, the counterbalance to MDM2-mediated degradation of TP53 (Lee et al. 2013). There are several potentially important connotations for the interaction between RUNX3 and point mutant TP53 (Fig. 21.2). First, the RUNX3-TP53 interaction very likely affects the stability of RUNX3 protein (Whittle et al. 2015). WT TP53 may enhance RUNX3 proteolytic degradation if it shunts RUNX3 to the proteasome. Conversely, in the context of accumulated mutant TP53 protein, the $TP53^{R175H}$-RUNX3 complexes may aberrantly stabilize RUNX3 protein by sequestration. We have also found that point mutant $Trp53^{R172H}$ forms stable complexes with Runx3 (unpublished observations). Additionally, RUNX3 itself is ubiquitinated by MDM2, the same E3 ubiquitin ligase that facilitates TP53 degradation; $TP53^{R175H}$ may instead sequester MDM2 or sterically hinder RUNX3-MDM2 interactions within a TP53-RUNX3 complex to limit RUNX3 degradation (Peng et al. 2001). In support of the latter, deletion of the C-terminal TP53-binding region of RUNX3 enhances RUNX3-MDM2 association, suggesting that TP53 binding to this region may affect RUNX3-MDM2 interaction (Chi et al. 2009; Yamada et al. 2010).

Secondly, the $TP53^{R175H}$-RUNX3 interaction also likely influences the transcriptional regulatory properties of each factor. The binding of TP53 to RUNX3 can directly affect gene regulation by inducing conformational shifts, recruiting additional, non-canonical protein factors and/or influencing DNA binding. In this way, novel complexes involving mutant TP53 and RUNX3 may shift the position and repertoire of transcriptional regulators that are recruited to the promoters of known or new RUNX3/$TP53^{R175H}$ target genes. Point mutation of TP53 has been shown to reverse the direction of regulation of other binding partners, including NF-Y, SP1 and VDR on their normal target genes (Freed-Pastor and

Prives 2012). That co-expression of Runx3 and Trp53^{R172H} occurs only in adenocarcinomas of the pancreas and not in the normal gland or inflammatory injury suggests that this complex may have profound effects on cellular homeostasis.

Finally, the interaction between RUNX3 and TP53 also suggests that the effects of acquired RUNX3 expression may be highly context-specific, particularly when considering the diverse mechanisms of *TP53* inactivation. The R175H mutation represents a fraction of the spontaneous *TP53* mutations that occur in human PDA, and distinct *TP53* mutations may produce different cancer cell phenotypes and patient outcomes (Olivier et al. 2006). In particular, it will be important to identify the unique influences on RUNX3 expression and function in PDA (or other cancers) where *TP53* bears a conformational mutation (such as R175H), a DNA-binding mutation, is homozygous deleted or remains intact.

21.5 RUNX3 and *SMAD4*

The tumor suppressor gene *SMAD4* is also commonly disrupted in pancreas cancer resulting in complete loss of expression in approximately 55 % of cases (Wilentz et al. 2000b). Complete inactivation of *SMAD4* is thought to be a late event in pancreas cancer progression but this may obscure the occurrence of early heterozygous mutation of the locus, which diverts PanIN towards the less aggressive MCN precursor lesion in certain contexts (Izeradjene et al. 2007; Wilentz et al. 2000a). SMAD4 mediates canonical TGFβ pathway signaling by forming a heterotrimeric complex with SMAD2/3 after phosphorylation of the latter by TGFβ receptors. The activated SMAD complex translocates to the nucleus to regulate gene expression by directly binding DNA and/or interacting with other transcription factor complexes. We and others have shown that RUNX3 is a SMAD cofactor that physically interacts with nuclear SMAD complexes (Hanai et al. 1999; Chi et al. 2005) and mediates TGFβ signaling output to modulate cell proliferation and differentiation (Chi et al. 2005; Shi and Stavnezer 1998; Fainaru et al. 2004; Hasegawa et al. 2007). Additionally, we see that the levels of *Smad4* expression can influence the expression and stability of Runx3 in GEMM of PDA: heterozygous deletion of *Smad4* attenuates Runx3 levels compared to *Smad4*$^{+/+}$ and *Smad4*$^{-/-}$ through an intricate and incompletely understood balance of *Runx3* transcription and protein stability (Fig. 21.2) (Whittle et al. 2015). As with the RUNX3-TP53 interaction, the RUNX3-SMAD

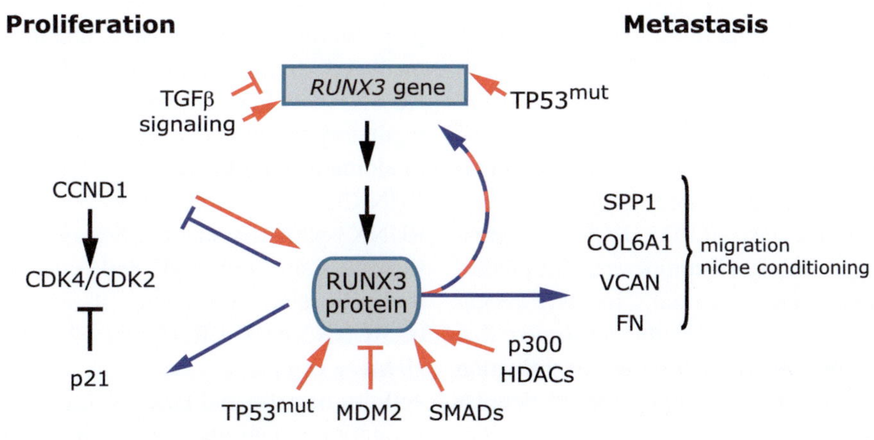

Fig. 21.2 Regulation and function of *RUNX3*. Expression of *RUNX3* is modulated by TGFβ signaling, *TP53* mutational status and RUNX3 itself. RUNX3 function is also affected by interactions with mutant TP53, receptor SMADs, acetyltransferases and deacetylases that combine to determine transcriptional regulation of cell cycle proteins such as p21 and secreted pro-metastatic factors such as SPP1. Inputs to RUNX3 levels and/or function are shown in *red* and RUNX3 outputs are highlighted in *blue*

interaction is likely to have widespread consequences on cell signaling and pancreas cancer progression. This interaction may similarly alter the association of RUNX3 and SMADs with the promoter regions of their respective target genes and thereby broadly influence gene expression. In this respect, the differences in RUNX3 function between PDA patients who have intact, heterozygous deleted and homozygous deleted *SMAD4* could be significant.

The TGFβ pathway has been intricately linked to the program of epithelial to mesenchymal transition (EMT), a cellular process that is defined by a loss of epithelial and gain in mesenchymal characteristics and functionalities (Lamouille et al. 2014). In reality, EMT is less categorical and more gradated than it is often portrayed in the literature, with an eclectic variety of expressed proteins and cellular phenotypes used as evidence for EMT. Canonical signaling through TGFβ receptors and SMAD proteins is commonly thought to induce EMT by stimulation of transcription factors (such as *SNAI1*, *ZEB1* and others) that regulate expression of a wide array of proteins contributing to cell morphology, adhesion, motility and invasiveness. Whereas others have shown a role for RUNX3 in inhibiting EMT, we have found that *RUNX3* expression in PDA does not factor into the ability of a cell to undergo this phenotypic transition (Whittle et al. 2015; Tanaka et al. 2012; Voon et al. 2012; Liu et al. 2014; Whittle and Hingorani 2015). Conversely, by comparing the *KPC* (EMT-competent, RUNX3-positive), *KPDC* (EMT-competent, RUNX3-negative) and *KPDDC* models (EMT–incompetent, RUNX3-positive) we have found that EMT competency is neither necessary nor sufficient for metastasis of PDA; RUNX3 expression in PDA trumps EMT competency for prediction of metastatic potential. Indeed, recent publications have emphatically confirmed that EMT is dispensable for metastatic dissemination of cancer, contrary to a long-espoused popular conception of the metastatic process (Fischer et al. 2015; Zheng et al. 2015). These data suggest the ability of RUNX3 to contribute to both EMT-dependent and EMT-independent modes of metastasis. Whereas tumor cells from *Smad4*$^{+/+}$ mice respond to TGFβ to invade and migrate efficiently (albeit as single cells), *Smad4*$^{-/-}$ cells are less motile but may seed secondary tumors more efficiently as disseminated cell clusters (Whittle and Hingorani 2015; Aceto et al. 2014). In both cases, the secretory program induced by RUNX3 likely contributes greatly to metastatic niche preparation in distant organs, regardless of the ability to undergo EMT.

21.6 Additional Factors and Phenotypes Associated with *RUNX3*

Several other proteins can associate with RUNX3, including p300, HDACs, CCND1 and BRD2 (Lee et al. 2013; Aceto et al. 2014; Yagi et al. 1999; Jin et al. 2004; Chung et al. 2007; Iwatani et al. 2010). The histone acetyltransferase p300 interacts with RUNX3 in a TGFβ-dependent manner and acetylates several RUNX3 lysines to protect them from ubiquitination by SMURFs and subsequent proteolysis (Jin et al. 2004). In opposition to this, HDACs 1, 2, 4 and 5 deacetylate these residues, leaving RUNX3 vulnerable to degradation. In transfected HEK293 cells, CCND1/Cyclin D1 was shown to antagonize the p300-RUNX3 interaction by physically competing with p300 for RUNX3 binding (Iwatani et al. 2010). This alternative RUNX3-CCND1 complex recruits HDAC4 and inhibits RUNX3-mediated transcription of *p21* by inducing RUNX3 degradation and/or interfering with its regulatory function (Lee et al. 2013). BRD2 bromodomains bind to acetylated lysine residues of RUNX3, an interaction enhanced by RUNX3-p300 association. Serum stimulation of starved cells revealed cell cycle-dependent, inversely related rises and falls in the formation of RUNX3-p300-BRD2 and RUNX3-CCND1-HDAC4 complexes (Lee et al. 2013). With the abilities of p300 and HDACs to broadly and disparately regulate gene expression by modifying histone acetylation states, differential recruitment of these proteins by RUNX3 to its target genes undoubtedly causes important changes to the transcriptome.

Binding to BRD2 facilitates upregulation of *p21* and *ARF* by RUNX3, suppressing primary tumor growth. In addition, Runx3 was lost or mislocalized in lung adenocarcinomas and targeted deletion of *Runx3* alone induced the formation of lung adenomas (Lee et al. 2013). Together, these observations suggest a primarily tumor suppressive role for *Runx3* in the lung. However, his particular lung cancer GEMM does not metastasize appreciably and concomitant mutation or deletion of *Trp53* was not incorporated in the model, so the effects on these aspects of the malignant phenotype remain unknown (Jackson et al. 2001, 2005). Analyzing whether *Runx3* influences the ability of cells to disseminate in the setting of a more metastatic lung cancer model would clarify whether the functional duality of *Runx3* observed in PDA is cancer-specific or more generalizable. Applying the mechanisms deduced in transfected HEK293 and human lung cancer cell lines to PDA would suggest that when TGFβ signaling is robust (as in *KPC* mice), RUNX3-p300-BRD2 complexes may predominate and suppress proliferation. In *KPDDC* mice, however, where mitogenic signaling significantly outweighs canonical TGFβ pathway activity, RUNX3-CCND1-HDAC4 complexes may instead predominate, enabling RUNX3 to induce *p21* expression. We would surmise that regulation of other target genes, such as *Spp1*, that support metastatic dissemination are not influenced by RUNX3-CCND1-HDAC complexes, as *KPDDC* cells also secrete Spp1 and disseminate widely. These are purely speculative hypotheses and have not been formally tested in pancreas cancer. The potential diversity and sequential acquisition of alterations in *TP53* and *SMAD4* (not to mention other major players in PDA such as *KRAS* and $p16^{INK4A}$) and the promiscuous interactions of RUNX3 with other impactful proteins add additional layers of complexity to RUNX3 regulation and function (Fig. 21.2).

21.7 *RUNX3* May Inform Treatment Decisions

Our data suggest that *Runx3* acts as a fulcrum in the impulse of PDA cells to divide or metastasize, shunting the cell towards a migratory phenotype at the expense of cell proliferation (Fig. 21.3). This duality recalls and likely reflects the well-characterized and counterpoised phenotypes associated with TGFβ signaling in which Runx3 serves as a downstream mediator (Hasegawa et al. 2007). With the substantial energy requirements for both division and dissemination, not to mention the physical challenges

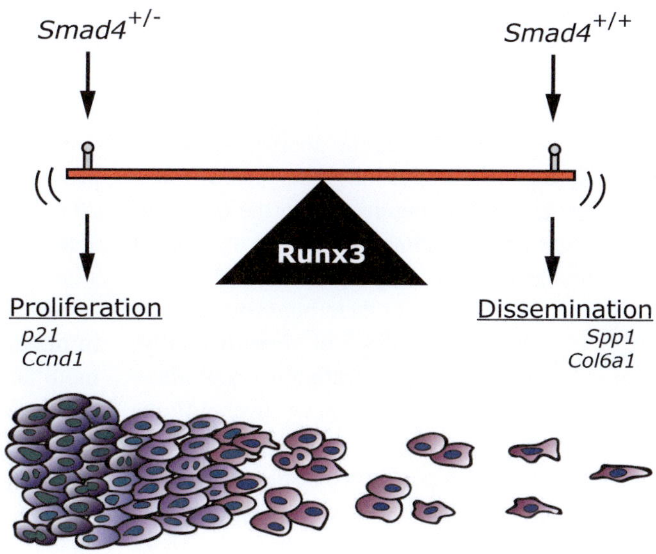

Fig. 21.3 ***Runx3*** **balances the cellular decision to proliferate or divide.** *Smad4* genomic status influences *Runx3* expression and function, committing cancer cells towards proliferation or dissemination through the regulated expression of cell cycle proteins or secreted factors

of performing both simultaneously, it is not surprising that commitment of cancer cells toward either proliferation or metastasis may occur. Not only do these diverse cell functions require the concerted activation of distinct, non-redundant metabolic and signaling pathways, the biosynthetic and physical outcomes of these processes are also highly divergent (LeBleu et al. 2014). Indeed, multiple studies suggest that invasion and division are inversely correlated (Hoek et al. 2008; Yano et al. 2014). That *Runx3* contributes critically to these cell fate decisions underscores its importance in modifying cancer behavior and alludes to its potential to guide clinical decision-making.

How exactly can information about RUNX3 expression be used to make treatment decisions in oncology? The first practical requirement is an efficient and accurate method to measure RUNX3 expression. The currently most promising technique in this respect is immunohistochemistry of tumor biopsies, which allows assessment in discrete cell populations. This obviates difficulties in distinguishing the origins of *RUNX3* transcripts from bulk tumor samples. Most solid tumors, and especially PDA, evolve into complex neo-organs that include the immigration of fibroblasts and immune cells; immune cells robustly express *RUNX3* and can confound an accurate assessment of *RUNX3* transcript levels in cancerous cells. Immunohistochemistry on intact tissue also provides additional information, such as the heterogeneity of expression levels across subpopulations and differences in subcellular localization. Such an approach is admittedly more labor-intensive than molecular analyses of whole tissue lysates and implies a greater need to test multiple regions and depths within the tissue, as false negatives will sometimes occur as a result of stochastic, spatially-restricted RUNX3-positive clones within a heterogeneous tumor. Advances in the quality of commercially-available RUNX3-specific antibodies now allow sensitive and specific detection of diverse tissue and cell types, an advantage that may have benefitted earlier studies performed with perhaps less robust reagents (Ogasawara et al. 2009; Soong et al. 2009; Bai et al. 2013; Xue et al. 2014). Moving forward with rigorously validated antibodies to enable clinical-grade testing for RUNX3 expression may additionally help to unify dissenters regarding tissue-specific levels of RUNX3 expression.

Intratumoral RUNX3 expression can predict survival after resection of pancreas tumors (where subsequent metastatic relapse is a leading cause of death) (Whittle et al. 2015). Given the dearth of effective treatment options in PDA and the limited window of opportunity in which to administer them, prognostic information regarding the metastatic potential of a given tumor could help prioritize therapies that preferentially target either local or distant relapse. For example, patients with high *RUNX3*-expressing tumors would be expected to benefit more from systemic therapy, whereas upfront surgery might be beneficial to patients whose tumors lacked *RUNX3* expression. Although RUNX3 levels may correlate with prognosis in PDA and help weigh the urgency with which to address microscopically disseminated disease, much more work will be needed to characterize the implications of RUNX3 expression across different cancers and among distinct cell types within a given cancer, with respect to likelihood and site of relapse, stage, grade, proliferation rate, drug responsiveness and so on.

Direct or indirect targeting of RUNX3 may be useful clinically, though inhibiting the functional activity of transcription factors is not trivial. Directly targeting RUNX3 is complicated by the lack of a catalytic pocket, in contrast to readily druggable kinases for example, but one can envision peptide and/or small molecule inhibitors that disrupt crucial RUNX3 interactions with other proteins (Lin et al. 2013; Rettenmaier et al. 2014). In principle, gene-based therapies may also prove effective at inhibiting RUNX3 expression and activity. However, with any of these approaches, one must tread cautiously and bear in mind the dichotomous nature of RUNX3 as a modifier of tumorigenicity. RUNX3 inhibition may prevent metastatic dissemination at the expense of enhancing local growth. This potential complication may be overcome by simultaneously inhibiting components of the cell cycle machinery while targeting RUNX3. Even if such dichotomous effects are observed, it may be beneficial to apply RUNX3-targeted therapies in circumstances where metastasis is the major risk, or as an adjuvant to surgical resection when mechanical disruption of the tumor can foment entry of cancer cells into the bloodstream. Conversely, an acceleration of proliferation by RUNX3 inhibition may be also capitalized upon to enhance the efficacy of conventional chemotherapeutics, such as gemcitabine, that rely on cell division to induce genetic damage and apoptosis. Should a significant acceleration of tumor growth be observed after direct inhibition of RUNX3, targeting downstream RUNX3 effectors, such as SPP1, may additionally help to specifically inhibit the metastasis-promoting sequelae. We stress that these possibilities remain highly speculative and are as yet untested and cannot be applied clinically without rigorous assessment of their merit.

21.8 Conclusions

RUNX3 both responds to and modulates the activity of a number of critical developmental programs and cell cycle pathways, and also exhibits complex effects on neoplasia. *RUNX3* has been described as an "oncogene" or "tumor suppressor" gene but it may be that neither designation alone suffices; instead, "tumor modifier" may be more appropriate. Exploring *RUNX3* biology through the lens of specific cell fate decisions may help to clarify its most important functions and allow more incisive clinical interpretations. The complexities and context-dependent uncertainties notwithstanding, *RUNX3* is undeniably an important regulator of oncogenesis and warrants deeper investigation as we work to advance strategies for eradicating cancer.

References

Aceto, N., Bardia, A., Miyamoto, D. T., Donaldson, M. C., Wittner, B. S., Spencer, J. A., et al. (2014). Circulating tumor cell clusters are oligoclonal precursors of breast cancer metastasis. *Cell, 158*(5), 1110–1122. doi:10.1016/j.cell.2014.07.013.

Aguirre, A. J., Bardeesy, N., Sinha, M., Lopez, L., Tuveson, D. A., Horner, J., et al. (2003). Activated Kras and Ink4a/Arf deficiency cooperate to produce metastatic pancreatic ductal adenocarcinoma. *Genes & Development, 17*(24), 3112–3126. doi:10.1101/gad.1158703.

Allison, D. C., Piantadosi, S., Hruban, R. H., Dooley, W. C., Fishman, E. K., Yeo, C. J., et al. (1998). DNA content and other factors associated with ten-year survival after resection of pancreatic carcinoma. *Journal of Surgical Oncology, 67*(3), 151–159.

Bae, S. C., Takahashi, E., Zhang, Y. W., Ogawa, E., Shigesada, K., Namba, Y., et al. (1995). Cloning, mapping and expression of PEBP2 alpha C, a third gene encoding the mammalian Runt domain. *Gene, 159*(2), 245–248.

Bagchi, A., Papazoglu, C., Wu, Y., Capurso, D., Brodt, M., Francis, D., et al. (2007). CHD5 is a tumor suppressor at human 1p36. *Cell, 128*(3), 459–475. doi:10.1016/j.cell.2006.11.052.

Bai, J., Yong, H. M., Chen, F. F., Song, W. B., Li, C., Liu, H., & Zheng, J. N. (2013). RUNX3 is a prognostic marker and potential therapeutic target in human breast cancer. *Journal of Cancer Research and Clinical Oncology, 139*(11), 1813–1823. doi:10.1007/s00432-013-1498-x.

Baker, S. J., Preisinger, A. C., Jessup, J. M., Paraskeva, C., Markowitz, S., Willson, J. K., et al. (1990). p53 gene mutations occur in combination with 17p allelic deletions as late events in colorectal tumorigenesis. *Cancer Research, 50*(23), 7717–7722.

Bangsow, C., Rubins, N., Glusman, G., Bernstein, Y., Negreanu, V., Goldenberg, D., et al. (2001). The RUNX3 gene--sequence, structure and regulated expression. *Gene, 279*(2), 221–232.

Barghout, S. H., Zepeda, N., Vincent, K., Azad, A. K., Xu, Z., Yang, C., et al. (2015). RUNX3 contributes to carboplatin resistance in epithelial ovarian cancer cells. *Gynecologic Oncology, 138*(3), 647–655. doi:10.1016/j.ygyno.2015.07.009.

Barton, C. M., Staddon, S. L., Hughes, C. M., Hall, P. A., O'Sullivan, C., Kloppel, G., et al. (1991). Abnormalities of the p53 tumour suppressor gene in human pancreatic cancer. *British Journal of Cancer, 64*(6), 1076–1082.

Bauer, O., Hantisteanu, S., Lotem, J., & Groner, Y. (2014). Carcinogen-induced skin tumor development requires leukocytic expression of the transcription factor Runx3. *Cancer Prevention Research, 7*(9), 913–926. doi:10.1158/1940-6207.CAPR-14-0098-T.

Bauer, O., Sharir, A., Kimura, A., Hantisteanu, S., Takeda, S., & Groner, Y. (2015). Loss of osteoblast Runx3 produces severe congenital osteopenia. *Molecular and Cellular Biology, 35*(7), 1097–1109. doi:10.1128/MCB.01106-14.

Bergamaschi, D., Gasco, M., Hiller, L., Sullivan, A., Syed, N., Trigiante, G., et al. (2003). p53 polymorphism influences response in cancer chemotherapy via modulation of p73-dependent apoptosis. *Cancer Cell, 3*(4), 387–402.

Bledsoe, K. L., McGee-Lawrence, M. E., Camilleri, E. T., Wang, X., Riester, S. M., van Wijnen, A. J., et al. (2014). RUNX3 facilitates growth of Ewing sarcoma cells.

Journal of Cellular Physiology, 229(12), 2049–2056. doi:10.1002/jcp.24663.

Bossi, G., Marampon, F., Maor-Aloni, R., Zani, B., Rotter, V., Oren, M., et al. (2008). Conditional RNA interference in vivo to study mutant p53 oncogenic gain of function on tumor malignancy. *Cell Cycle, 7*(12), 1870–1879.

Caldas, C., Hahn, S. A., da Costa, L. T., Redston, M. S., Schutte, M., Seymour, A. B., et al. (1994). Frequent somatic mutations and homozygous deletions of the p16 (MTS1) gene in pancreatic adenocarcinoma. *Nature Genetics, 8*(1), 27–32. doi:10.1038/ng0994-27.

Carvalho, R., Milne, A. N., Polak, M., Corver, W. E., Offerhaus, G. J., & Weterman, M. A. (2005). Exclusion of RUNX3 as a tumour-suppressor gene in early-onset gastric carcinomas. *Oncogene, 24*(56), 8252–8258. doi:10.1038/sj.onc.1208963.

Chambers, A. F., Wilson, S. M., Kerkvliet, N., O'Malley, F. P., Harris, J. F., & Casson, A. G. (1996). Osteopontin expression in lung cancer. *Lung Cancer, 15*(3), 311–323.

Chen, F., Bai, J., Li, W., Mei, P., Liu, H., Li, L., et al. (2013). RUNX3 suppresses migration, invasion and angiogenesis of human renal cell carcinoma. *PloS One, 8*(2), e56241. doi:10.1371/journal.pone.0056241.

Chen, F., Wang, M., Bai, J., Liu, Q., Xi, Y., Li, W., & Zheng, J. (2014). Role of RUNX3 in suppressing metastasis and angiogenesis of human prostate cancer. *PloS One, 9*(1), e86917. doi:10.1371/journal.pone.0086917.

Cheng, C. K., Li, L., Cheng, S. H., Lau, K. M., Chan, N. P., Wong, R. S., et al. (2008). Transcriptional repression of the RUNX3/AML2 gene by the t(8;21) and inv(16) fusion proteins in acute myeloid leukemia. *Blood, 112*(8), 3391–3402. doi:10.1182/blood-2008-02-137083.

Chi, X. Z., Yang, J. O., Lee, K. Y., Ito, K., Sakakura, C., Li, Q. L., et al. (2005). RUNX3 suppresses gastric epithelial cell growth by inducing p21(WAF1/Cip1) expression in cooperation with transforming growth factor {beta}-activated SMAD. *Molecular and Cellular Biology, 25*(18), 8097–8107. doi:10.1128/MCB.25.18.8097-8107.2005.

Chi, X. Z., Kim, J., Lee, Y. H., Lee, J. W., Lee, K. S., Wee, H., et al. (2009). Runt-related transcription factor RUNX3 is a target of MDM2-mediated ubiquitination. *Cancer Research, 69*(20), 8111–8119. doi:10.1158/0008-5472.CAN-09-1057.

Chung, D. D., Honda, K., Cafuir, L., McDuffie, M., & Wotton, D. (2007). The Runx3 distal transcript encodes an additional transcriptional activation domain. *The FEBS Journal, 274*(13), 3429–3439. doi:10.1111/j.1742-4658.2007.05875.x.

Colla, S., Morandi, F., Lazzaretti, M., Rizzato, R., Lunghi, P., Bonomini, S., et al. (2005). Human myeloma cells express the bone regulating gene Runx2/Cbfa1 and produce osteopontin that is involved in angiogenesis in multiple myeloma patients. *Leukemia, 19*(12), 2166–2176. doi:10.1038/sj.leu.2403976.

Courtin, A., Richards, F. M., Bapiro, T. E., Bramhall, J. L., Neesse, A., Cook, N., et al. (2013). Anti-tumour efficacy of capecitabine in a genetically engineered mouse model of pancreatic cancer. *PloS One, 8*(6), e67330. doi:10.1371/journal.pone.0067330.

Crippa, S., Salvia, R., Warshaw, A. L., Dominguez, I., Bassi, C., Falconi, M., et al. (2008). Mucinous cystic neoplasm of the pancreas is not an aggressive entity: lessons from 163 resected patients. *Annals of Surgery, 247*(4), 571–579. doi:10.1097/SLA.0b013e31811f4449.

Damdinsuren, A., Matsushita, H., Ito, M., Tanaka, M., Jin, G., Tsukamoto, H., et al. (2015). FLT3-ITD drives Ara-C resistance in leukemic cells via the induction of RUNX3. *Leukemia Research, 39*(12), 1405–1413. doi:10.1016/j.leukres.2015.09.009.

DeLeo, A. B., Jay, G., Appella, E., Dubois, G. C., Law, L. W., & Old, L. J. (1979). Detection of a transformation-related antigen in chemically induced sarcomas and other transformed cells of the mouse. *Proceedings of the National Academy of Sciences of the United States of America, 76*(5), 2420–2424.

Der, C. J., Krontiris, T. G., & Cooper, G. M. (1982). Transforming genes of human bladder and lung carcinoma cell lines are homologous to the ras genes of Harvey and Kirsten sarcoma viruses. *Proceedings of the National Academy of Sciences of the United States of America, 79*(11), 3637–3640.

Ducy, P., Zhang, R., Geoffroy, V., Ridall, A. L., & Karsenty, G. (1997). Osf2/Cbfa1: a transcriptional activator of osteoblast differentiation. *Cell, 89*(5), 747–754.

Estecio, M. R., Maddipoti, S., Bueso-Ramos, C., DiNardo, C. D., Yang, H., Wei, Y., et al. (2015). RUNX3 promoter hypermethylation is frequent in leukaemia cell lines and associated with acute myeloid leukaemia inv(16) subtype. *British Journal of Haematology, 169*(3), 344–351. doi:10.1111/bjh.13299.

Fainaru, O., Woolf, E., Lotem, J., Yarmus, M., Brenner, O., Goldenberg, D., et al. (2004). Runx3 regulates mouse TGF-beta-mediated dendritic cell function and its absence results in airway inflammation. *The EMBO Journal, 23*(4), 969–979. doi:10.1038/sj.emboj.7600085.

Fidler, I. J. (2003). The pathogenesis of cancer metastasis: the 'seed and soil' hypothesis revisited. *Nature Reviews Cancer, 3*(6), 453–458. doi:10.1038/nrc1098.

Finlay, C. A., Hinds, P. W., Tan, T. H., Eliyahu, D., Oren, M., & Levine, A. J. (1988). Activating mutations for transformation by p53 produce a gene product that forms an hsc70-p53 complex with an altered half-life. *Molecular and Cellular Biology, 8*(2), 531–539.

Fischer, K. R., Durrans, A., Lee, S., Sheng, J., Li, F., Wong, S. T., et al. (2015). Epithelial-to-mesenchymal transition is not required for lung metastasis but contributes to chemoresistance. *Nature, 527*(7579), 472–476. doi:10.1038/nature15748.

Fredika, M., Chu, K., Fernandez, S., Mu, Z., Zhang, X., Liu, H., et al. (2012). Genomic profiling of pre-clinical models of inflammatory breast cancer identifies a signature of epithelial plasticity and suppression of TGFß signaling. *Journal of Clinical and Experimental Pathology, 2*(5), 1000119.

Fredriksson, S., Horecka, J., Brustugun, O. T., Schlingemann, J., Koong, A. C., Tibshirani, R., & Davis, R. W. (2008). Multiplexed proximity ligation assays to profile putative plasma biomarkers relevant to pancreatic and ovarian cancer. *Clinical Chemistry, 54*(3), 582–589. doi:10.1373/clinchem.2007.093195.

Freed-Pastor, W. A., & Prives, C. (2012). Mutant p53: one name, many proteins. *Genes & Development, 26*(12), 1268–1286. doi:10.1101/gad.190678.112.

Furger, K. A., Allan, A. L., Wilson, S. M., Hota, C., Vantyghem, S. A., Postenka, C. O., et al. (2003). Beta(3) integrin expression increases breast carcinoma cell responsiveness to the malignancy-enhancing effects of osteopontin. *Molecular Cancer Research: MCR, 1*(11), 810–819.

Goel, A., Arnold, C. N., Tassone, P., Chang, D. K., Niedzwiecki, D., Dowell, J. M., et al. (2004). Epigenetic inactivation of RUNX3 in microsatellite unstable sporadic colon cancers. *International Journal of Cancer Journal international du cancer, 112*(5), 754–759. doi:10.1002/ijc.20472.

Gohler, T., Jager, S., Warnecke, G., Yasuda, H., Kim, E., & Deppert, W. (2005). Mutant p53 proteins bind DNA in a DNA structure-selective mode. *Nucleic Acids Research, 33*(3), 1087–1100. doi:10.1093/nar/gki252.

Guerra, C., & Barbacid, M. (2013). Genetically engineered mouse models of pancreatic adenocarcinoma. *Molecular Oncology, 7*(2), 232–247. doi:10.1016/j.molonc.2013.02.002.

Guo, W. H., Weng, L. Q., Ito, K., Chen, L. F., Nakanishi, H., Tatematsu, M., & Ito, Y. (2002). Inhibition of growth of mouse gastric cancer cells by Runx3, a novel tumor suppressor. *Oncogene, 21*(54), 8351–8355. doi:10.1038/sj.onc.1206037.

Hahn, S. A., Schutte, M., Hoque, A. T., Moskaluk, C. A., da Costa, L. T., Rozenblum, E., et al. (1996). DPC4, a candidate tumor suppressor gene at human chromosome 18q21.1. *Science, 271*(5247), 350–353.

Haley, J. A., Haughney, E., Ullman, E., Bean, J., Haley, J. D., & Fink, M. Y. (2014). Altered Transcriptional Control Networks with Trans-Differentiation of Isogenic Mutant-KRas NSCLC Models. *Frontiers in Oncology, 4*, 344. doi:10.3389/fonc.2014.00344.

Hanahan, D., & Weinberg, R. A. (2000). The hallmarks of cancer. *Cell, 100*(1), 57–70.

Hanahan, D., & Weinberg, R. A. (2011). Hallmarks of cancer: the next generation. *Cell, 144*(5), 646–674. doi:10.1016/j.cell.2011.02.013.

Hanai, J., Chen, L. F., Kanno, T., Ohtani-Fujita, N., Kim, W. Y., Guo, W. H., et al. (1999). Interaction and functional cooperation of PEBP2/CBF with Smads. Synergistic induction of the immunoglobulin germline Calpha promoter. *The Journal of Biological Chemistry, 274*(44), 31577–31582.

Hasegawa, K., Yazumi, S., Wada, M., Sakurai, T., Kida, M., Yamauchi, J., et al. (2007). Restoration of RUNX3 enhances transforming growth factor-beta-dependent p21 expression in a biliary tract cancer cell line. *Cancer Science, 98*(6), 838–843. doi:10.1111/j.1349-7006.2007.00460.x.

Higashiyama, M., Ito, T., Tanaka, E., & Shimada, Y. (2007). Prognostic significance of osteopontin expression in human gastric carcinoma. *Annals of Surgical Oncology, 14*(12), 3419–3427. doi:10.1245/s10434-007-9564-8.

Hingorani, S. R., Petricoin, E. F., Maitra, A., Rajapakse, V., King, C., Jacobetz, M. A., et al. (2003). Preinvasive and invasive ductal pancreatic cancer and its early detection in the mouse. *Cancer Cell, 4*(6), 437–450.

Hingorani, S. R., Wang, L., Multani, A. S., Combs, C., Deramaudt, T. B., Hruban, R. H., et al. (2005). Trp53R172H and KrasG12D cooperate to promote chromosomal instability and widely metastatic pancreatic ductal adenocarcinoma in mice. *Cancer Cell, 7*(5), 469–483. doi:10.1016/j.ccr.2005.04.023.

Hirama, M., Takahashi, F., Takahashi, K., Akutagawa, S., Shimizu, K., Soma, S., et al. (2003). Osteopontin overproduced by tumor cells acts as a potent angiogenic factor contributing to tumor growth. *Cancer Letters, 198*(1), 107–117.

Hoek, K. S., Eichhoff, O. M., Schlegel, N. C., Dobbeling, U., Kobert, N., Schaerer, L., et al. (2008). In vivo switching of human melanoma cells between proliferative and invasive states. *Cancer Research, 68*(3), 650–656. doi:10.1158/0008-5472.CAN-07-2491.

Horiguchi, S., Shiraha, H., Nagahara, T., Kataoka, J., Iwamuro, M., Matsubara, M., et al. (2013). Loss of runt-related transcription factor 3 induces gemcitabine resistance in pancreatic cancer. *Molecular Oncology, 7*(4), 840–849. doi:10.1016/j.molonc.2013.04.004.

Hruban, R. H., van Mansfeld, A. D., Offerhaus, G. J., van Weering, D. H., Allison, D. C., Goodman, S. N., et al. (1993). K-ras oncogene activation in adenocarcinoma of the human pancreas. A study of 82 carcinomas using a combination of mutant-enriched polymerase chain reaction analysis and allele-specific oligonucleotide hybridization. *The American Journal of Pathology, 143*(2), 545–554.

Hruban, R. H., Adsay, N. V., Albores-Saavedra, J., Anver, M. R., Biankin, A. V., Boivin, G. P., et al. (2006). Pathology of genetically engineered mouse models of pancreatic exocrine cancer: consensus report and recommendations. *Cancer Research, 66*(1), 95–106. doi:10.1158/0008-5472.CAN-05-2168.

Huang, B., Qu, Z., Ong, C. W., Tsang, Y. H., Xiao, G., Shapiro, D., et al. (2012). RUNX3 acts as a tumor suppressor in breast cancer by targeting estrogen receptor alpha. *Oncogene, 31*(4), 527–534. doi:10.1038/onc.2011.252.

Inman, C. K., & Shore, P. (2003). The osteoblast transcription factor Runx2 is expressed in mammary epithelial cells and mediates osteopontin expression. *The Journal of Biological Chemistry, 278*(49), 48684–48689. doi:10.1074/jbc.M308001200.

Irwin, M. S., Kondo, K., Marin, M. C., Cheng, L. S., Hahn, W. C., & Kaelin Jr., W. G. (2003). Chemosensitivity linked to p73 function. *Cancer Cell, 3*(4), 403–410.

Ito, Y. (2012). RUNX3 is expressed in the epithelium of the gastrointestinal tract. *EMBO Molecular Medicine, 4*(7), 541–542 author reply 543-544. doi:10.1002/emmm.201100203.

Iwatani, K., Fujimoto, T., & Ito, T. (2010). Cyclin D1 blocks the anti-proliferative function of RUNX3 by interfering with RUNX3-p300 interaction. *Biochemical and Biophysical Research Communications, 400*(3), 426–431. doi:10.1016/j.bbrc.2010.08.094.

Izeradjene, K., Combs, C., Best, M., Gopinathan, A., Wagner, A., Grady, W. M., et al. (2007). Kras(G12D) and Smad4/Dpc4 haploinsufficiency cooperate to induce mucinous cystic neoplasms and invasive adenocarcinoma of the pancreas. *Cancer Cell, 11*(3), 229–243. doi:10.1016/j.ccr.2007.01.017.

Jackson, E. L., Willis, N., Mercer, K., Bronson, R. T., Crowley, D., Montoya, R., et al. (2001). Analysis of lung tumor initiation and progression using conditional expression of oncogenic K-ras. *Genes & Development, 15*(24), 3243–3248. doi:10.1101/gad.943001.

Jackson, E. L., Olive, K. P., Tuveson, D. A., Bronson, R., Crowley, D., Brown, M., & Jacks, T. (2005). The differential effects of mutant p53 alleles on advanced murine lung cancer. *Cancer Research, 65*(22), 10280–10288. doi:10.1158/0008-5472.CAN-05-2193.

Jin, Y. H., Jeon, E. J., Li, Q. L., Lee, Y. H., Choi, J. K., Kim, W. J., et al. (2004). Transforming growth factor-beta stimulates p300-dependent RUNX3 acetylation, which inhibits ubiquitination-mediated degradation. *The Journal of Biological Chemistry, 279*(28), 29409–29417. doi:10.1074/jbc.M313120200.

Jin, Z., Han, Y. X., & Han, X. R. (2013). Loss of RUNX3 expression may contribute to poor prognosis in patients with chondrosarcoma. *Journal of Molecular Histology, 44*(6), 645–652. doi:10.1007/s10735-013-9511-x.

Ju, X., Ishikawa TO, Naka, K., Ito, K., Ito, Y., & Oshima, M. (2014). Context-dependent activation of Wnt signaling by tumor suppressor RUNX3 in gastric cancer cells. *Cancer Science, 105*(4), 418–424. doi:10.1111/cas.12356.

Katagiri, Y. U., Sleeman, J., Fujii, H., Herrlich, P., Hotta, H., Tanaka, K., et al. (1999). CD44 variants but not CD44s cooperate with beta1-containing integrins to permit cells to bind to osteopontin independently of arginine-glycine-aspartic acid, thereby stimulating cell motility and chemotaxis. *Cancer Research, 59*(1), 219–226.

Kim, W. Y., & Sharpless, N. E. (2006). The regulation of INK4/ARF in cancer and aging. *Cell, 127*(2), 265–275. doi:10.1016/j.cell.2006.10.003.

Kim, W. J., Kim, E. J., Jeong, P., Quan, C., Kim, J., Li, Q. L., et al. (2005). RUNX3 inactivation by point mutations and aberrant DNA methylation in bladder tumors. *Cancer Research, 65*(20), 9347–9354. doi:10.1158/0008-5472.CAN-05-1647.

Kim, E. K., Jeon, I., Seo, H., Park, Y. J., Song, B., Lee, K. A., et al. (2014). Tumor-derived osteopontin suppresses antitumor immunity by promoting extramedullary myelopoiesis. *Cancer Research, 74*(22), 6705–6716. doi:10.1158/0008-5472.CAN-14-1482.

Kinzler, K. W., & Vogelstein, B. (1997). Cancer-susceptibility genes. Gatekeepers and caretakers. *Nature, 386*(6627), 761–763. doi:10.1038/386761a0.

Kitago, M., Martinez, S. R., Nakamura, T., Sim, M. S., & Hoon, D. S. (2009). Regulation of RUNX3 tumor suppressor gene expression in cutaneous melanoma. *Clinical Cancer Research: An Official Journal of the American Association for Cancer Research, 15*(9), 2988–2994. doi:10.1158/1078-0432.CCR-08-3172.

Knudson Jr., A. G. (1971). Mutation and cancer: statistical study of retinoblastoma. *Proceedings of the National Academy of Sciences of the United States of America, 68*(4), 820–823.

Kolb, A., Kleeff, J., Guweidhi, A., Esposito, I., Giese, N. A., Adwan, H., et al. (2005). Osteopontin influences the invasiveness of pancreatic cancer cells and is increased in neoplastic and inflammatory conditions. *Cancer Biology & Therapy, 4*(7), 740–746.

Koopmann, J., Fedarko, N. S., Jain, A., Maitra, A., Iacobuzio-Donahue, C., Rahman, A., et al. (2004). Evaluation of osteopontin as biomarker for pancreatic adenocarcinoma. *Cancer Epidemiology, Biomarkers and Prevention: A Publication of the American Association for Cancer Research, Cosponsored by the American Society of Preventive Oncology, 13*(3), 487–491.

Kurklu, B., Whitehead, R. H., Ong, E. K., Minamoto, T., Fox, J. G., Mann, J. R., et al. (2015). Lineage-specific RUNX3 hypomethylation marks the preneoplastic immune component of gastric cancer. *Oncogene, 34*(22), 2856–2866. doi:10.1038/onc.2014.233.

Lacayo, N. J., Meshinchi, S., Kinnunen, P., Yu, R., Wang, Y., Stuber, C. M., et al. (2004). Gene expression profiles at diagnosis in de novo childhood AML patients identify FLT3 mutations with good clinical outcomes. *Blood, 104*(9), 2646–2654. doi:10.1182/blood-2003-12-4449.

Lamouille, S., Xu, J., & Derynck, R. (2014). Molecular mechanisms of epithelial-mesenchymal transition. *Nature Reviews Molecular Cell Biology, 15*(3), 178–196. doi:10.1038/nrm3758.

Lang, G. A., Iwakuma, T., Suh, Y. A., Liu, G., Rao, V. A., Parant, J. M., et al. (2004). Gain of function of a p53 hot spot mutation in a mouse model of Li-Fraumeni syndrome. *Cell, 119*(6), 861–872. doi:10.1016/j.cell.2004.11.006.

Lau, Q. C., Raja, E., Salto-Tellez, M., Liu, Q., Ito, K., Inoue, M., et al. (2006). RUNX3 is frequently inactivated by dual mechanisms of protein mislocalization and promoter hypermethylation in breast cancer. *Cancer Research, 66*(13), 6512–6520. doi:10.1158/0008-5472.CAN-06-0369.

LeBleu, V. S., O'Connell, J. T., Gonzalez Herrera, K. N., Wikman, H., Pantel, K., Haigis, M. C., et al. (2014). PGC-1alpha mediates mitochondrial biogenesis and oxidative phosphorylation in cancer cells to promote metastasis. *Nature Cell Biology, 16*(10), 992–1003 1001-1015. doi:10.1038/ncb3039.

Lee, K. S., Lee, Y. S., Lee, J. M., Ito, K., Cinghu, S., Kim, J. H., et al. (2010). Runx3 is required for the differentiation of lung epithelial cells and suppression of lung cancer. *Oncogene, 29*(23), 3349–3361. doi:10.1038/onc.2010.79.

Lee, C. W., Chuang, L. S., Kimura, S., Lai, S. K., Ong, C. W., Yan, B., et al. (2011a). RUNX3 functions as an oncogene in ovarian cancer. *Gynecologic Oncology, 122*(2), 410–417. doi:10.1016/j.ygyno.2011.04.044.

Lee, J. H., Pyon, J. K., Kim, D. W., Lee, S. H., Nam, H. S., Kang, S. G., et al. (2011b). Expression of RUNX3 in skin cancers. *Clinical and Experimental Dermatology, 36*(7), 769–774. doi:10.1111/j.1365-2230.2011.04069.x.

Lee, Y. S., Lee, J. W., Jang, J. W., Chi, X. Z., Kim, J. H., Li, Y. H., et al. (2013). Runx3 inactivation is a crucial early event in the development of lung adenocarcinoma. *Cancer Cell, 24*(5), 603–616. doi:10.1016/j.ccr.2013.10.003.

Levanon, D., & Groner, Y. (2004). Structure and regulated expression of mammalian RUNX genes. *Oncogene, 23*(24), 4211–4219. doi:10.1038/sj.onc.1207670.

Levanon, D., Brenner, O., Negreanu, V., Bettoun, D., Woolf, E., Eilam, R., et al. (2001). Spatial and temporal expression pattern of Runx3 (Aml2) and Runx1 (Aml1) indicates non-redundant functions during mouse embryogenesis. *Mechanisms of Development, 109*(2), 413–417.

Levanon, D., Bettoun, D., Harris-Cerruti, C., Woolf, E., Negreanu, V., Eilam, R., et al. (2002). The Runx3 transcription factor regulates development and survival of TrkC dorsal root ganglia neurons. *The EMBO Journal, 21*(13), 3454–3463. doi:10.1093/emboj/cdf370.

Levanon, D., Brenner, O., Otto, F., & Groner, Y. (2003). Runx3 knockouts and stomach cancer. *EMBO Reports, 4*(6), 560–564. doi:10.1038/sj.embor.embor868.

Levanon, D., Bernstein, Y., Negreanu, V., Bone, K. R., Pozner, A., Eilam, R., et al. (2011). Absence of Runx3 expression in normal gastrointestinal epithelium calls into question its tumour suppressor function. *EMBO Molecular Medicine, 3*(10), 593–604. doi:10.1002/emmm.201100168.

Li, Q. L., Ito, K., Sakakura, C., Fukamachi, H., Inoue, K., Chi, X. Z., et al. (2002). Causal relationship between the loss of RUNX3 expression and gastric cancer. *Cell, 109*(1), 113–124.

Li, J., Kleeff, J., Guweidhi, A., Esposito, I., Berberat, P. O., Giese, T., et al. (2004). RUNX3 expression in primary and metastatic pancreatic cancer. *Journal of Clinical Pathology, 57*(3), 294–299.

Lin, C., Lu, W., Zhang, W., Londono-Joshi, A. I., Buchsbaum, D. J., Bu, G., & Li, Y. (2013). The C-terminal region Mesd peptide mimics full-length Mesd and acts as an inhibitor of Wnt/beta-catenin signaling in cancer cells. *PloS One, 8*(2), e58102. doi:10.1371/journal.pone.0058102.

Linzer, D. I., & Levine, A. J. (1979). Characterization of a 54 K dalton cellular SV40 tumor antigen present in SV40-transformed cells and uninfected embryonal carcinoma cells. *Cell, 17*(1), 43–52.

Liu, Z., Chen, L., Zhang, X., Xu, X., Xing, H., Zhang, Y., et al. (2014). RUNX3 regulates vimentin expression via miR-30a during epithelial-mesenchymal transition in gastric cancer cells. *Journal of Cellular and Molecular Medicine, 18*(4), 610–623. doi:10.1111/jcmm.12209.

Lotem, J., Levanon, D., Negreanu, V., Bauer, O., Hantisteanu, S., Dicken, J., & Groner, Y. (2015). Runx3 at the interface of immunity, inflammation and cancer. *Biochimica et Biophysica Acta, 1855*(2), 131–143. doi:10.1016/j.bbcan.2015.01.004.

Mazur, P. K., & Siveke, J. T. (2012). Genetically engineered mouse models of pancreatic cancer: unravelling tumour biology and progressing translational oncology. *Gut, 61*(10), 1488–1500. doi:10.1136/gutjnl-2011-300756.

Michor, F., Iwasa, Y., & Nowak, M. A. (2004). Dynamics of cancer progression. *Nature Reviews Cancer, 4*(3), 197–205. doi:10.1038/nrc1295.

Miething, C., Grundler, R., Mugler, C., Brero, S., Hoepfl, J., Geigl, J., et al. (2007). Retroviral insertional mutagenesis identifies RUNX genes involved in chronic myeloid leukemia disease persistence under imatinib treatment. *Proceedings of the National Academy of Sciences of the United States of America, 104*(11), 4594–4599. doi:10.1073/pnas.0604716104.

Min, B., Kim, M. K., Zhang, J. W., Kim, J., Chung, K. C., Oh, B. C., et al. (2012). Identification of RUNX3 as a component of the MST/Hpo signaling pathway. *Journal of Cellular Physiology, 227*(2), 839–849. doi:10.1002/jcp.22887.

Mori, T., Nomoto, S., Koshikawa, K., Fujii, T., Sakai, M., Nishikawa, Y., et al. (2005). Decreased expression and frequent allelic inactivation of the RUNX3 gene at 1p36 in human hepatocellular carcinoma. *Liver International: Official Journal of the International Association for the Study of the Liver, 25*(2), 380–388. doi:10.1111/j.1478-3231.2005.1059.x.

Nakanishi, Y., Shiraha, H., Nishina, S., Tanaka, S., Matsubara, M., Horiguchi, S., et al. (2011). Loss of runt-related transcription factor 3 expression leads hepatocellular carcinoma cells to escape apoptosis. *BMC Cancer, 11*, 3. doi:10.1186/1471-2407-11-3.

Neesse, A., Frese, K. K., Bapiro, T. E., Nakagawa, T., Sternlicht, M. D., Seeley, T. W., et al. (2013). CTGF antagonism with mAb FG-3019 enhances chemotherapy response without increasing drug delivery in murine ductal pancreas cancer. *Proceedings of the National Academy of Sciences of the United States of America, 110*(30), 12325–12330. doi:10.1073/pnas.1300415110.

Nevadunsky, N. S., Barbieri, J. S., Kwong, J., Merritt, M. A., Welch, W. R., Berkowitz, R. S., & Mok, S. C. (2009). RUNX3 protein is overexpressed in human

epithelial ovarian cancer. *Gynecologic Oncology, 112*(2), 325–330. doi:10.1016/j.ygyno.2008.09.006.
Ng, L., Wan, T. M., Lam, C. S., Chow, A. K., Wong, S. K., Man, J. H., et al. (2015). Post-operative plasma osteopontin predicts distant metastasis in human colorectal cancer. *PloS One, 10*(5), e0126219. doi:10.1371/journal.pone.0126219.
Ogasawara, N., Tsukamoto, T., Mizoshita, T., Inada, K. I., Ban, H., Kondo, S., et al. (2009). RUNX3 expression correlates with chief cell differentiation in human gastric cancers. *Histology and Histopathology, 24*(1), 31–40.
Okawa, E. R., Gotoh, T., Manne, J., Igarashi, J., Fujita, T., Silverman, K. A., et al. (2008). Expression and sequence analysis of candidates for the 1p36.31 tumor suppressor gene deleted in neuroblastomas. *Oncogene, 27*(6), 803–810. doi:10.1038/sj.onc.1210675.
Olive, K. P., Tuveson, D. A., Ruhe, Z. C., Yin, B., Willis, N. A., Bronson, R. T., et al. (2004). Mutant p53 gain of function in two mouse models of Li-Fraumeni syndrome. *Cell, 119*(6), 847–860. doi:10.1016/j.cell.2004.11.004.
Olive, K. P., Jacobetz, M. A., Davidson, C. J., Gopinathan, A., McIntyre, D., Honess, D., et al. (2009). Inhibition of Hedgehog signaling enhances delivery of chemotherapy in a mouse model of pancreatic cancer. *Science, 324*(5933), 1457–1461. doi:10.1126/science.1171362.
Olivier, M., Langerod, A., Carrieri, P., Bergh, J., Klaar, S., Eyfjord, J., et al. (2006). The clinical value of somatic TP53 gene mutations in 1,794 patients with breast cancer. *Clinical Cancer Research: An Official Journal of the American Association for Cancer Research, 12*(4), 1157–1167. doi:10.1158/1078-0432.CCR-05-1029.
Omar, M. F., Ito, K., Nga, M. E., Soo, R., Peh, B. K., Ismail, T. M., et al. (2012). RUNX3 downregulation in human lung adenocarcinoma is independent of p53, EGFR or KRAS status. *Pathology Oncology Research: POR, 18*(4), 783–792. doi:10.1007/s12253-011-9485-5.
Paget, S. (1889). The distribution of secondary growths in cancer of the breast. *Lancet, 1*, 571–573.
Pan, H. W., Ou, Y. H., Peng, S. Y., Liu, S. H., Lai, P. L., Lee, P. H., et al. (2003). Overexpression of osteopontin is associated with intrahepatic metastasis, early recurrence, and poorer prognosis of surgically resected hepatocellular carcinoma. *Cancer, 98*(1), 119–127. doi:10.1002/cncr.11487.
Peng, Y., Chen, L., Li, C., Lu, W., Agrawal, S., & Chen, J. (2001). Stabilization of the MDM2 oncoprotein by mutant p53. *The Journal of Biological Chemistry, 276*(9), 6874–6878. doi:10.1074/jbc.C000781200.
Perez-Mancera, P. A., Guerra, C., Barbacid, M., & Tuveson, D. A. (2012). What we have learned about pancreatic cancer from mouse models. *Gastroenterology, 142*(5), 1079–1092. doi:10.1053/j.gastro.2012.03.002.
Poruk, K. E., Firpo, M. A., Scaife, C. L., Adler, D. G., Emerson, L. L., Boucher, K. M., & Mulvihill, S. J. (2013). Serum osteopontin and tissue inhibitor of metalloproteinase 1 as diagnostic and prognostic biomarkers for pancreatic adenocarcinoma. *Pancreas, 42*(2), 193–197. doi:10.1097/MPA.0b013e31825e354d.
Pratap, J., Lian, J. B., Javed, A., Barnes, G. L., van Wijnen, A. J., Stein, J. L., & Stein, G. S. (2006). Regulatory roles of Runx2 in metastatic tumor and cancer cell interactions with bone. *Cancer Metastasis Reviews, 25*(4), 589–600. doi:10.1007/s10555-006-9032-0.
Prosser, J., Thompson, A. M., Cranston, G., & Evans, H. J. (1990). Evidence that p53 behaves as a tumour suppressor gene in sporadic breast tumours. *Oncogene, 5*(10), 1573–1579.
Provenzano, P. P., Cuevas, C., Chang, A. E., Goel, V. K., Von Hoff, D. D., & Hingorani, S. R. (2012). Enzymatic targeting of the stroma ablates physical barriers to treatment of pancreatic ductal adenocarcinoma. *Cancer Cell, 21*(3), 418–429. doi:10.1016/j.ccr.2012.01.007.
Pulciani, S., Santos, E., Lauver, A. V., Long, L. K., Robbins, K. C., & Barbacid, M. (1982). Oncogenes in human tumor cell lines: molecular cloning of a transforming gene from human bladder carcinoma cells. *Proceedings of the National Academy of Sciences of the United States of America, 79*(9), 2845–2849.
Rettenmaier, T. J., Sadowsky, J. D., Thomsen, N. D., Chen, S. C., Doak, A. K., Arkin, M. R., & Wells, J. A. (2014). A small-molecule mimic of a peptide docking motif inhibits the protein kinase PDK1. *Proceedings of the National Academy of Sciences of the United States of America, 111*(52), 18590–18595. doi:10.1073/pnas.1415365112.
Rowe, G. C., Raghuram, S., Jang, C., Nagy, J. A., Patten, I. S., Goyal, A., et al. (2014). PGC-1alpha induces SPP1 to activate macrophages and orchestrate functional angiogenesis in skeletal muscle. *Circulation Research, 115*(5), 504–517. doi:10.1161/CIRCRESAHA.115.303829.
Rozenblum, E., Schutte, M., Goggins, M., Hahn, S. A., Panzer, S., Zahurak, M., et al. (1997). Tumor-suppressive pathways in pancreatic carcinoma. *Cancer Research, 57*(9), 1731–1734.
Rud, A. K., Boye, K., Oijordsbakken, M., Lund-Iversen, M., Halvorsen, A. R., Solberg, S. K., et al. (2013). Osteopontin is a prognostic biomarker in non-small cell lung cancer. *BMC Cancer, 13*, 540. doi:10.1186/1471-2407-13-540.
Salto-Tellez, M., Peh, B. K., Ito, K., Tan, S. H., Chong, P. Y., Han, H. C., et al. (2006). RUNX3 protein is overexpressed in human basal cell carcinomas. *Oncogene, 25*(58), 7646–7649. doi:10.1038/sj.onc.1209739.
Sangaletti, S., Tripodo, C., Sandri, S., Torselli, I., Vitali, C., Ratti, C., et al. (2014). Osteopontin shapes immunosuppression in the metastatic niche. *Cancer*

Research, 74(17), 4706–4719. doi:10.1158/0008-5472.CAN-13-3334.

Sato, K., Tomizawa, Y., Iijima, H., Saito, R., Ishizuka, T., Nakajima, T., & Mori, M. (2006). Epigenetic inactivation of the RUNX3 gene in lung cancer. *Oncology Reports, 15*(1), 129–135.

Sherman, M. H., Yu, R. T., Engle, D. D., Ding, N., Atkins, A. R., Tiriac, H., et al. (2014). Vitamin D receptor-mediated stromal reprogramming suppresses pancreatitis and enhances pancreatic cancer therapy. *Cell, 159*(1), 80–93. doi:10.1016/j.cell.2014.08.007.

Sherr, C. J. (2004). Principles of tumor suppression. *Cell, 116*(2), 235–246.

Shi, M. J., & Stavnezer, J. (1998). CBF alpha3 (AML2) is induced by TGF-beta1 to bind and activate the mouse germline Ig alpha promoter. *Journal of Immunology, 161*(12), 6751–6760.

Soong, R., Shah, N., Peh, B. K., Chong, P. Y., Ng, S. S., Zeps, N., et al. (2009). The expression of RUNX3 in colorectal cancer is associated with disease stage and patient outcome. *British Journal of Cancer, 100*(5), 676–679. doi:10.1038/sj.bjc.6604899.

Stromnes, I. M., Brockenbrough, J. S., Izeradjene, K., Carlson, M. A., Cuevas, C., Simmons, R. M., et al. (2014). Targeted depletion of an MDSC subset unmasks pancreatic ductal adenocarcinoma to adaptive immunity. *Gut, 63*(11), 1769–1781. doi:10.1136/gutjnl-2013-306271.

Stromnes, I. M., Schmitt, T. M., Hulbert, A., Brockenbrough, J. S., Nguyen, H. N., Cuevas, C., et al. (2015). T Cells Engineered against a Native Antigen Can Surmount Immunologic and Physical Barriers to Treat Pancreatic Ductal Adenocarcinoma. *Cancer Cell, 28*(5), 638–652. doi:10.1016/j.ccell.2015.09.022.

Subramaniam, M. M., Chan, J. Y., Soong, R., Ito, K., Ito, Y., Yeoh, K. G., et al. (2009). RUNX3 inactivation by frequent promoter hypermethylation and protein mislocalization constitute an early event in breast cancer progression. *Breast Cancer Research and Treatment, 113*(1), 113–121. doi:10.1007/s10549-008-9917-4.

Tabin, C. J., Bradley, S. M., Bargmann, C. I., Weinberg, R. A., Papageorge, A. G., Scolnick, E. M., et al. (1982). Mechanism of activation of a human oncogene. *Nature, 300*(5888), 143–149.

Takahashi, T., Nau, M. M., Chiba, I., Birrer, M. J., Rosenberg, R. K., Vinocour, M., et al. (1989). p53: a frequent target for genetic abnormalities in lung cancer. *Science, 246*(4929), 491–494.

Tanaka, S., Shiraha, H., Nakanishi, Y., Nishina, S., Matsubara, M., Horiguchi, S., et al. (2012). Runt-related transcription factor 3 reverses epithelial-mesenchymal transition in hepatocellular carcinoma. *International journal of cancer Journal international du cancer, 131*(11), 2537–2546. doi:10.1002/ijc.27575.

Tsunematsu, T., Kudo, Y., Iizuka, S., Ogawa, I., Fujita, T., Kurihara, H., et al. (2009). RUNX3 has an oncogenic role in head and neck cancer. *PloS One, 4*(6), e5892. doi:10.1371/journal.pone.0005892.

Tucker, R. W., Sanford, K. K., Handleman, S. L., & Jones, G. M. (1977). Colony morphology and growth in agarose as tests for spontaneous neoplastic transformation in vitro. *Cancer Research, 37*(5), 1571–1579.

Urquidi, V., Sloan, D., Kawai, K., Agarwal, D., Woodman, A. C., Tarin, D., & Goodison, S. (2002). Contrasting expression of thrombospondin-1 and osteopontin correlates with absence or presence of metastatic phenotype in an isogenic model of spontaneous human breast cancer metastasis. *Clinical Cancer Research: An Official Journal of the American Association for Cancer Research, 8*(1), 61–74.

van Es, J. M., Polak, M. M., van den Berg, F. M., Ramsoekh, T. B., Craanen, M. E., Hruban, R. H., & Offerhaus, G. J. (1995). Molecular markers for diagnostic cytology of neoplasms in the head region of the pancreas: mutation of K-ras and overexpression of the p53 protein product. *Journal of Clinical Pathology, 48*(3), 218–222.

Voon, D. C., Wang, H., Koo, J. K., Nguyen, T. A., Hor, Y. T., Chu, Y. S., et al. (2012). Runx3 protects gastric epithelial cells against epithelial-mesenchymal transition-induced cellular plasticity and tumorigenicity. *Stem Cells, 30*(10), 2088–2099. doi:10.1002/stem.1183.

Wada, M., Yazumi, S., Takaishi, S., Hasegawa, K., Sawada, M., Tanaka, H., et al. (2004). Frequent loss of RUNX3 gene expression in human bile duct and pancreatic cancer cell lines. *Oncogene, 23*(13), 2401–2407. doi:10.1038/sj.onc.1207395.

Wai, P. Y., & Kuo, P. C. (2008). Osteopontin: regulation in tumor metastasis. *Cancer Metastasis Reviews, 27*(1), 103–118. doi:10.1007/s10555-007-9104-9.

Wei, D., Gong, W., Oh, S. C., Li, Q., Kim, W. D., Wang, L., et al. (2005). Loss of RUNX3 expression significantly affects the clinical outcome of gastric cancer patients and its restoration causes drastic suppression of tumor growth and metastasis. *Cancer Research, 65*(11), 4809–4816. doi:10.1158/0008-5472.CAN-04-3741.

Whittle, M. C., & Hingorani, S. R. (2015). Disconnect between EMT and metastasis in pancreas cancer. *Oncotarget, 6*(31), 30445–30446. doi:10.18632/oncotarget.5720.

Whittle, M. C., Izeradjene, K., Rani, P. G., Feng, L., Carlson, M. A., DelGiorno, K. E., et al. (2015). RUNX3 Controls a Metastatic Switch in Pancreatic Ductal Adenocarcinoma. *Cell, 161*(6), 1345–1360. doi:10.1016/j.cell.2015.04.048.

Wilentz, R. E., Iacobuzio-Donahue, C. A., Argani, P., McCarthy, D. M., Parsons, J. L., Yeo, C. J., et al. (2000a). Loss of expression of Dpc4 in pancreatic intraepithelial neoplasia: evidence that DPC4 inactivation occurs late in neoplastic progression. *Cancer Research, 60*(7), 2002–2006.

Wilentz, R. E., Su, G. H., Dai, J. L., Sparks, A. B., Argani, P., Sohn, T. A., et al. (2000b). Immunohistochemical labeling for dpc4 mirrors genetic status in pancreatic adenocarcinomas : a new marker of DPC4 inactivation. *The American Journal of Pathology, 156*(1), 37–43. doi:10.1016/S0002-9440(10)64703-7.

Wu, M., Li, C., Zhu, G., Wang, Y., Jules, J., Lu, Y., et al. (2014). Deletion of core-binding factor beta (Cbfbeta) in mesenchymal progenitor cells provides new insights into Cbfbeta/Runxs complex function in cartilage and bone development. *Bone, 65*, 49–59. doi:10.1016/j.bone.2014.04.031.

Xue, L. N., Bai, F. H., Wang, X. Y., Lin, M., Tan, Y., Yao, X. Y., & Xu, K. Q. (2014). Expression of RUNX3 gene in pancreatic adenocarcinoma and its clinical significance. *Genetics and Molecular Research: GMR, 13*(2), 3940–3946. doi:10.4238/2014.May.23.4.

Yagi, R., Chen, L. F., Shigesada, K., Murakami, Y., & Ito, Y. (1999). A WW domain-containing yes-associated protein (YAP) is a novel transcriptional co-activator. *The EMBO Journal, 18*(9), 2551–2562. doi:10.1093/emboj/18.9.2551.

Yamada, C., Ozaki, T., Ando, K., Suenaga, Y., Inoue, K., Ito, Y., et al. (2010). RUNX3 modulates DNA damage-mediated phosphorylation of tumor suppressor p53 at Ser-15 and acts as a co-activator for p53. *The Journal of Biological Chemistry, 285*(22), 16693–16703. doi:10.1074/jbc.M109.055525.

Yamano, M., Fujii, H., Takagaki, T., Kadowaki, N., Watanabe, H., & Shirai, T. (2000). Genetic progression and divergence in pancreatic carcinoma. *The American Journal of Pathology, 156*(6), 2123–2133. doi:10.1016/S0002-9440(10)65083-3.

Yamashiro, T., Aberg, T., Levanon, D., Groner, Y., & Thesleff, I. (2002). Expression of Runx1, −2 and −3 during tooth, palate and craniofacial bone development. *Mechanisms of Development, 119*(Suppl 1), S107–S110.

Yang, Y., Ye, Z., Zou, Z., Xiao, G., Luo, G., & Yang, H. (2014). Clinicopathological significance of RUNX3 gene hypermethylation in hepatocellular carcinoma. *Tumour Biology: The Journal of the International Society for Oncodevelopmental Biology and Medicine, 35*(10), 10333–10340. doi:10.1007/s13277-014-2329-7.

Yano, S., Miwa, S., Mii, S., Hiroshima, Y., Uehara, F., Yamamoto, M., et al. (2014). Invading cancer cells are predominantly in G0/G1 resulting in chemoresistance demonstrated by real-time FUCCI imaging. *Cell Cycle, 13*(6), 953–960. doi:10.4161/cc.27818.

Zhang, Z., Chen, G., Cheng, Y., Martinka, M., & Li, G. (2011). Prognostic significance of RUNX3 expression in human melanoma. *Cancer, 117*(12), 2719–2727. doi:10.1002/cncr.25838.

Zhang, X., He, H., Zhang, X., Guo, W., & Wang, Y. (2015). RUNX3 Promoter Methylation Is Associated with Hepatocellular Carcinoma Risk: A Meta-Analysis. *Cancer Investigation, 33*(4), 121–125. doi:10.3109/07357907.2014.1003934.

Zheng, X., Carstens, J. L., Kim, J., Scheible, M., Kaye, J., Sugimoto, H., et al. (2015). Epithelial-to-mesenchymal transition is dispensable for metastasis but induces chemoresistance in pancreatic cancer. *Nature, 527*(7579), 525–530. doi:10.1038/nature16064.

Runx Genes in Breast Cancer and the Mammary Lineage

22

Nicholas Rooney, Alessandra I. Riggio, Daniel Mendoza-Villanueva, Paul Shore, Ewan R. Cameron, and Karen Blyth

Abstract

A full understanding of *RUNX* gene function in different epithelial lineages has been thwarted by the lethal phenotypes observed when constitutively knocking out these mammalian genes. However temporal expression of the *Runx* genes throughout the different phases of mammary gland development is indicative of a functional role in this tissue. A few studies have emerged describing how these genes impact on the fate of mammary epithelial cells by regulating lineage differentiation and stem/progenitor cell potential, with implications for the transformed state. The importance of the RUNX/CBFβ core factor binding complex in breast cancer has very recently been highlighted with both *RUNX1* and *CBFβ* appearing in a comprehensive gene list of predicted breast cancer driver mutations. Nonetheless, the evidence to date shows that the *RUNX* genes can have dualistic outputs with respect to promoting or constraining breast cancer phenotypes, and that this may be aligned to individual subtypes of the clinical disease. We take this opportunity to review the current literature on *RUNX* and *CBFβ* in the normal and neoplastic mammary lineage while appreciating that this is likely to be the tip of the iceberg in our knowledge.

Keywords

RUNX • CBFβ • Breast cancer • Mammary gland • Mammary epithelium

Nicholas Rooney and Alessandra I. Riggio contributed equally to this work.

N. Rooney • A.I. Riggio • K. Blyth (✉)
Beatson Institute for Cancer Research, Bearsden, Glasgow G61 1BD, UK
e-mail: K.Blyth@beatson.gla.ac.uk

D. Mendoza-Villanueva • P. Shore
Faculty of Life Sciences, University of Manchester, Manchester M13 9PT, UK

E.R. Cameron
School of Veterinary Medicine, University of Glasgow, Bearsden, Glasgow G61 1QH, UK

Y. Groner et al. (eds.), *RUNX Proteins in Development and Cancer*, Advances in Experimental Medicine and Biology 962, DOI 10.1007/978-981-10-3233-2_22

22.1 The RUNX Transcription Factors

The RUNX proteins (RUNX1, RUNX2, RUNX3) are three closely related transcription factors capable of participating in the regulation of a vast number of elements across the genome. In addition, indirect regulation, either as part of multi-factor complexes or through their capacity to control the expression of non-coding RNAs, further enhances their regulatory influence. Given the very large number of direct and indirect potential targets, and the capacity to both positively and negatively regulate gene transcription, simplistic characterization of their generic functions are challenging (Ito et al. 2015). This helps to explain their pleiotropic and apparently contradictory effects at different times or in different tissues. Constitutive knock-out models have led to signature phenotypes for each of the three *Runx* genes. *Runx1* was shown to have a critical role in hematopoietic development and vascular integrity, *Runx2* appeared to be essential for bone development, whereas the *Runx3* knock out mouse presented with ataxia secondary to abnormal neurogenesis (Blyth et al. 2005). More recently tissue specific gene deletion has permitted the investigation of RUNX function in a wider range of tissue types, including epithelial structures such as the mammary gland.

22.2 The Mammary Gland

The mammalian mammary gland is a highly plastic organ that undergoes profound changes in its structure and function through the different physiological stages of the reproductive cycle and pregnancy (Fig. 22.1a). The mouse mammary gland develops during puberty from an embryonic epithelial placode to form a branched network of collecting ducts and tubules which are composed of two distinct types of cell lineages: the luminal, or secretory cells, and the basal/myoepithelial cells. This arboreal structure develops within the mammary fat pad, a supporting stroma consisting of a variety of mesenchymal cells, such as adipocytes, fibroblasts and inflammatory cells, that together provide a physical and mechanical support to the rudimental epithelial tree (Hennighausen and Robinson 2005; Macias and Hinck 2012). During pregnancy, changes in progesterone and prolactin levels cause further growth, tertiary branching and the formation of lobulo-alveolar units containing the terminally differentiated cells capable of milk production. Following lactation, the gland returns to a virgin-like state through a process called involution, usually characterized by extensive tissue remodeling and epithelial cell death (Richert et al. 2000; Watson and Khaled 2008; Inman et al. 2015). Importantly, these cyclic changes acting on a complex structure require the integrated regulation of proliferation, differentiation and cell death, and have helped establish the murine mammary gland as a particularly useful model system for addressing key developmental questions.

22.3 Expression of the *Runx* Genes in the Normal Mammary Gland

The landmark publication that defined the osteogenic function of RUNX2 also gave an early hint that *Runx* genes may be involved in mammary gland development, with expression being noted in the early mammary placode of the developing foetus (Otto et al. 1997; Ferrari et al. 2015). Further evidence that *Runx* genes may have a role in normal mammary function came when it was demonstrated that RUNX2 was expressed in mammary epithelial cells (Inman and Shore 2003) and could bind the promoter region to regulate the mammary specific gene *Beta-casein* (Inman et al. 2005). Subsequent studies established that all three *Runx* genes are expressed at the mRNA level in whole mammary glands (Blyth et al. 2010). Crucially, and consistent with an important developmental function, *Runx* gene expression appeared to correlate with different physiological stages of mammary gland development (Fig. 22.1). Levels are highest in the virgin and early-pregnant glands, and at involution, but decrease through pregnancy reaching a nadir dur-

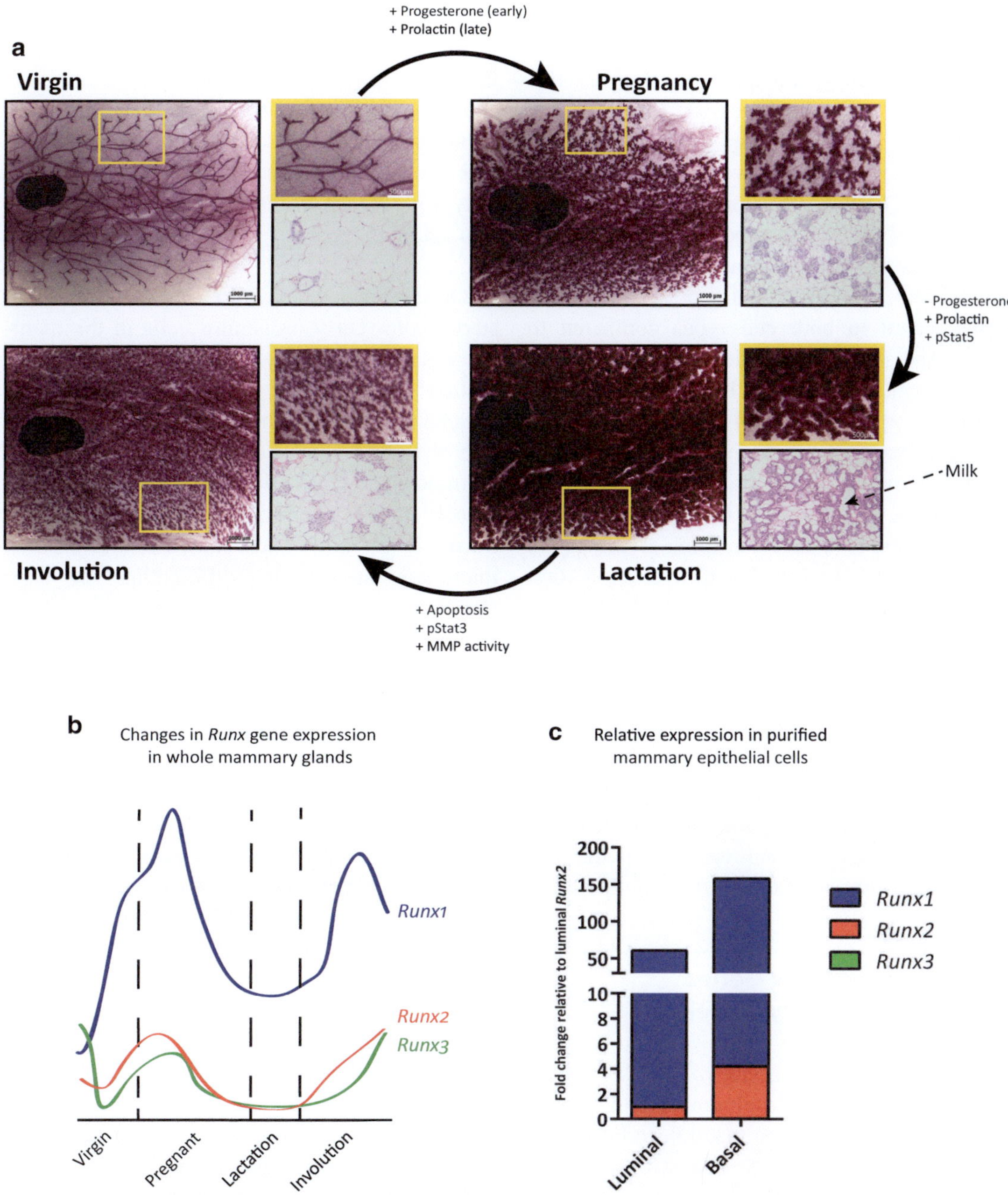

Fig. 22.1 Expression of the *Runx* genes during normal mammary gland development. (**a**) Whole mount and Hematoxylin and Eosin stained images of murine mammary glands depict the phenotypic changes at the indicated developmental stages (Virgin at 8 weeks of age; Pregnancy at day 14.5; Lactation at day 1 post-birth; Involution at 6 days post-wean). *Yellow boxed areas* are high power images of whole mount region. The mature glandular structure of the murine mammary gland is shown in the virgin panel. Progesterone drives proliferative expansion in early pregnancy leading to increased ductal branching and density, while Prolactin controls differentiation in late pregnancy by stimulating phosphoStat5. At parturition, Progesterone levels fall and milk accumulates in the hollow lumen of differentiated acini (*dashed arrow*). After weaning during the process of involution, the gland is returned to a pre-pregnancy like state by regulated apoptosis. The gland re-enters this developmental cycle on each subsequent pregnancy. (**b**) Schematic representation of the trend of *Runx* gene expression as assessed by RT-qPCR from whole mammary gland tissue. *Runx* gene expression fluctuates throughout the cycle of murine gland development and is highest during early pregnancy and rises again in involution (Adapted from Blyth et al. 2010). Note that *Runx1* is expressed at higher levels than *Runx2* and *Runx3*. (**c**) Stacked bar chart showing the relative proportion of *Runx* gene expression in flow cytometry sorted luminal and basal mammary epithelial cells. *Runx1* is the predominant gene expressed in mammary epithelial cells while *Runx3* was below detectable levels. Both *Runx1* and *Runx2* are more highly expressed in basal than luminal cells (Figure adapted from Mcdonald et al. 2014)

ing late pregnancy and lactation (Fig. 22.1b). Importantly, protein expression analysis for RUNX1 and RUNX2 mirrors that of the mRNA (McDonald et al. 2014; Owens et al. 2014; van Bragt et al. 2014). Expression of both *Runx1* and *Runx2* genes was shown to be highest in fractionated murine mammary basal cells compared to luminal cells, with *Runx1* being present at a much higher extent in both cell types compared to *Runx2*, and *Runx3* barely detectable in purified epithelial cells (Fig. 22.1c) (Kendrick et al. 2008; McDonald et al. 2014; van Bragt et al. 2014). *Runx* gene expression may in fact be incompatible with terminal differentiation and maturation of secretory units as *Runx1* expression is lost from differentiated alveolar luminal cells (van Bragt et al. 2014) while *Runx2* over expression leads to a delay in differentiation at late stages of pregnancy with a failure of lactation (McDonald et al. 2014). Consistent with this phenotypic observation is the finding that endogenous *Runx2* expression may be inhibited by prolactin signalling (Owens et al. 2014), an intriguing observation given that enforced *Runx2* expression may inhibit the prolactin receptor (McDonald et al. 2014), hinting at reciprocal regulation in the control of pregnancy associated differentiation.

22.4 Role of *Runx* Genes in Mammary Development

To date investigations into the role of *Runx* genes in the regulation of development and normal function of the mammary gland have been relatively limited. RUNX1 was shown to be fundamental for the 3D growth of basal-like MCF10A cells, where its loss correlated with increased proliferation and abnormal morphogenesis of their acinar structures (Wang et al. 2011). Interestingly, this observation was recently corroborated in a study demonstrating the requirement for RUNX1 to exit a bipotent state and differentiate into mature lobules and ducts when MCF10A cells were cultured in collagen (Sokol et al. 2015). Additionally, mammary specific deletion of *Runx1* led to a reduction in the proportion of luminal cells, with a particular deficit in mature luminal cells. *Runx1* expression was shown to be essential for maintaining and regulating the fate of the oestrogen receptor positive (ER-positive) luminal subpopulation, particularly with regard to directing ductal differentiation through repression of the alveolar Elf5 transcription factor (van Bragt et al. 2014).

On the other hand, *Runx2* has been linked to stem and/or early progenitor cells in the mammary gland, potentially downstream of WNT signalling (Ferrari et al. 2013). *Runx2* levels are enriched in mammospheres, whilst loss or inhibition of *Runx2* expression appears to curtail both mammosphere formation and mammary regeneration *in vivo* (Ferrari et al. 2015). Moreover, *Runx2* levels have been shown to decrease as mammary stem cells differentiate *in vitro* (Williams et al. 2009; Owens et al. 2014). Intriguingly, *Runx2* expression was also reported in terminal end buds (Kouros-Mehr and Werb 2006), pubertal structures representing the leading edge of the developing mammary epithelium and characterised by a high level of epithelial proliferation in the cap cells, which are thought to have stem cell like properties (Inman et al. 2015). Both *Runx2* over expression and deletion models have reported a delay in pubertal outgrowth and side branching of the mammary epithelial tree, suggesting that alterations in RUNX2 regulation may perturb normal development (McDonald et al. 2014; Owens et al. 2014). However, deletion of *Runx2* in transplanted embryonic buds did not affect the development of mature glands, suggesting that *Runx2* may be ultimately dispensable, at least after the initial formation of primordial tissue (Owens et al. 2014). Nevertheless, compensation from other *Runx* genes cannot be ruled out.

22.5 The *RUNX* Genes in Breast Cancer

Breast cancer is the most common female cancer worldwide (eg. 1.7 million estimated cases in 2012) and the leading cause of cancer-related death in women (eg. 521,900 female deaths in 2012) (Torre et al. 2016). In recent years a growing number of studies have suggested a role for *RUNX* genes in breast cancer; indeed 12.7% of breast invasive cancers have genetic alterations in *RUNX* and/or *CBFβ* genes according to the TCGA gene set (Fig. 22.2). However at the time of writing much of the available evidence to what that role may be is contradictory, certainly for *RUNX1*. As a result, obtaining a clear understanding of how *RUNX* gene dysregulation promotes or protects from this complex disease has been a challenge. In part this is due to breast cancer encompassing a set of very diverse diseases that differ in their cell of origin and the molecular changes that underpin the transition to the malignant state (Curtis et al. 2012). Clinically, the presence or absence of specific hormone or

A

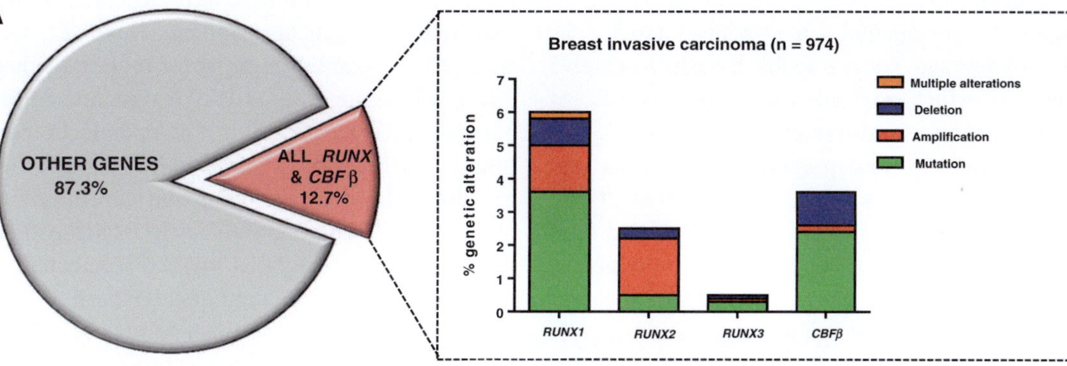

B

	All breast cancers (n=974)	ER-positive (n=594)	ER-negative (n=174)
RUNX1	6%	5.7%	6.9%
RUNX2	2.6%	0.8%	6.3%
RUNX3	0.5%	0.7%	0%
***CBF*β**	3.6%	3.9%	2.3%
ALL *RUNX* & *CBF* β	**12.7%**	**11.1%**	**15.5%**

C

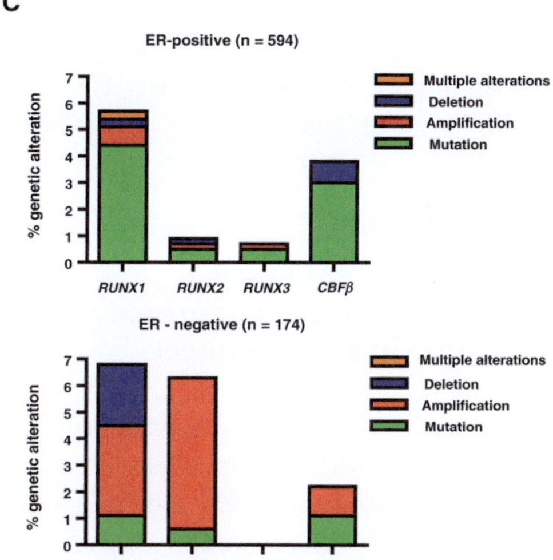

Fig. 22.2 ***RUNX* & *CBFβ* genetic alterations in breast cancer.** (**a**) In a study of breast invasive carcinoma, the *RUNX* & *CBFβ* gene set is altered in 124 (12.7%) of 974 cases (tumour samples with sequencing and copy number alteration (CNA) data), as reported in the TCGA dataset (TCGA, Cell 2015). The *RUNX1* gene accounts for the majority of these alterations (6%), followed by *CBFβ* (3.6%) & *RUNX2* (2.6%). (**b**) Table reporting the percentage of genetic alterations broken down for each gene or collectively for *'ALL RUNX&CBFβ'*, in all breast cancers (n = 974, as depicted in Figure (**a**), as well as in the ER-positive (n = 594) and ER-negative (n = 174) subtypes (TCGA, Cell 2015). (**c**) Stacked bar charts showing the relative proportion of the different genetic alterations, such as mutation, amplification, deletion and multiple alterations (i.e. coexistence of two genetic alterations), for *RUNX1, RUNX2, RUNX3* and CBFβ in the ER-positive (n = 594) and ER-negative (n = 174) subsets of the breast invasive carcinoma data set. Note that whilst mutations of the whole gene set predominate in the ER-positive subtype, *RUNX1* & *RUNX2* are frequently amplified in the ER-negative subgroup (Data obtained through cBioportal http://www.cbioportal.org/ (Cerami et al. 2012; Gao et al. 2013)

growth factor receptors, such as the oestrogen receptor (ER), the epidermal growth factor receptor 2 (HER2) and the progesterone receptor (PR), are essential in defining disease management. For example, the ER-positive group, the HER2-positive group and the triple negative group differ significantly in terms of their mutational landscape, genomic complexity and clinical prognosis. As such, while the ER-positive and HER2-positive groups are characterized by considerable clinical success, predominately due to tailored therapeutic treatments (such as Tamoxifen and Trastuzumab), the triple negative subtype, which accounts for almost 15–20% of all breast cancers, still remains an aggressive disease with very few prognostic indicators and identified molecular targets (Foulkes et al. 2010). Cell of origin descriptions can be overlaid onto this classification strategy such as luminal-like and basal-like, as well as pathological descriptions useful for marking the progress and aggressive nature of these diseases. More recently, molecular profiling has assisted in developing a higher level of sub-classification and a better understanding of disease complexity (Colombo et al. 2011; Cancer Genome Atlas 2012; Ciriello et al. 2015).

22.6 RUNX1: Evidence for a Tumour Suppressor Role in Breast Cancer

A plethora of evidence has suggested a tumour suppressor role for RUNX1 in at least some subtypes of breast cancer. *RUNX1* downregulation was identified as contributing to a 17-gene expression signature that would predict the propensity of a primary tumour to metastasize to distant loci. Retrospectively applying the same gene signature to an independent cohort successfully identified those primary tumours with a significantly poorer outcome (Ramaswamy et al. 2003). Later studies employing both RT-qPCR analyses on primary samples, as well as data mining on multiple independent Oncomine datasets, clearly showed a significant reduction of *RUNX1* expression in high histological grade tumours compared to low/mid-grade tumours (Kadota et al. 2010). The identification of recurrent genetic changes that drive breast cancer has been greatly aided by the advent of high throughput technologies and bioinformatics approaches, including next-generation sequencing (Kan et al. 2010). One such study performed by Ellis and colleagues reported a list of 18 significantly mutated genes in luminal breast cancer, with some previously identified targets, such as *PIK3CA*, *TP53* or *GATA3*, while some others were observed for the first time in clinical samples. Here, *RUNX1* emerged as a candidate target for mutation, clustering within the luminal-B-like signature of the disease (Ellis et al. 2012). An independent report found a further two mutations in *RUNX1* and four in *CBFβ*, the gene encoding the core-binding-factor-beta subunit required for RUNX function. Of note, these mutations were all detected in luminal ER-positive tumours (Banerji et al. 2012). Around the same time, following whole-exome sequencing of 510 breast tumours, *RUNX1* was also reported by The Cancer Genome Atlas (TCGA) as a significantly mutated gene. Again all 19 identified mutations of this study were found in the luminal A/B (ER-positive) and HER2-enriched subtypes, while none were present in the basal-like subgroup, thus highlighting a potential context dependent role (Cancer Genome Atlas 2012). The skew of *RUNX1* gene mutation in ER-positive disease is nicely illustrated in Fig. 22.2c.

However, other than the finding that *RUNX1* mutations are relevant to luminal tumours, there is little evidence relating to how *RUNX1* loss might impact on the disease process. In this regard, work from the Frenkel lab has recently demonstrated that loss of RUNX1 in ER-positive breast cancer cells would facilitate oestrogen induced β-catenin activation through suppression of the scaffold protein AXIN1. Furthermore, these authors showed that low AXIN1 expression correlated with low RUNX1 expression only in ER-positive patient samples, thus providing a mechanism for RUNX1-mediated tumour suppression in this subtype of the disease (Chimge et al. 2016). Although the majority of *RUNX1* somatic alterations reported so far, including point mutations, frame-shift mutations and deletions, seem to point towards a tumour suppressive function for the gene (Cornen et al. 2014; van Bragt et al. 2014), some caution must be applied when interpreting these reports as it is possible that not all these genetic changes would result in loss of function. It is of note nonetheless that in a high profile paper describing a comprehensive compendium of breast cancer mutations *RUNX1*, as well as *CBFβ*, featured in the top 50 list of protein-coding genes carrying probable driver mutations in breast cancer (Nik-Zainal et al. 2016).

22.7 RUNX1 as an Oncogenic Driver in Breast Cancer

In contrast to the putative loss of function discussed above, *RUNX1* was among the top most highly expressed genes in tumours of various cancers including breast cancer (Scheitz et al. 2012), and a number of transcriptomic papers have reported *RUNX1* to be overexpressed specifically in the triple negative subtype of the disease (Rody et al. 2011; Karn et al. 2011; Lehmann et al. 2011). This would be compatible with the observed gene amplification observed in ER-negative cancers (Fig. 22.2c). *RUNX1* was also found to be associated with super-enhancers in an ER-negative breast cancer cell line. Given the importance of these elements in directing the expression profile of cancers and their interaction with known oncogenic transcription factors (e.g. MYC), this represents an intriguing finding for this breast cancer subtype (Hnisz et al. 2013). A tissue microarray (TMA) study has confirmed a connection between RUNX1 and triple negative breast cancer (Ferrari et al. 2014). Although RUNX1 protein expression did not appear to be specifically enriched in any of the clinical subtypes of the disease, survival analysis showed a positive correlation between RUNX1 expression and poor outcome specifically in the ER-negative and triple negative subgroups. Indeed, multivariate analysis indicated that RUNX1 expression could act as an independent prognostic marker in this patient group (Ferrari et al. 2014). TMAs examined by Browne and colleagues also suggested that RUNX1 expression was elevated in a subset of ductal carcinomas *in situ* and invasive ductal carcinomas; a finding that was supported by the observation that RUNX1 expression positively correlated with advancing disease in the *MMTV-PyMT* mouse model of breast cancer (Browne et al. 2016; Browne et al. 2015). Moreover, inhibition of RUNX1 expression in cells derived from mouse tumours or in the triple negative MDA-MB-231 human breast cancer cell line resulted in decreased ability to proliferate, as well as to migrate and invade. For this pro-oncogenic phenotype, RUNX1 has been shown to bind to, and regulate expression of genes previously linked to breast cancer (Recouvreux et al. 2016).

22.8 RUNX1: Lessons from Cell Line Studies

Clues regarding the phenotypic consequences of altered RUNX1 function, as well as some of its critical downstream targets, have been derived from works in established breast cell lines. As aforementioned, deletion of *RUNX1* in MCF10A cells resulted in hyper-proliferation and abnormal morphogenesis in a 3D organotypic assay (Wang et al. 2011). In an effort to better elucidate the genetic alterations associated with breast cancer progression, Kadota and colleagues reported that *RUNX1* intragenic deletion was associated with a MCF10A subline capable of forming poorly-differentiated malignant xenografts. This discovery highlighted the propensity of breast cancer cells to lose RUNX1 expression and acquire a more malignant phenotype (Kadota et al. 2010). The identification of RUNX1 binding sites in the promoter of the *CDH1* gene and the possibility that RUNX1 positively regulates E-cadherin (Liu et al. 2005) could potentially contribute to this phenotype, as reducing E-cadherin levels facilitates epithelial-to-mesenchymal transition (EMT). However, RUNX1 expression was shown to vary widely in different breast cancer cell lines, consistent with the view that *RUNX1* may have diverse effects in different subtypes of breast cancer (Lau et al. 2006; Ferrari et al. 2014).

An intriguing aspect of both RUNX1 and RUNX2 function relevant to hormone responsive cancers is their capacity to interact with the ER signalling pathway, potentially modulating ER regulation across the genome (Chimge and Frenkel 2013). Given the importance of this pathway as a key driver of ER-positive breast tumours, the possibility that *RUNX* genes could redirect the ER signal to a subset of ERα-dependent genes makes them credible candidates for modulating tumour development, either positively or negatively (Stender et al. 2010). In a further twist to this complexity is the relationship between *cohesion*, *runx1* and ER target genes. The Cohesion subunit Rad21 has been reported to regulate *runx1* in Zebrafish (Horsfield et al. 2007), whilst *RAD21* and *RUNX1* expression are tightly correlated in endometrial cancer (Supernat et al. 2012). RAD21 itself has been implicated in aggressive breast cancer (van 't Veer et al. 2003; Xu et al. 2011) and is reported to co-localise with ER on oestrogen responsive genes with the potential to confer resistance to anti-oestrogen therapy (Schmidt et al. 2010; van Agthoven et al. 2010; McEwan et al. 2012). In view of this converging evidence the relationship between RAD21, RUNX1 and ER interdependency warrants further investigation. In another example of combinatorial gene regulation, the interaction between RUNX1 and FOXO illustrates the duality of RUNX1 effects on tumour cells (Janes 2011). In the presence of continued FOXO expression, RUNX1 down regulation results in abnormal proliferation, however in the absence of FOXO, loss of RUNX1 induces an oxidative stress that mediates growth arrest. Intriguingly these single cell based studies were echoed by the observation that *RUNX1* and *FOXO1* expression are negatively correlated in a cohort of triple negative breast cancer samples (Wang et al. 2011). Similarly, the interplay between RUNX1 and FOXP3 has been shown to regulate the expression of breast cancer associated genes. While physically inhibiting RUNX1 transcriptional activity in normal mammary epithelial cells, FOXP3 deregulation in breast cancer would lead to RUNX1-induced activation of the *RSPO3* oncogene and repression of the *GJA1* tumour suppressor gene (Recouvreux et al. 2016).

22.9 RUNX1: Insights from Animal Models

To date, relatively few studies have utilized mouse models to explore the role of *Runx1* in normal mammary development or breast cancer. As mentioned above, Van Bragt and colleagues reported that *Runx1* deletion resulted in a reduced proportion of the luminal lineage, with a particular deficit in the mature luminal subpopulation. Interestingly, simultaneous loss of the tumour suppressors *Trp53* or *Rb1* rescued this loss and restored the luminal population (van Bragt et al. 2014). However, the authors did not report that combined *Runx1/Trp53* or *Runx1/Rb1* loss resulted in abnormal proliferation or the development of a pre-neoplastic phenotype.

22.10 Expression of RUNX2 in Human Breast Tumours

RUNX2 expression, detected both by gene expression and protein analyses, has been reported in a proportion of breast cancer samples, although these studies have not consistently aligned the gene to a specific subtype (Ferrari et al. 2013). For example, small scale tumour microarray studies showed that RUNX2 expression was associated with ER-positive and HER2-positive tumours (Das et al. 2009; Onodera et al. 2010). However, RUNX2 target gene expression was observed to be negatively correlated with ER, arguing against a role for RUNX2 in ER-positive tumours (Khalid et al. 2008). A gene expression analysis of breast tumour samples from a large patient cohort, stratified into different phenotypic subtypes, revealed that *RUNX2* highly correlated with basal-like ER-negative tumours. Moreover, a subset of ER-negative tumours that had high RUNX2 expression was found to have a worse clinical outcome (Onodera et al. 2010). A similar association with poor survival has also been reported by El-Gendi and Mostafa, although here RUNX2 expression was linked to tumours of the HER2-positive subtype (El-Gendi and Mostafa 2015). Consistent with *RUNX2* gene amplification more commonly observed in ER-negative disease in the TCGA dataset (Fig. 22.2c), work from our lab using tissue microarray analysis of 416 human breast tumours showed that where RUNX2 was highly expressed, it significantly correlated with ER-negative and the triple negative subsets of breast cancer (McDonald et al. 2014). Triple negative breast cancer patients with high RUNX2 expression had a lower overall survival rate. An association between poor survival and RUNX2 has also been noted by Yang and colleagues, who further reported that RUNX2 expression was positively correlated with its downstream target SNAIL and that samples displaying strong co-expression had the worst outcome (Yang et al. 2015). Expression of the SNAIL related factor *SNAI2* also correlated with *RUNX2* in primary ER-negative breast cancers and a pro-metastatic phenotype (Chimge et al. 2011). Altogether, these studies propose that RUNX2 is associated with the more aggressive ER-negative and triple negative subtypes of breast cancer, in many cases correlating with a worse prognosis.

22.11 RUNX2: Studies on Breast Cancer Cell Lines

With regard to the role of RUNX2 in oestrogen responsive cancers, the work of Frenkel and colleagues has yielded some interesting insights (Chimge and Frenkel 2013). Their studies demonstrated a mutual antagonism between RUNX2 and ER (also seen for RUNX1 and RUNX3). This may result in restricting the growth of ER-positive cancers since RUNX2 inhibited colony formation in soft agar of oestrogen treated MCF7 cells (Chimge et al. 2012). In contrast, a wealth of data has emerged from breast cancer cell lines suggesting a role for RUNX2 in promoting the acquisition of an aggressive and metastatic phenotype (Barnes et al. 2004; Javed et al. 2005; Lau et al. 2006; Nagaraja et al. 2006; Pratap et al. 2008). This line of investigation stems from the observation that a number of known RUNX2 targets (such as MMP9, MMP13, Snail and Bone Sialoprotein amongst others) were also associated with motility, invasion and the transition to a more mesenchymal phenotype (Pratap et al. 2006; Pratap et al. 2011). Indeed, a series of overexpression studies have shown that RUNX2 can confer a transformed phenotype on mammary epithelial cells, as MCF10A cells can form larger and more dysplastic acini when cultured in 3D assays; HC11 cells acquire a more enhanced migratory ability; and the gene induces invasive properties in some breast cancer cell lines such as MCF7 cells (Pratap et al. 2009; Leong et al. 2010; Owens et al. 2014; Chimge et al. 2011). These alterations in cell behaviour were also observed in xenotransplantation studies. For example, inhibition of RUNX2 in MDA-MB-231 cells resulted in reduced bone lysis following direct intra-tibial injection of these aggressive cancer cells (Barnes et al. 2004; Javed et al. 2005; Pratap et al. 2008). Manipulation of the bone environment in cancer cells expressing RUNX2 may be explained by enhanced osteoclast activity resulting from increased levels of RANKL, PTHrP (via IHH signalling), IL-11, GM-CSF and ITGBL1, and by inhibiting osteoblast differentiation through the secretion of Sclerostin (Enomoto et al. 2003; Pratap et al. 2008; Mendoza-Villanueva et al. 2011; Li et al. 2015a). Of note, the role of RUNX2 in bone metastasis has been extensively reviewed elsewhere (Shore 2005; Pratap et al. 2006; Pratap et al. 2011; Ferrari et al. 2013; Chimge and Frenkel 2013). Altogether, these pieces of evidence point to a positive association between RUNX2 expression and invasiveness *in vitro* and aggressiveness *in vivo*, thus identifying RUNX2 as a potential target for pharmacological intervention in late stage breast tumours.

Given that RUNX2 appears to be upregulated in a proportion of patient samples and may provide an advantage to the growth and dissemination of tumour cells *in vitro*, then it follows that deregulation of RUNX2 expression would be associated with cancer progression. For example, two microRNAs (*miR135* and *miR203*) able to inhibit *RUNX2* were shown to reduce

MDA-MB-231 tumour growth and metastasis to the bone *in vivo*. Interestingly, these miRNAs are highly expressed in normal mammary epithelial cells, but their expression is curtailed in metastatic cell lines, implicating their loss in the genesis of breast cancer (Taipaleenmaki et al. 2015). Breast cancer patients with clinical depression have an increased risk of bone metastasis and it has been postulated that this may involve altered serotonin levels inducing increased levels of RUNX2 (Zong et al. 2016). RUNX2 is also regulated by its association with various binding partners, for instance RUNX2 is phosphorylated by AKT and can act downstream of the PI3K/AKT pathway in breast cancer cells (Pande et al. 2013; Cohen-Solal et al. 2015). In addition, the association of RUNX2 with the transcription factor TAZ has been shown to increase shedding of soluble E-Cadherin, which activates HER2 and contributes to oncogenesis in luminal breast cancers (Brusgard et al. 2015).

22.12 RUNX2: Insights from Animal Models

Compared to the body of work on cell lines, relatively few studies other than cell line xenografts have utilised animal models to investigate the role of *Runx2* in breast cancer. Enforced *Runx2* expression driven by the MMTV promoter revealed that RUNX2 overexpression was not sufficient to cause breast cancer. Ultimately, however, it did perturb epithelial homeostasis with over half of the mice in an aged cohort exhibiting epithelial hyperplasia, in some cases progressing to lesions resembling non-invasive ductal carcinoma *in situ* (DCIS) (McDonald et al. 2014). Germline deletion of *Runx2* was used to assess the contribution of RUNX2 in tumours induced by the well characterised *MMTV-PyMT* transgenic model. This study revealed that *Runx2* deletion caused a significant delay in tumour onset and an increase in survival. Time point analysis of the *Runx2* null glands showed that they were less neoplastic and had lower proliferation than their wild type counterparts (Owens et al. 2014). Other cancer models have also hinted at a pro-oncogenic role for RUNX2, for instance tumours from WNT driven models of breast cancer *(*APC^{1572T}*and BLG-Cre/*$APC^{fl/fl}$$Pten^{fl/fl}$*)* show high levels of RUNX2 protein (Ferrari et al. 2015). Altogether, genetically modified mouse models have shown that overexpression of *Runx2* causes early neoplastic changes in the mammary glands of aged mice, while *Runx2* deletion is able to delay tumourigenesis.

22.13 RUNX3 as a Putative Tumour Suppressor in Breast Cancer

Located on human chromosome 1p36, *RUNX3* resides in a region that has long been suspected to contain several tumour suppressor genes (Schwab et al. 1996; Weith et al. 1996). In line with this observation is a large body of work reporting evidence of *RUNX3* hyper-methylation (Kim et al. 2004; Suzuki et al. 2005; Lau et al. 2006; Hwang et al. 2007; Subramaniam et al. 2009; Park et al. 2011; Yu et al. 2014; Li et al. 2015b), genetic deletion (Chen et al. 2007; Hwang et al. 2007), protein mislocalisation (Lau et al. 2006; Subramaniam et al. 2009; Kim et al. 2009; Goh et al. 2010), and naturally occurring polymorphisms (Boone et al. 2013), in breast cancer; reviewed in more detail elsewhere (Chen 2012). Nevertheless, the role of RUNX3 in epithelial tumours has perplexed and divided the field for many years. The relevance of hyper-methylation and loss of function mutations as evidence for a tumour suppressor function, critically depends on the vexed question of whether RUNX3 is expressed in the normal epithelium (Levanon et al. 2011) and then downregulated during tumour genesis and progression (Hwang et al. 2007; Bai et al. 2013; Jiang et al. 2008; Huang et al. 2012). Whilst some reports have demonstrated RUNX3 expression in mammary tissue (Chen et al. 2007; Huang et al. 2012), others have failed to identify significant levels of *Runx3* in mammary epithelial cells (Wang et al. 2011; McDonald et al. 2014; van Bragt et al. 2014). That RUNX3 might not be expressed in this tissue compartment is also supported by data from the human protein atlas (http:www.proteinatlas.

org). Underlying some of the contradictory results obtained by different groups is the reliability of different antibodies, a frustration that also extends to RUNX1 and RUNX2 studies.

Regardless, ectopic expression of *RUNX3* suppresses the proliferative, invasive and tumourigenic properties of breast cancer cells (Lau et al. 2006; Chen et al. 2007; Huang et al. 2012). Furthermore, mice hemizygote for *Runx3* were shown to develop late onset mammary ductal carcinoma (Huang et al. 2012), perhaps due to increased ER stability and signalling in the absence of RUNX3. However, an increased incidence of mammary tumours has not been observed in an independently derived knock-out mouse model maintained on a different genetic background and it remains possible that changes in non-epithelial components could influence the incidence of spontaneous tumour development (Lotem et al. 2015). In light of this, and counter intuitively, it is interesting that *RUNX3* was identified as part of a tumour stromal gene expression signature that positively correlated with poor prognosis in human breast cancer (Finak et al. 2008), and is highly upregulated in breast cancer associated fibroblasts (McDonald & Rooney, unpublished). Therefore, a wealth of data suggests that RUNX3 could act as a tumour suppressor in mammary neoplasia however inconsistent findings on its expression and function in normal mammary epithelium continues to hamper a more complete understanding of this member of the family. Certainly *RUNX3* is the least genetically mutated within the TCGA dataset (Fig. 22.2) arguing for epigenetic mechanisms of alteration.

22.14 The RUNX Binding Factor CBFβ in Metastatic Breast Cancer Cells

Cancer cell migration is a critical feature of breast cancer metastasis, so understanding the underlying molecular mechanisms might reveal new ways to prevent cancer cell dissemination. Interestingly, the RUNX co-activator CBFβ is expressed in metastatic breast cancer cells, where it is mainly required for cell migration (Mendoza-Villanueva et al. 2010). In fact, inhibition of *CBFβ* expression in MDA-MB-231 cells via siRNA was able to reduce the ability of these cells to migrate in invasion assays, demonstrating that this critical feature of metastatic breast cancer cells is dependent on the presence of *CBFβ*. The main role of CBFβ is to enhance the capacity of RUNX proteins to bind to DNA and regulate gene transcription, by inducing allosteric changes in the Runt DNA binding domain (Tahirov et al. 2001). In MDA-MB-231 cells CBFβ associates with RUNX2 to regulate expression of several genes known to contribute to bone metastasis. Indeed, CBFβ is required for the expression of two important matrix metalloproteinase proteins, MMP-13 and MMP-9, which contribute to the proteolytic degradation of the extracellular matrix by metastatic cancer cells (Mendoza-Villanueva et al. 2010). Analysis of gene expression in MDA-MB-231 cells, in which either *RUNX2* or *CBFβ* expression was abrogated by siRNA-mediated knock-down, also revealed a number of genes that are involved in regulation of bone-cell function (Mendoza-Villanueva et al. 2011). The secreted osteoclastogenic factors *Osteopontin* (*OPN*), *IL-11* and *GM-CSF* were shown to be CBFβ targets in metastatic breast cancer cells. The dependence on CBFβ of these potent activators of osteoclast-mediated bone resorption suggests that the RUNX co-activator is necessary for the formation of osteolytic lesions, a key feature of late stage metastatic breast cancers. The same study also revealed a new role for RUNX2/CBFβ in mediating osteoblast inhibition by activating expression of the WNT-antagonist Sclerostin. When Sclerostin is secreted by osteocytes, it binds to the co-receptors LRP5 and LRP6, expressed on pre-osteoblasts, to inhibit WNT signalling and therefore osteoblast maturation (Bonewald and Johnson 2008). Inhibition of osteoblast maturation significantly contributes to osteolysis, and while current therapies target osteoclasts and bone degradation, complete healing of bone lesions is not achieved, possibly due to inhibited osteoblast maturation and new bone synthesis (Krawetz et al. 2009). Inhibitors of Sclerostin may therefore be of potential benefit in the treatment of bone metastases by enabling osteoblast maturation and thus improving healing of the osteolytic lesions. Interestingly, the study of RUNX2/CBFβ in breast cancer is revealing potential new avenues for the treatment of metastasis in patients. Future work aimed at developing new therapeutic strategies will focus on modulating the function of RUNX2/CBFβ effector target genes, and on specifically inhibiting the interaction between RUNX2 and CBFβ.

22.15 Conclusions

Since genetic alterations currently represent the most reliable indicator for predicting the importance of any gene in human neoplasia (Vogelstein and Kinzler 2004), it is not surprising to find the *RUNX-CBFβ* gene set is frequently altered in breast invasive carcinomas, albeit the contribution to disease phenotype is likely to be divergent depending on the ER status of tumours (Fig. 22.2). We are now entering an exciting era where additional *in vitro* and *in vivo* studies are underway in order to clarify the functional importance of each member of the family in the initiation and progression of breast cancer, and to better elucidate the dualistic roles played by these genes as tumour suppressors or as oncogenes in the different subtypes of this heterogeneous disease (Janes 2011).

Acknowledgements We thank Laura McDonald and Nicola Ferrari for providing material used in Fig. 22.1; all *in vivo* work in the Blyth lab is ethically approved (University of Glasgow) and carried out under Home Office licence in accordance with the EU Directive 2010. We also thank Catherine Winchester and Jim Neil for helpful comments on the manuscript. The Blyth lab is core funded by Cancer Research UK (grant code A17196).

References

Bai, J., Yong, H. M., Chen, F. F., Song, W. B., Li, C., Liu, H., & Zheng, J. N. (2013). RUNX3 is a prognostic marker and potential therapeutic target in human breast cancer. *Journal of Cancer Research and Clinical Oncology, 139*, 1813–1823.

Banerji, S., Cibulskis, K., Rangel-Escareno, C., Brown, K. K., Carter, S. L., Frederick, A. M., et al. (2012). Sequence analysis of mutations and translocations across breast cancer subtypes. *Nature, 486*, 405–409.

Barnes, G. L., Hebert, K. E., Kamal, M., Javed, A., Einhorn, T. A., Lian, J. B., et al. (2004). Fidelity of Runx2 activity in breast cancer cells is required for the generation of metastases-associated osteolytic disease. *Cancer Research, 64*, 4506–4513.

Blyth, K., Cameron, E. R., & Neil, J. C. (2005). The RUNX genes: Gain or loss of function in cancer. *Nature Reviews. Cancer, 5*, 376–387.

Blyth, K., Vaillant, F., Jenkins, A., Mcdonald, L., Pringle, M. A., Huser, C., et al. (2010). Runx2 in normal tissues and cancer cells: A developing story. *Blood Cells, Molecules & Diseases, 45*, 117–123.

Bonewald, L. F., & Johnson, M. L. (2008). Osteocytes, mechanosensing and Wnt signaling. *Bone, 42*, 606–615.

Boone, S. D., Baumgartner, K. B., Baumgartner, R. N., Connor, A. E., Pinkston, C. M., John, E. M., et al. (2013). Associations between genetic variants in the TGF-beta signaling pathway and breast cancer risk among Hispanic and non-Hispanic white women. *Breast Cancer Research and Treatment, 141*, 287–297.

Browne, G., Dragon, J. A., Hong, D., Messier, T. L., Gordon, J. A., Farina, N. H., et al. (2016). MicroRNA-378-mediated suppression of Runx1 alleviates the aggressive phenotype of triple-negative MDA-MB-231 human breast cancer cells. *Tumour Biology, 37*(7), 8825–8839.

Browne, G., Taipaleenmaki, H., Bishop, N. M., Madasu, S. C., Shaw, L. M., Van Wijnen, A. J., et al. (2015). Runx1 is associated with breast cancer progression in MMTV-PyMT transgenic mice and its depletion in vitro inhibits migration and invasion. *Journal of Cellular Physiology, 230*, 2522–2532.

Brusgard, J. L., Choe, M., Chumsri, S., Renoud, K., Mackerell Jr., A. D., Sudol, M., & Passaniti, A. (2015). RUNX2 and TAZ-dependent signaling pathways regulate soluble E-Cadherin levels and tumorsphere formation in breast cancer cells. *Oncotarget, 6*, 28132–28150.

Cancer Genome Atlas, N. (2012). Comprehensive molecular portraits of human breast tumours. *Nature, 490*, 61–70.

Cerami, E., Gao, J., Dogrusoz, U., Gross, B. E., Sumer, S. O., Aksoy, B. A., et al. (2012). The cBio cancer genomics portal: An open platform for exploring multidimensional cancer genomics data. *Cancer Discovery, 2*, 401–404.

Chen, L. F. (2012). Tumor suppressor function of RUNX3 in breast cancer. *Journal of Cellular Biochemistry, 113*, 1470–1477.

Chen, W., Salto-Tellez, M., Palanisamy, N., Ganesan, K., Hou, Q., Tan, L. K., et al. (2007). Targets of genome copy number reduction in primary breast cancers identified by integrative genomics. *Genes, Chromosomes & Cancer, 46*, 288–301.

Chimge, N. O., Baniwal, S. K., Little, G. H., Chen, Y. B., Kahn, M., Tripathy, D., et al. (2011). Regulation of breast cancer metastasis by Runx2 and estrogen signaling: the role of SNAI2. *Breast Cancer Research, 13*, R127.

Chimge, N. O., Baniwal, S. K., Luo, J., Coetzee, S., Khalid, O., Berman, B. P., et al. (2012). Opposing effects of Runx2 and estradiol on breast cancer cell proliferation: in vitro identification of reciprocally regulated gene signature related to clinical letrozole responsiveness. *Clinical Cancer Research, 18*, 901–911.

Chimge, N. O., & Frenkel, B. (2013). The RUNX family in breast cancer: Relationships with estrogen signaling. *Oncogene, 32*, 2121–2130.

Chimge, N. O., Little, G. H., Baniwal, S. K., Adisetiyo, H., Xie, Y., Zhang, T., et al. (2016). RUNX1 prevents oestrogen-mediated AXIN1 suppression and beta-catenin activation in ER-positive breast cancer. *Nature Communications, 7*, 10751.

Ciriello, G., Gatza, M. L., Beck, A. H., Wilkerson, M. D., Rhie, S. K., Pastore, A., et al. (2015). Comprehensive molecular portraits of invasive lobular breast cancer. *Cell, 163*, 506–519.

Cohen-Solal, K. A., Boregowda, R. K., & Lasfar, A. (2015). RUNX2 and the PI3K/AKT axis reciprocal activation as a driving force for tumor progression. *Molecular Cancer, 14*, 137.

Colombo, P. E., Milanezi, F., Weigelt, B., & Reis-Filho, J. S. (2011). Microarrays in the 2010s: the contribution of microarray-based gene expression profiling to breast cancer classification, prognostication and prediction. *Breast Cancer Research, 13*, 212.

Cornen, S., Guille, A., Adelaide, J., Addou-Klouche, L., Finetti, P., Saade, M. R., et al. (2014). Candidate luminal B breast cancer genes identified by genome, gene expression and DNA methylation profiling. *PloS One, 9*, e81843.

Curtis, C., Shah, S. P., Chin, S. F., Turashvili, G., Rueda, O. M., Dunning, M. J., et al. (2012). The genomic and transcriptomic architecture of 2,000 breast tumours reveals novel subgroups. *Nature, 486*, 346–352.

Das, K., Leong, D. T., Gupta, A., Shen, L., Putti, T., Stein, G. S., et al. (2009). Positive association between nuclear Runx2 and oestrogen-progesterone receptor gene expression characterises a biological subtype of breast cancer. *European Journal of Cancer, 45*, 2239–2248.

El-Gendi, S. M., & Mostafa, M. F. (2015). Runx2 expression as a potential prognostic marker in invasive ductal breast carcinoma. *Pathology Oncology Research, 22*(3), 461–470.

Ellis, M. J., Ding, L., Shen, D., Luo, J., Suman, V. J., Wallis, J. W., et al. (2012). Whole-genome analysis informs breast cancer response to aromatase inhibition. *Nature, 486*, 353–360.

Enomoto, H., Shiojiri, S., Hoshi, K., Furuichi, T., Fukuyama, R., Yoshida, C. A., et al. (2003). Induction of osteoclast differentiation by Runx2 through receptor activator of nuclear factor-kappa B ligand (RANKL) and osteoprotegerin regulation and partial rescue of osteoclastogenesis in Runx2−/− mice by RANKL transgene. *The Journal of Biological Chemistry, 278*, 23971–23977.

Ferrari, N., Mcdonald, L., Morris, J. S., Cameron, E. R., & Blyth, K. (2013). RUNX2 in mammary gland development and breast cancer. *Journal of Cellular Physiology, 228*, 1137–1142.

Ferrari, N., Mohammed, Z. M., Nixon, C., Mason, S. M., Mallon, E., Mcmillan, D. C., et al. (2014). Expression of RUNX1 correlates with poor patient prognosis in triple negative breast cancer. *PloS One, 9*, e100759.

Ferrari, N., Riggio, A. I., Mason, S., Mcdonald, L., King, A., Higgins, T., et al. (2015). Runx2 contributes to the regenerative potential of the mammary epithelium. *Scientific Reports, 5*, 15658.

Finak, G., Bertos, N., Pepin, F., Sadekova, S., Souleimanova, M., Zhao, H., et al. (2008). Stromal gene expression predicts clinical outcome in breast cancer. *Nature Medicine, 14*, 518–527.

Foulkes, W. D., Smith, I. E., & Reis-Filho, J. S. (2010). Triple-negative breast cancer. *The New England Journal of Medicine, 363*, 1938–1948.

Gao, J., Aksoy, B. A., Dogrusoz, U., Dresdner, G., Gross, B., Sumer, S. O., et al. (2013). Integrative analysis of complex cancer genomics and clinical profiles using the cBioPortal. *Science Signaling, 6*, l1.

Goh, Y. M., Cinghu, S., Hong, E. T., Lee, Y. S., Kim, J. H., Jang, J. W., et al. (2010). Src kinase phosphorylates RUNX3 at tyrosine residues and localizes the protein in the cytoplasm. *The Journal of Biological Chemistry, 285*, 10122–10129.

Hennighausen, L., & Robinson, G. W. (2005). Information networks in the mammary gland. *Nature Reviews. Molecular Cell Biology, 6*, 715–725.

Hnisz, D., Abraham, B. J., Lee, T. I., Lau, A., Saint-Andre, V., Sigova, A. A., et al. (2013). Super-enhancers in the control of cell identity and disease. *Cell, 155*, 934–947.

Horsfield, J. A., Anagnostou, S. H., Hu, J. K., Cho, K. H., Geisler, R., Lieschke, G., et al. (2007). Cohesin-dependent regulation of Runx genes. *Development, 134*, 2639–2649.

Huang, B., Qu, Z., Ong, C. W., Tsang, Y. H., Xiao, G., Shapiro, D., et al. (2012). RUNX3 acts as a tumor suppressor in breast cancer by targeting estrogen receptor alpha. *Oncogene, 31*, 527–534.

Hwang, K. T., Han, W., Bae, J. Y., Hwang, S. E., Shin, H. J., Lee, J. E., et al. (2007). Downregulation of the RUNX3 gene by promoter hypermethylation and hemizygous deletion in breast cancer. *Journal of Korean Medical Science, 22*, S24–S31.

Inman, C. K., Li, N., & Shore, P. (2005). Oct-1 counteracts autoinhibition of Runx2 DNA binding to form a novel Runx2/Oct-1 complex on the promoter of the mammary gland-specific gene beta-casein. *Molecular Cell. Biology, 25*, 3182–3193.

Inman, C. K., & Shore, P. (2003). The osteoblast transcription factor Runx2 is expressed in mammary epithelial cells and mediates osteopontin expression. *The Journal of Biological Chemistry, 278*, 48684–48689.

Inman, J. L., Robertson, C., Mott, J. D., & Bissell, M. J. (2015). Mammary gland development: Cell fate specification, stem cells and the microenvironment. *Development, 142*, 1028–1042.

Ito, Y., Bae, S. C., & Chuang, L. S. (2015). The RUNX family: Developmental regulators in cancer. *Nature Reviews. Cancer, 15*, 81–95.

Janes, K. A. (2011). RUNX1 and its understudied role in breast cancer. *Cell Cycle, 10*, 3461–3465.

Javed, A., Barnes, G. L., Pratap, J., Antkowiak, T., Gerstenfeld, L. C., Van Wijnen, A. J., et al. (2005). Impaired intranuclear trafficking of Runx2 (AML3/CBFA1) transcription factors in breast cancer cells inhibits osteolysis in vivo. *Proceedings of the National Academy of Sciences of the United States of America, 102*, 1454–1459.

Jiang, Y., Tong, D., Lou, G., Zhang, Y., & Geng, J. (2008). Expression of RUNX3 gene, methylation status and clinicopathological significance in breast cancer and breast cancer cell lines. *Pathobiology, 75*, 244–251.

Kadota, M., Yang, H. H., Gomez, B., Sato, M., Clifford, R. J., Meerzaman, D., et al. (2010). Delineating genetic alterations for tumor progression in the MCF10A series of breast cancer cell lines. *PloS One, 5*, e9201.

Kan, Z., Jaiswal, B. S., Stinson, J., Janakiraman, V., Bhatt, D., Stern, H. M., et al. (2010). Diverse somatic mutation patterns and pathway alterations in human cancers. *Nature, 466*, 869–873.

Karn, T., Pusztai, L., Holtrich, U., Iwamoto, T., Shiang, C. Y., Schmidt, M., et al. (2011). Homogeneous datasets of triple negative breast cancers enable the identification of novel prognostic and predictive signatures. *PloS One, 6*, e28403.

Kendrick, H., Regan, J. L., Magnay, F. A., Grigoriadis, A., Mitsopoulos, C., Zvelebil, M., & Smalley, M. J. (2008). Transcriptome analysis of mammary epithelial subpopulations identifies novel determinants of lineage commitment and cell fate. *BMC Genomics, 9*, 591.

Khalid, O., Baniwal, S. K., Purcell, D. J., Leclerc, N., Gabet, Y., Stallcup, M. R., et al. (2008). Modulation of Runx2 activity by estrogen receptor-alpha: implications for osteoporosis and breast cancer. *Endocrinology, 149*, 5984–5995.

Kim, T. Y., Lee, H. J., Hwang, K. S., Lee, M., Kim, J. W., Bang, Y. J., & Kang, G. H. (2004). Methylation of RUNX3 in various types of human cancers and premalignant stages of gastric carcinoma. *Laboratory Investigation, 84*, 479–484.

Kim, J. H., Choi, J. K., Cinghu, S., Jang, J. W., Lee, Y. S., Li, Y. H., et al. (2009). Jab1/CSN5 induces the cytoplasmic localization and degradation of RUNX3. *Journal of Cellular Biochemistry, 107*, 557–565.

Kouros-Mehr, H., & Werb, Z. (2006). Candidate regulators of mammary branching morphogenesis identified by genome-wide transcript analysis. *Developmental Dynamics, 235*, 3404–3412.

Krawetz, R., Wu, Y. E., Rancourt, D. E., & Matyas, J. (2009). Osteoblasts suppress high bone turnover caused by osteolytic breast cancer in-vitro. *Experimental Cell Research, 315*, 2333–2342.

Lau, Q. C., Raja, E., Salto-Tellez, M., Liu, Q., Ito, K., Inoue, M., et al. (2006). RUNX3 is frequently inactivated by dual mechanisms of protein mislocalization and promoter hypermethylation in breast cancer. *Cancer Research, 66*, 6512–6520.

Lehmann, B. D., Bauer, J. A., Chen, X., Sanders, M. E., Chakravarthy, A. B., Shyr, Y., & Pietenpol, J. A. (2011). Identification of human triple-negative breast cancer subtypes and preclinical models for selection of targeted therapies. *The Journal of Clinical Investigation, 121*, 2750–2767.

Leong, D. T., Lim, J., Goh, X., Pratap, J., Pereira, B. P., Kwok, H. S., et al. (2010). Cancer-related ectopic expression of the bone-related transcription factor RUNX2 in non-osseous metastatic tumor cells is linked to cell proliferation and motility. *Breast Cancer Research, 12*, R89.

Levanon, D., Bernstein, Y., Negreanu, V., Bone, K. R., Pozner, A., Eilam, R., et al. (2011). Absence of Runx3 expression in normal gastrointestinal epithelium calls into question its tumour suppressor function. *EMBO Molecular Medicine, 3*, 593–604.

Li, X. Q., Du, X., Li, D. M., Kong, P. Z., Sun, Y., Liu, P. F., et al. (2015a). ITGBL1 Is a Runx2 transcriptional target and promotes breast cancer bone metastasis by activating the TGFbeta signaling pathway. *Cancer Research, 75*, 3302–3313.

Li, Z., Guo, X., Wu, Y., Li, S., Yan, J., Peng, L., et al. (2015b). Methylation profiling of 48 candidate genes in tumor and matched normal tissues from breast cancer patients. *Breast Cancer Research and Treatment, 149*, 767–779.

Liu, Y. N., Lee, W. W., Wang, C. Y., Chao, T. H., Chen, Y., & Chen, J. H. (2005). Regulatory mechanisms controlling human E-cadherin gene expression. *Oncogene, 24*, 8277–8290.

Lotem, J., Levanon, D., Negreanu, V., Bauer, O., Hantisteanu, S., Dicken, J., & Groner, Y. (2015). Runx3 at the interface of immunity, inflammation and cancer. *Biochimica et Biophysica Acta, 1855*, 131–143.

Macias, H., & Hinck, L. (2012). Mammary gland development. *Wiley Interdiscip Rev Dev Biol, 1*, 533–557.

Mcdonald, L., Ferrari, N., Terry, A., Bell, M., Mohammed, Z. M., Orange, C., et al. (2014). RUNX2 correlates with subtype-specific breast cancer in a human tissue microarray, and ectopic expression of Runx2 perturbs differentiation in the mouse mammary gland. *Disease Models & Mechanisms, 7*, 525–534.

Mcewan, M. V., Eccles, M. R., & Horsfield, J. A. (2012). Cohesin is required for activation of MYC by estradiol. *PloS One, 7*, e49160.

Mendoza-Villanueva, D., Deng, W., Lopez-Camacho, C., & Shore, P. (2010). The Runx transcriptional co-activator, CBFbeta, is essential for invasion of breast cancer cells. *Molecular Cancer, 9*, 171.

Mendoza-Villanueva, D., Zeef, L., & Shore, P. (2011). Metastatic breast cancer cells inhibit osteoblast differentiation through the Runx2/CBFbeta-dependent expression of the Wnt antagonist, sclerostin. *Breast Cancer Research, 13*, R106.

Nagaraja, G. M., Othman, M., Fox, B. P., Alsaber, R., Pellegrino, C. M., Zeng, Y., et al. (2006). Gene expression signatures and biomarkers of noninvasive and invasive breast cancer cells: comprehensive profiles by representational difference analysis, microarrays and proteomics. *Oncogene, 25*, 2328–2338.

Nik-Zainal, S., Davies, H., Staaf, J., Ramakrishna, M., Glodzik, D., Zou, X., Martincorena, I., Alexandrov, L. B., Martin, S., Wedge, D. C., Van Loo, P., Ju, Y. S., Smid, M., Brinkman, A. B., Morganella, S., Aure, M. R., Lingjærde, O. C., Langerød, A., Ringnér, M., Ahn, S.-M., Boyault, S., Brock, J. E., Broeks, A., Butler, A., Desmedt, C., Dirix, L., Dronov, S., Fatima, A., Foekens, J. A., Gerstung, M., Hooijer, G. K. J., Jang, S. J., Jones, D. R., Kim, H.-Y., King, T. A., Krishnamurthy, S., Lee, H. J., Lee, J.-Y., Li, Y., Mclaren, S., Menzies, A., Mustonen, V., O'meara, S., Pauporté, I., Pivot, X., Purdie, C. A., Raine, K., Ramakrishnan, K., Rodríguez-González, F. G., Romieu, G., Sieuwerts, A. M., Simpson, P. T., Shepherd, R., Stebbings, L., Stefansson, O. A., Teague, J., Tommasi, S., Treilleux, I., Van Den Eynden, G. G., Vermeulen, P., Vincent-Salomon, A., Yates, L., Caldas, C., Veer, L. V. T., Tutt, A., Knappskog, S., Tan, B. K. T., Jonkers, J., Borg, Å., Ueno, N. T., Sotiriou, C., Viari, A., Futreal, P. A., Campbell, P. J., Span, P. N., Van Laere, S., Lakhani, S. R., Eyfjord, J. E., Thompson, A. M., Birney, E., Stunnenberg, H. G., Van De Vijver, M. J., Martens, J. W. M., Børresen-Dale, A.-L., Richardson, A. L., Kong, G., Thomas, G. & Stratton, M. R. 2016. Landscape of somatic mutations in 560 breast cancer whole-genome sequences. *Nature, 534*(7605), 47–54. Advance online publication. doi: 10.1038/nature17676.

Onodera, Y., Miki, Y., Suzuki, T., Takagi, K., Akahira, J., Sakyu, T., et al. (2010). Runx2 in human breast carcinoma: Its potential roles in cancer progression. *Cancer Science, 101*, 2670–2675.

Otto, F., Thornell, A. P., Crompton, T., Denzel, A., Gilmour, K. C., Rosewell, I. R., et al. (1997). Cbfa1, a candidate gene for cleidocranial dysplasia syndrome, is essential for osteoblast differentiation and bone development. *Cell, 89*, 765–771.

Owens, T. W., Rogers, R. L., Best, S. A., Ledger, A., Mooney, A. M., Ferguson, A., et al. (2014). Runx2 is a novel regulator of mammary epithelial cell fate in development and breast cancer. *Cancer Research, 74*, 5277–5286.

Pande, S., Browne, G., Padmanabhan, S., Zaidi, S. K., Lian, J. B., Van Wijnen, A. J., et al. (2013). Oncogenic cooperation between PI3K/Akt signaling and transcription factor Runx2 promotes the invasive properties of metastatic breast cancer cells. *Journal of Cellular Physiology, 228*, 1784–1792.

Park, S. Y., Kwon, H. J., Lee, H. E., Ryu, H. S., Kim, S. W., Kim, J. H., et al. (2011). Promoter CpG island hypermethylation during breast cancer progression. *Virchows Archiv, 458*, 73–84.

Pratap, J., Lian, J. B., Javed, A., Barnes, G. L., Van Wijnen, A. J., Stein, J. L., & Stein, G. S. (2006). Regulatory roles of Runx2 in metastatic tumor and cancer cell interactions with bone. *Cancer Metastasis Reviews, 25*, 589–600.

Pratap, J., Wixted, J. J., Gaur, T., Zaidi, S. K., Dobson, J., Gokul, K. D., et al. (2008). Runx2 transcriptional activation of Indian Hedgehog and a downstream bone metastatic pathway in breast cancer cells. *Cancer Research, 68*, 7795–7802.

Pratap, J., Imbalzano, K. M., Underwood, J. M., Cohet, N., Gokul, K., Akech, J., et al. (2009). Ectopic runx2 expression in mammary epithelial cells disrupts formation of normal acini structure: Implications for breast cancer progression. *Cancer Research, 69*, 6807–6814.

Pratap, J., Lian, J. B., & Stein, G. S. (2011). Metastatic bone disease: Role of transcription factors and future targets. *Bone, 48*, 30–36.

Ramaswamy, S., Ross, K. N., Lander, E. S., & Golub, T. R. (2003). A molecular signature of metastasis in primary solid tumors. *Nature Genetics, 33*, 49–54.

Recouvreux, M. S., Grasso, E. N., Echeverria, P. C., Rocha-Viegas, L., Castilla, L. H., Schere-Levy, C., et al. (2016). RUNX1 and FOXP3 interplay regulates expression of breast cancer related genes. *Oncotarget, 7*, 6552–6565.

Richert, M. M., Schwertfeger, K. L., Ryder, J. W., & Anderson, S. M. (2000). An atlas of mouse mammary gland development. *Journal of Mammary Gland Biology and Neoplasia, 5*, 227–241.

Rody, A., Karn, T., Liedtke, C., Pusztai, L., Ruckhaeberle, E., Hanker, L., et al. (2011). A clinically relevant gene signature in triple negative and basal-like breast cancer. *Breast Cancer Research, 13*, R97.

Scheitz, C. J., Lee, T. S., Mcdermitt, D. J., & Tumbar, T. (2012). Defining a tissue stem cell-driven Runx1/Stat3 signalling axis in epithelial cancer. *The EMBO Journal, 31*, 4124–4139.

Schmidt, D., Schwalie, P. C., Ross-Innes, C. S., Hurtado, A., Brown, G. D., Carroll, J. S., et al. (2010). A CTCF-independent role for cohesin in tissue-specific transcription. *Genome Research, 20*, 578–588.

Schwab, M., Praml, C., & Amler, L. C. (1996). Genomic instability in 1p and human malignancies. *Genes, Chromosomes & Cancer, 16*, 211–229.

Shore, P. (2005). A role for Runx2 in normal mammary gland and breast cancer bone metastasis. *Journal of Cellular Biochemistry, 96*, 484–489.

Sokol, E. S., Sanduja, S., Jin, D. X., Miller, D. H., Mathis, R. A., & Gupta, P. B. (2015). Perturbation-expression analysis identifies RUNX1 as a regulator of human mammary stem cell differentiation. *PLoS Computational Biology, 11*, e1004161.

Stender, J. D., Kim, K., Charn, T. H., Komm, B., Chang, K. C., Kraus, W. L., et al. (2010). Genome-wide analysis of estrogen receptor alpha DNA binding and tethering mechanisms identifies Runx1 as a novel tethering factor in receptor-mediated transcriptional activation. *Molecular Cell. Biology, 30*, 3943–3955.

Subramaniam, M. M., Chan, J. Y., Soong, R., Ito, K., Ito, Y., Yeoh, K. G., et al. (2009). RUNX3 inactivation by frequent promoter hypermethylation and protein mislocalization constitute an early event in breast cancer progression. *Breast Cancer Research and Treatment, 113*, 113–121.

Supernat, A., Lapinska-Szumczyk, S., Sawicki, S., Wydra, D., Biernat, W., & Zaczek, A. J. (2012). Deregulation of RAD21 and RUNX1 expression in endometrial cancer. *Oncology Letters, 4*, 727–732.

Suzuki, M., Shigematsu, H., Shames, D. S., Sunaga, N., Takahashi, T., Shivapurkar, N., et al. (2005). DNA methylation-associated inactivation of TGFbeta-related genes DRM/Gremlin, RUNX3, and HPP1 in human cancers. *British Journal of Cancer, 93*, 1029–1037.

Tahirov, T. H., Inoue-Bungo, T., Morii, H., Fujikawa, A., Sasaki, M., Kimura, K., et al. (2001). Structural analyses of DNA recognition by the AML1/Runx-1 Runt domain and its allosteric control by CBFbeta. *Cell, 104*, 755–767.

Taipaleenmaki, H., Browne, G., Akech, J., Zustin, J., Van Wijnen, A. J., Stein, J. L., et al. (2015). Targeting of Runx2 by miR-135 and miR-203 impairs progression of breast cancer and metastatic bone disease. *Cancer Research, 75*, 1433–1444.

Torre, L. A., Siegel, R. L., Ward, E. M., & Jemal, A. (2016). Global cancer incidence and mortality rates and trends--An update. *Cancer Epidemiology, Biomarkers & Prevention, 25*, 16–27.

Van 'T Veer, L. J., Dai, H., Van de Vijver, M. J., He, Y. D., Hart, A. A., Bernards, R., & Friend, S. H. (2003). Expression profiling predicts outcome in breast cancer. *Breast Cancer Research, 5*, 57–58.

Van Agthoven, T., Sieuwerts, A. M., Meijer, D., Meijer-Van Gelder, M. E., Van Agthoven, T. L., Sarwari, R., et al. (2010). Selective recruitment of breast cancer anti-estrogen resistance genes and relevance for breast cancer progression and tamoxifen therapy response. *Endocrine-Related Cancer, 17*, 215–230.

Van Bragt, M. P., Hu, X., Xie, Y., & Li, Z. (2014). RUNX1, a transcription factor mutated in breast cancer, controls the fate of ER-positive mammary luminal cells. *eLife, 3*, e03881.

Vogelstein, B., & Kinzler, K. W. (2004). Cancer genes and the pathways they control. *Nature Medicine, 10*, 789–799.

Wang, L., Brugge, J. S., & Janes, K. A. (2011). Intersection of FOXO- and RUNX1-mediated gene expression programs in single breast epithelial cells during morphogenesis and tumor progression. *Proceedings of the National Academy of Sciences of the United States of America, 108*, E803–E812.

Watson, C. J., & Khaled, W. T. (2008). Mammary development in the embryo and adult: A journey of morphogenesis and commitment. *Development, 135*, 995–1003.

Weith, A., Brodeur, G. M., Bruns, G. A., Matise, T. C., Mischke, D., Nizetic, D., et al. (1996). Report of the second international workshop on human chromosome 1 mapping 1995. *Cytogenetics and Cell Genetics, 72*, 114–144.

Williams, C., Helguero, L., Edvardsson, K., Haldosen, L. A., & Gustafsson, J. A. (2009). Gene expression in murine mammary epithelial stem cell-like cells shows similarities to human breast cancer gene expression. *Breast Cancer Research, 11*, R26.

Xu, H., Yan, M., Patra, J., Natrajan, R., Yan, Y., Swagemakers, S., et al. (2011). Enhanced RAD21 cohesin expression confers poor prognosis and resistance to chemotherapy in high grade luminal, basal and HER2 breast cancers. *Breast Cancer Research, 13*, R9.

Yang, Z., Zhang, B., Liu, B., Xie, Y., & Cao, X. (2015). Combined Runx2 and Snail overexpression is associated with a poor prognosis in breast cancer. *Tumour Biology, 36*, 4565–4573.

Yu, Y. Y., Chen, C., Kong, F. F., & Zhang, W. (2014). Clinicopathological significance and potential drug target of RUNX3 in breast cancer. *Drug Design, Development and Therapy, 8*, 2423–2430.

Zong, J. C., Wang, X., Zhou, X., Wang, C., Chen, L., Yin, L. J., et al. (2016). Gut-derived serotonin induced by depression promotes breast cancer bone metastasis through the RUNX2/PTHrP/RANKL pathway in mice. *Oncology Reports, 35*, 739–748.

Runx3 in Immunity, Inflammation and Cancer

23

Joseph Lotem, Ditsa Levanon, Varda Negreanu, Omri Bauer, Shay Hantisteanu, Joseph Dicken, and Yoram Groner

Abstract

In this chapter we summarize the pros and cons of the notion that Runx3 is a major tumor suppressor gene (TSG). Inactivation of TSGs in normal cells provides a viability/growth advantage that contributes cell-autonomously to cancer. More than a decade ago it was suggested that *RUNX3* is involved in gastric cancer development, a postulate extended later to other epithelial cancers portraying *RUNX3* as a major TSG. However, evidence that Runx3 is not expressed in normal gastric and other epithelia has challenged the *RUNX3*-TSG paradigm. In contrast, RUNX3 is overexpressed in a significant fraction of tumor cells in various human epithelial cancers and its overexpression in pancreatic cancer cells promotes their migration, anchorage-independent growth and metastatic potential. Moreover, recent high-throughput quantitative genome-wide studies on thousands of human samples of various tumors and new investigations of the role of *Runx3* in mouse cancer models have unequivocally demonstrated that *RUNX3* is not a *bona fide* cell-autonomous TSG. Importantly, accumulating data demonstrated that RUNX3 functions in control of immunity and inflammation, thereby indirectly influencing epithelial tumor development.

Keywords

Cancer • Immunity • Inflammation • RUNX3 • Tumor suppressor gene

J. Lotem • D. Levanon • V. Negreanu • O. Bauer
S. Hantisteanu • Y. Groner (✉)
Department of Molecular Genetics, Weizmann Institute of Science, Rehovot 76100, Israel
e-mail: yoram.groner@weizmann.ac.il

J. Dicken
Department of Molecular Genetics, Weizmann Institute of Science, Rehovot 76100, Israel

Compugen Ltd, Holon, Israel

23.1 Introduction

As alluded to in the Overview section by Ito and Speck, RUNX3 is one of the three mammalian Runt-domain transcription factors (TFs) (Levanon and Groner 2004). *Runx3* was originally cloned based on its similarity to *Runx1* (Levanon et al. 1994) and subsequently localized

Y. Groner et al. (eds.), *RUNX Proteins in Development and Cancer*, Advances in Experimental Medicine and Biology 962, DOI 10.1007/978-981-10-3233-2_23

on human and mouse chromosomes 1 and 4, respectively (Levanon et al. 1994; Avraham et al. 1995). *Runx3*$^{-/-}$ mice phenotypes reflect its expression pattern and vitality to the proper function of several important organs. These mice exhibit ataxia due to loss of TrkC proprioceptive neurons in the dorsal root ganglia (DRG) (Levanon et al. 2002), delayed chondrocyte maturation during embryogenesis (Yoshida et al. 2004), severe congenital osteopenia (Bauer et al. 2015), altered hair shape (Raveh et al. 2005), and a multitude of defects in the adaptive and innate immunity system including: defective proliferation and differentiation of activated cytotoxic CD8^{+} T cells (Cruz-Guilloty et al. 2009, Taniuchi et al. 2002; Woolf et al. 2003), helper Th1 cells (Djuretic et al. 2007) and natural killer (NK) cells (Levanon et al. 2014); defective development of intestinal innate lymphoid cells type 3 (ILC3) resulting in defective control of *C. Rodentium* infection (Ebihara et al. 2015); spontaneous development of colitis (Brenner et al. 2004), lung inflammation associated with accumulation of hyper-activated dendritic cells (DCs) (Fainaru et al. 2004) and lack of skin Langerhans cells (Fainaru et al. 2004) and dendritic epithelial T cells (Woolf et al. 2007). The important functions of Runx3 in controlling immunity and inflammation (see also Taniuchi et al. in this section) and the impact these phenomena have on cancer development (Hanahan and Weinberg 2011; Elinav et al. 2013), raises the possibility that Runx3 might be directly involved in development of hematopoietic malignancies or indirectly involved in epithelial malignancies in a cell-autonomous and non-cell-autonomous mechanisms, respectively.

More than a decade ago, the notion that *RUNX3* is a novel TSG and its inactivation plays a major role in gastric cancer (GC) development was put forward (Li et al. 2002). This contention was based on the following observations: *(i)* high expression of *Runx3* in normal mouse gastrointestinal tract (GIT) epithelium; *(ii)* gastric hyperplasia in *Runx3*$^{-/-}$ newborn mice; *(iii)* detection of a *RUNX3* point mutation in a single GC patient; *(iv)* *RUNX3* loss of heterozygosity (LOH) in 30 % of GC patients; *(v)* *RUNX3 P2* promoter DNA hypermethylation in GC; and *(vi)* reduced *RUNX3* mRNA in GC versus normal gastric epithelium in 60 % of GC patients assayed by RNA *in-situ* hybridization (RISH).

In view of the importance of TSGs in cancer development, hundreds of subsequent studies involving thousands of patients have attempted to verify and extend the suggested *RUNX3*-TSG paradigm in GC and other GIT cancers (Subramaniam et al. 2009). Many of these efforts (references included in Table S1 of Levanon et al. 2011) focused on the issue of *RUNX3 P2* promoter DNA hypermethylation in cancer, but have failed to confirm whether *RUNX3* is indeed expressed in normal GIT epithelial cells and to establish a quantitative relationship between *RUNX3 P2* hypermethylation and its expression.

For a gene to qualify as a cell-autonomous TSG in a given cell type, it has to be expressed in that cell type and its loss or inactivation should prove advantageous for viability or growth, thus promoting the malignancy. Compelling evidence from several laboratories (Levanon et al. 2001; Brenner et al. 2004; Li et al. 2004; Raveh et al. 2005; Carvalho et al. 2005; Friedrich et al. 2006; Levanon and Groner 2009; Levanon et al. 2011; Munoz et al. 2012; Lotem et al. 2013a; Stange et al. 2013; Bauer et al. 2014; Kim et al. 2014; McDonald et al. 2014; Treutlein et al. 2014; van Bragt et al. 2014; Kurklu et al. 2015; Na et al. 2015; Whittle et al. 2015; Llorca-Cardenosa et al. 2016) (Table 23.1) and the human proteome atlas portal (http:www.proteinatlas.org), demonstrating the lack of RUNX3 expression in normal GIT epithelium and other epithelial tissues, has challenged the possibility that *RUNX3* is a *bona fide* TSG whose inactivation leads cell autonomously to GC and other cancers. The large body of data that on one hand call into question the notion that *RUNX3* is a *bona fide* cell-autonomous TSG and on the other hand highlight its functions in immunity and inflammation is summarized in this chapter.

Table 23.1 Evidence for lack of Runx3 expression in various normal epithelia

Organ[a]	Cells	Detection method	References
Stomach (mouse)	Ep	LacZ, IHC	Brenner et al. (2004) and Levanon et al. (2001, 2011)
	Ep	RISH	Brenner et al. (2004)
	Ep	IHC	Kurklu et al. (2015) and Na et al. (2015)
	Chief-Troy$^+$ SC	Microarray[b]	Stange et al. (2013)
	Parietal	Microarray	Stange et al. (2013)
	Organoids from SC	Microarray	Stange et al. (2013)
Stomach (human)	Ep	IHC	Carvalho et al. (2005)
	Ep- microdissected	IHC, RT-PCR	Friedrich et al. (2006)
	Ep	IHC	Kurklu et al. (2015) and Llorca-Cardenosa et al. (2016)
Intestine (mouse)	Ep	IHC, LacZ	Brenner et al. (2004) and Levanon et al. (2001, 2011)
	Epcam$^+$	qRT-PCR	Levanon et al. (2011)
	Crypt-Lgr5$^+$ SC	Microarray	Munoz et al. (2012) and Kim et al. (2014)
	Crypt-Lgr5^{dim}	Microarray	Munoz et al. (2012)
	Crypt- progenitors	Microarray	Kim et al. (2014)
	Villi- enterocytes	Microarray	Kim et al. (2014)
Lung (mouse)	Ep	IHC	Levanon et al. (2001)
	Epcam$^+$	RNA-seq[b]	Treutlein et al. (2014)
Mammary gland (mouse)	Epcam$^+$	qRT-PCR	McDonald et al. (2014)
	Lin$^-$CD24$^+$CD29$^{low, high}$	qRT-PCR	Van Bragt et al. (2014)
Skin (mouse)	Keratinocytes	IHC	Raveh et al. (2005) and Bauer et al. (2014)
	Hair follicle matrix Ep	IHC	Levanon et al. (2001)
Pancreas	Ductal Ep	IHC	Li et al. (2004) and Whittle et al. (2015)

[a]Runx3 is also absent in esophagus, nasal olfactory and salivary gland ducts epithelia (Levanon et al. 2001). *Ep* epithelium, *IHC* immunohistochemistry, *SC* stem cells
[b]*Runx3* expression levels were obtained from the following microarray and RNA-seq data sets deposited in Gene Expression Omnibus (GEO) corresponding to the cited references: GSE51398 (Kim et al. 2014), GSE33948 (Munoz et al. 2012), GSE44060 (Stange et al. 2013) and GSE52583 (Treutlin et al. 2014). In all these data sets *Runx3* expression was below background level

23.2 *RUNX3* Involvement in Immune-Related Inflammatory Disorders and Its Implications for Cancer Development

As indicated above Runx3 is involved in many important immune system functions and its loss in *Runx3*$^{-/-}$ mice results in a variety of immunological defects including inflammation in the gut and lung. Several genetic studies in humans have implicated *RUNX3* in immune-related diseases in GIT as well as several other organs. Two studies associated *RUNX3* SNP rs2236851 with increased risk for ulcerative colitis (UC) (Weersma et al. 2008; Guo et al. 2010) and a genome wide association study (GWAS) of a Japanese population revealed a connection between certain *RUNX3*

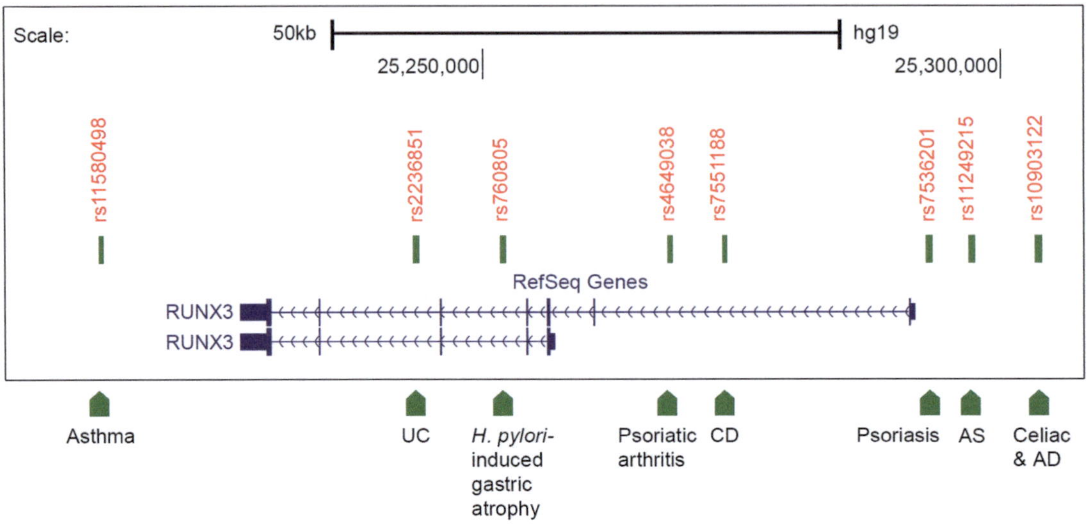

Fig. 23.1 *Location of RUNX3 polymorphic markers associated with human immune-related diseases.* UCSC Genome Browser map showing the positions of *RUNX3* SNPs associated with immune-related diseases in GIT and other tissues. *AD* atopic dermatitis, *AS* ankylosing spondylitis, *CD* Crohn's disease, *UC* ulcerative colitis. References are given in the text. The SNP rs7551188 in *RUNX3* intron 1 gave a strong association signal with CD in a Japanese population, but it did not reach the required GWAS significant level (Yamazaki et al. 2013)

SNPs and an increased risk for two other GIT diseases, celiac (Dubois et al. 2010) and Crohn's disease (CD) (Yamazaki et al. 2013), although the latter did not reach GWAS significance level (Fig. 23.1). GWAS also revealed an association of several SNPs located ~2-12 kb upstream of the *RUNX3 P1* promoter (rs4265380, rs4648889, rs7529070, rs11249215 and rs6600247) with ankylosing spondylitis (AS) (Evans et al. 2011; Cortes et al. 2013; Vecellio et al. 2016), rs7536201 with psoriasis (Tsoi et al. 2012) and rs10903122 with both celiac disease (Dubois et al. 2010) and atopic dermatitis (Esparza-Gordillo et al. 2013) (Fig. 23.1). Other relevant examples include the association of rs4649038 in *RUNX3* intron 1 with psoriatic arthritis (Apel et al. 2013); and rs11580498, located 13 kb downstream of *RUNX3*, with asthma in a Canadian population (Laprise 2014) (Fig. 23.1). A recent heterogeneous network edge prediction method based on an integration approach to prioritize disease-associated genes, predicted that RUNX3 is also associated with multiple sclerosis (Himmelstein and Baranzini 2015), another immune-related disease. It is interesting to note that several SNPs in other genes associated with immune-related diseases, including Lupus erythematosus, psoriasis and rheumatoid arthritis, disrupt or alter the canonical RUNX-binding sites within these susceptibility loci (Prokunina et al. 2002; Helms et al. 2003; Tokuhiro et al. 2003).

As indicated earlier, Runx3-deficient mice spontaneously develop early-onset colitis (Brenner et al. 2004), which can be transferred to immune-suppressed mice by fetal liver (FL) hematopoietic precursors from *Runx3*$^{-/-}$ embryos (Levanon et al. 2009; Sugai et al. 2011). It is thus possible that certain immune related diseases in GIT and other organs are associated with particular *RUNX3* SNPs, which might affect RUNX3 functions in leukocytes, such as CD8$^+$ T cells, NK, DCs and ILCs, all of which display distinct phenotypes when Runx3 is absent (Fainaru et al. 2004; Dicken et al. 2013; Lotem et al. 2013b; Levanon et al. 2014; Ebihara et al. 2015). This possibility is supported by a recent study showing that the AS-associated SNP rs4648889 located in the vicinity of the *RUNX3 P1* promoter resides in an enhancer region and the AA risk allele significantly reduces IRF4 recruitment and the level of *RUNX3* mRNA in CD8$^+$ T cells (Vecellio et al. 2016). Significantly, many of the genes reported to be associated with increased risk for inflammatory GIT diseases, including celiac (Trynka et al. 2010), Crohn's

Table 23.2 Runx3 target genes in CD8⁺ T, NK and/or DC that are associated with human immune-related diseases

Disease	Susceptibility genes that are Runx3-targets in CD8⁺ T, NK and/or DC[a]
Celiac	*Aldh2, Bach2, Ccr2, Ccr3, Ccr5, Ccrl2, Cd247, Cxcr6, Ctla4, Ets1, Fasl, Fli1, Icos, Il18rap, Irf4, Itga4, Park7, Plek, Ptpn2, Ptprk, Rgs1, Sh2b3, Tlr8, Tnfrsf9, Ube2e3, Xcr1, Zmiz1*
Crohn's	*Atg4b, Bach2, Cd19, Cd244, Cd27, Cpeb4, Crem, Fasl, Galc, Icoslg, Ifnar1, Ifngr2, Ikzf1, Ikzf3, Il18rap, Il2ra, Irf1, Irf4, Itln1, Jak2, Lat, Lta, Nfatc1, Plcl1, Prdm1, Ptpn2, Ptpn22, Rasgrp1, Ripk2, Sh2b1, Smad3, Spred1, Tnfsf8, Vamp3, Zfp36l1, Zmiz1*
UC	*Ahr, Calm3, Card11, Ccr1, Ccr2, Ccr3, Ccr5, Cd28, Ctla4, Dap, Exoc3, Fcgr2a, Gmppb, Gna12, Icos, Icosl, Ifng, Ikzf3, Il1r2, Il7r, Inpp5e, Itgal, Jak2, Nfkbiz, Pim3, Ormdl3, Prdm1, Ptger4, Serinc3, Smad3, Tnfrsf9, Tnfsf8, Tnpo3*
IBD	*Aldh2, Adcy3, Atxn2, Bre, Cd226, Crem, Crtc3, Dap, Dok3, Dusp1, Fcgr2a, Fyb, Galc, Gpr18, Gpr183, Icosl, Ifng, Ikzf1, Il18rap, Il2ra, Irf8, Jak2, Litaf, Loh12cr1, Lpxn, Nfil3, Ormdl3, Osm, Ptger4, Rorc, Rps6kb1, Sell, Sh2b3, Smad3, Socs1, Spred2, Spry4, Stat1, Tmem180, Tnfrsf18, Tnfrsf4, Tnfrsf9, Tnfsf8, Traf3ip2, Tspan14, Zfp36l1*
Psoriasis	*Ets1, Irf4, Rps6ka4, Ube2e3* and *Zmiz1*

[a]The lists of susceptibility genes for celiac, Crohn's, ulcerative colitis, IBD and psoriasis (Franke et al. 2010; Trynka et al. 2010; Anderson et al. 2011; Jostins et al. 2012; Tsoi et al. 2012; Liu et al. 2015) were intersected with the lists of Runx3-target genes in CD8⁺ T, NK and/or DCs (Dicken et al. 2013; Lotem et al. 2013b; Levanon et al. 2014). Common genes are indicated

(Franke et al. 2010; Liu et al. 2015), UC (Anderson et al. 2011; Liu et al. 2015) and inflammatory bowel disease (IBD) (Jostins et al. 2012; Liu et al. 2015), were identified as Runx3 targets in CD8⁺ T, NK and/or DCs (Dicken et al. 2013; Lotem et al. 2013a, b; Levanon et al. 2014) (Table 23.2). The Runx3 targets *Ets1, Ube2e3* and *Zmiz1* are also susceptibility genes for psoriasis (Tsoi et al. 2012) (Table 23.2). Together, these results implicate RUNX3 itself and many of its downstream target genes in immune and inflammatory cells as protectors against a range of immune-related diseases in various organs, including the GIT.

Several studies have underscored the role of various genes including TSGs in modulating inflammation in the tumor microenvironment and thereby affecting tumorigenesis. For example, deletion of the TSG *Smad4* in T cells, but not in epithelial cells, results in formation of epithelial GIT tumors, possibly mediated by enhanced expression of interleukins (IL) 5, 6 and 13 by Smad4-deficient T cells. Consequently, Stat3 is hyper-activated in the epithelium (Kim et al. 2006). Similarly, deletion of *Tgfbr2* in stromal fibroblasts increases expression of inflammatory mediators and cytokines and the development of squamous cell carcinoma in the forestomach (Bhowmick et al. 2004; Achyut et al. 2013). Deletion of *Pten* in mammary stromal fibroblasts results in massive remodeling of extracellular matrix, recruitment of macrophages and dramatically increases the incidence of Erbb2-driven mammary epithelial tumors (Trimboli et al. 2009). Deletion of *p53* in mesenchymal hepatic stellate cells promotes Ccl4-induced liver cirrhosis and accelerates carcinogen-induced liver epithelial tumorigenesis by secreting elevated levels of factors that stimulate polarization of macrophages into a tumor-promoting M2 state and their increased accumulation within the tumor microenvironment (Lujambio et al. 2013). Similarly, deletion of p53 in myeloid cells, enhances intestinal inflammation and promotes adenoma development in the mouse $APC^{min/+}$ model (He et al. 2015). Deletion of *Sqstm1/p62* in the whole organism, or specifically in stromal fibroblasts, causes IL-6 levels to rise, induces inflammation and increases tumorigenesis of prostate epithelial cancer (Valencia et al. 2014). Interestingly, *Sqstm1/p62* is highly-expressed in prostate cancer epithelial cells (Valencia et al. 2014) and is required for their proliferation *in vitro* and in xenografts via the mTORC1 pathway (Duran et al. 2011), indicating it is not an epithelium cell-autonomous TSG. Rather, the ensuing inflammation and IL-6 production following *Sqstm1/p62* deletion in stromal fibroblasts circumvents the requirement of this gene for cancer cell proliferation. Similar results were recently reported for a non-cell-autonomous TSG function of *Sqstm1/p62* whereby its loss in hepatic stellate cells enhances liver inflammation and fibrosis, thereby promoting liver cancer (Duran et al. 2016). It is thus tempting to speculate that loss of *Runx3* in immune cells, which results in colonic and lung inflammation in mice, and the association of *RUNX3* with several human inflammatory diseases might, under certain conditions, also indirectly promote the growth of epithelial cancer cells. Under these circumstances *RUNX3* acts as a non-cell autonomous TSG.

23.3 Loss of RUNX3 Affects Epithelial Tumors in a Non-cell Autonomous Manner

23.3.1 GIT Tumors

The GIT was the first organ in which *RUNX3* was argued to serve as a TSG (Li et al. 2002) although the absence of RUNX3 expression in normal GIT epithelium (Table 23.1) made it highly unlikely that *RUNX3* functions as a cell-autonomous TSG in this tissue. Nevertheless, it has been shown that loss of a TSG in stromal cells (Bhowmick et al. 2004; Katajisto et al. 2008; Lujambio et al. 2013) or T cells (Kim et al. 2006) can lead to epithelial tumorigenesis, raising the valid possibility that Runx3 activity in GIT leukocytes might protect GIT epithelium against tumorigenesis by a non-cell autonomous mechanism. Indeed, Runx3 is expressed in GIT leukocytes (Brenner et al. 2004; Levanon et al. 2011) including TCRγδCD8αα (E. Woolf, Ph.D. thesis, 2005) and TCRαβCD4⁺CD8αα (Reis et al. 2013, 2014) intraepithelial lymphocytes (IEL), whose development requires Runx3 (E. Woolf, Ph.D. thesis 2005, Reis et al. 2013, 2014). Significantly, proper function of IEL is important for protecting the epithelium against pathogens and inflammation (Cheroutre et al. 2011). Runx3 is also expressed in the 3 subtypes of intestinal innate lymphoid cells (ILC1–3) (Ebihara et al. 2015; Guri-BenAri et al. 2016) and is required for the development of ILC1 and ILC3 cells (Ebihara et al. 2015). Accordingly, loss of Runx3 in these cells results in impaired control of *C. Rodentium* infection (Ebihara et al. 2015). Moreover, ILC3 cells play an important role in maintaining intestinal integrity and their loss is associated with development of colitis (Zenewicz et al. 2008). As indicated above, absence of Runx3 in *Runx3*$^{-/-}$ mice is associated with early-onset colonic inflammation, epithelial hyperplasia, enlarged mesenteric lymph nodes and late-onset gastric hyperplasia (Brenner et al. 2004). Moreover, early-onset colitis also develops in *Runx3*$^{fl/fl}$/*Cd11c-Cre* mice, in which *Runx3* is specifically deleted in DCs (Levanon et al. 2009), and can be transferred to immune-suppressed mice by fetal liver (FL) hematopoietic precursors from *Runx3*$^{-/-}$ embryos (Levanon et al. 2009; Sugai et al. 2011). However, although inflammation can support tumor formation (Hanahan and Weinberg 2011; Elinav et al. 2013) and *H. pylori*-induced inflammation is a major contributing factor to GC (Nagini 2012), none of the *Runx3*$^{-/-}$ mice showed an increased incidence of GIT tumors or any other tumor (Brenner et al. 2004; Ito et al. 2008). Of note, transferred *Runx3*$^{-/-}$ FL cells induced colitis but not tumors in irradiated RAG2$^{-/-}$ mice housed under pathogen-free conditions but when housed under conventional conditions, these mice did develop colon and cecum tumors (Sugai et al. 2011). These results suggest that loss of Runx3 function in leukocytes can, under certain conditions, promote the development of colonic tumors derived from Runx3 non-expressing epithelium, presumably in a colitis-dependent mechanism influenced by the colonic flora. This scenario resembles the development of spontaneous colitis and outgrowth of colonic tumors in some of the mice in which the TF *Stat3* is specifically deleted in *Csf1r*-expressing myeloid cells (Deng et al. 2010). In contrast, conditional deletion of *Stat3* in the intestinal epithelium using *Villin-Cre*, actually inhibited growth of colonic tumors in a mouse inflammation-associated colon cancer model (Waldner et al. 2010). These results are clearly consistent with the idea that *Runx3* functions as a non-cell-autonomous TSG in GIT epithelium.

Using a carcinogen-induced GC model in Balb/c mice, it was reported that while ~10 % of wild-type (WT) and *Runx3*$^{+/-}$ mice treated with the carcinogen N-methyl-N-nitrosourea (MNU) developed gastric tumors, a prominent 70 % of MNU-treated *Runx3*$^{-/-}$ mice developed gastric tumors (Ito et al. 2011). As indicated earlier, normal gastric epithelial cells do not express Runx3 (Brenner et al. 2004; Stange et al. 2013; Kurklu et al. 2015) and it has been reported that MNU-induced gastric tumors in mice are associated with chronic infiltration of inflammatory cells and that inflammation promotes MNU-induced gastric tumors (Leung et al. 2008). Moreover, it has been shown that gastric cancer development in transgenic for *Wnt1, Ptgs2 and Ptges* (*Gan* mice) is dependent on TNFα and that tumor formation could be rescued in TNFα$^{-/-}$ mice by transfer of bone marrow-derived DCs from TNFα$^{+/+}$ mice, which infiltrated the tumors (Oshima et al. 2014). It has also been shown that macrophage-derived TNF is an important component of Wnt/β-catenin activation in gastric epithelium following *H. pylori* infection (Oguma et al. 2008). Taken together with the fact that *H. pylori*-induced inflammation is a major contributing factor to GC (Nagini 2012), the data is consistent with the notion that the higher incidence of MNU-induced gastric tumors in *Runx3*$^{-/-}$ Balb/c is not due to loss of a cell-autonomous Runx3 function in gastric epithelium but is rather due to a non-cell autonomous mechanism mediated by loss of leukocytic Runx3 function. It is well established that GIT epithelium-specific deletion of the TSG *Klf4*, using *Villin-Cre* transgenic mice, induces gastric tumors and accelerates MNU-induced gastric tumor development (Li et al. 2012a). Similarly, *Villin-Cre*-mediated deletion of both *Fbw7* and *Tp53*, but not each of these TSGs alone, induced adenocarcinoma in the small intestine, cecum and colon (Grim et al. 2012). In addition, targeting the TSG *APC* in epithelial stem cells using *Lgr5-Cre* or *Krt19-Cre* also resulted in development of colonic tumors (Asfaha et al. 2015). It would thus be very interesting to directly determine whether conditional deletion of *Runx3* specifically in GIT epithelium, or in inflammatory cells (using *MX1-Cre, Lck-Cre, CsflR-Cre, CD11c-Cre or Cx3cr1-Cre*) recapitulates the high incidence of MNU-induced gastric tumors reported to occur in *Runx3*$^{-/-}$ Balb/c mice and whether it also results in development of intestinal adenocarcinoma.

It is important to note that although *Runx3*$^{-/-}$ Balb/c mice survived for more than a year without developing GIT tumors, peculiarly ~50 % of mono-allelic inactivation of *Runx3* (i.e. *Runx3*$^{+/-}$) mice were reported to develop late-onset (15 months) small adenomas in the small intestine (Ito et al. 2008). In contrast, while ~20 % of homozygote *Runx3*$^{-/-}$ ICR mice developed inflammation of the small intestine, no intestinal adenomas were detected in *Runx3*$^{+/-}$ or *Runx3*$^{-/-}$ ICR mice even in 2-year old animals (Brenner et al. 2004). The reported formation of intestinal adenomas in *Runx3*$^{+/-}$ but not in *Runx3*$^{-/-}$ Balb/c mice is even more puzzling, because none of the *Runx3*$^{-/-}$ phenotypic features, including ataxia due to loss of TrkC neurons in DRG (Levanon et al. 2002), colitis (Brenner et al. 2004), asthma-like lung inflammation (Fainaru et al. 2004) and defective silencing of CD4 expression in CD8$^+$ T cells (Taniuchi et al. 2002; Woolf et al. 2003), is observed in *Runx3*$^{+/-}$ mice. These findings indicate that a sufficient amount of Runx3 is present in *Runx3*$^{+/-}$ mice to maintain homeostasis. It is thus unlikely that *RUNX3* is a *bona fide* TSG, since loss of only one TSG allele is generally not enough to induce tumors, and even in cases where tumor development does occur following the loss of one TSG allele (haploinsufficient TSG), the manifestation of the disease is less severe than when both alleles are lost (Berger et al. 2011).

23.3.2 Mammary Gland Tumors

Runx3 was suggested to function as a novel TSG also in the mammary gland by targeting estrogen receptors (ER) for degradation (Huang et al. 2012), implying that the presumed reduced level of Runx3 in *Runx3*$^{+/-}$ Balb/c mice is responsible for an increased expression of ER and the spontaneous development of breast ductal carcinoma in ~20 % of 15-months old female mice. Strangely enough, no data was provided for the incidence of mammary tumors in aged *Runx3*$^{-/-}$ mice (Huang et al. 2012). The fact that *Runx3* is not expressed in normal mammary gland epithelium (McDonald et al. 2014; van Bragt et al. 2014), makes it highly unlikely that the mammary tumors reported to develop in some of the aged *Runx3*$^{+/-}$ Balb/c mice reflect a cell-autonomous loss of Runx3 in the mammary epithelium itself. Nonetheless, unequivocally demonstrating that these *Runx3*$^{+/-}$ Balb/c mice mammary tumors are in fact the product of a cell-autonomous reduced expression of Runx3 in mammary epithelium requires that such tumors also develop following deletion of *Runx3* specifically in that tissue. It will therefore be interesting to determine whether conditional deletion of *Runx3* in mammary gland luminal (using *Wap-Cre* or *K8-Cre*) or basal (*K5-Cre*) epithelial cells or in immune cells will give rise to mammary adenocarcinoma in aged Balb/c mice. Notwithstanding, it should be stressed that the lack of mammary tumors or any other tumor in both *Runx3*$^{+/-}$ and *Runx3*$^{-/-}$ ICR mice, makes a Runx3-TSG scenario highly unlikely (Brenner et al. 2004).

23.3.3 Lung Tumors

Runx3 was also suggested to function as a TSG in the lung. *Runx3*$^{-/-}$ C57Bl/6 mouse embryos were reported to exhibit impaired lung development with alveolar epithelial hyperplasia, that newborn mice die within 24 h due to breathing abnormalities, and that 85% of 18-months old *Runx3*$^{+/-}$ mice develop lung adenomas, compared to only ~5% in WT mice (Lee et al. 2010). Treatment with urethane, a carcinogen that induces inflammation-dependent lung tumors, induced lung adenomas within 3 months in 85 % of *Runx3*$^{+/-}$ mice, whereas this outcome occurred only in ~15% of WT mice (Lee et al. 2010). In contrast, no lung adenomas were recorded in aged *Runx3*$^{+/-}$ or *Runx3*$^{-/-}$ ICR mice. In a more recent study, it was reported that post-natal deletion of *Runx3* in lung tissue of *Runx3*$^{fl/fl}$ mice by intranasal administration of Adeno-Cre (Ad-Cre) induced lung adenomas within 4 months and accelerated adenocarcinoma development driven by oncogenic K-Ras activity in *K-Ras*$^{LSL-G12D/+}$/*Runx3*$^{fl/fl}$ mice (Lee et al. 2013). However, because genuine Runx3 is not actually expressed in normal mouse lung alveolar or bronchiolar epithelium and in isolated lung Epcam^{+} cells (Levanon et al. 2011; Treutlein et al. 2014), an implied cell-autonomous loss of Runx3 in lung epithelium leading to the reported development of lung adenomas is an unlikely scenario (also see Whittle and Hingorani chapter in this section). Even so, to unequivocally affirm that deletion of *Runx3* specifically in lung bronchiolar or alveolar epithelium can induce adenomas and accelerate oncogenic K-Ras-induced adenocarcinoma would require conditional targeting of *Runx3* specifically in lung epithelium using *CC10-Cre* or *SPC-Cre* deleting strains, respectively. Alternatively, the pLV-CA2-Cre lentivirus system, which deletes specifically in the lung epithelium, could be used. This approach is suggested because it has been shown that activating oncogenic K-Ras in lung epithelium using either of these deleting strains effectively induces lung adenocarcinomas (Ji et al. 2006; Lin et al. 2012; Yeddula et al. 2015). Of note, a lentivirus-based shRNA screen for TSGs whose knockdown cooperates with oncogenic K-Ras in lung epithelium revealed several such TSGs but Runx3 was not among them (Yeddula et al. 2015).

Unlike its absence in normal lung epithelium, Runx3 is expressed in lung macrophages/DCs (Fainaru et al. 2004), raising the possibility that intranasal Ad-Cre-mediated deletion of *Runx3* in these leukocytes within lung airways and alveoli of *Runx3*$^{fl/fl}$ mice may induce lung inflammation, thereby leading indirectly to epithelial hyperplasia and, possibly also to adenoma. This putative scenario is compatible with the reported rapid accumulation of myeloid cells in the lungs following intranasal Ad-Cre (Wilderman et al. 2006), the finding that 40 % of lung cells with activated K-Ras in *K-Ras*$^{LSL-G12D/+}$ mice following intranasal Ad-Cre are CD45^{+} leukocytes (Choi et al. 2015) including lung macrophages (Lauber 2012) and accumulation of hyperactivated DCs and other inflammatory cells in the lungs of *Runx3*$^{-/-}$ (Fainaru et al. 2004) and lung-activated *K-Ras* mice (Ji et al. 2006; Lauber 2012). It is also in line with the formation of lung adenomas in ~60 % of *K-Ras*$^{LSL-G12D/+}$/*Mx1-Cre* mice, in which oncogenic K-Ras was induced specifically in hematopoietic cells (Chan et al. 2004). Moreover, development of lung tumors in *K-Ras*$^{LSL-G12D/+}$/

Ad-Cre mice depends heavily on macrophage influx into the lung (Cortez-Retamozo et al. 2012). These and other myeloid cells express and secrete increased levels of inflammatory cytokines including IL-1, IL-6, TNFα and insulin-like growth factor 1, which activate NF-κB and Stat3 in lung epithelial cells, leading to their enhanced proliferation (Takahashi et al. 2010; Fritz et al. 2011; Elinav et al. 2013; Houghton 2013). For example, deletion of *Ikkβ*, which is required for NF-κB activation, in myeloid cells (*LysM-Cre/Ikkβ$^{fl/fl}$*), but not in bronchiolar epithelial Clara cells (*CC10-Cre/Ikkb$^{fl/fl}$*), impairs myeloid cell accumulation, cytokine secretion in lungs of mice exposed to tobacco smoke and the proliferation of lung adenoma cells in chemically-induced lung tumors (Takahashi et al. 2010). Moreover, transplantation of bone marrow from *LysM-Cre/Ikkβ$^{fl/fl}$* mice into *K-Ras$^{LSL\text{-}G12D/+}$/Ad-Cre* mice or into urethane-treated mice, enhanced lung infiltration of IL-1β-producing neutrophils and lung tumor development as compared to transplantation of bone marrow from WT mice (McLeod et al. 2016). Together, these results strongly implicate Runx3 function in lung myeloid cells as an important regulator of lung inflammation, so its loss may contribute indirectly to lung adenoma development. It will be of great interest to determine whether deletion of *Runx3* specifically in myeloid cells using *Cd11c-Cre/Runx3$^{fl/fl}$* and/or *LysM-Cre/Runx3$^{fl/fl}$*, recapitulates the development of lung adenomas and the accelerated formation of K-Ras-induced lung adenocarcinomas, observed in *Runx3$^{fl/fl}$* and *K-Ras$^{LSL\text{-}G12D/+}$/Runx3$^{fl/fl}$* mice, respectively, following intranasal administration of *Ad-Cre*.

23.3.4 Skin Tumors

The two-stage skin carcinogenesis model in mice involving initiation/promotion by DMBA/TPA revealed that skin tumor formation is an inflammation-dependent process (reviewed in Rundhaug and Fischer 2010). It was shown that expression of Cxcr3 in T cells (Winkler et al. 2011) and the receptor for advanced glycation end-products RAGE (Gebhardt et al. 2008) in immune cells, but not in keratinocytes, is required for sustaining TPA-induced infiltration of inflammatory cells, epidermal hyperplasia and tumor promotion. Analysis of Runx3 role in this model has shown that while 100 % of DMBA/TPA treated WT ICR mice developed multiple skin papilloma tumors, the degree of inflammation-associated epithelial hyperplasia and frequency of papilloma-bearing mice were severely reduced in *Runx3$^{-/-}$* mice (Bauer et al. 2014). *Runx3$^{-/-}$* mice were also highly resistant to TPA-induced tumor promotion in skin already pre-initiated by an oncogenic Ha-Ras (Bauer et al. 2014). The reduced susceptibility of Runx3-deficient mice to skin carcinogenesis was associated with a marked reduction in TPA-recruited skin CD11b^{+} DCs and γδT cells and with an altered cytokine milieu in their skin, which together generated an anti-tumor environment (Bauer et al. 2014). Moreover, mice in which *Runx3* was specifically deleted in both DCs and T cells (using *Runx3$^{fl/fl::CD11c\text{-}Cre::Lck\text{-}Cre}$* mice), but not in keratinocytes (using *Runx3$^{fl/fl::K14\text{-}Cre}$* mice), fully simulated the resistance of *Runx3$^{-/-}$* mice to skin carcinogenesis (Bauer et al. 2014). These results directly show that *Runx3$^{-/-}$* mice resistance to inflammation-dependent skin carcinogenesis is due to loss of Runx3 activity in immune cells and not in keratinocytes (Fig. 23.2).

Unlike *Runx3*, deletion of *Runx1* specifically in epithelial cells, using *K19-Cre*, did inhibit DMBA/TPA-induced skin tumor formation (Scheitz et al. 2012). This finding indicates that although both Runx1 and Runx3 are required for skin tumor development in this model, they are required in different cell types, Runx1 in epithelial cells and Runx3 in immune cells. The two Runx members also differ in the carcinogenesis stage at which they are required, with Runx1 function in epithelium required at the initiation phase (Scheitz et al. 2012), while Runx3 function in immune cells is required at the promotion phase (Bauer et al. 2014). Hence, Runx3 function in immune cells is strongly implicated in TPA-induced skin tumor promotion, which is highly inconsistent with *Runx3* being a cell-autonomous TSG in epithelium. Moreover, the observation that RUNX3 is overexpressed in the nuclei of human skin basal cell carcinoma cells (Salto-

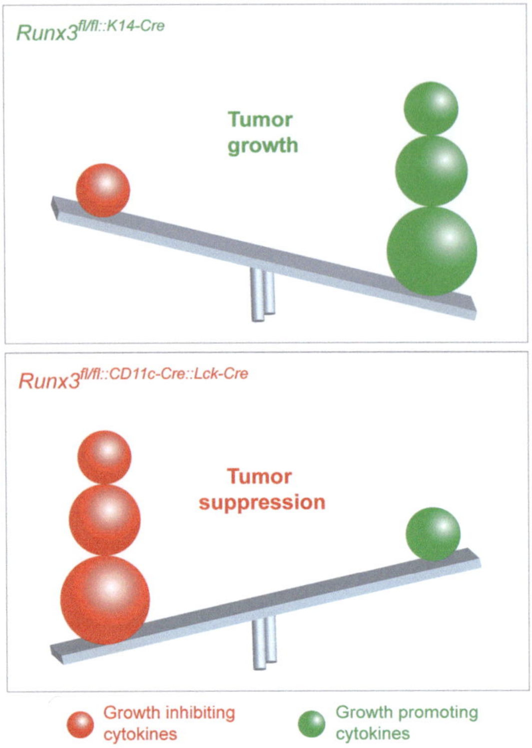

Fig. 23.2 *Leukocytic Runx3 regulates the balance of cytokines and tumor promotion in the TPA-treated inflamed skin microenvironment.* Deletion of *Runx3* in both dendritic and T cells (*bottom*) but not in keratinocytes (*top*) inhibits TPA-mediated skin inflammation and tumor promotion (Bauer et al. 2014)

Tellez et al. 2006) suggests that its presence in epithelial cells does not prevent carcinoma development. In fact, it raises the possibility that RUNX3 overexpression in epithelium rather than its absence may promote human skin cancer by a cell-autonomous mechanism.

23.4 Runx3 Promotes Epithelial Pancreatic Tumor Metastasis in a Cell-Autonomous Manner

Analysis of the molecular mechanism that contributes to development of pancreatic adenocarcinoma tumors in mice transgenic for activated K-Ras and mutant p53 revealed strong Runx3 nuclear expression in the tumor cells, but no expression in normal pancreatic ductal epithelium (Whittle et al. 2015). Moreover, overexpression of Runx3 increased the migration and anchorage-independent growth in agar and enhanced the metastatic potential of these tumors, whereas silencing Runx3 resulted in the opposite effects (Whittle et al. 2015) (see also Whittle and Hingorani in this section). Similarly, overexpression of RUNX3 in the human MiaPaCa-2 prostate cancer cell line strongly enhanced their migration, anchorage-independent growth and metastatic potential in immune-deficient mice while RUNX3 knockdown in two other cell lines inhibited these properties. Furthermore, high RUNX3 expression in the primary tumors of human pancreatic adenocarcinoma patients correlated with poorer survival following primary tumor resection (Whittle et al. 2015). Taken together, these results provide direct evidence that cancer cells can tolerate high levels of RUNX3 and aberrant induction of RUNX3 expression in pancreatic epithelial cells by activated oncogenes promotes, cell-autonomously, their ability to migrate away from the primary tumor and to proliferate at distant locations as metastases. These results again speak against a TSG role for RUNX3. It will be interesting to determine whether the reported overexpression of RUNX3 in a significant fraction of tumor cells in human gastric, head and neck and ovarian cancers (Carvalho et al. 2005; Kudo et al. 2011; Lee et al. 2011; Na et al. 2015; Llorca-Cardenosa et al. 2016) also plays a promoting role in their tumorigenic and metastatic potential. In addition, because chronic pancreatitis plays an essential role in K-Ras-induced pancreatic adenocarcinoma (Guerra et al. 2007) it will be interesting to determine the effect of conditional deletion of *Runx3* in immune cells on development of pancreatic adenocarcinoma.

23.5 Cancer Mutations and GWAS Do Not Support Involvement of *RUNX3* in GC or Other Cancers

Recurrent loss-of-function gene mutations in cancer are one of the characteristics of TSGs. The discovery of a *RUNX3* point mutation in tumor cells of a single GC patient was taken as evidence that *RUNX3* is a TSG (Li et al. 2002). In a later study, 2 out of 124 bladder cancer patients were shown to carry monoallelic *RUNX3* mutations, one of which had 4 different mutations on the same allele (Kim et al. 2005). However, while non-synonymous mutations in more than 600 genes, including the TSG *TP53* and several genes encoding chromatin-remodeling proteins were detected in a whole-exome sequencing (WES) study of 15 GC samples, no *RUNX3* mutations were detected in this study (Zang et al. 2012). Interestingly, one of these mutated chromatin-remodeling genes, *ARID1A*, resides at the same 1p36.11 chromosomal region as *RUNX3* (Fig. 23.3). Moreover, recent WES and whole genome sequencing studies on 100 and 295 GC samples (TCGR network 2014; Wang et al. 2014) and of thousands of samples from other cancers (Jones et al. 2010; Wiegand et al. 2010; Liang et al. 2012; Stephens et al. 2012; Dulak et al. 2013; Kandoth et al. 2013; Vogelstein et al. 2013; Wu and Roberts 2013; Lawrence et al. 2014) did not find *RUNX3* to be a significantly mutated cancer gene, whereas *ARID1A* was, thus solidifying *ARID1A* role as a *bona fide* TSG.

Cancer-prone human pedigrees often harbor germ-line mutations in the same genes that are frequently mutated in sporadic cancers. Approximately 2–10 % of gastric, breast and colorectal cancer cases have a familial predisposition, some of of which can be attributed to germ-line mutations in different TSGs (Jasperson et al. 2011; Chun and Ford 2012; Melchor and Benitez 2013) (Table 23.3). However, as shown in Table 23.3, not a single germ-line *RUNX3* mutation was detected in GC-prone pedigrees (Keller et al. 2004) and *RUNX3* was also not

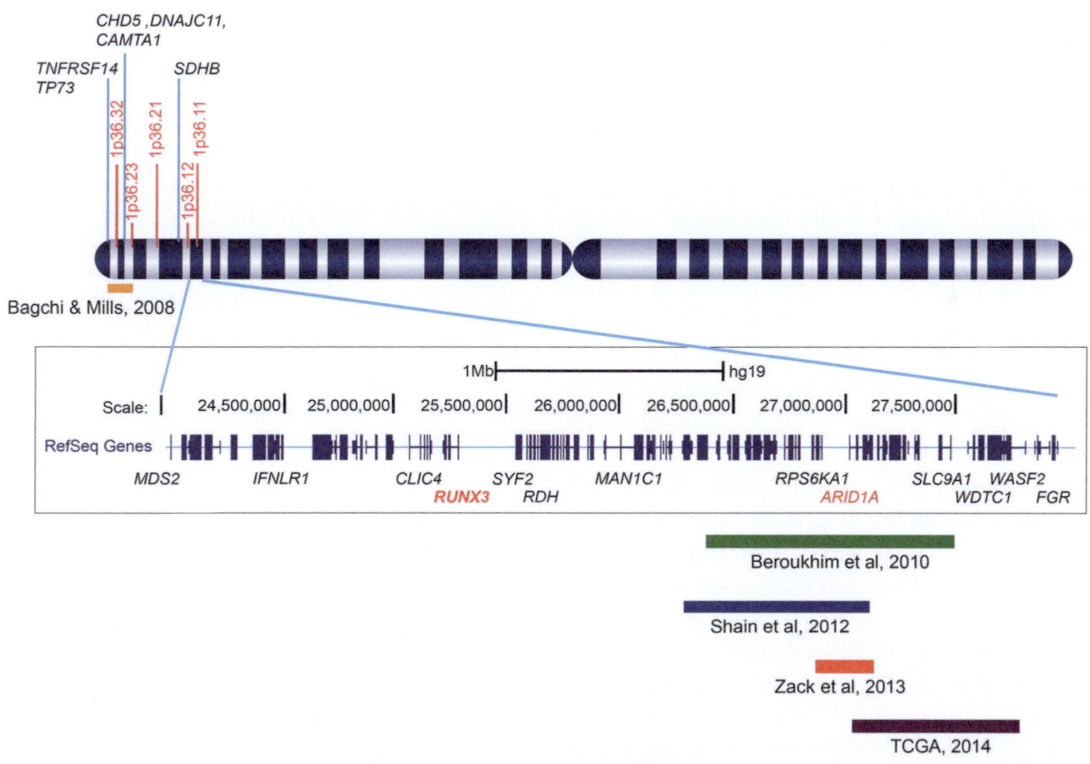

Fig. 23.3 *Focal recurrently deleted regions in cancer on human chromosome 1p36 region.* Schematic representation of human chromosome 1 and the UCSC Genome Browser blown-up map of the 4.1 Mb-long 1p36.11 region indicating some of its 86 RefSeq genes. *RUNX3* and *ARID1A* are marked in red. The colored bars below the genomic map mark the focal recurrently deleted 1.15 Mb, 850 kb, 250 kb and 746 kb 1p36.11 segments in cancers. These deletions include *ARID1A* but exclude *RUNX3* (Beroukhim et al. 2010; Shain et al. 2012; Zack et al. 2013; TCGR network 2014). The orange bar at the far left below the chromosome scheme marks the 5.7 Mb 1p36.33-1p36.23 segment recurrently deleted in multiple cancers (Bagchi and Mills 2008)

Table 23.3 Germ-line mutations predisposing to cancer

Cancer	Genes mutated in familial cancer[a]	References
GC	*APC, BRCA1, BRCA2, CDH1, SMAD4, STK11, TP53*	Chun and Ford (2012)
BC	*ATM, BRCA1, BRCA2, CDH1, PTEN, RAD51B RAD51C, RAD51D, STK11, TP53 and others*	Melchor and Benitez (2013)
CRC	*APC, MLH1, MSH2, MSH6, MYH, PMS2, PTEN, SMAD4, STK11*	Jasperson et al. (2011)

[a]No *RUNX3* germ-line mutations were detected in familial GC (Keller et al. 2004) and *RUNX3* is not a frequent cancer mutated gene in sporadic GC (Zang et al. 2012) or other cancers (Vogelstein et al. 2013)

found to be a cancer-predisposition gene across different cancer types (Vogelstein et al. 2013).

Analysis of the frequency of single nucleotide polymorphisms (SNPs) in control versus disease populations can identify genomic susceptibility loci associated with increased risk for various diseases. One such study with 10 *RUNX3* SNPs and ~300 GC and control Chinese patients reported that three *RUNX3* SNPs located in introns 1, 3 and 4, were associated with an increased risk of GC (Wu et al. 2009). Another study with two *RUNX3* SNPs located in exon 1 and intron 3 in a larger cohort of 583 GC and 1637 control patients, reported that the intron 3 SNP was associated with *H. pylori*-induced gastric atrophy but neither SNP was associated with late/tumorigenic stages of GC (Hishida et al. 2009). Interestingly, while the same intron 3 SNP rs760805 was analyzed in both studies (Hishida et al. 2009; Wu et al. 2009), it was reported to be associated with increased risk for GC only in the smaller population size study (Wu et al. 2009). Importantly, none of these *RUNX3* SNPs nor any other SNP in the whole 1p36.11 chromosomal region for that matter, was found to be a GC susceptibility locus in three GWAS on Chinese and Japanese populations, which did identify several GC susceptibility loci on other chromosomes (Sakamoto et al. 2008; Abnet et al. 2010; Shi et al. 2011) (Table 23.4). GWAS also revealed many genomic loci associated with increased risk for breast (Easton et al. 2007; Turnbull et al. 2010;

Table 23.4 Association of genomic loci with increased cancer risk

Type of study	Cancer associated loci	References
Individual SNPs GC	1p36.11 (3 SNPs, *RUNX3* introns 1, 3, 4) [a]	Wu et al. (2009)
GWAS GC	1q22, 3q13.31, 5p13.1, 8q24.3, 10q23	Sakamoto et al. (2008), Abnet et al. (2010), and Shi et al. (2011)
BC	1p11.2, 2p24.1, 2q33, 2q35, 3p24, 4q31.22, 5p12, 5q11.2, 5p15.2, 6q25.1, 8q24.21, 8q24.1, 10q26.13, 11p15.5, 12p11, 12q24, 14q24.1, 16q12.1, 16q23.2, 17q23.2, 19q13.41, 21q21	Easton et al. (2007), Turnbull et al. (2010), Fanale et al. (2012), Ghoussaini et al. (2012), and Sapkota et al. (2013)
CRC	2p22.1, 5p15.31, 5p15.33, 7q35, 8q23.3–24.11, 8q24.21, 10p14, 10q22.3, 10q25.2, 11q12.2, 11q23.1, 12p13.31, 12q24.21, 14q22.2, 15q13.3, 16q22.1, 17p13.3, 18q21.1, 19q13.1, 19q13.2, 20p12.3, 20q13.33	Tenesa and Dunlop (2009), Peters et al. (2012), and Zhang et al. (2014)
CC	4q12, 6p21.32, 17q12	Shi et al. (2013)
LC	3q28, 5p15.33, 6p21.32, 6q22.2, 10q25.2, 17q24.3	Wang et al. (2008), Landi et al. (2009), and Lan et al. (2012)
OC	3q25, 8q21, 10p12, 17q12, 17q21	Pharoah et al. (2013)
PC	1q32.1, 2p13.3, 3q29, 5p15.33, 7p13, 7q32.3, 9q34.2, 13q12.2, 13q22.1, 16q23.1, 17q25.1, 22q12.1	Petersen et al. (2010), Wolpin et al. (2014), and Childs et al. (2015)
PrC	1q21.3, 1q32.1, 2p25.1, 2q37.3, 3q13.2, 4q13.3, 5q35.2, 6p21.32, 6p21, 6q25.2, 7p15.3, 8p21.2, 10q24.32, 11q22.2, 12q24.21, 14q22.1, 14q24.1, 17p13.3, 17q21.32, 18q23, 19q13, 20p13, 20q13.33, 22q13, Xp22.2	Al Olama et al. (2013) and Eeles et al. (2013)

[a]The *RUNX3* intron-3 SNP, rs760805, was used in 2 studies (Hishida et al. 2009; Wu et al. 2009) but association with GC was found only in the smaller study (Hishida et al. 2009). *RUNX3* was not identified as a cancer predisposition gene across different cancer types (Vogelstein et al. 2013)
BC breast cancer, *CRC* colorectal cancer, *CC* cervical cancer, *LC* lung cancer, *OC* ovarian cancer, *PC* pancreatic cancer, *PrC* prostate cancer

Fanale et al. 2012; Ghoussaini et al. 2012; Sapkota et al. 2013), colorectal (Tenesa and Dunlop 2009; Peters et al. 2012; Zhang et al. 2014), cervical (Shi et al. 2013), lung (Wang et al. 2008; Landi et al. 2009; Lan et al. 2012), ovarian (Pharoah et al. 2013), pancreatic (Petersen et al. 2010; Wolpin et al. 2014; Childs et al. 2015) and prostate (Al Olama et al. 2013; Eeles et al. 2013) cancers, but no such association with increased cancer risk was detected on chromosome 1p36 (Table 23.4).

Taken together, these results indicate that although some *RUNX3* polymorphic markers may be associated with *H. pylori*-induced gastric changes, possibly due to an altered RUNX3 function in the infiltrating inflammatory immune cells, the absence of recurrent *RUNX3* mutations in cancer and the lack of association of *RUNX3* with increased cancer risk do not support the *RUNX3*-TSG paradigm.

23.6 Focal Genomic Copy-Number Alterations in Cancer do not Involve the *RUNX3* Locus

The presence of large genomic copy-number amplifications and deletions, often of an entire chromosomal arm, is one of the hallmarks of cancer. Such extensive genomic alterations, involving hundreds of genes, make it extremely difficult to identify the key genes within these regions that drive and/or contribute to cancer development. Using fluorescent *in-situ* hybridization (FISH), it was initially reported (Li et al. 2002) that ~30 % of GC samples had a hemizygous *RUNX3* deletion, based on a lower than 1:1 ratio between *RUNX3* and chromosome 1 centromere-specific FISH signals. However, the finding that in all GC samples there were at least two *RUNX3*-specific FISH signals per cell (Carvalho et al. 2005) indicates that there is no loss of *RUNX3* in GC but rather amplification of regions encompassing the chromosome 1 centromere (Carvalho et al. 2005). Furthermore, two recent high-resolution comparative genomic hybridization (CGH) studies of GC biopsies from 193 and 64 patients, respectively, revealed that the chromosome 1p36 region, which harbors *RUNX3*, was not frequently deleted in GC (Deng et al. 2012; Fan et al. 2012). An earlier study on 40 GC patients, likewise did not find chromosome 1p36 region to be a frequently deleted region in GC (Tsukamoto et al. 2008). The genomic regions that were recurrently deleted in GC harbored various known TSGs, while the frequently amplified ones housed known oncogenes (Deng et al. 2012) (Table 23.5). High-resolution analysis of the landscape of somatic copy number alterations across more than 3000 cancer samples representing 26 cancer types including GC (Beroukhim et al. 2010), revealed many focal alterations with an average of ~40 focal amplifications and ~35 focal deletions per GC sample. Interestingly, while focal deletions spanning *RUNX3* were not detected, a 1.15 Mb region on 1p36.11 containing 24 genes, including *ARID1A*, was recurrently deleted in multiple cancer types (Beroukhim et al. 2010) (Fig. 23.3). Thus, the lack of convincing evidence for fre-

Table 23.5 GC associated changes in gene copy number

Type of study	Genes	Chr. regions	References
CGH Amplified	*BLK, CDK6, CCND1, CCNE1,EGFR, ERBB2, FGFR2, FGF4, FGF19, GATA6, KLF5, KRAS, MET, MYC and others*	1q, 3q, 5p, 6p, 7pq, 8q, 12pq, 13q, 18pq, 19p, 20p, 21p	Deng et al. (2012) and Fan et al. (2012)
CGH Deleted[a]	*CDKN2A/B, CSMD1, FHIT, GMDS, PARK2, PDE4D, PTPRD, RB1, SMAD4, SMAD7, WWOX and others*	3p, 4pq, 5q, 6q, 8p, 9q, 11q, 14q, 16q, 17p, 18p, 18q, 19p, 21q, 22q	Tsukamoto et al. (2008), Deng et al. (2012), and Fan et al. (2012)

[a]Chromosomal region 1p36.1, in which *RUNX3* resides, is not deleted frequently in GC (Tsukamoto et al. 2008; Deng et al. 2012; Fan et al. 2012)

quent 1p36.11 focal genomic deletions encompassing *RUNX3* in cancer further puts into question the concept of *RUNX3* serving as a TSG.

23.7 *RUNX3 P2* Promoter Hypermethylation Is Not a Driver of GC or Other Cancers

GC is associated with DNA hypermethylation of many gene promoters (Zhao and Bu 2012), especially in EBV-positive GC (TCGR network 2014). The finding that the CpG-island adjacent to *RUNX3 P2* promoter is hypermethylated in GC, was interpreted as evidence for *RUNX3* silencing and presented as yet another *RUNX3* inactivation mechanism supporting a TSG role for *RUNX3* in GC (Li et al. 2002). In many subsequent studies on *RUNX3* hypermethylation in GC, summarized previously by Subramaniam (Subramaniam et al. 2009) and in a more recent study on 123 GC and 111 healthy patients (Hu et al. 2011), ~55% of GC samples were reported as having *RUNX3 P2* promoter hypermethylation, measured by the bisulfite-modified methylation-specific PCR (MSP) method. Yet, MSP is not a quantitative technique, and a subsequent analysis of GC samples using a quantitative method, i.e. a methyl-specific DNA microarray, revealed that merely 23% of GC samples showed *RUNX3 P2* hypermethylation as compared to 46% with the conventional MSP technique (Tamura et al. 2009). Moreover, several more recent studies employing high-density DNA methylation microarrays of gastric, colorectal and breast cancer biopsies all gave the *RUNX3 P2* methylation ratio a low ranking among the DNA hypermethylated genes (Van der Auwera et al. 2010; Dedeurwaerder et al. 2011; Kibriya et al. 2011; Park et al. 2011; Zouridis et al. 2012; Naumov et al. 2013). Taken together, it can be concluded that the role ascribed to *RUNX3 P2* hypermethylation in cancer has been greatly overrated.

Even though promoter DNA methylation prevents gene expression, finding such hypermethylation in tumor versus healthy control samples is not synonymous with its gene silencing causative role in the tumor cells. In fact, most hypermethylated gene promoters in cancer are already silent in the normal tissue of origin of these cancers (Keshet et al. 2006; Gal-Yam et al. 2008), being preferential targets for the polycomb complex that confers histone H3 lysine 27 methylation induced gene silencing (reviewed in Klutstein et al. 2016). Therefore, a parallel analysis of gene expression must be carried out alongside the DNA methylation assay in each tumor and control sample in order to establish the extent to which promoter methylation of a gene is causatively associated with its silencing in the tumor. However, in most of the studies summarized in (Subramaniam et al. 2009), cancer samples were analyzed either only for *RUNX3 P2* methylation by MSP or only for RUNX3 by IHC staining, precluding the ability to determine the extent to which *RUNX3* methylation actually reduced its expression in each of the GC samples tested. In fact, the study that analyzed both *RUNX3* methylation and RUNX3 expression by IHC or RT-PCR revealed a poor association between the percentage of GC samples with *RUNX3* promoter methylation and those expressing RUNX3 (Hu et al. 2011).

Since RT-PCR and IHC are not quantitative techniques, they cannot accurately measure the level of RUNX3 expression at the RNA and protein levels. Recent genome-wide quantitative analyses of both DNA methylation and gene expression in GC and other cancer types revealed that *RUNX3* was not listed among the genes that displayed both DNA hypermethylation and reduced expression (Van der Auwera et al. 2010; Dedeurwaerder et al. 2011; Kibriya et al. 2011; Park et al. 2011; Zouridis et al. 2012; Naumov et al. 2013; Wang et al. 2014). Hence, *RUNX3 P2* hypermethylation in cancer is not associated with its silencing. Of note, *RUNX3 P2* methylation in GC was found to be mostly monoallelic (Song et al. 2008), so even if one allele is silenced, the other allele could still potentially express *RUNX3*. Interestingly, unlike *P2*, the *RUNX3 P1* promoter is heavily methylated in gastric epithelial cells from *H. pylori* uninfected humans that do not express RUNX3, but is completely unmethylated in RUNX3-expressing immune cells (Kurklu et al. 2015). It is thus possible that this *P1* promoter methylation ensures *RUNX3* silencing in normal gastric epithelium, providing an explanation for the lack of association between *RUNX3 P2* hypermethylation in GC and its expression.

Lastly, analysis of the temporal relationship between accumulation of promoter DNA hypermethylation and gene expression in newly transformed human fibroblasts grown in culture revealed that while thousands of transcriptionally active genes were never methylated, hundreds of already silent genes including *RUNX3* gradually accumulated promoter methylation but only ~60 genes showed concomitant accumulation of promoter methylation and reduced expression (Landan et al. 2012). Similarly, proliferation history-dependent promoter DNA methylation events underlie hematopoietic stem cell (HSC) aging and often occur at polycomb PRC-2 complex-silenced target genes expressed in various other cell types, but not in HSC (Beerman et al. 2013). It is interesting to note that following *H. pylori* infection in mice, *Runx3 P2* remains unmethylated in gastric tissue, even at the precancerous gastric intestinal metaplasia stage, but becomes mildly hypermethylated in GC tissue from *gp130*$^{F/F}$ mice (Kurklu et al. 2015). In addition, immortalization of mouse gastric epithelial cells from *gp130*$^{F/F}$ mice by serial passage in culture is associated with increased *Runx3 P2* methylation, and similarly the low methylation ratio of *RUNX3 P2* in primary human GC samples (<15%) is strongly increased to ~90% in GC cell lines (Kurklu et al. 2015). In summary, *RUNX3 P2* methylation in GC is (1) significantly lower than originally proposed; (2) does not correlate with *RUNX3* expression level; and (3) is not an early driving event in the development of gastric or other cancers. Thus, promoter hypermethylation in cancer appears to be mostly a reflection of the proliferative history of the tumor cells, not the cause of early gene silencing that might drive cancer progression. Rather, gene silencing has been shown to correlate well with DNA hypermethylation at distal enhancer regulatory elements of many genes, including established TSGs (Aran et al. 2013).

23.8 Other *Bona Fide* TSGs Reside in 1p36 Chromosomal Region

Large deletions involving human chromosome 1p36 occur in a variety of cancers, including neural, epithelial and hematopoietic malignancies, raising the possibility that one or more TSGs reside at this region (Bagchi and Mills 2008). The part of 1p36 commonly deleted in these cancers was narrowed down to a DNA segment spanning 4.3 Mb in the syntenic mouse chromosome 4. This region does not include *RUNX3* (Fig. 23.3). An extra copy of this segment suppressed proliferation and enhanced apoptosis and senescence in cultured cells, whereas a heterozygous deficiency of the segment enhanced proliferation and suppressed senescence (Bagchi et al. 2007), consistent with the possibility that it harbors a TSG. Knocking down *CHD5*, but not 10 other candidate genes located in this segment, reenacted the outcome effect of deleting one copy of the 4.3 Mb segment. Accordingly, *Chd5*$^{+/-}$ mice spontaneously developed certain solid tumors and lymphoma (Bagchi et al. 2007) but no GC or other GIT tumors were observed.

Large heterozygous deletions spanning chromosome regions 1p36-p34 were also frequently found in pancreatic cancer (Birnbaum et al. 2011; Shain et al. 2012). Interestingly, the recurrent focal deletion that occurred in nearly half of pancreatic cancer patients spans 850 kb within chromosome 1p36.11 and contains 25 genes, including *ARID1A* but not *RUNX3* (Fig. 23.3). This region overlaps the 1.15 Mb segment on 1p36.11 that is recurrently deleted in multiple cancer types (Beroukhim et al. 2010). Analysis of somatic copy-number alterations in the framework of the Cancer Genome Atlas (TCGA) project, involving nearly 5000 cancer patients across 11 cancer types, revealed an even smaller recurrently deleted 1p36.11 region, spanning only ~250 kb and harboring just 2 genes, *ARID1A* and *PIGV* (Zack et al. 2013) (Fig. 23.3). Finally, a recent TCGA analysis of 295 GC patients detected another overlapping recurrent ~750 kb focally deleted region in 1p36.11 that contains 20 genes including *ARID1A* but excluding *RUNX3*

(TCGR network 2014) (Fig. 23.3). Together, these findings strongly suggest that while several known TSGs and TSG candidates are located on human chromosome 1p36, including *TNFRSF14*, *TP73, CHD5, DNAJC11, CAMTA1, SDHB* and *ARID1A* (Fig. 23.3), the real TSG on 1p36.11 is *ARID1A*, not *RUNX3*. This notion is supported by the observation noted earlier that *ARID1A* is significantly mutated in various cancer types (Jones et al. 2010; Wiegand et al. 2010; Liang et al. 2012; Stephens et al. 2012; Dulak et al. 2013; Kandoth et al. 2013; Vogelstein et al. 2013; Wu and Roberts 2013; Lawrence et al. 2014; TCGR network 2014; Wang et al. 2014).

23.9 Conclusions and Future Directions

Cancer is a major cause of mortality worldwide, highlighting the importance of studies aimed at identifying cancer-driving genes and promising gene targets for potential new therapies. More than a decade ago, *RUNX3* was suggested to be a major TSG in GIT epithelium, thus preventing the development of GC. Hundreds of following studies involving thousands of GC and other cancer patients have invested great effort in the attempt to verify and extend this *RUNX3*-TSG paradigm. Yet, all have failed to validate RUNX3 expression in normal GIT and other epithelia from which the cancer has arisen, which constitutes the first and foremost premise for a cell-autonomous TSG.

A second elementary criterion for proving that cell-autonomous loss of RUNX3 in the epithelium drives the formation of various epithelial cancers is that tumors should develop following *Runx3* targeting specifically in the epithelium, but such evidence is yet to be presented. Moreover, other premises inferring a *RUNX3*-TSG function, such as cancer-associated *RUNX3* inactivation by point mutations or focal loss of the 1p36.11 region that harbors *RUNX3*, are not supported by a string of genome-wide analyses, including ones carried out in the framework of TCGA. No germ-line *RUNX3* mutations have been found in cancer-prone pedigrees, and GWAS have likewise failed to show any association of *RUNX3* with increased cancer susceptibility. Finally, quantitative analysis of *RUNX3 P2* hypermethylation in cancer (the most studied aspect of its putative TSG function) reveals that *RUNX3 P2* promoter is not highly methylated in cancer and that its methylation does not impact its expression. Taken together, these findings do not support the paradigm of *RUNX3* being a cell-autonomous TSG in epithelial tissues. Accordingly, *RUNX3* is not mentioned in "Cancer Gene Census" repository (www.sanger.ac.uk/genetics/CGP/Census/), which lists 522 established cancer genes or in the "Network for Cancer Genes (NCG 4.0)" repository (ncg.kcl.ac.uk/), which contains 2000 known and candidate cancer genes based on 77 WES or whole-genome sequencing of more than 3000 cancer patients and 23 cancer types. It is interesting to note that unlike RUNX3, RUNX1 is expressed in normal epithelia (Levanon et al. 2001, 2011; McDonald et al. 2014; van Bragt et al. 2014) and *RUNX1* mutations are associated with human breast cancer (Banerji et al. 2012; Taniuchi et al. 2012). Furthermore, epithelium-specific conditional deletion of *Runx1* in mice induces the development of adenomas in the duodenum and significantly enhances development of GIT tumors in the colon, cecum and intestine in APC^{min} mice (Fijneman et al. 2012). Accordingly, unlike *RUNX3*, *RUNX1* is listed in the "Cancer Gene Census" and "Network for Cancer Genes (NCG 4.0)" repositories as a cancer gene.

RUNX3 has important functions in innate and adaptive immune cell types and has been associated with several immune-related diseases (see also Ebihara et al. in this section). Since chronic inflammatory reactions/diseases can promote epithelial cancer development by providing a tumor-promoting microenvironment, it is possible that the ensuing inflammatory reactions requiring RUNX3 (in TPA-treated skin) or those occurring in its absence (in lung and GIT) might support epithelial tumor development in a non-cell autonomous manner. In fact, loss of *Runx3* in DCs induces colitis, and transfer of *Runx3*$^{-/-}$ FL cells into irradiated RAG2$^{-/-}$ mice induces colitis that can even lead to the rise of colonic tumors when mice are housed under conventional conditions, but not under pathogen-free conditions. Tumor-infiltrating leukocytes can also kill inflammation-independent tumor cells, so loss of RUNX3 in immune cells might also augment epithelial tumor growth by weakening immune-surveillance. On the other hand, deletion of *Runx3* in immune cells, but not in keratinocytes, actually inhibits the outgrowth of inflammation-dependent skin tumors in a chemical carcinogenesis mouse model, indicating that Runx3 function in immune cells can also indirectly promote epithelial tumor formation.

The recent finding that RUNX3 is overexpressed in pancreatic cancer cells and enhances their migration, anchorage-independent growth and metastatic potential provides direct evidence that aberrant RUNX3 overexpression can also promote epithelial tumor development in a cell-autonomous mechanism (see Hingorani et al. in this section). The finding that RUNX3 is overexpressed in a significant fraction of tumor cells in human gastric cancer, skin basal cell carcinoma, head and neck and ovarian cancers (Carvalho et al. 2005; Kudo et al. 2011; Lee et al. 2011; Scheitz et al. 2012; Na et al. 2015; Llorca-Cardenosa et al. 2016) raises the possibility that RUNX3 overexpression may also contribute cell-autonomously to promote tumor formation and metastasis in these epithelial cancers as well. Similarly, RUNX1 overexpression acts cell-autonomously in epithelial cells to promote skin, oral and ovarian tumor formation (Scheitz et al. 2012). RUNX1 is also overexpressed in human breast cancer cells and in MMTV-driven mammary tumors including their lung metastases in mice (Browne et al. 2015), as well as in endometrial and epithelial ovarian tumors and their metastases (Doll et al. 2009; Keita et al. 2013). Moreover, overexpressed Runx1 is required for the increased migration and invasion of the MMTV-driven mammary tumors (Browne et al. 2015). Likewise, overexpressed *RUNX2* promotes ovarian cancer progression (Li et al. 2012b), breast and prostate cancer metastasis to bone (Pratap et al. 2011; Ferrari et al. 2013; McDonald et al. 2014; Li et al. 2016) and gastric cancer migration, invasion and metastasis (Guo et al. 2016). In addition, overexpressed RUNX2 in hepatocellular carcinoma promotes cell viability, growth, migration and invasion (Emma et al. 2016). It thus appears that all three RUNX members possess the ability to cell autonomously promote tumor metastasis when overexpressed. Such cell-autonomous cancer-promoting effects of RUNX members are manifested not only in epithelial cancers but also in various types of leukemia and lymphoma (Blyth et al. 2009; Brady et al. 2009; Ben-Ami et al. 2013).

Finally, the high frequency in tumors of *TP53* and *ARID1A* mutations and of deletions of regions harboring known TSGs, including *ARID1A*, which resides on chromosome 1p36.11, make them and not *RUNX3*, the relevant TSGs in cancer biology. The recurrent amplification of chromosomal regions harboring known oncogenes and activating mutations in these genes, especially the RAS/RTK family, makes them and not *RUNX3* the most promising targets for cancer therapy.

Acknowledgement The work was supported by the Israel Science Foundation (ISF) individual grant to YG and DL. We thank Elsevier publishers for permission to adapt and re-use in this review data we have previously published included in doi:10.1016/j.bbcan.2015.01.004 (Lotem et al. 2015).

References

Abnet, C. C., Freedman, N. D., Hu, N., Wang, Z., Yu, K., Shu, X. O., et al. (2010). A shared susceptibility locus in PLCE1 at 10q23 for gastric adenocarcinoma and esophageal squamous cell carcinoma. *Nature Genetics, 42*, 764–767.

Achyut, B. R., Bader, D. A., Robles, A. I., Wangsa, D., Harris, C. C., Ried, T., & Yang, L. (2013). Inflammation-mediated genetic and epigenetic alterations drive cancer development in the neighboring epithelium upon stromal abrogation of TGF-beta signaling. *PLoS Genetics, 9*, e1003251.

Al Olama, A. A., Kote-Jarai, Z., Schumacher, F. R., Wiklund, F., Berndt, S. I., Benlloch, S., et al. (2013). A meta-analysis of genome-wide association studies to identify prostate cancer susceptibility loci associated with aggressive and non-aggressive disease. *Human Molecular Genetics, 22*, 408–415.

Anderson, C. A., Boucher, G., Lees, C. W., Franke, A., D'Amato, M., Taylor, K. D., et al. (2011). Meta-analysis identifies 29 additional ulcerative colitis risk loci, increasing the number of confirmed associations to 47. *Nature Genetics, 43*, 246–252.

Apel, M., Uebe, S., Bowes, J., Giardina, E., Korendowych, E., Juneblad, K., et al. (2013). Variants in RUNX3 contribute to susceptibility to Psoriatic Arthritis, exhibiting further common ground with Ankylosing Spondylitis. *Arthritis and Rheumatism, 65*, 1224–1231.

Aran, D., Sabato, S., & Hellman, A. (2013). DNA methylation of distal regulatory sites characterizes dysregulation of cancer genes. *Genome Biology, 14*, R21.

Asfaha, S., Hayakawa, Y., Muley, A., Stokes, S., Graham, T. A., Ericksen, R. E., et al. (2015). Krt19(+)/Lgr5(-) cells are radioresistant cancer-initiating stem cells in the colon and intestine. *Cell Stem Cell, 16*, 627–638.

Avraham, K. B., Levanon, D., Negreanu, V., Bernstein, Y., Groner, Y., Copeland, N. G., & Jenkins, N. A. (1995). Mapping of the mouse homolog of the human runt domain gene, AML2, to the distal region of mouse chromosome 4. *Genomics, 25*, 603–605.

Bagchi, A., & Mills, A. A. (2008). The quest for the 1p36 tumor suppressor. *Cancer Research, 68*, 2551–2556.

Bagchi, A., Papazoglu, C., Wu, Y., Capurso, D., Brodt, M., Francis, D., et al. (2007). CHD5 is a tumor suppressor at human 1p36. *Cell, 128*, 459–475.

Banerji, S., Cibulskis, K., Rangel-Escareno, C., Brown, K. K., Carter, S. L., Frederick, A. M., et al. (2012). Sequence analysis of mutations and translocations across breast cancer subtypes. *Nature, 486*, 405–409.

Bauer, O., Hantisteanu, S., Lotem, J., & Groner, Y. (2014). Carcinogen-induced skin tumor development requires leukocytic expression of the transcription factor Runx3. *Cancer Prevention Research, 7*, 913–926.

Bauer, O., Sharir, A., Kimura, A., Hantisteanu, S., Takeda, S., & Groner, Y. (2015). Loss of osteoblast Runx3 produces severe congenital osteopenia. *Molecular and Cellular Biology, 35*, 1097–1109.

Beerman, I., Bock, C., Garrison, B. S., Smith, Z. D., Gu, H., Meissner, A., & Rossi, D. J. (2013). Proliferation-dependent alterations of the DNA methylation landscape underlie hematopoietic stem cell aging. *Cell Stem Cell, 12*, 413–425.

Ben-Ami, O., Friedman, D., Leshkowitz, D., Goldenberg, D., Orlovsky, K., Pencovich, N., et al. (2013). Addiction of t(8;21) and inv(16) acute myeloid leukemia to native RUNX1. *Cell Reports, 4*, 1131–1143.

Berger, A. H., Knudson, A. G., & Pandolfi, P. P. (2011). A continuum model for tumour suppression. *Nature, 476*, 163–169.

Beroukhim, R., Mermel, C. H., Porter, D., Wei, G., Raychaudhuri, S., Donovan, J., et al. (2010). The landscape of somatic copy-number alteration across human cancers. *Nature, 463*, 899–905.

Bhowmick, N. A., Neilson, E. G., & Moses, H.L. (2004). Stromal fibroblasts in cancer initiation and progression. *Nature, 432*, 332–337.

Birnbaum, D. J., Adelaide, J., Mamessier, E., Finetti, P., Lagarde, A., Monges, G., et al. (2011). Genome profiling of pancreatic adenocarcinoma. *Genes, Chromosomes & Cancer, 50*, 456–465.

Blyth, K., Slater, N., Hanlon, L., Bell, M., Mackay, N., Stewart, M., et al. (2009). Runx1 promotes B-cell survival and lymphoma development. *Blood Cells, Molecules & Diseases, 43*, 12–19.

Brady, G., Whiteman, H. J., Spender, L. C., & Farrell, P. J. (2009). Downregulation of RUNX1 by RUNX3 requires the RUNX3 VWRPY sequence and is essential for Epstein-Barr virus-driven B-cell proliferation. *Journal of Virology, 83*, 6909–6916.

Brenner, O., Levanon, D., Negreanu, V., Golubkov, O., Fainaru, O., Woolf, E., & Groner, Y. (2004). Loss of Runx3 function in leukocytes is associated with spontaneously developed colitis and gastric mucosal hyperplasia. *Proceedings of the National Academy of Sciences of the United States of America, 101*, 16016–16021.

Browne, G., Taipaleenmaki, H., Bishop, N. M., Madasu, S. C., Shaw, L. M., van Wijnen, A. J., et al. (2015). Runx1 is associated with breast cancer progression in MMTV-PyMT transgenic mice and its depletion in vitro inhibits migration and invasion. *Journal of Cellular Physiology, 230*, 2522–2532.

Carvalho, R., Milne, A. N., Polak, M., Corver, W. E., Offerhaus, G. J., & Weterman, M. A. (2005). Exclusion of RUNX3 as a tumour-suppressor gene in early-onset gastric carcinomas. *Oncogene, 24*, 8252–8258.

Chan, I. T., Kutok, J. L., Williams, I. R., Cohen, S., Kelly, L., Shigematsu, H., et al. (2004). Conditional expression of oncogenic K-ras from its endogenous promoter induces a myeloproliferative disease. *The Journal of Clinical Investigation, 113*(4), 528–538.

Cheroutre, H., Lambolez, F., & Mucida, D. (2011). The light and dark sides of intestinal intraepithelial lymphocytes. *Nature Reviews. Immunology, 11*, 445–456.

Childs, E. J., Mocci, E., Campa, D., Bracci, P. M., Gallinger, S., Goggins, M., et al. (2015). Common variation at 2p13.3, 3q29, 7p13 and 17q25.1 associated with susceptibility to pancreatic cancer. *Nature Genetics, 47*, 911–916.

Choi, H., Sheng, J., Gao, D., Li, F., Durrans, A., Ryu, S., et al. (2015). Transcriptome analysis of individual stromal cell populations identifies stroma-tumor crosstalk in mouse lung cancer model. *Cell Reports, 10*, 1187–1201.

Chun, N., & Ford, J. M. (2012). Genetic testing by cancer site: Stomach. *Cancer Journal, 18*, 355–363.

Cortes, A., Hadler, J., Pointon, J. P., Robinson, P. C., Karaderi, T., Leo, P., et al. (2013). Identification of multiple risk variants for ankylosing spondylitis through high-density genotyping of immune-related loci. *Nature Genetics, 45*, 730–738.

Cortez-Retamozo, V., Etzrodt, M., Newton, A., Rauch, P. J., Chudnovskiy, A., Berger, C., et al. (2012). Origins of tumor-associated macrophages and neutrophils. *Proceedings of the National Academy of Sciences of the United States of America, 109*, 2491–2496.

Cruz-Guilloty, F., Pipkin, M. E., Djuretic, I. M., Levanon, D., Lotem, J., Lichtenheld, M. G., et al. (2009). Runx3 and T-box proteins cooperate to establish the transcriptional program of effector CTLs. *The Journal of Experimental Medicine, 206*, 51–59.

Dedeurwaerder, S., Desmedt, C., Calonne, E., Singhal, S. K., Haibe-Kains, B., Defrance, M., et al. (2011). DNA methylation profiling reveals a predominant immune component in breast cancers. *EMBO Molecular Medicine, 3*, 726–741.

Deng, L., Zhou, J. F., Sellers, R. S., Li, J. F., Nguyen, A. V., Wang, Y., et al. (2010). A novel mouse model of inflammatory bowel disease links mammalian target of rapamycin-dependent hyperproliferation of colonic epithelium to inflammation-associated tumorigenesis. *The American Journal of Pathology, 176*, 952–967.

Deng, N., Goh, L. K., Wang, H., Das, K., Tao, J., Tan, I. B., et al. (2012). A comprehensive survey of genomic alterations in gastric cancer reveals systematic patterns of molecular exclusivity and co-occurrence among distinct therapeutic targets. *Gut, 61*, 673–684.

Dicken, J., Mildner, A., Leshkowitz, D., Touw, I. V., Hantisteanu, S., Jung, S., & Groner, Y. (2013). Transcriptional reprogramming of CD11b+Esamhi dendritic cell identity and function by loss of Runx3. *PloS One, 8*, e77490.

Djuretic, I. M., Levanon, D., Negreanu, V., Groner, Y., Rao, A., & Ansel, K. M. (2007). Transcription factors T-bet and Runx3 cooperate to activate Ifng and silence Il4 in T helper type 1 cells. *Nature Immunology, 8*, 145–153.

Doll, A., Gonzalez, M., Abal, M., Llaurado, M., Rigau, M., Colas, E., et al. (2009). An orthotopic endometrial cancer mouse model demonstrates a role for RUNX1 in distant metastasis. *International Journal of Cancer, 125*, 257–263.

Dubois, P. C., Trynka, G., Franke, L., Hunt, K. A., Romanos, J., Curtotti, A., et al. (2010). Multiple common variants for celiac disease influencing immune gene expression. *Nature Genetics, 42*, 295–302.

Dulak, A. M., Stojanov, P., Peng, S., Lawrence, M. S., Fox, C., Stewart, C., et al. (2013). Exome and whole-genome sequencing of esophageal adenocarcinoma identifies recurrent driver events and mutational complexity. *Nature Genetics, 45*, 478–486.

Duran, A., Amanchy, R., Linares, J. F., Joshi, J., Abu-Baker, S., Porollo, A., et al. (2011). p62 is a key regulator of nutrient sensing in the mTORC1 pathway. *Molecular Cell, 44*, 134–146.

Duran, A., Hernandez, E. D., Reina-Campos, M., Castilla, E. A., Subramaniam, S., Raghunandan, S., et al. (2016). *Cancer Cell, 30*, 595–609.

Easton, D. F., Pooley, K. A., Dunning, A. M., Pharoah, P. D., Thompson, D., Ballinger, D. G., et al. (2007). Genome-wide association study identifies novel breast cancer susceptibility loci. *Nature, 447*, 1087–1093.

Ebihara, T., Song, C., Ryu, S. H., Plougastel-Douglas, B., Yang, L., Levanon, D., et al. (2015). Runx3 specifies lineage commitment of innate lymphoid cells. *Nature Immunology, 16*, 1124–1134.

Eeles, R. A., Olama, A. A., Benlloch, S., Saunders, E. J., Leongamornlert, D. A., Tymrakiewicz, M., et al. (2013). Identification of 23 new prostate cancer susceptibility loci using the iCOGS custom genotyping array. *Nature Genetics, 45*, 385–391.

Elinav, E., Nowarski, R., Thaiss, C. A., Hu, B., Jin, C., & Flavell, R. A. (2013). Inflammation-induced cancer: Crosstalk between tumours, immune cells and microorganisms. *Nature Reviews. Cancer, 13*, 759–771.

Emma, M. R., Iovanna, J. L., Bachvarov, D., Puleio, R., Loria, G. R., Augello, G., et al. (2016). NUPR1, a new target in liver cancer: Implication in controlling cell growth, migration, invasion and sorafenib resistance. *Cell Death & Disease, 7*, e2269.

Esparza-Gordillo, J., Schaarschmidt, H., Liang, L., Cookson, W., Bauerfeind, A., Lee-Kirsch, M. A., et al. (2013). A functional IL-6 receptor (IL6R) variant is a risk factor for persistent atopic dermatitis. *The Journal of Allergy and Clinical Immunology, 132*, 371–377.

Evans, D. M., Spencer, C. C., Pointon, J. J., Su, Z., Harvey, D., Kochan, G., et al. (2011). Interaction between ERAP1 and HLA-B27 in ankylosing spondylitis implicates peptide handling in the mechanism for HLA-B27 in disease susceptibility. *Nature Genetics, 43*, 761–767.

Fainaru, O., Woolf, E., Lotem, J., Yarmus, M., Brenner, O., Goldenberg, D., et al. (2004). Runx3 regulates mouse TGF-beta-mediated dendritic cell function and its absence results in airway inflammation. *The EMBO Journal, 23*, 969–979.

Fan, B., Dachrut, S., Coral, H., Yuen, S. T., Chu, K. M., Law, S., et al. (2012). Integration of DNA copy number alterations and transcriptional expression analysis in human gastric cancer. *PloS One, 7*, e29824.

Fanale, D., Amodeo, V., Corsini, L. R., Rizzo, S., Bazan, V., & Russo, A. (2012). Breast cancer genome-wide association studies: There is strength in numbers. *Oncogene, 31*, 2121–2128.

Ferrari, N., McDonald, L., Morris, J. S., Cameron, E. R., & Blyth, K. (2013). RUNX2 in mammary gland development and breast cancer. *Journal of Cellular Physiology, 228*, 1137–1142.

Fijneman, R. J., Anderson, R. A., Richards, E., Liu, J., Tijssen, M., Meijer, G. A., et al. (2012). Runx1 is a tumor suppressor gene in the mouse gastrointestinal tract. *Cancer Science, 103*, 593–599.

Franke, A., McGovern, D. P., Barrett, J. C., Wang, K., Radford-Smith, G. L., Ahmad, T., et al. (2010). Genome-wide meta-analysis increases to 71 the number of confirmed Crohn's disease susceptibility loci. *Nature Genetics, 42*, 1118–1125.

Friedrich, M. J., Rad, R., Langer, R., Voland, P., Hoefler, H., Schmid, R. M., et al. (2006). Lack of RUNX3 regulation in human gastric cancer. *The Journal of Pathology, 210*, 141–146.

Fritz, J. M., Dwyer-Nield, L. D., & Malkinson, A. M. (2011). Stimulation of neoplastic mouse lung cell proliferation by alveolar macrophage-derived, insulin-like growth factor-1 can be blocked by inhibiting MEK and PI3K activation. *Molecular Cancer, 10*, 76.

Gal-Yam, E. N., Egger, G., Iniguez, L., Holster, H., Einarsson, S., Zhang, X., et al. (2008). Frequent switching of Polycomb repressive marks and DNA hypermethylation in the PC3 prostate cancer cell line. *Proceedings of the National Academy of Sciences of the United States of America, 105*, 12979–12984.

Gebhardt, C., Riehl, A., Durchdewald, M., Nemeth, J., Furstenberger, G., Muller-Decker, K., et al. (2008). RAGE signaling sustains inflammation and promotes tumor development. *The Journal of Experimental Medicine, 205*, 275–285.

Ghoussaini, M., Fletcher, O., Michailidou, K., Turnbull, C., Schmidt, M. K., Dicks, E., et al. (2012). Genome-wide association analysis identifies three new breast cancer susceptibility loci. *Nature Genetics, 44*, 312–318.

Grim, J. E., Knoblaugh, S. E., Guthrie, K. A., Hagar, A., Swanger, J., Hespelt, J., et al. (2012). Fbw7 and p53 cooperatively suppress advanced and chromosomally unstable intestinal cancer. *Molecular and Cellular Biology, 32*, 2160–2167.

Guerra, C., Schuhmacher, A. J., Canamero, M., Grippo, P. J., Verdaguer, L., Perez-Gallego, L., et al. (2007). Chronic pancreatitis is essential for induction of pancreatic ductal adenocarcinoma by K-Ras oncogenes in adult mice. *Cancer Cell, 11*, 291–302.

Guo, C., Yao, F., Wu, K., Yang, L., Zhang, X., & Ding, J. (2010). Chromatin immunoprecipitation and association study revealed a possible role of Runt-related transcription factor 3 in the ulcerative colitis of Chinese population. *Clinical Immunology, 135*, 483–489.

Guo, Z., Yang, L., Qian, F., Wang, Y., Yu, X., Ji, C.-D., et al. (2016). Transcription factor RUNX2 up-regulates chemokine receptor CXCR4 to promote invasive and metastatic potentials of human gastric cancer. *Oncotarget, 7*, 20999–21012.

Guri-BenAri, M., Thais, C. A., Serafini, N., Winter, D. R., Giladi, A., Lara-Astiaso, D., et al. (2016). The spectrum and regulatory landscape of intestinal innate lymphoid cells are shaped by the microbiome. *Cell, 166*, 1231–1246.

Hanahan, D., & Weinberg, R. A. (2011). Hallmarks of cancer: The next generation. *Cell, 144*, 646–674.

He, X., Xiang, C., Zhang, C., Xie, Y., Chen, L., Zhang, G., & Liu, G. (2015). p53 in the myeloid lineage modulates an inflammatory microenvironment limiting initiation and invasion of intestinal tumors. *Cell Reports, 13*, 888–897.

Helms, C., Cao, L., Krueger, J. G., Wijsman, E. M., Chamian, F., Gordon, D., et al. (2003). A putative RUNX1 binding site variant between SLC9A3R1 and NAT9 is associated with susceptibility to psoriasis. *Nature Genetics, 35*, 349–356.

Himmelstein, D. S., & Baranzini, S. E. (2015). Heterogeneous network edge prediction: A data integration approach to prioritize disease-associated genes. *PLoS Computational Biology, 11*, e1004259.

Hishida, A., Matsuo, K., Goto, Y., Mitsuda, Y., Naito, M., Wakai, K., et al. (2009). Significant association of RUNX3 T/A polymorphism at intron 3 (rs760805) with the risk of gastric atrophy in Helicobacter pylori seropositive Japanese. *Journal of Gastroenterology, 44*, 1165–1171.

Houghton, A. M. (2013). Mechanistic links between COPD and lung cancer. *Nature Reviews. Cancer, 13*, 233–245.

Hu, S. L., Huang, D. B., Sun, Y. B., Wu, L., Xu, W. P., Yin, S., et al. (2011). Pathobiologic implications of methylation and expression status of Runx3 and CHFR genes in gastric cancer. *Medical Oncology, 28*, 447–454.

Huang, B., Qu, Z., Ong, C. W., Tsang, Y. H. N., Xiao, G., Shapiro, D., et al. (2012). RUNX3 acts as a tumor suppressor in breast cancer by targeting estrogen receptor alpha. *Oncogene, 31*, 527–534.

Ito, K., Lim, A. C., Salto-Tellez, M., Motoda, L., Osato, M., Chuang, L. S., et al. (2008). RUNX3 attenuates beta-catenin/T cell factors in intestinal tumorigenesis. *Cancer Cell, 14*, 226–237.

Ito, K., Chuang, L. S., Ito, T., Chang, T. L., Fukamachi, H., Salto-Tellez, M., & Ito, Y. (2011). Loss of Runx3 is a key event in inducing precancerous state of the stomach. *Gastroenterology, 140,* 1536–1546.

Jasperson, J. W., Tuohy, T. M., Neklason, D. W., & Burt, R. W. (2011). Hereditary and familial colon cancer. *Gastroenterology, 138*, 2044–2058.

Ji, H., Houghton, A. M., Mariani, T. J., Perera, S., Kim, C. B., Padera, R., et al. (2006). K-ras activation generates an inflammatory response in lung tumors. *Oncogene, 25*, 2105–2112.

Jones, S., Wang, T. L., Shih Ie-M., Mao, T. L., Nakayama, K., Roden, R., et al. (2010). Frequent mutations of chromatin remodeling gene ARID1A in ovarian clear cell carcinoma. *Science, 330*, 228–231.

Jostins, L., Ripke, S., Weersma, R. K., Duerr, R. H., McGovern, D. P., Hui, K. Y., et al. (2012). Host-microbe interactions have shaped the genetic architecture of inflammatory bowel disease. *Nature, 491*, 119–124.

Kandoth, C., McLellan, M. D., Vandin, F., Ye, K., Niu, B., Lu, C., et al. (2013). Mutational landscape and significance across 12 major cancer types. *Nature, 502*, 333–339.

Katajisto, P., Vaahtomeri, K., Ekman, N., Ventela, E., Ristimaki, A., Bardeesy, N., et al. (2008). LKB1 signaling in mesenchymal cells required for suppression of gastrointestinal polyposis. *Nature Genetics, 40*, 455–459.

Keita, M., Bachvarova, M., Morin, C., Plante, M., Gregoire, J., Renaud, M. C., et al. (2013). The RUNX1 transcription factor is expressed in serous epithelial ovarian carcinoma and contributes to cell proliferation, migration and invasion. *Cell Cycle, 12*, 972–986.

Keller, G., Vogelsang, H., Becker, I., Plaschke, S., Ott, K., Suriano, G., et al. (2004). Germline mutations of the E-cadherin(CDH1) and TP53 genes, rather than of RUNX3 and HPP1, contribute to genetic predisposition in German gastric cancer patients. *Journal of Medical Genetics, 41*, e89.

Keshet, I., Schlesinger, Y., Farkash, S., Rand, E., Hecht, M., Segal, E., et al. (2006). Evidence for an instructive mechanism of de novo methylation in cancer cells. *Nature Genetics, 38*, 149–153.

Kibriya, M. G., Raza, M., Jasmine, F., Roy, S., Paul-Brutus, R., Rahaman, R., et al. (2011). A genome-wide DNA methylation study in colorectal carcinoma. *BMC Medical Genomics, 4*, 50.

Kim, W. J., Kim, E. J., Jeong, P., Quan, C., Kim, J., Li, Q. L., et al. (2005). RUNX3 inactivation by point mutations and aberrant DNA methylation in bladder tumors. *Cancer Research, 65*, 9347–9354.

Kim, B. G., Li, C., Qiao, W., Mamura, M., Kasprzak, B., Anver, M., et al. (2006). Smad4 signalling in T cells is required for suppression of gastrointestinal cancer. *Nature, 441*, 1015–1019.

Kim, T. H., Li, F., Ferreiro-Neira, I., Ho, L. L., Luyten, A., Nalapareddy, K., et al. (2014). Broadly permissive intestinal chromatin underlies lateral inhibition and cell plasticity. *Nature, 506*, 511–515.

Klutstein, M., Nerjman, D., Greenfield, R., & Cedar, H. (2016). DNA methylation in cancer and aging. *Cancer Research, 76*, 3446–3450.

Kudo, Y., Tsunematsu, T., & Takata, T. (2011). Oncogenic role of RUNX3 in head and neck cancer. *Journal of Cellular Biochemistry, 112*, 387–393.

Kurklu, B., Whitehead, R. H., Ong, E. K., Minamoto, T., Fox, J. G., Mann, J. R., et al. (2015). Lineage-specific RUNX3 hypomethylation marks the preneoplastic immune component of gastric cancer. *Oncogene, 34*, 2856–2866.

Lan, Q., Hsiung, C. A., Matsuo, K., Hong, Y. C., Seow, A., Wang, Z. M., et al. (2012). Genome-wide association analysis identifies new lung cancer susceptibility loci in never-smoking women in Asia. *Nature Genetics, 44*, 1330–1335.

Landan, G., Cohen, N. M., Mukamel, Z., Bar, A., Molchadsky, A., Brosh, R., et al. (2012). Epigenetic polymorphism and the stochastic formation of differentially methylated regions in normal and cancerous tissues. *Nature Genetics, 44*, 1207–1214.

Landi, M. T., Chatterjee, N., Yu, K., Goldin, L. R., Goldstein, A. M., Rotunno, M., et al. (2009). A genome-wide association study of lung cancer identifies a region of chromosome 5p15 associated with risk for Adenocarcinoma. *American Journal of Human Genetics, 85*, 679–691.

Laprise, C. (2014). The Saguenay-Lac-Saint-Jean asthma familial collection: The genetics of asthma in a young founder population. *Genes and Immunity, 15*, 247–255.

Lauber, S. (2012) The role of gp130 cytokines IL-6 and OSM on tumor development in mouse models for lung adenocarcinoma. Open AcessDissertations and Theses 7325: M.Sc., McMaster University, Hamilton, Ontario.

Lawrence, M. S., Stojanov, P., Mermel, C. H., Robinson, J. T., Garraway, L. A., Golub, T. R., et al. (2014). Discovery and saturation analysis of cancer genes across 21 tumour types. *Nature, 505*, 495–501.

Lee, K. S., Lee, Y. S., Lee, J. M., Ito, K., Cinghu, S., Kim, J. H., et al. (2010). Runx3 is required for the differentiation of lung epithelial cells and suppression of lung cancer. *Oncogene, 29*, 3349–3361.

Lee, C. W., Chuang, L. S., Kimura, S., Lai, S. K., Ong, C. W., Yan, B., et al. (2011). RUNX3 functions as an oncogene in ovarian cancer. *Gynecologic Oncology, 122*, 410–417.

Lee, Y. S., Lee, J. W., Jang, J. W., Chi, X. Z., Kim, J. H., Li, Y. H., et al. (2013). Runx3 inactivation is a crucial early event in the development of lung adenocarcinoma. *Cancer Cell, 24*, 603–616.

Leung, W. K., Wu, K. C., Wong, C. Y., Cheng, A. S., Ching, A. K., Chan, A. W., et al. (2008). Transgenic cyclooxygenase-2 expression and high salt enhanced susceptibility to chemical-induced gastric cancer development in mice. *Carcinogenesis, 29*, 1648–1654.

Levanon, D., & Groner, Y. (2004). Structure and regulated expression of mammalian RUNX genes. *Oncogene, 23*, 4211–4219.

Levanon, D., & Groner, Y. (2009). Runx3-deficient mouse strains circa 2008: Resemblance and dissimilarity. *Blood Cells, Molecules & Diseases, 43*, 1–5.

Levanon, D., Negreanu, V., Bernstein, Y., Bar-Am, I., Avivi, L., & Groner, Y. (1994). AML1, AML2, and AML3, the human members of the runt domain gene-family: cDNA structure, expression, and chromosomal localization. *Genomics, 23*, 425–432.

Levanon, D., Brenner, O., Negreanu, V., Bettoun, D., Woolf, E., Eilam, R., et al. (2001). Spatial and temporal expression pattern of Runx3 (Aml2) and Runx1 (Aml1) indicates non-redundant functions during mouse embryogenesis. *Mechanisms of Development, 109*, 413–417.

Levanon, D., Bettoun, D., Harris-Cerruti, C., Woolf, E., Negreanu, V., Eilam, R., et al. (2002). The Runx3 transcription factor regulates development and survival of TrkC dorsal root ganglia neurons. *The EMBO Journal, 21*, 3454–3463.

Levanon, D., Hantisteanu, S., Negreanu, V., Brenner, O., & Groner, Y. (2009). New insights into the mechanism of inflammatory bowel disease (IBD) etiology in Runx3-deficient mice. EMBO Worksop, RUNX Transcription Factors in Development & Disease 20.

Levanon, D., Bernstein, Y., Negreanu, V., Bone, K. R., Pozner, A., Eilam, R., et al. (2011). Absence of Runx3 expression in normal gastrointestinal epithelium calls into question its tumour suppressor function. *EMBO Molecular Medicine, 3*, 593–604.

Levanon, D., Negreanu, V., Lotem, J., Rae-Bone, K., Brenner, O., Leshkowitz, D., & Groner, Y. (2014). Runx3 regulates interleukin-15 dependent natural killer cell activation. *Molecular and Cellular Biology, 34*, 1158–1169.

Li, Q. L., Ito, K., Sakakura, C., Fukamachi, H., Inoue, K., Chi, X. Z., et al. (2002). Causal relationship between the loss of RUNX3 expression and gastric cancer. *Cell, 109*, 113–124.

Li, J., Kleeff, J., Guweidhi, A., Esposito, I., Berberat, P. O., Giese, T., et al. (2004). RUNX3 expression in primary and metastatic pancreatic cancer. *Journal of Clinical Pathology, 57*, 294–299.

Li, Q., Jia, Z., Wang, L., Kong, X., Li, Q., Guo, K., et al. (2012a). Disruption of Klf4 in villin-positive gastric progenitor cells promotes formation and progression of tumors of the antrum in mice. *Gastroenterology, 142*, 531–542.

Li, W., Xu, S., Lin, S., & Zhao, W. (2012b). Overexpression of runt-related transcription factor-2 is associated with advanced tumor progression and poor prognosis in epithelial ovarian cancer. *Journal of Biomedicine & Biotechnology, 2012*, 456534.

Li, X., Lu, J., Tan, C., Wang, Q., & Feng, Y. (2016). RUNX2 promotes breast cancer bone metastasis by increasing integrin a5-mediated colonization. *Cancer Letters, 380*, 78–86.

Liang, H., Cheung, L. W., Li, J., Ju, Z., Yu, S., Stemke-Hale, K., et al. (2012). Whole-exome sequencing combined with functional genomics reveals novel candidate driver cancer genes in endometrial cancer. *Genome Research, 22*, 2120–2129.

Lin, C., Song, H., Huang, C., Yao, E., Gacayan, R., Xu, S. M., & Chuang, P. T. (2012). Alveolar type II cells possess the capability of initiating lung tumor development. *PloS One, 7*, e53817.

Liu, J. Z., van Sommeren, S., Huang, H., Ng, S. C., Alberts, R., Takahashi, A., et al. (2015). Association analyses identify 38 susceptibility loci for inflammatory bowel disease and highlight shared genetic risk across populations. *Nature Genetics, 47*, 979–986.

Llorca-Cardenosa, M.J., Fleitas, T., Ibarrola-Villava, M., Pena-Chilet, M., Mongort, C., Martinez-Ciarpaglini, C., et al. (2016). Epigenetic changes in localized gastric cancer: The role of RUNX3 in tumor progression and the immune microenvironment. *Oncotarget* . doi:10.18632/Oncotarget, in press

Lotem, J., Levanon, D., Negreanu, V., & Groner, Y. (2013a). The false paradigm of Runx3 function as tumor suppressor in gastric cancer. *Journal of Cancer Therapy, 4*, 16–25.

Lotem, J., Levanon, D., Negreanu, V., Leshkowitz, D., Friedlander, G., & Groner, Y. (2013b). Runx3-mediated transcriptional program in cytotoxic lymphocytes. *PLOS One, 8*, e80467.

Lotem, J., Levanon, D., Negreanu, V., Bauer, O., Hantisteanu, S., Dicken, Y., & Groner, Y. (2015). Runx3 at the interface of immunity, inflammation and cancer. *Biochimica et Biophysica Acta, 1855*, 131–143.

Lujambio, A., Akkari, L., Simon, J., Grace, D., Tschaharganeh, D. F., Bolden, J. E., et al. (2013). Non-cell-autonomous tumor suppression by p53. *Cell, 153*, 449–460.

McDonald, L., Ferrari, N., Terry, A., Bell, M., Mohammed, Z. M., Orange, C., et al. (2014). RUNX2 in subtype specific breast cancer and mammary gland differentiation. *Disease Models & Mechanisms, 7*, 525–534.

McLeod, A. G., Sherrill, T. P., Cheng, D., Han, W., Saxon, J. A., Gleaves, L. A., et al. (2016). Neutrophil-derived IL-1β impairs the efficacy of NF-κB inhibitors against lung cancer. *Cell Reports, 16*, 120–132.

Melchor, L., & Benitez, J. (2013). The complex genetic landscape of familial breast cancer. *Human Genetics, 132*, 845–863.

Munoz, J., Stange, D. E., Schepers, A. G., van de Wetering, M., Koo, B. K., Itzkovitz, S., et al. (2012). The Lgr5 intestinal stem cell signature: Robust expression of proposed quiescent '+4' cell markers. *The EMBO Journal, 31*, 3079–3091.

Na, Y. J., Shim, K. N., Joo, Y. H., Kim, S. E., Jung, H. K., Jung, S. A., & Cho, M. S. (2015). RUNX3 methylation, loss of RUNX3 expression and clinicopathologic findings according to Helicobacter pylori CagA in gastric carcinoma. *The Korean Journal of Gastroenterology, 66*, 75–84.

Nagini, S. (2012). Carcinoma of the stomach: A review of epidemiology, pathogenesis, molecular genetics and chemoprevention. *World Journal of Gastrointestinal Oncology, 4*, 156–169.

Naumov, V. A., Generozov, E. V., Zaharjevskaya, N. B., Matushkina, D. S., Larin, A. K., Chernyshov, S. V., et al. (2013). Genome-scale analysis of DNA methylation in colorectal cancer using Infinium HumanMethylation450 BeadChips. *Epigenetics, 8*, 921–934.

Oguma, K., Oshima, H., Aoki, M., Uchio, R., Naka, K., Nakamura, S., et al. (2008). Activated macrophages promote Wnt signalling through tumour necrosis factor-alpha in gastric tumour cells. *The EMBO Journal, 27*, 1671–1681.

Oshima, H., Ishikawa, T., Yoshida, G. J., Naoi, K., Maeda, Y., Naka, K., et al. (2014). TNF-alpha/TNFR1 signaling promotes gastric tumorigenesis through induction of Noxo1 and Gna14 in tumor cells. *Oncogene, 33*, 3820–3829.

Park, J. H., Park, J., Choi, J. K., Lyu, J., Bae, M. G., Lee, Y. G., et al. (2011). Identification of DNA methylation changes associated with human gastric cancer. *BMC Medical Genomics, 4*, 82.

Peters, U., Hutter, C. M., Hsu, L., Schumacher, F. R., Conti, D. V., Carlson, C. S., et al. (2012). Meta-analysis of new genome-wide association studies of colorectal cancer risk. *Human Genetics, 131*, 217–234.

Petersen, G. M., Amundadottir, L., Fuchs, C. S., Kraft, P., Stolzenberg-Solomon, R. Z., Jacobs, K. B., et al. (2010). A genome-wide association study identifies pancreatic cancer susceptibility loci on chromosomes 13q22.1, 1q32.1 and 5p15.33. *Nature Genetics, 42*, 224–228.

Pharoah, P. D., Tsai, Y. Y., Ramus, S. J., Phelan, C. M., Goode, E. L., Lawrenson, K., et al. (2013). GWAS meta-analysis and replication identifies three new susceptibility loci for ovarian cancer. *Nature Genetics, 45*, 362–370.

Pratap, J., Lian, J. B., & Stein, G. S. (2011). Metastatic bone disease: Role of transcription factors and future targets. *Bone, 48*, 30–36.

Prokunina, L., Castillejo-Lopez, C., Oberg, F., Gunnarsson, I., Berg, L., Magnusson, V., et al. (2002). A regulatory polymorphism in PDCD1 is associated with susceptibility to systemic lupus erythematosus in humans. *Nature Genetics, 32*, 666–669.

Raveh, E., Cohen, S., Levanon, D., Groner, Y., & Gat, U. (2005). Runx3 is involved in hair shape determination. *Developmental Dynamics, 233*, 1478–1487.

Reis, B. S., Rogoz, A., Costa-Pinto, F. A., Taniuchi, I., & Mucida, D. (2013). Mutual expression of the transcription factors Runx3 and ThPOK regulates intestinal CD4(+) T cell immunity. *Nature Immunology, 14*, 271–280.

Reis, B. S., Hoytema van Konijnenburg, D. P., Grivennikov, S. I., & Mucida, D. (2014). Transcription factor T-bet regulates intraepithelial lymphocyte functional maturation. *Immunity, 41*, 244–256.

Rundhaug, J. E., & Fischer, S. M. (2010). Molecular mechanisms of mouse skin tumor promotion. *Cancers (Basel), 2*, 436–482.

Sakamoto, H., Yoshimura, K., Saeki, N., Katai, H., Shimoda, T., Matsuno, Y., et al. (2008). Genetic variation in PSCA is associated with susceptibility to diffuse-type gastric cancer. *Nature Genetics, 40*, 730–740.

Salto-Tellez, M., Peh, B. K., Ito, K., Tan, S. H., Chong, P. Y., Han, H. C., et al. (2006). RUNX3 protein is overexpressed in human basal cell carcinomas. *Oncogene, 25*, 7646–7649.

Sapkota, Y., Yasui, Y., Lai, R., Sridharan, M., Robson, P. J., Cass, C. E., et al. (2013). Identification of a breast cancer susceptibility locus at 4q31.22 using a genome-wide association study paradigm. *PloS One, 8*, e62550.

Scheitz, C. J., Lee, T. S., McDermitt, D. J., & Tumbar, T. (2012). Defining a tissue stem cell-driven Runx1/Stat3 signalling axis in epithelial cancer. *The EMBO Journal, 31*, 4124–4139.

Shain, A. H., Giacomini, C. P., Matsukuma, K., Karikari, C. A., Bashyam, M. D., Hidalgo, M., et al. (2012). Convergent structural alterations define SWItch/Sucrose NonFermentable (SWI/SNF) chromatin remodeler as a central tumor suppressive complex in pancreatic cancer. *Proceedings of the National Academy of Sciences of the United States of America, 109*, E252–E259.

Shi, Y., Hu, Z., Wu, C., Dai, J., Li, H., Dong, J., et al. (2011). A genome-wide association study identifies new susceptibility loci for non-cardia gastric cancer at 3q13.31 and 5p13.1. *Nature Genetics, 43*, 1215–1218.

Shi, Y., Li, L., Hu, Z., Li, S., Wang, S., Liu, J., et al. (2013). A genome-wide association study identifies two new cervical cancer susceptibility loci at 4q12 and 17q12. *Nature Genetics, 45*, 918–922.

Song, H. J., Shim, K. N., Joo, Y. H., Kim, S. E., Jung, S. A., & Yoo, K. (2008). Methylation of the tumor suppressor gene RUNX3 in human gastric carcinoma. *Gut Liver, 2*, 119–125.

Stange, D. E., Koo, B. K., Huch, M., Sibbel, G., Basak, O., Lyubimova, A., et al. (2013). Differentiated Troy+ chief cells act as reserve stem cells to generate all lineages of the stomach epithelium. *Cell, 155*, 357–368.

Stephens, P. J., Tarpey, P. S., Davies, H., Van Loo, P., Greenman, C., Wedge, D. C., et al. (2012). The landscape of cancer genes and mutational processes in breast cancer. *Nature, 486*, 400–404.

Subramaniam, M. M., Chan, J. Y., Yeoh, K. G., Quek, T., Ito, K., & Salto-Tellez, M. (2009). Molecular pathology of RUNX3 in human carcinogenesis. *Biochimica et Biophysica Acta, 1796*, 315–331.

Sugai, M., Aoki, K., Osato, M., Nambu, Y., Ito, K., Taketo, M. M., & Shimizu, A. (2011). Runx3 is required for full activation of regulatory T cells to prevent colitis-associated tumor formation. *Journal of Immunology, 186*, 6515–6520.

Takahashi, H., Ogata, H., Nishigaki, R., Broide, D. H., & Karin, M. (2010). Tobacco smoke promotes lung tumorigenesis by triggering IKKbeta- and JNK1-dependent inflammation. *Cancer Cell, 17*, 89–97.

Tamura, G., So, K., Miyoshi, H., Honda, T., Nishizuka, S., & Motoyama, T. (2009). Quantitative assessment of gene methylation in neoplastic and non-neoplastic gastric epithelia using methylation-specific DNA microarray. *Pathology International, 59*, 895–899.

Taniuchi, I., Osato, M., Egawa, T., Sunshine, M. J., Bae, S. C., Komori, T., et al. (2002). Differential requirements for Runx proteins in CD4 repression and epigenetic silencing during T lymphocyte development. *Cell, 111*, 621–633.

Taniuchi, I., Osato, M., & Ito, Y. (2012). Runx1: No longer just for leukemia. *The EMBO Journal, 31*, 4098–4099.

TCGAR network. (2014). Comprehensive molecular characterization of gastric adenocarcinoma. *Nature, 513*, 202–209.

Tenesa, A., & Dunlop, M. G. (2009). New insights into the aetiology of colorectal cancer from genome-wide association studies. *Nature Reviews. Genetics, 10*, 353–358.

Tokuhiro, S., Yamada, R., Chang, X., Suzuki, A., Kochi, Y., Sawada, T., et al. (2003). An intronic SNP in a RUNX1 binding site of SLC22A4, encoding an organic cation transporter, is associated with rheumatoid arthritis. *Nature Genetics, 35*, 341–348.

Treutlein, B., Brownfield, D. G., Wu, A. R., Neff, N. F., Mantalas, G. L., Espinoza, F. H., et al. (2014). Reconstructing lineage hierarchies of the distal lung epithelium using single-cell RNA-seq. *Nature, 509*, 371–375.

Trimboli, A. J., Cantemir-Stone, C. Z., Li, F., Wallace, J. A., Merchant, A., Creasap, N., et al. (2009). Pten in stromal fibroblasts suppresses mammary epithelial tumours. *Nature, 461*, 1084–1091.

Trynka, G., Wijmenga, C., & van Heel, D. A. (2010). A genetic perspective on coeliac disease. *Trends in Molecular Medicine, 16*, 537–550.

Tsoi, L. C., Spain, S. L., Knight, J., Ellinghaus, E., Stuart, P. E., Capon, F., et al. (2012). Identification of 15 new psoriasis susceptibility loci highlights the role of innate immunity. *Nature Genetics, 44*, 1341–1348.

Tsukamoto, Y., Uchida, T., Karnan, S., Noguchi, T., Nguyen, L. T., Tanigawa, M., et al. (2008). Genome-wide analysis of DNA copy number alterations and gene expression in gastric cancer. *The Journal of Pathology, 216*, 471–482.

Turnbull, C., Ahmed, S., Morrison, J., Pernet, D., Renwick, A., Maranian, M., et al. (2010). Genome-wide association study identifies five new breast cancer susceptibility loci. *Nature Genetics, 42*, 504–507.

Valencia, T., Kim, J. Y., Abu-Baker, S., Moscat-Pardos, J., Ahn, C. S., Reina-Campos, M., et al. (2014). Metabolic reprogramming of stromal fibroblasts through p62-mTORC1 signaling promotes inflammation and tumorigenesis. *Cancer Cell, 26*, 121–135.

van Bragt, M. P., Hu, X., Xie, Y., & Li, Z. (2014). RUNX1 a transcription factor mutated in breast cancer, controls the fate of ER-positive mammary luminal cells. *eLlife, 3*, 03881.

Van der Auwera, I., Yu, W., Suo, L., Van Neste, L., van Dam, P., Van Marck, E. A., et al. (2010). Array-based DNA methylation profiling for breast cancer subtype discrimination. *PloS One, 5*, e12616.

Vecellio, M., Roberts, A. R., Cohen, C. J., Cortes, A., Knight, J. C., Bowness, P., et al. (2016). The genetic association of RUNX3 with ankylosing spondylitis can be explained by allele-specific effects on IRF4 recruitment that alter gene expression. *Annals of the Rheumatic Diseases, 75*, 1534–1540.

Vogelstein, B., Papadopoulos, N., Velculescu, V. E., Zhou, S., Diaz Jr., L. A., & Kinzler, K. W. (2013). Cancer genome landscapes. *Science, 339*, 1546–1558.

Waldner, M. J., Wirtz, S., Jefremow, A., Warntjen, M., Neufert, C., Atreya, R., et al. (2010). VEGF receptor signaling links inflammation and tumorigenesis in colitis-associated cancer. *The Journal of Experimental Medicine, 207*, 2855–2868.

Wang, Y. F., Broderick, P., Webb, E., Wu, X. F., Vijayakrishnan, J., Matakidou, A., et al. (2008). Common 5p15.33 and 6p21.33 variants influence lung cancer risk. *Nature Genetics, 40*, 1407–1409.

Wang, K., Yuen, S. T., Xu, J., Lee, S. P., Yan, H. H., Shi, S. T., et al. (2014). Whole-genome sequencing and comprehensive molecular profiling identify new driver mutations in gastric cancer. *Nature Genetics, 46*, 573–582.

Weersma, R. K., Zhou, L., Nolte, I. M., van der Steege, G., van Dullemen, H. M., Oosterom, E., et al. (2008). Runt-related transcription factor 3 is associated with ulcerative colitis and shows epistasis with solute carrier family 22, members 4 and 5. *Inflammatory Bowel Diseases, 14*, 1615–1622.

Whittle, M. C., Izeradjene, K., Rani, P. G., Feng, L., Carlson, M. A., DelGiorno, K. E., et al. (2015). RUNX3 controls a metastatic switch in pancreatic ductal adenocarcinoma. *Cell, 161*, 1345–1360.

Wiegand, K. C., Shah, S. P., Al-Agha, O. M., Zhao, Y., Tse, K., Zeng, T., et al. (2010). ARID1A mutations in endometriosis-associated ovarian carcinomas. *The New England Journal of Medicine, 363*, 1532–1543.

Wilderman, M. J., Kim, S., Gillespie, C. T., Sun, J., Kapoor, V., Vachani, A., et al. (2006). Blockade of TNF-alpha decreases both inflammation and efficacy of intrapulmonary Ad.IFNbeta immunotherapy in an orthotopic model of bronchogenic lung cancer. *Molecular Therapy, 13*, 910–917.

Winkler, A. E., Brotman, J. J., Pittman, M. E., Judd, N. P., Lewis Jr., J. S., Schreiber, R. D., & Uppaluri, R. (2011). CXCR3 enhances a T-cell-dependent epidermal proliferative response and promotes skin tumorigenesis. *Cancer Research, 71*, 5707–5716.

Wolpin, B. M., Rizzato, C., Kraft, P., Kooperberg, C., Petersen, G. M., Wang, Z., et al. (2014). Genome-wide association study identifies multiple susceptibility loci for pancreatic cancer. *Nature Genetics, 46*, 994–1000.

Woolf, E., Xiao, C., Fainaru, O., Lotem, J., Rosen, D., Negreanu, V., et al. (2003). Runx3 and Runx1 are required for CD8 T cell development during thymopoiesis. *Proceedings of the National Academy of Sciences of the United States of America, 100*, 7731–7736.

Woolf, E., Brenner, O., Goldenberg, D., Levanon, D., & Groner, Y. (2007). Runx3 regulates dendritic epidermal T cell development. *Developmental Biology, 303*, 703–714.

Wu, J. N., & Roberts, C. W. (2013). ARID1A mutations in cancer: Another epigenetic tumor suppressor? *Cancer Discovery, 3*, 35–43.

Wu, D., Tian, Y., Gong, W., Zhu, H., Zhang, Z., Wang, M., et al. (2009). Genetic variants in the Runt-related transcription factor 3 gene contribute to gastric cancer risk in a Chinese population. *Cancer Science, 100*, 1688–1694.

Yamazaki, K., Umeno, J., Takahashi, A., Hirano, A., Johnson, T. A., Kumasaka, N., et al. (2013). A genome-wide association study identifies 2 susceptibility Loci for Crohn's disease in a Japanese population. *Gastroenterology, 144*, 781–788.

Yeddula, N., Xia, Y., Ke, E., Beumer, J., & Verma, I. M. (2015). Screening for tumor suppressors: Loss of ephrin receptor A2 cooperates with oncogenic KRas in promoting lung adenocarcinoma. *Proceedings of the National Academy of Sciences of the United States of America, 112*, E6476–E6485.

Yoshida, C. A., Yamamoto, H., Fujita, T., Furuichi, T., Ito, K., Inoue, K., et al. (2004). Runx2 and Runx3 are essential for chondrocyte maturation, and Runx2 regulates limb growth through induction of Indian hedgehog. *Genes & Development, 18*, 952–963.

Zack, T. I., Schumacher, S. E., Carter, S. L., Cherniack, A. D., Saksena, G., Tabak, B., et al. (2013). Pan-cancer patterns of somatic copy number alteration. *Nature Genetics, 45*, 1134–1140.

Zang, Z. J., Cutcutache, I., Poon, S. L., Zhang, S. L., McPherson, J. R., Tao, J., et al. (2012). Exome sequencing of gastric adenocarcinoma identifies recurrent somatic mutations in cell adhesion and chromatin remodeling genes. *Nature Genetics, 44*, 570–574.

Zenewicz, L. A., Yancopoulos, G. D., Valenzuela, D. M., Murphy, A. J., Stevens, S., & Flavell, R. A. (2008). Innate and adaptive interleukin-22 protects mice from inflammatory bowel disease. *Immunity, 29*, 947–957.

Zhang, B., Jia, W. H., Matsuda, K., Kweon, S. S., Matsuo, K., Xiang, Y. B., et al. (2014). Large-scale genetic study in East Asians identifies six new loci associated with colorectal cancer risk. *Nature Genetics, 46*, 533–542.

Zhao, C., & Bu, X. (2012). Promoter methylation of tumor-related genes in gastric carcinogenesis. *Histology and Histopathology, 27*, 1271–1282.

Zouridis, H., Deng, N., Ivanova, T., Zhu, Y., Wong, B., Huang, D., et al. (2012). Methylation subtypes and large-scale epigenetic alterations in gastric cancer. *Science Translational Medicine, 4*, 156ra140.

Roles of RUNX Complexes in Immune Cell Development

24

Takashi Ebihara, Wooseok Seo, and Ichiro Taniuchi

Abstract

During hematopoiesis, a variety of cells are generated from stem cells through successive rounds of cell fate determination processes. Studies in the last two decades have demonstrated the involvement of Runx transcription factor family members in differentiation of multiple types of hematopoietic cells. Along with evolutionary conservation, the Runx

family is considered to be one of the ancestral regulators of hematopoiesis. It is conceivable that the Runx family is involved in shaping the immune system, which is then comprised of innate and acquired lymphoid cells in vertebrates. In this chapter, we will first summarize roles of Runx proteins during the development of T- and B-lymphocytes, which appeared later during evolution and express antigen specific receptors as a result of DNA recombination processes. We also discuss the recent findings that have unraveled the functions of Runx during differentiation of innate lymphoid cells (ILCs).

Keywords

Runx family • Hematopoiesis • Lymphocytes • ILCs • Transcription factor • Development

24.1 Introduction

As expected from many studies showing multiple roles of Runx family in the control of development of many types of cells, genetic ablation of either Runx1 or Cbfβ resulted in lack of definitive hematopoiesis (Okuda et al. 1996; Wang et al. 1996a, b), placing Runx1/Cbfβ as one of the top regulators in the development of hematopoietic cells. Studies in invertebrates such as *Drosophila* also

T. Ebihara
Howard Hughes Medical Institute, Department of Medicine, Washington University School of Medicine, 660 South Euclid Avenue, St. Louis, MO 63110-1093, USA

W. Seo • I. Taniuchi (✉)
Laboratory for Transcriptional Regulation, RIKEN Center for Integrative Medical Sciences (IMS), 1-7-22 Suehiro-cho, Tsurumi-ku, Yokohama 230-0045, Japan
e-mail: ichiro.taniuchi@riken.jp

Y. Groner et al. (eds.), *RUNX Proteins in Development and Cancer*, Advances in Experimental Medicine and Biology 962, DOI 10.1007/978-981-10-3233-2_24

found that Runx family proteins play essential roles in hematopoiesis (Fossett and Schulz 2001; Lebestky et al. 2000), suggesting the ancestral function of Runx as an important regulator that controls the differentiation of hematopoietic cells (Braun and Woollard 2009). In this chapter, we summarize the roles of the Runx family in development of immune cells including lymphocytes and innate lymphoid cells (ILCs) in vertebrates.

24.2 Roles of Runx Complex in T Lymphocyte Development

24.2.1 Early Thymocyte Differentiation

After T-cell homing, precursors migrate from the fetal liver to the thymus rudiments around embryonic day 11.5–12.5 in mice, they colonize there and receive environmental cues to begin development into T-lymphoid cells. The expression of Notch ligands such as Delta-like 4 by thymic epithelial cells (TEC) is known to be a critical environmental signal for early thymocyte development (Hozumi et al. 2008; Koch et al. 2008). These early T cell progenitors (ETP) undergo sequential processes, which are controlled by transcription factor networks, in order to fully commit to the T cell lineage (Fig. 24.1) (Rothenberg et al. 2008). Since ETPs do not express CD4 and CD8 co-receptors, they are referred to as $CD4^-CD8^-$ double negative (DN) thymocytes, which are further divided into four subsets by the expression of CD25 and CD44 or CD117 (c-Kit). $CD25^-CD44^+$ DN1 subset is relatively heterogeneous (Porritt et al. 2004) and contains ETPs that differentiate into $CD25^+CD44^+$ DN2 cells. Recent studies addressing the transcriptional regulation of early T cell development have proposed that the DN2 stage is further divided into DN2a and DN2b, and that complete commitment to the T cell lineage occurs at the DN2a and DN2b transition in a Bcl11b dependent manner, based on the observation that T cell development was arrested at the DN2a stage in Bcl11b-deficient mouse (Li et al. 2010; Ikawa et al. 2010). Importantly, in host mice where T cell development is reconstituted from Runx1-deficient hematopoietic stem cells (HSCs), a similar developmental block at the DN2 stage was observed (Fig. 24.1) (Ichikawa et al. 2004). Although the precise mechanism of how lack of Runx1 is involved in DN2a arrest is not clear, it is possible that Runx1 is necessary for induction of *Bcl11b* expression. Indeed association of Runx with a 3′ downstream enhancer that plays an essential role to drive Bcl11b gene expression in T-cells was recently reported (Li et al. 2013).

There are two types of T-lymphocytes that express distinct T-cell antigen receptors, αβTCR and γδTCR. Although how and when the progenitors decide to become αβT cells or γδT cells remains controversial, there are ample studies addressing how the genes encoding α, β, δ and γ chains of TCR are activated and undergo DNA recombination, known as V(D)J recombination mediated by RAG-1/2 recombinase complexes at the antigen receptor loci. V(D)J recombination is a unique property endowed only to lymphocytes among somatic cells and is strictly controlled. Since there are other excellent reviews discussing the mechanisms of V(D)J recombination (Majumder et al. 2015), we simply focus on the effects of enhancer inactivation of each T-cell antigen gene. It has been shown that the Eα enhancer, Eβ enhancer, Eδ enhancer, and Eγ enhancer at the *Tcra*, *Tcrb*, *Tcrd*, and *Tcrg* genes, respectively, are essential for this recombination. Interestingly, each enhancer contains Runx recognition motifs (Hsiang et al. 1993; Redondo et al. 1991; Takeda et al. 1990; Hernandez-Munain and Krangel 1995) and is bound by Runx complexes (Tani-Ichi et al. 2011; Oestreich et al. 2006; Hollenhorst et al. 2007). Among these enhancers, Runx sites often locate closely to Ets binding sites (Wotton et al. 1994; Hernandez-Munain et al. 1998). For instance, in the Eβ enhancer, two ETS-Runx elements have been identified. Targeted mutations abrogating two Runx sites within the Eβ enhancers eliminated the enhancer function (Majumder et al. 2015), suggesting that Runx binding is essential for enhancer activation (Fig. 24.1). However, due to the arrest of DN2 stage through lack of Runx1 activity, the role of Runx1 function in the regula-

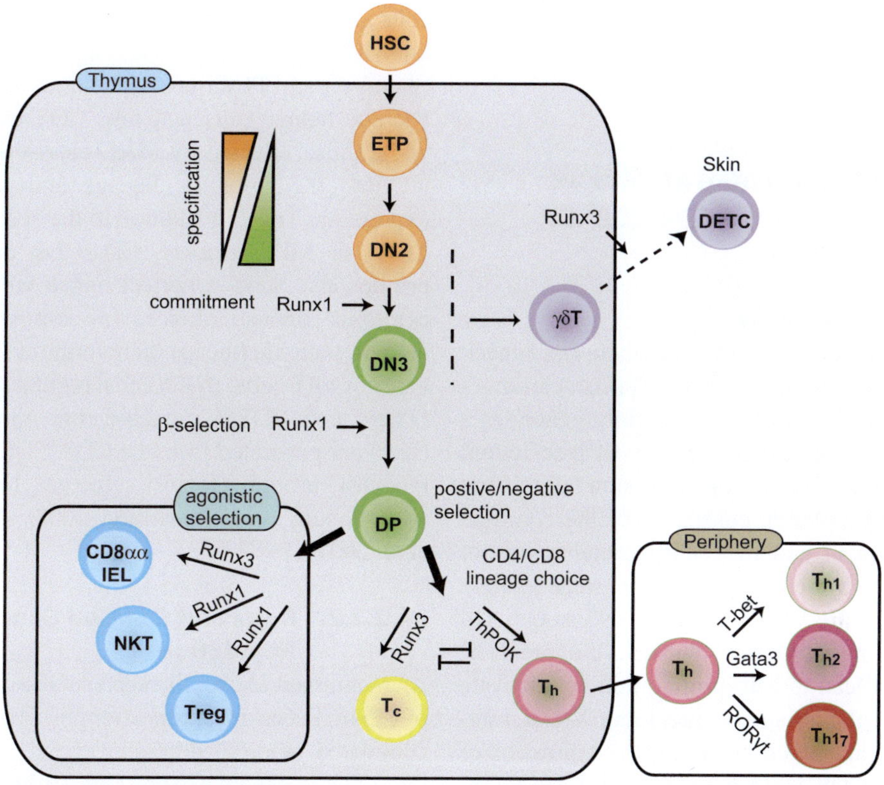

Fig. 24.1 Runx and T cell development. Upon prethymic acquisition of thymus homing property at some point during differentiation of hematopoietic stem cells (HSC), early thymocyte progenitors (ETP) colonize at the thymus and begin to develop into T lymphocyte-lineage by cascading activation of T cell programs as well as erasing developmental potency to alternative lineages, referred to as commitment process. Full commitment to T-lineage occurs at transition from DN2 to DN3 stage, and Runx1 is essential for this transition. γδT cells are differentiated from DN2/DN3 cells and development of skin-specific γδT cells, known as dendritic epidermal T cells (DETC), requires Runx3. Runx1 is important to pass β-selection. $CD4^{+}CD8^{+}$ DP thymocyte precursors are selected according to affinity of TCR with self-peptide (positive/negative selection) and positively selected thymocytes differentiate into CD4 helper (Th) or CD8 cytotoxic (Tc) T cells (CD4/CD8 lineage choice). Cross-antagonism between Runx3 and ThPOK plays a central role to fine separation of two fates as well as to couple MHC specificity of TCRs with appropriate fate. Some DP precursors develop into regulatory T cells (Treg), natural killer T cells (NKT) and CD8aa intra epithelial cells (IEL) through process known as agonistic selection, during which Runx1 and Runx3 play important roles for their differentiation. In the periphery, upon encountering antigens and depending on environmental cues, Th cells differentiate into distinct types of effector cells such as Th1, Th2 and Th17, each of which secrete signature cytokine, IFNγ, IL-4 and IL-17, respectively. Roles of Runx proteins during effector Th subset differentiation are discussed in the Sect. 24.2.3.

tion of *Tcrb* gene activation remains unclear. It is noteworthy, however, that in addition to enhancers in the T- cell antigen receptor loci, a μ enhancer in the *Igh* locus that encodes the B-cell antigen receptor also contains ETS-Runx composite elements (Erman et al. 1998). Thus, there might be a common regulatory mechanism by which antigen receptor loci are activated through co-binding of ETS and Runx to essential enhancers in these loci. Consistent with this note, skin specific γδT cells, referred to as DETC (dendritic epidermal T cells) expressing the invariant Vγ3 chain, were absent in the epidermis of Runx3-deficient mice (Fig. 24.1) (Woolf et al. 2007). However, fewer precursors expressing the Vγ3 chain were observed in the fetal thymus of Runx3-deficient mice, suggesting that Runx3 is also involved in the processes of expansion driven by interleukin 2 (IL-2) and migration guided by integrin CD103 for DETC differentiation after γδTCR expression.

24.2.2 Differentiation of αβT Cell Subsets from CD4⁺CD8⁺ DP Precursors

24.2.2.1 Overview

During αβT cell differentiation, there is a "check point" stage to monitor whether V(D)J rearrangement at the *Tcrb* locus successfully generates a functional TCRβ chain, known as β-selection. Cells lacking TCRβ chain expression fail to form pre-TCR complexes, consisting of the common pre-Tα protein and the TCRβ protein, are not allowed to differentiate to the next stage and are arrested at the $CD25^+CD44^-$ DN3 stage. On the other hand, signals through the pre-TCR complexes lead to cell proliferation and activate developmental programs including V(D)J rearrangement at the *Tcra* locus and the expression of CD4 and CD8 co-receptors. Thus $CD4^-CD8^-$ DN thymocytes become $CD4^+CD8^+$ DP thymocytes expressing the mature αβTCR with diverse antigen-specificity and face another selection process, known as positive/negative selection, which evaluates the quality of αβTCR in terms of their affinity to self-peptides presented on the major histocompatibility complex (MHC). Lack of TCR-mediated signals due to insufficient affinity to self-peptide–MHC complexes causes apoptosis, designated as 'death by neglect'. Cells with TCRs that react too strongly to self-peptide–MHC complexes, are thereby thought to be potentially self-reactive lymphocytes and are eliminated through a 'negative selection' process, to reduce the risk of auto-immunity. Only a few DP cells that express TCRs of appropriate affinity with self-peptide–MHC complexes are selected in a process known as 'positive selection', and proceed to develop into mature thymocytes. The CD4 and CD8 co-receptors help the TCRs to recognize self-peptides on MHC-class-II and MHC-class-I molecules via specific interaction with class-II and class-I, respectively. After positive selection, two major αβT-lineages, helper cells and cytotoxic cells are generated. It is well known that cells selected through MHC-class-II (thereby their TCRs are MHC-class-II specific) become helper cells and lose CD8 expression, while those selected by MHC-class-I differentiate into the cytotoxic lineage and lose CD4 expression. Thus, in addition to the specificity of TCRs to MHC classes, CD4/CD8 expression profiles also show a perfect match with helper/cytotoxic lineage choice. In addition to the helper/cytotoxic lineage dichotomy, at least three types of αβT cells, iNKT cells, regulatory T cells (Treg), and CD8aa intraepithelial lymphocytes (IEL) are generated from the $CD4^+CD8^+$ DP thymocytes through 'agonist selection' (Fig. 24.1) (Kronenberg and Rudensky 2005; Lambolez et al. 2007).

24.2.2.2 Runx and *Cd4/Cd8* Gene Regulation

As mentioned above, lineage specific expression of CD4/CD8 co-receptors after positive selection stimulated studies addressing the regulation of this lineage specific expression (Ellmeier et al. 1999). The search for critical *cis*-regulatory elements in the *Cd4* and *Cd8* loci first resulted in the identification of a transcriptional silencer in the *Cd4* locus as a critical regulatory region for helper-lineage specific CD4 expression (Sawada et al. 1994; Siu et al. 1994). A 434 bp transcriptional silencer, located at the first intron in the *Cd4* gene, is sufficient to repress reporter transgene expression in DN thymocytes and cytotoxic-lineage cells. Furthermore, following studies that removed the *Cd4* silencer from the mouse genome confirmed its relevance by revealing de-repression of CD4 in $CD8^+$ T cells at a comparable level to that in $CD4^+$ T cells (Leung et al. 2001; Zou et al. 2001). These results from mouse genetics clearly indicated that a single transcriptional silencer is responsible for helper-lineage specific CD4 expression by repressing *Cd4* in alternative cytotoxic-lineage T cells.

On the other hand, functionally negative regulatory elements have not been identified in the *Cd8* locus. Instead, at least six enhancer elements, termed as E8I to E8VI, have been isolated (Hostert et al. 1998; Ellmeier et al. 1998;

Sakaguchi et al. 2015). The effects of each enhancer deletion on *Cd8* expression revealed redundant function within the enhancers. Along with the CD8-lineage specific activity of E8I in a reporter transgene expression assay (Ellmeier et al. 1997), a combinatory regulation of enhancers rather than active repression by silencer(s) is supposed to control the lineage specificity of *Cd8* expression.

Runx proteins are known to directly regulate both *Cd4* and *Cd8* expression. Runx1 protein was isolated from a search for a *Cd4* silencer binding protein using the yeast one-hybrid screen (Taniuchi et al. 2002). Along with the essential requirement for the Runx recognition site for *Cd4* silencer activity, conditional inactivation of Runx1 in DN2/3 thymocytes by the *Lck-Cre* transgene resulted in CD4 expression in DN3 cells as was observed by lack of *Cd4* silencer (Taniuchi et al. 2002). On the other hand, loss of Runx3 but not Runx1, severely affected *Cd4* silencer activity in CD8-lineage cells (Taniuchi et al. 2002) as was manifested by de-repression of CD4 in mature Runx3-deficient $CD8^+$ T cells. Such a distinct role of two Runx proteins in *Cd4* gene repression at two stages may reflect different expression patterns of Runx1 and Runx3. While Runx1 is most highly expressed in immature DN and DP thymocytes, expression of Runx3 was nearly specific to CD8 SP thymocytes (Egawa et al. 2007). In addition to the involvement of Runx proteins in *Cd4* gene regulation, the binding of Runx3 to some enhancers in *Cd8* gene was also reported (Sato et al. 2005). Recently, the functional contribution of Runx to *Cd8* gene expression was demonstrated by the inefficient maintenance of CD8 expression in the absence of Runx3 (Hassan et al. 2011). Such dual roles of Runx3 in co-receptor gene expression, *Cd4* silencing, and *Cd8* reactivation, which comprise a key feature of cytotoxic-lineage cells predicted that induction of Runx3 expression might be a key event to activate programs that induce a cytotoxic fate. It should be noteworthy that transgene-mediated ectopic expression of Runx3 in immature thymocytes only partially redirected MHC-class-II-specific T cells to $CD8^+$ T-cells (Kohu et al. 2005; Grueter et al. 2005), at least in part via low CD4 expression on precursor cells as a result of prolonged *Cd4* silencer activity in the DN stage.

24.2.2.3 Antagonistic Interplay Between Runx and ThPOK

ThPOK is a member of the BTB/POZ transcription factor family; many members of this family have been shown to play essential roles in immune cell development (Ellmeier and Taniuchi 2014). Gain- and loss-of-function studies of ThPOK in mice have revealed that ThPOK is a master transcription factor for $CD4^+$ helper T cell development (Kappes and He 2006). For instance, a natural mutation referred to as the *hd* mutation that results in substitution of glycine for arginine in the putative second zinc-finger domain of the ThPOK protein, led to a loss of $CD4^+$ T cell development through fate conversion of MHC-class-II specific cells into $CD4^-CD8^+$ T cells (He et al. 2005). On the contrary, ectopic expression of ThPOK from DP precursors onwards resulted in lack of $CD8^+$ T cells due to redirection of MHC-class-I-specific thymocytes to $CD4^+CD8^-$ T cells (He et al. 2005).

This striking finding that the presence or absence of a single transcription factor, ThPOK, serves as a major determinant for CD4 helper versus CD8 cytotoxic lineage separation, raised the profound question of how helper-lineage-specific expression of ThPOK is controlled. The answer was revealed during characterization of Runx mutant mice. One issue to be considered while interpreting phenotypes caused by single Runx protein ablation is the redundancy between Runx family proteins. In particular, cross-regulation between Runx1 and Runx3 sometimes underestimates the effect caused by the lack of either protein. Given that the product of a single gene *Cbfβ* is the sole common subunit for all Runx proteins in mammals, inactivation of *Cbfβ* has the advantage in terms of avoiding redundancy between Runx proteins. It should be noted that the Cre/loxP-mediated recombination system occasionally suffers from the expansion of a leaky population that escapes Cre-mediated recombination. In addition, protein stability is another factor to be considered in Cre/loxP mediated conditional gene inactivation. Compared to Runx proteins that could be rapidly degraded by proteasomes, the Cbfβ protein seemed to be present longer after the inactivation of its gene. Thus, in thymocyte differentiation, inactivation of *Cbfβ* gene at the DN stage by *Lck-Cre* nearly recapitulated the compound inactivation of both *Runx1* and *Runx3* at the DP stage by *Cd4-Cre*. Analyses of T-cell development in these mice showed that loss of the Runx complex function in DP thymocytes led to severe reduction of mature thymocyte generation. Most importantly, in the remaining mature T-cell pool, $CD8^+$ T cells were almost absent (Setoguchi et al. 2008). Further analyses using mice in which *Runx3* inactivation was combined over *Runx*1 mutation, causing the deletion of the VWRPY motif at C-terminal end, revealed that redirection of MHC-class-I-specific T cells to $CD4^+$ T-cells was the reason for the loss of $CD8^+$ T cells. This was a phenocopy of the ThPOK transgenic mice, and prompted analyses of $Cbf\beta^{F/F}$*: Cd4-Cre* mice that retain a substantial number of $CD8^+$ T cells that abnormally de-repress CD4. These $CD4^+CD8^+$ cells developed under the gradual loss of Cbfβ protein after positive selection, expressed *Thpok* gene, providing supportive evidence that redirection of MHC-class-I-specific thymocytes to the $CD4^+$ T-cell lineage in mice lacking functional Runx proteins is due to inappropriate de-repression of *Thpok*. These observations clearly indicate that Runx proteins are involved in *Thpok* repression.

Naturally, the next question would be how Runx proteins are involved in *Thpok* regulation and whether Runx proteins directly regulate the regulatory regions of the *Thpok* gene. By using a ChIP-on-chip approach, two regions in the *Thpok* gene were identified as Runx binding sequences (RBS-1 and RBS-2). Further functional analysis of RBS-1 in a reporter transgene expression assay identified transcriptional silencer activity in this sequence. At the same time, a group led by Dr. Kappes also characterized the regulatory regions in *Thpok* and identified a silencer activity in a their distal regulatory element (DRE), that perfectly overlapped with RBS-1 (He et al. 2008). These observations indicate that silencer activity in DRE/RBS-1, hereafter referred to as *Thpok* silencer, is responsible for helper lineage specific expression via repression of *Thpok* expression in cytotoxic-lineage cells. The physiological relevance of *Thpok* silencer was confirmed by full *Thpok* de-repression as well as by the loss of $CD8^+$ T cells upon its removal from mouse genomes (Setoguchi et al. 2008). Importantly, the analytical ChIP assay detected Runx binding to *Thpok* silencer in both ThPOK-expressing helper and non-expressing cytotoxic cells (Setoguchi et al. 2008). This observation indicates that Runx binding is essential but not sufficient for *Thpok* silencer activity and that uncharacterized mechanisms beyond Runx binding may serve as the switch that controls the specificity of *Thpok* silencer activity.

Thus two silencers at different loci, the *Cd4* silencer and the *Thpok* silencer, require Runx protein binding to exert their silencer activity. Interestingly, albeit this common feature, both silencers show distinct dependency for VWRPY motifs at the C-terminal end of Runx proteins. While *Cd4* silencer activity depends completely on the VWRPY motif, *Thpok* silencer still represses the *Thpok* gene to a significant extent without the VWRPY motif (Seo et al. 2012b).

Collectively, one of the most important functions of Runx proteins during cell fate decision by $CD4^+CD8^+$ DP precursors is the repression of *Thpok* as well as *Cd4* genes. On the other hand, characterization of ThPOK function revealed that ThPOK acts to restrain Runx3 expression in $CD4^+$ T cells. Thus, ThPOK and Runx3 repress the expression of each other, forming an antagonistic interplay. Since such antagonism between two transcription factors, which play a central role in the development of alternative lineages, is often observed at developmental branch point (Laiosa et al. 2006), the cross antagonism between ThPOK and Runx3 serves as a central mechanism in CD4 helper/CD8 cytotoxic dichotomy (Fig. 24.1). Given that CD8-lineage specific expression of Runx3 reflects a CD8-lineage specific expression from the distal P1 promoter-driven *Runx3* transcript (Egawa et al. 2007), ThPOK should be involved in the regulatory mechanism that controls the lineage specific activity of the distal P1 promoter. IL-7 signals are known to be important for the generation of CD8-lineage cells in the thymus and can activate *Runx3* (Park et al. 2010). In line with this finding, it was proposed that ThPOK is indirectly involved in Runx3 repression through induction of the SOCS protein family, a strong inhibitor of the IL7 cytokine signals, thereby preventing Runx3 induction (Luckey et al. 2014). Unfortunately, at this moment, little is known about the regulatory regions that control *Runx3* expression in T cells. Further studies are necessary to precisely understand the regulation of CD8-lineage specific Runx3 expression.

24.2.2.4 Roles of Runx in Differentiation of Treg and iNKT Cells

Beyond their role in regulating the CD4/CD8 lineage separation, Runx proteins are known to have other important roles in development of T cell subsets. Results of Runx1 inactivation alone showed that the efficiency of β-selection and positive selection was impaired (Egawa et al. 2007). In addition, generation of invariant natural killer T (NKT) cells expressing the invariant Vα14 chain and the reactive lipid antigen on MHC class –I related CD1d was lost by lack of Runx1 (Egawa et al. 2005). Given that Runx3 is also expressed in iNKT cells, Runx3 is not likely to possess a compensatory function to support iNKT cell differentiation.

Another αβT cell subset generated from $CD4^+CD8^+$ DP precursors includes the regulatory T (Treg) cells (Ohkura et al. 2013; Josefowicz et al. 2012). Treg cells have a suppressive function and play an essential role in immune tolerance. FoxP3, a member of the forkhead box transcription factor family, is essential for the generation and function of Treg cells. Mutation of X chromosome-linked human *FOXP3* gene result in the IPEX (immunodysregulation, polyendocrinopathy, enteropathy, X-linked) syndrome, showing multi-organ autoimmune inflammatory disease, whereas mutations in the murine *Foxp3* gene, for instance, a natural mutation in the scurfy strain or engineered loss of the functional mutation, are known to cause severe postnatal lethal autoimmune disorders (Josefowicz et al. 2012). Thus, how the expression and functions of Foxp3 are regulated, is an essential and profound question in immunology. A study by the Sakaguchi's group has shown that RUNX1 can interact with FOXP3 protein by the immunoprecipitation assay (Ono et al. 2007). RUNX1 seemed to bind to the IL2 promoter and enhance IL-2 production upon TCR stimulation, which is repressed by FOXP3. In addition to regulation at the protein level, Runx proteins were shown to regulate *Foxp3* gene expression through binding to promoters and CNS2 enhancers (Bruno et al. 2009; Kitoh et al. 2009; Rudra et al. 2009; Klunker et al. 2009). A CNS2 enhancer undergoes Treg specific DNA demethylation processes and contributes to stable expression of the *Foxp3* gene (Zheng et al. 2010). Interestingly, CNS2 remained methylated in Treg cells lacking Cbfβ and *Foxp3* expression level was reduced (Rudra et al. 2009; Kitoh et al. 2009). Results of Treg specific conditional inactivation of *Runx1* or *Runx3* genes showed that Runx1/Cbfβ complexes are responsible for maintaining Foxp3 expression levels and thereby preventing immunological disorders such as gastritis, serum IgE elevation, and lymphadenopathy (Kitoh et al. 2009).

The third αβT cells subset generated through agonistic selection of $CD4^+CD8^+$ DP thymocytes include the CD8αα IEL cells due to their unique expression profile of CD8αα homodimers but not CD8αβ heterodimers as well as the tissue localization in the space between gut epithelial cells. Distinct from the CD8αβ heterodimer, CD8αα homodimers interact with the MHC class I like TL molecule (Leishman et al. 2001). By using a soluble TL tetramer, the sole reagent that is useful to separately detect CD8αα from CD8αβ on cells expressing both forms, it was shown that a proportion of DP thymocytes already expressed the CD8aa homodimer, thereby referred to as $CD4^+CD8\alpha\beta^+CD8\alpha\alpha^+$ triple positive (TP) thymocytes (Gangadharan et al. 2006). TP thymocytes are supposed to be precursors for CD8αα IEL, and they become $CD4^-CD8^-$ DN thymocytes after agonistic selection. Runx3 is essential for the re-expression of CD8αα (Pobezinsky et al. 2012), presumably through direct activation of E8I enhancers.

24.2.3 Role of Runx in Differentiation of Effector T Cells

Advances in the last decade have identified and characterized novel effector CD4$^+$ T cell subsets beside the classical Th1 and Th2 cells. In addition, these subsets retain a plasticity that allows them to occasionally modulate their identity (Nakayamada et al. 2012). A characteristic difference in the functions of these subsets is the pattern of cytokine secretion conferred by the induction of specific transcription factors, as is known in the case of classical Th1/Th2 subsets, these signature cytokines and transcription factors are IFNγ/Il-4 and T-bet/Gata3, respectively. Runx proteins have also been shown to regulate effector T cell subsets and cytokine expression. During Th1 cell differentiation, expression of Runx1 and Runx3 proteins exhibited unique reciprocal expression kinetics. Runx1 expression is downregulated while Runx3 expression is induced. Importantly, in differentiated Th1 cells, Runx3 represses *Il4* gene transcription through binding to an *Il4* silencer (Naoe et al. 2007; Djuretic et al. 2007).

Discovery of Th17 cells, whose characteristic cytokine is IL17, led to a renewed view of helper T cell differentiation as well as pathogenesis of autoimmune diseases (Harrington et al. 2005; Park et al. 2005). Detailed characterization of this IL17 producing cell subset including a comparison of gene expression profiles with other immune cells, combined with mouse genetics, identified RORγt (retinoic-acid-receptor-related orphan receptor-γt) as the master driver of Th17 differentiation (Ivanov et al. 2006). Enforced expression of Runx1 has been shown to induce IL17 expression, and vice versa, diminished Runx1expression by shRNA based knockdown showed an inhibitory effect against Th17 differentiation (Zhang et al. 2008). This Runx1 activity was reported to be mediated through direct regulation of the promoter/enhancer after protein complex formation with RORγt.

24.3 Roles of Runx Complexes in Early B Lymphocyte Development

B cell development occurs in the bone marrow in adult mice and initiates from the common lymphoid progenitors (CLPs) (Matthias and Rolink 2005). When CLPs adopt the B cell fate by starting to express the B cell lineage marker B220 and Rag recombinases, they are referred to as pre-pro-B cells. Once V(D)J fragments of immunoglobulin heavy chain locus (*Igh*) are successfully rearranged during the pro-B cell stage, the pre-B cell receptor is formed with the surrogate light chain and drives the transition from pro-B cells to pre-B cells. Like other lineages of blood cell types, it is no exception that early B cell development requires the expression of a series of transcription factors to activate B cell programming as well as to erase the potentials for alternative lineages. A widely-accepted model for transcription factor networks for B cells involves E2A, EBF, and Pax5, and a cascading cross-regulation among these three factors is central to the network. From a simplistic point of view, E2A activates EBF (Kee and Murre 1998), and EBF activates Pax5 (Decker et al. 2009).

Disruption of E2A or EBF (or both) results in developmental arrest at the pre-pro-B cell stage (Zhuang et al. 1994; Bain et al. 1994; Lin and Grosschedl 1995). Since E2A and EBF directly regulate genes specific to early B cell progenitors such as Rag1/2, Igαβ, VpreB, and λ5, these two factors have been considered as the main specification factors. On the other hand, inactivation of Pax5 results in blockade of B cell development at the pro-B-cell stage and such pro-B cells show the potential to follow other lineages after *in vivo* transplantation (Nutt et al. 1999; Rolink et al. 1999). Combined with the fact that Pax5 actively represses non-B-lineage genes (Mikkola et al. 2002), Pax5 has been considered to be the main commitment factor for B lymphoid-lineages. However, during recent years, other transcription factors important for B cell development in addition to E2A, EBF, and Pax5 have been identified, which include but are not limited to Bcl11a, Runx1, and Foxo1.

By using inducible-targeting strategies to overcome embryonic lethality, Runx1 was shown to have a pivotal role mainly in priming the lymphoid lineage (both B and T cells) during hematopoiesis (Ichikawa et al. 2004; Growney et al. 2005). More specifically, these studies clearly demonstrated that Runx1 is involved in the regulation of B cell generation since almost no B cells were developed in the absence of Runx1. A clue to the molecular mechanisms underlying these observations was obtained from a study showing that E2A and EBF cooperate with Runx1 to activate B cell programs before the involvement of Pax5 (Maier et al. 2004). Specifically, progressive demethylation of the *mb-1* gene, which encodes a critical signaling component of the pre-B cell receptor CD79a, requires synergistic activities of both EBF and Runx1. This poised epigenetic change in the *mb-1* gene is necessary for Pax5 to activate the *mb-1* promoter.

A further attempt was made to decipher the precise relationship between Runx1 and EBF during B lymphopoiesis by conditionally inactivating *Runx1*, *Runx3*, and *Cbfβ* genes by *mb1-cre* transgene (Seo et al. 2012a). Consistent with previous reports, inactivation of *Runx1* and *Cbfβ* (but not *Runx3*) resulted in defective B cell generation from early B cell progenitors. An interesting observation from this study was that Runx1 seems to directly activate *Ebf* transcription, since the amounts of *Ebf* mRNA was dramatically reduced from the progenitors in the absence of Runx1. Importantly, B cell development of Runx1-deficient cells was rescued by over-expression of EBF at least *in vitro*, suggesting that the major reason for developmental arrest in *Runx1*$^{F/F}$*: mb-1Cre* mice could be the reduced EBF expression. Therefore, it is possible that Runx1, together with E2A, is an upstream factor for EBF even though they seem to function together once expressed at the later stages (Lukin et al. 2010; Lin et al. 2010). Further studies will be required to dissect the extent to which E2A, Runx1, and EBF cooperate and the unique functions endowed to each transcription factor.

24.4 Roles of Runx Complexes in Development of Innate Lymphoid Cells (ILCs) and Conventional NK Cells

24.4.1 Overview of ILCs

Innate lymphoid cells (ILCs) are lymphocytes that reside in the mucosa and produce innate cytokines independently of antigen specificity upon infection or allergic stimulation to maintain the epithelial barrier (Sonnenberg and Artis 2015; Eberl et al. 2015; Cortez et al. 2015). ILCs do not have rearranged antigen-specific receptors but are dependent on the common γ chain of the cytokine receptor IL-2R (γc) for their differentiation. Based on their cytokine production and transcription factor requirements, ILCs are comprised of three groups, type I ILC (ILC1 and conventional NK: cNK), type II ILC (ILC2) and type III ILC (ILC3) populations. ILC populations generally express a dimeric IL-7 receptor α chain (CD127)/γc complex with some exceptions as described below. However, IL-7 is required only for ILC2 and ILC3 but not for the type I ILC population which expresses an IL-2/IL-15R β chain (CD122)/γc complex and is dependent on IL-15 instead of IL-7 for differentiation (Sonnenberg and Artis 2015; Fuchs et al. 2013; Klose et al. 2014).

The type I ILC population expresses the transcription factor T-bet and produces the type I cytokine IFNγ in response to IL-12, IL-15, and IL-18 (Sonnenberg and Artis 2015; Fuchs et al. 2013; Klose et al. 2013, 2014). In mice, $CD3^-$ $NK1.1^+$ NKp46 $(NCR)^+$ cells in tissues are recognized as type I ILC populations. Given that the type I ILC population is defined by IFNγ-producing cells, cNK cells in the spleen could also be categorized as type I ILC populations. However, recent studies suggest that cNK cells are a distinct population from ILC1 cells that are resident in mucosal tissues (Sojka et al. 2014; Klose et al. 2014; Gasteiger et al. 2015; Constantinides et al. 2014). The transcription factor Eomes is expressed only in cNK cells and is required for their differentiation while ILC1 cells in mucosal tissues are negative for Eomes (Gordon et al. 2012; Daussy et al. 2014). cNK cells do not express CD127 which is one of the markers for ILC1 cells (Cella et al. 2014). Thus, we hereafter term cNK cells as the Eomes$^+$ tissue-nonresident type I ILC population and ILC1 cells as the Eomes$^-$ tissue-resident type I ILC population.

ILC2 cells respond to IL-25, IL-33 and TSLP derived from epithelial cells and produce type II cytokines, IL-5, IL-9, and IL-13 (Sonnenberg and Artis 2015; Moro et al. 2010; Neill et al. 2010). ILC2 cells are characterized by high GATA-3 expression, which is required for their differentiation and secretion of type II cytokines (Hoyler et al. 2012). In contrast, ILC1 and ILC3 cells express GATA-3 at intermediate levels (Serafini et al. 2014; Klose et al. 2014). Thus, GATA-3 deficiency affects all ILC populations. Other transcription factors such as RORα, Gfi1, and Bcl11b are also necessary for ILC2 differentiation (Wong et al. 2012; Spooner et al. 2013; Califano et al. 2015).

ILC3 cells react to IL-1β and IL-23 from dendritic cells and produce IL-17 and IL-22 (Sonnenberg and Artis 2015; Cella et al. 2009; Satoh-Takayama et al. 2008). RORγt and AHR regulate cytokine production by ILC3 cells and their differentiation (Sawa et al. 2010, 2011; Sanos et al. 2009; Lee et al. 2012; Kiss et al. 2011). According to CD4 and NCR expression, ILC3 cells in the adult intestine were originally sub-grouped into NCR$^+$ ILC3 (NK-22, ILC22), CD4$^+$ ILC3 and CD4$^-$ NCR$^-$ ILC3 cells (Sonnenberg and Artis 2015). More recently, it has been reported that ILC3 cells are comprised of two distinct populations, T-bet$^-$ CCR6$^+$ ILC3 (Lymphoid tissue inducer-like cells: LTi-like cells) and T-bet$^+$ CCR6$^-$ ILC3, because T-bet$^-$ CCR6$^+$ ILC3 cells do not give rise to T-bet$^+$ CCR6$^-$ ILC3 cells, and vice versa (Klose et al. 2013). While T-bet$^-$ CCR6$^+$ ILC3s are CD4$^+$ or CD4$^-$, T-bet$^+$ CCR6$^-$ ILC3 cells are mostly NCR$^+$. LTi cells in the fetus highly express RORγt and contribute to lymphoid tissue formation including Peyer's patches and lymph nodes. LTi cells are also recognized as members of ILC3 cells (Sonnenberg and Artis 2015).

24.4.2 Developmental Pathway of ILCs

All ILC subsets develop from CLPs in the bone marrow and fetal liver (Sonnenberg and Artis 2015; Possot et al. 2011). CLPs also give rise to T- and B-lymphocytes as well as ILCs including cNK cells (Fig. 24.2). Some ILC progenitors downstream of CLPs were identified in the Lin$^-$ α4β7$^+$ fraction of bone marrow cells. CXCR6$^+$α4β7$^+$ lymphoid progenitors (αLP) and early ILC progenitors (EILPs) lose the potential to become lymphocytes but maintain the capacity to differentiate into cNK, ILC1, ILC2, and ILC3 cells (Yang et al. 2015; Yu et al. 2014). EILPs are marked by the expression of the transcription factor TCF-1. Deficiency of TCF-1 leads to absence of all ILCs. Downstream of EILPs, two ILC-specific progenitor cells are designated as common helper ILC progenitors (CHILPs) expressing the transcription factor Id2, and the common precursor to ILCs (ILCPs) expressing the transcription factor PLZF (Klose et al. 2014; Constantinides et al. 2014). CHILPs give rise to ILC1, ILC2, and ILC3 cells but not cNK cells. Along with the expression of PLZF in a small proportion of CHILPs, CHILPs are supposed to be upstream of ILCPs. In line with this model, ILCPs have a limited capacity to differentiate into ILCs because LTi-like cells as well as cNK cells do not develop from ILCPs. ILC2 progenitors (ILC2Ps) are ILC2-specific progenitor cells that are downstream of ILCPs in the bone marrow. ILC1- and ILC3-specific ILC progenitor cells are ILC-lineage negative (ILCLN) cells that are characterized as Lin$^-$ CD127$^+$ lacking ILC markers (ILC1: NK1.1, ILC2: KLRG1, ILC3: RORγt) and reside in the adult intestine. ILCLN cells were suggested to be the progenitor cells that migrate to the gut and lose their homing receptor α4β7. Apparently, ILCLN cells are likely to be a heterogeneous population because they differentiate into NCR$^+$ ILC3 and LTi-like cells, each of which arises from a different developmental pathway (Fig. 24.2).

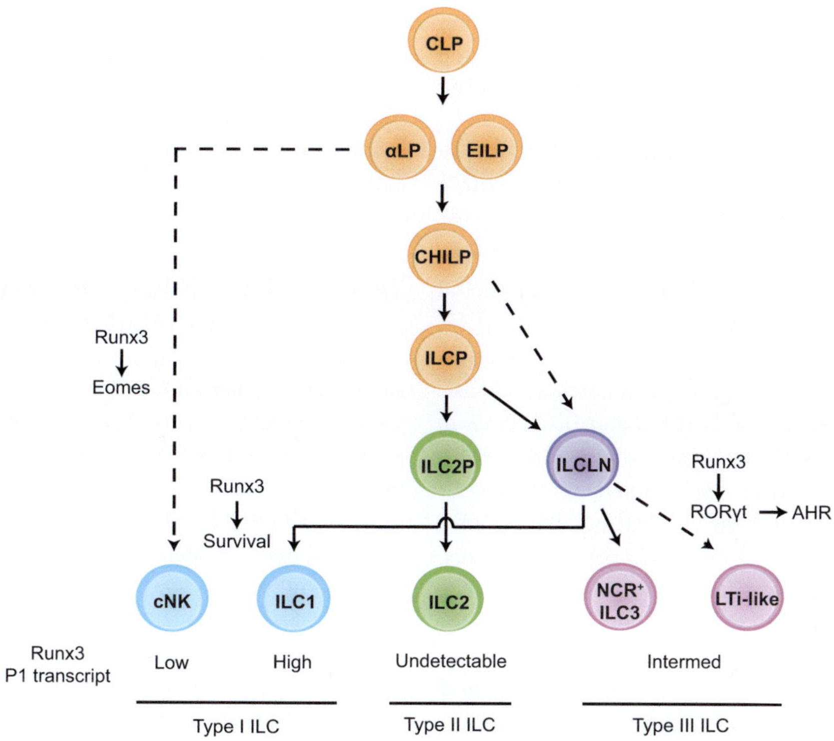

Fig. 24.2 Runx and ILC development. All ILC populations are the progeny of common lymphoid progenitors (CLPs) in the fetal liver and bone marrow. The ILC-committed precursor cells are CXCR6$^+$α4β7$^+$ lymphoid progenitors (αLPs) and early ILC progenitors (EILPs) that give rise to all ILC populations, but not T or B cells. The common helper ILC progenitors (CHILPs) do not differentiate into conventional NK (cNK) cells in spleen, but maintain a capacity for other tissue-resident ILC populations. The common precursor to ILCs (ILCPs) is downstream of CHILPs and do not differentiate into LTi-like cells. ILC-lineage negative (ILCLN) cells in the intestine are the progenitor cells specific to ILC1 and type III ILC3. All ILCs are differentially characterized by the levels of *P1-Runx3* transcripts. Roles of Runx3 during ILC specification are discussed in the Sect. 24.4.4

24.4.3 Runx Expression in ILC Progenitor Cells and ILC Subsets

Expression of *Runx1*, *Runx2*, and *Runx3* is relatively low in CLPs, αLPs, and CHILPs. However, both *Runx1* and *Runx3* are expressed at extremely high levels in PLZF$^+$ ILCPs (Ebihara et al. 2015). Runx3 transcript expression from the P1 promoter (*P1-Runx3*) is also very high in ILCPs and is correlated to PLZF expression during differentiation into CHILP, suggesting a possible cross-regulation between Runx3 and PLZF expression. However, the functions of Runx complexes in ILCPs remain to be elucidated. Given that ILC2Ps in the bone marrow and ILC2 cells in the intestine do not express *Runx3* from the P1 transcripts, it is possible that downregulation of *Runx3* expression is important for ILC2 differentiation. For ILC1 and ILC3 differentiation from ILCPs, *P1-Runx3* transcripts are reduced to intermediate level in ILCLN cells in the adult intestine. Then, upregulation of *P1-Runx3* occurs for ILC1 cells, while ILC3 cells maintain the intermediate level of it (Ebihara et al. 2015).

cNK cells develop from αLP and EILPs through NK progenitor cells (Lin$^-$ CD122$^+$ NK1.1$^-$ DX5$^-$). *P1-Runx3* transcripts are generally low throughout cNK cell differentiation compared to those in ILC1 cells in the intestine and liver (Ebihara et al. 2015; Ohno et al. 2008; Levanon et al. 2014).

ILC subsets in the intestine and cNK cells in the spleen have been well studied regarding Runx expression and function. In the intestine, cNK cells are relatively rare whereas other ILC subsets are dominant. The intestinal intraepithelial layer is enriched with ILC1 cells. ILC1, ILC2 and ILC3 cells are evenly distributed in the intestinal lamina propria and Peyer's patches. All ILC subsets including ILC1, ILC2 and ILC3 cells in the intestine predominantly express *Runx3* transcripts (Ebihara et al. 2015). However, ILC2 cells in the intestine use the P2, but not the P1 promoter although ILC1 and ILC3 cells in the intestine express only *P1-Runx3* transcripts. cNK cells express *Runx3* from both promoters, but the major transcript is the *P2-Runx3* (Ebihara et al. 2015; Levanon et al. 2014). When *P1-Runx3* transcripts are deleted in $CD8^+$ T cells, Runx3 protein is barely detected in αβT cells (Egawa et al. 2007; Egawa and Littman 2008) due to the inefficient activity of Kozak sequences to initiate translation for P2-Runx3 protein (Kim et al. 2015). When *P1-Runx3* transcript level was examined by a reporter allele that specifically reflects Runx3 P1-promoter activity, reporter expression was found to be very high in ILC1 cells, intermediate in ILC3 cells, low in cNK cells, and undetected in ILC2 cells. Given that *P1-Runx3* transcripts are correlated to Runx3 protein expression at least in T cells, differential expression pattern of *P1-Runx3* mRNA should be associated with Runx3 protein expression in ILCs. However, *P2-Runx3* transcripts in ILC2 cells might be translated into protein at least to some extent because high level of Runx3 protein can be detected in cNK cells when *P2-Runx3* transcripts are abundant (Levanon et al. 2014).

Runx3 is also expressed in ILC1 or cNK cells in other tissues. Liver-resident NK cells turned out to be ILC1 cells in which the *P1-Runx3 mRNA* amount was as high as those in the intestinal ILC1 cells (Ebihara et al. 2015). *Runx3* expression in skin-resident NK cells is as low as that in skin-nonresident NK cells. In the uterus, DBA^+ NK cells also express *Runx3* mainly from the P1 promoter (Levanon et al. 2014). Thus, among all Runx family members, Runx3 is the dominant Runx protein in all ILC subsets and is expressed highly in most type I ILC population and intermediately in ILC3 cells.

24.4.4 Roles of Runx Complexes in ILCs Development

24.4.4.1 Type I ILC

Several genetic approaches have clarified that Runx complexes are involved in cNK cell function and differentiation. Mice harboring the hypomorphic allele of Cbfβ exhibited absence of NKPs and cNK cells. When Cbfβ is deleted in hematopoietic cells with *Vav1-Cre* mice, LMPPs and CLPs do not emerge (Satpathy et al. 2014), suggesting the requirement of Cbfβ for early differentiation of lymphocytes before commitment to cNK cells. Recently, cNK cell-specific Cbfβ function has been examined by conditional *Cbfβ* gene inactivation using *NCR-iCre* mice. Cbfβ deletion in cNK cells leads to great reduction of NK cells in the spleen and an immature NK cell phenotype including low expression of Ly49, low DX5, and Eomes, and inefficient IFNγ production in response to IL-12 and IL-18 stimulation (Ebihara et al. 2015). Conditional deletion of Runx3 in cNK cells recapitulates the phenotypes of Cbfβ-deficiency in cNK cells. However, probably due to compensation by Runx1, the phenotypes of Runx3-deficient cNK cells are generally milder than those of Cbfβ-deficient cNK cells (Ebihara et al. 2015; Levanon et al. 2014). ChIP-seq and transcriptome analysis showed that products of Runx3-bound genes seemed to be associated with survival, proliferation, maturation, and migration of cNK cells (Levanon et al. 2014). Runx3 appears to function downstream of IL-15 signaling and contributes to cNK cell survival. However, the precise mechanism of Runx3 induction through IL-15 signaling is still unclear. Runx3 also positively regulates CD96 and Crtam, both of which are involved in cNK cell activity (Levanon et al. 2014). Thus, the Runx3/Cbfβ complex regulates cNK cell survival and functions.

ILC1 cells in the intestine and liver also require Runx3/Cbfβ complexes for their survival and IFNγ response to IL-12 (Ebihara et al. 2015). Normal levels of T-bet expression in Cbfβ-deficient ILC1 cells in the intestine suggested that T-bet might be upstream regulator for *Runx3* expression in ILC1 cells as was observed in $CD8^+$ T cells. Type I ILC populations in the skin and salivary gland were reduced in the absence of Runx3 or Cbfβ. Taken together, the Runx3/Cbfβ complex is indispensable to all type I ILC populations.

24.4.4.2 Type II ILC

ILC2 cells in the intestine express only the *P2-Runx3* transcript. Runx3 is dispensable for ILC2 differentiation in the intestine (Ebihara et al. 2015). During the differentiation of effector $CD4^+$ T cell subsets, GATA-3 antagonizes Runx3 through protein-protein interactions to promote T_H2 skewing (Yagi et al. 2010), whereas Runx3 blocks GATA-3 activity for T_H1 differentiation (Kohu et al. 2009). This balance between Runx3 and GATA-3 seems to be one of the determinant factors for ILC1 and ILC2 function and differentiation as well. Runx3 expression might overwhelm GATA3 expression in ILC1 cells and GATA-3 could suppress Runx3 in ILC2 cells. Further studies will be necessary to clarify the physiological roles of Runx complexes in ILC2s.

24.4.4.3 Type III ILC

An early study showed that LTi cells in the fetal gut are reduced in mice lacking *P1-Runx1* transcripts or Cbfβ2 variants, resulting in a severe deficit in secondary lymphoid organ formation (Tachibana et al. 2011). LTi cells express less RORγt in the fetal gut of Cbfβ2-deficient mice. However, the counterparts of LTi cells in the adult intestine normally express RORγt in those mice, suggesting the possible association of RORγt with Runx complexes in ILC3s. Recently, LTi cells turned out to be the progeny of CLPs which require the Runx1/Cbfβ complex for differentiation (Possot et al. 2011; Sonnenberg and Artis 2015; Constantinides et al. 2014; Klose et al. 2014; Cherrier et al. 2012; Satpathy et al. 2014). Therefore, it should be considered that reduction of LTi cells in Cbfβ2-deficient mice might reflect impaired CLP differentiation. Recently, as the ILC differentiation process has become more characterized, the roles of Runx complexes in ILC3s have been revealed (Ebihara et al. 2015). Among the Runx family members, Runx3 is predominantly expressed by all ILC3 subsets at intermediate levels, which is less than in ILC1 cells and more than in cNK cells and ILC2 cells. Runx3 ablation in all hematopoietic cells showed normal differentiation from CLPs to ILCPs stages, but resulted in accumulation of ILCLN cells in the intestine, reduction of ILC1 cells, and absence of ILC3 cells in the intestine. Runx3-deficient ILCLN cells are not apoptotic and cannot give rise to ILC3 cells *in vivo* when transferred into alymphoid mice. Fewer ILC1 cells were also developed from Runx3-deficient ILCLN cells than Runx3-competent ILCLN cells. Thus, Runx3 is necessary for ILCLN cells to differentiate into ILC1 and ILC3 cells.

Mechanistically, Runx3 directly contributes to RORγt expression in ILC3 cells (Ebihara et al. 2015). A reporter assay using the human NK cell line exhibited that Runx3 enhances RORγt promoter activity through the Runx binding site in the RORγt promoter. In addition, Runx3 binding to the RORγt promoter in ILC3 cells was confirmed by the ChIP assay, indicating that Runx3 regulates RORγt in ILC3 cells (Ebihara et al. 2015). The aryl hydrocarbon receptor (AHR) is another ILC3-related transcription factor that was shown to regulate ILC3 differentiation and IL-22 production together with RORγt. In the absence of Runx3, expression of AHR was also undetected in ILC3 cells (Ebihara et al. 2015). AHR expression in ILC3 cells was reduced to half in mice harboring half dosage of the RORγt gene, suggesting that RORγt is also involved in the mechanisms that regulate AHR expression in ILC3 cells. Although direct binding of RORγt to the AHR promoter in ILC3 cells has not been examined yet, RORγt binds to enhancer regions in the AHR promoter of T_H17 cells, counterpart αβT cells that share many features with ILC3 cells (Ebihara et al. 2015; Ciofani et al. 2012). These data support that Runx3 regulates RORγt and its downstream AHR. Collectively, Runx3 is indispensable for ILCLN cells to acquire two ILC3 transcription factors, RORγt and AHR, for final differentiation into ILC3 cells.

References

Bain, G., Maandag, E. C., Izon, D. J., Amsen, D., Kruisbeek, A. M., Weintraub, B. C., et al. (1994). E2A proteins are required for proper B cell development and initiation of immunoglobulin gene rearrangements. *Cell, 79*(5), 885–892.

Braun, T., & Woollard, A. (2009). RUNX factors in development: Lessons from invertebrate model systems. *Blood Cells, Molecules & Diseases, 43*(1), 43–48. doi:10.1016/j.bcmd.2009.05.001.

Bruno, L., Mazzarella, L., Hoogenkamp, M., Hertweck, A., Cobb, B. S., Sauer, S., et al. (2009). Runx proteins regulate Foxp3 expression. *The Journal of Experimental Medicine, 206*(11), 2329–2337. doi:10.1084/jem.20090226.

Califano, D., Cho, J. J., Uddin, M. N., Lorentsen, K. J., Yang, Q., Bhandoola, A., et al. (2015). Transcription factor Bcl11b controls identity and function of mature Type 2 innate lymphoid cells. *Immunity, 43*(2), 354–368. doi:10.1016/j.immuni.2015.07.005.

Cella, M., Fuchs, A., Vermi, W., Facchetti, F., Otero, K., Lennerz, J. K., et al. (2009). A human natural killer cell subset provides an innate source of IL-22 for mucosal immunity. *Nature, 457*(7230), 722–725. doi:10.1038/nature07537.

Cella, M., Miller, H., & Song, C. (2014). Beyond NK cells: The expanding universe of innate lymphoid cells. *Frontiers in Immunology, 5*, 282. doi:10.3389/fimmu.2014.00282.

Cherrier, M., Sawa, S., & Eberl, G. (2012). Notch, Id2, and RORgammat sequentially orchestrate the fetal development of lymphoid tissue inducer cells. *The Journal of Experimental Medicine, 209*(4), 729–740. doi:10.1084/jem.20111594.

Ciofani, M., Madar, A., Galan, C., Sellars, M., Mace, K., Pauli, F., et al. (2012). A validated regulatory network for Th17 cell specification. *Cell, 151*(2), 289–303. doi:10.1016/j.cell.2012.09.016.

Constantinides, M. G., McDonald, B. D., Verhoef, P. A., & Bendelac, A. (2014). A committed precursor to innate lymphoid cells. *Nature, 508*(7496), 397–401. doi:10.1038/nature13047.

Cortez, V. S., Robinette, M. L., & Colonna, M. (2015). Innate lymphoid cells: New insights into function and development. *Current Opinion in Immunology, 32*, 71–77. doi:10.1016/j.coi.2015.01.004.

Daussy, C., Faure, F., Mayol, K., Viel, S., Gasteiger, G., Charrier, E., et al. (2014). T-bet and Eomes instruct the development of two distinct natural killer cell lineages in the liver and in the bone marrow. *The Journal of Experimental Medicine, 211*(3), 563–577. doi:10.1084/jem.20131560.

Decker, T., Pasca di Magliano, M., McManus, S., Sun, Q., Bonifer, C., Tagoh, H., & Busslinger, M. (2009). Stepwise activation of enhancer and promoter regions of the B cell commitment gene Pax5 in early lymphopoiesis. *Immunity, 30*(4), 508–520. doi:10.1016/j.immuni.2009.01.012.

Djuretic, I. M., Levanon, D., Negreanu, V., Groner, Y., Rao, A., & Ansel, K. M. (2007). Transcription factors T-bet and Runx3 cooperate to activate Ifng and silence Il4 in T helper type 1 cells. *Nature Immunology, 8*(2), 145–153.

Eberl, G., Colonna, M., Di Santo, J. P., & McKenzie, A. N. (2015). Innate lymphoid cells. Innate lymphoid cells: A new paradigm in immunology. *Science, 348*(6237), aaa6566. doi:10.1126/science.aaa6566.

Ebihara, T., Song, C., Ryu, S. H., Plougastel-Douglas, B., Yang, L., Levanon, D., et al. (2015). Runx3 specifies lineage commitment of innate lymphoid cells. *Nature Immunology, 16*(11), 1124–1133. doi:10.1038/ni.3272.

Egawa, T., & Littman, D. R. (2008). ThPOK acts late in specification of the helper T cell lineage and suppresses Runx-mediated commitment to the cytotoxic T cell lineage. *Nature Immunology, 9*(10), 1131–1139. doi:10.1038/ni.1652.

Egawa, T., Eberl, G., Taniuchi, I., Benlagha, K., Geissmann, F., Hennighausen, L., et al. (2005). Genetic evidence supporting selection of the Valpha14i NKT cell lineage from double-positive thymocyte precursors. *Immunity, 22*(6), 705–716. doi:10.1016/j.immuni.2005.03.011.

Egawa, T., Tillman, R. E., Naoe, Y., Taniuchi, I., & Littman, D. R. (2007). The role of the Runx transcription factors in thymocyte differentiation and in homeostasis of naive T cells. *The Journal of Experimental Medicine, 204*(8), 1945–1957. doi:10.1084/jem.20070133.

Ellmeier, W., & Taniuchi, I. (2014). The role of BTB-zinc finger transcription factors during T cell development and in the regulation of T cell-mediated immunity. *Current Topics in Microbiology and Immunology, 381*, 21–49. doi:10.1007/82_2014_374.

Ellmeier, W., Sunshine, M. J., Losos, K., Hatam, F., & Littman, D. R. (1997). An enhancer that directs lineage-specific expression of CD8 in positively selected thymocytes and mature T cells. *Immunity, 7*(4), 537–547.

Ellmeier, W., Sunshine, M. J., Losos, K., & Littman, D. R. (1998). Multiple developmental stage-specific enhancers regulate CD8 expression in developing thymocytes and in thymus-independent T cells. *Immunity, 9*(4), 485–496.

Ellmeier, W., Sawada, S., & Littman, D. R. (1999). The regulation of CD4 and CD8 coreceptor gene expression during T cell development. *Annual Review of Immunology, 17*, 523–554.

Erman, B., Cortes, M., Nikolajczyk, B. S., Speck, N. A., & Sen, R. (1998). ETS-core binding factor: A common composite motif in antigen receptor gene enhancers. *Molecular and Cellular Biology, 18*(3), 1322–1330.

Fossett, N., & Schulz, R. A. (2001). Functional conservation of hematopoietic factors in Drosophila and vertebrates. *Differentiation; Research in Biological Diversity, 69*(2–3), 83–90. doi:10.1046/j.1432-0436.2001.690202.x.

Fuchs, A., Vermi, W., Lee, J. S., Lonardi, S., Gilfillan, S., Newberry, R. D., et al. (2013). Intraepithelial type 1 innate lymphoid cells are a unique subset of IL-12- and IL-15-responsive IFN-gamma-producing cells. *Immunity, 38*(4), 769–781. doi:10.1016/j.immuni.2013.02.010.

Gangadharan, D., Lambolez, F., Attinger, A., Wang-Zhu, Y., Sullivan, B. A., & Cheroutre, H. (2006). Identification of pre- and postselection TCRalphabeta+ intraepithelial lymphocyte precursors in the thymus. *Immunity, 25*(4), 631–641. doi:10.1016/j.immuni.2006.08.018.

Gasteiger, G., Fan, X., Dikiy, S., Lee, S. Y., & Rudensky, A. Y. (2015). Tissue residency of innate lymphoid cells in lymphoid and nonlymphoid organs. *Science, 350*(6263), 981–985. doi:10.1126/science.aac9593.

Gordon, S. M., Chaix, J., Rupp, L. J., Wu, J., Madera, S., Sun, J. C., et al. (2012). The transcription factors T-bet and Eomes control key checkpoints of natural killer cell maturation. *Immunity, 36*(1), 55–67. doi:10.1016/j.immuni.2011.11.016.

Growney, J. D., Shigematsu, H., Li, Z., Lee, B. H., Adelsperger, J., Rowan, R., et al. (2005). Loss of Runx1 perturbs adult hematopoiesis and is associated with a myeloproliferative phenotype. *Blood, 106*(2), 494–504. doi:10.1182/blood-2004-08-3280.

Grueter, B., Petter, M., Egawa, T., Laule-Kilian, K., Aldrian, C. J., Wuerch, A., Ludwig, Y., Fukuyama, H., Wardemann, H., Waldschuetz, R., Moroy, T., Taniuchi, I., Steimle, V., Littman, D. R., & Ehlers, M. (2005). Runx3 regulates integrin alpha E/CD103 and CD4 expression during development of CD4−/CD8+ T cells. *Journal of Immunology (Baltimore, Md: 1950), 175*(3), 1694–1705. doi:175/3/1694 [pii].

Harrington, L. E., Hatton, R. D., Mangan, P. R., Turner, H., Murphy, T. L., Murphy, K. M., & Weaver, C. T. (2005). Interleukin 17-producing CD4+ effector T cells develop via a lineage distinct from the T helper type 1 and 2 lineages. *Nature Immunology, 6*(11), 1123–1132. doi:10.1038/ni1254.

Hassan, H., Sakaguchi, S., Tenno, M., Kopf, A., Boucheron, N., Carpenter, A. C., Egawa, T., Taniuchi I., & Ellmeier, W. (2011). Cd8 enhancer E8I and Runx factors regulate CD8alpha expression in activated CD8+ T cells. *Proceedings of the National Academy of Sciences of the United States of America, 108*(45), 18330–18335. doi:1105835108 [pii] 10.1073/pnas.1105835108.

He, X., Dave, V. P., Zhang, Y., Hua, X., Nicolas, E., Xu, W., Roe, B. A., & Kappes, D. J. (2005). The zinc finger transcription factor Th-POK regulates CD4 versus CD8 T-cell lineage commitment. *Nature, 433*(7028), 826–833. doi:nature03338 [pii] 10.1038/nature03338.

He, X., Park, K., Wang, H., He, X., Zhang, Y., Hua, X., et al. (2008). CD4-CD8 lineage commitment is regulated by a silencer element at the ThPOK transcription-factor locus. *Immunity, 28*(3), 346–358.

Hernandez-Munain, C., & Krangel, M. S. (1995). c-Myb and core-binding factor/PEBP2 display functional synergy but bind independently to adjacent sites in the T-cell receptor delta enhancer. *Molecular and Cellular Biology, 15*(6), 3090–3099.

Hernandez-Munain, C., Roberts, J. L., & Krangel, M. S. (1998). Cooperation among multiple transcription factors is required for access to minimal T-cell receptor alpha-enhancer chromatin in vivo. *Molecular and Cellular Biology, 18*(6), 3223–3233.

Hollenhorst, P. C., Shah, A. A., Hopkins, C., & Graves, B. J. (2007). Genome-wide analyses reveal properties of redundant and specific promoter occupancy within the ETS gene family. *Genes & Development, 21*(15), 1882–1894. doi:10.1101/gad.1561707.

Hostert, A., Garefalaki, A., Mavria, G., Tolaini, M., Roderick, K., Norton, T., et al. (1998). Hierarchical interactions of control elements determine CD8alpha gene expression in subsets of thymocytes and peripheral T cells. *Immunity, 9*(4), 497–508.

Hoyler, T., Klose, C. S., Souabni, A., Turqueti-Neves, A., Pfeifer, D., Rawlins, E. L., et al. (2012). The transcription factor GATA-3 controls cell fate and maintenance of Type 2 innate lymphoid cells. *Immunity, 37*(4), 634–648. doi:10.1016/j.immuni.2012.06.020.

Hozumi, K., Mailhos, C., Negishi, N., Hirano, K., Yahata, T., Ando, K., et al. (2008). Delta-like 4 is indispensable in thymic environment specific for T cell development. *Journal of Experimental Medicine, 205*(11), 2507–2513. doi:10.1084/jem.20080134.

Hsiang, Y. H., Spencer, D., Wang, S., Speck, N. A., & Raulet, D. H. (1993). The role of viral enhancer "core" motif-related sequences in regulating T cell receptor-gamma and -delta gene expression. *Journal of Immunology (Baltimore, Md: 1950), 150*(9), 3905–3916.

Ichikawa, M., Asai, T., Saito, T., Seo, S., Yamazaki, I., Yamagata, T., et al. (2004). AML-1 is required for megakaryocytic maturation and lymphocytic differentiation, but not for maintenance of hematopoietic stem cells in adult hematopoiesis. *Nature Medicine, 10*(3), 299–304. doi:10.1038/nm997.

Ikawa, T., Hirose, S., Masuda, K., Kakugawa, K., Satoh, R., Shibano-Satoh, A., et al. (2010). An essential developmental checkpoint for production of the T cell lineage. *Science, 329*(5987), 93–96. doi:10.1126/science.1188995.

Ivanov, I. I., McKenzie, B. S., Zhou, L., Tadokoro, C. E., Lepelley, A., Lafaille, J. J., et al. (2006). The orphan nuclear receptor RORgammat directs the differentiation program of proinflammatory IL-17+ T helper cells. *Cell, 126*(6), 1121–1133. doi:10.1016/j.cell.2006.07.035.

Josefowicz, S. Z., Lu, L. F., & Rudensky, A. Y. (2012). Regulatory T cells: Mechanisms of differentiation and function. *Annual Review of Immunology, 30*, 531–564. doi:10.1146/annurev.immunol.25.022106.141623.

Kappes, D. J., & He, X. (2006). Role of the transcription factor Th-POK in CD4:CD8 lineage commitment. *Immunological Reviews*, *209*, 237–252. doi:IMR344 [pii] 10.1111/j.0105-2896.2006.00344.x.

Kee, B. L., & Murre, C. (1998). Induction of early B cell factor (EBF) and multiple B lineage genes by the basic helix-loop-helix transcription factor E12. *The Journal of Experimental Medicine, 188*(4), 699–713.

Kim, B., Sasaki, Y., & Egawa, T. (2015). Restriction of nonpermissive RUNX3 protein expression in T lymphocytes by the Kozak sequence. *Journal of Immunology, 195*(4), 1517–1523. doi:10.4049/jimmunol.1501039.

Kiss, E. A., Vonarbourg, C., Kopfmann, S., Hobeika, E., Finke, D., Esser, C., & Diefenbach, A. (2011). Natural aryl hydrocarbon receptor ligands control organogenesis of intestinal lymphoid follicles. *Science, 334*(6062), 1561–1565. doi:10.1126/science.1214914.

Kitoh, A., Ono, M., Naoe, Y., Ohkura, N., Yamaguchi, T., Yaguchi, H., et al. (2009). Indispensable role of the Runx1-Cbfbeta transcription complex for in vivo-suppressive function of FoxP3+ regulatory T cells. *Immunity, 31*(4), 609–620.

Klose, C. S., Kiss, E. A., Schwierzeck, V., Ebert, K., Hoyler, T., d'Hargues, Y., et al. (2013). A T-bet gradient controls the fate and function of CCR6-RORgammat+ innate lymphoid cells. *Nature, 494*(7436), 261–265. doi:10.1038/nature11813.

Klose, C. S., Flach, M., Mohle, L., Rogell, L., Hoyler, T., Ebert, K., et al. (2014). Differentiation of type 1 ILCs from a common progenitor to all helper-like innate lymphoid cell lineages. *Cell, 157*(2), 340–356. doi:10.1016/j.cell.2014.03.030.

Klunker, S., Chong, M. M., Mantel, P. Y., Palomares, O., Bassin, C., Ziegler, M., et al. (2009). Transcription factors RUNX1 and RUNX3 in the induction and suppressive function of Foxp3+ inducible regulatory T cells. *The Journal of Experimental Medicine, 206*(12), 2701–2715. doi:10.1084/jem.20090596.

Koch, U., Fiorini, E., Benedito, R., Besseyrias, V., Schuster-Gossler, K., Pierres, M., et al. (2008). Delta-like 4 is the essential, nonredundant ligand for Notch1 during thymic T cell lineage commitment. *The Journal of Experimental Medicine, 205*(11), 2515–2523. doi:10.1084/jem.20080829.

Kohu, K., Sato, T., Ohno, S., Hayashi, K., Uchino, R., Abe, N., Nakazato, M., Yoshida, N., Kikuchi, T., Iwakura, Y., Inoue, Y., Watanabe, T., Habu, S., & Satake, M. (2005). Overexpression of the Runx3 transcription factor increases the proportion of mature thymocytes of the CD8 single-positive lineage. *Journal of Immunology (Baltimore, Md: 1950), 174*(5), 2627–2636. doi:174/5/2627 [pii].

Kohu, K., Ohmori, H., Wong, W. F., Onda, D., Wakoh, T., Kon, S., et al. (2009). The Runx3 transcription factor augments Th1 and down-modulates Th2 phenotypes by interacting with and attenuating GATA3. *Journal of Immunology, 183*(12), 7817–7824. doi:10.4049/jimmunol.0802527.

Kronenberg, M., & Rudensky, A. (2005). Regulation of immunity by self-reactive T cells. *Nature, 435*(7042), 598–604. doi:10.1038/nature03725.

Laiosa, C. V., Stadtfeld, M., & Graf, T. (2006). Determinants of lymphoid-myeloid lineage diversification. *Annual Review of Immunology, 24*, 705–738. doi:10.1146/annurev.immunol.24.021605.090742.

Lambolez, F., Kronenberg, M., & Cheroutre, H. (2007). Thymic differentiation of TCR alpha beta(+) CD8 alpha alpha(+) IELs. *Immunological Reviews, 215*, 178–188. doi:10.1111/j.1600-065X.2006.00488.x.

Lebestky, T., Chang, T., Hartenstein, V., & Banerjee, U. (2000). Specification of Drosophila hematopoietic lineage by conserved transcription factors. *Science, 288*(5463), 146–149.

Lee, J. S., Cella, M., McDonald, K. G., Garlanda, C., Kennedy, G. D., Nukaya, M., et al. (2012). AHR drives the development of gut ILC22 cells and postnatal lymphoid tissues via pathways dependent on and independent of Notch. *Nature Immunology, 13*(2), 144–151. doi:10.1038/ni.2187.

Leishman, A. J., Naidenko, O. V., Attinger, A., Koning, F., Lena, C. J., Xiong, Y., et al. (2001). T cell responses modulated through interaction between CD8alphaalpha and the nonclassical MHC class I molecule, TL. *Science, 294*(5548), 1936–1939. doi:10.1126/science.1063564.

Leung, R. K., Thomson, K., Gallimore, A., Jones, E., Van den Broek, M., Sierro, S., et al. (2001). Deletion of the CD4 silencer element supports a stochastic mechanism of thymocyte lineage commitment. *Nature Immunology, 2*(12), 1167–1173.

Levanon, D., Negreanu, V., Lotem, J., Bone, K. R., Brenner, O., Leshkowitz, D., & Groner, Y. (2014). Transcription factor Runx3 regulates interleukin-15-dependent natural killer cell activation. *Molecular and Cellular Biology, 34*(6), 1158–1169. doi:10.1128/MCB.01202-13.

Li, L., Leid, M., & Rothenberg, E. V. (2010). An early T cell lineage commitment checkpoint dependent on the transcription factor Bcl11b. *Science, 329*(5987), 89–93. doi:10.1126/science.1188989.

Li, L., Zhang, J. A., Dose, M., Kueh, H. Y., Mosadeghi, R., Gounari, F., & Rothenberg, E. V. (2013). A far downstream enhancer for murine Bcl11b controls its T-cell specific expression. *Blood, 122*(6), 902–911. doi:10.1182/blood-2012-08-447839.

Lin, H., & Grosschedl, R. (1995). Failure of B-cell differentiation in mice lacking the transcription factor EBF. *Nature, 376*(6537), 263–267. doi:10.1038/376263a0.

Lin, Y. C., Jhunjhunwala, S., Benner, C., Heinz, S., Welinder, E., Mansson, R., et al. (2010). A global network of transcription factors, involving E2A, EBF1 and Foxo1, that orchestrates B cell fate. *Nature Immunology, 11*(7), 635–643. doi:10.1038/ni.1891.

Luckey, M. A., Kimura, M. Y., Waickman, A. T., Feigenbaum, L., Singer, A., & Park, J. H. (2014). The transcription factor ThPOK suppresses Runx3 and imposes CD4(+) lineage fate by inducing the SOCS suppressors of cytokine signaling. *Nature Immunology, 15*(7), 638–645. doi:10.1038/ni.2917.

Lukin, K., Fields, S., Lopez, D., Cherrier, M., Ternyak, K., Ramirez, J., et al. (2010). Compound haploinsufficiencies of Ebf1 and Runx1 genes impede B cell lineage progression. *Proceedings of the National Academy of Sciences of the United States of America, 107*(17), 7869–7874. doi:10.1073/pnas.1003525107.

Maier, H., Ostraat, R., Gao, H., Fields, S., Shinton, S. A., Medina, K. L., et al. (2004). Early B cell factor cooperates with Runx1 and mediates epigenetic changes associated with mb-1 transcription. *Nature Immunology, 5*(10), 1069–1077. doi:10.1038/ni1119.

Majumder, K., Bassing, C. H., & Oltz, E. M. (2015). Regulation of Tcrb gene assembly by genetic, epigenetic, and topological mechanisms. *Advances in Immunology, 128*, 273–306. doi:10.1016/bs.ai.2015.07.001.

Matthias, P., & Rolink, A. G. (2005). Transcriptional networks in developing and mature B cells. *Nature Reviews Immunology, 5*(6), 497–508. doi:10.1038/nri1633.

Mikkola, I., Heavey, B., Horcher, M., & Busslinger, M. (2002). Reversion of B cell commitment upon loss of Pax5 expression. *Science, 297*(5578), 110–113. doi:10.1126/science.1067518.

Moro, K., Yamada, T., Tanabe, M., Takeuchi, T., Ikawa, T., Kawamoto, H., et al. (2010). Innate production of T(H)2 cytokines by adipose tissue-associated c-Kit(+) Sca-1(+) lymphoid cells. *Nature, 463*(7280), 540–544. doi:10.1038/nature08636.

Nakayamada, S., Takahashi, H., Kanno, Y., & O'Shea, J. J. (2012). Helper T cell diversity and plasticity. *Current Opinion in Immunology, 24*(3), 297–302. doi:10.1016/j.coi.2012.01.014.

Naoe, Y., Setoguchi, R., Akiyama, K., Muroi, S., Kuroda, M., Hatam, F., et al. (2007). Repression of interleukin-4 in T helper type 1 cells by Runx/Cbf beta binding to the Il4 silencer. *The Journal of Experimental Medicine, 204*(8), 1749–1755.

Neill, D. R., Wong, S. H., Bellosi, A., Flynn, R. J., Daly, M., Langford, T. K., et al. (2010). Nuocytes represent a new innate effector leukocyte that mediates type-2 immunity. *Nature, 464*(7293), 1367–1370. doi:10.1038/nature08900.

Nutt, S. L., Heavey, B., Rolink, A. G., & Busslinger, M. (1999). Commitment to the B-lymphoid lineage depends on the transcription factor Pax5. *Nature, 401*(6753), 556–562. doi:10.1038/44076.

Oestreich, K. J., Cobb, R. M., Pierce, S., Chen, J., Ferrier, P., & Oltz, E. M. (2006). Regulation of TCRbeta gene assembly by a promoter/enhancer holocomplex. *Immunity, 24*(4), 381–391. doi:10.1016/j.immuni.2006.02.009.

Ohkura, N., Kitagawa, Y., & Sakaguchi, S. (2013). Development and maintenance of regulatory T cells. *Immunity, 38*(3), 414–423. doi:10.1016/j.immuni.2013.03.002.

Ohno, S., Sato, T., Kohu, K., Takeda, K., Okumura, K., Satake, M., & Habu, S. (2008). Runx proteins are involved in regulation of CD122, Ly49 family and IFN-gamma expression during NK cell differentiation. *International Immunology, 20*(1), 71–79. doi:10.1093/intimm/dxm120.

Okuda, T., van Deursen, J., Hiebert, S. W., Grosveld, G., & Downing, J. R. (1996). AML1, the target of multiple chromosomal translocations in human leukemia, is essential for normal fetal liver hematopoiesis. *Cell, 84*(2), 321–330.

Ono, M., Yaguchi, H., Ohkura, N., Kitabayashi, I., Nagamura, Y., Nomura, T., et al. (2007). Foxp3 controls regulatory T-cell function by interacting with AML1/Runx1. *Nature, 446*(7136), 685–689. doi:10.1038/nature05673.

Park, H., Li, Z., Yang, X. O., Chang, S. H., Nurieva, R., Wang, Y. H., et al. (2005). A distinct lineage of CD4 T cells regulates tissue inflammation by producing interleukin 17. *Nature Immunology, 6*(11), 1133–1141. doi:10.1038/ni1261.

Park, J. H., Adoro, S., Guinter, T., Erman, B., Alag, A. S., Catalfamo, M., Kimura, M. Y., Cui, Y., Lucas, P. J., Gress, R. E., Kubo, M., Hennighausen, L., Feigenbaum, L., & Singer, A. (2010). Signaling by intrathymic cytokines, not T cell antigen receptors, specifies CD8 lineage choice and promotes the differentiation of cytotoxic-lineage T cells. *Nature Immunology, 11*(3), 257–264. doi:ni.1840 [pii] 10.1038/ni.1840.

Pobezinsky, L. A., Angelov, G. S., Tai, X., Jeurling, S., Van Laethem, F., Feigenbaum, L., et al. (2012). Clonal deletion and the fate of autoreactive thymocytes that survive negative selection. *Nature Immunology, 13*(6), 569–578. doi:10.1038/ni.2292.

Porritt, H. E., Rumfelt, L. L., Tabrizifard, S., Schmitt, T. M., Zuniga-Pflucker, J. C., & Petrie, H. T. (2004). Heterogeneity among DN1 prothymocytes reveals multiple progenitors with different capacities to generate T cell and non-T cell lineages. *Immunity, 20*(6), 735–745. doi:10.1016/j.immuni.2004.05.004.

Possot, C., Schmutz, S., Chea, S., Boucontet, L., Louise, A., Cumano, A., & Golub, R. (2011). Notch signaling is necessary for adult, but not fetal, development of RORgammat(+) innate lymphoid cells. *Nature Immunology, 12*(10), 949–958. doi:10.1038/ni.2105.

Redondo, J. M., Pfohl, J. L., & Krangel, M. S. (1991). Identification of an essential site for transcriptional activation within the human T-cell receptor delta enhancer. *Molecular and Cellular Biology, 11*(11), 5671–5680.

Rolink, A. G., Nutt, S. L., Melchers, F., & Busslinger, M. (1999). Long-term in vivo reconstitution of T-cell development by Pax5-deficient B-cell progenitors. *Nature, 401*(6753), 603–606. doi:10.1038/44164.

Rothenberg, E. V., Moore, J. E., & Yui, M. A. (2008). Launching the T-cell-lineage developmental programme. *Nature Reviews Immunology, 8*(1), 9–21. doi:10.1038/nri2232.

Rudra, D., Egawa, T., Chong, M. M., Treuting, P., Littman, D. R., & Rudensky, A. Y. (2009). Runx-CBFbeta complexes control expression of the transcription factor Foxp3 in regulatory T cells. *Nature Immunology, 10*(11), 1170–1177. doi:10.1038/ni.1795.

Sakaguchi, S., Hombauer, M., Hassan, H., Tanaka, H., Yasmin, N., Naoe, Y., et al. (2015). A novel Cd8-cis-regulatory element preferentially directs expression in CD44hiCD62L+ CD8+ T cells and in CD8alphaalpha+ dendritic cells. *Journal of Leukocyte Biology, 97*(4), 635–644. doi:10.1189/jlb.1HI1113-597RR.

Sanos, S. L., Bui, V. L., Mortha, A., Oberle, K., Heners, C., Johner, C., & Diefenbach, A. (2009). RORgammat and commensal microflora are required for the differentiation of mucosal interleukin 22-producing NKp46+ cells. *Nature Immunology, 10*(1), 83–91. doi:10.1038/ni.1684.

Sato, T., Ohno, S., Hayashi, T., Sato, C., Kohu, K., Satake, M., & Habu, S. (2005). Dual functions of Runx proteins for reactivating CD8 and silencing CD4 at the commitment process into CD8 thymocytes. *Immunity, 22*(3), 317–328.

Satoh-Takayama, N., Vosshenrich, C. A., Lesjean-Pottier, S., Sawa, S., Lochner, M., Rattis, F., et al. (2008). Microbial flora drives interleukin 22 production in intestinal NKp46+ cells that provide innate mucosal immune defense. *Immunity, 29*(6), 958–970. doi:10.1016/j.immuni.2008.11.001.

Satpathy, A. T., Briseno, C. G., Cai, X., Michael, D. G., Chou, C., Hsiung, S., et al. (2014). Runx1 and Cbfbeta regulate the development of Flt3+ dendritic cell progenitors and restrict myeloproliferative disorder. *Blood, 123*(19), 2968–2977. doi:10.1182/blood-2013-11-539643.

Sawa, S., Cherrier, M., Lochner, M., Satoh-Takayama, N., Fehling, H. J., Langa, F., et al. (2010). Lineage relationship analysis of RORgammat+ innate lymphoid cells. *Science, 330*(6004), 665–669. doi:10.1126/science.1194597.

Sawa, S., Lochner, M., Satoh-Takayama, N., Dulauroy, S., Berard, M., Kleinschek, M., et al. (2011). RORgammat+ innate lymphoid cells regulate intestinal homeostasis by integrating negative signals from the symbiotic microbiota. *Nature Immunology, 12*(4), 320–326. doi:10.1038/ni.2002.

Sawada, S., Scarborough, J. D., Killeen, N., & Littman, D. R. (1994). A lineage-specific transcriptional silencer regulates CD4 gene expression during T lymphocyte development. *Cell, 77*(6), 917–929.

Seo, W., Ikawa, T., Kawamoto, H., & Taniuchi, I. (2012a). Runx1-Cbfbeta facilitates early B lymphocyte development by regulating expression of Ebf1. *The Journal of Experimental Medicine, 209*(7), 1255–1262. doi:10.1084/jem.20112745.

Seo, W., Tanaka, H., Miyamoto, C., Levanon, D., Groner, Y., & Taniuchi, I. (2012b). Roles of VWRPY motif-mediated gene repression by Runx proteins during T-cell development. *Immunology and Cell Biology, 90*(8), 827–830. doi:10.1038/icb.2012.6 icb20126 [pii].

Serafini, N., Klein Wolterink, R. G., Satoh-Takayama, N., Xu, W., Vosshenrich, C. A., Hendriks, R. W., & Di Santo, J. P. (2014). Gata3 drives development of RORgammat+ group 3 innate lymphoid cells. *The Journal of Experimental Medicine, 211*(2), 199–208. doi:10.1084/jem.20131038.

Setoguchi, R., Tachibana, M., Naoe, Y., Muroi, S., Akiyama, K., Tezuka, C., Okuda, T., & Taniuchi, I. (2008). Repression of the transcription factor Th-POK by Runx complexes in cytotoxic T cell development. Science, *319*(5864), 822–825. doi:319/5864/822 [pii] 10.1126/science.1151844.

Siu, G., Wurster, A. L., Duncan, D. D., Soliman, T. M., & Hedrick, S. M. (1994). A transcriptional silencer controls the developmental expression of the CD4 gene. *The EMBO Journal, 13*(15), 3570–3579.

Sojka, D. K., Plougastel-Douglas, B., Yang, L., Pak-Wittel, M. A., Artyomov, M. N., Ivanova, Y., et al. (2014). Tissue-resident natural killer (NK) cells are cell lineages distinct from thymic and conventional splenic NK cells. *eLife, 3*, e01659. doi:10.7554/eLife.01659.

Sonnenberg, G. F., & Artis, D. (2015). Innate lymphoid cells in the initiation, regulation and resolution of inflammation. *Nature Medicine, 21*(7), 698–708. doi:10.1038/nm.3892.

Spooner, C. J., Lesch, J., Yan, D., Khan, A. A., Abbas, A., Ramirez-Carrozzi, V., et al. (2013). Specification of type 2 innate lymphocytes by the transcriptional determinant Gfi1. *Nature Immunology, 14*(12), 1229–1236. doi:10.1038/ni.2743.

Tachibana, M., Tenno, M., Tezuka, C., Sugiyama, M., Yoshida, H., & Taniuchi, I. (2011). Runx1/Cbfbeta2 complexes are required for lymphoid tissue inducer cell differentiation at two developmental stages. *Journal of Immunology, 186*(3), 1450–1457. doi:10.4049/jimmunol.1000162.

Takeda, J., Cheng, A., Mauxion, F., Nelson, C. A., Newberry, R. D., Sha, W. C., et al. (1990). Functional analysis of the murine T-cell receptor beta enhancer and characteristics of its DNA-binding proteins. *Molecular and Cellular Biology, 10*(10), 5027–5035.

Tani-Ichi, S., Satake, M., & Ikuta, K. (2011). The pre-TCR signal induces transcriptional silencing of the TCRgamma locus by reducing the recruitment of STAT5 and Runx to transcriptional enhancers. *International Immunology, 23*(9), 553–563. doi:10.1093/intimm/dxr055.

Taniuchi, I., Osato, M., Egawa, T., Sunshine, M. J., Bae, S. C., Komori, T., et al. (2002). Differential requirements for Runx proteins in CD4 repression and epigenetic silencing during T lymphocyte development. *Cell, 111*(5), 621–633.

Wang, Q., Stacy, T., Binder, M., Marin-Padilla, M., Sharpe, A. H., & Speck, N. A. (1996a). Disruption of the Cbfa2 gene causes necrosis and hemorrhaging in the central nervous system and blocks definitive hematopoiesis. *Proceedings of the National Academy of Sciences of the United States of America, 93*(8), 3444–3449.

Wang, Q., Stacy, T., Miller, J. D., Lewis, A. F., Gu, T. L., Huang, X., et al. (1996b). The CBFbeta subunit is essential for CBFalpha2 (AML1) function in vivo. *Cell, 87*(4), 697–708.

Wong, S. H., Walker, J. A., Jolin, H. E., Drynan, L. F., Hams, E., Camelo, A., et al. (2012). Transcription factor RORalpha is critical for nuocyte development. *Nature Immunology, 13*(3), 229–236. doi:10.1038/ni.2208.

Woolf, E., Brenner, O., Goldenberg, D., Levanon, D., & Groner, Y. (2007). Runx3 regulates dendritic epidermal T cell development. *Developmental Biology, 303*(2), 703–714. doi:10.1016/j.ydbio.2006.12.005.

Wotton, D., Ghysdael, J., Wang, S., Speck, N. A., & Owen, M. J. (1994). Cooperative binding of Ets-1 and core binding factor to DNA. *Molecular and Cellular Biology, 14*(1), 840–850.

Yagi, R., Junttila, I. S., Wei, G., Urban Jr., J. F., Zhao, K., Paul, W. E., & Zhu, J. (2010). The transcription factor GATA3 actively represses RUNX3 protein-regulated production of interferon-gamma. *Immunity, 32*(4), 507–517. doi:10.1016/j.immuni.2010.04.004.

Yang, Q., Li, F., Harly, C., Xing, S., Ye, L., Xia, X., et al. (2015). TCF-1 upregulation identifies early innate lymphoid progenitors in the bone marrow. *Nature Immunology, 16*(10), 1044–1050. doi:10.1038/ni.3248.

Yu, X., Wang, Y., Deng, M., Li, Y., Ruhn, K. A., Zhang, C. C., & Hooper, L. V. (2014). The basic leucine zipper transcription factor NFIL3 directs the development of a common innate lymphoid cell precursor. *eLife, 3*. doi:10.7554/eLife.04406.

Zhang, F., Meng, G., & Strober, W. (2008). Interactions among the transcription factors Runx1, RORgammat and Foxp3 regulate the differentiation of interleukin 17-producing T cells. *Nature Immunology, 9*(11), 1297–1306. doi:10.1038/ni.1663.

Zheng, Y., Josefowicz, S., Chaudhry, A., Peng, X. P., Forbush, K., & Rudensky, A. Y. (2010). Role of conserved non-coding DNA elements in the Foxp3 gene in regulatory T-cell fate. *Nature, 463*(7282), 808–812. doi:10.1038/nature08750.

Zhuang, Y., Soriano, P., & Weintraub, H. (1994). The helix-loop-helix gene E2A is required for B cell formation. *Cell, 79*(5), 875–884.

Zou, Y. R., Sunshine, M. J., Taniuchi, I., Hatam, F., Killeen, N., & Littman, D. R. (2001). Epigenetic silencing of CD4 in T cells committed to the cytotoxic lineage. *Nature Genetics, 29*(3), 332–336. doi:10.1038/ng750.

CBFß and HIV Infection

25

Dong Young Kim and John D. Gross

Abstract

In order to achieve a persistent infection, viruses must overcome the host immune system. Host restriction factors dominantly block virus transmission, but are subject to down regulation by viral accessory proteins. HIV encodes several accessory factors that overcome different cellular restriction factors. For example, the HIV-1 protein Vif down regulates the human APOBEC3 family of restriction factors by targeting them for proteolysis by the ubiquitin-proteasome pathway. Recently, this function was shown to require the transcription cofactor CBFβ, which acts as a template to assist in Vif folding and allow for assembly of an APOBEC3-targeting E3 ligase complex. In uninfected cells, CBFβ is an essential binding partner of RUNX transcription factors. By binding CBFβ, Vif has also been shown to perturb transcription of genes regulated by the RUNX proteins, including restrictive APOBEC3 family members. Here we review how the link between CBFβ and Vif supports transcriptional and post-transcriptional repression of innate immunity. The ability of a single viral protein to coopt multiple host pathways is an economical strategy for a pathogen with limited protein coding capacity to achieve a productive infection.

Keywords

HIV • Transcription • RUNX • Restriction factors • Innate immunity • APOBEC3 • Vif

25.1 Introduction

Many viral pathogens have co-evolved with their host organism, often times resulting in a molecular arms race against host immune defenses (Daugherty and Malik 2012). Research designed to elucidate the intermolecular interactions at the

D.Y. Kim
College of Pharmacy, Yeungnam University, Gyeongsan 38541, South Korea
e-mail: dyokim@ynu.ac.kr

J.D. Gross (✉)
Department of Pharmaceutical Chemistry, University of California, San Francisco, CA 94158, USA
e-mail: jdgross@cgl.ucsf.edu

Y. Groner et al. (eds.), *RUNX Proteins in Development and Cancer*, Advances in Experimental Medicine and Biology 962, DOI 10.1007/978-981-10-3233-2_25

viral-host interface will therefore be critical to our understanding of both host immunity and viral pathogenesis. As our knowledge of the viral-host interface improves so will our efforts towards the development of new and innovative therapeutic agents to counter viral infections. In this chapter we will focus on HIV related research and how it has informed aspects of host innate immunity- specifically the identification of APOBEC3 host restriction factors and novel roles for CBFβ in viral restriction.

Nearly 25 years ago, binding sites for the core binding factor (CBF) were identified in the enhancer of the Moloney murine leukemia retrovirus (MoMLV) (S. Wang et al. 1993). Parallel studies identified key CBF binding sites in the polyoma virus enhancer (reviewed by Ito) (Ito 2008). These early investigations in viral systems paved the way for what we now know: that CBF is a heterodimeric complex consisting of a DNA binding subunit (RUNX 1, 2 or 3) and a non-DNA binding subunit termed CBFβ (de Bruijn and Speck 2004). CBFβ is required for all RUNX protein function, probably by allosterically enhancing their interactions with DNA and protecting RUNX proteins from degradation (Ogawa et al. 1993; Gu et al. 2000; Tang et al. 2000; Yan et al. 2004; Tahirov et al. 2001; Q. Wang et al. 1996; G. Huang et al. 2001). The diverse biological roles of RUNX proteins include hematopoiesis, neurogenesis and osteogenesis, which are reviewed in other chapters of this book. It is well appreciated that RUNX1 and 3 are important in T-cell differentiation, and have broad roles in immunity (de Bruijn and Speck 2004; S. Wang et al. 1993; Voon et al. 2015). Just like early work, where viruses were used to probe function of CBFβ during development, recent studies with HIV-1 and related lentiviruses reveal an emerging role of RUNX/CBFβ in the regulation of innate immunity (Ito 2008; Jager et al. 2012; W. Zhang et al. 2012; Hultquist et al. 2012; Anderson and Harris 2015).

HIV-1, the pathogen that causes AIDS, causes severe immunodeficiency by the depletion of CD4+ T-cells (Muro-Cacho et al. 1995; Finkel et al. 1995; Doitsh et al. 2015). In order to achieve a productive infection, HIV-1 must counteract several human restriction factors, which act as dominant blocks to viral replication in the absence of accessory proteins (Harris et al. 2012; Malim and Emerman 2008). The APOBEC3 family of restriction factors block the replication of retroviruses and retroelements by binding and enzymatically hypermutating newly transcribed cDNA prior to integration (Harris and Liddament 2004). In order to counteract the restrictive potential of APOBEC3 proteins, lentiviruses encode for the viral infectivity factor (Vif) accessory protein that promotes degradation of APOBEC3 proteins by hijacking the ubiquitin proteasome pathway (Yu et al. 2003). Several years ago, it was reported that CBFβ was required for this effect (Ogawa et al. 1993; W. Zhang et al. 2012; Gu et al. 2000; Jager et al. 2012; Tang et al. 2000; Yan et al. 2004; Tahirov et al. 2001; Q. Wang et al. 1996; G. Huang et al. 2001). This was surprising given that CBFβ has not been documented to play a direct role in the ubiquitin proteasome pathway. Here we review studies linking CBFβ to Vif function, including recent observations that Vif can perturb RUNX mediated transcription and the structural basis for these effects.

25.2 HIV-1 Vif Inhibits Human APOBEC3 to Promote Viral Replication

It has long been appreciated that the Vif protein is essential for viral replication in primary T-cells and natural infection (Fisher et al. 1987). Initially it was unclear what role Vif played in viral replication as it was only required in certain CD4+ T cell lines and not in others. CD4+ T cell lines that are able to support growth of Vif-deficient viruses are termed permissive cells (e.g. SupT1, CEM-SS and Jurkat cells), whereas those that do not support Vif-deficient viruses are termed non-permissive cells (e.g. HuT78, H9 and peripheral blood lymphocytes) (Fig. 25.1). Similar to non-permissive cells, transient heterokaryons formed by the fusion of non-permissive and permissive cells also restrict the spread of Vif-deficient HIV but allow the spread of wild-type HIV-1 (Simon et al. 1998). Characterization of permissive and

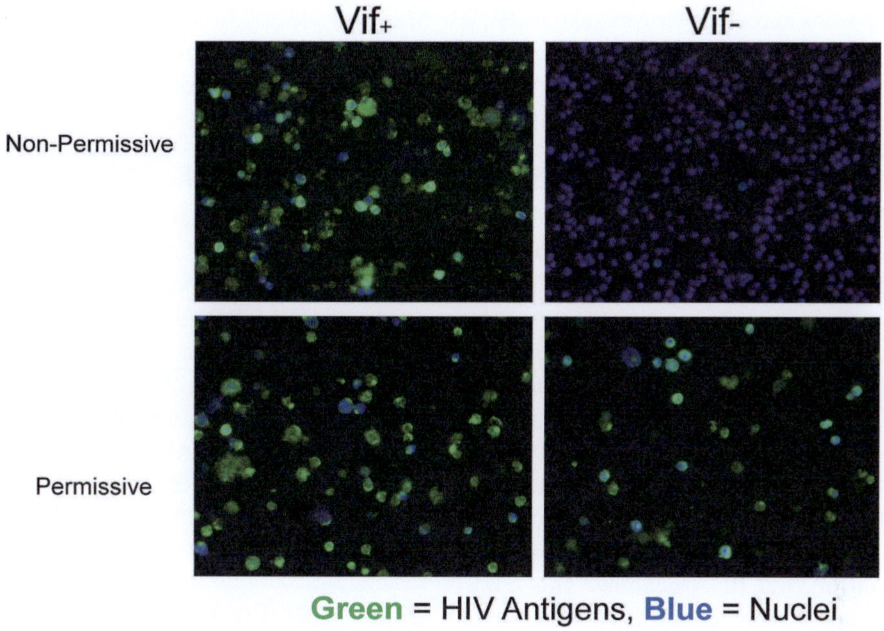

Fig. 25.1 Vif is required for the spread of HIV-1 in non-permissive and primary CD4+ T lymphocytes. Immunofluorescence showing HIV (*green*) and cell nuclei (*blue*). *Top row*, HIV spread in a non-permissive CD4+ T-lymphocyte cell line requires Vif. *Bottom row*, Vif is dispensable for spread in a permissive cell line. APOBEC3 family members are either expressed or not expressed in nonpermissive or permissive cell lines respectively

non-permissive cell lines, as well as cell fusion experiments, suggested that there was a host factor found in non-permissive cells that protected them against HIV infection in the absence of Vif (Simon et al. 1998), and that the primary role of Vif was to counteract this innate antiviral activity of non-permissive cells (Muro-Cacho et al. 1995; Fisher et al. 1987; Finkel et al. 1995; Gabuzda et al. 1992; Doitsh et al. 2015; Sakai et al. 1993; Sova and Volsky 1993; von Schwedler et al. 1993; Bouyac et al. 1997). By comparing the pattern of mRNA expression between permissive and non-permissive cells, APOBEC3G (A3G) was first identified as the antiviral factor that restricts the spread of Vif-deficient HIV-1 in non-permissive cells (Harris et al. 2012; Sheehy et al. 2002; Malim and Emerman 2008). Indeed, transient expression of A3G in permissive cells confers non-permissive phenotype (Harris and Liddament 2004; Sheehy et al. 2002).

In humans, seven members of the A3 protein family- A3A, A3B, A3C, A3D, A3F, A3G and A3H- are encoded in a tandem array on chromosome 22 and their expression levels vary in different tissues and cell types (Yu et al. 2003; Jarmuz et al. 2002; Koning et al. 2009). Although A3G displays the most potent antiviral activities against Vif-deficient HIV-1, other A3 proteins (A3D, A3F and A3H) are also expressed in non-permissive cells and contribute to the restriction of HIV-1 when the *Vif* gene is absent (Dang et al. 2006; Zheng et al. 2004; Mulder et al. 2010; Chaipan et al. 2013) and reviewed in (Desimmie et al. 2014).

Further support indicating the viruses need to counteract A3 proteins stems from the fact that the *vif* gene is found in all known lentiviruses except EIAV (Equine infectious anemia virus). These include HIV-2, simian immunodeficiency virus (SIV) and non-primate lentiviruses such as BIV, MVV, CAEV and FIV. In the absence of Vif, these viruses are all restricted by their respective host APOBEC3 family members. In addition to inhibiting lentiviruses, A3 proteins have also been reported to inhibit the infectivity of diverse retroviruses (HTLV-1, MLV and EIAV), retro-transposons and other viruses such as HBV and AAV (Holmes et al. 2007). Due to space limitations our review will be focused on HIV-1, the most extensively studied primate lentivirus.

25.3 Molecular Mechanisms of APOBEC3 Family Members and Primate HIV-1 Vif

The proteins of the A3 family contain either one (A3A, A3C and A3H) or two cytidine deaminase domains (A3B, A3D, A3F and A3G), suggesting that A3 proteins are able to restrict viral spread through DNA editing. Indeed, A3G mutates deoxy-cytidine to deoxy-uridine in the minus strand of HIV-1 DNA synthesized by reverse transcription of viral RNA genome. This enzymatic mutation results in the accumulation of non-functional proviruses by G-to-A hypermutation in the viral DNA (Harris et al. 2003; Lecossier et al. 2003; Mangeat et al. 2003; H. Zhang et al. 2003). However, catalytically inactive A3G can also display significant anti-retroviral activity when overexpressed, indicating that the deaminase-independent activity of A3G can contribute to HIV-1 inhibition. In this non-editing mode, A3G appears to inhibit the synthesis of viral DNA and its integration into human genome (Newman et al. 2005; F. Guo et al. 2006, 2007; Iwatani et al. 2007; X. Y. Li et al. 2007; Luo et al. 2007; Mbisa et al. 2007; X. Wang et al. 2012). In either case, it is widely accepted that in order for APOBEC3 family members to restrict retroviruses, they must be packed into the viral core to have access to the viral genetic material during reverse transcription in the target cells (Mangeat et al. 2003). The requirement for viral packaging of the APOBEC3 family of restriction factors is referred to as the Trojan Horse model (Fig. 25.2).

HIV-1 Vif plays a critical role in counteracting the APOBEC3 proteins. To this end, Vif reduces the steady-state level of A3G, A3C, A3D, A3F and A3H haplotype II in producer cells, targeting them for degradation by the ubiquitin-proteasome system and preventing their packaging into virions (Conticello et al. 2003; Mariani et al. 2003; Marin et al. 2003; Sheehy et al. 2003; Stopak et al. 2003; Yu et al. 2003). Cytoplasmic localization of HIV-1 Vif is required for its ability to neutralize APOBEC3G and probably all Vif susceptible APOBEC3 family members (Farrow et al. 2005; Goncalves et al. 1994; Wichroski 2004). In order to target A3 proteins for degradation, Vif hijacks a cellular E3 ubiquitin ligase of the Cullin-RING super family (CRL5), comprised of CUL5/RBX2 and the adaptor subunit ELOB/ELOC (Yu et al. 2003; D. J. Stanley et al. 2012; Kamura 2004). Within the context of the E3 ligase, Vif functions as the substrate receptor and directly binds APOBEC3 proteins, thus recruiting them to CRL5 for polyubiquitination (Fig. 25.2) (Yu et al. 2003).

25.4 CBFβ Acts as a Chaperone for HIV-1 Vif

Even though the functional roles of Vif in HIV-1 infectivity have been well-described, the purification of homogeneous recombinant Vif protein has been extremely difficult and for a long while limited research efforts in biochemistry and structure biology. Using novel proteomics approaches, CBFβ was discovered as a potential HIV-1 Vif-binding factor (Jager et al. 2012; W. Zhang et al. 2012). These studies employed affinity chromatography with mass spectrometry and found that over-expressed HIV-1 Vif in human cell lines was co-eluted with several cellular factors including CRL5 and CBFβ. The identification of CBFβ represented a major breakthrough in Vif biochemical studies, finally allowing researchers to reconstitute an active CRL5-Vif-CBFβ ubiquitin E3 ligase from recombinant purified components. This complex can polyubiquitinate A3G *in vitro* and recapitulates the known APOBEC3 substrate specificity in cells (Jager et al. 2012; D. Y. Kim et al. 2013). The role of CBFβ was further validated in cellular assays demonstrating that CBFβ is required for HIV-1 Vif to degrade all Vif-sensitive APOBEC3 family members (Hultquist et al. 2012). Specifically, knockdown of CBFβ reduces polyubiquitination of tran-

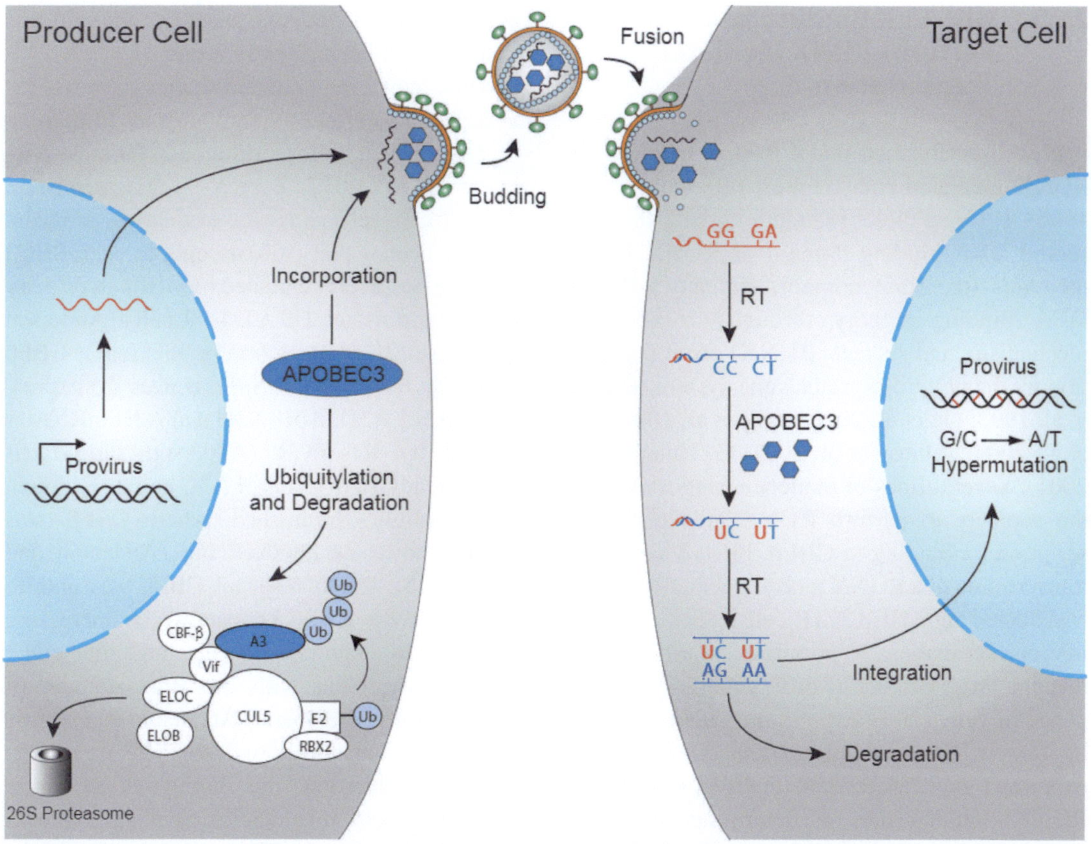

Fig. 25.2 Overview of restriction by APOBEC3 and suppression by HIV Vif. In the absence of Vif, APOBEC3 family members -A3D, A3F, A3G and A3H-are packed into budding virions and restrict HIV by acting as cytidine deaminases resulting in hypermutation, which leads to genetic catastrophe for the virus. Vif promotes infectivity by targeting A3 family members for degradation by the 26S proteasome. To do so, Vif hijacks a cellular E3 ligase (CRL5, or Cul5/RBX2, EloBC) and the transcription cofactor CBFß (Figure adapted from reference Harris et al. 2012)

siently expressed APOBEC3G and subsequently inhibits infectivity of HIV-1(Jager et al. 2012; Miyagi et al. 2014). These studies also showed that a reduction in CBFβ protein levels correlates with a decrease in steady-state levels of Vif protein in a variety of cell types, including CD4+ T-cell lines (Jager et al. 2012; W. Zhang et al. 2012; Miyagi et al. 2014; Han et al. 2014; Anderson and Harris 2015). With this in mind, reduced levels of Vif were compensated for by overexpression, however this does not restore the A3G degradation defect when CBFβ is knocked down (Miyagi et al. 2014). Knockdown of CBFβ by RNAi correlates with a loss of Vif binding to CUL5 in cells (W. Zhang et al. 2012; Han et al. 2014). These studies, together with the biochemical and structural data, indicate that CBFβ acts to 'chaperone' Vif by stabilizing its fold so that it can specifically engage CRL5 and A3 substrates to promote ubiquitination (*vide infra*). This function of CBFβ is likely conserved in all primate lentiviruses, based on sequence similarity and the fact that CBFβ is required for SIV Vif to degrade Vif-sensitive Rhesus Macaque APOBEC3 proteins (Hultquist et al. 2012). In contrast, CBFβ is dispensable for Vif function in several non-primate lentiviruses, consistent with the fact that the sequence of non-primate lentiviral Vif proteins diverges considerably from their primate counterparts (W. Zhang et al. 2014; Ai et al. 2014; Kane et al. 2015).

25.5 HIV-1 Vif Binding to CBFβ Perturbs RUNX Mediated Transcription

It is well established that CBFβ is required for RUNX mediated transcription (Q. Wang et al. 1996). It forms direct interactions with the conserved DNA binding domain of RUNX family members (the Runt domain), enhances RUNX DNA binding activity through an allosteric mechanism, and protects RUNX1 from degradation by the ubiquitin-proteasome system (Ogawa et al. 1993; Gu et al. 2000; Tang et al. 2000; Yan et al. 2004; Tahirov et al. 2001; G. Huang et al. 2001). Several lines of evidence suggest Vif has the capacity to perturb RUNX mediated transcription by binding to CBFβ. First, Vif reduces transcription of a RUNX reporter gene transiently transfected into HEK293T cells. Second, DNA microarray analysis in permissive Jurkat T-cells reveal a large number of differentially expressed genes in cells stably expressing Vif (Fig. 25.3a, b) (D. Y. Kim et al. 2013). Differentially expressed genes were statistically enriched with RUNX1 binding sites, as determined by ChIP-Seq studies using a RUNX1 antibody (D. Y. Kim et al. 2013). RT-qPCR of candidate genes confirmed that Vif had the capacity to repress transcription of known RUNX sensitive genes. For example, *Tbx21* (also known as *Tbet*) was identified in the ChIP-Seq and microarray analysis (D. Y. Kim et al. 2013). Consistent with prior studies, reduction in *Tbx21* expression correlated with enhanced production of IL-2(Szabo et al. 2000). Likewise, infection of a permissive CD4+ T-cell line with HIV-1 reduces expression of *Tbx21* in a Vif-dependent manner (D. Y. Kim et al. 2013). Third, co-IP, *in vitro* binding, and mutagenesis studies suggest CBFβ binds Vif in a manner that is mutually exclusive with RUNX transcription factors (D. Y. Kim et al. 2013). This notion is further supported by a high-resolution crystal structure of Vif-bound to CBFβ (*vide infra*) (Y. Guo et al. 2014). A parsimonious explanation for the ability of Vif to perturb transcription is that it can scavenge CBFβ so that it cannot be incorporated into transcription complexes with RUNX proteins, though this model has not been directly tested.

While the aforementioned studies show Vif has the capacity to affect numerous genes in permissive cells, a clear biological function for these phenomena in APOBEC3 expressing non-permissive CD4+ T-cell cells was not evaluated. A recent study indicates that Vif-susceptible APOBEC3 genes are positively regulated by CBFβ in primary and non-permissive H9 CD4+ T-cells (Anderson and Harris 2015). Reduction or ablation of CBFβ mRNA by RNAi or CRISPR reduces the expression of A3C, A3D, A3F, A3G and A3H mRNA as detected by RT-qPCR (Anderson and Harris 2015). Steady-state levels of A3G and A3F protein were severely diminished when CBFβ was knocked down or knocked out (Anderson and Harris 2015). RNAi resistant CBFβ was able to compliment the CBFβ knockdown by increasing A3G protein expression levels, and this effect required interaction with the Runt domain of RUNX family members (Anderson and Harris 2015). ChIP-Seq studies indicate there are numerous RUNX3 binding sites throughout the entire APOBEC3 locus (Fig. 25.3c). Strikingly, knockout or knockdown of CBFβ rendered non-permissive H9 cells permissive for infection with Vif-deficient virus: the restrictive potential of these cells provided by the APOBEC3 repertoire was nearly completely suppressed (Anderson and Harris 2015). These findings provide a compelling explanation for why HIV-1 Vif hijacks CBFβ: it allows Vif to interfere with RUNX mediated transcription of APOBEC3 family members. It has been suggested that the expanded APOBEC3 repertoire of primates compared to ancestral placental mammals has driven the evolution of primate lentiviral Vif to acquire CBFβ as a binding partner, endowing the primate lentiviruses with the ability to downregulate A3 transcriptionally, in addition to the well established post-transcriptional mechanism of ubiquitin mediated proteolysis (Anderson and Harris 2015).

It is worth mentioning that studies in non-permissive and permissive cells have allowed the unambiguous separation of function of CBFβ as it pertains to viral infectivity at the transcriptional

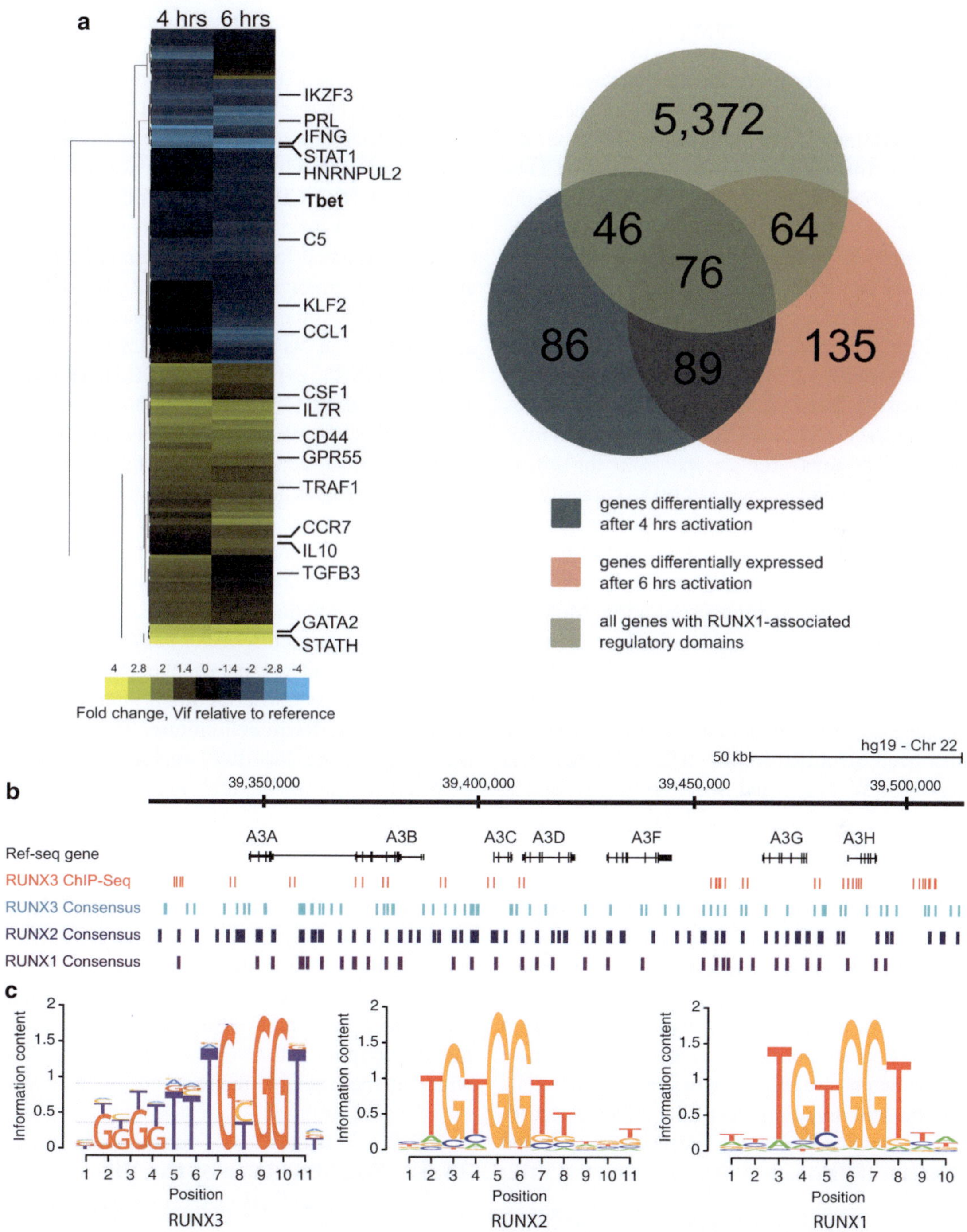

Fig. 25.3 Vif has the capacity to perturb transcription of RUNX genes in host cells. (**a**) Stable expression of Vif in permissive Jurkat T-cell lines perturbs gene expression. Shown is the differential expression of genes 4 and 6 h after activation with the phorbol ester PMA and the lectin PHA in cells either lacking or containing stably expressed Vif. In this study, there was a statistically significant enrichment of RUNX1 sites associated with differentially expressed genes. (**b**) Non-permissive T-cells have RUNX sites associated with APOBEC3 loci. Results from experimental ChIP-Seq data are shown for RUNX3-binding sites demonstrated by ENCODE ChIP-sequencing of the lymphoblastoid cell line GM12878 (ENCSR000BRI) (Consortium et al. 2013). (**c**) Motif for RUNX3 DNA binding site

and post-transcriptional levels. Early work where CBFβ was knocked down in permissive HEK293T or HeLa cells showed that degradation of heterologous expressed A3 family members by Vif required CBFβ, likely due to its chaperone function. In contrast, in non-permissive CD4+ T-cells, ablation of CBFβ nearly bypasses the requirement of the post-transcriptional degradative step, since steady state mRNA and protein levels of APOBEC3 are reduced from what is apparently a transcriptional defect, obviating the need for a functional Vif E3 ligase. The regulation of APOBEC3 expression at transcriptional and post-transcriptional steps by interactions between HIV-1 Vif and CBFβ is a molecular two-step that ensures viral escape from the innate immunity provided by APOBEC3 family of restriction factors (Anderson and Harris 2015).

Some studies indicate CBFβ could also regulate retroviral transcription. For example, this was initially observed in the retrovirus MoMLV, where CBF binding sites in the long-terminal repeat (LTR) were discovered as an enhancer of viral replication (S. W. Wang and Speck 1992). It is interesting to note that CBF binding sites have also been identified in the LTR of SIVmac and HIV-1(S. W. Wang and Speck 1992). Recently, CBFβ/RUNX1 was reported to repress HIV-1 transcription and suggested to be important for viral latency (Klase et al. 2014). Though combination therapy for HIV-1 infection can reduce plasma levels of virus to undetectable levels, HIV-1 can persist in a latent form in resting memory CD4+ T-cells (Finzi et al. 1997, 1999). In these cells, there is minimal transcription from the LTR because of the absence of necessary host factors which are present only in activated T-cells and the additional presence of a putative restriction factor that blocks viral reverse transcription (Nabel and Baltimore 1990; Baldauf et al. 2012; Laguette et al. 2011; Hrecka et al. 2011). Knockdown of CBFβ or RUNX1 in cell culture models of latency results in reactivation of virus (Klase et al. 2014). This observation is consistent with ChIP-qPCR showing RUNX and CBFβ associated with the HIV-1 LTR in latently infected cells (Klase et al. 2014). Likewise, a pharmacologic inhibitor of RUNX1/CBFβ (Ro5–3335) was also capable of reactivating latent cells (Klase et al. 2014; Cunningham et al. 2012). Synergy between Ro5–3335 and the HDAC inhibitor SAHA was observed in HIV-1 reactivation in cell-culture and in PBMCs isolated from patients infected with HIV-1. The available data suggest RUNX1/CBFβ is a potential target to reactive latent reservoirs, a strategy currently being investigated for curative treatments (Klase et al. 2014; Cunningham et al. 2012). In addition to Ro5–3335, other inhibitors RUNX1/CBFβ await testing in this regard (Gorczynski et al. 2007). RUNX1 expression in CD4+ memory T-cells of viremic HIV-1 patients correlates negatively with viral load and positively with CD4+ T-cell count, suggesting RUNX1 may be associated with progression of HIV-1 in the clinic (Klase et al. 2014).

Additional support for a role of RUNX1 and CBFβ in HIV-1 transcription is provided by the following observations (Klase et al. 2014). There are several conserved RUNX1 sites in the group B HIV-1 LTR. Overexpression of RUNX1 and CBFβ can repress transcription off integrated and unintegrated LTRs in cell culture. This effect is abrogated if the last two nucleotides of the RUNX1 consensus-binding motif (TGYGGT) are mutated within one of the predicted RUNX1 binding sites. Over expression of Vif partially restores transcription of the viral LTR and redistributes CBFβ from cell wide to cytoplasmic localization, a result that is consistent with Vif sequestering CBFβ in the cytoplasm. Therefore, CBFβ/RUNX has the capacity to repress transcription off the viral UTR in addition to promoting transcription of APOBEC3 genes, both of which are beneficial to the host.

25.6 The Structure of the Vif-CBFβ-ELOBC-CUL5 Complex

Insights into the multifunctional nature of HIV-1 Vif-CBFβ interactions are provided by the recent crystal structure of the HIV-1 Vif, CBFβ, ELOB and ELOC in complex with the CUL5 N-terminal domain (CUL5NTD) (Y. Guo et al. 2014) (Fig. 25.4). This structure sheds light on how Vif

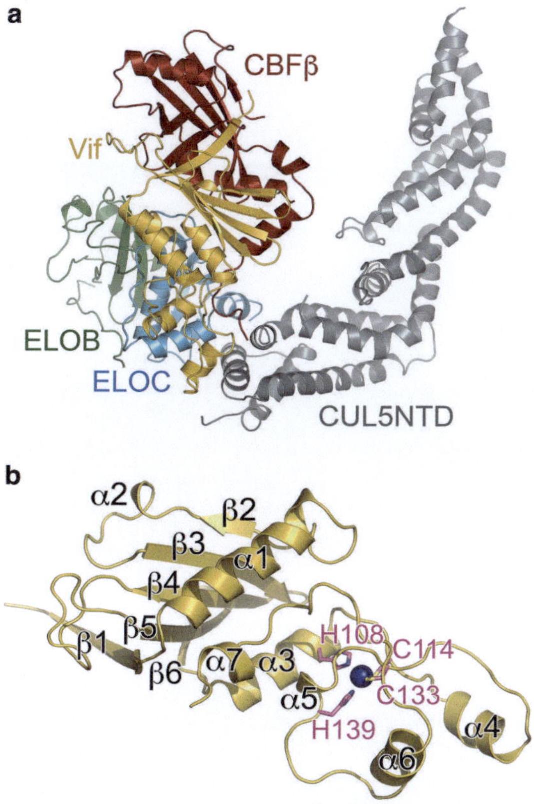

Fig. 25.4 Overview of Vif structure (**a**) Crystal structure of Vif-CBFß-ELOBC in complex with the N-terminal domain of CUL5 (CUL5NTD). The crystal contains 12 pentameric complexes in asymmetrical units, one of which is drawn as ribbon diagram (PDB id 4N9F; chain C, D, E, F and G). CUL5NTD, Vif, ELOB, ELOC and CBFβ are labeled and colored *grey*, *yellow*, *green*, *cyan* and maroon, respectively. Vif mediates direct interactions with CBFβ, CUL5 and ELOC. (**b**) The structure of the Vif monomer in the pentameric complex. Vif structure is drawn as a ribbon diagram and the secondary structure is labeled. The residues that coordinate a zinc atom, termed the HCCH motif, are depicted as sticks and the residue numbers are labeled. The α-helices (α1-α7) and β-strands (β1-β5) are also indicated

folding is promoted by host factors resulting in the formation of a functional CRL5-Vif-CBFβ holoenzyme, how Vif binding to CBFβ could perturb RUNX-mediated transcription, and how Vif recognizes A3 substrates. Below we will discuss each of these functions from a structural perspective.

The Vif subunit (residues 3–171) in the complex maintains a conical shape formed by the arrangement of two globular folds, a large α/β fold and small α fold (Fig. 25.4b). The large α/β fold consists of an antiparallel beta-sheet (β1–β5) and four alpha-helices (α1–α3 and α7) that are aligned along the convex side of the beta-sheet. The α fold (residues 112–161) inserted between α3 and α7 in the α/β fold is composed of three helices (α4, α5 and α6). Additionally, the HCCH motif in Vif coordinates a zinc atom between α/β and α folds. The coordination of the zinc atom is mediated by H108 in the α/β fold and by C114, C133 and H139 in the α fold and appears to stabilize the tertiary structure by reducing the flexibility between two globular folds of Vif (Fig. 25.4b). In the surface structure of Vif, most of the charged surface is exposed to solvent even in the pentameric complex. This indicates that the Vif surface for the binding of CUL5, ELOC and CBFβ is composed of mainly hydrophobic patches (Y. Guo et al. 2014).

The complete structure reveals that Vif mediates the formation of hetero-pentameric Vif-CBFβ-ELOBC-CUL5NTD complex by directly interacting with all affiliated subunits (Fig. 25.4a) (Y. Guo et al. 2014). As a central part of the assembly of CRL5-Vif E3 ligase, α3 in HIV-1 Vif (residues 121–127) interacts with the LWDD motif in CUL5 (residues 52–55), which is not conserved among other Cullin family proteins, indicating that the LWDD motif is required for the selective recruitment of CUL5 to HIV-1 Vif (Y. Guo et al. 2014). This mode of interaction between Vif and CUL5 is quite different than the binding in cellular SOCS2-CUL5, wherein the CUL box in SOCS2 (residues 182–186) interacts with residues L52, W53, Q113 and I116 in CUL5(Y. K. Kim et al. 2013; Y. Guo et al. 2014). For the binding of the ELOB/ELOC heterodimer, the BC box in HIV-1 Vif (residues 141–153 of α4) interacts with the surface residues distributed at the C-terminus of ELOC (B. J. Stanley et al. 2008). Vif seems to mimic the BC box in cellular SOCS proteins for ELOC binding (Kamura et al. 1998; Yu et al. 2003; Mehle 2004). As explained above, the BC box contains the one motif conserved among all lentiviral Vif proteins, the SLQxLA motif (Kane et al. 2015). Thus, Vif homologues may recruit ELOB/ELOC for the CRL-Vif assembly in a common manner.

25.7 The Binding Mode of CBFβ with HIV-1 Vif and Its Implications for Transcription

In contrast to the recruitment of CUL5 and ELOBC into CRL5-Vif, which is mediated by interactions between small binding motifs, the interaction between Vif and CBFβ is mediated by a large surface area, burying a total of 4800 Å^2 (Fig. 25.5) (Y. Guo et al. 2014). The binding interface is widely dispersed through residues 1–120 in HIV-1 Vif and encompasses the entire length of CBFβ. For example, a N-terminal beta strand in HIV-1 Vif (residues 2–11) pairs with a beta strand in CBFβ (residues 63–69) forming an intermolecular beta sheet through a network of hydrogen bonds whereas the C-terminal tail of CBFβ (residues 135–157), containing an alpha helix and a flexible loop, binds to a surface crevice near a zinc atom coordinated by the HCCH motif of Vif. Both interactions appear to support tight binding between HIV-1 Vif and CBFβ. The extensive buried surface area lining the interface between CBFβ and Vif may explain the susceptibility of Vif to proteasomal degradation when CBFβ is knocked down (Jager et al. 2012; W. Zhang et al. 2012; Miyagi et al. 2014; Anderson and Harris 2015). This effect is mirrored by an increase to protease susceptibility *in vitro* when CBFβ is absent from purification (D. Y. Kim et al. 2013). Therefore, CBFβ templates the folding of Vif to allow it to function as a substrate receptor for the CRL5-Vif-CBFβ E3 ligase. In this way, it functions to "chaperone" Vif into a conformation that allows proper folding and function.

CBFβ regulates transcriptional activities of RUNX proteins in human cells by binding to Runt domains conserved among the RUNX proteins and enhancing the association of the Runt domain with DNA (Ogawa et al. 1993; Gu et al. 2000; Tang et al. 2000; Tahirov et al. 2001; Q. Wang et al. 1996). It also protects RUNX family members from ubiquitination and proteasomal degradation (G. Huang et al. 2001). Even though there is no structural similarity between HIV-1 Vif and the Runt domain, available structures of both

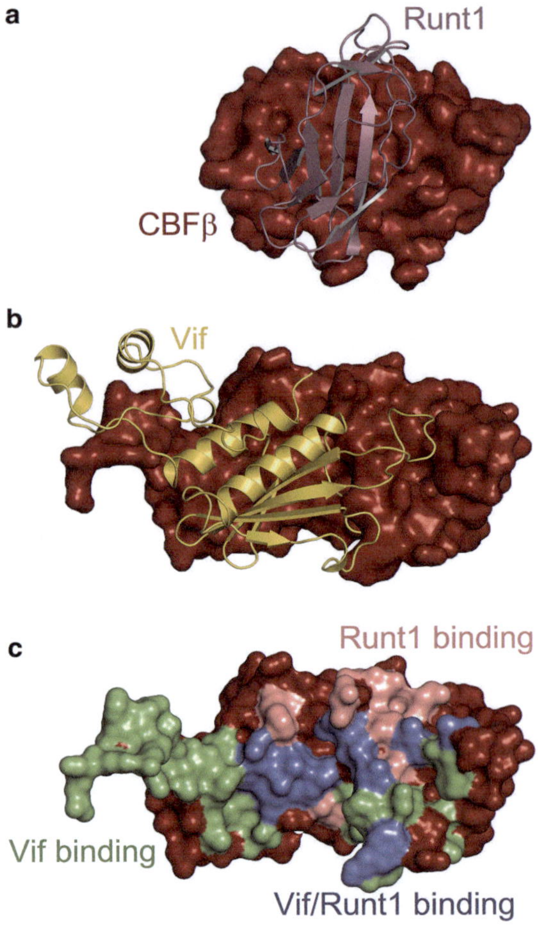

Fig. 25.5 Mutually exclusive interaction of Vif and RUNX proteins with CBFβ. (**a**) Binding interface between the Runt domain of RUNX1 (Runt1) and CBFβ. Complex structure of Runt1 and CBFβ is drawn as a ribbon diagram and surface model (PDB id 1E50; Chain G and H). The structure contains residues 2–135 in CBFβ and residues 57–175 in Runt1. (**b**) Binding interface between Vif (residues 3–172) and CBFβ (residues 3–156). (**c**) Surface patches on CBFβ for Vif and Runt1 binding. The residues specific for Vif and Runt1 binding are colored on the CBFβ surface model. Green, pink and blue indicate the residues for the binding of Vif, CBFβ and both, respectively

Runt1-CBFβ and Vif-CBFβ reveal insight into how each of these proteins engage CBFβ (Y. Guo et al. 2014; Tahirov et al. 2001; X. Huang et al. 1999). Based on the structural data it is apparent that Runt1 and Vif bind CBFβ with overlapping, but not identical, binding sites (Fig. 25.5). Therefore, it is not surprising that Vif binding to CBFβ partially occludes Runt1 binding (Warren et al. 2000; Y. Guo et al. 2014). Despite having overlapping binding site on CBFβ, the CBFβ binding modes for Vif and Runt1 are different (D. Y. Kim et al. 2013; Y. Guo et al. 2014). For example, HIV-1 Vif binds residues 63–69 on CBFβ to form a continuous beta sheet that is mediated by hydrogen bonds, whereas Runt1 domain binds the residues on CBFβ mainly through hydrophobic interactions. Moreover, an additional HIV-1 Vif binding motif on CBFβ (residues 135–157) reinforces the Vif-CBFβ interaction. Together these data establish that Vif and RUNX bind CBFβ in a mutually exclusive manner, and supports the hypothesis that HIV-1 Vif perturbs RUNX transcription activities by scavenging CBFβ from RUNX proteins.

25.8 Recognition of A3 Family Members by Vif

While there are no structures of a Vif-A3 complex, functional studies of mutants combined with structural studies of single domain A3C or the individual Vif interaction domains of A3F and A3G have provided insights into critical hotspots for Vif recognition (Kitamura et al. 2012; Bohn et al. 2013; Siu et al. n.d.; Nakashima et al. 2016; Kouno et al. 2015). This information is covered in several recent reviews, so only the key features will be summarized (Aydin et al. 2014; Chelico 2014; Desimmie et al. 2014). These experiments have revealed that the interaction modes between HIV-1 Vif and A3 family members differ even though the A3 proteins share a highly conserved tertiary structure. Binding of A3 family to Vif can be classified by different interaction surfaces. For example A3C, A3F and A3D form interactions with Vif using a shallow hydrophobic pocket and surface exposed acidic residues distributed across α-helices 2, 3 and 4, which form one face of the cytosine deaminase fold (Russell et al. 2009; Smith and Pathak 2010; Kitamura et al. 2012; Bohn et al. 2013; Siu et al. n.d.; Nakashima et al. 2016). In contrast, A3G forms interactions with Vif using hydrophobic and acidic residues within a loop between β4 and α4 (also known as the L7 loop). These include F126 and 128-DPD-130 (Bogerd et al. 2004; Mangeat et al. 2004; Schröfelbauer et al. 2004; Huthoff and Malim 2007; Russell et al. 2009; W. Zhang et al. 2008; Letko et al. 2015; Kouno et al. 2015). Interactions of Vif with A3HhapII are less well understood. Based on homology modeling, a residue (D121) on helix α4 is implicated in binding Vif and the charged character of the surface formed by α-helices 2, 3 and 4 is different from the A3C/D/F surface suggesting a different mode of recognition (M. M. H. Li et al. 2009; Zhen et al. 2010). Together these data suggest that A3C, A3F and A3D share a common Vif binding mode, which is distinct from that of A3G and A3HhapII.

Further support for different binding modes of A3 family members is provided by the crystal structure of the HIV-1 Vif-CBFβ-ELOBC-CUL5NTD complex and prior functional studies of mutants. Critical interaction residues have been mapped onto the crystal structure of Vif-CBFβ-ELOBC-CUL5NTD (Fig. 25.6) (Y. Guo et al. 2014). The A3HhapII binding motif (F39 and H48) (Binka et al. 2011; Ooms et al. 2013a, b), the A3G binding motif (40-YRHHY-44) (Russell and Pathak 2007; Yamashita et al. 2008), the A3F binding motifs (14-DRMR-17, 74-TGERxW-79 and 171-EDRW-174) (Russell and Pathak 2007; Z. He et al. 2008; Dang et al. 2010) and the shared A3F/A3G binding motifs (21-WxSLVK-26 and 55-VxIPLx$_4$L-64) (Chen et al. 2009; Dang et al. 2009) are surface-exposed Vif residues, indicating that these motifs may mediate direct binding between Vif and A3 proteins (Fig. 25.6). In addition, viral adaptation experiments and functional assays with patient derived Vif variants have allowed researchers to build models Vif-A3 domain complexes (Letko et al. 2015; Richards et al. 2015). While these models are informative, ultimately a structure of the Vif-A3 interface will be instrumental in improving our understanding of Vif-A3 interactions and the molecular details that drive specific Vif-A3 binding.

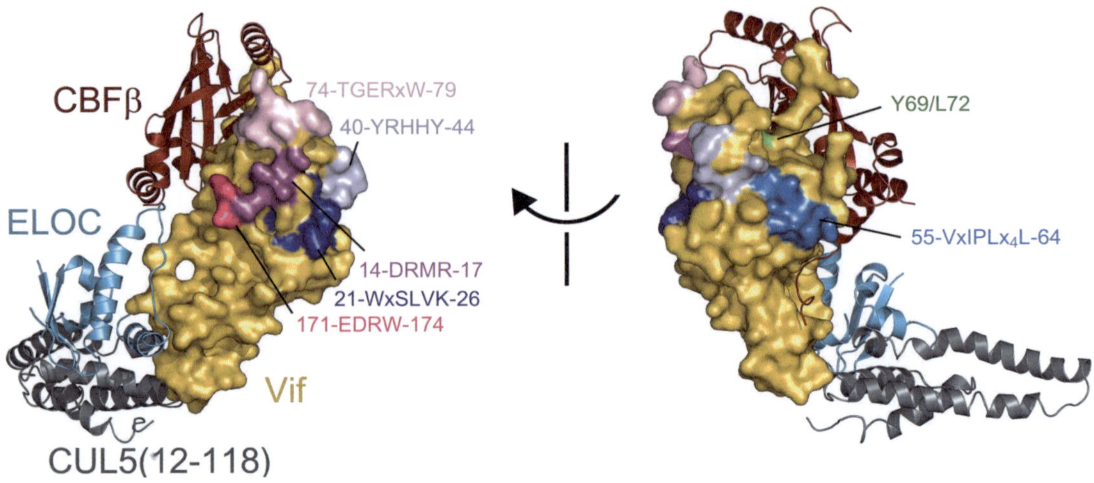

Fig. 25.6 Residues of Vif required for A3G and A3F neutralization. Clusters of surface exposed residues required for neutralization of A3F, A3G or both are indicated on a surface representation of Vif (*yellow*). Residues required for A3F, but not A3G, neutralization reside within the 14–17, 74–79 and 171–174 motifs colored *purple*, *fuchsia* and *light pink*. Residues required for A3G, but not A3F, neutralization are contained within the 40–44 motif colored light blue. Residues required for neutralization of both A3F and A3G reside within the 21–26 and 55–64 motif and are colored *blue*. Residues 69 and 72 are required for A3F and A3G neutralization (Pery et al. 2009), but probably because they are buried by CBFβ and stabilize the Vif fold (Y. Guo et al. 2014)

25.9 Outlook and Conclusion

The discovery of CBFβ as a Vif interaction partner is just one example of how systematic, unbiased studies of virus-host interactions have revealed new connections of how viruses co-opt host-cell biology (N. He et al. 2010). For the case of Vif, biochemical, structural and cell biological studies are consistent with the notion that CBFβ, along with other factors in the CRL5 E3, act to template Vif folding so that is can promote A3 polyubiquitination and degradation (Jager et al. 2012; D. Y. Kim et al. 2013). At the same time, Vif interacts with CBFβ in a manner that is competitive with RUNX binding, preventing formation of the CBFβ/RUNX heterodimer and perturbing RUNX transcriptional regulation (D. Y. Kim et al. 2013). CBFβ positively regulates expression of the full armament of A3 restriction factors, and Vif may repress transcription of these genes through interaction with CBFβ (Anderson and Harris 2015). Therefore, the interaction between Vif and CBFβ illustrates an economical strategy for a virus with limited protein coding capacity to perturb multiple host pathways, specifically by reducing the steady-state levels of A3 restriction factors by perturbing transcriptional and post-transcriptional steps of gene expression (Fig. 25.7). This phenomenon of dual hijacking is probably conserved in all primate lentiviruses, since CBFβ -Vif interactions are similarly conserved (Hultquist et al. 2012). It seems likely that other viruses with limited protein coding capacity may also use dual hijack mechanisms, but prevalence of this phenomenon is unclear as systematic, unbiased studies of virus-host interactions are in their infancy.

There are many remaining questions about the interaction of primate lentiviral Vif with CBFβ. First, though the simplest explanation for the Vif- CBFβ interaction is that it has evolved to reduce A3 gene expression, the effects on other host genes and their biological significance is unknown (Anderson and Harris 2015). It is tempting to speculate that the misregulation of additional host genes could be important for chronic development of infection in an animal model but this has not been tested. In this regard,

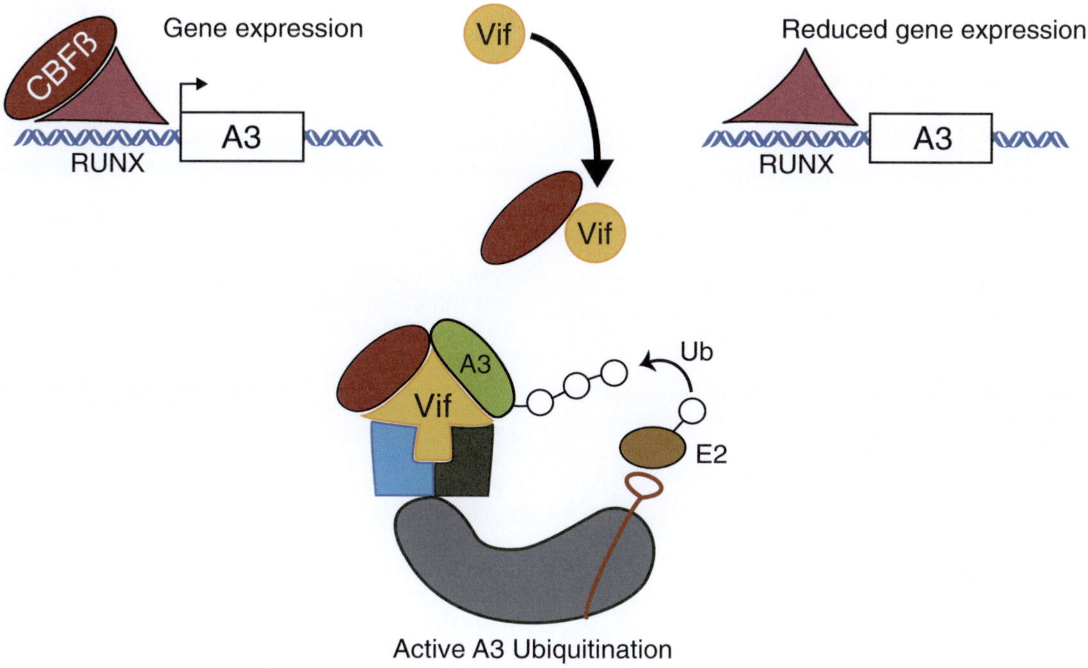

Fig. 25.7 Dual hijack model for Vif neutralization of APOBEC3. Vif binds CBFβ in a manner that is mutually exclusive with RUNX proteins, effectively preventing CBFβ from activating transcription of genes such as APOBEC3. In addition, Vif promotes the polyubiquitination and degradation of A3 family members by the 26S proteasome. This activity requires CBFβ, which promotes Vif folding. Thus, a single viral protein can perturb multiple host pathways by recruiting host factors to a common complex (Figure adapted from reference D. Y. Kim et al. 2013)

Vif could be used as a tool to study RUNX/CBFβ transcription in mouse models, since the murine CBFβ can complement the knockdown of the human counterpart (Han et al. 2014). Second, for Vif to effectively sequester CBFβ, its cellular concentrations during infection have to higher than CBFβ, which is relatively abundant in CD4+ T cells (GCID16P067063). Are Vif levels in infected cells high enough to sequester CBFβ and shutdown RUNX transcription, or might a catalytic mechanism be employed to promote CBFβ cytoplasmic retention? Third, binding of CBFβ to Vif limits its surface area for interaction with APOBEC3 family members, so does CBFβ/Vif form a composite surface for APOBEC3 binding? Models based on mutational analysis and viral adaptation have provided insights into how A3G and A3F are bound to Vif, but high-resolution structures of Vif bound to A3 enzymes will be required to address this question (Letko et al. 2015; Richards et al. 2015). Fourth, could small molecule inhibitors of the Vif- CBFβ interaction be discovered in order to unleash the restriction potential of APOBEC3 enzymes? The Vif-CBFβ interface is large, so finding a potent inhibitor of this protein-protein interaction could be challenging (Y. Guo et al. 2014). Allosteric inhibitors of Runt/CBFβ have been described, suggesting it may be feasible to inhibit Vif-CBFβ in a similar manner (Gorczynski et al. 2007). In sum, there are many exciting directions to explore between CBFβ, immunity and HIV infection.

Acknowledgements We thank Drs. Jennifer Binning, Serena Sanulli and Judd Hultquist for useful feedback on the manuscript. JDG acknowledges grant support from the National Institutes of Health (P50 GM082250). DYK was supported by the National Research Foundation of Korea (NRF) grant funded by the Korea government (MSIP) (NRF-2014R1A4A1071040).

References

Ai, Y., et al. (2014). Core-binding factor subunit beta is not required for non-primate lentiviral Vif-mediated APOBEC3 degradation. *Journal of Virology, 88*(20), 12112–12122.

Anderson, B. D., & Harris, R. S. (2015). Transcriptional regulation of APOBEC3 antiviral immunity through the CBF-/RUNX axis. *Science Advances, 1*(8), e1500296–e1500296.

Aydin, H., Taylor, M. W., & Lee, J. E. (2014). Structure-guided analysis of the human APOBEC3-HIV restrictome. *Structure/Folding and Design, 22*(5), 668–684.

Baldauf, H.-M., et al. (2012). SAMHD1 restricts HIV-1 infection in resting CD4+ T cells. *Nature Medicine, 18*(11), 1682–1689.

Binka, M., et al. (2011). The activity spectrum of Vif from multiple HIV-1 subtypes against APOBEC3G, APOBEC3F, and APOBEC3H. *Journal of Virology, 86*(1), 49–59.

Bogerd, H. P., et al. (2004). A single amino acid difference in the host APOBEC3G protein controls the primate species specificity of HIV type 1 virion infectivity factor. *Proceedings of the National Academy of Sciences of the United States of America, 101*(11), 3770–3774.

Bohn, M.-F., et al. (2013). Crystal structure of the DNA cytosine deaminase APOBEC3F: The catalytically active and HIV-1 Vif-binding domain. *Structure/Folding and Design, 21*, 1042–1050.

Bouyac, M., et al. (1997). Phenotypically Vif- human immunodeficiency virus type 1 is produced by chronically infected restrictive cells. *Journal of Virology, 71*(3), 2473–2477.

Chaipan, C., et al. (2013). APOBEC3G restricts HIV-1 to a greater extent than APOBEC3F and APOBEC3DE in human primary CD4+ T cells and macrophages. *Journal of Virology, 87*(1), 444–453.

Chelico, L. (2014). Suppression of APOBEC3-mediated restriction of HIV-1 by Vif. *Front Microbiol, 5*, 450.

Chen, G., et al. (2009). A patch of positively charged amino acids surrounding the human immunodeficiency virus type 1 Vif SLVx4Yx9Y motif influences its interaction with APOBEC3G. *Journal of Virology, 83*(17), 8674–8682.

Consortium, T. E. P., et al. (2013). An integrated encyclopedia of DNA elements in the human genome. *Nature, 488*(7414), 57–74.

Conticello, S. G., Harris, R. S., & Neuberger, M. S. (2003). The Vif protein of HIV triggers degradation of the human antiretroviral DNA deaminase APOBEC3G. *Current Biology: CB, 13*(22), 2009–2013.

Cunningham, L., et al. (2012). Identification of benzodiazepine Ro5-3335 as an inhibitor of CBF leukemia through quantitative high throughput screen against RUNX1-CBFβ interaction. *Proceedings of the National Academy of Sciences, 109*(36), 14592–14597.

Dang, Y., et al. (2006). Identification of APOBEC3DE as another antiretroviral factor from the human APOBEC family. *Journal of Virology, 80*(21), 10522–10533.

Dang, Y., et al. (2009). Identification of a novel WxSLVK motif in the N terminus of human immunodeficiency virus and simian immunodeficiency virus Vif that is critical for APOBEC3G and APOBEC3F neutralization. *Journal of Virology, 83*(17), 8544–8552.

Dang, Y., et al. (2010). Identification of 81LGxGxxIxW89 and 171EDRW174 domains from human immunodeficiency virus type 1 Vif that regulate APOBEC3G and APOBEC3F neutralizing activity. *Journal of Virology, 84*(11), 5741–5750.

Daugherty, M. D., & Malik, H. S. (2012). Rules of engagement: Molecular insights from host-virus arms races. *Annual Review of Genetics, 46*(1), 677–700.

de Bruijn, M. F., & Speck, N. A. (2004). Core-binding factors in hematopoiesis and immune function. *Oncogene, 23*(24), 4238–4248.

Desimmie, B. A., et al. (2014). Multiple APOBEC3 restriction factors for HIV-1 and one Vif to rule them all. *Journal of Molecular Biology, 426*(6), 1220–1245.

Doitsh, G., et al. (2015). Cell death by pyroptosis drives CD4 T-cell depletion in HIV-1 infection. *Nature, 505*(7484), 509–514.

Farrow, M. A., et al. (2005). Nuclear localization of HIV type 1 Vif isolated from a long-term asymptomatic individual and potential role in virus attenuation. *AIDS Research and Human Retroviruses, 21*(6), 565–574.

Finkel, T. H., et al. (1995). Apoptosis occurs predominantly in bystander cells and not in productively infected cells of HIV- and SIV-infected lymph nodes. *Nature Medicine, 1*(2), 129–134.

Finzi, D., et al. (1997). Identification of a reservoir for HIV-1 in patients on highly active antiretroviral therapy. *Science, 278*(5341), 1295–1300.

Finzi, D., et al. (1999). Latent infection of CD4+ T cells provides a mechanism for lifelong persistence of HIV-1, even in patients on effective combination therapy. *Nature Medicine, 5*(5), 512–517.

Fisher, A. G., et al. (1987). The sor gene of HIV-1 is required for efficient virus transmission in vitro. *Science, 237*(4817), 888–893.

Gabuzda, D. H., et al. (1992). Role of vif in replication of human immunodeficiency virus type 1 in CD4+ T lymphocytes. *Journal of Virology, 66*(11), 6489–6495.

Goncalves, J., Jallepalli, P., & Gabuzda, D. H. (1994). Subcellular localization of the Vif protein of human immunodeficiency virus type 1. *Journal of Virology, 68*(2), 704–712.

Gorczynski, M. J., et al. (2007). Allosteric inhibition of the protein-protein interaction between the leukemia-associated proteins Runx 1 and CBFβ. *Chemistry & Biology, 14*(10), 1186–1197.

Gu, T. L., et al. (2000). Auto-inhibition and partner proteins, core-binding factor beta (CBFbeta) and Ets-1, modulate DNA binding by CBFalpha2 (AML1). *Molecular and Cellular Biology, 20*(1), 91–103.

Guo, F., et al. (2006). Inhibition of formula-primed reverse transcription by human APOBEC3G during human immunodeficiency virus type 1 replication. *Journal of Virology, 80*(23), 11710–11722.

Guo, F., et al. (2007). The interaction of APOBEC3G with human immunodeficiency virus type 1 nucleocapsid inhibits tRNA3Lys annealing to viral RNA. *Journal of Virology, 81*(20), 11322–11331.

Guo, Y., et al. (2014). Structural basis for hijacking CBF-β and CUL5 E3 ligase complex by HIV-1 Vif. *Nature, 505*(7482), 229–233.

Han, X., et al. (2014). Evolutionarily conserved requirement for CBF- in the assembly of the HIV/SIV Vif-Cullin 5-RING E3 ubiquitin ligase. *Journal of Virology, 88*, 3320–3328.

Harris, R. S., & Liddament, M. T. (2004). Retroviral restriction by APOBEC proteins. *Nature Reviews Immunology, 4*(11), 868–877.

Harris, R. S., et al. (2003). DNA deamination mediates innate immunity to retroviral infection. *Cell, 113*(6), 803–809.

Harris, R. S., Hultquist, J. F., & Evans, D. T. (2012). The restriction factors of human immunodeficiency virus. *Journal of Biological Chemistry, 287*(49), 40875–40883.

He, Z., et al. (2008). Characterization of conserved motifs in HIV-1 Vif required for APOBEC3G and APOBEC3F interaction. *Journal of Molecular Biology, 381*(4), 1000–1011.

He, N., et al. (2010). HIV-1 Tat and host AFF4 recruit two transcription elongation factors into a bifunctional complex for coordinated activation of HIV-1 transcription. *Molecular Cell, 38*(3), 428–438.

Holmes, R. K., Malim, M. H., & Bishop, K. N. (2007). APOBEC-mediated viral restriction: Not simply editing? *Trends in Biochemical Sciences, 32*(3), 118–128.

Hrecka, K., et al. (2011). Vpx relieves inhibition of HIV-1 infection of macrophages mediated by the SAMHD1 protein. *Nature, 474*(7353), 658–661.

Huang, X., et al. (1999). Solution structure of core binding factor beta and map of the CBF alpha binding site. *Nature Structural Biology, 6*(7), 624–627.

Huang, G., et al. (2001). Dimerization with PEBP2beta protects RUNX1/AML1 from ubiquitin-proteasome-mediated degradation. *The EMBO Journal, 20*(4), 723–733.

Hultquist, J. F., et al. (2012). Vif proteins of human and simian immunodeficiency viruses require cellular CBFβ to degrade APOBEC3 restriction factors. *Journal of Virology, 86*(5), 2874–2877.

Huthoff, H., & Malim, M. H. (2007). Identification of amino acid residues in APOBEC3G required for regulation by human immunodeficiency virus type 1 Vif and Virion encapsidation. *Journal of Virology, 81*(8), 3807–3815.

Ito, Y. (2008). RUNX genes in development and cancer: Regulation of viral gene expression and the discovery of RUNX family genes. *Advances in Cancer Research, 99*, 33–76.

Iwatani, Y., et al. (2007). Deaminase-independent inhibition of HIV-1 reverse transcription by APOBEC3G. *Nucleic Acids Research, 35*(21), 7096–7108.

Jager, S., et al. (2012). Vif hijacks CBF-beta to degrade APOBEC3G and promote HIV-1 infection. *Nature, 481*(7381), 371–375.

Jarmuz, A., et al. (2002). An anthropoid-specific locus of orphan C to U RNA-editing enzymes on chromosome 22. *Genomics, 79*(3), 285–296.

Kamura, T. (2004). VHL-box and SOCS-box domains determine binding specificity for Cul2-Rbx1 and Cul5-Rbx2 modules of ubiquitin ligases. *Genes & Development, 18*(24), 3055–3065.

Kamura, T., et al. (1998). The Elongin BC complex interacts with the conserved SOCS-box motif present in members of the SOCS, ras, WD-40 repeat, and ankyrin repeat families. *Genes & Development, 12*(24), 3872–3881.

Kane, J. R., et al. (2015). Lineage-specific viral hijacking of non-canonical E3 ubiquitin ligase cofactors in the evolution of Vif anti-APOBEC3 activity. *Cell Reports, 11*, 1236–1250.

Kim, D. Y., et al. (2013). CBFβ stabilizes HIV Vif to counteract APOBEC3 at the expense of RUNX1 target gene expression. *Mol Cell, 49*(4), 632–644.

Kim, Y. K., et al. (2013). Structural basis of intersubunit recognition in elongin BC-cullin 5-SOCS box ubiquitin-protein ligase complexes. *Acta Crystallographica, Section D: Biological Crystallography, 69*(Pt 8), 1587–1597.

Kitamura, S., et al. (2012). The APOBEC3C crystal structure and the interface for HIV-1 Vif binding. *Nature Structural & Molecular Biology, 19*(10), 1005–1010.

Klase, Z., et al. (2014). Activation of HIV-1 from latent infection via synergy of RUNX1 inhibitor Ro5-3335 and SAHA J. Luban, ed. *PLoS Pathogens, 10*(3), e1003997.

Koning, F. A., et al. (2009). Defining APOBEC3 expression patterns in human tissues and hematopoietic cell subsets. *Journal of Virology, 83*(18), 9474–9485.

Kouno, T., et al. (2015). Structure of the Vif-binding domain of the antiviral enzyme APOBEC3G. *Nature Structural & Molecular Biology, 22*(6), 485–491.

Laguette, N., et al. (2011). SAMHD1 is the dendritic- and myeloid-cell-specific HIV-1 restriction factor counteracted by Vpx. *Nature, 474*, 654–657.

Lecossier, D., et al. (2003). Hypermutation of HIV-1 DNA in the absence of the Vif protein. *Science, 300*(5622), 1112.

Letko, M., et al. (2015). Identification of the HIV-1 Vif and human APOBEC3G protein interface. *Cell Reports, 13*(9), 1789–1799.

Li, X. Y., et al. (2007). APOBEC3G inhibits DNA strand transfer during HIV-1 reverse transcription. *The Journal of Biological Chemistry, 282*(44), 32065–32074.

Li, M. M. H., Wu, L. I., & Emerman, M. (2009). The range of human APOBEC3H sensitivity to lentiviral Vif proteins. *Journal of Virology, 84*(1), 88–95.

Luo, K., et al. (2007). Cytidine deaminases APOBEC3G and APOBEC3F interact with human immunodeficiency virus type 1 integrase and inhibit proviral DNA formation. *Journal of Virology, 81*(13), 7238–7248.

Malim, M. H., & Emerman, M. (2008). HIV-1 accessory proteins—Ensuring viral survival in a hostile environment. *Cell Host and Microbe, 3*(6), 388–398.

Mangeat, B., et al. (2003). Broad antiretroviral defence by human APOBEC3G through lethal editing of nascent reverse transcripts. *Nature, 424*(6944), 99–103.

Mangeat, B., et al. (2004). A single amino acid determinant governs the species-specific sensitivity of APOBEC3G to Vif action. *The Journal of Biological Chemistry, 279*(15), 14481–14483.

Mariani, R., et al. (2003). Species-specific exclusion of APOBEC3G from HIV-1 Virions by Vif. *Cell, 114*(1), 21–31.

Marin, M., et al. (2003). HIV-1 Vif protein binds the editing enzyme APOBEC3G and induces its degradation. *Nature Medicine, 9*(11), 1398–1403.

Mbisa, J. L., et al. (2007). Human immunodeficiency virus type 1 cDNAs produced in the presence of APOBEC3G exhibit defects in plus-strand DNA transfer and integration. *Journal of Virology, 81*(13), 7099–7110.

Mehle, A. (2004). Phosphorylation of a novel SOCS-box regulates assembly of the HIV-1 Vif-Cul5 complex that promotes APOBEC3G degradation. *Genes & Development, 18*(23), 2861–2866.

Miyagi, E., et al. (2014). CBFβ enhances de novo protein biosynthesis of its binding partners HIV-1 Vif and RUNX1 and potentiates the Vif-induced degradation of APOBEC3G. *Journal of Virology, 88*(9), 4839–4852.

Mulder, L. C., et al. (2010). Moderate influence of human APOBEC3F on HIV-1 replication in primary lymphocytes. *Journal of Virology, 84*(18), 9613–9617.

Muro-Cacho, C. A., Pantaleo, G., & Fauci, A. S. (1995). Analysis of apoptosis in lymph nodes of HIV-infected persons. Intensity of apoptosis correlates with the general state of activation of the lymphoid tissue and not with stage of disease or viral burden. *Journal of Immunology (Baltimore, Md.: 1950), 154*(10), 5555–5566.

Nabel, G., & Baltimore, D. (1990). An inducible transcription factor activates expression of human immunodeficiency virus in T cells. Nature 344(6262):178

Nakashima, M., et al. (2016). Structural insights into HIV-1 Vif-APOBEC3F interaction. *Journal of Virology, 90*(2), 1034–1047.

Newman, E. N., et al. (2005). Antiviral function of APOBEC3G can be dissociated from cytidine deaminase activity. *Current Biology: CB, 15*(2), 166–170.

Ogawa, E., et al. (1993). Molecular cloning and characterization of PEBP2 beta, the heterodimeric partner of a novel Drosophila runt-related DNA binding protein PEBP2 alpha. *Virology, 194*(1), 314–331.

Ooms, M., Brayton, B., et al. (2013a). HIV-1 Vif adaptation to human APOBEC3H haplotypes. *Cell Host and Microbe, 14*(4), 411–421.

Ooms, M., Letko, M., et al. (2013b). The resistance of human APOBEC3H to HIV-1 NL4-3 molecular clone is determined by a single amino acid in Vif L. Menéndez-Arias, ed. *PloS One, 8*(2), e57744.

Pery, E., et al. (2009). Regulation of APOBEC3 proteins by a novel YXXL motif in human immunodeficiency virus type 1 Vif and simian immunodeficiency virus SIVagm Vif. *Journal of Virology, 83*(5), 2374–2381.

Richards, C., et al. (2015). The binding interface between human APOBEC3F and HIV-1 Vif elucidated by genetic and computational approaches. *Cell Reports, 13*, 1781–1788.

Russell, R. A., & Pathak, V. K. (2007). Identification of two distinct human immunodeficiency virus type 1 Vif determinants critical for interactions with human APOBEC3G and APOBEC3F. *Journal of Virology, 81*(15), 8201–8210.

Russell, R. A., Smith, J., Barr, R., Bhattacharyya, D., & Pathak, V. K. (2009). Distinct domains within APOBEC3G and APOBEC3F interact with separate regions of human immunodeficiency virus type 1 Vif. *Journal of Virology, 83*(4), 1992–2003.

Sakai, H., et al. (1993). Cell-dependent requirement of human immunodeficiency virus type 1 Vif protein for maturation of virus particles. *Journal of Virology, 67*(3), 1663–1666.

Schröfelbauer, B., Chen, D., & Landau, N. R. (2004). A single amino acid of APOBEC3G controls its species-specific interaction with virion infectivity factor (Vif). *Proceedings of the National Academy of Sciences of the United States of America, 101*(11), 3927–3932.

Sheehy, A. M., et al. (2002). Isolation of a human gene that inhibits HIV-1 infection and is suppressed by the viral Vif protein. *Nature, 418*(6898), 646–650.

Sheehy, A. M., Gaddis, N. C., & Malim, M. H. (2003). The antiretroviral enzyme APOBEC3G is degraded by the proteasome in response to HIV-1 Vif. *Nature Medicine, 9*(11), 1404–1407.

Simon, J. H., et al. (1998). Evidence for a newly discovered cellular anti-HIV-1 phenotype. *Nature Medicine, 4*(12), 1397–1400.

Siu, K. K., Sultana, A., Azimi, F. C., & Lee, J. E. (n.d.). Structural determinants of HIV-1 Vif susceptibility and DNA binding in APOBEC3F. *Nature Communications, 4*, 1–10.

Smith, J. L., & Pathak, V. K. (2010). Identification of specific determinants of human APOBEC3F, APOBEC3C, and APOBEC3DE and African green monkey APOBEC3F that interact with HIV-1 Vif. *Journal of Virology, 84*(24), 12599–12608.

Sova, P., & Volsky, D. J. (1993). Efficiency of viral DNA synthesis during infection of permissive and nonpermissive cells with vif-negative human immunodeficiency virus type 1. *Journal of Virology, 67*(10), 6322–6326.

Stanley, B. J., et al. (2008). Structural insight into the human immunodeficiency virus Vif SOCS box and its role in human E3 ubiquitin ligase assembly. *Journal of Virology, 82*(17), 8656–8663.

Stanley, D. J., et al. (2012). Inhibition of a NEDD8 cascade restores restriction of HIV by APOBEC3G B. R. Cullen, ed. *PLoS Pathogens, 8*(12), e1003085.

Stopak, K., et al. (2003). HIV-1 Vif blocks the antiviral activity of APOBEC3G by impairing both its translation and intracellular stability. *Mol Cell, 12*(3), 591–601.

Szabo, S. J., et al. (2000). A novel transcription factor, T-bet, directs Th1 lineage commitment. *Cell, 100*(6), 655–669.

Tahirov, T. H., et al. (2001). Structural analyses of DNA recognition by the AML1/Runx-1 Runt domain and its allosteric control by CBFbeta. *Cell, 104*(5), 755–767.

Tang, Y. Y., et al. (2000). Biophysical characterization of interactions between the core binding factor alpha and beta subunits and DNA. *FEBS Letters, 470*(2), 167–172.

von Schwedler, U., et al. (1993). Vif is crucial for human immunodeficiency virus type 1 proviral DNA synthesis in infected cells. *Journal of Virology, 67*(8), 4945–4955.

Voon, D. C.-C., Hor, Y. T., & Ito, Y. (2015). The RUNX complex: Reaching beyond haematopoiesis into immunity. *Immunology, 146*(4), 523–536.

Wang, S. W., & Speck, N. A. (1992). Purification of core-binding factor, a protein that binds the conserved core site in murine leukemia virus enhancers. *Molecular and Cellular Biology, 12*(1), 89–102.

Wang, S., et al. (1993). Cloning and characterization of subunits of the T-cell receptor and murine leukemia virus enhancer core-binding factor. *Molecular and Cellular Biology, 13*(6), 3324–3339.

Wang, Q., et al. (1996). The CBFbeta subunit is essential for CBFalpha2 (AML1) function in vivo. *Cell, 87*(4), 697–708.

Wang, X., et al. (2012). The cellular antiviral protein APOBEC3G interacts with HIV-1 reverse transcriptase and inhibits its function during viral replication. *Journal of Virology, 86*(7), 3777–3786.

Warren, A. J., et al. (2000). Structural basis for the heterodimeric interaction between the acute leukaemia-associated transcription factors AML1 and CBFbeta. *The EMBO Journal, 19*(12), 3004–3015.

Wichroski, M. J. (2004). Analysis of HIV-1 viral infectivity factor-mediated proteasome-dependent depletion of APOBEC3G: Correlating function and subcellular localization. *Journal of Biological Chemistry, 280*(9), 8387–8396.

Yamashita, T., et al. (2008). Identification of amino acid residues in HIV-1 Vif critical for binding and exclusion of APOBEC3G/F. *Microbes and Infection, 10*(10–11), 1142–1149.

Yan, J., et al. (2004). CBFbeta allosterically regulates the Runx1 Runt domain via a dynamic conformational equilibrium. *Nature Structural & Molecular Biology, 11*(9), 901–906.

Yu, X., et al. (2003). Induction of APOBEC3G ubiquitination and degradation by an HIV-1 Vif-Cul5-SCF complex. *Science, 302*(5647), 1056–1060.

Zhang, H., et al. (2003). The cytidine deaminase CEM15 induces hypermutation in newly synthesized HIV-1 DNA. *Nature, 424*(6944), 94–98.

Zhang, W., et al. (2008). Distinct determinants in HIV-1 Vif and human APOBEC3 proteins are required for the suppression of diverse host anti-viral proteins. *PloS One, 3*(12), e3963.

Zhang, W., et al. (2012). T-cell differentiation factor CBF-β regulates HIV-1 Vif-mediated evasion of host restriction. *Nature, 481*(7381), 376–379.

Zhang, W., et al. (2014). Cellular requirements for BIV Vif-Mediated Inactivation of Bovine APOBEC3 Proteins. Journal of Virology. 88(21):12528–12540

Zhen, A., et al. (2010). A single amino acid difference in human APOBEC3H variants determines HIV-1 Vif sensitivity. *Journal of Virology, 84*(4), 1902–1911.

Zheng, Y.-H., et al. (2004). Human APOBEC3F is another host factor that blocks human immunodeficiency virus type 1 replication. *Journal of Virology, 78*(11), 6073–6076.

Part VI

Emerging Roles for RUNX

Roles of RUNX in Hippo Pathway Signaling

26

Antonino Passaniti, Jessica L. Brusgard, Yiting Qiao, Marius Sudol, and Megan Finch-Edmondson

Abstract

The Runt-domain (RD) transcription factors (*RUNX* genes) are an important family of transcriptional mediators that interact with a variety of proteins including the Hippo pathway effector proteins, YAP and TAZ. In this chapter we focus on two examples of RUNX-TAZ/YAP interactions that have particular significance in human cancer. Specifically, recent evidence has found that RUNX2 cooperates with TAZ to promote epithelial to mesenchymal transition mediated by the soluble N-terminal ectodomain of E-Cadherin, sE-Cad. Contrastingly, in gastric cancer, RUNX3 acts as a tumor suppressor via inhibition of the YAP-TEAD complex and disruption of downstream YAP-mediated gene transcription and the oncogenic phenotype. The reports highlighted in this chapter add to the growing repertoire of instances of Hippo pathway crosstalk that have been identified in cancer. Elucidation of these increasingly complex interactions may help to identify novel strategies to target Hippo pathway dysregulation in human cancer.

Keywords

RUNX proteins • Hippo pathway • YAP • TAZ • TEAD • sE-Cad • Breast cancer • Gastric cancer

A. Passaniti (✉) • J.L. Brusgard
Department of Pathology and Marlene & Stewart Greenebaum Comprehensive Cancer Center, University of Maryland School of Medicine, and the Veterans Administration Health Service, Baltimore, MD, USA
e-mail: TPassaniti@som.umaryland.edu

Y. Qiao • M. Finch-Edmondson
The Mechanobiology Institute (MBI) and the NUS Yong Loo Lin School of Medicine, National University of Singapore, Singapore, Republic of Singapore

M. Sudol
The Mechanobiology Institute (MBI) and the NUS Yong Loo Lin School of Medicine, National University of Singapore, Singapore, Republic of Singapore

Institute of Molecular and Cell Biology A*STAR, Singapore, Republic of Singapore

Y. Groner et al. (eds.), *RUNX Proteins in Development and Cancer*, Advances in Experimental Medicine and Biology 962, DOI 10.1007/978-981-10-3233-2_26

26.1 Introduction – *RUNX* Genes and Hippo Signaling

Identified initially in *Drosophila melanogaster* using genetic mosaic screens to identify novel tumor suppressors, the Hippo signaling pathway is a crucial regulator of organ size that is frequently dysregulated in human cancers (reviewed in Liu et al. 2012). The canonical Hippo pathway consists of the serine/threonine kinases Hippo (Hpo) and Warts, with their mammalian sterile 20-like kinases 1 and 2 (MST1/2) and large tumor suppressors 1 and 2 (LATS1/2) orthologs.

The respective adaptor proteins include Salvador (Sav; SAV1 ortholog) and Mob as tumor suppressor (Mats; Mps one binder kinase activator-like 1, MOB1 ortholog). These kinases center on Yorkie (Yki), a potent transcriptional coactivator that associates with the DNA-binding protein Scalloped (Sd) to drive transcription of genes involved in cell proliferation and survival (Justice et al. 1995; Tapon et al. 2002; Harvey et al. 2003; Pantalacci et al. 2003; Wu et al. 2003). The Yes-associated protein (YAP) and Transcriptional coactivator with PDZ-binding motif (TAZ/WWTR1) are both orthologs of Yki, and TEA domain proteins 1–4 (TEAD1–4) are the Sd mammalian orthologs. Activated by upstream signals, Hpo (MST) phosphorylates and activates Warts (LATS1/2), which in turn phosphorylates Yki (YAP/TAZ) on a specific serine residue to generate a 14-3-3 binding site, resulting in Yki (YAP/TAZ) cytoplasmic sequestration (Huang et al. 2005; Oh and Irvine 2008). For simplicity, only the mammalian nomenclature for Hippo components is referred to hereafter, unless otherwise specified.

In this chapter we explore the interplay between the Hippo signaling pathway and RUNX2 (oncogenic) and RUNX3 (tumor suppressive) proteins in carcinogenesis. We cover both the transcriptional and non-transcriptional interactions between RUNX and TAZ/YAP-TEAD, and in particular we focus on their contribution to breast and gastric cancers (Brusgard et al. 2015). The findings highlighted here may lead to the development of useful paradigms that integrate novel data with our current understanding of RUNX-Hippo crosstalk to better understand mechanisms driving cancer progression.

26.2 Basic Principles

26.2.1 Oncogenic and Tumor Suppressor Functions of the RUNX Genes

In addition to their '*classic*' role as transcriptional regulators during development and tumorigenesis, transcription-independent functions for RUNX proteins have been identified. Indeed RUNX has been implicated in the regulation of numerous physiological processes including DNA damage and cellular stress response, mitosis, autophagy, stem cell differentiation, and chromatin remodeling (Ito et al. 2015). Several studies have identified diverse roles for RUNX proteins via their interaction with numerous oncogenic and tumor suppressor mediators such as TGFβ, p53, Wnt and YAP/TAZ. One unifying principle in RUNX crosstalk with other cellular components is the ability of RUNX proteins to antagonize or enhance tumor suppressor or oncogenic functions. Recent evidence suggests RUNX proteins compete with each other to direct specific and opposing functions in part because they share identical DNA-recognition domains on target gene promoters (Chuang et al. 2013). For example, interaction of tumor suppressors RUNX1 or RUNX3 with p53 up-regulates BAX and PUMA to drive apoptosis following DNA damage (Ozaki et al. 2013a). Conversely, RUNX2 acts as a negative regulator of p53-dependent apoptosis via formation of a RUNX2/HDAC6/p53 transcriptional complex that represses BAX and PUMA (Ozaki et al. 2013b) (for detailed depiction of RUNX-p53 interaction refer to Fig. 26.1).

The oncogenic properties of RUNX2 are well established. In addition to its anti-apoptotic interaction with p53, RUNX2 attenuates the pro-apoptotic signaling of TAp73 to confer drug resistance (Ozaki et al. 2015), and negatively

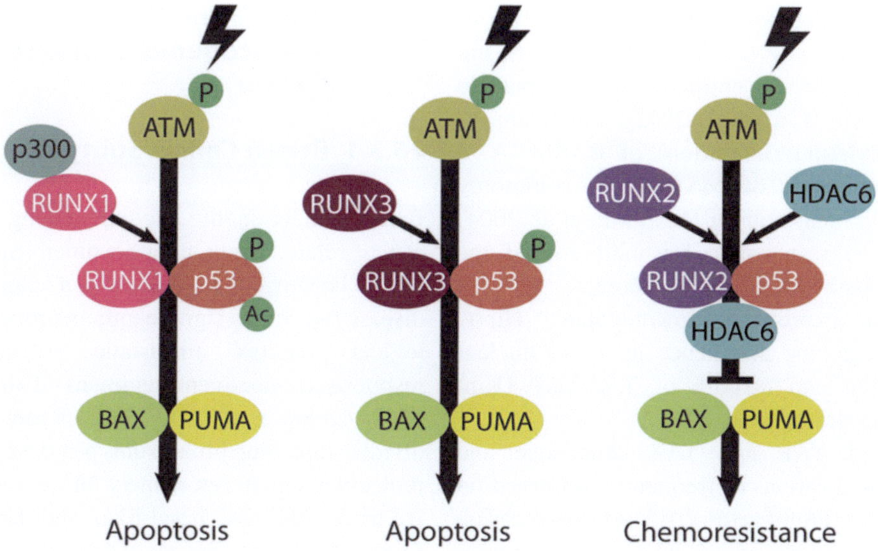

Fig. 26.1 Disparate regulatory functions of RUNX family members on p53 in the DNA damage response. RUNX1 and RUNX3 act as positive regulators of p53 in response to DNA damage. In contrast, the DNA damage-induced proapoptotic activity of p53 is inhibited by RUNX2 (Refer to Refs. (Ozaki et al. 2013a, b) for more details)

regulates the long non-coding RNA, MT1DP, a known tumor suppressor (Yu et al. 2014b). RUNX2 is upregulated during epithelial-mesenchymal transition (EMT) in breast and prostate cancer (Chimge et al. 2011; Baniwal et al. 2010) and increased RUNX2 abundance correlates with poor prognosis in luminal and triple-negative subtypes of breast cancer (Mcdonald et al. 2014; Brusgard et al. 2015). In animal models, RUNX2 mediates breast cancer metastasis (Barnes et al. 2004; Javed et al. 2005; Pratap et al. 2011) and was shown to promote drug resistance and escape from apoptosis (Ozaki et al. 2013b). Furthermore, RUNX2 negatively regulates mitochondrial SIRT6 and pyruvate dehydrogenase (PDH) and increases breast cancer cell glucose metabolism, which is a hallmark of cancer (Choe et al. 2015). Intriguingly, tumor suppressive functions for RUNX2 have also been described. RUNX2 promotes mammary epithelial cell differentiation (Inman and Shore 2003) and in a subset of breast cancers RUNX2 antagonizes estrogen receptor growth-stimulation (Chimge et al. 2012; Chimge and Frenkel 2013). Increased RUNX2 expression may also promote a more differentiated phenotype in osteosarcomas, providing the "brakes" against further tumor progression (Pratap et al. 2003).

26.2.2 The Hippo Signaling Pathway in Cancer

Increased activity of the Hippo pathway effectors YAP and TAZ has been reported in the majority of solid cancer types (Liu et al. 2012; Harvey et al. 2013; Plouffe et al. 2015; Zanconato et al. 2016). Studies in mice revealed that knockout of the upstream regulator Neurofibromin 2 (NF2), as well as LATS, MST, SAV1, and MOB1 frequently leads to cancer development (reviewed in Harvey et al. 2013). Common mechanisms of pathway dysregulation in humans include gene amplification of YAP/TAZ and epigenetic silencing of Hippo components, particularly by promoter hypermethylation. Interestingly, with the exception of NF2, somatic mutations within Hippo components are relatively rare. However numerous regulators of the core components of the Hippo pathway (MATS, SAV1, LATS and MOB1) have been identified that contribute to tumorigenesis (Liu et al. 2012). For example, hypermethylation (inhibition) of RASSF1A, a positive regulator of MST1/2, is commonly observed in breast cancer (Mehrotra et al. 2004) and may be responsible for inhibition of the Hippo pathway. Furthermore, reduced E-Cadherin expression downregulates Hippo pathway signaling and hence increases nuclear translocation and activity of TAZ/YAP (Kim et al. 2011; Harvey et al. 2013).

Increased YAP and TAZ abundance and nuclear localization is frequently observed in breast cancer (Plouffe et al. 2015). Overexpression of YAP in breast cancer cell lines promotes tumor formation in mouse xenograft models, which can be blocked by YAP knockdown (Wang et al. 2012; Chen et al. 2014). Likewise, increased TAZ abundance promotes cell transformation and EMT and correlates with a more invasive breast cancer phenotype (Lei et al. 2008; Chan et al. 2008). Mechanistically, LIFR (Leukemia inhibitory factor receptor), a suppressor of metastasis that is frequently lost in breast cancer, inactivates YAP via regulation of Hippo pathway signaling (Chen et al. 2012). Similar to RUNX2, YAP has also been reported to exhibit tumor suppressive functions. YAP knockdown in breast cancer cells increased tumor cell invasion and growth in nude mice (Yuan et al. 2008). Notably, hyperactivation of YAP alone is insufficient to give rise to tumors in normal mammary epithelial cells (Chen et al. 2014). From this study the authors hypothesize that other genetic disruptions are required to promote YAP-induced oncogenesis. Dysregulation of YAP activity was also reported to produce dysplasia (YAP overexpression) and hyperplasia (SAV1 conditional knockout) of the gastrointestinal epithelium (Harvey et al. 2013). Importantly, inactivation of the Hippo pathway does not induce gastric carcinoma, though the pathway is reported to promote development of pancreatic and colorectal cancers (Plouffe et al. 2015).

26.3 RUNX2 and TAZ as Oncogenes in Breast Cancer

26.3.1 Breast Cancer Subtypes

Breast cancer is the second leading cause of cancer-related death among women (Siegel et al. 2013). However breast cancer is a heterogeneous disease that varies significantly in terms of pathological features, metastatic potential, and response to treatment regimens (Eroles et al. 2012; Cadoo et al. 2013). Breast cancer can be divided into four broad subtypes based on their molecular signatures, namely luminal A, luminal B, triple negative basal-like, and HER2-type. Luminal subtypes are more common and generally have a better prognostic outcome compared to basal-like tumors, which tend to be more aggressive. As the name would suggest, HER2-type tumors are typically HER2 receptor-positive and thus can be treated with HER2-targeting drugs such as Herceptin or lapatinib.

26.3.2 RUNX2 and TAZ Expression in Breast Cancer

RUNX2 is normally expressed in developing breast epithelial cells and in the mammary stem cell population where it promotes terminal end bud differentiation (Ferrari et al. 2013; Mcdonald et al. 2014). In breast cancer cell lines however, RUNX2 promotes an osteomimetic phenotype and metastasis to bone through transcriptional activation of osteopontin, matrix metalloproteinases (MMPs), and VEGF (Barnes et al. 2004; Pratap et al. 2005, 2006). This is important since luminal breast cancers relapse predominantly to the bone microenvironment (Eroles et al. 2012; Foley et al. 2010) and account for 50 % of all metastasis-related breast cancer deaths (Ganapathy et al. 2012). As introduced above, overexpression of TAZ is observed in breast cancer patient samples (Chan et al. 2008) and cell lines (Hiemer et al. 2014), correlating with increased cell migration, tumorigenesis, invasiveness, and drug resistance (Lei et al. 2008). Notably, RUNX2 can bind YAP (Yagi et al. 1999) and TAZ (Cui et al. 2003) via interaction of the PPxY motif within its C-terminal transactivation domain with the WW domain/s of YAP/TAZ. Cooperation between RUNX and YAP/TAZ has been shown to promote cell transformation (Vitolo et al. 2007), osteoblast differentiation (Cui et al. 2003) and stem cell renewal (Varelas et al. 2008; Cordenonsi et al. 2011).

26.3.3 sE-Cad-Mediated EMT

EMT is typically characterized by downregulation of E-Cadherin and upregulation of vimentin (Lee et al. 2006; Thiery et al. 2009; Valastyan and Weinberg 2011). Whilst this '*classical*' EMT is usually required for cancer progression, cells may also metastasize from the primary tumor via an alternate mechanism involving proteolytic cleavage of E-Cadherin (120 kDa) to release the soluble, N-terminal ectodomain (sE-Cad; 80 kDa) (David and Rajasekaran 2012). MMP2 and -9 and ADAM (A Disintegrin and Metalloproteinase) -15 mediate cleavage of E-Cadherin to sE-Cad (David and Rajasekaran 2012; Najy et al. 2008; Davies et al. 2001; Huguenin et al. 2008; Noe et al. 2001; Symowicz et al. 2007; Zuo et al. 2011). sE-Cad exhibits autocrine and/or paracrine activity by binding HER2 (David and Rajasekaran 2012; Inge et al. 2011; Najy et al. 2008; Brouxhon et al. 2013, 2014) and interacts with full length E-Cadherin to destabilize adherens junctions (David and Rajasekaran 2012). The effect of sE-Cad signaling is promotion of migration, invasion, and proliferation while maintaining an epithelial morphology (David and Rajasekaran 2012; Grabowska and Day 2012; Chunthapong et al. 2004; Inge et al. 2011; Kuefer et al. 2003; Najy et al. 2008). Hence sE-Cad is a useful functional metastatic biomarker for numerous cancers, including breast cancer (David and Rajasekaran 2012; Chunthapong et al. 2004; Kuefer et al. 2003; Hofmann et al. 2013; Kuefer et al. 2005).

26.3.4 RUNX2 Cooperates with TAZ to Promote sE-Cad-Mediated EMT

Recently, our group discovered that cooperation between RUNX2 and TAZ increases shedding of sE-Cad to promote a tumorigenic phenotype characterized by anchorage-independent growth (tumorsphere formation) in breast cancer cells (Brusgard et al. 2015). RUNX2 promotes nuclear localization of TAZ, which is a driver for tumorigenesis since TAZ knockdown reduces tumorsphere growth. Intriguingly, given that TAZ and YAP are similarly regulated, expression and localization of YAP was not affected by RUNX2 expression in these cells (Brusgard et al. 2015). MMP expression (including MMP2, which can cleave E-Cadherin) was significantly elevated in RUNX overexpressing breast cancer cells and could be inhibited by MMP inhibitors. Treatment with E-Cadherin neutralizing antibody reduced the level of sE-Cad and inhibited tumorsphere formation. Binding of sE-Cad to HER2, which is expressed in a subset of luminal breast cancers (Ithimakin et al. 2013), promotes tumorgenicity (Brouxhon et al. 2013). Treatment of RUNX2 overexpressing cells with the HER2-targeting drugs Herceptin or lapatinib inhibited tumorsphere proliferation (Brusgard et al. 2015). Taken together, these results suggest that RUNX2 and TAZ cooperate to upregulate MMP expression in breast cancer and promote an sE-Cad/HER2-mediated EMT. Our working model is therefore as such: E-Cadherin, via its influence on Hippo pathway activity (Kim et al. 2011), maintains TAZ in a cytoplasmic (inactive), tumor suppressive state. RUNX2-induced cleavage of E-Cadherin to sE-Cad inactivates the Hippo pathway, resulting in nuclear localisation of TAZ and oncogenic transformation (Fig. 26.2).

26.3.5 Outstanding Questions and Future Directions

Correlation between RUNX2 signaling and increased TAZ nuclear localization in breast cancer cells suggests that factors which inhibit

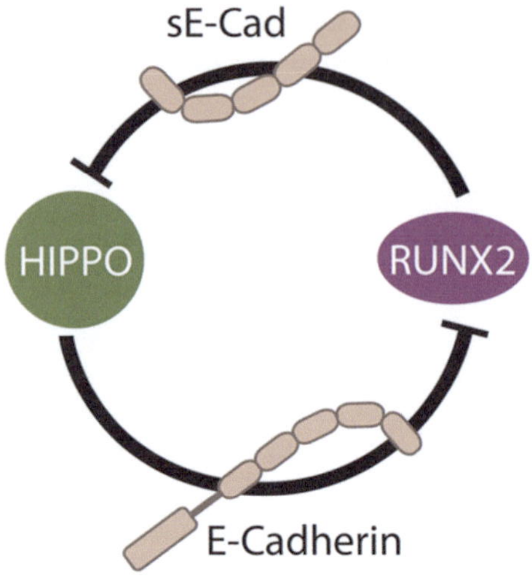

Fig. 26.2 RUNX2 manifests its oncogenic activity through upregulation of a soluble form of E-Cadherin (sE-Cad) that inactivates the Hippo tumor suppressor pathway. Conversely, full-length membrane bound E-Cadherin positively regulates the Hippo signaling pathway to keep RUNX2 oncogenic function in check

RUNX2 may restore Hippo signaling and block breast cancer progression. Mechanistically, we hypothesize that RUNX2 oncogenic activity is mediated, at least in part, by increased production of sE-Cad. However, whether sE-Cad is oncogenic, independent of RUNX2 overexpression, remains to be determined.

Furthermore, whilst RUNX2-induced TAZ nuclear localization suggests attenuation of Hippo signaling, a role for the Hippo pathway kinases MST1/2 and LATS1/2 in mediating RUNX2 oncogenic function has not been reported. To address these outstanding questions, cells could be treated directly with recombinant sE-Cad *in vitro* and assess the effect on Hippo signaling and tumorigenic properties.

Preliminary unpublished data from our lab indicate that treatment of breast cancer cells with recombinant sE-Cad reduces the abundance of active phosphorylated (phospho-) LATS1/2. Moreover, treatment of cells with the RUNX2 small molecule inhibitor $CADD_{522}$ increased phospho-LATS1/2 as well as the total level of LATS1 protein. This is consistent with our data showing significant reduction of TAZ abundance in the nucleus upon RUNX inhibition (Brusgard et al. 2015). Though these data support a role for RUNX2 in controlling Hippo pathway activity, this does not explain why YAP is not similarly regulated by RUNX2 overexpression in these cells. This is a curious observation that should be addressed in subsequent studies. Furthermore, data from the recombinant sE-Cad experiments would suggest that TAZ activation (nuclear localization) lies downstream of RUNX2 overexpression, MMP production and increased sE-Cad shedding. Therefore the mechanism linking TAZ activation and tumorigenic transformation of breast cancer cells should be determined.

Identification of TAZ-specific oncogenic target genes may reveal novel cancer biomarkers and therapeutic targets. Since RUNX2 inactivates several tumor suppressor pathways including p53, E-Cadherin, and SIRT6/PDH metabolic regulators (Choe et al. 2015), in addition to the Hippo pathway as discussed in detail here, RUNX2 inhibition could prove very effective as a novel cancer targeting strategy. Future effort should be employed to determine whether a combination of oncogene/tumor suppressor targeting and metabolic reprogramming strategies would be effective for other tumorigenic events where RUNX2 is a driving factor.

26.4 RUNX3 and TEAD-YAP Regulation in Gastric Cancer

26.4.1 RUNX3 Is a Tumor Suppressor in Gastric Cancer

Gastric cancer is the second leading cause of cancer-related mortality worldwide and is characterized by tumor heterogeneity driven by various signaling pathways (Shah and Ajani 2010). Consistent with gastric hyperplasia observed in *Runx3* knockout mice (Ito et al. 2011), loss of RUNX3 expression, typically due to hemizygous deletion or promoter hypermethylation, is observed in 60 % of human gastric cancers. Furthermore, reduced RUNX3 is causally linked to the initiation and progression of gastric cancer (Li et al. 2002; Fan et al. 2011). In gastric epithelial cells, RUNX3 cooperates with Smad/TGFβ signaling to drive expression of p21 (CIP1) (Chi et al. 2005) and BIM (Yano et al. 2006) to inhibit cell division and promote apoptosis, respectively. In the colon, RUNX3 attenuates oncogenic WNT signaling via inhibitory binding to the TCF4-β-catenin complex (Ito et al. 2008), and in the mouse lung Runx3 inhibits cellular transformation via upregulation of $p19^{Arf}$ and p21 in response to oncogenic K-Ras signaling (Lee et al. 2013).

26.4.2 The TEAD-YAP Complex Is Oncogenic in Gastric Cancer

YAP, in association with TEAD (TEAD-YAP complex), promotes cell proliferation by upregulation of target genes including connective tissue growth factor (*CTGF*) and Cysteine-rich angiogenic inducer 61 (*CYR61*) (Lai et al. 2011). YAP (Zhang et al. 2012; Lam-Himlin et al. 2006) and TEAD4 (Lim et al. 2013) have both been reported to be upregulated in gastric cancer patient samples, and expression of YAP target genes positively correlates with gastric carcinoma progression (Jiang et al. 2011; Lin et al. 2005), and patient outcome (Qiao et al. 2015). In gastric cancer cells, overexpression of a TEAD-YAP fusion protein increases anchorage-independent growth (Qiao et al. 2015), whilst YAP knockdown inhibits proliferation and metastasis (Zhang et al. 2012) and in some instances induces apoptosis (Zhou et al. 2011). Use of a YAP antagonist (Super-TDU: an inhibitor peptide mimicking the TDU region of VGLL4 that blocks YAP-TEAD binding) suppresses gastric cancer cell growth *in vitro* and is proposed as a therapeutic strategy to treat gastric cancer (Jiao et al. 2014).

Notably, crosstalk between RUNX and the Hippo signaling pathway, independent of direct interaction with YAP/TAZ, has been reported. Facilitated by MST2, RUNX3 and SAV1 form a complex to promote Hippo pathway-mediated cell death (Min et al. 2012). Consistent with this, expression of RUNX1 and RUNX3 inversely correlate with YAP abundance in cultured gastric cancer cells and patient samples (Qiao et al. 2015). Thus RUNX3 is a negative regulator of YAP activity and 'low RUNX/high YAP' expression might be a useful marker of gastric cancer progression.

26.4.3 RUNX3, TEAD and YAP Form a Ternary Complex

Recently, we showed that RUNX3 is a novel regulator of the TEAD-YAP complex in gastric carcinogenesis whereby RUNX3 physically interacts with TEAD, reducing its DNA-binding ability and effectively inhibiting downstream YAP signaling (Qiao et al. 2015). Mapping of the TEAD-RUNX3 interaction revealed that the Runt (DNA-binding) domain (RD) of RUNX3 is essential. We also discovered that the TEAD-RUNX3 interface overlaps with the TEAD DNA-recognition helix (Qiao et al. 2015). Even though RUNX family members share a high degree of sequence identity in their Runt domains, interaction between TEAD and RUNX2 was significantly weaker than that of RUNX1 and RUNX3, despite strong interaction between RUNX2 and YAP or TAZ (Qiao et al. 2015). Further experiments revealed that RUNX3, TEAD and YAP form a ternary complex, in which distinct domains mediate direct interaction of RUNX3 and YAP with TEAD (Qiao et al. 2015) (Fig. 26.3).

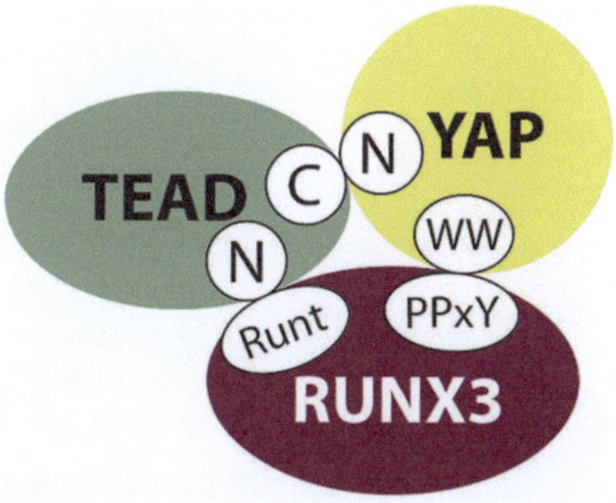

Fig. 26.3 RUNX3, TEAD and YAP form a tripartite protein complex
RUNX3 binds YAP via interaction of its carboxy terminal (C) PPxY motif with the WW domain(s) of YAP. YAP's amino terminal (N) TEAD-binding domain contacts the C-terminal region of TEAD. The Runt DNA-binding domain of RUNX3 completes the complex, associating with the N-terminus of TEAD, overlapping with TEAD's DNA-recognition helix

26.4.4 RUNX3 Binding Abolishes TEAD-YAP Oncogenic Activity in Gastric Cancer

The overlap of TEAD's DNA-recognition helix and RUNX binding interface led us to hypothesize that RUNX interaction would significantly reduce the DNA-binding ability of TEAD (see Fig. 26.4). Accordingly, RUNX3 overexpression in gastric cancer tissues correlates with downregulation of TEAD-YAP targets including CTGF, CYR61, and GLI2 (Qiao et al. 2015). Further analysis in gastric cancer cell lines revealed that RUNX3 abundance negatively correlates with YAP-induced CTGF expression, and this is associated with decreased binding of TEAD to CTGF promoters (Qiao et al. 2015). Screening of several RUNX3 mutants that are frequently observed in gastric cancer revealed that mutation of Leucine121 to Histidine (L121H) abolished its interaction with TEAD and significantly reduced the expression of CTGF (Qiao et al. 2015). In terms of biological significance, increased expression of wild-type RUNX3 (but not the L121H mutant) reduced anchorage-independent growth of gastric cancer cells *in vitro*, and tumor growth using nude mouse assay *in vivo* (Qiao et al. 2015). Moreover, overexpression of RUNX3 could block TEAD-YAP-induction of colony formation, providing evidence of a direct tumor suppressor role for RUNX3 in gastric carcinoma.

26.4.5 Future Perspectives and Potential for Therapeutic Application

Our recent report elucidates a novel mechanism of RUNX3 tumor suppressor activity in gastric cancer that has great potential for application in a range of human cancers driven by aberrant TEAD-YAP activity. Given the dual role of RUNX3 in regulating YAP activity via complex formation with SAV1/MST2 (Min et al. 2012) and TEAD-YAP (Qiao et al. 2015) it would be interesting to ascertain the relative contribution of these two mechanisms to RUNX3 anti-oncogenic activity in gastric cancer.

Precise control of RUNX3 and YAP expression is important during embryonic development

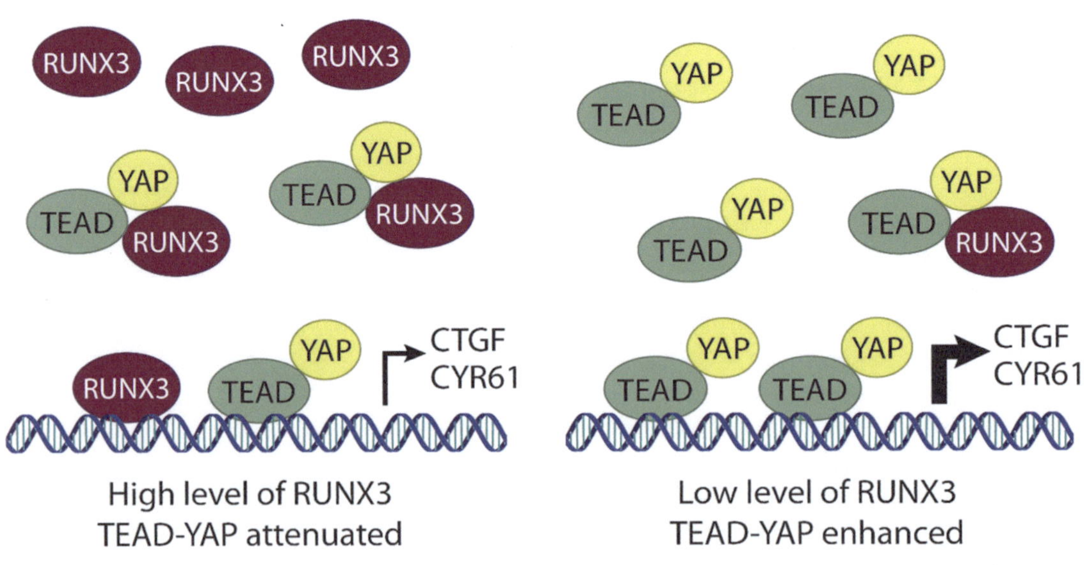

Fig. 26.4 RUNX3 inhibits TEAD/YAP-mediated gene transcription
Under conditions of high RUNX3 (*left panel*) DNA-binding ability of TEAD is inhibited and transcription of TEAD/YAP target genes (e.g., CTGF and CYR61) is attenuated. When RUNX3 is inactivated or expressed at relatively low levels (*right panel*) TEAD/YAP drives transcription of oncogenic target genes

and differentiation of the gastrointestinal tract; aberrant expression of RUNX3 (knockout) (Ito et al. 2011) or YAP (activation) (Camargo et al. 2007) promotes gastrointestinal dysplasia. The potent inhibition of TEAD-YAP by RUNX3 raises interesting questions regarding their roles and possible interaction during development. Preliminary data suggest that transcriptional activity of RUNX3 inversely correlates with TEAD4 expression (unpublished data). This mutual regulation between RUNX3 and TEAD4 might be a way to fine-tune the balance of proliferation and differentiation both during development and tumorigenesis.

Interestingly, only 35 % of TEAD-targeted genes were suppressed by RUNX3 overexpression in gastric cancer cells (Qiao et al. 2015). This suggests that inhibition of these targets might be due to adjacent RUNX3 and TEAD binding sites in their promoters, even though DNA binding does not seem to be required for RUNX3-TEAD interaction in our assays (Qiao et al. 2015). It is possible however that cognate DNA stabilizes and enhances the multicomponent RUNX3-TEAD-YAP complex, priming it for biological activity. A genome wide ChIP-seq experiment analyzing all TEAD-binding sites for proximity to RUNX motifs in gastric tissues could test this hypothesis.

In support of this, members of the Piccolo laboratory recently conducted a comprehensive ChIP-seq analysis to identify DNA-binding platforms for YAP and TAZ in breast cancer cells (Zanconato et al. 2015). Unsurprisingly, TEAD binding motifs were present in the majority of YAP/TAZ peaks. Encouragingly however, of the various DNA-binding factors proposed to cooperate with YAP/TAZ, RUNX-binding sites were the only other prominent motif identified. Moreover, for some YAP/TAZ target genes, there was a physical proximity of TEAD and RUNX binding sites in the cells analyzed (Zanconato et al. 2015).

Since RUNX3 is such a potent inhibitor of YAP in gastric cancer, these findings could lead to the development of novel RUNX3 mimicking compounds to target TEAD-YAP activity *in vivo*. Support for this proposal comes from studies demonstrating the efficacy of using YAP-TEAD inhibitors such as verteporfin (Liu-Chittenden et al. 2012; Yu et al. 2014a) and Super-TDU (Jiao et al. 2014) to suppress the oncogenic activity of YAP. Moreover, a publication showing forced overexpression of YAP in hematopoietic stem cells, in which RUNX1 and RUNX3 were highly expressed, did not lead to malignant cell growth (Jansson and Larsson 2012).

26.5 Concluding Remarks

The Hippo signaling pathway and its role in controlling the mammalian effector proteins YAP and TAZ was elucidated nearly ten years ago, yet we are still discovering novel regulators of this important signaling pathway. Recent findings from our laboratories linking RUNX2/TAZ with sE-Cad expression, and RUNX3 with TEAD-YAP in different models of cancer highlight the potential for development of effective targeting strategies for Hippo pathway dysregulation in various human pathologies. That being said, questions still remain regarding the mechanisms of YAP/TAZ regulation by the RUNX protein family.

Notably, apparent differences exist between the regulation of TAZ and YAP by RUNX2 and sE-Cad signaling in breast cancer. The Hippo pathway similarly regulates YAP and TAZ in terms of phosphorylation and nuclear localization (Hao et al. 2008; Kanai et al. 2000). Recently, we reported that YAP is a negative regulator of TAZ protein abundance in mammalian cells (Finch-Edmondson et al. 2015). This is relevant since it demonstrates that YAP and TAZ are subjected to discrete forms of regulation. Whether this direct relationship between YAP and TAZ abundance has implications for RUNX-mediated YAP/TAZ regulation remains to be determined.

Multiple isoforms of YAP harboring single (YAP1-1) or tandem (YAP1-2) WW domains are expressed in mammals (Gaffney et al. 2012). Because RUNX bind to YAP/TAZ via this key protein interaction domain, differences in the binding efficiency of RUNX to YAP1-1 or YAP1-2 isoforms may influence the signaling outcome. Especially in gastric cancer, where RUNX acts to inhibit TEAD-YAP activity, YAP isoforms that exhibit weaker binding to RUNX have the potential to be more oncogenic. Elucidation of the protein "interactome" of individual YAP isoforms may reveal striking differences in RUNX3 binding. Furthermore, since TAZ has only one WW domain, whether the number of WW domains influences RUNX interaction would be interesting to assess.

Finally, the development of CRISPR/Cas9 technology for efficient gene editing *in vitro* and *in vivo* has provided great opportunity for analyzing the effect of point mutations on protein-protein interactions. By taking advantage of clinical data signposting common mutants detected in cancer (e.g., RUNX3 mutant L121H) we can measure their effect using a biologically, and translationally relevant approach. This will enable us to better understand how mutations in critical proteins can drive cancer formation and progression, and may even pave the way for genetic engineering to combat cancer in humans.

Acknowledgments The research of MS, YQ, and MFE has been supported by generous "Seed Grants" from NUS-MBI-IMCB-A*STAR of the Republic of Singapore. AP was supported by a VA Health Services Research & Development award and a Pilot Grant project from the Marlene & Stewart Greenebaum Cancer Center. The authors would also like to acknowledge the assistance of Chun Xi Wong in preparation of the figure illustrations.

References

Baniwal, S. K., Khalid, O., Gabet, Y., Shah, R. R., Purcell, D. J., Mav, D., et al. (2010). Runx2 transcriptome of prostate cancer cells: Insights into invasiveness and bone metastasis. *Molecular Cancer, 9*, 258.

Barnes, G. L., Hebert, K. E., Kamal, M., Javed, A., Einhorn, T. A., Lian, J. B., et al. (2004). Fidelity of Runx2 activity in breast cancer cells is required for the generation of metastases-associated osteolytic disease. *Cancer Research, 64*, 4506–4513.

Brouxhon, S. M., Kyrkanides, S., Teng, X., Raja, V., O'banion, M. K., Clarke, R., et al. (2013). Monoclonal antibody against the ectodomain of E-cadherin (DECMA-1) suppresses breast carcinogenesis: Involvement of the HER/PI3K/Akt/mTOR and IAP pathways. *Clinical Cancer Research, 19*, 3234–3246.

Brouxhon, S. M., Kyrkanides, S., Teng, X., O'banion, M. K., Clarke, R., Byers, S., & Ma, L. (2014). Soluble-E-cadherin activates HER and IAP family members in HER2+ and TNBC human breast cancers. *Molecular Carcinogenesis, 53*, 893–906.

Brusgard, J. L., Choe, M., Chumsri, S., Renoud, K., Mackerell Jr., A. D., Sudol, M., & Passaniti, A. (2015). RUNX2 and TAZ-dependent signaling pathways regulate soluble E-Cadherin levels and tumorsphere formation in breast cancer cells. *Oncotarget, 6*, 28132–28150.

Cadoo, K. A., Fornier, M. N., & Morris, P. G. (2013). Biological subtypes of breast cancer: Current concepts and implications for recurrence patterns. *The Quarterly Journal of Nuclear Medicine and Molecular Imaging, 57*, 312–321.

Camargo, F. D., Gokhale, S., Johnnidis, J. B., Fu, D., Bell, G. W., Jaenisch, R., & Brummelkamp, T. R. (2007). YAP1 increases organ size and expands undifferentiated progenitor cells. *Current Biology, 17*, 2054–2060.

Chan, S. W., Lim, C. J., Guo, K., Ng, C. P., Lee, I., Hunziker, W., et al. (2008). A role for TAZ in migration, invasion, and tumorigenesis of breast cancer cells. *Cancer Research, 68*, 2592–2598.

Chen, D., Sun, Y., Wei, Y., Zhang, P., Rezaeian, A. H., Teruya-Feldstein, J., et al. (2012). LIFR is a breast cancer metastasis suppressor upstream of the Hippo-YAP pathway and a prognostic marker. *Nature Medicine, 18*, 1511–1517.

Chen, Q., Zhang, N., GRAY, R. S., Li, H., Ewald, A. J., Zahnow, C. A., & Pan, D. (2014). A temporal requirement for Hippo signaling in mammary gland differentiation, growth, and tumorigenesis. *Genes & Development, 28*, 432–437.

Chi, X. Z., Yang, J. O., Lee, K. Y., Ito, K., Sakakura, C., Li, Q. L., et al. (2005). RUNX3 suppresses gastric epithelial cell growth by inducing p21(WAF1/Cip1) expression in cooperation with transforming growth factor β-activated SMAD. *Molecular and Cellular Biology, 25*, 8097–8107.

Chimge, N. O., & Frenkel, B. (2013). The RUNX family in breast cancer: Relationships with estrogen signaling. *Oncogene, 32*, 2121–2130.

Chimge, N. O., Baniwal, S. K., Little, G. H., Chen, Y. B., Kahn, M., Tripathy, D., et al. (2011). Regulation of breast cancer metastasis by Runx2 and estrogen signaling: The role of SNAI2. *Breast Cancer Research, 13*, R127.

Chimge, N. O., Baniwal, S. K., Luo, J., Coetzee, S., Khalid, O., Berman, B. P., et al. (2012). Opposing effects of Runx2 and estradiol on breast cancer cell proliferation: In vitro identification of reciprocally regulated gene signature related to clinical letrozole responsiveness. *Clinical Cancer Research, 18*, 901–911.

Choe, M., Brusgard, J. L., Chumsri, S., Bhandary, L., Zhao, X. F., Lu, S., et al. (2015). The RUNX2 transcription factor negatively regulates SIRT6 expression to alter glucose metabolism in breast cancer cells. *Journal of Cellular Biochemistry, 116*, 2210–2226.

Chuang, L. S., Ito, K., & Ito, Y. (2013). RUNX family: Regulation and diversification of roles through interacting proteins. *International Journal of Cancer, 132*, 1260–1271.

Chunthapong, J., Seftor, E. A., Khalkhali-Ellis, Z., Seftor, R. E., Amir, S., Lubaroff, D. M., et al. (2004). Dual roles of E-cadherin in prostate cancer invasion. *Journal of Cellular Biochemistry, 91*, 649–661.

Cordenonsi, M., Zanconato, F., Azzolin, L., Forcato, M., Rosato, A., Frasson, C., et al. (2011). The Hippo transducer TAZ confers cancer stem cell-related traits on breast cancer cells. *Cell, 147*, 759–772.

Cui, C. B., Cooper, L. F., Yang, X., Karsenty, G., & Aukhil, I. (2003). Transcriptional coactivation of bone-specific transcription factor Cbfa1 by TAZ. *Molecular and Cellular Biology, 23*, 1004–1013.

David, J. M., & Rajasekaran, A. K. (2012). Dishonorable discharge: The oncogenic roles of cleaved E-cadherin fragments. *Cancer Research, 72*, 2917–2923.

Davies, G., Jiang, W. G., & Mason, M. D. (2001). Matrilysin mediates extracellular cleavage of E-cadherin from prostate cancer cells: A key mechanism in hepatocyte growth factor/scatter factor-induced cell-cell dissociation and in vitro invasion. *Clinical Cancer Research, 7*, 3289–3297.

Eroles, P., Bosch, A., Perez-Fidalgo, J. A., & Lluch, A. (2012). Molecular biology in breast cancer: Intrinsic subtypes and signaling pathways. *Cancer Treatment Reviews, 38*, 698–707.

Fan, X. Y., Hu, X. L., Han, T. M., Wang, N. N., Zhu, Y. M., Hu, W., et al. (2011). Association between RUNX3 promoter methylation and gastric cancer: A meta-analysis. *BMC Gastroenterology, 11*, 92.

Ferrari, N., Mcdonald, L., Morris, J. S., Cameron, E. R., & Blyth, K. (2013). RUNX2 in mammary gland development and breast cancer. *Journal of Cellular Physiology, 228*, 1137–1142.

Finch-Edmondson, M. L., Strauss, R. P., Passman, A., Sudol, M., Yeoh, G. C., & Callus, B. A. (2015). TAZ protein accumulation is negatively regulated by YAP abundance in mammalian cells. *The Journal of Biological Chemistry, 290*(46), 27928–27938.

Foley, J., Nickerson, N. K., Nam, S., Allen, K. T., Gilmore, J. L., Nephew, K. P., et al. (2010). EGFR signaling in breast cancer: Bad to the bone. *Seminars in Cell & Developmental Biology, 21*, 951–960.

Gaffney, C. J., Oka, T., Mazack, V., Hilman, D., Gat, U., Muramatsu, T., et al. (2012). Identification, basic characterization and evolutionary analysis of differentially spliced mRNA isoforms of human YAP1 gene. *Gene, 509*, 215–222.

Ganapathy, V., Banach-Petrosky, W., Xie, W., Kareddula, A., Nienhuis, H., Miles, G., & Reiss, M. (2012). Luminal breast cancer metastasis is dependent on estrogen signaling. *Clinical & Experimental Metastasis, 29*, 493–509.

Grabowska, M. M., & Day, M. L. (2012). Soluble E-cadherin: More than a symptom of disease. *Frontiers in Bioscience (Landmark Edition), 17*, 1948–1964.

Hao, Y., Chun, A., Cheung, K., Rashidi, B., & Yang, X. (2008). Tumor suppressor LATS1 is a negative regulator of oncogene *YAP*. *The Journal of Biological Chemistry, 283*, 5496–5509.

Harvey, K. F., Pfleger, C. M., & Hariharan, I. K. (2003). The *Drosophila* Mst ortholog, hippo, restricts growth and cell proliferation and promotes apoptosis. *Cell, 114*, 457–467.

Harvey, K. F., Zhang, X., & Thomas, D. M. (2013). The Hippo pathway and human cancer. *Nature Reviews. Cancer, 13*, 246–257.

Hiemer, S. E., Szymaniak, A. D., & Varelas, X. (2014). The transcriptional regulators TAZ and YAP direct transforming growth factor β-induced tumorigenic phenotypes in breast cancer cells. *The Journal of Biological Chemistry, 289*, 13461–13474.

Hofmann, G., Balic, M., Dandachi, N., Resel, M., Schippinger, W., Regitnig, P., et al. (2013). The predictive value of serum soluble E-cadherin levels in breast cancer patients undergoing preoperative systemic chemotherapy. *Clinical Biochemistry, 46*, 1585–1589.

Huang, J., Wu, S., Barrera, J., Matthews, K., & Pan, D. (2005). The hippo signaling pathway coordinately regulates cell proliferation and apoptosis by inactivating yorkie, the *Drosophila* homolog of YAP. *Cell, 122*, 421–434.

Huguenin, M., Muller, E. J., Trachsel-Rosmann, S., Oneda, B., Ambort, D., Sterchi, E. E., & Lottaz, D. (2008). The metalloprotease meprinβ processes E-cadherin and weakens intercellular adhesion. *PloS One, 3*, e2153.

Inge, L. J., Barwe, S. P., D'ambrosio, J., Gopal, J., Lu, K., Ryazantsev, S., et al. (2011). Soluble E-cadherin promotes cell survival by activating epidermal growth factor receptor. *Experimental Cell Research, 317*, 838–848.

Inman, C. K., & Shore, P. (2003). The osteoblast transcription factor Runx2 is expressed in mammary epithelial cells and mediates osteopontin expression. *The Journal of Biological Chemistry, 278*, 48684–48689.

Ithimakin, S., Day, K. C., Malik, F., Zen, Q., Dawsey, S. J., Bersano-Begey, T. F., et al. (2013). HER2 drives luminal breast cancer stem cells in the absence of HER2 amplification: Implications for efficacy of adjuvant trastuzumab. *Cancer Research, 73*, 1635–1646.

Ito, K., Lim, A. C., Salto-Tellez, M., Motoda, L., Osato, M., Chuang, L. S., et al. (2008). RUNX3 attenuates β-catenin/T cell factors in intestinal tumorigenesis. *Cancer Cell, 14*, 226–237.

Ito, K., Chuang, L. S., Ito, T., Chang, T. L., Fukamachi, H., Salto-Tellez, M., & Ito, Y. (2011). Loss of Runx3 is a key event in inducing precancerous state of the stomach. *Gastroenterology, 140*, 1536–1546.

Ito, Y., Bae, S. C., & Chuang, L. S. (2015). The RUNX family: Developmental regulators in cancer. *Nature Reviews. Cancer, 15*, 81–95.

Jansson, L., & Larsson, J. (2012). Normal hematopoietic stem cell function in mice with enforced expression of the Hippo signaling effector YAP1. *PloS One, 7*, e32013.

Javed, A., Barnes, G. L., Pratap, J., Antkowiak, T., Gerstenfeld, L. C., Van Wijnen, A. J., et al. (2005). Impaired intranuclear trafficking of Runx2 (AML3/CBFA1) transcription factors in breast cancer cells inhibits osteolysis *in vivo*. *Proceedings of the National Academy of Sciences of the United States of America, 102*, 1454–1459.

Jiang, C. G., Lv, L., Liu, F. R., Wang, Z. N., Liu, F. N., Li, Y. S., et al. (2011). Downregulation of connective tissue growth factor inhibits the growth and invasion of gastric cancer cells and attenuates peritoneal dissemination. *Molecular Cancer, 10*, 122.

Jiao, S., Wang, H., Shi, Z., Dong, A., Zhang, W., Song, X., et al. (2014). A peptide mimicking VGLL4 function acts as a YAP antagonist therapy against gastric cancer. *Cancer Cell, 25*, 166–180.

Justice, R. W., Zilian, O., Woods, D. F., Noll, M., & Bryant, P. J. (1995). The *Drosophila* tumor suppressor gene warts encodes a homolog of human myotonic dystrophy kinase and is required for the control of cell shape and proliferation. *Genes & Development, 9*, 534–546.

Kanai, F., Marignani, P. A., Sarbassova, D., Yagi, R., Hall, R. A., Donowitz, M., et al. (2000). TAZ: A novel transcriptional co-activator regulated by interactions with 14-3-3 and PDZ domain proteins. *The EMBO Journal, 19*, 6778–6791.

Kim, N. G., Koh, E., Chen, X., & Gumbiner, B. M. (2011). E-cadherin mediates contact inhibition of proliferation through Hippo signaling-pathway components. *Proceedings of the National Academy of Sciences of the United States of America, 108*, 11930–11935.

Kuefer, R., Hofer, M. D., Gschwend, J. E., Pienta, K. J., Sanda, M. G., Chinnaiyan, A. M., et al. (2003). The role of an 80 kDa fragment of E-cadherin in the metastatic progression of prostate cancer. *Clinical Cancer Research, 9*, 6447–6452.

Kuefer, R., Hofer, M. D., Zorn, C. S., Engel, O., Volkmer, B. G., Juarez-Brito, M. A., et al. (2005). Assessment of a fragment of e-cadherin as a serum biomarker with predictive value for prostate cancer. *British Journal of Cancer, 92*, 2018–2023.

Lai, D., Ho, K. C., Hao, Y., & Yang, X. (2011). Taxol resistance in breast cancer cells is mediated by the hippo pathway component TAZ and its downstream transcriptional targets Cyr61 and CTGF. *Cancer Research, 71*, 2728–2738.

Lam-Himlin, D. M., Daniels, J. A., Gayyed, M. F., Dong, J., Maitra, A., Pan, D., et al. (2006). The hippo pathway in human upper gastrointestinal dysplasia and carcinoma: A novel oncogenic pathway. *International Journal of Gastrointestinal Cancer, 37*, 103–109.

Lee, J. M., Dedhar, S., Kalluri, R., & Thompson, E. W. (2006). The epithelial-mesenchymal transition: New insights in signaling, development, and disease. *The Journal of Cell Biology, 172*, 973–981.

Lee, Y. S., Lee, J. W., Jang, J. W., Chi, X. Z., Kim, J. H., Li, Y. H., et al. (2013). Runx3 inactivation is a crucial early event in the development of lung adenocarcinoma. *Cancer Cell, 24*, 603–616.

Lei, Q. Y., Zhang, H., Zhao, B., Zha, Z. Y., Bai, F., Pei, X. H., et al. (2008). TAZ promotes cell proliferation and epithelial-mesenchymal transition and is inhibited by the hippo pathway. *Molecular and Cellular Biology, 28*, 2426–2436.

Li, Q. L., Ito, K., Sakakura, C., Fukamachi, H., Inoue, K., Chi, X. Z., et al. (2002). Causal relationship between the loss of RUNX3 expression and gastric cancer. *Cell, 109*, 113–124.

Lim, B., Park, J. L., Kim, H. J., Park, Y. K., Kim, J. H., Sohn, H. A., et al. (2013). Integrative genomics analysis reveals the multilevel dysregulation and oncogenic characteristics of TEAD4 in gastric cancer. *Carcinogenesis, 35*, 1020–1027.

Lin, M. T., Zuon, C. Y., Chang, C. C., Chen, S. T., Chen, C. P., Lin, B. R., et al. (2005). Cyr61 induces gastric cancer cell motility/invasion via activation of the integrin/nuclear factor-kappaB/cyclooxygenase-2 signaling pathway. *Clinical Cancer Research, 11*, 5809–5820.

Liu, A. M., Wong, K. F., Jiang, X., Qiao, Y., & Luk, J. M. (2012). Regulators of mammalian Hippo pathway in cancer. *Biochimica et Biophysica Acta, 1826*, 357–364.

Liu-Chittenden, Y., Huang, B., Shim, J. S., Chen, Q., Lee, S. J., Anders, R. A., et al. (2012). Genetic and pharmacological disruption of the TEAD–YAP complex suppresses the oncogenic activity of YAP. *Genes & Development, 26*, 1300–1305.

Mcdonald, L., Ferrari, N., Terry, A., BELL, M., Mohammed, Z. M., Orange, C., et al. (2014). RUNX2 correlates with subtype-specific breast cancer in a human tissue microarray, and ectopic expression of Runx2 perturbs differentiation in the mouse mammary gland. *Disease Models & Mechanisms, 7*, 525–534.

Mehrotra, J., Vali, M., Mcveigh, M., Kominsky, S. L., Fackler, M. J., Lahti-Domenici, J., et al. (2004). Very high frequency of hypermethylated genes in breast cancer metastasis to the bone, brain, and lung. *Clinical Cancer Research, 10*, 3104–3109.

Min, B., Kim, M. K., Zhang, J. W., Kim, J., Chung, K. C., Oh, B. C., et al. (2012). Identification of RUNX3 as a component of the MST/Hpo signaling pathway. *Journal of Cellular Physiology, 227*, 839–849.

Najy, A. J., Day, K. C., & Day, M. L. (2008). The ectodomain shedding of E-cadherin by ADAM15 supports ErbB receptor activation. *The Journal of Biological Chemistry, 283*, 18393–18401.

Noe, V., Fingleton, B., Jacobs, K., Crawford, H. C., Vermeulen, S., Steelant, W., et al. (2001). Release of an invasion promoter E-cadherin fragment by matrilysin and stromelysin-1. *Journal of Cell Science, 114*, 111–118.

Oh, H., & Irvine, K. D. (2008). In vivo regulation of Yorkie phosphorylation and localization. *Development, 135*, 1081–1088.

Ozaki, T., Nakagawara, A., & Nagase, H. (2013a). RUNX Family Participates in the Regulation of p53-Dependent DNA Damage Response. *International Journal of Genomics, 2013*, 271347.

Ozaki, T., Wu, D., Sugimoto, H., Nagase, H., & Nakagawara, A. (2013b). Runt-related transcription factor 2 (RUNX2) inhibits p53-dependent apoptosis through the collaboration with HDAC6 in response to DNA damage. *Cell Death & Disease, 4*, e610.

Ozaki, T., Sugimoto, H., Nakamura, M., Hiraoka, K., Yoda, H., Sang, M., et al. (2015). Runt-related transcription factor 2 attenuates the transcriptional activity as well as DNA damage-mediated induction of pro-apoptotic TAp73 to regulate chemosensitivity. *The FEBS Journal, 282*, 114–128.

Pantalacci, S., Tapon, N., & Leopold, P. (2003). The Salvador partner Hippo promotes apoptosis and cell-cycle exit in *Drosophila*. *Nature Cell Biology, 5*, 921–927.

Plouffe, S. W., Hong, A. W., & Guan, K. L. (2015). Disease implications of the Hippo/YAP pathway. *Trends in Molecular Medicine, 21*, 212–222.

Pratap, J., Galindo, M., Zaidi, S. K., Vradii, D., Bhat, B. M., Robinson, J. A., et al. (2003). Cell growth regulatory role of Runx2 during proliferative expansion of preosteoblasts. *Cancer Research, 63*, 5357–5362.

Pratap, J., Javed, A., Languino, L. R., Van Wijnen, A. J., Stein, J. L., Stein, G. S., & Lian, J. B. (2005). The Runx2 osteogenic transcription factor regulates matrix metalloproteinase 9 in bone metastatic cancer cells and controls cell invasion. *Molecular and Cellular Biology, 25*, 8581–8591.

Pratap, J., Lian, J. B., Javed, A., Barnes, G. L., Van Wijnen, A. J., Stein, J. L., & Stein, G. S. (2006). Regulatory roles of Runx2 in metastatic tumor and cancer cell interactions with bone. *Cancer Metastasis Reviews, 25*, 589–600.

Pratap, J., Lian, J. B., & Stein, G. S. (2011). Metastatic bone disease: Role of transcription factors and future targets. *Bone, 48*, 30–36.

Qiao, Y., Lin, S. J., Chen, Y., Voon, D. C., Zhu, F., Chuang, L. S., et al. (2015). RUNX3 is a novel negative regulator of oncogenic TEAD-YAP complex in gastric cancer. *Oncogene, 35*(20), 2664–2674.

Shah, M. A., & Ajani, J. A. (2010). Gastric cancer--an enigmatic and heterogeneous disease. *Journal of the American Medical Association, 303*, 1753–1754.

Siegel, R., Naishadham, D., & Jemal, A. (2013). Cancer statistics, 2013. *CA: A Cancer Journal for Clinicians, 63*, 11–30.

Symowicz, J., Adley, B. P., Gleason, K. J., Johnson, J. J., Ghosh, S., Fishman, D. A., et al. (2007). Engagement of collagen-binding integrins promotes matrix metalloproteinase-9-dependent E-cadherin ectodomain shedding in ovarian carcinoma cells. *Cancer Research, 67*, 2030–2039.

Tapon, N., Harvey, K. F., Bell, D. W., Wahrer, D. C., Schiripo, T. A., Haber, D., & Hariharan, I. K. (2002). salvador Promotes both cell cycle exit and apoptosis in *Drosophila* and is mutated in human cancer cell lines. *Cell, 110*, 467–478.

Thiery, J. P., Acloque, H., Huang, R. Y., & Nieto, M. A. (2009). Epithelial-mesenchymal transitions in development and disease. *Cell, 139*, 871–890.

Valastyan, S., & Weinberg, R. A. (2011). Tumor metastasis: Molecular insights and evolving paradigms. *Cell, 147*, 275–292.

Varelas, X., Sakuma, R., Samavarchi-Tehrani, P., Peerani, R., Rao, B. M., Dembowy, J., et al. (2008). TAZ controls Smad nucleocytoplasmic shuttling and regulates human embryonic stem-cell self-renewal. *Nature Cell Biology, 10*, 837–848.

Vitolo, M. I., Anglin, I. E., Mahoney Jr., W. M., Renoud, K. J., Gartenhaus, R. B., Bachman, K. E., & Passaniti, A. (2007). The RUNX2 transcription factor cooperates with the YES-associated protein, YAP65, to promote cell transformation. *Cancer Biology & Therapy, 6*, 856–863.

Wang, X., Su, L., & Ou, Q. (2012). Yes-associated protein promotes tumour development in luminal epithelial derived breast cancer. *European Journal of Cancer, 48*, 1227–1234.

Wu, S., Huang, J., Dong, J., & Pan, D. (2003). hippo encodes a Ste-20 family protein kinase that restricts cell proliferation and promotes apoptosis in conjunction with salvador and warts. *Cell, 114*, 445–456.

Yagi, R., Chen, L. F., Shigesada, K., Murakami, Y., & Ito, Y. (1999). A WW domain-containing yes-associated protein (YAP) is a novel transcriptional co-activator. *The EMBO Journal, 18*, 2551–2562.

Yano, T., Ito, K., Fukamachi, H., Chi, X. Z., Wee, H. J., Inoue, K., et al. (2006). The RUNX3 tumor suppressor upregulates Bim in gastric epithelial cells undergoing transforming growth factor beta-induced apoptosis. *Molecular and Cellular Biology, 26*, 4474–4488.

Yu, F. X., Luo, J., Mo, J. S., Liu, G., Kim, Y. C., Meng, Z., et al. (2014a). Mutant Gq/11 promote uveal melanoma tumorigenesis by activating YAP. *Cancer Cell, 25*, 822–830.

Yu, W., Qiao, Y., Tang, X., Ma, L., Wang, Y., Zhang, X., et al. (2014b). Tumor suppressor long non-coding RNA, MT1DP is negatively regulated by YAP and Runx2 to inhibit FoxA1 in liver cancer cells. *Cellular Signalling, 26*, 2961–2968.

Yuan, M., Tomlinson, V., Lara, R., Holliday, D., Chelala, C., Harada, T., et al. (2008). Yes-associated protein (YAP) functions as a tumor suppressor in breast. *Cell Death and Differentiation, 15*, 1752–1759.

Zanconato, F., Forcato, M., Battilana, G., Azzolin, L., Quaranta, E., Bodega, B., et al. (2015). Genome-wide association between YAP/TAZ/TEAD and AP-1 at enhancers drives oncogenic growth. *Nature Cell Biology, 17*, 1218–1227.

Zanconato, F., Cordenonsi, M., & Piccolo, S. (2016). YAP/TAZ at the roots of cancer. *Cancer Cell, 29*, 783–803.

Zhang, J., Xu, Z. P., Yang, Y. C., Zhu, J. S., Zhou, Z., & Chen, W. X. (2012). Expression of Yes-associated protein in gastric adenocarcinoma and inhibitory effects of its knockdown on gastric cancer cell proliferation and metastasis. *International Journal of Immunopathology and Pharmacology, 25*, 583–590.

Zhou, Z., Zhu, J. S., Xu, Z. P., & Zhang, Q. (2011). Lentiviral vector-mediated siRNA knockdown of the YAP gene inhibits growth and induces apoptosis in the SGC7901 gastric cancer cell line. *Molecular Medicine Reports, 4*, 1075–1082.

Zuo, J. H., Zhu, W., Li, M. Y., Li, X. H., Yi, H., Zeng, G. Q., et al. (2011). Activation of EGFR promotes squamous carcinoma SCC10A cell migration and invasion via inducing EMT-like phenotype change and MMP-9-mediated degradation of E-cadherin. *Journal of Cellular Biochemistry, 112*, 2508–2517.

Roles of RUNX in Hypoxia-Induced Responses and Angiogenesis

27

Sun Hee Lee, Sarala Manandhar, and You Mie Lee

Abstract

During the past two decades, Runt domain transcription factors (RUNX1, 2, and 3) have been investigated in regard to their function, structural elements, genetic variants, and roles in normal development and pathological conditions. The Runt family proteins are evolutionarily conserved from *Drosophila* to mammals, emphasizing their physiological importance. A hypoxic microenvironment caused by insufficient blood supply is frequently observed in developing organs, growing tumors, and tissues that become ischemic due to impairment or blockage of blood vessels. During embryonic development and tumor growth, hypoxia triggers a stress response that overcomes low-oxygen conditions by increasing erythropoiesis and angiogenesis and triggering metabolic changes. This review briefly introduces hypoxic conditions and cellular responses, as well as angiogenesis and its related signaling pathways, and then describes our current knowledge on the functions and molecular mechanisms of Runx family proteins in hypoxic responses, especially in angiogenesis.

Keywords

Hypoxia • HIF-1α • Angiogenesis • RUNX1 • RUNX2 • RUNX3 • VEGF

Abbreviations

VEGF	vascular endothelial growth factor
HIF	hypoxia-inducible factor
PTM	post-translational modification
CBFβ	core binding factor β
PHD	prolyl hydroxylase
pVHL	von Hippel Lindau

S.H. Lee • S. Manandhar • Y.M. Lee, Ph.D. (✉)
National Basic Research Laboratory of Vascular Homeostasis Regulation, BK21 Plus KNU Multi-Omics based Creative Drug Research Team, Research Institute of Pharmaceutical Sciences, College of Pharmacy, Kyungpook National University, Daegu 41566, South Korea
e-mail: lym@knu.ac.kr

Y. Groner et al. (eds.), *RUNX Proteins in Development and Cancer*, Advances in Experimental Medicine and Biology 962, DOI 10.1007/978-981-10-3233-2_27

FIH	factor inhibiting HIF
PI3K	phosphatidylinositol 3-kinase
MAPK	mitogen-activated protein kinase
PKB	protein kinase B
mTOR	mammalian target of rapamycin
ERK	extracellular signal-regulated kinase
S6 K	S6 kinase
eIF-4E	eukaryotic translational initiation factor 4E
4E–BP1	eIF-4E-binding protein
MNK	MAP kinase interacting kinase
VPF	vascular permeability factor
HRE	hypoxia-responsive element
PlGF	placental growth factor
VEGFR	VEGF-receptor
EC	endothelial cell
eNOS	endothelial nitric oxide
FAK	focal adhesion kinase
Ang	angiopoietin
PECAM	platelet-endothelial cell-adhesion molecule
bFGF	basic fibroblast growth factor
GM-CSF	granulocyte macrophage-colony stimulating factor
EPC	endothelial progenitor cell
CAC	circulating angiogenic cell
TIMP	tissue inhibitor of metalloproteinase
IGFBP-3	insulin-like growth factor-binding protein-3
AGM	aorta-gonad mesonephros
HSC	hematopoietic stem cell
HSPC	hematopoietic stem and progenitor cell
AML	acute myeloid leukemia
C/EBPα	CCAAT/enhancer-binding protein α
DNMT	DNA methyltransferases
ER	endoplasmic reticulum
UPR	unfolded protein response
TRAIL	tumor necrosis factor-related apoptosis-inducing ligand
HDAC	histone deacetylase
ODDD	oxygen-dependent degradation domain
HBME	bone marrow endothelial cell
MMP	matrix metalloproteinase
HMT	histone methyltransferase
BRD	bromodomain
MVD	microvascular density
vWF	von Willebrand factor
Dll4	Delta-like 4
Egr-3	early growth response-3

27.1 Introduction

Vertebrate organ development and the growth and metastasis of cancer rely on angiogenesis, the formation of new blood vessels from a pre-existing network of capillaries (Folkman 1997). Of the numerous angiogenic factors discovered thus far, vascular endothelial growth factor (VEGF) has been identified as a key mediator of angiogenesis (Senger et al. 1983). Elevated expression of VEGF in human tumor biopsy specimens has been detected in multiple types of cancers (Shi et al. 2001). Moreover, loss or inactivation of tumor suppressor genes and activation of oncogenes are associated with VEGF overexpression (Shi et al. 2001; Xie et al. 2004; Siewert et al. 1998). Both genetic and epigenetic alterations are involved in regulation of VEGF expression (Xie et al. 2004).

Hypoxic cellular microenvironments, which often arise in developing embryos, stem cell niches, solid tumors, and ischemic disease, trigger various adaptive responses in cells. Many of these responses are controlled by hypoxia-inducible factor (HIF). For example, the strong angiogenic factor VEGF is induced by HIF-1. HIF-1α and HIF-2α, central mediators of homeostatic responses that allow hypoxic cells to survive or differentiate (Giaccia et al. 2004), are primarily regulated at the level of proteasomal degradation, and other signaling pathways influence their stability under normoxic conditions (Lee et al. 2004). In particular, post-translational modifications (PTMs) such as phosphorylation, hydroxylation, acetylation, and nitrosylation are key mechanisms that regulate the stability of HIF-α. Molecules involved in the degradation process also play roles in tumor suppression, whereas those involved in activation or stabilization (including many growth factors) have oncogenic functions (Semenza 2001, 2010; Rey and Semenza 2010).

The RUNX family of transcription factors plays pivotal roles in normal development and neoplasia, discussed elsewhere in this book. The Runt domain transcription factors are composed of a larger DNA-binding subunit, α, and a smaller non-DNA-binding subunit, β (known as core binding factor β, CBFβ) (Ito 1999). Three mammalian genes, Runt-related genes 1, 2, and 3, encode the α subunits RUNX1, RUNX2, and RUNX3, respectively. RUNX1, which is mainly required for hematopoiesis, is the most frequent target of chromosomal translocations associated with human leukemia (Speck and Gilliland 2002). RUNX2 is essential for osteogenesis (Ducy et al. 1997), and *Runx2* knockout mice exhibit complete bone loss due to arrested osteoblast maturation (Komori et al. 1997). RUNX3 is required for development of CD8-lineage T cells (Woolf et al. 2003) and TrkC-dependent dorsal root ganglion neurons (Levanon et al. 2002), and also functions as a tumor suppressor in various cancers (Bae and Choi 2004). This review provides an update on our present understanding about the role of RUNX family proteins and their underlying molecular mechanisms in hypoxic microenvironments and angiogenic processes.

27.1.1 Hypoxic Microenvironment and Hypoxia-Inducible Factor (HIF)

Hypoxia refers to a condition in which tissues are not adequately oxygenated, usually due to an insufficient concentration of oxygen in the blood. Oxygen deprivation results in considerable stress in living cells. Actively growing cells, such as those in developing embryos or expanding tumors, respond to oxygen deprivation by instructing themselves and their microenvironment to engage in adaptations that maintain the supply of essential nutrients. These adaptive responses include changes to pathways that involved in glycolysis, apoptosis, cell-cycle arrest, survival, and angiogenesis (Semenza 2014).

Vertebrate cartilage is an avascular tissue and a well-known site of hypoxia. An intensive investigation of tissue-specific targeting of HIF-1α revealed that developmental growth plates are hypoxic (Schipani et al. 2001). An immunostaining technique employing an antibody against pimonidazole adduct (hypoxia marker), bound to DNA and protein in hypoxic regions, demonstrated substantial hypoxia at developmental growth plates and that hypoxia is an inducer of angiogenesis in various organs of developing embryos (Lee et al. 2001; Dunwoodie 2009). Similarly, the oxygen gradient in bone marrow renders the osteoblastic hematopoietic stem cell (HSC) niche at the endosteal surface hypoxic (Maes et al. 2012). The partial pressure of O_2 in human bone marrow is lower than that in peripheral blood, and medullary sinus architecture and arterial blood flow patterns generate an O_2 gradient. It has been proposed that HSCs and their proliferating progenitors are naturally distributed along this gradient, with the HSCs occupying the most hypoxic niches (Parmar et al. 2007). Moreover, the stem cell microenvironment (niche) in various organs, including the blood, epidermis, and intestine, is also hypoxic (Lane et al. 2014) and accumulating evidence has revealed that the oxygen level influences stem cell niches and that hypoxic conditions promote the differentiation of certain types of stem or progenitor cells *in vitro* (reviewed in (Simon and Keith 2008)). Thus, the molecular mechanisms and programs involved in vascular development and stem cell activity might be dependent on the hypoxic microenvironment both during development and in certain diseases (Simon and Keith 2008). In addition, hypoxia frequently occurs under pathological conditions, such as cancer and ischemic diseases (Maes et al. 2012).

Transcriptional responses to hypoxia are mediated for the most part by HIF, a heterodimer consisting of an α-subunit (HIF-α) and a β-subunit (HIF-β) (Harris 2002). HIF-1 consists of two basic helix-loop-helix proteins of the PER-ARNT-SIM subfamily, HIF-1α and HIF-1β, which heterodimerize and bind to the core responsive element RCGTG motif in target genes. HIF-1β is a stable subunit regulated in an oxygen-independent manner, whereas HIF-1α is labile with respect to oxygen level. Specifically, HIF-1α is constitutively expressed under nor-

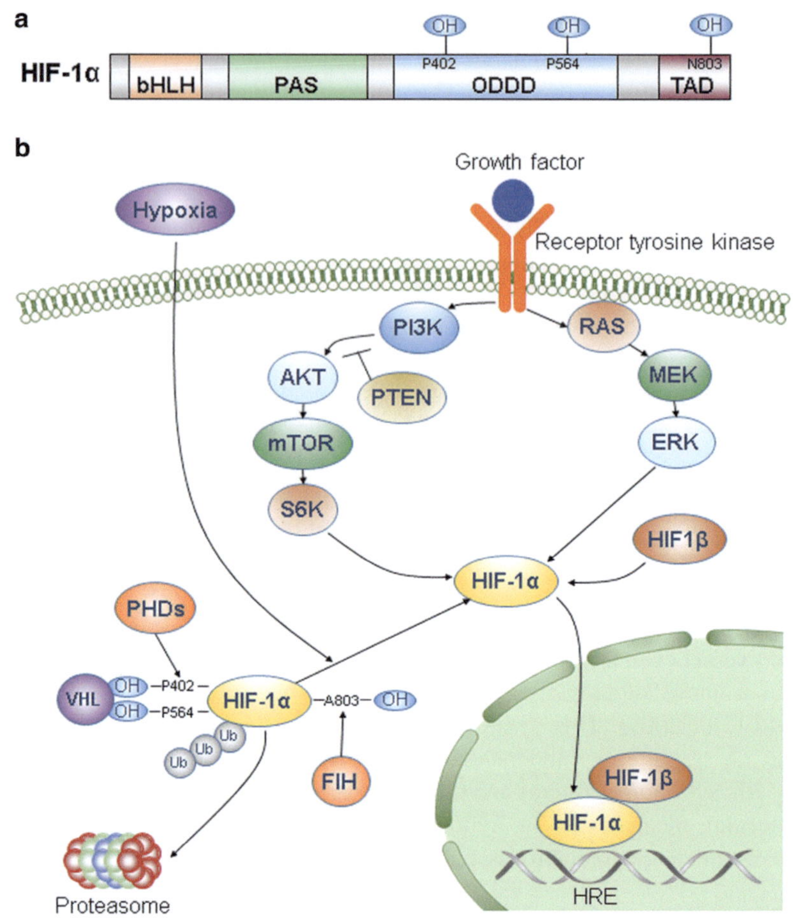

Fig. 27.1 Mechanisms of HIF-1α stabilization. (**a**) HIF-1α domain structure. Sites of proline hydroxylation are specified in the O_2-dependant degradation domain of the human protein as P402 and P564. N803 represents the position of the asparagine residue whose hydroxylation under normoxia interferes with p300/CBP binding. (**b**) The classical O_2 sensing pathway is represented by O_2-dependent enzymatic hydroxylation at P402 and/or P564 on HIF-1α. The hydroxylation reactions are carried out by various PHD enzymes that mediate recognition of the pVHL and are followed by ubiquitination (Ub) and subsequent proteasomal degradation of HIF-1α. In addition, growth factor mediated oncogenic activation, like activation of the Ras–RAF–MAPK, phosphoinositide 3-kinase PI3K, PTEN, or Akt pathways, can also cause HIF-1α accumulation. Stabilized HIF-1α associates with HIF-1β and binds to cognate HREs in target genes

moxic conditions, but post-translationally modified by a class of 2-oxoglutarate-dependent and Fe^{2+}-dependent prolyl hydroxylases (PHDs) at prolines 402 and 564; the modified protein is degraded after ubiquitination by von Hippel Lindau (pVHL), a tumor suppressor. HIF-1α is also regulated by transactivational inhibition of asparagine 803, which is hydroxylated by Factor Inhibiting HIF (FIH) (Fig. 27.1). The half-life of HIF-1α is very short (~5 min).

HIF-1α is also activated in a hypoxia-independent manner. Binding of growth factors to cognate receptor tyrosine kinases activates the phosphatidylinositol 3-kinase (PI3K) and mitogen-activated protein kinase (MAPK) pathways. PI3K in turn activates the downstream serine/threonine kinase AKT (protein kinase B, PKB) and mammalian target of rapamycin (mTOR). The extracellular signal-regulated kinase (ERK) and mTOR phosphorylate p70 S6 kinase (S6 K), which phosphorylates the ribosomal S6 protein and the eukaryotic translational initiation factor 4E (eIF-4E)-binding protein (4E-BP1). Phosphorylation of 4E-BP1 by MAP kinase interacting kinase (MNK) prevents its binding to eIF-4E and stimulates its activity directly. The result of growth factor signaling is an increase in the rate at which a subset of mRNAs within the cell (including HIF-1α mRNA) are translated into protein (Semenza 2003) (Fig. 27.1). Mammalian HIFα exists as three isoforms, HIF-1α, HIF-2α (or EPAS1), and HIF-3α (or IPAS). HIF-1α is ubiquitously expressed in all cells, but HIF-2α and HIF-3α are expressed in a tissue-specific manner (Majmundar et al. 2010).

27.1.2 Angiogenesis and Signaling Pathways

Hypoxia is the primary physiological stimulus that induces angiogenesis, a complex process by which new blood vessels are formed from existing vessels. Angiogenesis is orchestrated by endothelial cells, surrounding pericytes, smooth muscle cells, extracellular matrix (ECM), and angiogenic cytokines and growth factors. The steps of this process include degradation of the basement membrane surrounding an existing vessel; migration and proliferation of endothelial cells into the new space; maturation, differentiation, and adherence of the endothelial cells to each other; and lumen formation. Although many molecules and receptors have been investigated in the context of angiogenesis, VEGF-A (also known as vascular permeability factor, VPF) is one of the strongest angiogenic factors (Ferrara et al. 1996, 2003). VEGF-A, which is expressed in most cells, is secreted from inflammatory cells, mast cells, macrophages, and tumor cells in response to hypoxia. VEGF-A expression is activated by binding of HIF to hypoxia-responsive element (HRE) sites in the *VEGF* promoter. It acts as a chemoattractant and guides sprouting of new blood vessels into hypoxic regions of tissues (Carmeliet 2003; Gerhardt et al. 2003).

Among VEGF receptors (VEGFRs), VEGFR-2 is the major mediator of VEGF responses in ECs, including survival, proliferation, migration, and maturation. Phosphorylated VEGFR2 activates downstream signaling pathways, i.e., the ERK pathway (p42/44 MAPK) and PI3K, AKT/PKB pathway, via the activation of the small GTP-binding protein Rac, resulting in the regulation of proliferation, survival, and migration (Ferrara et al. 2003; Gerber et al. 1998). Activated endothelial nitric oxide (eNOS) is implicated in the vascular permeability and migration of ECs. Other MAPKs, p38 and focal adhesion kinase (FAK), along with the FAK substrate paxillin, regulate cytoskeletal regulation and migration (Mierke 2013) (Fig. 27.2).

Other angiogenic factors include the angiopoietins, Ang1–4. Ang1 and Ang4 function as agonists for their shared receptor, Tie2, whereas Ang2 and Ang3 act as competitive antagonists. Tie1 and Tie2 are receptor tyrosine kinases that are primarily expressed in ECs and early hematopoietic cells. Tie2 is the main receptor that mediates angiogenic responses, and Tie1 acts in complex with Tie2 (Fagiani and Christofori 2013). Ang/Tie signaling cascades are involved in fundamental events of angiogenesis, including vascular stabilization and remodeling, as well as recruitment of pericytes and smooth muscle cells (Fagiani and Christofori 2013). Ang1 is critical for vessel maturation, adhesion, migration, and survival. Ang2, on the other hand, promotes cell death and disrupts vascularization (Yuan and Rigor 2010). Ang1 and the junctional molecules VE-cadherin and platelet-endothelial cell-adhesion molecule (PECAM) tighten vessels, thereby counteracting active angiogenesis, which is associated with formation of leaky and immature vessels. Ang2 and proteinases mediate dissolution of the existing basement membrane and the interstitial matrix, and Ang1 and Ang2 work in conjunction with VEGF to modulate angiogenesis. Elevated levels of Ang2 promote tumor angiogenesis, metastasis, and inflammation (Eklund and Saharinen 2013) (Fig. 27.2).

In addition to angiogenesis (sprouting of preexisting vessels), secreted angiogenic cytokines such as VEGF, basic fibroblast growth factor (bFGF), granulocyte macrophage-colony stimulating factor (GM-CSF), IGF-1, and Ang(s) have been implicated in the mobilization of bone marrow – derived proangiogenic cells, such as endothelial progenitor cells (EPCs) or circulating

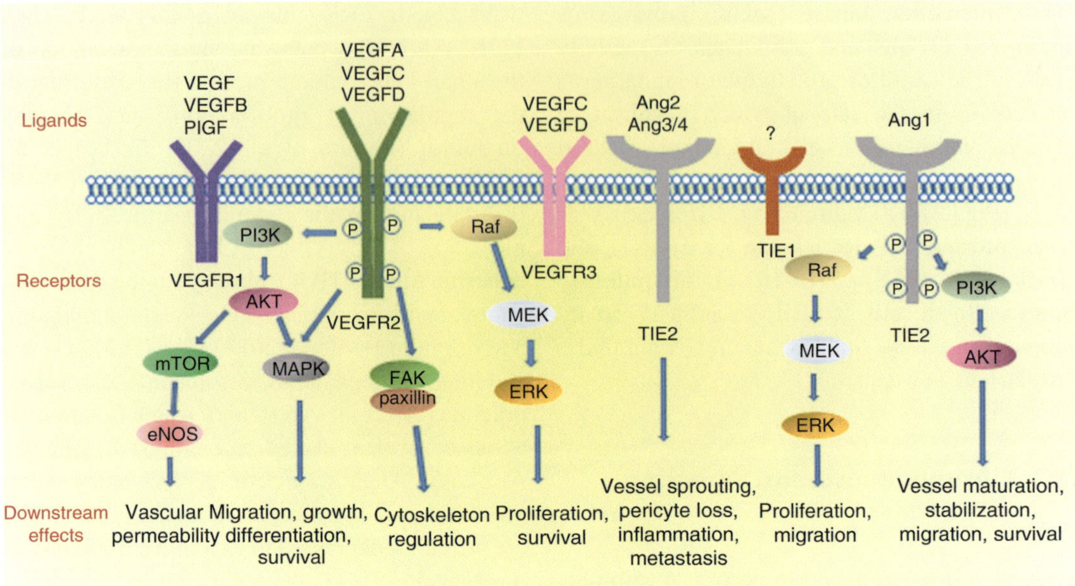

Fig. 27.2 Endothelial-cell receptors and growth factors mediated vasculogenesis and angiogenesis. Receptors are subgrouped into the VEGFR and Tie families. Specific ligand binding to the receptors is defined over each receptor. The VEGFR family consists of three transmembrane receptors, VEGFR-1, VEGFR-2, and VEGFR-3. VEGFR-2 is the main signal transducing receptor; it activates several downstream signaling molecules (*encircled*), and induces responses such as cell proliferation, migration, and survival. MEK/ERK mediated pathway mainly produces proliferation signals, while the key function of PI3K/Akt pathway is the regulation of cell survival. FAK and PI3K have also been implicated in cell migration via their ability to enhance the reorganization of actin and recruitment of actin-anchoring proteins to focal adhesions. Tie1 and Tie2 are two members of the Tie receptor family. The ligand, Ang1, is produced by non-endothelial cells, whereas the primary source of Ang2 is endothelial cells. Signaling through the Tie2 receptor increases survival, vascular permeability, and sprouting, and regulates pericyte/smooth muscle cell recruitment. Signaling molecules are encircled, and downstream effects are specified at the *bottom* with an *arrow*

angiogenic cells (CACs) (vasculogenesis) (Hirschi et al. 2008; Semenza 2010). The recruitment of CACs is critical for vessel formation in injured or ischemic tissues (Bosch-Marce et al. 2007).

27.1.3 Hypoxia and Tumor Angiogenesis

Due to the aberrant blood supply and unlimited growth of tumor cells, hypoxia is a hallmark of the tumor microenvironment. As a tumor grows, it rapidly outgrows its blood supply, leaving portions of the tumor with a significantly low oxygen tension. Hypoxic tumors are the result of available oxygen being consumed within a region 70–150 μm from the tumor vasculature by rapidly proliferating tumor cells, which limits the amount of oxygen available for diffusion into the tumor tissue. In practical terms, oxygen tension in cancer cells 100 μm from microvessels is approximately 2 mmHg (0.2–0.1 % oxygen, vs. end capillary levels of about 5 %), whereas at 200 μm the oxygen tension is approximately 0 mmHg (Helmlinger et al. 1997). Hypoxic tumor cells temporarily arrest the cell cycle and reduce energy consumption, but in order to proliferate they must further adapt to this stress condition by secreting survival and angiogenic factors. Solid tumors with hypoxic regions have a poorer prognosis than their well-oxygenated counterparts. This is a consequence of the genetic characteristics of viable hypoxic tumor cells, which invariably have a more aggressive phenotype (Hockel et al. 1996; Walenta et al. 2000).

Blood vessels in tumors exhibit abnormal features, e.g., structural abnormalities or chaotic and leaky blood flow, leading to local regions of hypoxia. This is due to the formation of immature and functionally imperfect blood vessels induced by VEGF-A. Upon normalization of tumor blood vessels, blood perfusion is improved and tumor hypoxic regions regress; consequently, normalization of tumor vessels has attracted a great deal of attention in the field of cancer treatment (reviewed in (Jain 2005)). Ang/Tie signaling is important for normalization of tumor vasculature. Maintenance of vascular homeostasis or normalization of tumor vasculature involves interplay between endogenous molecules such as soluble VEGFR1, Ang1, angiostatin, endostatin, tissue inhibitor of metalloproteinases (TIMPs, which inhibit MMPs), etc. (Bergers and Benjamin 2003).

27.2 RUNX1 in Hypoxic Responses and Angiogenesis

RUNX1 (also called AML1/PEBP2αB/CBFα2) is expressed in hematopoietic tissues, where it plays a critical role in hematopoietic cell differentiation (Tenen et al. 1997; Friedman 2002). Hematopoietic cells and ECs have common precursor cells called hemangioblasts (Murray 1932; Sabin 1920). In addition, hematopoietic cells directly differentiate from an endothelial precursor (Nishikawa et al. 1998). Budding of hematopoietic cells from hemogenic endothelium is facilitated by Runx1, and this process is abrogated in *Runx1*-deficient mice (North et al. 1999), suggesting that Runx1 is essential for the transition from ECs to hematopoietic cells (Chen et al. 2009). Furthermore, a recent report highlighted the importance of Runx1 phosphorylation at multiple sites during early hematopoiesis (Yoshimi et al. 2012).

Functional deregulation of RUNX1 occurs in leukemia (Ito 2004). In particular, this gene is located at the most frequent target breakpoint of chromosomal translocations t(8;21) resulting in human myeloid leukemia; these rearrangements generate fusion proteins like RUNX1/ETO (same as AML/ETO, t(8;21)), and RUNX1/Evi1 (t(3;21)) (Miyoshi et al. 1991; Look 1997; Okuda et al. 1996; Kurokawa and Hirai 2003). Site-specific acetylation of RUNX1/ETO promotes its leukemogenic activities (Wang et al. 2011). Mice heterozygous for Runx1/ETO die in midgestation due to central nervous system hemorrhage and profound inhibition of fetal liver hematopoiesis (Yergeau et al. 1997), similar to the phenotypes of mice harboring homozygous deletion of Runx1 (Wang et al. 1996; Okuda et al. 1996). In addition, yolk sac cells derived from Runx1/ETO heterozygotes can differentiate into macrophages in hematopoietic colony formation assays (Yergeau et al. 1997). These observations suggest differential roles for RUNX1 and its fusion gene(s) in the hematopoietic system.

In angiogenesis, Runx1 induces endothelial differentiation and maturation as well as vascular network formation activity by promoting expression of VE-cadherin (Iwatsuki et al. 2005) and Ang1 (Takakura et al. 2000). Insulin-like growth factor-binding protein-3 (IGFBP-3) has anti-angiogenic properties, inhibiting vascular EC survival but inducing tumor vasculature normalization (Delafontaine et al. 2004), and Runx1 promotes angiogenesis by repressing *IGFBP-3* expression in EPCs derived from the aorta-gonad mesonephros (AGM) region (Iwatsuki et al. 2005) (Table 27.1). Mice lacking *Runx1* exhibit defective hematopoiesis and massive central nervous system hemorrhage; the latter is due to a defect in erythrocyte differentiation (Okuda et al. 1996; Wang et al. 1996). These observations are supported by two studies, one showing that *Runx1*-deficient embryos have imperfect angiogenesis in head, pericardium, and liver (Suda and Takakura 2001), and the other showing that loss of Runx1 function in zebrafish embryo leads to defects in hematopoiesis and vasculogenesis (Kalev-Zylinska et al. 2002). In addition, impairment of angiogenesis is caused by a poor supply of Ang1 from HSCs, indicating that Runx1 regulates maturation of blood vessels via Ang1 (Takakura et al. 2000). Expression of Ang1 is also positively regulated by Runx1 in MSS31 ECs (Namba et al. 2000).

On the other hand, Runx1 reduces active angiogenesis in acute myeloid leukemia (AML).

Table 27.1 RUNX family-mediated angiogenesis

RUNX family	Molecule regulated	Angiogenesis status	System
RUNX1	Repressed IGFBP-3	Enhanced	Endothelial progenitor cell (Iwatsuki et al. 2005)
	Repressed VEGF	Reduced	Acute myeloid leukemia (Ter Elst et al. 2011; Suehiro et al. 2010)
RUNX2	Enhanced VEGF	Enhanced	Developing bone (Zelzer et al. 2001)
	Enhanced VEGF	Enhanced	Chondrosarcoma cells (Sun et al. 2009)
RUNX3	Repressed VEGF	Reduced	Human gastric cancer (Peng et al. 2006)
	Repressed VEGF	Reduced	Renal cell carcinoma (Chen et al. 2013)
	Repressed von Willebrand factor	Reduced	Human microvascular endothelial cells (Starke et al. 2011)
	Repressed VEGF	Reduced	Prostate cancer cells (Chen et al. 2014)

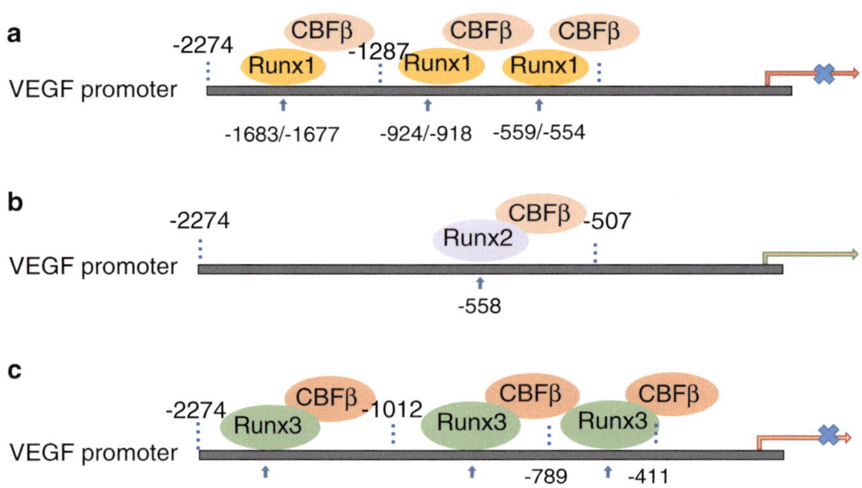

Fig. 27.3 RUNX mediates regulation of VEGF. The VEGF promotor comprises a total three RUNX1 (**a**), one RUNX2 (**b**) and three RUNX3 (**c**) binding sites. VEGF is negatively regulated when RUNX1 or RUNX3 bind to their corresponding sites on the VEGF promotor, while RUNX2 site specific binding to the VEGF promotor increases the expression of VEGF

Runx1 is involved in the transcriptional regulation of *VEGF-A* mRNA; specifically, it acts as a transcriptional repressor of *VEGF-A* by directly binding the promoter (Fig. 27.3a). By contrast, inhibition of RUNX1/ETO in RUNX1/ETO-positive Kasumi-1 AML cells decreases *VEGF-A* mRNA expression and VEGF-A protein secretion (Ter Elst et al. 2011). siRNA-mediated functional inhibition of RUNX1/ETO induces downregulation of the VEGF-A-induced Egr-3 in Kasumi-1 cells (Suehiro et al. 2010; Ter Elst et al. 2011) (Table 27.1). Collectively, these findings suggest that, in normal vascular formation, RUNX1 increases angiogenesis and the tightness of blood vessels via Ang1, whereas it inhibits active angiogenesis in AML, although some leukemogenic RUNX1 mutations can result in a dominant-negative effect on RUNX1 by stimulating *VEGF-A* expression, which induces vascular leakage.

VEGF is also a major mediator of angiogenesis, proliferation, migration, and survival of cancer cells, and RUNX1/ETO can aggravate AML by facilitating expression of VEGF-A (Ter Elst

Table 27.2 RUNX-mediated hypoxic responses

RUNX family	Molecule regulated	Hypoxic responses	System
RUNX1	Suppressed HIF-1α transcriptional activity	Suppressed	Hela, U937 transfectant (Peng et al. 2008)
	Enhanced RUNX1 transcriptional activity	Enhanced	Hela, U937 transfectant (Peng et al. 2008)
RUNX2	Enhanced HIF-1α stability	Enhanced	Hypertrophic chondrocytes (Lee et al. 2012)
RUNX3	Silenced RUNX3	Enhanced	Gastric cancer cells (Lee et al. 2009)
	Suppressed HIF-1 α	Suppressed	Gastric cancer cells (Lee et al. 2014b)

et al. 2011). A somatic mutation in RUNX1 has been detected in AML patients as well. RUNX1 somatic mutations are associated with poor prognosis in patients with *de novo* AML, possibly due to elevated VEGF-A levels in RUNX1-mutated blasts (Tang et al. 2009). These results indicate that inhibition of angiogenic factors such as VEGF-A in leukemia harboring RUNX1 mutations is a potential therapeutic target.

Experiments with forced expression of RUNX1 or HIF-1α under hypoxic conditions suggest that RUNX1 plays distinct functions in hematopoietic and leukemia cells during hypoxia-induced responses. Overexpression of RUNX1 suppresses HIF-1α transcriptional activity, whereas HIF-1α facilitates RUNX1 transcriptional activity under hypoxic conditions (Peng et al. 2008) (Table 27.2). Given that hypoxia/HIF-1α induces differentiation of AML cells and promotes the granulocytic differentiation of the normal hematopoietic cell line 32Dcl3 (Jiang et al. 2005; Huang et al. 2003), it is possible that RUNX1-mediated differentiation of hematopoietic cells under hypoxic conditions is mediated by HIF-1α (Fig. 27.4a). These findings suggest that RUNX1 inhibits angiogenic processes under hypoxic conditions, possibly by antagonizing the transcriptional activity of HIF-1 in hematopoietic cells. By contrast, in leukemic cells, promotion of transcriptional activity by Runx1 and CCAAT/enhancer-binding protein α (C/EBPα) through interaction with HIF-1α may promote differentiation (Peng et al. 2008; Jiang et al. 2005).

Runx1-binding sites in target gene promoters are frequently adjacent to binding sites of other hematopoietic transcription factors such as Ets-1, Myb, and C/EBPα. All of these factors cooperate with Runx1 to activate gene transcription (Kurokawa and Hirai 2003). However, high expression of HIF-1α in RUNX1/ETO-driven AML predicts inferior prognosis, possibly due to the epigenetic changes induced by RUNX1/ETO and HIF-1α. A recent study showed that RUNX1 and HIF-1α form a positive regulatory circuit and cooperate to transactivate the DNMT3a gene, leading to DNA hypermethylation. Runx1/ETO and HIF-1α, which are coexpressed and interact physically, act as a transcriptional repressor by recruiting co-repressors (histone deacetylases, DNMTs, nuclear receptor co-repressors) to target genes for silencing, particularly tumor suppressors such as $p14^{ARF}$ and $p15^{INK4b}$ (Linggi et al. 2002; Gao et al. 2015). Therefore, modulation of the Runx1/ETO–HIF-1α epigenetic machinery causes higher cell proliferation *in vitro* and more severe leukemic status in mice (Gao et al. 2015).

On the other hand, a growing body of evidence demonstrates the importance of RUNX1 in response to genotoxic and endoplasmic reticulum (ER) stress. Hematopoietic stem and progenitor cells (HSPCs) deficient in RUNX1 have reduced p53 levels and an attenuated unfolded protein response (UPR), making them less sensitive to apoptosis in response to genotoxic or ER stress (Cai et al. 2015). This phenomenon could explain the growth advantage of these cells over normal

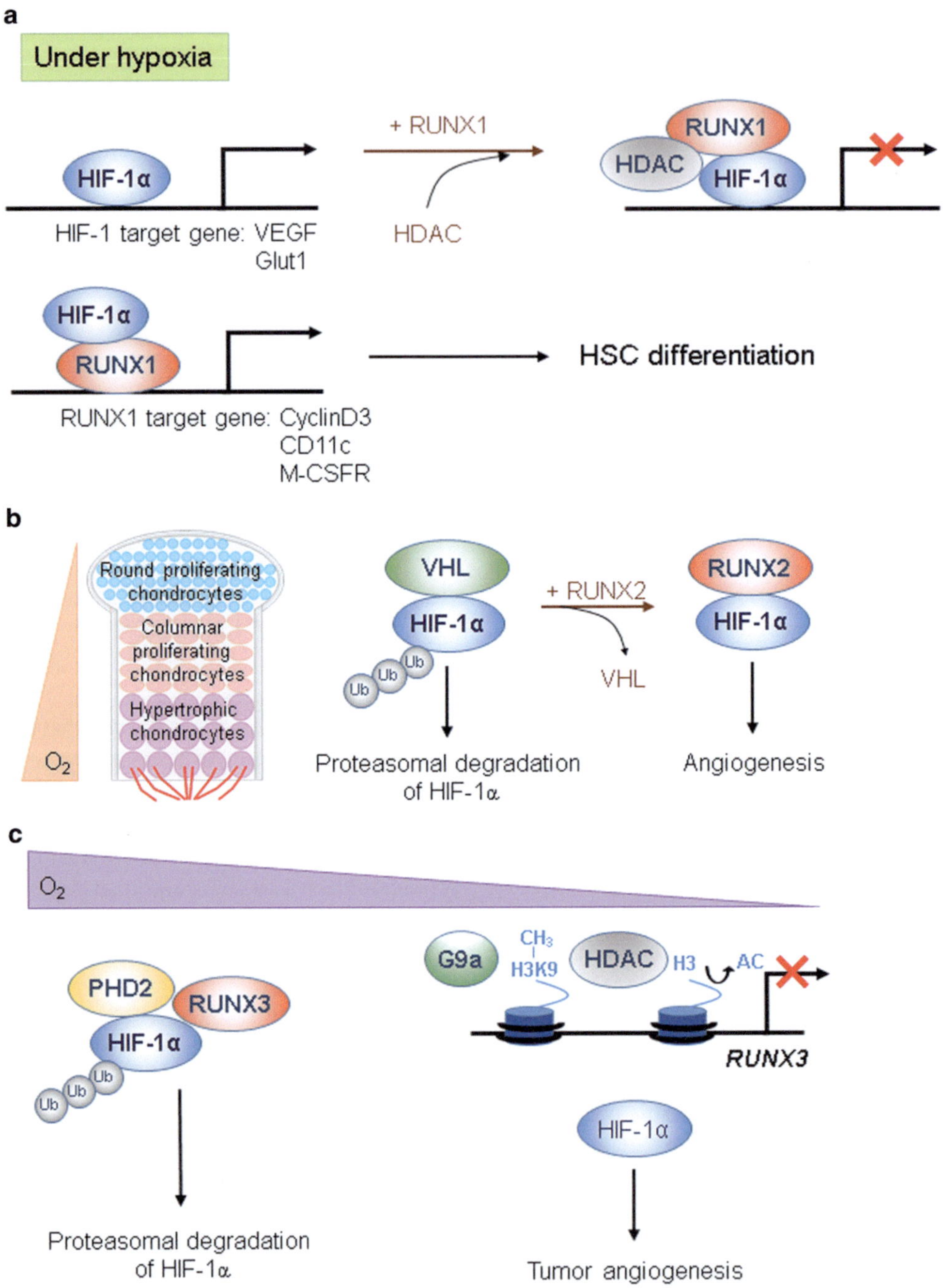

Fig. 27.4 Schematic representation of crosstalk between HIF-1 and RUNXs. (**a**) Under hypoxia, interactions between RUNX1 and HIF-1α represses transcription of HIF-1 targeted genes, such as VEGF and glucose transporter 1(Glut1). HIF-1α enhances the transcription of RUNX1 targeted genes, such as cyclin D3, CD11c, and macrophage colony stimulating factor receptor (M-CSFR). Interactions between RUNX1 and HIF-1α contribute to hypoxia-induced hematopoietic stem cell differentiation. (**b**) RUNX2 enhances HIF-1α stability through competition with VHL at the ODD domain and stimulates angiogenesis in hypertrophic chondrocytes at the growth plate. (**c**) RUNX3 destabilizes HIF-1α protein via PHD2 recruitment mediated by interactions with PHD2 and HIF-1α. Hypoxia silences *RUNX3* expression through the recruitment of G9a and HDAC1 to the *RUNX3* promoter, which in turn leads to H3K9 methylation and H3 deacetylation. Therefore, RUNX3 silencing stabilizes HIF-1 and enhances hypoxia-induced angiogenesis

HSPCs. By contrast, expression of the RUNX1 fusion oncogene RUNX1/ETO leads to growth arrest and apoptosis upon exposure to a severely hypoxic environment (Barbetti et al. 2013). Furthermore, in RUNX1/ETO-overexpressing cells, the pro-apoptotic effect of hypoxia is strengthened through direct binding of RUNX1/ETO to the tumor necrosis factor-related apoptosis-inducing ligand (TRAIL) promoter, thereby augmenting apoptosis (Barbetti et al. 2013). These observations suggest differential roles for RUNX1 and its fusion oncogene, dependent upon the cellular context and microenvironment.

27.3 RUNX2 in Hypoxic Responses and Angiogenesis

RUNX2 is a master regulator of osteoblast differentiation, bone formation, and mineralization (Komori et al. 1997; Otto et al. 1997; Ducy et al. 1997). Endochondral bone formation takes place in the growth plate, a highly specialized organ that generates practically all bone growth until adulthood (Erlebacher et al. 1995). Endochondral bone formation begins with aggregation of mesenchymal cells and their differentiation into chondrocytes. Chondrocyte hypertrophy initiates in the center of cartilaginous skeletal elements, and is followed by apoptosis; invasion of blood vessels, osteoclasts, and other mesenchymal cells from the perichondrium; and production of mature bone matrix, such as collagen type X (Ninomiya et al. 1986; Iyama et al. 1991).

Paradoxically, cartilage is highly resistant to vascularization, but longitudinal bone growth is highly dependent on angiogenic processes because hypertrophic cartilage is a target for capillary invasion and angiogenesis in endochondral ossification (Kuettner and Pauli 1983). Therefore, hypertrophic cartilage may produce angiogenic activators, whereas other types of cartilage produce angiogenic inhibitors (Descalzi Cancedda et al. 1995). Analysis of EF5 binding at different stages of fetal development clearly demonstrated the presence of a hypoxic central region in the fetal growth plate, in the round proliferative layer near the joint space, at the center of the columnar proliferative layer, and in the upper portion of the hypertrophic zone (Schipani et al. 2001). VEGF expressed by hypertrophic chondrocytes is required for chondrocyte survival and cartilage angiogenesis (Zelzer et al. 2004; Gerber et al. 1999). *Runx2* heterozygous knockout mice exhibit loss of vascularization and VEGF expression in hypertrophic chondrocytes, resulting in loss of endochondral ossification (Zelzer et al. 2001). The results demonstrate that RUNX2 is required for regulation of VEGF during endochondral bone formation in a hypoxia-independent manner (Zelzer et al. 2001) (Fig. 27.3b; Table 27.1).

An epigenetic mechanism is proposed to underlie Runx2-mediated endochondral ossification and angiogenesis. Histone deacetylase (HDAC) 4 homozygous knockout mice exhibit premature mineralization of endochondral bones, similar to the Runx2 gain-of-function phenotype (Vega et al. 2004). HDAC4 binds to Runx2 and inhibits Runx2 transcriptional activity. As HDAC4 decreases total and acetylated Runx2 through its deacetylation and transcriptional repressor activities, reduced expression of HDAC4 results in higher levels of Runx2, thereby increasing transcription of VEGF and its angiogenic activity on chondrosarcoma cells (Sun et al. 2009) (Table 27.1).

Osteoblasts express both HIF-1α and HIF-2α, which may play a role in modulating bone development, homeostasis, and angiogenesis; some of the effects of HIFs on bone and angiogenesis are mediated by VEGF (Schipani et al. 2001). RUNX2 interacts physically with HIF-1α in the nuclei of osteoblasts and on the chromatin of the VEGF gene. This interaction stimulates *VEGF* expression in mesenchymal osteogenic cells by direct and indirect binding of these two factors to regulatory regions of the *VEGF* gene (Kwon et al. 2011). A link between Runx2 and HIF-1α in hypertrophic chondrocytes has been identified: coexpression of Runx2 and HIF-1α, as well as higher vascular density, is observed in hypertrophic chondrocytes of the wild type, but expression of HIF-1α and vascular formation are not observed in growing tibial bones of *Runx2* knockouts. Further investigation of the role of RUNX2 in HIF-1α stability revealed that RUNX2 protects HIF-1α from degradation by blocking the interaction between HIF-1α and pVHL, thereby stimulating angiogenesis in hypertrophic chondrocytes that are not hypoxic (Lee et al. 2012) (Fig. 27.4b; Table 27.2). That report also showed that RUNX2 competes with pVHL for binding to the oxygen-dependent degradation domain (ODDD) of HIF-1α and increases HIF-1α nuclear translocation and stability, suggesting that RUNX2 is an upstream stabilizer of HIF-1α under normoxic states. This is another key mechanism that regulates vessel formation in growing long bones (Lee et al. 2012).

Stricker et al. used *in situ* hybridization to show that Runx2 and Runx3 transcripts overlap in immature and mature chondrocytes (Stricker et al. 2002). Simultaneous loss of Runx2 and Runx3 functions induces a dramatic delay in cartilage formation in axial and appendicular skeleton relative to Runx2 knockout alone (Yoshida et al. 2004). This reduction in cartilage maturation can be attributed to the arrest of chondrocyte differentiation before hypertrophy (Yoshida et al. 2004). RUNX1 is also involved in the formation of non-hematopoietic tissues during embryogenesis, including the development of skeleton (Lian et al. 2003; Wang et al. 2005). Furthermore, Runx1 is expressed in mesenchymal condensation together with Runx2, leading to the development of skeletal units (Kimura et al. 2010; Smith et al. 2005). However, it remains unclear how RUNX family proteins perform distinct functions in angiogenic processes during endochondral bone formation.

Expression of Runx2 in rat aorta – derived cells stimulates vascular sprouts (Sun et al. 2004), whereas introduction of a dominant-negative form of RUNX2 in human bone marrow endothelial (HBME)-1 cells inhibits their migration, invasion, and vascular tube formation on Matrigel (Sun et al. 2001), suggesting that it plays a positive role in angiogenesis (Sun et al. 2001; Bronckers et al. 2005). As expected from the positive effect of Runx2 on angiogenesis, RUNX2 is expressed at higher levels in tumors such as osteosarcoma and colon, prostate, and thyroid cancers (Brubaker et al. 2003; Kayed et al. 2007; Endo et al. 2008), highlighting its role as an oncogenic factor. Moreover, RUNX2 transactivates genes related to tumor progression, invasion, and metastasis, including survivin, MMP-2, MMP-9, and VEGF (Barnes et al. 2004; Pratap et al. 2005, 2006, 2008; Wai et al. 2006; Mendoza-Villanueva et al. 2010; Chimge and Frenkel 2013; Lim et al. 2010; Altieri 2008). These observations point toward the unfavorable outcomes of overexpressed RUNX2 in cancer progression, vascularization, and metastasis.

27.4 RUNX3 in Hypoxic Responses and Angiogenesis

RUNX3 is a well-established candidate tumor suppressor in human gastric cancer (Li et al. 2002). In addition, hemizygous deletion of *RUNX3* and resultant reduced expression of RUNX3 protein is common in other tumors, including bile duct, lung, and pancreatic cancer (Wada et al. 2004; Yanada et al. 2005). Although RUNX3 is silenced by promoter DNA hypermethylation in various cancer patient tissues and cell lines, it remains unknown whether the hypoxic tumor microenvironment regulates RUNX3. Lee et al. demonstrated that hypoxic status changes *RUNX3* expression at the transcriptional level not by DNA promoter hypermethylation, but instead by histone methylation and deacetylation mediated histone methyltransferase (HMT) G9a and HDAC1 (Lee et al. 2009), respectively (Fig. 27.4c; Table 27.2). HMT G9a is upregulated by hypoxia (Chen et al. 2006) and is associated with metastasis and poor prognosis of human cancers (Chen et al. 2010; Liu et al. 2015; Dong et al. 2012; Yuan et al. 2013). HDAC1 is also activated under hypoxic conditions (Kim et al. 2001), and histone deacetylation plays a major role in RUNX3 inactivation: acetylation of RUNX3 by p300 (Iwatani et al. 2010) and BRD (Lee et al. 2013b) is a key regulator of protein stability and cell-cycle arrest. Furthermore, HDAC inhibitors restore RUNX3 expression and tumor-suppressive function in cancer cells (Huang et al. 2007; Lee et al. 2009; Shio et al. 2011). Therefore, under hypoxic conditions, compounds that rescue epigenetic loss of RUNX expression could be utilized for prevention and treatment of cancer.

In the clinic, RUNX3 expression is inversely correlated with VEGF expression and microvascular density (MVD) status, suggesting that RUNX3 expression negatively regulates angiogenic phenotypes of human gastric cancer tissue (Peng et al. 2006). RUNX3 binds to putative RUNX3-binding sites of the *VEGF* promoter and directly suppresses its expression via transcriptional repression (Peng et al. 2006) (Fig. 27.3c; Table 27.1). Moreover, overexpression of RUNX3 reduces the mRNA expression level of von Willebrand factor (vWF), a regulator of angiogenesis, in human microvascular endothelial cells (Fu et al. 2011; Starke et al. 2011) (Table 27.1). In mice, following loss of *Runx3*, angiogenesis markers such as VEGF and vWF are remarkably elevated in developing liver at postnatal day 1. The level of CD31 in liver differs significantly between the *Runx3* knockout and wild-type mice. Thus, Runx3 may regulate angiogenesis during liver development, and in particular it may inhibit excessive angiogenesis (Lee et al. 2013a). Runx3 is also a critical regulator of normal lung development, including pulmonary vasculogenesis and angiogenesis (Lee et al. 2014a). RUNX3 overexpression significantly inhibits the expression and bioactivities of MMP-9 in renal cell carcinoma (Chen et al. 2013). RUNX3 suppresses metastasis by inducing an imbalance between MMP-2 and TIMP-2 in prostate cancer (Chen et al. 2014), as well as in gastric cancer (Chen et al. 2011). The inhibition of VEGF following restoration of RUNX3 greatly suppresses tumor angiogenesis by prostate cancer cells *in vitro* and *in vivo* (Chen et al. 2014), suggesting that RUNX3 reduces the availability of VEGF in the cancer microenvironment.

Like RUNX1, RUNX3 is prominently expressed in hematopoietic cells and different subsets of neurons (Marmigere et al. 2006; Le et al. 1999). Human CD34$^+$ HSCs and several hematopoietic cell lines, both normal and malignant, express RUNX3 (Le et al. 1999; Gomes et al. 2002). During mouse embryogenesis, blood cells in the liver and thymus express Runx3 in a pattern that overlaps with Runx1, suggesting cross-regulation or complementation by other Runx molecules (Levanon et al. 1994). Furthermore, RUNX3 is involved in myeloid differentiation through the retinoic acid receptor signaling pathway (Le et al. 1999), and it also plays a role in hematopoietic cells in zebrafish (Kalev-Zylinska et al. 2003). The involvement of RUNX3 in hematopoiesis suggest that it regulates aspects of the tumor microenvironment associated with inflammatory cells, such as macrophages, T cells, monocytes, and dendritic cells (Balkwill et al. 2012), as well as hematological malignancies, as described above in the section on Runx1.

The anti-angiogenic role of RUNX3 and hypoxia-induced silencing of RUNX3 raised the question of whether RUNX3 can control hypoxia-induced HIF-1α. Recent studies showed that RUNX3 decreases the half-life of HIF-1α, as well as its nuclear localization under hypoxia. Moreover, RUNX3 directly interacts with the C-terminal activation domain of HIF-1α and PHD2, promoting their interaction. Subsequently, it induces hydroxylation at prolines 402 and 564 in the ODDD, promoting the degradation of HIF-1α, suggesting that RUNX3 is essential for PHD2-mediated binding and hydroxylation of HIF-1α (Lee et al. 2014b). RUNX3 inhibits HIF-1α stability and downregulates HIF-1α transactivation activity and VEGF secretion under hypoxia (Fig. 27.4c). Furthermore, RUNX3 overexpression significantly inhibits hypoxia-induced angiogenesis, and siRNA against PHD2 restores the RUNX3-mediated inhibition of angiogenesis, suggesting that the interaction between RUNX3 and PHD2 is significant for the regulation of HIF-1α (Lee et al. 2014b) (Table 27.2). However, it remains unknown whether the interaction between PHD2 and HIF-1α occurs in the absence of RUNX3. If this interaction requires RUNX3, it would suggest a novel and critical role for RUNX3 in hypoxic responses, such as cancer progression, angiogenesis, stem cell maintenance, and ischemic diseases. Inflammation is a putative initiator of carcinogenesis (Lu et al. 2006), and inflammatory lesions are substantially hypoxic (Sitkovsky and Lukashev 2005). Some authors argue that

RUNX3 plays an important function in inflammatory cells, rather than acting as a tumor suppressor in epithelial cells (Lotem et al. 2015). However, regardless of the tumor suppressor function of RUNX3, hypoxia decreases RUNX3 expression and function, and RUNX3 facilitates degradation of oncogenic HIF-1α; therefore, RUNX3 must be important for regulation of active angiogenesis in tumor growth and inflammatory responses.

27.5 Perspectives

Vascular networks in adults that develop during physiological and pathological processes, including tumor vasculature, are formed by sprouting, intussusception of pre-existing vessels (angiogenesis), or incorporation of bone marrow – derived endothelial progenitors such as angioblasts or CACs (vasculogenesis) (Carmeliet and Jain 2000; Isner and Asahara 1999). Over the past few decades, a great deal of effort has been devoted to blocking VEGF signaling pathways, with the goal of developing a new approach for cancer treatment inhibiting angiogenesis, thereby starving cancer cells with nutrients and oxygen (Bridges and Harris 2011). The first clinically used angiogenic inhibitor in the US, the anti-VEGF-A antibody bevacizumab (Avastin®), was approved by the US FDA in 2003 for use in combination with standard chemotherapy against metastatic colon cancer. This antibody is also used to treat certain angiogenic diseases of the eye. However, in cancer patients, angiogenic therapy is challenging due to recurrent tumor growth and induction of resistance. Reducing the growth of tumor vessels induces hypoxia (Mehta et al. 2011; Yopp et al. 2011) and activates growth factor signaling and cytokines, such as hepatocyte growth factor receptor, c-MET, which in turn promote the growth of cancer cells and induce angiogenesis all over again (Gacche 2015). These results have frustrated many scientists and clinicians, leading them to seek alternative approaches to normalizing tumor blood vessels. As previously mentioned, because tumor vessels are leaky and chaotic, blood perfusion is reduced, significantly increasing hypoxia (Mehta et al. 2011; Yopp et al. 2011); consequently, chemotherapeutic agents are not efficiently delivered to the tumor. Early treatment with an anti-angiogenic therapy normalizes the tumor vessels and increases perfusion, resulting in increased delivery of chemotherapeutic agents and improved patient survival (Sorensen et al. 2012; Batchelor et al. 2013). Breakthroughs in anti-angiogenic therapy have highlighted the importance of research into cancer metabolism and ECs (McIntyre and Harris 2015). Anti-angiogenic therapy may be adopted based on the frequency of different responses, according to tissue types and metabolic profiles. Therefore, personalized therapeutic combinations, such as those that combine radiotherapy with chemotherapy in the scope of vascular normalization window, could be developed.

Another approach is to try to distinguish and characterize specific EC types in angiogenic processes. Tip cells, the cells that lead during migration, guide other cells via secretion of chemotactic molecules, whereas the stalk cells are the neighboring cells that follow and proliferate to make a hollow tube. Tip cell fate is induced by the VEGF–VEGFR2 signaling axis. Activation of the VEGFR2 signaling pathway in one cell induces Notch – Delta signaling that leads to Delta-like 4 (Dll4) binding to the Notch receptor in the neighboring cell, in which VEGFR2 and Dll4 expression is inhibited and VEGFR1 and Notch expression is induced to create a stalk cell fate (Eilken and Adams 2010; Hellstrom et al. 2007; Carmeliet et al. 2009). As tip cells produce filopodia that extend into the environment, they undergo chemotaxis toward angiogenic factors, or a hypoxic microenvironment. If tip cells are inactivated or eradicated, the active angiogenic process can be blocked at an early step without disturbing other types of ECs (Carmeliet et al. 2009). Conversely, if tip cell fate could be stimulated in a specialized area, organs or ischemic tissues could be specifically regenerated. Although RUNX family proteins all contain the conserved Runt homology domain, they have different expression patterns and function in tissue-specific and context-dependent ways. Therefore, RUNX family proteins might play distinctive roles in regulation of tip cell identity, or in the signaling pathways and metabolism of these cells, as well as in normalization of tumor vasculature. Therefore, manipulation of Runx family proteins in the hypoxic microenvironment represents a promising research direction and therapeutic approach for pathological angiogenic conditions such as cancer.

Acknowledgments This work was supported by NRF and KSEF grants funded by the Korean government (MSIP) (2012R1A4A1028835 and 2013R1A2A2A01068868).

References

Altieri, D. C. (2008). Survivin, cancer networks and pathway-directed drug discovery. *Nature Reviews Cancer, 8*(1), 61–70. doi:10.1038/nrc2293.

Bae, S. C., & Choi, J. K. (2004). Tumor suppressor activity of RUNX3. *Oncogene, 23*(24), 4336–4340. doi:10.1038/sj.onc.1207286.

Balkwill, F. R., Capasso, M., & Hagemann, T. (2012). The tumor microenvironment at a glance. *Journal of Cell Science, 125*(Pt 23), 5591–5596. doi:10.1242/jcs.116392.

Barbetti, V., Tusa, I., Cipolleschi, M. G., Rovida, E., & Dello Sbarba, P. (2013). AML1/ETO sensitizes via TRAIL acute myeloid leukemia cells to the pro-apoptotic effects of hypoxia. *Cell Death & Disease, 4*, e536. doi:10.1038/cddis.2013.49.

Barnes, G. L., Hebert, K. E., Kamal, M., Javed, A., Einhorn, T. A., Lian, J. B., et al. (2004). Fidelity of Runx2 activity in breast cancer cells is required for the generation of metastases-associated osteolytic disease. *Cancer Research, 64*(13), 4506–4513. doi:10.1158/0008-5472.can-03-3851.

Batchelor, T. T., Gerstner, E. R., Emblem, K. E., Duda, D. G., Kalpathy-Cramer, J., Snuderl, M., et al. (2013). Improved tumor oxygenation and survival in glioblastoma patients who show increased blood perfusion after cediranib and chemoradiation. *Proceedings of the National Academy of Sciences of the United States of America, 110*(47), 19059–19064. doi:10.1073/pnas.1318022110.

Bergers, G., & Benjamin, L. E. (2003). Tumorigenesis and the angiogenic switch. *Nature Reviews. Cancer, 3*(6), 401–410. doi:10.1038/nrc1093.

Bosch-Marce, M., Okuyama, H., Wesley, J. B., Sarkar, K., Kimura, H., Liu, Y. V., et al. (2007). Effects of aging and hypoxia-inducible factor-1 activity on angiogenic cell mobilization and recovery of perfusion after limb ischemia. *Circulation Research, 101*(12), 1310–1318. doi:10.1161/CIRCRESAHA.107.153346.

Bridges, E. M., & Harris, A. L. (2011). The angiogenic process as a therapeutic target in cancer. *Biochemical Pharmacology, 81*(10), 1183–1191. doi:10.1016/j.bcp.2011.02.016.

Bronckers, A. L., Sasaguri, K., Cavender, A. C., D'Souza, R. N., & Engelse, M. A. (2005). Expression of Runx2/Cbfa1/Pebp2alphaA during angiogenesis in postnatal rodent and fetal human orofacial tissues. *Journal of Bone and Mineral Research: the Official Journal of the American Society for Bone and Mineral Research, 20*(3), 428–437. doi:10.1359/JBMR.041118.

Brubaker, K. D., Vessella, R. L., Brown, L. G., & Corey, E. (2003). Prostate cancer expression of runt-domain transcription factor Runx2, a key regulator of osteoblast differentiation and function. *The Prostate, 56*(1), 13–22. doi:10.1002/pros.10233.

Cai, X., Gao, L., Teng, L., Ge, J., Oo, Z. M., Kumar, A. R., et al. (2015). Runx1 deficiency decreases ribosome biogenesis and confers stress resistance to hematopoietic stem and progenitor cells. *Cell Stem Cell, 17*(2), 165–177. doi:10.1016/j.stem.2015.06.002.

Carmeliet, P. (2003). Blood vessels and nerves: Common signals, pathways and diseases. *Nature Reviews. Genetics, 4*(9), 710–720. doi:10.1038/nrg1158.

Carmeliet, P., & Jain, R. K. (2000). Angiogenesis in cancer and other diseases. *Nature, 407*(6801), 249–257. doi:10.1038/35025220.

Carmeliet, P., De Smet, F., Loges, S., & Mazzone, M. (2009). Branching morphogenesis and antiangiogenesis candidates: Tip cells lead the way. *Nature Reviews. Clinical Oncology, 6*(6), 315–326. doi:10.1038/nrclinonc.2009.64.

Chen, H., Yan, Y., Davidson, T. L., Shinkai, Y., & Costa, M. (2006). Hypoxic stress induces dimethylated histone H3 lysine 9 through histone methyltransferase G9a in mammalian cells. *Cancer Research, 66*(18), 9009–9016. doi:10.1158/0008-5472.CAN-06-0101.

Chen, M. J., Yokomizo, T., Zeigler, B. M., Dzierzak, E., & Speck, N. A. (2009). Runx1 is required for the endothelial to haematopoietic cell transition but not thereafter. *Nature, 457*(7231), 887–891. doi:10.1038/nature07619.

Chen, M. W., Hua, K. T., Kao, H. J., Chi, C. C., Wei, L. H., Johansson, G., et al. (2010). H3K9 histone methyltransferase G9a promotes lung cancer invasion and metastasis by silencing the cell adhesion molecule Ep-CAM. *Cancer Research, 70*(20), 7830–7840. doi:10.1158/0008-5472.CAN-10-0833.

Chen, Y., Wei, X., Guo, C., Jin, H., Han, Z., Han, Y., et al. (2011). Runx3 suppresses gastric cancer metastasis through inactivation of MMP9 by upregulation of TIMP-1. *International Journal of Cancer Journal international du cancer, 129*(7), 1586–1598. doi:10.1002/ijc.25831.

Chen, F., Bai, J., Li, W., Mei, P., Liu, H., Li, L., et al. (2013). RUNX3 suppresses migration, invasion and angiogenesis of human renal cell carcinoma. *PloS One, 8*(2), e56241. doi:10.1371/journal.pone.0056241.

Chen, F., Wang, M., Bai, J., Liu, Q., Xi, Y., Li, W., & Zheng, J. (2014). Role of RUNX3 in suppressing metastasis and angiogenesis of human prostate cancer. *PloS One, 9*(1), e86917. doi:10.1371/journal.pone.0086917.

Chimge, N. O., & Frenkel, B. (2013). The RUNX family in breast cancer: Relationships with estrogen signaling. *Oncogene, 32*(17), 2121–2130. doi:10.1038/onc.2012.328.

Delafontaine, P., Song, Y. H., & Li, Y. (2004). Expression, regulation, and function of IGF-1, IGF-1R, and IGF-1 binding proteins in blood vessels. *Arteriosclerosis, Thrombosis, and Vascular Biology, 24*(3), 435–444. doi:10.1161/01.ATV.0000105902.89459.09.

Descalzi Cancedda, F., Melchiori, A., Benelli, R., Gentili, C., Masiello, L., Campanile, G., et al. (1995). Production of angiogenesis inhibitors and stimulators is modulated by cultured growth plate chondrocytes during in vitro differentiation: Dependence on extracellular matrix assembly. *European Journal of Cell Biology, 66*(1), 60–68.

Dong, C., Wu, Y., Yao, J., Wang, Y., Yu, Y., Rychahou, P. G., et al. (2012). G9a interacts with Snail and is critical for Snail-mediated E-cadherin repression in human breast cancer. *The Journal of Clinical Investigation, 122*(4), 1469–1486. doi:10.1172/JCI57349.

Ducy, P., Zhang, R., Geoffroy, V., Ridall, A. L., & Karsenty, G. (1997). Osf2/Cbfa1: A transcriptional activator of osteoblast differentiation. *Cell, 89*(5), 747–754.

Dunwoodie, S. L. (2009). The role of hypoxia in development of the Mammalian embryo. *Developmental Cell, 17*(6), 755–773. doi:10.1016/j.devcel.2009.11.008.

Eilken, H. M., & Adams, R. H. (2010). Dynamics of endothelial cell behavior in sprouting angiogenesis. *Current Opinion in Cell Biology, 22*(5), 617–625. doi:10.1016/j.ceb.2010.08.010.

Eklund, L., & Saharinen, P. (2013). Angiopoietin signaling in the vasculature. *Experimental Cell Research, 319*(9), 1271–1280. doi:10.1016/j.yexcr.2013.03.011.

Endo, T., Ohta, K., & Kobayashi, T. (2008). Expression and function of Cbfa-1/Runx2 in thyroid papillary carcinoma cells. *The Journal of Clinical Endocrinology and Metabolism, 93*(6), 2409–2412. doi:10.1210/jc.2007-2805.

Erlebacher, A., Filvaroff, E. H., Gitelman, S. E., & Derynck, R. (1995). Toward a molecular understanding of skeletal development. *Cell, 80*(3), 371–378.

Fagiani, E., & Christofori, G. (2013). Angiopoietins in angiogenesis. *Cancer Letters, 328*(1), 18–26. doi:10.1016/j.canlet.2012.08.018.

Ferrara, N., Carver-Moore, K., Chen, H., Dowd, M., Lu, L., O'Shea, K. S., et al. (1996). Heterozygous embryonic lethality induced by targeted inactivation of the VEGF gene. *Nature, 380*(6573), 439–442. doi:10.1038/380439a0.

Ferrara, N., Gerber, H. P., & LeCouter, J. (2003). The biology of VEGF and its receptors. *Nature Medicine, 9*(6), 669–676. doi:10.1038/nm0603-669.

Folkman, J. (1997). Angiogenesis and angiogenesis inhibition: An overview. *EXS, 79*, 1–8.

Friedman, A. D. (2002). Runx1, c-Myb, and C/EBPalpha couple differentiation to proliferation or growth arrest during hematopoiesis. *Journal of Cellular Biochemistry, 86*(4), 624–629. doi:10.1002/jcb.10271.

Fu, Y., Chang, A. C., Fournier, M., Chang, L., Niessen, K., & Karsan, A. (2011). RUNX3 maintains the mesenchymal phenotype after termination of the Notch signal. *The Journal of Biological Chemistry, 286*(13), 11803–11813. doi:10.1074/jbc.M111.222331.

Gacche, R. N. (2015). Compensatory angiogenesis and tumor refractoriness. *Oncogenesis, 4*, e153. doi:10.1038/oncsis.2015.14.

Gao, X. N., Yan, F., Lin, J., Gao, L., Lu, X. L., Wei, S. C., et al. (2015). AML1/ETO cooperates with HIF1alpha to promote leukemogenesis through DNMT3a transactivation. *Leukemia, 29*(8), 1730–1740. doi:10.1038/leu.2015.56.

Gerber, H. P., McMurtrey, A., Kowalski, J., Yan, M., Keyt, B. A., Dixit, V., & Ferrara, N. (1998). Vascular endothelial growth factor regulates endothelial cell survival through the phosphatidylinositol 3′-kinase/Akt signal transduction pathway. Requirement for Flk-1/KDR activation. *The Journal of Biological Chemistry, 273*(46), 30336–30343.

Gerber, H. P., Vu, T. H., Ryan, A. M., Kowalski, J., Werb, Z., & Ferrara, N. (1999). VEGF couples hypertrophic cartilage remodeling, ossification and angiogenesis during endochondral bone formation. *Nature Medicine, 5*(6), 623–628. doi:10.1038/9467.

Gerhardt, H., Golding, M., Fruttiger, M., Ruhrberg, C., Lundkvist, A., Abramsson, A., et al. (2003). VEGF guides angiogenic sprouting utilizing endothelial tip cell filopodia. *The Journal of Cell Biology, 161*(6), 1163–1177. doi:10.1083/jcb.200302047.

Giaccia, A. J., Simon, M. C., & Johnson, R. (2004). The biology of hypoxia: The role of oxygen sensing in development, normal function, and disease. *Genes & Development, 18*(18), 2183–2194. doi:10.1101/gad.1243304.

Gomes, I., Sharma, T. T., Edassery, S., Fulton, N., Mar, B. G., & Westbrook, C. A. (2002). Novel transcription factors in human CD34 antigen-positive hematopoietic cells. *Blood, 100*(1), 107–119.

Harris, A. L. (2002). Hypoxia – a key regulatory factor in tumour growth. *Nature Reviews. Cancer, 2*(1), 38–47. doi:10.1038/nrc704.

Hellstrom, M., Phng, L. K., Hofmann, J. J., Wallgard, E., Coultas, L., Lindblom, P., et al. (2007). Dll4 signalling through Notch1 regulates formation of tip cells during angiogenesis. *Nature, 445*(7129), 776–780. doi:10.1038/nature05571.

Helmlinger, G., Yuan, F., Dellian, M., & Jain, R. K. (1997). Interstitial pH and pO2 gradients in solid tumors in vivo: High-resolution measurements reveal a lack of correlation. *Nature Medicine, 3*(2), 177–182.

Hirschi, K. K., Ingram, D. A., & Yoder, M. C. (2008). Assessing identity, phenotype, and fate of endothelial progenitor cells. *Arteriosclerosis, Thrombosis, and Vascular Biology, 28*(9), 1584–1595. doi:10.1161/ATVBAHA.107.155960.

Hockel, M., Schlenger, K., Aral, B., Mitze, M., Schaffer, U., & Vaupel, P. (1996). Association between tumor hypoxia and malignant progression in advanced cancer of the uterine cervix. *Cancer Research, 56*(19), 4509–4515.

Huang, Y., Du, K. M., Xue, Z. H., Yan, H., Li, D., Liu, W., et al. (2003). Cobalt chloride and low oxygen tension trigger differentiation of acute myeloid leukemic cells: Possible mediation of hypoxia-inducible factor-1alpha. *Leukemia, 17*(11), 2065–2073. doi:10.1038/sj.leu.2403141.

Huang, C., Ida, H., Ito, K., Zhang, H., & Ito, Y. (2007). Contribution of reactivated RUNX3 to inhibition of gastric cancer cell growth following suberoylanilide hydroxamic acid (vorinostat) treatment. *Biochemical Pharmacology, 73*(7), 990–1000. doi:10.1016/j.bcp.2006.12.013.

Isner, J. M., & Asahara, T. (1999). Angiogenesis and vasculogenesis as therapeutic strategies for postnatal neovascularization. *The Journal of Clinical Investigation, 103*(9), 1231–1236. doi:10.1172/JCI6889.

Ito, Y. (1999). Molecular basis of tissue-specific gene expression mediated by the runt domain transcription factor PEBP2/CBF. *Genes to Cells: Devoted to Molecular & Cellular Mechanisms, 4*(12), 685–696.

Ito, Y. (2004). Oncogenic potential of the RUNX gene family: 'Overview'. *Oncogene, 23*(24), 4198–4208. doi:10.1038/sj.onc.1207755.

Iwatani, K., Fujimoto, T., & Ito, T. (2010). Cyclin D1 blocks the anti-proliferative function of RUNX3 by interfering with RUNX3-p300 interaction. *Biochemical and Biophysical Research Communications, 400*(3), 426–431. doi:10.1016/j.bbrc.2010.08.094.

Iwatsuki, K., Tanaka, K., Kaneko, T., Kazama, R., Okamoto, S., Nakayama, Y., et al. (2005). Runx1 promotes angiogenesis by downregulation of insulin-like growth factor-binding protein-3. *Oncogene, 24*(7), 1129–1137. doi:10.1038/sj.onc.1208287.

Iyama, K., Ninomiya, Y., Olsen, B. R., Linsenmayer, T. F., Trelstad, R. L., & Hayashi, M. (1991). Spatiotemporal pattern of type X collagen gene expression and collagen deposition in embryonic chick vertebrae undergoing endochondral ossification. *The Anatomical Record, 229*(4), 462–472. doi:10.1002/ar.1092290405.

Jain, R. K. (2005). Normalization of tumor vasculature: An emerging concept in antiangiogenic therapy. *Science, 307*(5706), 58–62. doi:10.1126/science.1104819.

Jiang, Y., Xue, Z. H., Shen, W. Z., Du, K. M., Yan, H., Yu, Y., et al. (2005). Desferrioxamine induces leukemic cell differentiation potentially by hypoxia-inducible factor-1 alpha that augments transcriptional activity of CCAAT/enhancer-binding protein-alpha. *Leukemia, 19*(7), 1239–1247. doi:10.1038/sj.leu.2403734.

Kalev-Zylinska, M. L., Horsfield, J. A., Flores, M. V., Postlethwait, J. H., Vitas, M. R., Baas, A. M., et al. (2002). Runx1 is required for zebrafish blood and vessel development and expression of a human RUNX1-CBF2T1 transgene advances a model for studies of leukemogenesis. *Development (Cambridge, England), 129*(8), 2015–2030.

Kalev-Zylinska, M. L., Horsfield, J. A., Flores, M. V., Postlethwait, J. H., Chau, J. Y., Cattin, P. M., et al. (2003). Runx3 is required for hematopoietic development in zebrafish. *Developmental Dynamics: An Official Publication of the American Association of the Anatomists, 228*(3), 323–336. doi:10.1002/dvdy.10388.

Kayed, H., Jiang, X., Keleg, S., Jesnowski, R., Giese, T., Berger, M. R., et al. (2007). Regulation and functional role of the Runt-related transcription factor-2 in pancreatic cancer. *British Journal of Cancer, 97*(8), 1106–1115. doi:10.1038/sj.bjc.6603984.

Kim, M. S., Kwon, H. J., Lee, Y. M., Baek, J. H., Jang, J. E., Lee, S. W., et al. (2001). Histone deacetylases induce angiogenesis by negative regulation of tumor suppressor genes. *Nature Medicine, 7*(4), 437–443. doi:10.1038/86507.

Kimura, A., Inose, H., Yano, F., Fujita, K., Ikeda, T., Sato, S., et al. (2010). Runx1 and Runx2 cooperate during sternal morphogenesis. *Development (Cambridge, England), 137*(7), 1159–1167. doi:10.1242/dev.045005.

Komori, T., Yagi, H., Nomura, S., Yamaguchi, A., Sasaki, K., Deguchi, K., et al. (1997). Targeted disruption of Cbfa1 results in a complete lack of bone formation owing to maturational arrest of osteoblasts. *Cell, 89*(5), 755–764.

Kuettner, K. E., & Pauli, B. U. (1983). Inhibition of neovascularization by a cartilage factor. *Ciba Foundation Symposium, 100*, 163–173.

Kurokawa, M., & Hirai, H. (2003). Role of AML1/Runx1 in the pathogenesis of hematological malignancies. *Cancer Science, 94*(10), 841–846.

Kwon, T. G., Zhao, X., Yang, Q., Li, Y., Ge, C., Zhao, G., & Franceschi, R. T. (2011). Physical and functional interactions between Runx2 and HIF-1alpha induce vascular endothelial growth factor gene expression. *Journal of Cellular Biochemistry, 112*(12), 3582–3593. doi:10.1002/jcb.23289.

Lane, S. W., Williams, D. A., & Watt, F. M. (2014). Modulating the stem cell niche for tissue regeneration. *Nature Biotechnology, 32*(8), 795–803. doi:10.1038/nbt.2978.

Le, X. F., Groner, Y., Kornblau, S. M., Gu, Y., Hittelman, W. N., Levanon, D., et al. (1999). Regulation of AML2/CBFA3 in hematopoietic cells through the retinoic acid receptor alpha-dependent signaling pathway. *The Journal of Biological Chemistry, 274*(31), 21651–21658.

Lee, Y. M., Jeong, C. H., Koo, S. Y., Son, M. J., Song, H. S., Bae, S. K., et al. (2001). Determination of hypoxic region by hypoxia marker in developing mouse embryos in vivo: A possible signal for vessel development. *Developmental Dynamics, 220*(2), 175–186. doi:10.1002/1097-0177(20010201)220:2<175::AID-DVDY1101>3.0.CO;2-F.

Lee, J. W., Bae, S. H., Jeong, J. W., Kim, S. H., & Kim, K. W. (2004). Hypoxia-inducible factor (HIF-1)alpha: Its protein stability and biological functions. *Experimental & Molecular Medicine, 36*(1), 1–12. doi:10.1038/emm.2004.1.

Lee, S. H., Kim, J., Kim, W. H., & Lee, Y. M. (2009). Hypoxic silencing of tumor suppressor RUNX3 by histone modification in gastric cancer cells. *Oncogene, 28*(2), 184–194. doi:10.1038/onc.2008.377.

Lee, S. H., Che, X., Jeong, J. H., Choi, J. Y., Lee, Y. J., Lee, Y. H., et al. (2012). Runx2 protein stabilizes hypoxia-inducible factor-1alpha through competition with von Hippel-Lindau protein (pVHL) and stimulates angiogenesis in growth plate hypertrophic chondrocytes. *The Journal of Biological Chemistry, 287*(18), 14760–14771. doi:10.1074/jbc.M112.340232.

Lee, J. M., Lee, D. J., Bae, S. C., & Jung, H. S. (2013a). Abnormal liver differentiation and excessive angiogenesis in mice lacking Runx3. *Histochemistry and Cell Biology, 139*(5), 751–758. doi:10.1007/s00418-013-1077-x.

Lee, Y. S., Lee, J. W., Jang, J. W., Chi, X. Z., Kim, J. H., Li, Y. H., et al. (2013b). Runx3 inactivation is a crucial early event in the development of lung adenocarcinoma. *Cancer Cell, 24*(5), 603–616. doi:10.1016/j.ccr.2013.10.003.

Lee, J. M., Kwon, H. J., Lai, W. F., & Jung, H. S. (2014a). Requirement of Runx3 in pulmonary vasculogenesis. *Cell and Tissue Research, 356*(2), 445–449. doi:10.1007/s00441-014-1816-x.

Lee, S. H., Bae, S. C., Kim, K. W., & Lee, Y. M. (2014b). RUNX3 inhibits hypoxia-inducible factor-1alpha protein stability by interacting with prolyl hydroxylases in gastric cancer cells. *Oncogene, 33*(11), 1458–1467. doi:10.1038/onc.2013.76.

Levanon, D., Negreanu, V., Bernstein, Y., Bar-Am, I., Avivi, L., & Groner, Y. (1994). AML1, AML2, and AML3, the human members of the runt domain gene-family: cDNA structure, expression, and chromosomal localization. *Genomics, 23*(2), 425–432. doi:10.1006/geno.1994.1519.

Levanon, D., Bettoun, D., Harris-Cerruti, C., Woolf, E., Negreanu, V., Eilam, R., et al. (2002). The Runx3 transcription factor regulates development and survival of TrkC dorsal root ganglia neurons. *The EMBO Journal, 21*(13), 3454–3463. doi:10.1093/emboj/cdf370.

Li, Q. L., Ito, K., Sakakura, C., Fukamachi, H., Inoue, K., Chi, X. Z., et al. (2002). Causal relationship between the loss of RUNX3 expression and gastric cancer. *Cell, 109*(1), 113–124.

Lian, J. B., Balint, E., Javed, A., Drissi, H., Vitti, R., Quinlan, E. J., et al. (2003). Runx1/AML1 hematopoietic transcription factor contributes to skeletal development in vivo. *Journal of Cellular Physiology, 196*(2), 301–311. doi:10.1002/jcp.10316.

Lim, M., Zhong, C., Yang, S., Bell, A. M., Cohen, M. B., & Roy-Burman, P. (2010). Runx2 regulates survivin expression in prostate cancer cells. *Laboratory Investigation; A Journal of Technical Methods and Pathology, 90*(2), 222–233. doi:10.1038/labinvest.2009.128.

Linggi, B., Muller-Tidow, C., van de Locht, L., Hu, M., Nip, J., Serve, H., et al. (2002). The t(8;21) fusion protein, AML1 ETO, specifically represses the transcription of the p14(ARF) tumor suppressor in acute myeloid leukemia. *Nature Medicine, 8*(7), 743–750. doi:10.1038/nm726.

Liu, S., Ye, D., Guo, W., Yu, W., He, Y., Hu, J., et al. (2015). G9a is essential for EMT-mediated metastasis and maintenance of cancer stem cell-like characters in head and neck squamous cell carcinoma. *Oncotarget, 6*(9), 6887–6901. doi:10.18632/oncotarget.3159.

Look, A. T. (1997). Oncogenic transcription factors in the human acute leukemias. *Science (New York, N.Y.), 278*(5340), 1059–1064.

Lotem, J., Levanon, D., Negreanu, V., Bauer, O., Hantisteanu, S., Dicken, J., & Groner, Y. (2015). Runx3 at the interface of immunity, inflammation and cancer. *Biochimica et Biophysica Acta, 1855*(2), 131–143. doi:10.1016/j.bbcan.2015.01.004.

Lu, H., Ouyang, W., & Huang, C. (2006). Inflammation, a key event in cancer development. *Molecular Cancer Research: MCR, 4*(4), 221–233. doi:10.1158/1541-7786.MCR-05-0261.

Maes, C., Carmeliet, G., & Schipani, E. (2012). Hypoxia-driven pathways in bone development, regeneration and disease. *Nature Reviews. Rheumatology, 8*(6), 358–366. doi:10.1038/nrrheum.2012.36.

Majmundar, A. J., Wong, W. J., & Simon, M. C. (2010). Hypoxia-inducible factors and the response to hypoxic stress. *Molecular Cell, 40*(2), 294–309. doi:10.1016/j.molcel.2010.09.022.

Marmigere, F., Montelius, A., Wegner, M., Groner, Y., Reichardt, L. F., & Ernfors, P. (2006). The Runx1/AML1 transcription factor selectively regulates development and survival of TrkA nociceptive sensory neurons. *Nature Neuroscience, 9*(2), 180–187. doi:10.1038/nn1631.

McIntyre, A., & Harris, A. L. (2015). Metabolic and hypoxic adaptation to anti-angiogenic therapy: a target for induced essentiality. *EMBO Molecular Medicine, 7*(4), 368–379. doi:10.15252/emmm.201404271.

Mehta, S., Hughes, N. P., Buffa, F. M., Li, S. P., Adams, R. F., Adwani, A., et al. (2011). Assessing early therapeutic response to bevacizumab in primary breast cancer using magnetic resonance imaging and gene expression profiles. *Journal of the National Cancer Institute. Monographs, 2011*(43), 71–74. doi:10.1093/jncimonographs/lgr027.

Mendoza-Villanueva, D., Deng, W., Lopez-Camacho, C., & Shore, P. (2010). The Runx transcriptional co-activator, CBFbeta, is essential for invasion of breast cancer cells. *Molecular Cancer, 9*, 171. doi:10.1186/1476-4598-9-171.

Mierke, C. T. (2013). The role of focal adhesion kinase in the regulation of cellular mechanical properties. *Physical Biology, 10*(6), 065005. doi:10.1088/1478-3975/10/6/065005.

Miyoshi, H., Shimizu, K., Kozu, T., Maseki, N., Kaneko, Y., & Ohki, M. (1991). t(8;21) breakpoints on chromosome 21 in acute myeloid leukemia are clustered within a limited region of a single gene, AML1. *Proceedings of the National Academy of Sciences of the United States of America, 88*(23), 10431–10434.

Murray, P. D. F. (1932). The development in vitro of the blood of the eearly chick embryo. *Proceedings of the Royal Society of London Series B, Containing Papers of a Biological Character, 111*(773), 497–521.

Namba, K., Abe, M., Saito, S., Satake, M., Ohmoto, T., Watanabe, T., & Sato, Y. (2000). Indispensable role of the transcription factor PEBP2/CBF in angiogenic activity of a murine endothelial cell MSS31. *Oncogene, 19*(1), 106–114. doi:10.1038/sj.onc.1203257.

Ninomiya, Y., Gordon, M., van der Rest, M., Schmid, T., Linsenmayer, T., & Olsen, B. R. (1986). The developmentally regulated type X collagen gene contains a long open reading frame without introns. *The Journal of Biological Chemistry, 261*(11), 5041–5050.

Nishikawa, S. I., Nishikawa, S., Kawamoto, H., Yoshida, H., Kizumoto, M., Kataoka, H., & Katsura, Y. (1998). In vitro generation of lymphohematopoietic cells from endothelial cells purified from murine embryos. *Immunity, 8*(6), 761–769.

North, T., Gu, T. L., Stacy, T., Wang, Q., Howard, L., Binder, M., et al. (1999). Cbfa2 is required for the formation of intra-aortic hematopoietic clusters. *Development (Cambridge, England), 126*(11), 2563–2575.

Okuda, T., van Deursen, J., Hiebert, S. W., Grosveld, G., & Downing, J. R. (1996). AML1, the target of multiple chromosomal translocations in human leukemia, is essential for normal fetal liver hematopoiesis. *Cell, 84*(2), 321–330.

Otto, F., Thornell, A. P., Crompton, T., Denzel, A., Gilmour, K. C., Rosewell, I. R., et al. (1997). Cbfa1, a candidate gene for cleidocranial dysplasia syndrome, is essential for osteoblast differentiation and bone development. *Cell, 89*(5), 765–771.

Parmar, K., Mauch, P., Vergilio, J. A., Sackstein, R., & Down, J. D. (2007). Distribution of hematopoietic stem cells in the bone marrow according to regional hypoxia. *Proceedings of the National Academy of Sciences of the United States of America, 104*(13), 5431–5436. doi:10.1073/pnas.0701152104.

Peng, Z., Wei, D., Wang, L., Tang, H., Zhang, J., Le, X., et al. (2006). RUNX3 inhibits the expression of vascular endothelial growth factor and reduces the angiogenesis, growth, and metastasis of human gastric cancer. *Clinical Cancer Research: An Official Journal of the American Association for Cancer Research, 12*(21), 6386–6394. doi:10.1158/1078-0432.CCR-05-2359.

Peng, Z. G., Zhou, M. Y., Huang, Y., Qiu, J. H., Wang, L. S., Liao, S. H., et al. (2008). Physical and functional interaction of Runt-related protein 1 with hypoxia-inducible factor-1alpha. *Oncogene, 27*(6), 839–847. doi:10.1038/sj.onc.1210676.

Pratap, J., Javed, A., Languino, L. R., van Wijnen, A. J., Stein, J. L., Stein, G. S., & Lian, J. B. (2005). The Runx2 osteogenic transcription factor regulates matrix metalloproteinase 9 in bone metastatic cancer cells and controls cell invasion. *Molecular and Cellular Biology, 25*(19), 8581–8591. doi:10.1128/mcb.25.19.8581-8591.2005.

Pratap, J., Lian, J. B., Javed, A., Barnes, G. L., van Wijnen, A. J., Stein, J. L., & Stein, G. S. (2006). Regulatory roles of Runx2 in metastatic tumor and cancer cell interactions with bone. *Cancer Metastasis Reviews, 25*(4), 589–600. doi:10.1007/s10555-006-9032-0.

Pratap, J., Wixted, J. J., Gaur, T., Zaidi, S. K., Dobson, J., Gokul, K. D., et al. (2008). Runx2 transcriptional activation of Indian Hedgehog and a downstream bone metastatic pathway in breast cancer cells. *Cancer Research, 68*(19), 7795–7802. doi:10.1158/0008-5472.can-08-1078.

Rey, S., & Semenza, G. L. (2010). Hypoxia-inducible factor-1-dependent mechanisms of vascularization and vascular remodelling. *Cardiovascular Research, 86*(2), 236–242. doi:10.1093/cvr/cvq045.

Sabin, F. R. (1920). Studies on the origin of blood vessels and of red corpuscles as seen in the living blastoderm of the chick during the second day of incubation. *Contributions to Embryology, 9*, 213–262.

Schipani, E., Ryan, H. E., Didrickson, S., Kobayashi, T., Knight, M., & Johnson, R. S. (2001). Hypoxia in cartilage: HIF-1alpha is essential for chondrocyte growth arrest and survival. *Genes & Development, 15*(21), 2865–2876. doi:10.1101/gad.934301.

Semenza, G. L. (2001). Hypoxia-inducible factor 1: Oxygen homeostasis and disease pathophysiology. *Trends in Molecular Medicine, 7*(8), 345–350.

Semenza, G. L. (2003). Targeting HIF-1 for cancer therapy. *Nature Reviews. Cancer, 3*(10), 721–732. doi:10.1038/nrc1187.

Semenza, G. L. (2010). Defining the role of hypoxia-inducible factor 1 in cancer biology and therapeutics. *Oncogene, 29*(5), 625–634. doi:10.1038/onc.2009.441.

Semenza, G. L. (2014). Oxygen sensing, hypoxia-inducible factors, and disease pathophysiology. *Annual Review of Pathology, 9*, 47–71. doi:10.1146/annurev-pathol-012513-104720.

Senger, D. R., Galli, S. J., Dvorak, A. M., Perruzzi, C. A., Harvey, V. S., & Dvorak, H. F. (1983). Tumor cells secrete a vascular permeability factor that promotes accumulation of ascites fluid. *Science, 219*(4587), 983–985.

Shi, Q., Le, X., Abbruzzese, J. L., Peng, Z., Qian, C. N., Tang, H., et al. (2001). Constitutive Sp1 activity is essential for differential constitutive expression of vascular endothelial growth factor in human pancreatic adenocarcinoma. *Cancer Research, 61*(10), 4143–4154.

Shio, S., Kodama, Y., Ida, H., Shiokawa, M., Kitamura, K., Hatano, E., et al. (2011). Loss of RUNX3 expression by histone deacetylation is associated with biliary tract carcinogenesis. *Cancer Science, 102*(4), 776–783. doi:10.1111/j.1349-7006.2011.01848.x.

Siewert, J. R., Bottcher, K., Stein, H. J., & Roder, J. D. (1998). Relevant prognostic factors in gastric cancer: Ten-year results of the German gastric cancer study. *Annals of Surgery, 228*(4), 449–461.

Simon, M. C., & Keith, B. (2008). The role of oxygen availability in embryonic development and stem cell function. *Nature Reviews. Molecular Cell Biology, 9*(4), 285–296. doi:10.1038/nrm2354.

Sitkovsky, M., & Lukashev, D. (2005). Regulation of immune cells by local-tissue oxygen tension: HIF1 alpha and adenosine receptors. *Nature Reviews Immunology, 5*(9), 712–721. doi:10.1038/nri1685.

Smith, N., Dong, Y., Lian, J. B., Pratap, J., Kingsley, P. D., van Wijnen, A. J., et al. (2005). Overlapping expression of Runx1(Cbfa2) and Runx2(Cbfa1) transcription factors supports cooperative induction of skeletal development. *Journal of Cellular Physiology, 203*(1), 133–143. doi:10.1002/jcp.20210.

Sorensen, A. G., Emblem, K. E., Polaskova, P., Jennings, D., Kim, H., Ancukiewicz, M., et al. (2012). Increased survival of glioblastoma patients who respond to antiangiogenic therapy with elevated blood perfusion. *Cancer Research, 72*(2), 402–407. doi:10.1158/0008-5472.CAN-11-2464.

Speck, N. A., & Gilliland, D. G. (2002). Core-binding factors in haematopoiesis and leukaemia. *Nature Reviews Cancer, 2*(7), 502–513. doi:10.1038/nrc840.

Starke, R. D., Ferraro, F., Paschalaki, K. E., Dryden, N. H., McKinnon, T. A., Sutton, R. E., et al. (2011). Endothelial von Willebrand factor regulates angiogenesis. *Blood, 117*(3), 1071–1080. doi:10.1182/blood-2010-01-264507.

Stricker, S., Fundele, R., Vortkamp, A., & Mundlos, S. (2002). Role of Runx genes in chondrocyte differentiation. *Developmental Biology, 245*(1), 95–108. doi:10.1006/dbio.2002.0640.

Suda, T., & Takakura, N. (2001). Role of hematopoietic stem cells in angiogenesis. *International Journal of Hematology, 74*(3), 266–271.

Suehiro, J., Hamakubo, T., Kodama, T., Aird, W. C., & Minami, T. (2010). Vascular endothelial growth factor activation of endothelial cells is mediated by early growth response-3. *Blood, 115*(12), 2520–2532. doi:10.1182/blood-2009-07-233478.

Sun, L., Vitolo, M., & Passaniti, A. (2001). Runt-related gene 2 in endothelial cells: Inducible expression and specific regulation of cell migration and invasion. *Cancer Research, 61*(13), 4994–5001.

Sun, L., Vitolo, M. I., Qiao, M., Anglin, I. E., & Passaniti, A. (2004). Regulation of TGFbeta1-mediated growth inhibition and apoptosis by RUNX2 isoforms in endothelial cells. *Oncogene, 23*(27), 4722–4734. doi:10.1038/sj.onc.1207589.

Sun, X., Wei, L., Chen, Q., & Terek, R. M. (2009). HDAC4 represses vascular endothelial growth factor expression in chondrosarcoma by modulating RUNX2 activity. *The Journal of Biological Chemistry, 284*(33), 21881–21890. doi:10.1074/jbc.M109.019091.

Takakura, N., Watanabe, T., Suenobu, S., Yamada, Y., Noda, T., Ito, Y., et al. (2000). A role for hematopoietic stem cells in promoting angiogenesis. *Cell, 102*(2), 199–209.

Tang, J. L., Hou, H. A., Chen, C. Y., Liu, C. Y., Chou, W. C., Tseng, M. H., et al. (2009). AML1/RUNX1 mutations in 470 adult patients with de novo acute myeloid leukemia: Prognostic implication and interaction with other gene alterations. *Blood, 114*(26), 5352–5361. doi:10.1182/blood-2009-05-223784.

Tenen, D. G., Hromas, R., Licht, J. D., & Zhang, D. E. (1997). Transcription factors, normal myeloid development, and leukemia. *Blood, 90*(2), 489–519.

Ter Elst, A., Ma, B., Scherpen, F. J., de Jonge, H. J., Douwes, J., Wierenga, A. T., et al. (2011). Repression of vascular endothelial growth factor expression by the runt-related transcription factor 1 in acute myeloid leukemia. *Cancer Research, 71*(7), 2761–2771. doi:10.1158/0008-5472.CAN-10-0402.

Vega, R. B., Matsuda, K., Oh, J., Barbosa, A. C., Yang, X., Meadows, E., et al. (2004). Histone deacetylase 4 controls chondrocyte hypertrophy during skeletogenesis. *Cell, 119*(4), 555–566. doi:10.1016/j.cell.2004.10.024.

Wada, M., Yazumi, S., Takaishi, S., Hasegawa, K., Sawada, M., Tanaka, H., et al. (2004). Frequent loss of RUNX3 gene expression in human bile duct and pancreatic cancer cell lines. *Oncogene, 23*(13), 2401–2407. doi:10.1038/sj.onc.1207395.

Wai, P. Y., Mi, Z., Gao, C., Guo, H., Marroquin, C., & Kuo, P. C. (2006). Ets-1 and runx2 regulate transcription of a metastatic gene, osteopontin, in murine colorectal cancer cells. *The Journal of Biological Chemistry, 281*(28), 18973–18982. doi:10.1074/jbc.M511962200.

Walenta, S., Wetterling, M., Lehrke, M., Schwickert, G., Sundfor, K., Rofstad, E. K., & Mueller-Klieser, W. (2000). High lactate levels predict likelihood of metastases, tumor recurrence, and restricted patient survival in human cervical cancers. *Cancer Research, 60*(4), 916–921.

Wang, Q., Stacy, T., Binder, M., Marin-Padilla, M., Sharpe, A. H., & Speck, N. A. (1996). Disruption of the Cbfa2 gene causes necrosis and hemorrhaging in the central nervous system and blocks definitive hematopoiesis. *Proceedings of the National Academy of Sciences of the United States of America, 93*(8), 3444–3449.

Wang, Y., Belflower, R. M., Dong, Y. F., Schwarz, E. M., O'Keefe, R. J., & Drissi, H. (2005). Runx1/AML1/Cbfa2 mediates onset of mesenchymal cell differentiation toward chondrogenesis. *Journal of Bone and Mineral Research: the Official Journal of the American Society for Bone and Mineral Research, 20*(9), 1624–1636. doi:10.1359/jbmr.050516.

Wang, L., Gural, A., Sun, X. J., Zhao, X., Perna, F., Huang, G., et al. (2011). The leukemogenicity of AML1-ETO is dependent on site-specific lysine acetylation. *Science (New York, N.Y.), 333*(6043), 765–769. doi:10.1126/science.1201662.

Woolf, E., Xiao, C., Fainaru, O., Lotem, J., Rosen, D., Negreanu, V., et al. (2003). Runx3 and Runx1 are required for CD8 T cell development during thymopoiesis. *Proceedings of the National Academy of Sciences of the United States of America, 100*(13), 7731–7736. doi:10.1073/pnas.1232420100.

Xie, K., Wei, D., Shi, Q., & Huang, S. (2004). Constitutive and inducible expression and regulation of vascular endothelial growth factor. *Cytokine & Growth Factor Reviews, 15*(5), 297–324. doi:10.1016/j.cytogfr.2004.04.003.

Yanada, M., Yaoi, T., Shimada, J., Sakakura, C., Nishimura, M., Ito, K., et al. (2005). Frequent hemizygous deletion at 1p36 and hypermethylation downregulate RUNX3 expression in human lung cancer cell lines. *Oncology Reports, 14*(4), 817–822.

Yergeau, D. A., Hetherington, C. J., Wang, Q., Zhang, P., Sharpe, A. H., Binder, M., et al. (1997). Embryonic lethality and impairment of haematopoiesis in mice heterozygous for an AML1-ETO fusion gene. *Nature Genetics, 15*(3), 303–306. doi:10.1038/ng0397-303.

Yopp, A. C., Schwartz, L. H., Kemeny, N., Gultekin, D. H., Gonen, M., Bamboat, Z., et al. (2011). Antiangiogenic therapy for primary liver cancer: Correlation of changes in dynamic contrast-enhanced magnetic resonance imaging with tissue hypoxia markers and clinical response. *Annals of Surgical Oncology, 18*(8), 2192–2199. doi:10.1245/s10434-011-1570-1.

Yoshida, C. A., Yamamoto, H., Fujita, T., Furuichi, T., Ito, K., Inoue, K., et al. (2004). Runx2 and Runx3 are essential for chondrocyte maturation, and Runx2 regulates limb growth through induction of Indian hedgehog. *Genes & Development, 18*(8), 952–963. doi:10.1101/gad.1174704.

Yoshimi, M., Goyama, S., Kawazu, M., Nakagawa, M., Ichikawa, M., Imai, Y., et al. (2012). Multiple phosphorylation sites are important for RUNX1 activity in early hematopoiesis and T-cell differentiation. *European Journal of Immunology, 42*(4), 1044–1050. doi:10.1002/eji.201040746.

Yuan, S. Y., & Rigor, R. R. (2010). *Regulation of endothelial barrier function, Integrated systems physiology: From molecule to function to disease*. San Rafael: Morgan & Claypool Life Sciences.

Yuan, Y., Tang, A. J., Castoreno, A. B., Kuo, S. Y., Wang, Q., Kuballa, P., et al. (2013). Gossypol and an HMT G9a inhibitor act in synergy to induce cell death in pancreatic cancer cells. *Cell Death & Disease, 4*, e690. doi:10.1038/cddis.2013.191.

Zelzer, E., Glotzer, D. J., Hartmann, C., Thomas, D., Fukai, N., Soker, S., & Olsen, B. R. (2001). Tissue specific regulation of VEGF expression during bone development requires Cbfa1/Runx2. *Mechanisms of Development, 106*(1–2), 97–106.

Zelzer, E., Mamluk, R., Ferrara, N., Johnson, R. S., Schipani, E., & Olsen, B. R. (2004). VEGFA is necessary for chondrocyte survival during bone development. *Development, 131*(9), 2161–2171. doi:10.1242/dev.01053.

The Emerging Roles of RUNX Transcription Factors in Epithelial-Mesenchymal Transition

28

Dominic Chih-Cheng Voon and Jean Paul Thiery

Abstract

Epithelial-mesenchymal transition (EMT) is an evolutionary conserved morphogenetic program necessary for the shaping of the body plan during development. It is guided precisely by growth factor signaling and a dedicated network of specialised transcription factors. These are supported by other transcription factor families serving auxiliary functions during EMT, beyond their general roles as effectors of major signaling pathways. EMT transiently induces in epithelial cells mesenchymal properties, such as the loss of cell-cell adhesion and a gain in cell motility. Together, these newly acquired properties enable their migration to distant sites where they eventually give rise to adult epithelia. However, it is now recognized that EMT contributes to the pathogenesis of several human diseases, notably in tissue fibrosis and cancer metastasis. The RUNX family of transcription factors are important players in cell fate determination during development, where their spatio-temporal expression often overlaps with the occurrence of EMT. Furthermore, the dysregulation of RUNX expression and functions are increasingly linked to the aberrant induction of EMT in cancer. The present chapter reviews the current knowledge of this emerging field and the common themes of RUNX involvement during EMT, with the intention of fostering future research.

D.C.-C. Voon (✉)
Institute for Frontier Science Initiative, Kanazawa University, Kanazawa, Ishikawa, Japan

Division of Genetics, Cancer Research Institute, Kanazawa University, Kanazawa, Ishikawa, Japan
e-mail: dvoon@staff.kanazawa-u.ac.jp

J.P. Thiery
Cancer Science Institute of Singapore, National University of Singapore, Singapore 117599, Singapore

Department of Biochemistry, Yong Loo Lin School of Medicine, National University of Singapore, Singapore 117596, Singapore

Institute of Molecular and Cell Biology, A-STAR, Singapore 138673, Singapore
e-mail: bchtjp@nus.edu.sg

Y. Groner et al. (eds.), *RUNX Proteins in Development and Cancer*, Advances in Experimental Medicine and Biology 962, DOI 10.1007/978-981-10-3233-2_28

Keywords

Epithelial-mesenchymal transition (EMT) • RUNX • Auxiliary transcription factors • Atrioventricular valve • Mammary gland development • Lacrimal gland repair • Cancer metastasis • Osteomimicry

28.1 Introduction

Metazoa are composed of three major cell types: epithelial, mesenchymal and neural. Epithelial cells are in close contact with one another and organized in sheets with apico–basal polarity. Within this arrangement, epithelial cells communicate with one another through an ordered series of cell–cell junctional complexes: adherens junctions, desmosomes and tight junctions. Through these contacts, the epithelium serves its function as a selective barrier. In contrast, mesenchymal cells are loosely organized within a three-dimensional extracellular matrix. Their main purpose is to act as connective tissues that provide structural support to the epithelia. Unlike epithelial cells, mesenchymal cells can function independently and are capable of migration, especially during development. The process Epithelial-Mesenchymal Transition (EMT) refers to a cellular program during normal development in which epithelial cells lose their junctional complexes to acquire mesenchymal properties, such as migratory and invasive capabilities. Following EMT, cells will often undergo the reverse process, referred to as Mesenchymal-Epithelial Transition (MET). EMT is a highly conserved morphogenetic process in evolution for the shaping of the body plan in metazoans. EMT and MET operate sequentially throughout embryogenesis and during organogenesis. A round of EMT–MET is reactivated in the adult stage during wound healing and EMT in the kidney epithelium induces fibrosis in the stroma. An EMT–MET round has also been proposed to be activated during the progression of carcinoma.

There is commonality between the process of EMT and the RUNX transcription factor family. Like the EMT process, RUNX proteins are also regulators of development that have been conserved through evolution and are frequently targeted during human carcinogenesis. The EMT process and RUNX participate in both vertebrate and invertebrate development. However, this chapter is devoted to describing the many instances where they intersect in vertebrate development and human disease.

28.2 Hallmarks and Drivers of EMT

28.2.1 The EMT Phenotype

Epithelial cells in contact are characterized by a set of specialized junctions including adherens junctions, which are established through the assembly of protein complexes at the site of E-cadherin clusters. These adhesion structures are the main targets in the execution of the EMT program (Huang et al. 2012).

In epithelial cells, the extracellular region of E-cadherin forms an interface between two adjacent epithelial cells, through *cis* and *trans* interactions. The E-cadherin adhesome consists of numerous proteins interacting directly or indirectly with the actin cytoskeleton. β-catenin bound to the carboxy-terminal region of the E-cadherin cytoplasmic domain anchors α-catenin, a crucial protein acting as a mechanosensor. P120-catenins also plays a major role in the dynamics of the junctional complexes (Engl et al. 2014; Yonemura et al. 2010). Although EMT occurs in different tissue context during development, these occurrences share a core set of common features. A principal mark of EMT is the reduction or complete loss of E-cadherin at the cell surface leading to the remodelling or disappearance of adherens junctions. From the onset of EMT to the full-blown mesenchymal stage, the disassembly of the adherens complex is timed with the dissolution of other epithelial junctional complexes, namely tight junctions and the desmosomes (Yilmaz and Christofori 2009; Huang et al. 2012; Lamouille et al. 2014). The initial stages of EMT include the transcriptional repression of E-cadherin and protein degradation (Cano et al. 2000; Batlle et al. 2000; van Roy and Berx 2008). As the cell takes on a more mesenchymal phenotype, E-cadherin is typically replaced by other cadherins, such as N-cadherin, cadherin-7 and cadherin 11 (Nakagawa and Takeichi 1995; Vallin et al. 1998). These cadherins sustain weaker intercellular adhesion (Chu et al. 2006). Collectively, the release from rigid epithelial junctional network and the switch to weaker cell-cell adhesion enables more temporal contacts suitable for greater cell motility (Theveneau and Mayor 2012). Furthermore, a cell that has undergone EMT commonly expresses vimentin, a cytoskeletal protein necessary for migration and the condensation of actin stress fibres that power its movement (Mendez et al. 2010). The migratory cell expresses metalloproteinases (e.g. MMP-2, -9 and -13) for the breaking down of the extracellular matrix as it migrates (Nistico et al. 2012).

The detachment of epithelial cells would trigger a form of program cell death, termed anoikis, which is a critical mechanism that safeguards them against anchorage-independent cell growth (Paoli et al. 2013). During EMT, this program is momentarily silenced through the activation of a pro-survival genetic program, rendering cells resistant to anoikis and apoptosis (De Craene and Berx 2013).

28.2.2 Coordinators of EMT

EMT is controlled by several evolutionarily conserved signaling pathways during development, which collaborate to achieve precise spatio-temporal coordination. The best characterized among these are the transforming growth factor-beta/bone morphogenic protein (TGF-β/BMP), Notch, wingless-Int (Wnt), Hedgehog (Hh), epidermal growth factor (EGF), Fibroblast growth factor (FGF) pathways. The importance of these cardinal developmental signals in the regulation of EMT has been extensively studied using *in vitro* and *in vivo* models and reviewed in depth in dedicated reviews (Thiery et al. 2009; Lamouille et al. 2014). Reflecting their central roles, mutations or disruptions of these pathways would result in the aberrant activation of EMT in human diseases, such as cancer (De Craene and Berx 2013).

Acting downstream of these principal growth signals is an integrated network of specialized transcription factors, which executes the transcriptomic and epigenetic changes as the cells passage through EMT. Like the upstream signaling cascades, molecular mechanisms employed by these EMT drivers have been studied in detail (reviewed in (Lamouille et al. 2014)). A concise summary is provided here for the purpose of the current discussion.

28.2.2.1 SNAIL

In the vertebrate, the SNAIL family of zinc finger transcription factors are represented by SNAI1 (also called SNAIL), SNAI2 (also called SLUG) and SNAI3. SNAIL transcription factors repress epithelial genes through the binding of an E-box binding sequence, which has been extensively characterised on the E-cadherin (*CDH1*) promoter (Cano et al. 2000; Batlle et al. 2000). In additional to their transcriptional effects, SNAIL proteins recruit the polycomb repressor complex 2 to target gene loci to repress gene expression epigenetically via histone modification (Herranz et al. 2008; Lamouille et al. 2014). Importantly, this would leave epithelial genes actively repressed but poised for reactivation. It is speculated that the maintenance of a bivalent state is key to the prompt reversion to an epithelial phenotype during MET, and the inherent plasticity associated with cells undergoing EMT (Lamouille et al. 2014).

28.2.2.2 TWIST

A number of basic helix-loop-helix (bHLH) transcription factors participate in EMT, including TWIST1, TWIST2, E12, E47 and inhibitor of differentiation (ID). These activating and deactivating bHLH proteins form homo- and heterodimers with one another to mediate signals from diverse pathways (Peinado et al. 2007). This includes the HIF-1α pathway, which promotes EMT in a hypoxic environment, such as in a tumor (Yang et al. 2008). Of the bHLH proteins, TWIST1/2 are best studied and mediate the repression of epithelial genes while activating mesenchymal ones (reviewed in (Castanon and Baylies 2002; Thiery et al. 2009; Lamouille et al. 2014)). TWIST proteins are highly conserved in evolution and are essential for the initiation of mesoderm development during gastrulation (Castanon and Baylies 2002). TWIST1 recruits epigenetic modifiers such as SET8 and BMI-1 to silence the E-cadherin while activating N-cadherin promoters (Yang et al. 2010, 2012). A further important function of TWIST proteins is their repression of the *INK4A* tumor suppressor gene, which is part of the pro-survival effects of EMT (Ansieau et al. 2008; Yang et al. 2010). Importantly, this pro-survival mechanism is often hijacked by cancer cells during tumorigenesis.

28.2.2.3 ZEB

The zinc finger E-box-binding homeobox (ZEB) transcription factors, ZEB1 and 2, also target the E-boxes in gene promoters to repress epithelial and activate mesenchymal genes (Peinado et al. 2007). In addition, ZEB1/2 remodel the chromatin structure of target loci by interacting with co-repressor CtBP, which recruits the chromatin remodelling complex SNFI/SWI to silence their target genes (Sanchez-Tillo et al. 2010). Conversely, ZEB1/2 interact with co-activators such as p300 and p300/CBP associated factors to activate transcription (Postigo et al. 2003).

Functionally, an important feature of SNAIL, TWIST and ZEB proteins is the sharing of a core set of key target genes, which allows these transcription factors to individually initiate EMT. In addition to functional redundancy, complex cross regulations exist between these transcription factors, for example *ZEB1* is directly regulated by SNAI1 and this could be further enhanced in partnership with TWIST1 (Dave et al. 2011). Lastly, it bears highlighting that the EMT inducers are not functionally identical, as was revealed in the distinct EMT programs driven by Snai2 and Snai1 in normal mammary stem cells and tumor-initiating cells, respectively (Ye et al. 2015).

28.2.2.4 Secondary Regulators of EMT

The primary EMT inducing transcription factors are supported by several evolutionarily conserved transcription factor families functioning as auxiliary regulators of EMT. Typically, these supporting transcription factors serve broader functions downstream of major signaling pathways, such as Smad (BMP/TGF-β pathway) and TCF (canonical Wnt pathway) (Derynck et al. 2014; Nawshad et al. 2007). However, the relationship between their general and EMT-promoting roles is often obscure, as is whether these roles are demarcated by spatio-temporal parameters or directed by alternative signaling routes. In this regard, a considerable body of evidence points to the involvement of the RUNX transcription factors during EMT, beyond their established roles in cell fate determination (Fig. 28.1). In the ensuing sections, these evidences are examined in light of normal development and human cancers, where common traits of RUNX involvement in EMT will be discussed.

28.3 RUNX and EMT During Development and Morphogenesis

28.3.1 Formation of Atrioventricular Valve in the Cardiac Canal

A clear example of RUNX's participation in EMT during normal development is that of RUNX2 during the formation of cardiac valves in the embryonic heart. The heart is unique amongst organs in that it is derived from three successive rounds of EMT–MET (reviewed in (Thiery et al. 2009; Kovacic et al. 2012)). Cardiac mesodermal cells are specified during gastrulation, and these cells undergo MET to become a transient, two-layered epithelium called the splanchnopleure. During the second round of EMT, the splanchnopleure folds on itself, while an endothelial lining is formed from the dissociated mesenchymal

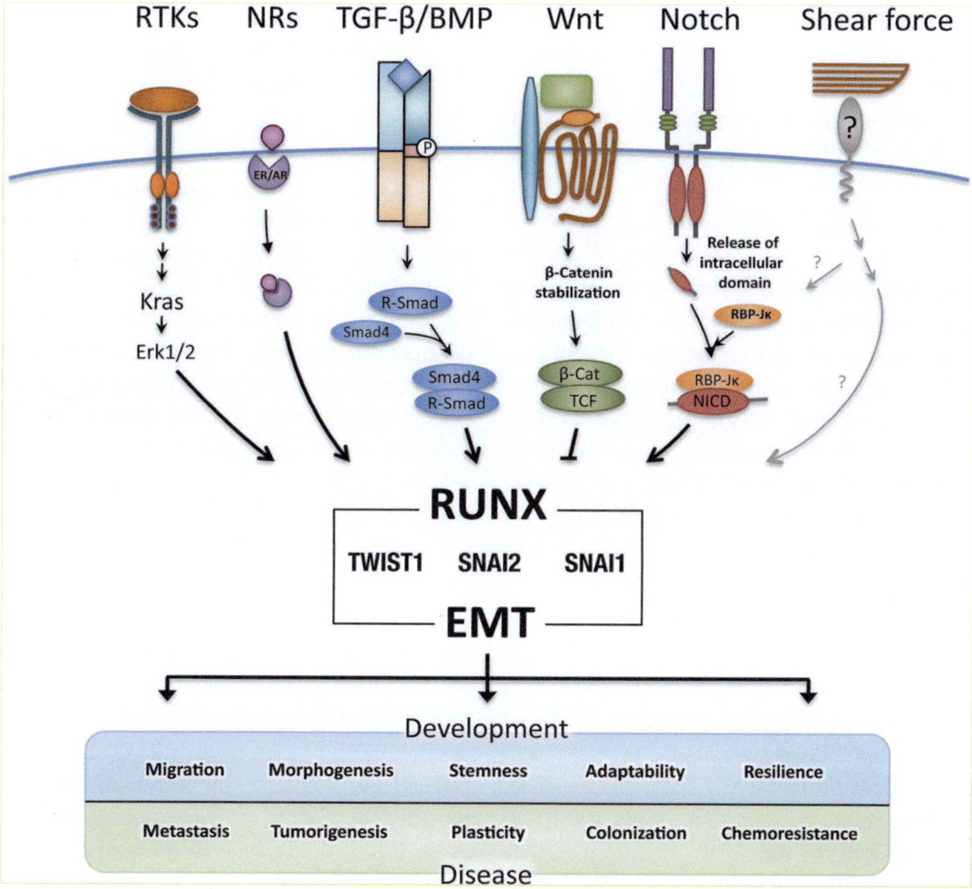

Fig. 28.1 RUNX proteins as a focal point of key signaling pathways and their partnership with key EMT-inducers to confer the attributes of EMT in normal development and cancer. *RTKs* receptor tyrosine kinases, *Erk1/2* extracellular response kinase 1 and 2, *NRs* nuclear receptors, *R-Smad* receptor Smad, *NICD* Notch intracellular domain

cells. The subsequent MET gives rise to two concentric tubes within which the primordial atrioventricular compartments are formed. During tertiary EMT, cells from the endocardial cushion undergo EMT to invade the underlying cardiac jelly. These mesenchymal cells then proliferate, thicken the endocardiac cushion, and give rise to the atrioventricular (AV) and ventricular outflow tract (OT) valves. This process is dependent on the interplay among Notch (Garg et al. 2005; Nigam and Srivastava 2009), TGFβ2/β3 (Brown et al. 1999), ALK (Mercado-Pimentel et al. 2007), RhoA (Tavares et al. 2006), ErbB (Erickson et al. 1997; Camenisch et al. 2002), and PDGF (Schatteman et al. 1992; Van Den Akker et al. 2005) pathways.

Akin to developmental EMT elsewhere, acting downstream of these pathways is a network of transcription factors that include Snai1/2, RBP-Jκ/CBF1 and Runx2. In the embryonic mouse heart, *Runx2* is detectable in the endocardial cells and the mesenchyme of both the AV and OT cushions from 9.5 days post-coitum (d.p.c.; E9.5), and is maintained throughout the mesenchymal transformation (Gitler et al. 2003). The role of Runx2 in the development of the AV canal was investigated in the context of bone morphogenetic protein (BMP) signaling using chicken heart explants (Tavares et al. 2006; Mercado-Pimentel et al. 2007). Runx2 is regulated by Alk2/5 and endoglin early in the development of the AV canal. The knockdown of endoglin by

RNAi downregulates Runx2 in the AV cushion but has an opposite effect in the OT cushion, indicative of a complex regulatory mechanism of multiple modulators (Mercado-Pimentel et al. 2007). Other members of the BMP/TGF-β family, namely BMP2 and TGF-β, also participate in the EMT of cardiac endothelial cells (Brown et al. 1999; Boyer et al. 1999; Ma et al. 2005). RUNX2 collaborates with these two signaling pathways in pluripotent mesenchymal precursors and therefore may act as focal points for multiple BMP/TGF-β signals in the development of cardiac valves (Lee et al. 2000). However, despite the dynamic and prominent expression of Runx2 in this tissue, cardiac valvular defects have not been observed in various Runx2-deficient mouse models. It is possible that functional compensation exists with the other Runx family members. Indeed, Runx3 expression is detected in the mouse AV cushion at E10.5. Runx3 is transcriptionally regulated by the Notch pathway and its nuclear effectors, CBF-1/Suppressor of Hairless/Lag-1 (CSL) and mastermind-like protein-1 (MAML-1) (Fu et al. 2011). Once induced, Runx3 sustains the long-term expression of Snai2 to maintain EMT-transformed endothelial cells in a mesenchymal state (Fu et al. 2011). Therefore, Runx3 is enlisted by the Notch pathway to prolong the expression of Snai2 during Notch-induced EMT in the AV cushion.

Interestingly, RUNX2 is also a positive regulator of Snai2 in multiple tissues and may also support Snai2 expression in the AV canal (Lambertini et al. 2010; Niu et al. 2012). However, in the cardiac valve, Notch represses RUNX2 activities via its target genes, Hes1 and Hrt2/Hey2 (Garg et al. 2005; Nigam and Srivastava 2009; McLarren et al. 2000). These Notch effectors physically interact with RUNX2 to prevent the differentiation of valvular cells into osteoblast-like cells, which is causal to calcification in aortic valves (Rajamannan et al. 2003; Garg et al. 2005; Nigam and Srivastava 2009). Collectively, these observations suggest a complex involvement of the Runx proteins in the overall development of the cardiac valves.

28.3.2 Emergence of Hematopoietic Stem Cells During Definitive Hematopoiesis

Closely related to the EMT of the cardiac endothelial cells is the dissociation of endothelial cells during their transition into hematopoietic stem cells (HSCs) (Kovacic et al. 2012; Kissa and Herbomel 2010). This process, termed Endothelial–Hematopoietic Transition (EHT), occurs in a specialized subpopulation of hemogenic endothelial cells, best characterized at the dorsal aortic floor in the embryonic aortic-gonadal-mesonephric (AGM) region (Medvinsky and Dzierzak 1996; Cumano et al. 1996; Taoudi and Medvinsky 2007). The emergence of HSCs is similarly observed in the vitelline and umbilical arteries (Chen et al. 2009).

Although the regulatory mechanisms underlying EHT are not fully understood, the process nevertheless bears strong resemblance to EMT, wherein is the dissolution of tight junctions (Yue et al. 2012; Zhang et al. 2014), a loss of cell polarity (Wilkinson et al. 2009), and a gain of cell motility and stem cell-like properties (Eilken et al. 2009; Kissa and Herbomel 2010; Boisset et al. 2010; Yue et al. 2012; Kovacic et al. 2012). Furthermore, a similar array of developmental signals is involved in coordinating EHT, including Notch (Hadland et al. 2004; Burns et al. 2005; Richard et al. 2013), TGF-β/BMP (Durand et al. 2007; Wilkinson et al. 2009; Zhang et al. 2014), Wnt (Clements et al. 2011), FGF (Pouget et al. 2014; Lee et al. 2014), ERK (Lan et al. 2014; Zhang et al. 2014), F2r-RhoA/ROCK (Yue et al. 2012) and Hedgehog (Gering and Patient 2005; Wilkinson et al. 2009).

In the mouse, the appearance of adult-repopulating HSCs during definitive hematopoiesis coincides with the spatial-temporal expression of Runx1, which begins at 9.5–10.5 d.p.c. (E9.5–10.5). Runx1 is necessary for the proper maintenance of adult HSCs and is a master regulator of adult hematopoiesis (Voon et al. 2015). Genetic ablation of Runx1 in mouse results in embryonic lethality at E11.5–12.5, characterized by a complete loss of HSC-populated definitive hematopoiesis, and the impairment of vascularization (Okuda et al. 1996; Wang et al. 1996). Numerous studies have shown that Runx1 is indispensable for the emergence of HSCs from the hemogenic endothelium of the dorsal aorta via EHT (North et al. 1999; Yokomizo et al. 2001; Chen et al. 2009; Kissa and Herbomel 2010; Lancrin et al. 2009; North et al. 2009; Adamo et al. 2009; Lam et al. 2010). Here, Runx1 is induced in hemogenic endothelial cells by the Notch signal from sub-aortic mesenchymal cells via Gata2, in order to act in tandem with Notch (Burns et al. 2005; Richard et al. 2013; Kobayashi et al. 2014; Gao et al. 2013). In addition, a number of other signals induce Runx1 *in vivo* through yet undetermined mechanisms, including BMP (Wilkinson et al. 2009) and mechanical shear force (Adamo et al. 2009; North et al. 2009). The induction of *Runx1* in hemogenic endothelial cells coincides strictly with their EHT. In the absence of *Runx1*, nascent HSCs initially emerge but succumb to a sudden death while exiting the aortic floor (Kissa and Herbomel 2010). Despite its apparent importance in EHT, the precise transcriptional program maintained by Runx1 remains to be elucidated, as is its relationship with the EMT regulators, such as Snail, Twist and Zeb transcription factors. Of note, Snai2 protects hematopoietic stem/progenitors cells against radiation-induced apoptosis by negating the effects of p53/Puma. It is also necessary for modulating the proliferation of HSCs during stress and regeneration (Inoue et al. 2002; Wu et al. 2005). It would be of interest to determine if these EMT-like protective roles trace back to an involvement of Snai2 during EHT in cooperation with Runx1.

28.3.3 RUNX and EMT in Other Tissue Contexts

RUNX proteins and EMT feature prominently in a number of other tissues both during development and in the adult. Although direct evidence is still lacking, RUNX proteins are likely part of the EMT genetic program in these tissue contexts. Here we describe two examples.

28.3.3.1 Mammary Gland

In recent years, compelling evidence points to the involvement of RUNX proteins in embryonic mammary development, postnatal mammary gland morphogenesis, and human breast cancer (Ferrari et al. 2013). Notably, this is a tissue in which EMT is prominent in organogenesis and disease. The mammary epithelium is unique in that much of its development occurs postnatally and its embryonic development is halted with rudimentary glands present at birth. During embryonic development, Runx2 is expressed transiently in the nascent mouse mammary buds at E12–12.5, during lineage segregation (Otto et al. 1997; Ferrari et al. 2015). This is replaced by the expression of Runx1 at E16 before their co-expression in the adult mammary glands (Ferrari et al. 2013).

Postnatal, mammary development resumes through branching morphogenesis and ductal elongation to form the mature mammary gland for milk production. During this period, the temporal and reversible activation of EMT, also termed "partial EMT", endows stem/progenitor cells at the terminal end buds (TEB) with transient motility to drive ductal elongation (Ewald et al. 2012; Nakaya and Sheng 2013). This is an precisely controlled process that is promoted by EMT inducers, such as Snail and Snai2; and restricted by EMT inhibitors, such as Ovol2 (Guo et al. 2012; Ye et al. 2015; Watanabe et al. 2014). In addition to the potential regulation of *Snai2*, a functional contribution by Runx2 in the "partial EMT" in the TEB is hinted by the co-expression of *Runx2* with *Snail, Twist1, Twist2* and other components of the EMT program (Kouros-Mehr and Werb 2006; Lambertini et al. 2010; Niu et al. 2012). Indeed, recent studies have demonstrated that the strict regulation of Runx2 is necessary for proper mammary gland development (Otto et al. 1997; McDonald et al. 2014; Owens et al. 2014; Ferrari et al. 2015). The targeting of *Runx2* in mouse mammary glands impaired ductal outgrowth at puberty and disrupted progenitor cell differentiation during pregnancy (Owens et al. 2014). On the other hand, the ectopic expression of Runx2 altered differentiation and led to a gain of EMT phenotype (McDonald et al. 2014;

Owens et al. 2014; Ferrari et al. 2015). Although the signaling pathways acting via Runx2 during osteoblast differentiation are well understood, little is known of the cooperating signals in this tissue that could explain these paradoxical phenotypes (Franceschi and Xiao 2003). Nevertheless, these observations are consistent with the need for a strict control of EMT during mammary morphogenesis.

Runx1 is likewise important in the maintenance of mammary gland homeostasis, which is reflected in its recurrent deletion in human luminal breast cancer (Banerji et al. 2012; Network 2012; Ellis et al. 2012). In the mouse, *Runx1* is the most highly expressed amongst Runx genes and is present in all subpopulations of mammary epithelial cells except secretary alveolar luminal cells (McDonald et al. 2014; van Bragt et al. 2014). The mammary-specific deletion of *Runx1* by *MMTV*-Cre reduces the proportion of estrogen receptor (ER)-positive luminal cells in virgin adult mice, and this is likely mediated via the interaction between Runx1 and ERα (van Bragt et al. 2014; Stender et al. 2010). Furthermore, Runx1 promotes a mature luminal phenotype while suppressing alveolar luminal cell differentiation through its repression of *Elf5* (van Bragt et al. 2014). In addition to being a key driver of alveolar luminal cell differentiation during lactation, Elf5 is an important regulator of *Snai2* and EMT of mammary epithelial cells (Chakrabarti et al. 2012). Therefore, it is likely that Runx1 also exerts an influence in partial EMT during mammary gland morphogenesis. Collectively, the associations described herein provide clear reasons to precisely eluciate the roles of Runx proteins during mammary development, especially with respect to "partial EMT", in future studies.

28.3.3.2 Lacrimal Gland

The lacrimal gland (LG) is a tubuloacinar exocrine gland that generates most of the aqueous preocular tear film. It is composed of three main cell types: acinar, ductal, and myoepithelial cells. The development of the LG shares common features with that of the mammary gland, with an analogous branching morphogenesis that is governed by factors such as Sox9, fibroblast growth factor (FGF) and BMP signaling (Chen et al. 2014; Dean et al. 2004; Lu et al. 2006). Furthermore, it is likely that LG development progresses by way of a "partial EMT" similar to that seen in the mammary gland. Indeed, EMT is observed during LG tissue repair following acute inflammation, giving rise to nestin-positive mesenchymal stem-like cells at the sites of the injury, which also expressed *vimentin* and *Snail* (Zoukhri et al. 2007, 2008; You et al. 2012).

Runx proteins are also similarly involved during LG development and regeneration (Voronov et al. 2013). All Runx proteins are expressed in the developing LG epithelium and their down-regulation in organoid culture has been shown to reduce LG growth, branching and proliferation (Voronov et al. 2013). Of the Runx proteins, Runx1 is highly expressed and its targeting *in vivo* can impair the timing of LG bud outgrowth during development (Voronov et al. 2013). Of particular note, Runx1 and Runx3 are induced during tissue regeneration and are correlated with epithelial cell proliferation following injury (Voronov et al. 2013); however, it is currently unclear if Runx proteins contribute to the EMT program during LG regeneration. Future studies on this topic will serve as helpful comparisons to understand the involvement of Runx during EMT in exocrine tissues.

28.4 RUNX and EMT in Disease

28.4.1 RUNX2 and Bone Metastasis of Cancers

Parallel to its involvement in EMT during development, RUNX proteins have been implicated in the aberrant activation of EMT in cancer in different tissues. The best characterized of these phenomena is the involvement of RUNX2 in the metastases of breast and prostate cancers (Fig. 28.1).

Metastatic breast and prostate cancers are typified by their tropism for the bone, with skeletal colonization accounting for approximately 70% of metastases (Roodman 2004). During dissemination, metastatic carcinoma cells acquire an EMT-like phenotype characterized by increased cell migration and the expression of mesenchymal markers and EMT-promoting transcription factors. Metastatic breast cancer (BCa) and prostate cancer (PCa) cells display osteocyte-like properties that enhance their survival at distant osseous niches, such as the expression of osteopontin (OPN), osteocalcin, bone sialoprotein (BSP) and, importantly, RUNX2 (Koeneman et al. 1999; Barnes et al. 2003; Brubaker et al. 2003; Javed et al. 2005) (Table 28.1).

RUNX2 is a key regulator of osteogenesis (Lian and Stein 2003; Ito et al. 2015). Therefore, its over-expression in breast cancer (BCa) cells offers a mechanistic explanation to the skeletal tropism and osteomimicry of BCa cells (Barnes et al. 2003; Inman and Shore 2003; Khalid et al. 2008; Selvamurugan et al. 2004; Javed et al. 2005; Lau et al. 2006). Indeed, RUNX2 expression is required for the metastasis of BCa *in vivo* and correlates with the increased grade and poor prognosis in the clinic (Barnes et al. 2004; Javed et al. 2005; Das et al. 2009; Onodera et al. 2010; Ferrari et al. 2013). RUNX2 and its binding partner CBFβ promote BCa cell migration *in vitro* (Pratap et al. 2005; Chimge et al. 2011; Mendoza-Villanueva et al. 2010), likely via its regulation of invasion-associated genes, such as matrix metalloproteinase (MMP)-13 (Selvamurugan et al. 2004; Mendoza-Villanueva et al. 2010), MMP-9 (Pratap et al. 2005; Mendoza-Villanueva et al. 2010), vascular endothelial growth factor (VEGF) (Zelzer et al. 2001) and BSP (Barnes et al. 2003). In particular, RUNX2 is necessary for the induction of *SNAI2*, a central coordinator of mammary epithelial stemness and EMT during branch morphogenesis (Chimge et al. 2011; Guo et al. 2012). In metastatic BCa cells, RUNX2 also regulates S100A4, a well-established marker of motility and invasion implicated in cancer EMT and exosome-mediated metastasis (Xu et al. 2015; Hoshino et al. 2015).

Analogous to these observations, RUNX2 enables the osteomimicry of PCa cells, promotes the formation of a pro-surviving osseous niche and is similarly associated with the EMT metastatic phenotype (Yang et al. 2001; Zayzafoon et al. 2004; Baniwal et al. 2009; Lim et al. 2010; Akech et al. 2010; Baniwal et al. 2010; Little et al. 2012). RUNX2 positively regulates EMT drivers such as *SMAD3*, *SNAI2* and *SOX9* (Baniwal et al. 2010; Little et al. 2012, 2014). Notably, RUNX2 and androgen receptor cooperatively activate *SNAI2* expression to induce EMT-like properties in PCa cells *in vitro* (Little et al. 2014). The association of RUNX2 with cancer metastasis and aberrant EMT is also observed in thyroid cancers. In this context, the RNAi targeting of RUNX2 suppressed known target genes, such as *SNAI2* and *VEGF*, as well as other EMT-related genes, including *TWIST1* and *MMP2* (Niu et al. 2012). Interestingly, Twist1 and 2 are shown to functionally antagonize Runx2 during osteoblast differentiation, raising the possibility of intricate cross regulation between Runx2 and the EMT transcription factors during bone metastasis (Bialek et al. 2004).

Collectively, the major lesson from these studies is that the aberrant expression of RUNX2 is key to the concurrent activation of EMT and osteomimicry, especially in BCa and PCa. These insights could be instructive in interpreting the involvement of RUNX proteins in other cancer types.

28.4.2 RUNX3 and EMT – Suppression or Promotion?

In contrast, other RUNX proteins appear to exert an opposite effect to RUNX2. In hepatocellular carcinoma (HCC), RUNX3 is frequently silenced due to promoter methylation, hemizygous deletion and loss of heterozygosity (Mori et al. 2005; Nakanishi et al. 2011; Xiao and Liu 2004). The re-introduction of RUNX3 in EMT-prone HCC cell lines would promote an epithelial-like cell morphology and increased E-cadherin, as well as suppressing mesenchymal markers, N-cadherin and vimentin (Tanaka et al. 2012). In a Runx3-knockout mouse model, increased EMT features and ERK1/2 phosphorylation were associated with defects in the intra-alveolar septa. These phenotypes were partially ameliorated when mice were treated with a pharmacological inhibitor of ERK1/2 (Lee et al. 2011).

Further evidence of a protective role for RUNX3 against aberrant EMT was reported in

Table 28.1 Known and potential involvement of RUNX proteins in EMT during development and disease

Tissue context	RUNX	Signal	Description of involvement	References
I. Involvement of RUNX in EMT during development and organogenesis				
Atrioventricular Canal	Runx2	Alk2/Alk5	Invasion of mesenchymal cells into cardiac cushions.	Tavares et al. (2006) and Mercado-Pimentel et al. (2007)
	Runx3	Notch	Sustain the expression of Snai2 to maintain mesenchymal state	Fu et al. (2011)
Hemogenic endothelial cells of the dorsal aorta	Runx1		The emergence of HSC via endothelial-hematopoietic transition	North et al. (1999), Yokomizo et al. (2001), Chen et al. (2009), Kissa and Herbomel (2010) and Lam et al. (2010)
		Shear force	Induced by blood flow together with Myb to induce NO and direct EHT	Adamo et al. (2009), North et al. (2009) and Gao et al. (2013)
		Notch	Induced in endothelial cells by Notch signal from sub-aortic mesenchymal cells via Gata2, to partner Notch.	Burns et al. (2005), Richard et al. (2013) and Gao et al. (2013)
Mammary gland development	Runx2		Necessary for ductal outgrown at puberty and progenitor cell differentiation during pregnancy	Otto et al. (1997), McDonald et al. (2014), Owens et al. (2014) and Ferrari et al. (2015)
	Runx1	Estrogen receptor	Support ERa$^+$ luminal cells and suppress alveolar luminal cells via repressing *Elf5*	van Bragt et al. (2014)
Lacrimal gland regeneration	Runx1		Control timing of bud outgrowth. Needed for regeneration and LG branching.	Voronov et al. (2013)
	Runx3		Induced during regeneration and needed for LG branching	
	Runx2		Needed for LG branching in organoid cultures	
II. Involvement of RUNX in EMT in cancer				
Breast cancer metastasis	RUNX2/ CBFβ	ERα, Wnt	Promotes bone metastasis of breast carcinoma cells via *IBSP* and sclerostin	Barnes et al. (2003), Javed et al. (2005), Khalid et al. (2008) and Mendoza-Villanueva et al. (2010)
		Wnt, TGFβ	Promotes cell migration via *SNAI2*, *OPN*, *MMP-13*, *MMP-9*, *VEGF* & *S100A4*	Inman and Shore (2003), Selvamurugan et al. (2004), Pratap et al. (2005), Zelzer et al. (2001) and Chimge et al. (2011)

(continued)

Table 28.1 (continued)

Tissue context	RUNX	Signal	Description of involvement	References
Prostate cancer metastasis	RUNX2	TGFβ/BMP; Androgen receptor	Induces invasive phenotype via *SNAI2*, *SMAD3*, and *SOX9*.	Baniwal et al. (2009, 2010) and Little et al. (2012, 2014)
Thyroid cancer	RUNX2	Unclear	Postive regulation of *SNAI2*, *TWIST1*, *VEGF*, *MMP2*	Niu et al. (2012)
Hepatocellular carcinoma	RUNX3		Reverts HCC cells to an epithelial-like phenotype	Tanaka et al. (2012)
Lung carcinogenesis	RUNX3	Kras/ERK	Gain of EMT-like features following the loss of *Runx3 in vivo*	Lee et al. (2011)
Gastric carcinogenesis	Runx3	TGFβ, Wnt	Protects cells against EMT-induced plasticity and tumorigenicity.	Voon et al. (2013) and Voon et al. (2012)
	RUNX3		Inhibits cell migration by suppressing *OPN*	Cheng et al. (2013)
Pancreatic ductal adenocarcinoma metastasis	Runx3	Kras, TGFβ/BMP	Inhibits cell proliferation via suppression of *p21* at low dose. Promotes metastasis via induction of *Opn* and *Col6a1* at high dose	Whittle et al. (2015)

two independent studies in gastric carcinogenesis. The first reported an inverse correlation between RUNX3 and OPN expression in gastric tumors and cell lines. Concordant with the pro-metastatic role of OPN, RUNX3 expression was associated with better clinical outcome (Cheng et al. 2013). This relationship was confirmed *in vitro* where RUNX3 suppressed *OPN* expression and gastric cancer cell migration (Cheng et al. 2013). In the second study, immortalized *p53*-null/*Runx3*-deficient gastric cells were found prone to spontaneous EMT due to increased sensitivity to TGF-β, Wnt and EGFR/Ras signaling (Voon et al. 2012; 2013). This led to increased epithelial plasticity, reflected in the expression of the gastric stem cell marker Lgr5, which amplifies canonical Wnt signal (Li et al. 2002; Voon et al. 2012; de Lau et al. 2014). Interestingly, the *Runx3*-deficient cells were refractory to TGF-β-induced apoptosis but sensitive to TGF-β-mediated EMT, highlighting the importance of RUNX proteins in the context-specific interpretation of TGF-β signaling (Voon et al. 2012).

Overall, the current evidence in the literature supports the notion that RUNX2 promotes cancer metastasis by inducing EMT and osteomimicry, whereas RUNX3 exerts a protective effect. However, there are exceptions to this general observation, underscoring the complexity of cancer biology. For example, a recent study has found that RUNX3 acts as a "metastatic switch" in an oncogenic *Kras/p53* mutant mouse model of pancreatic ductal adenocarcinoma (PDA/PDAC) (Whittle et al. 2015). Here, a high expression of Runx3 was correlated with lung and liver metastases and promoted PDA cell migration through its positive regulation of *Opn*. Intriguingly, Runx3 had the opposite effect on *Opn* in gastric carcinoma (Cheng et al. 2013; Whittle et al. 2015). This is likely a result of the context-specific nature of Runx protein functions, which stems from their ability to partner a diverse range of transcription factors and co-factors (reviewed in (Blyth et al. 2005; Ito et al. 2015)). These permutations are further compounded by the interplay and cross-regulations between co-expressed Runx family members. Lastly, the proper execution of Runx function is often dependent on gene dosage (Osato and Ito 2005; Ben-Ami et al. 2013; McDonald et al. 2014; Owens et al. 2014; Whittle et al. 2015). Indeed, Whittle et al. observed that Runx3 could serve tumor suppressive or tumor promoting functions in PDAC cells depending on its dosage and the availability of its binding partner, Smad4 (Whittle et al. 2015). In this regard, the dose-dependency of Runx3 function in PDAC is similar to the varied levels of Twist1 required to drive distinct stages of skin carcinogenesis (Beck et al. 2015).

In the clinic, the challenge of combating cancer metastasis is intimately linked to tumor chemoresistance. It is becoming clear that these seemingly distinct phenomena are in fact driven by the same cellular plasticity afforded by aberrant EMT. Furthermore, recent studies have raised the possibility that EMT promotes the spread and survival of cancer cells via a combination of mechanisms, some of which become prominent in the context of medical intervention, such as chemoresistance. As such, it would be profitable to determine the contribution of Runx proteins to these additional EMT-enabled capabilities in future research.

28.5 Concluding Remarks

The RUNX family of transcription factors and the process of EMT are often regarded as being at the "crossroads of development and disease". This Chapter undertakes a collation of the published literature where the two topics have overlapped to reveal an intimate connection between RUNX and EMT. In particular, as nuclear effectors of key developmental signals, RUNX proteins function in partnership with EMT initiators, notably SNAI2, to modulate tissue morphogenesis. Consequently, a disruption to normal RUNX functionality often accompanies the aberrant activation of EMT in disease. It remains to be seen if this hypothesis is supported by future findings. In any case, these early impressions provide a framework upon which more definitive studies can be designed to show that RUNX and EMT are in fact working in arms, on the *same* intersection between development and disease.

Acknowledgement The authors thank Rebecca Jackson for her generous help in the preparation of the manuscript. We are also grateful to Jormay Lim and Isao Kobayashi for their valuable input.

References

Adamo, L., Naveiras, O., Wenzel, P. L., McKinney-Freeman, S., Mack, P. J., Gracia-Sancho, J., et al. (2009). Biomechanical forces promote embryonic haematopoiesis. *Nature, 459*(7250), 1131–1135. doi:10.1038/nature08073.

Akech, J., Wixted, J. J., Bedard, K., van der Deen, M., Hussain, S., Guise, T. A., et al. (2010). Runx2 association with progression of prostate cancer in patients: Mechanisms mediating bone osteolysis and osteoblastic metastatic lesions. *Oncogene, 29*(6), 811–821. doi:10.1038/onc.2009.389.

Ansieau, S., Bastid, J., Doreau, A., Morel, A. P., Bouchet, B. P., Thomas, C., et al. (2008). Induction of EMT by twist proteins as a collateral effect of tumor-promoting inactivation of premature senescence. *Cancer Cell, 14*(1), 79–89. doi:10.1016/j.ccr.2008.06.005.

Banerji, S., Cibulskis, K., Rangel-Escareno, C., Brown, K. K., Carter, S. L., Frederick, A. M., et al. (2012). Sequence analysis of mutations and translocations across breast cancer subtypes. *Nature, 486*(7403), 405–409. doi:10.1038/nature11154.

Baniwal, S. K., Khalid, O., Sir, D., Buchanan, G., Coetzee, G. A., & Frenkel, B. (2009). Repression of Runx2 by androgen receptor (AR) in osteoblasts and prostate cancer cells: AR binds Runx2 and abrogates its recruitment to DNA. *Molecular Endocrinology (Baltimore, Md), 23*(8), 1203–1214. doi:10.1210/me.2008-0470.

Baniwal, S. K., Khalid, O., Gabet, Y., Shah, R. R., Purcell, D. J., Mav, D., et al. (2010). Runx2 transcriptome of prostate cancer cells: Insights into invasiveness and bone metastasis. *Molecular Cancer, 9*, 258. doi:10.1186/1476-4598-9-258.

Barnes, G. L., Javed, A., Waller, S. M., Kamal, M. H., Hebert, K. E., Hassan, M. Q., et al. (2003). Osteoblast-related transcription factors Runx2 (Cbfa1/AML3) and MSX2 mediate the expression of bone sialoprotein in human metastatic breast cancer cells. *Cancer Research, 63*(10), 2631–2637.

Barnes, G. L., Hebert, K. E., Kamal, M., Javed, A., Einhorn, T. A., Lian, J. B., et al. (2004). Fidelity of Runx2 activity in breast cancer cells is required for the generation of metastases-associated osteolytic disease. *Cancer Research, 64*(13), 4506–4513. doi:10.1158/0008-5472.can-03-3851.

Batlle, E., Sancho, E., Franci, C., Dominguez, D., Monfar, M., Baulida, J., & Garcia De Herreros, A. (2000). The transcription factor snail is a repressor of E-cadherin gene expression in epithelial tumour cells. *Nature Cell Biology, 2*(2), 84–89. doi:10.1038/35000034.

Beck, B., Lapouge, G., Rorive, S., Drogat, B., Desaedelaere, K., Delafaille, S., et al. (2015). Different levels of Twist1 regulate skin tumor initiation, stemness, and progression. *Cell Stem Cell, 16*(1), 67–79. doi:10.1016/j.stem.2014.12.002.

Ben-Ami, O., Friedman, D., Leshkowitz, D., Goldenberg, D., Orlovsky, K., Pencovich, N., et al. (2013).

Addiction of t(8;21) and inv(16) acute myeloid leukemia to native RUNX1. *Cell Reports, 4*(6), 1131–1143. doi:10.1016/j.celrep.2013.08.020.

Bialek, P., Kern, B., Yang, X., Schrock, M., Sosic, D., Hong, N., et al. (2004). A twist code determines the onset of osteoblast differentiation. *Developmental Cell, 6*(3), 423–435.

Blyth, K., Cameron, E. R., & Neil, J. C. (2005). The RUNX genes: Gain or loss of function in cancer. *Nature Reviews Cancer, 5*(5), 376–387. doi:10.1038/nrc1607.

Boisset, J. C., van Cappellen, W., Andrieu-Soler, C., Galjart, N., Dzierzak, E., & Robin, C. (2010). In vivo imaging of haematopoietic cells emerging from the mouse aortic endothelium. *Nature, 464*(7285), 116–120. doi:10.1038/nature08764.

Boyer, A. S., Ayerinskas, I. I., Vincent, E. B., McKinney, L. A., Weeks, D. L., & Runyan, R. B. (1999). TGFbeta2 and TGFbeta3 have separate and sequential activities during epithelial-mesenchymal cell transformation in the embryonic heart. *Developmental Biology, 208*(2), 530–545. doi:10.1006/dbio.1999.9211.

Brown, C. B., Boyer, A. S., Runyan, R. B., & Barnett, J. V. (1999). Requirement of type III TGF-beta receptor for endocardial cell transformation in the heart. *Science (New York, N.Y.), 283*(5410), 2080–2082.

Brubaker, K. D., Vessella, R. L., Brown, L. G., & Corey, E. (2003). Prostate cancer expression of runt-domain transcription factor Runx2, a key regulator of osteoblast differentiation and function. *The Prostate, 56*(1), 13–22. doi:10.1002/pros.10233.

Burns, C. E., Traver, D., Mayhall, E., Shepard, J. L., & Zon, L. I. (2005). Hematopoietic stem cell fate is established by the Notch-Runx pathway. *Genes & Development, 19*(19), 2331–2342. doi:10.1101/gad.1337005.

Camenisch, T. D., Schroeder, J. A., Bradley, J., Klewer, S. E., & McDonald, J. A. (2002). Heart-valve mesenchyme formation is dependent on hyaluronan-augmented activation of ErbB2-ErbB3 receptors. *Nature Medicine, 8*(8), 850–855. doi:10.1038/nm742.

Cano, A., Perez-Moreno, M. A., Rodrigo, I., Locascio, A., Blanco, M. J., del Barrio, M. G., et al. (2000). The transcription factor snail controls epithelial-mesenchymal transitions by repressing E-cadherin expression. *Nature Cell Biology, 2*(2), 76–83. doi:10.1038/35000025.

Castanon, I., & Baylies, M. K. (2002). A Twist in fate: Evolutionary comparison of Twist structure and function. *Gene, 287*(1–2), 11–22.

Chakrabarti, R., Hwang, J., Andres Blanco, M., Wei, Y., Lukacisin, M., Romano, R. A., et al. (2012). Elf5 inhibits the epithelial-mesenchymal transition in mammary gland development and breast cancer metastasis by transcriptionally repressing Snail2. *Nature Cell Biology, 14*(11), 1212–1222. doi:10.1038/ncb2607.

Chen, M. J., Yokomizo, T., Zeigler, B. M., Dzierzak, E., & Speck, N. A. (2009). Runx1 is required for the endothelial to haematopoietic cell transition but not thereafter. *Nature, 457*(7231), 887–891. doi:10.1038/nature07619.

Chen, Z., Huang, J., Liu, Y., Dattilo, L. K., Huh, S. H., Ornitz, D., & Beebe, D. C. (2014). FGF signaling activates a Sox9-Sox10 pathway for the formation and branching morphogenesis of mouse ocular glands. *Development (Cambridge, England), 141*(13), 2691–2701. doi:10.1242/dev.108944.

Cheng, H. C., Liu, Y. P., Shan, Y. S., Huang, C. Y., Lin, F. C., Lin, L. C., et al. (2013). Loss of RUNX3 increases osteopontin expression and promotes cell migration in gastric cancer. *Carcinogenesis, 34*(11), 2452–2459. doi:10.1093/carcin/bgt218.

Chimge, N. O., Baniwal, S. K., Little, G. H., Chen, Y. B., Kahn, M., Tripathy, D., et al. (2011). Regulation of breast cancer metastasis by Runx2 and estrogen signaling: the role of SNAI2. *Breast Cancer Research : BCR, 13*(6), R127. doi:10.1186/bcr3073.

Chu, Y. S., Eder, O., Thomas, W. A., Simcha, I., Pincet, F., Ben-Ze'ev, A., et al. (2006). Prototypical type I E-cadherin and type II cadherin-7 mediate very distinct adhesiveness through their extracellular domains. *The Journal of Biological Chemistry, 281*(5), 2901–2910. doi:10.1074/jbc.M506185200.

Clements, W. K., Kim, A. D., Ong, K. G., Moore, J. C., Lawson, N. D., & Traver, D. (2011). A somitic Wnt16/Notch pathway specifies haematopoietic stem cells. *Nature, 474*(7350), 220–224. doi:10.1038/nature10107.

Cumano, A., Dieterlen-Lievre, F., & Godin, I. (1996). Lymphoid potential, probed before circulation in mouse, is restricted to caudal intraembryonic splanchnopleura. *Cell, 86*(6), 907–916.

Das, K., Leong, D. T., Gupta, A., Shen, L., Putti, T., Stein, G. S., et al. (2009). Positive association between nuclear Runx2 and oestrogen-progesterone receptor gene expression characterises a biological subtype of breast cancer. *European Journal of Cancer (Oxford, England : 1990), 45*(13), 2239–2248. doi:10.1016/j.ejca.2009.06.021.

Dave, N., Guaita-Esteruelas, S., Gutarra, S., Frias, A., Beltran, M., Peiro, S., & de Herreros, A. G. (2011). Functional cooperation between Snail1 and twist in the regulation of ZEB1 expression during epithelial to mesenchymal transition. *The Journal of Biological Chemistry, 286*(14), 12024–12032. doi:10.1074/jbc.M110.168625.

De Craene, B., & Berx, G. (2013). Regulatory networks defining EMT during cancer initiation and progression. *Nature Reviews Cancer, 13*(2), 97–110. doi:10.1038/nrc3447.

de Lau, W., Peng, W. C., Gros, P., & Clevers, H. (2014). The R-spondin/Lgr5/Rnf43 module: Regulator of Wnt signal strength. *Genes & Development, 28*(4), 305–316. doi:10.1101/gad.235473.113.

Dean, C., Ito, M., Makarenkova, H. P., Faber, S. C., & Lang, R. A. (2004). Bmp7 regulates branching morphogenesis of the lacrimal gland by promoting mesenchymal proliferation and condensation. *Development (Cambridge, England), 131*(17), 4155–4165. doi:10.1242/dev.01285.

Derynck, R., Muthusamy, B. P., & Saeteurn, K. Y. (2014). Signaling pathway cooperation in TGF-beta-induced epithelial-mesenchymal transition. *Current Opinion in Cell Biology, 31*, 56–66. doi:10.1016/j.ceb.2014.09.001.

Durand, C., Robin, C., Bollerot, K., Baron, M. H., Ottersbach, K., & Dzierzak, E. (2007). Embryonic stromal clones reveal developmental regulators of definitive hematopoietic stem cells. *Proceedings of the National Academy of Sciences of the United States of America, 104*(52), 20838–20843. doi:10.1073/pnas.0706923105.

Eilken, H. M., Nishikawa, S., & Schroeder, T. (2009). Continuous single-cell imaging of blood generation from haemogenic endothelium. *Nature, 457*(7231), 896–900. doi:10.1038/nature07760.

Ellis, M. J., Ding, L., Shen, D., Luo, J., Suman, V. J., Wallis, J. W., et al. (2012). Whole-genome analysis informs breast cancer response to aromatase inhibition. *Nature, 486*(7403), 353–360. doi:10.1038/nature11143.

Engl, W., Arasi, B., Yap, L. L., Thiery, J. P., & Viasnoff, V. (2014). Actin dynamics modulate mechanosensitive immobilization of E-cadherin at adherens junctions. *Nature Cell Biology, 16*(6), 587–594. doi:10.1038/ncb2973.

Erickson, S. L., O'Shea, K. S., Ghaboosi, N., Loverro, L., Frantz, G., Bauer, M., et al. (1997). ErbB3 is required for normal cerebellar and cardiac development: A comparison with ErbB2-and heregulin-deficient mice. *Development (Cambridge, England), 124*(24), 4999–5011.

Ewald, A. J., Huebner, R. J., Palsdottir, H., Lee, J. K., Perez, M. J., Jorgens, D. M., et al. (2012). Mammary collective cell migration involves transient loss of epithelial features and individual cell migration within the epithelium. *Journal of Cell Science, 125*(Pt 11), 2638–2654. doi:10.1242/jcs.096875.

Ferrari, N., McDonald, L., Morris, J. S., Cameron, E. R., & Blyth, K. (2013). RUNX2 in mammary gland development and breast cancer. *Journal of Cellular Physiology, 228*(6), 1137–1142. doi:10.1002/jcp.24285.

Ferrari, N., Riggio, A. I., Mason, S., McDonald, L., King, A., Higgins, T., et al. (2015). Runx2 contributes to the regenerative potential of the mammary epithelium. *Scientific Reports, 5*, 15658. doi:10.1038/srep15658.

Franceschi, R. T., & Xiao, G. (2003). Regulation of the osteoblast-specific transcription factor, Runx2: Responsiveness to multiple signal transduction pathways. *Journal of Cellular Biochemistry, 88*(3), 446–454. doi:10.1002/jcb.10369.

Fu, Y., Chang, A. C., Fournier, M., Chang, L., Niessen, K., & Karsan, A. (2011). RUNX3 maintains the mesenchymal phenotype after termination of the Notch signal. *The Journal of Biological Chemistry, 286*(13), 11803–11813. doi:10.1074/jbc.M111.222331.

Gao, X., Johnson, K. D., Chang, Y. I., Boyer, M. E., Dewey, C. N., Zhang, J., & Bresnick, E. H. (2013). Gata2 cis-element is required for hematopoietic stem cell generation in the mammalian embryo. *The Journal of Experimental Medicine, 210*(13), 2833–2842. doi:10.1084/jem.20130733.

Garg, V., Muth, A. N., Ransom, J. F., Schluterman, M. K., Barnes, R., King, I. N., et al. (2005). Mutations in NOTCH1 cause aortic valve disease. *Nature, 437*(7056), 270–274. doi:10.1038/nature03940.

Gering, M., & Patient, R. (2005). Hedgehog signaling is required for adult blood stem cell formation in zebrafish embryos. *Developmental Cell, 8*(3), 389–400. doi:10.1016/j.devcel.2005.01.010.

Gitler, A. D., Lu, M. M., Jiang, Y. Q., Epstein, J. A., & Gruber, P. J. (2003). Molecular markers of cardiac endocardial cushion development. *Developmental Dynamics: An Official Publication of the American Association of the Anatomists, 228*(4), 643–650. doi:10.1002/dvdy.10418.

Guo, W., Keckesova, Z., Donaher, J. L., Shibue, T., Tischler, V., Reinhardt, F., et al. (2012). Slug and Sox9 cooperatively determine the mammary stem cell state. *Cell, 148*(5), 1015–1028. doi:10.1016/j.cell.2012.02.008.

Hadland, B. K., Huppert, S. S., Kanungo, J., Xue, Y., Jiang, R., Gridley, T., et al. (2004). A requirement for Notch1 distinguishes 2 phases of definitive hematopoiesis during development. *Blood, 104*(10), 3097–3105. doi:10.1182/blood-2004-03-1224.

Herranz, N., Pasini, D., Diaz, V. M., Franci, C., Gutierrez, A., Dave, N., et al. (2008). Polycomb complex 2 is required for E-cadherin repression by the Snail1 transcription factor. *Molecular and Cellular Biology, 28*(15), 4772–4781. doi:10.1128/mcb.00323-08.

Hoshino, A., Costa-Silva, B., Shen, T. L., Rodrigues, G., Hashimoto, A., Tesic Mark, M., et al. (2015). Tumour exosome integrins determine organotropic metastasis. *Nature*. doi:10.1038/nature15756.

Huang, R. Y., Guilford, P., & Thiery, J. P. (2012). Early events in cell adhesion and polarity during epithelial-mesenchymal transition. *Journal of Cell Science, 125*(Pt 19), 4417–4422. doi:10.1242/jcs.099697.

Inman, C. K., & Shore, P. (2003). The osteoblast transcription factor Runx2 is expressed in mammary epithelial cells and mediates osteopontin expression. *The Journal of Biological Chemistry, 278*(49), 48684–48689. doi:10.1074/jbc.M308001200.

Inoue, A., Seidel, M. G., Wu, W., Kamizono, S., Ferrando, A. A., Bronson, R. T., et al. (2002). Slug, a highly conserved zinc finger transcriptional repressor, protects hematopoietic progenitor cells from radiation-induced apoptosis in vivo. *Cancer Cell, 2*(4), 279–288.

Ito, Y., Bae, S. C., & Chuang, L. S. (2015). The RUNX family: Developmental regulators in cancer. *Nature Reviews Cancer, 15*(2), 81–95. doi:10.1038/nrc3877.

Javed, A., Barnes, G. L., Pratap, J., Antkowiak, T., Gerstenfeld, L. C., van Wijnen, A. J., et al. (2005). Impaired intranuclear trafficking of Runx2 (AML3/CBFA1) transcription factors in breast cancer cells inhibits osteolysis in vivo. *Proceedings of the National Academy of Sciences of the United States of America, 102*(5), 1454–1459. doi:10.1073/pnas.0409121102.

Khalid, O., Baniwal, S. K., Purcell, D. J., Leclerc, N., Gabet, Y., Stallcup, M. R., et al. (2008). Modulation of Runx2 activity by estrogen receptor-alpha: Implications for osteoporosis and breast cancer. *Endocrinology, 149*(12), 5984–5995. doi:10.1210/en.2008-0680.

Kissa, K., & Herbomel, P. (2010). Blood stem cells emerge from aortic endothelium by a novel type of cell transition. *Nature, 464*(7285), 112–115. doi:10.1038/nature08761.

Kobayashi, I., Kobayashi-Sun, J., Kim, A. D., Pouget, C., Fujita, N., Suda, T., & Traver, D. (2014). Jam1a-Jam2a interactions regulate haematopoietic stem cell fate through Notch signalling. *Nature, 512*(7514), 319–323. doi:10.1038/nature13623.

Koeneman, K. S., Yeung, F., & Chung, L. W. (1999). Osteomimetic properties of prostate cancer cells: A hypothesis supporting the predilection of prostate cancer metastasis and growth in the bone environment. *The Prostate, 39*(4), 246–261.

Kouros-Mehr, H., & Werb, Z. (2006). Candidate regulators of mammary branching morphogenesis identified by genome-wide transcript analysis. *Developmental Dynamics: An Official Publication of the American Association of the Anatomists, 235*(12), 3404–3412. doi:10.1002/dvdy.20978.

Kovacic, J. C., Mercader, N., Torres, M., Boehm, M., & Fuster, V. (2012). Epithelial-to-mesenchymal and endothelial-to-mesenchymal transition: From cardiovascular development to disease. *Circulation, 125*(14), 1795–1808. doi:10.1161/circulationaha.111.040352.

Lam, E. Y., Hall, C. J., Crosier, P. S., Crosier, K. E., & Flores, M. V. (2010). Live imaging of Runx1 expression in the dorsal aorta tracks the emergence of blood progenitors from endothelial cells. *Blood, 116*(6), 909–914. doi:10.1182/blood-2010-01-264382.

Lambertini, E., Franceschetti, T., Torreggiani, E., Penolazzi, L., Pastore, A., Pelucchi, S., et al. (2010). SLUG: A new target of lymphoid enhancer factor-1 in human osteoblasts. *BMC Molecular Biology, 11*, 13. doi:10.1186/1471-2199-11-13.

Lamouille, S., Xu, J., & Derynck, R. (2014). Molecular mechanisms of epithelial-mesenchymal transition. *Nature Reviews Molecular Cell Biology, 15*(3), 178–196. doi:10.1038/nrm3758.

Lan, Y., He, W., Li, Z., Wang, Y., Wang, J., Gao, J., et al. (2014). Endothelial Smad4 restrains the transition to hematopoietic progenitors via suppression of ERK activation. *Blood, 123*(14), 2161–2171. doi:10.1182/blood-2013-09-526053.

Lancrin, C., Sroczynska, P., Stephenson, C., Allen, T., Kouskoff, V., & Lacaud, G. (2009). The haemangioblast generates haematopoietic cells through a haemogenic endothelium stage. *Nature, 457*(7231), 892–895. doi:10.1038/nature07679.

Lau, Q. C., Raja, E., Salto-Tellez, M., Liu, Q., Ito, K., Inoue, M., et al. (2006). RUNX3 is frequently inactivated by dual mechanisms of protein mislocalization and promoter hypermethylation in breast cancer. *Cancer Research, 66*(13), 6512–6520. doi:10.1158/0008-5472.can-06-0369.

Lee, K. S., Kim, H. J., Li, Q. L., Chi, X. Z., Ueta, C., Komori, T., et al. (2000). Runx2 is a common target of transforming growth factor beta1 and bone morphogenetic protein 2, and cooperation between Runx2 and Smad5 induces osteoblast-specific gene expression in the pluripotent mesenchymal precursor cell line C2C12. *Molecular and Cellular Biology, 20*(23), 8783–8792.

Lee, J. M., Shin, J. O., Cho, K. W., Hosoya, A., Cho, S. W., Lee, Y. S., et al. (2011). Runx3 is a crucial regulator of alveolar differentiation and lung tumorigenesis in mice. *Differentiation; research in biological diversity, 81*(4), 261–268. doi:10.1016/j.diff.2011.02.001.

Lee, Y., Manegold, J. E., Kim, A. D., Pouget, C., Stachura, D. L., Clements, W. K., & Traver, D. (2014). FGF signalling specifies haematopoietic stem cells through its regulation of somitic Notch signalling. *Nature Communications, 5*, 5583. doi:10.1038/ncomms6583.

Li, Q. L., Ito, K., Sakakura, C., Fukamachi, H., Inoue, K., Chi, X. Z., et al. (2002). Causal relationship between the loss of RUNX3 expression and gastric cancer. *Cell, 109*(1), 113–124.

Lian, J. B., & Stein, G. S. (2003). Runx2/Cbfa1: A multifunctional regulator of bone formation. *Current Pharmaceutical Design, 9*(32), 2677–2685.

Lim, M., Zhong, C., Yang, S., Bell, A. M., Cohen, M. B., & Roy-Burman, P. (2010). Runx2 regulates survivin expression in prostate cancer cells. *Laboratory Investigation; a journal of technical methods and pathology, 90*(2), 222–233. doi:10.1038/labinvest.2009.128.

Little, G. H., Noushmehr, H., Baniwal, S. K., Berman, B. P., Coetzee, G. A., & Frenkel, B. (2012). Genome-wide Runx2 occupancy in prostate cancer cells suggests a role in regulating secretion. *Nucleic Acids Research, 40*(8), 3538–3547. doi:10.1093/nar/gkr1219.

Little, G. H., Baniwal, S. K., Adisetiyo, H., Groshen, S., Chimge, N. O., Kim, S. Y., et al. (2014). Differential effects of RUNX2 on the androgen receptor in prostate cancer: Synergistic stimulation of a gene set exemplified by SNAI2 and subsequent invasiveness. *Cancer Research, 74*(10), 2857–2868. doi:10.1158/0008-5472.can-13-2003.

Lu, P., Sternlicht, M. D., & Werb, Z. (2006). Comparative mechanisms of branching morphogenesis in diverse systems. *Journal of Mammary Gland Biology and Neoplasia, 11*(3–4), 213–228. doi:10.1007/s10911-006-9027-z.

Ma, L., Lu, M. F., Schwartz, R. J., & Martin, J. F. (2005). Bmp2 is essential for cardiac cushion epithelial-mesenchymal transition and myocardial patterning. *Development (Cambridge, England), 132*(24), 5601–5611. doi:10.1242/dev.02156.

McDonald, L., Ferrari, N., Terry, A., Bell, M., Mohammed, Z. M., Orange, C., et al. (2014). RUNX2 correlates with subtype-specific breast cancer in a human tissue microarray, and ectopic expression of Runx2 perturbs differentiation in the mouse mammary gland. *Disease Models & Mechanisms, 7*(5), 525–534. doi:10.1242/dmm.015040.

McLarren, K. W., Lo, R., Grbavec, D., Thirunavukkarasu, K., Karsenty, G., & Stifani, S. (2000). The mammalian basic helix loop helix protein HES-1 binds to and modulates the transactivating function of the runt-related factor Cbfa1. *The Journal of Biological Chemistry, 275*(1), 530–538.

Medvinsky, A., & Dzierzak, E. (1996). Definitive hematopoiesis is autonomously initiated by the AGM region. *Cell, 86*(6), 897–906.

Mendez, M. G., Kojima, S., & Goldman, R. D. (2010). Vimentin induces changes in cell shape, motility, and adhesion during the epithelial to mesenchymal transition. *FASEB Journal : official publication of the Federation of American Societies for Experimental Biology, 24*(6), 1838–1851. doi:10.1096/fj.09-151639.

Mendoza-Villanueva, D., Deng, W., Lopez-Camacho, C., & Shore, P. (2010). The Runx transcriptional co-activator, CBFbeta, is essential for invasion of breast cancer cells. *Molecular Cancer, 9*, 171. doi:10.1186/1476-4598-9-171.

Mercado-Pimentel, M. E., Hubbard, A. D., & Runyan, R. B. (2007). Endoglin and Alk5 regulate epithelial-mesenchymal transformation during cardiac valve formation. *Developmental Biology, 304*(1), 420–432. doi:10.1016/j.ydbio.2006.12.038.

Mori, T., Nomoto, S., Koshikawa, K., Fujii, T., Sakai, M., Nishikawa, Y., et al. (2005). Decreased expression and frequent allelic inactivation of the RUNX3 gene at 1p36 in human hepatocellular carcinoma. *Liver International : official journal of the International Association for the Study of the Liver, 25*(2), 380–388. doi:10.1111/j.1478-3231.2005.1059.x.

Nakagawa, S., & Takeichi, M. (1995). Neural crest cell-cell adhesion controlled by sequential and subpopulation-specific expression of novel cadherins. *Development (Cambridge, England), 121*(5), 1321–1332.

Nakanishi, Y., Shiraha, H., Nishina, S., Tanaka, S., Matsubara, M., Horiguchi, S., et al. (2011). Loss of runt-related transcription factor 3 expression leads hepatocellular carcinoma cells to escape apoptosis. *BMC Cancer, 11*, 3. doi:10.1186/1471-2407-11-3.

Nakaya, Y., & Sheng, G. (2013). EMT in developmental morphogenesis. *Cancer Letters, 341*(1), 9–15. doi:10.1016/j.canlet.2013.02.037.

Nawshad, A., Medici, D., Liu, C. C., & Hay, E. D. (2007). TGFbeta3 inhibits E-cadherin gene expression in palate medial-edge epithelial cells through a Smad2-Smad4-LEF1 transcription complex. *Journal of Cell Science, 120*(Pt 9), 1646–1653. doi:10.1242/jcs.003129.

Network CGA. (2012). Comprehensive molecular portraits of human breast tumours. *Nature, 490*(7418), 61–70. doi:10.1038/nature11412.

Nigam, V., & Srivastava, D. (2009). Notch1 represses osteogenic pathways in aortic valve cells. *Journal of Molecular and Cellular Cardiology, 47*(6), 828–834. doi:10.1016/j.yjmcc.2009.08.008.

Nistico, P., Bissell, M. J., & Radisky, D. C. (2012). Epithelial-mesenchymal transition: general principles and pathological relevance with special emphasis on the role of matrix metalloproteinases. *Cold Spring Harbor Perspectives in Biology, 4*(2). doi:10.1101/cshperspect.a011908.

Niu, D. F., Kondo, T., Nakazawa, T., Oishi, N., Kawasaki, T., Mochizuki, K., et al. (2012). Transcription factor Runx2 is a regulator of epithelial-mesenchymal transition and invasion in thyroid carcinomas. *Laboratory Investigation; a journal of technical methods and pathology, 92*(8), 1181–1190. doi:10.1038/labinvest.2012.84.

North, T., Gu, T. L., Stacy, T., Wang, Q., Howard, L., Binder, M., et al. (1999). Cbfa2 is required for the formation of intra-aortic hematopoietic clusters. *Development (Cambridge, England), 126*(11), 2563–2575.

North, T. E., Goessling, W., Peeters, M., Li, P., Ceol, C., Lord, A. M., et al. (2009). Hematopoietic stem cell development is dependent on blood flow. *Cell, 137*(4), 736–748. doi:10.1016/j.cell.2009.04.023.

Okuda, T., van Deursen, J., Hiebert, S. W., Grosveld, G., & Downing, J. R. (1996). AML1, the target of multiple chromosomal translocations in human leukemia, is essential for normal fetal liver hematopoiesis. *Cell, 84*(2), 321–330.

Onodera, Y., Miki, Y., Suzuki, T., Takagi, K., Akahira, J., Sakyu, T., et al. (2010). Runx2 in human breast carcinoma: its potential roles in cancer progression. *Cancer Science, 101*(12), 2670–2675. doi:10.1111/j.1349-7006.2010.01742.x.

Osato, M., & Ito, Y. (2005). Increased dosage of the RUNX1/AML1 gene: A third mode of RUNX leukemia? *Critical Reviews in Eukaryotic Gene Expression, 15*(3), 217–228.

Otto, F., Thornell, A. P., Crompton, T., Denzel, A., Gilmour, K. C., Rosewell, I. R., et al. (1997). Cbfa1, a candidate gene for cleidocranial dysplasia syndrome, is essential for osteoblast differentiation and bone development. *Cell, 89*(5), 765–771.

Owens, T. W., Rogers, R. L., Best, S. A., Ledger, A., Mooney, A. M., Ferguson, A., et al. (2014). Runx2 is a novel regulator of mammary epithelial cell fate in development and breast cancer. *Cancer Research, 74*(18), 5277–5286. doi:10.1158/0008-5472.can-14-0053.

Paoli, P., Giannoni, E., & Chiarugi, P. (2013). Anoikis molecular pathways and its role in cancer progression. *Biochimica et Biophysica Acta, 1833*(12), 3481–3498. doi:10.1016/j.bbamcr.2013.06.026.

Peinado, H., Olmeda, D., & Cano, A. (2007). Snail, Zeb and bHLH factors in tumour progression: an alliance against the epithelial phenotype? *Nature Reviews Cancer, 7*(6), 415–428. doi:10.1038/nrc2131.

Postigo, A. A., Depp, J. L., Taylor, J. J., & Kroll, K. L. (2003). Regulation of Smad signaling through a differential recruitment of coactivators and corepressors by ZEB proteins. *The EMBO Journal, 22*(10), 2453–2462. doi:10.1093/emboj/cdg226.

Pouget, C., Peterkin, T., Simoes, F. C., Lee, Y., Traver, D., & Patient, R. (2014). FGF signalling restricts haematopoietic stem cell specification via modulation of the BMP pathway. *Nature Communications, 5*, 5588. doi:10.1038/ncomms6588.

Pratap, J., Javed, A., Languino, L. R., van Wijnen, A. J., Stein, J. L., Stein, G. S., & Lian, J. B. (2005). The Runx2 osteogenic transcription factor regulates matrix metalloproteinase 9 in bone metastatic cancer cells and controls cell invasion. *Molecular and Cellular Biology, 25*(19), 8581–8591. doi:10.1128/mcb.25.19.8581-8591.2005.

Rajamannan, N. M., Gersh, B., & Bonow, R. O. (2003). Calcific aortic stenosis: From bench to the bedside – emerging clinical and cellular concepts. *Heart (British Cardiac Society)., 89*(7), 801–805.

Richard, C., Drevon, C., Canto, P. Y., Villain, G., Bollerot, K., Lempereur, A., et al. (2013). Endothelio-mesenchymal interaction controls runx1 expression and modulates the notch pathway to initiate aortic hematopoiesis. *Developmental Cell, 24*(6), 600–611. doi:10.1016/j.devcel.2013.02.011.

Roodman, G. D. (2004). Mechanisms of bone metastasis. *The New England Journal of Medicine, 350*(16), 1655–1664. doi:10.1056/NEJMra030831.

Sanchez-Tillo, E., Lazaro, A., Torrent, R., Cuatrecasas, M., Vaquero, E. C., Castells, A., et al. (2010). ZEB1 represses E-cadherin and induces an EMT by recruiting the SWI/SNF chromatin-remodeling protein BRG1. *Oncogene, 29*(24), 3490–3500. doi:10.1038/onc.2010.102.

Schatteman, G. C., Morrison-Graham, K., van Koppen, A., Weston, J. A., & Bowen-Pope, D. F. (1992). Regulation and role of PDGF receptor alpha-subunit expression during embryogenesis. *Development (Cambridge, England), 115*(1), 123–131.

Selvamurugan, N., Kwok, S., & Partridge, N. C. (2004). Smad3 interacts with JunB and Cbfa1/Runx2 for transforming growth factor-beta1-stimulated collagenase-3 expression in human breast cancer cells. *The Journal of Biological Chemistry, 279*(26), 27764–27773. doi:10.1074/jbc.M312870200.

Stender, J. D., Kim, K., Charn, T. H., Komm, B., Chang, K. C., Kraus, W. L., et al. (2010). Genome-wide analysis of estrogen receptor alpha DNA binding and tethering mechanisms identifies Runx1 as a novel tethering factor in receptor-mediated transcriptional activation. *Molecular and Cellular Biology, 30*(16), 3943–3955. doi:10.1128/mcb.00118-10.

Tanaka, S., Shiraha, H., Nakanishi, Y., Nishina, S., Matsubara, M., Horiguchi, S., et al. (2012). Runt-related transcription factor 3 reverses epithelial-mesenchymal transition in hepatocellular carcinoma. *International Journal of Cancer/Journal international du cancer, 131*(11), 2537–2546. doi:10.1002/ijc.27575.

Taoudi, S., & Medvinsky, A. (2007). Functional identification of the hematopoietic stem cell niche in the ventral domain of the embryonic dorsal aorta. *Proceedings of the National Academy of Sciences of the United States of America, 104*(22), 9399–9403. doi:10.1073/pnas.0700984104.

Tavares, A. L., Mercado-Pimentel, M. E., Runyan, R. B., & Kitten, G. T. (2006). TGF beta-mediated RhoA expression is necessary for epithelial-mesenchymal transition in the embryonic chick heart. *Developmental Dynamics: An Official Publication of the American Association of the Anatomists, 235*(6), 1589–1598. doi:10.1002/dvdy.20771.

Theveneau, E., & Mayor, R. (2012). Cadherins in collective cell migration of mesenchymal cells. *Current Opinion in Cell Biology, 24*(5), 677–684. doi:10.1016/j.ceb.2012.08.002.

Thiery, J. P., Acloque, H., Huang, R. Y., & Nieto, M. A. (2009). Epithelial-mesenchymal transitions in development and disease. *Cell, 139*(5), 871–890. doi:10.1016/j.cell.2009.11.007.

Vallin, J., Girault, J. M., Thiery, J. P., & Broders, F. (1998). Xenopus cadherin-11 is expressed in different populations of migrating neural crest cells. *Mechanisms of Development, 75*(1–2), 171–174.

van Bragt, M. P., Hu, X., Xie, Y., & Li, Z. (2014). RUNX1, a transcription factor mutated in breast cancer, controls the fate of ER-positive mammary luminal cells. *eLife, 3*, e03881. doi:10.7554/eLife.03881.

Van Den Akker, N. M., Lie-Venema, H., Maas, S., Eralp, I., DeRuiter, M. C., Poelmann, R. E., & Gittenberger-De Groot, A. C. (2005). Platelet-derived growth factors in the developing avian heart and maturating coronary vasculature. *Developmental Dynamics: An Official Publication of the American Association of the Anatomists, 233*(4), 1579–1588. doi:10.1002/dvdy.20476.

van Roy, F., & Berx, G. (2008). The cell-cell adhesion molecule E-cadherin. *Cellular and Molecular Life Sciences: CMLS, 65*(23), 3756–3788. doi:10.1007/s00018-008-8281-1.

Voon, D. C., Wang, H., Koo, J. K., Nguyen, T. A., Hor, Y. T., Chu, Y. S., et al. (2012). Runx3 protects gastric epithelial cells against epithelial-mesenchymal transition-induced cellular plasticity and tumorigenicity.

Stem Cells (Dayton, Ohio), 30(10), 2088–2099. doi:10.1002/stem.1183.

Voon, D. C., Wang, H., Koo, J. K., Chai, J. H., Hor, Y. T., Tan, T. Z., et al. (2013). EMT-induced stemness and tumorigenicity are fueled by the EGFR/Ras pathway. *PloS One, 8*(8), e70427. doi:10.1371/journal.pone.0070427.

Voon, D. C., Hor, Y. T., & Ito, Y. (2015). The RUNX complex: Reaching beyond haematopoiesis into immunity. *Immunology*. doi:10.1111/imm.12535.

Voronov, D., Gromova, A., Liu, D., Zoukhri, D., Medvinsky, A., Meech, R., & Makarenkova, H. P. (2013). Transcription factors Runx1 to 3 are expressed in the lacrimal gland epithelium and are involved in regulation of gland morphogenesis and regeneration. *Investigative Ophthalmology & Visual Science, 54*(5), 3115–3125. doi:10.1167/iovs.13-11791.

Wang, Q., Stacy, T., Binder, M., Marin-Padilla, M., Sharpe, A. H., & Speck, N. A. (1996). Disruption of the Cbfa2 gene causes necrosis and hemorrhaging in the central nervous system and blocks definitive hematopoiesis. *Proceedings of the National Academy of Sciences of the United States of America, 93*(8), 3444–3449.

Watanabe, K., Villarreal-Ponce, A., Sun, P., Salmans, M. L., Fallahi, M., Andersen, B., & Dai, X. (2014). Mammary morphogenesis and regeneration require the inhibition of EMT at terminal end buds by Ovol2 transcriptional repressor. *Developmental Cell, 29*(1), 59–74. doi:10.1016/j.devcel.2014.03.006.

Whittle, M. C., Izeradjene, K., Rani, P. G., Feng, L., Carlson, M. A., DelGiorno, K. E., et al. (2015). RUNX3 controls a metastatic switch in pancreatic ductal adenocarcinoma. *Cell, 161*(6), 1345–1360. doi:10.1016/j.cell.2015.04.048.

Wilkinson, R. N., Pouget, C., Gering, M., Russell, A. J., Davies, S. G., Kimelman, D., & Patient, R. (2009). Hedgehog and Bmp polarize hematopoietic stem cell emergence in the zebrafish dorsal aorta. *Developmental Cell, 16*(6), 909–916. doi:10.1016/j.devcel.2009.04.014.

Wu, W. S., Heinrichs, S., Xu, D., Garrison, S. P., Zambetti, G. P., Adams, J. M., & Look, A. T. (2005). Slug antagonizes p53-mediated apoptosis of hematopoietic progenitors by repressing puma. *Cell, 123*(4), 641–653. doi:10.1016/j.cell.2005.09.029.

Xiao, W. H., & Liu, W. W. (2004). Hemizygous deletion and hypermethylation of RUNX3 gene in hepatocellular carcinoma. *World Journal of Gastroenterology, 10*(3), 376–380.

Xu, H., Li, M., Zhou, Y., Wang, F., Li, X., Wang, L., & Fan, Q. (2015). S100A4 participates in epithelial-mesenchymal transition in breast cancer via targeting MMP2. *Tumour Biology : the journal of the International Society for Oncodevelopmental Biology and Medicine*. doi:10.1007/s13277-015-3709-3.

Yang, J., Fizazi, K., Peleg, S., Sikes, C. R., Raymond, A. K., Jamal, N., et al. (2001). Prostate cancer cells induce osteoblast differentiation through a Cbfa1-dependent pathway. *Cancer Research, 61*(14), 5652–5659.

Yang, M. H., Wu, M. Z., Chiou, S. H., Chen, P. M., Chang, S. Y., Liu, C. J., et al. (2008). Direct regulation of TWIST by HIF-1alpha promotes metastasis. *Nature Cell Biology, 10*(3), 295–305. doi:10.1038/ncb1691.

Yang, M. H., Hsu, D. S., Wang, H. W., Wang, H. J., Lan, H. Y., Yang, W. H., et al. (2010). Bmi1 is essential in Twist1-induced epithelial-mesenchymal transition. *Nature Cell Biology, 12*(10), 982–992. doi:10.1038/ncb2099.

Yang, F., Sun, L., Li, Q., Han, X., Lei, L., Zhang, H., & Shang, Y. (2012). SET8 promotes epithelial-mesenchymal transition and confers TWIST dual transcriptional activities. *The EMBO Journal, 31*(1), 110–123. doi:10.1038/emboj.2011.364.

Ye, X., Tam, W. L., Shibue, T., Kaygusuz, Y., Reinhardt, F., Ng Eaton, E., & Weinberg, R. A. (2015). Distinct EMT programs control normal mammary stem cells and tumour-initiating cells. *Nature, 525*(7568), 256–260. doi:10.1038/nature14897.

Yilmaz, M., & Christofori, G. (2009). EMT, the cytoskeleton, and cancer cell invasion. *Cancer Metastasis Reviews, 28*(1–2), 15–33. doi:10.1007/s10555-008-9169-0.

Yokomizo, T., Ogawa, M., Osato, M., Kanno, T., Yoshida, H., Fujimoto, T., et al. (2001). Requirement of Runx1/AML1/PEBP2alphaB for the generation of haematopoietic cells from endothelial cells. *Genes to Cells: Devoted to Molecular & Cellular Mechanisms, 6*(1), 13–23.

Yonemura, S., Wada, Y., Watanabe, T., Nagafuchi, A., & Shibata, M. (2010). alpha-Catenin as a tension transducer that induces adherens junction development. *Nature Cell Biology, 12*(6), 533–542. doi:10.1038/ncb2055.

You, S., Avidan, O., Tariq, A., Ahluwalia, I., Stark, P. C., Kublin, C. L., & Zoukhri, D. (2012). Role of epithelial-mesenchymal transition in repair of the lacrimal gland after experimentally induced injury. *Investigative Ophthalmology & Visual Science, 53*(1), 126–135. doi:10.1167/iovs.11-7893.

Yue, R., Li, H., Liu, H., Li, Y., Wei, B., Gao, G., et al. (2012). Thrombin receptor regulates hematopoiesis and endothelial-to-hematopoietic transition. *Developmental Cell, 22*(5), 1092–1100. doi:10.1016/j.devcel.2012.01.025.

Zayzafoon, M., Abdulkadir, S. A., & McDonald, J. M. (2004). Notch signaling and ERK activation are important for the osteomimetic properties of prostate cancer bone metastatic cell lines. *The Journal of Biological Chemistry, 279*(5), 3662–3670. doi:10.1074/jbc.M308158200.

Zelzer, E., Glotzer, D. J., Hartmann, C., Thomas, D., Fukai, N., Soker, S., & Olsen, B. R. (2001). Tissue specific regulation of VEGF expression during bone development requires Cbfa1/Runx2. *Mechanisms of Development, 106*(1–2), 97–106.

Zhang, C., Lv, J., He, Q., Wang, S., Gao, Y., Meng, A., et al. (2014). Inhibition of endothelial ERK signalling by Smad1/5 is essential for haematopoietic stem cell emergence. *Nature Communications, 5*, 3431. doi:10.1038/ncomms4431.

Zoukhri, D., Macari, E., & Kublin, C. L. (2007). A single injection of interleukin-1 induces reversible aqueous-tear deficiency, lacrimal gland inflammation, and acinar and ductal cell proliferation. *Experimental Eye Research, 84*(5), 894–904. doi:10.1016/j.exer.2007.01.015.

Zoukhri, D., Fix, A., Alroy, J., & Kublin, C. L. (2008). Mechanisms of murine lacrimal gland repair after experimentally induced inflammation. *Investigative Ophthalmology & Visual Science, 49*(10), 4399–4406. doi:10.1167/iovs.08-1730.

A Regulatory Role for RUNX1, RUNX3 in the Maintenance of Genomic Integrity

29

Vaidehi Krishnan and Yoshiaki Ito

Abstract

All human cells are constantly attacked by endogenous and exogenous agents that damage the integrity of their genomes. Yet, the ensuing damage is mostly fixed and very rarely gives rise to genomic defects that promote cancer formation. This is due to the co-ordinated functioning of DNA repair proteins and checkpoint mechanisms that accurately detect and repair DNA damage to ensure genomic fitness. According to accumulating evidence, the RUNX family of transcription factors participate in the maintenance of genomic stability through transcriptional and non-transcriptional mechanisms. *RUNX1* and *RUNX3* maintain genomic integrity in a transcriptional manner by regulating the transactivation of apoptotic genes following DNA damage via complex formation with p53. RUNX1 and RUNX3 also maintain genomic integrity in a non-transcriptional manner during interstand crosslink repair by promoting the recruitment of FANCD2 to sites of DNA damage. Since *RUNX* genes are frequently aberrant in human cancer, here, we argue that one of the major modes by which *RUNX* inactivation promotes neoplastic transformation is through the loss of genomic integrity. In particular, there exists strong evidence that leukemic RUNX1-fusions such as RUNX1-ETO disrupt genomic integrity and induce a "mutator" phenotype during the early stages of leukemogenesis. Consistent with increased DNA damage accumulation induced by RUNX1-ETO, PARP inhibition has been shown to be an effective synthetic-lethal therapeutic approach against RUNX1-ETO expressing leukemias. Here, in this chapter we will examine current evidence suggesting that the tumor suppressor potential of RUNX proteins can be at least partly attributed to their ability to ensure high-fidelity DNA repair and thus prevent mutational accumulation during cancer progression.

V. Krishnan (✉) • Y. Ito
Cancer Science Institute of Singapore, National University of Singapore,
Singapore 117599, Singapore
e-mail: csivk@nus.edu.sg; csiitoy@nus.edu.sg

Y. Groner et al. (eds.), *RUNX Proteins in Development and Cancer*, Advances in Experimental Medicine and Biology 962, DOI 10.1007/978-981-10-3233-2_29

Keywords

RUNX • DNA repair • DNA damage response • Fanconi anemia • Genomic instability • Cancer

29.1 Introduction

Cancer was nothing more than a mystery before the discovery of DNA as the genetic material. Yet, as early as in 1902, Theodor Boveri accurately concluded that the cancer cell was a changed normal cell in that tiny microscopic bodies called chromosomes were abnormally distributed within cancer (Boveri 2008). By the 1930s, it was possible to perform karyotypic analysis of cancer cells. Through these studies came the realization that whereas a normal human cell had 46 chromosomes, cancer cells often had an abnormal number of chromosomes, often in excess of 46. Following the discovery of DNA as the genetic material, genes were identified as the constituting units of chromosomes and that mutation in DNA involved a change in the chemical structure of DNA. Later, many chemicals that were demonstrated to mutate DNA by forming adducts turned out to be human carcinogens (Jeggo et al. 2016). These and other pioneering work carried out over several decades created the paradigm that "genomic instability", comprising of abnormalities in chromosome number and/or aberrations in chromosome structure like deletions, duplication, insertions or mutations, is an important hallmark for most human cancers and generates the genetic diversity required to expedite the acquisition of the cancer phenotype (Hanahan and Weinberg 2011; Negrini et al. 2010).

A century since Theodor Boveri's original observation, the advent of next-generation sequencing has revolutionized our understanding of the genomic landscape of human cancer. Using sophisticated genomic tools and bioinformatic approaches to examine the mutational patterns of 7042 cancers, more than 20 distinct so-called "mutational signatures" have been extracted (Alexandrov et al. 2013a, b). These mutational signatures are shown to be dictated by mutagen exposure, altered DNA repair processes, the chronological age of patient and proliferation rate of the tissue and can hence be used as a retrospective blueprint to trace the history of cancer genesis (Helleday et al. 2014; Nik-Zainal et al. 2012, 2015). One such mutational signature, the 'CC→TT' mutations are an outcome of the failure to repair cyclobutane pyrimidine dimers generated by ultraviolet light in skin cancers. Hence, the presence of C>T signature would implicate UV as the root cause of carcinogenesis in such a patient. Cancers also carry another mutational signature called as Kataegis, which is a processive clustering of cytosine mutations due to the hyperactivity of DNA editing enzymes. However, the underlying aetiology for a vast majority of mutational signatures remain unknown (Helleday et al. 2014). For instance, chromoplexy (balanced chains of rearrangements across many chromosomes), or chromoanasynthesis (localized gains or losses of chromosomal material by replication based-mechanisms) and other types genomic instability arise by as-yet-unknown cancer specific defects and/or exposure to mutagenic agents (Baca et al. 2013; Holland and Cleveland 2012).

In the following sections, we will examine the three principal underlying reasons for genomic instability in human cancer. As described in the schematic shown in figure (Fig. 29.1), genomic instability accumulates during cancer progression as a function of three main factors (a) Heightened exposure to endogenous and exogenous DNA damaging agents in a chronic manner (b) Reduced fidelity of DNA repair due to downregulation of DNA repair proteins or enzymes (c) Loss of DNA damage checkpoint signalling allows a cell with damaged genome to persist in the population and propagate its mutation to the progeny during cell division. Apart from discussions on mechanisms inducing genomic instability in cancers, we will

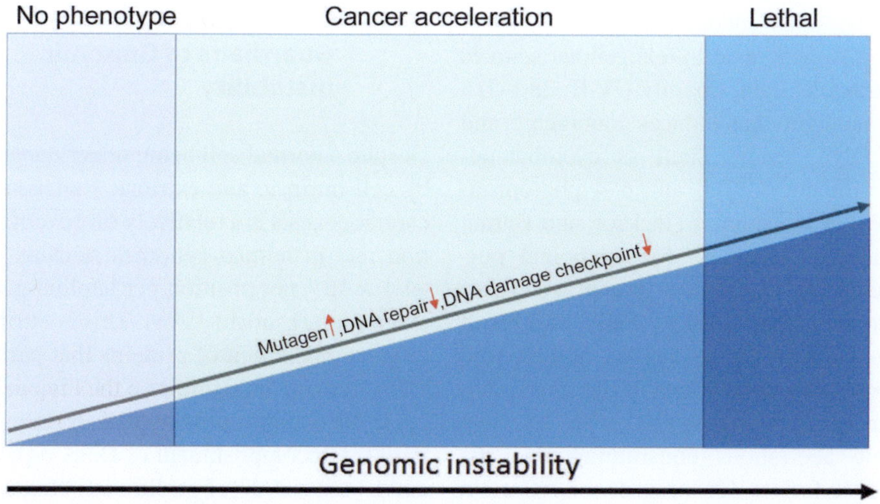

Fig. 29.1 The interplay between exposure to mutagenic agents, DNA repair competence and robust DNA damage checkpoints determines whether a cell accumulates genomic instability during cancer progression. In the "no phenotype" status, either cells are not facing high loads of extrinsic or intrinsic DNA damage, or have robust DNA repair and intact cell cycle checkpoint machinery. Hence, this "low range of genomic instability has little, if any phenotypic consequences. In the "permissive-state" for accelerated carcinogenesis, genomic instability levels are intermediate and hence most suitable for promoting a mutator phenotype in tumors. This stage can result either due to DNA repair downregulation or/and checkpoint loss. In the "lethal stage", genomic instability levels are too high and hence such cells are eliminated from the population with no consequence to cancer progression. This concept was further supported by the genomic analysis of copy-number alterations (CNA) as a measure of genomic instability across 12 cancer types (Andor et al. 2016). It was found that CNA affecting either <25 % or >75 % of a tumour's genome predicted for reduced risk of mortality, whereas cancers bearing a CNA abundance of 50–75 % were identified as the highest risk group amongst patients among six of the cancers studied

also examine current evidence that the tumor suppressor potential of Runx proteins can be partly attributed to their ability to maintain genomic integrity. We will discuss current evidence that *RUNX* loss can predict for heightened sensitivity to DNA repair inhibitors, providing a novel therapeutic avenue that exploits the principal of synthetic lethality to target *RUNX*-deficient tumours.

29.2 Genomic Instability in Human Cancer

29.2.1 Sources of Exogenous and Endogenous DNA Damaging Agents

An average eukaryotic cell encounters about 100,000 lesions per cell per day. Endogenous by-products of metabolism like reactive oxygen species (ROS) and their highly reactive intermediates account for a vast majority of endogenous DNA damage in normal cells (Jackson and Bartek 2009). DNA itself is sensitive to spontaneous hydrolysis resulting in abasic sites and base deamination (1000 per cell per day). Impaired dNTP incorporation during DNA replication can also account for a significant proportion of DNA alterations. More recently, aldehydes, an intermediate various metabolic pathways or alcohol metabolism have been proved to be potent DNA damaging agents (Pontel et al. 2015). Similarly, certain metabolic by-products like deoxycholic acid released by intestinal bacteria can cause DNA damage and liver cancer (Yoshimoto et al. 2013). Endogenous stresses like endoplasmic reticulum stress also generates DNA damage through the expression of inflammatory mediators like leukotriene C4 (Dvash et al. 2015). Intriguingly, the hormone estrogen and estrogen-induced metabolites have also shown to form adducts with DNA and induce gene mutations in breast cancer (Santen et al. 2015).

Apart from such endogenous sources, DNA damage is also generated by exogenous sources such as ultraviolet light, (mainly UV-B: 280–315 nm) from sunlight that induces mutagenic and cytotoxic DNA lesions such as cyclobutane-pyrimidine dimers (CPDs) and 6–4 photoproducts (64PPs) in melanomas (Jackson and Bartek 2009). Radiation exposure for medicinal purposes or accidental exposure to radiation can give to DNA double strand breaks (DSBs) and result in greater incidence of leukemias and thyroid cancers. Tobacco smoke is one of the most powerful exogenous carcinogens and causes a variety of genomic aberrations constituting one the strongest risk factors for lung cancer. Certain herbal remedies such as Aristolochic acid are powerful mutagens and cause upper urinary-tract urothelial cancer by mediating A>T mutations (Poon et al. 2013).

Some viruses and bacteria also pose a genomic threat by causing DNA damage and such pathways are implicated in the carcinogenesis initiated by these pathogens. For example, the human papillomavirus induces DNA double strand breaks through the viral proteins E1 and E7 (Duensing and Munger 2002). Infection with *Helicobacter.pylori*, a strong predisposition factor for gastric cancers, generates high levels of oxidative stress due to the upregulation of the pro-oxidant, spermine oxidase (Xu et al. 2004). Many cancer-associated genes like activated RAS and MYC increase DNA damage through heightened ROS production, which facilitates tumorigenic signaling and metabolic reprogramming in cancer cells (Maya-Mendoza et al. 2015). The overexpression of MYC and RAS can also lead to DNA replication stress, a form of genomic instability associated with the depletion of the cellular pool of nucleotides (dNTPs) due to the increased firing of replication origins (Lecona and Fernandez-Capetillo 2014). Chronic inflammation has also been recognised as an important factor for carcinogenesis, because inflammatory cells release ROS and reactive nitrogen species (RNS) that result in oxidative and nitrative DNA damage through the formation of 8-oxo-7,8-dihydro-2′-deoxyguanosine (8-oxodG) adducts and 8-nitroguanine, respectively.

29.2.2 DNA Repair Pathways Guardians of Genomic Instability

Despite a normal cell being under constant attack by cell-intrinsic and extrinsic mutagenic agents, carcinogenesis is a relatively rare event and mutation rate in human genomes remains at rate of $\sim 1.1 \times 10^{-8}$ per position per haploid genome per generation (Lindahl 1993). This is attributed to a large compendium of proteins that participate in DNA damage recognition, DNA repair or in cell cycle checkpoint to delay cell cycle progression until the accomplishment of DNA repair (Harper and Elledge 2007). For different classes of DNA damaging lesions, cells utilise biochemically distinct sets of DNA repair proteins with some degree of overlap thus constituting a comprehensive toolkit for genome maintenance (Fig. 29.2).

For example, DNA replication errors or errors resulting from polymerases result in single base mismatches or insertion-deletion loops and these are corrected by the DNA **MisMatch repair** (MMR) pathway (Kunkel and Erie 2005). While most such DNA replication-associated errors are corrected by the proofreading activity of DNA pol-δ, some may escape correction especially in regions of DNA carrying repetitive mono or dinucleotide repeats. In such genomic regions, the action of the MMR pathway becomes very critical. The "Mut" proteins are the main active components of the MMR pathway and they scan the newly synthesised DNA for bulges and loops arising from mismatches. The two main components of the MMR apparatus, MutS and MutL correct mismatched DNA.

Small chemical alterations that do not cause any distortion of the double helix are repaired by the **Base excision repair** (BER) pathway (Wilson and Bohr 2007). This repair pathway handles mutagenic lesions that arise by spontaneous hydrolytic non-enzymatic DNA alkylation reactions or oxidised bases like 8-oxo-G. During BER, DNA glycosylases accomplish repair by cleaving the *N*-glycosyl bond between the damaged base and sugar. The APE1 lyase cuts out the base-free deoxyribose to generate 5′-deoxyribose phosphate termini (dRP) which is further

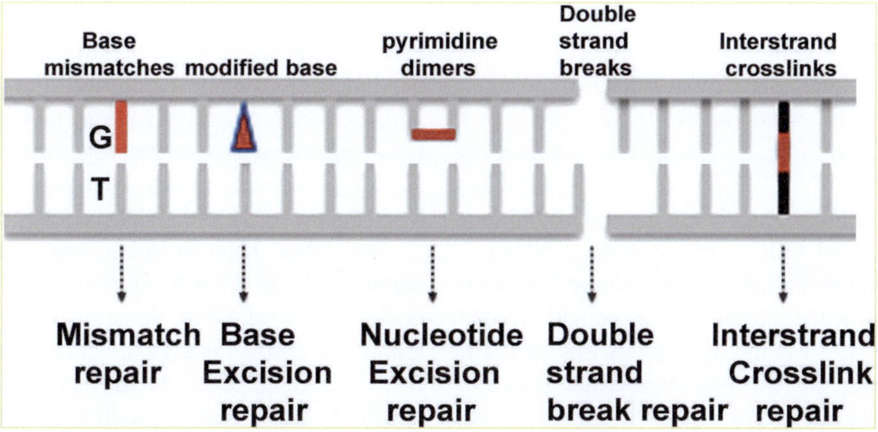

Fig. 29.2 In response to different classes of DNA damaging lesions, cells activate distinct sets of repair pathways such as Mismatch repair, Base excision repair, Nucleotide excision repair, Double strand break repair and Interstrand crosslink repair to cope with each type of DNA damaging lesion

removed by the action of the DNA polymerase β (pol-β) in the short-patch BER pathway. When the 5′ terminals are refractory to pol-β activity, strand displacement synthesis incorporates multiple nucleotides by the long patch BER pathway. In this case, several enzymes such as PCNA (proliferating cell nuclear antigen), FEN1 and pol-β and/or pol-δ/ε act together to remove the blocking terminus. The final step of BER consists of ligation of the remaining nick, by either DNA Lig1 alone or the DNA Lig3–XRCC1 complex.

When single strand breaks arise by ROS-induced disintegration of oxidized deoxyribose, genotoxic stresses such as IR, by the erroneous activity of DNA topoisomerase 1 (TOP1) or as a BER intermediate, the **Single Strand Break Repair (SSBR)** pathway is utilized. SSBR utilizes many of the same proteins as BER, such as APE1, pol-β, and DNA Lig3, but along with the scaffold proteins, PARP1 and XRCC1. SSBR involves the activation of PARP (poly-ADP ribose polymerase) which act as sensors of SSBs, through their two zinc finger motifs (Caldecott 2008). PARP activation results in the synthesis of poly (ADP-ribose) chains which are then assembled onto target proteins like histones H1/H2B and PARP1 itself (Schreiber et al. 2006).

Lesions that cause a structural distortion of DNA such as pyrimidine dimers induced by ultraviolet light or DNA intrastrand crosslinks are repaired by the **Nucleotide Excision Repair** (NER) pathway (Guo et al. 2010; Nouspikel 2009). Depending on the molecular mechanism used to identify the damaged base, NER can be further subdivided into two types- **Global genome repair** (GGR) which repairs damage throughout the genome or **Transcription-coupled repair** (TCR) which repairs DNA lesions in genomic regions undergoing transcription. During **GGR**, the DDB1/DDB2 heterodimer binds to UV-damaged sites, recruits XPC, a constituent of the XPC/HR23B/Centrin heterodimer (Sugasawa et al. 2005). In a sequential reaction, XPC, in turn recruits the TFIIH (transcription factor IIH) complex, followed by recruitment of XPA and RPA. The two helicases within the TFIIH complex (XPB and XPD) unwind the DNA by about 20 base pair around the damage, followed by exit of XPC-HR23 and the entry of the endonucleases XPG and XPF-ERCC1. The resulting gap is filled in by the combined actions of DNA pol-δ or ε, proliferating cell nuclear antigen (PCNA), RPA, and DNA ligase I. **Transcription-coupled repair** is initiated at genomic regions undergoing transcription where bulky lesions lead to RNA polymerase II stalling (Hanawalt and Spivak 2008). In a process regulated by the Cockayne Syndrome proteins, CSA and CSB, stalled RNA polymerase is evicted and additional factors like the histone acetytransferase p300, the CSA-DDB1 E3 ubiquitin/COP9 signalosome complex are recruited. The latter steps of TCR are very similar to GGR.

DNA double strand breaks (DSBs) are one of the most deleterious lesions within cells and are repaired either by utilising the **non-homologous end joining** (NHEJ) pathway or the **homologous recombination** (HR) repair pathway. NHEJ is error-prone DNA repair process, where DSB ends that do not carry any homology are joined together by a repair process that is requires the DNA-PK (DNA protein kinase) complex and the Lig4 complex (Lieber 2010). The DNA-PK complex comprises of the KU70/Ku80 heterodimer, which bind to DSBs and recruit the enzyme DNA-PKcs (PRKDC), the catalytic subunit of DNA-PK. This results in DNA-PKcs auto phosphorylation and destabilization of DNA-PKcs interaction with DNA, providing access to enzymes such as Artemis that are involved in end processing. End processing is also carried out by APLF nucleases and the PNK kinase/phosphatase. If the end processing results in gap generation, then these gaps are filled in by DNA pol-μ, DNA pol-λ and terminal deoxyribonucleotidyltransferase. The final step involves the ligation of DNA end by X4-L4 (a complex containing XRCC4, DNA ligase IV and XLF) (Hartlerode and Scully 2009). When classical NHEJ pathway is non-functional, cells utilise a back-up pathway called as **alternative-NHEJ** or **MMEJ** (microhomology-mediated end joining) (Simsek and Jasin 2010). Here, cells utilise microhomology sequences of 1–10 base pairs that often appear at end junctions to mediate end joining. Alt-NHEJ is a more error-prone process than NHEJ and often is the reason for small deletions and insertions at the site of double strand breaks in human cancers.

In contrast to NHEJ, **homologous recombination** (HR) is an error-free process that involves the use of the intact homologues sister chromatid as the template for DNA repair (Moynahan and Jasin 2010). HR involves the detection of the double strand break (DSB) by the MRN complex (Mre11-Rad5-Nbs1) and ATM activation by autophosphorylation. MRE11 along with CtIP generate single stranded DNA (ssDNA) at the DSB site by carrying out end resection. Next, Rad51 is assimilated to ssDNA in a BRCA2/PALB2 dependent manner, stimulating Rad51-dependent homology search and strand invasion and the formation of the Holliday junction. DNA-pol δ-mediated DNA synthesis ensues and the Holliday junction is resolved via the Rad54/Mus81/Emc1 and Rad51C/Xrcc3 complexes.

The DNA **Interstrand Cross Link (ICL) repair** pathway is involved in the repair of a specialized type of DNA damage, where the two strand of DNA undergo covalent crosslinking (Ceccaldi et al. 2016; Moldovan and D'Andrea 2009). At least 18 gene products are involved in the successful repair of DNA ICLs and mutation in any one of them can give rise to the human cancer predisposition syndrome, Fanconi Anemia (FA). The FA pathway of ICL repair is activated when DNA replication forks encounter an interstrand crosslinked DNA and are hence stalled in the S phase of cell cycle. The repair process is initiated by the FA core complex (FANCA, FANCB, FANCC, FANCE, FANCF, FANCG, FANCL and FANCM) which is a multi-subunit ubiquitin ligase that monoubiquitinates the central repair proteins FANCD2 at lysine 561 and FANCI at lysine 523. Monoubiquitinated FANCD2 in turn, recruits the nucleases FAN1 and SLX4 (FANCP), which together with XPF/ERCC4 (FANCQ) incise on both sides of the crosslink. Finally, the original DNA double helix is restored by HR repair by the DNA repair proteins FANCD1 (BRCA2), FANCN (PALB2), FANCJ (BRIP1) and FANCO (RAD51C) together with translesion repair proteins. There are also several other proteins like FAAP20, FAAP24, FAAP100, MHF1, MHF2, USP1, that actively participate in the ICL repair process. Defects in any of these proteins can also result in ICL repair defect, although mutations in these proteins have not been uncovered in FA patients as yet.

29.2.3 Cancer Predisposition Syndromes, Mutator Phenotype and Human Cancer

The best evidence for a causal relationship between loss of DNA repair and increased predisposition towards malignancies emerged through the study of familial cancer-predisposition syndromes. In one of the earliest such descriptions, Xeroderma Pigmentosa patients defective for NER fail to repair UV-induced DNA damage and exhibit a ~1000-fold greater risk for skin cancers and a ~100,000-fold greater risk for squamous cell carcinoma of the tongue, as compared to the general population (Guo et al. 2010). Similarly, FA patients fail to repair ICL type of DNA damage and hence their risk for getting all types of cancers including leukaemia and solid tumours is ~50-fold higher than the non-FA population (Alter 2003). The loss of MMR proteins such as MSH2 and MLH1 results in the Lynch syndrome, also called as the hereditary non-polyposis colorectal cancer (HNPCC), an inherited disorder that increases the risk of many types of cancer, particularly colorectal cancers in such families (Leach et al. 1993). Interestingly, most cases of familial cancers are attributed to defects in genes encoding for DNA repair or DNA damage signaling (summarised in Table 29.1), such as Xeroderma pigmentosa (NER defects), Fanconi Anemia (ICL repair defect), Ataxia-Telangiectasia (DNA damage signalling defect) and Li-Fraumeni syndrome (p53 mutation).

In early 1990s, it was discovered that ~15 % of sporadic gastric, endometrial and colon tumours had defect in MMR due to promoter hypermethylation and the consequent silencing of MLH1. MMR defect gave rise to microsatellite instability, a phenomenon that caused concomitant mutations in hundreds of downstream genes some of which may confer a proliferative advantage to incipient cancer cells and drive tumor progression (Shibata et al. 1994). These observations in turn gave rise to the idea of a 'mutator phenotype' associated with cancer progression, wherein the mutation of a single DNA repair gene accelerates carcinogenesis by dramatically increasing the mutability of the entire

Table 29.1 Table depicting familial DNA repair syndromes and the genes and cancers associated with them

Human syndrome	Mutated gene	Cancer phenotypes	Disrupted DNA repair pathway
Hereditary non-polyposis colorectal cancer (HNPCC)	*MSH2, MSH3, MSH6, MLH1, PMS2*	Colon and gynaecological cancers	MMR
Rothmund Thomsun syndrome (RTS)	*RECQL4*	Osteosarcomas and skin cancers	BER
Xeroderma pigmentosum	*XPA-XPG, POL H*	Skin cancers	NER
Ataxia Telangiectasia (AT)	*ATM*	Increased risk for leukemias and lymphoid malignancies	DNA damage signalling
Nijmegan break syndrome (NBS)	*NBS1*	B cell lymphoma	DNA damage signalling
Li-Fraumeni syndrome	*TP53*	Brain, breast cancer, sarcomas,leukemias, melanomas and gastrointestinal cancers	DNA damage signalling
Bloom syndrome	*BLM*	Elevated predisposition to all cancers	HR
Werner syndrome	*WRN*	Cancer predisposition	HR, BER, telomere maintenance
Early onset breast cancer	*BRCA1*	Breast and ovarian cancer	HR
Early onset breast cancer	*BRCA2*	Breast and ovarian cancer	HR
Fanconi anemia	*FANCA-FANCL, BRCA2 (FANCD1)*	Greater predisposition to acute myeloid leukemias, myelodysplasia, head and neck squamous cell carcinomas	ICL repair and HR

genome. However, the concept of a "mutator phenotype" as the basis for sporadic tumours languished for the next 20 years since the time because apart from MMR gene mutations, pervasive DNA repair gene mutations were seldom retrieved in spontaneous human cancers by small-scale sequencing efforts.

More recently, the advent of more sophisticated next-generation sequencing technologies have brought the mutator phenotype hypothesis back to the fore. For example, in a study published by the TCGA (The Cancer Genome Atlas) consortium on ovarian cancers, pathway analysis revealed that ~50 % of ovarian cancers harboured defects in DSB repair by homologous recombination repair (*BRCA1* and *BRCA2* aberrations, *RAD51C* hypermethylation, *ATM/ATR* mutations, *FANCD2* mutations). A cross-cancer alteration summary for ten commonly mutated DNA repair genes *ATM*, *ATR*, *BRCA1*, *BRCA2*, *FANCD2*, *MDC1*, *MLH1*, *MSH2*, *PRKDC*, *RAD51* (126 cancer studies/ at least 50 cases per study/at least 5 % mutation) is shown in Fig. 29.3.

Based on these analyses, it is currently accepted that widespread mutations in DNA repair proteins can be observed in bladder, ovarian, gastric, breast, lung, melanoma and other solid tumours, although future studies are required to understand the functional relevance of these mutations.

29.2.4 DNA Damage Checkpoint Pathways and Oncogene-Induced Senescence as Barriers to Carcinogenesis

In the preceding sections, it was discussed that heightened exposure to mutagens or DNA repair down regulation can result in genomic instability during cancer progression. However, cells have an additional tier of protection in the form of DNA damage checkpoint, which has to be breached for cells bearing DNA double strand breaks or replications stress to be clonally selected for during cancer evolution.

The DNA damage checkpoint is activated following the successful detection of DNA damage by the sensor proteins like ATM and ATR to delay cell cycle progression until the accomplishment of DNA repair (Bartek and Lukas 2007). The DNA damage checkpoints prevent cell cycle transitions by decreasing cyclin-dependent kinase (CDK) activity through ATM and ATR signaling. For instance, ATM activation by DSBs in G1 leads to Chk2 phosphorylation which, in turn phosphorylates the phosphatase CDC25A. The resultant CDC25A degradation in turn prevents the Thr14/Tyr15 dephosphorylation-mediated activation of CDK2, preventing cell cycle transition from G1 to S phase. Through an independent mechanism, ATM or CHK2-mediated stabilization of p53 activates the transcription of the cell cycle inhibitor, p21 resulting in G1/S arrest. During the G2/M checkpoint, Chk1 and Chk2 phosphorylate and inactivate CDC25C phosphatase by promoting its inhibitory sequestration by 14-3-3 proteins. Hence, the CDC25C dependent activation of CDK1-Cyclin B complex is prevented and cells are arrest at the G2/M phase of cell cycle. Similarly, during the S phase checkpoint, ATR is activated in response to stalled replication forks. ATR phosphorylates CHK1 which in turn phosphorylates CDC25A. As a result, CDC25A undergoes degradation and prevents the activation of CDK2/Cyclin E, a kinase required for the initiation of new replication origins during DNA replication.

The concept that **DNA damage signaling posed an anti-cancer barrier** was elegantly illustrated through a path-breaking work by Jiri Bartek and colleagues (Bartkova et al. 2005, 2006). By tracing cancer progression of skin, lung and breast cancers, it was shown that precancerous lesions experience high amounts of DNA damage, but such lesions did not progress to frank malignancy unless they lost checkpoint signaling by developing mutations in checkpoint proteins such as p53. According to this model, when normal cells are exposed to endogenous or exogenous mutagens generating DSBs (intracellular ROS) or experience oncogene-induced DNA replication stress, there are three possible outcomes. The cell cycle checkpoint machinery either stops the proliferation of cells with damaged genomes to allow DNA repair or triggers a state of irreversible cell cycle arrest called senes-

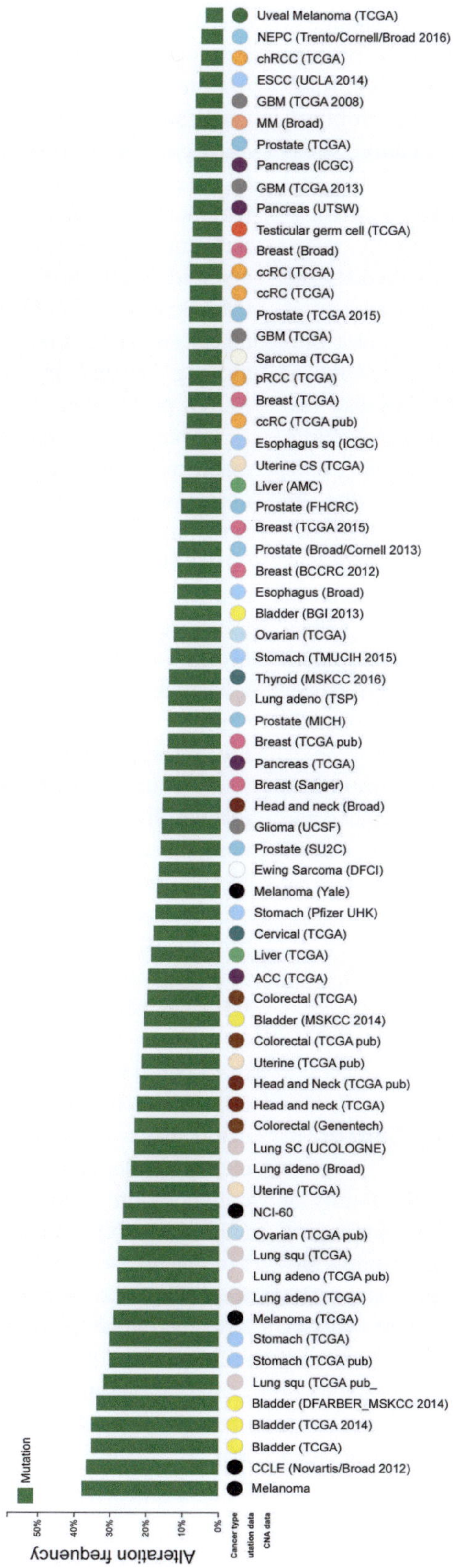

Fig. 29.3 A cross-cancer alteration summary for 10 commonly mutated DNA repair genes *ATM*, *ATR*, *BRCA1*, *BRCA2*, *FANCD2*, *MDC1*, *MLH1*, *MSH2*, *PRKDC*, *RAD51* (126 cancer studies/ at least 50 cases per study/at least 5 % mutation) is shown

cence or mediates apoptosis when the damage is too severe to be repaired. Consistently, overexpression of oncogenes such as Ras, Myc, Stat3 and E2F1 was shown to increase cell proliferation rate and generate DNA replication stress leading to persistent ATM/ATR activation and senescence (Gorgoulis and Halazonetis 2010).

The senescence pathway evoked by oncogenes upon hyper-activation was called as the **oncogene-induced senescence** pathway and is recognised as an important barrier to tumor progression. Consistently, precancerous lesions of the skin, lung and colon show both apoptosis and senescence at the early stages of tumor development, whereas these processes are actively suppressed during cancer progression due to breaches to this anti-cancer barrier (Bartkova et al. 2005). Taken together, tumor clones harbouring mutational or epigenetic inactivation of DNA damage checkpoint proteins are subsequently selected for during tumor development because they help in the evasion of anti-cancer barriers such as cell cycle checkpoint/senescence/apoptosis.

In the next section, we will review current data that *RUNX* misregulation in human cancer can disrupt the genomic integrity of incipient cancer cells to promote tumor progression.

29.3 *RUNX* and Regulation of Genomic Integrity

29.3.1 Evidence for RUNX1 and RUNX3 as Tumor Suppressors in Human Cancer

Soon after their discovery, the RUNX family of genes, *RUNX1*, *RUNX2* and *RUNX3* were mainly being studied as DNA binding transcription factors that during key developmental steps (Ito et al. 2015). It turned out that all three RUNX family members have important roles during lineage-specifying steps at the various stages of development. Indeed, *RUNX1* is essential in haematopoiesis, *RUNX2* indispensable for skeletal development and *RUNX3* is required for T cell development and for regulating the axonal projections of a specific subpopulation of dorsal root ganglion population. Gene-knockouts of each *Runx* gave rise to very well-defined developmental defects. However, at around the same time, it was identified that the *AML1* gene that undergoes frequent translocation, t(8;21), in human acute myeloid leukemia (AML) was in fact *RUNX1* (Maseki et al. 1993; Miyoshi et al. 1991). In turn, this observation spurred a lot of excitement in moving beyond the developmental roles of RUNX proteins and foray into cancer biology.

Perhaps the strongest links between *RUNX* inactivation and cancer progression have been made for *RUNX1* and *RUNX3*, whereas *RUNX2* is mainly implicated as an oncogene (Blyth et al. 2010). Hence, we focus below mainly on *RUNX1* and *RUNX3*. *RUNX1* aberrations in the form of translocations and mutations are found in acute myeloid leukemia (AML), acute lymphocytic leukemia, therapy-related myeloid leukemia, chronic myelomonocytic leukemia, and in myelodysplastic syndrome. Monoallelic germline mutations in *RUNX1* also result in familial platelet disorder predisposed to acute myelogenous leukemia (FPD/AML). *RUNX1* is also one of the most frequent targets of chromosomal translocations in AML, generating the fusion protein, RUNX1-ETO. The dominant inhibition of endogenous RUNX1 and RUNX3 by RUNX1-ETO has been considered as a key pathological factor that drives leukemic transformation in these cells. The tumor suppressor role of RUNX1 in leukemia has been convincingly demonstrated using many mouse models where *RUNX1* deficiency accelerated AML development alone or in collaboration with other oncogenes such as MLL-ENL, NRAS and EVI5 (Jacob et al. 2010; Motoda et al. 2007; Nishimoto et al. 2011). Recently, next generation sequencing studies have also unveiled point mutation of *RUNX1* in solid tumors, such as deletions in oesophageal cancers (7 %) and inactivating mutations in breast cancers (4 %) (Gao et al. 2013; Taniuchi et al. 2012).

Similarly, the inactivation of *RUNX3* has been unequivocally demonstrated across several human tumors. According to the first such report, *RUNX3* is rendered inactive due to a combination of hemizygous deletion and DNA hypermethylation of the *RUNX3* promoter region in 60 % of primary gastric cancer samples (Li et al. 2002). *RUNX3* promoter methylation was also observed in precursor lesions such as chronic gastritis, gastric adenomas, and colonic polyps, suggesting that this event is an early event during cancer progression. *RUNX3* deficiency has been reported in solid tumors of the lung, liver, bladder, breast, colon, laryngeal cancers and prostate amongst others.

The most convincing evidence for *RUNX3* as a tumor suppressor during tumor progression was obtained by the analysis of mice deficient for *Runx3*. *Runx3*-null mice examined on the first day of birth show hyperplasia of the lung and stomach while *Runx3+/−* mice have increased predisposition to tumor development across many tissues after a long latency (Lee et al. 2010, 2013; Li et al. 2002). By 18 months of age, about 80 % of *Runx3+/−* mice develop lung adenomas and about 20 % of *Runx3+/−* mice develop ductal adenocarcinoma. Tellingly, Runx3 expression in these spontaneously developed tumors was markedly reduced as compared to corresponding normal tissue of wild type mice (Huang et al. 2012; Lee et al. 2010). By 16 months of age, *Runx3+/−* mice also develop spontaneous colonic adenomas, which progressed into adenocarcinomas when *Runx3+/−* mice are crossed to ApcMin/+ mice (Ito et al. 2008). Recent studies also revealed *Runx3+/−* stomach to be precancerous, and administration of the carcinogen MNU resulted in adenocarcinomas more frequently and earlier in fundic and pyloric glands of *Runx3*-deficient mice than WT mice (Ito et al. 2011). It can be concluded that the tumorigenicity imparted by the loss of *Runx3* has to cooperate with additional hits such as carcinogen exposure or oncogenic activation (Wnt) to trigger complete malignant transformation.

29.3.2 DNA Damaging Agents Induce *RUNX3* Promoter Hypermethylation and Inactivation in Early Precancerous Lesions

RUNX3 inactivation mainly occurs by CpG island DNA methylation of *RUNX3* promoter, EZH2-dependent H3K27 trimethylation of *RUNX3* promoter or cytoplasmic mislocalization (gastric and breast cancers) and more rarely by mutation (gastric and bladder cancer) (Chuang and Ito 2010). Interestingly, several studies have shown a close integration between RUNX3 levels and environmental stresses such as exposure to genotoxic chemicals, attack by infectious agents or exposure to hormones. For example, in a study that correlated smoking with bladder cancer, it was found that *RUNX3* promoter methylation was not present in normal urothelial cells, but occurred early during tumorigenesis in response to smoking leading to the suggestion that *RUNX3* methylation can be used as a molecular clock to determine the age of a bladder tumor (Wolff et al. 2008). In another study using colorectal cancer cells, it was found that *RUNX3* was silenced by epigenetic regulation in a ROS-dependent manner (Kang et al. 2012). *RUNX3* inactivation in gastric cancer is brought about by *H.pylori* infection again through promoter hypermethylation. The CagA oncoprotein of *H.pylori* was also shown to directly associate with RUNX3 through a specific recognition of the PY motif of RUNX3 by a WW domain of CagA and mediate RUNX3 ubiquitination and degradation (Tsang et al. 2010). *H.pylori* induced *RUNX3* inactivation by promoter hypermethylation was also shown to be influenced by dietary factors. In a more comprehensive study on the correlation between *RUNX3* inactivation and *H.pylori* infection, methylation status of the *RUNX3* promoter was correlated with a spectrum of gastric lesions, including chronic atrophic gastritis, intestinal metaplasia, gastric adenoma, dysplasia, and gastric adenocarcinoma with paired noncancerous mucosa tissues. *RUNX3* promoter methylation was correlated with distinct stages of GC progression and was present in a very early precancerous lesion, atrophic gastritis (Chung et al. 2013; Lu et al. 2012). *RUNX3* promoter hypermethylation was estrogen-dependent in breast cancer mammospheres (Cheng et al. 2008). However, the relationship between the genotoxic role of estrogen (Estrogen DNA adduct formation) and *RUNX3* promoter methylation was not investigated. In summary, the above studies clear outline a regulatory mechanism through which endogenous and exogenous DNA damaging agents (described in Sect. 29.2.1 of this chapter) such as smoking, ROS, estrogen and *H.pylori* directly act at the level of *RUNX3* promoter hypermethylation to epigenetically silence *RUNX3* at the very initial stages of carcinogenesis. Based on these data, it can be speculated that RUNX3 imposes a barrier against genotoxic stresses, which, has to be alleviated to promote genomic instability and cancer progression.

29.3.3 RUNX and p53 Signaling

A direct regulatory function for RUNX proteins during DNA damage signaling was revealed through the study of the p53 in the DNA damage response pathway. In the well-characterised DNA damage response pathway, the exposure of cells to DNA damage insults results in the activation of p53, and p53 in turn, transcriptionally regulates the expression of proteins mediating cell cycle arrest or apoptosis by binding to p53 consensus at the target promoters. In the study by Wu et al., RUNX1 was found to participate in p53-dependent DNA damage responses, through protein-protein interactions between RUNX1 and p53. RUNX1 when recruited together with p53 to p53 target promoters facilitated higher transactivation of the p53 target genes BAX, PUMA, NOXA and p21(Wu et al. 2013). Consistently, *RUNX1* depletion significantly attenuated p53 transcriptional response after doxorubicin treatment and reduced doxorubicin-dependent apoptosis. RUNX1 depletion also reduced p53 acetylation at Lys-373/382, a site that is acetylated by the p300 histone acetyltransferase.

In another study by the same group, RUNX3 was found to regulate p53-mediated DNA damage-dependent responses following the exposure to doxorubicin (Yamada et al. 2010). This work was prompted by the observation that *RUNX3* knockdown inhibited DNA damage-dependent apoptosis in p53 wild-type cells but not in p53-deficient cells. RUNX3 co-immunoprecipitated in the nucleus with p53 and regulated ATM-dependent phosphorylation of p53 at ser15. Intriguingly, RUNX3 also formed a complex with DNA damage-activated phosphorylated ATM (ser1981). It was proposed that RUNX3 assisted ATM in the doxorubicin mediate phosphorylation of p53 at ser15 position during DNA damage response (Satoh et al. 2012). Here, when the C-terminal truncated form of RUNX1, RUNX1-dc was expressed in hematopoietic stem/progenitor cells, robust accumulation of the double strand break marker γH2AX accumulation was observed. Mechanistic studies revealed that RUNX1 and p53 synergistically activate the transcription of Gadd45a, a sensor of DNA damage. Taken together, RUNX1/RUNX3 and p53 form a transcriptionally active complex that upon recruitment to p53 target promoters influenced the outcome of the DNA damage response.

29.3.4 RUNX and Oncogene-Induced Senescence

As described earlier under Sect. 29.2.4, cells activate the oncogene-induced senescence (OIS) programme as a fail-safe mechanism to protect themselves against hyper-proliferative signals elicited by deregulated oncogenes (Bartek et al. 2007). In the context of Ras-induced OIS, Ras activates OIS by increasing the firing of DNA replication origins causing DNA replication stress. According to a recent study, following Ras activation, RUNX proteins cooperate with the p19 alternative reading frame (ARF)/p53 pathway to induce a senescence-like growth arrest in primary mouse embryonic fibroblasts. These findings were extended to *in vivo* mouse models, where *RUNX3* loss accelerated K-Ras induced lung adenocarcinoma formation because of impaired activation of the p14(ARF)-p53 pathway (Lee et al. 2013). Functional analysis revealed that Runx3 forms a complex with BRD2 in a K-Ras-dependent manner in the early phase of the cell cycle and this complex induces the expression of p14(ARF)/p19(Arf) providing a missing link between oncogenic K-Ras and the p14(ARF)-p53 pathway. Indeed, given that RUNX proteins were originally identified as regulator of polyoma virus induced DNA replication it is possible that RUNX proteins "sense" Ras induced DNA replication stress and activate BRD2 dependent p14(Arf) transcription. Future studies are required to provide a clearer understanding of Ras-induced DNA replication stress, RUNX status and OIS.

29.3.5 RUNX1, RUNX3 and Oxidative Stress Response

The increased production of ROS is known to be a hallmark of many cancers and cancer cell lines. Early studies had demonstrated that increased ROS can have a damaging effect on several biomolecules such as proteins, lipids, and DNA. ROS species can bind to a variety of DNA nucleotides to form oxidised bases, which are potently mutagenic when left unrepaired by the BER pathway. ROS can damage DNA also by causing strand breaks, which at high levels or upon collision with the DNA replication or transcription apparatus, can give rise to DNA double strand breaks. Based on these evidences, it is clear that one of the mechanisms through which tumor suppressor genes can influence tumorigenesis is by the regulation of ROS levels.

Intriguingly, it was found that RNT-1, the *C.elegans* homolog of RUNX is stabilized by oxidative stress through the MAPK pathway. It was shown that RNT-1 is constantly expressed and degraded in the *C.elegans* intestine (Lee et al. 2012). But, in response to oxidative stress, RNT-1 is immediately stabilised. The MAP kinase pathway was required for RNT-1 stabilization, and RNT-1 was phosphorylated by SEK-1/PMK-1 *in vitro*. It was proposed RNT-1 stabilisation in the intestine facilitated rapid response to environmental stress challenges in the intestine.

The relationship between RUNX proteins and maintenance of redox balance was further demonstrated in a study that investigated mechanisms that restrained high-ROS accumulation in leukemia-initiating cells (LIC) isolated from T cell acute lymphoblastic leukemia (T-ALL) (Giambra et al. 2015). In this model, low ROS accumulation was dependent on downregulation of the gene encoding for protein kinase C θ (PKC-θ). It was demonstrated that PKC-θ was regulated by NOTCH1 in a *RUNX*-dependent manner. Here, NOTCH1 induces *RUNX3* and RUNX3, in turn, represses *RUNX1* and RUNX1 induces PKC-θ. Hence, RUNX1 and RUNX3 interplay was a critical determining factor of ROS production and leukemic transformation in T-ALL. In this model, reciprocal increased *RUNX1* and decreased *RUNX3* resulted in heightened ROS production. These results are consistent with findings that the overexpression of RUNX1 in home foreskin fibroblasts (Hs68) resulted in heightened oxidative-stress, p38 MPAK activation, and increases p53 stabilization and caused premature senescence. In contrast, in another study on the role of Forkhead box O (FOXO) transcription factors during 3D breast epithelial acinar morphogenesis , single-cell gene expression profiling led to the identification that a subset of FOXO target genes was jointly regulated by *RUNX1* (Wang et al. 2011). Here, in this model, inhibition of RUNX1 and FOXO1 synergistically causes widespread oxidative stress during 3D morphogenesis *in vitro*. Taken together, the above results suggest that the relationship between RUNX and oxidative stress is context-dependent and it is possible that RUNX1 and RUNX3 have contrasting roles in the maintenance of redox balance.

29.3.6 RUNX1, RUNX3 and DNA Repair

According to accumulating evidence RUNX proteins may have a direct role to play in DNA repair and genome maintenance. In one study, it was found that RUNX3 directs associated with the Ku70/Ku80 heterodimer, an important upstream regulator of the NHEJ pathway (Tanaka et al. 2007). Here, RUNX3 association with Ku70 was found to increase RUNX3-mediated transactivation of p21. However, whether RUNX3-KU70 interaction can modulate NHEJ efficacy was not investigated.

In a more recent work, the relationship between RUNX proteins and ICL repair by the Fanconi anemia pathway of DNA repair was investigated (Wang et al. 2014). This study was prompted by the observation that single conditional knockout of *Runx1* or *Runx3* generally resulted in mild phenotypes in mice. For example, *Runx1*-deficient mice showed hematopoietic stem cell expansion followed by subsequent exhaustion. Similarly, *Runx3*-deficient mice developed a myeloproliferative disorder (MPD) only after 18 months of age. It was hence hypothesized that the mild phenotypes of *Runx1* or *Runx3* deficient mice could be because these two RUNX family proteins compensate in the absence of each other.

Hence, mice concurrently deficient for *Runx1* and *Runx3* were created under the control of the Mx1-cre promoter (*Mx1-Cre+*, *Runx1fl/fl* and *Runx3fl/fl* mice). By depleting *Runx1* and *Runx3* in the hematopoietic cells, a severe phenotype was seen in mice. About 82 % of the mice developed bone marrow failure (BMF) and the remaining 18 % developed MPD (predisposition to leukemia). The co-occurrence of BMF and MPD, supposedly contradictory phenotypes, is usually observed in the DNA repair deficiency-associated syndrome, FA. Hence, it was hypothesized that RUNX proteins may have a critical role in ICL repair by the FA pathway. According to several lines of evidence, *Runx1/Runx3* double knockout mice showed signs of genomic instability. *Runx1/Runx3* double knock-out mice showed increased sensitivity to DNA damaging agents like radiation and mitomycin C. Also, hematopoietic stem/progenitor cells (HSPCs), thec-Kit+Sca-1+Lineage (KSL) fraction from *Runx1/Runx3* double knock-out had persistent γH2AX foci 12 h after *in vitro* irradiation, indicating irreparable DNA damage.

Mechanistic studies revealed that RUNX proteins have a non-transcriptional role in the ICL repair pathway. Co-immunoprecipitation assay showed increased interaction between RUNX1 and FANCI/FANCD2, central mediators of ICL repair. RUNX3 was also shown to co-immunoprecipitate with FANCI and FANCD2. The RUNX/FANCI/FANCD2 complex formation was independent of the canonical transcription RUNX/CBFβ complex. More importantly, using biochemical methods and immunofluorescence studies, it was demonstrated that *RUNX1* and *RUNX3* co-depletion resulted in diminished recruitment of FANCI and FANCD2 to DNA damage sites. Taken together, the above study emphasises a critical and unexpected role for RUNX1 and RUNX3 in the FA pathway of DNA repair. Consistently, at least two human FA patients without any mutations in the known FA genes had genomic deletion the region encompassing RUNX1 (Byrd et al. 2011; Click et al. 2011).

Recent studies indicate that the FA pathway is critical for the suppression tumorigenesis in response to endogenous DNA damage in the form of aldehydes such as acetaldehyde, malonilaldehydes generated as intermediates in metabolic pathways (Ceccaldi et al. 2016). In summary, RUNX1 and RUNX3 impose a very critical barrier against carcinogenesis by preventing genomic instability in response to intrinsic DNA damage in the form of DNA interstrand crosslinks.

29.3.7 RUNX1-ETO Induces a "Mutator" Phenotype

Perhaps the best evidence that RUNX misregulation can generate genomic instability exists for the leukemogenic RUNX fusion protein, RUNX1-ETO. The t(8;21) translocation that fuses the N terminal of *RUNX1* with almost the entire gene of *ETO* is a leukemia-initiating event, and fusion gene sequences may be found long before the onset of leukemia, yet, the induction of fully developed AML has a very long latency. Moreover, many strains of RUNX1-ETO-expressing transgenic mice have been generated but none of these mice develop leukemia spontaneously, but are highly susceptible to *N*-ethyl-*N*-nitrosourea (ENU)-induced leukemia (Yuan et al. 2001). These studies have given rise to the idea that although RUNX1-ETO is a leukemia initiating event it has to be co-operate with secondary mutations to cause AML.

However, the mechanisms driving the acquisition of co-operating mutations remain unclear. According to one view, RUNX1-ETO can promote mutagenesis of the genome, promoting the acquisition of additional mutations. In a genome-wide study, the leukemogenic proteins, AML1/ETO, PML/RAR, and PLZF/RAR were expressed in U937 hematopoietic precursor cells and global gene expression was measured using oligonucleotide chips. Here, RUNX1-ETO expression repressed DNA repair genes, particularly of the base excision repair pathway such as *OGG1* and compromised OGG1 DNA glycosylase activity (Alcalay et al. 2003). Similarly, ectopic expression of RUNX1-ETO has been shown to increase the basal levels of γH2AX across multiple studies (Wajapeyee et al. 2010; Wolyniec et al. 2009). RUNX1-ETO was also shown to downregulate the expression of DNA repair genes such as *BRCA2* and *ATM* (Krejci et al. 2008). In a study on the role of RUNX proteins in ICL repair, it has been shown that RUNX1-ETO suppresses the FA pathway of DNA repair, possibly also contributing to heightened basal DNA damage levels (Wang et al. 2014). In turn, due to increased DNA damage, RUNX1-ETO over expressing cells have an activated p53 pathway which increases basal apoptosis and enhances the sensitivity to DNA damaging agents (Krejci et al. 2008).

Recently, using two elegant *in vitro* models, it has been convincingly demonstrated that the expression of RUNX1-ETO imparts a mutator phenotype to cells (Forster et al. 2016). RUNX1-ETO was expressed in the non-transformed TK6 lymphoblastoid cell line and the acquisition of mutations at the *PIGA* reporter gene was measured both at basal levels as well as after DNA damage (Doxorubicin and irradiation). RUNX1-ETO expression predisposed cells to the acquisition of mutations, both spontaneously and after treatment with genotoxic agents. In this model, RUNX1-ETO dependent downregulation of OGG1 expression was proposed as the main driver of genomic instability. Importantly, the strength of the mutator phenotype was related to the expression level of the fusion protein. Taken together, the above studies convincingly demonstrate a direct role for RUNX1-ETO in the generation of genomic instability in human leukemia.

29.3.8 PARP Inhibition as a Synthetic Lethal Approach as RUNX1-ETO Expressing Leukaemia

The observation that RUNX1-ETO engenders a mutator phenotype makes t(8;21) leukemic cells as good candidates for synthetic lethal approaches such as PARP inhibition therapy. According to the concept of synthetic lethality, cells treated with PARP inhibitors become highly dependent on alternative or back-up DNA repair pathways. Hence, down regulating those back-up DNA repair pathways will create massive genomic instability and cell death. Consistently, RUNX1-ETO expressing Kasumi-1 leukemic cells were shown to be more sensitive to PARP inhibition (Wang et al. 2014). In this study, increased PARP inhibitor sensitivity was attributed to the ability of RUNX1-ETO to curtail the FA pathway of ICL repair by impairing FANCD2 localization to sites of DNA damage. Consistently, using a variety of *in vitro* and *in vivo* models, it has been demonstrated by Esposito et al. that the transduction of RUNX1-ETO into hematopoietic progenitor cells imparted heightened sensitivity to olaparib, a clinically approved PARP inhibitor (Esposito et al. 2015; Wang et al. 2015). The putative proposed mechanism being that RUNX1-ETO imparts "BRCAness" by downregulation of BRCA1 and thus makes cells sensitive to PARP inhibitor. While the precise mechanism underlying PARP inhibitor sensitivity is not clear, the above studies unveil a new therapeutic approach to treat t (8;21) leukaemia.

29.4 Concluding Remarks and Future Directions

Multiple studies have demonstrated a strong role for RUNX proteins in the maintenance of genomic integrity. We propose that intact RUNX1 and RUNX3 impose a critical barrier that needs to be breached for cells to accumulate further secondary hits during carcinogenesis (Fig. 29.4). At least four lines of evidence support this assumption-firstly, *RUNX3* inactivation occurs even in pre-neoplastic lesions and its promoter hypermethylation and inactivation are closely linked to exposure to multiple genotoxic exogenous agents such factors like smoking, ROS and *H.pylori* infection. Second, *Runx1* and *Runx3* double-knockout mice show a variety of DNA repair deficiency related phenotypes. Thirdly, both RUNX1 and RUNX3 associate with p53 to form a complex that modulates the transactivation potential of p53 during the DNA damage response. Last but not the least, ectopic expression of the leukemogenic protein, RUNX1-ETO, directly endows cells with a potent "mutator" phenotype, either by impaired the FA pathway of DNA repair or/and downregulation of BER repair genes and HR genes like OGG1 and BRCA1, thereby promoting the acquisition of secondary hits that drive carcinogenesis. Further affirmative data can now be obtained for all RUNX-deficient cancers through the use next generation sequencing and genomic analysis. In-depth genomic analysis of human tumors stratified based on RUNX expression followed by the detection of any of the 33 so-far reported mutational signatures, specifically in RUNX-deficient human cancers, will reinforce the exciting idea that RUNX dysfunction accelerates carcinogenesis by promoting genomic instability. A second potentially exciting area of research is to identify DNA repair proteins that form "synthetic-lethal" pairs with RUNX deficiency. A case in point is the utility of PARP inhibitors for the targeting of RUNX1-ETO expressing leukemias. Then, akin

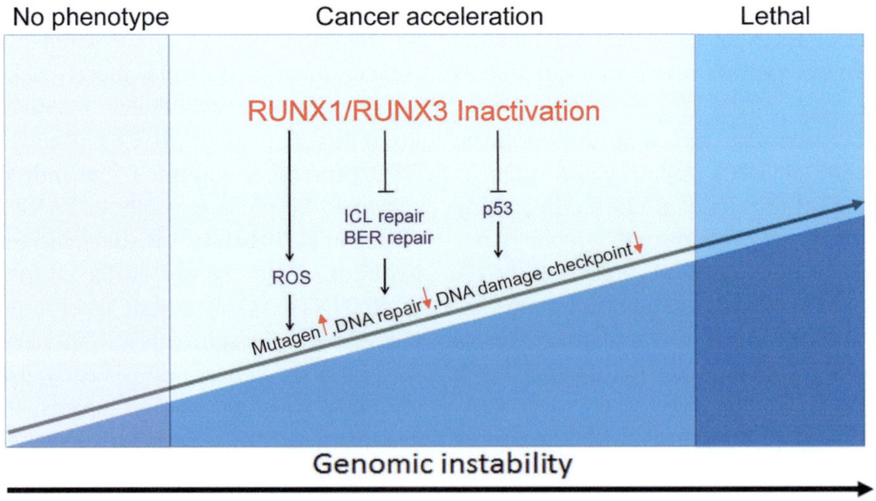

Fig. 29.4 Induction of genomic instability by RUNX1 and RUNX3 during cancer progression – We hypothesise that the inactivation of RUNX1and RUNX3 results in an intermediate level of genomic instability, the state that is more suitable for acquisition of mutations for the purpose of cancer progression. RUNX-deficiency increases mutation load by increasing ROS levels, reduces the efficiency of ICL repair and BER repair and impairs DNA damage checkpoint by misregulation of p53 function. In summary, RUNX deficiency might promote cancer progression by imparting a mutator phenotype

to PARP inhibitor therapy for BRCA1-deficient tumors, RUNX-deficiency can be targeted using the currently available DNA repair inhibitors as a therapeutic strategy.

Acknowledgement The authors are supported by the National Medical Research Council (NMRC), Singapore, the National Research Foundation (NRF) under the Singapore Ministry of Education under its Research Centres of Excellence initiative and by the NRF under its Translational and Clinical Research Flagship Programme.

References

Alcalay, M., Meani, N., Gelmetti, V., Fantozzi, A., Fagioli, M., Orleth, A., et al. (2003). Acute myeloid leukemia fusion proteins deregulate genes involved in stem cell maintenance and DNA repair. *The Journal of Clinical Investigation, 112*, 1751–1761.

Alexandrov, L. B., Nik-Zainal, S., Wedge, D. C., Aparicio, S. A., Behjati, S., Biankin, A. V., et al. (2013a). Signatures of mutational processes in human cancer. *Nature, 500*, 415–421.

Alexandrov, L. B., Nik-Zainal, S., Wedge, D. C., Campbell, P. J., & Stratton, M. R. (2013b). Deciphering signatures of mutational processes operative in human cancer. *Cell Reports, 3*, 246–259.

Alter, B. P. (2003). Cancer in Fanconi anemia, 1927–2001. *Cancer, 97*, 425–440.

Andor, N., Graham, T. A., Jansen, M., Xia, L. C., Aktipis, C. A., Petritsch, C., et al. (2016). Pan-cancer analysis of the extent and consequences of intratumor heterogeneity. *Nature Medicine, 22*, 105–113.

Baca, S. C., Prandi, D., Lawrence, M. S., Mosquera, J. M., Romanel, A., Drier, Y., et al. (2013). Punctuated evolution of prostate cancer genomes. *Cell, 153*, 666–677.

Bartek, J., & Lukas, J. (2007). DNA damage checkpoints: From initiation to recovery or adaptation. *Current Opinion in Cell Biology, 19*, 238–245.

Bartek, J., Bartkova, J., & Lukas, J. (2007). DNA damage signalling guards against activated oncogenes and tumour progression. *Oncogene, 26*, 7773–7779.

Bartkova, J., Horejsi, Z., Koed, K., Kramer, A., Tort, F., Zieger, K., et al. (2005). DNA damage response as a candidate anti-cancer barrier in early human tumorigenesis. *Nature, 434*, 864–870.

Bartkova, J., Rezaei, N., Liontos, M., Karakaidos, P., Kletsas, D., Issaeva, N., et al. (2006). Oncogene-induced senescence is part of the tumorigenesis barrier imposed by DNA damage checkpoints. *Nature, 444*, 633–637.

Blyth, K., Vaillant, F., Jenkins, A., McDonald, L., Pringle, M. A., Huser, C., et al. (2010). Runx2 in normal tissues and cancer cells: A developing story. *Blood Cells, Molecules, and Diseases, 45*, 117–123.

Boveri, T. (2008). Concerning the origin of malignant tumours by Theodor Boveri. Translated and annotated by Henry Harris. *Journal of Cell Science, 121*(Suppl 1), 1–84.

Byrd, R. S., Zwerdling, T., Moghaddam, B., Pinter, J. D., & Steinfeld, M. B. (2011). Monosomy 21q22.11-q22.13 presenting as a Fanconi anemia phenotype. *American Journal of Medical Genetics. Part A, 155A*, 120–125.

Caldecott, K. W. (2008). Single-strand break repair and genetic disease. *Nature Reviews. Genetics, 9*, 619–631.

Ceccaldi, R., Sarangi, P., & D'Andrea, A. D. (2016). The Fanconi anaemia pathway: New players and new functions. *Nature Reviews. Molecular Cell Biology, 17*, 337–349.

Cheng, A. S., Culhane, A. C., Chan, M. W., Venkataramu, C. R., Ehrich, M., Nasir, A., et al. (2008). Epithelial progeny of estrogen-exposed breast progenitor cells display a cancer-like methylome. *Cancer Research, 68*, 1786–1796.

Chuang, L. S., & Ito, Y. (2010). RUNX3 is multifunctional in carcinogenesis of multiple solid tumors. *Oncogene, 29*, 2605–2615.

Chung, W. C., Jung, S. H., Joo, K. R., Kim, M. J., Youn, G. J., Kim, Y., et al. (2013). An inverse relationship between the expression of the gastric tumor suppressor RUNX3 and infection with Helicobacter pylori in gastric epithelial dysplasia. *Gut Liver, 7*, 688–695.

Click, E. S., Cox, B., Olson, S. B., Grompe, M., Akkari, Y., Moreau, L. A., et al. (2011). Fanconi anemia-like presentation in an infant with constitutional deletion of 21q including the RUNX1 gene. *American Journal of Medical Genetics. Part A, 155A*, 1673–1679.

Duensing, S., & Munger, K. (2002). The human papillomavirus type 16 E6 and E7 oncoproteins independently induce numerical and structural chromosome instability. *Cancer Research, 62*, 7075–7082.

Dvash, E., Har-Tal, M., Barak, S., Meir, O., & Rubinstein, M. (2015). Leukotriene C4 is the major trigger of stress-induced oxidative DNA damage. *Nature Communications, 6*, 10112.

Esposito, M. T., Zhao, L., Fung, T. K., Rane, J. K., Wilson, A., Martin, N., et al. (2015). Synthetic lethal targeting of oncogenic transcription factors in acute leukemia by PARP inhibitors. *Nature Medicine, 21*, 1481–1490.

Forster, V. J., Nahari, M. H., Martinez-Soria, N., Bradburn, A. K., Ptasinska, A., Assi, S. A., et al. (2016). The leukemia-associated RUNX1/ETO oncoprotein confers a mutator phenotype. *Leukemia, 30*, 250–253.

Gao, J., Aksoy, B. A., Dogrusoz, U., Dresdner, G., Gross, B., Sumer, S. O., et al. (2013). Integrative analysis of complex cancer genomics and clinical profiles using the cBioPortal. *Science Signal, 6*, 11.

Giambra, V., Jenkins, C. E., Lam, S. H., Hoofd, C., Belmonte, M., Wang, X., et al. (2015). Leukemia stem cells in T-ALL require active Hif1alpha and Wnt signaling. *Blood, 125*, 3917–3927.

Gorgoulis, V. G., & Halazonetis, T. D. (2010). Oncogene-induced senescence: The bright and dark side of the response. *Current Opinion in Cell Biology, 22*, 816–827.

Guo, C., Tang, T. S., & Friedberg, E. C. (2010). SnapShot: Nucleotide excision repair. *Cell, 140*(754–754), e751.

Hanahan, D., & Weinberg, R. A. (2011). Hallmarks of cancer: The next generation. *Cell, 144*, 646–674.

Hanawalt, P. C., & Spivak, G. (2008). Transcription-coupled DNA repair: Two decades of progress and surprises. *Nature Reviews. Molecular Cell Biology, 9*, 958–970.

Harper, J. W., & Elledge, S. J. (2007). The DNA damage response: Ten years after. *Molecular Cell, 28*, 739–745.

Hartlerode, A. J., & Scully, R. (2009). Mechanisms of double-strand break repair in somatic mammalian cells. *The Biochemical Journal, 423*, 157–168.

Helleday, T., Eshtad, S., & Nik-Zainal, S. (2014). Mechanisms underlying mutational signatures in human cancers. *Nature Reviews. Genetics, 15*, 585–598.

Holland, A. J., & Cleveland, D. W. (2012). Chromoanagenesis and cancer: Mechanisms and consequences of localized, complex chromosomal rearrangements. *Nature Medicine, 18*, 1630–1638.

Huang, B., Qu, Z., Ong, C. W., Tsang, Y. H., Xiao, G., Shapiro, D., et al. (2012). RUNX3 acts as a tumor suppressor in breast cancer by targeting estrogen receptor alpha. *Oncogene, 31*, 527–534.

Ito, K., Lim, A. C., Salto-Tellez, M., Motoda, L., Osato, M., Chuang, L. S., et al. (2008). RUNX3 attenuates beta-catenin/T cell factors in intestinal tumorigenesis. *Cancer Cell, 14*, 226–237.

Ito, K., Chuang, L. S., Ito, T., Chang, T. L., Fukamachi, H., Salto-Tellez, M., & Ito, Y. (2011). Loss of Runx3 is a key event in inducing precancerous state of the stomach. *Gastroenterology, 140*(1536–1546), e1538.

Ito, Y., Bae, S. C., & Chuang, L. S. (2015). The RUNX family: Developmental regulators in cancer. *Nature Reviews. Cancer, 15*, 81–95.

Jackson, S. P., & Bartek, J. (2009). The DNA-damage response in human biology and disease. *Nature, 461*, 1071–1078.

Jacob, B., Osato, M., Yamashita, N., Wang, C. Q., Taniuchi, I., Littman, D. R., et al. (2010). Stem cell exhaustion due to Runx1 deficiency is prevented by Evi5 activation in leukemogenesis. *Blood, 115*, 1610–1620.

Jeggo, P. A., Pearl, L. H., & Carr, A. M. (2016). DNA repair, genome stability and cancer: A historical perspective. *Nature Reviews. Cancer, 16*, 35–42.

Kang, K. A., Zhang, R., Kim, G. Y., Bae, S. C., & Hyun, J. W. (2012). Epigenetic changes induced by oxidative stress in colorectal cancer cells: Methylation of tumor suppressor RUNX3. *Tumour Biology, 33*, 403–412.

Krejci, O., Wunderlich, M., Geiger, H., Chou, F. S., Schleimer, D., Jansen, M., et al. (2008). p53 signaling in response to increased DNA damage sensitizes AML1-ETO cells to stress-induced death. *Blood, 111*, 2190–2199.

Kunkel, T. A., & Erie, D. A. (2005). DNA mismatch repair. *Annual Review of Biochemistry, 74*, 681–710.

Leach, F. S., Nicolaides, N. C., Papadopoulos, N., Liu, B., Jen, J., Parsons, R., et al. (1993). Mutations of a mutS homolog in hereditary nonpolyposis colorectal cancer. *Cell, 75*, 1215–1225.

Lecona, E., & Fernandez-Capetillo, O. (2014). Replication stress and cancer: It takes two to tango. *Experimental Cell Research, 329*, 26–34.

Lee, K. S., Lee, Y. S., Lee, J. M., Ito, K., Cinghu, S., Kim, J. H., et al. (2010). Runx3 is required for the differentiation of lung epithelial cells and suppression of lung cancer. *Oncogene, 29*, 3349–3361.

Lee, K., Shim, J., Bae, J., Kim, Y. J., & Lee, J. (2012). Stabilization of RNT-1 protein, runt-related transcription factor (RUNX) protein homolog of *Caenorhabditis elegans*, by oxidative stress through mitogen-activated protein kinase pathway. *The Journal of Biological Chemistry, 287*, 10444–10452.

Lee, Y. S., Lee, J. W., Jang, J. W., Chi, X. Z., Kim, J. H., Li, Y. H., et al. (2013). Runx3 inactivation is a crucial early event in the development of lung adenocarcinoma. *Cancer Cell, 24*, 603–616.

Li, Q. L., Ito, K., Sakakura, C., Fukamachi, H., Inoue, K., Chi, X. Z., et al. (2002). Causal relationship between the loss of RUNX3 expression and gastric cancer. *Cell, 109*, 113–124.

Lieber, M. R. (2010). The mechanism of double-strand DNA break repair by the nonhomologous DNA end-joining pathway. *Annual Review of Biochemistry, 79*, 181–211.

Lindahl, T. (1993). Instability and decay of the primary structure of DNA. *Nature, 362*, 709–715.

Lu, X. X., Yu, J. L., Ying, L. S., Han, J., Wang, S., Yu, Q. M., et al. (2012). Stepwise cumulation of RUNX3 methylation mediated by Helicobacter pylori infection contributes to gastric carcinoma progression. *Cancer, 118*, 5507–5517.

Maseki, N., Miyoshi, H., Shimizu, K., Homma, C., Ohki, M., Sakurai, M., & Kaneko, Y. (1993). The 8;21 chromosome translocation in acute myeloid leukemia is always detectable by molecular analysis using AML1. *Blood, 81*, 1573–1579.

Maya-Mendoza, A., Ostrakova, J., Kosar, M., Hall, A., Duskova, P., Mistrik, M., et al. (2015). Myc and Ras oncogenes engage different energy metabolism programs and evoke distinct patterns of oxidative and DNA replication stress. *Molecular Oncology, 9*, 601–616.

Miyoshi, H., Shimizu, K., Kozu, T., Maseki, N., Kaneko, Y., & Ohki, M. (1991). t(8;21) breakpoints on chromosome 21 in acute myeloid leukemia are clustered within a limited region of a single gene, AML1. *Proceedings of the National Academy of Sciences of the United States of America, 88*, 10431–10434.

Moldovan, G. L., & D'Andrea, A. D. (2009). FANCD2 hurdles the DNA interstrand crosslink. *Cell, 139*, 1222–1224.

Motoda, L., Osato, M., Yamashita, N., Jacob, B., Chen, L. Q., Yanagida, M., et al. (2007). Runx1 protects hematopoietic stem/progenitor cells from oncogenic insult. *Stem Cells, 25*, 2976–2986.

Moynahan, M. E., & Jasin, M. (2010). Mitotic homologous recombination maintains genomic stability and suppresses tumorigenesis. *Nature Reviews. Molecular Cell Biology, 11*, 196–207.

Negrini, S., Gorgoulis, V. G., & Halazonetis, T. D. (2010). Genomic instability – An evolving hallmark of cancer. *Nature Reviews. Molecular Cell Biology, 11*, 220–228.

Nik-Zainal, S., Alexandrov, L. B., Wedge, D. C., Van Loo, P., Greenman, C. D., Raine, K., et al. (2012). Mutational processes molding the genomes of 21 breast cancers. *Cell, 149*, 979–993.

Nik-Zainal, S., Kucab, J. E., Morganella, S., Glodzik, D., Alexandrov, L. B., Arlt, V. M., et al. (2015). The genome as a record of environmental exposure. *Mutagenesis, 30*, 763–770.

Nishimoto, N., Arai, S., Ichikawa, M., Nakagawa, M., Goyama, S., Kumano, K., et al. (2011). Loss of AML1/Runx1 accelerates the development of MLL-ENL leukemia through down-regulation of p19ARF. *Blood, 118*, 2541–2550.

Nouspikel, T. (2009). DNA repair in mammalian cells: Nucleotide excision repair: Variations on versatility. *Cellular and Molecular Life Sciences, 66*, 994–1009.

Pontel, L. B., Rosado, I. V., Burgos-Barragan, G., Garaycoechea, J. I., Yu, R., Arends, M. J., et al. (2015). Endogenous formaldehyde is a hematopoietic stem cell genotoxin and metabolic carcinogen. *Molecular Cell, 60*, 177–188.

Poon, S. L., Pang, S. T., McPherson, J. R., Yu, W., Huang, K. K., Guan, P., et al. (2013). Genome-wide mutational signatures of aristolochic acid and its application as a screening tool. *Science Translational Medicine, 5*, 197ra101.

Santen, R. J., Yue, W., & Wang, J. P. (2015). Estrogen metabolites and breast cancer. *Steroids, 99*, 61–66.

Satoh, Y., Matsumura, I., Tanaka, H., Harada, H., Harada, Y., Matsui, K., et al. (2012). C-terminal mutation of RUNX1 attenuates the DNA-damage repair response in hematopoietic stem cells. *Leukemia, 26*, 303–311.

Schreiber, V., Dantzer, F., Ame, J. C., & de Murcia, G. (2006). Poly(ADP-ribose): novel functions for an old molecule. *Nature Reviews. Molecular Cell Biology, 7*, 517–528.

Shibata, D., Peinado, M. A., Ionov, Y., Malkhosyan, S., & Perucho, M. (1994). Genomic instability in repeated sequences is an early somatic event in colorectal tumorigenesis that persists after transformation. *Nature Genetics, 6*, 273–281.

Simsek, D., & Jasin, M. (2010). Alternative end-joining is suppressed by the canonical NHEJ component Xrcc4-ligase IV during chromosomal translocation formation. *Nature Structural & Molecular Biology, 17*, 410–416.

Sugasawa, K., Okuda, Y., Saijo, M., Nishi, R., Matsuda, N., Chu, G., et al. (2005). UV-induced ubiquitylation of XPC protein mediated by UV-DDB-ubiquitin ligase complex. *Cell, 121*, 387–400.

Tanaka, Y., Imamura, J., Kanai, F., Ichimura, T., Isobe, T., Koike, M., et al. (2007). Runx3 interacts with DNA repair protein Ku70. *Experimental Cell Research, 313*, 3251–3260.

Taniuchi, I., Osato, M., & Ito, Y. (2012). Runx1: No longer just for leukemia. *The EMBO Journal, 31*, 4098–4099.

Tsang, Y. H., Lamb, A., Romero-Gallo, J., Huang, B., Ito, K., Peek Jr., R. M., et al. (2010). Helicobacter pylori CagA targets gastric tumor suppressor RUNX3 for proteasome-mediated degradation. *Oncogene, 29*, 5643–5650.

Wajapeyee, N., Wang, S. Z., Serra, R. W., Solomon, P. D., Nagarajan, A., Zhu, X., & Green, M. R. (2010). Senescence induction in human fibroblasts and hematopoietic progenitors by leukemogenic fusion proteins. *Blood, 115*, 5057–5060.

Wang, L., Brugge, J. S., & Janes, K. A. (2011). Intersection of FOXO- and RUNX1-mediated gene expression programs in single breast epithelial cells during morphogenesis and tumor progression. *Proceedings of the National Academy of Sciences of the United States of America, 108*, E803–E812.

Wang, C. Q., Krishnan, V., Tay, L. S., Chin, D. W., Koh, C. P., Chooi, J. Y., et al. (2014). Disruption of Runx1 and Runx3 leads to bone marrow failure and leukemia predisposition due to transcriptional and DNA repair defects. *Cell Reports, 8*, 767–782.

Wang, L., Hamard, P. J., & Nimer, S. D. (2015). PARP inhibitors: A treatment option for AML? *Nature Medicine, 21*, 1393–1394.

Wilson 3rd, D. M., & Bohr, V. A. (2007). The mechanics of base excision repair, and its relationship to aging and disease. *DNA Repair (Amst), 6*, 544–559.

Wolff, E. M., Liang, G., Cortez, C. C., Tsai, Y. C., Castelao, J. E., Cortessis, V. K., et al. (2008). RUNX3 methylation reveals that bladder tumors are older in patients with a history of smoking. *Cancer Research, 68*, 6208–6214.

Wolyniec, K., Wotton, S., Kilbey, A., Jenkins, A., Terry, A., Peters, G., et al. (2009). RUNX1 and its fusion oncoprotein derivative, RUNX1-ETO, induce senescence-like growth arrest independently of replicative stress. *Oncogene, 28*, 2502–2512.

Wu, D., Ozaki, T., Yoshihara, Y., Kubo, N., & Nakagawara, A. (2013). Runt-related transcription factor 1 (RUNX1) stimulates tumor suppressor p53 protein in response to DNA damage through complex formation and acetylation. *The Journal of Biological Chemistry, 288*, 1353–1364.

Xu, H., Chaturvedi, R., Cheng, Y., Bussiere, F. I., Asim, M., Yao, M. D., et al. (2004). Spermine oxidation induced by Helicobacter pylori results in apoptosis and DNA damage: Implications for gastric carcinogenesis. *Cancer Research, 64*, 8521–8525.

Yamada, C., Ozaki, T., Ando, K., Suenaga, Y., Inoue, K., Ito, Y., et al. (2010). RUNX3 modulates DNA damage-mediated phosphorylation of tumor suppressor p53 at Ser-15 and acts as a co-activator for p53. *The Journal of Biological Chemistry, 285*, 16693–16703.

Yoshimoto, S., Loo, T. M., Atarashi, K., Kanda, H., Sato, S., Oyadomari, S., et al. (2013). Obesity-induced gut microbial metabolite promotes liver cancer through senescence secretome. *Nature, 499*, 97–101.

Yuan, Y., Zhou, L., Miyamoto, T., Iwasaki, H., Harakawa, N., Hetherington, C. J., et al. (2001). AML1-ETO expression is directly involved in the development of acute myeloid leukemia in the presence of additional mutations. *Proceedings of the National Academy of Sciences of the United States of America, 98*, 10398–10403.

Index

Y. Groner et al. (eds.), *RUNX Proteins in Development and Cancer*, Advances in Experimental Medicine and Biology 962, DOI 10.1007/978-981-10-3233-2

J

K

L

M

N

O

MIX
Papier aus verantwortungsvollen Quellen
Paper from responsible sources
FSC® C105338

If you have any concerns about our products,
you can contact us on
ProductSafety@springernature.com

In case Publisher is established outside the EU,
the EU authorized representative is:
Springer Nature Customer Service Center GmbH
Europaplatz 3, 69115 Heidelberg, Germany

Printed by Libri Plureos GmbH
in Hamburg, Germany